KIRK-OTHMER

ENCYCLOPEDIA OF
CHEMICAL
TECHNOLOGY

FOURTH EDITION

VOLUME **8**

DEUTERIUM AND TRITIUM
TO
ELASTOMERS, POLYETHERS

EXECUTIVE EDITOR
Jacqueline I. Kroschwitz

EDITOR
Mary Howe-Grant

KIRK-OTHMER

ENCYCLOPEDIA OF CHEMICAL TECHNOLOGY

FOURTH EDITION

VOLUME **8**

DEUTERIUM AND TRITIUM
TO
ELASTOMERS, POLYETHERS

A Wiley-Interscience Publication
JOHN WILEY & SONS

New York • Chichester • Brisbane • Toronto • Singapore

This text is printed on acid-free paper.

Library of Congress Cataloging-in-Publication Data

Encyclopedia of chemical technology / executive editor, Jacqueline
  I. Kroschwitz; editor, Mary Howe-Grant.—4th ed.
    p. cm.
  At head of title: Kirk-Othmer.
  "A Wiley-Interscience publication."
  Includes index.
  Contents: v. 8. Deuterium and tritium to elastomers, polyethers.
  ISBN 0-471-52676-2 (v. 8)
  1. Chemistry, Technical—Encyclopedias. I. Kirk, Raymond E.
(Raymond Eller), 1890–1957. II. Othmer, Donald F. (Donald
Frederick), 1904– . III. Kroschwitz, Jacqueline I., 1942– .
IV. Howe-Grant, Mary, 1943– . V. Title: Kirk-Othmer encyclopedia
of chemical technology.
TP9.E685 1992   91-16789
660′.03—dc20

CONTENTS

EDITORIAL STAFF
FOR VOLUME 8

Executive Editor: **Jacqueline I. Kroschwitz**
Editor: **Mary Howe-Grant**
Editorial Supervisor: **Lindy J. Humphreys**
Assistant Editor: **Cathleen A. Treacy**
Copy Editor: **Christine Punzo**

CONTRIBUTORS
TO VOLUME 8

Ulrich Baurmeister, *Enka AG, Ohder, Germany,* Dialysis
Dwight H. Bergquist, *Henningsen Foods, Inc., Omaha, Nebraska,* Eggs
Bret Berner, *CIBA-GEIGY Corporation, Ardsley, New York,* Drug delivery systems
Seymour S. Block, *University of Florida, Gainesville,* Disinfectants and antiseptics
Thomas E. Breuer, *Humko Chemical, Memphis, Tennessee,* Dimer acids

Jeffrey T. Books, *Ethyl Corporation, Baton Rouge, Louisiana,* Phosphazenes (under Elastomers, synthetic)

Robert R. Cantrell, *Union College, Jackson, Tennessee,* Dicarboxylic acids

Peter Cervoni, *American Cyanamid Company, Pearl River, New York,* Diuretic agents

Peter S. Chan, *American Cyanamid Company, Pearl River, New York,* Diuretic agents

Wai-kai Chen, *University of Illinois, Chicago,* Dimensional analysis

A. J. Cofrancesco, *Consultant, Delanson, New York,* Dyes, natural

Ann R. Comfort, *CIBA-GEIGY Corporation, Ardsley, New York,* Drug delivery systems

Ernesto de Guzman, *Sybron Chemicals Inc., Wellford, South Carolina,* Dye carriers

Steven M. Dinh, *CIBA-GEIGY Corporation, Ardsley, New York,* Drug delivery systems

Theresa M. Dobel, *Du Pont Chemical Company, Stow, Ohio,* Ethylene–acrylic elastomers (under Elastomers, synthetic)

Michael Doherty, *University of Massachusetts, Amherst,* Distillation, azeotropic and extractive

Kenneth R. Engh, *CELITE Corporation, Lompac, California,* Diatomite

Royce Ennis, *Du Pont-Beaumont Works, Beaumont, Texas,* Chlorosulfonated polyethylene (under Elastomers, synthetic)

James R. Fair, *The University of Texas at Austin, Austin,* Distillation

Brian Glover, *Zeneca Colours, Manchester, UK,* Dyes, application and evaluation

Peter Gregory, *Zeneca Specialties, Manchester, UK,* Dyes and dye intermediates

Werner M. Grootaert, *3M Company, St. Paul, Minnesota,* Fluorocarbon elastomers (under Elastomers, synthetic)

Adel F. Halasa, *Goodyear Tire & Rubber Company, Akron, Ohio,* Polybutadiene (under Elastomers, synthetic)

William M. Hann, *Rohm & Haas Company, Spring House, Pennsylvania,* Dispersants

Makoto Hattori, *Sumitomo Chemical Company, Tokyo,* Dyes, anthraquinone

Robert W. Johnson, *Union Camp Corporation, Savannah, Georgia,* Dicarboxylic acids

August H. Jorgensen, *Zeon Chemicals, USA, Louisvllle, Kentucky,* Nitrile rubber (under Elastomers, synthetic)

Steven W. Kaiser, *Union Carbide Chemicals and Plastics Company Inc., South Charleston, West Virginia,* Diamines and higher amines, aliphatic

Joseph J. Katz, *Argonne National Laboratories, Argonne, Illinois,* Deuterium and tritium

Jeffrey P. Knapp, *E. I. du Pont de Nemours & Co., Inc.,* Distillation, azeotropic and extractive

Edward Kresge, *Exxon Chemical Company, Linden, New Jersey,* Butyl rubber (under Elastomers, synthetic)

Vernon L. Kyllingstad, *Zeon Chemicals, Louisville, Kentucky,* Polyethers (under Elastomers, synthetic)

Marvin Landau, *Huls America, Inc., Piscataway, New Jersey,* Driers and metallic soaps

B. A. Lewis, *Cornell University, Ithaca, New York,* Dietary fiber

Michael J. Lysaght, *Cyto Therapeutics, Inc., Providence, Rhode Island,* Dialysis

Paul H. McCormick, *Drying Unincorporated, Newark, Delaware,* Drying

Donald Mackey, *Zeon Chemicals, USA, Louisville, Kentucky,* Nitrile rubber (under Elastomers, synthetic)

J. M. Massie, *Goodyear Tire & Rubber Company, Akron Ohio,* Polybutadiene (under Elastomers, synthetic)

George H. Millet, *3M Company, St. Paul, Minnesota,* Fluorocarbon elastomers (under Elastomers, synthetic)

Booker Morey, *SRI International, Menlo Park, California,* Dewatering

Maurice Morton, *Consultant, Beachwood, Ohio,* Survey (under Elastomers, synthetic)

Jacobus W. N. Noordermeer, *DSM Elastomers Europe BV, Geleen, The Netherlands,* Ethylene–propylene–diene rubber (under Elastomers, synthetic)
mers, synthetic)

Charles M. Pollock, *Union Camp Corporation, Savannah, Georgia,* Dicarboxylic acids

Elizabeth Quadros, *CIBA-GEIGY Corporation, Ardsley, New York,* Drug delivery systems

Abraham Reife, *CIBA-GEIGY Corporation, Tom's River, New Jersey,* Dyes, environmental chemistry

J. Shacter, *Martin Marietta Energy Systems, Oak Ridge, Tennessee,* Diffusion separation methods

Roy E. Smith, *CIBA-GEIGY Corporation, Greensboro, North Carolina,* Dyes, reactive

A. L. Spelta, *EniChem Elastomer, Milano, Italy,* Acrylic elastomers (under Elastomers, synthetic)

David M. Sturmer, *Eastman Kodak Company, Rochester, New York,* Dyes, sensitizing

Boyce Sutton, Jr., *Sybron Chemicals Inc., Wellford, South Carolina,* Dye carriers

E. Von Halle, *Martin Marietta Energy Systems, Oak Ridge, Tennessee,* Diffusion separation methods

H. C. Wang, *Exxon Chemical Company, Linden, New Jersey,* Butyl rubber (under Elastomers, synthetic)

Thomas J. Ward, *Clarkson University, Potsdam, New York,* Economic evaluation

Zeno Wicks, Jr., *Consultant, Las Cruces, New Mexico,* Drying oils

W. K. Witsiepe, *Consultant, Louisville, Kentucky,* Polychloroprene (under Elastomers, synthetic)

Allen R. Worm, *3M Company, St. Paul, Minnesota,* Fluorocarbon elastomers (under Elastomers, synthetic)

NOTE ON CHEMICAL ABSTRACTS SERVICE REGISTRY NUMBERS AND NOMENCLATURE

Chemical Abstracts Service (CAS) Registry Numbers are unique numerical identifiers assigned to substances recorded in the CAS Registry System. They appear in brackets in the *Chemical Abstracts* (CA) substance and formula indexes following the names of compounds. A single compound may have synonyms in the chemical literature. A simple compound like phenethylamine can be named β-phenylethylamine or, as in *Chemical Abstracts*, benzeneethanamine. The usefulness of the *Encyclopedia* depends on accessibility through the most common correct name of a substance. Because of this diversity in nomenclature careful attention has been given to the problem in order to assist the reader as much as possible, especially in locating the systematic CA index name by means of the Registry Number. For this purpose, the reader may refer to the CAS Registry Handbook—Number Section which lists in numerical order the Registry Number with the *Chemical Abstracts* index name and the molecular formula; eg, **458-88-8**, Piperidine, 2-propyl-, (*S*)-, $C_8H_{17}N$; in the *Encyclopedia* this compound would be found under its common name, coniine [*458-88-8*]. Alternatively, this information can be retrieved electronically from CAS Online. In many cases molecular formulas have also been provided in the *Encyclopedia* text to facilitate electronic searching. The Registry Number is a valuable link for the reader in retrieving additional published information on substances and also as a point of access for on-line data bases.

In all cases, the CAS Registry Numbers have been given for title compounds in articles and for all compounds in the index. All specific substances indexed in *Chemical Abstracts* since 1965 are included in the CAS Registry System as are a large number of substances derived from a variety of reference works. The CAS Registry System identifies a substance on the basis of an unambiguous computer-language description of its molecular structure including stereochemical detail. The Registry Number is a machine-checkable number (like a Social Security number) assigned in sequential order to each substance as it enters the registry system. The value of the number lies in the fact that it is a concise and unique means of substance identification, which is independent of, and therefore bridges, many systems of chemical nomenclature. For polymers, one Registry Number may be used for the entire family; eg, polyoxyethylene (20) sorbitan monolaurate has the same number as all of its polyoxyethylene homologues.

Cross-references are inserted in the index for many common names and for some systematic names. Trademark names appear in the index. Names that are incorrect, misleading, or ambiguous are avoided. Formulas are given very frequently in the text to help in identifying compounds. The spelling and form used, even for industrial names, follow American chemical usage, but not always the usage of *Chemical Abstracts* (eg, *coniine* is used instead of *(S)-2-propylpiperidine, aniline* instead of *benzenamine*, and *acrylic acid* instead of *2-propenoic acid*).

There are variations in representation of rings in different disciplines. The dye industry does not designate aromaticity or double bonds in rings. All double bonds and aromaticity are shown in the *Encyclopedia* as a matter of course. For example, tetralin has an aromatic ring and a saturated ring and its structure

appears in the *Encyclopedia* with its common name, Registry Number enclosed in brackets, and parenthetical CA index name, ie, tetralin [*119-64-2*] (1,2,3,4-tetrahydronaphthalene). With names and structural formulas, and especially with CAS Registry Numbers, the aim is to help the reader have a concise means of substance identification.

CONVERSION FACTORS, ABBREVIATIONS, AND UNIT SYMBOLS

SI Units (Adopted 1960)

The International System of Units (abbreviated SI), is being implemented throughout the world. This measurement system is a modernized version of the MKSA (meter, kilogram, second, ampere) system, and its details are published and controlled by an international treaty organization (The International Bureau of Weights and Measures) (1).

SI units are divided into three classes:

BASE UNITS

length	meter[†] (m)
mass	kilogram (kg)
time	second (s)
electric current	ampere (A)
thermodynamic temperature[‡]	kelvin (K)
amount of substance	mole (mol)
luminous intensity	candela (cd)

SUPPLEMENTARY UNITS

plane angle	radian (rad)
solid angle	steradian (sr)

[†]The spellings "metre" and "litre" are preferred by ASTM; however, "-er" is used in the *Encyclopedia*.

[‡]Wide use is made of Celsius temperature (t) defined by

$$t = T - T_0$$

where T is the thermodynamic temperature, expressed in kelvin, and $T_0 = 273.15$ K by definition. A temperature interval may be expressed in degrees Celsius as well as in kelvin.

DERIVED UNITS AND OTHER ACCEPTABLE UNITS

These units are formed by combining base units, supplementary units, and other derived units (2–4). Those derived units having special names and symbols are marked with an asterisk in the list below.

Quantity	Unit	Symbol	Acceptable equivalent
*absorbed dose	gray	Gy	J/kg
acceleration	meter per second squared	m/s^2	
*activity (of a radionuclide)	becquerel	Bq	1/s
area	square kilometer	km^2	
	square hectometer	hm^2	ha (hectare)
	square meter	m^2	
concentration (of amount of substance)	mole per cubic meter	mol/m^3	
current density	ampere per square meter	$A//m^2$	
density, mass density	kilogram per cubic meter	kg/m^3	g/L; mg/cm^3
dipole moment (quantity)	coulomb meter	C·m	
*dose equivalent	sievert	Sv	J/kg
*electric capacitance	farad	F	C/V
*electric charge, quantity of electricity	coulomb	C	A·s
electric charge density	coulomb per cubic meter	C/m^3	
*electric conductance	siemens	S	A/V
electric field strength	volt per meter	V/m	
electric flux density	coulomb per square meter	C/m^2	
*electric potential, potential difference, electromotive force	volt	V	W/A
*electric resistance	ohm	Ω	V/A
*energy, work, quantity of heat	megajoule	MJ	
	kilojoule	kJ	
	joule	J	N·m
	electronvolt[†]	eV[†]	
	kilowatt-hour[†]	kW·h[†]	
energy density	joule per cubic meter	J/m^3	
*force	kilonewton	kN	
	newton	N	$kg·m/s^2$

[†]This non-SI unit is recognized by the CIPM as having to be retained because of practical importance or use in specialized fields (1).

Quantity	Unit	Symbol	Acceptable equivalent
*frequency	megahertz	MHz	
	hertz	Hz	1/s
heat capacity, entropy	joule per kelvin	J/K	
heat capacity (specific), specific entropy	joule per kilogram kelvin	J/(kg·K)	
heat transfer coefficient	watt per square meter kelvin	W/(m²·K)	
*illuminance	lux	lx	lm/m²
*inductance	henry	H	Wb/A
linear density	kilogram per meter	kg/m	
luminance	candela per square meter	cd/m²	
*luminous flux	lumen	lm	cd·sr
magnetic field strength	ampere per meter	A/m	
*magnetic flux	weber	Wb	V·s
*magnetic flux density	tesla	T	Wb/m²
molar energy	joule per mole	J/mol	
molar entropy, molar heat capacity	joule per mole kelvin	J/(mol·K)	
moment of force, torque	newton meter	N·m	
momentum	kilogram meter per second	kg·m/s	
permeability	henry per meter	H/m	
permittivity	farad per meter	F/m	
*power, heat flow rate, radiant flux	kilowatt	kW	
	watt	W	J/s
power density, heat flux density, irradiance	watt per square meter	W/m²	
*pressure, stress	megapascal	MPa	
	kilopascal	kPa	
	pascal	Pa	N/m²
sound level	decibel	dB	
specific energy	joule per kilogram	J/kg	
specific volume	cubic meter per kilogram	m³/kg	
surface tension	newton per meter	N/m	
thermal conductivity	watt per meter kelvin	W/(m·K)	
velocity	meter per second	m/s	
	kilometer per hour	km/h	
viscosity, dynamic	pascal second	Pa·s	
	millipascal second	mPa·s	
viscosity, kinematic	square meter per second	m²/s	
	square millimeter per second	mm²/s	

Quantity	Unit	Symbol	Acceptable equivalent
volume	cubic meter	m^3	
	cubic decimeter	dm^3	L (liter) (5)
	cubic centimeter	cm^3	mL
wave number	1 per meter	m^{-1}	
	1 per centimeter	cm^{-1}	

In addition, there are 16 prefixes used to indicate order of magnitude, as follows:

Multiplication factor	Prefix	Symbol	Note
10^{18}	exa	E	
10^{15}	peta	P	
10^{12}	tera	T	
10^9	giga	G	
10^6	mega	M	
10^3	kilo	k	
10^2	hecto	h^a	[a]Although hecto, deka, deci, and centi
10	deka	da^a	are SI prefixes, their use should be
10^{-1}	deci	d^a	avoided except for SI unit-multiples
10^{-2}	centi	c^a	for area and volume and nontech-
10^{-3}	milli	m	nical use of centimeter, as for body
10^{-6}	micro	μ	and clothing measurement.
10^{-9}	nano	n	
10^{-12}	pico	p	
10^{-15}	femto	f	
10^{-18}	atto	a	

For a complete description of SI and its use the reader is referred to ASTM E 380 (4) and the article UNITS AND CONVERSION FACTORS which appears in Vol. 24.

A representative list of conversion factors from non-SI to SI units is presented herewith. Factors are given to four significant figures. Exact relationships are followed by a dagger. A more complete list is given in the latest editions of ASTM E 380 (4) and ANSI Z210.1 (6).

Conversion Factors to SI Units

To convert from	To	Multiply by
acre	square meter (m^2)	4.047×10^3
angstrom	meter (m)	1.0×10^{-10}[†]
are	square meter (m^2)	1.0×10^{2}[†]

[†]Exact.

To convert from	To	Multiply by
astronomical unit	meter (m)	1.496×10^{11}
atmosphere, standard	pascal (Pa)	1.013×10^{5}
bar	pascal (Pa)	$1.0 \times 10^{5\dagger}$
barn	square meter (m^2)	$1.0 \times 10^{-28\dagger}$
barrel (42 U.S. liquid gallons)	cubic meter (m^3)	0.1590
Bohr magneton (μ_B)	J/T	9.274×10^{-24}
Btu (International Table)	joule (J)	1.055×10^{3}
Btu (mean)	joule (J)	1.056×10^{3}
Btu (thermochemical)	joule (J)	1.054×10^{3}
bushel	cubic meter (m^3)	3.524×10^{-2}
calorie (International Table)	joule (J)	4.187
calorie (mean)	joule (J)	4.190
calorie (thermochemical)	joule (J)	$4.184^{\dagger}$
centipoise	pascal second (Pa·s)	$1.0 \times 10^{-3\dagger}$
centistokes	square millimeter per second (mm^2/s)	$1.0^{\dagger}$
cfm (cubic foot per minute)	cubic meter per second (m^3/s)	4.72×10^{-4}
cubic inch	cubic meter (m^3)	1.639×10^{-5}
cubic foot	cubic meter (m^3)	2.832×10^{-2}
cubic yard	cubic meter (m^3)	0.7646
curie	becquerel (Bq)	$3.70 \times 10^{10\dagger}$
debye	coulomb meter (C·m)	3.336×10^{-30}
degree (angle)	radian (rad)	1.745×10^{-2}
denier (international)	kilogram per meter (kg/m)	1.111×10^{-7}
	tex‡	0.1111
dram (apothecaries')	kilogram (kg)	3.888×10^{-3}
dram (avoirdupois)	kilogram (kg)	1.772×10^{-3}
dram (U.S. fluid)	cubic meter (m^3)	3.697×10^{-6}
dyne	newton (N)	$1.0 \times 10^{-5\dagger}$
dyne/cm	newton per meter (N/m)	$1.0 \times 10^{-3\dagger}$
electronvolt	joule (J)	1.602×10^{-19}
erg	joule (J)	$1.0 \times 10^{-7\dagger}$
fathom	meter (m)	1.829
fluid ounce (U.S.)	cubic meter (m^3)	2.957×10^{-5}
foot	meter (m)	$0.3048^{\dagger}$
footcandle	lux (lx)	10.76
furlong	meter (m)	2.012×10^{-2}
gal	meter per second squared (m/s^2)	$1.0 \times 10^{-2\dagger}$
gallon (U.S. dry)	cubic meter (m^3)	4.405×10^{-3}
gallon (U.S. liquid)	cubic meter (m^3)	3.785×10^{-3}
gallon per minute (gpm)	cubic meter per second (m^3/s)	6.309×10^{-5}
	cubic meter per hour (m^3/h)	0.2271

†Exact.
‡See footnote on p. xiv.

To convert from	To	Multiply by
gauss	tesla (T)	1.0×10^{-4}
gilbert	ampere (A)	0.7958
gill (U.S.)	cubic meter (m^3)	1.183×10^{-4}
grade	radian	1.571×10^{-2}
grain	kilogram (kg)	6.480×10^{-5}
gram force per denier	newton per tex (N/tex)	8.826×10^{-2}
hectare	square meter (m^2)	$1.0 \times 10^{4\dagger}$
horsepower (550 ft·lbf/s)	watt (W)	7.457×10^2
horespower (boiler)	watt (W)	9.810×10^3
horsepower (electric)	watt (W)	$7.46 \times 10^{2\dagger}$
hundredweight (long)	kilogram (kg)	50.80
hundredweight (short)	kilogram (kg)	45.36
inch	meter (m)	$2.54 \times 10^{-2\dagger}$
inch of mercury (32°F)	pascal (Pa)	3.386×10^3
inch of water (39.2°F)	pascal (Pa)	2.491×10^2
kilogram-force	newton (N)	9.807
kilowatt hour	megajoule (MJ)	$3.6^\dagger$
kip	newton(N)	4.448×10^3
knot (international)	meter per second (m/S)	0.5144
lambert	candela per square meter (cd/m^3)	3.183×10^3
league (British nautical)	meter (m)	5.559×10^3
league (statute)	meter (m)	4.828×10^3
light year	meter (m)	9.461×10^{15}
liter (for fluids only)	cubic meter (m^3)	$1.0 \times 10^{-3\dagger}$
maxwell	weber (Wb)	$1.0 \times 10^{-8\dagger}$
micron	meter (m)	$1.0 \times 10^{-6\dagger}$
mil	meter (m)	$2.54 \times 10^{-5\dagger}$
mile (statute)	meter (m)	1.609×10^3
mile (U.S. nautical)	meter (m)	$1.852 \times 10^{3\dagger}$
mile per hour	meter per second (m/s)	0.4470
millibar	pascal (Pa)	1.0×10^2
millimeter of mercury (0°C)	pascal (Pa)	$1.333 \times 10^{2\dagger}$
minute (angular)	radian	2.909×10^{-4}
myriagram	kilogram (kg)	10
myriameter	kilometer (km)	10
oersted	ampere per meter (A/m)	79.58
ounce (avoirdupois)	kilogram (kg)	2.835×10^{-2}
ounce (troy)	kilogram (kg)	3.110×10^{-2}
ounce (U.S. fluid)	cubic meter (m^3)	2.957×10^{-5}
ounce-force	newton (N)	0.2780
peck (U.S.)	cubic meter (m^3)	8.810×10^{-3}
pennyweight	kilogram (kg)	1.555×10^{-3}
pint (U.S. dry)	cubic meter (m^3)	5.506×10^{-4}

†Exact.

To convert from	To	Multiply by
pint (U.S. liquid)	cubic meter (m³)	4.732×10^{-4}
poise (absolute viscosity)	pascal second (Pa·s)	$0.10^{\dagger}$
pound (avoirdupois)	kilogram (kg)	0.4536
pound (troy)	kilogram (kg)	0.3732
poundal	newton (N)	0.1383
pound-force	newton (N)	4.448
pound force per square inch (psi)	pascal (Pa)	6.895×10^3
quart (U.S. dry)	cubic meter (m³)	1.101×10^{-3}
quart (U.S. liquid)	cubic meter (m³)	9.464×10^{-4}
quintal	kilogram (kg)	$1.0 \times 10^{2\dagger}$
rad	gray (Gy)	$1.0 \times 10^{-2\dagger}$
rod	meter (m)	5.029
roentgen	coulomb per kilogram (C/kg)	2.58×10^{-4}
second (angle)	radian (rad)	$4.848 \times 10^{-6\dagger}$
section	square meter (m²)	2.590×10^6
slug	kilogram (kg)	14.59
spherical candle power	lumen (lm)	12.57
square inch	square meter (m²)	6.452×10^{-4}
square foot	square meter (m²)	9.290×10^{-2}
square mile	square meter (m²)	2.590×10^6
square yard	square meter (m²)	0.8361
stere	cubic meter (m³)	$1.0^{\dagger}$
stokes (kinematic viscosity)	square meter per second (m²/s)	$1.0 \times 10^{-4\dagger}$
tex	kilogram per meter (kg/m)	$1.0 \times 10^{-6\dagger}$
ton (long, 2240 pounds)	kilogram (kg)	1.016×10^3
ton (metric) (tonne)	kilogram (kg)	$1.0 \times 10^{3\dagger}$
ton (short, 2000 pounds)	kilogram (kg)	9.072×10^2
torr	pascal (Pa)	1.333×10^2
unit pole	weber (Wb)	1.257×10^{-7}
yard	meter (m)	$0.9144^{\dagger}$

†Exact.

Abbreviations and Unit Symbols

Following is a list of common abbreviations and unit symbols used in the *Encyclopedia*. In general they agree with those listed in *American National Standard Abbreviations for Use on Drawings and in Text (ANSI Y1.1)* (6) and *American National Standard Letter Symbols for Units in Science and Technology (ANSI Y10)* (6). Also included is a list of acronyms for a number of private and government organizations as well as common industrial solvents, polymers, and other chemicals.

Rules for Writing Unit Symbols (4):

1. Unit symbols are printed in upright letters (roman) regardless of the type style used in the surrounding text.
2. Unit symbols are unaltered in the plural.
3. Unit symbols are not followed by a period except when used at the end of a sentence.
4. Letter unit symbols are generally printed lower-case (for example, cd for candela) unless the unit name has been derived from a proper name, in which case the first letter of the symbol is capitalized (W, Pa). Prefixes and unit symbols retain their prescribed form regardless of the surrounding typography.
5. In the complete expression for a quantity, a space should be left between the numerical value and the unit symbol. For example, write 2.37 lm, *not* 2.37lm, and 35 mm, *not* 35mm. When the quantity is used in an adjectival sense, a hyphen is often used, for example, 35-mm film. *Exception:* No space is left between the numerical value and the symbols for degree, minute, and second of plane angle, degree Celsius, and the percent sign.
6. No space is used between the prefix and unit symbol (for example, kg).
7. Symbols, not abbreviations, should be used for units. For example, use "A," not "amp," for ampere.
8. When multiplying unit symbols, use a raised dot:

$$N \cdot m \quad \text{for} \quad \text{newton meter}$$

   In the case of W·h, the dot may be omitted, thus:

$$Wh$$

   An exception to this practice is made for computer printouts, automatic typewriter work, etc, where the raised dot is not possible, and a dot on the line may be used.
9. When dividing unit symbols, use one of the following forms:

$$m/s \quad or \quad m \cdot s^{-1} \quad or \quad \frac{m}{s}$$

   In no case should more than one slash be used in the same expression unless parentheses are inserted to avoid ambiguity. For example: write:

$$J/(mol \cdot K) \quad or \quad J \cdot mol^{-1} \cdot K^{-1} \quad or \quad (J/mol)/K$$

   but *not*

$$J/mol/K$$

10. Do not mix symbols and unit names in the same expression. Write:

$$\text{joules per kilogram} \quad or \quad \text{J/kg} \quad or \quad \text{J·kg}^{-1}$$

but *not*

$$\text{joules/kilogram} \quad nor \quad \text{joules/kg} \quad nor \quad \text{joules·kg}^{-1}$$

ABBREVIATIONS AND UNITS

A	ampere	AOAC	Association of Official Analytical Chemists
A	anion (eg, HA)		
A	mass number	AOCS	Americal Oil Chemists' Society
a	atto (prefix for 10^{-18})		
AATCC	American Association of Textile Chemists and Colorists	APHA	American Public Health Association
		API	American Petroleum Institute
ABS	acrylonitrile–butadiene–styrene	aq	aqueous
abs	absolute	Ar	aryl
ac	alternating current, *n.*	*ar-*	aromatic
a-c	alternating current, *adj.*	*as-*	asymmetric(al)
ac-	alicyclic	ASHRAE	American Society of Heating, Refrigerating, and Air Conditioning Engineers
acac	acetylacetonate		
ACGIH	American Conference of Governmental Industrial Hygienists		
		ASM	American Society for Metals
ACS	American Chemical Society	ASME	American Society of Mechanical Engineers
AGA	American Gas Association		
Ah	ampere hour	ASTM	American Society for Testing and Materials
AIChE	American Institute of Chemical Engineers		
		at no.	atomic number
AIME	American Institute of Mining, Metallurgical, and Petroleum Engineers	at wt	atomic weight
		av(g)	average
		AWS	American Welding Society
		b	bonding orbital
AIP	American Institute of Physics	bbl	barrel
		bcc	body-centered cubic
AISI	American Iron and Steel Institute	BCT	body-centered tetragonal
		Bé	Baumé
alc	alcohol(ic)	BET	Brunauer-Emmett-Teller (adsorption equation)
Alk	alkyl		
alk	alkaline (not alkali)	bid	twice daily
amt	amount	Boc	*t*-butyloxycarbonyl
amu	atomic mass unit	BOD	biochemical (biological) oxygen demand
ANSI	American National Standards Institute		
		bp	boiling point
AO	atomic orbital	Bq	becquerel

C	coulomb	DIN	Deutsche Industrie Normen
°C	degree Celsius		
C-	denoting attachment to carbon	*dl*-; DL-	racemic
		DMA	dimethylacetamide
c	centi (prefix for 10^{-2})	DMF	dimethylformamide
c	critical	DMG	dimethyl glyoxime
ca	circa (approximately)	DMSO	dimethyl sulfoxide
cd	candela; current density; circular dichroism	DOD	Department of Defense
		DOE	Department of Energy
CFR	Code of Federal Regulations	DOT	Department of Transportation
		DP	degree of polymerization
cgs	centimeter-gram-second	dp	dew point
CI	Color Index	DPH	diamond pyramid hardness
cis-	isomer in which substituted groups are on same side of double bond between C atoms	dstl(d)	distill(ed)
		dta	differential thermal analysis
cl	carload	(*E*)-	entgegen; opposed
cm	centimeter	ϵ	dielectric constant (unitless number)
cmil	circular mil		
cmpd	compound	*e*	electron
CNS	central nervous system	ECU	electrochemical unit
CoA	coenzyme A	ed.	edited, edition, editor
COD	chemical oxygen demand	ED	effective dose
coml	commercial(ly)	EDTA	ethylenediaminetetra-acetic acid
cp	chemically pure		
cph	close-packed hexagonal	emf	electromotive force
CPSC	Consumer Product Safety Commission	emi	electromagnetic interference
		emu	electromagnetic unit
cryst	crystalline	en	ethylene diamine
cub	cubic	EPA	Environmental Protection Agency
D	Debye		
D-	denoting configurational relationship	epr	electron paramagnetic resonance
d	differential operator	eq.	equation
d	day; deci (prefix for 10^{-1})	esca	electron spectroscopy for chemical analysis
d-	*dextro*-, dextrorotatory		
da	deka (prefix for 10^1)	esp	especially
dB	decibel	esr	electron-spin resonance
dc	direct current, *n.*	est(d)	estimate(d)
d-c	direct current, *adj.*	estn	estimation
dec	decompose	esu	electrostatic unit
detd	determined	exp	experiment, experimental
detn	determination	ext(d)	extract(ed)
Di	didymium, a mixture of all lanthanons	F	farad (capacitance)
		F	faraday (96,487 C)
dia	diameter	f	femto (prefix for 10^{-15})

FAO	Food and Agriculture Organization (United Nations)	hyd	hydrated, hydrous
		hyg	hygroscopic
		Hz	hertz
fcc	face-centered cubic	^{i} (eg, Pr^i)	iso (eg, isopropyl)
FDA	Food and Drug Administration	i-	inactive (eg, i-methionine)
		IACS	International Annealed Copper Standard
FEA	Federal Energy Administration	ibp	initial boiling point
FHSA	Federal Hazardous Substances Act	IC	integrated circuit
		ICC	Interstate Commerce Commission
fob	free on board		
fp	freezing point	ICT	International Critical Table
FPC	Federal Power Commission		
		ID	inside diameter; infective dose
FRB	Federal Reserve Board		
frz	freezing	ip	intraperitoneal
G	giga (prefix for 10^9)	IPS	iron pipe size
G	gravitational constant = 6.67×10^{11} N·m^2/kg^2	ir	infrared
		IRLG	Interagency Regulatory Liaison Group
g	gram		
(g)	gas, only as in H_2O(g)	ISO	International Organization Standardization
g	gravitational acceleration		
gc	gas chromatography	ITS-90	International Temperature Scale (NIST)
gem-	geminal		
glc	gas–liquid chromatography	IU	International Unit
		IUPAC	International Union of Pure and Applied Chemistry
g-mol wt; gmw	gram-molecular weight		
GNP	gross national product	IV	iodine value
gpc	gel-permeation chromatography	iv	intravenous
		J	joule
GRAS	Generally Recognized as Safe	K	kelvin
		k	kilo (prefix for 10^3)
grd	ground	kg	kilogram
Gy	gray	L	denoting configurational relationship
H	henry		
h	hour; hecto (prefix for 10^2)	L	liter (for fluids only) (5)
ha	hectare	l-	levo-, levorotatory
HB	Brinell hardness number	(l)	liquid, only as in NH_3(l)
Hb	hemoglobin	LC$_{50}$	conc lethal to 50% of the animals tests
hcp	hexagonal close-packed		
hex	hexagonal	LCAO	linear combination of atomic orbitals
HK	Knoop hardness number		
hplc	high performance liquid chromatography	lc	liquid chromatography
		LCD	liquid crystal display
HRC	Rockwell hardness (C scale)	lcl	less than carload lots
		LD$_{50}$	dose lethal to 50% of the animals tested
HV	Vickers hardness number		

LED	light-emitting diode	N-	denoting attachment to nitrogen
liq	liquid		
lm	lumen	n (as n_D^{20})	index of refraction (for 20°C and sodium light)
ln	logarithm (natural)		
LNG	liquefied natural gas	^{n} (as Bu^n),	
log	logarithm (common)	n-	normal (straight-chain structure)
LPG	liquefied petroleum gas		
ltl	less than truckload lots	n	neutron
lx	lux	n	nano (prefix for 10^9)
M	mega (prefix for 10^6); metal (as in MA)	na	not available
		NAS	National Academy of Sciences
M	molar; actual mass		
$\overline{M}_w$	weight-average mol wt	NASA	National Aeronautics and Space Administration
$\overline{M}_n$	number-average mol wt		
m	meter; milli (prefix for 10^{-3})	nat	natural
		ndt	nondestructive testing
m	molal	neg	negative
m-	meta	NF	*National Formulary*
max	maximum	NIH	National Institutes of Health
MCA	Chemical Manufacturers' Association (was Manufacturing Chemists Association)	NIOSH	National Institute of Occupational Safety and Health
MEK	methyl ethyl ketone	NIST	National Institute of Standards and Technology (formerly National Bureau of Standards)
meq	milliequivalent		
mfd	manufactured		
mfg	manufacturing		
mfr	manufacturer		
MIBC	methyl isobutyl carbinol		
MIBK	methyl isobutyl ketone	nmr	nuclear magnetic resonance
MIC	minimum inhibiting concentration	NND	New and Nonofficial Drugs (AMA)
min	minute; minimum	no.	number
mL	milliliter	NOI-(BN)	not otherwise indexed (by name)
MLD	minimum lethal dose		
MO	molecular orbital	NOS	not otherwise specified
mo	month	nqr	nuclear quadruple resonance
mol	mole		
mol wt	molecular weight	NRC	Nuclear Regulatory Commission; National Research Council
mp	melting point		
MR	molar refraction		
ms	mass spectrometry	NRI	New Ring Index
MSDS	material safety data sheet	NSF	National Science Foundation
mxt	mixture		
μ	micro (prefix for 10^{-6})	NTA	nitrilotriacetic acid
N	newton (force)	NTP	normal temperature and pressure (25°C and 101.3 kPa or 1 atm)
N	normal (concentration); neutron number		

NTSB	National Transportation Safety Board	qv	quod vide (which see)
O-	denoting attachment to oxygen	R	univalent hydrocarbon radical
o-	ortho	(*R*)-	rectus (clockwise configuration)
OD	outside diameter	*r*	precision of data
OPEC	Organization of Petroleum Exporting Countries	rad	radian; radius
o-phen	*o*-phenanthridine	RCRA	Resource Conservation and Recovery Act
OSHA	Occupational Safety and Health Administration	rds	rate-determining step
		ref.	reference
owf	on weight of fiber	rf	radio frequency, *n*.
Ω	ohm	r-f	radio frequency, *adj*.
P	peta (prefix for 10^{15})	rh	relative humidity
p	pico (prefix for 10^{-12})	RI	Ring Index
p-	para	rms	root-mean square
p	proton	rpm	rotations per minute
p.	page	rps	revolutions per second
Pa	pascal (pressure)	RT	room temperature
PEL	personal exposure limit based on an 8-h exposure	RTECS	Registry of Toxic Effects of Chemical Substances
pd	potential difference	^s (eg, Bu^s); *sec*-	secondary (eg, secondary butyl)
pH	negative logarithm of the effective hydrogen ion concentration	S	siemens
		(*S*)-	sinister (counterclockwise configuration)
phr	parts per hundred of resin (rubber)	*S*-	denoting attachment to sulfur
p-i-n	positive-intrinsic-negative		
pmr	proton magnetic resonance	*s*-	symmetric(al)
p-n	positive-negative	s	second
po	per os (oral)	(s)	solid, only as in $H_2O(s)$
POP	polyoxypropylene	SAE	Society of Automotive Engineers
pos	positive		
pp.	pages	SAN	styrene-acrylonitrile
ppb	parts per billion (10^9)	sat(d)	saturate(d)
ppm	parts per million (10^6)	satn	saturation
ppmv	parts per million by volume	SBS	styrene–butadiene–styrene
		sc	subcutaneous
ppmwt	parts per million by weight	SCF	self-consistent field; standard cubic feet
PPO	poly(phenyl oxide)		
ppt(d)	precipitate(d)		
pptn	precipitation	Sch	Schultz number
Pr (no.)	foreign prototype (number)	sem	scanning electron microscope(y)
pt	point; part		
PVC	poly(vinyl chloride)	SFs	Saybolt Furol seconds
pwd	powder	sl sol	slightly soluble
py	pyridine	sol	soluble

soln	solution	*trans-*	isomer in which
soly	solubility		substituted groups are
sp	specific; species		on opposite sides of
sp gr	specific gravity		double bond between C
sr	steradian		atoms
std	standard	TSCA	Toxic Substances Control
STP	standard temperature and		Act
	pressure (0°C and 101.3	TWA	time-weighted average
	kPa)	Twad	Twaddell
sub	sublime(s)	UL	Underwriters' Laboratory
SUs	Saybolt Universal seconds	USDA	United States Department
syn	synthetic		of Agriculture
^{t} (eg, Bu^t),		USP	*United States*
t-, tert-	tertiary (eg, tertiary		*Pharmacopeia*
	butyl)	uv	ultraviolet
T	tera (prefix for 10^{12}); tesla	V	volt (emf)
	(magnetic flux density)	var	variable
t	metric ton (tonne)	*vic-*	vicinal
t	temperature	vol	volume (not volatile)
TAPPI	Technical Association of	vs	versus
	the Pulp and Paper	v sol	very soluble
	Industry	W	watt
TCC	Tagliabue closed up	Wb	weber
tex	tex (linear density)	Wh	watt hour
T_g	glass-transition	WHO	World Health
	temperature		Organization (United
tga	thermogravimetric		Nations)
	analysis	wk	week
THF	tetrahydrofuran	yr	year
tlc	thin layer chromatography	(*Z*)-	zusammen; together;
TLV	threshold limit value		atomic number

Non-SI (Unacceptable and Obsolete) Units		Use
Å	angstrom	nm
at	atmosphere, technical	Pa
atm	atmosphere, standard	Pa
b	barn	cm^2
bar†	bar	Pa
bbl	barrel	m^3
bhp	brake horsepower	W
Btu	British thermal unit	J
bu	bushel	m^3; L
cal	calorie	J
cfm	cubic foot per minute	m^3/s
Ci	curie	Bq
cSt	centistokes	mm^2/s
c/s	cycle per second	Hz

†Do not use bar (10^5Pa) or millibar (10^2Pa) because they are not SI units, and are accepted internationally only for a limited time in special fields because of existing usage.

Non-SI (Unacceptable and Obsolete) Units		Use
cu	cubic	exponential form
D	debye	$C \cdot m$
den	denier	tex
dr	dram	kg
dyn	dyne	N
dyn/cm	dyne per centimeter	mN/m
erg	erg	J
eu	entropy unit	J/K
°F	degree Fahrenheit	°C; K
fc	footcandle	lx
fl	footlambert	lx
fl oz	fluid ounce	m^3; L
ft	foot	m
ft·lbf	foot pound-force	J
gf den	gram-force per denier	N/tex
G	gauss	T
Gal	gal	m/s^2
gal	gallon	m^3; L
Gb	gilbert	A
gpm	gallon per minute	(m^3/s); (m^3/h)
gr	grain	kg
hp	horsepower	W
ihp	indicated horsepower	W
in.	inch	m
in. Hg	inch of mercury	Pa
in. H_2O	inch of water	Pa
in.-lbf	inch pound-force	J
kcal	kilo-calorie	J
kgf	kilogram-force	N
kilo	for kilogram	kg
L	lambert	lx
lb	pound	kg
lbf	pound-force	N
mho	mho	S
mi	mile	m
MM	million	M
mm Hg	millimeter of mercury	Pa
mμ	millimicron	nm
mph	miles per hour	km/h
μ	micron	μm
Oe	oersted	A/m
oz	ounce	kg
ozf	ounce-force	N
η	poise	$Pa \cdot s$
P	poise	$Pa \cdot s$
ph	phot	lx
psi	pounds-force per square inch	Pa
psia	pounds-force per square inch absolute	Pa
psig	pounds-force per square inch gage	Pa
qt	quart	m^3; L
°R	degree Rankine	K
rd	rad	Gy
sb	stilb	lx
SCF	standard cubic foot	m^3
sq	square	exponential form
thm	therm	J
yd	yard	m

BIBLIOGRAPHY

1. The International Bureau of Weights and Measures, BIPM (Parc de Saint-Cloud, France) is described in Appendix X2 of Ref. 4. This bureau operates under the exclusive supervision of the International Committee for Weights and Measures (CIPM).
2. *Metric Editorial Guide (ANMC-78-1)*, latest ed., American National Metric Council, 5410 Grosvenor Lane, Bethesda, Md. 20814, 1981.
3. *SI Units and Recommendations for the Use of Their Multiples and of Certain Other Units (ISO 1000-1981)*, American National Standards Institute, 1430 Broadway, New York, N.Y. 10018, 1981.
4. Based on *ASTM E 380-89a (Standard Practice for Use of the International System of Units (SI))*, American Society for Testing and Materials, 1916 Race Street, Philadelphia, Pa. 19103, 1989.
5. *Fed. Regist.*, Dec. 10, 1976 (41 FR 36414).
6. For ANSI address, see Ref. 3.

R. P. LUKENS
ASTM Committee E-43 on SI Practice

D

Continued

DEUTERIUM AND TRITIUM

Deuterium, **1**
Tritium, **17**

The element hydrogen [*12385-13-6*] (qv) has three known isotopes. These hydrogen species are identical in atomic number, ie, they have identical extranuclear electronic configurations ($1s^1$), but they differ in nuclear mass. Over 99.98% of the hydrogen in nature has a nucleus consisting of a single proton and, therefore, has mass 1 (symbol ^1H). Two heavier isotopes of hydrogen, present in small amounts in nature, of mass 2 and 3 are also known; these have nuclei consisting of one proton and one neutron (deuterium) or two neutrons (tritium). The three isotopes of hydrogen resemble each other closely in chemical and physical properties, but because the ratios of their masses are the largest for any set of isotopes in the periodic table, differences in chemical and physical properties exist that are larger than those encountered in any other set of isotopes. Whereas ordinary hydrogen and deuterium are stable isotopes, tritium is unstable and its nucleus undergoes radioactive decay (see RADIOACTIVE TRACERS; RADIOISOTOPES). The two heavy isotopes of hydrogen are always present to a small extent in any compound or substance containing the light isotope of hydrogen. Because both isotopically pure deuterium and tritum can be manufactured on a large industrial scale, a large variety of isotopically pure deuterium and tritium compounds is readily available. Heavy water, D_2O, is the most important compound of deuterium and the only form of deuterium produced and used on a large scale.

DEUTERIUM

Deuterium [*16873-17-9*] (symbol ^2H or D) occurs in nature in all hydrogen-containing compounds to the extent of about 0.0145 atom %. Small but real differences in the deuterium content of water from various sources (rain, snow, glaciers, freshwater, seawater from different oceans) can readily be detected, and variations in the natural abundance of deuterium resulting from evaporation, precipitation, and molecular exchange make it possible to draw far-reaching conclusions about the genesis and geological history of natural waters.

Molecular deuterium [7782-39-0], D_2, was first isolated in relatively pure form by Urey and co-workers at Columbia University in 1931 (1), and nearly pure D_2O was prepared by G. N. Lewis shortly thereafter by electrolysis (2). Subsequently applications of deuterium as a tracer for the path of hydrogen in biological systems were developed and became widely used. In physical organic chemistry, the differences in rates of reaction between corresponding ^1H and D compounds became an important tool for the elucidation of organic reaction mechanisms. The significance of deuterium isotope effects in a biological context attracted attention very early (3,4). The discovery in 1959 that it was possible to grow fully deuterated organisms (5,7) opened new areas of isotope chemistry and biology for exploration.

The recognition in 1940 that deuterium as heavy water [7789-20-0] has nuclear properties that make it a highly desirable moderator and coolant for nuclear reactors (qv) (8,9) fueled by uranium (qv) of natural isotopic composition stimulated the development of industrial processes for the manufacture of heavy water. Between 1940 and 1945 four heavy water production plants were operated by the United States Government, one in Canada at Trail, British Columbia, and three at the U.S. Army Ordinance Works operated by the DuPont Company at Morgantown, West Virginia; Childersburg, Alabama; and Dana, Indiana. The plant at Trail used chemical exchange between hydrogen gas and steam for the initial isotope separation followed by electrolysis for final concentration. The three plants in the United States used vacuum distillation of water for the initial separation followed by electrolysis. Details of these plants and their operations may be found in the literature (10).

In 1950 construction of a large-scale heavy water facility was initiated at the Savannah River site near Aiken, South Carolina, for the nuclear reactors operated there. The dual temperature exchange of deuterium between hydrogen sulfide and water (the GS process) was selected (11,12) for isotopic enrichment of deuterium. This plant was designed to produce 450 metric tons of heavy water per year, and a plant of similar capacity was constructed at Dana, Indiana. As of this writing, the plant at Savannah River is the only heavy water plant in operation in the United States, producing about 160 t of heavy water annually. Canadian D_2O capacity is about 1450 t/yr (13). France, Norway, Switzerland, and India also have small production plants. The continued production of heavy water is closely coupled to the use of heavy water in nuclear reactor technology. The general decline in the construction of new nuclear reactors has dimmed the prospects for the large-scale production of deuterium.

Physical Properties

Although the chemical and physical properties of all isotopes of an element are qualitatively the same, there are quantitative differences among them. The physical and chemical differences between the hydrogen isotopes are relatively much greater than those among the isotopes of all other elements because of the large relative differences in mass, ie, H:D:T = 1:2:3.

As in the case of hydrogen and tritium, deuterium exhibits nuclear spin isomerism (see MAGNETIC SPIN RESONANCE) (14). However, the spin of the deuteron [12597-73-8] is 1 instead of ½ as in the case of hydrogen and tritium. As a consequence, and in contrast to hydrogen, the ortho form of deuterium is more stable than the para form at low temperatures, and at normal temperatures the ratio of ortho- to para-deuterium is 2:1 in contrast to the 3:1 ratio for hydrogen.

The physical and thermodynamic properties of elemental hydrogen and deuterium and of their respective oxides illustrate the effect of isotopic mass differences.

Properties of Light and Heavy Hydrogen. Vapor pressures from the triple point to the critical point for hydrogen, deuterium, tritium, and the various diatomic combinations are listed in Table 1 (15). Data are presented for the equilibrium and normal states. The equilibrium state for these substances is the low temperature ortho–para composition existing at 20.39 K, the normal boiling point of normal hydrogen. The normal state is the high (above 200 K) temperature ortho–para composition, which remains essentially constant.

Thermodynamic data on H_2, the mixed hydrogen–deuterium molecule [13983-20-5], HD, and D_2, including values for entropy, enthalpy, free energy, and specific heat have been tabulated (16). Extensive PVT data are also presented in Reference 16 as are data on the equilibrium–temperature behavior of the ortho and para forms of H_2 and D_2. Some physical properties of liquid H_2 and D_2 at 20.4 K are presented in Table 2.

Properties of Light and Heavy Water. Selected physical properties of light and heavy water are listed in Table 3 (17). Thermodynamic properties are given in Table 4. The liquid plus vapor critical-temperature curve for $(xD_2O - (1 - x)H_2O)$ mixtures over the entire concentration range has been reported (28).

Table 1. Vapor Pressures and Triple and Critical Points of Hydrogen Isotopes[a,b]

Factor	e-H_2	n-H_2	HD	HT	e-D_2	n-D_2	DT	n-T_2
triple point, T_{tr}, K	13.81	13.96	16.60	17.62	18.69	18.73	21.71	20.62
vapor pressure, kPa[c]								
T_{tr}	7.04	7.20	12.37	14.59	17.13	17.14	19.43	21.60
20.0 K	93.22	90.31	51.02	38.50	29.66	29.41	21.98	
22.5 K	185.1	179.1	112.3	89.49	71.82	71.17	56.46	44.80
25.0 K	327.65	316.63	214.42	177.66	147.19	145.79	120.82	100.03
27.5 K	534.07	515.13	369.98	315.36	288.02	265.28	227.11	194.10
30.0 K	819.56	788.49	592.74	515.65	447.55	442.56	388.42	339.88
32.5 K	1203.4	1153.8	898.50	792.91	700.58	691.91	619.04	548.47
35.0 K			1307.4	1164.0	1044.7	1030.0	935.27	843.06
T_{cr}	1293.9	1298.0	1484.41	1570.53	1649.59	1664.79	1773.18	1850.24
critical point, T_{cr}, K	32.99	33.24	35.91	37.13	38.26	38.35	39.42	40.44

[a]Adapted from Ref. 15.
[b]The prefixes e- and n- refer to equilibrium and normal states. For T, data are available only for the normal state. The equilibrium state for these substances is the low temperature ortho-para composition existing at 20.39 K, the normal boiling point of normal hydrogen. The normal state is the ortho-para composition above 200 K. See text.
[c]To convert kPa to mm Hg, multiply by 7.5.

Table 3. Physical Properties of Light and Heavy Water

Property	H_2O	Ref.[a]	D_2O	Ref.[b]
molecular weight	18.015		20.028	
melting point, T_m, °C	0.00		3.81	
triple point, T_{tr}, °C	0.01		3.82	
temp of max density, °C	3.98		11.23	
normal boiling point, T_b, °C	100.00		101.42	
critical constants		19		20
temperature, °C	374.1		371.1	
pressure, MPa[c]	22.12		21.88	
volume, cm^3/mol	55.3		55.0	
density at 25°C, g/cm^3	0.99701		1.1044	
vapor pressure, liquid at 25°C, kPa[d]	3.166		2.734	
molar volumes, cm^3/mol				
solid at T_m	19.65		18.679	
liquid at T_m	18.018		18.118	
liquid at T_{max} density	18.016		18.110	
liquid at 25°C	18.069		18.134	
coefficients of thermal expansion, °C^{-1}				
solid at T_m	1.39×10^{-4}		1.39×10^{-4}	
liquid at T_m	-5.9×10^{-5}		-3.2×10^{-5}	
liquid at 25°C	26.2×10^{-5}		21.8×10^{-5}	
compressibility at 20°C, Pa^{-1d}	4.45	19	4.59	19
crystallographic parameters at 0°C, nm		21		21
a	0.45228		0.45258	
c	0.73673		0.73689	
c/a	0.1629		0.1628	
length of the hydrogen bond, nm	0.2765		0.2760	
dipole moment, C·m[e]				
benzene soln at 25°C	5.87×10^{-30}		5.94×10^{-30}	
vapor at 100–200°C	6.14×10^{-30}		6.14×10^{-30}	
dielectric constant at 25°C	78.304		77.937	
refractive index, n_D^{20}	1.3330		1.3283	
polarizability of vapor near 100°C, cm^3/mol	58.5		61.7	
viscosity at 25°C, mPa·s($=$cP)	0.8903	22	1.107	
surface tension at 25°C, mN/m ($=$dyn/cm)	71.97		71.93	
moments of inertia, 10^{40} g·cm^2				
I_e^A	1.0224		1.833	
I_e^B	1.9180		3.841	
I_e^C	2.9404		5.674	
vibrational fundamentals, cm^{-1}		23		23
v_1	3657.05		2671.69	
v_2	1594.59		1178.33	
v_3	3755.79		2788.03	
ion product constant at 25°C	1.01×10^{-14}	24	1.11×10^{-15}	24
heat of ion product at 25°C, kJ/mol[f]	56.27	24	60.33	24

[a]Unless otherwise indicated, data are from Ref. 17.
[b]Unless otherwise indicated, data are from Ref. 18.
[c]To convert MPa to atm, multiply by 10.1.
[d]To convert kPa to mm Hg, multiply by 7.5.
[e]To convert C·m to debye, multiply by 3.00×10^{29}.
[f]To convert Joules to calories, divide by 4.184.

Table 4. Thermodynamic Properties of Light and Heavy Water

Parameter[a]	H_2O	Ref.[b]	D_2O	Ref.[c]
ΔH_{fus} at T_m, kJ/mol	−6.008		6.339	16
ΔS_{fus} at T_m, J/K	22.0		22.6_7	
ΔH_{vap} at 3.82°C, kJ/mol	44.77		46.8	
ΔE_{vap} at 3.82°C, kJ/mol	42.47	d	44.18	d
ΔS_{vap} at 3.82°C, J/K	161.62		167.8	
ΔH_{subl} at T_{tr}, kJ/mol	50.91		52.84	
ΔE_{subl} at T_{tr}, jJ/mol	48.66	d	50.54	
ΔS_{subl} at T_{tr}, J/K	186.4		190.7	
ΔH^*_{vap} at 25°C, kJ/mol	44.02	25	45.39	
ΔG^*_{vap} at 25°C, kJ/mol	8.62	25	8.95	19
ΔS^*_{vap} at 25°C, J/K	118.8	d	122.25	d
ΔE^*_{vap} at 25°C, kJ/mol	41.55	d	42.93	d
$(\Delta H^\circ_f)_{298}$, liquid, kJ/mol	−285.9	26	−294.6	25
$(\Delta G^\circ)_{298}$, liquid, kJ/mol	−237.2	26	−243.5	25
S°_{298}, liquid, J/K	70.08	26	76.11	25
$(\Delta H^\circ_f)_{298}$, gas, kJ/mol	−241.8	26	−249.2	25
$(\Delta G^\circ_f)_{298}$, gas, kJ/mol	−228.6	26	−234.6	25
S°_{298}, gas, J/K	188.8	26	198.3	27
$H^\circ_{298} - H^\circ_0$ gas, kJ/mol	9.908	26	9.954	25
C_p, liquid at 25°C, J/(mol·K)	75.27		84.35	
C_v, liquid at 25°C, J/(mol·K)	74.48		83.7	

[a]To convert J to cal, divide by 4.184.
[b]Unless otherwise indicated, data are from Ref. 17.
[c]Unless otherwise indicated, data are from Ref. 18.
[d]Calculated from data given in Table 3.

Table 5. Properties of Neutron Moderators

Moderator	Slowing-down power $\xi \times \Sigma_s$, cm^{-1}	Macroscopic absorption cross section Σ_a, cm^{-1}	Moderating ratio, $\xi \times \Sigma_a/\Sigma_a$
water	1.28	2.2×10^{-2}	58
heavy water	0.18	8.5×10^{-6}	21,000
helium	10^{-5}	2.2×10^{-7}	45
beryllium	0.16	1.2×10^{-3}	130
graphite	0.065	3.3×10^{-4}	200

greater amount of energy is required to activate a C—D bond for reaction, and general reactions involving the rupture of a C—D bond proceed considerably more slowly than for a comparable C—H bond. The most important factor contributing to the difference in bond energy and the kinetic isotope effect is the lower (5.021–5.275 kJ/mol (1.2–1.5 kcal/mol)) zero-point vibrational energy for D bonds. A comprehensive quantum and statistical mechanical theory of kinetic isotope effects has been developed (35). Kinetic isotope effects are particularly important in the elucidation of organic reaction mechanisms, and are the subject of a large literature (36).

Biological Effects of Deuterium. Replacement of more than one third of the hydrogen by deuterium in the body fluids of mammals or two thirds of the hydrogen in higher green plants has catastrophic consequences for the organisms. At lower deuterium levels in higher plants, growth is markedly slowed. In mice and rats low levels of deuterium result in sterility, and at higher concentrations neuromuscular disturbances, fine muscle tremors, and a tendency to convulsions can be noted (37,38). Impairment of kidney function, anemia, disturbed carbohydrate metabolism, central nervous system disturbances, and altered adrenal function have been found in deuterated mice (39,40). Hemoglobin and red blood cell count, serum glucose, and cholesterol all decrease in deuterated dogs (38).

Extensive replacement of H by D in living organisms is, however, not invariably fatal to the organisms. Numerous green and blue-green algae have been grown in which >99.5% of the hydrogen has been replaced by D. These fully deuterated organisms can be used to start a food chain to provide fully deuterated nutrients for organisms that have more demanding nutritional requirements. Numerous varieties of bacteria, molds, fungi, and even a protozoan have been successfully grown in fully deuterated form. These organisms of unnatural isotopic composition and the deuterated compounds that can be extracted from them have found uses in many areas of biological research (41).

Kinetic isotope effects are an important factor in the biology of deuterium. Isotopic fractionation of hydrogen and deuterium in plants occurs in photosynthesis. The lighter isotope ^1H is preferentially incorporated from water into carbohydrates and lipids formed by photosynthesis. Hydrogen isotopic fractionation has thus become a valuable tool in the elucidation of plant biosynthetic pathways (42,43).

Deuterium isotope effects provide information about the mechanisms of enzyme action (44–46). A bacterial alkaline phosphatase and a plant alkaline phosphatase have been prepared from fully deuterated organisms grown in 99.6% D_2 (47). The enzymatic behavior of the H and D enzymes were quite similar. Fully deuterated griseofulvin (48) (see ANTIBIOTICS, PEPTIDES) and benzylpenicillin (49) (see ANTIBIOTICS, β-LACTAMS–PENICILLIN AND OTHERS) have antifungal and antibiotic potencies at least as great as their ordinary hydrogen analogues, and are probably metabolized *in vivo* more slowly because of the enhanced stability of C—D relative to C—H bonds. Synthetic deuterated drugs have also been considered as therapeutic agents (50). Administration of D_2O has been reported to prevent hypertension in hypertensive rats (51), and this has even been the subject of patent applications. Considering the well-documented toxic effects of deuterium in mammals, long-term administration of D_2O to humans is not likely to become a routine procedure. However, D_2O confers significant protection to yeast against hydrostatic pressure damage (52), and pretreatment with D_2O has been claimed to protect cultured cells against x-ray damage (53). Protein structure is affected by dissolution in D_2O. One manifestation is increased resistance to thermal denaturation. Few systematic studies of fully deuterated proteins or the effects of D_2O on the structures of proteins of natural isotopic composition are available.

Isotope Effects on Superconductivity. Substitution of hydrogen by deuterium affects the superconducting transition temperature of palladium hydride [26929-60-2], PdH_2 (54,55), palladium silver hydride, $Pd_{1-y}Ag_yH_zD_x$ (56), and vanadium–zirconium–hydride, $V_2ZrH_x(D_x,T_x)$ (57).

Production of Heavy Water

Because of the low natural abundance of deuterium, very large amounts of starting material, which is water, must be processed to produce relatively small amounts of highly enriched deuterium. No water or other hydrogen compound has been found either in nature or as a by-product of an industrial operation that is significantly enriched in deuterium. The cost of subsequent enrichment to 99% is negligible compared to the costs incurred in the initial enrichment from natural abundance to 1%. For small-scale preparations, a highly efficient but very expensive process such as electrolysis can be used. For large-scale use, however, a high enrichment factor per stage is of only secondary importance to the overall costs of operation both in power and in capital investment. The isotope separation methods that have attracted the greatest interest include chemical exchange between water and hydrogen sulfide, hydrogen and water, and hydrogen and ammonia; distillation of water or hydrogen; and electrolysis of water in combination with other procedures (58).

Chemical Exchange Processes. Isotope exchange reactions (20) between hydrogen gas and water or ammonia, and between water and hydrogen sulfide provide the basis for the most efficient large-scale methods known for the concentration of deuterium. Equilibrium constants for these reactions are shown in Table 6. These equilibria are temperature dependent. The efficiency of chemical exchange processes can be increased by taking advantage of the difference in the equilibrium constants as a function of temperature in the form of a dual-temperature exchange process.

In the dual-temperature H_2O/H_2S process (61,62), exchange of deuterium between $H_2O(l)$ and $H_2S(g)$ is carried out at pressures of ca 2 MPa (20 atm). At elevated temperatures deuterium tends to displace hydrogen in the hydrogen sulfide and thus concentrates in the gas. At lower temperatures the driving force is reversed and the deuterium concentrates in H_2S in contact with water on the liquid phase.

Table 6. Chemical Exchange Reactions

Reaction	Equilibrium constant[a]	Reference
$H_2O(l) + HDS(g) \rightleftharpoons HDO(l) + H_2S(g)$	$K_{30} = 2.18$ $K_{130} = 1.83$	59
$NH_3(l) + HD(g) \overset{catalyst}{\rightleftharpoons} NH_2D(l) + H_2(g)$	$K^*_{-50} = 6.60$ $K^*_0 = 4.42$	60
$H_2O(g) + HD(g) \overset{catalyst}{\rightleftharpoons} HDO(g) + H_2(g)$	$K_{25} = 3.62$ $K_{125} = 2.43$	59

[a] Subscripts indicate temperature in °C.

The deuterium exchange reactions in the H_2S/H_2O process (the GS process) occur in the liquid phase without the necessity for a catalyst. The dual-temperature feature of the process is illustrated in Figure 1a. Dual-temperature operation avoids the necessity for an expensive chemical reflux operation that is essential in a single-temperature process (11,163) (Fig. 1b).

As shown in Figure 1a, the basic element of the H_2S/H_2O process is a pair of gas–liquid contacting towers each containing a number of sieve or bubble-cap plates (see DISTILLATION). The cold tower operates at a temperature of 30°C and the hot tower at 120–140°C. Water entering this system flows downward through the cold tower and then through the hot tower countercurrent to a stream of hydrogen sulfide gas at 1896 kPa (275 psig). The water is progressively enriched in deuterium as it passes through the cold tower, and progressively depleted as it passes through the hot tower, eventually leaving the hot tower at a concentration below that at which it entered the system. The HDO and HDS that build up within the process are withdrawn from the base of the cold tower and top of the hot tower, respectively, by withdrawing a fraction of the water and gas flow. These enriched fractions are fed to a succeeding stage for further concentration. The hydrogen sulfide gas, which acts as a transport medium for the deuterium, circulates in a closed loop within the several stages of the process. About 20% of the deuterium in natural water can be economically extracted in this manner.

In producing tonnage quantities of heavy water, large equipment must be used to perform the initial separation because of the low feed concentration and the high throughput. To produce one metric ton of D_2O the plant must process 41,000 t of water, and must cycle 135,000 t of hydrogen sulfide. By appropriate process staging, the amount of materials handled and the size of equipment can be reduced almost in proportion to the increase in concentration of heavy water

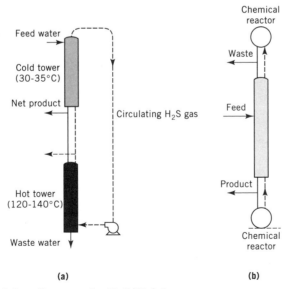

(a)   (b)

Fig. 1. Simplified flow diagrams for H_2S/H_2O heavy water processes. (**a**) Dual-temperature system where the pressure is 1.90 MPa; (**b**) single-temperature system. To convert MPa to atm, divide by 0.100. Courtesy of George Newnes, Ltd., London.

through the plant. This staging may be done using any of the possible heavy-water processes, or the process may be changed from stage to stage. In a properly staged plant most of the capital investment is incurred in the first hundredfold concentration gain, ie, from about 0.015 to 1.5 mol % D_2O.

In the heavy-water plants constructed at Savannah River and at Dana, these considerations led to designs in which the relatively economical GS process was used to concentrate the deuterium content of natural water to about 15 mol %. Vacuum distillation of water was selected (because there is little likelihood of product loss) for the additional concentration of the GS product from 15 to 90% D_2O, and an electrolytic process was used to produce the final reactor-grade concentrate of 99.75% D_2O.

In addition to the large contacting towers, a large amount of heat-recovery equipment was required to improve thermal efficiency. There were three principal heat-recovery circuits at the Savannah River production units: (*1*) the humidifier, comprising the bottom ten plates of the hot tower and the primary gas coolers; (*2*) the liquid heaters; and (*3*) the waste-stripper feed preheaters. The most important is the humidifier circuit, in which about half of the heat contained in the gas stream leaving the top of the hot tower is exchanged with a recirculating water stream that in turn heats and humidifies the gas entering the base of the hot tower. To make up for heat losses, steam is added to the hot tower after first being used in the stripper to remove the dissolved H_2S from the waste.

A dual-temperature system requires twice the number of separating stages (hot and cold columns) required for a single temperature system having the same effective separation factor. However, the requirement for a refluxing step is eliminated. The heat energy needed to maintain the temperature difference between the hot and cold columns can be minimized by heat exchange at appropriate points to preheat the gas and liquid streams entering the hot column and to precool the gas entering the cold column (64). All of the installations producing highly enriched deuterium in quantities greater than 20 metric tons D_2O annually use the H_2S/H_2O dual-temperature exchange system.

The fundamental parameters for the ammonia–hydrogen exchange (Table 6) are much more favorable than the corresponding factors in the H_2S/H_2O system, but the exchange reaction must be catalyzed to achieve a useful rate of exchange. The discovery (65) that the amide ion, NH_2^-, produced by addition of alkali metal to liquid ammonia, is an efficient catalyst for the NH_3/H_2 exchange stimulated intensive interest in this system (66,67). A dual-temperature system operating with a hot column at 70°C (single stage separation factor of 2.9) and a cold column at -40°C (single stage separation factor of 5.9) would have an effective separation factor of 2.0, which would permit extraction of 50% of the deuterium from the ammonia feed. Catalysis of the exchange by potassium amide is sufficiently effective even at -40°C to attain equilibrium in reasonably sized exchange columns. A single-temperature plant has been operated in France to produce about 20 t/yr D_2O (68). The primary limitation on the use of this process has been the availability of sufficient quantities of ammonia for plant feed. Even an ammonia plant producing 1000 t/d would provide sufficient feed only to permit production of 60–70 t/yr D_2O. A concept using an enrichment stripping system with a regeneration column has been developed (59,60) in which water is the deuterium feed via a hydrogen–water exchange step. This would, in principle, allow H_2/NH_3 chemical exchange plants of unlimited production capacity.

A variant of the H_2/NH_3 chemical exchange process uses alkyl amines in place of ammonia. Hydrogen exchange catalyzed by NH_2^- is generally faster using alkyl amines than ammonia, and a dual-temperature flow sheet for a H_2/CH_3NH_2 process has been developed (69).

Chemical exchange between hydrogen and steam (catalyzed by nickel–chromia, platinum, or supported nickel catalysts) has served as a pre-enrichment step in an electrolytic separation plant (10,70). If the $H_2(g)/H_2O(l)$ exchange could be operated as a dual-temperature process, it very likely would displace the H_2S/H_2O process. However, suitable catalysts for liquid-phase use have not been reported.

Distillation. Vacuum distillation (qv) of water, which contains the three molecular species H_2O, HDO, and D_2O, was the first method used for the large-scale extraction of deuterium (10,58) (Fig. 2). From the equilibrium constant in the liquid phase it is evident that the distribution of ^1H and D is not statistical. The differences in vapor pressure between H_2O and D_2O are significant, and a fractionation factor (see Table 7) of 1.05 can be obtained at about 50°C. As the fractionation factor decreases with increasing temperature, vacuum distillation is used that can be carried out at low temperatures. However, low pressure distillation requires larger towers to handle a given mass flow rate of vapor at lower pressure. In practice, tower top pressures between 6.7–16.7 kPa (50–125 mm Hg) were used, corresponding to temperatures between 30–50°C and separation factors between 1.06–1.05. Maximum deuterium recovery in a water distillation plant was considerably less than 5%, and the deuterium was concentrated from an initial value of 0.0143 atom % to 87–91 atom %. Further concentration to 99.8%

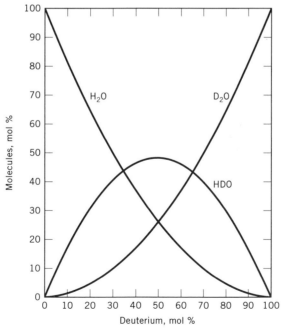

Fig. 2. Equilibrium concentrations of H_2O, HDO, and D_2O vs overall concentration of deuterium in water. $K_{25°C} = 3.80$ for $H_2O + D_2O \rightleftharpoons 2 HDO$.

Table 7. Distillation Process Requirements for the Production of Deuterium[a]

Requirement	Water distillation	Hydrogen distillation
deuterium content of feed, %	0.0149	0.0149
separation factor, α	1.05[b]	1.52[c]
temperature, K	323	23
pressure, kPa[d]	ca 13.3 kPa	ca 160 kPa
min no. stages	308	41
min reflux ratio	141,000	19,600
recovery, %	ca 5	90
mol feed per mol product	133,940	7,442
ratio of operating costs	11	1

[a]Ref. 61.
[b]$\alpha = (p_{H_2O}/p_{D_2O})^{1/2}$.
[c]$\alpha = (p_{H_2}/p_{D_2})^{1/2}$.
[d]To convert kPa to psi, multiply by a 0.145.

was achieved by electrolysis. Operating parameters for a water distillation plant for deuterium recovery are given in Table 7. Distillation of water is now only used for final enrichment of D_2O (71).

Distillation of liquid hydrogen as a method for separating deuterium received early consideration (10,58) because of the excellent fractionation factor that can be attained and the relatively modest power requirements. The cryogenic temperatures, and the requirement that the necessarily large hydrogen feed be extremely pure (traces of air, carbon monoxide, etc, are solids at liquid hydrogen temperature) have been deterrents to the use of this process (see CRYOGENICS).

In the separation of deuterium by distillation of liquid hydrogen, the presence of nuclear spin isomers must be taken into account. Ordinary hydrogen gas consists of three parts of ortho-H_2 (nuclear spins parallel) and one part para-H_2 (nuclear spins antiparallel)(see HYDROGEN). The antiparallel orientation is the lowest energy state ($J = 0$). It is the usual practice in the production of liquid hydrogen to include chromium oxide, Cr_2O_3, a paramagnetic species that catalyzes the conversion of the parallel to the antiparallel para spin orientation. A comparison of the operating characteristics of a water distillation plant with those of a liquid hydrogen distillation plant are given in Table 7. Hydrogen distillation may become important to deuterium production (see also HYDROGEN ENERGY).

Electrolysis.   For reasons not fully understood (76), the isotope separation factor commonly observed in the electrolysis of water is between 7 and 8. Because of the high separation factor and the ease with which it can be operated on the small scale, electrolysis has been the method of choice for the further enrichment of moderately enriched H_2O–D_2O mixtures. Its usefulness for the production of heavy water from natural water is limited by the large amounts of water that must be handled, the relatively high unit costs of electrolysis, and the low recovery.

Economic Aspects.   The principal market for deuterium has been as a moderator for nuclear fission reactors fueled by unenriched uranium. The decline in nuclear reactor construction has sharply reduced the demand for heavy water. The United States has stopped large-scale production of D_2O, and Canada is the only supplier of heavy water at this time. Heavy water is priced as a fine chemical, and its price is not subject to market forces.

Methods of Analysis

The principal methods for determination of the deuterium content of hydrogen and water are based upon measurements of density, mass, or infrared spectra. Other methods are based on proton magnetic resonance techniques (77,78), ^{19}F nuclear magnetic resonance (79), interferometry (80), osmometry (81), nuclear reaction (82), combustion (83), and falling drop methods (84).

Density.   Measurement of the density of water by pycnometry is the classical method (30) for establishing deuterium concentrations in heavy water. Very precise measurements can be made by this method, provided the sample is prepared free of suspended or dissolved impurities and the concentration of oxygen-18 in water is about 0.2 mol %. However, in nearly all heavy water manufactured since 1950 in the United States, the concentration of oxygen-18 is about 0.4 mol % as the result of its concentration in the water-distillation portion of the production facilities. Using a correction for oxygen-18, the sensitivity of the densimetric method is $\pm 0.2\%$ D_2O, but the accuracy is not better than 0.03% D_2O because of an uncertainty of 30 ppm in the density of pure heavy water.

Mass Spectrometry.   Mass spectrometric methods (85) are used for both hydrogen gas and water (see MASS SPECTROMETRY). These are capable of determining directly the concentration of different isotopic species in samples of H_2O, HDO, and D_2O, eg, by measuring the relative abundance of ions of mass 18 and 19 or mass 19 and 20. The principal limitation of the method is the contamination of the sample-introduction system with water vapor. This contamination causes significant memory effects on succeeding samples. To obtain accurate analyses it is necessary to use a comparative standard having an isotopic concentration within 7% of that of the unknown. The precision of this method is 0.005–0.1 mol % over the range 0.05–15% and is 0.2 mol % in the range 99–100% D_2O.

Infrared Spectrophotometry.   The isotope effect on the vibrational spectrum of D_2O makes infrared spectrophotometry the method of choice for deuterium analysis. It is as rapid as mass spectrometry, does not suffer from memory effects, and requires less expensive laboratory equipment. Measurement at either the O–H fundamental vibration at 2.94 µm (O–H) or 3.82 µm (O–D) can be used. This method is equally applicable to low concentrations of D_2O in H_2O, or the reverse (86,87). Absorption in the near infrared can also be used (88,89) and this procedure is particularly useful (see INFRARED AND RAMAN SPECTROSCOPY; SPECTROSCOPY). The D/H ratio in the nonexchangeable positions in organic compounds can be determined by a combination of exchange and spectrophotometric methods (90).

Raman Spectroscopy.   Raman spectroscopy is an excellent method for the analysis of deuterium containing mixtures, particularly for any of the diatomic H–D–T molecules. For these, it is possible to predict absolute light scattering intensities for the rotational Raman lines. Hence, absolute analyses are possible, at least in principle. The scattering intensities for the diatomic hydrogen isotope species is comparable to that of dinitrogen, N_2, and thus easily observed.

Gas Chromatography.   Gas chromatography is a well recognized method for the analysis of H–D–T mixtures. The substrate is alumina, Al_2O_3, coated with ferric oxide, Fe_2O_3. Neon is used as the carrier gas. Detectors are usually both thermal conductivity (caratherometer) and ion chamber detectors when tritium is involved (see CHROMATOGRAPHY).

Uses

The only large-scale use of deuterium in industry is as a moderator, in the form of D_2O, for nuclear reactors. Because of its favorable slowing-down properties and its small capture cross section for neutrons, deuterium moderation permits the use of uranium containing the natural abundance of uranium-235, thus avoiding an isotope enrichment step in the preparation of reactor fuel. Heavy water-moderated thermal neutron reactors fueled with uranium-233 and surrounded with a natural thorium blanket offer the prospect of successful fuel breeding, ie, production of greater amounts of ^{233}U (by neutron capture in thorium) than are consumed by nuclear fission in the operation of the reactor. The advantages of heavy water-moderated reactors are difficult to assess.

Deuterium Fusion.   At sufficiently high temperatures, deuterium undergoes nuclear fusion with the production of large amounts of energy:

$$^2H + {}^2H \longrightarrow {}^3H + {}^1H + 4.0 \text{ MeV}$$

$$^2H + {}^2H \longrightarrow {}^3He + n + 3.3 \text{ MeV}$$

The temperature required to surmount the barrier to fusion is considerably reduced for the $^2H(^3H,n)^4He$ reaction, and the efforts to develop a nuclear fusion reactor are all based on the D–T fusion reaction. Many different pathways have been explored to reach the temperature at which a self-sustaining D–T fusion reaction occurs. These include magnetic confinement of a D–T plasma, laser implosion, and particle acceleration. The prospect of an energy source that meets global energy demand and mitigates environmental concerns about global warming, coupled with the depletion of fossil fuel energy sources, provide powerful incentives to pursue nuclear fusion (see FUSION ENERGY).

It has been claimed that the D–D fusion reaction occurs when D_2O is electrolyzed with a metal cathode, preferably palladium, at ambient temperatures. This claim for a cold nuclear fusion reaction that evolves heat has created great interest, and has engendered a voluminous literature filled with claims for and against. The proponents of cold fusion report the formation of tritium and neutrons by electrolysis of D_2O, the expected stigmata of a nuclear reaction. Some workers have even claimed to observe cold fusion by electrolysis of ordinary water (see, for example, Ref. 91). The claim has also been made for the formation of tritium by electrolysis of water (92). On the other hand, there are many experimental results that cast serious doubts on the reality of cold fusion (93–96). Theoretical calculations indicate that cold fusions of D may indeed occur, but at the vanishingly small rate of 10^{-23} events per second (97). As of this writing the cold fusion controversy has not been entirely resolved.

BIBLIOGRAPHY

"Deuterium and Tritium" in *ECT* 2nd ed., Vol. 6, pp. 895–910, by J. F. Proctor, E. I. du Pont de Nemours & Co., Inc.; in *ECT* 3rd ed., Vol. 7, pp. 539–564, by Joseph J. Katz, Argonne National Laboratories.

1. H. C. Urey, *Science* 78, 566 (1933).
2. G. N. Lewis and R. T. MacDonald, *J. Chem. Phys.* 1, 341 (1933).
3. G. N. Lewis, *Science* 79, 151 (1934).
4. J. H. Morowitz and L. M. Brown, *U.S. Nat. Bur. Stand. Rep.* 2179, (1953).
5. H. L. Crespi, S. M. Archer, and J. J. Katz, *Nature* 184, 729 (1959).
6. W. Chorney and co-workers, *Biochim. Biophys. Acta* 37, 280 (1960).
7. D. Kritchevsky, ed., *Ann. N.Y. Acad. Sci.* 84, 573 (1960).
8. W. H. Zinn and C. A. Trilling, *Proc. Int. Conf. Peaceful Uses At. Energy, 3rd Geneva* 28, 209 (1964).
9. W. B. Lewis and T. G. Church in Ref. 8, p. 1.
10. G. M. Murphy, ed., *Production of Heavy Water*, McGraw-Hill Book Co, Inc., New York, 1955, p. 4.
11. W. P. Bebbington and V. R. Thayer, *Chem. Eng. Progr.* 55(9), 70 (1959).
12. *Technica Ed Economia Della Produzione Di Acqua Pesante*, Comitato Nazionale Energia Nucleare Symposium, Turin, Italy, Sept. 30, 1970.
13. *U.S. Atomic Energy Commission Document 1174-71*, U.S. Government Printing Office, Washington, D.C., 1971.
14. A. Farkas, *Orthohydrogen, Parahydrogen, and Heavy Hydrogen*, Cambridge University Press, Cambridge, 1935.
15. H. M. Mittelhauser and G. Thodos, *Cryogenics* 4, 368 (1963).
16. H. W. Wooley, R. B. Scott, and F. G. Brickwedde, *J. Res. Natl. Bur. Stand.* 41, 379 (1948).
17. E. C. Arnett and D. R. McKelvey in J. F. Coetzee and C. D. Ritchie, eds., *Solute–Solvent Interactions*, Marcel Dekker, Inc., New York, 1969, pp. 343–398.
18. G. Nemethy and H. A. Scheraga, *J. Chem. Phys.* 41, 680 (1964).
19. E. Whalley, *Proc. Conf. Thermodynamics Transport Properties of Fluids, 10–12 July 1957*, Institute of Mechanical Eng., London, 1958, p. 15.
20. F. T. Barr and W. P. Drews, *Chem. Eng. Progr.* 56, 49 (1960).
21. H. D. Megaw, *Nature* 134, 900 (1934).
22. J. R. Coe and T. B. Godfrey, *J. Appl. Phys.* 15, 625 (1944).
23. A. S. Friedman and L. Haar, *J. Chem. Phys.* 22, 2051 (1954).
24. A. K. Covington, R. A. Robinson, and R. G. Bates, *J. Phys. Chem.* 70, 3820 (1966).
25. F. D. Rossini, J. W. Knowlton, and H. L. Johnston, *J. Res. Natl. Bur. Stand.* 24, 369 (1940).
26. F. D. Rossini, *J. Res. Natl. Bur. Stand.* 22, 407 (1939).
27. E. A. Long and J. D. Kemp, *J. Am. Chem. Soc.* 58, 1829 (1936).
28. W. L. Marshall and J. M. Simonson, *J. Chem. Thermodyn.* 23, 613–616 (1991).
29. E. C. Noonan, *J. Am. Chem. Soc.* 70, 2915 (1948).
30. I. Kirschenbaum, *Physical Properties and Analysis of Heavy Water*, McGraw-Hill Book Co., Inc., New York, 1951, p. 54.
31. A. K. Covington, R. A. Robinson, and R. G. Bates, *J. Chem. Ed.* 44, 635 (1967).
32. H. L. Clever, *J. Chem. Ed.* 45, 231 (1968).
33. A. K. Covington and co-workers, *Anal. Chem.* 40, 700 (1968).
34. S. Glasstone and A. Sesonske, *Nuclear Reactor Engineering*, D. Van Nostrand Co., Inc., Princeton, N.J., 1963, p. 134.
35. J. Bigeleisen in P. A. Rock, ed., *Isotopes and Chemical Principles*, American Chemical Society Symposium Series No. 11, 1975, pp. 1–43.

36. C. J. Collins and N. S. Bowman, eds., *Isotope Effects in Chemical Reactions, American Chemical Society Monograph 167*, Van Nostrand Reinhold Co., New York, 1971.
37. J. J. Katz and co-workers, *Am. J. Physiol* **203**, 357 (1961).
38. D. M. Czajka and co-workers, *Am. J. Physiol.* **201**, 357 (1961).
39. J. F. Thomson, *Ann. N.Y. Acad. Sci.* **84**, 736 (1960).
40. J. F. Thomson, *Biological Effects of Deuterium*, Pergamon Press, Oxford, 1963.
41. J. J. Katz and H. L. Crespi in Ref. 36, Chapt. 5, pp. 286–363.
42. C. R. Hutchinson, *J. Nat. Prod.* **45**, 27–37 (1982).
43. B. N. Smith and H. Ziegler, *Bot. Acta* **103**, 335–342 (1990).
44. W. J. Malaisse and co-workers, *Mol. Cell. Biochem.* **93**, 153–165, 1990.
45. X. Guo and co-workers, *Biochem. J.* **278**, 487–491 (1991).
46. L. D. Sutton, J. S. Stout, and D. M. Quinn, *J. Am. Chem. Soc.* **112**, 8398–8403 (1990).
47. S. Rokop and co-workers, *Biochim. Biophys. Acta* **191**, 707–715 (1969).
48. D. A. Nona and co-workers, *J. Pharm. Sci.* **57**, 1993–1995 (1968).
49. B. C. Carlstedt and co-workers, *J. Pharm. Sci.* **35**, 856–857 (1973).
50. P. Dumont, *Rev. IRE* **6**, 2–10 (1982).
51. S. Vasdev and co-workers, *Hypertension (Dallas)* **18**, 550–557 (1991).
52. Y. Komatsu and co-workers, *Biochem. Biophys. Res. Commun.* **174**, 1141–1147 (1991).
53. R. H. Laeng and co-workers, *Int. J. Radiat. Biol.* **59**, 165–173 (1991).
54. R. Griessen and D. G. De Groot, *Helv. Phys. Acta* **55**, 699–710 (1982).
55. S. L. Drechsler and A. P. Zhernov, *Phys. Status Solidi B.* **130**, 381–386 (1985).
56. E. Matsushita and T. Matsubara, *J. Magn. Magn. Mater.*, 31–34, 517–518 (1983).
57. K. Mori and co-workers, *Kenkyu Hokoku-Toyama Daigaku Torichumu Kagaku Senta* **9**, 39–46 (1989).
58. M. Benedict in *Progress in Nuclear Energy, Series IV, Technology and Engineering*, Pergamon Press, Ltd., Oxford, 1956, pp. 3–56.
59. U. Schindewolf and G. Lang in Ref. 12, pp. 77–88.
60. S. Walter and E. Nitschke in Ref. 12, pp. 175–193.
61. J. Spevack in Ref. 12, pp. 25–43.
62. U.S. Pats. 2,787,526 (Apr. 2, 1957), 2,895,803 (Sept. 29, 1950), 3,142,540 (Sept. 29, 1950), 4,008,046 (Mar. 21, 1971), J. Spevack (to U.S. Atomic Energy Commission).
63. H. London, ed., *Separation of Isotopes*, George Newnes, Ltd., London, 1961, p. 7.
64. W. Spindel in P. A. Rick, ed., *Isotopes and Chemical Principles*, American Chemical Society Symposium Series No. 11, 1975, pp. 77–100.
65. Y. Claeys, J. C. Dayton, and W. K. Wilmarth, *J. Chem. Phys.* **18**, 759 (1950).
66. M. Perlman, J. Bigeleisen, and N. Elliot, *J. Chem. Phys.* **21**, 70 (1953).
67. J. Bigeleisen in J. Kistenmaker, J. Bigeleisen, and A. O. C. Nier, eds., *Proceedings of the International Symposium on Isotope Separation*, North Holland Publishing Co., Amsterdam, 1958, p. 121.
68. E. Roth and M. Rostain in Ref. 12, pp. 69–74.
69. A. R. Bancroft and H. K. Ral in Ref. 12, pp. 49–60.
70. Ref. 50, pp. 32–52.
71. P. Baertschi and W. Kuhn in Ref. 58, pp. 57–61.
72. M. Kinoshita, *Fusion Technol.* **6**, 574–583 (1984):
73. R. H. Sherman, J. R. Bartlit, and D. K. Veirs, *Fusion Technol.* **6**, 625–628 (1984).
74. H. Gutowski in Ref. 12, pp. 93–100.
75. H. Gutowski, *Kerntechnik* **12**(5/6), (1970).
76. B. E. Conway, *Proc. Royal Soc. London* **A247**, 400 (1958).
77. D. E. Leyden and C. N. Reilley, *Anal. Chem.* **37**, 1333 (1965).
78. W. Johnson and R. A. Keller, *Anal. Lett.* **2**, 99 (1969).
79. C. Deverell and K. Schaumberg, *Anal. Chem.* **39**, 1879 (1967).
80. J. Mercea, *Chem. Ing. Tech.* **41**, 508 (1969).
81. E. Lazzarini, *Nature* **204**, 875 (1964).

82. S. Amiel and M. Peisach, *Anal. Chem.* **34,** 1305 (1962).

83. R. N. Jones and M. A. MacKenzie, *Talanta* **7,** 124 (1960).

84. M. F. Clarke, *Anal. Biochem.* **31,** 81 (1969).

85. P. Chastagner, H. L. Daves, and W. B. Hess, *Anal. Chem. Nucl. Technol., Proc. Conf. Anal. Chem. Energy Technol, 25th,* 1982, pp. 153–160.

86. W. H. Stevens and W. Thurston, *Atomic Energy of Canada, Ltd., Report No. 295,* Chalk River, Ontario, Canada, 1954.

87. M. D. Turner and co-workers, *Microchem. J.* **8,** 395 (1964).

88. H. L. Crespi and J. J. Katz, *Anal. Biochem.* **2,** 274 (1961).

89. M. Kobayashi, *Kyoto Daigaku Genshiro Jikkensho Gakujutsu Koenkai Koen Yoshishu* **24,** 109–114 (1990).

90. A. Schimmellmann, *Anal. Chem.* **63,** 2456–2459.

91. T. Matsumoto, *Fusion Technol.* **17,** 490–492 (1990).

92. E. Storms and C. Talcott, *Fusion Technol.* **17,** 680–695 (1990).

93. J. A. Knapp and co-workers, *J. Fusion Energy* **9,** 371–375 (1990).

94. G. M. McCracken and co-workers, *J. Phys. D:* **23,** 469–475 (1990).

95. M. Gai and co-workers, *J. Fusion Energy* **9,** 217 (1990).

96. I. I. Astakhov and co-workers, *Electrochim. Acta* **36,** 1127–1128 (1991).

97. R. H. Parmenter and W. E. Lamb, Jr., *Proc. Natl. Acad. Sci. U.S.A.* **87,** 8652–8654 (1990).

Joseph J. Katz
Argonne National Laboratory

TRITIUM

Tritium [*15086-10-9*], the name given to the hydrogen isotope of mass 3, has symbol ^3H or more commonly T. Its isotopic mass is 3.0160497 (1). Molecular tritium [*10028-17-8*], T_2, is analogous to the other hydrogen isotopes. The tritium nucleus is energetically unstable and decays radioactively by the emission of a low-energy β particle. The half-life is relatively short ($\sim$12 yr), and therefore tritium occurs in nature only in equilibrium with amounts produced by cosmic rays or man-made nuclear devices.

Tritium was first prepared in the Cavendish Laboratory by Rutherford, Oliphant, and Harteck in 1934 (2,3) by the bombardment of deuterophosphoric acid using fast deuterons. The D–D nuclear reaction produced tritium ($^2D + {}^2D \longrightarrow {}^3T + {}^1H +$ energy), but also produced some ^3He by a second reaction ($^2D + {}^2D \longrightarrow {}^3He + {}^1n +$ energy). It was not immediately known which of the two mass-3 isotopes was radioactive. In 1939 it was established (4) that ^3He occurred in nature and was stable. Tritium was later proved to be radioactive. A vivid history of the unravelling of the complex relationships between tritium, deuterium, and helium-3 is available (5).

Physical Properties

Tritium is the subject of various reviews (6–8), and a book (9) provides a comprehensive survey of the preparation, properties, and uses of tritium compounds. Selected physical properties for molecular tritium, T_2, are given in Table 1.

Table 1. Physical Properties of Molecular Tritium[a]

Property	Value
melting point, at 21.6 kPa[b], K	20.62
boiling point, K	25.04
critical temperature, K	40.44[c]
critical pressure, MPa[d]	1.850
critical volume, cm³/mol	57.1[e]
heat of sublimation, J/mol[f]	1640
heat of vaporization, J/mol[f]	1390
entropy of vaporization, J/(mol·K)[f]	54.0
molar density of liquid, mol/L	
20.62 K[b]	45.35
25 K	42.65
29 K	39.66

[a]Values are from Ref. 11 unless otherwise indicated.
[b]Value represents the triple point (162 mm Hg).
[c]From Ref. 10.
[d]To convert MPa to psi, multiply by 145.
[e]Value is calculated.
[f]To convert J to cal, divide by 4.184.

Calculated vapor pressure relationships of T_2, HT, and DT have been reported (10) (see DEUTERIUM AND TRITIUM, DEUTERIUM). An equation for the vapor pressure of solid tritium in units of kPa, T in Kelvin, has been given (11):

$$\log p = 5.6023 - 88.002/T$$

The three-phase region of D_2–DT–T_2 has been studied (12). Relative volatilities for the isotopic system deuterium–deuterium tritide–tritium have been found (13) to be 5–6% below the values predicted for ideal mixtures.

All components appear miscible in both liquid and solid phases from 17 to 22 K. For a 50–50 mol % mixture of liquid D–T at ca 19.7 K, the gas phase contains ca 42% T and the solid phase 52%.

The T–T bond energy has been estimated at 4.5881 eV (14).

The entropy of T_2 at 298.15 K is 164.8562 kJ/mol (39.4016 kcal/mol), the specific heat is 29.1997 J/(mol·°C) (6.9789 cal/(mol·°C)), and the Gibbs free energy is 135.9083 kJ/mol (32.4829 kcal/mol). These values were derived from extensive *ab initio* calculations for T_2 at spin equilibrium, ortho and para T_2, and the isotopomers HT and DT (15).

Ortho-Para Tritium. As in the case of molecular hydrogen, molecular tritium exhibits nuclear spin isomerism. The spin of the tritium nucleus is ½, the same as that for the hydrogen nucleus, and therefore H_2 and T_2 obey the same nuclear isomeric statistics (16). Below 5 K, molecular tritium is 100% para at equilibrium. At high (100°C) temperatures the equilibrium concentration is 25% para and 75% ortho. The kinetic parameters of conversion for T_2 at low temperatures are faster than rates at corresponding temperatures for H_2. In the solid phase the conversion of molecular titrium to a state of ortho-para equilibrium is 210 times as fast as that for molecular hydrogen.

The experimental and theoretical aspects of the radiation and self-induced conversion kinetics and equilibria between the ortho and para forms of hydrogen, deuterium, and tritium have been correlated (17). In general, the radiation-induced transitions are faster than the self-induced transitions.

Properties of T_2O. Some important physical properties of T_2O are listed in Table 2. Tritium oxide [14940-65-9] can be prepared by catalytic oxidation of T_2 or by reduction of copper oxide using tritium gas. T_2O, even of low (2–19% T) isotopic abundance, undergoes radiation decomposition to form HT and O_2. Decomposition continues, even at 77 K, when the water is frozen. Pure tritiated water irradiates itself at the rate of 10 MGy/d (10^9 rad/d). A stationary concentration of tritium peroxide, T_2O_2, is always present (9). All of these factors must be taken into account in evaluating the physical constants of a particular sample of T_2O.

Table 2. Physical Properties of T_2O

Property	Value	Reference
mol wt	22.032	
triple point, °C	4.49	18
temperature of maximum density, °C	13.4	
boiling point, °C	101.51	18
density at 25°C, g/mL	1.2138	19
$\Delta H°_{vap}$ at 25°C, kJ/mol^a	ca 45.81	18
liquid vapor pressure at 25°C, kPa^b	2.64	
vibrational fundamentals, cm^{-1}	1017, 2438	20
ionization constant at 25°C	ca 6×10^{-16}	21

^{a} To convert J to cal, divide by 4.184.
^{b} To convert kPa to mm Hg, multiply by 7.5.

Nuclear Properties

Radioactivity. Tritium decays by β emission, $^3T \longrightarrow {}^3He + \beta^-$. A summary of the radioactive properties of T, adapted from Ref. 22, is given in Table 3.

Nuclear Fusion Reactions. Tritium reacts with deuterium or protons (at sufficiently high temperatures) to undergo nuclear fusion:

Table 3. Radioactive Properties of Tritium

Property	Value
half-life, yr	12.43^a
decay constant, s^{-1}	1.7824×10^{-9}
mean β energy, keV	5.7
molar activity, TBq/mol^b	2128

^{a} Courtesy of the National Institute of Standards and Technology.
^{b} To convert Bq to Ci, divide by 3.7×10^{10}

$$^3T + {}^2D \longrightarrow {}^4He + {}^1n + 17.6 \text{ MeV}$$

$$^3T + {}^1H \longrightarrow {}^4He + 19.6 \text{ MeV}$$

The first of these nuclear fusion reactions produces a neutron that can be used to form a new atom of tritium as well as evolving a very large amount of energy in the form of extremely hot helium. Nuclear fusion using tritium can be initiated and sustained at the lowest temperature (at least in principle) of any nuclear fusion reaction known. Tritium thus becomes the key element both for controlled thermonuclear energy sources and in the uncontrolled release of thermonuclear energy in the hydrogen bomb (see FUSION ENERGY).

Nuclear Magnetic Resonance.   All three hydrogen isotopes have nuclear spins, $I \neq 0$, and consequently can all be used in nmr spectroscopy (Table 4) (see MAGNETIC SPIN RESONANCE). Tritium is an even more favorable nucleus for nmr than is ^1H, which is by far the most widely used nucleus in nmr spectroscopy. The radioactivity of T and the ensuing handling problems are a deterrent to widespread use for nmr. Considerable progress has been made in the applications of tritium nmr (23,24).

Table 4. Nuclear Magnetic Resonance Properties of Hydrogen Isotopes

Isotope	Nuclear spin	Resonance frequency,[a] MHz	Relative sensitivity[b]	Magnetic moment, 10^{-4} J/T[c]
H	1/2	100.56	1.000	25.8995
D	1	15.360	9.64×10^{-3}	7.9513
T	1/2	104.68	1.21	27.625

[a]At a field of 2.35 T (23.5 kilogauss).
[b]At constant field.
[c]To convert J/T to μ_B (nuclear Bohr magnetons), divide by 9.274×10^{-24}.

Chemical Properties

Most of the chemical properties of tritium are common to those of the other hydrogen isotopes. However, notable deviations in chemical behavior result from isotope effects and from enhanced reaction kinetics induced by the β-emission in tritium systems. Isotope exchange between tritium and other hydrogen isotopes is an interesting manifestation of the special chemical properties of tritium.

Isotope Effects.   Any difference in the chemical or physical properties of two substances that differ only in isotopic composition constitutes an isotope effect. Isotope effects are usually largest when the isotope is directly involved in the rate-determining step of a reaction. The greater the mass of an atom, the lower its zero-point bond energy, and the greater the activation energy required to cleave a chemical bond. In general, therefore, reactions involving rupture of a —C—T bond may proceed at a markedly slower rate than those involving cleavage of a corresponding —C—H bond. Kinetic isotope effects arising from bond cleavage are termed primary isotope effects. Secondary isotope effects occur as a result of the presence of the isotope in nearby molecular sites. Bonds involving the isotope are neither broken nor formed in the reaction. Secondary isotope effects are generally smaller than primary effects. The latter may easily be 10 to 100 times greater. Solvent isotope effects include many important primary effects, eg, solvolyses and acid-base reactions, as well as secondary effects such as solvation. Although primary and secondary kinetic isotope effects are the most extensively studied, numerous other isotope effects have been observed with tritium. Thus, an H/T separation factor of about 14 occurs in the electrolysis of HOT (25,26). Other tritium isotope effects of significant magnitude have been observed in ion exchange (qv) (27) and gas chromatography (qv) (28,29). Many other examples have been described (9).

Enhanced Reaction Kinetics. For reactions involving tritium, the reaction rates are frequently larger than expected because of the ionizing effects of the tritium β-decay. For example, the uncatalyzed reaction $2\,T_2 + O_2 \longrightarrow 2\,T_2O$ can be observed under conditions (25°C) for which the analogous reaction of H_2 or D_2 would be too slow for detection (30).

Isotopic Exchange Reactions. Exchange reactions between the isotopes of hydrogen are well known and well substantiated. The equilibrium constants for exchange between the various hydrogen molecular species have been documented (18). Kinetics of the radiation-induced exchange reactions of hydrogen, deuterium, and tritium have been critically and authoritatively reviewed (31). The reaction $T_2 + H_2 \longrightarrow 2\,HT$ equilibrates at room temperature even without a catalyst (30).

In 1957, it was demonstrated (32) that tritium could be introduced into organic compounds by merely exposing them to tritium gas. Since that time hundreds of compounds, of types as simple as methane and as complex as insulin, have been labeled with tritium by this basic method of isotope exchange. Much work has been done to optimize conditions for the exchange technique through control of operating variables such as temperature, pressure, and addition of noble gases to facilitate energy transfer (33). Exchange of the hydrogen of organic compounds with tritium gas has been facilitated by activating the gas by ultraviolet light, γ- and x-ray irradiation, and microwave discharge. Although high chemical yields are obtained by the original, ie, Wilzbach, technique, highly tritiated impurities of structures similar to but not identical with that of the starting material are formed by direct irradiation damage of the target material and by decomposition of the tritiated products by self-irradiation. Separation of radioactive impurities produced in Wilzbach labeling may prove difficult, as is the task of proving that the radioactive T introduced into the substrate compound is actually in the compound of interest.

Catalytic exchange in solution between an organic compound and tritium gas or T_2O is a general procedure for introducing tritium with high specificity. Much higher molar specific activities (greater than 1.85 GBq/mmol (50 mCi/mmol)) can be attained than with the Wilzbach method. Both homogeneous and heterogeneous catalysts, as well as acid-base catalysis, can be used.

Hot atom reactions have also been used to label organic compounds with T. Irradiation of helium-3 with neutrons according to the nuclear reaction $^3He(n,p)^3H$ produces very energetic tritium atoms that can displace ordinary hydrogen in organic compounds. This procedure is not very selective, and the labeling pattern must be determined to enable the tritiated product to be used effectively as a tracer (34).

Natural Production and Occurrence

Tritium arises in nature by the action of primary cosmic rays (high-energy protons) or cosmic-ray neutrons on a number of elements. Natural tritium was first
detected in the atmosphere (35) and was later shown to be present in rainwater
(36). Because of its relatively short radioactive half-life, naturally produced tritium does not accumulate indefinitely and the amount of tritium found in nature
is very small. The unit for measurement of natural tritium, the tritium unit (TU)
signifies a ratio of 1 atom of tritium per 10^{18} atoms of hydrogen. A very excellent
worldwide survey of tritium levels in natural waters has been compiled (37). Values range from less than one TU for water in certain deep wells and at extreme
sea depths to several hundred TU in samples of rainwater taken during periods
of active thermonuclear weapons testing. The level of tritium in atmospheric hydrogen increased from 3800 TU in 1948–1949 to 490,000 TU in 1959 (38).

The principal source of natural tritium is the nuclear reactions induced by
cosmic radiation in the upper atmosphere, where fast neutrons, protons, and deuterons collide with components of the stratosphere to produce tritium:

$$^{14}N + {}^{1}n \longrightarrow {}^{3}T + {}^{12}C$$

$$^{14}N + {}^{1}H \longrightarrow {}^{3}T + \text{fragments}$$

$$^{2}D + {}^{2}D \longrightarrow {}^{3}T + {}^{1}H$$

The most important of these reactions by far is the $^{14}N(n,{}^{12}C)^{3}T$ reaction (39). The
energetic tritons so produced are incorporated into water molecules by exchange
or oxidation, and the tritium reaches the earth's surface as rainwater.

Large-scale detonation of hydrogen bombs in the atmosphere during the
1950s and 1960s, and inadvertent tritium losses have had a permanent impact
on the tritium content of the atmosphere. For example, some 20,000 TBq (500,000
Ci) of tritium was released by accident into the atmosphere at Aiken, South Carolina in September, 1984 (40). The deposition of tritium in rainfall (mainly as
HTO) is not uniform, and water in different localities can have very different
tritium contents. The total deposition of tritium on the continental United States
for the period 1953–1983 is estimated (41) at 12 ± 2 kg. In the United States,
tritium in ground water is smallest in the Southwest, and highest in the Midwest.
These differences may be associated with annual rainfall, the introduction of
water from the Pacific Ocean, weather patterns, and the like. The transport, dispersion, cycling, and analysis of tritium in the environment receives considerable
attention (42) because of the possible large-scale use of tritium in nuclear fusion.

Tritium has also been observed in meteorites and material recovered from
satellites (see also EXTRATERRESTRIAL MATERIALS). The tritium activity in meteorites can be reasonably well explained by the interaction of cosmic-ray particles
and meteoritic material. The tritium contents of recovered satellite materials
have not in general agreed with predictions based on cosmic-ray exposure. For
observations higher than those predicted (Discoverer XVII and satellites), a theory of exposure to incident tritium flux in solar flares has been proposed. For
observations lower than predicted (Sputnik 4), the suggested explanation is a
diffusive loss of tritium during heating up on reentry.

Production

Nuclear Reactions.   The primary reaction for the production of tritium is

$$^6\text{Li} + {}^1n \longrightarrow {}^3\text{T} + {}^4\text{He} + 4.78 \text{ MeV}$$

The capture cross section of ^{6}Li for this reaction using thermal neutrons is 930×10^{-28} m^2 (930 b) (43). All of the experimental data available to the end of 1986 on the cross sections for the nuclear reaction $^7\text{Li} + n \longrightarrow {}^4\text{He} + n + \text{T} - 2.47$ MeV, which utilizes the much more abundant isotope ^{7}Li, are collected in a review (44).

A second, more favorable reaction is

$$^3\text{He} + {}^1n \longrightarrow {}^3\text{T} + {}^1\text{H}$$

for which the capture cross section of ^{3}He to thermal neutrons is 5200×10^{-28} m^2 (5200 b). The limited availability of ^{3}He (0.00013–0.00017% natural abundance) (45) restricts the practical importance of this mode of production (see also HELIUM-GROUP GASES). Tritium is also produced by the action of high energy protons (such as primary cosmic rays) on a number of elements, and by the reaction of cosmic ray neutrons and ^{14}N. Reaction cross sections are small, generally a few to a few hundred m$^2 \times 10^{-31}$ (millibarns).

Production in Target Elements.   Tritium is produced on a large scale by neutron irradiation of ^{6}Li. The principal U.S. site of production is the Savannah River plant near Aiken, South Carolina where tritium is produced in large heavy-water moderated, uranium-fueled reactors. The tritium may be produced either as a primary product by placing target elements of Li–Al alloy in the reactor, or as a secondary product by using Li–Al elements as an absorber for control of the neutron flux.

The confinement region in which nuclear fusion proceeds is surrounded by a blanket in which the neutrons produced by the fusion reaction are captured to produce tritium. Because of its favorable cross section for neutron capture, lithium is the favored blanket material. Various lithium blanket materials have been considered, ranging from liquid lithium metal, lithium-lead alloys, or lithium dioxide to aqueous solutions of lithium salts. Lithium ceramics such as lithium aluminate, LiAlO_2, and lithium zirconates, Li_2ZrO_3, $\text{Li}_6\text{Zr}_2\text{O}_7$, and Li_8ZrO_6, continue to show promise as candidate breeder materials (46). Extraction of tritium from the breeder blanket also poses many problems.

Production in Heavy Water Moderator.   A small quantity of tritium is produced through neutron capture by deuterium in the heavy water used as moderator in the reactors. The thermal neutron capture cross section for deuterium is extremely small (about 6×10^{-32} m^2), and consequently the tritium produced in heavy water moderated reactors is generally significant only as a potential health hazard. However, in a high-flux reactor such as that at the Institute Max von Laue-Paul Langevin (Grenoble, France), the heavy water moderator is a useful source of tritium (39).

Production in Fission of Heavy Elements.   Tritium is produced as a minor product of nuclear fission (47). The yield of tritium is one to two atoms in 10,000 fissions of natural uranium, enriched uranium, or a mixture of transuranium nuclides (see ACTINIDES AND TRANSACTINIDES; URANIUM).

Production-Scale Processing.   The tritium produced by neutron irradiation of ^6Li must be recovered and purified after target elements are discharged from nuclear reactors. The targets contain tritium and ^4He as direct products of the nuclear reaction, a small amount of ^3He from decay of the tritium and a small amount of other hydrogen isotopes present as surface or metal contaminants.

In the recovery process the gaseous constituents of the target are evolved, and the hydrogen isotopes separated from other components of the gas mixture. A number of methods that can be applied to a process mixture or to naturally occurring sources are available for separating the tritium from the hydrogen and deuterium. Because of the military importance of tritium, details of the large-scale production of this isotope have not been published. A report, however, is available that describes the large-scale production of tritium and its uses in France (48).

Isotopic Concentration.   A number of techniques have been reported for concentrating tritium from naturally occurring sources. For example, separation factors (H/T) of 6.6 to 29 were observed (49) for the concentration of tritium by electrolysis of tritiated water. Tritium is concentrated in the undecomposed water.

Low (20–25 K) temperature distillation has been widely used to separate hydrogen and deuterium and has also been successfully applied to the separation of tritium from the other hydrogen isotopes (50,51). At Los Alamos National Laboratory, a system of four interlinked cryogenic fractionation columns has been designed for the separation of an approximately equal mixture of deuterium and tritium containing a small amount of ordinary hydrogen into a tritium-free stream of HD for waste disposal, and streams of high-purity D_2, DT, and T_2 (52) (see CRYOGENICS). Mathematical models that are in good agreement with experimental results on the separation of deuterium–tritium by cryogenic distillation have been developed (53).

Concentration by gas chromatography has also been demonstrated. Elution chromatography has been used on an activated alumina column to resolve the molecular species H_2, HT, and T_2, thereby indicating a technique for separation or concentration of tritium (54). This method was extended (55) to include deuterium components. The technique was first demonstrated in 1964 using macro quantities of all six hydrogen molecular species (56).

Successful separations of tritium from hydrogen and deuterium have been achieved by a cryogenic thermal diffusion column (57), by diffusion through a palladium–silver–nickel membrane (58), and by chromatography on coated molecular sieves (59). Laser separation of tritium also appears to be competitive with more conventional isotope separation methods (60) (see LASERS). Large separation coefficients have been reported for zirconium chromium hydride ($ZrCr_2H_x$)–hydrogen isotope systems (61). A number of processing options thus are available for processing fuel and product streams containing H, D, and T in nuclear fusion reactors.

Analysis and Detection

Tritium is readily detectable because of its radioactivity. Under certain conditions concentrations as low as 370 μBq/mL (10^{-8} μCi/mL) can be detected. Most detection devices and many analytical techniques exploit the ionizing effect of the tritium β-decay as a principle of operation (62,63).

Ionization Chamber. The ionization chamber is a simple, sensitive, and sturdy device filled with gas and containing two electrodes between which a potential difference is maintained. When gas containing tritium is admitted to the chamber, the radiation ionizes the gas, the ions are drawn to the electrodes, and a flow of current results that is proportional to the tritium concentration. Because of the ability to detect very small currents, this technique is very sensitive and can be adapted to either static or flow systems. Ionization chambers are used as process-stream monitors, leak detectors, stack monitors, breathing-air monitors, detectors for surface contamination, and detectors instrumented with vibrating-reed electrometers. For the last instrument, concentrations of tritiated water vapor in air of 370 μBq/mL (10^{-8} μCi/mL) (STP) can be measured. Developments in very low background proportional counters and chambers to be used for low-level internal gas counting of β-particles emitted by tritium are the subject of a comprehensive review (64). The combination of a proportional counter and a computer significantly enhances the measurement of very low levels of tritium in air.

Mass Spectrometer. The mass spectrometer is the principal analytical tool of direct process control for the estimation of tritium. Gas samples are taken from several process points and analyzed rapidly and continually to ensure proper operation of the system. Mass spectrometry is particularly useful in the detection of diatomic hydrogen species such as HD, HT, and DT. Mass spectrometric detection of helium-3 formed by radioactive decay of tritium is still another way to detect low levels of tritium (65). Accelerator mass spectroscopy (ams) has also been used for the detection of tritium and carbon-14 at extremely low levels. The principal application of ams as of this writing has been in archeology and the geosciences, but this technique is expected to facilitate the use of tritium in biomedical research, various clinical applications, and in environmental investigations (66).

Thermal-Conductivity Analyzer. The thermal-conductivity analyzer operates on the principle that the loss of heat from a hot wire by gaseous conduction to a surface at a lower temperature varies with the thermal conductivity of the gas, and is virtually independent of pressure between 1.3 kPa (10 mm Hg) and 101 kPa (1 atm). This technique is frequently used in continuous monitors for tritium in binary gas mixtures for immediate detection of process change.

Calorimeter. The β-decay energy of tritium is very precisely known (9). The thermal energy generated by the decay can thus be used with a specially designed calorimeter to measure the quantity of tritium in a system of known heat capacity (see THERMAL, GRAVIMETRIC, AND VOLUMETRIC ANALYSIS).

Liquid Scintillation Counter. The rapid and sensitive technique of liquid scintillation counting is applied for the determination of tritium in liquid systems. The tritiated sample is dissolved in a solvent that contains an organic scintillator. Because many samples, particularly those of a biological nature, are not soluble in water or organic solvents, emulsions are usually used in scintillation counting (67,68). The accurate determination of tritium in biosystems is a particularly difficult problem because of the sizeable isotope effect encountered in the removal of tissue water. Methods for minimizing tritium fractionation during water removal have been described (69).

The emitted β particles excite the organic molecules which, in returning to normal energy levels, emit light pulses that are detected by a photomultiplier tube, amplified, and electronically counted. Liquid scintillation counting is by far the most widely used technique in tritium tracer studies and has superseded most other analytical techniques for general use (70).

Analysis of H–D–T Mixtures.   Raman spectroscopy is a very useful and practical method for the analysis of diatomic molecules containing tritium and deuterium or hydrogen. Absolute light scattering intensities for the rotational Raman lines can be predicted, thus making absolute analyses possible. Gas chromatography is another excellent method for the analysis of H–D–T mixtures. The stationary phase consists of alumina, Al_2O_3, coated with ferric oxide, Fe_2O_3; neon is used as the carrier gas. Detectors are usually both thermal conductivity (caratherometer) and ion chamber detectors.

Health Physics Aspects

Hazards.   Because tritium decays with emission of low-energy radiation ($E_{av} = 5.7$ keV), it does not constitute an external radiation hazard. However, tritium presents a serious hazard through ingestion and subsequent exposure of vital body tissue to internal radiation. The body assimilates tritiated water and distributes it throughout body fluids with remarkable speed and efficiency. When exposed to tritiated water vapor via inhalation, people absorb 98–99% of the activity inspired through the respiratory system (71). Uniform distribution throughout body fluids occurs within 90 minutes. Also, when exposed to such an atmosphere, tritium entering the body through the total skin area approximately equals that entering the lungs. Molecular or elemental tritium (T_2) or HT is much less readily assimilated. Approximately 0.004% of such activity inspired is absorbed, apparently after preliminary oxidation in the lungs. Negligible amounts of elemental tritium are absorbed through the skin.

It is generally assumed that ingested tritiated water is rapidly absorbed and uniformly distributed in the body fluids, with the result that the entire organism is uniformly irradiated. This may not necessarily be the case. In experiments using mice, the ingestion of low-level tritiated water causes liver damage (72). The ingestion of tritium used as a tracer in organic molecules that are metabolized in specific pathways can concentrate tritium, resulting in the possibility of localized radiation damage. For example, tritiated thymidine concentrates selectively in the DNA of cells, resulting in selective damage to cell nuclei (73). Environmental assessments of tritium must take these factors into consideration (74).

The body excretes tritium with a biological half-life of 8–14 d (10.5 d average) (75), which can be reduced significantly with forced fluid intake. For humans, the estimated maximum permissible total body burden is 37 MBq (1 mCi). The median lethal dose (LD$_{50}$) of tritium assimilated by the body is estimated to be 370

GBq (10 Ci). Higher doses can be tolerated with forced fluid intake to reduce the biological half-life.

Monitoring and Control. Detailed descriptions of methods used for handling and monitoring tritium at Savannah River (76,77) and the European Tritium Handling Program (78) have been published.

A widely used instrument for air monitoring is a type of ionization chamber called a Kanné chamber. Surface contamination is normally detected by means of smears, which are simply disks of filter paper wiped over the suspected surface and counted in a windowless proportional-flow counter. Uptake of tritium by personnel is most effectively monitored by urinalyses normally made by liquid scintillation counting on a routine or special basis. Environmental monitoring includes surveillance for tritium content of samples of air, rainwater, river water, and milk.

The radiological hazard of tritium to operating personnel and the general population is controlled by limiting the rates of exposure and release of material. Maximum permissible concentrations (MPC) of radionuclides were specified in 1959 by the International Commission on Radiological Protection (79). For purposes of control all tritium is assumed to be tritiated water, the most readily assimilated form. The MPC of tritium in breathing air (continuous exposure for 40 h/wk) is specified as 185 kBq/mL (5 μCi/mL) and the MPC for tritium in drinking water is set at 3.7 GBq/mL (0.1 Ci/mL) (79). The maximum permitted body burden is 37 MBq (one millicurie). Whenever bioassay indicates this value has been exceeded, the individual is withdrawn from further work with tritium until the level of tritium is reduced.

Personnel are protected in working with tritium primarily by containment of all active material. Containment devices such as process lines and storage media are normally placed in well-ventilated secondary enclosures (hoods or process rooms). The ventilating air is monitored and released through tall stacks; environmental tritium is limited to safe levels by atmospheric dilution of the stack effluent. Tritium can be efficiently removed from air streams by catalytic oxidation followed by water adsorption on a microporous solid absorbent (80) (see ABSORPTION).

Several new technologies are in the process of development at the Savannah River plant that would considerably enhance safety in handling large amounts of tritium. Metal hydride technology has been developed to store, purify, pump, and compress hydrogen isotopes. Conversion to or extraction from metal triteride would offer flexibility and size advantages compared to conventional processing methods that use gas tanks and mechanical compressors, and should considerably reduce the risk of tritium gas leaks (see HYDRIDES).

Personnel who must work in areas in which tritium contamination exceeds permitted levels are safeguarded by protective clothing, such as ventilated plastic suits. Detailed descriptions of laboratories suitable for manipulation of tritium can be found in Reference 9.

Uses

Nuclear fusion is an approach to the ever increasing global demands for energy. All nuclear fusion reactions require very high temperatures for initiation. The thermal threshold is lowest for light ions, and the nuclear reaction $D(T,n)^4He$, involving the fusion of deuterium and tritium nuclei is considered to be the most practical approach to the realization of nuclear fusion energy (qv). Whereas the technology for large-scale production of deuterium exists, the production and handling of tritium is one of the key problems in the achievement of practical nuclear fusion.

The development of a tritium fuel cycle for fusion reactors is likely to be the focus of tritium chemical research into the twenty-first century.

Tritium is widely used as a tracer in molecular biology (see RADIOACTIVE TRACERS).

BIBLIOGRAPHY

"Deuterium and Tritium" in *ECT* 2nd ed., Vol. 6, pp. 895–910, by J. F. Proctor, E. I. du Pont de Nemours & Co., Inc.; in *ECT* 3rd ed., Vol. 7, pp. 539–564, by Joseph J. Katz, Argonne National Laboratory.

1. A. H. Wapstra, *Nat. Bur. Stds (U.S.) Spec. Publ.* **343**, 151 (1971).
2. M. L. E. Oliphant, P. Harteck, and E. Rutherford, *Prac. Roy. Soc.* **A144**, 692 (1934).
3. T. W. Bonner, *Phys. Rev.* **53**, 711 (1938).
4. L. W. Alvarez and R. Cornog, *Phys. Rev.* **56**, 613 (1939).
5. J. L. Heilbron and R. W. Seidel, *Lawrence and His Laboratory: A History of the Lawrence Berkeley Laboratory*, Vol. 1, University of California Press, Berkeley, 1989, pp. 368–373.
6. R. Viallard in P. Pascal, ed., *Nouveau Traite de Chimie Mineral*, Tome 1, Masson et Cie, Paris, 1956, p. 911.
7. E. L. Meutterties, *Transition Metal Hydrides*, Marcel Dekker, Inc., New York, 1971, Chapt. 1, pp 1–7.
8. K. M. MacKay and M. F. A. Dove, in *Comprehensive Inorganic Chemistry*. Pergamon Press, Oxford, 1973, Chapt. 3, pp. 77–116.
9. E. A. Evans, *Tritium and Its Compounds*, 2nd ed., John Wiley & Sons, Inc., New York, 1974.
10. H. M. Mittelhauser and G. Thodos, *Cryogenics* **4**, 368 (1964).
11. E. R. Grilly, *J. Am. Chem. Soc.* **73**, 843 and 5307 (1951).
12. P. C. Souers and co-workers, *Three-Phase Region of D_2-DT-T_2, Lawrence Livermore Laboratory Report UCRL-79036*, 1977.
13. R. H. Sherman, J. R. Bartlit, and R. A. Briesmeister, *Cryogenics* **16**, 611 (1976).
14. F. D. Rossini and co-workers, *Natl. Bur. Stds. (U.S.) Circular 500*, U.S. Government Printing Office, Washington, D.C., 1950.
15. R. J. Le Roy, S. G. Chapman, and F. R. W. McCourt, *J. Phys. Chem.* **94**, 923–929 (1990).
16. E. W. Albers, P. Harteck, and R. R. Reeves, *J. Am. Chem. Soc.* **86**, 204 (1964).
17. J. W. Pyper and C. K. Briggs, *The Ortho-Para Forms of Hydrogen, Deuterium and Tritium: Radiation and Self-induced Conversion Kinetics and Equilibrium, Lawrence Livermore Laboratory Report UCRL-52278*, 1977.
18. W. M. Jones, *J. Chem. Phys.* **48**, 207 (1968).
19. M. Goldblatt, *J. Phys. Chem.* **68**, 147 (1964).
20. P. A. Staats, H. W. Morgan, and J. H. Goldstein, *J. Chem. Phys.* **24**, 916 (1956).
21. M. Goldblatt and W. M. Jones, *J. Chem. Phys.* **51**, 1881 (1969).
22. Ref. 9, p. 9.

23. P. Diehl in T. Axenrod and G. Webb, eds. *Nuclear Resonance Spectroscopy of Nuclei Other Than Protons*, John Wiley & Sons, Inc., New York, 1974, pp. 275–285.
24. J. P. Bloxsidge and co-workers, *J. Chem. Res. (Part S)*, 42 (1977).
25. M. L. Eidenoff, *J. Am. Chem. Soc.* **69**, 977 (1947).
26. *Tritium in the Physical and Biological Sciences*, Vol. I, International Atomic Energy Agency, Vienna, 1962, p. 162.
27. H. Gottschling and E. Freese, *Nature* **196**, 829 (1962).
28. K. E. Wilzbach and P. Riesz, *Science* **126**, 748 (1957).
29. P. D. Klein, *Adv. Chromatog.* **3**, 3 (1966).
30. L. M. Dorfman and B. A. Hemmer, *Phys. Rev.* **94**, 754 (1954).
31. J. W. Pyper and C. K. Briggs, *Kinetics of the Radiation-induced Exchange Reactions of Hydrogen, Deuterium, and Tritium, Lawrence Livermore Laboratory Report UCRL-52380*, 1978.
32. K. Wilzbach, *J. Am. Chem. Soc.* **79**, 1013 (1957).
33. M. Wenzel and P. E. Schulze, *Tritium Markierung, Preparation, Measurement and Uses of Wilzbach Labelled Compounds*, Berlin, Walter de Gruzter, 1962.
34. G. R. Choppin and J. Rydberg, *Nuclear Chemistry*, Pergamon Press, Oxford, UK, 1980, p. 186.
35. V. Fallings and P. Harteck, *Z. Naturforsch.* **5a**, 438 (1950).
36. A. V. Grosse and co-workers, *Science* **113**, 1 (1951).
37. Ref. 24, pp. 5–32.
38. Ref. 26, pp. 56–67.
39. P. Pautrot and J. P. Arnauld, *Trans. Am. Nucl. Soc.* **20**, 202 (1975).
40. D. D. Hoel, R. J. Kurzeja, and A. G. Evans, *Energy Res. Abstr.* **16**, abstr. 5101 (1991).
41. R. L. Michel, *IAHS Publ.* **179**, 109–115 (1989); C. E. Murphy, Jr., *Energy Abstr.* **16**, abstr. 5100 (1991).
42. B. Brigoli and co-workers, *Health Phys.* **61**, 105–110 (1991).
43. G. Friedlander and J. W. Kennedy, *Nuclear and Radiochemistry*, John Wiley & Sons, Inc., New York, 1955, p. 404.
44. B. Yu and D. Cai, *Chin. J. Nucl. Phys.* **12**, 107–116 (1990).
45. Ref. 43, p. 415.
46. C. E. Johnson, *Ceram. Int.* **17**, 253–258 (1991).
47. E. L. Albenesius, *Phys. Rev. Lett.* **3**, 274 (1959).
48. *Le Tritium*, Commissariat a l'Energie Atomique, Bull. Inform. Scientifiques et Techniques, No. 178, Feb. 1973.
49. S. Kaufman and W. F. Libby, *Phys. Rev.* **93**, 1337 (1954).
50. T. M. Flynn and co-workers, *Proceedings of the 1957 Cryogenic Engineering Conference*, U.S. National Bureau of Standards, Boulder, Col., 1958, p. 58.
51. M. Damiani, R. Getraud, and A. Senn, *Sulzer Techn. Rev.* **4**, 41 (1972).
52. J. R. Bartlet, W. H. Denton, and R. H. Sherman, "Hydrogen Isotope Distillation for the Tritium Systems Test Assembly," *American Nuclear Society Conference on the Technology of Controlled Nuclear Fusion, May 9–11, 1978*, Santa Fe, N. M.
53. T. Yamanishi and co-workers, *Nippon Genshiryoku Kenkyusho*, (1988); R. H. Sherman, "Fusion Technology," *Second National Topical Meeting on Tritium in Fission, Fusion, and Isotopic Applications, Apr.–May, 1985*, Dayton, Ohio.
54. Ref. 26, pp. 121–133.
55. J. King, *J. Phys. Chem.* **67**, 1397 (1963).
56. D. L. West and A. L. Marston, *J. Am. Chem. Soc.* **86**, 4731 (1964).
57. I. Yamamoto and A. Kanagawa, *J. Nucl. Sci. Technol.* **27**, 250–255 (1990).
58. V. M. Bystritskii and co-workers, *Prib. Tekh. Eksp.*, 216–219 (1991).
59. K. K. Pushpa, K. A. Rao, and R. M. Iyer, *J. Chromatogr. Sci.* **28**, 441–444 (1990).
60. I. P. Herman, K. Takeuchi, and Y. Makide, *Opt. Eng. (N.Y.)* **20**, 173–220 (1989).

61. I. L. Vedernikova and co-workers, *Zh. Fiz. Khim.* **65,** 1657–1660 (1991).
62. E. L. Albenesius and L. H. Meyer, *DP-771*, Savannah River Plant, E. I. du Pont de Nemours & Co., Inc., Aiken, South Carolina, 1962.
63. D. B. Hoisington, *Nucleonics Fundamentals*, McGraw-Hill Book Co., Inc., New York, 1959, pp. 316–319.
64. M. Garcia-Leon and G. Madurga, eds., *Low-Level Meas. Man-Made Radionuclides Environ., Proc. Inc. Summer Sch., 2nd*, World Science, Singapore, 1991, pp. 38–71.
65. S. Halas, *Zesz. Nauk. Politech. Slask., Mat.-Fiz* **47,** 9–21 (1986).
66. M. L. Roberts and co-workers, *Nucl. Instrum. Methods Phys. Res., Sect. B, (Pt. 2)* **B56-B57** 882–885 (1991).
67. A. Dyer and J. C. J. Dean, *J. Radioanal. Nucl. Chem.* **141,** 139–154 (1990).
68. G. Rauret and co-workers, *Analyst (London)* **115,** 1097–1101 (1990).
69. M. A. Kim, F. Baumgaertner, and C. Schulze, *Radiochim. Acta* **55,** 101–106 (1991).
70. K. D. Neame and C. A. Homewood, *Liquid Scintillation Counting*, John Wiley & Sons, Inc., New York, 1974.
71. E. A. Pinson and W. H. Langham, *J. Appl. Physiol.* **10,** 108 (1957).
72. S. Pareek and A. L. Bhatia, *Radiobiol. Radiother.* **30,** 167–172 (1989).
73. J. Schapiro, *Radiation Protection*, 3rd ed., Harvard University Press, Cambridge, Mass., 1990, pp. 147–148.
74. *Ibid.*, pp. 396–397.
75. H. L. Butler and R. W. Van Wyck, *DP-329*, Savannah River Plant, E. I. du Pont de Nemours & Co., Inc., Aiken, South Carolina, 1962.
76. W. C. Reinig and E. L. Albenesius, *Am. Ind. Hyg. Assoc. J.* **24,** 276 (1963).
77. M. S. Ortman and co-workers, *J. Vac. Sci. Technol.* **8,** 2881–2889 (1990).
78. B. Hircq, *Fusion Eng. Des.* **14,** 161–170 (1991).
79. International Commission on Radiological Protection Publication 2, *Report of Committee II on Permissible Dose for Internal Radiation*, Pergamon press, Oxford, 1959.
80. A. E. Sherwood, *Tritium Removal from Air Streams by Catalytic Oxidation and Water Adsorption, Lawrence Livermore Laboratory Report UCRL-78173, 1976.*

JOSEPH J. KATZ
Argonne National Laboratory

DEWATERING

Dewatering is the last process applied to separate water from a solid, unless thermal drying (qv) is used. Dewatering is usually a mechanical process that presses residual water from solids or displaces the water with a gas, and the energy required is negligible compared with the heat required for drying (1). Thus there is a significant incentive for adding a dewatering step to a process. Whereas the broader term deliquoring includes the separation of nonaqueous liquids from solids, the discussion herein is specific to water removal.

In municipal wastewater treatment, dewatering is regarded as the final process used to achieve a water content of ≤85% in the sludge (2,3) or to change the behavior of sludge to that of a solid (4) (see WATER, SEWAGE). Many other industries regard 85% water content as *feed* to a primary liquid–solid separation device, such as a thickener, and consider 15% moisture (85% solids) an appropriately dewatered product. Dewatering, then, is not defined by moisture content or by the use of a type of equipment (5).

Maximizing mechanical dewatering is a matter of selecting the most appropriate equipment, optimizing the variables that affect dewatering on the equipment, and optimizing the feed. Optimizing the feed is usually most important. However, if the feed contains a product the properties of which should not be changed or if dewatering is easy, then equipment selection is the most important variable. Selection of equipment is dependent on the particular feed stream, for example, cotton (qv), fruit juices (qv), or sewage, and the field is in constant development. Aids to selecting equipment are found in the literature (see General References). Once the choice of equipment has been narrowed, eg, centrifuges, belt presses, or the like, testing on the equipment is extremely important.

Interaction of Water and Solids

Water associates with solids in a range of energies (6). The energy needed for water removal is indicated in Figure 1. The highest dewatering energies are associated with a monolayer or less of water, which is not generally considered moisture content.

Dewatering processes, normally concerned with water bound in capillaries, affect the water and solids by changing the size distribution of capillary radii,

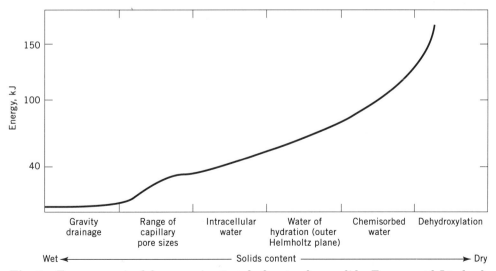

Fig. 1. Energy required for removing 1 mol of water from solids. To convert kJ to kcal, divide by 4.184.

reducing adhesion of water to the solids, displacing water from the capillaries, and reducing the energy required to cause flow in the capillaries. These effects are achieved by three methods (1). (*1*) The particulate matrix may be compacted by applying stress, that is, forces can be induced by frictional drag of the liquid as it flows through the pores. Body forces can arise from gravity or centrifugal motion, and boundary stresses can be applied with rolls, membranes, pistons or screw presses, or acoustic energy (see MEMBRANE TECHNOLOGY; ULTRASONICS). (*2*) The water may be displaced, usually with a gas, by application of vacuum or pressure. Centrifugal force in a pusher centrifuge also results in the water being displaced by a gas. (*3*) An electrical field may be applied to a slurry of charged particles (6).

Although all the techniques are effective, in industrial applications there is rarely time to achieve an equilibrium reduced saturation state (see FILTRATION), so variables that affect only the kinetics of dewatering and not the equilibrium and residual moisture are also very important. The most important kinetic variables in displacing the liquid from the solid are increases in pressure differentials and viscosity reduction.

Cake Dewatering

The most important function of filtration (qv) is the formation of a filter or centrifuge cake from a slurry (see also SEPARATION, CENTRIFUGAL). The most important function of dewatering is removal of the liquid from the resulting cake. Theories of cake dewatering are covered elsewhere (7–12). Although theory lags behind practice, the principal variables controlling dewatering have been identified and both dewatering rate as well as equilibrium and residual moisture in fully defined cakes can be predicted. The behavior of relatively incompressible cakes can be modeled using changes in porosity with pressure. For more compressible cakes, empirical models have been developed (13). Theoretical models of expression dewatering of compressible cakes exist (12). Very few industrial applications of dewatering have consistent or fully characterized cakes, and no process in the design stage can fully characterize a filter cake.

Ways to reduce the final moisture content of a centrifuge cake include the use of steam (qv), surfactants (qv), or flocculants (see FLOCCULATING AGENTS), as well as pretreatments by pelletizing, oil agglomeration, thermal treatment, and freeze–thaw processes. The main dewatering variable in the centrifuge itself is centrifugal force sufficient to expel the liquid from the pores in the cake through the filter medium. In some newer centrifuges the pitch of the screw is shortened, applying compression to the cake to further reduce cake pore volume and, with it, entrained water (14). Other basket centrifuge designs incorporate baffles to prevent materials from slipping out of the basket before dewatering (35).

Cake dewatering is related to cake formation; in a filter cake, the final moisture content is dependent on many variables that also control cake formation. Pretreatment processes, for example, affect both cake formation and cake dewatering. Cake dewatering is achieved by compacting the solids; displacing the residual liquid in the cake with another phase, usually a gas; or applying an electrical current to remove the liquid. Each process relies on different properties of the cake.

Compaction, Compression, and Expression. Compaction is a newer term for compression and is used to describe the movement of particles relative to one another within a device until the matrix of particles gains enough strength to resist further consolidation (16). Compaction occurs in a plate and frame filter both while the chamber is filling and at the end of the cycle when the chamber is nearly full and the pressure rises steeply. Compactibility (or compressibility) describes the reduction in volume of the particle matrix. Compaction also takes place in the bed of a thickener as the solids continuously deposit on the top of the bed and a thickened slurry is withdrawn from the bottom.

Compression virtually always reduces the permeability of the filter cake. The reduced permeability in highly compressible cakes is greatest at the surface of the filter medium, where the pressure drop is greatest. Further dewatering of these cakes in a filter press can be achieved by reversing the flow on a plate-and-frame filter and using the compressed "skin" of filter cake as a membrane to squeeze water from the center of the cake, where pressure differentials are lower (17). Another method of applying compression is by delayed cake formation using continuous pressure filters. These filters easily remove large quantities of water and produce a highly thickened discharge, which sometimes forms a rope as it discharges. Continuous pressure filters can be very effective for thickening dilute slurries and for washing in slurry form (18).

Expression is the application of mechanical stress to a matrix of particles in the fully formed cake (12). Expression, which refers to squeezing the solids rather than the slurry (19), is used to further reduce void space in order to separate more liquid from the solids. Expression dewatering is most effective with compressible cakes (12,20) of particle sizes below 50 μm (7), superflocculated cakes or cakes otherwise containing loosely entrapped liquid, and organic solids containing liquid-filled cells that must be ruptured for effective dewatering. Filters that use expression include variable-chamber plate-and-frame filters, belt-filter presses, very high pressure belt presses (21,22), tube presses, and screw presses. Tube presses can exert pressures up to 14 MPa (2000 psi) (23), the highest pressure of any of these expression devices. After pressing, an air-blow step can be added to further reduce the entrained moisture (23). Screw presses are used primarily with fibrous or polymeric materials containing entrapped liquid. Pressures up to 110 MPa (16,000 psi) have been measured in these devices at the points between the screw and screen when solids are present (24).

Significant improvements were made in the 1980s and early 1990s in high capacity, automated variable volume filters that incorporate automatic pressure filtration, expression, washing, and air displacement. Some of the large plate-and-frame automatic presses can operate at up to 2 MPa (ca 285 psig), with up to 100 chambers (25,26).

If expression is effective, it reduces the permeability of the cake being compacted and, as a consequence, the resistance to flow of the liquid increases considerably (27). The effectiveness of expression is governed by cake thickness, specific resistance, consolidation properties, and shear forces.

Displacement Dewatering. Replacement of the liquid in the voids of a cake by another liquid or a gas is termed displacement dewatering. Air displacement can be accomplished by using a pressure difference to force the liquid from the pores in the cake. The types of filters that provide displacement dewatering include virtually all vacuum filters, rotary pressure disk and drum filters, the Lasta (25), Larox (28), and Vertipress (29) automatic filter presses, hyperbaric filters, tube press filters (23), and a hybrid continuous pressure filter–expression press (30). In a centrifuge, displacement dewatering is accomplished by applying a body force directly to the liquid by the spinning motion. The factors that control displacement dewatering are

Property	Variable
cake	particle size and size distribution, shape, packing, dimensions of the cake
fluid	density, viscosity
interfacial forces	surface tension (gas–liquid), interfacial tension (solid–liquid, gas–solid)
other	temperature, pressure gradient, rate of displacement

In a study on dewatering methods for peat, displacement dewatering was done using acetone, a polar solvent having a lower heat of vaporization than water. Dewatering was improved in terms of both the pressure filtering step and the quantity of heat required. Less heat was required to dry the cake and recover the acetone from the filtrate by distillation (31).

The rate of displacement dewatering increases by increasing the driving force, bed permeability, and filter area, and decreasing viscosity or cake thickness. The dewatered cake becomes drier by increasing the driving force, bed permeability, and the contact angle or decreasing the surface tension. There are a number of techniques to achieve these results.

Expression Dewatering of Fibrous Materials. Fibrous materials are frequently dewatered in belt-filter, screw, disk, and roll presses and in batch pot and cage presses. Table 1 lists applications of screw, roll, and pot presses. Screw and high pressure belt presses are continuous and have replaced batch pot and cage presses in most applications. Traditionally, however, batch presses have been used for squeezing cocoa butter from cocoa beans, which require pressures up to 41 MPa (6000 psi) (39). A description of many types of batch presses is included in Reference 40.

Screw presses (Fig. 2) do not produce a clear liquid product. Frequently, the product is further filtered in a filter press to give a clear liquid product. Press aids are added to feed materials containing fine particles or particles that can deform and plug the slots in the cage of a screw press. Typical press aids include sawdust, rice hulls, perlite, and diatomaceous earth (see DIATOMITE). A vertical screw press is a continuous press that has been used for dewatering sewage sludge (2).

A disk press, shown in Figure 3, can achieve a compression ratio of about 4:1 and produce a paper pulp having a consistency of 45–50% solids. It has also been used on brewers' spent grains and coffee grounds. The two surfaces of the rotating, converging press disks have a screen backing that retains and presses the solids while letting liquid pass through the screen. Slurry is fed at the wide

Table 1. Applications of Screw, Roll, and Pot Presses[a]

Material	Liquid, %	
	In feed	In product
paper pulp	97	50
	90	65
wood chips	85	50
sugar cane	68	43
oilseeds		
high oil content[b]	>30	3–7
low oil content[c]	<30	3–6
cocoa (separation of cocoa butter)	53	12
food[d]	60–90	10–30
polymers, elastomeric and thermoplastic[e]	60	5–8
rendered tissue	20[f]	6–10
sewage sludge[g]	98	85

[a]Refs. 32–37.
[b]Includes copra, cottonseed, corn germ, peanuts, flax, safflower, sunflower, sesame, palm kernels, and linseed.
[c]Includes soybean, rice bran, and dry-process corn germ.
[d]Includes apples, carrots, coffee grounds, fish, grapes, pineapples, and tomatoes.
[e]Includes ABS, nitriles, styrene–butadiene rubber (SBR), natural rubber, and ethylene–propylene–diene rubber (EPDM).
[f]Fat.
[g]The sludge was steam-heated in the press and treated with CaO and polymer. About 95% of the solids were retained in the cake (38).

part of the space between the disks, and the slurry is carried through the maximum compression zone before being released as the disks diverge (41). A similar device, called the shoe rotary press, has been tested on both fine coal and fine coal refuse (42).

A different type of press is the Vari-Nip shown in Figure 4 (43). A slurry is forced into the vat at up to 240 kPa (20 psig). Two porous rolls rotate in a pressurized vat of pulp slurry. As the rolls rotate together into the slurry, the differential pressure across the face of the rolls forces dewatering of the pulp and deposits fibers on the roll faces, forming fiber mats. The pressate, which has passed through the roll perforations, drains to the discharge ports. As the rotation continues, the mat enters the nip areas, and the rolls press the mat together, forcing dewatering to a high consistency. The resulting mat is then guided to a breaker conveyer (43). This press is somewhat similar to drilled press rolls used in paper making (see PAPER).

Improving Cake Dewatering. *Viscosity Reduction.* Equations relating the rate of liquid flow through a filter cake can be simplified to

$$\frac{V}{A} = \frac{K\Delta P}{\mu l}$$

where V in units of m³/s is the flow rate through the cake, A in m², is the area of the cake, K, m², is the cake permeability, ΔP, Pa, is the pressure drop across the

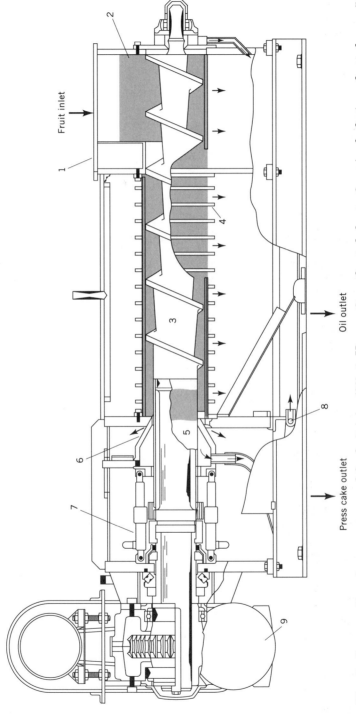

Fig. 2. Cross section of screw press used for fruit juice (32). 1, Hopper; 2, perforated sheets; 3, main shaft; 4, perforated cage; 5, draining cylinder; 6, cone; 7, hydraulic cylinder; 8, draining cylinder oil; 9, gear box. Courtesy of the French Oil Mill Machinery Co.

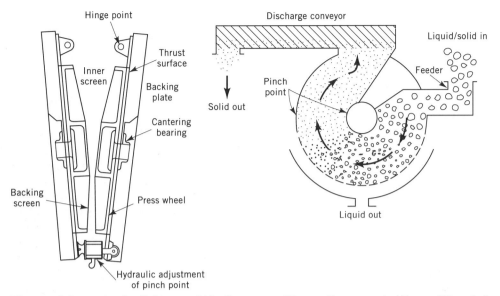

Fig. 3. Schematic of a disk press (41). Courtesy of Bepex Corp., a subsidiary of Berwind Corp.

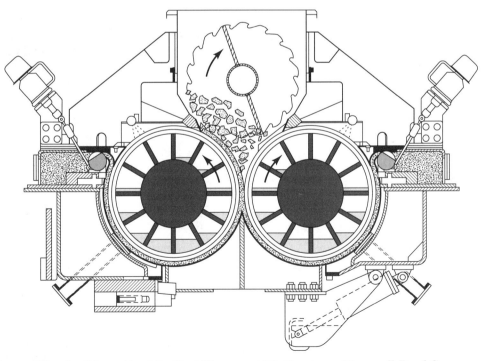

Fig. 4. Schematic of the Vari-Nip press (43). Courtesy of Ingersoll-Rand Co.

cake, l in m, is the thickness of the cake, and μ is the cake viscosity (Darcy's law). Viscosity, a kinetic variable, does not appear in equations describing reduced saturation levels in a cake (44). Because most practical filtration is limited by the time available for the steps of cake forming and dewatering, an increase in the flow rate of filtrate during dewatering translates directly to lower cake moisture levels. If the liquid is water, the viscosity can drop by a factor of three as the temperature rises from 15°C to 80°C, and consequently the flow rate of water through the cake is tripled. The addition of moderate quantities of salts, polymers, or small amounts of less viscous miscible liquids, such as alcohols, has very little effect on the viscosity of water. Temperature is the most important control of this variable.

The usual method of heating to improve dewatering is to apply low pressure steam to the filter cake as it is in the dewatering phase of the filter cycle. Steam has been used on rotary vacuum filters to dewater fine coal and on horizontal belt filters to dewater pipeline coal. Steaming typically removes an additional 0.7–1.5 kg of water per kilogram of steam applied and reduces moisture levels in coal filter cakes by about 5% (45–47). The effectiveness of steam depends directly on the permeability of the cake. In highly permeable cakes, up to 90% of the contained moisture can be removed. Generally, steam is ineffective on filter cakes in which particles smaller than 10 μm predominate (48). However, permeability is the critical factor: one of the largest installations of steam-assisted filtration is on <20 μm nonmagnetic taconites (49).

Use of Surfactants. Although the use of steam to improve dewatering is consistently beneficial, the effects of surfactants on residual moisture are highly inconsistent. Additions of anionic, nonionic, or sometimes cationic surfactants of a few hundredths weight percent of the slurry, 0.02–0.5 kg/t of solids (50), are as effective as viscosity reduction in removing water from a number of filter cakes, including froth-floated coal, metal sulfide concentrates, and fine iron ores (Table 2). A few studies have used both steam and a surfactant on coal and iron ore and found that the effects are additive, giving twice the moisture reduction of either treatment alone (44–46,49).

Surfactants aid dewatering of filter cakes after the cakes have formed and have very little observed effect on the rate of cake formation. Equations describing

Table 2. Effect of Surfactants on Residual Moisture in Filter Cakes

	Moisture content, %	
Material	Without surfactant	With surfactant
sulfide flotation concentrates[a]	15	12
	12	9
iron ore	17	15
	21	17
coal	20–22	16
	36–40	30–34
	6	9
silica sand	12	8

[a]CuFeS$_2$, MoS$_2$, and ZnS.

the effect of a surfactant show that dewatering is enhanced by lowering the capillary pressure of water in the cake rather than by a kinetic effect. The amount of residual water in a filter cake is related to the capillary forces holding the liquids in the cake. Laplace's equation relates the capillary pressure (P_c) to surface tension (σ), contact angle of air and liquid on the solid (θ) which is a measure of wettability, and capillary radius (r_c), or a similar measure applicable to filter cakes.

$$P_c = \frac{2\,\sigma\,\cos\theta}{r_c}$$

Surfactants lower the surface tension of water, typically from 72 to ca 30–35 mN/m ($\doteq$ dyn/cm), and many surfactants have a strong effect on the contact angle when used at low concentrations. Both changes help dewatering. Too much surfactant, near or above the critical micelle concentration (CMC), reverses the effect that the surfactant has on contact angle at lower concentrations, and at or above the CMC there is no further lowering of surface tension. At the higher concentrations, the surfactant loses some of its beneficial effect on dewatering, as shown in Figure 5. The beneficial effects of surfactants on dewatering are most pronounced in cakes that have been partially deslimed or in cakes of partially hydrophobic particles (eg, flotation concentrates) that are adsorbed onto each other. Surfactants at or above CMC have little practical effect on extremely fine cakes, where pores are small and the cake has no further opportunity to consolidate. A number of filter cakes do not respond to surfactant addition at any level.

pH Adjustment. Virtually all solids become charged in water, either by reaction with the water to form surface hydroxyl groups that can ionize or by adsorption of ions from the water. The charges on the particles affect how much water is bound to the particle. Reducing the charge on the particle increases the amount of dewatering possible in conventional dewatering equipment. For example, adjusting the pH of peat to approximately 3 increases dewatering, and

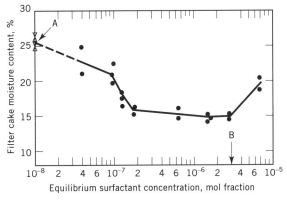

Fig. 5. Effect of surfactant concentration on moisture content of <500 μm coal filter cake (51). Point A represents zero surfactant concentration; Point B, the critical micelle concentration (CMC).

reducing the charge on coal using pH and metal ions improves the results of pressure and vacuum filtration (2,52,53). When an electrical field is externally applied for dewatering, the effect of the charge changes.

Use of Flocculants.   In the minerals industry and in water treatment (see WATER, MUNICIPAL WATER TREATMENT), the primary purpose of flocculants is to improve sedimentation rates and overflow clarity in thickening operations. Flocculants can also have a beneficial effect on dewatering in a filter or centrifuge. Generally, the flocculants that work best on sedimentation are not the best for filtration. For example, large, loose flocs are effective in causing rapid settling but trap water in the filter cake and can deform easily and block the filter medium. Any excess polymer may stick to the filter medium, causing further blinding (54).

Other flocculants are capable of improving filtration rates up to 100 times, especially of fine clay, sludges, or tailings (55). One of the main uses of flocculants in filtration is to make extremely slow filtering slurries filterable at reasonable rates. In addition, flocculants are critical in making belt-filter presses and dewatering centrifuges effective. There are extensive and helpful reviews on the selection of flocculants and the effects on the performance of dewatering devices (1,56–59). In municipal sludge processing, where often no flocculant is added to the primary thickening devices, flocculants are subsequently chosen to improve dewatering rates.

Although filtration rates can be much faster with flocculants, the final cake moisture is often higher in a flocculated cake (60–63). In contrast, using flocculants optimized for filtration, coal, and other mineral slurries can be dewatered to moisture contents significantly lower than the untreated cake (64–68). The advantages of rapid filtration rates can also be preserved. Flocculants that provide better filtration tend to form flocs having the following characteristics (65):

Floc characteristics	Beneficial effects
small	reduces intrafloccular water in the late stages of filtration; reduces pickup problems resulting from gravity settling in the rotary filter chamber
strong	prevents floc breakdown owing to suspending agitation in the filter tank; resists collapse and premature loss of cake permeability in the early interfloccular stage of filtration
equisized	prevents localized breakthrough of air, cake shrinkage, cracking, and early loss of vacuum
good fines capture (into floc structure)	provides good filtrate clarity; prevents cloth binding and poor discharge

Pumping a polymer-flocculated slurry to a filter degrades or destroys the floccules. To repair the damage, other flocculants, chosen for their optimum filtration characteristics (68), can be added. For example, to filter froth-floated coal (nominally <0.5-mm particle size), a medium molecular-weight anionic flocculant (average molecular weight of 10^7) is used. For sedimentation, much higher molecular-weight polymers are more effective (64,65).

In addition to specifying molecular weight, the chemical structure of poly-acrylamide flocculants has a significant effect on final moisture content. Two references (59,65) show the marked effects of both chemical structure and molecular weight on filtration rates and are useful guides to flocculant selection for coal- and clay-containing fine slurries (Fig. 6). Further dewatering of a flocculated filter cake can be achieved by using a surfactant dewatering aid, as described. The effectiveness of surfactants as dewatering aids seems neither to impair nor be impaired by the flocculant (65). Additions of 1–5 kg/t of an insoluble but highly water-absorbable polyacrylamide superabsorbent to a very sticky coal fines filter cake convert the material from a glue to a friable material that can be handled in normal material handling equipment (66) (see also WATER-SOLUBLE POLYMERS).

Sludge Conditioning. Sludge conditioning is the chemical, physical, or heat treatment of wastewater sludges to improve dewatering (2,3,69–71). Because sludge handling costs can be 25–50% of the total cost of wastewater treatment, dewatering is critical to cost control. A number of substances have been added to thickened sludge to increase the permeability of the filter or centrifuge cake. In addition to polymer flocculants, coagulants such as ferric and ferrous chloride, alum, and lime have been added, which chemically react in the sludge and improve dewatering. Diatomaceous earth, fly ash, and ash derived from incinerating dewatered sludge or bark have been added as body feeds to improve sludge dewatering.

Biological processing of sludge reduces the amount of sludge needing dewatering and changes the dewatering properties. The changes are not always helpful. A comparison of high biological oxidation of wastewater sludges, to minimize the amount of sludge that needed disposal, showed higher overall disposal costs compared with the costs of less oxidation and twice as much "wasting" (removal and disposal) of sludge (72). In an unusual application of biological conditioning, enzymes have been used to aid in the dewatering of phosphatic clay ponds (73).

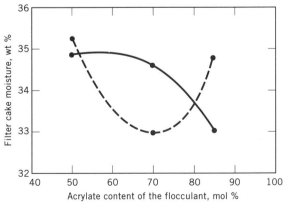

Fig. 6. Effect of acrylate content on coal refuse filter-cake moisture (59): (——), the effect on a longer chain flocculant (mol wt of 8×10^5); (– – –), shorter chain flocculant (mol wt of 2.5×10^6). Concentration of flocculant is 150 g/t.

The Hi-Compact mechanical dewatering process of Humboldt Wedag takes advantage of sludge conditioning to achieve a high degree of secondary dewatering of highly compressible sludges. The sludge is first flocculated and settled. The thickened sludge is then chopped into pellets that are coated with an incompressible filter aid, such as ash, fine coal, or another material. The coated pellets are then compressed in a batch press at about 5000 kPa (50 atm) to remove 60% of the remaining water. The incompressible coating on the pellets provides a network of channels for the expressed water to follow out of the pressed cake (74).

Comparisons are available on the relative performance and costs for dewatering municipal sludges (2). The relative performance of different filters and conditioners on waste sludges is shown in Table 3. The same sludge was treated on two belt-filter presses, two different centrifuges, and rotary vacuum filter (75). In another study, a variable chamber filter press, fixed-volume filter press, continuous belt-filter press, and rotary vacuum filter were compared for performance, capacity, and capital and operating costs (69).

Figure 7 shows the ranges of solids content achieved by various dewatering methods. The high solids centrifuges can achieve the same or higher solids content achieved by belt presses on municipal sludges. For recommended test procedures and expected results, consult References 2, 3, and 69–71. Particle size is the most important variable in dewatering municipal sludges (25) and is directly related to many of the other variables correlated with moisture content (76). The amount of $<5\ \mu$ particles is the largest variable, other than the nature of the sludge itself, in determining dewatering behavior (2).

Use of Mechanical Vibration. Vibration of the sludge or cake can further release entrapped moisture. Pretreatment of organic sludges and highly hydrated inorganic sludges using ultrasonic energy reduced by half the amount of polyelectrolyte polymer needed to achieve the same moisture content of the dewatered cake (77). Vibration of thixotropic slurries improved dewatering on a vacuum filter by breaking up trapped air and improving capillary channels (78). Other devices use ultrasonics to break up loose agglomerates, often of cosmetics (qv), so that these pass through a strainer (79).

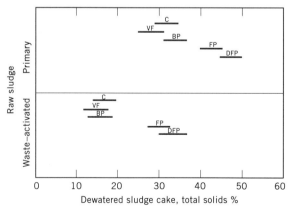

Fig. 7. Solids content of sludges dewatered by C, solid bowl centrifuge; VF, vacuum filter; BP, belt press; FP, filter press, and DFP, diaphragm filter press (2).

Table 3. Comparison of Filters on Aerobically Digested Sludges[a]

Property	Belt-filter press		Filter press		Rotary vacuum filter	Centrifuge
	Standard	Vacuum-assisted	Plates[b]	Cloth[b]		
cake produced, % solids	12–20	14–25	35–50	35–50	3–6	
operating pressure, MPa[e]	[c]	[c]	0.7–1.4	0.7–1.4	0.1[d]	2000–3000 g^f
relative space requirements	2.6	2.6	3.0	3.0	1.0	
relative cost, $[d]	2.2	2.8	3.0	5.2	1.0	
maximum size[d], m^2	60	60	400	60	100	
usual sludge conditioning	polymer	polymer	chemical and filter aid	chemical and filter aid	chemical and filter aid	polymer, 0–10 kg/t

[a] Ref. 69.
[b] Movable parts.
[c] Gravity plus pressure rolls.
[d] Values given are approximate.
[e] MPa unless otherwise noted. To convert MPa to psi, multiply by 145.
[f] Values are in units of gravitational acceleration.

43

Less Common Commercial Dewatering Processes

When solids dewatering is known to be a problem early in the process-design stage of a plant or is serious enough to warrant consideration of a range of dewatering alternatives, two approaches are available, and both can be used together. First, processes that begin the dewatering process while in the original suspension may be used. A second approach is to extract water or apply unusual desaturating forces to water present in sludges and cakes.

Agglomeration of Suspended Solids. *Pelletizing Precipitation.* Typical, cold, soda lime water softening generates a 5–15% soupy calcium carbonate sludge, which has been dumped into lagoons for disposal. By controlling reaction conditions of lime and hard water and providing for recirculating seeds of sand or calcium carbonate, precipitation of the carbonate can be controlled to form on the sand and to grow to 1.5-mm tight pellets. The pellets dewater by draining to less than 10% moisture (80,81). Control of crystal growth and crystal habit is used to improve dewatering in the production of phosphoric acid and in scrubbing of flue gas with lime (82,83). In each process, a precipitate of gypsum is formed. The most frequent applications of crystal growth regulation are to prevent scaling and to control freezing, for example, of trainloads of coal. The chemical principles are similar.

Additions of new flocculants after conventional thickening produce further dewatering of mineral slimes. A clay flocculated with polyacrylamides and rotated in a drum can produce a growth of compact kaolin pellets (84), which can easily be wet-screened and dewatered. A device called a Dehydrum, which flocculates and pelletizes thickened sludges into round, 3-mm pellets, was developed for this purpose. Several units reported in commercial operation in Japan thicken fine refuse from coal-preparation plants. The product contains 50% moisture, compared with 3% solids fed into the Dehydrum from the thickener underflow (85). In Poland, commercial use of the process to treat coal fines has been reported (86), and is said to compare favorably both economically and technically to thickening and vacuum filtration.

The U.S. Bureau of Mines has run large-scale tests on a similar process for treating <600-μm coal tailings. The tailings of a flotation cell are treated at a concentration of 3–7% solids. After mixing the slurry with 0.14 kg of polyethylene oxide flocculant per metric ton of dry solids and working for less than 30 s, the solids dewater to 50–70% moisture (87,88). The process has been demonstrated at a rate of 2300 L/min of waste slurry. Waste phosphate slimes have been similarly successfully consolidated from 4% to 40% solids (87,89). Mechanisms of flocculation in concentrated suspensions have been studied relatively little; however, conditions favoring pelletizing flocculation are described for colloidal latex suspensions (90).

Other applications include dewatering extremely fine (0.1 μm) laterite leach tailings (91). These pelletizing processes should be compared in flocculant consumption and operating and capital costs with belt-filter presses.

Oil Agglomeration. Pelletizing can also be accomplished using chemicals which make hydrophilic particles hydrophobic and thus agglomerate the particles into tight hydrophobic clusters (92,93). These processes, eg, Cattermole and Murex, were first used to make selective separations of a relatively minor amount of sulfide minerals (1–2%) in a slurry containing mostly silicates and carbonates. A more recent agglomeration process that was selective for coal added Freon to a slurry of <30 μm coal and mineral matter. The Freon caused the fine coal to agglomerate into large pellets that were recovered from the slurry on coarse static screens. The only remaining water was occluded in the pellets. The Freon, with its low heat capacity and boiling point (24°C), was recovered in a dryer (94).

A successful variation of oil agglomeration was used for removal and dewatering of soot from a 1–3% solids suspension consisting of <5-μm particles in refinery process waters (Fig. 8). Heavy oil was added to the dilute slurry and intensely agitated in a multistage mixer. The soot agglomerated with the oil to form 3–5 mm pellets that were easily screened from the water (95). The pellets contained only 5–10% water. The process was modified to recover very fine clean coal, and it produced highly uniform, hard, spherical pellets 1–2 mm in diameter.

This process has been applied to certain mineral oxides. Examples include recovery of cassiterite, SnO_2, from silicates (96), gold from ores (97), and ilmenite, $FeTiO_2$, from silicates (98). In each case, the normally hydrophilic mineral was treated with a surfactant (often a fatty acid) under conditions that would selectively coat the desired mineral. Additional oil is then added to agglomerate the treated minerals. Because of the cost of added reagent, the particles should have some intrinsic value. If no excess oil is added, but enough reagent is added and highly agitated, the pelletizing process is known as shear flocculation (99–101).

Another modification of dewatering methods using oil-agglomeration techniques combines the water exclusion of agglomeration with the ability of froth flotation to thicken. Dissolved air flotation (DAF) or induced air flotation (102) is used in over 300 municipalities for thickening wastewater solids and is also used at a much larger number of oil-well sites, refineries, and food-processing and rendering plants for removal of oil from wastewater. For oily wastes, or for selective removal of oily solids, DAF works very effectively. Typically, however, the thickened product contains only 2–4% solids. Because of poor performance, high energy consumption, and problems with controlling volatile organic gases and odors in the excess air, use as a thickening device is declining in favor of centrifuges and

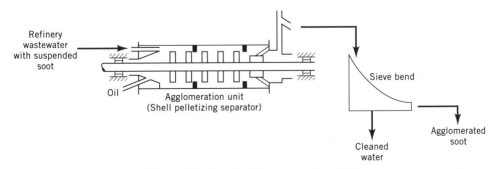

Fig. 8. Shell pelletizing separator (92).

belt presses. On food and industrial wastes, as a separator of oily wastes, DAF usage is strong. Preceding the operation with oil agglomeration or shear flocculation is particularly useful for inorganic particles finer than 15 μm.

Extraction Processes. *Oil-Phase Extraction.* In processes of dewatering by wetting a solid with an immiscible phase, another step in water displacement is possible. The use of very large quantities of an immiscible liquid allows extraction or transfer of the particles from one phase into the other, and the particles remain in a dispersed state in the new phase. To extract solids (as small as 0.1 μm) from water, a hydrophobic surface on the solid is needed. This surface is usually provided by using flotation reagents such as long-chain fatty acids, alkyl sulfonates, amines, or xanthates. Conversely, water has been used to agglomerate and extract hydrophilic solids that are dispersed in inorganic liquids (103). Crud, a concentration of solids at the liquid–liquid interface in normal solvent extraction, represents partial extraction (104). Pigment flushing is a technique used in paint (qv) and ink (qv) manufacture for transferring paint particles from an aqueous solution, where formation takes place, to a dispersed state in a nonaqueous carrier (105,106). Because the pigments (qv) are first filtered to remove soluble salts but are not dried before dispersion in the oil phase, the process function includes dewatering. Similarly, a process for manufacturing ferrofluids (stable dispersions of submicrometer magnetite in kerosene) consists of precipitating magnetite from water, adding oleic acid to coat the fine precipitate, and then contacting the wet filter cake with kerosene. The magnetite transfers into the kerosene and forms a stable suspension (107).

Solvent Extraction of the Liquid. Water contained in a cake or slurry can be extracted from the solids by dissolving the water in a solvent that is less expensive to evaporate than the contained water alone. The Institute of Gas Technology (IGT) has developed a laboratory process called solvent dewatering based on the principle that the solubility of water in selected solvents changes significantly with a change in temperature. In one example, hot solvent is mixed with wet peat and the water–solvent solution is then decanted. Upon cooling, the water precipitates as a separate phase (108). As of this writing, the process has not proved economical. The Resource Conservation Company has used chilled triethylamine (TEA) to dissolve water from sludge, and then warmed the extracted liquid to form an immiscible water phase. The effect of temperature on water solubility is opposite to the IGT solvent. TEA requires only 309 kJ/kg (129 Btu/lb) to evaporate, less than water (93,109). Similar processes have been considered for desalination (see WATER, SUPPLY AND DESALINATION).

Thermal Processes. *Thermal Drying.* The solvent-extraction processes discussed have progressively included evaporation (qv). In this aspect, the Carver-Greenfield process of multiple-effect evaporation of water from sludges is an important alternative to very late stage dewatering. Organic sludges of about 20% solids are fed to the unit, along with enough oil to keep the sludge moving in the processing equipment. Using three effects in the evaporators, only 700–900 kJ/kg (300–387 Btu/lb) of evaporated water are required, compared with the 2.3 MJ/kg (1000 Btu/lb) needed in a single-effect evaporator (38 + 110) (see DRYING). A plant using this process to treat municipal wastewater sludges began operations in 1991 in Ocean County, New Jersey.

Thermal Treatment. A number of dewatering processes alter the interaction of solid with liquid. Most depend on making a hydrophilic solid hydrophobic by adding small quantities of surfactants and oils. Many biological sludges cannot be economically treated with reagents to provide hydrophobicity, and such treatments would have no effect on water bound inside the mycelium. Thermal treatment is intended for these organic sludges. Partial wet-air oxidation lowers the specific cake resistance of many biological gels and colloidal sludges by 50–100 times. For example, if a municipal sludge that normally thickens to 5% solids is successfully thermally treated, it thickens to 10–15% solids. On filtering the thickened sludge, the cake formed from the untreated sludge has about 15% solids. The thermally treated sludge can be filtered to 30–50% solids (2,111).

Thermal treatment consists of heating the sludge under a pressure of about 2.4 MPa (350 psi) and to temperatures of 150–225°C (111) for 15–40 min either with (low pressure oxidation) or without (heat treatment) additional air (2). Reactions, including partial oxidation, occur that change the nature and consume 1–5% of the solids. Unfortunately, this process produces significant quantities of acetic acid and other short-chain, soluble organics. Whereas in 1979 there were over 100 installations operating on sewage sludge, in 1992, because of improvements in mechanical dewatering processes, only 30 to 35 remain.

At least five related dewatering processes have been applied to peat and lignite. Peat and lignite have a high absorbed-moisture content (90% in peat and 40–50% water in lignite) and have a tendency to break down to undesirable fines and to become pyrophoric when dried (see LIGNITE AND BROWN COAL). Steam drying comprises a family of processes very different from steam dewatering of filter cakes. These processes involve heating the peat or lignite containing the initial water content in an autoclave to temperatures of 150–200°C under pressures of about 1.3 MPa (189 psi) for about 15 min. This treatment causes the solids to shrink, eliminating water from pores and removing carboxylic acid and its salts from the surface. The steam treatment itself, considered separately from subsequent evaporative flashing, allows 30–50% of the initial water to drain or be pressure filtered from the product (112). Higher temperature and pressure lead to greater dewatering, and pressures up to 10 MPa (1500 psi) have been tested (113). These high pressures produce a completely dewatered lignite having high stability and little tendency to reabsorb moisture.

The only commercially used process in this group is the Fleissner process, developed in 1927 for drying lignite. One plant, operating in Austria between 1927 and 1960, achieved a capacity of 1700 t/d. In 1982, a number of other plants licensed by Fleissner were operating (114). There were also related processes in the pilot-plant stage (95,113–116), including one for dewatering peat with a capacity of 50,000 metric tons per year (117).

By raising the pressure, temperature, and available oxygen, virtually all the organic solids can be oxidized to CO_2 and H_2O. Ignition occurs at 200–225°C and wet-air oxidation is then autogenous. At 250–300°C, reactions occur rapidly. The vapor pressure of water at 300°C is about 9 MPa (1300 psi). Rather than needing 25–30% solids to achieve autogenous combustion of a sludge in air, only 0.5–1% organics in water is needed for autogenous wet-air oxidation (118). In supercritical water oxidation (SCWO), the temperature and pressure are raised still further to the critical point of water, 374°C and 22 MPa (3190 psi), which completely burns sewage sludge and toxic organics (119–121).

Freeze–Thaw Dewatering.   Slow freezing of hydroxide, clay, and municipal sludges affects the water-retention properties of the solids when the frozen slurry is thawed. Two plants used this process on water-treatment sludges. The sludge is gelatinous aluminum hydroxide with organic and inorganic matter that typically thickens to 1–3% solids. In tests in the United Kingdom, the sludge, after a cycle of freezing and thawing, became a sandy, granular material that drained without needing a filter. A few small plants use freeze–thaw dewatering for municipal wastewater sludges (3–7% solids) and achieve a solids content of 25% with natural drainage after thawing the frozen sludge (2). The probable mechanisms of dewatering by freezing have been described (12). The ability to withstand freeze–thaw cycling is an important consideration in latex-paint manufacture (106). Similar considerations must be important in some frozen-food formulations (see FOOD PROCESSING).

Freeze Crystallization.   Freezing may be used to form pure ice crystals, which are then removed from the slurry by screens sized to pass the fine solids but to catch the crystals and leave behind a more concentrated slurry. The process has been considered mostly for solutions, not suspensions. However, freeze crystallization has been tested for concentrating orange juice where solids are present (see FRUIT JUICES). Commercial applications include fruit juices, coffee, beer, wine (qv), and vinegar (qv). A test on milk was begun in 1989 (123). Freeze crystallization has concentrated pulp and paper black liquor from 6% to 30% dissolved solids and showed energy savings of over 75% compared with multiple-effect evaporation. Only 35–46 kJ/kg (15–20 Btu/lb) of water removed was consumed in the process (124).

Clathrate Freezing.   Clathrate freezing uses methane or ethane under pressure, where 1 mol of ethane traps 18 mol of water. Methane clathrates apparently can form in natural gas pipelines at room temperature. The process has been studied for dewatering wastewater treatment sludge, using Freon 11 (125). It has also been considered for removing water from the black liquor derived in the Kraft pulping process for making paper fiber.

Capillary Suction Processes.   The force needed to remove water from capillaries increases proportionately with a decrease in capillary radius, exceeding 1400 kPa (200 psi) in a 1-μm-diameter capillary. Some attempts have been made to use this force as a way to dewater sludges and cakes by providing smaller dry capillaries to suck up the water (27). Sectors of a vacuum filter have been made of microporous ceramic, which conducts the moisture from the cake into the sector and removes the water on the inside by vacuum. Pore size is sufficiently small that the difference in pressure during vacuum is insufficient to displace water from the sector material, thus allowing a smaller vacuum pump to be effective (126).

Electromagnetic Processes.   *Electrical Enhancement of Dewatering.*   Electrophoresis (qv) can be used to prevent a filter cake from forming on a filter medium while allowing water to pass through the medium from the slurry. Electrophoresis is used to move the particles upstream, opposite to the liquid movement, in order to prevent blinding of the medium.

Once a matrix of particles is formed, whether filter cake, thickened under-flow, or soil, applying a current to the fluid causes a movement of ions in the water and, with the ions, water of hydration. The phenomenon is called electroosmosis. The pressure generated on the fluid is given by (127):

$$P = \frac{2 \, \zeta ED}{\pi r^2}$$

where P = pressure in Pa; ζ = zeta potential in V; E = electric field in V/m; D = dielectric constant; r = radius of capillary in m. The amount of water moved is proportional to the intensity and time that power is applied, proportional to the zeta potential of the solid, and inversely proportional to the conductivity of the fluid (128). Results are often measured in kWh/t of dewatered product or in kWh/t of water removed.

High pressures are generated in the small capillary openings. Unlike pressure generated on a fluid by an externally applied force, however, the largest forces are generated at the shear plane of the liquid and the solid in the pores. The effect of a particle size on capillary retention force is shown in Table 4 (26). To calculate the pore radius used in the table, it is assumed that the pore is a cylinder that can just pass between three monosized spheres. The entries following P_E are the pressures developed by electroosmosis in those same pores, assuming a field of 3000 V/m. To generate that field in an electrolyte, a current of 600 A/m^2 must flow, and, in this case, it is assumed to be created in a 10^{-4} M, 1:1 electrolyte (0.2 S/m conductivity) (129).

In most applications, far less current and lower voltages are used. For example, in dewatering clay soils to stabilize dams, foundations, or dredged spoil, 20–100 V/m are commonly applied (130,131). In soil stabilization (qv), power is applied for weeks to months.

The effectiveness and costs of electroosmotic dewatering on a large number of clay-containing tailings from metallic, nonmetallic, and coal mines has been shown (132,133). The process can be used *in situ* or in a batch dewatering cell. One large test dewatered a very old, stable, 50% solid slime generated by a coal-washing plant. Applying 37 kWh/t of the final product, moisture was reduced to 19% in 24 h. Using a batch dewatering cell, 1100 t/week of slimes were dewatered (132). There is interest in using the technique to dewater hazardous waste sludge

Table 4. Effect of Particle Size on Capillary Retention Force

Parameter	Diameter, μm				
	50	20	15	12	1
r_c = 0.165 d/2, μm	4.125	1.65	1.24	0.99	0.0825
$P_c{}^a$, kPa^b	35.5	88.2	117	147	1760
$P_E{}^c$, kPa^b				17.2	2430

$^a P_c$ = 2 σ cos θ/r_c, capillary pressure.
^{b}To convert kPa to psi, multiply by 0.145.
$^c P_E$ = electrical pressure.

ponds before excavating. In the Netherlands, electroosmosis is used for *in situ* washing of metals from the soil.

Electroosmosis with Vibration.   A commercially available electroacoustic dewatering (EAD) filter combines ultrasonic vibration of the cake with electroosmosis of water in the cake to achieve greater dewatering. Figure 9 shows a 2-m wide commercial machine developed for processing the cakes produced by conventional dewatering methods. The electrical current provides the force to move the liquid; an appropriate level of ultrasonic agitation helps to consolidate the cake, releases trapped gases and liquids, and maintains a liquid continuum for current to low. Typically, half the remaining water is removed from the dewatered cake fed to the EAD filter (135).

Magnetic Enhancement of Dewatering.   *Liquid–Solid Separation.*   When magnetic forces are considered for liquid–solid separations, it is usually for thickening and filtration rather than for dewatering. The Frantz Ferrofilter, commonly used to remove suspended ferromagnetic impurities from liquids, is somewhat analogous to a depth filter in its use of multiple collection sites and lack of a definable porous filter medium surface (135). The Ferrofilter principle has been extended to very large magnetized volumes of 1.7 m^3 and at high fields, from 0.15 T (1500 gauss) for the Ferrofilter to 2.0 T (2×10^4 gauss) for the new machines. The large, high gradient magnetic filters use a depth filter medium of 430 stainless-steel wool to form high gradients in the high field. Commercially, these are widely used for removing paramagnetic impurities from kaolin (136), and new superconducting magnets are now also commercially used (137). Less intense versions are used to separate and dewater magnetic iron ores and nonferromagnetic

Fig. 9.   Two-meter wide electroacoustic dewatering press (134). Courtesy of Ashbrook-Simon-Hartley (134).

iron ores, for example, itabirite, and other ores containing specular hematite (138,139).

Large-volume magnetic separators have been used to remove 90% of the suspended mill scale solids in steel rolling-mill wastewater, and for cleanup of steam-boiler water (see SEPARATION, MAGNETIC). A much wider range of potential uses of magnetic separation for clarifying, filtering, and dewatering is possible when nonmagnetic impurities are made magnetic. In Japan, plating wastes with dissolved Cr, Mn, Cu, Zn, Cd, and so on, have been precipitated as magnetic ferrites (qv) and recovered from the wastewater with a simple magnetic separator. Details of this commercial process have been reported (140,141).

A simple separator used to recover the magnetic particles consists of a series of disks mounted on a shaft. Each disk has a number of permanent magnets mounted flush with the surface at its perimeter. The disks rotate into and out of the liquid containing the suspended magnetic material and lift the magnetic particles out of the stream. The magnets are then scraped clean (Fig. 10). Very low residence times are needed for removal of the particles compared to settling or flotation (142).

Adsorption of nonmagnetic suspended materials onto magnetic seeds has been proposed and tested for removal of suspended solids from drinking water (143) and of bacteria from municipal wastewater effluent. Using aluminum sulfate as a coagulant and 100–1000 ppm of magnetite, 80–90% of the suspended solids were removed (144). In laboratory experiments, 0.1% of magnetite was added to sewage, and the magnetite quickly became coated with biological flocs. These settled under the influence of a fairly weak magnetic field of 0.04 T in 4% of the time required for settling the untreated feed. The authors estimated that this would allow a fourfold increase in flow rate, a fivefold decrease in sludge volume, and an effluent with half the normal BOD content over normal wastewater treatment (146).

At least one study has specifically evaluated the use of a high gradient magnetic separator for dewatering a paramagnetic mineral slurry, malachite, $CuCO_3 \cdot Cu(OH)_2$, of an average particle size of 4 μm. Initial slurries of 2–12% solids were passed through the magnetic matrix and allowed to drain. When the field was turned off, the mineral could be successfully washed off the matrix to give a 40% solid slurry (146).

Finally, selective separation and dewatering of one suspended substance in a slurry containing different minerals or precipitates is possible by selectively

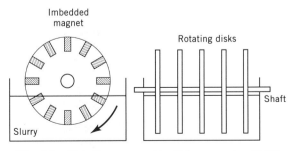

Fig. 10.   Rotating magnetic-disk separator (142).

adsorbing a magnetic material (usually hydrophobic) onto a solid that is also naturally or chemically conditioned to a hydrophobic state. This process (Murex) was used on both sulfide ores and some oxides (145). More recently, hydrocarbon-based ferrofluids were tested and shown to selectively adsorb on coal from slurries of coal and mineral matter, allowing magnetic recovery (147). Copper and zinc sulfides were similarly recoverable as a dewatered product from waste-rock slurries (148).

Economic Aspects

Dewatering, a part of the liquid–solid separation equipment and supplies market, is not well segmented. The same equipment is often used for both separation and dewatering. The larger market for U.S. industrial and municipal liquid–solid separation equipment, not including disposable cartridges or membranes or high purity electronic or pharmaceuticals filtration, has been estimated at $1 billion for 1991 (149). Earlier estimates of the European market were $1.2 billion in 1988. Municipal water and wastewater treatment accounts for about $300 million, the largest segment, followed by the chemicals and allied industries segment, at $160 million. Other market segments are pulp and paper, general manufacturing, food and beverages, mineral processing, and oil and gas exploration and production, and electric utilities.

The market for filter equipment alone was about $580 million in 1991 (149), and pressure filters are on the order of $115–$160 million of the total filter market. About 500 U.S. companies manufacture liquid–solid separation equipment. Some of the manufacturers in this worldwide business include Ametek, Anderson International Corp., Arus-Andritz Inc., Ashbrook-Simon-Hartley, Bird Machine Co., Inc., Bepex Corp., Black Clawson, Centrifugal and Mechanical Industries, Denver Process Equipment Co., Dorr-Oliver Inc., Ebara Infilco, Envirex Inc., Eimco Process Equipment Co., Hitachi Plant Engineering and Construction Co., Ltd., Infilco-Degremont Inc., Ingersoll Rand Co., JWI Inc., KHD Humboldt Wedag, Krauss-Maffei Corp., Komline-Sanderson Engineering Corp., Kurita Machinery Manufacturing Co., Ltd., Larox Oy, Mitsubishi Kakoki Kaisha, Ltd., Rosenmund Inc., Sharples Division, Alfa-Laval Separation Inc., Sparkler Filters Inc., U.S. Filter, Wemco, and Zimpro-Passavant Environmental Systems Inc.

Flocculants and coagulants are sometimes used as pretreatments before dewatering. The market for flocculants and coagulants for water and wastewater treatment in the United States for 1989 was about $250 million, or 68,000 t, provided by 18 companies. In Europe, 22,000 t of flocculants and coagulants, made by 15 companies, had a market value of $115 million. In Japan, 12 companies made 23,000 t, valued at $184 million (150).

BIBLIOGRAPHY

"Dewatering" in *ECT* 3rd ed., Supplement, pp. 310–339, by B. Morey, Telic Technical Services.

1. F. M. Tiller and C. S. Yeh, *Filtr. Sep.* **27**, 129 (1990).

2. O. E. Albertson and co-workers, *Dewatering Municipal Wastewater Sludges Design Manual*, EPA/625/1-87/014, Sept. 1987.

3. W. E. Stanley, *Sludge Dewatering, Manual of Practice 20*, Water Pollution Control Federation, Washington, D.C., 1969, p. 101.

4. P. A. Vesilind, *Treatment and Disposal of Wastewater Sludges*, Ann Arbor Science Publishers, Ann Arbor, Mich., 1974, Chapt. 6.

5. P. W. Thrush, ed., *Dictionary of Mining, Metallurgical and Related Terms*, U.S. Government Printing Office, Washington, D.C., 1968, p. 319; T. C. Collocott, ed., *Chambers Dictionary of Science and Technology*, Barnes and Noble, New York, 1971; D. N. Lapedes, *McGraw-Hill Dictionary of Scientific and Technical Terms*, 2nd ed., McGraw-Hill Book Co., New York, 1978.

6. H. Sato and co-workers, *Filtr. Sep.* **19,** 492 (1982).

7. H. B. Gala and S. H. Chiang, *Filtration and Dewatering; Review of Literature*, Report #DOE/ET/14291-1, U.S. Department of Energy, Washington, D.C., 1980.

8. L. Svarofsky, *Solid–Liquid Separations*, 3d ed., Butterworths, London, 1990.

9. R. J. Wakeman, *Filtration Post-Treatment Processes*, Elsevier Publishing Co., Amsterdam, 1975.

10. D. B. Purchas, *Solid/Liquid Separation*, Uplands Press, Croyden, UK, 1981.

11. F. M. Tiller, C. S. Yeh, and W. F. Leu, *Sep. Sci. Tech.* **22,** 1037 (1987).

12. F. M. Tiller and C. S. Yeh, *AIChE J.* **33**(8), 1241 (Aug. 1987).

13. A. Rushton and M. A. A. Arab, *Filtr. Sep.* **26,** 181 (1989).

14. *Humboldt Centripress ADS (Advanced Dewatering System)*, Humboldt Decanter, Inc., Norcross, Ga., 1990.

15. R. J. Wakeman and Fan Deshun, *Chemical Engineering Design Research* **69**(A5), 403 (1991).

16. F. M. Tiller, in B. M. Moudgil and B. J. Scheiner, eds., *Flocculation and Dewatering*, Engineering Foundation, New York, 1989, p. 89.

17. F. M. Tiller and L.-L. Horng, *AIChE J.* **29**(2), 297 (Mar. 1983).

18. *Continuous Pressure Filter for the Process Industries*, Ingersoll-Rand, Nashua, N.H., 1992.

19. M. Shirato and co-workers, *Filtr. Sep.* **7,** 277 (1970).

20. H. G. Schwartzberg, J. R. Rosenau, and G. Richardson, *AIChE Symp. Ser.* **73**(163), 177 (1977).

21. *Eimco Expressor Press*, Eimco Process Equipment Co., Salt Lake City, Utah, 1990.

22. *Magnum Press*, Bulletin MP-201, Parkson Corp., Fort Lauderdale, Fla., 1987.

23. *High Pressure Filtration Using the Tube Filter Press: A Technical and Economic Review*, Alfa-Dyne Inc., Cleveland, Ohio, 1991; J. Quilter, *Indust. Miner. Mag., Energy Suppl.*, 29 (Mar. 1983).

24. D. K. Bredeson, *J. Am. Oil Chem. Soc.* **60,** 163A (1983).

25. *Lasta Automatic Filterpress*, Ingersoll-Rand, Nashua, N.H., 1991.

26. A. F. Westergard, *Eng. Min. J.*, 60 (June 1983).

27. R. J. Wakeman and A. Rushton, *Filtr. Sep.* **13,** 450 (1976).

28. *Filtr. News*, 10 (Nov. 1989).

29. S. A. Bratten and S. V. Tracy, "Improved Concentrate Dewatering Utilizing Variable Volume Pressure Filters," paper presented at *Annual Meeting Society for Mining, Metallurgy and Exploration*, Denver, Colo., 1991.

30. *Filtr. News*, 32 (Jan. 1992).

31. M. Münter and U. Grén, *Filtr. Sep.* **27,** 264 (1990).

32. *Elastomer and Polymer Processing Systems*, Bulletin MPR 76, 1976, *Pre-Press*, Bulletin FO 175; and *French Dual Cage Screw Press*, Bulletin OP8130, 1981, The French Oil Mill Machinery Co., Piqua, Ohio.

33. L. H. Tindale and S. R. Hill-Haas, *J. Am. Oil Chem. Soc.* **53,** 265 (1976).

34. J. A. Ward, *J. Am. Oil Chem. Soc.* **53,** 261 (1976).

35. D. K. Bredeson, *J. Am. Oil Chem. Soc.* **55,** 762 (1978).

36. *Anderson Duo Crackling Expeller Presses,* Bulletin Duo 375-2, 1981; *Anderson Rubber and Plastic Polymer Dewatering and Drying Equipment,* Bulletin RDD 73, 1980; and *Anderson Expeller Presses,* Bulletin 359 R, 1980, Anderson International Corp., Cleveland, Ohio.

37. *Pressmaster Press,* Bulletin SB82-002B, Beloit Corp., Jones Division, Dalton, Mass., 1982.

38. K. Ohmiya and S. Takahashi, *J. Water Pollut. Control Fed.* **52,** 943 (1980).

39. *Carver Cocoa Presses,* Bulletins HV-A and FP-1, Fred S. Carver, Inc., Menomonee Falls, Wisc., 1981.

40. R. H. Perry and C. H. Chilton, *Chemical Engineers Handbook,* 5th ed., McGraw-Hill Book Co., New York, 1973, pp. 19-101–19-104.

41. *V-Press,* Bulletin 64-5, Bepex Corp., Rietz Division, Santa Rosa, Calif., 1982; Ref. 14, p. 516.

42. B. K. Parekh and J. P. Matoney, in J. W. Leonard, III, ed., *Coal Preparation,* 5th ed., Society for Mining, Metallurgy and Exploration, Inc., Littleton, Colo., 1991, Chapt. 8.

43. *Technical Bulletin 2-2-16/1-B* and *Vari-Nip Technical Discussion,* Ingersoll-Rand, Nashua, N.H.

44. C. E. Silverblatt and D. A. Dahlstrom, *Ind. Eng. Chem.* **46,** 1201 (1954).

45. C. S. Simons and D. A. Dahlstrom, *Chem. Eng. Prog.* **62**(1), 75 (1966).

46. A. F. Baker and A. W. Duerbrouck, in A. C. Partridge, ed., *Proceedings of the International Coal Preparation Congress,* 1977.

47. J. H. Brown, *Can. Min. Metall. Bull.* **58,** 315 (1965); *Transactions Can. Inst. Min. Met.* **68,** 105 (1965).

48. F. M. Tiller and J. R. Crump, *Chem. Eng. Prog.* **74,** 65 (Oct. 1977).

49. U.S. Pat. 4,107,028 (Aug. 15, 1978), R. K. Emmett, S. D. Heden, and R. A. Summerhays (to Envirotech Corp.).

50. S. M. Moos and R. E. Dugger, *Min. Eng.* **31,** 1479 (1979).

51. H. B. Gala, S. H. Chiang, and W. W. Wen, *Proceedings World Filtration Congress III,* The Filtration Society, Downington, Pa., 1982.

52. B. Herath, P. Geladi, and C. Albano, *Filtr. Sep.* **26,** 53 (1989).

53. J. G. Groppo and B. K. Parekh, *Effect of Metal Ions on Vacuum Filtration of Coal,* Society for Mining, Metallurgy and Exploration, Annual Meeting, Denver, Colo., 1991.

54. A. Rushton, *Filtr. Sep.* **13,** 573 (1976).

55. P. J. Lafforgue and co-workers, paper presented at *Society of SME-AIME Annual Meeting,* Feb. 1982, preprint 82-22, available from the United Engineering Society Library, New York.

56. *Proceedings of the Consolidation and Dewatering of Fine Particles Conference,* University of Alabama, Aug. 1982, available from U.S. Bureau of Mines, Tuscaloosa, Ala.

57. *Proceedings of the Progress in the Dewatering of Fine Particles Conference,* University of Alabama, Apr. 1981, available from the U.S. Bureau of Mines, Tuscaloosa, Ala.

58. F. N. Kemmer and J. McCallion, eds., *Nalco Water Handbook,* McGraw-Hill Book Co., New York, 1979, Chapts. 8–9.

59. M. E. Lewellyn and S. S. Wang, in R. B. Seymour and G. A. Stahl, eds., *Macromolecular Solutions Solvent-Property Relationships in Polymers,* Pergamon Press, New York, 1982, pp. 134–150.

60. S. K. Nicol, *Proc. Australas Inst. Min. Metall.,* 37 (Dec. 1976).

61. M. J. Pearse and T. Barnett, *Filtr. Sep.* **17,** 460 (1980).

62. Ref. 14, p. 53.

63. R. Leutz and M. Clement, *Filtr. Sep.* **7,** 193 (1970).
64. S. K. Mishra, in B. M. Moudgil and B. J. Scheiner, eds., *Flocculation and Dewatering,* 1989, Engineering Foundation, New York, p. 89.
65. M. J. Pearse, in Ref. 56, pp. 41–89.
66. G. M. Moody, *Trans. Inst. Min. Metall.* **99,** C137 (1990).
67. V. P. Mehrotra and co-workers, *Filtr. Sep.* **19,** 197, (1982).
68. R. J. Schwartz, *Sludge Dewatering, Manual of Practice 20,* Water Pollution Control Federation, Washington, D.C., 1969, pp. 13–40.
69. A. F. Cassel and B. P. Johnson, *Evaluation of Devices for Producing High Solids Sludge Cake,* NTIS Report No. PB80-111503, National Technical Information Service, Washington, D.C., 1980.
70. D. DiGregorio and J. F. Zievers, in W. W. Eckenfelder, Jr., and C. J. Santhanam, eds., *Sludge Treatment,* Marcel Dekker, Inc., New York, 1981, Chapt. 6, pp. 142–207.
71. *Wastewater Engineering,* 2nd ed., McGraw-Hill Book Co., New York, 1979, Chapt. 11.
72. G. Smith, "Optimizing Operation of Low-Load Aeration Systems: Wasting More . . . and Paying Less", presented at *15th Annual Conference of the Alabama Association of Water Pollution Control,* Orange Beach, Ala., Nov. 1991, available from Envirex Inc.
73. M. Anazia, in *Mining Eng.,* 485 (May 1990).
74. *Hi-Compact Method—A Purely Mechanical Process for Maximum Secondary Dewatering of Sludges,* Bulletin 5-400e, KHD Humboldt Wedag AG, Cologne, Germany, 1988.
75. B. Sawyer, R. Watkins, and C. Lue-Hing, *Proceedings 31st Industrial Waste Conference,* Purdue University, Lafayette, Ind., 1976, p. 537.
76. P. R. Karr and T. M. Keinath, *J. Water Pollut. Control Fed.* **50,** 1911 (1978).
77. J. Bien, *Filtr. Sep.* **25,** 425 (1988).
78. *Filtr. Sep.* **27,** 163 (1990).
79. *Fuji Micro-Sonic Filter,* Fuji Filter Manuf. Co., Tokyo, 1988.
80. *Spiractor,* Bulletin 5852, Permutit Co., Inc., Paramus, N.J., 1979.
81. *SWA/KW Reactor,* Esmil Water Systems Ltd., Buckinghamshire, UK, 1976, 1987.
82. D. A. Dahlstrom, in M. P. Freeman and J. A. FitzPatrick, eds., *Theory, Practice and Process Principles for Physical Separations,* Engineering Foundation, New York, 1977, pp. 261–273; *EPRI Report F.P. 937,* Electric Power Research Institute, Palo Alto, Calif., 1979.
83. A. D. Randolph and D. Etherton, *Study of Gypsum Crystal Nucleation and Growth Rates in Simulated Flue Gas Desulfurization Liquors,* EPRI Report CS1885, Electric Power Research Institute, Palo Alto, Calif., 1981.
84. M. Yusa and A. M. Gaudin, *Am. Ceram. Soc. Bull.* **43,** 402, (1964).
85. M. Yusa and co-workers, in A. C. Partridge, ed., *Proceedings of the 7th International Coal Preparation Congress,* Australian National Committee, Sydney, 1976.
86. J. Szczpya, in P. Somasundaran, ed., *Fine Particles Processing,* Society of Mining Engineers of AIME, Littleton, Colo., 1980, p. 1676.
87. B. J. Scheiner and A. G. Smelley, *Dewatering of Thickened Phosphate Clay Waste from Disposal Ponds,* Paper A81-6, The Metallurgical Society of AIME, Warrendale, Pa., 1981; J. R. Pederson, ed., *U.S. Bureau of Mines Research 81,* U.S. Government Printing Office, Washington, D.C., 1981, p. 83.
88. B. J. Scheiner and co-workers, "New Dewatering Techniques for Fine Particle Waste," *Proc. 16 Int. Min. Proc. Cong.,* 1951, Elsevier Publishing Co., Amsterdam, 1988.
89. B. J. Scheiner and M. M. Ragin, *Society for Mining Metallurgy and Exploration Transactions* **284,** 1801 (1988).
90. K. Higashitani and T. Kubota, *Powder Technol.* **51,** 61 (1987).

91. R. M. Hoover and P. V. Avotins, *Development of Polymer Pelletization for Enhancing Solid Liquid Separation of Leached Laterite Residue*, Paper A78-13, The Metallurgical Society of AIME, Warrendale, Pa., 1978.
92. V. P. Mehrotra and co-workers, *Int. J. Miner. Process.* **11,** 175 (1983).
93. V. P. Mehrotra and co-workers, *Min. Eng. (NY)* **32,** 1230 (1980).
94. D. V. Keller, Jr., in Ref. 56, pp. 152–171.
95. F. J. Zuiderweg and co-workers, *Chem. Engineer (London)*, 223 (July 1968).
96. F. W. Meadus and co-workers, *Can. Min. Metall. Bull.*, 968 (1966).
97. F. W. Meadus and co-workers, *Can. Min. Metall. Bull.*, 1326 (1969).
98. I. E. Puddington and B. D. Sparks, *Miner. Sci. Eng.* **7,** 282 (1975).
99. P. T. L. Koh and L. T. Warren, *13th International Mineral Processing Congress*, Warsaw, Poland, 1979.
100. A. M. Gaudin and P. Malozemoff, *J. Phys. Chem.* **37,** 599 (1933).
101. A. M. Gaudin and P. Malozemoff, *Trans. Am. Inst. Min. Metall. Engrs.* **112,** 303 (1934).
102. O. E. Albertson, *Sludge Thickening, Manual of Practice FD1*, Task Force on Sludge Thickening, Water Pollution Control Federation, Washington, D.C., 1980, p. 33.
103. H. M. Smith and I. E. Puddington, *Can. J. Chem.* **38,** 1911 (1960).
104. G. M. Ritcey and A. W. Ashbrook, *Solvent Extraction Principles and Applications to Process Metallurgy*, Elsevier Publishing Co., Amsterdam, 1979, Part II, p. 669.
105. R. Stratton Crawley, in P. Somasundaran and M. Arbiter, eds., *Beneficiation of Mineral Fines*, National Science Foundation, Society of Mining Engineers, AIME, Littleton, Colo., 1979, p. 317.
106. D. Bass, *Paint Manuf.*, 5 (Jan. 1957).
107. G. W. Reimers and S. E. Khalafalla, *Preparing Magnetic Fluids by a Peptizing Method*, U.S. Bureau of Mines Technical Progress Report 59, U.S. Bureau of Mines, Washington, D.C., Sept. 1972; U.S. Pat. 3,843,540 (Oct. 22, 1974), G. W. Reimers and S. E. Khalafalla (to U.S. Department of the Interior).
108. C. L. Tsaros, in J. W. White and B. F. Feingold, eds., *Peat Energy Alternatives*, Institute of Gas Technology, Chicago, 1980.
109. *Chem. Eng.*, 82 (June 4, 1979).
110. S. A. Raksit, *Carver-Greenfield Pilot Demonstration*, LA-OMA Project Los Angeles Department of Public Works, Los Angeles, 1978.
111. J. Jacknow, *Sludge* **2**(4), 26, (July 1979).
112. Can. Pat. 1,010,477 (Nov. 8, 1977), E. J. Wasp (to Bechtel International Corp.).
113. W. H. Oppelt and co-workers, *Drying North Dakota Lignite to 1500 Psi by the Fleissner Process*, Report of Investigations 5527, U.S. Bureau of Mines, Washington, D.C., 1959.
114. B. Stanmore, D. N. Boria, and L. E. Paulson, *Steam Drying of Lignite: A Review of Processes and Performance*, DOE/GFETC/R1-82/1 (DE82007849), U.S. Department of Energy, available from National Technical Information Service, Washington, D.C., 1982.
115. J. B. Murray and D. G. Evans, *Fuel* **51,** 290 (1972).
116. U.S. Pats. 4,052,168; 4,129,420 (1977), E. Koppelman; G. Parkinson, *Chem. Eng.*, 77 (Mar. 27, 1978).
117. J. Rohr, Wheelabrator-Frye, Hampton, N.H., personal communication, 1992.
118. D. F. Othmer, *Mech. Eng.*, 30 (Dec. 1979).
119. J. Josephson, *Environ. Sci. Technol* **16**, 548A (1982).
120. R. W. Shaw and co-workers, *Chem. Eng. News* **69**(51), 26 (Dec. 1991).
121. *Supercritical Water Oxidation Engineering Bulletin*, U.S. Environmental Protection Agency, EPA 540/S-92/006, 1992.
122. G. S. Logsdon and E. Edgerley, Jr., *J. Am. Water Works Assoc.* **63,** 734 (Nov. 1971).

123. J. Douglas and A. Amarnath, *Freeze Concentration: an Energy-Efficient Separation Process*, EPRI Journal, p. 17, 1989.

124. H. E. Davis and C. J. Egan, *AIChE Symp. Ser. 207* **77,** 50 (1981).

125. B. Molayem and T. Bardakci, *Dewatering Wastewater Treatment Sludge by Clathrate Freezing: a Bench-Scale Study*, EPA, NTIS PB 86-239779/AS, 1986.

126. *Filtr. Sep.* **28,** 238 (1991).

127. A. W. Adamson, *Physical Chemistry of Surfaces*, 3rd ed., John Wiley & Sons, Inc., New York, 1974, p. 212.

128. N. C. Lockhart, in Ref. 51, pp. 325–332.

129. M. P. Freeman, in G. Hetsrom, ed., *Handbook of Multiphase Systems*, Hemisphere, New York, 1982, Chapt. 9.3, pp. 9-9–9-115.

130. B. A. Segall and co-workers, *ASCE Geotech Engineering Division J. GT* **106,** 1148 (1980).

131. C. A. Fetzer, *Proceedings ASCE, Journal of Soil Mechanics and Foundations Division* **93 SM4,** 85 (1967).

132. R. H. Sprute and D. J. Kelsh, "Dewatering Fine Particle Waste Suspensions with Direct Current," *Encyclopedia of Fluid Mechanics*, Gulf Publishing, Houston, Tex., 1986, Chapt. 27.

133. R. H. Sprute and D. J. Kelsh, *Electrokinetic Densification of Solids in a Coal Mine Sediment Pond—A Feasibility Study*, Bureau of Mines Report of Investigations 9137, 1988.

134. T. Schiene, *ElectroAcoustic Dewatering*, Ashbrook-Simon-Hartley, Houston, Tex., 1990.

135. *Frantz Ferrofilter Magnetic and Electromagnetic Separators*, Bulletins EM and PM, S. G. Frantz Co., Inc., Trenton, N.J., 1980.

136. C. Mills, *Ind. Miner. (London)*, 41 (Aug. 1977).

137. *Superconducting High Gradient Magnetic Separator System*, Eriez Magnetics, Erie, Pa., 1991.

138. J. E. Lawver and D. M. Hopstock, *Miner. Sci. Eng.* **6,** 154 (July 1974).

139. D. M. Thayer and P. B. Linkson, *Trans. AIME* **270,** 1897 (1981).

140. N. Nojiri and co-workers, *J. Water Pollut. Control Fed.*, **52,** 1898 (1980).

141. Y. Tamaura and co-workers, *Water Res.* **13,** 21 (1979).

142. M. Miura and T. M. Williams, *Chem. Eng. Prog.*, 66 (Apr. 1978).

143. B. A. Bolto and co-workers, *J. Polym. Sci. Polym. Symp.*, 211 (1975).

144. R. R. Oder and B. I. Horst, *Filtr. Sep.* **13,** 363 (1976).

145. A. Faseur and co-workers, *Filtr. Sep.* **25,** 344 (1988).

146. P. Chakrabarti and co-workers, *Filtr. Sep.* **19,** 105 (1982).

147. T. A. Sladek and C. H. Cox, *Coal Preparation Using Magnetic Separation*, Vol. 4, *Evaluation of Magnetic Fluids for Coal Beneficiation*, EPRI Report CS1517, Energy Electric Power Research Institute, Palo Alto, Calif., July 1980.

148. U.S. Pats. 1,043,851 (Nov. 1912); 1,043,850 (Nov. 1912); 996,491 (Aug. 1911); 993,717 (June 1911), A. A. Lockwood.

149. *Filtr. Sep.* **29,** 278 (1992).

150. *SRI International Estimates*, SRI International, Menlo Park, Calif., 1991.

General References

O. E. Albertson and co-workers, *Dewatering Municipal Wastewater, Sludges Design Manual*, EPA/625/1-87/014, Sept. 1987.

H. B. Gala and S. H. Chiang, *Filtration and Dewatering: Review of Literature*, DOE/ET/14291-1, 1980.

Liquid Filtration Manual, Sedimentation and Centrifugation Manual, McIlvaine Co., Northbrook, Ill.

D. B. Purchas, *Solid-Liquid Separation Technology*, Uplands Press, 1981.
L. Svarovsky, *Solid–Liquid Separation*, 3rd ed., Butterworths, 1990.

Equipment Selection

Liquid Filtration Newsletter, Sedimentation and Centrifugation Newsletter, McIlvaine Co., Northbrook, Il.
E. Mayer, *Filtration News*, 24 (May 1988).
A. Ruston, *Selection and Use of Liquid/Solid Separation Equipment*, Institution of Chemical Engineers, 1982.

Solid/Liquid Separation Scaleup

O. E. Albertson and co-workers, *Dewatering Municipal Wastewater, Sludges Design Manual*, EPA/625/1-87/014, Sept. 1987.
D. B. Purchas and R. J. Wakeman, Uplands Press, 1986.
F. M. Tiller and C. S. Yeh, *Filtr. Sep.* **27,** 129 (1990).

Equipment Optimization

D. A. Dahlstrom, *Coal* **95** (4), 52 (1990).
B. J. Scheiner, *Fluid/Particle Separation Journal* **1,** 46 (1990).

Pretreatment

B. M. Moudgil and B. J. Scheiner, eds., *Flocculation and Dewatering*, Engineering Foundation, New York, 1989.
R. J. Wakeman, *Filtration Post Treatment Processes*, Elsevier, Amsterdam, 1975.

BOOKER MOREY
SRI International

DEXTROSE AND STARCH SYRUPS. See SYRUPS.

DIAGNOSTICS. See MEDICAL DIAGNOSTIC REAGENTS.

DIALYSIS

Dialysis is a membrane separation process in which one or more dissolved species flow across a selective barrier in response to a difference in concentration. It is the earliest molecularly separative membrane process to be identified and described (1). The mode of transport is diffusion, and separation occurs because small molecules diffuse more rapidly than larger ones, and also because the degree to which membranes restrict solute transport usually increases with permeant size. The basic principles are illustrated in Figure 1. Solute c is present at

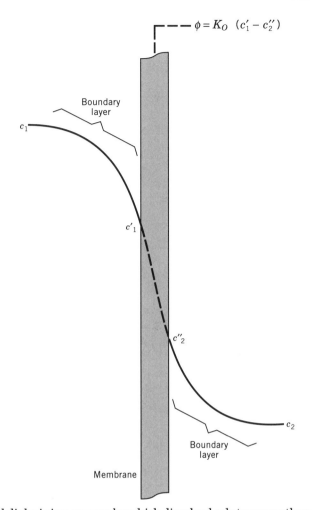

$$\phi = K_O \ (c'_1 - c''_2)$$

Fig. 1. General dialysis is a process by which dissolved solutes move through a membrane in response to a difference in concentration and in the absence of differences in pressure, temperature, and electrical potential. The rate of mass transport or solute flux, ϕ, is directly proportional to the difference in concentration at the membrane surfaces (eq. 1). Boundary layer effects, the difference between local and wall concentrations, are important in most practical applications.

concentrations c' and c'' on opposite sides of a membrane. In the absence of differences in pressure, temperature, or electrical potential, Fick's phenomenological first-order description of diffusion, published in 1855 (2), states that solute will move from region of greater to lesser concentration and at a rate proportional to the difference. In equation 1, ϕ = unit solute flux in $g/cm^2 \cdot s$; D = diffusion coefficient, cm^2/s; c = concentration in g/cm^3; x = distance in cm; and the minus sign accounts for the convention that flux is considered positive in the direction of decreasing concentration.

$$\phi = -D \frac{\partial c}{\partial x} \tag{1}$$

Diffusion coefficients decrease roughly in proportion to the square root of molecular weight, are widely tabulated for aqueous solutions, or may be estimated from the Stokes Einstein equation (3). Ignoring boundary layer effects for the moment, and by assuming that diffusion within the membrane is analogous to that in free solution, equation 1 can be integrated across a homogeneous membrane of thickness d to yield the following equation, where S represents the dimensionless solute partition coefficient, ie, the ratio of solute concentration in external solution to that at the membrane surface, and D_M represents solute diffusion within the membrane and is assumed independent of solute concentration.

$$\phi = \frac{SD_M \, \Delta c}{d} \tag{2}$$

The product SD_M is often termed permeability; if two or more solutes are dialysing at the same time, the degree of separation or enrichment is proportional to the ratio of their permeabilities. The closer the permeability of a membrane is to that of an equivalent thickness of free solution, the more rapid is the resultant dialytic transport. Considerable effort has been devoted to understanding how the physical and chemical properties of a membrane determine its permeability. The simplest approaches are geometric and consider the membrane to comprise a series of parallel pores that provide a topographic obstacle to hard noninteracting permeant molecules (4); far more complex analyses are also available (5,6). As a general rule, permeability for a particular species increases with porosity (solute content) of the membrane and with the diameter of its pores. Equation 2 also states that the mass flow rate of solute is inversely proportional to membrane thickness, but the degree of separation (selectivity) is independent of thickness. For this reason, membranes are always made as thin as possible consistent with the requirements of mechanical strength and reliability. Equation 2 is often further simplified to the following expression for flux per unit of membrane area, where thickness is incorporated into an overall membrane mass-transfer coefficient, K_M, with units of cm/s.

$$\phi = K_M \, \Delta_C \tag{3}$$

Dialysis transport relations need not start with Fickian diffusion; they may also be derived by integration of the basic transport equation (7) or from the phenomenological relationships of irreversible thermodynamics (8,9).

Solutions adjacent to the membranes are rarely well mixed, and the resistance to transport resides not just in the membrane but also in the fluid regions, termed boundary layers, on both the dialysate and feed side. Boundary layer effects typically account for from 25 to 75% of overall resistance. They are minimized by rapid convective flow tangential to the surface of the dialysing membrane. When fluid pathways are thin, juxtamembrane flow is laminar, and boundary layer resistance decreases with increasing wall shear rate. Where geometry permits higher Reynolds numbers, flow becomes turbulent and resistance varies with net tangential velocity. Geometric turbulence promoters are often employed. All tactics to reduce boundary layer result in higher energy utilization. Quantitatively, the membrane resistance becomes part of an overall coefficient K_O which, for conceptual purposes, is broken down into three independent and reciprocally additive components:

$$\frac{1}{K_O} = \frac{1}{K_B} + \frac{1}{K_M} + \frac{1}{K_D} \tag{4}$$

$$R_O = R_B + R_M + R_D \tag{5}$$

where K is device-averaged mass-transfer coefficient (or permeability) in cm/s, R is device-averaged resistance in s/cm, and the subscripts B, M, and D respectively denote the feedstream, membrane, and dialysate. Note that K_M in equation 4 is identical to that in equation 3. K_B can be estimated for many relevant conditions of geometry and flow using mass transport analysis based on wall Sherwood Numbers (10). K_M is best obtained by measurements employing special test fixtures in which boundary layer resistances are negligible or known (11,12). K_D is more problematic, and is usually obtained by extrapolations based on Wilson plots (13). Boundary layer theory, as well as techniques for correlation, estimation, and prediction of the constituent mass-transfer coefficients have been reviewed in two particularly lucid monographs (14,15). Overall solute transport is obtained from local flux by mass balance and integration; for the most common case of countercurrent flow:

$$\phi = (c_i' - c_i') \frac{Q_B}{A} \frac{\exp\left[\frac{K_O A}{Q_B}\left(1 - \frac{Q_B}{Q_D}\right)\right] - 1}{\exp\left[\frac{K_O A}{Q_B}\left(1 - \frac{Q_B}{Q_D}\right)\right] - \frac{Q_B}{Q_D}} \tag{6}$$

where c_i' and c_i'' represent inlet concentrations in the feed and dialysate streams in g/cm^3, A represents membrane surface area in cm^2, Q_B and Q_D are feed and dialysate flow rates in cm^3/min, and ϕ and K_O are as defined in equations 3 and 4. Derivations of this relationship and similar expressions for cocurrent or crossflow geometries can be found in the literature (14,16,17) (see MASS TRANSPORT).

Dialysis is a highly constrained process. Molecular diffusion is slow in the context of industrial dimensions. The driving force is set by the system itself, decreases in the course of purification, and is not amenable to extrinsic augmentation. The permeant species is not recovered in pure form, and is necessarily more dilute in the dialysate than in the starting stream. Low energy utilization is offset by high capital costs. For these reasons, dialysis has been largely limited to laboratory separations or specialized *in vivo* pharmacological investigations, and has enjoyed very limited success as a broad-based commercial unit operation. But the slow and gentle nature of dialysis has a special appeal for biologic applications, particularly when partial purification of the feed stream, rather than recovery of a product, is intended. Commercially significant examples include the adjustment of alcohol content of beverages and the removal of salts from solutions of proteins or other biologic macromolecules. However, the most successful and widespread application of dialysis—or for that matter of any membrane process— is the support of patients with kidney failure by repeated intermittent blood cleansing. In 1992 nearly half a million patients were maintained on dialysis, and the worldwide commercial aspects of this enterprise exceeded 15 billion U.S. dollars. Dialysis is closely related to membrane gas separation, pervaporation, ultrafiltration, and controlled release of pharmaceuticals discussed in separate sections of this *Encyclopedia* (see CONTROLLED RELEASE TECHNOLOGY, PHARMACEUTICALS). Particularly common is diafiltration, combined simultaneous dialysis, and ultrafiltration (qv) (see MEMBRANE TECHNOLOGY).

Industrial Dialysis

The recovery of caustic from hemicellulose (qv) in the rayon process was well established in the 1930s (18), and is still used in modern times (19) (see PULP). Very few new industrial applications of dialysis emerged during the 1940–1980 period. More recently, interest has reawakened in isobaric dialysis as a unit operation for the removal of alcohol from beverages (20,21) and in the production of products derived from biotechnology (22,23).

Although to many an oxymoron, alcohol-free beer has grown in popularity over the past decade in response to changing life-styles and legislative restraints on alcohol consumption; markets are also developing for alcohol-free wine. By the end of 1992, 40 key beer breweries worldwide had installed dialysis plants with an annual capacity of more than 189,000 m^3 (5 × 10^7 gal) of beer (qv). The process is illustrated in Figure 2. Alcohol is removed from beer by dialysis, the dialysate is distilled to remove alcohol, and the raffinate is recycled as a dialysate stream. The combination of dialysis and distillation preserves the flavor of the product

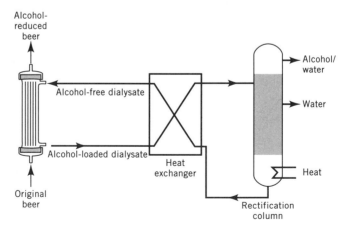

Fig. 2. Schematic of alcohol reduction in beverages. Countercurrent dialysis is combined with distillation. The separation process is isothermal, and high boiling ingredients, present in the dialysate, are preserved. In this fashion, alcohol removal is accomplished with minimal perturbation in flavor.

(24); dialysis is isothermal so the beer need not be heated. Higher boiling alcohols, esters, and carbohydrates that impart the special flavor to the beverage are already present in the dialysate and thus are not removed from the feed stream. A typical commercial installation is shown in Figure 3.

Fig. 3. A commercial dialysis facility showing the dialysis section of a German brewery where alcohol is removed from beer. Technical dialysis modules contain up to 50,000 capillaries and around 23 m^2 (250 ft^2) of membrane surface area. Typical plants might contain between 50 and 100 modules. Courtesy of Holstein and Kappert Processtechnik GmBH, Dortmund, Germany.

Dialysis plays an important role in the expanding biotechnology industry, but rarely as a stand-alone unit operation. It is applicable to the removal of salts from heat-sensitive or mechanically labile compounds such as vaccines, hormones, enzymes, and other bioactive cell secretions. In these instances, process efficiency is almost always increased by combining dialysis with ultrafiltration in the process known as diafiltration. Dialysis provides a simple means to control media and extracellular environment in bioreactors. Dialyzers can also offer the basis for a novel bioreactor design: the extraluminal region of a hollow fiber dialyzer provides an excellent growth environment for mammalian cells when the lumen is perfused with oxygen and nutrients (25). In the production of monoclonal antibodies, for example, a small benchtop bioreactor can readily equal the antibody production of several thousand mice. This technology is still in its early stages and considerable evolution can be anticipated in the future (see FERMENTATION; REACTOR TECHNOLOGY).

Laboratory Dialysis

Until the early 1960s, laboratory investigators relied on dialysis for the separation, concentration, and purification of a wide variety of biologic fluids. Examples include removal of a buffer from a protein solution or concentrating a polypeptide with hyperosmotic dialysate. Specialized fixtures were sometimes employed;

alternatively, dialysis tubes, ie, cylinders of membrane about the size of a test tube and sealed at both ends, were simply suspended in a dialysate bath. In recent years, dialysis as a laboratory operation has been replaced largely by ultrafiltration and diafiltration.

Microdialysis is a highly specialized application of the technique (26–28). In its simplest form, a U-shaped dialysis capillary is surgically implanted into the tissue of a living animal. Isotonic dialysate is pumped through the tubing at a flow rate low enough to allow equilibration with small solutes in the host's extracellular fluid. Concentration of solutes in the exiting fluid thus approaches those in the extracellular portion of the tissue. This technique is extremely useful because it permits uninterrupted sampling of the chemistry of individual tissues or body compartments without drawing blood. Once the implant is established, a microdialysis probe is capable of sampling continuously for days or even weeks. The procedure is most widely used in rodent studies, and is most popular for direct implantation into the brain by standard animal neurosurgical techniques. Microprobe designs range from straightforward U tubes to complex concentric capillaries. Perfusate flow rate is extremely low, around 10^{-6} mL/min. Microdialysis is performed on anesthetized animals usually under microprocessor control with on-line analysis of the eluate. Between 500 and 1000 articles have appeared in the literature describing microdialysis experiments in animals. The technique is likely to increase in popularity in the future, though therapeutic application seems a remote possibility.

Hemodialysis

Serious kidney disease is surprisingly uncommon in relation to the complexity of the organ, striking between 1 in 5,000 and 1 in 10,000 of the population per year. The origin of kidney disease may be genetic, traumatic, metabolic, vascular, or immunologic, and the response of the kidney, although essentially sclerotic, may be reversible or permanent, local or systemic, rapid or slow, or any combination thereof (29,30). Kidney failure, as distinct from kidney disease, occurs when renal function has declined to the point that the kidney can no longer satisfactorily perform its homeostatic and excretory functions. Since nature has provided kidneys with an abundance of overcapacity, patients become overtly symptomatic and identifiably diseased only after about 90% of function has been lost. When kidneys decline further and loss of capacity exceeds about 95%, some form of renal replacement therapy is required. Current alternatives include kidney transplantation and dialysis.

Despite widespread consensus that a successful transplant is the most satisfactory form of therapy for end stage renal disease, a chronic shortage of donor organs limits the number of patients receiving transplantation to about 18,000 per year (31,32). The remainder of renal failure patients require maintenance dialysis. About 12% elect continuous ambulatory peritoneal dialysis (CAPD), the remaining 88% hemodialysis. In CAPD, approximately 2 liters of a sterile, nonpyrogenic and hypertonic solution of glucose and electrolyte are instilled via gravity flow into the peritoneal cavity through an indwelling catheter four times per day. Intraperitoneal fluid partially equilibrates with solutes in the plasma, and plasma water is ultrafiltered due to osmotic gradients. After 4–5 h, except at night when the exchange is lengthened to 9–11 h to accommodate sleep, the peritoneal fluid is drained, and the process repeated. Patients perform the exchanges themselves in 20–30 min, at home or in the work environment, after a training cycle which lasts only 1–2 weeks. The literature on CAPD is abundant, but is well summarized in reference texts (33,34) and review articles (35,36).

The remaining 88% of untransplanted patients with kidney failure receive hemodialysis. This is an intermittent therapy with patients typically having thrice-weekly treatments of from 2.5 to 4 hours. Although most hemodialysis is performed in free-standing treatment centers, it may also be provided in a hospital or performed by the patient at home. The hemodialysis circuit consits of two fluid pathways. The blood side is entirely disposable, though many centers re-use some or all circuit components in order to reduce costs. It comprises a 16-gauge needle for access to the circulation (usually through a fistula created in the patient's forearm), lengths of dioctyl phthalate plasticized poly(vinyl chloride) tubing including a special tubing segment adapted to fit into a peristaltic blood pump, the hemodialyzer itself, a venous bubble trap and an open mesh screen filter, various ports for samples and gauge connections, and a return cannula. Components of the blood-side circuit are supplied in sterile and nonpyrogenic condition; ethylene oxide is the most common sterilant, although both radiation and steam sterilization are rapidly gaining favor. The dialysate side is essentially a machine capable of proportioning out glucose and electrolyte concentrates with water to provide dialysate of appropriate composition, pumping dialysate past a restrictor valve and through the hemodialyzer at subatmospheric pressure, and monitoring temperature, circuit pressures, and flow rates. During treatment the patient's blood is anticoagulated with heparin. Typical blood flow rates are 200–350 mL/min; dialysate flow rates are usually 500 mL/min. Straightforward techniques have been developed to prime the blood side with sterile saline prior to use and to rinse back nearly all the formed elements after treatment. Although most mass transport occurs by diffusion, circuits are operated with pressure on the blood side controlled to 13.3 to 66.7 kPa (100 to 500 mm Hg) higher than on the dialysate side. This provides an opportunity to remove 2 to 4 liters of fluid along with the solute; higher rates of fluid removal are technically possible but physiologically unacceptable. Hemodialyzers must be designed with high enough hydraulic permeabilities to provide adequate fluid removal at the upper pressure range, but not so high that excessive dewatering will occur at the lower pressure ranges.

Figure 4 is a schematic of a typical hemodialyzer. Although other geometries are still employed, the preferred format is a hollow fiber hemodialyzer about 25 cm in length and 5 cm in diameter. Devices typically contain 6,000 to 10,000 capillaries, each with an inner diameter of 200 μm and a wall thickness of around 10 μm. Mean total membrane surface area is 1.1 ± 0.4 m^2. Well over 60 million hemodialyzers were produced in 1992. Because of economies of scale, unit price was on the order of $10 per unit, much lower than would be anticipated from the complexity of the device or by comparison with other membrane products. Interestingly, the hemodialyzer rarely represents more than 10% of the cost of a treatment session. This therapy is extensively described in the literature; by mid-1992,

Med Line contained over 29,000 citations on hemodialysis. Several excellent reference texts provide concise and comprehensive coverage of all aspects of hemodialysis (37–40).

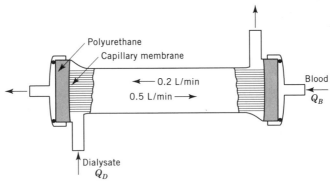

Fig. 4. Schematic of a hemodialyzer. The design of a dialyzer is close to that of a shell and tube heat exchanger. Blood enters through an inlet manifold, is distributed to a parallel bundle of fibers, and exits into a collection manifold. Dialysate flows countercurrent in an external chamber; the blood and dialysate are separated from the fibers by a polyurethane potting material. Housings are typically prepared from acrylate or polycarbonate. Production volume is greater than 50 million units per year and cost is very low, around $10 U.S. in 1992.

Engineering Aspects of Hemodialysis.   Engineering interest in hemodialysis is concentrated on the optimization of the hemodialysis membrane (4,41), the dependency of solute removal on membrane and device characteristics (14,15), and quantitation of hemodialysis therapy through urea pharmacokinetics (42–44).

Hemodialysis membranes vary from one another in chemical composition, transport properties, and in their biocompatibility, defined here as the capacity of a material to avoid recognition and response by various host defense mechanisms (Table 1). Table 2 divides hemodialysis membranes into three classes: cellulosics, modified cellulosics, and synthetics. Cellulosics are prepared from regenerated cellulose by the cupramonium process; these extremely hydrophilic

Table 1. Contemporary Hemodialysis Membrane Characteristics

Hydraulic permeability[a]	Solute clearance		Market share[b]	Absolute growth
	mol wt 250	mol wt >1000		
low-flux				
KUFR = 2–6	high	low	70%	steady
middle-flux				
KUFR = 5–12	high	medium	20%	growing
high-flux/high-performance				
KUFR = 10–200	high	high	10%	growing

[a]KUFR = ultrafiltration coefficient in mL/h·m²·mm Hg.
[b]Estimated 1992.

Table 2. Polymeric Materials for Dialysis Membranes

Material	Manufacturer
Regenerated cellulosics	
Cuprophan	Akzo
cuprammonium cellulose	Asahi
	Terumo
SCE[a]	Teijin
	Althin
Synthetically modified cellulose	
Hemophan	Akzo
cellulose acetate	Akzo
	Toyobo
	Althin
	Teijin
cellulose triacetate	Toyobo
SMC[b]	Akzo
Synthetics	
polysulfone	Akzo
	Fresenius
	NMC
	Kurary
	Kawasumi
polycarbonate	Gambro
polyamide	Gambro
polyacrylonitrile	Hospal
	Asahi
SPAN[c]	Akzo
EVAL[d]	Kawasumi/Kuraray
PMMA[e]	Torray

[a] SCE = saponified cellulose ester.
[b] SMC = specially modified cellulose.
[c] SPAN = sulfonated polyacrylonitrile.
[d] EVAL is a poly(vinyl alcohol), a copolymer of ethylene and vinyl alcohol.
[e] PMMA = poly(methyl methacrylate).

structures sorb water, bind it tightly, and form true hydrogels as is illustrated in the left hand panels of Figure 5. Their principle advantage is low unit cost; this is complemented by the strength of the highly crystalline cellulose which allows membranes to be made very thin and thus provides effective small-solute transport in relatively small hemodialyzers. The vulnerabilities of regenerated cellulose are its limited permeability to larger molecules, and the presence of labile nucleophilic groups that trigger complement activation and transient leukopenia during the first hour of a hemodialysis session. The advantages appear to outweigh the disadvantages: over 70% of current hemodialyzers are prepared from

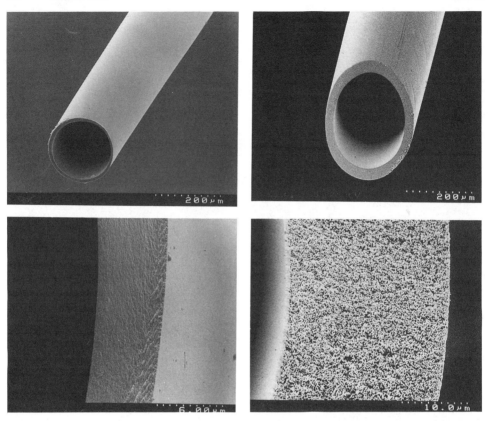

Fig. 5. Scanning electron micrographs of hollow fiber dialysis membranes. Membranes in left panels are prepared from regenerated cellulose (Cuprophan) and those on the right from a copolymer of polyacrylonitrile. The cellulosic materials are hydrogels and the synthetic thermoplastic forms a microreticulated open cell foam with a tight skin on the inner wall. Pictures at top are membrane cross sections; those below are of the wall region. Dimensions as indicated.

cellulosics, most of which are supplied by Akzo Faser AG under the trade name Cuprophan. At the opposite end of the spectrum are membranes prepared from synthetic, engineering thermoplastics, such as polysulfone and polyamide. These materials form anisotropic membranes with foamlike or trebacular cross sections (see the right hand panel in Fig. 5). They appear less active to the complement cascade and other physiologic identifiable defense mechanisms. In addition to this improved biocompatibility, these membranes are the least restrictive in transport to larger molecules. Drawbacks are increased cost and sufficiently high hydraulic permeability to require specialty control mechanisms and to raise concerns over the biologic quality of dialysate fluid. Roughly 10% of hemodialyzers are produced from such hydrophobic membranes. A middle group, also accounting for 10–15% of total hemodialyzer production, comprises both derivatized cellulosics, eg, cellulose diacetate, and synthetic hydrophilic polymers. Because of regulatory vigilence, all hemodialysis membranes in use are both safe and effective; there is no sound epidemiologic evidence that selection of one membrane over another will alter a patient's morbidity, mortality, or quality of life.

The clinical performance of a hemodialyzer is usually described in terms of clearance, a term having its roots in renal physiology, which is defined as the rate of solute removal divided by the inlet flow concentration as shown in equation 7, where Cl is clearance in mL/min and all other terms are as defined previously except that, in deference to convention, flow rates are now expressed in minutes rather than seconds and feed side (c') is now synonymous with blood flow on the luminal side.

$$Cl = \frac{\phi A}{c_i'} = \frac{Q_B(c_i' - c_o')}{c_i'} \tag{7}$$

Note that the numerator in each of the ratios in equation 7 represents the rate of solute removal from the patient. By mass balance, clearance is related to mass-transfer coefficient K_O as defined earlier in equations 3, 4, and 5, and where each of the three expressions equal rate of mass removal in g/s.

$$K_O A \, \Delta c = \phi A = Cl c_i' \tag{8}$$

For consistency, clearance here is expressed in cm^3/s although the more common clinical units, and those used later in this chapter, are mL/min. Combination and rearrangement of equations 6–8 allows clearance to be estimated from mass-transfer coefficient and vice versa; the conditions of countercurrent flow with no dialysate recycling are shown below.

$$Cl = Q_B \frac{\exp\left[\frac{K_O A}{Q_B}\left(1 - \frac{Q_B}{Q_D}\right)\right] - 1}{\exp\left[\frac{K_O A}{Q_B}\left(1 - \frac{Q_B}{Q_D}\right)\right] - \frac{Q_B}{Q_D}} \tag{9}$$

$$K_O = \frac{Q_B}{A\left(1 - \frac{Q_B}{Q_D}\right)} \ln\left[\frac{1 - \frac{Cl}{Q_D}}{1 - \frac{Cl}{Q_B}}\right] \tag{10}$$

Similar expressions for other conditions of geometry and flow are found in References 14 and 15.

Clearance decreases with increasing permeant molecular weight and depends in complex fashion upon blood and dialysate flow rate and upon device geometry. Detailed engineering analyses are available in References 14–17. As a general rule in most contemporary dialyzers, the clearance of small solutes such as urea (mol wt = 58), creatinine (mol wt = 113) has either approached a maximum (clearance can never exceed blood flow rate) or is limited by boundary layers adjacent to the membrane; for these solutes changes in membrane permeability or membrane surface area will not significantly affect clearance whereas increases in blood flow will lead to increased clearance. In contrast, larger solutes such as inulin (mol wt ~ 5200 daltons) or beta-2-microglobulin (mol wt = 11,118 daltons), are membrane limited. Their clearance will increase, often linearly, with increasing membrane surface area, but will be largely unaffected by changes in blood or dialysate flow rate. These relationships are illustrated in Figure 6 and summarized in Table 3.

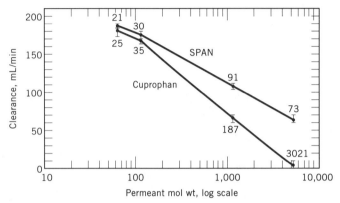

Fig. 6. Solute transport in hemodialysis. Clearance vs solute mol wt for dialyzers prepared from the two different membranes illustrated in Figure 5. Numbers next to points represent R_O in min/cm calculated from equations 10 and 5. Data is *in vitro* at 37°C with saline as the perfusion fluid. Lumen flow, dialysate flow, and transmembrane pressure were 200 mL/min, 500 mL/min, and 13.3 kPa (100 mm Hg); area = 1.6 m^2. Inulin clearance of the SPAN fiber was elevated by inulin transported by the filtering fluid.

Table 3. Effects of Changes in Conditions of Geometry and Flow on Hemodialyser Clearance

	Effect on clearance of	
Parameter increased	Low mol wt solutes[a]	High mol wt solutes[b]
blood flow rate	increases	little or no effect
dialysate flow rate	little effect	little or no effect
membrane surface area	little effect	almost linear increase
membrane permeability	little effect	almost linear increase

[a]Mol wts of <200 daltons, eg, urea (60), creatinine (113), or uric acid (158).
[b]Mol wts of >1000 daltons, vitamin B$_{12}$ (1355) or inulin (~5200).

Urea Pharmacokinetics. Pharmacokinetics summarizes the relationships between solute generation, solute removal, and concentration in a patient's blood stream. In the context of hemodialysis, this analysis is most readily applied to urea, which has, as a consequence, become a surrogate for other uremic toxins in the quantitation of therapy and in attempts to describe its adequacy. In the simplest case, a patient is assumed to have no residual renal function. Urea is generated from the breakdown of dietary protein, accumulates in a single pool equivalent to the patient's fluid volume, and is removed uniformly from that pool during hemodialysis. A mass balance around the patient yields the following differential equation:

$$\frac{d\,(cV)}{dt} = G - Clc \tag{11}$$

where c = whole blood urea concentration normally expressed as mg % (mg/100 mL), V = urea distribution volume in the patient in mL, G = urea generation rate in mg/min, t = time from onset of hemodialysis in minutes, and Cl = urea clearance in mL/min.

Urea concentration in the United States medical literature is often reported as BUN (blood urea nitrogen), which is urea concentration, usually in mg/dL, multiplied by a factor of 0.47. V, in equation 11, can be measured by dilution studies, but is often estimated in kinetic modeling studies as 58% of patient weight. Generation is calculated from a knowledge or an estimate of patient protein intake (each gram of protein consumed produces about 250 mg of urea; see References 43 and 44 for more exact correlations based on metabolic studies of uremic patients). Thus a 70 kg patient, consuming a typical 1.0 g of protein per kg of body weight per day, would produce 28 g of urea distributed over a fluid volume of 40.6 L and, in the absence of any clearance, urea concentration would increase by 70 mg % (mg/100 mL) every 24 hours. The reduction of urea concentration during hemodialysis is readily obtained from equation 11 by neglecting intradialytic generation and changes in volume where c^o and c^t represent the urea concentrations in blood at the beginning and during the course of treatment.

$$c^t = c^o \exp\left(-\frac{Cl\,t}{V}\right) \tag{12}$$

A 3.5 h treatment of a 70 kg patient (V = 40.6 liters) with a urea clearance of 200 mL/min should result in a 64% reduction in urea concentration or a value of 0.36 for the ratio c^t/c^o; this parameter almost always falls between 0.30 and 0.45. The increase in urea concentration between hemodialysis treatments is obtained from equation 13, again assuming a constant V, where c^o is the urea concentration in the patient's blood at the end of the hemodialysis, and c^t the concentration at time t during the intradialytic interval.

$$c^t = c^o + \frac{G}{V}\,t \tag{13}$$

Urea concentration typically increases by about 50 to 100 mg/100 mL/24 h. Even a small residual clearance will prove numerically significant and, for oliguric patients, the slightly more complex formulas given in References 43 and 44 should be employed. The exponential decay constant in equation 12, $Cl\,t/V$, is the net normalized quantity of hemodialysis therapy. It is calculated simply by multiplying the urea clearance in mL/min by the duration of hemodialysis, also in minutes, and dividing by the distribution volume in mL, which, in the absence of a better estimate, is taken as 0.58 × the patient weight. This parameter provides an index of the adequacy of hemodialysis (45) and based on retrospective analysis of various therapy formats, a value of 1.0 or greater for urea proposedly provides an adequate amount of hemodialysis for most patients. Although not without its critics, this approach has found nearly universal clinical acceptance, and represents the current prescriptive norm to hemodialysis therapy.

Maintenance hemodialysis has grown and expanded beyond the expectations of even the most enthusiastic of its earliest proponents. Figure 7 is a plot of the overall estimated dialysis population by year since 1970. The population at the end of 1992 exceeded 475,000; another 500,000 patients or so have received therapy at one time but have since died or had transplants. Maintenance dialysis is now available to some extent in all but the poorest nations; in economically advanced countries, excepting the United Kingdom, it is rendered as a virtual entitlement. The current worldwide mean cost of a single dialysis patient is about $30,000 per year (47); the aggregate economic magnitude of the medical application of hemodialysis thus approaches $15 billion.

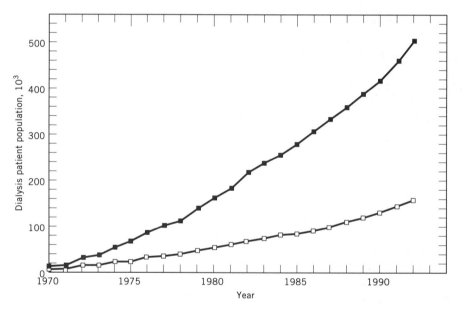

Fig. 7. Estimate of the total number of patients receiving maintenance dialysis over the past 20 years. Totals include both hemodialysis and peritoneal dialysis, but exclude transplant recipients. The fraction of patients receiving peritoneal dialysis has grown steadily from 0% in 1978 to about 12% in 1992. These data were combined from various regional registries and industry sources; demographic estimates of this ilk are accurate to within 5% (46). □, United States; ■, worldwide.

BIBLIOGRAPHY

"Dialysis and Electrodialysis" in *ECT* 1st ed., Vol. 5, pp. 1–20, by F. K. Daniel, Chemical Consultant; "Dialysis" in *ECT* 2nd ed., Vol. 7, pp. 1–21, by E. F. Leonard, Columbia University; in *ECT* 3rd ed., Vol. 7, pp. 564–579, by E. F. Leonard, Columbia University.

1. T. Graham, *Phil. Mag.* **49,** 337 (1866).
2. A. Fick, *Pogg. Ann.* **94,** 59–86 (1855).
3. A. Einstein, *Ann. Physik.* **19,** 289 (1906).
4. M. J. Lysaght, *Contrib. Nephrol.* **61,** 1–17 (1988).
5. W. Pusch, *Desalination* **59,** 105–198 (1986).
6. H. Strathmann, *Trennung von Molekularan Mischungen mit Hilfe Synthetischer Membranen,* Steinkopff, Darmstadt, FRG, 1959, pp. 15–55.
7. R. Schloegl, *Stofftransport durch Membranen,* Steinkopff, Darmstadt, FRG, 1966.
8. O. Kedem and A. Katachalsky, *Biochem. Biophys. Acta* **27,** 229–246 (1961).
9. A. Katachalsky and P. F. Curran, *Nonequilibrium Thermodynamics in Biophysics,* Harvard University Press, Cambridge, Mass., 1967.
10. C. K. Colton and co-workers, *AIChE J.* **17,** 772–780 (1971).
11. C. K. Colton and co-workers, *J. Biomed. Mater. Res.* **5,** 459–488 (1971).
12. E. Klein and co-workers, *J. Membr. Sci.* **2,** 349–364 (1977).
13. E. F. Leonard and W. Bluemle, *Trans. NY Acad. Sci.,* 585–598 (1959).
14. C. K. Colton and E. G. Lowrie, in B. M. Brenner and F. C. Rector, eds., *The Kidney,* 2nd ed., Saunders Publishing Co., Philadelphia, 1981, pp. 2425–2489.
15. C. K. Colton, *Blood Purif.* **5,** 202–251 (1987).
16. A. S. Michaels, *Trans. Am. Soc. Artif. Intern. Organs* **12,** 387–392 (1966).
17. F. A. Gotch and co-workers, *The Evaluation of Hemodialyzers,* DHEW publication NIH 72-103, Washington, D.C. 1972.
18. H. B. Volrath, *Chem. Met. Eng.* **43,** 303 (1936).
19. T. Nishiwaki and S. Itoi, *Jpn. Chem. Q.* **41,** 36 (1982).
20. H. Moonen and H. J. Niefind, *Desalination* **41,** 327–335 (1982).
21. F. Jonaaon, in P. M. Bungay, H. K. Lonsdale, and M. N. de Pinho, eds., *Synthetic Membranes: Science, Engineering, Applications,* Reidel, Dordrecht, 327–335, 1982.
22. C. Heath and G. Belfort, *Int. J. Biochem.* **22,** 823–835 (1990).
23. J. M. Piret and C. L. Cooney, *Biotechnol. Adv.* **8,** 763–783 (1990).
24. H. G. Tilgner and F. J. Schmitz, *Eur. Pat* 36,175 (1981).
25. P. M. Knazek and co-workers, *Science* **178,** 65–67 (1974).
26. U. Ungerstedt, *J. Int. Med.* **230,** 365–373 (1991).
27. N. Lindefors, G. Amberg, and U. Ungerstedt, *Pharm. Methods.* **22,** 141–156 (1989).
28. G. Amberg and N. Lindefors, *Pharm. Methods* **22,** 157–183 (1989).
29. H. Smith, *From Fish to Philosopher,* Doubleday, Garden City, N.J., 1961.
30. B. M. Brenner and F. C. Rector, eds., *The Kidney,* 3rd ed., Saunders Publishing Co., Philadelphia, 1986.
31. P. A. Keown and C. R. Stiller, *Kidney Int. Suppl* **24,** S145-9 (1988).
32. T. B. Strom and N. L. Tilney, in Ref. 30, pp. 1941–1985.
33. R. Gokal, *Continuous Ambulatory Peritoneal Dialysis,* Churchill Livingston, Edinburgh, 1987.
34. K. D. Nolph, *Peritoneal Dialysis,* 3rd ed., Martinus Nihjoff, The Hague, 1988.
35. K. D. Nolph, A. S. Lindblad, and W. J. Novak, *N. Engl. J. Med.* **318,** 1595–1600 (1988).
36. M. J. Lysaght and P. C. Farrell, *J. Membr. Sci.* **44,** 5–33 (1989).
37. J. F. Maher, *Replacement of Renal Function by Dialysis,* 3rd ed., Kluwer, Boston, 1989.
38. H. J. Gurland, *Uremia Therapy,* Springer-Verlag, Berlin, 1987.
39. T. H. Frost, *Technical Aspects of Renal Dialysis,* Pittman Press, Kent, UK, 1977.
40. A. R. Nissenson, R. Fine, and D. Gentile, *Clinical Dialysis,* Appleton & Lange Century Crofts, Nowalk, Conn., 1984.
41. H. Strathmann and H. Goehl, *Contrib. Nephrol.* **78,** 119–141 (1990).
42. E. G. Lowrie, ed., *Kidney Int.* **23,** suppl. 13 (1983).
43. J. A. Sargeant and F. Gotch, in Ref. 37, pp. 87–143.

44. P. C. Farrell, *Dialysis Kinetics: ASAIO Primers in Artificial Organs 4*, J. B. Lippincott, Philadelphia, 1988.
45. F. Gotch and J. A. Sargent, *Kidney Int.* **23,** S103-6 (1983).
46. M. J. Lysaght, Ph.D. dissertation, University of New South Wales, Australia, 1989, Chapt. 2.
47. P. W. Eggers, *N. Engl. J. Med.* **318,** 223–229 (1988).

MICHAEL J. LYSAGHT
CytoTherapeutics, Inc.

ULRICH BAURMEISTER
Akzo Faser AG

DIAMINES AND HIGHER AMINES, ALIPHATIC

The aliphatic diamine and polyamine family encompasses a wide range of multifunctional, multireactive compounds. This family includes ethylenediamine (EDA) and its homologues, the polyethylene polyamines (commonly referred to as ethyleneamines), the diaminopropanes and several specific alkanediamines, and analogous polyamines. The molecular structures of these compounds may be linear, branched or cyclic, or combinations of these.

The ethyleneamines have found the broadest commercial application and are the primary focus of this article. The lower molecular weight ethylenediamines, ie, EDA, diethylenetriamine (DETA), piperazine (PIP), and *N*-(2-aminoethyl)-piperazine (AEP), are available commercially as industrially pure products. The tetramine (TETA), pentamine (TEPA), hexamine (PEHA), and higher polyamine products are commercially available as boiling point fractions consisting of natural mixtures of linear, branched, and cyclic compounds. Their compositions are largely determined by the chemical processes used in their production. The individual components in these higher ethyleneamines are generally not available in industrial quantities.

The predominant commercial diaminopropanes are 1,2-propylenediamine (1,2-PDA), 1,3-diaminopropane (1,3-PDA), iminobispropylamine (IBPA), and dimethylaminopropylamine (DMAPA). Other commercially important products include other higher alkylenediamines, such as hexamethylenediamine (HMDA); certain cyclic amines, such as triethylenediamine (TEDA); and various alkyl- and hydroxyalkyl-derivatives. These compounds have commercial significance in other specific areas (see also AMINES, AROMATIC AMINES-DIARYL AMINES; AMINES, CYCLOALIPHATIC AMINES; AMIDES, FATTY ACID; POLYAMIDES).

Physical Properties

Physical properties of some commercially available polyamines appear in Table 1. Generally, they are slightly to moderately viscous, water-soluble liquids with mild to strong ammoniacal odors. Although completely soluble in water initially, hydrates may form with time, particularly with the heavy ethyleneamines (TETA, TEPA, PEHA, and higher polyamines), to the point that gels may form or the total solution may solidify under ambient conditions. The amines are also completely miscible with alcohols, acetone, benzene, toluene and ethyl ether, but only slightly soluble in heptane. Piperazine, the lowest mol wt cyclic diamine, freezes above room temperature. As such, it is available commercially as either the anhydrous solid or an aqueous solution.

EDA forms the following binary azeotropes (at atmospheric pressure):

Secondary component	Bp, °C	Amine, %
water	119	81.6
isobutyl alcohol	120.5	50.0
n-butanol	124.7	35.7
toluene	103	30.8

Chemical Properties

The aliphatic alkyleneamines are strong bases exhibiting behavior typical of simple aliphatic amines. Additionally, dependent on the location of the primary or secondary amino groups in the alkyleneamines, ring formation with various reactants can occur. This same feature allows for metal ion complexation or chelation (1). The alkyleneamines are somewhat weaker bases than aliphatic amines and much stronger bases than ammonia as the pK_b values indicate (Table 2).

Inorganic Acids. Alkyleneamines react vigorously with commonly available inorganic acids forming crystalline, water soluble salts. The free alkyleneamines can be regenerated by reaction of their salts with aqueous caustic.

Ethylenedinitramine [505-71-5], an explosive compound, is made by reaction of two moles of nitric acid [7697-37-2] per mole of EDA, splitting out two moles of water from the salt at elevated temperatures (6).

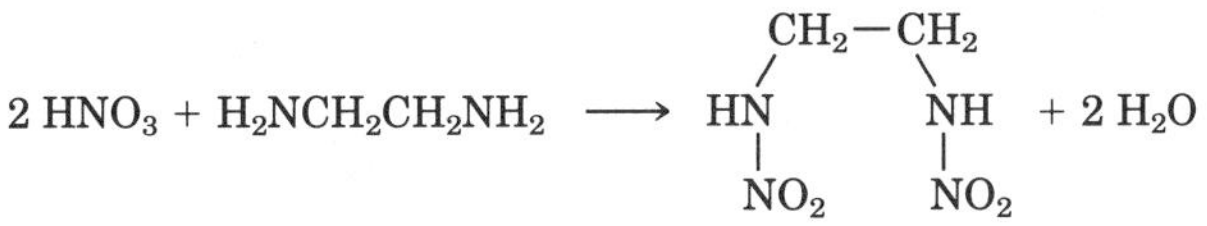

$$2\ HNO_3 + H_2NCH_2CH_2NH_2 \longrightarrow \underset{\underset{NO_2}{|}}{HN}\overset{CH_2-CH_2}{\diagup\diagdown}\underset{\underset{NO_2}{|}}{NH} + 2\ H_2O$$

Alkylene Oxides and Aziridines. Alkyleneamines react readily with epoxides, such as ethylene oxide [75-21-8] (EO) or propylene oxide [75-56-9] (PO), to form mixtures of hydroxyalkyl derivatives. Product distribution is controlled by the amine to epoxide mole ratio. If EDA, which has four reactive amine hydrogens, reacts at an EDA to EO mole ratio which is greater than 1:4, a mixture of mono-,

Table 1. Properties of Commercial Diamines and Higher Amines

Commercial name	CAS Registry Number	Structural formula	Molecular weight	Freezing point, °C	Bp^a, °C	ΔH^a_{vap}, kJ/mol	Refractive index, n_D^{20}	Viscosity at 20°C, mPa·s (= cP)
ethylenediamine	[107-15-3]	H$_2$N(CH$_2$)$_2$NH$_2$	60.1	10.8	117.0	40.7	1.4565	1.8
diethylenetriamine	[111-40-0]	H$_2$N(CH$_2$)$_2$NH(CH$_2$)$_2$NH$_2$	103.2	−39	206.9	54.0	1.4859	7.2
triethylenetetramine	[112-24-3]	H$_2$N(CH$_2$CH$_2$NH)$_2$CH$_2$CH$_2$NH$_2$^{b}	146.2^b	−35	277.4	56.4	1.4986	26
tetraethylenepentamine	[112-57-2]	H$_2$N(CH$_2$CH$_2$NH)$_3$CH$_2$CH$_2$NH$_2$^{b}	189.3^b	<−40	315		1.5067	76
pentaethylenehexamine^c	[4067-16-7]	H$_2$N(CH$_2$CH$_2$NH)$_4$CH$_2$CH$_2$NH$_2$^{b}	232.4^b	−30	180–280^d			100–300
aminoethylpiperazine	[140-31-8]	NH$_2$CH$_2$CH$_2$N(CH$_2$CH$_2$)$_2$NH	129.6	−17	221	41.2	1.5003	15
piperazine	[110-85-0]	HN(CH$_2$CH$_2$)$_2$NH	86.1	109.6	144.1			
1,2-propylenediamine	[78-90-0]	H$_2$NCH(CH$_3$)CH$_2$NH$_2$	74.1	−27	120–123	38.2	1.4455	1.6
1,3-diaminopropane	[109-76-2]	H$_2$NCH$_2$CH$_2$CH$_2$NH$_2$	74.1	−12	137–140	46.4^e	1.4555	2.0
iminobispropylamine	[56-18-8]	H$_2$N(CH$_2$)$_3$NH(CH$_2$)$_3$NH$_2$	131.2	−16	110–120	76.2^e	1.4791	9.6
N-(2-aminoethyl)-1,3-propylenediamine	[13531-52-7]	H$_2$N(CH$_2$)$_2$NH(CH$_2$)$_3$NH$_2$	117.2		80			
N,N'-bis-(3-aminopropyl)-ethylenediamine	[10563-26-5]	H$_2$N(CH$_2$)$_3$NH(CH$_2$)$_2$NH(CH$_2$)$_3$NH$_2$	174.3		170			
dimethylaminopropylamine	[109-55-7]	(CH$_3$)$_2$NCH$_2$CH$_2$CH$_2$NH$_2$	102.2	−56	134.9	35.6	1.4350	1.1
menthanediamine	[80-52-4]	(see structure)	170.3	−45	107–126^f		1.479	17.5^g
triethylenediamine	[280-57-9]	N(CH$_2$CH$_2$)$_3$N	112.2	158	174	61.9^h		
hexamethylenediamine^i	[124-09-4]	H$_2$N(CH$_2$)$_6$NH$_2$	116.2	41	204.0	51.0		

Structure for menthanediamine:

$$\text{C(CH}_3\text{)}_2\text{—NH}_2 \text{ (ring)}, \quad \text{CH}_3, \quad \text{H}_2\text{N}$$

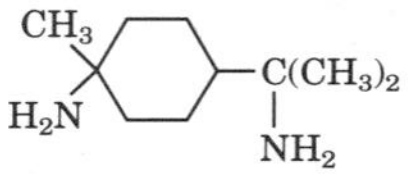

^{a}At 101.3 kPa = 1 atm unless otherwise noted.
^{b}Linear component. Commercial product consists of a mixture of linear, branched, and cyclic structures with the same number of nitrogen atoms.
^{c}Commercial higher polyamine products contain up to about 40% PEHA.
^{d}At 0.67 kPa, 10–60% distills in this range.
^{e}At 93.3°C.
^{f}At 1.3 kPa.
^{g}At 25°C.
^{h}Heat of sublimation, below 78°C.
^{i}For manufacture of HMDA in preparation of nylon-6,6, see POLYAMIDES.

Table 2. Alkyleneamine pK Values

Amine	pK_1	pK_2	Reference
EDA	3.83	6.56	2
1,2-PDA	4.03	6.90	2
1,3-PDA	3.70	5.71	3
DETA	3.89	4.60	4
TETA	3.84	4.47	4
NH_3	4.76		5
$CH_3CH_2NH_2$	3.25		5

di-, tri-, and tetrahydroxyethyl derivatives of EDA are formed. A 10:1 EDA:EO feed mole ratio gives predominantly 2-hydroxyethylethylenediamine [111-41-1]; the remainder is a mixture of bis-(2-hydroxyethyl)ethylenediamines (7). If the reactive NH to epoxide feed mole ratio is less than one and, additionally, a strong basic catalyst is used, then oxyalkyl derivatives, like those shown for EDA and excess PO result (8,9).

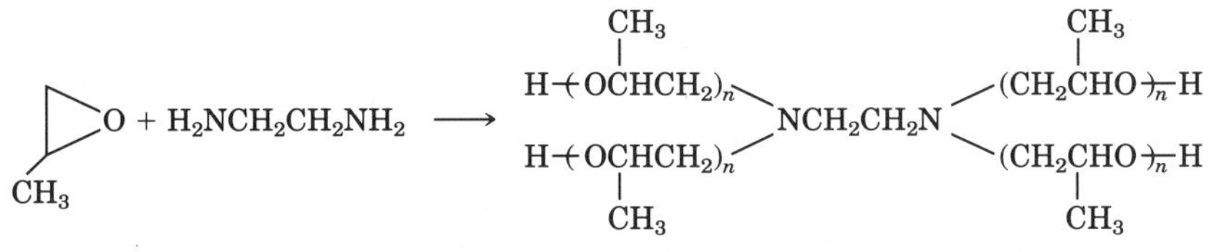

Aziridines react with alkyleneamines in an analogous fashion to epoxides (10,11). Product distribution is controlled by the alkyleneamine-to-aziridine mole ratio.

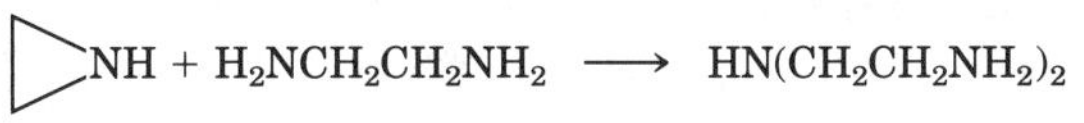

Aliphatic Alcohols and Alkylene Glycols. Simple aliphatic alcohols, such as methanol [67-56-1], can be used to alkylate alkyleneamines. For example, piperazine reacts with methanol over a reductive amination catalyst to yield a mixture of 1-methyl- [109-01-3] and 1,4-dimethylpiperazine [106-58-1] (12).

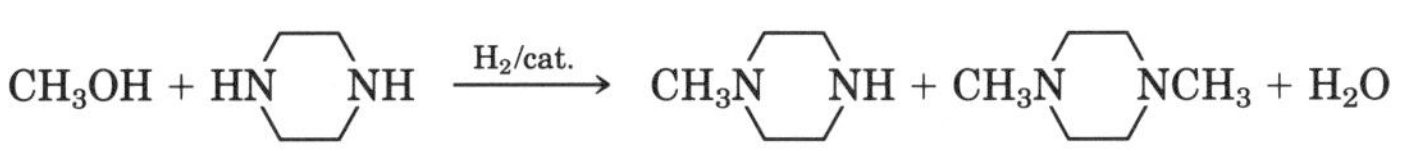

Either gas- or liquid-phase reactions of ethyleneamines with glycols in the presence of several different metal oxide catalysts leads to predominantly cyclic ethyleneamine products (13). At temperatures exceeding 400°C, in the vapor phase, pyrazine [290-37-9] formation is favored (14). Ethyleneamines bearing 2-hydroxyalkyl substituents can undergo a similar reaction (15).

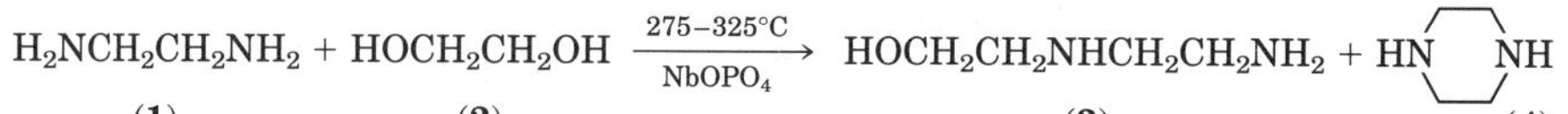

$$H_2NCH_2CH_2NH_2 + HOCH_2CH_2OH \xrightarrow[\text{NbOPO}_4]{275-325°C} HOCH_2CH_2NHCH_2CH_2NH_2 + HN\underset{}{\overset{}{\bigcirc}} NH$$

(1)   **(2)**   **(3)**   **(4)**

Reaction of either **(1)** + **(2)** or **(3)** + **(4)** at 400°C over ZnO gives pyrazine **(5)**.

$$N\underset{}{\overset{}{\bigcirc}} N$$

(5)

Organic Halides. Alkyl halides and aryl halides, activated by electron withdrawing groups (such as NO_2) in the ortho or para positions, react with alkyleneamines to form mono- or disubstituted derivatives. Product distribution is controlled by reactant ratio, metal complexation or choice of solvent (16,17). Mixing methylene chloride [75-09-2] and EDA reportedly causes a runaway reaction (18).

$$RX + H_2NCH_2CH_2NH_2 \longrightarrow RNHCH_2CH_2NH_2 \cdot HX$$

$$2\,RX + H_2NCH_2CH_2NH_2 \longrightarrow RNHCH_2CH_2NHR \cdot 2HX$$

Aliphatic dihalides, like ethylene dichloride [107-06-2], react with alkyleneamines to form various polymeric, cross-linked, water soluble cationic products (19) or higher alkyleneamine products depending on the reactant ratio.

$$2n\,X{-}R{-}X + n\,HN(CH_2CH_2NH_2)_2 \longrightarrow {-}(R{-}NHCH_2CH_2\overset{\overset{\displaystyle R}{|}}{N}CH_2CH_2NH)_{\overline{n}} \cdot 2nHX$$

Aldehydes. Alkyleneamines react exothermically with aliphatic aldehydes. The products depend on stoichiometry, reaction conditions, and structure of the alkyleneamine. Reactions of aldehydes with ethyleneamines like EDA or DETA give mono- and disubstituted imidazolidines via cyclization of the intermediate Schiff base (20).

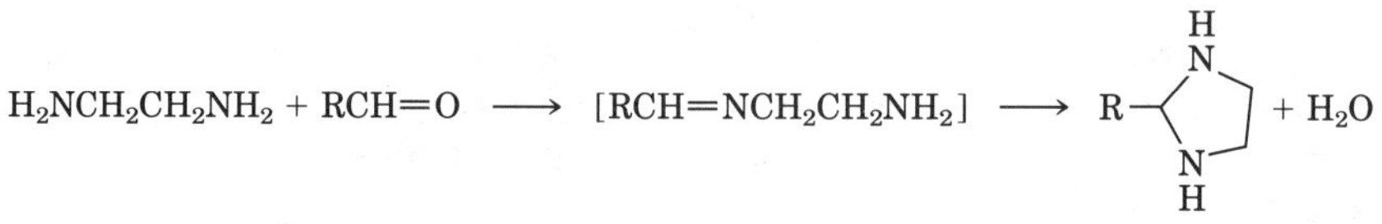

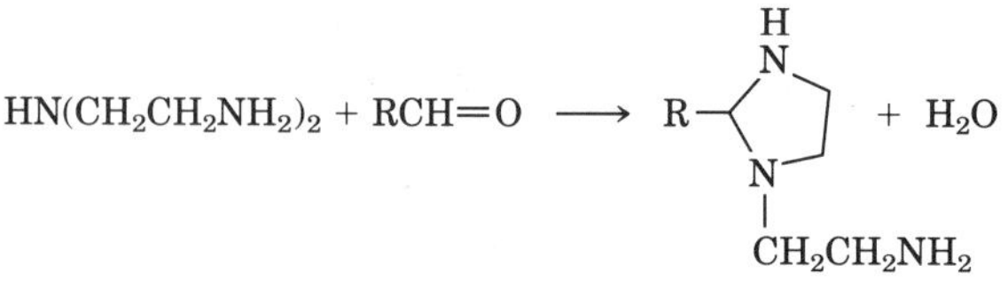

$$H_2NCH_2CH_2NH_2 + RCH{=}O \longrightarrow [RCH{=}NCH_2CH_2NH_2] \longrightarrow R{-}\underset{\underset{H}{N}}{\overset{\overset{H}{N}}{\diagup\!\!\!\diagdown}} + H_2O$$

$$HN(CH_2CH_2NH_2)_2 + RCH{=}O \longrightarrow R{-}\underset{\underset{CH_2CH_2NH_2}{N}}{\overset{\overset{H}{N}}{\diagup\!\!\!\diagdown}} + H_2O$$

Reaction of $H_2N(CH_2)_nNH_2$ with excess formaldehyde [50-00-0] is reversible and gives products of widely varying types which depend on the alkane chain length. For example, reaction of formaldehyde with EDA yields both 1,3,6,8-tetraazatricyclo[4.4.1.1^{3,8}]dodecane and 3-oxa-1,5-diazabicyclo[3.2.1]octane; the simple imidazolidine is not observed. However, 1,4-butanediamine [*110-60-1*] and diamines with longer chains give two-dimensional polymers (21).

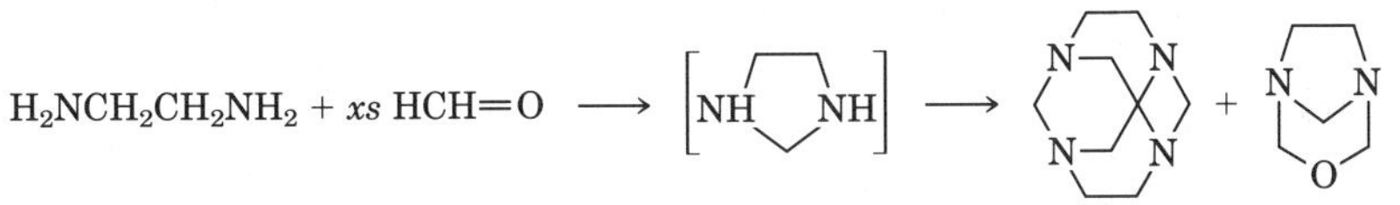

Substituted methyl- or dimethylamines are prepared by treating the appropriate primary or secondary amine with formaldehyde in the presence of hydrogenation catalysts (22,23).

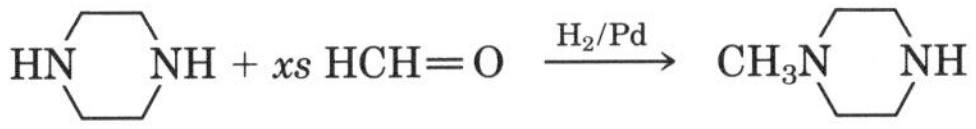

EDA reacts with formaldehyde and sodium cyanide under the appropriate alkaline conditions to yield the tetrasodium salt of ethylenediaminetetraacetic acid (24). By-product ammonia is removed at elevated temperatures under a partial vacuum. The free acid or its mono-, di-, or trisodium salts can be produced by the appropriate neutralization using a strong mineral acid. This same reaction with other amines is used to produce polyamino acetic acids and their salts. These products are used widely as chelating agents.

Organic Acids and Their Derivatives (Anhydrides, Nitriles, Ureas). Alkyleneamines react with acids, esters, acid anhydrides or acyl halides to form amidoamines and polyamides. Various diamides of EDA are prepared from the appropriate methyl ester or acid at moderate temperatures (25,26).

$$H_2NCH_2CH_2NH_2 + 2\ CH_3COOH \longrightarrow CH_3CONHCH_2CH_2NHCOCH_3 + 2\ H_2O$$

Diacids or carbonates are used with alkyleneamines to make amide polymers (27,28).

$$n\ H_2NCH_2CH_2NH_2 + n\ HOOC(R)COOH \longrightarrow HO\ \text{-[}CO(R)CONHCH_2CH_2NH\text{]}_{\overline{n}}\ H + n\ H_2O$$

Under more forcing conditions with acid anhydrides, EDA can form tetraacyl derivatives (29). However, much milder conditions or less active acylating agents are needed to obtain the monoamide essentially free of the diamide (30–32).

$$H_2NCH_2CH_2NH_2 + CH_3(CH_2)_{10}COOCH_3 \xrightarrow[75°C]{CH_3OH} CH_3(CH_2)_{10}\overset{\overset{O}{\|}}{C}NHCH_2CH_2NH_2 + CH_3OH$$

The monoamides can be cyclized to form imidazolines. This can be done without isolation of the monoamide by a variety of techniques including distilling the alcohol and water away as formed (33), heating under vacuum (34), heating in a hydrocarbon solvent under vacuum (35), or heating in the presence of catalysts such as alumina treated with H_3PO_4 (36) or CaO (37).

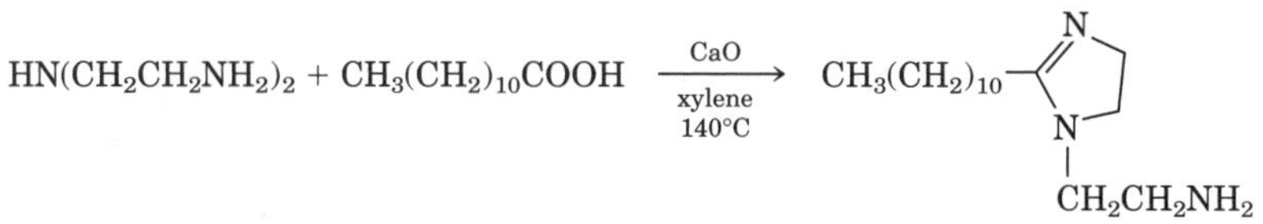

$$HN(CH_2CH_2NH_2)_2 + CH_3(CH_2)_{10}COOH \xrightarrow[\substack{xylene \\ 140°C}]{CaO} CH_3(CH_2)_{10}\text{—imidazoline ring}$$

Hydrogen cyanide (HCN) and aliphatic nitriles (RCN) can be used to form imidazolines. For example, EDA and HCN form 2-imidazoline (38). In the presence of sulfur or polysulfides as catalysts, 2-alkyl-2-imidazolines can be prepared from aliphatic nitriles and EDA (39,40).

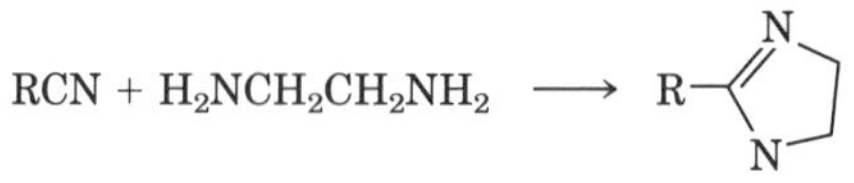

$$RCN + H_2NCH_2CH_2NH_2 \longrightarrow R\text{—imidazoline ring}$$

Alkyleneamines, like AEP, react with urea [57-13-6] to form substituted ureas and ammonia (41).

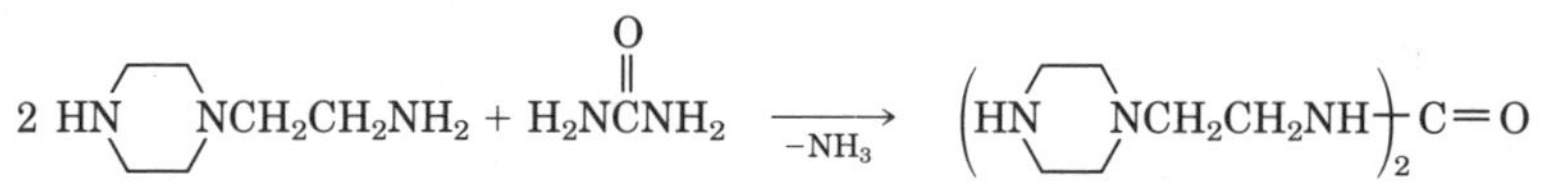

$$2\ HN\underset{}{\diagup\!\!\!\!\!\!\diagdown}NCH_2CH_2NH_2 + H_2NCNH_2 \xrightarrow[-NH_3]{} \left(HN\underset{}{\diagup\!\!\!\!\!\!\diagdown}NCH_2CH_2NH\right)_2\!\!C=O$$

DETA has been produced by reaction of monoethanolamine [141-43-5] (MEA) and EDA with urea (42). In this process, MEA reacts with urea to form 2-oxazolidinone which then reacts with EDA to form the cyclic urea of diethylenetriamine. Hydrolysis of the urea liberates the free amine.

Alkyl-substituted succinimides are prepared by reaction of alkyleneamines such as TETA or TEPA with the corresponding alkyl substituted succinic anhydride (43).

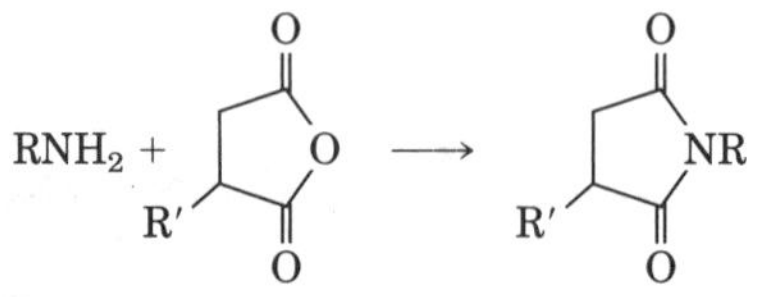

$$RNH_2 + \text{(anhydride)} \longrightarrow \text{(succinimide)}$$

Amines and other bases catalyze the exothermic decomposition of molten maleic anhydride [108-31-6] at temperatures above 150°C, accompanied by the rapid evolution of gaseous products (44,45). The rate of reaction reportedly increases with the basicity of the amine and higher initial temperatures. The reaction mixture can become explosive.

The imidazolines can be dehydrogenated at high temperatures over metal oxide catalysts to give the corresponding imidazoles (46).

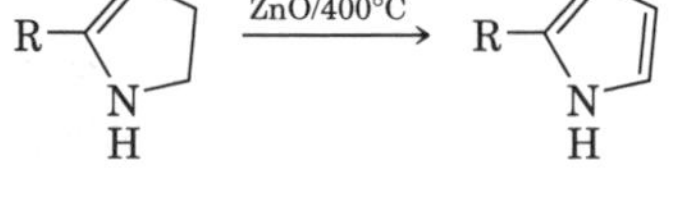

$$R\text{—imidazoline} \xrightarrow{ZnO/400°C} R\text{—imidazole}$$

Sulfur Compounds.   Ethylene thiourea [96-45-7], a suspected human carcinogen, is prepared by reaction of carbon disulfide [75-15-0] (CS_2) in aqueous EDA (47).

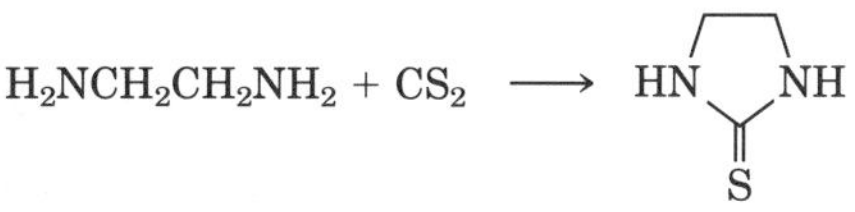

$$H_2NCH_2CH_2NH_2 + CS_2 \longrightarrow \underset{\underset{S}{\|}}{HN\diagdown\diagup NH}$$

EDA reacts readily with two moles of CS_2 in aqueous sodium hydroxide to form the bis sodium dithiocarbamate. When aqueous ammonia and zinc oxide (or manganese oxide or its hydrate) is used with a basic catalyst, the zinc (or manganese) dithiocarbamate salt is isolated. Alternatively, the disodium salt can react with $ZnSO_4$ or $MnSO_4$ followed by dehydration in an organic solvent to yield the same salts (48–50).

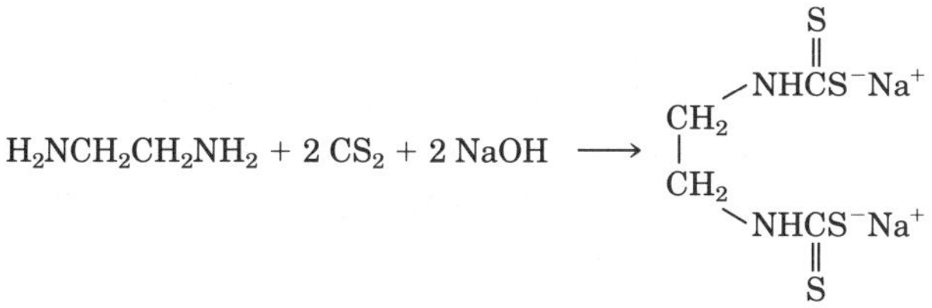

$$H_2NCH_2CH_2NH_2 + 2\ CS_2 + 2\ NaOH \longrightarrow \begin{array}{l} \overset{\displaystyle S}{\underset{\displaystyle \|}{}} \\ CH_2{\diagup}^{NHCS^-Na^+} \\ | \\ CH_2{\diagdown}_{NHCS^-Na^+} \\ \qquad\quad \underset{\|}{\overset{\|}{S}} \end{array}$$

Olefins.   EDA reacts with isobutylene [115-11-7] over a borosilicate zeolite containing 3.2% Cr at 300°C to give the monoalkylation product (51).

$$(CH_3)_2CH{=}CH_2 + H_2NCH_2CH_2NH_2 \longrightarrow (CH_3)_3CNHCH_2CH_2NH_2$$

EDA and other alkyleneamines react readily with acrylonitrile or acrylate esters. EDA reacts with acrylonitrile to give tetrakis(2-cyanoethyl)-ethylenediamine which is reduced over Raney nickel to give tetrakis(3-aminopropyl)-ethylenediamine (52). With methyl acrylate and EDA under controlled conditions, a new class of starburst dendritic macromolecules forms (53,54).

$$H_2NCH_2CH_2NH_2 + CH_2{=}CHCN \xrightarrow[\text{2. Ra–Ni}]{\text{1. }\Delta} [H_2N(CH_2)_3]_2NCH_2CH_2N[(CH_2)_3NH_2]_2$$

Olefins can be aminomethylated with carbon monoxide [630-08-0] (CO) and amines in the presence of rhodium-based catalysts. For example, piperazine reacts with cyclohexene [110-83-8] to form N,N'-di-(1-cyclohexylmethyl)-piperazine [79952-94-6] (55).

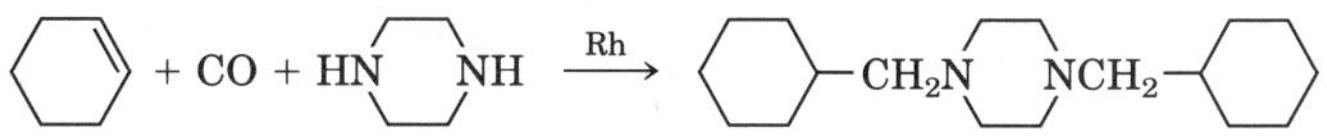

$$\langle\text{hexagon}\rangle + CO + HN\diagdown\diagup NH \xrightarrow{Rh} \langle\text{hexagon}\rangle{-}CH_2N\diagdown\diagup NCH_2{-}\langle\text{hexagon}\rangle$$

Environmentally Available Reactants.   Under normal conditions ethyleneamines are considered to be thermally stable molecules. However, they are sufficiently reactive that upon exposure to adventitious water, carbon dioxide, nitrogen oxides, and oxygen, trace levels of by-products can form and increased color usually results.

Carbon dioxide reacts readily with EDA in methanol to form crystalline *N*-(2-aminoethyl)carbamate [*109-58-0*] (56).

$$H_2NCH_2CH_2NH_2 + CO_2 \longrightarrow \overset{+}{H_3N}CH_2CH_2NHCOO^-$$

EDA, when heated with carbon dioxide under pressure (57), or treated with urea (58), or ethylene carbonate [*96-49-1*] (59), or CO and oxygen using a selenium catalyst (60), produces ethyleneurea.

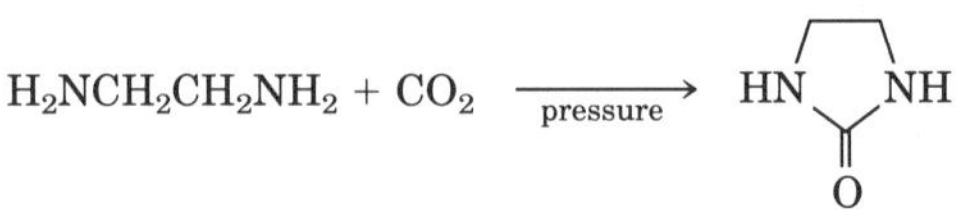

$$H_2NCH_2CH_2NH_2 + CO_2 \xrightarrow[\text{pressure}]{} \quad HN \underset{O}{\overset{\frown}{}} NH$$

N-Alkylpiperazines and PIP can react with nitrosating agents such as nitrogen oxides, nitrites or nitrous acid to form nitrosamine derivatives (61,62). Piperazine dihydrochloride [*142-64-3*] reacts with aqueous sodium nitrite and HCl to give the dinitrosamine that melts at 156–158°C (61).

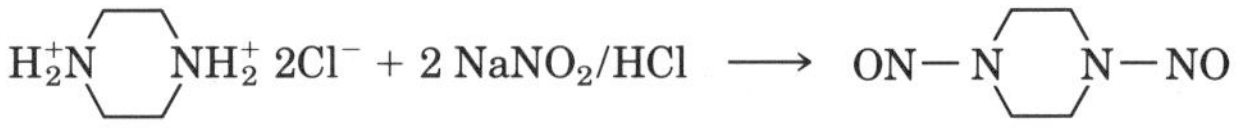

$$\overset{+}{H_2N} \underset{}{\overset{\frown}{}} NH_2^+ \, 2Cl^- + 2\,NaNO_2/HCl \longrightarrow ON-N \underset{}{\overset{\frown}{}} N-NO$$

PIP, if ingested (as the acid salt) or inhaled, is nitrosated *in vivo*, and is excreted partially as the mono- and dinitrosamines (63,64).

Ethyleneamines are soluble in water. However, in concentrated aqueous solutions, amine hydrates may form; the reaction is mildly exothermic. The hydrates of linear TETA and PIP melt around 50°C (65,66).

Manufacture

Ethyleneamine Processes.   Present industrial processes are based on ethylene and ammonia. The sixty year old ethylene dichloride (EDC) process is still the most widely practiced industrial route for producing ethyleneamines (67). In this process, aqueous ammonia reacts with EDC at high temperature and pressure to produce the total spectrum of ethyleneamine products as a mixture of amine hydrochloride salts. These salts are neutralized, usually with aqueous caustic, generating the free amines and forming inorganic salts, normally sodium chloride (68–70). Product distribution can be controlled by varying the NH_3:EDC mole ratio (71,72), product recycle (73–75), pH (72,76–78), and reactor geometry (70,71,77). A typical product distribution obtained at a 20:1 mole ratio of NH_3:EDC is EDA 55%, PIP 1.9%, DETA 23%, AEP 3.5%, TETA 9.9%, TEPA 3.9%, and higher polyamines 2.3% (72). Raising the NH_3:EDC mole ratio increases EDA, whereas lowering this ratio leads to the formation of higher mol wt polyethylene-amines. The higher mol wt ethyleneamines are made also by reaction of EDC with EDA, DETA, or mixtures of these (75,79–81). The amine products can be separated from the aqueous salt solution by evaporative crystallization (82,83), dehydration (84,85), solvent extraction (86–90), or combinations of these. In one process, the amines can be extracted from a neutralized reaction mixture with a solvent system comprised of a reactive ketone and an alcohol, followed by back extraction with CO_2 and water, and thermally cracking the so-formed carbamates (91,92). After the amines are separated from their salts, they are purified by conventional fractional distillation.

Alternative processes for the manufacture of ethyleneamines have been actively sought since the late 1960s. The catalytic reductive amination of monoethanolamine (MEA), which was the first such process to appear, produces the lighter ethyleneamines (EDA, DETA, PIP, and AEP) and coproduct aminoethylethanolamine [114-41-1] (AEEA), and hydroxyethylpiperazine [103-76-4] (HEP). Catalyst development has progressed from simple nickel catalysts to catalysts containing nickel with up to three additional metals such as Co, Cu, Cr, Fe, Re, Ru, and B or copper–cobalt catalysts associated with Re, W, and Zr (93–102). Aluminas, silicas, aluminosilicates, titania and zirconia are used as support materials. Most catalysts exhibit high selectivities to EDA, but produce more PIP than characteristically observed in the older EDC process. Product distribution can be controlled by temperature (103), NH_3:MEA mole ratio, hydrogen partial pressure (104), and conversion. A 9.3:1 NH_3:MEA mole ratio (96) gives the following selectivity to products at a 60% MEA conversion: EDA 71%, DETA 10.2%, AEEA 6.9%, PIP 9.2%, AEP 1.4%, and HEP 1.3%. An integrated two-step process produces MEA from a supercritical mixture of ethylene oxide and ammonia in the first stage and reductively aminates the ethanolamines in the second stage (105). The reductive amination of EDA or MEA without ammonia can be used to produce DETA or AEEA, respectively (106–108).

The condensation of MEA with EDA over heterogeneous catalysts to form primarily DETA represents the newest commercial technology for making ethyleneamines. Prior to 1977, this type of condensation was used primarily to make cyclic ethyleneamines (109), not the more linear alkyleneamines which are now being produced (110–112). This process is operated at high temperatures (200–350°C) and in either the vapor phase (113–115) or liquid phase (116). Catalysts for this condensation include Group 4 (IVB), 5 (VB), and 6 (VIB) metal oxides (117,118); Group IVB and Group VB supported phosphates (119–123); and supported tungstates (124,125). Product distribution can be controlled by the EDA:MEA mole ratio (110,111,126,127), temperature (120), product recycle (116), and conversion. The TETA, TEPA, and higher polyamines produced by this technology have a higher linear content than similar products produced by the EDC-based technology. In a typical example (120), an EDA:MEA mole ratio of 2:1 yields the following product selectivities at 51% MEA conversion: DETA 77.2%, PIP 2.7%, AEEA 0.8%, AEP 2.0%, TETA 16.6%, TEPA 0.5%. The TETA fraction is comprised of over 99% acyclic ethyleneamines.

A process for the production of ethylenimine [151-56-4], a suspect carcinogen, by the vapor phase dehydration of monoethanolamine has been developed (128–132). By using an alkyleneamine co-feed with the alkanolamine, higher alkyleneamines are made *in situ* (133). The catalysts are tungsten-, niobium-, or phosphate-based.

At one time, alkyleneamines were made primarily from raw materials derived from methane or methanol. This technology is still being practiced selectively to form branched ethyleneamines, for example. The reaction of formaldehyde and hydrogen cyanide in the presence of water yields glycolnitrile [107-16-4] (134,135) or in the presence of ammonia yields aminoacetonitrile and its condensation products (136–138). Reduction of the glycolnitrile with hydrogen in the presence of ammonia or the reduction of the aminoacetonitrile with hydrogen yields alkyleneamines (139–141).

Diaminopropane Processes. 1,2-Propylenediamine can be produced by the reductive amination of propylene oxide (142), 1,2-propylene glycol [57-55-6] (143), or monoisopropanolamine [78-96-6] (144). 1,3-Propanediol [504-63-2] can be used to make 1,3-diaminopropane (143). Various propaneamines are produced by reducing the appropriate acrylonitrile–amine adducts (145–147). Polypropaneamines can be obtained by the oligomerization of 1,3-diaminopropane (148,149).

Economic Aspects

Ethyleneamine production capacities for 1992, broken down by geographical area and main producers in each area, are given in Table 3. Announced expansions in

Table 3. Ethyleneamine Production Capacities, 1992

Company	Process technology[a]	Capacity, 10^3 t/yr
United States		
Union Carbide	EDC, MEA	82
Dow	EDC	45
Texaco	MEA	32
Total		*159*
Western Europe		
Dow	EDC	27
BASF	MEA	30
Delamine	EDC	27
Berol Kemi	MEA	25
Bayer	EDC	18
Total		*127*
Japan		
Tosoh	EDC	15
Others		
	EDC	6
Approximate world capacity		*307*

[a]EDC = ethylene dichloride-based, MEA = monoethanolamine-based.

1993 will increase U.S. production by 29,500 t. Several small regional producers are expected to come on-stream in the next several years. Worldwide growth for most ethyleneamines is expected to parallel GDP. Some regional and certain applications demands will show somewhat higher growth rates.

Of the worldwide ethyleneamines capacity, over 50% is EDC-based; the balance is monoethanolamine-derived. A complete breakdown of the ethyleneamines capacity by product is not feasible since most manufacturers can vary production mix to meet market demand. A rough estimate is that EDA represents about 40% of the total production of the family of ethyleneamines. In Europe, the product mix is skewed somewhat more toward EDA.

Prices of the principal commercial polyamines in 1992, in the United States, are given in Table 4.

Table 4. Prices of Some Commercially Available Polyamines

Product	Price, U.S. \$/kg[a]
ethylenediamine	3.06
diethylenetriamine	3.74
triethylenetetramine	3.33
tetraethylenepentamine	3.90
higher polyamines[b]	3.26
aminoethylpiperazine	2.97
piperazine (65% aqueous soln)[c]	2.97
1,2-propylenediamine	5.95
1,3-diaminopropane	7.11
iminobispropylamine	5.77
dimethylaminopropylamine	2.71
hexamethylenediamine	2.56
menthanediamine	55.07

[a]List price for shipment of bulk quantities as of July, 1992.
[b]Product marketed under a variety of commercial names.
[c]Contained PIP basis.

Specifications and Test Methods

Typical specifications for the commercially available ethyleneamines are given in Table 5. For more detailed specifications on these products and for specifications of other polyamines not listed, the individual manufacturers should be consulted.

The assay of ethyleneamines is usually done by gas chromatography. Compared to packed columns, in which severe tailing is often encountered due to the high polarity of the ethyleneamines, capillary columns provide better component separation and quantification. Typically, amines can be analyzed using fused silica capillary columns with dimethylsilicones, substituted dimethylsilicones or PEG Compound 20 M as the stationary phase (150).

Because the heavy ethyleneamines are very complex materials, assays by titration in aqueous and nonaqueous media are often performed (151). The result

Table 5. Typical Specifications of Some Aliphatic Polyamines

Commercial name	Assay, min %	Amine value, mg KOH/g	Specific gravity, 20°C/20°C	Color, APHA, max
ethylenediamine	99.0		0.896–0.904	15
diethylenetriamine	98.5		0.948–0.954	30
triethylenetetramine	95.0	1410–1480	0.976–0.983	50
tetraethylenepentamine	90.0	1290–1375	0.987–1.001	2[a]
higher polyamines[b]		1100–1300	1.005–1.025	
1,2-propylenediamine	99.0		0.855–0.885	15
1,3-diaminopropane	99.0		0.886–0.890	20
iminobispropylamine	99.0		0.926–0.936	30
dimethylaminopropylamine	99.0		0.815–0.820	15
hexamethylenediamine	99.9[c]		0.889–0.900	10

[a] 1963 Gardner Scale.
[b] Commercial product contains 30–40% PEHA, remainder higher amines.
[c] Anhydrous basis, commercial material contains 9–16% water.

is usually expressed as an amine number or amine value, a measure of the total basic nitrogen content of the product. Titrimetric procedures are also available to define primary, secondary, and tertiary amine content (152).

Determination of ethyleneamines in air can be accomplished by absorbing the amines on NITC (1-naphthyl isothiocyanate) treated XAD-2 resin, then desorbing the derivative from the treated tubes and quantifying the amount using high performance liquid chromatography (hplc). Sensitivity is reported as 0.37 and 0.016 mg/m^3 for EDA and DETA, respectively, per sample (153,154).

Storage and Handling

By virtue of their unique combination of reactivity and basicity, the polyamines react with, or catalyze the reaction of, many chemicals, sometimes rapidly and usually exothermically. Some reactions may produce derivatives that are explosives (eg, ethylenedinitramine). The amines can catalyze a runaway reaction with other compounds (eg, maleic anhydride, ethylene oxide, acrolein, and acrylates), sometimes resulting in an explosion.

As commercially pure materials, the ethyleneamines exhibit good temperature stability, but at elevated temperatures noticeable product breakdown may result in the formation of ammonia and lower and higher mol wt species. This degradation becomes more pronounced at higher temperature and over longer time periods. Certain contaminants, such as mineral acids, can lower the onset temperature for rapid thermal decomposition. The manufacturer should be contacted and thermal stability testing conducted whenever ethyleneamines are mixed with other materials.

Like many other combustible liquids, self-heating of ethyleneamines may occur by slow oxidation in absorbent or high-surface-area media, eg, dumped filter cake, thermal insulation, spill absorbents, and metal wire mesh (such as that used in vapor mist eliminators). In some cases, this may lead to spontaneous combustion; either smoldering or a flame may be observed. These media should be washed with water to remove the ethyleneamines, or thoroughly wet prior to disposal in accordance with local and Federal regulations.

Since ethyleneamines react with many other chemicals, dedicated processing equipment is usually desirable. Proper selection of the materials of construction for ethyleneamine service is essential to ensure the integrity of the handling system and to maintain product quality. Amines slowly absorb water, carbon dioxide, nitrogen oxides, and oxygen from the atmosphere, which may result in the formation of low concentrations of by-products and generally increase color. Storage under an inert atmosphere minimizes this sort of degradation.

Galvanized steel, copper and copper-bearing alloys are unacceptable for all ethyleneamine service. Although corrosion of carbon steel is not extreme (<0.03 mm/yr) at typical storage temperatures, the lighter amines (particularly EDA and DETA) pick up iron and discolor badly. If this cannot be tolerated, then 300 series stainless steels or aluminum are recommended for the storage tanks. Carbon steel generally can be used for storage of the heavier ethyleneamines without noticeable impact on product quality if the storage temperature is modest (<60°C), nitrogen blankets are maintained to exclude air, and the material is anhydrous. A 300 series stainless steel is often specified for heating coils, transfer lines and small agitated tanks, because carbon steel can suffer enhanced corrosion due to the erosion of the passive film by the product velocity. Similar logic suggests cast 316 stainless steel for pumps and valves in ethyleneamine service.

Nonmetallic equipment normally is not used for ethyleneamine service. Ethyleneamines can permeate polyethylene and polypropylene, even at ambient temperature. However, certain grades of these materials may be acceptable in some storage applications. Baked phenolic-lined carbon steel is acceptable for storage of many pure ethyleneamines, except EDA.

Gaskets utilized in ethyleneamine service generally are made of Grafoil flexible graphite or polytetrafluoroethylene (TFE). There is no single elastomer that is acceptable for the entire product line, although TFE may be considered as an alternative to elastomers across the product line. However, because TFE is not a true elastomer, it may not always prove suitable as a replacement.

Most common thermal insulating materials are acceptable for ethyleneamine service. However, porous insulation may introduce the hazard of spontaneous combustion if saturated with ethyleneamines from a leak or external spill. Good housekeeping, maintenance of weather barriers, minimization of flanges and fittings, and training of personnel can reduce the potential for smoldering insulation fires. Closed cellular glass insulation is normally resistant to insulation fires because it is difficult for the ethyleneamines to saturate these materials.

Certain ethyleneamines require storage above ambient temperature to keep them above their freezing points (EDA and PIP) or to lower the viscosity (the heavy amines). As a result, the vapors "breathing" from the storage tank can contain significant concentrations of the product. Water scrubbers may be used to capture these vapors.

Solid ethyleneamine carbamates, formed by the reaction of the amines with carbon dioxide, can foul tank vents and pressure relief devices. Vent fouling can be minimized by using a nitrogen blanket that prevents atmospheric CO_2 from being drawn in, or by steam-tracing the vents (>160°C) to decompose the carbamates. Vents must be inspected regularly to ensure that the carbamate does not gradually accumulate and clog the lines.

Although the ethyleneamines are water soluble, solid amine hydrates may form at certain concentrations that may plug processing equipment, vent lines, and safety devices. Hydrate formation usually can be avoided by insulating and heat tracing equipment to maintain a temperature of at least 50°C. Water cleanup of ethyleneamine equipment can result in hydrate formation even in areas where routine processing is nonaqueous. Use of warm water can reduce the extent of the problem.

Health and Safety Factors

Ethyleneamine vapors are painful and irritating to the eyes, nose, throat, and respiratory system. Extremely high vapor concentration may cause lung damage. Prolonged or repeated inhalation may lead to kidney, liver, and respiratory system injury. Contact with the liquids will severely damage the eyes and may cause serious burns to the skin. When swallowed, the concentrated liquid materials may produce considerable local injury. Repeated oral exposures may cause kidney and liver changes. Both vapors and liquid can cause sensitization in some individuals, resulting in contact dermatitis and/or the development of an asthmatic respiratory response. This may occur in certain susceptible individuals following exposure to extremely low concentrations of ethyleneamines, even below the irritation threshold.

The ACGIH has adopted TLVs of 10 ppm (25 mg/m^3) and 1 ppm (4 mg/m^3) for EDA and DETA, respectively. Thus, for a normal 8 hr work day, the time weighted average concentrations of EDA and DETA in the air of the workplace should not exceed these levels.

Strict precautions should be observed to prevent direct contact with ethyleneamines, including eye, skin, and respiratory protection. If contact is made, medical treatment should be obtained immediately, in addition to flushing and washing with copious amounts of water. Vomiting is not to be induced following ingestion. Before handling any of these products it is important to contact the manufacturer to obtain complete toxicological information, safe-handling recommendations, and an MSDS.

Applications

Polyalkylene polyamines find use in a wide variety of applications by virtue of their unique combination of reactivity, basicity, and surface activity. With a few significant exceptions, they are used predominantly as intermediates in the production of functional products. End-use profiles for the various ethyleneamines are given in Table 6.

Fungicides. The ethylenebisdithiocarbamates (EBDCs) are a class of broad-spectrum, preventive, contact fungicides first used in the early 1940s. They

Table 6. World End-Use Profiles for Ethyleneamines, 1990, 10^3 t/yr

Application	EDA	DETA	PIP	Higher homologues	Total
fungicide	18				18
oil and fuel additives	6	7		29	42
polyamides/epoxy curing	7	5	1	12	25
paper resins		20		2	22
chelating agents	26	5		1	32
fabric softeners/surfactants		7		<1	7
petroleum production	2	1		2	5
bleach activator	15				15
anthelmintics/pharmaceuticals			4		4
other	10	7	4	13	34
Total					*204*

have found application on many fruits, vegetables, potatoes, and grains for prevention of mildew, scab, rust, and blight. Following a 1989 proposed ban on use of EBDCs on most crops, the cancer risk potential of these products was reviewed by the United States Environmental Protection Agency (EPA), which ruled in early 1992 that these products were within guidelines for safe use on 45 specific crops (155). However, in July 1992, a California Federal Appeals court ruled that the use of the manganese/zinc salt [8018-01-7] (mancozeb) could not be allowed.

The EBDCs are prepared by reaction of EDA with carbon disulfide in the presence of sodium or ammonium hydroxide initially, then with zinc and/or manganese salts, as appropriate (156–160). A continuous process has recently been reported (161). The common names of these salts are nabam [142-59-6] (Na salt), amobam (ammonium salt), zineb [12122-67-7] (Zn salt), maneb [12427-38-2] (Mn salt), and mancozeb.

Other materials based on EDA have also been suggested as fungicides. The most important of the imidazoline type (162) is 2-heptadecyl-2-imidazoline (163), prepared from EDA and stearic acid [57-11-4]. It is used as the acetate salt for control of apple scab and cherry leaf spot. A 2:1 EDA–copper sulfate complex has been suggested for control of aquatic fungi (164).

Lubricant and Fuel Additives. The preparation of ashless dispersants for motor oil and other lubricants, and of certain detergents for motor fuels, has become the largest application for polyamines in the past several years. The most widely used derivatives are the mono- and bis-polyisobutenylsuccinimides. They are most commonly prepared by the condensation reaction of polyisobutenylsuccinic anhydrides with polyethylene polyamines (165–169). TETA, TEPA, and higher ethyleneamines are generally used; however, all of the ethyleneamines and many other polyamines, and substituted diamines and polyamines, have been employed for various products. Polyamines are also suggested for use in dispersant-varnish inhibitors for lubricating oils for two-stroke engines (170). Lubricating oil compositions for marine diesel engines containing TEPA in microemulsion form have also been reported (171). Reduced attack of fluoroelastomers is claimed for succinimides based on TEPA that have reacted with adipic acid [124-04-9] (172).

Although connection of polyalkylene or poly(alkylene oxide) groups to the polyamine is most commonly by the succinimide linkage, a different linking group is employed in another important class of ashless dispersants—the Mannich bases. They are prepared on a commercial scale by reaction of an alkylphenol with formaldehyde and a polyamine (173–177). The alkyl and polyamine moieties are similar to those used in the succinimide products.

Additives for lubricating oils providing a combination of viscosity index improvement (VII) and dispersancy have also been reported. These additives are prepared from ethyleneamines by reaction with various VII-type polymers that have been chlorinated or modified in some other way to provide an ethyleneamine reaction site. Antirust additives for lubricating oils have been prepared by reaction of polyamines with fatty acids followed by reaction with polyalkylenesuccinic anhydrides (178,179).

Over-based and other metal-containing oil additives providing detergency have also been prepared from polyamines (180,181). A number of antiwear/antiscuff additives for lube oils and greases also use polyamines. The combination of 2,6-di-*t*-butylphenol [*128-39-2*] and AEP is said to be an effective antioxidant in lube oil formulations containing certain viscosity index improvers (182). The thermal oxidative stability of a lubricating composition containing zinc diisopropyl dithiophosphate is reportedly increased by addition of sulfur and ethylenediamine (183). Polyamines are also used in preparation of fatty acid/amine soaps and poly(urea) thickeners for lubricating greases. A complex tolylene polyurea thickener made with EDA has been reported (184).

In the fuel additives area, EDA and DETA, as well as *N*-(2-hydroxyethyl)ethylenediamine, have found significant commercial application as dispersant detergent additives for gasoline after reacting with chlorinated polybutylenes (185,186). Improved antirust properties are reported when these compounds are neutralized with carboxylic acids (187). Numerous similar products made by alkylating or acylating EDA or DETA (188–196) have also been suggested as fuel detergent and deposit control additives. Cetane improvement in diesel fuels has been reported for a mixture containing EDA (197); and antiknock properties of unleaded gasoline have been improved by addition of PIP or C-methyl substituted PIPs (198). The barium salt of a bis-nonylphenol (prepared by reaction of nonylphenol, DETA and formaldehyde) has been used as an antismoke additive in diesel fuels (199).

Epoxy Curing Agents. A variety of polyamines and their derivatives that contain primary and secondary amine functionality are used as epoxy resin hardeners in various functional coatings, adhesives, castings, laminates, grouts, etc (200). All of the commercial ethyleneamines have been used, DETA and TETA probably most commonly. There are numerous examples in the literature of adhesives using TETA (201) and DETA (202), coatings using EDA (203,204), tile adhesive and grout using EDA and DETA (205), casting or potting systems using DETA (206, 207) and PIP (208,209). PIP (210) and AEP (211) also function as epoxy cure accelerators in certain applications (see EPOXY RESINS).

The ethyleneamines are commonly modified in various ways to achieve desired performance changes. Amidoamines and reactive polyamides, prepared by condensation of TETA or higher polyamines with fatty acids (212–214) and dimer acids (215,216), respectively, are widely used for longer pot life and better flexibility, adhesion, and solvent resistance (217). An amidoamine prepared from EDA and lysine methyl ester reportedly gives good low temperature cure (218); and another prepared from TEPA and bis(salicylate)boric acid reportedly gives high reactivity, low toxicity, and unlimited solubility in epoxy resins (219). Most of the amidoamines and reactive polyamides contain significant imidazoline structure which provides improved compatibility with liquid epoxy resins (220–224).

Other modifications of the polyamines include limited addition of alkylene oxide to yield the corresponding hydroxyalkyl derivatives (225) and cyanoethylation of DETA or TETA, usually by reaction with acrylonitrile [107-13-1], to give derivatives providing longer pot life and better wetting of glass (226). Also included are ketimines, made by the reaction of EDA with acetone for example. These derivatives can also be hydrogenated, as in the case of the equimolar adducts of DETA and methyl isobutyl ketone [108-10-1] or methyl isoamyl ketone [110-12-3] (227), or used as is to provide moisture cure performance. Mannich bases prepared from a phenol, formaldehyde and a polyamine are also used, such as the hardener prepared from cresol, DETA, and formaldehyde (228). Other modifications of polyamines for use as epoxy hardeners include reaction with aldehydes (229), epoxidized fatty nitriles (230), aromatic monoisocyanates (231), or propylene sulfide [1072-43-1] (232).

Polyamide Resins. Another class of polyamide resins, in addition to the liquid resins used as epoxy hardeners, are the thermoplastic type, prepared generally by the condensation reaction of polyamines with polybasic fatty acids. These resins find use in certain hot-melt adhesives, coatings, and inks. Diamines, typically EDA (233), are the principal amine reactant; however, tri- and tetramines are sometimes used at low levels to achieve specific performance.

In the adhesives area, thermoplastic, fatty polyamides are used in hot-melt and heat-seal adhesives for leather, paper, plastic and metal. Blends of EDA- and DETA-based polyamides are suggested for use in metal can seam sealants with improved toughness (234); pressure sensitive adhesives have been formulated with DETA-based polyamides (235); and anionic and cationic suspensoid adhesives are used as heat-seal coatings in paper converting (236). PIP and certain PIP derivatives are used with EDA in some applications (237).

Thermoplastic polyamides are used in coatings to modify alkyd resins (qv) in thixotropic systems (238) and to plasticize nitrocellulose lacquers (239). DETA-tall oil fatty acid-based polyamides are suggested for use as corrosion inhibitors in alkyd paints (240). Printing inks for flexo-gravure application on certain paper, film and foil webs rely on EDA- and PDA-based polyamides for their specific performance (241).

Paper Pulping, Resins and Additives. Considerable interest has been generated in the sulfur-free delignification of wood chips with EDA–soda liquors since the late 1970s (242–244), with more recent interest in EDA–sulfide pulping (245,246). Improved rates of EDA recovery have been developed for the latter process (247).

Another significant end-use for polyamines is in preparation of paper wet-strength resins. These are polyamide, modified formaldehyde, and polyamine resins used to improve the physical strength of tissue, toweling, and packaging paper products. The cationic formaldehyde resins include both urea–formaldehyde and melamine–formaldehyde types (248,249). Cationic functionality is imparted by incorporation of DETA, TETA, and/or TEPA in the backbone of the resins. This is accomplished either by reaction of the amine with the formaldehyde resin (250), by coreaction of the amine with the resin intermediates (251), or by prereaction of the amine with formaldehyde, followed by reaction with urea (252). The cationic functionality provides substantivity to the cellulose fibers in paper. Modifications of polyamine-modified formaldehyde resins with epichlorohydrin [106-89-8] (253) or ε-caprolactam [105-60-2] (254) have also been reported.

The most widely used paper wet-strength resins are the modified polyamide variety commonly prepared from DETA (sometimes TETA), adipic acid, and epichlorohydrin (255–259). In contrast with the polyamine-modified formaldehyde resins that require acidic conditions, these resins can be used under less corrosive neutral or alkaline conditions to make nonacidic papers. Stability of these resins reportedly can be improved by addition of formaldehyde (256) or alkoxylation prior to addition of epichlorohydrin (258). Polyamides made from polymeric fatty acids and ethyleneamines and neutralized with an organic base reportedly provide waterproofing as well as wet-strength (260). Polyamine wet-strength resins, that contain no amide functionality and that provide improved dry strength, are made by the reaction of ethyleneamines with epichlorohydrin (261).

Pigment retention and drainage additives made with polyamines include polyamines made from ethyleneamines, ethanolamines, and epichlorohydrin (262); ethyleneamines combined with phosphorus-modified polyamines made by reaction of ethyleneamines with $POCl_3$ [10025-87-3] (263); and a DETA–glutaric acid polyamide crosslinked with PEG-bis(3-chloro-2-hydroxypropyl) ether (264). Polyamines made from ethyleneamines and EDC are useful flocculating agents (265).

Waterproofing and sizing agents made with polyamines include amidoamines from ethyleneamines and fatty acids, used as is (266) or quaternized (267) or polyamides from polymeric fatty acids and ethyleneamines (268); ethyleneamine-based polyamides which have reacted with formaldehyde and sodium bisulfite and also provide antistatic properties (269); ethyleneamine-modified rosin gums (270), polyamides made from ethyleneamines and maleic anhydride which then further react with chlorinated paraffin (271); and ethyleneamine-treated sodium polyacrylate emulsions to give a waxy surface on Kraft paper (272). TEPA is incorporated in a heat-curable, polyamide–polyurea resin used in paper coating systems (273).

Chelates and Chelating Agents. Poly(aminoacetic acid)s and their salts derived from polyamines are used in a variety of systems where metal ions, commonly iron, need to be inactivated, buffered, concentrated, or transported. These chelating agents operate by forming stoichiometric complexes, called chelates, with most polyvalent metals. The most important industrial chelating agent is made by reaction of EDA with formaldehyde and hydrogen cyanide or an alkali metal cyanide in the presence of excess sodium hydroxide to form the tetrasodium salt of ethylenediamine tetraacetic acid (Na_4EDTA) (274–276). The intermediate ethylenediaminetetraacetonitrile may also be isolated and hydrolyzed to Na_4EDTA separately (277). The free tetraacid can then be formed by neutralization with HCl and precipitation from solution. The free ethyleneamines themselves are good chelating agents (qv) for many metals, such as copper and zinc. They are also used in specialized applications such as electroplating and electroless metal coating with gold, silver, platinum, palladium, copper, zinc, nickel, and many alloys, where their metal complexing properties are required (278,279). This same property makes the ethyleneamines useful as ingredients in systems for etching, and for stripping nickel and nickel alloy coatings.

DETA is also used as the starting ethyleneamine in preparing analogous chelating agents. Diethylenetriaminepentaacetate [67-43-6] (DETPA) is used in chlorine-free paper pulp bleaching to prevent iron from decreasing the efficiency of the hydrogen peroxide (280).

Other developments in chelating resins include fibers made from poly(ethylene glycol) and poly(vinyl alcohol) to which EDA was attached with epichlorohydrin (281); and a styrene–divinylbenzene resin with pendant EDTA or DETPA groups (282).

Fabric Softeners, Surfactants and Bleach Activators. Mono- and bisamidoamines and their imidazoline counterparts are formed by the condensation reaction of one or two moles of a monobasic fatty acid (typically stearic or oleic) or their methyl esters with one mole of a polyamine. Imidazoline formation requires that the ethyleneamine have at least one segment in which a secondary amine group lies adjacent to a primary amine group. These amidoamines and imidazolines form the basis for a wide range of fabric softeners, surfactants, and emulsifiers. Commonly used amines are DETA, TETA, and DMAPA, although most of the polyethylene and polypropane polyamines can be used.

The most common alkyleneamine-based fabric softeners use bisamidoamines made from DETA, which are either cyclized by further dehydration to the corresponding imidazoline or lightly alkoxylated (283) to convert the central secondary amine group to a tertiary amine. The imidazoline or substituted bisamidoamine can then be quaternized with an alkylating agent, generally dimethyl sulfate, to yield softeners popular for household use (284–286). Bisimidazoline compounds are prepared by condensing TETA or TEPA with fatty acids (287) or triglycerides (288), followed by quaternization with dimethyl sulfate [77-78-1] to yield cationic fabric softeners. Another version uses sodium chloroacetate as the quaternizing agent to give amphoteric softeners (289). The products from the reaction of 1,3-PDA with alkyl isocyanates (290) or alkylene oxides (291) have also been suggested as softeners for use in detergent formulations (see DETERGENCY).

Many of the surfactants made from ethyleneamines contain the imidazoline structure or are prepared through an imidazoline intermediate. Various 2-alkylimidazolines and their salts prepared mainly from EDA or monoethoxylated EDA are reported to have good foaming properties (292–295). Ethyleneamine-based imidazolines are also important intermediates for surfactants used in shampoos by virtue of their mildness and good foaming characteristics. 2-Alkylimidazolines made from DETA or monoethoxylated EDA and fatty acids or their methyl esters are the principal commercial intermediates (296–298). They are converted into shampoo surfactants commonly by reaction with one or two moles of sodium chloroacetate to yield amphoteric surfactants (299–301). The ease with which the imidazoline intermediates are hydrolyzed leads to amidoamine-type structures when these derivatives are prepared under aqueous alkaline conditions. However, reaction of the imidazoline under anhydrous conditions with acrylic acid [79-10-7] to make salt-free, amphoteric products, leaves the imidazoline structure essentially intact. Certain polyamine derivatives also function as water-in-oil or oil-in-water emulsifiers. These include the products of a reaction between DETA, TETA, or TEPA and fatty acids (302) or oxidized hydrocarbon wax (303). The amidoamine made from lauric acid [143-07-7] and DETA mono- and bis(2-ethylhexyl) phosphate is a very effective water-in-oil emulsifier (304).

Several cleaning formulations for specific uses contain unreacted polyamines. Examples include mixtures of ammonium alkylbenzenesulfonate, solvents, and PIP which give good cleaning and shine performance on mirrors and other hard surfaces without rinsing (305), and a hard-surface cleaner composed of a water-soluble vinyl acetate–vinyl alcohol copolymer, EDA, cyclohexanone [108-94-1], dimethyl sulfoxide [67-68-5], a surfactant, and water (306). TEPA, to which an average of 17 moles of ethylene oxide are added, improves the clay soil removal and soil antiredeposition properties of certain liquid laundry detergents (307).

Tetraacetylethylenediamine [10543-57-4] (TAED) has been widely adopted for use in home laundry products as an activator for peroxygen bleaches.

Petroleum Production and Refining.   Specific polyamine derivatives are used in the petroleum production and refining industries as corrosion inhibitors, demulsifiers, neutralizers, and additives for certain operations.

The derivatives used in corrosion inhibitor formulations for down-hole use constitute a significant industrial application for polyamines. Again, mono- and bisamidoamines, imidazolines, and polyamides made from the higher polyamines are the popular choices. The products made from DETA and fatty acids have been widely used (308). A wide variety of other polyamine-based, corrosion inhibiting derivatives have been developed, generally incorporating some form of oil-soluble or oil-dispersible residue. Sulfur and its derivatives are also used in these polyamine-based corrosion inhibitors on occasion.

Other polyamine derivatives are used to break the oil/water emulsions produced at times by petroleum wells. Materials such as polyether polyols prepared by reaction of EDA with propylene and ethylene oxides (309); the products derived from various ethyleneamines reacting with isocyanate-capped polyols and quaternized with dimethyl sulfate (310); and mixtures of PEHA with oxyalkylated alkylphenol–formaldehyde resins (311) have been used.

In secondary operations, where chemicals are injected into hydrocarbon formations in conjunction with a chemical flooding process, polyamines are used to reduce the loss of injected chemicals to the formation by adsorption and precipitation (312). TEPA and other ethyleneamines are used with water-soluble polymeric thickeners in water–flood petroleum recovery operations to stabilize viscosity, mobility, and pH while imparting resistance to hydrolysis (313).

Ethyleneamines are used in certain petroleum refining operations as well. For example, an EDA solution of sodium 2-aminoethoxide is used to extract thiols from straight-run petroleum distillates (314); a combination of substituted phenol and AEP are used as an antioxidant to control fouling during processing of a hydrocarbon (315); AEP is used to separate alkenes from thermally cracked petroleum products (316); and TEPA is used to separate carbon disulfide from a C_5 pyrolysis fraction from ethylene production (317). EDA and DETA are used in the preparation and reprocessing of certain cracking and hydrotreating catalysts (318–321).

Asphalt Additives and Emulsifiers.   Mono- and bisamidoamines, imidazolines and their mixtures are commonly used in formulating antistrip additives used to promote adhesion between the asphalt and mineral aggregate in bitu-minous mixtures for road paving, surfacing, and patching. They are commonly prepared from the higher molecular weight polyamines and tall oil fatty acids or other readily available organic acids and acylating agents (322–328). Similar polyamine derivatives are used, generally as their HCl or acetic acid salts, to make stable asphalt-in-water emulsions (329–337).

Other Applications.   Polyamines and their derivatives are used in a variety of other industries besides those already mentioned. In the fibers and textiles area, EDA, DETA, TETA, or TEPA are used to treat rayon (338), flax (339), hemp (340), jute (341,342), ramie (343), polyacrylonitrile (344,345), wool (346), polyester (347), acetate (348), and cotton (349) fibers and/or fabrics for modification of var-ious properties (see TEXTILE FINISHING). Durable press properties are developed in cotton fabrics using additives based on EDA or DETA (350); and compounds based on EDA are used as light stabilizers for spandex fibers (351), for shrink-proofing wool (352), as antistats for polypropylene carpet backing (353), and as moth-proofing agents (354). Industrial softening agents based on DETA and TEPA are also used (355,356), as are dyeing assists based on EDA, DETA, TETA, and TEPA for polyester (357), nylon (358), and other synthetic fabrics (359). Ny-lon-6 is a polyamide fiber based on DETA, ε-caprolactam and adipic acid (360) (see CAPROLACTAM; FIBERS, POLYAMIDES).

Iron and phosphate ore beneficiation processes, involving enrichment of the raw ores, use amidoamines and imidazolines made from polyamines and fatty acids as flotation agents that selectively adhere to silica particles and permit their removal by flotation procedures (361,362). A bis (fatty alkyl DETA) imidazoline is a flotation agent for beneficiation of niobium and tantalum ores (363). Ethyl-eneamines are used for extraction of copper, chromium, and nickel from their sul-fide and oxide ores (364). EDA and DETA are used in lead ore processing and lead recovery processes (365–368). The principal use for PIP is as the active ingredient in anthelmintic preparations. These are inorganic or low mol wt organic acid salts of PIP that are used primarily in veterinary applications to combat intestinal worms, especially round worms (*Ascaris lumbricoides*) and pinworms [*Enterobi-um (Oxyuris) vermicularis*] (369) (see ANTIPARASITIC AGENTS, ANTHELMINTICS).

The polyamines also are used in several ways in the manufacture and proc-essing of elastomers, polymers, and related materials. A significant volume of ethyleneamines is used in the coagulation of SBR and other synthetic rubber latexes (370–373) (see ELASTOMERS, STYRENE–BUTADIENE RUBBER; LATEX TECHNOLOGY). They also are used in vulcanization processes for various rubbers including thiodiethanol (374), vinyl acetate copolymer (375), ethylene copolymer (376), EPDM terpolymer (377–379), SBR (380), modified diene (381,382), chlo-roprene telomer (383), and urethane (384). Ethyleneamines and certain of their derivatives are also used in a variety of urethane systems as curing agents, cat-alysts, polyol precursors, stabilizers, and internal parting agents—even in the recycling of rigid polyurethane waste. EDA also finds use as a curing catalyst for phenolic resins (qv), a stabilizer for urea resins, an antistatic treatment for poly-styrene foam, and as a polymerization inhibitor for isoprene. EDA is a key com-ponent of the polymer in spandex fiber (see FIBERS, ELASTOMERS). TETA is a popular curing agent for furfural resin binders for molded graphite structures and foundry molds; and EDA finds use in systems for etching polyimide films. EDA reacts with haloalkylalkoxysilanes and phthalocyaninatosilanols to make agents that improve the adhesion between inorganic surfaces and polymers. Scrubber solvents containing TETA are used to remove acrylate and other monomer vapors from exhaust streams generated during handling and processing operations, such as latex paint vehicle manufacturing. The bisamide made by reaction of 1 mole of EDA with 2 moles of stearic acid, [ethylenebis(stearamide)], is used as an external lubricant for ABS resin and PVC, parting material, viscosity regulator, preservative, and surface gloss enhancer. It is also useful as a defoamer in paper mill operations and certain detergent formulations (see AMIDES, FATTY ACID).

The polyamines are also used in a number of other industries, including ceramics, coal–oil emulsification and coal extraction, polymer concrete, concrete additives, stone and concrete coatings and plaster systems, other industrial corrosion inhibitors (metal pickling, steel rolling, etc), coatings and adhesives systems other than epoxies, explosives, fire retardants, functional fluids (hydraulic, metal working, etc), gas treating, water treating, ion exchange resins, and membranes for gas separation, water treating, ion exchange, ultrafiltration, and electrolysis.

BIBLIOGRAPHY

"Ethylene Amines" in *ECT* 1st ed., Vol. 5, pp. 898–905, by J. Conway, Carbide and Carbon Chemicals Division, Union Carbide and Carbon Corp.; "Diamines and Higher Amines, Aliphatic" in *ECT* 2nd ed., Vol. 7, pp. 22–39, by A. W. Hart, The Dow Chemical Company; in *ECT* 3rd ed., Vol. 7, pp. 580–602, by R. D. Spitz, Dow Chemical U.S.A.

1. S. Kobayashi and co-workers, *Macromolecules* **20**, 1496 (1987).
2. R. N. Keller and L. J. Edwards, *J. Am. Chem. Soc.* **74**, 2931 (1952).
3. A. Gero, *J. Am. Chem. Soc.* **76**, 5159 (1954).
4. H. B. Jonassen and co-workers, *J. Am. Chem. Soc.* **72**, 2430 (1950).
5. H. H. Willard and N. H. Furman, *Elementary Quantitative Analysis*, D. Van Nostrand, Inc., New York, 1940, p. 474.
6. U.S. Pat. 4,539,430 (Sept. 3, 1985), K. Y. Lee (to U.S. Dept. of Energy).
7. Ger. Offen. DE 2,716,946 (Oct. 19, 1978), R. Schubart (to Bayer, AG).
8. Czech. Pat. 222,448 (Sept. 15, 1985), to M. Capka, R. Rericha, Z. Sir, J. Vilim, and J. Hetflejs.
9. J. Plucinski and H. Prystasz, *Pol. J. Chem.* **54**, 2201 (1980).
10. J. A. Deyrup and A. Hassner, eds. *Chemistry of Heterocyclic Compounds*, Vol. 42 of *Small Ring Heterocycles. Pt. 1: Aziridines*, John Wiley & Sons, Inc., New York, 1983, pp. 1–214.
11. A. R. Dalin and co-workers, *Zh. Obshch. Khim.* **58**, 2098 (1988).
12. T. W. Geiger and H. F. Rase, *Ind. Eng. Chem. Prod. Res. Dev.* **20**, 688 (1981).
13. U.S. Pat. 4,983,735 (Jan. 8, 1991), G. E. Hartwell, R. G. Bowman, and D. C. Molzahn (to The Dow Chemical Co.).
14. Jpn. Kokai Tokkyo Koho JP 54 132,588 (Oct. 15, 1979), K. Sato (to Tokai Electro-Chemical Co., Ltd.).
15. Jpn. Kokai Tokkyo Koho JP 55 122,769 (Sept. 20, 1980), (to Sanwaka Junyaku Co., Ltd.).
16. Eur. Pat. Appl. EP 374,929 (June 27, 1990), W. J. Kruper, Jr. (to The Dow Chemical Co.).
17. J. J. Yaounanc and co-workers, *J. Chem. Soc. Chem. Commun.*, 206 (1991).

18. W. A. Heskey, *Chem. Eng. News* **64**(21), 2 (1986).
19. U.S. Pat. 3,523,892 (Aug. 11, 1970), D. L. Schiegg (to Calgon Corp.).
20. J. Hine and K. W. Narducy, *J. Am. Chem. Soc.* **95**, 3362 (1973).
21. J. Dale and T. Sigvartsen, *Acta Chem. Scand.* **45**, 1064 (1991).
22. Ger. Offen. DE 3,544,510 (June 19, 1987), J. Weber, V. Falk, and C. Kniep (to Ruhr-chemie AG).
23. Jpn. Kokai Tokkyo Koho JP 01 16,751 (Jan. 20, 1989), Y. Yokota, K. Matsutani, and K. Okabe (to Kao Corp.).
24. E. V. Anderson and J. A. Gaunt, *Ind. Eng. Chem.* **52**, 190 (1952).
25. I. K. Chernova and co-workers, *Plast. Massy*, (7), 88 (1989).
26. Ger. Offen. DE 2,941,023 (Apr. 23, 1981), W. Wellbrock (to Hoechst AG).
27. V. Kale and co-workers, *J. Appl. Polym. Sci.* **36**, 1517 (1988).
28. Ger. Offen. DE 3,542,230 (June 4, 1987), W. Heitz and R. Schwalm (to Bayer AG).
29. Brit. Pat. Appl. GB 2,096,133 (Oct. 13, 1982), K. Coupland (to Croda Chemicals, Ltd.).
30. A. R. Jacobson, A. N. Makris, and L. M. Sayre, *J. Org. Chem.* **52**, 2592 (1987).
31. Jpn. Kokai Tokkyo Koho JP 60 233,042 (Nov. 19, 1985), (to Johnson and Johnson Baby Products Co.).
32. Jpn. Kokai Tokkyo Koho JP 63 208,560 (Aug. 30, 1988), S. Fukuda, S. Kojima, Y. Kadoma, and H. Kobashi (to Nippon Oils and Fats Co., Ltd.).
33. Ger. Offen. DE 2,615,886 (Oct. 20, 1977), T. Dockner and A. Frank (to BASF AG).
34. Czech. Pat. CS 184,110 (July 15, 1980), Z. Vodak, F. Krsnak, and L. Bechtold.
35. Span. Pat. ES 540,640 (Nov. 16, 1985), J. P. Pujada.
36. Eur. Pat. Appl. EP 411,456 (Feb. 6, 1991), K. Ebel, J. Schroeder, D. Juergen, T. Dockner, and H. Krug (to BASF AG).
37. Jpn. Kokai Tokkyo Koho JP 54 63, 077 (May 21, 1979), Y. Kai and H. Hino (to Nisshin Oil Mills, Ltd.).
38. W. Jentzsch and M. Seefelder, *Chem. Ber.* **98**, 1342 (1960).
39. N. Sawa, *Nippon Kagaku Zasshi* **89**, 780 (1968).
40. Ger. Offen. 2,512,513 (Oct. 7, 1976), A. Frank and T. Dockner (to BASF AG).
41. U.S. Pat. 4,477,646 (Oct. 16, 1984), J. Myers (to The Dow Chemical Co.).
42. U.S. Pat. 4,387,249 (June 7, 1983), R. M. Harnden and D. W. Calvin (to The Dow Co.).
43. Eur. Pat. Appl. EP 417,990 (Mar. 20, 1991), D. J. Malfer (to Ethyl Petroleum Additives, Inc.).
44. W. R. Davie, *Chem. Eng. News* **42**, 41 (Feb. 24, 1964).
45. C. E. Vogler and co-workers, *J. Chem. Eng. Data* **8**, 620 (1963).
46. Ger. Offen. DE 2,728,976 (Jan. 18, 1979), T. Dockner, A. Frank, and H. Krug (to BASF AG).
47. Ger. Offen. DE 2,703,312 (Aug. 3, 1978), D. Cramm and C. D. Barnikel (to Bayer AG).
48. Ger. Offen. DE 3,534,245 (Mar. 26, 1987), M. Bergfeld and L. Eisenhuth (to AKZO GmbH).
49. Ger. Offen. DE 3,534,246 (Mar. 26, 1987), M. Bergfeld and L. Eisenhuth (to AKZO GmbH).
50. Rep. of China Pat. CN 1,040,583 A (Mar. 21, 1990), D. Wan and co-workers (to Xian Research Institute of Modern Chemistry).
51. Eur. Pat. Appl. EP 296,495 (Dec. 28, 1988), M. Hesse, W. Hoelderich, and M. Schwarzmann (to BASF AG).
52. Ger. Offen. 2,714,403 (Oct. 6, 1977), G. Soula and P. Duteurtre (to Société Orogil).
53. D. A. Tomalia and co-workers, *Polym. J. (Tokyo)* **17**, 117 (1985).
54. P. B. Smith and co-workers in J. Mitchell, ed., *Appl. Polym. Anal. Charact.*, Hanser, Munich, 1987, p. 357.
55. F. Jachimowicz and J. W. Raksis, *J. Org. Chem.* **47**, 445 (1982).

56. G. L. Gaines, Jr., *J. Org. Chem.* **50,** 410 (1985).
57. U.S. Pat. 2,497,309 (Feb. 14, 1950), A. T. Larson and A. G. Weber (to E. I. du Pont de Nemours & Co., Inc.).
58. U.S. Pat. 3,597,443 (Aug. 3, 1971), M. Crowther (to Procter Chemical Co., Inc.).
59. U.S. Pat. 2,892,843 (June 30, 1959), L. Levine (to The Dow Chemical Co.).
60. U.S. Pat. 3,737,428 (June 5, 1973), S. Tsutsumi and N. Sonoda (to Asahi Kasei Kogyo Kabushiki Kaisha).
61. J. B. Lanbert and co-workers, *J. Org. Chem.* **34,** 4147 (1969).
62. J. Casado, A. Castro, and M. A. Lopez-Quintela *Bull. Soc. Chim. Fr.*, (3), 401 (1987).
63. T. Bellander and co-workers, *Int. Arch. Occup. Environ. Health* **60,** 25 (1988).
64. S. S. Hecht, J. B. Morrison, and R. Young, *Carcinogenesis* **5,** 979 (1984).
65. Ger. Offen. DE 1,075,118 (Feb. 11, 1960), G. Spielberger (to Bayer AG).
66. U.S. Pat. 3,481,933 (Dec. 2, 1969), R. L. Mascioli (to Air Products and Chemicals, Inc.).
67. U.S. Pat. 1,832,534 (Nov. 17, 1932), G. O. Curme, Jr. and F. W. Lommen (to Carbide and Carbon Chemicals Corp.).
68. U.S. Pat. 2,769,841 (Nov. 6, 1956), S. W. Dylewski, H. G. Dulude, and G. W. Warren (to The Dow Chemical Co.).
69. Brit. Pat. GB1,147,984 (Apr. 10, 1969), J. G. Blears and P. Simpson (to Simon-Carves Ltd.).
70. U.S. Pat. 4,980,507 (Dec. 25, 1990), N. Mizui, K. Mitarai, and Y. Tsutsumi (to Tosoh Corp.).
71. Jpn. Pat. Appl. 42 31,994 (May 20, 1967), S. Wakiyama and Y. Kashida (to Toyo Soda Manufacturing Co., Ltd.).
72. Eur. Pat. Appl. EP222,934 (May 27, 1987), E. G. Ramirez (to The Dow Chemical Co.).
73. U.S. Pat. 3,462,493 (Aug. 19, 1969), W. P. Coker and G. E. Ham (to The Dow Chemical Co.).
74. U.S. Pat. 3,573,311 (Mar. 30, 1971), L. L. Valka (to The Dow Chemical Co.).
75. Jpn. Kokai Tokkyo Koho 54 32,600 (Mar. 9, 1979), H. Ohfuka, I. Miyanohara, and N. Nagai (to Toyo Soda Manufacturing Co., Ltd.).
76. C. M. P. V. Nunes, P. J. Garner, and M. Towhidi, *Rev. Part. Quim* **16,** 164 (1974).
77. U.S. Pat. 3,484,488 (Dec. 16, 1969), M. Lichtenwalter and T. H. Cour (to Jefferson Chemical Co.).
78. U.S. Pat. 4,324,724 (Apr. 13, 1982), H. Mueller, K. Wulz, K.-H. Beyer, and W. Streit, (to BASF AG).
79. Rom. Pat. RO 90,714 (Nov. 29, 1986), (to Comb. Chimic Rimnicu).
80. USSR Pat. 1,162,786 (June 23, 1985), A. M. Potapov and co-workers (to Ufa Petroleum Institute).
81. V. S. Borisenko, V. A. Bobylev, and I. V. Borisenko, *Zhur. Obshch. Khim.* **59,** 1131 (1989).
82. U.S. Pat. 3,202,713 (Aug. 24, 1965), G. Marullo, D. Costabello, G. Boffa, E. Fernasieri, and G. Maioranno, (to Montecatini Societa Generale per l'Industria Mineraria e Chimica).
83. U.S. Pat. 3,862,234 (Jan. 21, 1975), C. S. Steele (to Jefferson Chemical Co.).
84. U.S. Pat. 3,337,630 (Aug. 22, 1967), H. C. Moke and J. M. F. Leathers (to The Dow Chemical Co.).
85. U.S. Pat. 3,394,186 (July 23, 1968), H. G. Muhlbauer (to Jefferson Chemical Co.).
86. U.S. Pat. 3,433,788 (Mar. 18, 1969), G. S. Somekh and E. N. Hawkes (to Union Carbide Corp.).
87. Israel Pat. IL 57,019 (Sept. 30, 1983), J. Segall and L. M. Shorr (to Institute of Research and Development Ltd.).

88. U.S. Pat. 4,582,937 (Apr. 15, 1986), Y. Hiraga, T. Murakami, H. Saito, and O. Fujii (to Toyo Soda Manufacturing Co. Ltd.).

89. Jpn. Kokai Tokkyo Koho 58 213,737 (Dec. 12, 1983), Y. Hiraga, T. Murakami, H. Saito, and O. Fujii (to Toyo Soda Manufacturing Co. Ltd.).

90. Jpn. Kokai Tokkyo Koho 59 175,457 (Oct. 4, 1984), T. Murakami and T. Kawamoto (to Toyo Soda Manufacturing Co. Ltd.).

91. Eur. Pat. Appl. EP110,470 (June 16, 1984), F. J. Budde (to AKZO N. V.).

92. U.S. Pat. 4,650,906 (Mar. 17, 1987), T. Murakami and T. Kawamoto (to Toyo Soda Manufacturing Co., Ltd.).

93. U.S. Pat. 4,123,462 (Oct. 31, 1978), D. C. Best (to Union Carbide Corp.).

94. U.S. Pat. 4,912,260 (Mar. 27, 1990), I. D. Dobson, W. A. Lidy, and P. S. Williams (to BP Chemicals, Ltd.).

95. U.S. Pat. 4,863,890 (Sept. 5, 1989), J. Koll (to Berol Kemi AB).

96. U.S. Pat. 4,855,505 (Aug. 8, 1989), J. Koll (to Berol Kemi AB).

97. U.S. Pat. 4,772,750 (Sept. 20, 1988), C. E. Habermann (to The Dow Chemical Co.).

98. U.S. Pat. 4,642,303 (Feb. 10, 1987), T. L. Renken (to Texaco, Inc.).

99. Ger. Pat. DD 236,728 (June 18, 1986), D. Voigt, H. Haack, I. Bartosch, and R. Thaetner (to VEB Leuna Werke, Walter Ulbricht).

100. U.S. Pat. 5,002,922 (Mar. 26, 1991), M. Irgang, J. Schossig, W. Schroeder, and S. Winderl (to BASF AG).

101. U.S. Pat. 4,806,690 (Feb. 21, 1989), R. G. Bowman (to The Dow Chemical Co.).

102. U.S. Pat. 4,922,024 (May 1, 1990), R. G. Bowman, M. H. Tegen, and G. E. Hartwell (to The Dow Chemical Co.).

103. Fr. Pat. FR2,281,920 (Mar. 12, 1976), (to BASF AG).

104. U.S. Pat. 4,234,730 (Nov. 18, 1980), T. T. McConnell and T. H. Cour (Texaco Development Corp.).

105. U.S. Pat. 4,400,539 (Aug. 23, 1983), C. A. Gibson and J. R. Winters (to Union Carbide Corp.).

106. U.S. Pat. 4,568,746 (Feb. 4, 1986), F. G. Cowherd (to Union Carbide Corp.).

107. U.S. Pat. 5,068,329 (Nov. 26, 1991), L. M. Burgess and C. A. Gibson (to Union Carbide Corp.).

108. U.S. Pat. 5,068,330 (Nov. 26, 1991), L. M. Burgess and C. A. Gibson (to Union Carbide Corp.).

109. U.S. Pat. 2,073,671 (Mar. 16, 1937), C. E. Andrews (to Rohm and Haas Co.).

110. U.S. Pat. 4,036,881 (July 19, 1977), M. E. Brennen and E. L. Yeakey (to Texaco Development Co.).

111. U.S. Pat. 4,463,193 (July 31, 1984), T. A. Johnson and M. E. Ford (to Air Products and Chemicals, Inc.).

112. U.S. Pat. 4,394,524 (July 19, 1983), T. A. Johnson and M. E. Ford (to Air Products and Chemicals, Inc.).

113. U.S. Pat. 4,617,418 (Oct. 14, 1986), M. E. Ford and T. A. Johnson (to Air Products and Chemicals, Inc.).

114. U.S. Pat. 4,720,588 (Jan. 19, 1988), M. G. Turcotte, C. A. Cooper, M. E. Ford, and T. A. Johnson (to Air Products and Chemicals, Inc.).

115. U.S. Pat. 4,910,342 (Mar. 20, 1990), M. G. Turcotte and C. A. Cooper (to Air Products and Chemicals, Inc.).

116. Eur. Pat. Appl. EP115,138 (Aug. 8, 1984) S. H. Vanderpool, L. W. Watts, Jr., J. M. Larkin, and T. L. Renken (to Texaco Development Co.).

117. Eur. Pat. Appl. EP315,189 (May 10, 1989), Y. Hara, N. Suzuki, Y. Ito, and K. Sekizawa (to Tosoh Corp.).

118. Eur. Pat. Appl. EP412,611 (Feb. 13, 1991), A. R. Doumaux, Jr., D. J. Schreck, S. W. King, and G. A. Skoler, (to Union Carbide Chemicals and Plastics Co., Inc.).

119. U.S. Pat. 4,806,517 (Feb. 21, 1989), S. H. Vanderpool, L. W. Watts, Jr., J. M. Larkin, and T. L. Renken (to Texaco, Inc.).

120. U.S. Pat. 4,914,072 (Apr. 3, 1990), N. J. Grice, J. F. Knifton, and C.-H. Yang (to Texaco, Inc.).

121. U.S. Pat. 4,983,736 (Jan. 8, 1991), A. R. Doumaux, Jr. and D. J. Schreck (to Union Carbide Chemicals and Plastics Co., Inc.).

122. U.S. Pat. 4,927,931 (May 22, 1990), D. C. Molzahn, G. E. Hartwell, and R. G. Bowman (to The Dow Chemical Co.).

123. U.S. Pat. 5,011,999 (Apr. 30, 1991), R. G. Bowman, G. E. Hartwell, D. C. Molzahn, E. G. Ramirez, and J. E. Lastovica, Jr. (to The Dow Chemical Co.).

124. U.S. Pat. 4,983,565 (Jan. 8, 1991), J. F. Knifton and W.-Y. Su (to Texaco Chemical Co.).

125. U.S. Pat. 5,030,740 (July 9, 1991), R. G. Bowman, D. C. Molzahn, and G. E. Hartwell (to The Dow Chemical Co.).

126. Jpn. Kokai Tokkyo Koho JP 02 000735 (Jan. 5, 1990), Y. Hara, Y. Ito, and K. Sekizawa (to Tosoh Co., Ltd.).

127. U.S. Pat. 4,605,770 (Aug. 12, 1986), M. E. Ford and T. A. Johnson (to Air Products and Chemicals, Inc.).

128. U.S. Pat. 4,301,036 (Nov. 17, 1981), W. V. Haynes and D. L. Childress (to The Dow Chemical Co.).

129. U.S. Pat. 4,358,405 (Nov. 9, 1982), W. V. Haynes and D. L. Childress (to The Dow Chemical Co.).

130. U.S. Pat. 4,376,732 (Mar. 15, 1983), E. G. Ramirez (to The Dow Chemical Co.).

131. U.S. Pat. 4,841,061 (June 20, 1989), Y. Shimasaki, M. Ueshima, H. Tuneki, and K. Ariyoshi (to Nippon Shokubai Kagaku Kogyo Co., Ltd.).

132. Jpn. Kokai Tokkyo Koho JP 01 207,265 (Aug. 21, 1989), H. Tsuneki, T. Kamei, K. Yamamoto, Y. Morimoto, and T. Ueshima (to Nippon Shokubai Kagaku Kogyo Co., Ltd.).

133. Jpn. Kokai Tokkyo Koho JP 63 122,652 (May 26, 1988), S. Suzuki and M. Kitano (to Nippon Shokubai Kagaku Kogyo Co., Ltd.).

134. U.S. Pat. 3,167,582 (Jan. 26, 1965), K. W. Saunders, W. H. Montgomery, and J. C. French (to American Cyanamid Co.).

135. Eur. Pat. Appl. EP 426,394 (May 8, 1991), B. A. Cullen and B. A. Parker (to W. R. Grace & Co.).

136. U.S. Pat. 4,704,465 (Nov. 3, 1987), K. P. Lannert and S.-M. Lee (to Monsanto Co.).

137. U.S. Pat. 4,895,971 (Jan. 23, 1990), M. B. Sherwin and J.-L. Su (to W. R. Grace & Co.).

138. U.S. Pat. 5,008,428 (Apr. 16, 1991), M. B. Sherwin and J.-L. Su (to W. R. Grace & Co.).

139. Eur. Pat. Appl. EP212,986 (Mar. 4, 1987), M. B. Sherwin, S.-C. P. Wang, and S. R. Montgomery (to W. R. Grace & Co.).

140. Jpn. Kokai 62 201,848 (Sept. 5, 1987), M. Inomata, K. Fukayama, A. Yamauchi, and Y. Tanaka (to Mitsui Toatsu Chemical Co.).

141. Jpn. Kokai 62 285,921 (Dec. 11, 1987), M. Inomata, K. Miyama, and M. Kitagawa (to Mitsui Toatsu Chemical Co.).

142. U.S. Pat. 3,597,483 (Aug. 3, 1971), E. Haarer, H. Corr, and S. Winderl (to BASF AG).

143. U.S. Pat. 3,270,059 (Aug. 30, 1966), S. Winderl and co-workers (to BASF AG).

144. U.S. Pat. 2,519,560 (Aug. 22, 1950), G. W. Fowler (to Union Carbide & Carbon Corp.).

145. U.S. Pat. 4,552,862 (Nov. 12, 1985), J. M. Larkin (to Texaco, Inc.).

146. Eur. Pat. Appl. EP 135,725 (Apr. 3, 1985), S. Kumol, K. Mitaral, and Y. Tsutsumi (to Toyo Soda Manufacturing Co., Ltd.).

147. Ger. Pat. DD 238,043 (Aug. 6, 1986), D. Voigt, H. Voigt, and W. Wilfried (to VEB Leuna, Walter Ulbricht).

148. Ger. Offen. 2,540,871 (Mar. 24, 1977), H. Graefje and co-workers (to BASF AG).

149. Ger. Offen. 2,605,212 (Aug. 25, 1977), W. Mesch, H. Hoffmann, and D. Voges (to BASF AG).

150. V. A. Bobylev and co-workers, *Zh. Anal. Khim.* **41,** 324 (1986).

151. J. M. Weber in F. D. Snell and L. S. Ettre, eds., *Encyclopedia of Industrial Chemical Analysis*, Vol. 11, John Wiley & Sons, Inc., 1971, pp. 421–428.

152. E. D. Smith and R. D. Radford, *Anal. Chem.* **33,** 1160 (1961).

153. P. M. Eller, ed., *NIOSH Manual of Analytical Methods*, Vol. 1, U.S. Dept. of Health and Human Services, Washington, D.C., 1984, method no. 2540.

154. *OSHA method no. 60.* U.S. Department of Labor, Washington, D.C., 1986.

155. *Chem. Eng. News*, 6 (Feb. 24, 1992).

156. V. Vrabel and co-workers, *Proc. Conf. Coord. Chem., 11th*, 1987, pp. 473–474.

157. U.S. Pat. 2,504,404 (Apr. 18, 1950), A. L. Flenner (to E. I. du Pont de Nemours & Co., Inc.).

158. U.S. Pat. 2,844,623 (July 22, 1968), E. A. Fike (to Roberts Chemicals, Inc.).

159. U.S. Pat. 2,317,765 (Apr., 1943), W. F. Hester (to Rohm & Haas Co.).

160. U.S. Pat. 3,379,610 (Apr. 23, 1968), C. B. Lyon, J. W. Nemec, and V. H. Unger (to Rohm & Haas Co.).

161. Czech Patent CS 237,725 (Apr. 15, 1987), J. Teren, D. Lucansky, V. Kubala, and E. Hutar.

162. R. H. Wellman and S. E. A. McCallan, *Contrib. Boyce Thompson Inst.* **14,** 151 (1946).

163. H. S. Cunningham and E. G. Sharvelle, *Phytopathology* **30,** 4 (1940).

164. Ger. Pat. 2,506,431 (Sept. 18, 1975), M. D. Meyers and G. A. Stoner (to Sandox GmbH).

165. U.S. Pat. 3,623,985 (Nov. 30, 1971), Y. G. Hendrickson (to Chevron Research Co.).

166. U.S. Pat. 3,451,931 (June 24, 1969), D. J. Kahn and M. L. Robbins (to Esso Research and Engineering Co.).

167. U.S. Pat. 3,390,082 (June 25, 1968), W. M. Le Suer and G. R. Norman (to Lubrizol Corp.).

168. V. I. Karzhev, N. V. Goncharova, and E. A. Bulekova, *Sb. Nauchn. Tr. - Vses. Nauchno-Issled. Inst. Pererab. Nefti* **44,** Pt. 3, 64–69 (1983); *Chem. Abstr.* **100,** 106081a (1984).

169. V. V. Danilenko and co-workers, *Khim. Tekhnol. (Kiev)* **2,** 26–28 (1983); *Chem. Abstr.* **98,** 218422j (1983).

170. PCT Int. Appl. WO 8,603,220 (June 5, 1986), K. E. Davis (to Lubrizol Corp.).

171. Eur. Pat. Appl. EP 8,193 (Feb. 20, 1980), E. L. Neustadter (to British Petroleum Co., Ltd.).

172. Eur. Pat. Appl. EP 438,849 (July 31, 1991), R. Scattergood and D. K. Walters (to Ethyl Petroleum Additives Ltd.).

173. Netherlands Pat. 7,409,282 (Jan. 15, 1976), (to Toa Nenryo Kogyo).

174. U.S. Pat. 3,374,174 (Mar. 19, 1968), W. M. Le Seur (to Lubrizol).

175. U.S. Pat. 3,634,515 (Jan. 11, 1972), R. E. Karll and E. J. Piasek (to Standard Oil (Indiana)).

176. U.S. Pat. 3,725,277 (Apr. 3, 1973), C. Worrel (to Ethyl Corp.).

177. U.S. Pat. 3,798,165 (Mar. 19, 1974), R. Karll and E. Piasek (to Amoco Corp.).

178. U.S. Pat. 2,568,876 (Sept. 25, 1961), R. V. White and P. S. Landis (to Socony-Vacuum Oil Co., Inc.).

179. U.S. Pat. 2,794,782 (June 4, 1957), E. P. Cunningham and D. W. Dinsmore (to Monsanto Chemical Co.).

180. U.S. Pat. 4,219,431 (Aug. 26, 1980), S. Chibnik (to Mobil Oil Corp.).

181. Fr. Demande FR 2,616,441 (Dec. 16, 1988), P. Hoornaert and C. Rey (to Elf France).

182. U.S. Pat. 4,798,678 (Jan. 17, 1989), C. S. Liu, L. D. Grina, and M. M. Kapuscinski (to Texaco Inc.).

183. USSR Pat. SU 1,214,739 (Feb. 28, 1986), O. N. Grishina and co-workers (to A. E. Arbuzov, Institute of Organic and Physical Chemistry and Kazan Chemical–Technological Institute, USSR).

184. U.S. Pat. 5,011,617 (Apr. 30, 1991), G. L. Fagan (to Chevron Research and Technology Co.).

185. Fr. Pat. 1,410,400 (July 31, 1964), (to Shell International).

186. U.S. Pat. 3,864,270 (Feb. 4, 1975), H. Chafetz, W. P. Cullen, and E. F. Miller (to Texaco, Inc.).

187. U.S. Pat. 3,996,024 (Dec. 7, 1976), M. D. Coon (to Chevron Research Co.).

188. U.S. Pat. 4,294,714 (Oct. 13, 1981), R. A. Lewis and L. R. Honnen (to Chevron Research Co.).

189. U.S. Pat. 4,292,046 (Sept. 29, 1981), A. B. Piotrowski (to Mobil Oil Corp.).

190. Can. Pat. CA 1,096,381 (Feb. 24, 1981), W. H. Machleder and J. M. Bollinger (to Rohm & Haas Co.).

191. U.S. Pat. 4,234,321 (Nov. 18, 1980), J. E. Lilburn (to Chevron Research Co.).

192. U.S. Pat. 4,230,588 (Oct. 28, 1980), B. R. Bonazza and S. Schiff (to Phillips Petroleum Co.).

193. U.S. Pat. 4,185,965 (Jan. 29, 1980), R. C. Schlicht and W. M. Cummings (to Texaco Inc.).

194. U.S. Pat. 4,105,417 (Aug. 8, 1978), M. D. Coon and J. H. MacPherson.

195. U.S. Pat. 3,231,348 (Jan. 25, 1966), E. G. Lindstrom and W. L. Richardson (to Chevron Research Co.).

196. Ger. Offen. 2,401,930 (July 24, 1975), D. Wagnitz (to B. P. Benzin and Petroleum A.G.).

197. Span. Pat. ES 2,004,658 (Feb. 1, 1989), D. P. G. Nandia and J. Jose (to Inproven S. A.).

198. U.S. Pat. 4,321,061 (Mar. 23, 1982), R. M. Parlman, R. C. Lee, and L. D. Burns (to Phillips Petroleum Co.).

199. Rom. Pat. RO 83,008 (Jan. 30, 1984), T. I. Decean, G. Iordache, A. Dumitru, and S. Popescu (to Combinatul Petrochimic Teleajen).

200. I. Skeist and G. R. Somerville, *Epoxy Resins*, Reinhold Publishing Co., New York, 1958, pp. 21, 167, 185, 233.

201. F. J. Allen and W. M. Hunter, *J. Appl. Chem.* **7**, 86 (1957).

202. Jpn. Kokai Tokkyo Koho JP 03 045,683 (Feb. 27, 1991), N. Murata (to Nippon Telegraph and Telephone Corp.).

203. Czech. Pat. CS 260,717 (May 15, 1989), J. Klugar, M. Lidarik, J. Snuparek, B. Hajkova, and M. Sip.

204. Chin. Pat. CN 86,101,597 (Sept. 23, 1987), G. Fang (to Wuhan Bicycle Electric Plating Plant, Peoples' Republic of China).

205. U.S. Pat. 4,740,536 (Apr. 26, 1988), Y. Y. H. Chao (to Rohm & Haas Co.).

206. Czech. Pat. CS 212,066 (Feb. 26, 1982), I. Wiesner and J. Novak.

207. USSR Pat. SU 1,608,194 (Nov. 23, 1990), E. V. Amosova and co-workers.

208. Czech. Pat. CS 249,845 (Apr. 25, 1988), J. Hejlova.

209. Jpn. Kokai Tokkyo Koho JP 62 224,954 (Oct. 2, 1987), T. Shiraki, Y. Nakamura, and H. Tabata (to Nitto Electric Industrial Co., Ltd.).

210. Jpn. Kokai Tokkyo Koho JP 01 223,112 (Sept. 6, 1989), S. Kanazawa (to Kojima Press Co., Ltd.).

211. U.S. Pat. 4,800,222 (Jan. 24, 1989), H. G. Waddill (to Texaco Inc.).

212. U.S. Pat. 4,766,186 (Aug. 23, 1988), K. B. Sellstrom and H. G. Waddill (to Texaco Inc.).

213. Jpn. Kokai Tokkyo Koho JP 75 33,693 (Nov. 1, 1975), M. Nakajima and A. Yanaguchi (to Nippon Carbide Industries Co., Inc.).

214. Jpn. Kokai Tokkyo Koho JP 75 123,200 (Mar. 27, 1975), T. Hoki, T. Toyomoto, and H. Komoto (to Asahi Chemical Industry Co., Ltd.).

215. Jpn. Kokai Tokkyo Koho JP 75 151,998 (Dec. 6, 1975), Y. Nakamura, S. Aoyama, and T. Suzuki (to Toto Chemical Industry Co., Ltd.).

216. U.S. Pat. 2,450,940 (Oct. 12, 1948), J. C. Cowan, L. B. Falkenburg, and A. Lewis (to U.S. Dept. of Agriculture).

217. R. Anderson and D. H. Wheeler, *J. Am. Chem. Soc.* **70,** 760 (1948).

218. U.S. Pat 2,705,223 (Mar. 29, 1955), M. Renfrew and H. Wittcoff (to General Mills, Inc.).

219. Jpn. Kokai Tokkyo Koho JP 02 115,153 (Apr. 27, 1990), Y. Kimura and M. Honma (to Ajinomoto Co., Inc.).

220. Czech. Pat. CS 254519 (Sept. 15, 1988), I. Wiesner, L. Wiesnerova, K. Exnerova, and B. Exner.

221. Jpn. Kokai Tokkyo Koho JP 75 117,900 (Sept. 16, 1975), A. Kotone, T. Hori, and M. Hoda (to Sakai Chemical Industry Co., Ltd.).

222. Jpn. Kokai Tokkyo Koho JP 77 26,598 (Feb. 28, 1977), M. Magakura and Y. Higaki (to Nisshin Oil Mills, Ltd.).

223. Jpn. Kokai Tokkyo Koho JP 77 00,898 (Jan. 6, 1977), H. Suzuki and co-workers (to Asahi Denka Kogyo K. K.).

224. Ger. Offen. 2,405,111 (Aug. 14, 1975), G. Johannes and co-workers (to Hoechst A.G.).

225. Ger. Offen. 2,326,668 (Dec. 12, 1974), F. Schuelde, J. Obendorf, and V. Kulisch (to Veba-Chemie A.G.).

226. F. Pitt and M. N. Paul, *Mod. Plast.* **34,** 124 (1957).

227. U.S. Pat. 4,126,640 (Nov. 21, 1978), D. E. Floyd (to General Mills Chemicals, Inc.).

228. Czech. Pat. CS 260720 (May 15, 1989), J. Klugar, M. Lidarik, J. Snuparek, H. Jaromir, B. Hajkova, and M. Sip.

229. U.S. Pat. 3,026,285 (Mar. 20, 1962), F. N. Hirosawa and J. Delmonte.

230. U.S. Pat. 3,356,647 (Dec. 5, 1967), W. M. Bunde (to Ashland Oil & Refining).

231. U.S. Pat. 3,407,175 (Oct. 22, 1968), W. E. Presley and T. J. Hairston (to The Dow Chemical Co.).

232. U.S. Pat. 3,548,002 (Dec. 15, 1970), L. Levine (to The Dow Chemical Co.).

233. U.S. Pat. 3,595,816 (July 27, 1971), F. O. Barrett (to Emery Industries, Inc.).

234. U.S. Pat. 2,839,219 (June 17, 1958), J. H. Groves and G. G. Wilson (to General Mills, Inc.).

235. U.S. Pat. 3,462,284 (Aug. 19, 1969), L. R. Vertnik (to General Mills, Inc.).

236. U.S. Pat. 2,811,459 (Oct. 29, 1957), H. Wittcoff and W. A. Jordan (to General Mills, Inc.).

237. Ger. Offen. 2,361,486 (June 12, 1975), W. Imoehl and M. Drawert (to Schering A.G.).

238. U.S. Pat. 2,663,649 (Dec. 22, 1953), W. B. Winkler (to T. F. Washburn Co.).

239. U.S. Pt. 2,379,413 (July 3, 1945), T. F. Bradley (to American Cyanamid Co.).

240. Pol. Pat. 60,441 (Aug. 5, 1970), L. Chromy, J. Polaczy, and E. Smieszek (to Instytut Badawczo Projectowy Przemyslu Farb i Lakierow).

241. U.S. Pat. 3,412,115 (Nov. 19, 1968), D. E. Floyd and D. W. Glaser (General Mills, Inc.).

242. G. J. Kubes and co-workers, *Cellul. Chem. Technol.* **13,** 803–811 (1979).

243. N. Hartler, *EUCEPA Symp. 2, 11/1–11/29*, Finn. Pulp Pap. Res. Inst., Helsinki, 1980.

244. J. M. MacLeod and co-workers, *Pulping Conf. Proc.*, 25–30 (1980).

245. A. B. J. Du Plooy, *Gov. Rep. Announce. Index (U.S.) 1981* **81,** 5572 (1980).
246. R. G. Nayak and J. L. Wolfhagen, *J. Appl. Polym. Sci., Appl. Polym. Symp.* **37,** 955–965 (1983).
247. R. G. Nayak and J. L. Wolfhagen, *Biotechnol. Bioeng. Symp.* **13,** 657–662 (1984).
248. U.S. Pat. 2,769,799 (Nov. 6, 1956), T. Suen and Y. Jen (to American Cyanamid Co.).
249. U.S. Pat. 2,769,800 (Nov. 6, 1956), T. Suen and Y. Jen (to American Cyanamid Co.).
250. U.S. Pat. 2,554,475 (May 22, 1951), T. Suen and J. H. Daniel, Jr. (to American Cyanamid Co.).
251. U.S. Pat. 2,683,134 (July 6, 1954), J. B. Davidson and E. J. Romatowski (to Allied Chemical and Dye Corp.).
252. U.S. Pat. 2,742,450 (Apr. 17, 1956), R. S. Yost and R. W. Auten (to Rohm & Haas Co.).
253. Ger. Pat. 2,453,826 (May 15, 1975), (to Scott Paper).
254. U.S. Pat. 2,689,239 (Sept. 14, 1954), S. Melamed (to Rohm & Haas Co.).
255. U.S. Pat. 2,926,116 and 2,926,154 (Feb. 23, 1960), G. I. Keim (to Hercules Powder Co.).
256. U.S. Pat. 3,227,671 (Jan. 4, 1966), G. I. Keim (to Hercules Powder Co.).
257. U.S. Pat. 3,565,754 (Feb. 23, 1971), N. W. Dachs and G. M. Wagner (to Hooker Chemical Corp.).
258. U.S. Pat. 2,609,126 (Sept. 28, 1971), H. Asao, F. Yoshida, K. Tomihara, M. Akimoto, and G. Kubota (to Toho Kagaku Kogyo Kabushiki Kaisha).
259. U.S. Pat. 3,442,754 (May 6, 1969), H. H. Espy (to Hercules, Inc.).
260. U.S. Pat. 2,926,117 (Feb. 23, 1960), H. Wittcoff (to General Mills Inc.).
261. U.S. Pat. 2,595,935 (May 6, 1952), J. H. Daniel, Jr. and C. G. Landes (to American Cyanamid Co.).
262. U.S. Pat. 3,577,313 (May 4, 1971), J. C. Bolger, R. W. Hausslein, and H. E. McCollum (to Amicon Corp.).
263. U.S. Pat. 3,591,529 (July 6, 1971), J. Fertig and H. H. Stockmann (to National Starch & Chem.).
264. Ger. Pat. 2,434,816 (Feb. 5, 1976), E. Scharf, R. Fikentseher, W. Auhorn, and W. Streit (to BASF AG).
265. Ger. Pat. 2,351,754 (Apr. 17, 1975), G. Spielberger and K. Hammerstroem (to Bayer AG).
266. U.S. Pat. 2,772,969 (Dec. 4, 1956), L. A. Lundberg and W. F. Reynolds, Jr. (to American Cyanamid Co.).
267. Ger. Offen. DE 3,515,480 (Oct. 30, 1986), W. Von Bonin, U. Beck, J. Koenig, and H. Baeumgen (to Bayer AG).
268. U.S. Pat. 2,767,089 (Oct. 16, 1956), N. A. Kjelson, M. M. Renfrew, and H. Witcoff (to General Mills Inc.).
269. Swiss Pat. 571,474 (Jan. 15, 1976), R. Hochreuter (to Sandoz).
270. U.S. Pat. 2,772,966 (Dec. 4, 1956), J. H. Daniel, Jr. and S. T. Moore (to American Cyanamid Co.).
271. Jpn. Pat. 70 28,722 (Sept. 19, 1970), Y. Chiba and H. Adachi (to Kinkai Kagaku).
272. U.S. Pat. 3,902,958 (Sept. 2, 1975), D. L. Breen and A. J. Frisque (to Nalco Chemical Co.).
273. Jpn. Kokai Tokkyo Koho JP 83 126,394 (July 27, 1983) (to Sumitomo Chemical Co., Ltd.).
274. S. Chaberek and A. E. Martell, *Organic Sequestering Agents*, John Wiley & Sons, Inc., New York, 1959.
275. J. C. Bailar, Jr., ed., *The Chemistry of Coordination Compounds*, Reinhold Publishing Co., New York, 1956.
276. U.S. Pat. 2,461,519 (Feb. 15, 1949), F. C. Bersworth (to Martin Dennis Co.).

277. U.S. Pat. 2,860,164 (Nov. 11, 1958), H. Kroll and F. P. Butler (to Geigy Chemical Corp.).
278. U.S. Pat. 2,205,995 (June 25, 1940), H. Ulrich and E. Ploetz (to I. G. Farbenindustrie).
279. U.S. Pat. 2,855,428 (Oct. 7, 1958), J. J. Singer (to Hampshire Chemical Corp.).
280. U.S. Pat. 5,013,404 (May 7, 1991), S. H. Christiansen, T. Littleton, and R. T. Patton (to Dow Chemical Co.).
281. Jpn. Kokai Tokkyo Koho JP 55 071,814 (May 30, 1980) (to Nichibi Co., Ltd.).
282. K. Takeda, M. Akiyama, and T. Yamamizu, *React. Polym. Ion Exch. Sorbents* **4,** 11–20 (1985).
283. U.S. Pat. 3,933,871 (Jan. 20, 1976), L. J. Armstrong (to Armstrong Chemical Co., Inc.).
284. Ger. Offen. 2,165,947 (July 20, 1972), T. V. Kandathil (to S. C. Johnson and Son, Inc.).
285. U.S. Pat. 3,954,634 (May 4, 1976), J. A. Monson, W. L. Stewart, and H. F. Gruhn (to S. C. Johnson and Son, Inc.).
286. Ger. Offen. 2,520,150 (Jan. 2, 1976), P. Goullet (to Azote et Produits Chimiques SA).
287. U.S. Pat. 3,887,476 (June 3, 1975), R. B. McConnell (to Ashland Oil, Inc.).
288. U.S. Pat. 3,855,235 (Dec. 17, 1974), R. B. McConnell (to Ashland Oil, Inc.).
289. U.S. Pat. 3,898,244 (Aug. 5, 1975), R. B. McConnell (to Ashland Oil, Inc.).
290. U.S. Pat. 3,965,015 (June 22, 1976), R. A. Bauman (to Colgate-Palmolive Co.).
291. U.S. Pat. 4,049,557 (Sept. 20, 1977), H. E. Wixton (to Colgate-Palmolive Co.).
292. U.S. Pat. 2,155,877 (Apr. 25, 1939), E. Waldmann (to I.G. Farbenind, AG).
293. U.S. Pat. 2,215,863 (Sept. 24, 1941), E. Waldmann (to General Aniline & Film Corp.).
294. U.S. Pat. 2,215,864 (Sept. 24, 1941), E. Waldmann and A. Chwala (to General Aniline & Film Corp.).
295. V. I. Lysenko and co-workers, *Kolloida Zh.* **38,** 914–918 (1976).
296. Span. Pat. ES 540,640 (Nov. 16, 1985), J. Pomares Pujada (to Pulcra SA).
297. Czech. Pat. CS 184,110 (July 15, 1980), Z. Vodak, F. Krsnak, and L. Bechtold.
298. Jpn. Kokai Tokkyo Koho JP 61 039,939 (Sept. 6, 1986), M. Kusumi and S. Ando (to Lion Corp.).
299. U.S. Pat. 2,781,349 (Feb. 12, 1957), H. S. Mannheimer.
300. Jpn. Kokai Tokkyo Koho JP 75 137,917 (Nov. 1, 1975), Y. Nakamura and co-workers (to Toho Chemical Industry Co., Ltd.).
301. U.S. 4,189,593 (Feb. 19, 1980), J. R. Wechsler, T. G. Baker, G. T. Battaglini, and F. L. Skradski.
302. U.S. Pat. 2,622,067 (Dec. 15, 1952), (to Socony-Vacuum Oil Co.).
303. Brit. Pat. 1,224,440 (Nov. 10, 1971), (to Witco Chemical Corp.).
304. Jpn. Kokai Tokkyo Koho JP 03 008,427 (Jan. 16, 1991), F. Mayuzumi (to Daiichi Kogyo Seiyaku Co., Ltd.).
305. Eur. Pat. Appl. EP 442,251 (Aug. 21, 1991), J. K. E. De Waele and A. Koenig (to Procter and Gamble Co., U.S.A.).
306. USSR Pat. SU 939,511 (June 30, 1982), M. K. Nikitin and co-workers.
307. Eur. Pat. Appl. EP 112,593 (July 4, 1984), J. M. Vander Meer, D. N. Rubingh, and E. P. Gosselink (to Procter and Gamble Co.).
308. Brit. Pat. 1,177,134 (Jan. 7, 1970), C. O. Bundrant, C. R. Hainebach, and F. H. Mays (to Champion Chemicals, Inc.).
309. Ger. Offen. 1,944,569 (Mar. 11, 1971), H. J. Schuestler and R. Scharf (to Henkel and Cie., GmbH).
310. U.S. Pat. 3,993,615 (Nov. 23, 1976), S. B. Markofsky and L. L. Wood (to W. R. Grace).
311. U.S. Pat. 4,089,803 (May 16, 1978), D. U. Bessler (to Petrolite Corp.).
312. U.S. Pat. 4,444,262 (Apr. 24, 1984), H. K. Haskin and P. E. Figdore (to Texaco Inc.).
313. Ger. Offen. DE 3,131,461 (Apr. 22, 1982), J. E. Glass, Jr. (to Union Carbide Corp.).
314. L. A. Mel'nikova and N. K. Lyapina, *Neftekhimiya* **20,** 914–917 (1980).

315. U.S. Pat. 4,744,881 (May 17, 1988), D. K. Reid (to Betz Laboratories, Inc.).
316. Jpn. Kokai Tokkyo Koho JP 60 044,586 (Mar. 9, 1985), (to The Dow Chemical Co.).
317. Jpn. Kokai Tokkyo Koho JP 58 188,824 (Nov. 4, 1983), (to Japan Synthetic Rubber Co., Ltd.).
318. Y. Dai and S. Zhu, *Shiyou Huagong* **14,** 641–645 (1985).
319. U.S. Pat. 4,155,875 (May 22, 1979), K. Yamaguchi, K. Kawakami, and Y. Nakamoto (to Nippon Mining Co., Ltd.).
320. U.S. Pat. 4,902,404 (Feb. 20, 1990), T. C. Ho (to Exxon Research and Engineering Co.).
321. Eur. Pat. Appl. EP 181,035 (May 14, 1986), M. S. Thompson (to Shell Internationale Research Maatschappij BV).
322. O. K. Dobozy, *Egypt K. Chem.* **16,** 419 (1973).
323. U.S. Pat. 2,426,220 (Aug. 25, 1947), J. M. Johnson (to Nostrip, Inc.).
324. U.S. Pat. 2,812,339 (Nov. 5, 1957), M. L. Kalinowski and L. T. Crews (to Standard Oil Co.).
325. O. Dobozy, *Tenside* **7,** 83 (1970).
326. U.S. Pat. 2,766,132 (Oct. 9, 1956), C. M. Blair, Jr., W. Groves, and K. L. Lissant (to Petrolite Corp.).
327. V. G. Ostroverkhov and co-workers, *Neftepererab. Neftekhim. (Kiev)* **38,** 43–47 (1990); *Chem. Abstr.* **115,** 34390k (1991).
328. PCT Int. Appl. WO 8,807,066 (Sept. 22, 1988), D. L. Hopkins (to Lubrizol Corp., U.S.A.).
329. U.S. Pat. 3,097,292 (July 2, 1963), E. W. Mertens (to California Research Corp.).
330. U.S. Pat. 3,230,104 (Jan. 18, 1966), C. W. Falkenberg, R. A. Paley, and J. J. Patti (to Components Corp. of America).
331. Brit. Pat. 1,174,577 (Dec. 17, 1969), N. H. Greatorex and K. N. Shaw (to Swan, Thomas and Co., Ltd.).
332. Ger. Pat. 2,513,843 (Oct. 9, 1975), M. Fujita and S. Okada.
333. U.S. Pat. 3,249,451 (May 3, 1966), E. D. Evans and C. H. Hopkins (to Skelly Oil Co.).
334. U.S. Pat. 2,721,807 (Oct. 25, 1955), J. L. Rendall and D. R. Husted (to Minnesota Mining and Manufacturing Co.).
335. Ger. Pat. DD 275,470 (Jan. 24, 1990), J. Seupel and co-workers (to VEB Otto Grotewohl Boehlen).
336. Span. Pat. ES 554,977 (Sept. 1, 1987), J. Sanchez-Marcos Sanchez (to Elsamex SA).
337. U.S. Pat. 4,338,136 (July 6, 1982), P. Goullet and P. Scotte (to Azote et Produits Chimiques, SA).
338. C. I. Kim and J. C. Kim, *Choson Minjujuui Inmin Konghwaguk Kwahagwon Tongbo* **25,** 249–252 (1977); *Chem. Abstr.* **88,** 171601t (1978).
339. V. I. Lebedeva and S. Y. Shibashova, *Izv. Vyssh. Uchebn. Zaved., Tekhnol. Tekst. Prom-sti.* **3,** 66–69 (1989).
340. N. Chand and S. Verma, *Sci. Cult.* **55,** 137–139 (1989).
341. J. Das, A. K. Mohanty, and B. C. Singh, *Text. Res. J.* **59,** 525–529 (1989).
342. P. Ghosh and C. Datta, *Indian J. Technol.* **26,** 431–436 (1988).
343. M. Luo and F. Liang, *Fangzhi Xuebao* **10,** 57–60 (1989).
344. S. G. Abdurakhmanova, V. I. Shoshina, and G. V. Nikonovich, *Khim. Volokna* **2,** 34–36 (1989).
345. U.S. Pat. 2,758,003 (Aug. 7, 1956), O. Bayer and H. Kleiner (to Farbenfabriken Bayer, AG).
346. South Afr. Pat. ZA 7,901,957 (May 28, 1980), T. Jellinek (to Commonwealth Scientific and Industrial Research Organization, Australia).
347. Jpn. Kokai Tokkyo Koho JP 55 045,859 (Mar. 31, 1980), S. Fukuoka (to Toyobo Co., Ltd.).

348. USSR Pat. SU 1,420,090 (Aug. 30, 1988), J. Libonas, V. Stanevicius, and V. Paskevicius (Vilnius State University).
349. X. P. Lei and D. M. Lewis, *J. Soc. Dyers Colour.* **106,** 352–356 (1990).
350. N. R. Bertoniere and S. P. Rowland, *J. Appl. Polym. Sci.* **23,** 2567–2577 (1979).
351. Eur. Pat. Appl. EP 342,974 (Nov. 23, 1989), A. Kawaguchi (to E. I. du Pont de Nemours & Co.).
352. G. B. Guise and F. W. Jones, *Text. Res. J.* **48,** 705–709 (1978).
353. Ger. Pat. DE 2,824,614 (Dec. 13, 1979), K. Dahmen (to Chemische Fabrik Stockhausen und Cie).
354. Jpn. Kokai Tokkyo Koho JP 63 264,506 (Nov. 1, 1988), K. Saito, M. Fujino, and T. Okamoto (to Toray Industries, Inc.).
355. Jpn. Kokai Tokkyo Koho JP 53 065,494 (June 10, 1978), H. Ogawara, Y. Yokota, T. Yonezawa, and K. Onoda (to Miyoshi Oil and Fat Co., Ltd.).
356. Jpn. Kokai Tokkyo Koho JP 02 160,976 (June 10, 1990), H. Fukuda and K. Okada (to Nikka Chemical Industry Co., Ltd.).
357. Jpn. Kokai Tokkyo Koho JP 53 114,984 (Oct. 6, 1978), N. Sugioka and co-workers (to Sakai Textile Mfg. Co., Ltd. and Toho Chemical Industry Co., Ltd.).
358. Jpn. Kokai Tokkyo Koho JP 74 42,864 (Nov. 18, 1974), (to Unitika, Ltd.).
359. Ger. Pat. 2,249,610 (Apr. 24, 1974), (to Chemiische Fabrik Pfersee, GmbH).
360. Jpn. Kokai Tokkyo Koho JP 62 104,913 (May 15, 1987), T. Yarino and Y. Yamahara (to Teijin Ltd.).
361. U.S. Pat. 3,866,603 (Feb. 18, 1975), D. J. Wilpers (to Shell Oil Co.).
362. U.S. Pat. 2,857,331 (Oct. 21, 1958), C. A. Hollingsworth, K. F. Schilling, and J. L. Wester (to Smith-Douglass Co., Inc.).
363. Ger. Offen. 2,546,180 (Apr. 28, 1977) V. Cuntze, G. Edelmann, and L. Metza (to Hoechst, A.G.).
364. U.S. Pat. 2,322,201 (June 15, 1943), D. W. Jayne, Jr., J. M. Day, and S. E. Erickson (to American Cyanamid Co.).
365. U.S. Pat. 3,475,163 (Oct. 28, 1969), W. A. Mod, F. N. Teumac, and J. D. Watson, Sr. (to Dow Chemical).
366. A. Tonev and E. Staikova, *God. Nauchnoizszled, Proektno-Konstr. Inst. Tsventna Metal. (Plovdiv),* 7–12; (1974); *Chem. Abstr.* **83,** 104027u (1975).
367. U.S. Pat. 2,940,964 (Aug. 30, 1960), F. A. Forward, H. Veltman, and A. I. Vizsolyi (to Sherritt Gordon Mines Ltd.).
368. N. Lyakov and T. Nikolov, *Metalurgiya (Sofia)* **32,** 23 (1977).
369. L. S. Goodman and A. Gilman, eds., *The Pharmacological Basis of Therapeutics*, 5th ed., MacMillan Publishing Co., New York, 1975, pp. 1027–1028.
370. U.S. Pat. 3,751,474 (Aug. 7, 1973), K. G. Phillips and M. J. Geerts (to Nalco Chemical Co.).
371. Ger. Pat. DD 206,382 (Jan. 25, 1984), M. Bertram, H. G. Fuchs, R. Jost, and J. Stricker (to VEB Chemische Werke Buna).
372. U.S. Pt. 4,408,038 (Oct. 4, 1983), R. A. Covington, Jr. and O. M. Ekiner (to E. I. du Pont de Nemours & Co.).
373. Ger. Offen. DE 2,905,651 (Aug. 21, 1980), H. Perrey and M. Matner (to Bayer AG).
374. U.S. Pat. 4,218,559 (Aug. 19, 1980), R. A. Behrens and D. R. Maulding (to American Cyanamid Co.).
375. Jpn. Kokai Tokkyo Koho JP 55 000,726 (Jan. 7, 1980), K. Takahashi, T. Kondo, M. Koga, and T. Fukuda (to TDK Electronics Co., Ltd.).
376. U.S. Pat. 4,381,378 (Apr. 26, 1983), L. L. Harrell, Jr. (to E. I. du Pont de Nemours & Co.).
377. Jpn. Kokai Tokkyo Koho JP 60 108,447 (June 13, 1985), (to Japan Synthetic Rubber Co., Ltd.).

378. K. Tenchev and D. Dimitrov, *Khim. Ind. (Sofia)* **5**, 227–229 (1982).
379. K. Tenchev and D. Dimitrov, *Hem. Ind.* **37**, 44–46 (1983).
380. Jpn. Kokai Tokkyo Koho JP 55 031,846 (Mar. 6, 1980), H. Ikeda, K. Goto, and Y. Shimozato (to Japan Synthetic Rubber Co., Ltd.).
381. Jpn. Kokai Tokkyo Koho JP 54 148,043 (Nov. 19, 1979), H. Ikeda and Y. Shimozato (to Japan Synthetic Rubber Co., Ltd.).
382. S. Yamashita, J. Akiyama, and S. Kohjiya, *Asahi Garasu Kogyo Gijutsu Shoreikai Kenkyu Hokoku* **31**, 315–331, 1997; *Chem. Abstr.* **90**, 56067c (1979).
383. Jpn. Kokai Tokkyo Koho JP 52 127,953 (Oct. 27, 1977), K. Marubashi and M. Doi (to Denki Kagaku Kohyo K. K.).
384. Czech. Pat. CS 211,054 (May 15, 1984), P. Svoboda, P. Vanek, O. Vilim, Z. Smely, and M. Zajicek.

RICHARD G. CARTER
ARTHUR R. DOUMAUX, JR.
STEVEN W. KAISER
PAMLA R. UMBERGER
Union Carbide Chemicals and Plastics Company Inc.

DIAMOND. See CARBON.

DIARYLAMINES. See AMINES, AROMATIC AMINES-DIARYLAMINES.

DIASPORE. See ALUMINUM COMPOUNDS, ALUMINUM OXIDE.

DIASTASES. See ENZYME APPLICATIONS, INDUSTRIAL.

DIATOMITE

Diatomite is a naturally occurring, porous, high surface area form of hydrous silica that is used as a filter aid and as a mineral filler. Diatomite products may be classified according to manufacturing method into three categories: natural diatomite [7631-86-9], calcined diatomite [91053-39-3], and flux-calcined diatomite [68855-54-9]. Products from all three categories find widespread use in industrial filtration (qv) applications as a filter aid for achieving higher clarity, longer filter cycles, and removing high solids concentrations. Products from all three categories are also used as functional fillers (qv) where diatomite properties add to the performance of paints, plastics, rubber, catalysts, agricultural chemicals, pharmaceuticals (qv), toothpastes, polishes, and other chemicals.

Diatomite, also known as diatomaceous earth, or kieselguhr, consists mainly of accumulated shells or frustules of intricately structured amorphous hydrous silica secreted by diatoms, which are microscopic, one-celled golden brown algae of the class Bacillariophyceae. Diatoms exist in many different environments and are abundant in regions of oceanic upwelling: 12,000 to 16,000 species of diatoms live in fresh, brackish, or saline waters. Diatom species can live both solitarily and colonially; some are mobile, others stationary. All diatom species have an elaborately ornamented siliceous skeleton, which results in accumulations of uniquely porous particles.

Diatoms are single-celled photosynthetic plants consisting of two shells that fit together in the same manner as the two halves of a pill box (1). Reproduction is by division at such a rate that it is estimated that one diatom can produce 10^{10} descendants in 30 days under the most favorable conditions. The diatom plants extract silica from the water to form an encasing shell or exoskeleton. Ocean floor mapping has revealed that diatoms accumulate in areas of oceanic upwelling, where nutrient-rich waters circulate near the sunlit surface, such as along the West Coast and most continents and at the equator (2,3). Freshwater diatoms accumulate most commonly where silica-rich springs have contributed enough nutrients to foster diatom blooms.

At the end of a brief life, the diatom settles to the bottom of the body of water where the organic matter decomposes, leaving the siliceous skeleton. These fossil skeletons, or frustules, are in the shape of the original diatom plant and have designs as varied and intricate as snowflakes. Examples are shown in Figure 1.

In AD 532 the Roman emperor Justinian used diatomite bricks for lightweight construction of the dome when building the Church of St. Sophia in Constantinople (Istanbul). The names bergmehl, fossil flour, farine fossile, and mountain flour apparently originated when early poverty-driven peoples extended supplies of meal and flour by dilution with diatomaceous earth. Tripoli is a name

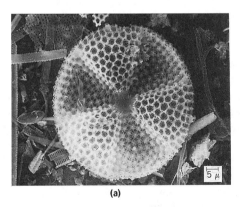
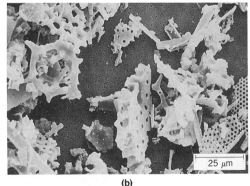

(a)   (b)

Fig. 1. (**a**) Scanning electron photomicrograph of a diatom of the genus *Actinoptycus*, illustrating the ornate and porous nature of the diatom frustule. (**b**) Photomicrograph of a diatomite product showing the variety of shapes and pores that give the products high surface area and ability to trap solids for separation from clear liquids.

given to diatomite formerly mined in Tripoli, North Africa. Kieselguhr, the name given to the diatomite first mined in Hanover, Germany, in 1836 or earlier, is still used as a general name for all diatomite products in Europe. An incorrect name that persisted for many years was infusorial earth; incorrect because *Infusoria* comprises a group of the animal kingdom (1). Nobel developed the first important industrial use of diatomite as an absorbent for liquid nitroglycerin in the making of dynamite late in the nineteenth century.

The first commercial shipment of diatomite in the United States was made in 1893 and consisted of material from a small quarry operation in the vast deposit near Lompoc, California. It went to San Francisco to be used for pipe insulation. Small-scale operation of parts of the Lompoc deposit continued until it was acquired by the Kieselguhr Co. of America, which later became the Celite Co. (4). Since that first work, the industry has grown immensely, and diatomite products are used in almost every country.

Origin of Deposits

Diatoms inhabit fresh, brackish, or sea waters. Environmental changes in the bodies of water where diatoms flourish are reflected by the different types of diatoms that may appear at different levels of the same deposit.

Most commercial marine diatomite deposits exploit accumulations resulting from large blooms of diatoms that occurred in the oceans during the Miocene geological epoch. Diatomite sediments older than the Jurassic period are rare in the fossil record. Commercial deposits of diatomite are accumulations of the fossil skeletons, which can occur in beds as thick as 900 m in some locations (5). Marine deposits must have been formed on the bottom of protected basins or other bodies of quiet water, undisturbed by strong currents, in an environment similar to the existing Santa Barbara Channel or Gulf of California (3,6).

The main deposits of freshwater diatomite were laid down in large lakes. Many of these deposits in the western United States formed during glacial times, when the local climate was wetter. Several tens of square kilometers in Nevada west of Tonopah are covered with diatomite as are other large areas in the Great Basin.

The principal marine deposits were formed during the Tertiary period and more particularly the upper Miocene epoch. Deposits of freshwater origin date from the Pliocene to Miocene epochs to more recent times, dating to as late as 100,000 years ago. U.S. commercial deposits are at or comparatively near the surface. Bog deposits are exploited and lake beds are dredged for use in other parts of the world.

Location of Deposits.   Deposits of diatomite are known to exist on every continent and in nearly every country. Over half of the states in the United States reportedly contain diatomaceous earth deposits. In some cases, deposits of marine and freshwater origin occur almost side by side as do deposits of widely varying ages (5). Most of the deposits are not large enough or sufficiently pure to have commercial value. Production figures show the location of the deposits that meet commercial standards in both respects (7).

In the United States the most extensive commercial deposits are located in California, Nevada, Oregon, and Washington. The U.S. Bureau of Mines also reports the commercial operation of diatomite deposits in Arizona (7).

California contains the largest formation of diatomite in the United States; the Monterey Formation extends from Point Arena in Mendocino County in the north to San Onofre in the south (7). The most extensive deposit is near Lompoc, Santa Barbara County, and is of marine origin. A freshwater deposit is being mined near Burney, Shasta County. Other important deposits (not mined as of this writing) are located in Monterey, Fresno, Inyo, Kern, Orange, San Bernadino, San Joaquin, Sonoma, and San Luis Obispo counties.

Oregon production of filter aids is from a deposit near Ontario, Oregon, and of pet litter from near Christmas Valley. An extensive deposit near Terrebonne, Oregon, was operated from 1936 to 1961. Four different companies operate at least five deposits in Nevada. Of the several comparatively large deposits in Washington, only one, near Quincy, is being operated on a commercial scale (7).

In Canada, large freshwater deposits are found in British Columbia, and small deposits are located in Nova Scotia, New Brunswick, Quebec, and Ontario. A small amount of diatomite production has been reported in Costa Rica, Chile, Brazil, and Argentina. Although deposits exist on other continents, the most important are in Europe, Africa, and Asia, primarily in Japan and Korea. Large deposits have been reported in the Caucasus Mountains of the former USSR. The leading producers in Europe operate in Romania, Germany, Italy, Iceland, and Denmark. Both Danish and Romanian diatomite production, although high in terms of tonnage, is of comparatively low value. Most Romanian diatomite is used as a binder in the construction industry and the Möler earth of Denmark is used as an absorbent, for bricks, and in agriculture (8). African deposits are located in Algeria and Kenya, and there are some small operations in the Republic of South Africa. Diatomite deposits also have been developed in Japan and China.

Physical and Chemical Properties

Chemically, diatomite consists primarily of silicon dioxide [7631-86-9], $SiO_2 \cdot nH_2O$, and is essentially inert. It is attacked by strong alkalies and by hydrofluoric acid but is virtually unaffected by other acids. The silicon dioxide has a unique structure, resulting from the intricate form of the diatom skeletons. The chemically combined water content varies from 2 to 10%. Impurities that are often found mixed with the diatomite are other aquatic fossils such as sponge residues, Radiolaria, silicoflagellata, sand, clay, volcanic ash, mineral aerosols, calcium carbonate, magnesium carbonate, soluble salts, and organic matter. The types and amounts of impurities are highly variable between the deposits and depend on the conditions of sedimentation at the time of diatom deposition. Variations also exist within deposits. Typical chemical analyses of diatomite products are given in Table 1.

The color of pure diatomite is white, or near white, but impurities such as carbonaceous matter, clay, iron oxide, volcanic ash, etc, may darken it. The refractive index ranges from 1.41 to 1.48, almost that of opaline silica. Diatomite is optically isotropic.

Table 1. Spectrograpic Analysis of Diatomite Products[a,b]

| Constituent, wt % | Diatomite | | |
	Natural	Calcined	Flux-calcined
Al_2O_3	4.06	3.54	3.63
Fe_2O_3	1.54	1.45	1.40
CaO	0.91	0.69	0.71
Na_2O	0.53	0.59	3.86
P_2O_5	0.27	0.18	0.17
MgO	0.67	0.54	0.60
K_2O	0.67	0.62	0.62
SiO_2	89.90	90.80	87.90
TiO_2	0.21	0.20	0.21
Total	*98.76*	*98.61*	*99.10*

[a]Refs. 9,10.
[b]On a dry basis.

Individual diatom frustules are porous. The diatoms are highly variable in shape and size, having particles that range in effective diameter from 0.75 to 1000 μm, but most are 50 to 100 μm in diameter. Diatom shapes can range from simple cylinders and disks to complex, highly variable, but always punctate, forms. The highly variable array of shapes gives marine diatomite an advantage in certain filtration applications over freshwater diatomite. The latter usually contains fewer genera and less variation in size and shape.

The bulk density of powdered diatomite varies from 112 to 320 kg/m^3. The true specific gravity of diatomite is 2.1 to 2.2, the same as for opaline silica, or opal (1). The thermal conductivity of bulk quantities of diatomite is low but increases with higher percentages of impurities and a higher density. The fusion point depends on the purity but averages about 1430°C for pure material, which is slightly less than for pure silica. The addition of chemical agents, such as soda ash, reduces the fusion point.

Diatomite has only weak adsorption (qv) powers but shows excellent absorption (qv) because of its structure and high surface area. Acids, liquid fertilizers (qv), alcohol, water, oils, and other fluids are absorbed by diatomite.

Mining and Processing

Diatomite deposits are usually discovered by observation of outcrop, and the value of the deposits is determined by geological prospecting and exploration. Samples are taken from the surface outcrops by digging or trenching; underground samples are secured from test holes, core drill holes, or tunnels. Samples are examined chemically, physically, and microscopically to determine the suitability of the diatomite for various uses.

Mining. Most diatomite is mined by open-pit methods. Layers of crude are cleaned by bulldozers or scrapers, and overburden is removed. Samples are obtained and analyzed. After the material has been assigned a grade, the diatomite is broken up using bulldozer scarifiers and then is loaded into trucks by belts, hydraulic excavators, or scrapers. Stockpiles are used at the mine or plant site for air drying, storing, or blending the crudes. Underground mining techniques were used in the United States for a short time and are still in use in countries where low cost labor is available.

Processing. Three general types of diatomite are produced, and a range of grades exists in each. Grade, as used herein, refers not to the quality of the product but designates one of the series of products made for specific uses. Producers often supply custom-made materials for specialized applications.

The crude diatomite, which may contain up to 60% moisture, is first milled in a method that preserves the intricate structure of the diatomite. This material is fed to dryers operating at relatively low temperatures, where virtually all of the moisture is removed (see DRYING). Coarse and gritty nondiatomaceous earth material is removed in separators and preliminary particle size separation is made in cyclones. For many producers, all of the manufacturing processes, with the exception of the calcination step, take place while the material is being pneumatically conveyed. The resultant material is termed natural product. This is the only type of diatomite made by some producers.

Calcined diatomite is produced from natural diatomite, which is then subjected to high temperature calcination in a rotary kiln at about 980°C. The calcined material is then again milled and classified to remove coarse agglomerates as well as extreme fines.

The third type of product, flux-calcined diatomite, is obtained by calcination of the natural product in the presence of a fluxing agent, generally soda ash, although sodium chloride can also be employed. Such processing has the effect of reducing the surface area of the particles, changing the color from the natural tan color to a true white, and rendering various impurities insoluble. Some of the diatomite is converted on calcination to cristobalite [*14464-46-1*]. Most diatomite contains a small percentage of quartz (less than 5%), calcined diatomite can contain up to 25% cristobalite, and flux-calcined diatomite can contain up to 65% cristobalite.

Economic Aspects

Processed diatomite powders and aggregates in the United States range in price from \$20 to \$225/t for products in carload quantities. Materials for specialized applications, which are sold in small volumes and require special processing, range up to \$900/t in carload quantities. All of these prices are fob the diatomite plant.

Owing to the low bulk density of diatomite, freight and trucking rates (on a weight basis) are high. Domestic finished products are packed and shipped in laminated kraft-paper bags, usually containing 22.5 kg, or the product is shipped in bulk or semibulk bags. Bagged products are shipped by truck or rail boxcar, normal boxcar loading being 27 to 36 metric tons. It may also be palletized and wrapped using a polyethylene shrink wrap. Bulk shipments are also made in pressure-differential trucks of 45 and 74 m^3 (1600 and 2600 ft^3). These carry 12 to 15 tons and 16 to 21 tons, respectively. Distribution by bulk truck is confined to the economical trucking distance from the producing plant. Semibulk shipments are made in bags holding approximately 1.27 m^3.

For longer distances, pressure-differential rail hopper cars can be used, which hold between 36 to 45 metric tons per car. In-plant storage is in conventional silos, usually having a 60° cone bottom. The silo may be pressurized for discharge, but this is exceptional. Normally, diatomite is moved from the silo to an adjacent small pressure vessel for transfer to user sites. The material may be aerated and pumped using a modified diaphragm pump. Bulk handling has the advantage of improved environmental conditions as well as lower handling cost in comparison with bagged material. Care must be taken in all types of conveying not to degrade the diatom structure.

Domestic Producers. A principal company mining diatomite and processing it into finished products is Celite Corp. (Lompoc, California), which has wholly owned mines and processing facilities in Lompoc, California; Quincy, Washington; Jalisco, Mexico; Murat, France; Alicante, Spain; and a joint venture mine in Iceland. Other companies are Grefco, Inc. in Lompoc and Burney, California, and Mina, Nevada, and Eagle-Picher Industries, Inc., in Sparks and Lovelock, Nevada, and Vale, Oregon. Production was also reported by the U.S. Bureau of Mines for 1990 by Whitecliff Industries, Mammoth, Arizona; Canyon Resources Minerals Corp., Fernley, Nevada; and Oil-Dri Production Co., Christmas Valley, Oregon (7,11).

Production. Annual diatomite production in the United States fluctuated in the 1980s from a high of 689,000 t (1980) to 570,000 t (1986). In 1990, production was 619,000 t. After the United States, Romania, the former USSR, and France are the largest producers of diatomite. Combined with the United States, these countries account for more than 75% of the world's production. The production totals for the 12 highest diatomite-producing countries are shown in Table 2.

Table 2. Annual World Production of Diatomite[a]

Country	Production, t	
	1989	1990
United States	617,000	619,000
Brazil	20,000	20,000
Denmark	72,000	70,000
France	260,000	260,000
Germany	48,000	50,000
Iceland	25,000	25,000
Italy	28,000	30,000
Korea, Republic of	70,000	70,000
Mexico	35,000	35,000
Romania	260,000	260,000
Spain	100,000	100,000
USSR	260,000	260,000
other countries	43,000	45,000
World Total	*1,838,000*	*1,844,000*

[a]Estimated values from Ref. 7.

Specifications, Standards, and Quality Control

Diatomite is tested from the mining bench through the production process to the bag. Methods depend in part on use. For use in pigments, there are ASTM (12) and military (13) specifications. Other specifications relate to use in the pulp (qv) and paper (qv) industry (14), in water (qv) purification (15), and for use in pharmaceuticals (qv) (16).

The diatomite is analyzed for principal element oxides by the wavelength-dispersive x-ray fluorescence spectrometer (xrf), for trace element analysis, and for compliance with *Food Chemical Codex* specifications (17) by inductively coupled plasma spectroscopy (icp) or by atomic absorption spectrophotometry. The material is tested for permeability by timing the passage of a known volume of liquid through a filter cake under a constant pressure differential. The diatomite is tested for wet cake density and for color using a photovolt meter or the Hunter colorimeter. Particle sizes are obtained by sieving, by using a monochromatic light-scattering particle size analyzer, or by using the Hegman fineness test.

Uses

Diatomite products have unique characteristics of high surface area, low bulk density, high water and oil absorption, and high permeability. Each fossil particle, because of its silica composition, is a rigid but porous and irregular shape. Table 3 lists useful properties of typical diatomite products. In general, diatomite products can be grouped according to use as filter aids, fillers (qv) or extenders, adsorbents, catalyst carriers, insecticide carriers or dilutents, fertilizer conditioners, thermal insulation, and miscellaneous. In physical form, powders make up by far the greatest proportion of diatomite products. Mean particle diameters range from 0.75 to 20 μm. Aggregates are available for special uses and range from 1.5-cm particles to fine powders.

There are two principal ways in which finished diatomite products are used in manufacturing plants: either as a filter aid (see FILTRATION), where the diatomite is expendable, or as a filler, where the diatomite becomes a component and remains as part of the manufactured product. As of 1990, the use of diatomite products was 71% filtration, 15% fillers, and 14% other (7).

Filtration. Diatomite is used as a filter aid for applications with difficult-to-filter solids to improve permeability of the filter cake, to prevent the blinding of filter elements, and where high clarity is required such as in the polish filtration of wine (qv) or beer (qv) before bottling. It is also used in sugar (qv) refining, water treatment, and in the production of fruit juices (qv) and industrial chemicals.

Typically, a filter cake or precoat is built up on the filter septa to prevent blinding, short filter cycle times, and costly cleaning of the septa. Then diatomite is added as body feed to the liquid to be filtered so that the permeability of the filter cake may be maintained. Filler aid permeability of diatomite ranges from 0.06 to 30 μm^2. At the end of the filter cycle the filtrate is clear and the solids are retained in the solid or semisolid diatomite filter cake. The type and amount of diatomite for precoat and body feed are normally determined by pilot studies (18,19).

Table 3. Property Ranges of Diatomite Products[a,b]

| Property | Filter aids | | | Fillers, all types |
	Natural	Calcined	Flux-calcined	
permeability range, μm^{2c}	0.06	0.5–2.0	1.0–29.6	
density, kg/m^3				
wet cake	240–350	270–350	290–380	
bulk	112	120–128	144–336	104–160
particle size distribution				
(granulometer), μm				
10% less than	1.5–3.6	2.5–4.4	7–11	2–4
50% less than	7.0–13.4	10.0–16.1	25–37	6–20
90% less than	25–44.5	30.0–58.9	65–97	14–30
approximate pressure differential[d], kPa[e]	36.5	2.33–4.56	1.11–0.058	
specific gravity	2.00	2.25	2.33	2.0–2.3
porosity, by volume, %	65–85	65–85	65–85	65–85
median pore size range, μm	1.5	3.5–5.0	7–22	f
surface area, m^2/g	10–20	4–6	1–4	0.7–30
pH	6.0–8.0	6.0–8.0	8.0–10.0	6.0–10.0
refractive index	1.42	1.44	1.48	1.40–1.49
oil absorption, %				100–210
particle size, retained %				
98 μm (150 mesh)	1–2	4–7	6–40	
44 μm (325 mesh)	0–12			0–14
Mohs' hardness	3.5–4.0	4.5	5.5–6.0	3.5–6.0

[a]Values given are typical or estimated values, not specifications.
[b]Refs. 9–11.
[c]To convert from μm^2 to d'Arcys, multiply by 1.013.
[d]Measurement at 0.034 cm/s and 0.1 g/m^2 precoat.
[e]To convert from kPa to psi, multiply by 0.145.
[f]The Hegman gauge readings, useful for paint manufacture, run from 0–55.

The first use of diatomite as a filter aid was for the filtration of sewage sludge (4). Further use developed in cane sugar refining (20). In this technology, diatomite filter aids are used in the filtration of various liquors in refining cane, beet, and corn sugars and in clarification of syrups (qv), molasses, etc. From the techniques developed in the sugar refining industry, diatomite filtration was applied to a wide range of separation problems involving beer, chemicals, water, solvents, antibiotics (qv), oils and fats, phosphoric acids, and many others. The necessity of producing clean plant effluent has spurred the application of diatomite filtration for solids removal to clarify the waste streams of manufacturing plants (see WASTES, INDUSTRIAL).

Fillers. Diatomite mineral fillers are used primarily (1) where bulk is needed with minimum weight increase, (2) as an extender where economy of more expensive ingredients is a factor, or (3) where the structure of the particle is important. In other applications, diatomite can add strength, toughness, and resistance to abrasion, or it can act as a mild abrasive and polishing agent (see ABRASIVES).

The paint (qv) and plastics industries are typical of those employing diatomite extensively as fillers and pigment extenders. Diatomite is useful as a filler in paint where it forms a rougher surface, provides a flatting effect, and because of its porous intricate structure, allows faster drying and may improve intercoat adhesion. Also diatomite is extensively used as a polyethylene antiblocking agent where the irregular shapes of the diatom act as a mechanical barrier that prevents sticking of hot polyethylene films during production.

Diatomaceous earth has been used as a chromatographic support since the inception of gas chromatography. A diverse line of products are available. The high surface area and structure of diatomite enables it to carry the liquid phase while an inert surface prevents interference with partitioning. Properly treated diatomite chromatographic supports make use of all its unique properties (10).

Insulation. Diatomite makes an efficient thermal insulator because of its high resistance to heat (fusion point at 1430°C) and its high porosity (see INSULATION, THERMAL). Materials in the form of powders, aggregates, and bricks are most commonly used. At one time, solid bricks were sawed directly from strata in a deposit, dried in a kiln, and then milled to size. In this form, the diatomite can withstand direct service temperatures up to 870°C without undue shrinkage. This type is no longer available, although it was widely used as a brick course in walls, bases, and tops of heated equipment. Diatomite insulating bricks for all temperatures are now formed by adding binder, molding the mixture to sizes and shapes desired, and then firing in a kiln. Diatomite powders and aggregates are often installed loose over tops or in hollow wall spaces of furnaces, kilns, ovens, etc. Calcined aggregates are supplied for mixture with water and Portland cement (qv) for casting bases, doors, baffles, etc, for various types of heated equipment.

Other Uses. There are many miscellaneous uses of diatomite; some are highly specialized and extensive. As a pozzolanic admixture in concrete mixes, diatomite improves the workability of the mix, permitting easier chuting and placement in intricate forms. Diatomite is also used as a catalyst support (see CATALYSIS). Diatomaceous earth powders are used as carriers for insecticides and as fluffing agents for heavier dusts. Certain diatomite powders act as a natural insecticide and are used to protect seeds and stored grain (see INSECT CONTROL TECHNOLOGY). The fertilizer industry uses large quantities of diatomite as an anticaking agent or conditioner, particularly for prilled ammonium nitrate. The diatomite greatly reduces absorption of moisture by the fertilizer, thus preventing caking in the bag and making spreading easier (see FERTILIZERS).

BIBLIOGRAPHY

"Diatomite" in *ECT* 1st ed., Vol. 5, pp. 3–37, by H. Mulray, Sierra Talc & Co.; in *ECT* 2nd ed., Vol. 7, pp. 53–63, by E. L. Neu, Great Lakes Carbon Corp.; in *ECT* 3rd ed., Vol. 7, pp. 603–614, by E. L. Neu and A. F. Alciatore, Grefco, Inc.

1. R. Calvert, *Diatomaceous Earth, American Chemical Society Monograph Series, Chemical Monograph Series*, J. J. Little & Ives, Co., New York, 1930.
2. A. P. Lisitsyn, *Int. Geol. Rev.* **9**, 631 (1967); **9**, 842 (1967); **9**, 980 (1967); **9**, 1114 (1967).

3. S. E. Calvert, *Geol. Soc. Am. Bull.* **77,** 569 (1966).

4. A. B. Cummins, *The Development of Diatomite Filter Aid Filtration and Separation,* Uplands Press, Ltd., Craydon, UK, 1973.

5. W. W. Wornardt, Jr., *Occasional Papers of the California Academy of Sciences, No. 63,* California Academy of Sciences, Los Angeles, 1967.

6. A. Soutar, S. R. Johnson, and T. R. Baumgartner, in C. Isaacs and J. Garison, eds., *The Monterey Formation and Related Siliceous Rocks of California,* Society of Economic Paleontologists and Mineralogists, 1981, p. 123.

7. L. L. Davis, in *Mineral Commodity Summaries, 1991,* U.S. Bureau of Mines, Dept. of the Interior, Washington, D.C., 1991, p. 50.

8. L. Pettifer, *Ind. Minerals* **1** (175), 47 (1982).

9. *Celite Filter Aids for Maximum Clarity at Lowest Cost,* Internal Publication No. FA-84, CELITE Corp., Lompoc, Calif., 1984.

10. *Functional Fillers for Industrial Applications,* Internal Publication No. FF-396, CELITE Corp., Lompoc, Calif., 1984.

11. F. L. Kadey, in S. J. Lefond, ed., *Industrial Minerals and Rocks,* 5th ed., AIME, New York, 1983, p. 677.

12. *ASTM D604-81; D719-86,* American Society for Testing and Materials, Philadelphia, Pa., 1989.

13. U.S. Military Specifications, *MIL-S-15191C, 1986; 52-MA-522a.*

14. TAPPI, *T658 OS-77,* Washington, D.C.

15. U.S. Military Specification, *MIL-F-52637A,* 1978.

16. *U.S. Pharmacopeia, The National Formulary,* USP XXII, NF XVII, 1990, p. 1738.

17. *U.S. Food Chemical Codex,* 3rd ed., 3rd suppl., 1992, p. 99.

18. C. W. Cain, Jr., in J. J. McKetta, ed., *Encyclopedia of Chemical Processing and Design,* Vol. 21, Marcel Dekker, Inc., New York, 1984, p. 348.

19. J. Kiefer, *Brauwelt Int.* **6,** 300 (1991).

20. H. S. Thatcher, *Sugar Filtration—Improved Methods Filtration,* Kieselguhr Co. of America, Lompoc, Calif., 1915.

KENNETH R. ENGH
Celite Corporation

DICARBOXYLIC ACIDS

The diacids are characterized by two carboxylic acid groups attached to a linear or branched hydrocarbon chain. Aliphatic, linear dicarboxylic acids of the general formula $HOOC(CH_2)_nCOOH$, and branched dicarboxylic acids are the subject of this article. The more common aliphatic diacids (oxalic, malonic, succinic, and adipic) as well as the common unsaturated diacids (maleic acid, fumaric acid), the dimer acids (qv), and the aromatic diacids (phthalic acids) are not discussed here (see ADIPIC ACID; MALEIC ANHYDRIDE, MALEIC ACID, AND FUMARIC ACID; MALONIC ACID AND DERIVATIVES; OXALIC ACID; PHTHALIC ACID AND OTHER

BENZENE-POLYCARBOXYLIC ACIDS; SUCCINIC ACID AND SUCCINIC ANHYDRIDE). The bifunctionality of the diacids makes them versatile materials, ideally suited for a variety of condensation polymerization reactions. Several diacids are commercially important chemicals that are produced in multimillion kg quantities and find application in a myriad of uses.

Nomenclature

Unsubstituted aliphatic dicarboxylic acids, $HOOC(CH_2)_nCOOH$, are most often referred to by their trivial names for $n = 2$ to 10 (see Table 1). Higher homologues are named using the IUPAC system by adding the suffix dioic to the parent hydrocarbon. Older literature may refer to compounds using a system whereby the name is formed by adding the suffix dicarboxylic acid to the name of the hydrocarbon skeleton. Note that carbons from the carboxyl groups are not included in formulating the name of the base hydrocarbon in this latter system. By way of illustration, the compound

$$
\begin{array}{cccccccccc}
1 & 2 & 3 & 4 & 5 & 6 & 7 & 8 & 9 & 10 \\
\end{array}
$$

$$HOOC\text{-}CH_2\text{-}CH_2\text{-}CH_2\text{-}CH_2\text{-}CH_2\text{-}CH_2\text{-}CH_2\text{-}CH_2\text{-}COOH$$

$$
\begin{array}{cccccccc}
\alpha & \beta & \tau & \delta & \delta' & \tau' & \beta' & \alpha'
\end{array}
$$

Table 1. Physical Properties[a] of C_2–C_{21} Aliphatic Dicarboxylic Acids

IUPAC name	CAS Registry Number	Common name	Mp, °C	Bp, °C[b]	Water solubility,[c] g/100 mL	Density, g/mL
ethanedioic	[144-62-7]	oxalic	187 (dec)		9.5	
propanedioic	[141-82-2]	malonic	134–136 (dec)		152	1.619
butanedioic	[110-15-6]	succinic	187.6–187.9		8.35	1.572
pentanedioic	[110-94-1]	glutaric	98–99	200[d]	130	1.424
hexanedioic	[124-04-9]	adipic	153.0–153.1	265[e]	3.08[f]	1.345
heptanedioic	[111-16-0]	pimelic	105.7–105.8	272	5.0	1.287
octanedioic	[505-48-6]	suberic	143.0–143.4	279	0.16	1.270
nonanedioic	[123-99-9]	azelaic	107–108	286.5[e]	0.214	1.235
decanedioic	[110-20-6]	sebacic	134.0–134.4	294.5[e]	0.10	1.231
undecanedioic	[1852-04-6]		110.5–112[g]		0.003	
dodecanedioic	[693-23-2]		128.7–129.0	254[h]	0.004	1.16
tridecanedioic	[505-52-2]	brassylic	114			
tetradecanedioic	[821-38-5]		126.5[i]			
pentadecanedioic	[1460-18-0]		114.7[i]			
hexadecanedioic	[505-54-4]	thapsic	125[i]			
heptadecanedioic	[2424-90-0]		117–118[j]			
octadecanedioic	[871-70-5]		124.6–124.8[k]			
nonadecanedioic	[6250-70-0]		118–119.5			
eicosanedioic	[2424-92-2]		124–125[l]			
heneicosanedioic	[505-55-5]	japanic	118–120[i]			

[a]Data from Ref. 1 except as noted.
[b]At 13.3 kPa = 100 mm Hg unless otherwise noted.
[c]At 20–25°C unless otherwise noted.
[d]At 2.7 kPa = 20 mm Hg (2).
[e]Ref. 3.
[f]At 34.1°C.
[g]Ref. 4.
[h]At 2.0 kPa = 15 mm Hg (5).
[i]Ref. 6.
[j]Ref. 7.
[k]Ref. 8.
[l]Ref. 9.

is designated sebacic acid, decanedioic acid, and 1,8-octanedicarboxylic in the respective systems. Locations of substituents are specified by numbers as shown in the illustration or by Greek letters where the carbons adjacent to the carboxyl groups are designated α and α' as illustrated.

Physical Properties

Detailed summaries of physical properties are given (1,2). The diacids are colorless, crystalline solids that melt somewhat higher than monoacids of the same molecular weight. For diacids of even carbon number, melting points decrease sharply for numbers 2–10 and remain relatively constant for numbers 12–20 (see Table 1). There is a marked alternation in melting point and other physical properties with changes in carbon number from even to odd within the series. Odd members exhibit lower melting points, and higher solubility. Theoretical treatments have been developed to correlate these physical properties (6,10). The alternating effects are the result of the inability of odd carbon number compounds to assume an in-plane orientation of both carboxyl groups with respect to the hydrocarbon chain (1). Other properties showing these alternations are decarboxylation temperature and index of refraction. Boiling point, heat of combustion, density, and dielectric constant do not show the alternating effect and vary, predictably, with the ratio of methylene to carboxyl groups within the molecule. Alternating effects can be quite pronounced as illustrated by glutaric acid which melts lower than succinic or adipic (see Table 1). Such alternation persists throughout the series with odd-numbered acids always melting lower than their neighbors; the effect diminishes as the number of carbons increases (Fig. 1). These effects have practical consequences in the selection of material for a given preparation since acid melting point, decarboxylation temperature, and solubility are often key considerations. The effects persist in derivatives based on the diacids, particularly polyamides (qv), polyurethanes, and polyesters (qv) (11–13) (see URETHANE POLYMERS).

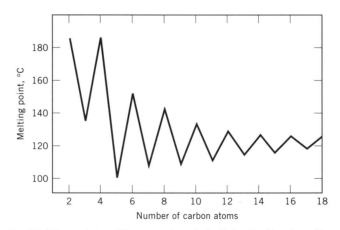

Fig. 1. Melting points of linear saturated aliphatic dicarboxylic acids.

The temperature at which decarboxylation occurs is of particular interest in manufacturing processes based on polymerization in the molten state where reaction temperatures may be near the point at which decomposition of the diacid occurs. Decarboxylation temperatures are tabulated in Table 2 along with molar heats of combustion. The diacids become more heat stable at carbon number four with even-numbered acids always more stable. Thermal decomposition is strongly influenced by trace constituents, surface effects, and other environmental factors; actual stabilities in reaction systems may therefore be lower.

Table 2. Decarboxylation Temperatures and Molar Heats of Combustion of Dicarboxylic Acids

Dicarboxylic acid	Decarboxylation temp[a], °C	Molar heat of combustion[b], kJ/mol[c]
oxalic	166–180	246
malonic	140–160	864
succinic	290–310	1492
glutaric	280–290	2151
adipic	300–320	2800
pimelic	290–310	3464
suberic	340–360	4115
azelaic	320–340	4778
sebacic	350–370	5429
dodecanedioic		6740

[a]Refs. 14–16.
[b]Refs. 17, 18.
[c]To convert J to cal, divide by 4.184.

Order of thermal stability as determined by differential thermal analysis is sebacic (330°C) > azelaic = pimelic (320°C) > suberic = adipic = glutaric (290°C) > succinic (255°C) > oxalic (200°C) > malonic (185°C) (19). This order is somewhat different than that in Table 2, and is the result of differences in test conditions. The energy of activation for decarboxylation has been estimated to be 251 kJ/mol (60 kcal/mol) for higher members of the series and 126 kJ/mol (30 kcal/mol) for malonic acid (1).

Lower members of the series are water soluble; solubility falls off sharply above adipic acid. Alternating effects are again expressed with acids of odd carbon numbers being the most soluble (see Table 1). Dibasic acids are ionized in aqueous solution to varying degree depending upon the proximity of the carboxyl groups within the individual structures. The carboxyl group, being electron-withdrawing, causes the neighboring carboxyl hydrogen to be more readily dissociated. With more than one interspersed methylene group the pK_a approaches that of the monocarboxylic acids of similar molecular weight. Ionization constants are tabulated in Table 3.

Table 3. Ionization Constants[a] of Dicarboxylic Acids

Dicarboxylic acid	K_1	K_2	References
oxalic	5.36×10^{-2}	5.42×10^{-5}	20
malonic	1.42×10^{-3}	2.01×10^{-6}	19
succinic	6.21×10^{-5}	2.31×10^{-6}	21
glutaric	4.58×10^{-5}	3.89×10^{-6}	22
adipic	3.85×10^{-5}	3.89×10^{-6}	23
pimelic[b]	3.19×10^{-5}	3.74×10^{-6}	23
suberic[b]	3.05×10^{-5}	3.85×10^{-6}	23–25
azelaic	2.88×10^{-5}	3.86×10^{-6}	23–25
sebacic	3.1×10^{-5}	3.6×10^{-6}	25
dodecanedioic[c]	2.0×10^{-6}	2.5×10^{-7}	25
tridecanedioic[c]	1.6×10^{-6}	2.9×10^{-7}	25

[a]In water at 25°C unless otherwise noted.
[b]At 18°C.
[c]In ethanol:water = 40:60.

Chemical Properties

The dibasic acids undergo the reactions typical of monocarboxylic acids (see CARBOXYLIC ACIDS). The dibasic acids respond to heat by either losing one carboxyl group, yielding a monocarboxylic acid, or dehydration to a cyclic or polymeric anhydride. A carbon suboxide, having a bis-ketene structure, may be formed under some conditions. Cyclic anhydrides are formed by glutaric acid on heating or dehydrating with acetic anhydride or acetyl chloride. Heating adipic acid results in a high molecular weight polymeric anhydride that can be vacuum distilled to an unstable cyclic anhydride. Polyanhydrides are formed from all of the diacids of six carbons or higher when these are heated with acetic anhydride (26).

Cyclization with loss of one carboxyl takes place in the presence of metal oxides, notably barium and thorium. Thus adipic acid yields cyclopentanone, carbon dioxide, and water (Dieckmann reaction).

Manufacture, Preparation, and Processes

Glutaric Acid. Until 1990–1991 glutaric acid was available commercially from Du Pont as a by-product in the production of adipic acid. It is no longer available, but Du Pont produces dimethyl glutarate and mixtures of dimethyl succinate and dimethyl glutarate, as well as mixtures of dimethyl glutarate and dimethyl adipate; these esters are known commercially as DBE (dibasic esters). Du Pont fractionates DBE into 99% dimethyl glutarate (DBE-5) and 98% dimethyl adipate (DBE-6) as well as mixtures of the two (DBE-2 and DBE-3). Also available is DBE-9, which contains 66% dimethyl glutarate and 33% dimethyl succinate. Physical properties of these methyl esters are published in Du Pont's technical bulletin on dibasic esters (DBE) (27).

Several procedures for making glutaric acid have been described in *Organic Syntheses* starting with trimethylene cyanide (28), methylene bis (malonic acid)

(29), γ-butyrolactone (30), and dihydropyran (31). Oxidation of cyclopentane with air at 140° and 2.7 MPa (400 psi) gives cyclopentanone and cyclopentanol, which when oxidized further with nitric acid at 65–75° gives mixtures of glutaric acid and succinic acid (32).

Pimelic Acid. This acid is manufactured by Tateyama Chemical Company in Japan in quantities of about 1000–2000 kg/yr, and by Heinrich Mock Nachf in Germany. The method or process they are using has not been disclosed. Pimelic acid is available in small quantities with purities of 98% from laboratory chemical supply companies. The preparation of pimelic acid has been described in *Organic Syntheses*; cyclohexanone condenses with diethyl oxalate, followed by decarboxylation to ethyl 2-keto-hexahydrobenzoate, and then cleavage of the β-keto ester with strong alkali (33). It has also been made from salicylic acid by reduction and subsequent cleavage with sodium in isoamyl alcohol (34). A potential commercial method has been described in which ε-caprolactone reacts with carbon monoxide and water in the presence of a catalyst such as palladium, and a co-catalyst such as hydriodic acid. Pimelic acid was obtained in yields as high as 46.7% (35). Other synthetic routes to pimelic acid have been described (36).

Suberic Acid. This acid is not produced commercially at this time. However, small quantities of high purity (98%) can be obtained from chemical supply houses. If a demand developed for suberic acid, the most economical method for its preparation would probably be based on one analogous to that developed for adipic and dodecanedioic acids; air oxidation of cyclooctane to a mixture of cyclooctanone and cyclooctanol. This mixture is then further oxidized with nitric acid to give suberic acid (37).

Another method that appears to have commercial potential is the ozonolysis of cyclooctene. Ozonolysis is carried out using a short chain carboxylic acid, preferably propanoic acid, as solvent. The resultant mixture is thermally decomposed in the presence of oxygen at about 100°C to give suberic acid in about 60–78% yield (38–40). Carboxylation of 1,6-hexanediol using nickel carbonyl as catalyst is reported to give suberic acid in 90% yield (41).

There are several laboratory methods useful for the preparation of suberic acid. One starting material is 1,6-hexanediol which can be converted to the dibromide with HBr. Reaction of the dibromide with NaCN gives the dinitrile which can be hydrolyzed to suberic acid. The overall yield is 76% (42). Another laboratory method is the condensation of 1,3-cyclohexanedione with ethyl bromoacetate followed by reductive cleavage to give suberic acid in 50% yield (43).

Azelaic Acid. This acid is produced by the Emery Group of Henkel Corporation in Cincinnati, Ohio in multimillion kg quantities. The process that is currently used is based on the ozonolysis of oleic acid (from grease or tallow) followed by the decomposition of the ozonide with oxygen. Oleic acid [*112-80-1*] and pelargonic acid [*112-05-0*] are fed into an ozone absorber countercurrent to a continuous flow of oxygen gas containing about 2% ozone. The pelargonic acid serves as a solvent or diluent to help moderate the reaction. Since the reaction is highly exothermic the ozone absorber is cooled and usually maintained at a temperature of 25–45°C. The reaction mixture is then fed into reactors maintained at about 100°C and sparged with oxygen gas where the ozonides are decomposed and oxidized rapidly. Manganese salts are used to catalyze the oxidation of any aldehydes that were formed to acids. The product at this stage contains pelargonic, azelaic, and palmitic and stearic acids that were present in the feed and high molecular weight materials such as esters and dimer that were formed during ozonization.

The mixed oxidation products are fed to a still where the pelargonic and other low boiling acids are removed as overhead while the heavy material, esters and dimer acids, are removed as residue. The side-stream contains predominately azelaic acid along with minor amounts of other dibasic acids and palmitic and stearic acids. The side-stream is then washed with hot water that dissolves the azelaic acid, and separation can then be made from the water-insoluble acids, palmitic and stearic acids. Water is removed from the aqueous solution by evaporators or through crystallization (44,45).

Chromic acid was used by Emery to oxidize oleic acid on a commercial scale in the late 1940s and early 1950s until it was replaced by the ozone route. The process was relatively expensive in spite of the fact that the spent chromic acid solutions were regenerated in electrolytic cells. Pelargonic acid was removed from the oxidation mixture by distillation at reduced pressure. Azelaic acid was recovered from the residue by recrystallization from hot water. The resulting azelaic acid contained lower molecular weight dibasic acids that resulted from degradation during the oxidation reaction (46,47).

A good laboratory method for preparing azelaic acid is the permanganate oxidation of oleic acid (48). Less degradation occurs as only small amounts of suberic acid are formed. When the oxidation is carried out in acetone, an azelaic acid yield of 83% is reported (49). Nitric acid oxidations of oleic acid have been studied by a number of investigators (50–52). The main disadvantage of using nitric acid is that it leads to degradation of azelaic acid, resulting in relatively large amounts of suberic acid being formed. Typically, oxidation products contain 65% azelaic acid and 35% suberic acid. Like nitric acid oxidation, air oxidation has been studied extensively, and the azelaic acid thus obtained contains other dibasic acids, mainly suberic acid. Generally, a cobalt catalyst is used and yields are usually no higher than 15% (53,54).

A U.S. patent describes the reaction of commercial oleic acid with hydrogen peroxide in acetic acid followed by air oxidation using a heavy metal compound and an inorganic bromine or chlorine compound to catalyze the oxidation. Excellent yields of dibasic acids are obtained (up to 99%) containing up to 72% azelaic acid (55).

A novel route to azelaic acid is based on butadiene. Butadiene is dimerized to 1,5-cyclooctadiene, which is carbonylated to the monoester in the presence of an alcohol. Hydrolysis of this ester followed by a caustic cleavage step produces azelaic acid in both high yield and purity (56).

Sebacic Acid.   This acid is produced commercially by Union Camp in Dover, Ohio, by Hokoku Oil Company in Japan, and by a state enterprise in the People's Republic of China (57). The process used in each case is based on the caustic oxidation of castor oil or ricinoleic acid [141-22-0] in either a batch or continuous process. The castor oil or ricinoleic acid and caustic are fed to a reactor (usually Monel or nickel) at a temperature of 180–270°C where the ricinoleic acid undergoes a series of reactions with evolution of hydrogen to give disodium sebacate and capryl alcohol (58). When the reaction is complete, the soaps are dissolved in water and acidified to a pH of about 6. At this pH, the soaps of the monobasic acids (C-16 to C-20 plus dimer acids) are converted to free acids that are insoluble in water. The disodium sebacate is only partially neutralized to the half acid salt which is water soluble. The oil and aqueous layers are separated. The aqueous layer containing the half salt is acidulated to a pH of about 2 causing the resulting sebacic acid to precipitate from the solution. It is then filtered, water washed, and finally dried (59).

A number of process improvements have been described, and include the use of white mineral oil having a boiling range of 300–400°C (60) or the use of a mixture of cresols (61). These materials act to reduce the reaction mixture's viscosity, thus improving mixing. Higher sebacic acid yields are claimed by the use of catalysts such as barium salts (62), cadmium salts (63), lead oxide, and salts (64).

An electrooxidation process was developed by Asahi Chemical Industry in Japan, and was also piloted by BASF in Germany. It produces high purity sebacic acid from readily available adipic acid. The process consists of 3 steps. Adipic acid is partially esterified to the monomethyl adipate. Electrolysis of the potassium salt of monomethyl adipate in a mixture of methanol and water gives dimethyl sebacate. The last step is the hydrolysis of dimethyl sebacate to sebacic acid. Overall yields are reported to be about 85% (65).

Another alternative method to produce sebacic acid involves a four-step process. First, butadiene [106-99-0] is oxycarbonylated to methyl pentadienoate which is then dimerized, using a palladium catalyst, to give a triply unsaturated dimethyl sebacate intermediate. This unsaturated intermediate is hydrogenated to dimethyl sebacate which can be hydrolyzed to sebacic acid. Small amounts of branched chain isomers are removed through solvent crystallizations giving sebacic acid purities of greater than 98% (66).

The electrochemical conversions of conjugated dienes into alkadienedioic acid have been known for some time. Butadiene has been converted into diethyl-3,7-decadiene-1,10,dioate by electrolysis in a methanol–water solvent (67). An improvement described in the patent literature (68) uses an anhydrous aprotic solvent and an electrolyte along with essentially equimolar amounts of carbon dioxide and butadiene; a mixture of decadienedioic acids is formed. This material can be hydrogenated to give sebacic acid.

In another method based on butadiene, it is dimerized in the presence of sodium to form an isomeric mixture of disodiooctadiene (69). Carbonation of the mixture using dry ice gives the unsaturated acids, 3,7-decadienedioic acid, 2-vinyl-5-octenoic acid, and 2,5-divinyladipic acid in a ratio of about 3.5:5:1, respectively. Hydrogenation of this mixture gives the saturated dibasic acid product consisting of approximately 51% 2-ethylsuberic, 38% sebacic, and 11% 2,5-diethyladipic. Sebacic acid is separated by fractional crystallization in about 35% yield. By-product acids, consisting of 72–80% 2-ethylsuberic, 12–18% diethyladipic, and 5–10% sebacic acids are obtained and referred to as isosebasic acid (70).

Dodecanedioic Acid. Dodecanedioic acid (DDDA) is produced commercially by Du Pont in Victoria, Texas, and by Chemische Werke Hüls in Germany. The starting material is butadiene which is converted to cyclododecatriene using a nickel catalyst. Hydrogenation of the triene gives cyclododecane, which is air oxidized to give cyclododecanone and cyclododecanol. Oxidation of this mixture with nitric acid gives dodecanedioic acid (71).

Other methods have been described to produce dodecanedioic acid. Cyclododecene is prepared from cyclododecatriene by partial hydrogenation. Ozonolysis of the cyclododecene followed by oxidation of the intermediate ozonides gives dodecanedioic acid (72). Hydrogenation of ricinoleic acid gives 12-hydroxystearic acid, which upon treatment with caustic at high temperatures, 325–330°C, gives a mixture of undecanedioic and dodecanedioic acids. The use of a catalyst such as cadmium oxide increases the yield of dibasic acids to about 51% of theoretical. The composition of the mixed acids is about 75% C-11 and 25% C-12 dibasic acids (73). Reaction of undecylenic acid with carbon monoxide using a triphenylphosphine–rhodium complex as catalyst gives 11-formylundecanoic acid, which, upon reaction with oxygen in the presence of Co(II) salts, gives 1,12-dodecanedioic acid in 70% yield (74).

Another process for the production of dodecanedioic acid is by oxidation of cyclododecene using a two-phase system in which ruthenium tetroxide serves as the oxidizing agent in the organic phase, and is regenerated in the second phase, an aqueous phase containing cerium(IV) ions (75).

Brassylic Acid. This acid is commercially available from Nippon Mining Company (Tokyo, Japan). It is made by a fermentation process (76). Several years ago, Emery Group, Henkel Corp. (Cincinnati, Ohio) produced brassylic acid via ozonization of erucic acid primarily for captive use in making dimethyl brassylate and ethylene brassylate. A pilot-scale preparation based on ozonization of erucic acid has been described in which brassylic acid yields of 72–82% were obtained in purities of 92–95%. Recrystallization from toluene gave purities of 99% (77).

C-19 Dicarboxylic Acids. The C-19 dicarboxylic acids are generally mixtures of isomers formed by the reaction of carbon monoxide on oleic acid. Since the reaction produces a mixture of isomers, no single chemical name can be used to describe them. Names that have been used include 2-nonyldecanedioic acid, 2-octylundecanedioic acid, 1,8-(9)-heptadecanedicarboxylic acid, and 9-(10)-carboxystearic acid. The name 9-(10)-carboxystearic acid can be used correctly if the product is made with no double bond isomerization (rhodium triphenylphosphine catalyst system).

There are currently no commercial producers of C-19 dicarboxylic acids. During the 1970s BASF and Union Camp Corporation offered developmental products, but they were never commercialized (78). The Northern Regional Research Laboratory (NRRL) carried out extensive studies on preparing C-19 dicarboxylic acids via hydroformylation using both cobalt catalyst and rhodium complexes as catalysts (78). In addition, the NRRL developed a simplified method to prepare 9-(10)-carboxystearic acid in high yields using a palladium catalyst (79).

C-19 dicarboxylic acid can be made from oleic acid or derivatives and carbon monoxide by hydroformylation, hydrocarboxylation, or carbonylation. In hydroformylation, ie, the Oxo reaction or Roelen reaction, the catalyst is usually cobalt carbonyl or a rhodium complex (see OXO PROCESS). When using a cobalt catalyst a mixture of isomeric C-19 compounds results due to isomerization of the double bond prior to carbon monoxide addition (80).

$$—CH{=}CH— \ + \ CO \ + \ H_2 \ \xrightarrow{\text{catalyst}} \ — \underset{\underset{CHO}{|}}{CH}CH_2—$$

Hydroformylation catalyzed by rhodium triphenylphospine results in only the 9 and 10 isomers in approximately equal amounts (79). A study of recycling the rhodium catalyst and a cost estimate for a batch process have been made (81).

In hydrocarboxylation, the Reppe reaction, the catalyst can be nickel or cobalt carbonyl or a palladium complex where R = H or alkyl.

$$-CH{=}CH- \ + \ CO \ + \ ROH \ \xrightarrow{\text{catalyst}} \ - \ \underset{\underset{COOR}{|}}{CH}CH_2-$$

The nickel or cobalt catalyst causes isomerization of the double bond resulting in a mixture of C-19 isomers. The palladium complex catalyst produces only the 9-(10)-carboxystearic acid. The advantage of the hydrocarboxylation over the hydroformylation reaction is it produces the carboxylic acids in a single step and obviates the oxidation of the aldehydes produced by hydroformylation.

Carbonylation, or the Koch reaction, can be represented by the same equation as for hydrocarboxylation. The catalyst is H_2SO_4. A mixture of C-19 dicarboxylic acids results due to extensive isomerization of the double bond. Methyl-branched isomers are formed by rearrangement of the intermediate carbonium ions. Reaction of oleic acid with carbon monoxide at 4.6 MPa (45 atm) using 97% sulfuric acid gives an 83% yield of the C-19 dicarboxylic acid (82). Further optimization of the reaction has been reported along with physical data of the various C-19 dibasic acids produced. The mixture of C-19 acids was found to contain approximately 25% secondary carboxyl and 75% tertiary carboxyl groups. As expected, the tertiary carboxyl was found to be very difficult to esterify (80,83).

C-20 Dicarboxylic Acids. These acids have been prepared from cyclohexanone via conversion to cyclohexanone peroxide followed by decomposition by ferrous ions in the presence of butadiene (84–87). Okamura Oil Mill (Japan) produces a series of commercial acids based on a modification of this reaction. For example, Okamura's modifications of the reaction results in the following composition of the reaction product: C-16 (linear) 4–9%, C-16 (branched) 2–4%, C-20 (linear) 35–52%, and C-20 (branched) 30–40%. Unsaturated methyl esters are first formed that are hydrogenated and then hydrolyzed to obtain the mixed acids. Relatively pure fractions of C-16 and C-20, both linear and branched, are obtained after solvent crystallization and fractionation (88).

C-21 Dicarboxylic Acids. C-21 dicarboxylic acids are a mixture of predominately 5-(6)-carboxy-4-hexyl-2-cyclohexene-1-octanoic acid, 5-isomer [42763-47-3] and 6-isomer [42763-46-2]. C-21 dicarboxylic acids were first described at the Northern Regional Research Laboratory in Peoria, Illinois (89).

C-21 dicarboxylic acids are produced by Westvaco Corporation in Charleston, South Carolina in multimillion kg quantities. The process involves reaction of tall oil fatty acids (TOFA) (containing about 50% oleic acid and 50% linoleic acid) with acrylic acid [79-10-7] and iodine at 220–250°C for about 2 hours (90). A yield of

C-21 as high as 42% was reported. The function of the iodine is apparently to conjugate the double bond in linoleic acid, after which the acrylic acid adds via a Diels-Alder type reaction to form the cyclic reaction product. Other catalysts have been described and include clay (91), palladium, and sulfur dioxide (92). After the reaction is complete, the unreacted oleic acid is removed by distillation, and the crude C-21 diacid can be further purified by thin film distillation or molecular distillation.

A 22 carbon atom adduct has been prepared in a similar reaction to that described above by replacing the acrylic acid with methacrylic acid. Somewhat lower yields are obtained when using methacrylic acid [79-41-4] vs acrylic acid (93).

Dicarboxylic Acids via Microorganisms

During the 1980s a number of patents were issued describing the preparation of dicarboxylic acids or esters using microorganisms. The α, ω-n-alkanedioic acids that have been prepared generally have 5–25 carbons. One of the first methods described the preparation of dimethyl 1,16-hexadecanedioate using a nutrient solution, n-hexadecyl bromide, and certain strains of *Torulopsis* (94). Other methods have been described to give dibasic acid of 8–22 carbons using n-alkanes or n-alcohols and various organisms (95–98). One particular yeast converts C12–18 n-alkanes not only to dicarboxylic acids but also to 3-hydroxydicarboxylic acids as well (99). The unsaturated dicarboxylic acid itaconic acid (2-methylenebutanedioic acid [97-65-4] is produced by fermentation (100). Finally, unsaturated dibasic acids having 14–22 carbons have been prepared from the corresponding unsaturated fatty acids and certain microorganisms. For example, 9-*trans*-octadecenedioic acid has been prepared from elaidic acid; 6,9-octadecadienedioic acid from linoleic acid, and 5-docosenedioic acid from erucic acid (101).

Derivatives and Uses

Diacids, owing to their ready incorporation into polymers, are components in a wide variety of materials (100–103). The diacids are important industrial intermediates for the manufacture of diesters, polyesters, and polyamides. These derivatives find application as plasticizing agents, lubricants, heat transfer fluids, dielectric fluids, fibers, copolymers, inks and coatings resins, surfactants, fungicides, insecticides, hot-melt coatings, and adhesives. Of the higher diacids, azelaic, sebacic, and dodecanoic find the greatest application. Derivatives of glutaric and C-21 diacids also enjoy significant commercial applications.

Diesters. Many of the diester derivatives are commercially important. The diesters are important plasticizers, polymer intermediates, and synthetic lubricants. The diesters of azelaic and sebacic acids are useful as monomeric plasticizing agents; these perform well at low temperatures and are less water-soluble and less volatile than are diesters of adipic acid. Azelate diesters, eg, di-n-hexyl, di(2-ethylhexyl), and dibutyl, are useful plasticizing agents for poly(vinyl chloride), synthetic rubbers, nitrocellulose, and other derivatized celluloses (104). The dihexyl azelates and dibutyl sebacate are sanctioned by the U.S. Food and Drug Administration for use in poly(vinyl chloride) films and in other plastics with direct contact to food. The di(2-ethylhexyl) and dibenzyl sebacates are also valuable plasticizers. Monomeric plasticizers have also been prepared from other diacids, notably dodecanedioic, brassylic, and 8-ethylhexadecanedioic (88), but these have not enjoyed the commercialization of the sebacic and azelaic diesters.

The dioctyl, didecyl, and di-(tridecyl) esters of azelaic, sebacic, dodecanedioic, and brassylic acids serve as synthetic lubricants that offer some advantages over petroleum-based lubricants. The dihexyl and dioctyl esters of 8-ethylhexadecanedioic acid also are useful lubricants (88). The viscosity/temperature relationships, low viscosity rise upon oxidation, high flash points, low peroxide value, good additive response, low coking characteristics, and overall excellent lubricity make these diesters desirable for demanding engine and equipment environments (104).

Other applications for the diesters of the diacids are known. Du Pont's Dibasic Ester (DBE) is an effective industrial cleaning solvent and paint stripper. DBE is marketed as an environmentally superior substitute to the chlorinated solvents (105).

Polyesters.   Azelaic and sebacic acids are commonly used for polyester applications. Typical glycols that react with the diacids to give polyesters include 1,2-propylene glycol, 1,3-butylene glycol, and 1,4 butanediol. Aliphatic polyesters, unsaturated polyesters, and copolyesters find application in fibers, films, casting and potting resins, lubricants, resins, adhesives, and laminates. The plasticizers offer greater resistance to solvents, lower volatility, and less migration, but are generally less effective and have poorer low temperature flex characteristics than those of the corresponding aliphatic monomerics. Polyester properties may be tuned by judicious choice of monofunctional additive, either carboxylic acid or alcohol. Diacid-based polyester fibers and films exhibit excellent low temperature flexibility and high tensile strength. The alkyd resins (qv) derived from sebacic and azelaic acids impart flexibility and are used as plasticizers (106).

Polyamides.   The production of aliphatic polyamides, or nylons, consumes a large portion of the total production of the diacids. These polyamides find application in apparel and carpet fibers, engineering plastics, nylon copolymers for monofilament, wire-coating, and molding resin applications. An interesting potential application for nylon-5,7, prepared from pimelic acid and 1,5-pentanediamine, is as a conducting polymer (107). The polyamide produced from suberic acid and 1,4-cyclohexanebis(methylamine) has an exceptionally high melting point (295°C) that enables high melt processing temperatures to be employed without concurrent thermal decomposition; it has properties well-suited for fiber, film, and molding plastic applications (108). The 6,9, 6,10, and 6,12 nylons, derived from azelaic, sebacic, and dodecanedioic acids respectively, find application in engineering plastics, bristles, and fibers. These nylons are more moisture-resistant than is the adipic based nylon-6,6. The alicyclic nylon, Du Pont's Quiana, derived from dodecanedioic acid and bis(4-aminocyclohexyl)methane, was introduced as a replacement for silk in 1968 (109). A recent patent describes the incorporation of dodecanedioic acid into nylon-6,6 to improve the fibers' dyeability (110). Brassylic acid based nylons 13, 6,13, and 13,13 are very moisture resistant. Azelaic and sebacic acids find application in dimer acid based polyamides and contribute to higher tensile strength and higher melting point resins.

Other Polymeric Derivatives. Hydroxyl-terminated polyesters, derived from reaction of azelaic, sebacic, or dodecanedioic acids with glycols, may be used in polyurethanes. These polymers offer several advantages over adipic acid based materials, including excellent low-temperature properties, water resistance, tear strength, and good elongation. These polymers may be utilized in specialty films, fabric coatings, and Spandex fibers (104). The same diacids may be used as modifiers in poly(ethylene terephthalate)/glycol copolyesters to contribute improved flexibility, melting point, and dyeability. Azelaic, sebacic, dodecanedioic, and brassylic acids may be used in copolyetheresteramides (111). Two patents describe additional applications for the C-9–C-40 diacids for the preparation of polyester carbonates (112), and the copolymerization of epoxides and carbon dioxide by reaction of either glutaric or adipic acids with zinc oxide (113).

Miscellaneous Derivatives. Pimelic acid is used as an intermediate in some pharmaceuticals and in aroma chemicals; ethylene brassylate is a synthetic musk (114). Salts of the diacids have shown utility as surfactants and as corrosion inhibitors. The alkaline, ammonium, or organoamine salts of glutaric acid (115) or C-5–C-16 diacids (116) are useful as noncorrosive components for antifreeze formulations, as are methylene azelaic acid and its alkali metal salt (117). Salts derived from C-21 diacids are used primarily as surfactants and find application in detergents, fabric softeners, metal working fluids, and lubricants (118). The salts of the unsaturated C-20 diacid also exhibit anticorrosion properties, and the sodium salts of the branched C-20 diacids have the ability to complex heavy metals from dilute aqueous solutions (88).

Table 4. Dicarboxylic Acid Prices, $/kg

Acid	Grade	Mode of shipment	1979	1991
adipic	resin grade	bags or hopper cars	0.91	1.43
pimelic			132	575
suberic			65	103
azelaic	E1110	multiwall	1.85	3.48
	E1144	bags		3.74
sebacic	purified	multiwall	3.48	4.49
	CP	bags or		4.51
	nylon	super-sacks		4.51
dodecanedioic		multiwall bags	2.65	4.36
eicosanedioic	ULB-20	steel drums		8.79
(C_{20} diacids)	SB20	steel drums		9.66
	SL20	poly-lined paper bags		12.41
C_{21} diacids	D1550	lined steel		1.54
	D1575	drums or tank trunks		3.74

Economic Aspects

The prices and mode of shipment for the various commercial grades of diacids are provided in Table 4. The price of adipic acid is included for comparison. In addition to these diacids, undecanedioic, brassylic, tetradecanedioic, hexadecanedioic, docosanedioic, and tetracosanedioic acids are available, expensive, and in limited quantity from research chemical supply houses.

Azelaic acid is available from Henkel's Emery Group in two grades; E1110 and E1144 are 80% and 90% azelaic acid, respectively. Union Camp Corporation offers three grades of sebacic acid: purified, CP, and nylon grades. The nylon grade is >95% sebacic acid with low ash content and low color. Oakamura Oil Mill Ltd.'s C-20 diacids are each unique products. ULB-20 is a mixture of straight-chain and branched unsaturated acids. SL-20 is 85–90% eicosanedioic acid. SB-20 is 80–90% 8-ethyloctadecanedioic acid. Westvaco's C-21 Diacid is offered in two grades. Diacid 1550 is darker in color and lower in acid value than Diacid 1575, with diacid contents of 88% and >97% respectively.

Health and Safety Factors

The acute oral toxicities of the diacids and some common derivatives are provided in Table 5. In general, the higher diacids are essentially nontoxic. There are no indications that the dicarboxylic acids detailed here are carcinogenic or teratogenic in animals or humans. It is generally recognized that these diacids are ocular irritants and that the inhalation of the dust of these diacids is irritating to the mucous membranes and the respiratory tract.

The water solubility of glutaric acid fosters its toxicity. Glutaric acid is a known nephrotoxin. Renal failure has been documented in rabbits administered sodium glutarate subcutaneously (124). Dibasic ester (Du Pont), which contains

Table 5. Acute Toxicities of Diacids and Derivatives

Compound	Oral LD$_{50}$, mg/kg		Dermal LD$_{50}$, mg/kg	
	Species	Dose	Species	Dose
glutaric	mouse	6000[a]		
dimethyl glutarate	rat	8191[b]	rabbit	>2250[b]
pimelic	rat	7000[c]		
azelaic	rat	>10000[d]	rat	>10000[d]
dihexyl azelate	rat	16000[e]		
sebacic	mouse	6000[f]		
diethyl sebacate	rat	14500[g]	guinea pig	7320[g]
dodecanedioic	rat	17000[h,i]	rabbit	>6000[i]
Diacid 1550 (C21)	rat	6176[j]		

[a]Ref. 119, p. 1478.   [d]Ref. 122.   [g]Ref. 121, p. 2337.   [j]Ref. 119.
[b]Ref. 120.   [e]Ref. 119, p. 334.   [h]Average lethal dose.
[c]Ref. 121, p. 4937.   [f]Ref. 119, p. 2338.   [i]Ref. 123.

primarily dimethyl glutarate, has low acute toxicity by inhalation and by inges-tion, and is moderately toxic via dermal absorption. The acid is both a dermal and ocular irritant of humans. The ester is a severe skin irritant and may cause a rash in humans (120).

The sodium salts of suberic and azelaic acids are mildly nephrotoxic to rab-bits when administered subcutaneously (125). Azelaic acid is both a dermal and ocular irritant, and is a sensitizing agent to guinea pigs (122). Interestingly, nei-ther dermal absorption or irritation are reported for guinea pigs exposed to pimelic acid (121). Sebacic and dodecanedioic acids are not dermal irritants, but are mild ocular irritants. Long-term exposure in white rats to sebacic acid has produced irritation of the respiratory tract and alteration of kidney and liver functions (126).

Of the higher diacids, the alicyclic, unsaturated Diacid 1550 (the Westvaco C-21 diacid) is significantly water soluble and is moderately irritating both der-mally and ocularly. The corresponding dipotassium salt, Diacid H-240, is sub-stantially more irritating (127).

Environmental Effects. In general, the higher diacids do not pose substan-tial environmental risk; however, releases of significant quantities into surface or ground waters may be reportable under the Clean Water Act. The low biotoxicity of the higher diacids results, in part, from their limited water solubility. Glutaric acid is significantly more water soluble than the other diacids described herein, and the aquatic biotoxicity of glutaric acid and dimethyl glutarate is established. This acid is toxic to both protozoa and fish, and its dimethyl ester is toxic to both aquatic invertebrates and to fish (128,129). Westvaco's Diacid 1550 (C-21) is rela-tively more water soluble than the higher, linear diacids, and this diacid is mildly toxic to aquatic invertebrates, fish, and algae (130).

BIBLIOGRAPHY

"Acids, Dicarboxylic" in *ECT* 1st ed., Vol. 1, pp. 152–157, by C. J. Knuth and P. F. Bruins, Polytechnic Institute of Brooklyn, and R. R. Umbdenstock, Chas. Pfizer and Co., Inc.; in *ECT* 2nd ed., Vol. 1, pp. 240–254, by W. M. Muir, Harris Research Laboratories, Inc.; "Dicarboxylic Acids" in *ECT* 3rd ed., Vol. 7, pp. 614–628, by Paul Morgan, Consultant.

1. E. H. Pryde and J. C. Cowan in J. K. Stille and T. W. Campbell, eds., *Condensation Monomers*, Wiley-Interscience, New York, 1972, pp. 1–153.
2. J. A. Dean, ed., *Lange's Handbook of Chemistry*, 11th ed., McGraw-Hill Book Co., Inc., New York, 1973.
3. F. Krafft and H. Noerdlinger, *Chem. Ber.* **22,** 816 (1889).
4. L. J. Durham, D. J. McLeod, and J. Cason, *Organic Synthesis Collective Volumes*, Vol. 4, Wiley, New York, 1963, p. 510.
5. H. Noerdlinger, *Chem. Ber.* **23,** 2356 (1890).
6. J. G. Erickson, *J. Am. Chem. Soc.* **71,** 307 (1949).
7. C. Hell and C. Jordanoff, *Chem. Ber.* **24,** 991 (1891).
8. L. Ruzicka, Pl. A. Plattner, and W. Widmer, *Helv. Chim. Acta.* **25,** 1086 (1942).
9. *Ibid.*, p. 604.
10. D. E. F. Armstead, *Sch. Sci. Rev.* **55,** 416 (1973).
11. R. Hill, ed., *Fibres from Synthetic Polymers*, Elsevier Science Publishing Co., Inc., New York, 1953, pp. 135, 164, 311.

12. P. W. Morgan, *Condensation Polymers by Interfacial and Solutions Methods*, Wiley-Interscience, New York, 1965, pp. 244, 382.

13. P.W. Morgan and S. L. Kwolek, *Macromolecules* **8**, 104 (1975).

14. V. V. Korshak and S. V. Rogozhin, *Izv. Akad. Nauk SSSR Otd. Khim, Nauk.*, 531 (1952).

15. V. V. Korshak and S. V. Rogozhin, *Dokl. Akad. Nauk. SSSR* **76**, 539 (1951).

16. W. Sweeny and J. Zimmerman in N. M. Bikales, ed., *Encyclopedia of Polymer Science and Technology*, Vol. 10, Wiley-Interscience, New York, 1969, pp. 483–507.

17. R. C. Wilhoit and D. Shiao, *J. Chem. Eng. Data* **9**, 595 (1964).

18. P. E. Verkade, H. Hartman, and J. Coops, *Recl. Trav. Chim. Pays-Bas* **45**, 380 (1926).

19. S. N. Das and D. L. G. Ives, *Proc. Chem. Soc.*, 373 (1961); W. J. Hamer, J. O. Burton, and S. F. Acree, *J. Res. Natl. Bur. Stand.* **24**, 269 (1940).

20. L. S. Darken, *J. Am. Chem. Soc.* **63**, 1007 (1941); G. D. Pinching and R. G. Bates, *J. Res. Natl. Bur. Stand.* **40**, 405 (1948).

21. G. D. Pinching and R. G. Bates, *J. Res. Natl. Bur. Stand.* **45**, 322 and 444 (1950).

22. I. Jones and F. G. Soper, *J. Chem. Soc.*, 133 (1936).

23. B. Adell, *Z. Phys. Chem. (Leipzig)* **185**, 161 (1939).

24. G. Kortum, W. Vogel, and K. Andrussow, *Pure Appl. Chem.* **1**, 190 (1960).

25. G. Bonhomme, *Bull. Soc. Chim. Fr.*, 60 (1968); *Chem. Abstr.* **68**, 117603 (1968); **82**, 72408 (1975).

26. F. L. Breusch and E. Ulusoy, *Fette, Seifen, Anstrichm.* **66**, 739 (1964).

27. *Dibasic Esters*, technical bulletin, Du Pont Chemicals, Wilmington, Del., 1991.

28. C. S. Marvel and W. F. Tuley, in Ref. 4, Vol. 1, rev. ed., 1946, p. 289.

29. T. J. Otterbocher, in Ref. 4, Vol. 1, 2nd ed., 1946, p. 290.

30. G. Paris, L. Berlinguet, and R. Gaudry, in Ref. 4, Vol. 4, 1963, p. 496.

31. J. English, Jr. and J. E. Dayan, in Ref. 4, Vol. 4, 1963, p. 499.

32. U.S. Pat. 2,452,741 (Nov. 2, 1948), H. W. Fleming (to Phillips Petroleum Co.).

33. H. R. Snyder, C. A. Brooks, and S. H. Shapino, in Ref. 4, Vol. 2, rev. ed., 1943, p. 531.

34. A. Muller, in Ref. 4, Vol. 2, rev. ed., 1943, p. 535.

35. U.S. Pat. 4,888,443 (Dec. 19, 1989), J. H. Murib and J. H. Katy (to National Distillers and Chemical Corp.).

36. Ref. 1, pp. 56–57.

37. Belgian Pat. 660,522 (Sept. 2, 1965), R. B. Judge, M. F. Levy, and J. H. Quinn (to Wallace and Tiernan).

38. U.S. Pat. 3,219,675 (Nov. 23, 1965), R. Seekerchen (to Columbia Carbon Co.).

39. U.S. Pat. 3,441,604 (Apr. 29, 1969), E. K. Baylis, W. Pickles, and K. D. Sparrow (to Gergy Chemical Corp.).

40. U.S. Pat. 3,280,183 (Oct. 18, 1966), A. Maggiolo (to Wallace and Tiernan, Inc.).

41. W. Reppe and co-workers, *Justus Liebigs Ann. Chem.* **560**, 1 (1948).

42. K. E. Miller and co-workers, *J. Chem. Eng. Data* **9**, 227 (1964).

43. H. Stetter in W. Forest, ed., *Newer Methods of Preparative Organic Chemistry*, Vol. 2, Academic Press, Inc., New York, 1963, p. 79.

44. U.S. Pat, 2,813,113 (Nov. 12, 1957), C. G. Goebel, A. C. Brown, H. F. Oehlschlaeger, and R. P. Rolfes (to Emery Industries, Inc.).

45. U.S. Pat. 3,402,108 (Sept. 17, 1968), H. F. Ohlschlaeger and H. G. Rodenberg (to Emery Industries, Inc.).

46. U.S. Pat. 2,450,858 (Oct. 5, 1948), J. D. Fitzpatrick and L. D. Meyers (to Emery Industries, Inc.).

47. U.S. Pat. 2,389,191 (Nov. 20, 1945), J. D. Fitzpatrick and L. D. Meyers (to Emery Industries, Inc.).

48. J. W. Hall and W. L. McEwen, in Ref. 4, Vol. 2, 1943, p. 53.

49. E. F. Armstrong and T. P. Hilditch, *J. Soc. Chem. Ind. Trans.* **44**, 43T (1925).

50. U.S. Pat. 2,203,680 (June 11, 1940), E. K. Ellingboe (to E. I. du Pont de Nemours & Co., Inc.).
51. U.S. Pat. 2,365,290 (Dec. 19, 1944), F. J. Sprules and R. Griffith (to Napco Chemical Co.).
52. U.S. Pat. 3,021,348 (Feb. 13, 1963), V. P. Kuceski (to C. P. Hall Co.).
53. U.S. Pat. 2,292,950 (Aug. 11, 1942), D. J. Loden and P. L. Salzberg (to E. I. du Pont de Nemours & Co., Inc.).
54. T. M. Patrick and W. S. Emerson, *Ind. Eng. Chem.* **41,** 636 (1949).
55. U.S. Pat. 4,606,863 (Aug. 19, 1986), M. Nakazawa, K. Fujitani, and H. Manomi (New Japan Chemical Co., Ltd.).
56. D. T. Thompson, *Platinum Met. Rev.* **19,** 88 (1975).
57. H. K. Schwitzer in T. H. Applewhite, ed., *World Conference on Oleochemicals into the 21st Century*, American Oil Chemists Society, Champaign, Ill., 1990.
58. D. D. Nanavat, *J. Sci. Ind. Rev.* **35,** 163 (1976).
59. U.S. Pat. 2,731,495 (Jan. 17, 1956), R. S. Emslie (to E. I. du Pont de Nemours & Co., Inc.).
60. U.S. Pat. 2,318,762 (May 11, 1943), G. D. Davis and B. A. Dombrow (to National Oil Products Co.).
61. U.S. Pat. 2,674,608 (Apr. 6, 1954), G. Dupont and O. Kostlitz (to Société Organico).
62. U.S. Pat. 2,734,916 (Feb. 14, 1956), D. S. Bolley and F. C. Naughton (to Baker Castor Oil Co.).
63. U.S. Pats. 2,696,500 and 2,696,501 (Dec. 7, 1954), W. Stein (to Henkle & Cie, GmbH).
64. U.S. Pat. 2,851,491 (Sept. 9, 1958), F. C. Naughton, and R. C. Daidone (to Baker Castor Oil Co.).
65. U.S. Pat. 4,237,317 (Dec. 2, 1980), K. Yamataka, Y. Matsuoka, and T. Isoya (to Asahi Kasei Kogyo Kabushiki Kaisha).
66. U.S. Pat. 4,299,976 (Nov. 10, 1981), C. Hsu and H. S. Kesling, Jr. (to Atlantic Richfield Co.).
67. R. V. Lindsey, *J. Am. Chem. Soc.* **81,** 2073 (1959).
68. U.S. Pat. 4,377,451 (Mar. 22, 1983), W. J. M. van Tilborg and C. J. Smit (to Shell Oil Co.).
69. C. E. Frank and W. E. Foster, *J. Org. Chem.* **26,** 303 (1961).
70. U.S. Pat. 2,858,337 (Oct. 28, 1958), S. A. Mednisk, R. Wynkoop, and J. Feldman (to National Distillers and Chemical Corp.).
71. Ger. Offen. 1,912,569 (Oct. 2, 1969), J. O. White and D. D. Daves (to E. I. du Pont de Nemours & Co., Inc.).
72. U.S. Pat. 3,280,183 (Oct. 18, 1966), A. Maggiolo (to Wallace and Tiernan, Inc.)
73. T. R. Steadman and J. O. Peterson, *Ind. Eng. Chem.* **50,** 59 (1958).
74. U.S. Pat. 4,733,007 (Mar. 22, 1988), J. Andrade, K. Koehler, and G. Prescher (to Degussa Aktiengesellschaft).
75. U.S. Pat. 5,026,461 (June 25, 1991), D. D. Davis and D. L. Sullivan (to E. I. du Pont de Nemours & Co., Inc.).
76. U.S. Pat. 4,339,536 (July 13, 1982), K. Kato and N. Vemura (to Nippon Mining Co., Ltd.).
77. E. J. Dufek, W. E. Parker, and R. E. Koos, *J. Am. Oil Chem. Soc.* **51,** 351 (1974).
78. E. N. Frankel and E. H. Pryde, *J. Am. Oil Chem. Soc.* **54,** 873A (1977).
79. E. N. Frankel and F. L. Thomas, *J. Am. Oil Chem. Soc.* **50,** 39 (1973).
80. E. T. Roe and D. Swern, *J. Am. Oil Chem. Soc.* **37,** 661 (1960).
81. J. P. Friedrich, G. R. List, and V. E. Sohns, *J. Am. Oil Chem. Soc.* **50,** 455 (1973).
82. H. Koch, *Fette, Seifen, Anstrichm.* **59,** 493 (1957).
83. N. E. Lawson, T. T. Chang, and F. B. Slezak, *J. Am. Oil Chem. Soc.* **54,** 215 (1977).
84. W. Cooper and W. H. T. Davison, *J. Chem. Soc.*, 1380 (1952).

85. U.S. Pat. 2,601,223 (June 24, 1952), M. J. Roedel (to E. I. du Pont de Nemours & Co., Inc.).
86. M. S. Kharasch and W. Nudenberg, *J. Org. Chem.* **19,** 1921 (1954),
87. D. D. Coffman and H. N. Cripps, *J. Am. Chem. Soc.* **80,** 2880 (1958).
88. *Higher Dibasic Acids and Their Derivatives*, technical bulletin, Okamura Oil Mill, Ltd., Osaka, Japan, 1985.
89. J. C. Cowan, W. C. Ault, and H. M. Teeter, *Ind. Eng. Chem.* **38,** 1138 (1946).
90. U.S. Pat. 3,753,968 (Aug. 21, 1973), B. F. Ward (to Westvaco Corp.).
91. U.S. Pat. 4,156,095 (May 22, 1979), A. H. Jevne and G. L. Schwebke (to Henkel Corp.).
92. Ger. Offen. 2,406,401, (1974), B. F. Ward (to Westvaco Corp.).
93. U.S. Pat. 3,899,476 (Aug. 12, 1975), B. F. Ward (to Westvaco Corp.).
94. U.S. Pat. 3,483,083 (Dec. 9, 1969), G. W. Elson, R. Howe, and D. F. Johes (to Imperial Chemical Industries, Ltd.).
95. U.S. Pat, 3,843,466 (Oct. 22, 1974), S. A. Akabori, I. Shiio, and R. Uchio (to Ajinomato Co., Inc.).
96. U.S. Pat. 3,975,234 (Aug. 17, 1976), D. O. Hitzman (to Phillips Petroleum Co.).
97. U.S. Pat. 4,220,720 (Sept. 2, 1980), A. Tooka and S. Uchida (to Bio Research Center Co., Ltd.).
98. U.S. Pat, 4,624,920 (Nov. 25, 1986), S. Inove, Y. Kimura, and S. Adachi (to Eiji Suzuki).
99. U.S. Pat. 4,827,030 (May 2, 1989), F. F. Hill (to Huels Akliengesellschaft).
100. Jpn. Kokai 8,463,190 (1984), Iwata Kagasku Kogyo K. K.; Jpn. Kokai 81,137,893 (1981), Banda Kagaku Kogyo K. K. (to Shizuoka Prefecture); N. Nakagawa and co-workers, *J. Ferment. Technol.* **62,** 201–203 (1984); Jpn. Kokai 8,034,017 (1980) (to Mitsubishi Chemical Industries Co. Ltd.); S. Ikeda, *Hakko Kogaku Kaishi* **60,** 208–210 (1982).
101. U.S. Pat. 4,474,882 (Oct. 2, 1984), E. Kuneshige and T. Morinaga (to Daicel Chemical Industries, Ltd.).
102. E. C. Leonard in E. H. Pryde, ed., *Fatty Acids*, American Oil Chemist's Society, Champaign, Ill., 1979, Chapt. 25.
103. R. W. Johnson in R. W. Johnson and E. Fritz, eds., *Fatty Acids in Industry*, Marcel Dekker, Inc., New York, 1989, Chapt. 13.
104. *Emerox Azelaic Acid and Polymer Intermediates*, technical bulletin, Emery Industries, Cincinnati, Ohio, 1981.
105. *DBE Solvent Applications*, technical bulletin, Du Pont Chemicals, Wilmington, Del., 1991.
106. H. J. Lansen in J. I. Kroschwitz, ed., *Encyclopedia of Plastics and Engineering*, Vol. 1, 1985, p. 653.
107. G. Froyer and co-workers, *J. Polymer Sci., Polymer Chem. Ed.* **19,** 165–174 (1981).
108. M. T. Watson and G. M. Armstrong, *SPE J.*, **21,** 475 (1965).
109. Ref. 102, p. 15.
110. U.S. Pat. 5,025,087 (June 18, 1991), F. P. Williams (to E. I. du Pont de Nemours & Co., Inc.).
111. U.S. Pat. 4,483,975 (Nov. 20, 1984), E. de Jong, K. H. Hapelt, and H. Knipp (to Plate Bonn Gesellschaft Haftung).
112. U.S. Pat. 5,025,081 (June 18, 1991), L. P. Fontana and P. W. Buckley (to General Electric Co.).
113. U.S. Pat. 5,026,676 (June 25, 1991), S. A. Montika, T. L. Pickering, A. Rokicki, and B. K. Stein (to Air Products and Chemicals, Inc., Arco Chemical Co., and Mitsui Petrochemical Industries, Ltd.).
114. *Chem. Week* **97,** 69 (1965).
115. U.S. Pat. 4,448,702 (May 15, 1984), G. Kaes (to Lang & Co.).

116. U.S. Pat. 4,647,392 (Mar. 3, 1987), J. W. Darden, C. A. Triebel, W. A. Van Neste, and J. P. Maes (to Texaco Inc. S. A. and Texaco Belgium N.V.).
117. U.S. Pat. 4,578,205 (Mar. 25, 1986), E. L. Yeakey, G. P. Speranza, C. A. Triebel, and D. R. McCoy (to Texaco Inc.).
118. *Diacid Surfactants*, technical bulletin, Westvaco Chemical Division, Charleston Heights, S.C.
119. N. I. Sax, *Dangerous Properties of Industrial Materials*, Van Nostrand Reinhold Co., New York, 1984.
120. MSDS 00000004, *Dibasic Esters*, Du Pont Chemicals, July 1991.
121. G. D. Clayton and F. E. Clayton, eds., *Patty's Industrial Hygiene and Toxicology*, 3rd ed., John Wiley & Sons, Inc., New York, 1981.
122. Emerox 1144, Quantum Chemical Corp., Jan. 1989.
123. MSDS 00000003, *Dodecanedioic Acid*, DuPont Chemicals, Mar. 6, 1990.
124. W. C. Rose, *J. Pharmacol.* **24**, 147 (1924).
125. W. C. Rose and co-workers, *J. Pharmacol.* **25,** 59 (1925).
126 MSDS 240,104,541, Union Camp Corp., Apr. 1989.
127. *Diacid Surfactants*, technical bulletin, Westvaco Chemical Division, Charleston Heights, S.C.
128. K. Verschuerren, *Handbook of Environmental Data on Organic Compounds*, Van Nostrand Reinhold Co., Inc., New York, 1977, p. 354.
129. MSDS E-77995-3, Du Pont Chemicals, Dec. 1987.
130. *Westvaco Diacid Toxicological and Environmental Testing,* technical bulletin, Westvaco Chemical Division, Charleston Heights, S.C.

ROBERT W. JOHNSON
CHARLES M. POLLOCK
Union Camp Corporation

ROBERT R. CANTRELL
Union University

DIENE POLYMERS.   See ELASTOMERS, SYNTHETIC.

DIESEL FUEL.   See GASOLINE AND OTHER MOTOR FUELS; FEEDSTOCKS.

DIETARY FIBER

Historically, dietary fiber referred to insoluble plant cell wall material, primarily polysaccharides, not digested by the endogenous enzymes of the human digestive tract. This definition has been extended to include other nondigestible polysaccharides, from plants and other sources, that are incorporated into processed foods. Cellulose [9004-34-6] (qv) is fibrous; however, lignin [9005-53-2] (qv) and many other polysaccharides in food do not have fiberlike structures (see also CARBOHYDRATES).

Cell-wall dietary fiber is a complex system composed of variable amounts of cellulose, other polysaccharides such as hemicellulose [9034-32-6] (qv) and pectin [9000-69-5], and lignin. The precise composition and proportion of polysaccharide types is related to the plant source, stage of maturity, and growing conditions. The composition and physical properties of dietary fiber also may be affected by both postharvest physiological changes and food processing (qv). Cereal grains, legumes, vegetables, and fruits are primary sources of dietary fiber (see WHEAT AND OTHER CEREAL GRAINS). A smaller proportion of total dietary fiber comes from polysaccharides (gums and mucilages) added for their functionality in processed foods.

The variation in water solubility among polysaccharides results in varied physiological roles. Plant cell-wall polysaccharides and lignin provide insoluble dietary fiber (IDF); nondigestible storage polysaccharides, some pectic polysaccharides, and most of the functional additives contribute soluble dietary fiber (SDF).

A resurgence of interest in dietary fiber has been stimulated by epidemiological evidence of differences in colonic disease patterns between cultures with diets containing large quantities of fiber, and Western cultures having more highly refined diets. Many African countries, for example, are relatively free of diverticular disease, ulcerative colitis, hemorrhoids, polyps, and cancer of the colon (1). Whereas most interest has focused on the beneficial role of dietary fiber, there is also concern that high fiber diets may cause disturbances in the absorption of nutrients such as minerals (see MINERAL NUTRIENTS) and vitamins (qv). The interrelationships between consumption of dietary fiber and health status have been obscured by inadequate knowledge of the quantity and structural chemistry of dietary fiber in various food sources and the physiological roles of the fiber components.

Terminology. Various names have been proposed for the nondigestible part of plant cells, names suggesting either the fibrous nature of some of the cell-wall polysaccharides or their resistance to digestive tract enzymes. Fiber values in food composition tables originally were based on crude fiber analyses, but this assay does not give a true picture of even the fibrous cell-wall material. The term crude fiber is not equivalent to dietary fiber and has little practical meaning. Likewise, the terms unavailable (or nonavailable) carbohydrate and nonnutritive fiber fail to recognize the presence of lignin in fiber or the microbial fermentation of polysaccharides that occurs in the colon. The term nonstarch polysaccharides (NSP) emphasizes the nondigestible character of fiber and is supported by appropriate analytical methodology, but excludes lignin.

Other names have been proposed based on specific methods of analysis. In the Van Soest detergent methods (2), the terms *neutral detergent fiber* (NDF) or *neutral detergent residue* (NDR), and *acid detergent fiber* (ADF) refer to the insoluble hemicellulose–cellulose–lignin complex and to the cellulose–lignin complex of the plant cell wall, respectively. Before the importance of soluble dietary fiber was recognized, NDF was accepted as a measure of dietary fiber even though it reflects insoluble fiber only.

Dietary fiber is the accepted terminology in the United States for nutritional labeling. Total dietary fiber (TDF) and its subfractions, insoluble dietary fiber (IDF) and soluble dietary fiber (SDF), are defined analytically by official methods (3–5).

Sources, Composition, and Structure of Dietary Fiber

Natural sources of fiber in the diet include fruits, vegetables, legumes, and cereal grain products. The insoluble fiber content of some processed foods and breads is supplemented by incorporation of purified cellulose, cereal brans, or other plant fiber preparations. Cellulose and its chemically modified derivatives; seaweed polysaccharides, alginates [9005-32-7] and carrageenans [9000-07-1]; seed mucilaginous polysaccharides, guar and locust bean galactomannans [11078-30-1]; highly complex plant exudate polysaccharides, gum arabic [9000-01-5], tragacanth [9000-65-1], and others; microbially synthesized polysaccharides, xanthan [11138-66-2] and gellan gum [71010-52-1]; pectins; and other plant polysaccharides are added to foods for a variety of purposes. Because these materials are also nondigestible, they contribute to the total effect of dietary fiber and are encompassed by its definition.

Dietary fiber is a mixture of simple and complex polysaccharides and lignin. In intact plant tissue these components are organized into a complex matrix, which is not completely understood. The physical and chemical interactions that sustain this matrix affect its physicochemical properties and probably its physiological effects. Several of the polysaccharides classified as soluble fiber are soluble only after they have been extracted under fairly rigorous conditions.

Lignin, a highly polymerized alkylaromatic substance, is associated with the decreased digestibility by colonic microbial enzymes of some cell-wall polysaccharides. Lignification also renders the structural polysaccharides less soluble and extractable. The degree of lignification is specific to plant type and increases with plant maturity as the lignin infiltrates the primary and secondary cell walls. Edible plant material has a relatively low content of lignin.

There is a great diversity of chemical structures in the polysaccharides of plant tissue. Table 1 lists some principal types of polysaccharides in edible land plants using simplified structures. Other reviews (6–8) provide a more complete description of the complex structures. Complete structures for many of the polysaccharides of edible sources have not yet been determined.

Any starch (qv) escaping digestion in the upper gastrointestinal tract also contributes to dietary fiber effects. Some food starches, and the amylose fraction in particular, are readily converted into a nondigestible or slowly digestible physical form under certain food processing conditions. These resistant starches are

Table 1. Polysaccharides of Land Plants

Polysaccharide	CAS Registry Number	Structure
Structural		
cellulose	[9004-34-6]	(1→4)-β-D-glucan
hemicelluloses	[9034-32-6]	
xylan	[9014-63-5]	(1→4)-β-D-xylan
arabinoxylan	[98513-12-3]	(1→4)-β-D-xylan with 3-linked α-L-arabinose branches
glucuronoarabinoxylans		(1→4)-β-D-xylan with 2-linked 4-*O*-methyl-α-D-glucuronic acid and arabinose branches
xyloglucan	[37294-28-3]	(1→4)-β-D-glucan with 6-linked α-D-xylose branches
β-glucans	[55965-23-6]	(1→3), (1 → 4)-β-D-glucan
pectic substances[a]	[9046-38-2]	(1→4)-α-D-galacturonan
	[39280-21-2]	(1→2)-L-rhamno-(1 → 4)-α-D-galacturonan
associated polysaccharides[a]		
arabinan	[9060-75-7]	(1→5)-α-L-arabinan with 3-linked α-L-arabinose branches
arabinogalactan	[9036-66-2]	(1→4)-β-D-galactan with α-L-ara-(1→5)-α-L-ara-(1→3)-L-arabinose branches
galactan	[9051-94-9]	(1→4)-β-D-galactan
Nonstructural		
fructan	[9005-80-5],	(2→1)-β-D-fructan
	[9013-95-0]	(2→6)-β-D-fructan
galactomannan	[11078-30-1]	(1→4)-β-D-mannan with single 6-linked α-D-galactose branches
glucomannan	[37230-82-3]	(1→4)-β-D-glucomannan

[a]The rhamnogalacturonans are associated physically and covalently with associated polysaccharides. Several less common sugars are also covalently linked to the polysaccharide complex.

readily fermented by colonic bacteria. Small amounts of waxes, cutin, and minerals in fruits and vegetables contribute to total dietary fiber values but may be physiologically inert.

Physiological Properties

The beneficial effects of dietary fiber, including both soluble and insoluble fiber, are generally recognized. Current recommendations are for daily intakes of 20–35 g in a balanced diet of cereal products, fruits, vegetables, and legumes. However, the specific preventive role of dietary fiber in certain diseases has been difficult to establish, in part because dietary risk factors such as high saturated fat and high protein levels are reduced as fiber levels increase.

Dietary fiber is important in the functioning of the entire gastrointestinal (GI) tract and affects the structure and morphology of the intestine. The process of chewing insoluble fiber-rich foods increases salivation and the flow of gastric secretions. Fiber components that increase viscosity or gel, for example, guar gum and pectin, delay gastric emptying. Fibers exert buffering action and may alter gastric pH by their effect on gastrointestinal hormones (qv). A reduced glycemic response, probably resulting from delayed absorption of glucose, is associated with various fiber fractions, particularly viscosity-enhancing fibers.

Dietary fiber has a pronounced effect on the characteristics of the fecal mass and on the rate of passage of digest through the GI tract. The particle size and shape, density, and water-holding capacity (WHC) of dietary fiber influence the flow rates. Because hydratability and WHC are determined by chemical structure, crystallinity, overall plant tissue morphology, and lignin content, food processing, such as grinding and heat processing, indirectly affects the rate of passage. Insoluble fiber in the diet increases the rate of passage, frequency of defecation, and the fecal mass, although there are large variations among individuals.

High fiber diets also play a role in the excretion of bile acids and cholesterol [57-88-5]. Insoluble fiber, particularly lignin, promotes excretion of bile acids and salts. Viscous soluble fiber components, such as pectin, guar, and soluble $(1\rightarrow3),(1\rightarrow4)$-$\beta$-D-glucan [55965-23-6], are associated with a lowering of serum cholesterol and triglycerides in hyperlipemic individuals by an unclear mechanism. There has been concern that high fiber diets might impair absorption of minerals and some vitamins but healthy adults on a balanced diet should not be affected. Epidemiological studies show a negative correlation between colon cancer and high fiber diets; however, a definitive protective role in humans has not been established experimentally.

Dietary fiber is degraded extensively in the colon by bacterial action, yielding short-chain fatty acids, primarily acetic acid [64-19-7], propionic acid [79-09-4], and butyric acid [107-92-6], and gases, such as hydrogen, carbon dioxide, and methane. Because the human digestive enzymes known to hydrolyze polysaccharides are amylases, that is, $(1 \rightarrow 4)$-α-D-glucanases [9000-90-2], acting in the upper GI tract, dietary fiber was assumed to pass into the colon largely unchanged. However, studies of ileostomy patients have provided evidence that not all of the dietary hemicelluloses are passed from the small intestine into the colon intact, suggesting that some bacterial activity also occurs in the lower small intestine. Colonic bacteria possess inducible enzymes that actively degrade and ferment much of the fiber polysaccharides as well as the mucosal polysaccharides from the host (9). Some individuals possess bacteria with cellulolytic activity, but hemicelluloses and soluble polysaccharides are more actively fermented.

Fiber components are the principal energy source for colonic bacteria with a further contribution from digestive tract mucosal polysaccharides. Rate of fermentation varies with the chemical nature of the fiber components. Short-chain fatty acids generated by bacterial action are partially absorbed through the colon wall and provide a supplementary energy source to the host. Therefore, dietary fiber is partially caloric. The short-chain fatty acids also promote reabsorption of sodium and water from the colon and stimulate colonic blood flow and pancreatic secretions. Butyrate has added health benefits. Butyric acid is the preferred energy source for the colonocytes and has been shown to promote normal colonic epithelial cell differentiation. Butyric acid may inhibit colonic polyps and tumors. The relationships of intestinal microflora to health and disease have been reviewed (10).

Physicochemical Properties

Several physicochemical properties of dietary fiber contribute to its physiological role. Water-holding capacity, ion-exchange capacity, solution viscosity, density, and molecular interactions are characteristics determined by the chemical structure of the component polysaccharides, their crystallinity, and surface area.

Water-Holding Capacity (WHC). All polysaccharides are hydrophilic and hydrogen bond to variable amounts of water. Hydratability is a function of the three-dimensional structure of the polymer (11) and is influenced by other components in the solvent. Fibrous polymers and porous fiber preparations also absorb water by entrapment. The more highly crystalline fiber components are more difficult to hydrate and have less tendency to swell. Structural features and other factors, including grinding, that decrease crystallinity or alter structure, may increase hydration capacity and solubility. However, fine grinding of insoluble dietary fiber such as bran reduces WHC. In general, branched polysaccharides are more soluble than are linear polysaccharides because close packing of molecular chains is precluded. WHC is strongly influenced by the pentosan components of cell-wall dietary fiber and varies with the structure and source of these hemicelluloses.

Soluble polysaccharides, such as agar [9002-18-0] and pectin, which form three-dimensional networks stabilized by physical or covalent interactions imbibe large quantities of water to form rigid gels (11). The gelling behavior and high viscosity of both the single-unit branched polysaccharides, such as the galactomannans, and the cereal $(1 \rightarrow 3)$, $(1 \rightarrow 4)$-β-D-glucans affect GI tract function. The moderation of blood glucose levels observed with diets containing the viscous soluble polysaccharides may be related to a decrease in intraluminal mixing and a consequent decrease in the rate of absorption of glucose.

Ion Exchange. Acidic polysaccharides containing uronic acids, sulfate, or phosphate groups are cation exchangers, binding metal ions. The type of cation bound to these groups influences the physical properties of the polysaccharide. For example, alginic acid [9005-32-7], which is relatively insoluble as the free acid, is soluble as the sodium salt and forms a gel, calcium alginate [9005-35-0], with calcium ion. The cation-exchange capacity of dietary fiber is primarily dependent on the pectic acid content and glucuronic acid-containing hemicelluloses. Ion-exchange capacities have been determined for several dietary fiber sources (12), and the evidence suggests that cooking may alter the exchange capacity.

Molecular Interactions. Various polysaccharides readily associate with other substances, including bile acids and cholesterol, proteins, small organic molecules, inorganic salts, and ions. Anionic polysaccharides form salts and chelate complexes with cations; some neutral polysaccharides form complexes with inorganic salts; and some interactions are structure specific. Starch amylose and the linear branches of amylopectin form inclusion complexes with several classes of polar molecules, including fatty acids, glycerides, alcohols, esters, ketones, and iodine/iodide. The absorbed molecule occupies the cavity of the amylose helix, which has the capacity to expand somewhat to accommodate larger molecules. The starch–lipid complex is important in food systems. Whether similar inclusion complexes can form with any of the dietary fiber components is not known.

Soluble polysaccharides interact indirectly with proteins (qv) by competing for water and decreasing the solubility of the protein. Structure-specific interactions are exemplified by the lectins. These proteins interact with carbohydrates having a specific structure; the consequence of such interactions is aggregation and precipitation of the insoluble complex. This interaction displays the characteristics of an antigen–antibody reaction because it is characterized by a strict structural requirement for both carbohydrate and protein, competitive inhibition by the monomer sugar, a reversible reaction, and dissolution of the complex at high carbohydrate concentrations. If such interactions occur in the gastrointestinal tract, the nutritional availability of the protein can be impaired.

Dietary fiber and fiber-rich food fractions bind bile acids and bile salts *in vitro*. This interaction is more pronounced for the lignin component.

Analysis of Dietary Fiber

Analytical methods suitable for routine assays measure groups of fiber components having similar solubility properties. Values may be expressed as total dietary fiber (TDF), insoluble dietary fiber (IDF), or soluble dietary fiber (SDF). Quantitative analysis of specific fiber components other than cellulose, $(1\rightarrow3)$, $(1\rightarrow4)$-β-D-glucans, $(1\rightarrow4)$-α-D-galacturonans, and lignin is time-consuming because it involves fractionation of complex, chemically similar polysaccharides. The more commonly used fiber analysis methods are detergent methods for cell-wall fiber, enzymatic gravimetric methods for soluble and insoluble polysaccharides, and methods that include a quantitative analysis of the component sugars in the fiber fractions. The method selected defines IDF and SDF, since solubility of polysaccharides is affected by temperature and pH.

Detergent Methods. The neutral detergent fiber (NDF) and acid detergent fiber (ADF) methods (2), later modified for human foods (13), measure total insoluble plant cell wall material (NDF) and the cellulose–lignin complex (ADF). The easily solubilized pectins and some associated polysaccharides, galactomannans of legume seeds, various plant gums, and seaweed polysaccharides are extracted away from the NDF. They cannot be recovered easily from the extract, and therefore the soluble fiber fraction is lost.

The detergent method for insoluble fiber superseded the crude fiber method and became the method of choice for insoluble fiber analysis until the 1980s, when methods were developed to recover soluble fiber as well. Some analysts still prefer the NDF procedure for insoluble fiber. The method is simple, inexpensive, reproducible, and amenable to routine assays. The disadvantage is the inability to recover the soluble fraction. See Reference 14 for more information on detergent methods.

Neutral Detergent Fiber (AACC Method 32-20). The ground sample is extracted for 1 h under reflux at pH 7.0 with a buffered detergent solution containing sodium lauryl sulfate [151-21-3] and ethylenediaminetetraacetic acid [60-00-4].

Protein, soluble polysaccharides, and low molecular-weight substances are dissolved. Starch in high concentration interferes and is removed by α-amylase [9000-90-2] digestion. Although pancreatic α-amylase was used in the original modified NDF method (13), the heat stable α-amylase is more convenient (14). The insoluble fiber residue is collected by filtration, dried, and corrected for ash to give the NDF value. Certain products containing highly viscous water-soluble gums may be difficult to analyze because of retarded filtration, but in general the method is convenient and reliable for insoluble fiber.

Acid Detergent Fiber. The ground sample is heated for 1 h under reflux in a solution of 2% cetyltrimethylammonium bromide [57-09-0] in 1N sulfuric acid [7664-93-9]. The acid hydrolyzes and dissolves the noncellulosic polysaccharides. The insoluble residue, relatively free of hemicelluloses and containing all the cellulose and lignin, is filtered, dried, and corrected for ash to give the ADF value.

Cellulose may be extracted from the ADF with 72% sulfuric acid (w/w) at 4°C for 24 h leaving an insoluble residue of lignin. The loss in mass of the ADF estimates the cellulose component. Alternatively, cellulose may be estimated by hydrolysis of the ADF and determination of glucose.

In some processed foods insoluble artifacts generated during heat processing remain in the ADF residue together with the true lignin. The permanganate method for lignin gives a correct value. Ligninlike artifacts are not measured by this lignin procedure.

Enzymatic Gravimetric Methods for TDF, SDF, and IDF. These methods use an α-amylase and protease to remove starch and reduce protein. They differ from each other in the conditions for gelatinization of starch. Elimination of detergent permits recovery of soluble fiber, which is not possible with the detergent methods.

AOAC Method 985.29 for TDF. This AOAC method (3), referred to as the method of Prosky and co-workers (4), was cited in the Nutritional Labeling and Education Act of 1990 as the general analytical approach for food labeling of dietary fiber content. The method has undergone several modifications for TDF and for the primary fractions, SDF and IDF.

The dry milled food sample is digested sequentually with heat-stable α-amylase in phosphate buffer at 95–100°C and then at 60°C with protease and amyloglucosidase [9032-08-0] to remove protein and starch. The soluble fiber is precipitated with ethanol, and the combined soluble and insoluble fiber is collected by filtration, dried, weighed, and corrected for residual protein and ash to give a value for TDF. This method has also received AOAC official first action for IDF. For this value the insoluble residue from the amyloglucosidase digestion is collected by filtration, dried, weighed, and corrected for protein and ash. Although SDF can be obtained from the supernatant by precipitation with ethenol, the accuracy is not acceptable for some products.

AACC Method 32-07 for TDF, SDF, and IDF. This approved method is a modification of the Prosky method. The phosphate buffer is replaced by a buffer of 2(*N*-morpholino)ethanesulfonic acid (MES) [4432-31-9] and tris(hydroxymethyl)aminomethane (TRIS) [77-86-1]. Precision is improved and analysis time is reduced. TDF may be determined on a separate sample or calculated by summing the SDF and IDF values. The MES/TRIS-modified method has received final approval by AACC and official first action by AOAC (5).

AACC 32-06 Rapid Dietary Fiber Method for TDF, SDF, and IDF. The dietary fiber is determined by separate assays for soluble and insoluble fiber using a modification of AOAC 985.29 for SDF and the pancreatic α-amylase/NDF procedure for IDF. The SDF and IDF values are added to give TDF. Determination of the SDF requires autoclaving at 120°C for starch gelatinization, a more rigorous treatment than in the other enzymatic gravimetric methods. Corrections for protein are not required. TDF values are generally comparable to those obtained by the other methods described and affords a considerable time savings (15). The procedure, with minor modifications, has been submitted for final approval by AACC.

Urea Enzymatic Dialysis Method. This method (16) uses 8 M urea [57-13-6] to gelatinize and facilitate removal of starch and promote extraction of the soluble fiber at mild (50°C) temperatures. Following digestion with heat-stable α-amylase and protease, IDF is isolated by filtration or TDF is obtained after ethanol precipitation. Values for TDF are comparable to those obtained by the methods described earlier, and this method is less time-consuming than are the two AOAC-approved methods. Corrections for protein are required as in the AOAC methods.

Methods Based on Constituent Sugar Analysis. Unlike the gravimetric methods, these methods (17,18) arrive at dietary fiber values (TDF, SDF, IDF) by chromatographic (glc or hplc) analysis of the constituent sugars after extraction and fractionation of the fiber. Uronic acids, determined by decarboxylation or colorimetric analysis, and lignin values are taken into account in the calculation of the fiber values to give values comparable to those obtained by the gravimetric methods. If the lignin value is not included in the calculation, then a value for nonstarch polysaccharides (NSP) (18) is obtained. These methods are more time-consuming but give additional information about the component polysaccharides in the fiber. Because of the heterogeneity of plant polysaccharides, it is virtually impossible to give precise quantitative information on the individual polysaccharides in dietary fiber.

Sources of Dietary Fiber for Processed Foods

An increasing number of fiber sources are available for food processing, representing a diversity of sources and technological advances. Commercial food-grade purified cellulose products and cereal brans are used in food products to enhance the fiber content or for other functional purposes. Many purified or partially purified nondigestible polysaccharides are used in food systems for their physicochemical properties, for example, for viscosity or as suspending agents. These polysaccharides contribute to the dietary fiber content even though they are not used for that purpose.

Commercial Cellulose Products. High quality purified cellulose is prepared from soft or hard woods by pulping methods that remove most of the associated hemicelluloses, lignin, waxes, and other constituents. The pulp (qv) is then bleached to give a white cellulose (α-cellulose) that is a relatively pure (>90%) (1 → 4)-β-D-glucan. Although the main intent is to degrade and remove the lignin and hemicelluloses, the cellulose, too, may be partially depolymerized. Purified fibrous cellulose is white, odorless, and nonabrasive. Although devoid of flavor, it possesses an undesirable texture in the mouth related to its fibrous nature, particle size, and insolubility. Other cellulosic products (Table 2) are made from the cellulose pulp or α-cellulose.

Powdered Cellulose. Solka-Floc, a purified fibrous powdered cellulose, is said to be 99% cellulose on a dry weight basis, virtually lignin free, free of fat and protein, and with low ash content. Food grades are available in fiber lengths averaging 20–25 to 100–140 μm.

Microcrystalline Cellulose. Limited hydrolysis of fibrous cellulose pulp yields a product more highly crystalline than native cellulose. Mineral acids hydrolyze glycosidic bonds in the less crystalline regions of the cellulose fibers. The resulting cellulose microcrystals can be mechanically dispersed in water to give a gel or emulsionlike colloidal suspension for use in low calorie whipped toppings as well as in low fat or fat-free emulsions that mimic oil-in-water emulsions at cellulose concentrations in the 0.75–3% range. These microcrystals, like the native cellulose, do not melt and can be used in food systems under heat-processing conditions that would destroy the emulsion stability of fat- and oil-based systems.

Microcrystalline celluloses are marketed under the trade name Avicel. The physical characteristics of microcrystalline celluloses differ markedly from those of the original cellulose. The free-flowing powders have particle sizes as small as 0.2–10 μm. Avicel celluloses coated with xanthan gum, guar gum, or carboxymethylcellulose to modify and stabilize their properties are also available. The Avicel products are promoted for use in low calorie whipped toppings and icings and in fat-reduced salad dressings and frozen desserts (see FAT SUBSTITUTES).

Cellulose Derivatives. Chemical modification markedly alters the physical properties of cellulose. Common derivatives include methylcellulose [*9004-67-5*], ethylcellulose [*9004-57-3*], propylcellulose [*9005-18-9*], hydroxyethylcellulose [*9004-62-0*], hydroxypropylcellulose [*9004-64-2*], and carboxymethylcellulose (CMC) [*9000-11-7*] as well as mixed ethers of cellulose (see CELLULOSE ETHERS). These derivatives are obtained by alkylation of the cellulose under basic conditions. The ether substituents disrupt the extensive hydrogen bonding in the cellulose and permit increased hydration. At certain degrees of substitution the insoluble cellulose is transformed into a water-soluble colloid. The nature and size of the substituent, the regularity of substitution, and the molecular weight of the cellulose determine the degree of substitution at which water solubility occurs. In addition to increased hydration, CMC displays increased ion-exchange capacity.

Table 2. Commercial Cellulose Products

Type	Trade name	Producer
powdered	Solka-Floc	James River Corp., Saddle Brook, N.J.
microcrystalline	Avicel	FMC Corp., Philadelphia, Pa.
derivatives	Methocel methylcellulose	Dow Chemical Co., Midland, Mich.
	Benecel methylcellulose	Aqualon/Hercules Inc., Wilmington, Del.
	hydroxypropylcellulose	Aqualon/Hercules Inc., Wilmington, Del.
	Ticalose CMC	TIC Gums, Belcamp, Md.
bacterial	Cellulon	Weyerhaeuser Co., Tacoma, Wash., with Cetus, Emeryville, Calif.

These cellulose ethers are used in food products for enhanced water retention, reduction of oil absorption in fried foods, and as thickeners and binders (see FOOD ADDITIVES). Consumer products include Methocel methylcellulose, Benecel methylcellulose, hydroxypropylcellulose, and Ticalose CMC.

Bacterial Cellulose. Development of a new strain of Acetobacter may lead to economical production of another novel cellulose. Cellulon fiber has a very fine fiber diameter and therefore a much larger surface area, which makes it physically distinct from wood cellulose. Its physical properties more closely resemble those of the microcrystalline celluloses; thus it feels smooth in the mouth, has a high water-binding capacity, and provides viscous aqueous dispersions at low concentration. It interacts synergistically with xanthan and CMC for enhanced viscosity and stability.

Bran. Bran fractions are more concentrated sources of fiber than are grain flours. As a fiber supplement in foods, cereal grain brans are more representative of natural dietary fiber than is cellulose because bran contains hemicelluloses, lignin, and cellulose. The composition and properties of bran are dependent on the cereal grain source, plant variety, and milling practices. Varying amounts of the endosperm components are present, and the particle size depends on the break roll from which the bran is taken. Wheat bran contains a large proportion of arabinoxylans, considerable lignin, and some starch in addition to cellulose. Wheat brans are readily available from any flour mill. Brans from corn, oats, barley, and rice are also available and differ from wheat bran in composition and properties. Oat bran and barley bran, for example, contain high proportions of $(1 \longrightarrow 3)$, $(1 \longrightarrow 4)$-β-D-glucans, part of which is soluble fiber, and little or no pectin. Oatrim products, developed at the National Center for Agricultural Research (ARS), Peoria, Illinois, contain up to 10% of the β-glucan. Rice brans are available in grades ranging from defatted bran (0.5–1.5% oil, 30–40% TDF) to stabilized full fat (18–22% oil, 25–30% TDF). Oat, barley, and rice brans contain significant amounts of soluble fiber, in contrast with wheat bran.

The American Association of Cereal Chemists has made available a certified food grade wheat bran as a reference standard for research purposes. This bran is a blend of soft white wheat (87.3%) and club white wheat (12.7%).

Other Insoluble Fiber Sources. Other insoluble fiber sources are commercially available as well, including fiber from sugar-beet pulp, a by-product of sugar production. Table 3 lists other insoluble fiber sources.

Sources of Functional Polysaccharides Contributing to SDF. Polysaccharides of diverse and frequently complex structure and origin are used as food

Table 3. Other Insoluble Fiber Sources

Source	Trade name	TDF, %	Producer
sugar-beet pulp	Fibrex	74[a]	Delta Fibre Foods, St. Louis Park, Minn.
soybean cotyledons	Fibrin	75[b]	Protein Technologies, Int., St. Louis, Mo.
pea fiber		90	Woodstone Foods Ltd., Ogilvie Mills, Winnipeg, Canada

[a]24% soluble fiber, primarily pectic polysaccharides.
[b]TDF primarily composed of pectin and associated polysaccharides.

additives for their functional properties, viscosity, emulsification, bodying and suspending, water binding, gelation, and fat-mimicking texture. They are used at relatively low levels (up to about 3%) and are regulated by the FDA for specific functional purposes. These polysaccharides tend to be water-soluble and contribute to SDF. They are usually classified by source, that is, seaweed or algal, microbial, plant exudate, legume seeds, and tubers. The more commonly used gums and mucilages are listed in Table 4 along with some commercial producers.

Table 4. Commercial Food Additive Sources of Soluble Polysaccharides

Source	Gum/mucilage	Polysaccharide composition[a]	Supplier[b]
algae (seaweed)			
	agar	anhydro-L-Gal, Gal	AEP, CIE, TIC, MEE
	alginates	L-GulA, ManA	CIE, TIC, MEE, KEL
	carrageenan	anhydro-Gal, Gal	AEP, CIE, TIC, MEE, FMC, CNI, SHC, SBI
microbial			
	gellan	Glc, GlcA, L-Rha	KEL
	xanthan	Glc, GlcA, Man	CIE, TIC, MEE, KEL, SBI, RPF
plants			
exudate gums			
	arabic	L-Ara, Gal, GlcA, 4-0-methyl-GlcA, L-Rha	AEP, CIE, TIC, MEE, CNI, RPF
	ghatti	L-Ara, Gal, GlcA, Man, Xyl	CIE, TIC, MEE, RPF
	karaya	Gal, GalA, GlcA, L-Rha	AEP, CIE, TIC, MEE, RPF
	tragacanth	arabinogalactan + tragacanthic acid (L-Fuc, Gal, GalA, Xyl)	AEP, CIE, TIC, MEE, RPF
legume seeds			
	guar gum	galactomannan	AEP, CIE, TIC, MEE, RPF, AQH
	locust bean (carob)	galactomannan	AEP, CIE, TIC, MEE, RPF, AQH
other			
	konjac flour	glucomannan	FMC
	pectin	rhamnogalacturonan	SBI, HIF
synthetic/chemically modified			
	polydextrose	Glc	PFZ
	modified celluloses	Glc	AQH, HIF, DOW

[a]Principal polysaccharide constituent listed. Sugars are D- unless otherwise noted. Ara = arabinose; Gal = galactose; GalA = galacturonic acid; Glc = glucose; GlcA = glucuronic acid; GulA = guluronic acid; Man = mannose; ManA = mannuronic acid; Rha = rhamnose; Xyl = xylose; Fuc = fucose. Chemical structures and properties are available (19).
[b]Principal producers and suppliers. AEP = AEP Colloids, Div. of Sarcom, Balston Spa, N.Y.; AQH = Aqualon/Hercules, Wilmington, Del.; CIE = Colony Import & Export Corp., Garden City, N.Y.; CNI = Colloides Naturels, Inc., Bridgewater, N.J.; DOW = Dow Chemical Co., Midland, Mich.; FMC = FMC Corp., Marine Colloids Div., Philadelphia, Pa.; HIF = Hercules Inc., Food Ingredients Group, Middletown, N.Y.; KEL = Kelco, Div. of Merck & Co., San Diego, Calif.; MEE = Meer Corp., North Bergen, N.J.; PFZ = Pfizer, New York, N.Y.; RPF = Rhone-Poulenc, Food Ingredients Div., Cranbury, N.J.; SBI = Sanofi BioIngredients, Inc., Waukesha, Wisc.; SHC = Shemberg Corp., Santa Anna, Calif.; TIC = TIC Gums, Inc., Belcamp, Md.

BIBLIOGRAPHY

"Dietary Fibers" in *ECT* 3rd ed., Vol. 7, pp. 628–638, by B. Lewis, Cornell University.
1. H. Trowell, *Nutr. Rev.* **35,** 6 (1977).
2. P. J. Van Soest and R. H. Wine, *J. Assoc. Off. Anal. Chem.* **50,** 50 (1967).
3. *AOAC Official Methods of Analysis*, 15th ed., Washington, D.C., 1990, pp. 1105–1106.
4. L. Prosky and co-workers, *J. Assoc. Off. Anal. Chem.* **68,** 677 (1985).
5. *AOAC Official Methods of Analysis*, 15th ed., 3rd supp., Washington, D.C., 1992, in press.
6. R. R. Selvendran, *Amr. J. Clin. Nutr.* **39,** 320 (1984).
7. A. Bacic, P. J Harris, and B. A. Stone, in J. Preiss, ed., *The Biochemistry of Plants*, Vol. 14, Academic Press, New York, 1988, pp. 297–370.
8. A. M. Stephen, in G. O. Aspinall, ed., *The Polysaccharides*, Vol. 2, Academic Press, New York, 1983, Chapt. 3.
9. A. A. Salyers, S. E. H. West, J. R. Vercellotti, and T. D. Wilkins, *Appl. Environ. Microbiol.* **34,** 529 (1977); **33,** 319 (1977).
10. D. F. Hentges, ed., *Human Intestinal Microflora in Health and Disease*, Academic Press, New York, 1983.
11. D. A. Rees, E. R. Morris, D. Thom, and J. K. Madden, in G. O. Aspinall, ed., *The Polysaccharides*, Vol. 1, Academic Press, New York, 1982, Chapt. 5.
12. A. A. McConnell, M. A. Eastwood, and W. D. Mitchell, *J. Sci. Food Agric.* **25,** 1457 (1974).
13. D. Schaller, *Food Prod. Develop.* **11,** 70 (1977).
14. P. J. Van Soest, J. B. Robertson, and B. A. Lewis, *J. Dairy Sci.* **74,** 3583 (1991).
15. R. Mongeau and R. Brassard, *J. Food Comp. Anal.* **2,** 189 (1989).
16. J. L. Jeraci, B. A. Lewis, P. J. Van Soest, and J. B. Robertson, *J. Assoc. Off. Anal. Chem.* **72,** 677 (1989).
17. O. Theander and E. A. Westerlund, *J. Agric. Food Chem.* **34,** 330 (1986).
18. H. N. Englyst and J. H. Cummings, *J. Assoc. Off. Anal. Chem.* **71,** 808 (1988).
19. G. O. Aspinall, ed., *The Polysaccharides*, Vol. 2, Academic Press, New York, 1983.

General References

J. W. Anderson, D. A. Deakins, T. L. Floore, B. M. Smith, and S. E. White, *Crit. Rev. Food Sci. Nutr.* **29,** 95 (1990).
M. L. Dreher, *Handbook of Dietary Fiber, An Applied Approach*, Marcel Dekker, New York, 1987.
I. Furda and C. J. Brine, eds., *New Developments in Dietary Fiber*, Plenum Press, New York, 1990.
D. Kritchevsky, C. Bonfield, and J. W. Anderson, eds., *Dietary Fiber, Chemistry, Physiology and Health Effects*, Plenum Press, New York, 1990.
S. M. Pilch, *Physiological Effects and Health Consequences of Dietary Fiber*, Life Sciences Research Office, Federation of American Societies for Experimental Biology, Bethesda, Md., 1987.
G. A. Spiller, ed., *Handbook of Dietary Fiber in Human Nutrition*, CRC Press, Boca Raton, Fl., 1986.
G. V. Vahouny and D. Kritchevsky, eds., *Dietary Fiber, Basic and Clinical Aspects*, Plenum Press, New York, 1986.

B. A. LEWIS
Cornell University

DIETHANOLAMINE.   See ALKANOLAMINES.

DIETHYLENE GLYCOL.   See GLYCOLS.

DIFFUSION.   See DIFFUSION SEPARATION METHODS; MASS TRANSFER.

DIFFUSION SEPARATION METHODS

Ordinary diffusion involves molecular mixing caused by the random motion of molecules. It is much more pronounced in gases and liquids than in solids. The effects of diffusion in fluids are also greatly affected by convection or turbulence. These phenomena are involved in mass-transfer processes, and therefore in separation processes (see MASS TRANSFER; SEPARATION SYSTEMS SYNTHESIS). In chemical engineering, the term diffusional unit operations normally refers to the separation processes in which mass is transferred from one phase to another, often across a fluid interface, and in which diffusion is considered to be the rate-controlling mechanism. Thus, the standard unit operations such as distillation (qv), drying (qv), and the sorption processes, as well as the less conventional separation processes, are usually classified under this heading (see ABSORPTION; ADSORPTION; ADSORPTION, GAS SEPARATION; ADSORPTION, LIQUID SEPARATION).

A number of special processes have been developed for difficult separations, such as the separation of the stable isotopes of uranium and those of other elements (see NUCLEAR REACTORS; URANIUM AND URANIUM COMPOUNDS). Two of these processes, gaseous diffusion and gas centrifugation, are used by several nations on a multibillion dollar scale to separate partially the uranium isotopes and to produce a much more valuable fuel for nuclear power reactors. Because separation in these special processes depends upon the different rates of diffusion of the components, the processes are often referred to collectively as diffusion separation methods. There is also a thermal diffusion process used on a modest scale for the separation of helium-group gases (qv) and on a laboratory scale for the separation of various other materials. Thermal diffusion is not discussed herein.

The most important industrial application of the diffusion separation methods has been for the enrichment of uranium-235 [*5117-96-1*], ^{235}U. Natural uranium consists mostly of ^{238}U and 0.711 wt % ^{235}U plus an inconsequential amount of ^{234}U. The United States was the first country to employ the gaseous diffusion process for the enrichment of ^{235}U, the fissionable natural uranium isotope. During the 1940s and 1950s, this enrichment application led to the investment of several billion dollars in process facilities. The original plants were built in 1943–1945 in Oak Ridge, Tennessee, as part of the Manhattan Project of World War II.

As of the early 1990s, diffusion separation methods are being employed or developed internationally in (1) Argentina, which has a gaseous diffusion project; (2) Brazil, where work is ongoing on gas centrifuges; (3) China, which has gaseous diffusion and gas centrifuges under development; (4) France, including Eurodif, owned by France, Italy, Spain and Belgium, which has a gaseous diffusion plant at Tricastin, plus a topping plant at Pierrelatte; (5) Germany, which has a large-scale centrifuge plant (Urenco); (6) India, which has gas centrifuges; (7) Japan, which has gas centrifuges; (8) the Netherlands, where there is a large-scale Urenco gas centrifuge plant; (9) Pakistan, which has gas centrifuges; (10) South Africa, which has a version of an advanced vortex tube process and has been working on the gas centrifuge process; (11) the CIS, which has large-scale gaseous diffusion and gas centrifuge plants; and (12) the UK, which has gaseous diffusion and Urenco gas centrifuge plants.

In the United States, a group of domestic investors, including Duke Power, and Urenco, have applied for permission to construct a new gas centrifuge plant in Louisiana.

General Process and Design Selection

For difficult separations, such as isotope separations that involve the separation of molecules having very similar physical and chemical properties, the enrichment that can be obtained in a single equilibrium stage or transfer unit of the process is quite small. Hence, an extremely large number of these elementary separating units must be connected to form a separation cascade in order to achieve most desired separations. Consequently, very large separation systems requiring large amounts of energy are needed, and the total energy requirement is one of the most important cost considerations.

The energy or power required by any separation process is related more or less directly to its thermodynamic classification. There are, broadly speaking, three general types of continuous separation processes: reversible, partially reversible, and irreversible.

Reversible Processes. Distillation is an example of a theoretically reversible separation process. In fractional distillation, heat is introduced at the bottom stillpot to produce the column upflow in the form of vapor which is then condensed and turned back down as liquid reflux or column downflow. This system is fed at some intermediate point, and product and waste are withdrawn at the ends. Except for losses through the column wall, etc, the heat energy spent at the bottom vaporizer can be recovered at the top condenser, but at a lower temperature. Ideally, the energy input of such a process is dependent only on the properties of feed, product, and waste. Among the diffusion separation methods discussed herein, the centrifuge process (pressure diffusion) constitutes a theoretically reversible separation process.

Partially Reversible Processes. In a partially reversible type of process, exemplified by chemical exchange, the reflux system is generally derived from a chemical process and involves the consumption of chemicals needed to transfer the components from the upflow into the downflow at the top of the cascade, and to accomplish the reverse at the bottom. Therefore, although the separation process itself may be reversible, the entire process is not, if the reflux is not accomplished reversibly.

Insofar as the consumption of chemicals is concerned, it is obvious that the total consumption of reflux-producing chemicals is proportional to the interstage flows, or width of the cascade, but independent of the number of stages in series, or length of the system.

Irreversible Processes.   Irreversible processes are among the most expensive continuous processes. These are used only in special situations, such as when the separation factors of more efficient processes (that is, processes that are theoretically more efficient from an energy point of view) are found to be uneconomically small. Except for pressure diffusion, the diffusion methods discussed herein are essentially irreversible processes. Thus, gaseous diffusion, in which gas expands from a region of high pressure to one of low pressure, mass diffusion, in which a vapor flows from a region of high partial pressure to one of low partial pressure, and thermal diffusion, in which heat flows from a high temperature source to a low temperature sink, are all irreversible processes. In contrast with reversible and partially reversible processes, the energy demand in an irreversible process is distributed over the whole cascade in direct proportion to the distribution flow.

In gaseous diffusion, the cascade consists of individual stages that are connected in series. In each stage part of the gaseous feed is forced through a diffusion membrane or barrier with holes smaller than the mean free path of the gas (see MEMBRANE TECHNOLOGY). Because of slightly greater mobility, the lighter components flow preferentially through the barrier. This enriched portion of the feed is transported to a neighboring stage, up the cascade, where the lighter components tend to concentrate. The other portion of the gas that does not pass through the barrier is rejected to a neighboring stage, down the cascade, where the heavier components tend to concentrate. The feed to each stage is thus composed of combined upflow and downflow from neighboring stages.

In pressure diffusion, a pressure gradient is established by gravity or in a centrifugal field. The lighter components tend to concentrate in the low pressure (center) portion of the fluid. Countercurrent flow and cascading extend the separation effect.

Irreversible processes are mainly applied for the separation of heavy stable isotopes, where the separation factors of the more reversible methods, eg, distillation, absorption, or chemical exchange, are so low that the diffusion separation methods become economically more attractive. Although application of these processes is presented in terms of isotope separation, the results are equally valid for the description of separation processes for any ideal mixture of very similar constituents such as close-cut petroleum fractions, members of a homologous series of organic compounds, isomeric chemical compounds, or biological materials.

Cascade Design

Less conventional diffusional separation operations are characterized by the relatively small separations that can be obtained by the elementary separation mechanism. That is, the changes in fluid composition attained in gaseous diffusion across the barrier, in thermal diffusion between the hot and cold walls, in mass diffusion between the inlet and the condensing surface for the sweep vapor, and in the centrifuge between the axis of the rotor and its periphery, are all quite small. Thus, a large number of separating units must be employed. Cascade is the term given to the aggregation of separating units that have been interconnected so as to be able to produce the desired material. The optimum arrangement of the separating units in a separation cascade generally minimizes the unit cost of product, and its design is a problem common to all separation processes. In a stagewise separation process such as gaseous diffusion, each unit of equipment consists of one separation stage.

The Separation Stage. A fundamental quantity, α, exists in all stochastic separation processes, and is an index of the steady-state separation that can be attained in an element of the process equipment. The numerical value of α is developed for each process under consideration in the subsequent sections. The separation stage, which in a continuous separation process is called the transfer unit or equivalent theoretical plate, may be considered as a device separating a feed stream, or streams, into two product streams, often called heads and tails, or product and waste, such that the concentrations of the components in the two effluent streams are related by the quantity, α. For the case of the separation of a binary mixture this relationship is

$$\left(\frac{y}{1-y}\right)\Big/\left(\frac{x}{1-x}\right) = \alpha \tag{1}$$

where y is the mol fraction of the desired component in the upflowing (heads) stream from the stage and x the mol fraction of the same component in the downflowing (tails) stream from the stage. The quantity α is usually called the stage separation factor.

For the case of separating a binary mixture, the following conventions are used. The concentrations of the streams are specified by the mol fraction of the desired component. The purpose of the separation process is usually to obtain one component of the mixture in an enriched form. If both components are desired, the choice of the desired component is an arbitrary one. The upflowing stream from the separation stage is the one in which the desired component is enriched, and by virtue of this convention, α is defined as a quantity the value of which is greater than unity. However, for the processes considered here, α exceeds unity by only a very small fraction, and the relationship between the concentrations leaving the stage can be written, without appreciable error, in the form

$$y - x = (\alpha - 1)x(1 - x) \tag{2}$$

A separation stage or transfer unit operating on a binary mixture is shown schematically in Figure 1. In a cascade of separating stages, the feed stream can be formed by mixing the downflowing stream from the stage above and the upflowing stream from the stage below. The quantity, θ, ie, the fraction of the combined stage feed that goes into the stage upflow stream, is termed the cut of the stage. In cascades ordinarily designed for difficult separations, the stage cut is

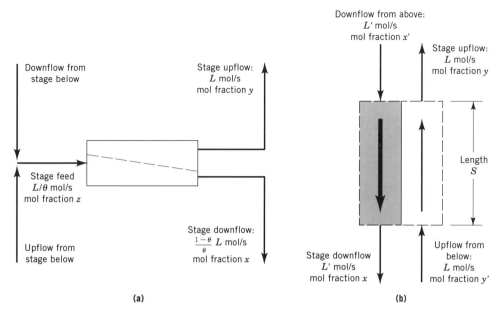

Fig. 1. The analogy between the separation stage and the transfer unit or equivalent theoretical plate: (**a**) in a stagewise process; (**b**) in a continuous-separation process. The terms are defined in text.

normally very nearly equal to one-half. In the case of a theoretical plate in a continuous process, the feed consists of two separate streams, one from above and one from below. In cascades for either stagewise or continuous processes the upflow rate L and the downflow rate L' (or $L(1 - \theta)/\theta$) are very nearly equal. For continuous process units, the length S is the length of equipment necessary to satisfy the requirement of equation 1, that the streams leaving the unit be related by α; it is usually called the height of a transfer unit (HTU) or the height equivalent to a theoretical plate (HETP). Although the HTU and HETP are defined differently and are not precisely equivalent to each other, the difference between them becomes negligible when the value of the quantity $\alpha - 1$ is small.

The Separative Capacity. The separation stage is characterized not only by the separation factor α but also by its capacity or throughput of which the upflow L is a measure, and in the case of the continuous process, also by the length S. It is therefore desirable to define and determine a quantity indicative of the amount of useful separative work that can be done per unit time by a single stage. Such a quantity is called the separative capacity of the stage. It is postulated that the separation stage does useful work on the streams it processes, hence increasing their net value. The value of a stream must be a function of its concentration; let this value function be designated by $v(x)$. Then the separative capacity of the stage by definition is set equal to the increase in value it creates. The separative capacity of a unit is a very useful concept and permits comparisons to be made between different separation processes.

The separative capacity of the stage, termed δU, is set equal to the net increase in value of the streams it processes (see Fig. 1):

$$\delta U = Lv(y) + \frac{1 - \theta}{\theta} Lv(x) - \frac{1}{\theta} Lv(z) \qquad (3)$$

The value functions appearing in equation 3 may be expanded in Taylor series about x and, because the concentration changes effected by a single stage are relatively small, only the first nonvanishing term is retained. When the value of z is replaced by its material balance equivalent, ie, equation 4:

$$z = (1 - \theta)x + \theta y \qquad (4)$$

the separative capacity of the stage is given by:

$$\delta U = \tfrac{1}{2} L (1 - \theta)(y - x)^2 v''(x) \qquad (5)$$

where $v''(x)$ is the second derivative of the value function. The concentrations y and x are related by equation 2; thus the separative capacity can also be written as:

$$\delta U = \tfrac{1}{2} L (1 - \theta)(\alpha - 1)^2 x^2 (1 - x)^2 v''(x) \qquad (6)$$

As it is desirable that the separative capacity of the stage be independent of the concentration of the material with which it is operating, the terms in the equation involving the concentration are set equal to a constant, taken for convenience to be unity, and the separative capacity of a single stage operating with a cut of one-half is seen to be:

$$\delta U = \tfrac{1}{4} L (\alpha - 1)^2 \qquad (7)$$

Thus, the separative capacity of a stage is directly proportional to the stage upflow as well as to the square of the separation effected.

Equivalent Theoretical Plate. The separative capacity of a theoretical plate in a continuous process can be obtained in the same manner. By equating the separative capacity of the unit to the net increase in value of the four streams handled (eq. 8):

$$\delta U = Lv(y) + L'v(x) - Lv(y') - L'v(x') \qquad (8)$$

After expansion in Taylor series about the concentration x and replacing the concentration y' by its material balance equivalent:

$$\delta U = L(y - x')(x' - x)v''(x) \qquad (9)$$

The separative capacity of the equivalent theoretical stage in the continuous process is seen to depend on the concentration difference between the countercurrent streams as well as on the concentration difference between the top and bottom of the stage. The separative capacity is zero when x' is equal to y or x' is equal to x;

inspection shows that it attains a maximum value when x' is equal to the arithmetic average of x and y and that this maximum value is:

$$(\delta U)_{\max} = \tfrac{1}{4} L(y - x)^2 v''(x) = \tfrac{1}{4} L (\alpha - 1)^2 \tag{10}$$

Thus, the maximum value of the separative capacity of a theoretical plate in a continuous process is equal to that of a single separation stage when both units have the same value of the $L(\alpha - 1)^2$ product. When the continuous process is operated so as to yield its maximum separative capacity, the concentrations y' and x' of the streams entering the unit are equal and the similarity between the separation stage and the theoretical plate is accentuated because, for this case, both may be considered to separate a single feed stream into two product streams having concentrations related by α. The definition of a theoretical plate in the continuous process is essentially arbitrary and not required; however, it is a useful concept, permitting both the stagewise and continuous processes to be treated with the same set of cascade equations.

The Value Function. The value function itself is defined, as has been indicated above, by the second-order differential:

$$v''(x) = 1/[x^2(1 - x)^2] \tag{11}$$

In the design of cascades, a tabulation of $v(x)$ and of $v'(x)$ is useful. The solution of the above differential equation contains two arbitrary constants. A simple form of this solution results when the constants are evaluated from the boundary conditions $v(0.5) = v'(0.5) = 0$. The expression for the value function is then:

$$v(x) = (2x - 1)\ln[x/(1 - x)] \tag{12}$$

and for the derivative of the value function (eq. 13):

$$v'(x) = \frac{2x - 1}{x(1 - x)} + 2 \ln \frac{x}{1 - x} \tag{13}$$

Therefore, $v(1 - x) = v(x)$ and $v'(1 - x) = -v'(x)$.

Application. In addition to providing a relatively simple means for estimating the production of separation cascades, the separative capacity is useful for solving some basic cascade design problems; for example, the problem of determining the optimum size of the stripping section.

It can be assumed that P, y_P, and x_F for the cascade have been specified, and that the cost of feed and the cost per unit of separative work, the product of separative capacity and time, are known. The basic assumption is that the unit cost of separative work remains essentially constant for small changes in the total plant size. The cost of the operation can then be expressed as the sum of the feed cost and cost of separative work:

$$C_{\text{total}} = (C_F)(F)(\Delta t) + C_{\Delta U}[Pv(y_P) + Wv(x_W) - Fv(x_F)]\Delta t \tag{14}$$

where C_{total} is the total cost of operation for the period of time Δt, and C_F and $C_{\Delta U}$ are the cost per unit of feed and the cost per unit of separative work, respectively. The optimum value of x_W is that which minimizes the total cost and can be found by differentiating the total cost with respect to x_W under the restrictions that P, y_P, and x_F remain constant, and setting the result equal to zero. The result of this procedure is that the optimum x_W is the solution to equation 15:

$$v(x_F) - v(x_W) - (x_F - x_W)v'(x_W) = C_F/C_{\Delta U} \tag{15}$$

When the cascade is operated using the optimum x_W, the cost of producing material at any other concentration, y_P, is given by:

$$C_{\text{total}} = C_{\Delta U}\, P[v(y_P) - v(x_W) - (y_P - x_w)v'(x_W)]\Delta t \tag{16}$$

obtained by combining equations 14 and 15. An equation of this form can be used to establish the value of material of different concentrations from separation cascades.

Cascade Gradient Equations.   An arrangement of separation stages to form a simple cascade is shown in Figure 2. A simple cascade is one that divides a single cascade feed stream into a product stream and a waste stream. Additional side streams, however, could easily be handled. To be consistent with the conventions given for the single stage, the desired component is assumed to be enriched in the product stream at the top of the cascade. The cascade feed is introduced at some intermediate stage between the top and bottom of the cascade. The portion of the cascade that lies above the feed point is termed the enriching section; that which lies below the feed point is termed the stripping section. The gradient equations for the cascade are obtained from a combination of the material balance equations, frequently called the operating-line equations, and the α relationship, usually called the equilibrium-line equation. From a material balance around the top of the cascade down to, but not including, stage n of the enriching section, is obtained the operating-line equation:

$$L_n y_n = (L_n - P)x_{n+1} + Py_P \tag{17}$$

which can be combined with the equilibrium-line (eq. 2) to give:

$$x_{n+1} - x_n = \frac{L_n}{L_n - P}[(\alpha - 1)x_n(1 - x_n) - (P/L_n)(y_P - x_n)] \tag{18}$$

For the case under consideration, where the value of $\alpha - 1$ is quite small, it follows that everywhere in the cascade, except possibly at the extreme ends, the stage upflow is many times greater than the product withdrawal rate. Thus $L/(L - P)$ can be set equal to unity. Furthermore when the value of $\alpha - 1$ is small, the stage enrichment $x_{n+1} - x_n$ can be approximated by the differential ratio dx/dn without appreciable error. The gradient equation for the enriching section of a simple cascade therefore takes the form

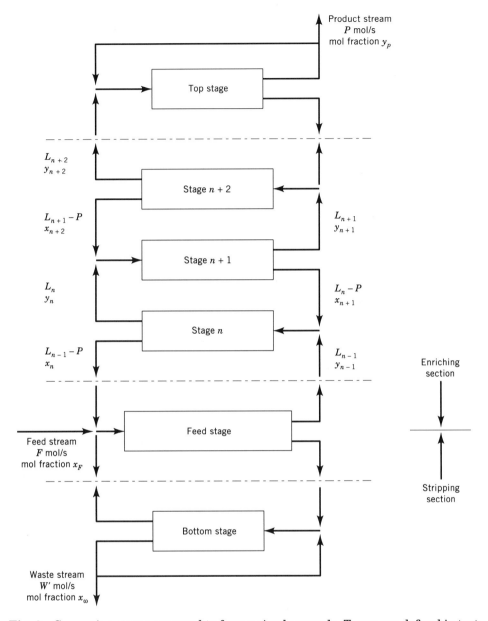

Fig. 2. Separation stages arranged to form a simple cascade. Terms are defined in text.

$$dx/dn = (\alpha - 1)x(1 - x) - (P/L)(y_P - x) \qquad (19)$$

Similarly, one obtains a gradient equation for the stripping section that has the form

$$dx/dn = (\alpha - 1)x(1 - x) - (W/L)(x - x_W) \qquad (20)$$

Equations 19 and 20 are the basic equations for cascade design. Although these equations were derived from a consideration of a cascade composed of discrete separation stages, equations of the same form are also obtained for cascade designs based on continuous or differential separation processes. For use in the case of continuous separation processes, however, the term dx/dn, which is the enrichment per stage, is usually replaced by the equivalent terms $S\, dx/dz$, where S is the stage length and dx/dz the enrichment per unit length of process equipment. These equations may then be used to calculate the output from a given cascade configuration, that is, from a cascade for which the variation of $\alpha - 1$ and L is known as a function of the stage number.

Minimum Length or Minimum Number of Stages.   It is evident from the gradient equations that the enrichment per stage decreases as the withdrawal rate increases. Thus the minimum number of stages required to span a given concentration difference is obtained when no material is withdrawn from the cascade. This mode of operation $(P = W = F = 0)$ is frequently called total reflux operation. Integration of the gradient equation for this case with $\alpha - 1$ taken to be constant gives:

$$N_{\min} = \frac{1}{\alpha - 1} \ln \left(\frac{x_T}{1 - x_T} \middle/ \frac{x_B}{1 - x_B} \right) \tag{21}$$

where the concentration range to be spanned is from the concentration x_B at the bottom to concentration x_T at the top. As an example of the magnitudes involved, consider the enrichment of ^{235}U by gaseous diffusion from $x_B = 0.005$ to $x_T = 0.90$. For a value of α equal to 1.0043 the minimum number of diffusion stages required is 1742.

Minimum Width or Minimum Stage Upflow.   It also follows directly from the gradient equations that if the withdrawal rates from the cascade are nonzero, it is necessary that the stage upflow from the stage at which the cascade concentration is x must exceed some critical value in order that there be any enrichment at that point in the cascade. This critical value is called the minimum stage upflow and is obtained by setting dx/dn equal to zero in the gradient equation. Thus, for any point in the enriching section the minimum stage upflow is given by

$$L_{\min} = P(y_P - x)/[(\alpha - 1)x(1 - x)] \tag{22}$$

For the case of enriching ^{235}U to 90 mol % product concentration, the stage upflow at the feed point $(x_F = 0.0072)$ must therefore exceed 29,046 times the product withdrawal rate. It can now be seen from a consideration of the minimum stage upflow that the approximation made in deriving the gradient equation, ie, taking the quantity $(1 - P/L)$ equal to unity, introduces negligible error except possibly in the immediate vicinity of the withdrawal points. The condition that arises in a cascade at points where the stage upflow approaches the value $L_{\min}$ is commonly called pinching.

Gradient Equations for a Square Section.   A section of a cascade composed of identical stages, that is, a number of stages having the same separation factor and the same stage upflow, is called a square section. For sections of this type the gradient equations are readily integrable. For a section in the enricher of a cascade producing material at rate P and concentration y_P, the solution can be written:

$$N_{\text{sect}} = [(\alpha - 1)(X_1 - X_0)]^{-1} \ln \frac{(X_1 - x_B)(x_T - X_0)}{(X_1 - x_T)(x_B - X_0)} \qquad (23)$$

where X_1 and X_0 are the roots of a quadratic equation and are given by:

$$X_1, X_0 = \frac{L(\alpha - 1) + P \pm \{[L(\alpha - 1) + P]^2 - 4 L(\alpha - 1)Py_P\}^{1/2}}{[2L(\alpha - 1)]} \qquad (24)$$

and N_{sect} is the number of stages necessary to span the concentration difference from x_B at the bottom of the section to x_T at the top of the section. Equation 23 is also obtained for a square section in the stripping section, but in the case of the stripper the value of X_1 and X_0 are given by:

$$X_1, X_0 = \frac{\{L(\alpha - 1) - W \pm \{[L(\alpha - 1) - W]^2 + 4 L(\alpha - 1)Wx_W\}^{1/2}\}}{[2L(\alpha - 1)]} \qquad (25)$$

Graphical Solution.   Some of the preceding concepts can be illustrated graphically by means of a McCabe-Thiele diagram, shown in Figure 3. In such a diagram the equilibrium line, equations 1 or 2, and the operating line, equation 17, are

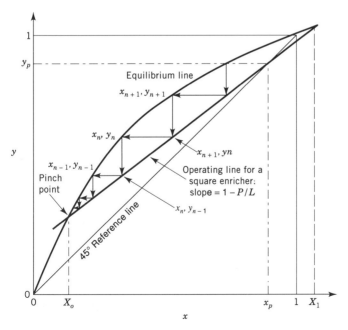

Fig. 3.   A McCabe-Thiele diagram for a hypothetical square cascade section illustrating pinching. Terms are defined in text.

plotted on a set of x, y coordinates. For the case of a square cascade section, the operating line is straight and has the slope $(1 - P/L)$; if the section lies at the product withdrawal end of the cascade, its opening line passes through the point $x = y = y_P$, as shown. The number of stages required to span a given concentration difference, or conversely the concentration difference obtained across a square section with a given number of stages can be illustrated in such a figure.

It is evident from the construction that the closer the operating line lies to the 45-degree reference line, the fewer the number of stages required to span a given concentration difference. The minimum number of stages is therefore required when the operating line coincides with the 45-degree line, that is, when P/L is equal to zero. It is also evident that in the neighborhood of a point of intersection of the operating and equilibrium lines the enrichment per stage becomes quite small and is equal to zero at the point of intersection itself. At such a point pinching is said to occur, and the origin of the term is made clear by the diagram. In order for a cascade section to span a concentration difference from x_B to y_P, it follows that the operating line may not intersect the equilibrium line before the concentration x_B is reached; the value of L for which the two curves intersect at x_B is the minimum upflow corresponding to the cascade concentration x_B. It is also noteworthy that the values X_0 and X_1 appearing in equations 23 through 25 are the x coordinates of the two points of intersection of the operating line of a square section with the equilibrium line. Although the graphical solution of the cascade gradient equation is simple in principle and exact in theory, it becomes quite cumbersome in practice when processes having separation factors close to unity and hence cascades with thousands of stages are under consideration. For this reason analytic solutions to the gradient equation are usually preferred.

The Ideal Cascade. A cascade of particular interest to design engineers is the ideal cascade: a continuously tapered cascade (ie, L is a continuously varying function of x or n) that has the property of minimizing the sum of the stage upflows of all the stages required to achieve a given separation task. Because, in general, the total volume of the equipment required and the total power requirement of the cascade are directly proportional to the sum of the stage upflows, a consideration of the ideal plant requirements often permits a good economic estimate of the unit cost of product to be made without having to resort to the much more painstaking labor of designing a real (as opposed to ideal) cascade to accomplish the separation job. A simple, intuitive approach to the ideal cascade concept in the case of a cascade composed of discrete stages follows. Again, the resulting equations are also valid for a cascade based on a continuous or differential separation process.

For the case of a stagewise enrichment process the ideal cascade may be defined as one in which there is no mixing of streams of unequal concentrations. Clearly, the mixing of streams of unequal concentrations in the cascade to form the feed to a separation stage constitutes an inefficiency because it is precisely the reverse of the process taking place in the stage itself. Figure 2 shows that the no-mixing condition at the entrance to stage $n + 1$ requires that $y_n = x_{n+2}$. If the enrichment per stage is essentially constant, x_{n+2} may be written as $x_n + 2(dx/dn)$. The concentration y_n is related to x_n by the α-relationship (eq. 2).

Thus the no-mixing condition leads directly to the gradient equation for the ideal plant:

$$\frac{dx}{dn} = \frac{\alpha - 1}{2} x(1 - x) \tag{26}$$

The number of stages required to span a given concentration difference in an ideal plant in which all stages have the same separation factor is therefore

$$N_{\text{ideal}} = \frac{2}{\alpha - 1} \ln \left(\frac{x_T}{1 - x_T} \bigg/ \frac{x_R}{1 - x_R} \right) = 2N_{\min} \tag{27}$$

and is twice the minimum number of stages required. The combination of equations 19 and 26 gives the equation for the stage upflow at any point in the enricher of an ideal cascade that is twice the minimum upflow

$$L_{\text{ideal}} = \frac{2P(y_P - x)}{(\alpha - 1)x(1 - x)} = 2L_{\min} \tag{28}$$

Equations 27 and 28 can be used in conjunction, along with the corresponding equations for the stripping section, to produce an ideal plant profile such as is shown in Figure 4 where L_{ideal} is plotted against N_{ideal} for the example of an ideal cascade to produce one mol of uranium per unit time enriched to 90 mol % in ^{235}U from natural feed containing 0.72 mol % ^{235}U, with a waste stream rejected at a concentration of 0.5% ^{235}U. The characteristic lozenge shape of the ideal cascade

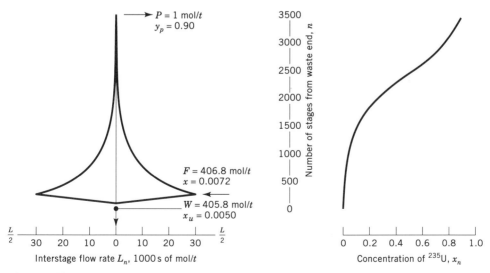

Fig. 4. Characteristics of an ideal separation cascade for uranium isotope separations. For this cascade: $\alpha = 1.0043$; $N_T = 3484$ stages; $\Delta U = 153.08$ mol/t; $\Sigma L_T = 33.116 \times 10^2$ mol/t, where t represents unit time.

is evident: no two stages in either the enricher or stripper are the same size. It can be deduced from the above statements that, except at the terminals, the operating line for an ideal cascade on a McCabe-Thiele diagram is a curved line lying midway between the equilibrium line and the 45° reference line.

Total Upflow in an Ideal Plant.   The sum of the upflows from all of the stages in the ideal plant, or more simply, the total upflow, is the area enclosed by the cascade shown in Figure 4. An analytical expression for this quantity is obtained as the summation of all the stage upflows in the enriching section expressed as an integral:

$$\sum^{\text{enr}} L_n = \int L \, dn = \int_{x_F}^{y_P} L \frac{dn}{dx} \, dx \tag{29}$$

For the ideal cascade, L is given by equation 28 and dx/dn by equation 26. Making these substitutions:

$$\sum^{\text{enr}} L_n = \int_{x_F}^{y_P} \frac{4 \, P}{(\alpha - 1)^2} \frac{(y_P - x)}{x^2(1 - x)^2} \, dx \tag{30}$$

However, recalling the definition of the value function, equation 11, and assuming that the value of α is the same for all stages, the integral may be written in the form:

$$\sum^{\text{enr}} L_n = \frac{4 \, P}{(\alpha - 1)^2} \int_{x_F}^{y_P} (y_P - x)v''(x)dx \tag{31}$$

which is readily integrated by parts to give:

$$(\text{total upflow})_{\text{enr}} = [4 \, P/(\alpha - 1)^2][v(y_P) - v(x_F) - (y_P - x_F)v'(x_F)] \tag{32}$$

The equation for the total flow in the stripping section is obtained in the same manner:

$$(\text{total upflow})_{\text{str}} = [4 \, W/(\alpha - 1)^2][v(x_W) - v(x_F) - (x_W - x_F)v'(x_F)] \tag{33}$$

The total flow in the cascade is then given by the sum of equations 32 and 33, which can be simplified with the use of the cascade material balances:

$$P + W = F \tag{34}$$

and

$$Py_P + Wx_W = Fx_F \tag{35}$$

to give the convenient form

$$(\text{total upflow})_{\text{cascade}} = [4/(\alpha - 1)^2][Pv(y_P) + Wv(x_W) - Fv(x_F)] \qquad (36)$$

For the example considered above, the total cascade upflow is found to be 33×10^6 mols per unit time.

The second term in brackets in equation 36 is the separative work produced per unit time, called the separative capacity of the cascade. It is a function only of the rates and concentrations of the separation task being performed, and its value can be calculated quite easily from a value balance about the cascade. The separative capacity, sometimes called the separative power, is a defined mathematical quantity. Its usefulness arises from the fact that it is directly proportional to the total flow in the cascade and, therefore, directly proportional to the amount of equipment required for the cascade, the power requirement of the cascade, and the cost of the cascade. The separative capacity can be calculated using either molar flows and mol fractions or mass flows and weight fractions. The common unit for measuring separative work is the separative work unit (SWU) which is obtained when the flows are measured in kilograms of uranium and the concentrations in weight fractions.

The great utility of the separative capacity concept lies in the fact that if the separative capacity of a single separation element can be determined, perhaps from equations 7 or 10, then the total number of such identical elements required in an ideal cascade to perform a desired separation job is simply the ratio of the separative capacity of the cascade to that of the element. The concept of an ideal plant is useful because moderate departures from ideality do not appreciably affect the results. For example, if the upflow in a cascade is everywhere a factor of m times the ideal upflow, the actual total upflow required to perform a separative task is $m^2/(2m - 1)$ times the ideal cascade total upflow. Thus, if the upflow is 20% greater than ideal at every point in that cascade ($m = 1.2$), the number of separation elements would be only 2.86% greater than that calculated from ideal cascade considerations.

Equations for Large Stage Separation Factors. The preceding results have been obtained with the use of equation 2 and by replacing the finite difference, $x_{n+1} - x_n$, by the differential, dx/dn, both of which are valid only when the quantity $(\alpha - 1)$ is very small compared with unity. However, there has been renewed interest, partly because of the development of the gas centrifuge process to commercial status, in the design of cascades composed of stages with large stage separation factors. When the stage separation factor is large, the number of stages required in an ideal cascade in which all stages have the same separation factor is given by

$$N_{\text{ideal}} = \frac{2}{\ln\alpha} \ln \left(\frac{y_P}{1 - y_P} \Big/ \frac{x_W}{1 - x_W} \right) - 1 \qquad (37)$$

instead of equation 27. When dealing with cascades composed of stages having large separation factors, it is somewhat more convenient to calculate the sum of all the stage feed flows in the cascade rather than the sum of all the stage upflows as was done in the case when $(\alpha - 1)$ is small. When $(\alpha - 1)$ is small with respect to unity, the stage feed flow is essentially just twice the stage upflow rate, and the stage feed flow rate in an ideal cascade (see eq. 28) is:

$$(L/\theta)_{ideal} = \frac{4P(y_P - x)}{(\alpha - 1)x(1 - x)} \tag{38}$$

However, when α is large, the corresponding equation for the stage feed rate takes the form

$$(L/\theta)_{ideal} = \frac{(\alpha)^{1/2} + 1}{(\alpha)^{1/2} - 1} \frac{P(y_P - z)}{z(1 - z)} \tag{39}$$

The sum of the stage feed flow rates of all of the stages in an ideal cascade is just twice the total cascade upflow rate when $(\alpha - 1)$ is small with respect to unity, or

$$(total\ stage\ feed)_{cascade} = \frac{8}{(\alpha - 1)^2} [Pv(y_P) + Wv(x_W) - F_V(x_F)] \tag{40}$$

but is given by

$$(total\ stage\ feed)_{cascade} = \frac{2}{\ln\alpha} \frac{(\alpha)^{1/2} + 1}{(\alpha)^{1/2} - 1} [Pv(y_P) + WV(x_W) - FV(x_F)] \tag{41}$$

when α is larger. It can be seen that when α is close to unity, equation 41 gives the same result as equation 40. However, if α is equal to 1.1, the total stage feed would be underestimated by 9.2%, and if α is equal to 2.0, the total stage feed would be underestimated by 42.4%, using equation 40 instead of equation 41. Some of the work in this area of cascade theory has dealt with the design of cascades using asymmetric isotope separation stages (1,2) with the analysis of two-up, one-down cascades, that is, cascades in which the upflow from the nth stage in the cascade bypasses the $(n + 1)^{th}$ stage and is reintroduced at the $(n + 2)^{th}$ stage.

Real Cascades. Although the ideal cascade minimizes the volume of equipment and the energy requirements, the cost of the cascade is generally not minimized because production economies are realized in the manufacture of the process equipment when a large number of identical units are produced. Thus, a minimum-cost cascade consists of a number of square cascade sections rather than uniformly tapered nonidentical stages. A first approximation to the optimum practical cascade, once the size (length and width) of the individual separating units available is known, is obtained by fitting the ideal plant shape with square sections in some intuitively appealing manner, as illustrated in Figure 5.

Figure 5 shows an ideally tapered enricher that has been replaced by three square cascade sections, a process called squaring-off the cascade (3–6). During the squaring-off process, two essential requirements must be kept in mind: The interstage flow in all-square sections must always exceed the local value of L_{min} at all points in the cascade, and the squared-off cascade must contain a total

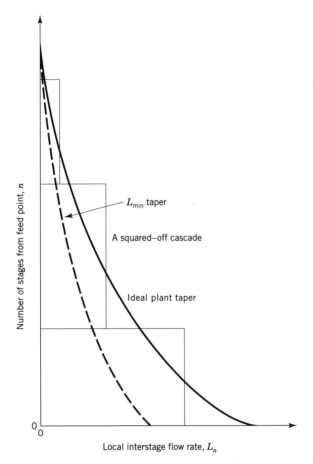

Fig. 5. Design of a real cascade obtained by squaring off an ideal enriching section.

number of stages which exceeds N_{min}. In order for the squared-off cascade to give a performance closely resembling that of the ideal cascade, the shape of the squared-off cascade should approximate the shape of the ideal cascade.

In the final analysis the problem of determining the optimum practical cascade is rather complex. Equipment performance and costs need to be related to the selected independent process variables. The main process equipment usually consists of a large number of separating units, pumping and heat exchange equipment, control devices, and connecting piping. The whole process is provided with services, and auxiliary systems and feed and withdrawal facilities. It is usually enclosed by a building and surrounded by land of the proper type. Sizes and cost of these important items must be related to the process variables. The details of procedures used for the optimization of real gaseous diffusion cascades are presented in Reference 7, and the problems of optimization of real gas centrifuge cascades are discussed in Reference 8.

Time-Dependent Cascade Behavior. The period of time during which a cascade must be operated from start-up until the desired product material can be withdrawn is called the equilibrium time of the cascade. The equilibrium time of cascades utilizing processes having small values of $\alpha - 1$ is a very important quantity. Often a cascade may prove to be quite impractical because of an excessively long equilibrium time. An estimate of the equilibrium time of a cascade can be obtained from the ratio of the enriched inventory of desired component at steady state, H, to the average net upward transport of desired component over the entire transient period from start-up to steady state, $\bar{\tau}$. In equation form this definition can be written as

$$T_{eq} = H/\bar{\tau} = \frac{1}{\bar{\tau}} \int_0^N h_n(x_n - x_F)dn \tag{42}$$

where h_n is the holdup of the nth stage. The average net upward transport for the entire transient period is not usually known; the initial and final values of the net transport, however, are known. At start-up the concentration gradient is flat, because the column is filled with material at feed concentration, and the transport is a maximum. Using this transport in equation 42 gives a lower limit for the equilibrium time

$$(T_{eq})_{min} = H/[L(\alpha - 1)x_F(1 - x_F)] \tag{43}$$

At steady state, with a fully developed gradient, the net transport is $P(y_P - x_F)$, which is a lower limit for net upward transport. Substituting this into equation 42, leads to an expression for an upper limit for the equilibrium time:

$$(T_{eq})_{max} = H/[P(y_P - x_F)] \tag{44}$$

Equations 43 and 44 thus yield a lower and upper limit, respectively, and used together usually give a satisfactory estimate for the equilibrium time of a cascade.

Examination of equation 42 shows that T_{eq} is directly proportional to the average stage holdup of process material. Thus, in conjunction with the fact that liquid densities are on the order of a thousand times larger than gas densities at normal conditions, the reason for the widespread use of gas-phase processes in preference to liquid-phase processes in cascades for achieving difficult separations becomes clear.

The unsteady-state behavior of a separation unit is, furthermore, of interest because it can be used for the experimental determination of the separation parameters of the unit. If the holdup of the separating unit is known, the separation factor, $\alpha - 1$, can be obtained from a knowledge of the transient behavior of the unit during start-up. Figure 6 shows the concentration gradient in a square column shortly after start-up. As long as the gradient is flat at the feed point as shown, an equation much like equation 44 for the maximum equilibrium time can be used to relate the enriched holdup to the elapsed time. From a knowledge of the gradient, the enriched holdup H can be computed, and with L known the separation factor $\alpha - 1$ can be computed from

$$(\alpha - 1) = H/[Lx_F(1 - x_F) \, \Delta t] \tag{45}$$

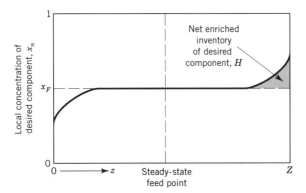

Fig. 6. Concentration gradient in a column or cascade a short time after start-up.

where Δt is the elapsed time since start-up when the column was filled with material at feed concentration.

The Gaseous Diffusion Process

The gaseous diffusion separation process depends on the separation effect arising from the phenomenon of molecular effusion (that is, the flow of gas through small orifices). When a mixture of two gases is confined in a vessel and is in thermal equilibrium with its surroundings, the molecules of the lighter gas strike the walls of the vessel more frequently, relative to its concentration, than the molecules of the heavier gas. This is caused by the greater average thermal velocity of the lighter molecules. If the walls of the container are porous with holes large enough to permit the escape of individual molecules, but sufficiently small so that bulk flow of the gas as a whole is prevented (that is, with pore diameters approaching mean-free-path dimensions of the gas), then the lighter molecules escape more readily than the heavier ones, and the escaping gas is enriched with respect to the lighter component of the mixture. The equation for the separation factor, α, for this process reflects the relative ease of light versus heavy molecules in escaping through the pores. Indeed, α^*, the ideal separation factor, is the ratio of the two molecular velocities. Because the kinetic energies, $\frac{1}{2}\,mv^2$, of the two species are the same, α^*, the ratio of the two velocities is equal also to the square root of the inverse ratio of the two molecular weights. In 1895 Rayleigh and Ramsey used this method to separate argon from nitrogen, and in 1920 it was employed to slightly enrich the concentration of the neon-22 isotope.

A primary improvement in diffusion separation technology was the development in 1932 of a cascade of diffusion stages for isotope separation by an arrangement similar to that shown in Figure 7. Using a 24-stage cascade an appreciable enrichment in the isotopes of neon was obtained. Subsequently (9), almost pure deuterium was obtained from a cascade of 48 stages, and the isotopes of nitrogen and of carbon were enriched (10) using a 34-stage cascade.

The large plants built for the separation of uranium isotopes following World War II are outstanding applications of the gaseous diffusion process. In the United

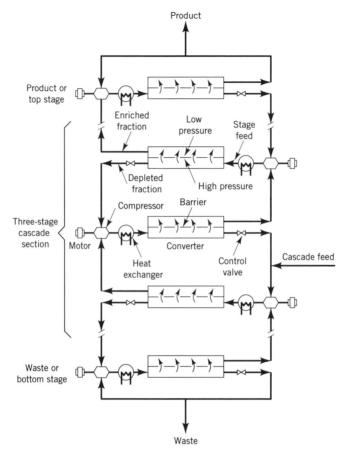

Fig. 7. A cascade of gaseous diffusion stages.

States this work culminated in the construction of the gaseous diffusion cascade called the K-25 plant at Oak Ridge, Tennessee. Other plants were built at Oak Ridge, Paducah, Kentucky, and Portsmouth, Ohio, in the United States (11–13) and at Capenhurst, England (14), and at Pierrelatte and Tricastin, France (15). Gaseous diffusion plants have also been reported to be in operation in the former USSR and in the People's Republic of China.

Process Description. The basic unit of a gaseous diffusion cascade is the gaseous diffusion stage. The main components are the converter holding the barrier in tubular form, motors, and compressors moving the gas between stages, a heat exchanger removing the heat of compression introduced by the stage compressors, the interstage piping, and special instruments and controls to maintain the desired pressures and temperatures.

Figure 7 is a schematic representation of a section of a cascade. The feed stream to a stage consists of the depleted stream from the stage above and the enriched stream from the stage below. This mixture is first compressed and then cooled so that it enters the diffusion chamber at some predetermined optimum temperature and pressure. In the case of uranium isotope separation the process gas is uranium hexafluoride [7783-81-5], UF_6. Within the diffusion chamber the gas flows along a porous membrane or diffusion barrier. Approximately one-half of the gas passes through the barrier into a region of lower pressure. This gas is enriched in the component of lower molecular weight, ^{235}U. The enriched fraction, upon leaving the diffusion chamber, is directed to the stage above where it is recompressed to the barrier high-side pressure. The gas that does not pass through the barrier is depleted with respect to the light component. This depleted fraction, upon leaving the chamber, passes through a control valve, and is directed to the stage below where it too is recompressed to the barrier high-side pressure. However, because it is necessary in this case to compensate only for the frictional losses and the control valve pressure drop, the compression ratio may not need to be as high as that for the enriched fraction. Thus there is some freedom in the design of the stage; that is, two compressors might be used, a larger one for the enriched fraction and a smaller one for the depleted fraction; or the pressure drop across the control valve may be made equal to the pressure drop across the barrier so that both streams can be recompressed by the same compressor (this scheme, although wasteful of power, might make up for it in savings in equipment costs), or some other compromise mode of operation might be used.

For operational efficiency a number of gaseous diffusion stages are operated together in units referred to as cells and buildings. Cells and buildings can be removed from operation for routine maintenance and bypassed without disturbing the diffusion cascade.

Successful operation of the gaseous diffusion process requires a special, fine-pored diffusion barrier, mechanically reliable and chemically resistant to corrosive attack by the process gas. For an effective separating barrier, the diameter of the pores must approach the range of the mean free path of the gas molecules, and in order to keep the total barrier area required as small as possible, the number of pores per unit area must be large. Seals are needed on the compressors to prevent both the escape of process gas and the inflow of harmful impurities. Some of the problems of cascade operation are discussed in Reference 16.

The need for a large number of stages and for the special equipment makes gaseous diffusion an expensive process. The three United States gaseous diffusion plants represent a capital expenditure of close to 2.5×10^9 dollars (17). However, the gaseous diffusion process is one of the more economical processes yet devised for the separation of uranium isotopes on a large scale.

Stage Design. The important parameters of a separation cascade employing gaseous diffusion stages are the stage separation factor and the size of a stage required to handle the desired stage flows. Both of these parameters depend on the characteristics of the barrier.

Barrier Characteristics. The barrier material must be fine-pored and have many pores per unit area. Preparation and characterization of such a material presents a difficult technological problem. The characteristics of a barrier suitable for the separation of isotopes by gaseous diffusion are discussed in Reference 18, including various effective pore sizes and pore size distributions. Experimental techniques used to evaluate barrier characteristics are presented in Reference 19. These techniques include adsorption methods, electron microscopy, x-ray analysis, porosity measurements with mercury, permeability measurements with liquids and gases, and measurements of separation effectiveness. Barrier materials have pore sizes in the range of 10–30 nm (20). One electrolytic technique leads to a thin sheet material having about 10^{10} pores per square centimeter and radii on the order of 15 nm (21). The separating performance of a barrier has been evaluated by means of a 12-stage pilot plant (22).

Barrier Flow.   An ideal separation barrier is one that permits flow only by effusion, as is the case when the diameter of the pores in the barrier is sufficiently small compared to the mean free path of the gas molecules. If the pores in the barrier are treated as a collection of straight circular capillaries, the rate of effusion through the barrier is governed by Knudsen's law (eq. 46):

$$N = \frac{4}{3} (2\pi MRT)^{-1/2} \frac{\phi d}{l} (p_f - p_b) \tag{46}$$

where N is the molar flow of gas per unit area through the barrier, M is its molecular weight, R is the gas constant, T is the absolute temperature, ϕ is the fraction of the barrier area open to flow, d is the effective pore diameter, l is the pore length or thickness of the barrier, and p_f and p_b are the high- and low-side pressures of the barrier, respectively. In practice not all of the flow through the barrier is effusive flow. Through those pores where the diameters are of the order of the mean free path or greater, a nonseparative Poiseuille flow occurs. The two types of flow are additive and the total flow can be represented by

$$N = \frac{a}{(M)^{1/2}} (p_f - p_b) + \frac{b}{\mu} (p_f^2 - p_b^2) \tag{47}$$

where a and b are functions of temperature for a particular barrier and μ is the viscosity of the gas. The first term on the right is the contribution to the total flow of the effusive flow, the second that of the nonseparative Poiseuille flow. Because the pressure dependence of each type of flow is different, the constants a and b can be evaluated from a series of measurements at different pressures (24).

The Fundamental Separation Effect.   An ideal-point separation factor can be defined on the basis of the separation obtained when a binary mixture flows through an ideal barrier into a region of zero back pressure. For this case an expression of the form of equation 46 can be written for each component. The flow of light component through the barrier is proportional to $p_f x'/(M_A)^{1/2}$, and the flow of heavy component is proportional to $p_f (1 - x')/(M_B)^{1/2}$ where x' is the mol fraction of the light component on the high-pressure side of the barrier and M_A and M_B are the mol wts of the light and heavy components, respectively. The concentration, y', of the effusing gas is therefore

$$y' = \frac{x'/(M_A)^{1/2}}{x'/(M_A)^{1/2} + (1 - x')/(M_B)^{1/2}} \tag{48}$$

and from the definition of α (eq. 1), it follows that the ideal-point separation factor is equal to

$$\alpha^* = \frac{y'/(1 - y')}{x'/(1 - x')} = (M_B/M_A)^{1/2} \tag{49}$$

which, for the case of uranium isotope separation using UF_6, is equal to 1.00429. As has been pointed out, this is also the expression for the ratio of the velocity of the light molecules to that of the heavy molecules.

The Stage Separation Factor.   The stage separation factor, in all probability, is appreciably different from the ideal-point separation factor because of the existence of four efficiency terms:

(*1*) A Barrier Efficiency Factor. In practice, diffusion plant barriers do not behave ideally; that is, a portion of the flow through the barrier is bulk or Poiseuille flow which is of a nonseparative nature. In addition, at finite pressure the Knudsen flow (25) is not separative to the ideal extent, that is, $(M_A/M_B)^{1/2}$. Instead, the degree of separation associated with the Knudsen flow is less separative by an amount that depends on the pressure of operation. To a first approximation, the barrier efficiency is equal to the Knudsen flow multiplied by a pressure-dependent term associated with its degree of separation, divided by the total flow.

(*2*) A Back-Pressure Efficiency Factor. Because a gaseous diffusion stage operates with a low-side pressure p_b which is not negligible with respect to p_f, there is also some tendency for the lighter component to effuse preferentially back through the barrier. To a first approximation the back-pressure efficiency factor is equal to $(1 - r)$, where r is the pressure ratio p_b/p_f.

(*3*) A Mixing Efficiency Factor. As the gas flows along the high-pressure side of the diffusion barrier, it becomes, as a result of the effusion process, preferentially depleted with respect to the lighter component in the neighborhood immediately adjacent to the barrier. As a result a concentration gradient perpendicular to the barrier is set up on the high-pressure side, and the average concentration x' of the light component in the bulk of the gas flowing past a point is greater than x'', the concentration of the light component at the surface of the barrier at that point. The mixing efficiency factor is equal to the ratio $(y' - x')/(y' - x'')$ as indicated in Figure 8. A value for the point mixing efficiency factor can be calculated from a consideration of diffusion through an effective film representing the resistance to diffusion. It is given by an expression of the form: $\exp(-Nl_f/\rho D)$, where l_f is the thickness of the effective film.

(*4*) A Cut-Correction Factor. The stage separation factor has been defined as relating the concentrations in the streams leaving the stage. Because the concentrations, x' and y', on each side of the barrier are changing continuously as the gas flows through the diffusion stage, the relationship between the concentrations of the streams differs from the point relationship. This difference is taken into account with the cut correction factor. If the gas on the high-pressure side flows through the stage with no appreciable mixing taking place in the direction of flow, and if the effused fraction is withdrawn from the stage directly upon passing through the barrier, the cut correction can be calculated from material balance considerations. For this case, the exit or stream concentration of the stage upflow is equal to the average concentration of the effused gas, whereas the exit or stream concentration of the downflow is equal to the terminal concentration of the uneffused gas which is, of course, at maximum and not average depletion. Consequently, the stage separation factor relating to exit concentration is greater

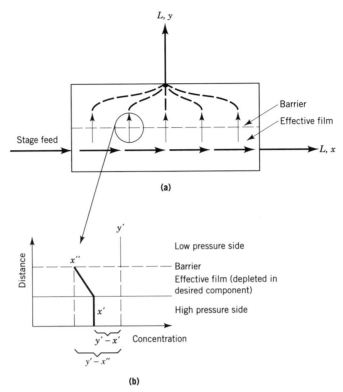

Fig. 8. Flow of process gas through a gaseous diffusion stage. (**a**) Gaseous diffusion stage; (**b**) local concentration profiles near the diffusion barrier; y' = point concentration of light component on low pressure side; x' = average concentration of light component in bulk of gas on high pressure side of barrier flowing past the specified point; x'' = point concentration of light component at surface of barrier on the high pressure side; $y' - x''$ = separation that would be obtained across barrier in the absence of the effective film; and $y' - x'$ = actual separation obtained across barrier, taking a film into account.

than the point separation factor, and the cut correction factor exceeds unity. (This phenomenon is analogous to cross-flow in a plate distillation column.)

The stage separation factor can therefore be related to the ideal-point separation factor by an equation of the form

$$(\alpha - 1) = (E_b)(E_p)(E_M)(E_c)(\alpha^* - 1) \tag{50}$$

where

$$E_b = \text{barrier efficiency} = \frac{\text{separative flow through barrier}}{\text{total flow through barrier}} \tag{51}$$

$$E_p = \text{back-pressure efficiency} = 1 - r \tag{52}$$

$$E_M = \text{mixing efficiency} = \exp^{-Nlf/\rho D} \tag{53}$$

and

$$E_c = \text{cut correction} = \frac{1}{\theta} \ln \frac{1}{1 - \theta} \tag{54}$$

For the usual case where approximately one-half of the gas entering the stage passes through the barrier, the value of the cut, θ, is equal to 0.5 and the cut correction takes on the value 1.386.

These efficiency factors are discussed in more detail in References 26–28. Actually, the barrier and back-pressure efficiencies are interrelated and cannot be formulated independently, except only as an approximation. A better formation that has been found to fit the experimental results is:

$$(E_b)(E_p) = (1 - r)/[1 + (1 - r)(p_f/p^*)] \tag{55}$$

where p^* is a constant, the value of which must be determined experimentally. It may be noted that for an ideal barrier, $E_b = 1$, p^* is equal to ∞, and the back-pressure efficiency is given by equation 52.

Separative Capacity. An expression for the separative capacity of a single gaseous diffusion stage where the upflow rate is L mols per unit time, given in equation 7, can be written as

$$\delta U = 1/4 \, L \left(\frac{1 - r}{1 + (1 - r)(p_f/p^*)} \right)^2 E_M^2 \left(\frac{1}{\theta} \ln \frac{1}{1 - \theta} \right)^2 (\alpha^* - 1)^2 \tag{56}$$

For a very high quality barrier ($p^* \longrightarrow \infty$), the separative capacity of a stage having a mixing efficiency of 100% and operating at a cut of one-half would be:

$$\delta U = \frac{1.921}{4} L(1 - r)^2(\alpha^* - 1)^2 \tag{57}$$

If the power requirement of the gaseous diffusion process were no greater than the power required to recompress the stage upflow from the pressure on the low-pressure side of the barrier to that on the high-pressure side, then the power requirement of the stage would be $LRT \ln (1/r)$ for the case where the compression is performed isothermally. The power requirement per unit of separative capacity would then be given simply by the ratio

$$\frac{\text{power requirement}}{\text{unit separative capacity}} = \frac{2.082 \, RT \ln (1/r)}{(1 - r)^2(\alpha^* - 1)^2} \tag{58}$$

This quantity is minimized when the stage is operated at a pressure ratio across the barrier corresponding to $r = 0.285$. Furthermore, if power were the only economic consideration, the stage would be operated at this pressure ratio. However, as the value of r is decreased from this optimum, although the cost of power is increased, the number of stages required and hence the capital cost of the plant is decreased. Thus, in practice a compromise between these factors is made.

The optimum pressure level for gaseous diffusion operation is also determined by comparison; at some pressure level the decrease in equipment size and volume to be expected from increasing the pressure and density is outweighed by the losses that occur in the barrier efficiency. Nevertheless, because it is well known that the cost of power constitutes a large part of the total cost of operation of gaseous diffusion plants, it can perhaps be assumed that a practical value of r does not differ greatly from the above optimum. Inclusion of this value in the preceding equations yields

$$\delta U = 0.246 \, L(\alpha^* - 1)^2 \tag{59}$$

and

$$\frac{\text{power requirement}}{\text{unit separative capacity}} = \frac{5.11 \, RT}{(\alpha^* - 1)^2} \tag{60}$$

The actual power requirement is greater than that given by equation 58 or 60 because of the occurrence of frictional losses in the cascade piping, compressor inefficiencies, and losses in the power distribution system.

Plant Operation and Costs.   The operation and economics of the three United States gaseous diffusion plants running in 1972 is discussed in References 29 and 30. These plants were operated as a single gaseous diffusion complex such that interplant shipments occurred so as to optimize the overall system. Independent operation of the plants would have resulted in about a 1% loss in separative work.

In 1985, owing to the declining demand by the nuclear power industry for enriched uranium, the Oak Ridge gaseous diffusion plant was taken out of operation and, subsequently, was shut down. The U.S. gaseous diffusion plants at Portsmouth, Ohio and Paducah, Kentucky remain in operation and have a separative capacity of 19.6 million SWU (separative work unit) per year which as of this writing is not fully utilized.

Information on the design of new gaseous diffusion plants is available (30). The shape of a gaseous diffusion cascade, based on 1970 U.S. technology, having 8.75 million SWU/yr separative capacity, designed to produce uranium containing 4% ^{235}U from natural feed having tails at 0.25% ^{235}U is shown in Figure 9.

In 1973, Eurodif, a multinational consortium organized under French leadership, decided to build a large gaseous diffusion plant at Tricastin in France. This plant was completed in 1982, has a separative capacity of 10.8 million SWU/yr, and is based on French gaseous diffusion technology. Some design and progress information regarding this cascade is available (31–33). The engineering design of the French gaseous diffusion stage, although functionally the same as the United States stage, differs appreciably in appearance, and the motor, compressor, and diffuser are arranged vertically and are contained in a single housing in the French plant. Some of the main features of the Tricastin plant are UF$_6$ rates in kg/h of product, feed, and waste are 600, 3025, and 2425, respectively; percentage ^{235}U in the product, feed, and waste is 1.35–3.15, 0.72, and 0.25, respectively; stages in the enricher are 220 small sizes at 0.6 MW, 280

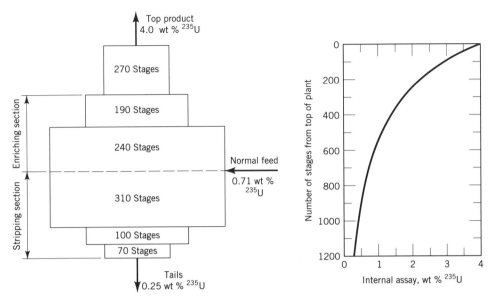

Fig. 9. Schematic diagram and internal gradient for an 8.75 million-SWU/yr plant.

medium sizes at 1.6 MW, and 320 large sizes at 3.3 MW; stages in the stripper are 60 small sizes, 120 medium, and 400 large at 0.6, 1.6, and 3.3 MW, respectively; the total cascade power is 3100 MW, and the total cascade separative capacity is 10.7×10^6 SWU/yr; and the specific power requirement is 2538 kWh/SWU.

From equation 60 one can obtain a theoretical power requirement of about 900 kWh/SWU for uranium isotope separation assuming a reasonable operating temperature. A comparison of this number with the specific power requirements of the United States (2433 kWh/SWU) or Eurodif plants (2538 kWh/SWU) indicates that real gaseous diffusion plants have an efficiency of about 37%. This represents not only the barrier efficiency, the value of which has not been reported, but also electrical distribution losses, motor and compressor efficiencies, and frictional losses in the process gas flow.

The cost of enriched material from a gaseous diffusion plant depends both on the cost of separative work and of feed material. It can be seen from equation 15 that if the optimum tails concentration from a gaseous diffusion plant is 0.25%, the ratio of the cost of a kg of normal uranium to the cost of a kg of separative work equal to 0.80 is implied. Because the cost of separative work in new gaseous diffusion plants is expected to be about $100/SWU, equation 16 gives the cost per kg of uranium containing 4% ^{235}U as about $1,240.

Pressure Diffusion Processes

The development of the kinetic theory of gases led to the conclusion that a partial separation of the components of a gaseous mixture results when the gas is subjected to a pressure gradient. Thus, a column of gas standing in the earth's gravity field should show a separation effect, the lighter components concentrating at the top of the column, the heavier components at the bottom. This is indeed the case, but the effect is too slight to be utilized in a practical separation process. In the case of the isotopes of uranium, a column about 0.4 km in height would be required to give an enrichment equal to that of a single gaseous diffusion stage. Therefore, in order to utilize the pressure diffusion phenomenon, steeper pressure gradients than are normally available are needed.

Several devices have been developed for the purpose of producing such pressure gradients. The best known is the gas centrifuge (see also SEPARATION, CENTRIFUGAL). High-speed centrifuges can develop gravitational fields equal to many thousand times that of the earth. Thus relatively large pressure gradients can exist between the axis and periphery of a centrifuge, giving rise to appreciable separation effects. By moving streams of gas at the periphery and at the axis countercurrently, the centrifuge can be made equivalent to a multistage separating column.

A second type of apparatus based on the pressure diffusion effect is the separation nozzle. Pressure gradients in a curved expanding jet produce an isotopic separation similar to that in a centrifuge. The separation effect obtained with a single jet is relatively small, and separation nozzle stages, similar to gaseous diffusion stages, must be used in a cascade to realize most of the desired separations.

A third device that utilizes pressure diffusion is the vortex chamber. Here, as in the centrifuge, angular acceleration effects in a rapidly rotating gas provide the pressure gradient. The vortex chamber may be considered as a centrifuge with a stationary outer wall. The mechanical difficulties of high-speed rotating machinery are avoided at the expense of friction effects between the gas and the stationary wall. The literature concerning the use of such a device for isotope separation is limited. Results of experimentation indicate the effect of some of the process variables and the separation factors in H_2–CO_2, H_2–HD and ^{36}Ar–^{40}Ar binary gas mixtures have been measured (34,35). A vortex tube has been used for isotope separation (36), and for the separation of gases in nuclear rocket or ramjet engines.

A vortex tube process has been developed in South Africa and is being used there for the enrichment of uranium (37). It appears that cascades of this type are characterized by an extremely high power consumption.

The Gas Centrifuge.   The first suggestion that centrifugal gravitational fields might be used to effect separation of isotopes was made in 1919 (38). Then in 1934 the convection-free vacuum ultracentrifuge was developed (39–41). Extensive information on the construction and operation of high-speed centrifuges and experimental data on the separation of the isotopes of argon, xenon, and uranium are available (39,42–47). Early work on centrifuge development and centrifuge theory is discussed in References 48–55. Simplified approximate models of the flow have been developed (56–58), as have more accurate approximations (59–70). Japanese researchers have made an appreciable contribution to the literature in this field (71–78). Other surveys can be found in References 28 and 79–84. An excellent source of information on gas centrifuge development and centrifuge theory can be found in the proceedings of the early workshops on gases in strong rotation and in the proceedings of the workshops on separation phenomena in gases and liquids which followed (85–91).

The Groth and Zippe Centrifuges.   A schematic drawing of the ZG 5 gas centrifuge, in Figure 10, is typical of the Groth centrifuge (39,45,46). It is suspended and driven from above directly by an electric motor. The rotor spins in a vacuum-tight casing. Gas is introduced through a central tube and removed through scoop tubes at the ends of the rotor shielded by baffles from the main part of the bowl to prevent disturbance of the internal gas flow pattern. The gas is caused to undergo countercurrent axial flow by maintaining a temperature difference between the ends of the rotor. The top end of the bowl is heated by eddy currents in an aluminum ring at the top end cap; the bottom end cap is cooled by a cooling

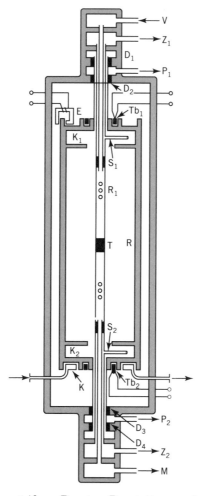

Fig. 10.   The Groth ZG 5 centrifuge. R, rotor; R_1, stationary shaft; T, Teflon seal; K_1, K_2, chambers for gas scoops; S_1, S_2, scoops; V, gas supply; M, manometer; Z_1, Z_2, tapping points for enriched and depleted gas; P_1, P_2, vacuum chambers; E, electromagnet for eddy current heating; Tb_1, Tb_2, temperature measuring devices; K, cooling coil; and D_1, D_2, D_3, D_4, labyrinth seals.

coil. Thermocouples are used to measure the end cap temperatures and the internal pressure at the centrifuge axis is measured by connections to the center tube. Labyrinth seals are used at the ends to maintain a gas seal.

The short bowl Zippe centrifuge (55) shown in Figure 11 is somewhat simpler. It is supported on a needle bearing at the base and driven by an electric motor, the armature of which is a steel plate rigidly attached to the bottom of the rotor. The stator consists of a flat winding on an iron core positioned so that the poles are separated from the armature by only a small gap (about 6 mm). Power is supplied by an alternator. Damping bearings are used to resist vibrations at both ends of the rotor. The centrifuge is completely closed at the bottom. The other end is connected with the top region of the outer vacuum casing only by a small annular gap around the feed tube. A small amount of gas that leaks from the interior of the bowl at the low pressure near the axis is confined to the region above the top of the rotor by a Holweck-type spiral groove molecular pump surrounding the rotor near the top,

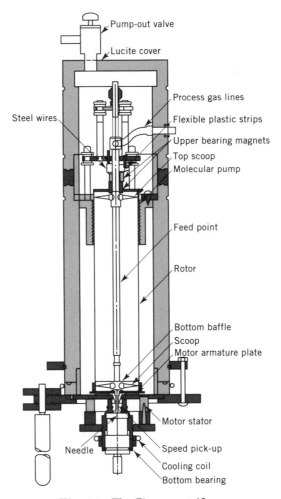

Fig. 11. The Zippe centrifuge.

and pumped out of this region to maintain the necessary vacuum. Dimensions of the two types of centrifuges are given in Table 1.

The maximum theoretical separative capacity of a centrifuge is proportional to its length and to the fourth power of its peripheral speed, putting a premium particularly on high peripheral speeds and to a lesser extent on long rotors. The allowable peripheral speed of a cylindrical rotor is limited by the ratio of the tensile strength of the material of construction to its density. The maximum peripheral speed, or the burst speed, of a centrifuge rotor is given by:

$$V_{\mathrm{max}} = \left(\frac{\sigma}{\rho}\right)^{1/2} \tag{61}$$

where σ is the tensile strength of the material and ρ is its density. The rotor must be resistant to the process gas, then only certain materials may be usable. Both Groth and Zippe centrifuges have used an aluminum alloy for rotors for use with UF_6. Table 2 gives the values of these properties for several materials

Table 1. Dimensions of Centrifuge Bowls[a]

Bowl	Length, cm	Radius, cm
UZ I	40	6.0
UZ IIIB	63.5	6.7
ZG 3	66.5	9.25
ZG 5	113.0	9.25
ZG 6[b]	240.0	20.0
ZG 7[b]	316.0	22.5
Zippe	30–38	3.81

[a]See Refs. 45 and 47.
[b]Proposed.

Table 2. Maximum Peripheral Speeds for Various Rotor Materials

Material	Density, ρ, kg/m³	Tensile strength, σ, GPa[a]	V_{max}, m/s
aluminum[b]	2800	0.448	400
aluminum[c]	2800	0.64	478
steel[b]	7800	1.381	421
maraging steel[b]	7800	1.932	498
maraging steel[c]	8100	3.0	608
glass fiber[b]	1800	0.49	522
carbon fiber[b]	1600	0.829	720
kevlar[c]	1334	2.17	1271
carbon fiber[c]	1560	3.50	1498

[a]To convert from GPa to psi, multiply by 145,000.
[b]From Ref. 82.
[c]From Ref. 90.

that could be used for fabricating centrifuge rotors and the value of the corresponding maximum peripheral speed. The allowable operating speed would be expected to be about 80% of the maximum speed in order to provide a margin of safety.

Mechanical Features. The construction and operation of a precision, high-speed centrifuge in a high vacuum environment presents some formidable mechanical problems. One difficulty with high-speed rotating machinery is the critical-speed phenomena. A long rod, or its equivalent, undergoes resonant vibrations at its fundamental and higher natural frequencies. This can cause large displacements from the axis of rotation unless the rod is properly restrained at high frequencies by damping devices capable of applying sufficient restraining forces. In the Zippe centrifuge the resonance frequency problem is avoided by limiting the ratio of length to diameter to less than four so that the customary operating speeds (300–350 m/s) are below the first fundamental flexural critical frequency, a so-called subcritical centrifuge. On the other hand Groth models, the ZG 6 and ZG 7, are long bowl or supercritical centrifuges. These run at rotational speeds above that corresponding to one or more flexural critical values and operate at a speed not too close to any of the critical values.

The principal power consumption of a centrifuge at operating speed occurs in friction in the bearing systems and in gas drag on the internal parts, particularly the scoops. A poorly balanced rotor results in high power consumption. Wide variations result from variations in the number, length, diameter of tubing, and tip design of the scoops (47). The long-term maintenance and lubrication of the bearing and support systems are a problem.

The scoop system in the Zippe centrifuge is used to control the internal circulation of the gas in the centrifuge, and in common with the Groth machines, must also extract a sufficient volume of gas at a pressure adequate to pump the gas to the feed point of the next centrifuge in a cascade. If this can be accomplished, no intermachine pumps are required in a cascade. This becomes an increasingly difficult problem at higher speeds because the scoop tips must be close to the centrifuge bowl wall in order to have access to the process gas at a higher pressure. The size and length of the piping in the scoops and the feed insertion tubing is critical because of limitations of their conductance for gas flow at low pressures. Other problems include long-term fatigue and creep of structural materials at high speeds and possibly stress corrosion in some systems. The Zippe centrifuge has been taken as the starting point for most of the modern centrifuge research and development programs.

The Urenco/Centec organization, formed in 1971 by British, Dutch, and German companies to carry on centrifuge research and development efforts, is a primary manufacturer and operator of centrifuge cascades for uranium enrichment. The research and development activities pursued by Urenco have succeeded roughly in doubling the separative capacity of individual centrifuges every five years since 1971. This has been accomplished primarily by increases in the length of the centrifuges and by increases in the peripheral speed of the centrifuges by the use of stronger and lighter materials of construction. Urenco currently operates commercial centrifuge enrichment facilities at Almelo in the Netherlands, at Capenhurst in England, and at Gronau in Germany. These plants have a combined separative capacity of 2.5 million SWU per year. A centrifuge plant to be built by Urenco having a separative capacity of 1.5 million SWU per year is planned for Homer, Louisiana.

An aggressive centrifuge development program culminating in the design, construction, and testing of large centrifuges that had a nominal separative capacity of 200 SWU per machine was carried out in the United States. This program was abandoned in 1985 before construction of the centrifuge facility in Portsmouth, Ohio, was completed. No gas centrifuge work has been done in the United States since that date. The former USSR has revealed that it has employed gas centrifuge technology for uranium enrichment since 1960 and that it operates centrifuge plants that have a combined separative capacity of 10 million SWU annually. Japan has also been developing gas centrifuges for a number of years. A 200,000 SWU uranium enrichment demonstration plant was placed in service at Ningyo-Toge in May 1989. A 1.5 million SWU/y gas centrifuge plant is under construction at Rokkashomura.

Design Principles. Although the separation of fluid mixtures can be accomplished using several different types of centrifuges, discussion of the centrifuge separation theory is herein confined to the consideration of the countercurrent gas centrifuge. In order to design separation cascades consisting of countercurrent gas centrifuges, it is necessary to know the separative performance of the individual units. Gas centrifuge theory serves fairly well for predicting the performance of a single centrifuge. However, the separation behavior of a particular gas centrifuge depends on the flow pattern of the gas circulating within it, which in turn depends on the geometry of any baffles and scoops within the centrifuge bowl as well as on any temperature gradients in the gas and on the method used to introduce feed to the centrifuge. Owing to the complexity of the general case, the equations for centrifuge performance are presented for only a few idealized circulation patterns.

Radial Density and Pressure Gradients. Consider a centrifuge of length Z and of radius r_2, the internal dimensions of the centrifuge bowl, that rotates at a constant angular velocity of ω radians per second. If the centrifuge contains a single pure gas rotating at the same angular velocity as the centrifuge bowl, each element of the gas has a force impressed on it by virtue of its angular acceleration. This force is directed outward in a cylindrical coordinate system, and can be expressed as $(\rho\omega^2 r)(rdrd\phi dz)$. At steady state this force must be balanced by a force resulting from the radial pressure gradient established in the centrifuge bowl. The inward force on an element of the gas owing to this pressure gradient is given by $(dp/dr)(rdrd\phi dz)$. Equating these two forces gives:

$$dp/dr = \rho\omega^2 r \qquad (62)$$

where p is the pressure, r is the spatial coordinate in the radial direction, ρ is the density of the gas, and ω is the angular velocity of the centrifuge.

The pressure and the density of a gas are related by an equation of state. If the maximum pressure permitted within the centrifuge bowl is not too high, the equation of state for an ideal gas will suffice. The relationship between the pressure and density of an ideal gas is given by the well-known equation:

$$p = \rho RT/M \qquad (63)$$

where T is the absolute temperature of the gas, K; M is the mol wt of the gas, and R is the gas constant, 8.3147 J/(mol·K). Elimination of the density from equations 62 and 63 yields the differential equation for the pressure gradient in the centrifuge (eq. 64),

$$\frac{dp}{dr} = \frac{Mp}{RT}\omega^2 r \tag{64}$$

which, for the case of an isothermal centrifuge, is readily integrated to yield

$$p(r) = p(0)\exp^{(M\omega^2 r^2/2RT)} \tag{65}$$

Equation 65 gives the pressure at any point within the centrifuge, $p(r)$, as a function of the coordinate r, the pressure at the axis $p(0)$, the angular velocity of the centrifuge, and the temperature and mol wt of the gas. Should the centrifuge contain not a single pure gas, but a gas mixture, equations of the above forms could be written for each species present. In particular for the case of a binary gas mixture, consisting of species A and B.

$$p_A(r) = p_A(0)\exp^{(M_A\omega^2 r^2/2RT)} \tag{66}$$

and

$$p_B(r) = 2p_B(0)\exp^{(M_B\omega^2 r^2/2RT)} \tag{67}$$

The ratio of these two equations gives the radial separation afforded by the gas centrifuge under equilibrium conditions, that is, for no internal gas circulation. An equilibrium separation factor between gas at the axis of the centrifuge and gas at the periphery is therefore given by:

$$\alpha_0 \equiv \frac{x_A(0)}{x_B(0)}\Big/\frac{x_A(r_2)}{x_B(r_2)} = \exp^{[(M_B-M_A)\omega^2 r_2^2/2RT]} \tag{68}$$

It should be noted that the separation factor for the centrifuge process is a function of the difference in the mol wts of the components being separated rather than, as is the case in gaseous diffusion, a function of their ratio. The gas centrifuge process would therefore be expected to be relatively more suitable for the separation of heavy molecules. As an example of the equilibrium separation factor of a gas centrifuge, consider the Zippe centrifuge, operating at 60°C with a peripheral velocity ωr_2 of 350 m/s. From equation 68, α_0 is calculated to be 1.0686 for uranium isotopes in the form of UF_6.

Mass Transport. An expression for the diffusive transport of the light component of a binary gas mixture in the radial direction in the gas centrifuge can be obtained directly from the general diffusion equation and an expression for the radial pressure gradient in the centrifuge. For diffusion in a binary system in the absence of temperature gradients and external forces, the general diffusion equation retains only the pressure diffusion and ordinary diffusion effects and takes the form

$$J_A = -cD_{AB} \left[\frac{M_A x_A}{RT} \left(\frac{\overline{V}_A}{M_A} - \frac{1}{\rho} \right) \frac{dp}{dr} + \frac{dx_A}{dr} \right] \qquad (69)$$

where cD_{AB} is the product of the molar density and binary diffusion coefficient of the process gas and $\overline{V}_a$ is the partial molal volume of component A.

Because the total pressure gradient is the sum of the partial pressure gradients, the following substitution can be made in equation 69

$$\frac{dp}{dr} = \frac{dp_A}{dr} + \frac{dp_B}{dr} = \frac{(M_A p_A + M_B p_B)}{RT} \omega^2 r \qquad (70)$$

and the equation for the radial flux of component A in a mixture of ideal gases is found to be

$$J_A = cD_{AB} \left[\frac{(M_B - M_A) x_A (1 - x_A)}{RT} \omega^2 r + \frac{dx_A}{dr} \right] \qquad (71)$$

Figure 12 is a schematic drawing of a section of a countercurrent gas centrifuge in which an arbitrary axial convective flow pattern is shown. It is assumed that the convective velocity v in the centrifuge can be expressed as a function of r only, and is independent of z. The convective velocity is assumed to be in the z direction only, and the regions at the ends of the centrifuge in which the direction of the flow is changed are neglected.

The net transport of component A in the $+z$ direction in the centrifuge τ_A is equal to the sum of the convective transport and the axial diffusive transport. At the steady state the net transport of component A toward the product withdrawal point must be equal to the rate at which component A is being withdrawn from the top of the centrifuge. Thus, the transport of component A is given by equation 72:

$$\tau_A = P x_P = \int_0^{r_2} 2\pi r c v x \, dr - \int_0^{r_2} 2\pi r c D_{AB} \frac{\partial x}{\partial z} \, dr \qquad (72)$$

where x is used in place of x_A for the mol fraction of component A, and the net transport of both components toward the product withdrawal point, τ, is given by

$$\tau = P = \int_0^{r_2} 2\pi r c v \, dr \qquad (73)$$

where P is the product withdrawal rate, mol/s, and x_P is the concentration of component A in the product material. The integrals appearing in equation 72 are evaluated using the flux equation 71 (92). Several approximations are involved and are most satisfactory for the case of relatively long units in which the axial

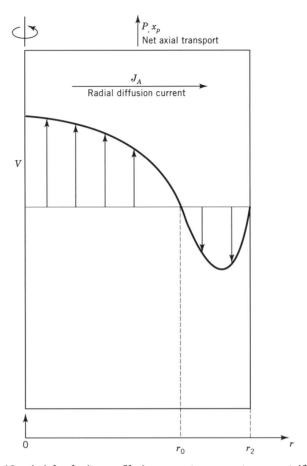

Fig. 12. Axial velocity profile in a countercurrent gas centrifuge.

concentration difference is large compared with the concentration differences in the radial direction, and in which the magnitudes of the feed and withdrawal rates are small with respect to the circulation rate of the internal convective flow. The results are not completely satisfactory for application to short-bowl centrifuges with relatively high throughput rates. For the case of the gas centrifuge, application of the method leads to a gradient equation for the enriching section of the centrifuge that can be written in the form

$$Sdx/dz = (\alpha - 1)x(1 - x) - (P/L)(x_P - x) \tag{74}$$

where S is the stage length in the centrifuge and α is the stage separation factor. The quantity L is a measure of the convective circulation rate of the gas in the centrifuge and may be evaluated from the integral

$$L = \int_0^{r_o} 2\pi r c v \, dr \tag{75}$$

where r_0 is the radius at which the convective velocity is equal to zero (see Fig. 12). The stage length S is the sum of two terms

$$S = \frac{\int_0^{r_2} \frac{dr}{rcD_{AB}} \left[\int_0^r r'cvdr'\right]^2}{\int_0^{r_0} rcvdr} + \frac{\int_0^{r_2} rcD_{AB}dr}{\int_0^{r_0} rcvdr} \tag{76}$$

The first term may be considered as the contribution of the internal circulation or convective flow to the stage length, the second term as the contribution of the axial diffusion to the stage length. The stage separation factor is given by

$$\alpha - 1 = \frac{(M_B - M_A)\,\omega^2}{RT} \frac{\int_0^{r_2} rdr \int_0^r r'cvdr'}{\int_0^{r_0} rcvdr} \tag{77}$$

From an inspection of the preceding three equations, it is evident that for the case of a given velocity profile in which v retains its functional dependence on r but is permitted to vary in magnitude by a factor, that is, $v = bf(r)$, the convective contribution to the stage length varies directly with the magnitude of L, whereas the diffusive contribution to the stage length varies inversely with the magnitude of the circulation rate L. Thus there exists a value of L for which the stage length for the separation process is a minimum. Designating this value of L by L_0, analysis of the expression for the stage lengths shows that

$$L_0 = 2\pi \int_0^{r_0} rcvdr \left[\frac{\int_0^{r_2} rcD_{AB}dr}{\int_0^{r_2} \frac{dr}{rcD_{AB}} \left(\int_0^r r'cvdr'\right)^2}\right]^{1/2} \tag{78}$$

The corresponding minimum value of the stage length, designated by S_0, is given by

$$S_0 = \frac{2\left[\int_0^{r_2} \frac{dr}{rcD_{AB}} \left(\int_0^r r'cvdr'\right)^2 \int_0^{r_2} rcD_{AB}dr\right]^{1/2}}{\int_0^{r_0} rcvdr} \tag{79}$$

The preceding two equations may be used to write the gradient equation for the countercurrent gas centrifuge in an alternative form. If the ratio of the actual gas circulation rate in the centrifuge to the circulation rate that minimizes the stage length L/L_0 is designated by m, then equation 74 may be rewritten

$$\left(\frac{1 + m^2}{2m}\right) S_0 \frac{dx}{dz} = (\alpha - 1)x(1 - x) - \frac{P}{mL_0}(x_P - x) \tag{80}$$

For the stripping section of the centrifuge, that is, the section between the point at which the feed is introduced and the end at which the waste stream is withdrawn, the gradient equation has the corresponding form

$$\left(\frac{1 + m^2}{2m}\right) S_0 \frac{dx}{dz} = (\alpha - 1)x(1 - x) - \frac{W}{mL_0}(x - x_W) \tag{81}$$

where W is the waste withdrawal rate, mol/s, and x_W is the concentration of component A in the waste material.

Maximum Separative Capacity and the Separative Efficiency. The separative efficiency of a gas centrifuge used for isotope separation is best defined in terms of separative work. Thus, the separative efficiency E is defined by

$$E = \frac{\delta U \text{ (experimental)}}{\delta U \text{ (max)}} \tag{82}$$

where δU (experimental) is the actual separative work produced per unit time by the centrifuge under consideration and δU (max) is the maximum theoretical separative capacity of the machine. The maximum separative capacity of a gas centrifuge (41) is given by

$$\delta U \text{ (max)} = \frac{\pi Z c D_{AB}}{2} \left(\frac{\Delta M V^2}{2 RT}\right)^2 \tag{83}$$

where δU is the separative capacity in mols per unit time, Z is the length of the rotor, ΔM is the difference in the mol wts of the components being separated, and V is the peripheral velocity of the centrifuge ($V = \omega r_2$). The expression for the maximum separative capacity of a centrifuge indicates a desirability for: (*1*) Low-temperature operation because the theoretical maximum separative capacity of a centrifuge varies inversely as the temperature; (*2*) Long centrifuge bowls because the theoretical maximum separative capacity varies directly as Z and that δU (max) is independent of the radius of the bowl; and (*3*) High peripheral velocity because the theoretical maximum separative capacity varies as the fourth power of the peripheral speed. At the higher speeds the predicted separative capacity increases with increasing peripheral speed much more slowly than the fourth-power relationship. Nevertheless, over the entire range of speeds investigated there is still an appreciable gain in separative capacity to be realized from an increase in speed.

Theoretical Formulation of the Separative Efficiency. The separative efficiency E of a countercurrent gas centrifuge may be considered to be the product of four factors, all but one of which can be evaluated on the basis of theoretical considerations. In this formulation the separative efficiency is defined by

$$E \equiv e_C e_I e_F e_E \tag{84}$$

where e_C designates the circulation efficiency, e_I designates the ideality efficiency, e_F designates the flow pattern efficiency, and e_E designates the experimental efficiency and includes all phenomena such as turbulence and end effects not taken into account by the preceding terms. The circulation efficiency for a countercurrent gas centrifuge is given by

$$e_C = m^2/(1 + m^2)$$ (85)

As has been previously noted, m is the ratio of the rate at which gas flows upward in a centrifuge to the quantity L_0 that depends on the geometry of the bowl, the physical properties of the gas, and the flow pattern. Thus, m is directly proportional to the upflow rate. It is evident from the definition of e_C that it approaches unity as m takes on increasingly larger values. This is understood when it is realized that the circulation efficiency is representative of the loss in separative capacity owing to axial diffusion against the axial concentration gradient established in the bowl, and is, in fact, equal to the ratio of the convective contribution to the stage length of the sum of the convective and diffusive contributions, that is, to the total stage length. As m increases, the convective contribution of the stage length increases proportionally, and the diffusive contribution decreases as m^{-1}; at high circulation rates the diffusive transport becomes negligible with respect to the convective transport within the centrifuge.

The ideality efficiency takes into account the difference between the shape of a centrifuge that may be regarded as a square cascade and that of an ideal cascade. As has been pointed out in the section on cascade theory, the separative capacity of an element of length in a centrifuge is the greatest when the circulation rate L through the element bears a certain relationship to the withdrawal rate, withdrawal concentration, and the concentration in the centrifuge at that point, as has been indicated by equation 28. When this condition is satisfied the cascade is termed ideal. In a square cascade, however, this condition cannot be satisfied at more than a single point in the enricher and in the stripping sections. Thus, one can associate an efficiency with each point in the cascade that is a function of the departure of the actual flow from the ideal flow. The ideality efficiency may be regarded as the average of these point efficiencies over the entire cascade.

Analysis of the gradient equations for a countercurrent gas centrifuge shows that when the withdrawal rates are optimized, the ideal efficiency assumes a maximum value of 81%. Curves of the ideal efficiency for both a stripping and enriching section of five stages ($Z/S = 5$) are shown in Figure 13. The flow model assumed in these calculations is also shown. In order to achieve this maximum value of 0.81 for the ideality efficiency, it is necessary that in addition to operating at the optimum withdrawal rates there be no mixing of gas of unlike concentrations at the feed point. The difference in the behavior of the curves for the efficiency of the enriching and stripping sections results primarily from the fact that in the model considered the feed is assumed to enter the downflowing stream, and therefore the flows in the enriching and stripping sections are not symmetric.

The flow pattern efficiency e_F depends solely upon the shape of the velocity profile in the circulating gas. In terms of the integrals appearing in the gradient equation, the flow pattern efficiency is given by equation 86.

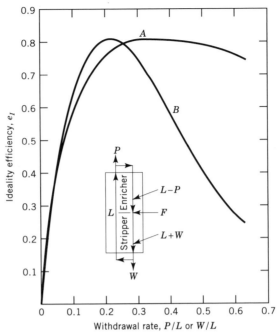

Fig. 13. The ideal efficiency of a five-stage enricher and stripper as a function of the product or waste withdrawal rate, where A represents stripping section efficiency, and B represents enriching section efficiency.

$$e_F = \frac{4 \left(\int_0^{r_2} r \, dr \int_0^r r' c v \, dr' \right)^2}{c D_{AB} r_2^4 \int_0^{r_2} \dfrac{dr}{r c D_{AB}} \left(\int_0^r r' c v \, dr' \right)^2} \tag{86}$$

To evaluate the flow pattern efficiency, a knowledge of the actual hydrodynamic behavior of the process gas circulating in the centrifuge is necessary. Primarily because of the lack of such knowledge, the flow pattern efficiency has been evaluated for a number of different assumed isothermal centrifuge velocity profiles.

The Optimum Velocity Profile.   The optimum velocity profile (41), that is the velocity profile that yields the maximum value for the flow pattern eficiency, is one in which the mass velocity pv is constant over the radius of the centrifuge except for a discontinuity at the wall of the centrifuge ($r = r_2$). This optimum velocity profile is shown in Figure 14a. For this case the following values for the separation parameters of the centrifuge are obtained

$$\alpha - 1 = \frac{1}{2} \frac{\Delta M V^2}{2 \, RT} \tag{87}$$

$$L_0 = 2 \, (2)^{1/2} \, \pi r_2 c D_{AB} \tag{88}$$

$$S_0 = r_2 (2)^{1/2} \tag{89}$$

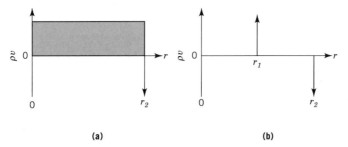

(a)   (b)

Fig. 14. Hypothetical velocity profile models for a countercurrent-flow gas centrifuge. (**a**) The optimum velocity profile in a countercurrent gas centrifuge. (**b**) The two-shell velocity profile.

and $e_F = 1.0$.

   The Two-Shell Velocity Profile. A second simple velocity profile (41) is shown in Figure 14**b** in which the flow consists of two thin streams, one situated at radius r_1, flowing upward, and the other situated at the wall ($r = r_2$), flowing downward. For this case the values of the separation parameters are

$$\alpha - 1 = \left[1 - \left(\frac{r_1}{r_2}\right)^2\right] \frac{\Delta M V^2}{2 \, RT} \tag{90}$$

$$L_0 = \left[\frac{2}{\ln(r_2/r_1)}\right]^{1/2} \pi r_2 c D_{AB} \tag{91}$$

$$S_0 = [2 \, ln(r_2/r_1)]^{1/2} \, r_2 \tag{92}$$

$$e_F = \left[1 - \left(\frac{r_1}{r_2}\right)^2\right]^2 \Bigg/ \ln\frac{r_2}{r_1} \tag{93}$$

The value of the flow pattern efficiency is shown as a function of the spacing between the streams in Figure 15**a**. It can be seen that the flow pattern efficiency is a maximum when the position of the upflowing stream is chosen such that r_1/r_2 is equal to 0.5335. For this particular case the flow pattern efficiency assumes the value $e_F = 0.8145$.

   These simple velocity profiles do not indicate directly any dependence of the flow pattern efficiency upon the rotational speed of the centrifuge. A dependence on speed is to be expected on the basis of the argument that at high speeds the gas in the centrifuge is crowded toward the periphery of the rotor and that the effective distance between the countercurrent streams is thereby reduced. It can be seen from the two-shell model that, as the position of upflowing stream approaches the periphery, the flow pattern efficiency drops off from its maximum value.

   The Martin Velocity Profile. It has been suggested (50) that the velocity profile in a gas centrifuge in which the countercurrent flow is caused by a temperature difference between the circulating gas and the end caps is given by

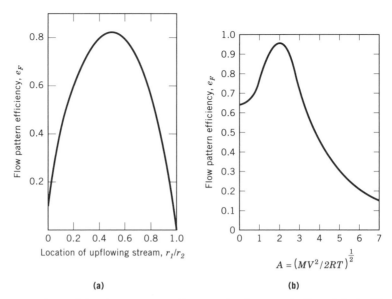

Fig. 15. (**a**) Values of the flow-pattern efficiency for the two-shell model. (**b**) The dependence of the flow-pattern efficiency on the dimensionless parameter A for the Martin profile.

$$\int_0^r r'v\,dr' = \frac{\Delta T}{\omega}\left(\frac{\lambda^3}{\eta T}\right)^{1/4}\left(\frac{2\,rp(r_2)}{MRT}\right)^{1/2}\exp^{(M\omega^2(r^2-r_2^2)/4RT)} \tag{94}$$

where λ is the thermal conductivity of the process gas, η is the viscosity of the process gas, and ΔT is the temperature difference between the gas and the end caps, one warmer, the other cooler than the gas. Equation 94 was derived by considering the flow along a heated plate in a strong gravity field. All other considerations such as coriolis forces and any resistance to the axial flow, were neglected.

The separation parameters have been calculated for a centrifuge in which the behavior of the circulating gas is described by Martin's equation. The flow pattern efficiency is shown in Figure 15(**b**) as a function of the dimensionless parameter A, where A is equal to $(MV^2/2RT)^{1/2}$. In this case the maximum flow pattern efficiency attainable is 0.956.

Cascade Design. The efficiency of a Zippe-type centrifuge, separating uranium isotopes when UF_6 is the process gas, operating at a peripheral speed of 350 m/s and at a temperature of 320 K ($A = 2.85$), would be expected to be

$$E = \frac{m^2}{1+m^2}(0.81)(0.76) \tag{95}$$

The observed efficiency of 35% could be interpreted to mean that the circulation efficiency of this machine is about 60%, corresponding to an m value of 1.2. According to the theory presented, if the centrifuge could be operated with $m = 3$, it should be possible to obtain a separative efficiency of about 55%. With this assumption, the separative capacity of a single machine would be 1.78×10^{-3} kg/d of uranium. A cascade for uranium isotope separation designed to produce 1 kg/d of enriched uranium containing 90% ^{235}U from natural uranium would therefore require approximately 116,000 Zippe-type gas centrifuges. Table 3 shows the size of a cascade consisting of Zippe-type centrifuges required for the production of 1 kg/d of UF$_6$. Modern centrifuges have attained much higher separative capacities than the original Zippe machine. Were the cascade described in Table 3 to be constructed using today's centrifuges, the number of centrifuges required would be lower by one to two orders of magnitude.

Table 3. Characteristics of an Ideal Centrifuge Cascade[a]

Characteristics	Value	Source of value
Centrifuge parameters		
length of rotor Z, cm	30.48	given[b]
diameter of rotor $2r_2$, cm	7.62	given[b]
peripheral speed $V = wr_2$, m/s	350	given
operating temperature T, K	320	given
Centrifuge separative capacity		
separative capacity δU, SWU/yr, maximum	1.17	eq. 83
separative capacity, δU, SWU/yr, actual	0.65	E U(max)
circulation efficiency, e_c	0.90	eq. 85
ideality efficiency, e_I	0.81	maximum, Fig. 13
flow pattern efficiency, e_F	0.76	Fig. 15**b**, $A = 2.85$
Cascade parameters		
product concentration y_P, mol fraction	0.90	given
feed concentration x_F, mol fraction	0.0072	given
waste concentration x_W, mol fraction	0.0025	given
product rate P, kg/d	1.0	given
feed rate F, kg/d	191.0	eqs. 34, 35
separative capacity, δU, SWU/yr	75,380	eq. 36
number of centrifuges required	115,970	

[a]Cascade to yield 1 kg/d UF$_6$ enriched to 0.90 mol fraction ^{235}U.
[b]Predetermined by equipment or operation.

The Separation Nozzle Process

The separation nozzle process, developed at the Karlsruhe Nuclear Research Center in Germany for the enrichment of the light uranium isotope ^{235}U, is also referred to as the jet diffusion method for the separation of gas mixtures. Isotopes were first separated (93) in a slit-type gas jet (94) in 1946. A device for separating gaseous mixtures by jet diffusion was patented in the United States in 1952. Soon thereafter this separation effect associated with high-speed gas flow through a nozzle was applied to the separation of isotopes (95–97). More recent work by the

German research group is described (98–121). Interest in the jet separation process also led to experimental and theoretical work in the United States (94,122–125), and in Japan (126–130).

Apparatus and Method of Operation. The separation nozzle process stage planned for commercial use differs appreciably from the stage used in early investigations. The basic features of the separation nozzle method are illustrated in Figure 16. Gaseous uranium hexafluoride mixed with a light auxiliary gas is expanded through a nozzle into a curved flow channel. At the end of the curved flow path, after turning 180°, the stream is divided by a knife edge into two parts, an interior fraction which is enriched in ^{235}U and a wall fraction which is depleted with respect to the ^{235}U. The light auxiliary gas present in a large molar excess increases the flow velocity of the UF_6 and, hence, it increases the centrifugal force determining the separation. In addition, the light gas delays the sedimentation of the two UF_6-isotopes in the centrifugal field slightly differently, which also has a favorable effect upon the separation of the isotopes.

Usually, a mixture of 2–5 mol % UF_6 and 95–98 mol % H_2 is used as a process gas; the expansion ratios range from 1.8–2.5. According to the gas kinetic scaling relations the optimum operating pressure of the nozzle is inversely proportional to its characteristic dimensions; for example, the optimum inlet pressure of a commercial separation nozzle system with a radius of curvature of 0.1 mm is on the order of tens of kPa (tenths of atmospheres). Figure 17 illustrates the design of a commercial separation nozzle element. The ten slit-shaped separation nozzles are mounted on the periphery of an extruded aluminum tube. Feed gas is introduced into the segments marked F and expands through the nozzles. The heavy fraction is pumped off through the segments marked H and the light fraction is pumped off from the space around the element.

Theory. A good understanding of the separation phenomenon of the separation nozzle process is obtained from a very simple model that treats the sepa-

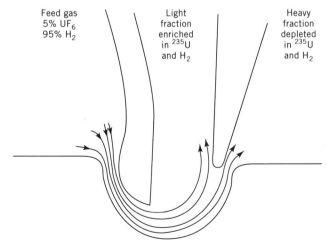

Fig. 16. Cross-section of the separation nozzle system used in the commercial implementation of the separation nozzle process.

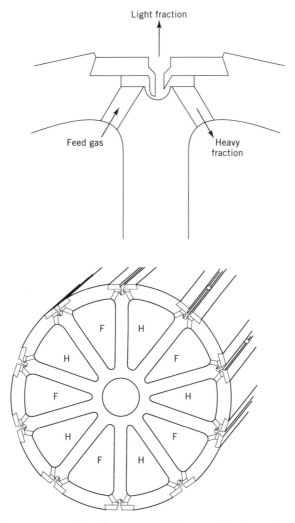

Fig. 17. Schematic representation of a commercial separation element tube manufactured by the Messerschmitt-Bölkow-Blohm Company, Munich. Terms are defined in text.

ration nozzle as a gas centrifuge at steady state. It is assumed that the feed mixture traverses the circular flow path at a constant and uniform angular velocity (wheel flow) and that the peripheral velocity of the flow is equal to the sonic velocity of the entering feed gas. The separation of the isotopes is effected, as in the gas centrifuge, by pressure diffusion in the pressure gradient resulting from the curved streamlines and the associated centrifugal forces. One important function served by the light auxiliary gas in the feed is to increase the flow velocity of the mixture and hence the magnitude of the separation factor attained.

When the gas speed is sufficiently high, the separation factor corresponding to a given value of the cut is essentially independent of the gas velocity and, hence, at high speeds, is given (104) to a good approximation as

$$(\alpha - 1) \cong \left(\frac{1}{1 - \theta} \ln \frac{1}{\theta} \right) \frac{\Delta M}{M} \tag{96}$$

where θ is the stage cut defined as the fraction of the uranium in the feed stream to the separation nozzle that is withdrawn in the light or enriched product stream and $\Delta M/M$ is the fractional difference in the mol wts of the isotopic species being separated. In the case of uranium isotopes where UF_6 is the process gas, $\Delta M/M = 3/352 = 0.0085$. The separative work produced by the stage is

$$\delta U = F \frac{\theta(1 - \theta)}{2} (\alpha - 1)^2 \tag{97}$$

where F is the feed rate of uranium to the separating unit and δU is the amount of separative work produced by the nozzle system per unit time. The separation performance of the separation nozzle in the limit of high gas speed, as described in equations 96 and 97, is shown in graphical form in Figure 18. An equilibrium separation nozzle produces its maximum separative work rate at a cut of about 0.2. The secondary enrichment effects caused by ternary diffusion involving the light auxiliary gas, are treated in some detail in References 131 and 132.

Cascade for Uranium Enrichment. The design for a 5 million SWU/yr plant producing uranium enriched to 3% ^{235}U and stripping the feed to 0.3% ^{235}U in the waste stream is shown in Figure 19. Data for such a plant are given in Table 4.

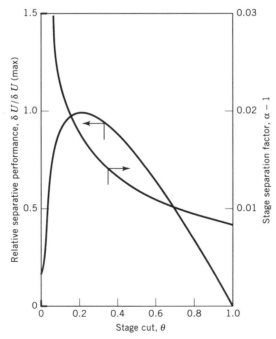

Fig. 18. High speed performance limit of an equilibrium separation nozzle.

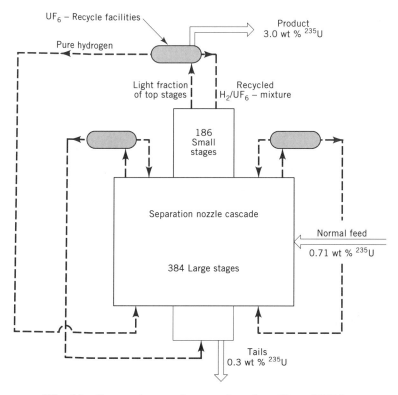

Fig. 19. Separation nozzle cascades of 5 million SWU/yr.

The 570 stages and 2520 MW required by the 5 million SWU/yr nozzle plant should be compared with the 1180 stages (to span the range from 0.25–4.0% ^{235}U) and 2430 MW required by the 8.75 million SWU/yr gaseous diffusion plant shown in Figure 9.

Three prototype separation nozzle stages of different sizes have been constructed at the Karlsruhe Nuclear Center. A small stage, designated the SR-33, has a compressor suction volume of 33,000 m^3/h, 7.5 m in height and 1.5 m in diameter. A stage of intermediate size, designated the SR-100, has a compressor suction volume of 100,000 cm^3/h, 10 m in height and 2.5 m in diameter. A large stage, designated the SR-300, has a compressor suction volume of 300,000 cm^3/h, 14 m in height and 4 m in diameter. When separation nozzle systems of an advanced design (double-deflecting nozzles having small radii of curvature of 20 μm and 10 μm) are installed in these stages, the separative capacity of the SR-33 stage is expected to be 2400 SWU annually; the separative capacity of the SR-100, 7200 SWU per year; and the separative capacity of the SR-300, 22,000 SWU per year. It was estimated in 1982 that a 3.8 million SWU per year cascade, designed to enrich uranium to 3.2 percent ^{235}U, would have a separative work cost of $120/SWU (75).

Table 4. Conceptual Design Data For a 5 Million SWU/yr Plant

Characteristics		Value	
separating element[a]			
N_0 (mol % UF_6)		4.2	
expansion ratio, π		2.1	
uranium cut, ϑ_u		1/4	
elementary separation effect E_A (%)		1.48	
separation nozzle inlet pressure, kPa[b]		38.7	
separation nozzle outlet pressure, kPa[b]		18.4	
separation stage	*Large*		*Small*
suction flow of compressor, m³/h	1,030,600		312,000
rated power of compressor motor, kW	5,500		1,720
separating element slit length, m	26,660		8,050
separative work capacity, SWU/yr	12,800		3,900
cascade			
mass flow, kg/yr of U			
product[c]		1,585,000	
waste[d]		9,452,000	
feed[e]		11,037,000	
number of large stages		384	
number of small stages		186	
net separative work capacity, kg/yr of U		5,045,000	
total energy requirement of plant, MW		2,520	

[a]Hydrogen is used as a carrier gas.
[b]To convert kPa to mm Hg, multiply by 7.5.
[c]3% ^{235}U.
[d]0.337% ^{235}U.
[e]Natural uranium.

Nomenclature

Some symbols that appear in the text only once and are clearly defined at that time are not listed. Dimensions are given in terms of the mass, M, length, L, time, t, temperature, T, and mols.

C_F = cost of feed material, eq. 14, $/M$ or $/mol.
$C_{\Delta U}$ = cost of separative work, eq. 14, $/M$ or $/mol.
C_{total} = total cost of enriched product, eq. 14, $/M$ or $/mol.
c = total molar concentration, mol/L³.
D_{AB} = binary diffusion coefficient for the pair $A - B$, L^2/t.
E = overall efficiency of a separation process, eq. 82, dimensionless.
E_b = barrier efficiency in gaseous diffusion, eq. 51, dimensionless.
E_c = cut correction in gaseous diffusion, eq. 54, dimensionless.
E_M = mixing efficiency in gaseous diffusion, eq. 53, dimensionless.
E_p = back-pressure efficiency in gaseous diffusion, eq. 52, dimensionless.
e_C = circulation efficiency of a centrifuge, eq. 85, dimensionless.
e_E = experimental efficiency of a centrifuge, eq. 84, dimensionless.

e_F = flow pattern efficiency of a centrifuge, eq. 86, dimensionless.

e_i = ideality efficiency of a centrifuge, eq. 84, dimensionless.

F = feed flow rate to a unit or cascade, eq. 34, M/t or mol/t.

H = steady-state enriched inventory of desired component in a cascade, eq. 42, M or mols.

h_n = total material holdup of the nth stage, eq. 42, M or mols.

J_A = molar flux of component A in a binary mixture relative to the molar average velocity, eq. 69, mol/tL^2.

L = upflow rate of process gas in a stage, eq. 3, M/t or mol/t.

L_{ideal} = stage upflow of process gas in an ideal cascade, eq. 28, M/t or mol/t.

L_{min} = minimum stage upflow of process gas, eq. 22, M/t or mol/t.

L_0 = value of interstage flow rate which minimizes the stage length, eq. 78, M/t or mol/t.

l_f = thickness of effective stagnant film on gaseous diffusion barrier, eq. 53, L.

M_i = molecular weight of component i.

m = ratio of actual interstage flow rate to interstage flow rate which minimizes the stage length, eq. 80, dimensionless.

N_A = molar flux of component A with respect to stationary coordinates, mol/L^2t.

N_{ideal} = number of stages required to span a given concentration span in an ideal cascade, eq. 27, dimensionless

N_{min} = minimum number of stages required to span a given range of concentrations, eq. 21, dimensionless.

N_{sect} = number of stages required to span the concentration range of a square section, eq. 23, dimensionless.

n = stage counting index, eq. 17, dimensionless.

P = product flow rate of a unit or cascade, eq. 34, M/t or mol/t.

p = fluid pressure, M/t^2L.

p_b = fluid pressure on the low pressure side of the gaseous diffusion barrier, eq. 46, M/t^2L.

p_f = fluid pressure on the high pressure side of the gaseous diffusion barrier, eq. 46, M/t^2L.

R = gas constant, ML^2/t^2T mol.

r = pressure ratio p_b/p_f across the gaseous diffusion barrier, eq. 52, dimensionless.

r_0 = transverse coordinate of the plane of zero axial velocity, eq. 75, L.

r_2 = radius of a centrifuge bowl, eq. 68, L.

S = stage length of a continuous process, L.

T = absolute temperature, T.

T_{eq} = equilibrium time of a cascade or separating unit, eq. 42, t.

δU = separative capacity of a unit. eq. 3, M/t or mol/t.

ΔU = separative capacity of a cascade, M/t or mol/t.

$\overline{V}$ = peripheral velocity of a centrifuge, eq. 83, L/t.

$\overline{V}_A$ = partial molal volume of a component A in a mixture, eq. 69, L^3/mol.

$v(x)$ = value function, eq. 12, dimensionless.

$v'(x)$ = first derivative of the value function with respect to concentration, eq. 13, dimensionless.

$v''(x)$ = second derivative of the value function with respect to concentration, eq. 11, dimensionless.

W = waste flow rate from a unit or cascade, eq. 34, M/t or mol/t.

x = mol fraction of desired component of a binary mixture in the downflow or depleted stream, dimensionless.

x_F = mol fraction of desired component of a binary mixture in the feed stream of a unit or cascade, dimensionless.

x_i = mol fraction of component i in a mixture, dimensionless.

x_W = mol fraction of desired component of a binary mixture in the waste stream of a unit or cascade, dimensionless.

y = mol fraction of desired component of a binary mixture in the upflow or enriched stream, dimensionless.

y_P = mol fraction of desired component of a binary mixture in the product stream of a unit or cascade, dimensionless.

Z = overall length of the separating column, L.

z = mol fraction of desired component of a binary mixture in the feed stream of a stage, eq. 4, dimensionless.

z = axial distance or length coordinate in a column, L.

α = stage separation factor, eq. 1, dimensionless.

α^* = ideal stage separation factor, eq. 49, dimensionless.

α_0 = equilibrium stage separation factor in a centrifuge, eq. 90, dimensionless.

θ = fraction of stage feed which goes into the stage upflow, "cut," eq. 3, dimensionless.

μ = fluid viscosity, $M/$Lt.

ρ = mass density, M/L^3.

$\bar{\tau}$ = average net upward transport of desired material in a cascade approaching equilibrium, eq. 42, M/t or mol/t.

ω = angular velocity, eq. 62, radians/t.

BIBLIOGRAPHY

"Diffusion Separation Methods" in *ECT* 1st ed., Vol. 5, pp. 76–133, by M. Benedict, Hydrocarbon Research, Inc.; "Diffusion Separation" in *ECT* 1st ed., 2nd Suppl., pp. 297–315, by K. B. McAfee, Jr., Bell Telephone Labs.; "Diffusion Separation Methods" in *ECT* 2nd ed., Vol. 7, pp. 91–175, by J. Shacter, E. Von Halle, and R. L. Hoglund, Union Carbide Corporation; in *ECT* 3rd ed., Vol. 7, pp. 639–723, by R. L. Hoglund, J. Shacter, and E. Von Halle, Union Carbide Nuclear Division.

1. A. Kanagawa, I. Yamamoto, and Y. Mizuno, *J. Nucl. Sci. Tech.* **14,** 892 (1977).
2. G. Jansen and J. L. Robertson, *Analysis of Nonideal Asymmetric Cascades*, paper presented at American Chemical Society Meeting, Montreal, May 1976.
3. G. A. Garrett and J. Shacter, *Proceedings of the International Symposium on Isotope Separatio*, Amsterdam, 1958, pp. 17–31.
4. G. R. H. Geoghegan in Ref. 3, pp. 518–523.
5. H. Barwich, *Ann. Physik* **20,** 70 (1957).
6. E. Oliveri, *Energia Nucl. Milan* **8,** 453 (1961).
7. J. C. Guais, *BNES Intern. Conference on Uranium Isotope Separation*, Paper 21, London, Mar. 5–7, 1975.

8. N. Ozaki and I. Harada, *BNES Intern. Conference on Uranium Isotope Separation*, Paper 22, London, Mar. 5–7, 1975.

9. H. Harmsen, G. Hertz, and W. Schütze, *Z. Physik* **90,** 703 (1934).

10. D. E. Wooldridge and W. R. Smythe, *Phys. Rev.* **50,** 233 (1936).

11. H. D. Smyth, *Atomic Energy for Military Purposes*, Princeton University Press, Princeton, N.J., 1945.

12. P. C. Keith, *Chem. Eng.* **53,** 112 (1946).

13. J. F. Hogerton, *Chem. Eng.* **52,** 98 (1945).

14. K. E. B. Jay, *Britain's Atomic Factories*, H. M. Stationery Office, London, 1954.

15. M. Molbert in J. R. Merriman and M. Benedict, eds., *Recent Developments in Uranium Enrichment, AIChE Symposium Series, No. 221,* Vol. 78, *The Eurodif Program*, AIChE, New York, 1982.

16. H. Albert, *Proceedings of the U.N. International Conference on Peaceful Uses of Atomic Energy, 2nd Geneva*, Vol. 4, P/1268, 1958, pp. 412–417.

17. *Major Activities in the Atomic Energy Programs*, Jan.–Dec., 1962. U.S. Gov't. Printing Office, Washington, D.C., 1963.

18. D. Massignon in Ref. 16, P/1266, pp. 388–394.

19. J. Charpin, P. Plurien, and S. Mommejac in Ref. 16, P/1265, pp. 380–387.

20. C. Frejacques and co-workers in Ref. 16, P/1262, pp. 418–421.

21. M. Martensson and co-workers in Ref. 16 P/181, pp. 395–404.

22. O. Bilous and G. Counas in Ref. 16, P/1263, pp. 405–411.

23. W. G. Pollard and R. D. Present, *Phys. Rev.* **73,** 762 (1948).

24. P. C. Carman, *Flow of Gases through Porous Media*, Butterworths Publications Ltd., London, 1956.

25. R. D. Present and A. J. de Bethune, *Phys. Rev.* **75,** 1050 (1949).

26. H. T. C. Pratt, *Countercurrent Separation Processes*, American Elsevier Publishing Company, Inc., New York, 1967.

27. C. Boorman in H. London, ed., *Separation of Isotopes*, George Newnes Ltd., London, 1961, Chapt. 8; D. Massignon, *Gaseous Diffusion*, in S. Villani, ed., *Topics in Applied Physics*, Vol. 35, Springer-Verlag, New York, 1979.

28. M. Benedict, T. Pigford, and H. Levi, *Nuclear Chemical Engineering*, 2nd ed., McGraw-Hill Book Co., New York, 1981, Chapt. 14.

29. *AEC Gaseous Diffusion Plant Operations*, USAEC Report No. ORO-684, U.S. Atomic Energy Commission, Washington, D.C., Jan. 1972.

30. *AEC Data on New Gaseous Diffusion Plants*, USAEC Report No. ORO-685, U.S. Atomic Energy Commission, Washington, D.C., Apr. 1972.

31. *CEA Bulletin d'Informations Scientifiques et Techniques*, No. 206, 3-134, Sept. 1975 (in French).

32. G. Besse, *BNES International Conference on Uranium Isotope Separation*, Paper 17, London, Mar. 5–7, 1975.

33. J. P. Gougeau, *Developments in Uranium Enrichment, AIChE Symposium Series 169,* Vol. 73, AIChE, New York, 1977, pp. 12–14.

34. H. J. Mürtz and H. G. Nöller, *Z. Naturforsch.* **16a,** 569 (1961).

35. Ger. Pat. 1,154,793 (Sept. 26, 1963), H. G. Nöller.

36. K. Bornkessel and J. Pilot, *Z. Physik. Chem.* **221,** 177 (1962).

37. A. J. A. Roux, W. L. Grant, R. A. Barbour, R. S. Loubser, and J. J. Wannenburg, *Development and Progress of the South African Enrichment Project, International Conference of Nuclear Power and its Fuel Cycle*, IAEA-CN-36/300, Salzburg, Austria, May 1977.

38. F. A. Lindemann and F. W. Aston, *Phil. Mag.* **37,** 523 (1919).

39. W. Groth in H. London, ed., *Separation of Isotopes*, George Newnes Ltd., London, 1961, Chapt. 6.

40. J. W. Beams, L. B. Snoddy, and A. R. Kuhlthau in Ref. 16, P/723, pp. 428–434.
41. K. Cohen, *The Theory of Isotope Separation as Applied to the Large-Scale Production of U-235*, Natl. Nuclear Energy Ser. Div. III, Vol. 1B, McGraw-Hill Book Co., New York, 1951, Chapt. 6.
42. W. Groth, E. Nann, and K. H. Welge, *Z. Naturforsch.* **12a,** 81 (1957).
43. W. Groth and K. H. Welge, *Z. Physik. Chem.* **19,** 1 (1959).
44. W. Buland and co-workers, *Z. Physik. Chem. Frankfurt* **24,** 249 (1960).
45. W. E. Groth and co-workers in Ref. 16, P/1807, pp. 439–446.
46. K. Beyerle and co-workers in Ref. 3, pp. 667–694.
47. G. Zippe, *The Development of Short Bowl Ultracentrifuges*, Rept. EP-4420-101-60U, Research Laboratories for the Engineering Sciences, University of Virginia, Charlottesville, 1960.
48. A. Bramley, *Science* **92,** 427 (1940).
49. H. Martin and W. Kuhn, *Z. Physik. Chem.* **189,** 219 (1941).
50. H. Martin, *Z. Elektrochem.* **54,** 120 (1950).
51. M. Steenbeck, *Kernenergie* **1,** 921 (1958).
52. J. Los and J. Kistemaker *Proceedings of the International Symposium on Isotope Separation*, Amsterdam, 1958, pp. 695–700.
53. A. Kanagawa and Y. Oyama, *J. At. Energy Soc. Jpn.* **3,** 868 (1961).
54. A. Kanagawa and Y. Oyama, *Nippon Genshiryoku Gakkaishi* **3,** 918 (1961).
55. S. Whitley, *Revs. Modern Physics* **56,** 41 (1984).
56. A. S. Berman, *A Theory of Isotope Separation in a Long Countercurrent Gas Centrifuge*, Rept. K-1536, Union Carbide Corp., Nuclear Div., 1962.
57. A. S. Berman, *A Simplified Model for the Axial Flow in a Long Countercurrent Gas Centrifuge*, Rept. K-1535, Union Carbide Corp., Nuclear Div., 1963.
58. H. M. Parker and T. T. Mayo, IV, *Countercurrent Flow in a Semi-Infinite Gas Centrifuge*, Rept. UVA-279-63R, Research Laboratories for the Engineering Sciences, University of Virginia, Charlottesville, 1963.
59. J. L. Ging, *Countercurrent Flow in a Semi-Infinite Gas Centrifuge: Axially Symmetric Solution in the Limit of High Angular Speed*, Rept. EP-4422-198-62S, Research Laboratories for the Engineering Sciences, University of Virginia, Charlottesville, 1962.
60. J. L. Ging, *The Nonexistence of Pure Imaginary Eigenvalues and the Uniqueness Theorem for the Linearized Gas Flow Equations*, Rept. EP-4422-245-62S, Research Laboratories for the Engineering Sciences, University of Virginia, Charlottesville, 1962.
61. J. L. Ging, *Onsager Minimum Principle for Stationary Flow in Axially Symmetric Rotating Systems*, Rept. EP-3912-321-64U, Research Laboratories for the Engineering Sciences, University of Virginia, Charlottesville, 1964.
62. J. L. Ging, *Eigenvalue Problem—Limit of Low Angular Speed*, Rept. EP-3912-64U, Research Laboratories for the Engineering Sciences, University of Virginia, Charlottesville, 1964.
63. J. L. Ging, *Modified Minimum Principle for Stationary Flow in a Gas Centrifuge*, Rept. EP-3912-325-64U, Research Laboratories for the Engineering Sciences, University of Virginia, Charlottesville, 1964.
64. J. L. Ging, *Onsager Minimum Principle for Axially Decaying Eigenmodes*, Rept. EP-3912-326-64U, Research Laboratories for the Engineering Sciences, University of Virginia, Charlottesville, 1964.
65. G. F. Carrier and S. H. Maslen, *Flow Phenomena in Rapidly Rotating Systems*, Rept. TID 18065, U.S.-D.O.E., 1962.
66. G. F. Carrier in H. Görtler, ed., *Proceedings of the Eleventh International Congress of Applied Mechanics*, Springer, Berlin, 1964.
67. Soubbaramayer in Ref. 27, Chapt. 4, Centrifugation.
68. H. G. Wood, *J. Fluid Mech.* **101,** 1 (1980).

69. R. J. Ribando, *A Finite Difference Solution of Onsager's Model for Flow in a Gas Centrifuge*, Rept. UVA-ER-822-83U, University of Virginia, Charlottesville, 1983.

70. J. J. H. Brouwers, *On the Motion of a Compressible Fluid in a Rotating Cylinder*, Doctoral Dissertation, The Technische Hogeschool, Twente, the Netherlands, June, 1976.

71. T. Sakurai and T. Matsuda, *J. Fluid Mech.* **62,** 727 (1974).

72. T. Sakurai, *J. Fluid Mech.* **72,** 321 (1975).

73. T. Matsuda, K. Hashimoto, and H. Takeda, *J. Fluid Mech.* **73,** 389 (1976).

74. K. Hashimoto, *J. Fluid Mech.* **76,** 289 (1976).

75. T. Matsuda and K. Hashimoto, *J. Fluid Mech.* **78,** 337 (1976).

76. T. Matsuda and K. Hashimoto, *J. Fluid Mech.* **85,** 433 (1978).

77. T. Kai, *J. Nucl. Sci. Technol.* **14,** 267 (1977).

78. T. Kai, *J. Nucl. Sci. Technol.* **14,** 506 (1977).

79. E. Von Halle, *The Countercurrent Gas Centrifuge for the Enrichment of U-235*, K/OA-4058, Union Carbide Nuclear Div., Oak Ridge, Tenn., Nov., 1977.

80. E. Krause and E. H. Hirschel, eds., *DFVLR—Colloquium*, Proz-Wahn, West Germany, 1970.

81. D. R. Olander, *Adv. Nucl. Sci. Technol.* **6,** 105 (1972).

82. D. G. Avery and E. Davies, *Uranium Enrichment by Gas Centrifuge*, Mills & Boon Ltd., London, 1973.

83. S. Villani, *Isotope Separation*, American Nuclear Society, 1976.

84. E. Rätz in *Aerodynamic Separation of Gases and Isotopes, Lecture Series 1978*, Von Karmen Institute for Fluid Dynamics, Belgium, 1978.

85. Soubbaramayer, ed., *Proceedings of the Second Workshop on Gases in Strong Rotation*, Cadarache, France, Apr. 1977.

86. G. B. Scuricini, ed., *Proceedings of the Third Workshop on Gases in Strong Rotation*, Rome, Mar. 1979.

87. E. Rätz, ed., *Proceedings of the Fourth Workshop on Gases in Strong Rotation*, Oxford, UK, Aug. 1981.

88. H. G. Wood, ed., *Proceedings of the Fifth Workshop on Gases in Strong Rotation*, Charlottesville, Va., June 1983.

89. Y. Takashima, ed., *Proceedings of the Sixth Workshop on Gases in Strong Rotation*, Tokyo, Aug. 1985.

90. K. G. Roesner and E. Rätz, eds., *Proceedings of the First Workshop on Separation Phenomena in Liquids and Gases*, Technische Hochschule Darmstadt, Darmstadt, Germany, July 1987.

91. P. Louvet, P. Noe, and Soubbaramayer, eds., *Proceedings of the Second Workshop on Separation Phenomena in Liquids and Gases*, Cite Scientifique Parcs et Technopoles Ile de France Sud, Versailles, France, July 1989.

92. W. H. Furry, R. C. Jones, and L. Onsager, *Phys. Rev.* **55,** 1083 (1939).

93. P. A. Tahourdin, *Final Report on the Jet Separation Methods*, Oxford Rept. No. 36, Br. 694, Clarendon Lab., Oxford, UK, 1946.

94. S. A. Stern, P. C. Waterman, and T. F. Sinclair, *J. Chem. Phys.* **33,** 805 (1960).

95. E. W. Becker and co-workers in Ref. 16, P/1002, pp. 455–457.

96. E. W. Becker in Ref. 3, pp. 560–578.

97. E. W. Becker in H. London, ed., *Separation of Isotopes*, George Newnes Ltd., London, 1961, Chapt. 9.

98. E. W. Becker and co-workers, *Angew. Chemie. Intern. Ed. (Engl.)* **6,** 507 (1967).

99. E. W. Becker and co-workers, *Atomwirtschaft* **18,** 524 (1973).

100. E. W. Becker and co-workers, *International Conference on Uranium Isotope Separation*, London, 1975.

101. E. W. Becker and co-workers, *European Nuclear Conference*, Paris, 1975.

102. E. W. Becker and co-workers, American Nuclear Society Meeting, *KFK-Bericht 2235*, Gesellschaft für Kernforschung, Karlsruhe, 1975.
103. H. Geppert and co-workers, *International Conference on Uranium Isotope Separation*, London, 1975.
104. E. W. Becker and co-workers, *Z. Naturforsch.* **26a,** 1377 (1971).
105. P. Bley and co-workers, *Z. Naturforsch.* **28a,** 1273 (1973).
106. K. Bier and co-workers, *KFK-Bericht 1440*, Gesellschaft für Kernforschung, Karlsruhe, 1971.
107. U. Ehrfeld and W. Ehrfeld, *KFK-Bericht 1634*, Gesellschaft für Kernforschung, Karlsruhe, 1972.
108. W. Ehrfeld and E. Schmid, *KFK-Bericht 2004*, Gesellschaft für Kernforschung, Karlsruhe, 1974.
109. Ger. Pat. 1,096,875 (Jan. 12, 1961), E. W. Becker (to Deutsche Gold-und Silber-Scheideanstalt vorm. Roessler).
110. W. Ehrfeld and U. Knapp, *KFK-Bericht 2138*, Gesellschaft für Kernforschung, Karlsruhe, 1975.
111. E. W. Becker and co-workers, *4th United Nations International Conference on the Peaceful Uses of Atomic Energy*, Geneva, 1971, paper 383.
112. H. J. Fritsch and R. Schütte, *KFK-Bericht 1437*, Gesellschaft für Kernforschung, Karlsruhe, 1971.
113. R. Schütte and co-workers, *Chemie-Ing. Technik* **44,** 1099 (1972).
114. W. Fritz and co-workers, *Chemie-Ing. Technik* **45,** 590 (1973).
115. R. Schütte, *KFK-Bericht 1986*, Gesellschaft für Kernforschung, Karlsruhe, 1974.
116. P. Bley and co-workers, *KFK-Bericht 2092*, Gesellschaft für Kernforschung, Karlsruhe, 1975.
117. W. Ehrfeld in Ref. 84.
118. U. Ehrfeld in Ref. 84.
119. E. W. Becker in Ref. 27.
120. E. W. Becker, P. Noguira Batista, and H. Volcker, *Nucl. Technol.* **52,** 105 (1981).
121. E. W. Becker and co-workers in Ref. 16.
122. P. C. Waterman and S. A. Stern, *J. Chem. Phys.* **31,** 405 (1959).
123. R. R. Chow, *On the Separation Phenomenon of Binary Gas Mixture in an Axisymmetric Jet*, Rept. HE-150-175, Institute of Engineering Research, University of California, Berkeley, 1959.
124. E. E. Gose, *Am. Inst. Chem. Engs. J.* **6,** 168 (1960).
125. V. H. Reis and J. B. Fenn, *J. Chem. Phys.* **39,** 3240 (1963).
126. H. Mikami, *J. Nucl. Sci. Technol.* **6,** 452 (1969).
127. H. Mikami, *I & EC Fundam.* **9,** 121 (1970).
128. H. Mikami and Y. Takashima, *Bull. Tokyo Inst. Technol.* **61,** 67 (1964).
129. H. Mikami and Y. Takashima, *J. Nucl. Sci. Technol.* **5,** 572 (1968).
130. H. Mikami and Y. Takashima, *Int. J. Heat Mass Transfer* **11,** 1597 (1968).
131. W. Berkahn, W. Ehrfeld, and G. Krieg, *Calculations of Uranium Isotope Separation in the Separation Nozzle for Small Mol Fractions of UF$_6$ in the Auxiliary Gas*, Institut für Kernverfahrenstechnik, Kernforschungszentrum Karlsruhe, West Germany, report KFK-2351, Nov., 1976.
132. G. F. Malling and E. Von Halle, *Aerodynamic Isotope Separation Processes for Uranium Enrichment: Process Requirement*, paper presented at the Symposium on New Advances in Isotope Separation, Div. of Nuclear Chemistry and Technology, American Chemical Society, San Francisco, Calif., Aug. 1976; *UCC-ND Report K/OA-2872*, Oak Ridge Gaseous Diffusion Plant, Oak Ridge, Tenn., Oct. 7, 1976.

General References

M. Benedict, T. Pigford, and H. Levi, *Nuclear Chemical Engineering*, 2nd ed., McGraw-Hill Book Co., New York, 1981, Chapts. 12 and 14.

K. P. Cohen, *The Theory of Isotope Separation as Applied to the Large-Scale Production of U-235, National Nuclear Energy Series Division III*, Vol. 1B, McGraw-Hill Book Co., New York, 1951.

H. London, ed., *Separation of Isotopes*, George Newnes Ltd., London, 1961.

H. R. C. Pratt, *Countercurrent Separation Processes*, American Elsevier Publishing Company, Inc., New York, 1967.

S. Villani, *Isotope Separation*, American Nuclear Society Monograph, ANS Publications, 1976.

S. Villani, ed., *Topics in Applied Physics*, Vol. 35, Springer-Verlag, New York, 1979.

J. Kistemaker, J. Bigeleisen, and A. O. Neir, eds., *Proceedings of the International Symposium on Isotope Separation*, Amsterdam, Apr. 23–27, 1957, North-Holland Publishing Company, Amsterdam, and Interscience Publishers, Inc., New York, 1958.

Proceedings of the Second United Nations International Conference on the Peaceful Uses of Atomic Energy, Geneva, Sept. 1–3, 1958, United Nations publication, Geneva, 1958, particularly Vol. 4: *Production of Nuclear Materials and Isotopes*.

Proceedings of the Third International Conference on the Peaceful Uses of Atomic Energy, Geneva, Aug. 31–Sept. 9, 1964, United Nations publication, New York, 1965, particularly Vol. 12: *Nuclear Fuels—III, Raw Materials*.

Proceedings of the Fourth International Conference on the Peaceful Uses of Atomic Energy, Geneva, Sept. 6–16, 1971, United Nations and the International Atomic Energy Agency, 1972, particularly Vol. 9, *Isotope Enrichment, Fuel Cycles and Safeguards*.

Proceedings of the International Conference on Uranium Isotope Separation, London, Mar. 5–7, 1975, British Nuclear Energy Society, 1975.

M. Benedict, ed., *Development in Uranium Enrichment, AIChE Symposium Series*, Vol. 73, No. 169, American Institute of Chemical Engineers, New York, 1977.

J. R. Merriman and M. Benedict, eds., *Recent Developments in Uranium Enrichment, AIChE Symposium Series*, Vol. 78, no. 221, American Institute of Chemical Engineers, New York, 1982.

E. Von Halle
Martin Marietta Energy Systems

J. Shacter
Consultant

DIGITAL DISPLAYS. See Optical displays.

DIGITAL OPTICAL RECORDING. See Information storage
materials.

DIKETENE. See Ketenes and related substances.

DIMENSIONAL ANALYSIS

Dimensional analysis is a technique that treats the general forms of equations governing natural phenomena. It provides procedures of judicious grouping of variables associated with a physical phenomenon to form dimensionless products of these variables; therefore, without destroying the generality of the relationship, the equation describing the physical phenomenon may be more easily determined experimentally. It guides the experimenter in the selection of experiments capable of yielding significant information and in the avoidance of redundant experiments, and makes possible the use of scale models for experiments (see also DESIGN OF EXPERIMENTS). The method is particularly valuable when the problems involve a large number of variables. On such occasions, dimensional analysis may reveal that, whatever the form of the inaccessible final solution, certain features of it are obligatory. The technique has been utilized effectively in engineering modeling (1–7).

The method of dimensional analysis is not new. It can be traced to Newton, who at the time was laying the foundation of mechanics as a fundamental branch of science. The validity of the method is based on the premise that any equation that correctly describes a physical phenomenon must be dimensionally homogeneous. This principle of dimensional homogeneity, which states in effect that quantities of different kinds cannot be added together, is of fundamental importance in dimensional analysis, and was first expressed by J. Fourier's classic work *La Théorie Analytique de la Chaleur*, published in 1822. He not only introduced the notion of dimensional homogeneity but also the conception of what is today termed the dimensional formula. In 1914, Buckingham (8) made a significant contribution with his famous Pi Theorem which possibly prompted Lord Rayleigh (9) shortly thereafter to observe, "It happens not infrequently that results in the form of 'laws' are put forward as novelties on the basis of elaborate experiments which might have been predicted *a priori* after a few minutes' consideration." Needless to say, Fourier's and Buckingham's works have been used, elaborated upon, and extended by many others (10–38). A measure of this interest is reflected in three comprehensive bibliographies (10–12) in which more than 600 research contributions are referenced.

Units and Dimensions

The concepts used to describe natural phenomena are based on the precise measurement of quantities. The quantitative measure of anything is a number that is found by comparing one magnitude with another of the same type. It is necessary to specify the magnitude of the quantity used in making the comparison if the number is to be meaningful. The statement that "the length of a car is 6 meters" implies that a length has been chosen, namely, one meter, and that the ratio of the length of the car to the chosen length is 6. The chosen magnitudes, such as the meter, are called *units* of measurement. The result of a measurement is represented by a number followed by the name of the unit that was used in making the measurement (see UNITS). The statement that the area of a room is 30 square meters indicates that the unit of measurement is one square meter. Thus, to each kind of physical quantity there corresponds an appropriate kind of unit. The physical concepts such as length, area, and time are referred to as *dimensions*, which are different from units. The length of a car is 6 m, which is equivalent to 19.7 ft or 6.56 yards.

Classical physics is built on the foundation of the laws of motion. It was felt at the time that the entire subject could be based on the laws of classical mechanics, and further work would undoubtedly make electromagnetism another branch of mechanics. Under these circumstances, it was natural to regard length l, mass m, and time t as the fundamental, primary, or reference dimensions. However, such designations lead to dimensional ambiguity in that two distinct concepts may possess the same dimensions. The system works fairy well for mechanics. The most notable ambiguity occurs with energy and torque. However, in electromagnetism the situation is bad. The classical arrangement employs electrostatic and electromagnetic systems, and the same concept may lead to different dimensions. For example, in the electrostatic system capacitance and length have the same dimensions, and in the electromagnetic system inductance and length are not dimensionally distinguished. On the subject of heat, ambiguities occur between entropy and mass (13), or temperature and reciprocal length (14), depending upon the assumptions made about the dimensionality of some constants. This does not mean that dimensional ambiguity is a fact of nature; it merely shows an imperfection in our human scheme of assigning primary concepts.

Over the years the number of reference dimensions in physics has evolved from the original three, to four, to five, and then gradually downwards to an absolutely necessary one, and then upwards again through an understanding that, though only one is absolutely necessary, a considerable convenience can stem from using three, to four, or five reference dimensions depending on the problem at hand (1,6,7,15–20). There is nothing sacrosanct about the number of reference dimensions, and dimensional analysis is merely a tool that may be manipulated at will (1). This principle of free choice of the reference dimensions has been widely accepted, although one still finds references to true dimensions. Thus, an important step in dimensional analysis is the selection of reference dimensions in such a way that the others, called the secondary or derived dimensions, can be expressed in terms of them. The relation between reference and derived dimensions is generally established either through the fundamental law or equation governing the phenomenon or through definitions. When length, mass, and time are taken to be the reference dimensions, the dimensions of velocity v, for example, are the dimensions of length divided by time, or expressed by symbols, $v = lt^{-1}$. Likewise, through Newton's law of motion, which relates force, mass, and acceleration by

$$\text{force} = \text{constant} \times \text{mass} \times \text{acceleration} \tag{1}$$

the dimensions of force f must be (mass·length)/time^2 or $f = mlt^{-2}$. The expressions like $v = lt^{-1}$ and $f = mlt^{-2}$ are referred to as dimensional formulas. The exponents of dimensions of a physical quantity are the powers of the reference dimensions in which it is expressed. Thus, the exponents of the dimensions of force are 1 in mass, 1 in length and -2 in time. If force, length, and time are chosen as the reference dimensions, then mass becomes secondary. In either of these two choices, the constant in Newton's law is dimensionless. However if force, mass, length, and time are chosen as the reference dimensions, the constant is no longer dimensionless and the units are generally selected so that the constant is numerically equal to the standard acceleration of gravity. Table 1 lists the exponents of dimensions of some common variables in mechanics with respect to these three choices of reference dimensions $[(m,l,t); (f,l,t); (f,m,l,t)]$ which give rise to the absolute, gravitational, and engineering systems of dimensions, respectively.

To eliminate the ambiguities in the subject of electricity and magnetism, it is convenient to add charge q to the traditional l, m and t dimensions of mechanics to form the reference dimensions. In many situations permittivity ϵ or permeability μ is used in lieu of charge. For thermal problems temperature T is considered as a reference dimension. Tables 2 and 3 list the exponents of dimensions of some common variables in the fields of electromagnetism and heat.

Other dimensional systems have been developed for special applications which can be found in the technical literature. In fact, to increase the power of dimensional analysis, it is advantageous to differentiate between the lengths in radial and tangential directions (13). In doing so, ambiguities for the concepts of energy and torque, as well as for normal stress and shear stress, are eliminated (see Ref. 13).

Table 1. Exponents of Dimensions for Mechanical Quantities in Absolute, Gravitational, and Engineering Systems

	Absolute			Gravitational			Engineering			
Quantity	m	l	t	f	l	t	f	m	l	t
acceleration	0	1	-2	0	1	-2	0	0	1	-2
angular acceleration	0	0	-2	0	0	-2	0	0	0	-2
angular velocity	0	0	-1	0	0	-1	0	0	0	-1
area	0	2	0	0	2	0	0	0	2	0
angular momentum	1	2	-1	1	1	1	0	1	2	-1
density, mass	1	-3	0	1	-4	2	0	1	-3	0
energy, work	1	2	-2	1	1	0	1	0	1	0
force	1	1	-2	1	0	0	1	0	0	0
frequency	0	0	-1	0	0	-1	0	0	0	-1
length	0	1	0	0	1	0	0	0	1	0
linear acceleration	0	1	-2	0	1	-2	0	0	1	-2
linear momentum	1	1	-1	1	0	1	0	1	1	-1
linear velocity	0	1	-1	0	1	-1	0	0	1	-1
mass	1	0	0	1	-1	2	0	1	0	0
moment of inertia	1	2	0	1	1	2	0	1	2	0
power	1	2	-3	1	1	-1	1	0	1	-1
pressure	1	-1	-2	1	-2	0	1	0	-2	0
stress	1	-1	-2	1	-2	0	1	0	-2	0
surface tension	1	0	-2	1	-1	0	1	0	-1	0
time	0	0	1	0	0	1	0	0	0	1
viscosity, absolute	1	-1	-1	1	-2	1	1	0	-2	1
viscosity, kinematic	0	2	-1	0	2	-1	0	0	2	-1
volume	0	3	0	0	3	0	0	0	3	0

Table 2. Exponents of Dimensions for Electromagnetic Quantities

Quantity	l	m	t	q	l	m	t	ϵ	l	m	t	μ
charge	0	0	0	1	³⁄₂	½	−1	½	½	½	0	−½
capacitance	−2	−1	2	2	1	0	0	1	−1	0	2	−1
current	0	0	−1	1	³⁄₂	½	−2	½	½	½	−1	−½
electric field intensity	1	1	−2	−1	−½	½	−1	−½	½	½	−2	½
electric potential difference	2	1	−2	−1	½	½	−1	−½	³⁄₂	½	−2	½
electric flux	0	0	0	1	³⁄₂	½	−1	½	½	½	0	−½
electric flux density	−2	0	0	1	−½	½	−1	½	−³⁄₂	½	0	−½
inductance	2	1	0	−2	−1	0	2	−1	1	0	0	1
magnetic field intensity	−1	0	−1	1	½	½	−2	½	−½	½	−1	−½
magnetic flux	2	1	−1	−1	½	½	0	−½	³⁄₂	½	−1	½
magnetic flux density	0	1	−1	−1	−³⁄₂	½	0	−½	−½	½	−1	½
magnetomotive force	0	0	−1	1	³⁄₂	½	−2	½	½	½	−1	−½
permeability	1	1	0	−2	−2	0	2	−1	0	0	0	1
permittivity	−3	−1	2	2	0	0	0	1	−2	0	2	−1
resistance	2	1	−1	−2	−1	0	1	−1	0	0	−1	1

Table 3. Exponents of Dimensions for Thermal Quantities

Quantity	l	m	t	T	f	l	t	T
coefficient of thermal expansion	0	0	0	−1	0	0	0	−1
entropy	2	1	−2	−1	1	1	0	−1
temperature	0	0	0	1	0	0	0	1
thermal energy (heat)	2	1	−2	0	1	1	0	0
thermal power	2	1	−3	0	1	1	−1	0
thermal conductivity	1	1	−3	−1	1	0	−1	−1
thermittivity	2	0	−2	−1	0	2	−2	−1

Dimensional Matrix and Dimensionless Products

An appropriate set of independent reference dimensions may be chosen so that the dimensions of each of the variables involved in a physical phenomenon can be expressed in terms of these reference dimensions. In order to utilize the algebraic approach to dimensional analysis, it is convenient to display the dimensions of the variables by a matrix. The matrix is referred to as the dimensional matrix of the variables and is denoted by the symbol D. Each column of D represents a variable under consideration, and each row of D represents a reference dimension. The ith row and jth column element of D denotes the exponent of the reference dimension corresponding to the ith row of D in the dimensional formula of the variable corresponding to the jth column. As an illustration, consider Newton's law of motion, which relates force F, mass M, and acceleration A by (eq. 2):

$$F = \text{constant} \times MA \tag{2}$$

If length l, mass m, and time t are chosen as the reference dimensions, from Table 1 the dimensional formulas for the variables F, M and A are as follows:

Variables	Dimensional formulas
F	$m^1 l^1 t^{-2}$
M	$m^1 l^0 t^0$
A	$m^0 l^1 t^{-2}$

The dimensional matrix associated with Newton's law of motion is obtained as (eq. 3)

$$
\begin{array}{c}
\quad\; F \;\; M \;\; A \\
\boldsymbol{D} = \begin{array}{c} m \\ l \\ t \end{array}
\left[\begin{array}{ccc}
1 & 1 & 0 \\
1 & 0 & 1 \\
-2 & 0 & -2
\end{array} \right]
\end{array}
\tag{3}
$$

In the example, the exponents of dimensions in the dimensional formula of the variable F are 1, 1 and -2, and hence the first column is $(1,1,-2)$. Likewise, the second and third columns of $\boldsymbol{D}$ correspond to the exponents of dimensions in the dimensional formulas of the variables M and A, respectively.

As indicated earlier, the validity of the method of dimensional analysis is based on the premise that any equation that correctly describes a physical phenomenon must be dimensionally homogeneous. An equation is said to be dimensionally homogeneous if each term has the same exponents of dimensions. Such an equation is of course independent of the systems of units employed provided the units are compatible with the dimensional system of the equation. It is convenient to represent the exponents of dimensions of a variable by a column vector called dimensional vector represented by the column corresponding to the variable in the dimensional matrix. In equation 3, the dimensional vector of force F is $[1,1,-2]'$ where the prime denotes the matrix transpose.

Suppose that there are n variables Q_1 Q_2, $\cdots$, Q_n that are involved in a physical phenomenon whose dimensional vectors are $\boldsymbol{D}_1, \boldsymbol{D}_2, \cdots, \boldsymbol{D}_n$, respectively. This phenomenon can generally be expressed by (eq. 4):

$$
f(Q_1, Q_2, \cdots, Q_n) = 0
\tag{4}
$$

When such a function is established or assumed, it will still exist even after the variables are intermultiplied in any manner whatsoever. This means that each variable in the equation can be combined with other variables of the equation to form dimensionless products whose dimensional vectors are the zero vector. Equation 4 can then be transformed into the nondimensional form as (eq. 5):

$$
f(\pi_1, \pi_2, \cdots, \pi_n) = 0
\tag{5}
$$

where the dimensionless products π_i ($i = 1, 2, \cdots, n$) can generally be expressed as the power products of the form (eq. 6):

$$\pi_i = Q_1^{x_{1i}} Q_2^{x_{2i}} \cdots Q_n^{x_{ni}} \tag{6}$$

Let $R_1, R_2, \cdots, R_m$ be a set of chosen reference dimensions. Then the dimensional formulas for the variables Q_i are given by (eq. 7):

$$R_1^{d_{1i}} R_2^{d_{2i}} \cdots R_m^{d_{mi}} \tag{7}$$

where the exponents of dimensions are represented by the dimensional vectors as (eq. 8):

$$D_i' = [d_{1i}, d_{2i}, \cdots, d_{mi}], i = 1, 2, \cdots, n \tag{8}$$

Using equation 7 the dimensional formulas for π_i of equation 6 can be written to give (eq. 9):

$$\left[R_1^{d_{11}} R_2^{d_{21}} \cdots R_m^{d_{m1}} \right]^{x_{1i}} \left[R_1^{d_{12}} R_2^{d_{22}} \cdots R_m^{d_{m2}} \right]^{x_{2i}} \cdots \left[R_1^{d_{1n}} R_2^{d_{2n}} \cdots R_m^{d_{mn}} \right]^{x_{ni}} \tag{9}$$

Since π_i are dimensionless products having dimensional vectors equal to the zero vector, the exponents of the R_j ($j = 1, 2, \ldots, m$) must add up to zero, giving (eq. 10):

$$\begin{aligned}
d_{11}x_{1i} + d_{12}x_{2i} + \cdots + d_{1n} x_{ni} &= 0 \\
d_{21}x_{1i} + d_{22}x_{2i} + \cdots + d_{2n} x_{ni} &= 0 \\
\cdots \qquad \cdots \qquad \cdots & \\
\cdots \qquad \cdots \qquad \cdots & \\
\cdots \qquad \cdots \qquad \cdots & \\
d_{m1}x_{1i} + d_{m2}x_{2i} + \cdots + d_{mn}x_{ni} &= 0
\end{aligned} \tag{10}$$

In terms of the dimensional vectors of equation 8, equation 10 can be rewritten as (eqs. 11–13):

$$[D_1, D_2, \cdots, D_n] X_i = 0, i = 1, 2, \cdots, n \tag{11}$$

where

$$X_i' = [x_{1i}, x_{2i}, \cdots, x_{ni}] \tag{12}$$

or more compactly

$$DX = 0 \tag{13}$$

where $X = X_i$, $i = 1, 2, \cdots, n$. Thus, the product of a set of variables is dimensionless if, and only if, the exponents of these variables are a solution of the homogeneous linear algebraic equation (13). A vector X is said to be a B-vector of D if it is a solution of equation 13. The corresponding dimensionless product associated with the variables of a B-vector is called a B-number (21,22). Frequently, the term pi-number is also used by many authors because it was first introduced by Buckingham (8) in 1914 who used the symbol π for a dimensionless product or group. In fact, the term pi was even attached to his contributions to dimensional analysis, and is known as Buckingham's Pi-Theorem. But this usage is deprecated because of possible confusion with the universal constant of $\pi = 3.14159$. Therefore, the choice of his initial B is preferred to that of the term π.

The following example illustrates the above procedure.

Example 1. The problem is to find the period P of a simple pendulum swinging in a vacuum under the influence of gravity. To write an equation for the period, the first step is to consider what physical quantities affect the period. This requires some prior knowledge upon which the intuitive judgment can be based. On this basis, it is clear that the period depends on the mass M of the bob, the length L of the string supporting the bob, and of course, the acceleration g owing to the force of gravity. As before, mass m, length l, and time t are chosen as the reference dimensions, ie, $R_1 = m$, $R_2 = l$, and $R_3 = t$. From Table 1 the dimensional formulas for the variables M, L, P, and g, together with their dimensional vectors, are as shown below:

Variables	Dimensional formulas	Dimensional vectors
$Q_1 = M$	$m^1 l^0 t^0$	$\boldsymbol{D}_1' = [1, 0, 0]$
$Q_2 = L$	$m^0 l^1 t^0$	$\boldsymbol{D}_2' = [0, 1, 0]$
$Q_3 = P$	$m^0 l^0 t^1$	$\boldsymbol{D}_3' = [0, 0, 1]$
$Q_4 = g$	$m^0 l^1 t^{-2}$	$\boldsymbol{D}_4' = [0, 1, -2]$

Thus, the dimensional formulas for π_i of equation 6 can be expressed as (eq. 14):

$$\left(m^1 l^0 t^0\right)^{x_{1i}} \left(m^0 l^1 t^0\right)^{x_{2i}} \left(m^0 l^0 t^1\right)^{x_{3i}} \left(m^0 l^1 t^{-2}\right)^{x_{4i}} \tag{14}$$

whose dimensional vector must be the zero vector requiring (eq. 15):

$$\begin{array}{c} \\ m \\ l \\ t \end{array} \begin{array}{cccc} M & L & P & g \\ \left[\begin{array}{cccc} 1 & 0 & 0 & 0 \\ 0 & 1 & 0 & 1 \\ 0 & 0 & 1 & -2 \end{array}\right] \end{array} \begin{bmatrix} x_{1i} \\ x_{2i} \\ x_{3i} \\ x_{4i} \end{bmatrix} = \begin{bmatrix} 0 \\ 0 \\ 0 \end{bmatrix} \tag{15}$$

the coefficient matrix of which is identified as the dimensional matrix $\boldsymbol{D}$ associated with the pendulum problem, and $i = 1, 2, 3, 4$. Solving x_{1i}, x_{2i} and x_{3i} in terms of x_{4i} yields the desired B-vectors of $\boldsymbol{D}$ as (eq. 16):

$$\boldsymbol{X}_i = [0, -x_{4i}, 2x_{4i}, x_{4i}], \quad i = 1, 2, 3, 4 \tag{16}$$

where $x_{4i} \neq 0$ are arbitrary constants. The corresponding B-numbers become (eq. 17).

$$\pi_i = M^0 \, L^{-x_{4i}} \, P^{2x_{4i}} \, g^{x_{4i}} \tag{17}$$

Since the solutions X_i are related to one another by a multiplicative constant, there is only one linearly independent solution and hence only one independent B-number. Choose, for simplicity, $x_{4i} = 1$. Equation 17 can be rewritten as (eq. 18):

$$P = \text{constant} \, \sqrt{L/g} \tag{18}$$

since π_i is a constant. The value of this constant, which is known to be 2π, cannot be determined by the method of dimensional analysis, and must be evaluated experimentally or analytically.

The example demonstrates that not all the B-numbers π_i of equation 5 are linearly independent. A set of linearly independent B-numbers is said to be complete if every B-number of D is a product of powers of the B-numbers of the set. To determine the number of elements in a complete set of B-numbers, it is only necessary to determine the number of linearly independent solutions of equation 13. The solution to the latter is well known and can be found in any text on matrix algebra (see, for example, (39) and (40)). Thus the following theorems can be stated.

Theorem 1. The number of products in a complete set of B-numbers associated with a physical phenomenon is equal to $n - r$, where n is the number of variables that are involved in the phenomenon and r is the rank of the associated dimensional matrix.

This result was first discussed by Buckingham (8) and stated in its present form by Langhaar (23). It states in effect that an equation is dimensionally homogeneous if and only if it can be reduced to a relationship among a complete set of B-numbers. Buckingham's result (8) was originally stated as Theorem 2.

Theorem 2. A necessary and sufficient condition for an equation $f(Q_1, Q_2, \cdots, Q_n) = 0$ to be dimensionally homogeneous is that it should be reducible to the form $g(\pi_1, \pi_2, \cdots, \pi_p) = 0$, where the πs are a complete set of B-numbers of the variables Qs.

This theorem does not specify how many products in a complete set of B-numbers can be expected from a given set of variables, but it does state that a physical phenomenon describable by n quantities can be rigorously and accurately described by a complete set of B-numbers. The number of products in a complete set of B-numbers may also be determined by another rule, which is equivalent to Theorem 1. The rule (Theorem 3) was first given by Van Driest (24).

Theorem 3. The number of products in a complete set of B-numbers is equal to the total number of variables minus the maximum number of these variables that will not form a dimensionless product.

To show the equivalence of Theorems 1 and 3, it is only necessary to demonstrate that the maximum number of the variables that will not form a dimensionless product is equal to the rank of the dimensional matrix D.

In terms of linear vector space, Buckingham's theorem (Theorem 2) simply states that the null space of the dimensional matrix has a fixed dimension, and Van Driest's rule (Theorem 3) then specifies the nullity of the dimensional matrix. The problem of finding a complete set of B-numbers is equivalent to that of computing a fundamental system of solutions of equation 13 called a complete set of B-vectors. For simplicity, the matrix formed by a complete set of B-vectors will be called a complete B-matrix. It can also be demonstrated that the choice of reference dimensions does not affect the B-numbers (22).

Theorem 4. The set of B-numbers associated with a physical phenomenon is invariant with respect to the choice of the reference dimensions provided that the reference dimensions are considered independent, and that the number of these reference dimensions is not altered.

The implication of this theorem is important in that in computing a complete set of dimensionless products or B-numbers associated with a physical phenomenon, it does not matter which set of dimensions are chosen as the reference dimensions as long as they are independent and their number is not altered.

In example 1, there are four variables that are involved in the pendulum problem. The associated dimensional matrix D is given in equation 15. Since the rank r of D is 3, according to Theorem 1 there are only $n - r = 4 - 3 = 1$ independent B-numbers, as expected.

Suppose now that force f, length l, and time t are chosen as the reference dimensions. From Table 1 the new dimensional matrix $\tilde{D}$ becomes (eq. 19)

$$\tilde{D} = \begin{matrix} f \\ l \\ t \end{matrix} \begin{bmatrix} \begin{matrix} M & L & P & g \end{matrix} \\ 1 & 0 & 0 & 0 \\ -1 & 1 & 0 & 1 \\ 2 & 0 & 1 & -2 \end{bmatrix} \tag{19}$$

The matrix $\hat{D}$ that will transform $\tilde{D}$ to D is the dimensional matrix of the variables force, length, and time with respect to the reference dimensions m, l and t. Again from Table 1 equation 20 is obtained.

$$\hat{D} = \begin{matrix} m \\ l \\ t \end{matrix} \begin{bmatrix} \begin{matrix} f & l & t \end{matrix} \\ 1 & 0 & 0 \\ 1 & 1 & 0 \\ -2 & 0 & 1 \end{bmatrix} \tag{20}$$

It is straightforward to confirm that $D = \hat{D}\tilde{D}$. Since $\hat{D}$ is nonsingular, $DX = 0$ and $\tilde{D}X = 0$ are equivalent, possessing the same set of B-numbers.

In applying dimensional analysis, it is first necessary to be able to identify the variables that govern a particular physical phenomenon. The naming of the governing variables requires some prior knowledge of a particular branch of physics involved. This may include analytical studies, experimental observations, or both. Whatever the source, there must be some prior knowledge upon which a selection can be made.

Systematic Calculation of a Complete B-Matrix

Once the dimensional matrix has been set up and the number of products in a complete set of B-numbers determined, a complete set of B-vectors must be computed. In the following, a systematic procedure for this purpose is presented.

Let D be the dimensional matrix of order m by n associated with a set of variables of a physical phenomenon, where m is the number of chosen reference dimensions and n the number of variables of the set. Without loss of generality, it may be assumed that $n \geq m$. Consider the augmented matrix (eq. 21):

$$[D' \quad I_n]$$

(21)

where, as before, the prime denotes the matrix transpose and I_n is the identity matrix of order n. Suppose that the rank of D is r. Then a finite sequence of elementary row operations of equation 21 yields an equivalent matrix of the following form [see, for example, (39)] (eq. 22):

$$\begin{bmatrix} D_{11} & D_{12} & C_{13} \\ 0 & 0 & C_{23} \end{bmatrix}$$

(22)

where D_{11} is a nonsingular upper triangular matrix of order r, and D_{12}, C_{13}, and C_{23} are matrices of orders $r \times (m - r)$, $r \times n$ and $(n - r) \times n$, respectively.

Theorem 5. The transpose C'_{23} of C_{23} is a complete B-matrix of equation 13. It is advantageous if the dependent variables or the variables that can be regulated each occur in only one dimensionless product, so that a functional relationship among these dimensionless products may be most easily determined (8). For example, if a velocity is easily varied experimentally, then the velocity should occur in only one of the independent dimensionless variables (products). In other words, it is sometimes desirable to have certain specified variables, each of which occurs in one and only one of the B-vectors. The following theorem gives a necessary and sufficient condition for the existence of such a complete B-matrix. This result can be used to enumerate such a B-matrix without the necessity of exhausting all possibilities by linear combinations.

Theorem 6. Let A_1 be a given complete B-matrix associated with a set of variables. Then there exists a complete B-matrix A_2 of these variables such that certain specified variables each occur in only one of the B-vectors of A_2 if, and only if, the rows corresponding to these specified variables in A_1 are linearly independent.

The foregoing procedures are illustrated by the following examples.

Example 2. A smooth spherical body of projected area A moves through a fluid of density ρ and viscosity μ with speed v. The total drag δ encountered by the sphere is to be determined. Clearly, the total drag δ is a function of v, A, ρ, and μ. As before, mass m, length l, and time t are chosen as the reference dimensions. From Table 1 the dimensional matrix is (eq. 23):

$$D = \begin{array}{c} \\ m \\ l \\ t \end{array} \begin{array}{c} \delta \quad v \quad A \quad \rho \quad \mu \\ \begin{bmatrix} 1 & 0 & 0 & 1 & 1 \\ 1 & 1 & 2 & -3 & -1 \\ -2 & -1 & 0 & 0 & -1 \end{bmatrix} \end{array}$$

(23)

To compute a complete B-matrix, the augmented matrix (eq. 24):

$$[\boldsymbol{D}'\ \ \boldsymbol{I}_5] \tag{24}$$

is considered, which is given by (eq. 25):

$$
\begin{bmatrix}
1 & 1 & -2 & 1 & 0 & 0 & 0 & 0 \\
0 & 1 & -1 & 0 & 1 & 0 & 0 & 0 \\
0 & 2 & 0 & 0 & 0 & 1 & 0 & 0 \\
1 & -3 & 0 & 0 & 0 & 0 & 1 & 0 \\
1 & -1 & -1 & 0 & 0 & 0 & 0 & 1
\end{bmatrix}
\tag{25}
$$

The objective is to apply a sequence of elementary row operations (39) to equation 25 to bring it to the form of equation 22. Since the rank of $\boldsymbol{D}$ is 3, the order of the matrix $\boldsymbol{C}_{23}$ is $(n-r) \times n = 2 \times 5$. The following sequence of elementary row operations will result in the desired form:

new row 4 = row 4 − row 1 ≡ (designated as) row 4′;

new row 5 = row 5 − row 1 ≡ row 5′;

new row 3 = row 3 − 2 × row 2 ≡ row 3′;

new row 4 = row 4′ + 4 × row 2 ≡ row 4″;

new row 5 = row 5′ + 2 × row 2 ≡ row 5″;

new row 4 = row 3′ + row 4″ ≡ row 4‴;

new row 5 = row 5″ + ½ × row 3′ ≡ 5‴.

The corresponding matrix in partitioned form is given by (eq. 26):

$$
\begin{bmatrix}
1 & 1 & -2 & 1 & 0 & 0 & 0 & 0 \\
0 & 1 & -1 & 0 & 1 & 0 & 0 & 0 \\
0 & 0 & 2 & 0 & -2 & 1 & 0 & 0 \\
0 & 0 & 0 & -1 & 2 & 1 & 1 & 0 \\
0 & 0 & 0 & -1 & 1 & \tfrac{1}{2} & 0 & 1
\end{bmatrix}
=
\begin{bmatrix}
\boldsymbol{D}_{11} & \boldsymbol{C}_{13} \\
\boldsymbol{0} & \boldsymbol{C}_{23}
\end{bmatrix}
\tag{26}
$$

where $\boldsymbol{D}_{12}$ is null. This gives (eq. 27):

$$
\boldsymbol{C}_{23} =
\begin{array}{ccccc}
\delta & v & A & \rho & \mu
\end{array}
\begin{bmatrix}
-1 & 2 & 1 & 1 & 0 \\
-1 & 1 & \tfrac{1}{2} & 0 & 1
\end{bmatrix}
\tag{27}
$$

According to Theorem 5, the transpose $\boldsymbol{C}'_{23}$ of $\boldsymbol{C}_{23}$ is a complete B-matrix. Since there are five variables and since the rank of $\boldsymbol{D}$ is 3, Theorem 1 reveals that there are two dimensionless products in a complete set of B-numbers, each of which corresponds to a row of C_{23}. This yields a functional relation between the two B-numbers as (eq. 28):

$$f(v^2 A\rho/\delta, \, v\mu A^{1/2}/\delta) = 0 \tag{28}$$

where $\pi_1 = v^2 A\rho/\delta$ and $\pi_2 = v\mu A^{1/2}/\delta$, or alternatively (eq. 29):

$$v^2 A\rho/\delta = f_1(v\mu A^{1/2}/\delta) \tag{29}$$

This relation is not in the best form for the calculation of the drag since δ appears in both products. Hence it is necessary to change the two independent B-numbers by requiring that δ occur in only one of them. To this end, we let M be a nonsingular submatrix of C'_{23} of order 2 containing the row corresponding to δ. Thus, row 1 and, eg, row 5 of C_{23} are chosen to give

$$M = \begin{bmatrix} -1 & -1 \\ 0 & 1 \end{bmatrix} \tag{30}$$

the adjoint matrix of which is given by (39) (eq. 31):

$$M_a = \begin{bmatrix} 1 & 1 \\ 0 & -1 \end{bmatrix} \tag{31}$$

Then the matrix product (eq. 32):

$$C'_{23}M_a = \begin{bmatrix} -1 & -1 \\ 2 & 1 \\ 1 & \frac{1}{2} \\ 1 & 0 \\ 0 & 1 \end{bmatrix} \begin{bmatrix} 1 & 1 \\ 0 & -1 \end{bmatrix} = \begin{matrix} \delta \\ v \\ A \\ \rho \\ \mu \end{matrix} \begin{bmatrix} -1 & 0 \\ 2 & 1 \\ 1 & \frac{1}{2} \\ 1 & 1 \\ 0 & -1 \end{bmatrix} \tag{32}$$

is a desired B-matrix. The associated B-numbers are obtained as $\pi_1 = A\rho v^2/\delta$ and $\pi_2 = v\rho A^{1/2}/\mu$, yielding a functional relation (eq. 33):

$$\delta/A\rho v^2 = f_2(v\rho A^{1/2}/\mu) \tag{33}$$

Let d be the diameter of the sphere. Then $A = \pi d^2/4$ and $\pi_2 = \pi^{1/2} v\rho d/2\mu$. The dimensionless product $v\rho d/\mu$, which was first derived by Osborne Reynolds, is the familiar Reynolds number, and is denoted by Re. Equation 33 can now be expressed as (eq. 34):

$$\delta = (A\rho v^2/2)f_3(Re) \tag{34}$$

Defining a drag coefficient C_δ by (eq. 35) leads to (eq. 36):

$$\delta = C_\delta(A\rho v^2/2) \tag{35}$$

$$C_\delta = f_3(Re) \tag{36}$$

Thus, the drag problem is reduced to an equation involving only two dimensionless products C_δ and Re. The plot of the drag coefficient C_δ as a function of the Reynolds number Re can be obtained from experimental data. Knowing the speed of the sphere, equation 34 together with the drag coefficient is now in the best form for the direct determination of the drag (see also RHEOLOGICAL MEASUREMENTS).

On the other hand, suppose that the speed is to be determined when the drag is given. Then equations 33 or 34 are not convenient, and the two independent B-numbers must be changed again, so that the speed v will occur only in one of the B-numbers. To this end, let $\boldsymbol{M}$ be a nonsingular submatrix of $\boldsymbol{C'_{23}}$ of order 2 containing the row corresponding to v. Thus, choosing row 2 and, again say, row 5 of the transpose of the matrix of equation 27 gives (eq. 37):

$$\boldsymbol{M} = \begin{bmatrix} 2 & 1 \\ 0 & 1 \end{bmatrix} \tag{37}$$

Then the matrix product (eq. 38):

$$-\boldsymbol{C'_{23}}\boldsymbol{M}_a = - \begin{bmatrix} -1 & -1 \\ 2 & 1 \\ 1 & \frac{1}{2} \\ 1 & 0 \\ 0 & 1 \end{bmatrix} \begin{bmatrix} 1 & -1 \\ 0 & 2 \end{bmatrix} = \begin{matrix} \delta \\ v \\ A \\ \rho \\ \mu \end{matrix} \begin{bmatrix} 1 & 1 \\ -2 & 0 \\ -1 & 0 \\ -1 & 1 \\ 0 & -2 \end{bmatrix} \tag{38}$$

is a desired B-matrix, where $\boldsymbol{M}_a$ is the adjoint matrix of $\boldsymbol{M}$. The associated B-numbers become $\pi_1 = \delta/A\rho v^2$ and $\pi_2 = \delta\rho/\mu^2$ yielding (eq. 39):

$$\delta/A\rho v^2 = f_4(\delta\rho/\mu^2) \tag{39}$$

From equation 35, it is simple to demonstrate that (eq. 40):

$$v\rho d/\mu = Re = (8\rho\delta/C_\delta\pi)^{1/2}/\mu \tag{40}$$

and that equation 39 can be expressed as (eq. 41):

$$C_\delta = f_5(\rho\delta/\mu^2) \tag{41}$$

The drag coefficient C_δ can be plotted as a function of the dimensionless product $\rho\,\delta/\mu^2$. Thus, equations 40 and 41 are in proper form for direct determination of the speed once the drag is given.

Suppose that an experiment were set up to determine the values of drag for various combinations of v, A, ρ, and μ. If each variable is to be tested at ten values, then it would require $10^4 = 10{,}000$ tests for all combinations of these values. On the other hand, as a result of dimensional analysis the drag can be calculated by means of the drag coefficient, which, being a function of the Reynolds number Re, can be uniquely determined by the values of Re. Thus, for data of equal accuracy, it now requires only 10 tests at ten different values of Re instead of 10,000, a remarkable saving in experiments.

In addition, dimensional analysis can be used in the design of scale experiments. For example, if a spherical storage tank of diameter d is to be constructed, the problem is to determine wind load at a velocity v. Equations 34 and 36 indicate that, once the drag coefficient C_δ is known, the drag can be calculated from C_δ immediately. But C_δ is uniquely determined by the value of the Reynolds number Re. Thus, a scale model can be set up to simulate the Reynolds number of the spherical tank. To this end, let a sphere of diameter $\hat{d}$ be immersed in a fluid of density $\hat{\rho}$ and viscosity $\hat{\mu}$ and towed at the speed of $\hat{v}$. Requiring that this model experiment have the same Reynolds number as the spherical storage tank gives

$$\frac{\hat{v}\hat{\rho}\hat{d}}{\hat{\mu}} = \frac{v\rho d}{\mu} \tag{42}$$

where ρ and μ are the air density and viscosity, respectively. Thus, a sphere of 1 m in diameter, immersed in water and towed at 32.5 kilometers per hour has the same Reynolds number as the spherical storage tank of 10 m in diameter with air flowing over it at 50 kilometers per hour. By towing a smaller sphere in a water tank and measuring its drag force $\hat{\delta}$, the drag coefficient is determined from equation 35 by the formula

$$C_\delta = \frac{8\hat{\delta}}{\pi\hat{\rho}\hat{d}^2\hat{v}^2} \tag{43}$$

This C_δ can then be used in the original full-scale spherical storage tank to calculate its wind load (eq. 35):

$$\delta = \frac{\pi\rho d^2 v^2 C_\delta}{8} \tag{44}$$

Example 3. The mean free path P_m of electrons scattered by a crystal lattice is known to involve temperature θ, energy E, the elastic constant C, the Planck's constant h, the Boltzmann constant k, and the electron mass M (see, for example, (25)). The problem is to derive a general equation among these variables.

Again length l, mass m, time t, and temperature T are chosen as the reference dimensions. Then the associated dimensional matrix $\boldsymbol{D}$ is obtained as (eq. 45):

$$\mathbf{D} = \begin{array}{c} \\ l \\ m \\ t \\ T \end{array} \begin{array}{c} P_m \quad C \quad E \quad h \quad k \quad \theta \quad M \\ \begin{bmatrix} 1 & -1 & 2 & 2 & 2 & 0 & 0 \\ 0 & 1 & 1 & 1 & 1 & 0 & 1 \\ 0 & -2 & -2 & -1 & -2 & 0 & 0 \\ 0 & 0 & 0 & 0 & -1 & 1 & 0 \end{bmatrix} \end{array} \tag{45}$$

To compute a complete B-matrix, the augmented matrix (eq. 46):

$$[\mathbf{D}' \quad \mathbf{I}_7] \tag{46}$$

is considered, which is given by (eq. 47):

$$\begin{bmatrix} 1 & 0 & 0 & 0 & 1 & 0 & 0 & 0 & 0 & 0 & 0 \\ -1 & 1 & -2 & 0 & 0 & 1 & 0 & 0 & 0 & 0 & 0 \\ 2 & 1 & -2 & 0 & 0 & 0 & 1 & 0 & 0 & 0 & 0 \\ 2 & 1 & -1 & 0 & 0 & 0 & 0 & 1 & 0 & 0 & 0 \\ 2 & 1 & -2 & -1 & 0 & 0 & 0 & 0 & 1 & 0 & 0 \\ 0 & 0 & 0 & 1 & 0 & 0 & 0 & 0 & 0 & 1 & 0 \\ 0 & 1 & 0 & 0 & 0 & 0 & 0 & 0 & 0 & 0 & 1 \end{bmatrix} \tag{47}$$

and which may be put in the form of equation 22 by a sequence of elementary row operations yielding (eq. 48):

$$\begin{bmatrix} 1 & 0 & 0 & 0 & 1 & 0 & 0 & 0 & 0 & 0 & 0 \\ 0 & 1 & -2 & 0 & 1 & 1 & 0 & 0 & 0 & 0 & 0 \\ 0 & 0 & 0 & 0 & 2 & 0 & 1 & -2 & 0 & 0 & 1 \\ 0 & 0 & -1 & 0 & -2 & 0 & 0 & 1 & 0 & 0 & -1 \\ 0 & 0 & 0 & -1 & 2 & 0 & 0 & -2 & 1 & 0 & 1 \\ 0 & 0 & 0 & 0 & 2 & 0 & 0 & -2 & 1 & 1 & 1 \\ 0 & 0 & 0 & 0 & -5 & -1 & 0 & 2 & 0 & 0 & -1 \end{bmatrix} \tag{48}$$

from which the appropriate submatrices can be identified: $\mathbf{D}_{12}$ is null and (eq. 49–51):

$$\mathbf{D}_{11} = \begin{array}{c} (\text{row 1}) \\ (\text{row 2}) \\ (\text{row 4}) \\ (\text{row 5}) \end{array} \begin{bmatrix} 1 & 0 & 0 & 0 \\ 0 & 1 & -2 & 0 \\ 0 & 0 & -1 & 0 \\ 0 & 0 & 0 & -1 \end{bmatrix} \tag{49}$$

$$
C_{13} = \begin{matrix} \text{(row 1)} \\ \text{(row 2)} \\ \text{(row 4)} \\ \text{(row 5)} \end{matrix} \begin{bmatrix} 1 & 0 & 0 & 0 & 0 & 0 & 0 \\ 1 & 1 & 0 & 0 & 0 & 0 & 0 \\ -2 & 0 & 0 & 1 & 0 & 0 & -1 \\ 2 & 0 & 0 & -2 & 1 & 0 & 1 \end{bmatrix} \tag{50}
$$

$$
C_{23} = \begin{matrix} \\ \text{(row 3)} \\ \text{(row 6)} \\ \text{(row 7)} \end{matrix} \begin{matrix} P_m & C & E & h & k & \theta & M \\ \begin{bmatrix} 2 & 0 & 1 & -2 & 0 & 0 & 1 \\ 2 & 0 & 0 & -2 & 1 & 1 & 1 \\ -5 & -1 & 0 & 2 & 0 & 0 & -1 \end{bmatrix} \end{matrix} \tag{51}
$$

Thus, the transpose of C_{23} is a complete B-matrix. Since there are seven variables involved in the phenomenon and the rank of D is 4, from Theorem 1 there are three dimensionless products in a complete set of B-numbers, each of which corresponds to a row of C_{23}.

Suppose that the problem is to find a B-matrix of D such that the variables P_m, C, and E each occur in one and only one of the B-vectors. Since the submatrix M of C'_{23} consisting of the first three rows corresponding to the variables P_m, C, and E is nonsingular, according to Theorem 6 there exists a B-matrix with the desired property. Let M_a be the adjoint matrix of M. Then (eq. 52):

$$
C'_{23}M_a = \begin{bmatrix} 2 & 2 & -5 \\ 0 & 0 & -1 \\ 1 & 0 & 0 \\ -2 & -2 & 2 \\ 0 & 1 & 0 \\ 0 & 1 & 0 \\ 1 & 1 & -1 \end{bmatrix} \begin{bmatrix} 0 & 0 & -2 \\ -1 & 5 & 2 \\ 0 & 2 & 0 \end{bmatrix} = \begin{bmatrix} -2 & 0 & 0 \\ 0 & -2 & 0 \\ 0 & 0 & -2 \\ 2 & -6 & 0 \\ -1 & 5 & 2 \\ -1 & 5 & 2 \\ -1 & 3 & 0 \end{bmatrix} \tag{52}
$$

Hence, the right-hand side of equation 52 is a desired complete B-matrix. A functional relationship among the associated B-numbers can be obtained and is given by (eq. 53):

$$
f(h^2/P_m^2 k\theta M,\ k^5\theta^5 M^3/C^2 h^6,\ k^2\theta^2/E^2) = 0 \tag{53}
$$

Observe that the variables P_m, C, and E each occur in only one dimensionless product. Alternatively, equation 53 can be written as (eq. 54):

$$
P_m^2 = (h^2/k\theta M)f_1(k^5\theta^5 M^3/C^2 h^6,\ k^2\theta^2/E^2) \tag{54}
$$

The functional relation in equation 53 or 54 cannot be determined by dimensional analysis alone; it must be supplied by experiments. The significance is that the mean-free-path problem is reduced from an original relation involving seven variables to an equation involving only three dimensionless products, a considerable saving in terms of the number of experiments required in determining the governing equation.

Optimization of the Complete B-Matrices

In the foregoing, the computation of a complete B-matrix from a given dimensional matrix has been indicated. With the exception that certain variables may each be required to occur in only one dimensionless product, the selection of a complete B-matrix is totally arbitrary. In order to simplify the formulas associated with a complete B-matrix and to provide a procedure for establishing an explicit set of B-numbers, it is necessary to impose additional constraints in the selection of these B-vectors in forming a complete B-matrix. To avoid the fractional exponents of the formulas, the elements of the matrices are restricted to integers. In addition, the following criteria are proposed for the optimization of the B-matrices: (1) maximize the number of zeros in a complete B-matrix, and (2) minimize the sum of the absolute values of all the integers of a complete B-matrix. These criteria (21) are chosen so that the formulas associated with a physical phenomenon are in their simplest form, otherwise they are completely arbitrary. Evidently the order of the two optimization criteria is important. For the purpose of this review the sequence consisting of criterion (1) followed by criterion (2) is assumed.

Since the columns of any complete B-matrix are a basis for the null space of the dimensional matrix, it follows that any two complete B-matrices are related by a nonsingular transformation. In other words, a complete B-matrix itself contains enough information as to which linear combinations should be formed to obtain the optimized ones. Based on this observation, an efficient algorithm for the generation of an optimized complete B-matrix has been presented (22). No attempt is made here to demonstrate the algorithm. Instead, an example is being used to illustrate the results.

Example 4. For a given lattice, a relationship is to be found between the lattice resistivity and temperature using the following variables: mean free path L, the mass of electron M, particle density N, charge Q, Planck's constant h, Boltzmann constant k, temperature θ, velocity v, and resistivity ρ. Suppose that length l, mass m, time t, charge q, and temperature T are chosen as the reference dimensions. The dimensional matrix $\boldsymbol{D}$ of the variables is given by (eq. 55):

$$
\boldsymbol{D} = \begin{array}{c} \\ l \\ m \\ t \\ q \\ T \end{array}
\begin{array}{c} \begin{array}{ccccccccc} L & M & N & Q & h & k & \theta & v & \rho \end{array} \\
\left[\begin{array}{ccccccccc}
1 & 0 & -3 & 0 & 2 & 2 & 0 & 1 & 3 \\
0 & 1 & 0 & 0 & 1 & 1 & 0 & 0 & 1 \\
0 & 0 & 0 & 0 & -1 & -2 & 0 & -1 & -1 \\
0 & 0 & 0 & 1 & 0 & 0 & 0 & 0 & -2 \\
0 & 0 & 0 & 0 & 0 & -1 & 1 & 0 & 0
\end{array}\right]
\end{array} \tag{55}
$$

Using the procedure outlined in the preceding section, a matrix similar to that of equation 22 is obtained as follows (eq. 56):

$$
\begin{bmatrix}
1 & 0 & 0 & 0 & 0 & 1 & 0 & 0 & 0 & 0 & 0 & 0 & 0 & 0 \\
0 & 1 & 0 & 0 & 0 & 0 & 1 & 0 & 0 & 0 & 0 & 0 & 0 & 0 \\
0 & 0 & 0 & 0 & 0 & 3 & 0 & 1 & 0 & 0 & 0 & 0 & 0 & 0 \\
0 & 0 & 0 & 1 & 0 & 0 & 0 & 0 & 1 & 0 & 0 & 0 & 0 & 0 \\
0 & 0 & -1 & 0 & 0 & -2 & -1 & 0 & 0 & 1 & 0 & 0 & 0 & 0 \\
0 & 0 & 0 & 0 & 0 & 2 & 1 & 0 & 0 & -2 & 1 & 1 & 0 & 0 \\
0 & 0 & 0 & 0 & 1 & 0 & 0 & 0 & 0 & 0 & 0 & 1 & 0 & 0 \\
0 & 0 & 0 & 0 & 0 & 1 & 1 & 0 & 0 & -1 & 0 & 0 & 1 & 0 \\
0 & 0 & 0 & 0 & 0 & -1 & 0 & 0 & 2 & -1 & 0 & 0 & 0 & 1
\end{bmatrix}
\tag{56}
$$

Thus, the transpose of a complete B-matrix is given by (eq. 57):

$$
C_{23} =
\begin{array}{c}
\\
\text{(row 3)} \\
\text{(row 6)} \\
\text{(row 8)} \\
\text{(row 9)}
\end{array}
\begin{array}{cccccccccc}
L & M & N & Q & h & k & \theta & v & \rho \\
\left[\begin{array}{ccccccccc}
3 & 0 & 1 & 0 & 0 & 0 & 0 & 0 & 0 \\
2 & 1 & 0 & 0 & -2 & 1 & 1 & 0 & 0 \\
1 & 1 & 0 & 0 & -1 & 0 & 0 & 1 & 0 \\
-1 & 0 & 0 & 2 & -1 & 0 & 0 & 0 & 1
\end{array}\right]
\end{array}
\tag{57}
$$

From this matrix, an optimized complete B-matrix can be generated by the algorithm proposed by Chen (22), which results in a matrix whose transpose is given by (eq. 58):

$$
\begin{bmatrix}
3 & 0 & 1 & 0 & 0 & 0 & 0 & 0 & 0 \\
1 & 1 & 0 & 0 & -1 & 0 & 0 & 1 & 0 \\
0 & 1 & 0 & 0 & 0 & -1 & -1 & 2 & 0 \\
-1 & 0 & 0 & 2 & -1 & 0 & 0 & 0 & 1
\end{bmatrix}
\tag{58}
$$

Thus, the general relationship between the lattice resistivity and temperature can be expressed as (eq. 59):

$$
f(L^3N,\ LMv/h,\ Mv^2/k\theta,\ Q^2\rho/Lh) = 0
\tag{59}
$$

The number of independent variables is reduced from the original nine to four. This is a great saving in terms of the number of experiments required to determine the desired function. For example, suppose that a decision is made to test only four values for each variable. Then it would require $4^9 = 262144$ experiments to test all combinations of these values in the original equation. As a result of equation 59, only $4^4 = 256$ tests are now required for four values each of the four B-numbers.

Nomenclature

$$
\begin{aligned}
A &= \text{area or acceleration} \\
C &= \text{elastic constant} \\
C_\delta &= \text{drag coefficient}
\end{aligned}
$$

$$\hat{d}, d \; = \; \text{diameter}$$
$$\boldsymbol{D}, \hat{\boldsymbol{D}}, \tilde{\boldsymbol{D}} \; = \; \text{dimensional matrix of variables}$$
$$E \; = \; \text{energy}$$
$$f, F \; = \; \text{force}$$
$$g \; = \; \text{acceleration of gravity}$$
$$h \; = \; \text{Planck's constant}$$
$$k \; = \; \text{Boltzmann constant}$$
$$L \; = \; \text{length or mean free path}$$
$$l \; = \; \text{length}$$
$$m, M \; = \; \text{mass}$$
$$N \; = \; \text{particle density}$$
$$P \; = \; \text{period}$$
$$P_m \; = \; \text{mean free path}$$
$$q, Q \; = \; \text{charge}$$
$$Re \; = \; \text{Reynolds number}$$
$$\theta, T \; = \; \text{temperature}$$
$$t \; = \; \text{time}$$
$$\hat{v}, v \; = \; \text{velocity or speed}$$
$$\delta \; = \; \text{drag}$$
$$\epsilon \; = \; \text{permittivity}$$
$$\hat{\mu}, \mu \; = \; \text{permeability or viscosity}$$
$$\hat{\rho}, \rho \; = \; \text{fluid density or resistivity}$$

BIBLIOGRAPHY

"Dimensional Analysis" in *ECT* 1st ed., Vol. 5, pp. 133–141, by D. Q. Kern, The Patterson Foundry & Machine Co.; in *ECT* 2nd ed., Vol. 7, pp. 176–190, by I. H. Silberberg, Texas Petroleum Research Committee, and J. J. McKetta, The University of Texas; in *ECT* 3rd ed., Vol. 7, pp. 752–767, by Wai-Kai Chen, Ohio University.

1. P. W. Bridgman, *Dimensional Analysis*, Yale University Press, New Haven, Conn., 1922.
2. G. Murphy, *Similitude in Engineering*, The Ronald Press Co., New York, 1950.
3. J. F. Douglas, *An Introduction to Dimensional Analysis for Engineers*, Sir Isaac Pitman & Sons, London, 1969.
4. H. L. Langhaar, *Dimensional Analysis and Theory of Models*, John Wiley & Sons, Inc., New York, 1951.
5. S. J. Kline, *Similitude and Approximation Theory*, McGraw-Hill Book Co., New York, 1965.
6. L. I. Sedov, *Similarity and Dimensional Methods in Mechanics*, Academic Press, Inc., New York, 1959.
7. H. E. Huntley, *Dimensional Analysis*, Dover Publications, Inc., New York, 1967.
8. E. Buckingham, *Phys. Rev.* **4,** 345 (1914).
9. Lord Rayleigh, *Nature* **95,** 66 (1915).
10. T. J. Higgins, *Appl. Mech. Rev.* **10,** 331 (1957).
11. *Ibid.*, p. 443.
12. A. D. Sloan and W. W. Happ, "Literature Search: Dimensional Analysis," *NASA Rept. ERC/CQD 68-631* (Aug. 1968).

13. P. Moon and D. E. Spencer, *J. Franklin Inst.* **248,** 495 (1949).
14. E. U. Condon, *Am. J. Phys.* **2,** 63 (1934).
15. R. C. Tolman, *Phys. Rev.* **9,** 237 (1917).
16. W. E. Duncanson, *Proc. Phys. Soc.* **53,** 432 (1941).
17. G. B. Brown, *Proc. Phys. Soc.* **53,** 418 (1941).
18. H. Dingle, *Phil. Mag.* **33,** 321 (1942).
19. E. A. Guggenheim, *Phil. Mag.* **33,** 479 (1942).
20. C. M. Focken, *Dimensional Methods and Their Applications*, Edward Arnold and Co., London, 1953.
21. W. W. Happ, *J. Appl. Phys.* **38,** 3918 (1967).
22. W. K. Chen, *J. Franklin Inst.* **292,** 403 (1971).
23. H. L. Langhaar, *J. Franklin Inst.* **242,** 459 (1946).
24. E. R. Van Driest, *J. Appl. Mech.* **13,** A-34 (1946).
25. W. Shockley, *Electrons and Holes in Semiconductors*, D. Van Nostrand Co., Princeton, N.J., 1950.
26. R. P. Kroon, *J. Franklin Inst.* **292,** 45 (1971).
27. A. Klinkenberg and H. H. Mooy, *Chem. Eng. Progr.* **44,** 17 (1948).
28. S. Corrsin, *Am. J. Phys.* **19,** 180 (1951).
29. J. Geertsma, G. A. Croes, and N. Schwarz, *Trans. AIME* **207,** 118 (1956).
30. L. Brand, *Am. Math. Month,* **59,** 516 (1952).
31. B. Leroy, *Am. J. Phys.* **52,** 230 (1984).
32. D. I. H. Barr, *J. Eng. Mech.* **110,** 1357 (1984); **113,** 1431 (1987).
33. J. M. Supplee, *Am. J. Phys.* **53,** 549 (1985).
34. J. Puretz, *J. Phys. D. Appl. Phys.* **19,** 1237 (1986).
35. M. Strasberg, *J. Acoust. Soc. Am.* **83,** 544 (1988).
36. J. J. Chen, *Can. J. Chem. Eng.* **66,** 701 (1988).
37. T. Szirtes, *Mach. Design* **61,** 113 (1989).
38. R. Bhaskar and A. Nigam, *Art. Intel.* **45,** 73 (1990).
39. F. E. Hohn, *Elementary Matrix Algebra*, The Macmillan Co., New York, 1958.
40. R. Bellman, *Introduction to Matrix Analysis*, McGraw-Hill Book Co., New York, 1960.

WAI-KAI CHEN
University of Illinois at Chicago

DIMER ACIDS

The dimer acids [*61788-89-4*], 9- and 10-carboxystearic acids, and C-21 dicarboxylic acids are products resulting from three different reactions of C-18 unsaturated fatty acids. These reactions are, respectively, self-condensation, reaction with carbon monoxide followed by oxidation of the resulting 9- or 10-formylstearic acid (or, alternatively, by hydrocarboxylation of the unsaturated fatty acid), and Diels-Alder reaction with acrylic acid. The starting materials for these reactions have been almost exclusively tall oil fatty acids or, to a lesser degree, oleic acid, although other unsaturated fatty acid feedstocks can be used (see CARBOXYLIC ACIDS, FATTY ACIDS FROM TALL OIL; TALL OIL).

The basic research that led to these products was done at the Northern Regional Research Center of the USDA: dimer acids research in the 1940s (1–3), C-21 dicarboxylic acid work in 1957 (4), and carboxystearic acid synthetic studies (5,6) in the 1970s (see DICARBOXYLIC ACIDS).

Physical Properties

The physical properties of polymerized fatty acids are influenced by the basestock, by the dimerization conditions and catalysis, and by the degree to which monomer, dimer, and higher oligomers are separated following the dimerization.

Dimer acids are relatively high mol wt (ca 560) and yet are liquid at 25°C. This liquidity is a consequence of the many isomers present, most with branching or cyclic structures.

Most of the products listed in Tables 1–3 are based on manufacture from tall oil fatty acids. Dimer acids based on other feedstocks (eg, oleic acid) may have different properties. A European manufacturer recently announced availability of a 44-carbon dimer acid, presumably made from an erucic acid feedstock (7).

Chemical Properties

Structure and Mechanism of Formation. Thermal dimerization of unsaturated fatty acids has been explained both by a Diels-Alder mechanism and by a

Table 1. Properties of Dimer Acid[a]

Physical property	Usual range
composition, %	
dimer acids	82–83
trimer acids (and higher)	14–16
monobasic acids	1–5
neutralization equivalent	285–297
acid number	189–197
saponification number	189–199
unsaponifiables, %	0.5–1.0
color, Gardner (1963)	7–9
viscosity at 25°C, mm^2/s (= cSt)	7500–9000
specific gravity,	
25/25°C	~0.95
100/25°C	~0.91
density, 25°C, kg/m^3	~955
refractive index, 25°C	~1.485
pour point, °C	~4 to ~10
flash point (COC) °C	~300

[a]Hystrene series of dimer acids, Humko Chemical Division of Witco Corporation.

Table 2. Properties of Distilled Dimer Acids

Physical property	Value	
	Hydrogenated[a]	Unhydrogenated[b]
Composition, %		
dimer	97	95
trimer	3	4
monomer	trace	1
neutralization equivalent	284–294	286–285
acid number	191–197	190–196
saponification number	193–200	190–202
unsaponifiables, %	0.1	0.5
color, Gardner (1963)	1	~5
iodine value	~20	
viscosity at 25°C, mm²/s (=cSt)	~5200	7000–8000
specific gravity, 25/25°	0.94	0.9
density at 25°C, kg/m³	~950	~950
refractive index, 25°C	1.475	1.483
pour point, °C	~12	~ −8

[a]Empol series of dimer acids, Henkel Corp., Emery Group (oleic-based, thus of lower viscosity).
[b]Hystrene series of dimer acids, Humko Chemical Div. of Witco Corporation.

Table 3. Properties of Trimer Acids[a]

Physical property	Value
composition, %	
dimer	40
trimer	60
monobasic acids	trace
neutralization equivalent	295–330
acid number	170–190
saponification number	170–202
unsaponifiables, %	~1.0
color, Gardner (1963)	dark
viscosity at 25°C, mm²/s (=cSt)	~30,000
density at 25°C, kg/m³	~955

[a]Hystrene 5460, Humko Chemical Div. of Witco Corp.

free-radical route involving hydrogen transfer. The Diels-Alder reaction appears to apply to starting materials high in linoleic acid content satisfactorily, but oleic acid oligomerization seems better rationalized by a free-radical reaction (8–10).

    Clay-catalyzed dimerization of unsaturated fatty acids appears to be a carbonium ion reaction, based on the observed double bond isomerization, acid catalysis, chain branching, and hydrogen transfer (8,9,11).

    It has been shown (12) that different precursors for dimer preparation give quite different structures (Table 4).

Table 4. Dimer Acid: Feedstock–Structure Relationship

Feedstock	Dimer structure		
	Acrylic	Monocyclic	Polycyclic
oleic[a] or elaidic[b] acid	40	55	5
tall oil fatty acids	15	70	15
linoleic acid[c]	5	55	40

[a]cis-9-octadecenoic acid [112-80-1]
[b]trans-9-octadecenoic acid [112-79-8]
[c]cis-9, cis-12-octadecadienoic acid [60-33-3]

The following are three possible structures of the methyl esters of dimer acids.

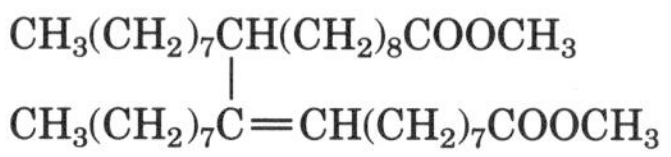

(1)

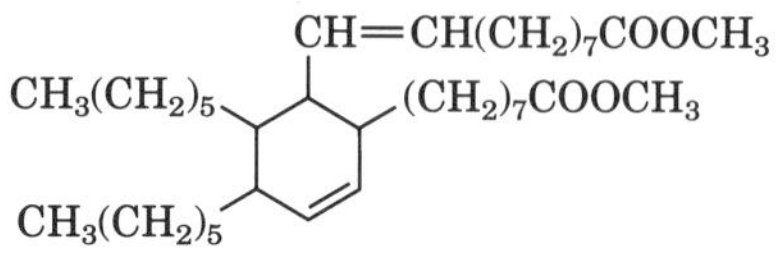

(2)

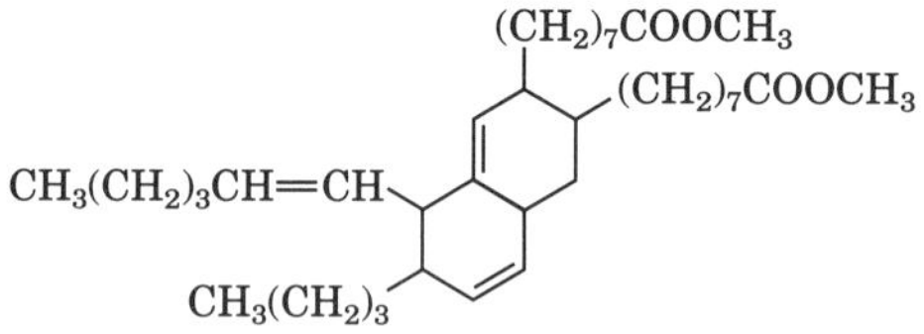

(3)

The C-18 acid source for the acyclic (**1**) (*E*)-, (*Z*)- [28923-98-0] is the Δ9 acid (oleic or elaidic); for monocyclic (**2**) (*E*)-, (*E*)- [26796-50-9], (*E*)-, (*Z*)- [56636-20-5] it is the Δ9,11-octadecadienoic acid; and for bicyclic (**3**) [32733-04-3] the source is Δ9,11,13-octadecatrienoic acid.

There are a myriad of possible isomers, including positional and geometrical isomers of the double bond(s) as well as structural isomers resulting from head-to-head or head-to-tail alignment of the reacting fatty acids.

Chemical Reactions.   The reactions of dimer acids were reviewed fully in 1975 (13). The most important is polymerization; the greatest quantities of dimer acids are incorporated into the non-nylon polyamides. Other reactions of dimer acids that are applied commercially include polyesterification, hydrogenation, esterification, and conversion of the carboxy groups to various nitrogen-containing functional groups. Table 5 summarizes the nonpolymeric chemical reactions of dimer acids. Polymerization reactions of dimer acids include polyamidation (34–36), polyesterification (37–39), reactions resulting in polymeric nitrogen derivatives other than polyamides (40–42), and reactions involving dimer diprimary amines (43,44).

Table 5. Nonpolymeric Chemical Reactions of Dimer Acids

Reactions at the double bond and at the α-carbon		Reactions of the carboxyl group	
Reaction	Refs.	Reaction	Refs.
halogenation	14	salt formation	15
epoxidation	14	esterification	16,17
sulfation	14	hydrogenolysis	18,19
sulfonation	14	ethoxylation	20–23
hydrogenation	24	amidation	25–29
sulfurization	14	ammonolysis, reduction	
		(dimer nitriles, dimer amines)	30–32
		isocyanate formation	33

A scan of the literature over the years 1980–1991 shows that most of the current dimer activity involves the reaction of dimer acids to form a huge variety of polyamide and polyester structures to modify their properties for a wide range of industries and uses. Many of these property modifications seem to make use of the flexibilizing properties or adhesion-promoting properties of the dimer structure.

Manufacture

The clay-catalyzed intermolecular condensation of oleic and/or linoleic acid mixtures on a commercial scale produces approximately a 60:40 mixture of dimer acids (C_{36} and higher polycarboxylic acids) and monomer acids (C_{18} isomerized fatty acids). The polycarboxylic acid and monomer fractions are usually separated by wiped-film evaporation. The monomer fraction, after hydrogenation, can be fed to a solvent separative process that produces commercial isostearic acid, a complex mixture of saturated fatty acids that is liquid at 10°C. Dimer acids can be further separated, also by wiped-film evaporation, into distilled dimer acids and trimer acids. A review of dimerization gives a comprehensive discussion of the subject (10).

Thermal Oligomerization. Commercial manufacture of dimer acids began in 1948 with Emery Industries use of a thermal process involving steam pressure. Patents were issued in 1949 (45) and 1953 (46) that describe this process. Earlier references to fatty acid oligomerization, antedating the USDA work of 1941–1948, occur in patents in 1918 and 1919 (47,48), and in papers written in 1929–1941 (49–51). There appears to still be some small use of this approach to making dimer products.

Clay-Catalyzed Oligomerization. Emery Industries modified its commercial process in 1953 and began producing dimer acids by using a combination of thermal- and Montmorillonite clay-catalyzed oligomerization. Such a process has been described in patents (52–55). In general, in present-day commercial practice, 100 parts of the fatty acid, 4 parts of clay, and 2 parts of water are heated in an autoclave under autogenous pressure at approximately 230°C for 4 hours with agitation. After cooling and removal of the clay by filtration, separation of monomeric and polymeric material can be carried out by wiped-film evaporation or molecular distillation. Most current production is based on this type of technology.

In 1991, in addition to Henkel (now including Emery), commercial producers of dimer acids in the U.S. were Arizona Chemical, Schering Berlin, Humko Chemical Division of Witco, and Union Camp Corporation. There are other producers throughout the world.

Process Modification.   Dimer acid process modifications have fallen into three categories (8); those claiming higher dimer:trimer ratios, those utilizing varying types of clays, and those purporting to result in improved yields. Higher dimer:trimer ratios are said to be obtained through use of alkali (56), ammonia or amines (57), aryl sulfohalides (58), 1-mercapto-2-naphthol (59), and a "Texas natural acid clay" (60). Natural or synthetic clay catalysts have included lithium salt–acetic anhydride stabilized clay (61), synthetic magnesium silicate (62), and neutral Alabama bentonite (63). Improved yields of dimer acids are claimed to result from a number of two-step clay-catalyzed procedures (64,65) and by the addition of low mol wt saturated alcohols to the reaction mixture (66) (see CLAYS). Clay, lithium salts and phosphoric acid treatment, in one or two stages, are said to yield 63–73% dimer and trimer (67).

Another aspect of process improvement is color improvement. For example, use of phosphoric acid and formaldehyde on the dimer product is said to improve color (68). Other treatments, both on the raw materials and on the finished product, have also been used. Most of these are variations of standard fatty material color reduction techniques.

Other Polymerization Methods.   Although none has achieved commercial success, there are a number of experimental alternatives to clay-catalyzed or thermal oligomerization of dimer acids. These include the use of peroxides (69), hydrogen fluoride (70), a sulfonic acid ion-exchange resin (71), and corona discharge (72) (see INITIATORS).

Energy Requirements.   The production of dimer acids is quite energy intensive. A standard operating sequence normally results in the expenditure of about 18.6 MJ (17,600 Btu) (equivalent to 0.67 kg coal or 0.33 kg natural gas or fuel oil) to produce each kg of crude dimer and to separate it into monomer, dimer, and trimer. Of this energy it is estimated that 10% is electrical, consumed by the pumps and agitators of the system. The other 90% is fuel-derived, and is consumed mainly for process thermal requirements with some usage for the steam necessary to obtain reduced pressures for certain operations. Energy requirements for the storage and handling of either the raw materials or the products are not included; thus the 18.6 MJ (17,600 Btu) used per kg represents only the energy consumption within the process units themselves.

Storage and Handling

Since dimer acids, monomer acids, and trimer acids are unsaturated, they are susceptible to oxidative and thermal attack, and under certain conditions they are slightly corrosive to metals. Special precautions are necessary, therefore, to prevent product color development and equipment deterioration. Type 304 stainless steel is recommended for storage tanks for dimer acids. For heating coils and for agitators 316 stainless steel is preferred (heating coils with about $4\frac{3}{4}$ m^2 (50 ft^2) of heat transfer surface in the form of a 5.1 cm schedule-10 U-bend scroll are recommended for a 37.9-m^3 (10,000-gal) tank. Dimer acid storage tanks should have an inert gas blanket.

316 Stainless steel centrifugal pumps may be used to transfer dimer acid stocks. Pipe, valves, and fittings may be of 304 stainless steel if the liquid temperature is maintained below 107°C. The recommended temperature ranges for transfer (pumping) of dimer acids and related stocks are as follows.

Stock	Range, °C
monomer acids	46–49
dimer acids	54–77 (71 optimum)
trimer acids	77–82

Specific gravities and viscosities for these chemicals, in the recommended temperature ranges for transfer, are shown in Table 6.

The temperature for handling dimers should never exceed 82°C. Even with an inert gas blanket, color deterioration of the products accelerates at higher temperatures. Dimer and trimer acids stored for longer times should be held below 50°C. Tank agitators should be left on or interlocked with the heating cycle of the steam coils so that the agitators will be moving the liquid to prevent discoloration of the material near the coils during heating.

Stainless steel or epoxy-lined tank cars and tank trucks are recommended for shipping. Aluminum also has been used. The tank can be flushed with carbon dioxide before loading and blanketed with nitrogen after loading. Drum shipments are recommended in epoxy-lined open-head drums fitted with a bung. Dimer acids and their by-products contaminated with iron or copper show accelerated color deterioration. Exposure to these metals or their salts should be minimized.

Table 6. Specific Gravities and Viscosities at Pumping Temperatures

Stock	Temperature, °C	Specific gravity, t/25°	Viscosity, mm^2/s (=cSt)
monomer acids	49	0.890	23
dimer acids	71	0.923	305
trimer acids	80	0.926	669

Economic and Market Aspects

According to one estimate (73), the current capacity for manufacturing dimer acids in the U.S. is around 55,000 t per year. Current demand is estimated at about 33,600 t per year, and is expected to grow at about 2–3% per year to 35,000 t in 1993. The historical growth rate for dimer acids (1980–1989) was 0.8% per year. Prices of tall oil fatty acids, the raw material for over 90% of dimers, currently fluctuates in the $0.55–0.66 per kg range. The dimer acids themselves are presently selling at about $1.10 per kg for the standard 75–80% dimer acids, and about $2.20 per kg for the distilled (90–95%) dimer acids.

The current market situation for dimer acids includes relatively high raw material costs, high energy costs, slow growth and relatively low prices. It is generally recognized as a mature market, with hopes for future growth hinging on factors such as increased polyamide use and a resurgence of oil drilling, where dimers are used for corrosion inhibition.

Analysis

The American Society for Testing and Materials (ASTM) and the American Oil Chemists Society (AOCS) provide standard methods for determining properties that are important in characterization of dimer acids. Characterization of dimer acids for acid and saponification values, unsaponifiables, and specific gravity are done by AOCS standard methods:

Property	Method
acid value	AOCS Te 1a-64
saponification number	AOCS Tl 1a-64
unsaponifiable	AOCS Tk 1a-64
specific gravity	AOCS To 1a-64

Flash and fire points, Gardner color, kinematic viscosity, and pour points are determined by ASTM methods:

Property	Method
flash and fire points	
open cup	ASTM D92-85
closed cup	ASTM D93-85
Gardner color	ASTM D1544-80
kinematic viscosity	ASTM D445-86
pour point	ASTM D97-85

The determination of iodine value (IV), AOCS Tg 1-64, is sometimes used to determine the extent of unsaturation. Because the tertiary allylic hydrogen in the compounds is capable of substitution by halogen atoms, this only approximates a value for the degree of unsaturation.

Currently, there is continuing work on an industry standard method for the direct determination of monomer, dimer, and trimer acids. Urea adduction (of the methyl esters) has been suggested as a means of determining monomer in distilled dimer (74). The method is tedious and the nonadducting branched-chain monomer is recovered with the polymeric fraction. A microsublimation procedure was developed as an improvement on urea adduction for estimation of the polymer fraction. Incomplete removal of monomer esters or loss of dimer during distillation can lead to error (75).

Thin-layer chromatography (76,77) has been used for the estimation of the amounts of dimer, trimer, and monomer in methyl esters. Both this method and paper chromatography are characterized by lack of precision (78,79) (see CHROMATOGRAPHY).

Microscale distillation techniques have also been used to determine monomer, dimer, and trimer in polymerized fatty acids. Such a method was used for some time, but has been largely replaced by various methods of high pressure liquid chromatography (hplc). Gas–liquid chromatography (glc) has been used for analysis of monomer–dimer–trimer content also. An older approach, packed column chromatography of the methyl ester of the dimer acid with thermal conductivity detection, based on a combination of molecular weight and boiling point separation, was used at one time. Currently, methods of high temperature capillary chromatography of the dimer acid using flame ionization detection and a combination of molecular weight and boiling point for separation find some use in the industry.

Various high pressure liquid chromatographic methods are being widely used for the analysis of dimer acids today. Gel permeation chromatography with separation according to molecular size and response to a refractive index detector as monomer, dimer, and trimer is currently utilized. Liquid chromatographic separation according to functionality and response to ultraviolet photometric absorption followed by component collection and gravimetric analysis as monobasic, dibasic, and polybasic segments is also currently in use. An AOCS committee is currently working on a liquid chromatography method with separation according to functionality and detection by response to a laser light-scattering mass detector as monobasic, dibasic, and polybasic components.

For production use, once an accurate determination of monomeric, dimeric, and higher oligomeric dimer acids has been established for a specific process and feedstock's oligomerization, subsequent reaction extent can be estimated rapidly by a viscometric method (80).

Health and Safety Aspects

The acute oral toxicity and the primary skin and acute eye irritative potentials of dimer acids, distilled dimer acids, trimer acids, and monomer acids have been evaluated based on the techniques specified in the Code of Federal Regulations (CFR) (81). The results of this evaluation are shown in Table 7. Based on these results, monomer acids, distilled dimer acids, dimer acids, and trimer acids are classified as nontoxic by ingestion, are not primary skin irritants or corrosive materials, and are not eye irritants as these terms are defined in the Federal regulations.

Food Additive and Food Packaging Regulations. Federal regulations do not permit direct use of dimer acids in food products. CFR, however, permits indirect use of dimer acids in packaging materials with incidental food contact. These permitted applications include use of dimer acids as components of polyamide, epoxy, and polyester resins for use in coating plastic films, paper, and paper-

Table 7. Toxicity Data for Dimer Acids and Related Products

Sample	Oral LD_{50}, g/kg	Primary irritation index	Eye irritation
monomer acids	>21.5	1.0	very slight erythema in four rabbits
distilled dimer acids	>21.5	0.50	very slight erythema in four rabbits
dimer acids	>21.5	0.75	very slight erythema in three rabbits
trimer acids	>10.0	0	very slight erythema in six rabbits

board. These regulations are 21 CFR 177.1200, dimer acids as a component of polyamide resins for coating cellophane; 21 CFR 175.300, dimer acids as a component of epoxy, polyester, or polyamide resins for "resinous and polymeric coatings" that come into contact with food; 21 CFR 175.390, dimer acids as a component of zinc–silicon dioxide matrix coating, which is the food-contact surface of articles intended for use in producing, manufacturing, packing, processing, preparing, treating, packaging, transporting, or holding food; 21 CFR 177.1210, dimer acids as components of "resinous and polymeric coatings" used in closures with sealing gaskets for food containers; 21 CFR 176.200, dimer and trimer acids as defoaming agents in coatings ultimately destined for food use; 21 CFR 175.380, dimer acids as components of "resinous and polymeric coatings", used as adjuvants for other resins used in food contact coatings; 21 CFR 175.320, dimer acids as components of polyamide resins used as coatings for food-contact polyolefin films; and 21 CFR 176.180, dimer acids as components of "resinous and polymeric coatings" for paper and paperboard in contact with dry food.

Flammability. Dimer and trimer acids, as well as monomer acids derived from dimer acid processing, are neither flammable nor combustible as defined by the Department of Transportation (DOT) and do not represent a fire hazard:

Product	Flash point, °C		Fire point, °C
	open cup	closed cup	open cup
monomer acids	193	154	216
dimer acids	279	246	318
trimer acids	329	299	352

Uses

Nonreactive Polyamide Resins. The non-nylon dimer based polyamide resins are characterized by lack of crystallinity, relatively low softening points, adhesiveness, hydrophobicity and, generally, relatively low transition-temperature ranges. These properties contrast sharply with the crystallinity and high melting temperatures of the nylon polyamides (qv) based on C_6 to C_{12} dibasic acids and are sufficiently unique to carve out large markets of their own (13,82). About 65% of dimer use is in this area, according to one recent estimate (71).

Dimer-based polyamide resin markets are divided into those for reactive polyamides and those for nonreactive polyamides. The largest volume commercial application of dimer acids is in nonreactive polyamide resins. These resins, solids at 25°C, are manufactured by the reaction of dimer acids (or trimer acids) or their esters with aliphatic diamines. Polyamide resins with a broad spectrum of properties can be obtained, with the transition temperatures (the phase change from a glass to a liquid) being determined by the diamine used, the stoichiometry of the reactants, and the amount and type of short-chain dibasic acids added to increase the melting range. Table 8 shows the melting ranges of typical neutral polyamide resins (34,36). Dimer acids impart flexibility, corrosion resistance, chemical resistance, moisture resistance, and adhesion to nonreactive polyamides. Hot melt adhesives (qv), the largest commercial application of nonreactive polyamide resins, are thermoplastics that have fairly sharp melting ranges. They are particularly useful in high-speed assembly operations such as packaging, can assembly, bookbinding, and shoe assembly because they can be applied in liquid form, eliminating the need for solvents.

The dimer acid-based polyamide resins appear best suited to specialized uses requiring high performance. A major application for the dimer acid-based polyamide hot-melt adhesives is in the shoe industry. Here good adhesion and high-temperature properties permit bonding of shoe soles to uppers without the need for stitching with thread. Metal bonding (eg, side-seam welding) and plastic and metal film and foil lamination are other hot-melt adhesive applications of dimer acid-based polyamides.

Flexographic printing inks utilize nonreactive polyamides from dimer acids as resin binders. Polyamide resins are especially well suited for flexographic printing on plastic films and metallic foil laminates because the resins adhere very well to the printed surface, give high-gloss surfaces, and accommodate to substrate deformation (see INKS).

The most important coating application for the nonreactive polyamide resins is in producing thixotropy. Typical coating resins such as alkyds, modified alkyds, natural and synthetic ester oils, varnishes, and natural vegetable oils can be made thixotropic by the addition of dimer acid-based polyamide resins (see ALKYD RESINS). Specialty high performance coating applications often require the properties imparted by dimer acid components.

Table 8. Melting Ranges for Typical Neutral Polyamides

Dimer:trimer mol ratio	Diamine	Polyamide melting range, °C
1.7:1	ethylenediamine	96–103
1.8:1	ethylenediamine	108–112
1.7:1	propylenediamine	53–59
1.7:1	hexamethylenediamine	70–80
1.7:1	dimer diprimary amine	liquid at 25°

Reactive Polyamide Resins.   Another significant commercial application of dimer acids is in reactive polyamide resins. These are formed by the reaction of dimer acids with polyamines such as diethylenetriamine to form polyamides containing reactive secondary amine groups (see DIAMINES AND HIGHER AMINES, ALIPHATIC). In contrast to nonreactive polyamides, these materials are generally liquids at 25°C.

They are used extensively to react with epoxy or phenolic resins, yielding adhesives that are useful in casting and laminating, in structural work, for patching and sealing compounds, and for protective coatings. The amount used in epoxy applications far exceeds the use with phenolic resins.

Miscellaneous Commercial Applications.   Dimer acids are components of "downwell" corrosion inhibitors for oil-drilling equipment (see PETROLEUM; CORROSION AND CORROSION INHIBITORS). This may account for 10% of current dimer acid use (71). The acids, alkyl esters, and polyoxyalkylene dimer esters are used commercially as components of metal-working lubricants (see LUBRICATION). Dimer esters have achieved some use in specialty lubricant applications such as gear oils and compressor lubricants. The dimer esters, compared to dibasic acid esters, polyol esters and poly(α-olefin)s, are higher in cost and of higher viscosity. The higher viscosity, however, is an advantage in some specialties, and the dimer esters are very stable thermally and can be made quite oxidatively stable by choice of proper additives.

Other dimer acid markets include intermediates for nitriles, amines and diisocyanates. Dimers are also used in polyurethanes, in corrosion inhibition uses other than for downwell equipment, as a "mildness" additive for metal-working lubricants, and in fiber glass manufacture.

BIBLIOGRAPHY

"Dimer Acids" in *ECT* 3rd ed., Vol. 7, pp. 768–782, by Edward C. Leonard, Humko Sheffield Chemical.

1. J. C. Cowan, W. C. Ault, and H. M. Teeter, *Ind. Eng. Chem.* **38,** 1138 (1946).
2. L. B. Falkenburg and co-workers, *Oil Soap* **22,** 143 (1945).
3. J. C. Cowan, A. J. Lewis, and L. B. Falkenburg, *Oil Soap* **21,** 101 (1944).
4. H. M. Teeter and co-workers, *J. Am. Oil Chem. Soc.* **22,** 512 (1957).
5. E. N. Frankel and E. H. Pryde, *J. Am. Oil Chem. Soc.* **54,** 873A (1977).
6. N. E. Lawson, T. T. Cheng, and F. B. Slezak, *J. Am. Oil Chem. Soc.* **54,** 215 (1977).
7. *Unichema International News Bulletin*, Vol. 3, Unichema North America, Chicago, Spring/Summer 1991.
8. R. W. Johnson in E. H. Pryde, ed., *Fatty Acids*, American Oil Chemists Society, Champaign, Ill., 1979.
9. M. J. A. M. den otter, *Fette Seifen Anstrichm.* **72,** 667, 875, 1056 (1970).
10. R. W. Johnson in R. W. Johnson and E. Fritz, eds., *Fatty Acids in Industry*, Marcel Dekker, Inc., New York, 1989.
11. R. W. Johnson, *Clay Catalyzed Dimerization of Fatty Acids*, American Oil Chemists Society Meeting, Chicago, 1991.
12. D. H. McMahon and E. P. Crowell, *J. Am. Oil Chem. Soc.* **51,** 522 (1974).
13. E. C. Leonard, ed., *The Dimer Acids*, Humko Sheffield Chemical, Memphis, Tenn., 1975.
14. *Empol Dimer and Trimer Acids*, technical bulletin, Emery Industries, Inc., Cincinnati, Ohio, 1971, p. 8.

15. J. Levy in E. S. Pattison, ed., *Fatty Acids and Their Industrial Applications*, Marcel Dekker, Inc., New York, 1968, p. 209.
16. U.S. Pat. 2,673,184 (Mar. 23, 1954), A. J. Morway, D. W. Young, and D. L. Cottle (to Standard Oil Development Co.).
17. U.S. Pat. 2,849,399 (Aug. 26, 1958), A. H. Matuzak and W. J. Craven (to Esso Research and Engineering Co.).
18. U.S. Pat. 2,347,562 (Apr. 25, 1944), W. B. Johnston (to American Cyanamid . Co.).
19. U.S. Pat. 2,413,612 (Dec. 31, 1946) E. W. Eckey and J. E. Taylor (to The Procter and Gamble Co.).
20. U.S. Pat. 3,173,887 (Mar. 16, 1965), T. E. Yeates and C. M. Thierfelder (to the U.S. Dept. of Agriculture).
21. U.S. Pat. 2,473,798 (June 21, 1949), R. E. Kienle and G. P. Whitcomb (to American Cyanamid Co.).
22. U.S. Pat. 2,758,976 (Aug. 14, 1956), G. E. Barker (to Atlas Powder Co.).
23. U.S. Pat. 3,429,817 (Feb. 25, 1969), M. J. Furey and A. F. Turbak (to Esso Research and Engineering Co.).
24. U.S. Pat. 3,595,887 (July 27, 1971), M. V. Kulkarni and R. L. Scheribel (to General Mills, Inc.).
25. U.S. Pat. 2,537,493 (Jan. 9, 1951), and U.S. Pat. 2,470,081 (May 10, 1949), J. T. Thurston and R. B. Warner (to American Cyanamid Co.).
26. U.S. Pat. 2,992,145 (July 11, 1961), C. Santangelo and B. H. Kress (to Quaker Chemical Products Corp.).
27. U.S. Pat. 3,256,182 (June 14, 1966), G. F. Scherer (to Rockwell Manufacturing . Co.).
28. U.S. Pat. 2,965,591 (Dec. 20, 1960), J. Dazzi (to Monsanto Chemical Co.).
29. U.S. Pat. 3,219,612 (Nov. 23, 1965), E. L. Skau, R. R. Mod, and F. C. Magne (to the U.S. Dept. of Agriculture).
30. U.S. Pat, 2,526,044 (Oct. 17, 1950), A. W. Ralston, O. Turinsky, and C. W. Christensen (to Armour & Co.).
31. U.S. Pat. 3,223,631 (Dec. 14, 1965), A. J. Morway and A. J. Rutkowski (to Esso Research and Engineering Co.).
32. U.S. Pat. 3,010,782 (Nov. 28, 1961), K. E. McCaleb, L. Vertnik, and D. L. Anderson (to General Mills, Inc.).
33. U.S. Pat. 3,481,959 (Dec. 2, 1969), G. Egle (to Henkel).
34. D. E. Floyd, *Polyamide Resins*, 2nd ed., Reinhold Publishing Corp., New York, 1966, p. 11.
35. L. B. Falkenburg and co-workers, *Oil Soap* **22,** 143 (1945).
36. U.S. Pat. 2,450,740 (Oct. 12, 1948), J. C. Cowan, L. B. Falkenburg, H. M. Teeter, and P. S. Skell (to the U.S. Dept. of Agriculture).
37. U.S. Pat. 2,411,178 (Nov. 19, 1946), D. W. Young and E. Lieber (to Standard Oil Development Co.); U.S. Pat. 2,424,588 (July 29, 1947), W. J. Sparks and D. W. Young (to Standard Oil Development Co.); U.S. Pat. 2,435,619 (Feb. 17, 1948), D. W. Young and W. J. Sparks (to Standard Oil Development Co.).
38. U.S. Pat. 3,492,232 (Jan. 27, 1970), M. Rosenberg (to The Cincinnati Milling Machine Co.).
39. U.S. Pat. 3,769,215 (Oct. 30, 1973), R. J. Sturwold and F. O. Barrett (to Emery Industries, Inc.).
40. U.S. Pat. 3,217,028 (Nov. 9, 1965), L. R. Vertnik (to General Mills, Inc.).
41. U.S. Pat. 3,281,470 (Oct. 25, 1966), L. R. Vertnik (to General Mills, Inc.).
42. U.S. Pat. 3,235,596 (Feb. 15, 1966), R. Nordgren, L. R. Vertnik, and H. Wittcoff (to General Mills, Inc.).
43. U.S. Pat. 3,231,545 (Jan. 25, 1966); U.S. Pat. 3,242,141 (Mar. 22, 1966), L. R. Vertnik and H. Witcoff (to General Mills, Inc.).

44. U.S. Pat. 3,483,237 (Dec. 9, 1969), D. E. Peerman and L. R. Vertnik (to General Mills, Inc.).
45. U.S. Pat. 2,483,761 (Sept. 27, 1949), C. G. Goebel (to Emery Industries, Inc.).
46. U.S. Pat. 2,664,429 (Dec. 29, 1953), C. G. Goebel (to Emery Industries, Inc.).
47. Brit. Pat. 121,777 (1918), J. Craven.
48. Brit. Pat. 127,814 (1919), (to De Nordiske Fabriker).
49. J. Scheiber, *Farben, Lacke, Anstrichst.*, 585 (1929).
50. C. P. A. Kappelmeier, *Farben Ztg.* **38**, 1018, 1077 (1983).
51. T. F. Bradley and W. B. Johnston, *Ind. Eng. Chem.* **33**, 86 (1941); T. F. Bradley and D. Richardson, *Ind. Eng. Chem.* **32**, 963, 802 (1940).
52. U.S. Pat. 2,347,562 (Apr. 25, 1944), W. B. Johnston (to American Cyanamid Co.).
53. U.S. Pat. 2,426,489 (Aug. 26, 1947), M. De Groote (to Petrolite Corp.).
54. U.S. Pat. 2,793,219 (May 21, 1957), F. O. Barrett, C. G. Goebel, and R. M. Peters (to Emery Industries, Inc.).
55. U.S. Pat. 2,793,220 (May 21, 1957), F. O. Barrett, C. G. Goebel, and R. M. Peters (to Emery Industries, Inc.).
56. U.S. Pat. 2,955,121 (Oct. 4, 1960), L. D. Myers, C. G. Goebel, and F. O. Barrett (to Emery Industries, Inc.).
57. U.S. Pat. 3,076,003 (Jan. 29, 1963), L. D. Myers, C. G. Goebel, and F. O. Barrett (to Emery Industries, Inc.).
58. U.S. Pat. 3,773,806 (Nov. 20, 1973), M. Morimoto, M. Saito, and A. Gouken (to Kao Soap Co.).
59. U.S. Pat. 3,925,342 (Dec. 9, 1975), R. P. F. Scharrer (to Arizona Chemical Co.).
60. U.S. Pat. 3,157,681 (Nov. 17, 1964), E. M. Fischer (to General Mills, Inc.).
61. U.S. Pat. 3,412,039 (Nov. 19, 1968), S. E. Miller (to General Mills, Inc.).
62. U.S. Pat. 3,444,220 (May 13, 1969), D. H. Wheeler (to General Mills, Inc.).
63. U.S. Pat. 3,732,263 (May 8, 1973), L. U. Berman (to Kraftco Corp.).
64. U.S. Pat. 3,110,784 (Aug. 13, 1963), C. G. Goebel (to Emery Industries, Inc.).
65. U.S. Pat. 3,632,822 (Jan. 4, 1972), N. H. Conroy (to Arizona Chemical Co.).
66. U.S. Pat. 3,507,890 (Apr. 21, 1970), G. Dieckelmann and H. Rutzen (to Henkel).
67. Brit. Pat. 2,172,597 (Sept. 24, 1986), K. S. Hayes (to Union Camp Corp.).
68. Jpn. Pat. 62,022,742 (Jan. 30, 1987), Arimoto and co-workers (to Harima Chemicals, Inc.).
69. U.S. Pat. 2,964,545 (Dec. 13, 1960), S. A. Harrison (to General Mills, Inc.).
70. U.S. Pat. 2,670,361 (Feb. 23, 1954), C. E. Croston, H. B. Teeter, and J. C. Cowan (to the U.S. Dept. of Agriculture).
71. U.S. Pat. 3,367,952 (Feb. 6, 1968), H. G. Arlt (to Arizona Chemical Co.).
72. U.S. Pat. 3,533,932 (Oct. 13, 1970), J. A. Coffman and W. R. Browne (to General Electric Co.).
73. "Chemical Profile: Dimer Acid", in *Chem. Mark. Rep.* (May 15, 1989).
74. D. Firestone and co-workers, *J. Am. Oil Chem. Soc.* **44**, 465 (1961); D. Firestone, S. Nesheim, and W. Horwitz, *J. Assoc. Off. Anal. Chem.* **38**, 253 (1961).
75. A. Huang and D. Firestone, *J. Assoc. Off. Anal. Chem.* **52**, 958 (1969).
76. A. K. Sen Gupta and H. Scharmann, *Fette, Seifen, Anstrichm.* **70**, 86 (1969).
77. G. Billek and O. Heisz, *Fette, Seifen, Anatrichm.* **71**, 189 (1969).
78. D. Firestone, *J. Am. Oil Chem. Soc.* **40**, 247 (1963).
79. H. E. Rost, *Fette, Seifen, Anstrichm.* **64**, 427 (1962); **65**, 463 (1963).
80. R. P. A. Sims, *Ind. Eng. Chem.* **47**, 1049 (1955).
81. J. M. Cox and S. E. Friberg, *JAOCS* **58**(6), 743–745 (1981).
82. R. W. Johnson and co-workers in J. L. Kroschwitz, ed., *Encyclopedia of Polymer Science and Engineering*, Vol. 11, 2nd ed., Wiley-Interscience, New York, 1988, pp. 476–489.

General References

R. W. Johnson in E. H. Pryde, ed., *Fatty Acids*, American Oil Chemists Society, Champaign, Ill., 1979.

E. C. Leonard, ed., *The Dimer Acids*, Humko Sheffield Chemical, Memphis, Tenn., 1975.

E. C. Leonard in E. H. Pryde, ed., *Fatty Acids*, American Oil Chemists Society, Champaign, Ill., 1979.

E. H. Pryde and J. C. Cowan, Wiley-Interscience, New York, 1972.

R. W. Johnson in R. W. Johnson and E. Fritz eds., *Fatty Acids in Industry*, Marcel Dekker, Inc., New York, 1989.

THOMAS E. BREUER
Humko Chemical Division of Witco Corporation

DIOXIN. See GROUNDWATER MONITORING; HAZARDOUS WASTE TREATMENT; HERBICIDES; SOIL CHEMISTRY AND PESTICIDES.

DIPHENYL AND TERPHENYLS. See BIPHENYL AND TERPHENYLS.

DIPHENYL ETHER, DIPHENYL OXIDE. See HEAT-EXCHANGE TECHNOLOGY, HEAT-TRANSFER MEDIA OTHER THAN WATER; ETHERS.

DISINFECTANTS AND ANTISEPTICS

Agents that served as disinfectants and antiseptics were known and utilized by ancient peoples. Soldiers disinfected equipment and clothing with fire or boiling water, houses were fumigated with burning sulfur, drinking water was purified by storing it in silver or copper vessels, and food was preserved by drying, salting, acidifying, and treating with spices (1). These methods worked, but it was not known why they worked.

In the 1500s, Girolamo Fracastoro, a physician in Italy who studied contagion, proposed that infection was caused by the passage from one person to another of minute bodies capable of self-multiplication (2). In the late seventeenth century van Leeuwenhoek, who made microscopes, was the first to observe bacteria. He reported seeing them become inactive when treated with ordinary vinegar (3). Chlorine and hypochlorite, both first prepared in the 1770s, were tried in France and England in the 1820s for water purification and general sanitation to destroy effluvia believed to cause disease. French doctors reported favorably on the use of hypochlorite in surgery, and Faraday showed that the cowpox virus was inactivated by chlorine (1). This information, however, was largely ignored or rejected.

In Austria in 1847, Semmelweis demonstrated that washing the hands of doctors with a solution of hypochlorite prevented the spread of childbed fever (4). Louis Pasteur published in 1861 (5) the results of his experiments proving that bacteria existed in air and were the cause of fermentations. Joseph Lister, a surgeon in Scotland, met Pasteur and became convinced that bacteria in the air produced infection of wounds. He demonstrated conclusively in 1865 that disinfection with phenol would greatly reduce infection during surgery (6); he later learned that not only the bacteria in the air but those on the hands and clothing were responsible for the infection.

Robert Koch, in Germany in 1881, did scientific laboratory tests on 70 different chemicals, at different concentrations and in different solvents, to assess their ability to kill spores of anthrax bacteria (7). Refinement of the testing methods were made in 1897, 1903, and 1908 (8). They continued to be improved, standardized, and published under the auspices of organizations like the Association of Official Analytical Chemists (AOAC) (now called AOAC International).

In the twentieth century, upon the development of the sulfa drugs (see ANTIBACTERIAL AGENTS, SYNTHETIC) and antibiotics (qv), it was expected that microbial diseases could be completely controlled. This did not occur. A survey of hospital infection rates shows that infection rates run as high as 34% in certain unit areas (9). There are numerous reasons for this: immunity-suppressing agents that make patients much more susceptible to infection are used for some treatments; treatments that control some organisms may allow competing organisms to proliferate; and strains of resistant organisms have developed that elude the treatments meant to control that species.

A new profession, that of infection control practitioner, has been developed to deal with hospital infection problems and at least one practitioner for every 250 hospital beds has been recommended (10). In the mid-1970s, the cost of nosocomial infections was said to be in excess of 1 billion dollars (11), and has continued to rise with health care costs.

The Environmental Protection Agency (EPA) and the Centers for Disease Control (CDC) reported in the early 1990s that diseases caused by viruses and parasites are on the increase. The cause may be the drinking water supply, because these organisms are not always destroyed by the water-treatment processes (12). Wider, intelligent use of disinfectants and antiseptics can greatly aid in removing microbes to limit the chance of infection. Disinfectants find additional use in preventing spoilage of products such as food (see FOOD ADDITIVES), pharmaceuticals (qv), cosmetics (qv), paints (see PAINT), wood (qv), cloth (see TEXTILES), and even in helping to keep office buildings from becoming uninhabitable (13).

Definitions.   Physical and chemical agents that combat pathogenic and nonpathogenic microorganisms are often referred to as disinfectants. Attempts to standardize terminology by health agencies in the United States, such as the CDC, the Food and Drug Administration (FDA), and the EPA, have resulted in the following definitions.

Disinfectant.   A disinfectant is a chemical or physical agent that frees from infection, that kills bacteria, fungi, viruses, and protozoa, but may not kill or inactivate bacterial spores, and is used only on inanimate objects, not on or in living tissue. A bactericide, fungicide, virucide, etc, is a disinfectant intended to kill the organisms indicated in the term. A germicide claims to kill pathogenic microorganisms, or germs.

Antiseptic. An antiseptic is a chemical substance that prevents or inhibits the action or growth of microorganisms but may not necessarily kill them, and is used topically on living tissue. The distinction between a disinfectant and an antiseptic is that the former is expected to kill all vegetative cells and is used only on inanimate objects, whereas the latter may not kill all cells and is used on the body.

Other Terms. *Antimicrobial agents* refers to all chemical and physical agents used to combat microorganisms.

Sterilant is a chemical or physical agent that destroys all forms of microbial life.

Biocide is an antimicrobial agent that kills or inhibits the growth of microorganisms.

Sporicide kills (inactivates) bacterial spores, and is therefore expected to kill all other microorganisms of less resistance. According to the AOAC International it may not kill 100% of the spores, and therefore may not be as powerful as a sterilant. However, according to the EPA, sporicide and sterilant are considered identical.

Germicide destroys microorganisms, with emphasis on pathogenic agents, called germs. It may be a disinfectant or an antiseptic. It does not claim to be a sporicide.

Bactericide, fungicide, virucide, protozoacide, and *algicide* are chemical agents that kill either bacteria, fungi (including yeasts), viruses, protozoa, or algae. They may kill organisms other than the name implies, but not necessarily so.

Bacteriostat, fungistat, etc, are agents that only claim to inhibit, not necessarily kill, microorganisms. In some cases, if they prevent proliferation of microorganisms for sufficient time, these agents may serve as effectively as if they killed the microorganisms.

Sanitizer is a chemical agent used on inanimate objects that reduces the number of microbial contaminants to safe limits as judged by public health requirements.

Antibiotic is an organic chemical substance produced by microorganisms that has the capacity in low concentration to selectively destroy or inhibit the growth of other microorganisms without injuring the host cells. It may be administered systemically and be an antimicrobial chemotherapeutic agent.

An antimicrobial preservative serves to protect materials and products from the deleterious consequences of microbial growth and activities.

Antimicrobial agents don't necessarily fall into only one category. For example, a sterilant under various conditions that could affect its action, such as time, temperature, pH, concentration, and presence of organic matter, might become less potent and act only as a disinfectant, a bactericide, or a sanitizer. Likewise, the reverse situation is also possible—a weaker agent under favorable conditions can exert greater activity and move up in category.

Use of Disinfectants and Antiseptics

From 1985 to 1987, the average annual amount of disinfectants sold in the U.S. was 537×10^3 kg of solid chemical products and 2.05×10^9 L of liquid product. About 4100 disinfectants were registered with the EPA: 44% were labeled as disinfectant, 25% as sanitizer, 14% as virucide, 14% as fungicide, 2% as tuberculocide, and 1% as sterilizer (14). The disinfectant market is about $1 billion annually at the retail level in the United States (14). The total manufacturer's cost for all biocides in 1989 was $850 million, with $60–90 million for disinfectants and sanitizers, $35 million for hospital and medical antiseptics, $150 million for swimming pools, $115 million for food and feed preservatives, and the rest for miscellaneous industrial uses including wood preservation, cosmetics and toiletries, paint, plastics, paper (qv), cooling water, petroleum production, metal working fluids, pharmaceuticals, adhesives (qv), slurries, leather (qv), and latex (15). These figures do not include the cost for chlorine and its compounds, ozone (qv), and uv light, which are used in the purification of drinking water and in treating sewage effluents.

Disinfectants are used in janitorial supplies for hospitals and the home to treat toilet bowls, floors and walls in sick rooms, operating rooms, and wherever infective microorganisms are a problem. Instruments such as scalpels, scissors, catheters, and endoscopes used to invade tissues are treated with disinfectants, as are dental instruments. Laws require that hospital waste must be disinfected so that bacteria and viruses, such as the hepatitis virus and the AIDS virus, do not infect hospital workers and people in the community.

Disinfecting chemicals also experience wide application in the food industry in the growth and production of plant and animal foods. Disinfectants give protection against plant and animal diseases, provide sanitation necessary in food processing (qv) plants, such as dairy, poultry, and seafood plants where human disease microorganisms and food spoilage organisms can be a problem, and assist in food preservation and in the preparation and service of foods. About 20% of the world's food supply is lost to microbial spoilage (16). Finally, disinfectants are employed in a range of industrial applications such as prevention of paint mildewing, microbial contamination of pharmaceuticals and cosmetics, bacterial clogging of oil wells, biocorrosion of airplane storage tanks, and decay of timber.

Antiseptics are used in the home for simple cuts and wounds, and in hospitals for treating patients' skin and surgeons' hands prior to operative procedures. They also are used for preparation of the skin prior to insertion of items such as intravascular lines, chest tubes, temporary pacemakers, and catheters of all kinds. About 80% of hospital patients undergo insertion of an intravascular catheter. Soaps, mouthwashes, lotions, ointments, nose drops, suppositories, and vaginal creams that contact the skin and mucous membranes are often treated with germ-killing antiseptics. Antiseptics are especially important in high risk situations when normal body defense mechanisms are compromised because of an invasive procedure like surgery or catheterization, and when there is an immune deficiency resulting from burns, wounds, extremes of age, cytotoxic drugs, and radiation therapy.

Dyes. No group of compounds is so intimately tied to the history and fabric of microbiology as are the dyes (17) (see DYES AND DYE INTERMEDIATES). Dyes have been used to selectively stain microorganisms for microscopic examination and identification, and also as antiseptics to inhibit growth. Dye preparations have been recommended for a wide variety of infections ranging from the top of the head (*tinea capitis*, ringworm of the scalp) to the bottom of the feet (*tinea pedia*, athletes foot). Dyes are generally no longer used, however, in part because of stains to clothing, bedding, and noninfected body parts.

Halogens

Chlorine, Hypochlorites, and Chlorine Dioxide. Chlorine [*7782-50-57*] and its compounds are not only among the oldest disinfectants, but are used in greatest amount because these materials are cheap and effective (see ALKALI AND CHLORINE PRODUCTS, CHLORINE). They find use in treatment of drinking water, wastewater, swimming pools, and general sanitation in commercial plants and the home (18). Chlorine has the greatest antimicrobial activity of the halogens. Fluorine (qv) is too reactive to be used as a disinfectant, but fluoride and silicofluoride salts have found use as mold-control agents (see ANTIPARASITIC AGENTS, ANTIMYCOTICS) (19) and wood preservatives (20) (see also BLEACHING AGENTS).

Elemental chlorine is used in industrial applications, but chlorine is more readily dispensed in the form of hypochlorites. Chlorine reacts with water to give hypochlorous acid and hydrochloric acid.

$$\text{Cl}_2 + \text{H}_2\text{O} \underset{\text{H}^+}{\overset{\text{OH}^-}{\rightleftharpoons}} \text{HOCl} + \text{H}^+ + \text{Cl}^- $$
$$\text{H}^+ \updownarrow \text{OH}^+$$
$$\text{OCl}^- + \text{H}^+$$

The dissociation of hypochlorous acid depends on the pH. The unionized acid is present in greater quantities in acid solution, although in strongly acid solution the reaction with water is reversed and chlorine is liberated. In alkaline solutions the hypochlorite ion OCl^- is increasingly liberated as the pH is increased. The pH is important because unionized hypochlorous acid is largely responsible for the antimicrobial action of chlorine in water. Chlorine compounds are therefore more active in the acid or neutral range. The hypochlorites most commonly employed are sodium hypochlorite [*7681-52-9*] or calcium hypochlorite [*7778-54-3*].

The hypochlorites have a broad spectrum of activity toward microorganisms. They are antimicrobial to most bacteria, fungi, viruses, protozoans, and algae in just a few parts per million (21). Higher concentrations of chlorine, however, are necessary to inactivate hepatitis A virus (22); even 5000 ppm were not effective against rotavirus (23) and cysts of the protozoan *Acanthamoeba castellanii* (24). The hypochlorites have some sporicidal activity, although it is less than against vegetative cells (25). Being highly chemical reactive, their antimicrobial activity is reduced by the presence of ammonia and organic compounds like proteins, amino acids, and lignin, as well as inorganic reducing agents such as sulfides, nitrites, and compounds of iron and manganese. Chlorine solutions are unstable to light, heat, and acidity. There are numerous reports that small additions of bromine (qv) enhance the bactericidal activity of chlorine (26,27). Bromine is not as active as chlorine although it has been proposed as a disinfectant for swimming pools (18).

A chlorine compound of increasing importance for water treatment is chlorine dioxide [10049-04-4]. It has greater oxidation capacity than hypochlorite and has greater germicidal potency, being sporicidal as well as bactericidal and virucidal. Chlorine compounds have come under scrutiny because of the reaction with organics in drinking water to produce carcinogenic trichloromethanes such as chloroform. Chlorine dioxide does not produce trichloromethanes and it does not react with ammonia nitrogen. It is, however, five times the cost of chlorine and hypochlorites and must be generated on site (18,21), although a stabilized form can be used for small applications (21).

N-Chloramines. N-Chloramines comprise the derivatives of amines in which one or two valences of trivalent nitrogen are taken up by chlorine (see CHLORAMINES AND BROMAMINES). When chlorine combines with ammonia in water, monochloramine [10599-90-3], NH_2Cl, is produced. It releases hypochlorous acid slowly, requiring a greater concentration and longer time to produce germicidal action. Organic chloramines are produced when hypochlorous acid reacts with an amine, amide, imine, or imide. Table 1 presents the structure of a few chloramines and their available chlorine, ie, a measurement of the oxidizing capacity. These compounds, like hypochlorites, are used in the cleaning and sanitizing of equipment and utensils in the food and dairy industries, for water and

Table 1. Available Chlorine Content of N-Chloramines

Name	CAS Registry Number	Structure	Available Cl, %
chloramine-T	[127-65-1]	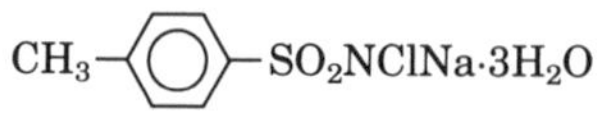 CH_3—◯—$SO_2NClNa \cdot 3H_2O$	23–26
dichloramine-T	[473-34-7]	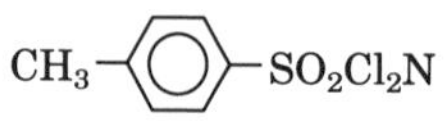 CH_3—◯—SO_2Cl_2N	56–60
halazone	[80-13-7]	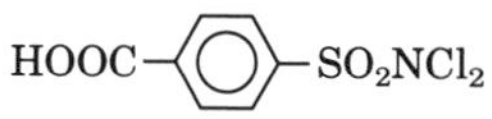 $HOOC$—◯—SO_2NCl_2	48–52.5
succinchlorimide	[128-09-6]	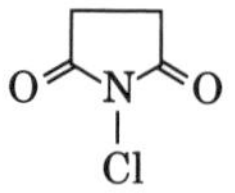	50–54
chloroazodin	[502-98-7]	$H_2NCN{=}NCNH_2$, $\underset{NCl}{\|} \quad \underset{NCl}{\|}$	75–79
trichloroisocyanuric acid	[87-90-1]	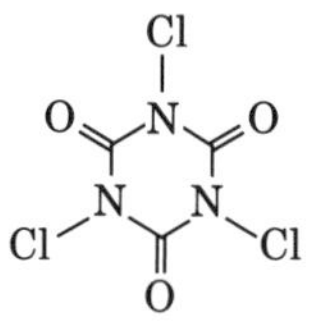	88–90

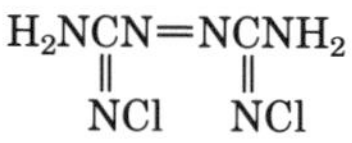

sewage treatment, and for bleaching and sanitizing in commercial laundries. Like chloramine, they are slower in effecting germicidal action than hypochlorous acid. In swimming pools this is desirable because the germicidal effect lasts longer where chlorine is lost from sunlight and aeration (28). As dry powders, chloramines are stable in storage.

Chloramine-T, sodium N-chloro-p-toluenesulfonamide [127-65-1], was widely used during World War I for the treatment of infected wounds, and subsequently for hygienic purposes such as mouthwashes, douches, etc. It can be used for sanitizing food-handling equipment, but its activity is considerably slower than that of hypochlorites. The Grade A Pasteurized Milk Ordinance (1985) and the Ordinance and Code for Restaurants (1948) of the U.S. Public Health Service permitted the use of chloramine-T.

Dichloramine-T, N,N-dichloro-p-toluenesulfonamide [473-34-7], is insoluble in water, but soluble in a number of organic solvents, including chlorinated paraffin. Its medical usage appears to have declined.

Halazone, N,N-dichloro-p-carboxybenzenesulfonamide [80-13-7], is suitable for the decontamination of water, as is also succinchlorimide, N-chlorosuccinimide [128-09-6], which is a white crystalline compound having a chlorine odor. Succinchlorimide is strongly bactericidal when compared to hypochlorites, and is less affected by organic matter than halazone. However, it is inferior to hypochlorites as a cysticide (29). Chloroazodin, also known as azochloramide and N,N-dichloro-azodicarbonamidine [502-98-7], is claimed to be relatively nontoxic to tissue. Applied to a wound it acts as a mild and slow oxidant (30).

Chlorinated cyanuric acid derivatives include dichloroisocyanuric acid [2782-57-2], the sodium [2893-78-9] and potassium [2244-21-5] salts, and trichloroisocyanuric acid [87-90-1]. These compounds can be used in sanitation as such or formulated into various products such as machine dishwashing compounds, scouring powders, industrial sanitizing compounds for food plants, dairies, and swimming pools. When used in the health field, 100–200 ppm available chlorine is recommended for average soiling, 1000–2500 ppm for organic soiling, and up to 10,000 ppm for spillages. Where blood is spilled, use of granules is preferred to limit the contaminated area by absorption and to yield a higher chlorine concentration (31) (see CYANURIC AND ISOCYANURIC ACIDS).

1,3-Dichloro-5,5-dimethyl hydantoin [118-52-5] is used in several commercial cleaner–sanitizer products as an antibacterial component. In a pH range of 5.8–7.0, it has similar antimicrobial activity to hypochlorite; it is less effective under alkaline conditions.

Chloromelamine [7673-09-8] is a chlorination product of 1,3,5-triaminotriazine. Formulated with a suitable anionic surfactant, it has been considered as a bactericidal rinse for mess kits, and also for the treatment of contaminated fruits and vegetables (32).

Iodine. Iodine has been important for many years, primarily as an antiseptic (see IODINE AND IODINE COMPOUNDS). In the American Civil War physicians used it to treat battle wounds. Elemental iodine is not very soluble in water, but dissolves readily when sodium iodide is added, forming triiodide: $I_2 + I^- \rightleftharpoons I_3^-$. Iodine may thus be used as an aqueous solution but it has generally been used as a tincture of 2% iodine in 70% alcohol. Tests on bacteria and spores demonstrate that the triiodide ion has a negligible effect on the organisms, and that the diatomic iodine is responsible for the antimicrobial action (33,34).

Chemically highly active, though not as active as chlorine, the degerming capacity of iodine as tinctures of 2–7% reduced the bacterial count of the skin by 97.5–100% by the serial basin test, making iodine superior to organomercurials and quaternary ammonium antiseptics (35). Iodine, like chlorine, is an oxidizing agent and has a broad spectrum of activity (36,37). Its activity is diminished by proteins and other organics, but less so than chlorine, most likely because of a much slower rate of reaction (38). Iodine shows greater activity at lower pH values. Its activity increases with increasing temperature, but not as greatly as chlorine. Iodine does not react with ammonia and amines to produce the iodine analogues of chloramines.

Tincture of iodine and aqueous iodine are not as popular as they used to be because they stain skin and clothes a brown color and also because of their toxicity. These problems have been considerably reduced, but not completely resolved, in the production of iodophors.

Iodophors are the product of the chemical reaction of iodine and surface-active agents or polymers to produce complexes that retain the germicidal activity but not the undesirable properties of iodine. These are water-soluble, nonstaining, less irritating to the skin and other tissues, nonirritating to the eyes and mucous membranes (39,40), and do not cause a burning sensation when applied to raw skin. For these reasons, iodophores have virtually replaced tincture of iodine. The organic carrier of the iodine may be polyvinylpyrrolidine (Povidone–iodine), a nonionic surfactant, or a cationic detergent. Povidone–iodine is a brown, water-soluble powder containing approximately 10% iodine. However, the amount of free iodine, which is responsible for the antimicrobial activity, is low in a concentrated solution, but is released as the solution is diluted (41). Concentrated solutions have actually been contaminated with bacteria (42). For use as an antiseptic, povidone–iodine is diluted with water or alcohol to a concentration of 1% iodine. Detergents are added if it is used as a surgical scrub. Iodophors are important as broad-spectrum antiseptics for the skin, although they do not have the persistent action of some other antiseptics. They are also used as disinfectants for clinical thermometers that have been used by tuberculous patients, for surface disinfection of tables, etc, and for clean equipment in hospitals, food plants, and dairies, much as chlorine disinfectants are used.

Alcohols

Alcohols, particularly ethanol [64-17-5] and 2-propanol [67-63-9], are important disinfectants and antiseptics. In the aliphatic series of straight-chain alcohols, the antimicrobial activity increases with increasing molecular weight up to a maximum, depending on the organism tested. For *Staphylococcus aureus*, the maximum activity occurs using amyl alcohol [71-41-0], $C_5H_{12}O$; for *Salmonella typhosa*, octyl alcohol [111-87-5], $C_8H_{18}O$ (43); for *Mycobacterium tuberculosis* and *Trichophyton gypseum*, cetyl alcohol [36653-82-4], $C_{16}H_{34}O$ (44); and for *Mycoplasma gallisepticum* and *M. pneumoniae*, alcohols of 16–19 carbons (45). The order of bactericidal activity for alcohols is primary > secondary > tertiary. Alcohols are bactericidal, fungicidal, and virucidal but not sporicidal (43).

Methanol has little or no antimicrobial activity (46) but ethanol, 2-propanol, and *n*-propanol [71-23-8] have high activity toward gram-positive and gram-negative bacteria and lipophilic viruses. Whereas *n*-propanol is the most active (47), it has had little application because of greater cost and odor. Ethanol and 2-propanol are similar in antimicrobial activity and physical properties, but ethanol is lower in toxicity. 2-Propanol, however, has the advantage of being available without the restriction of government regulation and taxation and is most widely used. Ethanol, which is generally not antimicrobial in less than 20% aqueous solution, is most effective when used at 80–95% by volume, 70–92% by weight (48), rather than at 100%. It is believed that alcohol functions by denaturation of proteins, which are not as readily denatured in the absence of water. In practice, ethanol is used as a 70% solution by weight because this percentage is less expensive and just as effective as greater strengths (Table 2). 2-Propanol has been reported to be somewhat more effective as an antimicrobial agent toward *Staphylococcus aureus* and *E. coli* than ethanol (64) but has about the same performance toward *M. tuberculosis* on dried sputum smears (54). 2-Propanol is usually used as an aqueous solution of 60–70% by volume, although at least one study has found it to be more active at full strength (99%) (65). Most results, however, show that water is necessary for alcohol germicidal action (66).

Ethanol and 2-propanol find application as skin antiseptics for personnel handwashing, surgical scrub, and preoperative skin preparations because these alcohols evaporate leaving no residue and are rapid in action. Washing the hands for 1 minute in 70% ethanol (by weight) has a degerming effect equivalent to 6–7 minutes of scrubbing (66,67). 2-Propanol in 70% solution degerms the skin about as well or a little better than ethanol (68). Alcohols also defat and tend to dry and roughen the skin; therefore, glycerol is sometimes added as an emollient. Newer degerming processes use alcohol foams. These apply smaller quantities of alcohol to the skin, which is then rubbed to dryness, allowing the natural fats and oils to be redeposited.

Ethanol and 2-propanol have also found use in disinfecting clinical thermometers, and as preservatives to prevent microbial deterioration of cosmetics and medicinals. They are sometimes combined with other disinfectants, namely formaldehyde (69), phenolics (70), chlorhexidine (71), hypochlorite (72), and phenols (70).

Other alcohols used as preservatives include propylene glycol [57-55-6], $C_3H_8O_2$; benzyl alcohol [100-51-6], C_7H_8O; chlorobutanol [57-15-8], $C_4H_7Cl_3O$; phenylethyl alcohol [60-12-8], $C_8H_{10}O$; phenoxyethanol [122-99-6], $C_8H_{10}O_2$; and Bronopol [52-51-7], $C_3H_6BrNO_4$. Propylene glycol and triethylene glycol [112-27-6], $C_6H_{14}O_4$, are sometimes used in air-conditioning systems to protect against airborne infective and contaminating microorganisms.

Phenolic Compounds

As a disinfectant or antiseptic, phenol [108-95-2] (carbolic acid) is mostly of historical interest. However, its extensive use continues in both investigative and

Table 2. Bactericidal and Virucidal Action of Ethanol

Organism	Effective conc, %	Time required, min	Reference
Salmonella typhosa	21.6	5	49
	17.3	10	49
	40–100	0.17	50
Staphylococcus aureus	34.5	5	49
	60–95	0.17	50
	100	0.85–1.53	50
Staphylococcus albus	70–80	5	51
	15	1440	52
Streptococcus pyogenes	100	0.85–1.53	50
	50–100	0.51	50
	40	4	50
	30	45	50
	20	60	50
Escherichia coli	60	1–5	43
	70	0.68	53
	80	0.51	53
	60	5	53
	40–100	0.17	50
Pseudomonas aeruginosa	30–100	0.17	50
	20	30	50
Mycobacterium tuberculosis			
aqueous suspension	95	0.26	54
	100	0.51	54
	70	1	54
in dry sputum			
thin smears	70	1–5	54
thick smears	95	30	54
Trichophyton mentagrophytes	50	5	55
	25	20	55
Trichophyton rubrum	50	5	55
	25	20	55
Microsporum canis	50	5	55
	25	20	55
Microsporum audouini	70	1440	55
fowl pox virus	70–95	10	56
	50	30	56
Newcastle virus	70–95	3	57
vaccinia virus	40,50	10,60	58
foot and mouth disease virus,	50	15–20	59
filtered (Berkefeld)	60	1–15	59
unfiltered	20–60	360	59
poliovirus, type 1	70	10	60
Coxsackie B-1	60	10	60
ECHO 6	50	10	60
adenovirus, type 2	50	10	60
herpes simplex	30	10	60
influenza, Asian	30	10	60
human immunodeficiency virus	70	1	61
hepatitus B virus	80	2	62
rotavirus	80	10	63

analytical microbiology, eg, as in the AOAC phenol coefficient and use-dilution methods.

Phenol Derivatives.   Derivatives of phenol have been investigated more thoroughly than any other group of disinfectants, and have found use as antiseptics, as the active ingredient in germicidal soaps and lotions, as hard-surface disinfectants, as preservatives for toiletries, and as antimicrobial products for institutions and the household. In industry they have varied uses, such as wood preservatives, mildewcides for leather, and to control algae, slime-forming bacteria, fungi, and sulfate-reducing bacteria in oil fields.

Studies of the effect of the structure of phenol derivatives on antimicrobial activity have generated a number of general observations. (*1*) In a homologous series of monoalkyl phenols, the potency against four common organisms, ie, *Salmonella typhosa*, *Staphylococcus aureus*, *Mycobacterium tuberculosis*, and *Candida albicans*, increases with the increase in molecular weight until the *n*-amyl derivative is reached (Table 3). Further increases in molecular weight result in a substantial increase in germicidal potency toward some microorganisms, and a decrease with respect to others. The term quasispecific has been proposed to describe the action of a disinfectant, such as *n*-heptylphenol, which is extremely effective against *Staphylococcus aureus* but only slightly so against *Salmonella typhosa* (*73*). The increased activity is presumed to be caused by the increasing surface activity with molecular weight and the ability to orient at an interface. Still further increase in molecular weight produces declined activity because of decreased water solubility. Straight-chain alkyl substituents confer greater activity than branched-chain substituents having the same number of carbon atoms. (*2*) Halogenation increases the antimicrobial activity of phenols. Introduction of aliphatic or aromatic groups into the aromatic nucleus of halogenated phenols

Table 3. Microbicidal Action of Phenol Derivatives[a]

Substituting radical	CAS Registry Number	*Salmonella typhosa*	*Staphylococcus aureus*	*Mycobacterium tuberculosis*	*Candida albicans*
none	[108-95-7]	1.0	1.0	1.0	1.0
2-methyl	[95-48-7]	2.3	2.3	2.0	2.0
3-methyl	[108-39-4]	2.3	2.3	2.0	2.0
4-methyl	[106-44-5]	2.3	2.3	2.0	2.0
4-ethyl	[123-07-9]	6.3	6.3	6.7	7.8
2,4-dimethyl	[105-67-9]	5.0	4.4	4.0	5.0
2,5-dimethyl	[95-87-4]	5.0	4.4	4.0	4.0
3,4-dimethyl	[95-65-8]	5.0	3.8	4.0	4.0
2,6-dimethyl	[576-26-1]	3.8	4.4	4.0	3.5
4-*n*-propyl	[645-56-7]	18.3	16.3	17.8	17.8
4-*n*-butyl	[99-71-8]	46.7	43.7	44.4	44.4
4-*n*-amyl	[1322-06-1]	53.3	125.0	133.0	156.0
4-*t*-amyl	[80-46-6]	30.0	93.8	111.1	100.0
4-*n*-hexyl	[61902-50-9]	33.3	313.0	389.0	333.0
4-*n*-heptyl	[26997-02-4]	16.7[b]	625.0	667.0	556.0

[a]Expressed as phenol coefficient at 37°C. See text.
[b]Approximate value.

increases the activity up to certain limits depending on the number of carbons in the substituting groups. (3) Greater activity is found where the alkyl group is ortho to the phenolic moiety, and the halogen is para, rather than the reverse. Pseudohalogens, SCN substituents, follow the same generalizations. (4) Nitro group substitution in phenols also increases antimicrobial activity, believed to result from uncoupling of oxidative phosphorylation. In the case of the higher homologues, the germicidal action manifests the quasispecific character as illustrated in Figure 1, which relates to the homologous series of o-alkyl derivatives of p-chlorophenol. Many instances of the same effect have been established also in the group of polyalkyl-, aryl-, and aralkylchlorophenol derivatives (see also CHLOROCARBONS). It is significant that the increase in molecular weight that accompanies the increase in the antimicrobial potential is also accompanied by a decrease in toxicity to animals (mice).

Because of lower toxicity and high antimicrobial activity, the phenols having the greatest use in disinfections are o-phenylphenol (Dowicide 1) [90-43-7], $C_{12}H_{10}O$; o-benzyl-p-chlorophenol (Santophen 1) [120-32-1], $C_{13}H_{11}ClO$; and p-tert-amylphenol [80-46-6], $C_{11}H_{16}O$. They possess similar general characteristics; ie, broad-spectrum antimicrobial activity toward gram-negative and gram-

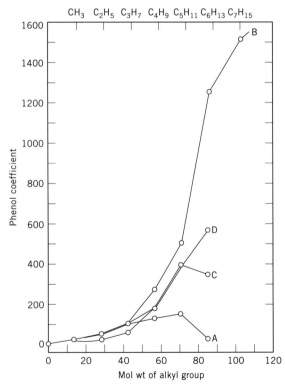

Fig. 1. The quasispecific effect in the homologous series of o-alkyl-p-chlorophenol derivatives against A = *Salmonella typhosa*; B = *Staphyloccus aureus*; C = *Mycobacterium tuberculosis*; D = *Candida albicans*. Phenol coefficient is the activity of the chemical tested compared to that of phenol.

positive bacteria, fungi, *Mycobacterium tuberculosis*, and protozoa; virucidal activity against lipophilic and intermediate but not hydrophilic viruses; tolerance for organic loading and hard water; residual activity; and biodegradability.

Phenols are considered to be low-to-intermediate level disinfectants, appropriate for general disinfection of noncritical and semicritical areas. They are not sporicidal and should not be used when sterilization is required. They are dispensed in aqueous formulations or in mixed water–alcohol solutions. Levels of 400–1300 ppm in the diluted formulation are typical.

Ortho-phenylphenol in ethanol has demonstrated extraordinary action against *Pseudomonas aeruginosa*, *Staphylococcus aureus*, *Salmonella choleraesuis*, *Mycobacterium tuberculosis*, and poliovirus within 3 minutes in the presence of 50% whole human blood (74). It also is effective in combating the growth of common molds, and is an active ingredient of commercial mildewcides. *O*-benzyl-*p*-chlorophenol and *p-tert*-amylphenol also have enjoyed wide acceptance in household and institutional disinfectant products. An example of a liquid concentrate formulation (Lysol, pine scent) is 4.5% o-benzyl-*p*-chlorophenol, 1.5% isopropyl alcohol, 16.5% soap, 5.0% pine oil, 0.7% tetrasodium EDTA. Standard dilution is 1 to 100 with water, but greater dilution is used in some applications (74). Pine oil is employed in disinfectant products mainly to supply a clean, woodsy odor, and whereas it has antibacterial activity, it lacks the ability to kill *Staphylococcus aureus*. *p*-Chloro-*m*-xylenol (PCMX, 4-chloro-3,5-dimethylphenol) [88-04-0], C_8H_9ClO, has important use in antibacterial soaps because of its safety and low toxicity profile. These applications include nonsurgical handscrub soaps, deodorant soaps, liquid soaps, and underarm deodorants. PCMX is also used as a preservative in printing ink (see INKS), cosmetics, adhesives, photographic emulsions, and shoe polishes.

Phenolic dispersions made using natural soaps are sensitive to hard water, but under proper formulation they can tolerate water of 400 ppm total hardness and maintain complete clarity and germicidal activity. The highest activity for phenols was found when using secondary alkane sulfonates as solubilizing agents (75).

Although resorcinol [108-46-3], C_6H_6O, a dihydric phenol, is a comparatively weak bactericide, a nuclear-substituted alkyl derivative, *n*-hexylresorcinol [136-77-6], $C_{13}H_{12}O$, has a phenol coefficient of 45, and has had considerable use as a topical antiseptic.

Bisphenols. As a group, bisphenols show outstanding antimicrobial properties, and two members have achieved considerable commercial importance (Fig. 2). Hexachlorophene [70-30-4], also known as hexachlorophane and methylene-bistrichlorophenol, has found intensive use because it does not lose activity in the presence of soap. The treated soap reduces bacterial skin flora to a small fraction of the original number (76,77). Hexachlorophene also has been incorporated in soapless detergent bases, eg, pHisohex, and in deodorant soaps where it controls the proliferation of cutaneous bacteria that cause perspiration malodor by the decomposition of apocrine sweat. It is successful in preventing the spread of staphylococcal infections when used as surgical scrubs and in nurseries. Extremely low concentrations of hexachlorophene suffice to produce inhibition of *Staphylococcus aureus*, namely 1–1.5 ppm, depending on the inoculum (78,79).

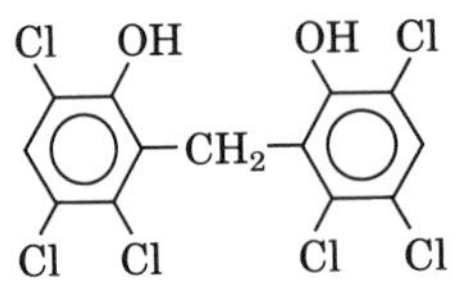

Fig. 2. Bisphenols having antimicrobial properties.

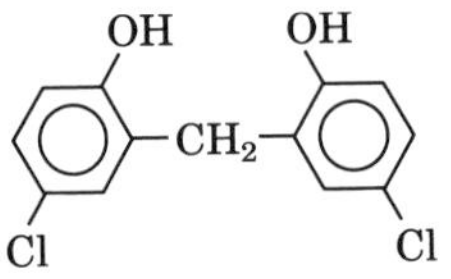

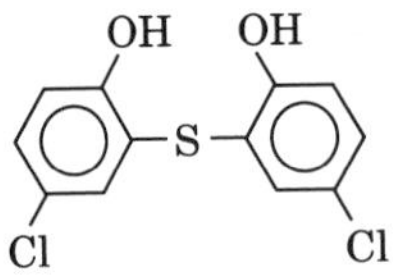

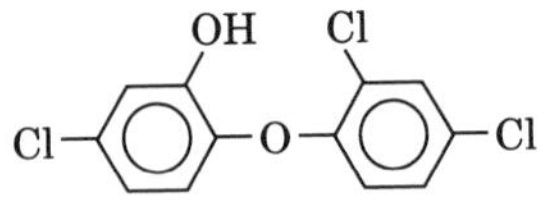

Against *Escherichia coli* and *Pseudomonas aeruginosa* the bacteriostatic range is 20–100 ppm (80).

The action of hexachlorophene is reversed upon contact with blood. This is deemed significant for any prophylactic application to broken skin against infection in injury, or for preoperative use (80,81). Skin that has been degermed using hexachlorophene is therefore not protected against subsequent contamination from transient pathogens or against lesions of bacterial origin (80,81). Although reports of the extensive use of hexachlorophene do not show toxic symptoms in humans (82,83), neurotoxicity has been demonstrated in rats using large doses (84,85), in monkeys (86), and in premature infants (87). The FDA (88) has banned over-the-counter sale of soaps, cosmetics, and drugs containing more than 0.1% hexachlorophene. All products of higher percentage have been put on a prescription basis. All products must be labeled to prevent use by pregnant women and children. The United Kingdom has initiated similar restrictions.

Another important bisphenol is dichlorophene (dichlorophane, methylene-bis(4-chlorophenol)) [97-23-4]. Whereas it is not as active against bacteria as hexachlorophane, it has found miscellaneous applications, eg, as a rot preservative for textiles (89); as a treatment for athlete's foot and for tapeworm in humans and domestic animals; as a slimicide in paper manufacture; and as an antibacterial agent in water-cooling systems (90).

Fentichlor (bithionol, thiobis-(4,6-dichlorophenol)) [97-18-7] is a bisphenol having a sulfur bridge rather than a methylene bridge. Like hexachlorophene, it is more inhibitory to gram-positive than gram-negative bacteria, but it is highly active against fungi and yeasts. In a series of eight thiobisphenols it exhibited the greatest fungistatic potency (91). Its chief application is in the treatment of dermatophytic conditions; however it has shown photosensitivity in humans in some cosmetic and soap applications (92). A comprehensive review of bisphenols is available (93).

Triclosan (Irgasan, trichlorohydroxydiphenyl ether) [66943-50-6] is a bisphenol with an oxygen bridge. According to a 1977 survey (94), it was the most widely used phenolic preservative, appearing in 52 formulations. It has been recommended for soaps and washing products (95), and many antibacterial soaps contain triclosan. It inhibits staphylococci at 0.03–0.1 ppm, and some *E. coli* in the same range, but *Pseudomonas aeruginosa* requires 100–1000 ppm. Molds are inhibited at 1–30 ppm.

Coal-Tar Disinfectants. Coal-tar disinfectants formerly constituted the most important category of disinfectants for general use. These are obtained from coal tar (see COAL; COAL CONVERSION PROCESSES; TAR AND PITCH) which, when fractionated, yields a group of chemicals including phenols, organic bases, and neutral hydrocarbon oils. The phenol fraction, or tar acids, is separated by distillation to give phenol, cresols, xylenols, and high boiling acids consisting of such higher homologues as propyl and butyl phenols and naphthols. The phenol coefficient, a measure of activity with respect to phenol (96), increases from 1 for phenol, 2 for cresols, and 5 for xylenols, to 25–60 for the higher homologues. The water solubility decreases with increasing molecular weight. The terms black fluid and white fluid refer to the way the fractions are diluted; the black fluids are crude phenol fractions solubilized with soaps, whereas white fluids are prepared by emulsifying the fractions. The term clear soluble fluid is used to describe the solubilized products lysol and sudol.

Lysol consists of a mixture of the three cresol isomers solubilized using a soap prepared from linseed oil and potassium hydroxide, to form a clear solution on dilution. Most vegetative pathogens, including mycobacteria, are killed in 15 minutes by 0.3–0.6% lysol. Lysol has a phenol coefficient of 2. Bacterial spores are very resistant. Lysol is also the name of a proprietary product, the formula of which has changed over the years; other phenols have been substituted for the cresols.

Sudol uses fractions of coal tar rich in xylenols and ethylphenols. It is much more active and less corrosive than lysol, and remains more active in the presence of organic matter. The phenol coefficients of sudol against *Mycobacterium tuberculosis*, *Staphylococcus aureus*, and *Pseudomonas aeruginosa* are 6.3, 6, and 4, respectively. It also is slowly sporicidal (97).

In addition to coal tar, petroleum has been a source of the same chemicals, and many of the individual phenols have been produced in the pure state by synthetic processes.

Acids

Hydroxybenzoic Acids and Esters. The phenolic hydroxyl group of benzoic acid derivatives shown in Figure 3 is largely responsible for their antimicrobial activity. Benzoic acid [65-85-0], itself effective as a preservative in foods and cosmetics, is relatively weak as a disinfectant as compared to phenols. *ortho*-Hydroxybenzoic acid [69-72-7] (salicylic acid) has its phenolic group masked by hydrogen bonding to the oxygens of the carboxyl group. It has keratinolytic activity, and is used with benzoic acid in a successful athlete's foot treatment, ie, Whitfield's ointment. The methyl, ethyl, propyl, and butyl esters of vanillic acid (4-hydroxy-3-methoxy benzoic acid) [121-34-6] have antifungal properties at 0.1–0.2% concentration, and have been used in the preservation of foods and food packaging (qv) materials.

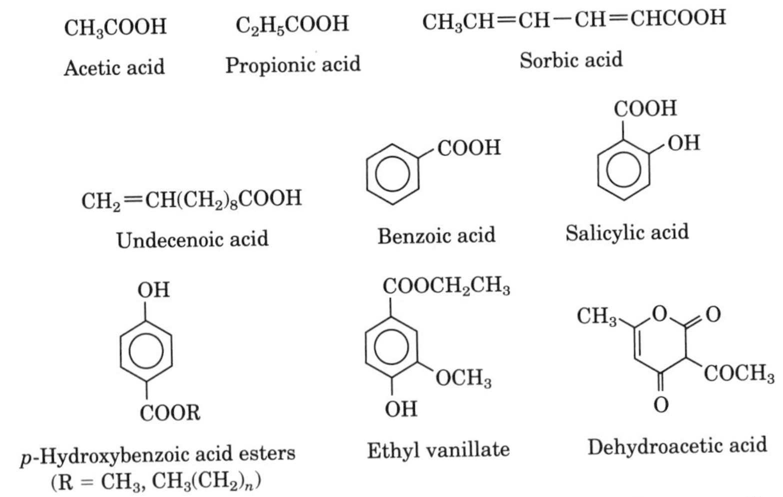

$$CH_3COOH \qquad C_2H_5COOH \qquad CH_3CH{=}CH{-}CH{=}CHCOOH$$

Acetic acid   Propionic acid   Sorbic acid

$$CH_2{=}CH(CH_2)_8COOH$$

Undecenoic acid   Benzoic acid   Salicylic acid

p-Hydroxybenzoic acid esters   Ethyl vanillate   Dehydroacetic acid
$(R = CH_3, CH_3(CH_2)_n)$

Fig. 3. Organic acids and esters useful as disinfectants and preservatives.

The most successful compounds in this group are the esters of p-hydroxy-benzoic acid, known as parabens, which have been in continuous use since the 1920s. Their antimicrobial activity increases from the methyl to the benzyl ester, but water solubility limits use mainly to the esters listed in Table 4. These compounds are active against gram-positive and gram-negative bacteria, yeasts, and fungi (98,99). The low order of toxicity (100,101), lack of irritation, and absorption and excretion characteristics in both humans and animals make these compounds well suited as preservatives for pharmaceuticals, cosmetics, and food (102–104). Their application in antimycotic therapy has been considered by several investigators (105) and tests on their use in a cosmetic lotion show that in combination with another antimicrobial preservative, imidazolindyl urea, they are ideal in providing a broad spectrum preservative system (106).

Inorganic Acids. Strong inorganic acids have little antimicrobial activity in themselves but inhibit microorganism growth by lowering the pH. Disinfectant toilet bowl cleaners that contain 9.5% HCl or more are antimicrobial. Carbonic acid [463-79-6] in soft drinks provides some antibacterial preservation. Sulfurous acid [7782-99-2] is an effective preservative used to preserve wines (see WINE), fruit juices (qv), and dried fruits.

Organic Acids. Among the organic acids (Fig. 3), acetic acid [64-19-7], as vinegar (qv), is an effective food preservative and has been used medically since ancient times for treating open wounds. Propionic acid [79-09-47] is a preservative for cheeses and food wrappers and prevents the growth of microbial rope and mold in bread. Lactic acid [598-82-3] preserves pickles, sauerkraut, and silage. Benzoic acid has been used for many years to control mold growth in ketchup and other foods. Sorbic acid (2,4-hexadienoic acid) [22500-92-1] inhibits yeasts (qv) and molds at 25–500 ppm, and finds application as a preservative in foods, cosmetics (qv), and medicines (107). Dehydroacetic acid [520-45-6] has similar activity for bacteria and fungi, and has had similar applications (108). Undecenoic acid (undecylenic acid) [112-38-9] alone, or with its zinc salt, is used in a popular treat-

Table 4. Physical and Antimicrobial Properties of *p*-Hydroxybenzoic Acid Esters[a]

Parameter	*p*-Hydroxybenzoic acid ester			
	Methyl	Ethyl	Propyl	Butyl
Properties				
CAS Registry Number	[99-76-3]	[120-47-8]	[94-13-3]	[94-26-8]
solubility at 15°C, %	0.16	0.08	0.023	0.005
oil/water partition coefficient	2.4	13.4	38.1	239.6
Minimum inhibitory concentration[b]				
Staphylococcus aureus	1000	500	125	63
Escherichia coli	800	560	350	160
Pseudomonas aeruginosa	1000	700	350	150
Proteus vulgaris	2000	1000	500	500
Penicillium sp.	500	250	125	200
Aspergillius niger	1000	400	200	200
Trichophyton interdigitale	160	80	40	20
Saccharomyces cerevisiae	1000	500	125	63
Candida albicans	1000	800	250	125
Fungicidal concentration[b]				
Candida albicans	5000	2500	625	625
Aspergillus niger	5000	5000	2500	1250
Penicillium chrysogenum	5000	2500	1250	1250

[a]See Figure 3.
[b]Concentration values are given in ppm.

ment for athlete's foot fungus and other superficial skin dermatophytoses (109). Glutaric acid [110-94-1] and glutaric acid analogues have been reported to be virucidal against rhinoviruses (110).

Aldehyde Antimicrobials

Two aldehydes (qv) have made their mark in the field of disinfection, namely formaldehyde [50-00-0] and glutaraldehyde [111-30-8]. Other aldehydes do not match these compounds in activity (111,112).

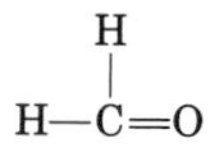

$$\begin{array}{cc} \overset{\text{H}}{\underset{|}{}} & \overset{\text{H}}{\underset{|}{}} \qquad \overset{\text{H}}{\underset{|}{}} \\ \text{H}-\text{C}=\text{O} & \text{O}=\text{CCH}_2\text{CH}_2\text{CH}_2\text{C}=\text{O} \\ \text{formaldehyde} & \text{glutaraldehyde} \end{array}$$

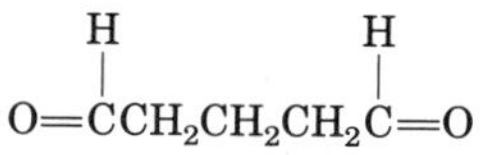

Formaldehyde (qv) was used for many years as a fumigant to treat premises, furniture, and objects exposed to patients with contagious illness. This practice has been virtually abandoned, and is considered of little value as a disinfection procedure because of the problems involved and the toxicity of the vapor (113). Gaseous formaldehyde is effective for the sterilization of heat-sensitive medical instruments and hospital supplies, such as bedding and blankets. The process employs low temperature steam and formaldehyde in a specially designed, gastight chamber. Low temperature, subatmospheric pressure steam destroys vegetative bacteria at 73–80°C; steam used with gaseous formaldehyde destroys the spores as well. In this process, air is removed and steam, at 73°C with a total of 10 mL formalin per cubic foot (357 mL/m^3) of space, is introduced in several stages to penetrate the materials. It is held for 2 h, followed by evacuation of the formaldehyde and admission of filtered air (114,115). Formaldehyde vapor also is used in the fumigation of poultry houses and hatcheries. A relative humidity of 75–100% is essential for effective microbiocidal action.

Formalin is an aqueous solution of 34–38% formaldehyde. As an alkylating agent, formaldehyde reacts with proteins and nucleic acids through the amino acid groups and the sulfhydryl, phenolic, or indole residues. Microbiologically, it is active against bacteria, fungi, bacterial spores, and many viruses. Formaldehyde has had varied uses in embalming fluids, as a preservative for laboratory specimens, for inactivating poliovirus, and for the disinfection of isolators, ion-exchange columns, and soils. Used at 8% concentration, formaldehyde is a much more active microbiocide in 70% isopropanol than in water. The alcohol solution is said to be rapidly bactericidal, tuberculocidal, and sporicidal. A 6–8% formaldehyde solution is rated as a sterilant, and at 1–8% as a low-to-high level disinfectant. Controversy in the early 1990s regarding formaldehyde as a potential occupational carcinogen has limited its use (116).

A number of disinfectants apparently owe their activity to formaldehyde, although there is argument on whether some of them function by other mechanisms. In this category, the drug with the longest history is hexamethylenetetramine (hexamine, urotropin) [100-97-0], which is a condensation product of formaldehyde and ammonia that breaks down by acid hydrolysis to produce formaldehyde. Hexamine was first used for urinary tract antisepsis. Other antimicrobials that are adducts of formaldehyde and amines have been made; others are based on methylolate derivations of nitroalkanes. The applications of these compounds are widespread, including inactivation of bacterial endotoxin; preservation of cosmetics, metal working fluids, and latex paint; and use in spin finishes, textile impregnation, and secondary oil recovery (117).

The other aldehyde of importance in disinfection is glutaraldehyde. It has two aldehyde groups in its five-carbon molecule, and is a more powerful germicide than other dialdehydes and formaldehyde. The 2% alkaline solution of glutaraldehyde is rapid in action, reportedly killing most bacteria in less than 1 min, the tubercle bacillus and viruses in less than 10 min, and bacterial spores in 3 h or less (118); however, EPA registration requires 45 min for the tubercle bacillus and 10 h for spores. At 2% concentration glutaraldehyde is classified as a high level germicide, capable of producing sterility. It has demonstrated high stability and continued activity in the presence of organic matter (119), and is stable to heat and light. It is not very effective at low temperatures, but increases in activity as the temperature is increased (120). Glutaraldehyde is sold as the 2% solution at pH 4, to which is added an activator to adjust the pH to 7.5–8.5. This makes it active and ready for use. The acid solution is relatively stable, but the activated solution may be used only for 2–4 wk and then discarded; this is because the alkalinized glutaraldehyde slowly polymerizes and decreases in activity. After 24 wk of storage, 2% alkaline glutaraldehyde has a concentration of only 0.3%, whereas the more stable acid glutaraldehyde is 1.5% (121). If the solution becomes contaminated with organic matter or diluted with water from wet or unclean instruments, it must be discarded and new solution used. Heating the acid solution or using it with ultrasonics has the effect of activating the acid solution, whereas heating the alkaline solution more rapidly inactivates it.

The primary application for glutaraldehyde is in the disinfection or sterilization of heat-sensitive medical and surgical instruments like endoscopes. This requires a minimum of 20 min for disinfection, and 10–12 h for sterilization (122). The instruments should be thoroughly cleaned, disassembled, put into the alkaline glutaraldehyde for the proper length of time, then rinsed three times with sterile water. A vertical jar may be used so that air is not entrapped in the tubes, which would prevent the glutaraldehyde solution from making contact with all interior surfaces. A 30 min immersion is recommended following use of an endoscope on a patient or carrier of hepatitis B or HIV, and a 1 h immersion following use in a tuberculosis patient, to prevent infection between patients (123). Glutaraldehyde also has been recommended for cold sterilization of hemostats, cystoscopes, food containers, anesthetic equipment, dental equipment, urological equipment, and gynecologic laproscopy equipment (124).

Caution should be taken when using glutaraldehyde. Gloves and aprons should be worn and adequate ventilation provided. It has been reported to produce contact dermatitis, eye irritation, nausea, headache, rashes, and asthmatic reaction (125).

Peroxygen Compounds

Hydrogen Peroxide. Hydrogen peroxide [7722-84-1], H_2O_2, a long-time favored antiseptic in the home and in medical circles, fell out of favor once it was discovered that peroxide is quickly decomposed by the catalase in tissues. There was also a problem with the stability of peroxide preparations, but this has been overcome and stable solutions having a long shelf-life are available. Peroxide has found important applications as a disinfectant and sporicide rather than as an antiseptic. Applications include purifying drinking water (126); treating contaminated water supplies in hospitals (127); treating raw milk (128); sterilizing spacecraft (129); and disinfecting contact lenses (qv) (130) and acrylic resin sections of surgical implants (131). These applications use peroxide in concentrations of 10–25% rather than the 3% solution used as an antiseptic. Another important application of peroxide is in the sterilization of the contact surfaces of food packaging (qv) for nonrefrigerated milk and fruit juices (qv). In this process, peroxide at a concentration of 15–35% is applied at a temperature of 70–140°C for a period of 1–2 seconds to 5 minutes, depending on the strength of the peroxide and the temperature (132).

At 70–140°C, peroxide is vaporized. Peroxide vapor has been reported to rapidly inactivate pathogenic bacteria, yeast, and bacterial spores in very low concentrations (133). Experiments using peroxide vapor for space decontamination of rooms and biologic safety cabinets hold promise (134). The use of peroxide vapor and a plasma generated by radio frequency energy releasing free radicals, ions, excited atoms, and excited molecules in a sterilizing chamber has been patented (135).

Peroxide is believed to produce the free hydroxyl radical said to be the strongest oxidant known. Cupric, ferric, cobaltous, manganous, and dichromate ions are found to enhance the bactericidal action of peroxide by causing the release of the hydroxyl radical (136,137). Heat sharply increases the rate of activity of peroxide, as does uv light (138) and ultrasonic energy (139). Peroxide has the environmental advantage over other disinfectants of leaving no toxic residue, yielding only oxygen and water.

Peracetic Acid. Peracetic acid (peroxyacetic acid) [79-21-0], $C_2H_4O_3$, the peroxide of acetic acid, is a disinfectant having the desirable properties of hydrogen peroxide, ie, broad-spectrum activity against microorganisms, lack of harmful decomposition products, and infinite water solubility. Peracetic acid also has greater lipid solubility and is free from deactivation by catalase and peroxidase enzymes. However, it is corrosive, and degradation products may have to be rinsed from the surface of disinfected materials. Peracetic acid is a more powerful antimicrobial agent than hydrogen peroxide and most other disinfectants (140–142). It has advantages for disinfection and sterilization not found in any other agent (143). Against spores of *Bacillus thermoacidurans*, it was reported to be the most active of 23 germicides (144); against a range of bacteria it is lethal at 6–250 ppm, toward yeasts 25–83 ppm, fungi 50–500 ppm, bacterial spores 100–500 ppm, and viruses 15–2000 ppm (145). The values obtained are determined by the medium employed and the time necessary for inactivation.

Table 5 gives a comparison of peracetic acid and two other disinfectants against food-poisoning bacteria (146). Peracetic retains its activity better than many disinfectants at refrigeration temperatures, and is more effective at lower pH values (141). Aqueous solutions are comprised of the acid in combination with hydrogen peroxide, acetic acid, sulfuric acid, water, and a stabilizing agent. All of these ingredients are necessary to keep it stable in storage, and the concentration

Table 5. Comparison of Disinfectants Against Food-Poisoning Bacteria[a,b]

Organism	Peracetic acid	Active chlorine	Benzalkonium chloride
At 20°C			
Listeria monocytogenes	45	100	200
Staphylococcus aureus ATCC 6538	90	860	500
Enterococcus faecium DSM 2918	45	300	250
At 5°C			
Listeria monocytogenes	90	860	500
Staphylococcus aureus ATCC 6538	90	1100	750
Enterococcus faecium DSM 2918	90	450	500

[a]Lethality in 5 min.
[b]Concentrations of disinfectants given in ppm.

of the hydrogen peroxide in some formulations may considerably exceed that of the peracetic acid. The vapor of peracetic acid, like that of hydrogen peroxide, is active against bacterial spores and has been found to be most effective at 80% relative humidity and to have little activity at 20% relative humidity (147).

The strong antimicrobial activity of peracetic acid makes it valuable in maintaining sterile conditions in the production of germ-free animals (148). It has been accepted worldwide in the food processing and beverage industries as an ideal for clean-in-place systems (149); it does not require rinsing where the breakdown product, acetic acid, is not objectionable in high dilution. Peracetic acid is more toxic than hydrogen peroxide and is a weak carcinogen (150) but can be used with safety when diluted. Like all peroxides, it is a powerful oxidizer and should be handled with proper safety precautions. It is more corrosive to metals and plastics than is hydrogen peroxide (149).

Other Peroxygen Compounds.   The magnesium salt of peroxyphthalate is a commercial product formulated as a sanitizer for diapers, bathrooms, and kitchens (151). Benzoyl peroxide is in general used in skin formulations for treating acne. Inorganic peroxides also have found their niche, eg, perborates have been used in dental preparations; permanganate has been used as an antiseptic and a treatment for fungal skin infections; and the potassium salt of monoperoxysulfuric acid is used in toilet bowl cleaners, denture cleansers, and as a swimming pool disinfectant.

Surface-Active Agents

Quaternary Ammonium Compounds.   Compared to most of the standard germicides, quaternary ammonium compounds (qv) (quats) are a more recent addition to the arsenal of microbe-fighting weaponry. Quats became important in 1935 with the announcement of the antibacterial activity of the long-chain quaternary ammonium salts (152). The quats may be regarded as long-chain alkyl relatives of ammonium salts, NH_4X, where the hydrogens are replaced by organic groups, R_4NX. They have been termed invert soaps, the counterpart of soaps. Like fatty acid soaps, these are salts containing a long-chain hydrocarbon group, and are surface-active substances. In soap the anion contributes the hydrophobic portion; in quats the cation is hydrophobic (see SURFACTANTS).

Many primary long-chain amine salts have appreciable antibacterial properties, but long-chain aliphatic amine salts derived from weakly basic aliphatic amines and their aqueous solutions require a pH low enough to counteract hydrolysis and partial liberation of the water-insoluble amine base. The quaternary ammonium compounds are salts of strong bases, and therefore remain in solution in acidic as well as basic media. This has led to the preference of quats in antimicrobial preparations. Quats, like other cationic-active compounds, are incompatible with anionic-active substances like soaps, and mutual precipitation occurs when they are brought together in aqueous solution. This limits use. Many nonionics also interfere with the activity of quats. In a germicidal quat, at least one of the four organic radicals must impart surface-activity, and because this activity resides in the cation, these compounds are referred to as cationic germicides.

The unique property of quats is the ability to produce bacteriostasis in very high dilution (Table 6). Because of this ability, and the carryover of the disinfectant in early test procedures, it was originally thought that these compounds were highly bactericidal. However, for bactericidal action, 10 to 20 times the concentration is required. The quats have a narrower antibacterial spectrum than the phenols, and are much more active against gram-positive bacteria than gram-negative. A concentration of 2–5 ppm inhibits many gram-positives, whereas it may take 300–500 ppm to inhibit gram-negatives (153). Acid-fast bacteria and bacterial spores are resistant to quats. Quats, moderately active against most fungi and lipophilic viruses but not against hydrophilic viruses, are very active against algae, which makes them useful in those industrial water processes where algae are a problem.

Quats are most effective on the alkaline side, pH 9–10, and are less effective as the pH decreases below 7, eg, using 8 ppm quat, it took 5 minutes to kill bacteria at pH 10, 15 minutes at pH 7, 30 minutes at pH 5, 100 minutes at pH 3 (154). Quats are adsorbed and inactivated by proteins. Cotton and polymers having a negative charge remove quats from solution, and cellulosic cleaning materials such as rags, mops, and sponges inactivate quats in this way (155). Hard water containing calcium, magnesium, and iron salts interferes with the action of quats and thereby necessitates higher concentrations of the quat to be used. In some formulations, a sequestering agent or chelating agent may be employed with the quat to remove the hardness. EDTA used in these formulations serves both to remove the metal ions and increase the activity against gram-negative bacteria.

Many quats were synthesized and tested as antimicrobial agents (153,156). To have appreciable activity, the sum of the carbon atoms in the four R groups should be more than 10 (see Fig. 4). The anion is not important except to help

Table 6. Bacteriostatic Dilutions of
Benzalkonium Chloride[a]

Microorganism	Concentration, ppm
Salmonella typhosa	3.91
Shigella dysenteriae	1.95
Escherichia coli	15.6
Aerobacter aerogenes	62.5
Salmonella paratyphi	15.6
Salmonella enteritidis	31.2
Proteus vulgaris	62.5
Pseudomonas aeruginosa	31.2
Vibrio cholerae	1.95
Staphylococcus aureus	1.25
Pneumococcus II	5.00
Streptococcus pyogenes	1.25
Clostridium welchii	5.00
Clostridium tetani	5.00
Clostridium histolyticum	5.00
Clostridium oedematiens	5.00

[a]See Figure 4.

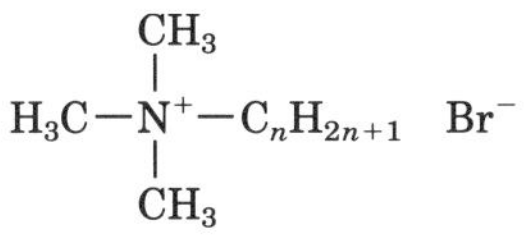

n = 12, 14, or 16
Cetrimide

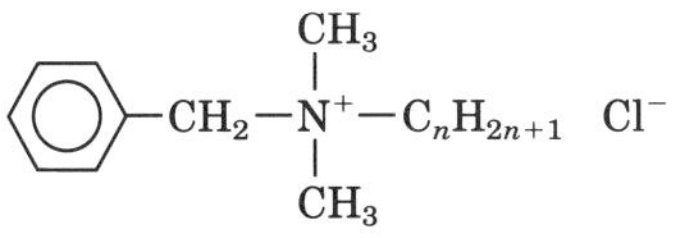

n = 8 to 18
Benzalkonium chloride

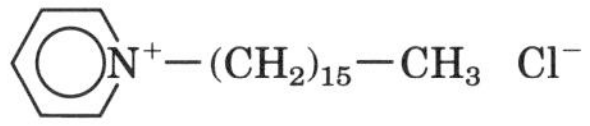

Cetylpyridinium chloride

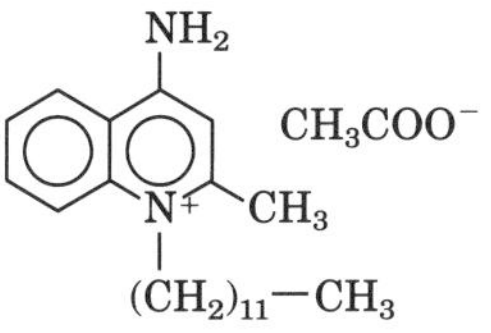

Laurolinium acetate

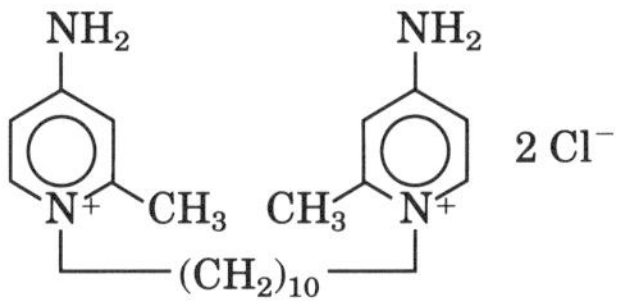

Dequalinium chloride

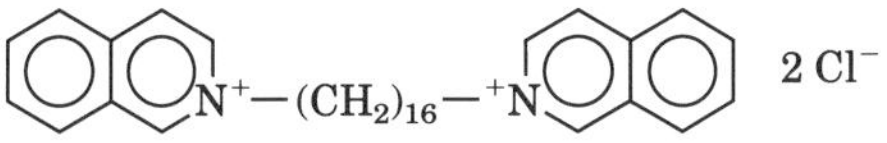

Hedaquinium chloride

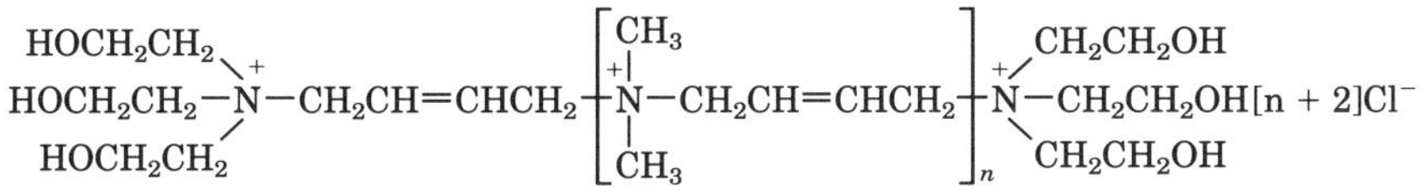

Polyquaternium 1

Fig. 4. Quaternary ammonium compounds (quats) tested as antimicrobial agents. Centrimide is a mixture of dodecyl, tetrodecyl-, and hexadecyltrimethylammonium bromide.

provide solubility. To have high activity, at least one of the R groups must have a chain length in the range of C_8 to C_{18}. A chain length of 14 or 16 is usually best. Three of the four covalent links may be satisfied by nitrogen in a pyridine ring, as in cetylpyridinium chloride [*123-03-5*] (1-hexadecylpyridinium chloride), a piperidinium, or other heterocyclic group (157). Laurolinium acetate [*146-37-2*], a derivative of 4-aminoquinaldine, is a powerful antimicrobial agent, as are bisquaternary compounds like dequalinium chloride [*522-51-0*] and hedaquinium chloride [*4310-89-8*]. Cetylpyridinium chloride is the active ingredient of a popular mouthwash (Cepacol); laurolinium acetate has been used as a preoperative skin disinfectant; dequalinium chloride is used in lozenges and paint for treating infections of the mouth and throat. Only a few of the many quats that were investigated have reached and retained commercial importance.

Further developments have brought forth polymeric quats having antimicrobial properties (158–160). Different kinds of polyquats have been described with molecular weight from 2,000 to 60,000 (153). Polymeric quats have two characteristics that make them uniquely different from the monomeric quats. One is the absence of foaming, even at high concentrations. The other is their remarkably low toxicity in skin and eye irritation tests and oral ingestion tests. This is indeed unique, for quats in general are noted for low toxicity. Acute oral LD_{50} for benzalkonium chloride [68391-01-5] is 0.2 g/kg, and for other quats are as high as 2.2 g/kg, but polyquaternium 1 [75345-27-6] (Onamer M) has an LD_{50} of 4.47 g/kg. Yet, even at this low toxicity, these polymer quats have significant antibacterial and antifungal activity (161). Polyquaternium 1 is considered a candidate preservative for ophthalmic preparations (162), and is registered with FDA for contact lens solutions; WSCP (Busan 77), LD_{50} of 2.77 g/kg, is registered with EPA as a swimming pool algaecide; and the biguanides, another group of polymers (161,163), are known under such trademarks as Vantocil and Cosmocil CQ. A further development to extend the use of quats in medicine is the so-called soft drugs or labile antimicrobial quats; ie, those that would be broken down *in vivo* to nontoxic fragments (164).

When first introduced, quats were used in surgery as hand scrubs and for treatment of instruments; however, these uses have been discouraged. Quats are used as surface disinfectants for floors, walls, and equipment in hospitals and nursing homes, breweries, food plants, and the home. They may be used as sanitizers in rinsewater for glasses and dishes in restaurants. In combination with compatible nonionic wetting agents, quats are employed in products giving one-step cleaning and disinfection for environmental surfaces. A popular commercial disinfectant cleaner contains 2.7% quat in the concentrate and is diluted to give a usage concentration of approximately 400 ppm. Quats effectively and economically curtail algae in industrial cooling water towers and outdoor swimming pools. An application that takes advantage of the charge attraction to negatively charged fabrics is in treatment of diapers to suppress bacterial diaper rash and to prevent the liberation of ammonia produced by bacterial breakdown of urea in urine (165,166). For laundry where hot water is not used, a final rinse with 200 ppm of quats based on the dry weight of the fabric provides residual bacteriostatic activity to the fabric (167). An antibacterial and antimildew treatment of carpeting, underwear, socks, mattress ticking, etc, was introduced as Slygard and Biogard; it resists washing, and results from an organic silicon quat. This treatment claimed 90% reduction in bacterial numbers on the treated fabric, and resisted 40 test wash cycles with commercial detergents (168).

Considerable work has been done to try to explain why quats are antimicrobial. The following sequence of steps is believed to occur in the attack by the quat on the microbial cell: (*1*) adsorption of the compound on the bacterial cell surface; (*2*) diffusion through the cell wall; (*3*) binding to the cytoplasmic membrane; (*4*) disruption of the cytoplasmic membrane; (*5*) release of cations and other cytoplasmic cell constituents; (*6*) precipitation of cell contents and death of the cell.

Acid–Anionic Sanitizers. This group, like the cationic disinfectants, are surface active agents but, like common soaps, these compounds have the hydrophobic principle in the anion rather than the cation. Whereas the cationic agents perform best in an alkaline environment, the anionics work best under acid conditions, usually between pH 1 and 3. The acidic groups donating the hydrophilic groups are carboxylic acid, sulfonic acid, sulfuric acid ester, phosphoric acid ester, or phosphonic acid. The hydrophobic portion is contributed by alkyl chains, which may be substituted with aromatic rings such as benzene or naphthalene. The acid–anionic sanitizers combine low levels of an anionic surface-active agent and an organic or inorganic acid, a solubilizer, and in some cases, a small amount of nonionic surfactant. One acid anionic product on the market (Per-Vad) contains 15% orthophosphoric acid and 2.6% sulfonated oleic acid–sodium salt as the active ingredients, with pH 1.0.

Many types of anionic surfactants have been investigated in combination with different acids (169). It was found that alkyl aryl sulfonates exhibit the best overall results. Food grade phosphoric acid was selected as the most suitable acid, based on low toxicity, corrosion considerations, and economic matters. Rapid bactericidal action toward both gram-positives and gram-negatives was obtained using about 200 ppm of the sulfonates at pH 2–3. Surfactants having alkyl chain lengths of 12 carbon atoms showed maximum biologic activity, eg, dodecylbenzene sulfonic acid (170,171). Benzene ring substitution has the effect of increasing the length of the alkyl chain by 3.5 ethylene residues. Because solubility is an important requirement of the anionic surfactant, anionic groups like sulfates and sulfonates are commonly used in acid–anionic sanitizers.

Whereas these preparations do not possess the high bacteriostatic activity of quaternary ammonium germicides, they have the alternate advantage of being rapidly functional in acid solution. In comparative experiments of several different disinfectants, the acid–anionic killed bacteria at lower concentration than five other disinfectants. Only sodium hypochlorite and an iodine product were effective at higher dilution than the acid–anionic. By the AOAC use dilution test, the acid–anionic killed *Pseudomonas aeruginosa* at 225 ppm, *Salmonella choleraesuis* at 175 ppm, and *Staphylococcus aureus* at 325 ppm (172).

The acid–anionics are particularly valuable in the dairy industry, where they have had the greatest acceptance. The acid helps remove calcium deposit from milk (milkstone), thus enabling the anionic to do a more efficient job of destroying the bacteria. The action in killing bacteria is rapid, approximately 30 seconds in the absence of protein. Protein increases the kill time several fold. The phosphoric acid surfactant sanitizer has been shown to have virucidal activity against *Streptococcus cremoris* bacteriophage, which destroys the lactic acid streptococci used as starter cultures in the dairy industry. The phage was inactivated in 15 seconds in hard water and distilled water (173). It is important to eliminate these phages as they interfere with the cheesemaking process. The detergency of acid–anionic surfactants is useful for cleaning equipment in dairies and food plants, and they are not corrosive to the stainless steel equipment used in these plants. Because they contain common detergents and food-grade acid, the sanitizers have very low toxicity, and are used to disinfect cows' udders prior to milking as a precaution against mastitis infection. Foaming may be a problem, but new low-foam acid–anionic surfactants have been introduced (174,175). Another use proposed for these sanitizers is in hard surface disinfection of walls, floors, and equipment in hospitals, nursing homes, schools, hotels, restaurants, etc, where the surfaces are not adversely affected by the acid solution.

Amphoteric Surfactant Disinfectants. Amphoteric surfactants (ampholytes) differ from cationic and anionic surfactants because they ionize in water to give zwitterions, ie, ions with both a positive and a negative charge in the same molecule, and, depending on pH, may act more like anions or cations. They are typically amino acids (qv) with a long-chain alkyl group, and are more basic than acidic in nature because they have two or three amine groups but only one carboxylic acid group. As in the case of the other surfactants, the long-chain hydrocarbon group donates the hydrophobic part of the molecule and the surface activity. In some disinfectant preparations the active ingredient is a mixture of two amphoteric compounds, or a mixture of an amphoteric and a long-chain amine. Increasing the amine nitrogens in the molecule from 1 to 3 increases the bactericidal activity in derivatives of dodecylglycine. With 1 amine nitrogen, bactericidal action against *Staphylococcus aureus* took 10 min; with 2 nitrogens, 5 min; and with 3 nitrogens, 1 minute. The optimum pH is also dependent on the basicity of the ampholyte. In the compound having one nitrogen, the optimum pH was 3, for 2 nitrogens it was 6, and for 3 nitrogens it was 9 (176).

Whereas the ampholyte germicides are more closely related to the quaternaries than they are to the acid–anionics, there are some distinct differences. The quats have greater activity toward gram-positive bacteria than gram-negative, but the ampholytes are less selective in this regard. The quats are more adversely affected in the presence of protein than are the ampholytes (177,178). On the other hand, quats are bacteriostatic at much lower concentrations than the ampholytes. In some cases, this disadvantage of the ampholytes can be an advantage, as in the case of processing foods like cheese, beer, and wine which are produced with the aid of microorganisms and the residual disinfectant in the equipment can deter the growth of the organisms. Using ampholytes, rinsing is not as important, because the toxicity is low, and there is no effect on the organisms when there is ordinary dilution.

Microbiological results for ampholytes tend to be varied. For example, in one study Tego 51, ie, 10% dodecyl di(aminoethyl)glycine chlorohydrate [*6843-97-6*], pH 7.7 (Table 7), has been found to kill a number of different gram-positive and gram-negative bacteria with a 1% solution in 30 min and a 10% solution in 1 minute. In another study (180) it was found that 1% Tego 51 killed bacteria in 10 minutes. Two other ampholytes at 1% did not kill most of the organisms tested in 10 min, and the study concluded by comparison with other disinfectants, that the bactericidal properties of the two ampholytes were slight. In other studies, another ampholyte (Anon 300) was found to be active both bacteriostatically and bactericidally (181,182), and Tego 51 killed *M. tuberculosis* in 2.5 min and *M. bovis* in 10 minutes (183).

Table 7. Composition of Tego Disinfectants

Tego	Active ingredient[a]	Composition[b], %	pH[c]
103S	$RNH(CH_2)_2NH(CH_2)_2NHCH_2COOH \cdot HCl$	15	7.7
103G	$RNH(CH_2)_2NH(CH_2)_2NHCH_2COOH \cdot HCl$ + $[RNH(CH_2)_2]_2NCH_2COOH \cdot HCl$	10	7.7
51	$RNH(CH_2)_2NH(CH_2)_2NHCH_2COOH$ + $RNH(CH_2)_3NHCH_2COOH$	9	8.2
51B	$RNH(CH_2)_2NH(CH_2)_2NHCH_2COOH$ + $RNH(CH_2)_3NHCH_2COOH$	22.5	8.2
2000	$RNH(CH_2)_3NHCH_2COOH$ + $RNH(CH_2)_3NH_2$	20	8.0

[a]R = $C_{12}H_{25}$ or $C_{14}H_{29}$.
[b]Aqueous solution.
[c]Values given are approximate.

The ampholytes, which must be used in higher concentrations than many other disinfectants when employed at room temperature, are greatly improved as the temperature is increased. At 20°C, 3250 ppm of Tego 51 killed *Pseudomonas aeruginosa* in 30 min, at 50°C it required only 100 ppm in the same time (184,185).

Whereas tests (186) indicated that ampholytes were effective in skin cleansing for preoperative use, for wound cleansing, and as an antiseptic in the oral cavity (187), as well as other medical applications, the food and beverage industries have proved to be the principal employers of these compounds. Ampholytes are used as sanitizers and disinfectants, not as food preservatives. Low toxicity, absence of skin irritation, and noncorrosiveness, along with antimicrobial activity, has given ampholytes acceptance in dairies, meat plants, and the brewing and soft drink industries. These disinfectants have been manufactured and distributed in Europe and Japan, but not in the United States.

Chelating Agents. 8-Hydroxyquinoline [*148-24-3*] (8-quinolinol, oxine) might be thought to function as a phenol, but of the 7 isomeric hydroxyquinolines only oxine exhibits significant antimicrobial activity, and is the only one to have the capacity to chelate metals. If the hydroxyl group is blocked so that the compound is unable to chelate, as in the methyl ether, the antimicrobial activity is destroyed. The relationship between chelation and activity of oxine has been investigated (188–190). Oxine itself is inactive, and exerts activity by virtue of the metal chelates produced in its reaction with metal ions in the medium (see CHELATING AGENTS) (Fig. 5). Used by itself or as the sulfate [*134-31-6*] (Chinosol) or

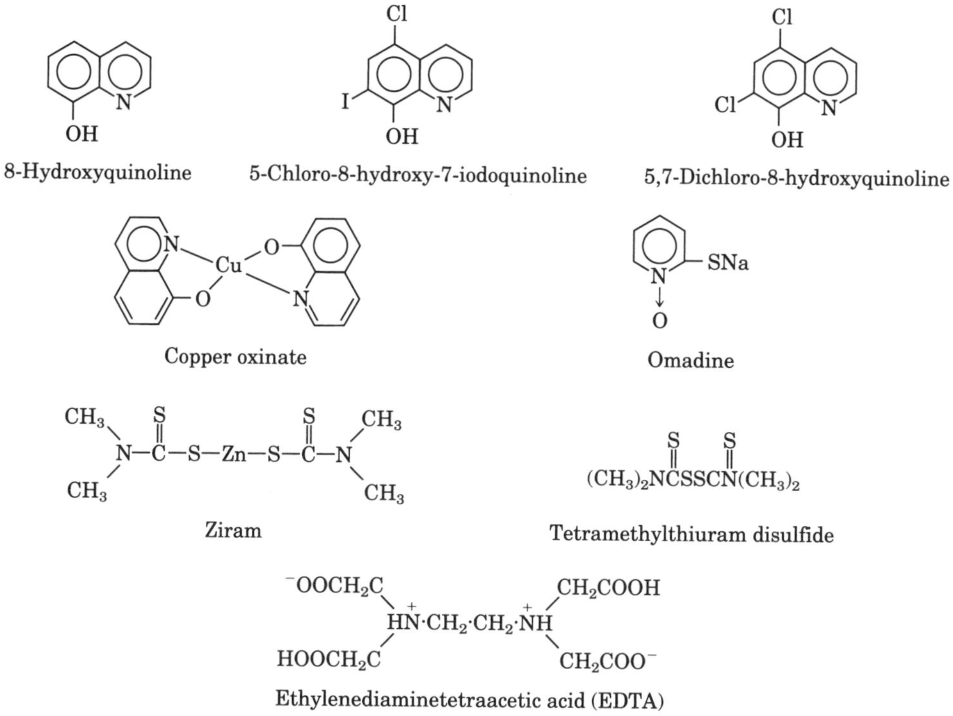

8-Hydroxyquinoline   5-Chloro-8-hydroxy-7-iodoquinoline   5,7-Dichloro-8-hydroxyquinoline

Copper oxinate   Omadine

Ziram   Tetramethylthiuram disulfide

Ethylenediaminetetraacetic acid (EDTA)

Fig. 5. Chelating agents useful as antimicrobials.

benzoate in antiseptics, the effect is bacteriostatic and fungistatic rather than microbiocidal. Inhibitory action is more pronounced upon gram-positive than gram-negative bacteria; the growth-preventing concentrations for staphylococci being 10 ppm; for streptococci 20 ppm; for *Salmonella typhosa* and for *E. coli* 100 ppm (191,192). However, a 1% solution requires at least 10 hours to kill staphylococci and 30 hours for *E. coli* bacilli. The oxine benzoate [*86-75-9*] was the most active antifungal agent in a series of 24 derivatives of quinoline tested. A 2.5% preparation of this compound was successful in treating dermatophytosis (193,194). Iron and cupric salts were found to prolong the antibacterial effect of oxine on teeth (195).

Certain halogen derivatives of 8-hydroxyquinoline have a record of therapeutic efficacy in the treatment of cutaneous fungus infections and also of amebic dysentery. Among these are 5-chloro-7-iodo-8-quinolinol [*130-26-7*] (iodochlorhydroxyquin, Vioform), 5,7-diiodo-8-hydroxyquinoline [*83-73-8*] (diiodohydroxyquin), and sodium 7-iodo-8-hydroxyquinoline-5-sulfonate [*885-04-1*] (chiniofon) (196–198).

Copper 8-quinolinolate [*10380-28-6*] (copper oxinate), the copper compound of 8-hydroxyquinoline, is employed as an industrial preservative for a variety of purposes, including the protection of wood and textiles against fungus-caused rotting, and interior paints for food plants. It has 25 times greater antifungal activity than oxine (199). More recent concern that hydroxyquinolines are carcinogenic has resulted in reduced interest in this group.

Another compound, the antimicrobial action of which is associated with chelation, is 2-pyridinethiol-*N*-oxide [*3811-73-2*] (Omadine). Activity has been shown to depend on coordinating property. The iron chelate is active, but not the free pyridine compound (200). In the form of its zinc chelate it is found in shampoos to control seborrheic dermatitis (201). Other applications of this useful chemical include preservation of adhesives, plastics, latex paints, polyurethane foam, and metal working fluids (202).

Many compounds capable of chelation have been tested for antimicrobial properties. Those showing positive results include salicylaldoxime [*94-67-7*], 1-nitroso-2-naphthol [*131-91-9*], mercaptobenzothiazol [*149-30-4*], dimethylglyoxime [*95-45-4*], salicylaldehyde [*90-02-8*], cupferron [*135-20-6*], phenanthroline [*66-71-7*], isoniazid [*54-85-3*], thiosemicarbazones, the sulfur analogue of oxine, and numerous antibiotics (qv) including tetracyclines. Whether these compounds function exclusively, partially, or at all by virtue of their ability to chelate is open to debate.

Whereas many active compounds have been found, their usefulness depends on many factors, such as toxicity, stability, solubility, color, odor, manufacturing problems, and cost. Among the compounds capable of chelation that have proved especially useful are the dithiocarbamates, which for years have controlled fungus diseases on tomatoes, potatoes, and other plants. They also have been used as industrial preservatives and sodium-*N*-methyldithiocarbamate [*137-42-8*] is an effective soil fumigant (see FUNGICIDES, AGRICULTURAL). Some dithiocarbamates have been found to exert powerful inhibitory action upon fungi pathogenic for man, ie, *Trichophyton mentagrophytes*, *Epidermophyton floccosum*, *Microsporum audouini*, and *Torula histolytica* (see ANTIPARASITIC AGENTS). Bacteria, especially those of the gram-positive variety, were also found to be susceptible, as illustrated by disodium bisdithiocarbamate [69943-51-7] which has an inhibitory concentration for *Staphylococcus aureus* of 1:100,000. The corresponding dilutions for the gram-negative *Salmonella typhosa* and *Escherichia coli* are 1:5,000 and 1:10,000, respectively (203). In another investigation correlating the chemical structure of dithiocarbamic acid derivatives with their *in vitro* antibacterial and antifungal activity, it was found that within each category of dithiocarbamates, thiuram monosulfides and methyl and ethyl esters were the most active, whereas the higher alkyl derivatives were comparatively inactive (204). Tetramethyl-thiuram disulfide [137-26-8] (TMTD, thiram) (see Fig. 5) was the most active of the compounds tested against pathogenic fungi as well as against bacteria; *Streptococcus pyogenes*, *Streptococcus faecalis*, and *Staphylococcus aureus* show a markedly greater susceptibility to its action than *Escherichia coli* and *Pseudomonas aeruginosa* (205). Thiram has been used in disinfectant soaps.

Ethylenediaminetetraacetic acid (EDTA), an important chelating agent, is not considered a bactericide in its own right, as it has generally no effect on gram-positive bacteria. However, it can act alone or with other agents to cause lysis of some gram-negative bacteria, eg, *Pseudomonas aeruginosa*. It potentiates the activity of chemically unrelated antibacterial compounds against gram-negative bacteria. It does so by chelating cations in the outer membrane of the bacteria, thus disrupting the membrane, and causing the release of lipopolysaccharide (206). It has no effect on yeast or mold. EDTA is reported to increase the activity of benzalkonium chloride to gram-negative organisms and prevent the growth of *Pseudomonas aeruginosa* resistant to this germicide (207). EDTA is also reported to similarly potentiate the action of chloroxylenol (208).

Biguanidine Compounds. *Chlorhexidine.* Chlorhexidine [55-56-1] (1,6-di(4-chlorophenyl-diguanidino)hexane, Hibitane), a leading skin antiseptic and a cationic bisguanide, is a strong base that reacts with acids to form salts, most of which are water-insoluble. However, the salt with gluconic acid is soluble and is dispensed as a 20% solution that is colorless and odorless. Chlorhexidine is moderately surface active, stable in the range of pH 5 to 8, and is generally compatible with other cationic germicides. Some disinfectant products incorporate both chlorhexidine and a quaternary, eg, Presept Liquid which has 1.25% chlorhexidine gluconate and 0.1% quaternary ammonium compounds in 70% alcohol. Chlorhexidine is incompatible with organic anions such as soaps, anionic detergents, and many dyes. Inorganic anions also precipitate this chemical in all but very dilute solutions. Solutions of less than 1% chlorhexidine are stable to autoclave sterilizing temperature of 121°C for 15 minutes.

chlorhexidine   Vantocil IB

Unlike the quats, chlorhexidine is highly active against both gram-positive and gram-negative bacteria; yeasts and fungi are also sensitive to this agent which has high activity against the lipid viruses, including many respiratory viruses, herpes, cytomegalovirus, and HIV, but not the hydrophilic viruses, such as poliovirus and the enteric viruses. Chlorhexidine is inactive against spores except at high temperatures. Acid-fast bacteria are inhibited but not killed by aqueous chlorhexidine solutions, but are killed by alcoholic solutions (209). Prolonged use of the drug has not led to the development of resistant bacteria (210). The bacteriostatic and bactericidal activities of chlorhexidine have been studied (Table 8). Virucidal activity toward 22 viruses found activity toward 17 at a concentration range of 0.001–0.25% (209). Regarding activity against spores, chlorhexidine inhibits spore germination, but at 25 ppm it is not sporicidal at 20–37°C after 2 hours exposure; at 70°C, there was a 5 log reduction in number of spores (211).

The optimum pH for chlorhexidine antimicrobial activity is in the range 5.5–7.0 but varies with the buffer used and the organism, having a range of pH 5–8. With *S. aureus* and *E. coli*, activity increases with increased pH; the reverse occurs with *P. aeruginosa*.

Chlorhexidine achieved its outstanding position as a skin disinfectant because of its apparent low toxicity to most body parts, absence of skin irritation, and ability to reduce hospital nosocomial skin infections (212,213). Its low toxicity was demonstrated when used on the intact skin of newborns (214), but in the U.S. restrictions are made for its use around ears. In early studies (215), 4% chlorhexidine handwash was found to be as effective as hexachlorophene in reducing numbers of bacteria on the hands, and in maintaining the low numbers for several hours under surgical gloves. Povidone iodine gave less reduction and allowed survivors to increase dramatically during operations. In a two minute wash, 4% chlorhexidine gave an 86% reduction in skin flora compared to 68% with iodophor and 47% with hexachlorophene. The benefit of a chlorhexidine handwash for hygienic handwashing by hospital personnel in preventing transfer of organisms between patients was demonstrated for 47 subjects whose hands were intentially contaminated. After washing with chlorhexidine skin cleanser for 15 seconds, a 99.9% reduction of transient contaminants was achieved (216). Chlorhexidine is sometimes combined with alcohol to achieve the rapid reduction obtained with alcohol and the residual effect of chlorhexidine. It is substantive to the skin, and repeated use results in lower levels of bacteria on the skin.

Chlorhexidine has found other medical applications, eg, in urology in preventing urinary tract infections (217), in obstetrics and gynecology (218), in controlling infection in burns and wounds (219), and in the prevention of oral disease

Table 8. Bacteriostatic and Bactericidal Activities of Chlorhexidine[a,b]

Number and type of organism	Bacteriostatic activity, ppm	Bactericidal activity, log reduction[c]
18 gram-positive bacteria	0.5–38 (128)[d]	1.1–5.8 (0.3)
28 gram-negative bacteria	0.5–64 (102 and 115)	2.9–6.4 (1.8)
24 fungi/yeast	1.0–16 (32 and 64)	1.6–4.4 (0.5)

[a]Ref. 209.
[b]Values in parentheses are extreme variances.
[c]10 min exposure.
[d]*Lactobacillus casei.*

(220). Hypersensitivity to chlorhexidine has been reported in Japan (221) but 0.05% concentration is considered to be safe.

Polyhexamethylene Biguanide. Another important biguanide disinfectant is polyhexamethylene biguanide [*76770-08-6*] (PHMB, Vantocil IB). This compound is a polymer containing a spread of polydispersed oligomers of molecular weight between 500 and 6000, for which the tetramer is the predominant species. Each oligomer can have an amine or a cyanoguanidine group at either end position.

Like chlorhexidine, PHMB is classed as a membrane-active antimicrobial agent, along with the quaternary ammonium compounds, phenols, alcohols, and anilides. They all induce leakage of low molecular-weight components from bacterial cells because of a change in permeability arising from a change in the integrity of the cytoplasmic membrane. The negative charge of the bacterial cell is neutralized and then reversed.

PHMB is one of the few biologically active synthetic polymers. It is active against most gram-positive and gram-negative bacteria, yeasts, many fungi, and aquatic algae, but is not active against Mycobacteria and bacterial spores (222). It is water-soluble, and is used as a solid surface disinfectant at concentrations of 1000–2000 ppm. It is registered with EPA for the control of pathogenic bacteria in bottle washwater, cannery cooling water, synthetic adhesives, leather processing liquors, and metalworking fluids, and has also been proposed for the preservation of food and cosmetics.

Anilides. Salicylanilide [*87-17-2*], the product of salicylic acid and aniline, was developed in 1930 under the name Shirlan for preventing mildew on stored and shipped woolen goods. It has been used as a mildewcide in plastics, paints, lacquers, leather, and paper (223,224). Later, halogenated salicylanilides (Fig. 6) were introduced as antimicrobial soap additives for degerming skin. Tetrachlorosalicylanilide [*1322-37-8*] (Anobial), 3,4′,5′-tribromosalicylanilide [*87-10-5*] (tribromosalan), and dibromofluoromethylsalicylanilide [*4776-06-1*] (fluorosalan) were effective as antimicrobials (225). However, these were found to cause primary dermatitis, with 53 cases of photosensitization (226–232), and were classed as not recommended as safe and effective by the FDA in 1974 (233). Whereas the salicylanilides contain a phenol moiety, this cannot entirely explain the activity of these compounds because carbanilide [*102-07-8*] (diphenylurea) and its halogenated derivatives, which have the anilide grouping without the phenol, are also

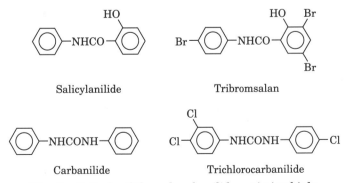

Fig. 6. Salicylanilide and carbanilide antimicrobials.

excellent antimicrobials (234). The trichlorocarbanilide [*101-20-2*] (triclocarban) was found to be effective *in vivo* against *Staphylococcus aureus* and *Cornebacterium minitissimus*. This agent was incorporated in a bar soap (235), the use of which was permitted by the FDA (233). Salicylanilides and carbanilides have been reviewed (225).

Nitrogen Heterocycles

Imidazole and imidazoline derivatives have provided some useful antimicrobial compounds (Fig. 7). Metronidazole [*433-48-1*] (2-methyl-5-nitroimidazole-1-ethanol) inhibits the growth of pathogenic protozoa, *Trichomonas vaginalis*, and *Eistamoeta histolyticum* in urogenital infections, and is effective against infections resulting from anaerobic bacteria and facultative anaerobes. Other derivatives, ie, clotrimazole [*23593-75-1*], miconazole [*22916-47-8*], econazole [*27220-47-9*], and ketoconazole [*65277-42-1*], demonstrate a broad antimycotic spectrum of activity (236). Glyodin [*556-22-9*] (2-heptadecyl-2-imidazoline acetate) protects apple and cherry trees from leaf spot disease. It functions as a cationic surfactant (237). Dantoin [*6440-58-0*] (DMDMH-55, Glydant (1,3-di(hydroxymethyl)-5,5-dimethylhydantoin)) has a wide spectrum of activity against bacteria and fungi at 250–500 ppm. It is water-soluble and active over a wide pH range. It can be used as a preservative for resins, adhesives, and emulsions (238). It may act through the slow release of formaldehyde. Germall 115, *N,N'''*-methylene bis(5'-(1-hydroxylmethyl)-2,5-dioxo-4-imidazolidinylurea) [*39236-46-9*], is an antimicrobial preservative tested for the cosmetic industry. It is more active against bacteria than fungi. It is claimed to be nontoxic, compatible with emulsions and proteins, and to act synergistically with other preservatives (239). Germall II, diazolidinyl urea [*78491-02-8*], is reported to have a superior antimicrobial spectrum to that of Germall 115, but only marginally better antifungal activity (240). Concentrations of 0.1–0.3% are recommended.

Some triazines and other nitrogen heterocycles are produced by reacting amines and formaldehyde and may exert antimicrobial activity by virtue of the slow release of formaldehyde. These compounds find use as preservatives for latex paints, adhesives, and cosmetic products. Grotan BK, Onyxide 200, Triadine 3, Busan 1060, and Bioban GK are trade names for 1,3,5-tris (2-hydroxyethyl) hexahydro-s-triazine [*4719-04-4*]. A mixed oxazole compound, Nuosept 95, is recommended for cutting oils, starch, and cellulose-based products. Figure 7 shows 5-hydroxymethyl-1-aza-3,7-dioxabicyclo(3.3.0)octane [*56709-13-8*], one of three cyclo-octanes in the commercial mixture. Dowicil 200, 1-(3-chloroallyl)-3,5,7-triaza-1-azonia-adamantane chloride [*4080-31-3*], is active against bacteria and fungi at 50–400 ppm, and is used to control their growth in shampoos and cosmetics (241).

Mercaptobenzothiazole [*149-30-4*] is an effective agent against antibacterial and antifungal properties. It is used in combination with dimethyldithiocarbamate (Vancide 51) as a general industrial preservative. A related compound, Busan 1030, 2-(thiocyanomethythio)benzothiazole, is a preservative for caulking compounds, vinyl acetate wallcovering adhesives, particle board, and various paints and protective coatings. It is also sold as a protectant for lumber against sapstain and mold.

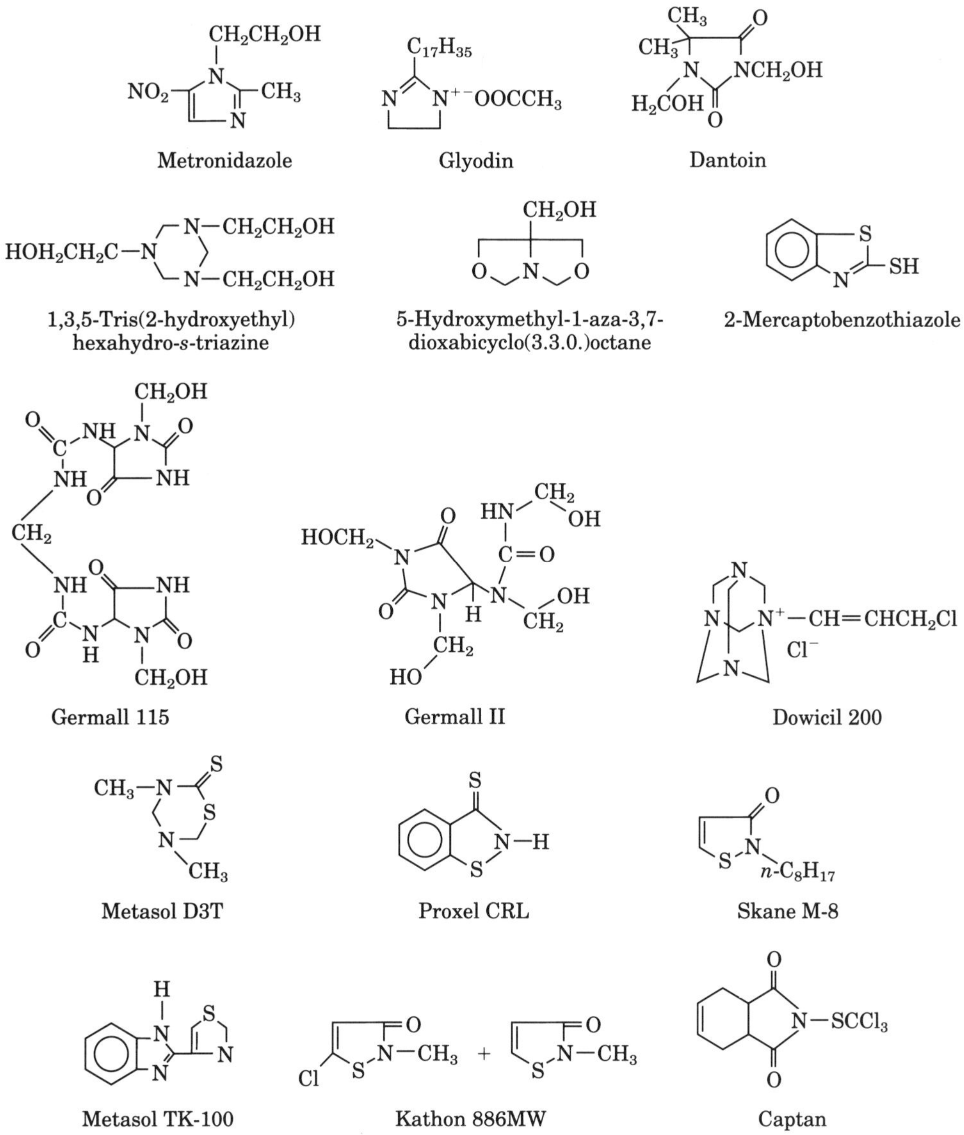

Fig. 7. Nitrogen heterocyclics.

Mylone [533-74-4] (tetrahydro-3,5,dimethyl-2H-1,3,5-thiadiazine-2-thione), a preservative with various trade names, ie, Metasol D3T, Busan 1058, and Biocide-N-521, has a range of suggested applications in leather, paint, glue, casein, starch, pigment slurries, and paper manufacture. Antimicrobial thiazole compounds include Proxel CRL, 1,2-benzoisothiazolin-3-one [2634-33-5]; Kathon 886 (Kathon CT, Kathon CG), a mixture of 5-chloro-2-methyl-3(2H)-isothiazalone and 2-methyl-3(2H)-isothiazalone [2682-20-4]; Skane M-8, 2-n-octyl-4-isthiazolin-3-one [26530-20-1]; and Metasol TK-100, (2-(4-thiazolyl)benzimidazole) [148-79-8] (see Fig. 7). These chemicals are used as preservatives for paint, leather, fabrics, hydraulic fluids, cooling tower water, etc (242).

Captan, N-trichloromethylthio-4-cyclohexene-1,2-carboximide [133-06-2] (Fig. 7), ranks second to the dithiocarbamates in importance as an organic fungicide in agriculture (see FUNGICIDES, AGRICULTURAL). It has a broad spectrum of action, and is safe to spray on many crops. Its principal use is against diseases of fruit trees, and it has also been used to prevent spoilage of stored fruit. In nonagricultural applications, captan has been used in the treatment of skin infections in humans and animals, and as an industrial preservative in paint, plasticizers, and lacquer. Related compounds are Folpet, N-(trichloromethylthio) phthalimide [133-07-3], and Difolatan, cis-N-[1,1,2,2-(tetrachloroethyl)thio]cyclohexene-1,2-dicarboximide [2425-06-1] (243).

Sulfur Compounds

Sulfur has long been known for its properties as a pesticide and a curative agent. Homer spoke of the pest-averting sulfur as far back as 800–1000 BC, Hippocrates (400 BC) considered sulfur as an antidote against plague, and Dioscorides (100 AD) used sulfur ointment in dermatology (244). In 1803, the use of a lime–sulfur protective treatment for fruit trees was reported, and in 1850 sulfur dust was used to protect foliage (245). In 1891 sulfur dust was used on soil to control onion smut (246).

Sulfur is selective in its antimicrobial activity, and there is debate whether it has any toxicity of its own or if it acts through chemical conversion to some other form. However, colloidal sulfur, a mixture of 70% sulfur and 30% polythionic acids, has a phenol coefficient against common gram-positive and gram-negative bacteria of 1.5–5, and against plant pathogens of 4–276 (247). Colloidal sulfur has been reported fungicidal to *Tinea corporis* and *Trichophyton interdigitale* at concentrations of 1%, but not to *Monilia tropicalis*, even at 5% (248).

Sulfur dioxide, sulfites, and metabisulfites have had extensive use as antimicrobial preservatives in the food industry. In pharmaceuticals they have had a dual role, acting as preservatives and antioxidants. The sulfa drugs, or sulfonamides, the first effective chemotherapeutic agents to be employed systemically for the prevention and cure of bacterial infections, have a wide range of activity against gram-positive and gram-negative organisms in a concentration of about 50–100 ppm (see ANTIBACTERIAL AGENTS, SYNTHETIC—SULFONAMIDES). There is a generally direct correlation between their efficacy *in vitro* and *in vivo*. The sulfas are bacteriostatic in the body; in some cases, at very high concentration, they are bactericidal. Unlike the more highly active antibiotics that have largely replaced them, however, sulfas are inhibited by blood, pus, and tissue breakdown products. Sulfas are employed to some degree in combination with antibiotics or other drugs. There have been thousands of sulfa derivatives synthesized and tested; a few of historical interest are sulfanilamide [16-74-1], sulfapyridine [144-83-2], sulfaguanidine, sulfathiazole [72-14-0], and sulfadiazine [68-35-9] (249).

The toxicity of sulfur to microorganisms is taken advantage of in several industrial biocides. Methylene bisthiocyanate [6317-18-6] (Metasol TK, Biosperse 284) has powerful action against slime-forming bacteria, spore formers, and fungi, and is used in paper mills and water cooling systems, where high activity and low cost are primary considerations. It is sometimes used in combination with 2-(thiocyanomethylthio)benzothiazole to broaden the microbial spectrum. Other sulfur-containing slimicides are disodium cyanodithioimidocarbonate [138-93-2], 2-hydroxypropyl methanethiosulfonate (Busan 1005) [30388-01-3], and methylenebis(butanethiolsulfonate) [16008-32-5]. Thiosulfonates, as a group, are active antimicrobial agents with the activity in the divalent sulfur (250), but the effect of strong electronegative groups, as in RSO_2SCCl_3 and RSO_2SCF_3, markedly increase the activity (251). The effect of the electron-withdrawing chlorines is also demonstrated in hexachlorodimethyl sulfone (Stauffer N-138b) [3064-70-8], which has been used to control corrosion-causing sulfate-reducing bacteria in oil well flooding systems (252). It should be noted that sulfones without the adjacent halogens are generally inactive toward microorganisms.

$$\underset{\text{methylene bisthiocyanate}}{NCS-\underset{\overset{\displaystyle |}{H}}{\overset{\overset{\displaystyle H}{|}}{C}}-SCN} \qquad\qquad \underset{\text{hexachlorodimethyl sulfone}}{Cl_3C-\underset{\overset{\displaystyle \|}{O}}{\overset{\overset{\displaystyle O}{\|}}{S}}-CCl_3}$$

Metal Compounds

Metal compounds, particularly compounds of the heavy metals, have a history of importance as antimicrobial agents. Because of regulations regarding economic poisons in the environment they are no longer widely used in this application. Mercury, lead, cadmium, uranium, and other metals have been implicated in cases of poisoning that resulted in government response. The metals whose compounds have been of primary interest as antimicrobials are mercury, silver, and copper.

Mercury.   Mercury is remarkable for its great bacteriostatic activity. An organic mercurial, phenylmercuric nitrate (merphenyl nitrate) [55-68-5] showed inhibition of S. aureus at 0.1 ppm and of E. coli at 3 ppm. If only a small inoculum is used, inhibition at 0.003 ppm for S. aureus and 0.25 ppm for E. coli is noted (253). Dilutions of 1–2 ppm inhibit the growth of common molds. But mercurials are less powerful than microbiocides; 5 ppm of a mercurial is reported to be lethal to S. aureus, and 20 ppm to E. coli (254). Against pathogenic spores, 750 ppm was not reliably lethal (255). Mercuric chloride [7487-94-7] was reported in 1881 to have killed anthrax spores, and was the only compound that killed them in such great dilution (256). Some years later, mercuric chloride was shown to be a poor germicide, although it had an almost unbelievable bacteriostatic property (257). The mercury-treated spores were actually only inhibited, and when treated with sulfide to neutralize the mercury, they revived and grew.

Organic mercurials with popular names like mercurochrome [129-16-8], metaphen [133-58-4], and merthiolate (thimerosal) [54-64-8], appeared in the decade after World War I. Mercurochrome was only one-tenth as active as mercuric chloride, but merthiolate was 2–10 times more active, and phenylmercuric nitrate was even greater in activity. Using Staphylococcus aureus and Pseudomonas aeru-

ginosa, merthiolate and three phenylmercurials were much more active against *S. aureus* than was mercuric chloride, but there was no significant difference in regard to *Ps. aeruginosa* (258).

mercurochrome   metaphen   merthiolate

Sulfides, thiols, and proteinacious organic matter, particularly plasma and whole blood, seriously depress and may even abolish the germicidal action of mercury compounds (qv). As of this writing approved uses for mercurials are limited to contact lens cleaning fluids, spoilage prevention of stored water-based paints, mildew control in finished paints, and a retardant for sapstain in lumber. It is uncertain, however, whether even these applications will be permitted to continue. A comprehensive review of mercurial antimicrobials is available (259).

Silver.   Silver [7440-22-4] vessels were used to store water dating back to 4000 BC. Workers in 1869 (260) began scientific investigation of the antimicrobial properties of silver. The outstanding bactericidal properties of highly diluted silver metal, as apart from silver ions in salts like silver nitrate [7761-88-8], was termed oligodynamic action. The prevention of bacterial eye infection by gonorrhea in newborns with the use of 1% silver nitrate became a medical milestone. And the use of silver nitrate (261) and later silver sulfadiazine [21548-73-2] to prevent infection in burn patients (262,263) was a major advance in burn therapy.

Silver nitrate is astringent and a protein precipitant, which is not medically desirable. Other forms of silver have been used to avoid this problem, including colloidal silver, silver-protein preparations, and finely divided silver metal called Katadyn silver.

Introduced in 1928, Katadyn silver consisted originally of metallic silver in spongy form with an added activating metal below silver in the electromotive series, such as gold or palladium. When deposited on filter elements, sand, etc, Katadyn silver slowly but continuously reduced the bacteria count of *Escherichia coli* from 500,000 to 0 within several hours. More intensive antibacterial performance was delivered by the Electro-Katadyn process, which has been recommended for the purification of water in swimming pools. This process operates by means of a battery of silver plates with a direct current induced across them (264). Unfortunately, some important pathogens, including *Staphylococci*, appear to resist the action of ionic silver. A more recent modification of Katadyn is Movidyn; it is claimed to possess a wider bactericidal spectrum that includes staphylocidal action (265).

Inhibitory concentration for electrically generated silver using groups of bacteria had a range of 0.03 to 1.03 ppm, and lethality ranged from 0.26 to 10.5. For different pathogenic yeasts, the mean inhibitory concentration (MIC) ranged from 0.5 to 4.7 ppm, and lethality from 1.9 to 13.8 ppm. For silver sulfadiazine, the MIC of a large number of different bacteria varied from <0.78 to 100 ppm (266).

Against *M. pyogenes*, the lethal concentrations of silver compounds were approximately: lactate 1%, citrate 0.2%, nitrate 1%, proteinate 6%. Against *S. typhosa*, the respective lethal concentrations were roughly: lactate 0.2%, citrate 0.02%, nitrate 0.25%, and proteinate 0.5–1% (267). Penicillin and other antibiotics have since replaced the caustic silver nitrate for anti-infection treatment of the eyes of newborn infants, but silver, which is effective against both *Pseudomonas aeruginosa* and other gram-negative bacteria associated with burned tissue, is still employed in that therapy. Silver also may be expected to continue to find use in water purification.

Copper.   Copper, like mercury and silver, has a long history of use as an antimicrobial agent. Its use has been mainly as a fungicide and algicide. Copper salts were used in agriculture in 1761 (268), and the salts and water in which metallic copper had been left were shown to prevent germination of the spores of the bunt fungus (see COPPER COMPOUNDS). In 1767 the use of copper sulfate as a wood preservative was recommended (269). A mixture of copper sulfate and lime was introduced in 1885, and became an important treatment of plants for the prevention of many devastating diseases such as late blight of potato, responsible for mass starvation in Ireland in 1845–1846. Copper fungicides for plants have been largely replaced by organic compounds like the dithiocarbamates. The cupric ion is algicidal at 0.5–2.9 ppm, and is used to prevent algae growth in swimming pools. Copper compounds were used by the military and industry during World War II to treat tents, ropes, tarpaulins, and gun covers to prevent rotting; copper is highly toxic to the fungi that degrade cotton cellulose. Copper naphthenate [*1338-02-9*] and copper-8-quinolinolate [*10380-28-6*], $C_{18}H_{12}CuN_2O_2$, are still used for these treatments (270). The latter has also proved to be an effective mildewcide with many applications (271).

Other Metals.   Other metals having antimicrobial properties include zinc, chromium, arsenic, boron, and tin. Zinc, as zinc oxide [*1314-13-2*], has been used in paint to enhance mildew resistance, and zinc chloride [*7646-85-7*] is a wood preservative. Chromated zinc chloride and copper chromate are wood preservatives that combine zinc and copper with chromium. Arsenic, in the form of lead arsenate, was formerly used as a fungicide–insecticide for protecting plants. Sodium arsenate is a wood preservative; an organic arsenical, oxybisphenoxyarsine [*58-36-6*] (Durotex) (272), is effective in combating fungi in adhesives, wall paper, paint, plastics, and emulsions (see ARSENIC COMPOUNDS).

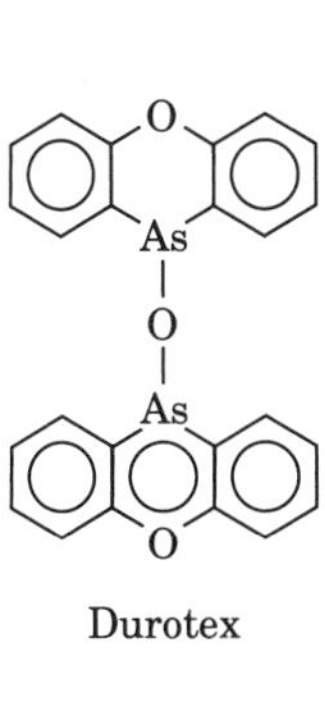

Durotex

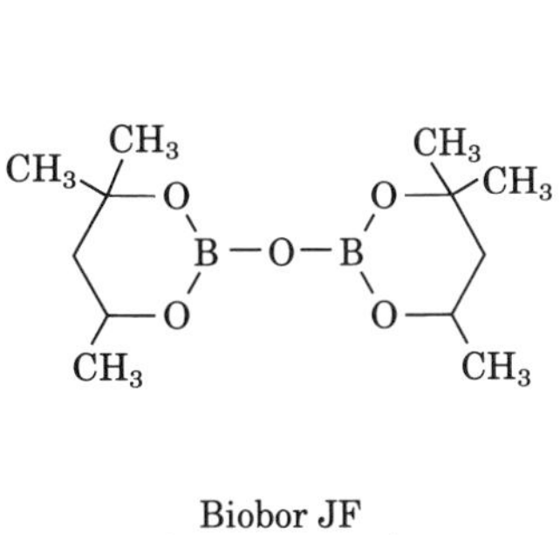

Biobor JF
(one structure)

Boron, as barium metaborate, is marketed as a mildew preventative for paints (273). Borax is used as a wood preservative, and an organic boron, 2,2′-(1-methyltrimethylenedioxy)-bis(4,4,6-trimethyl)-1,3,2-dioxaborinane (Biobor JF) [*14697-50-80*] is a biocide for jet fuel (274). Whereas tin metal is used to coat steel cans used as food containers, organic tin in the form of tributyl tin compounds have proven to be powerful antimicrobials, and have found use in antifouling coatings for ship bottoms, paints, and wood preservatives (275).

Table 9 gives a comparison of the antibacterial activity of a number of metal salts and nonmetals (276,277) in the form of germicidal concentrations.

Table 9. Minimum Germicidal Concentrations of Metal Salts and Nonmetals, 10 min

Compound	*Salmonella typhosa*	*Staphylococcus aureus*
	Metal salts	
$HgCl_2$		1:16,000
$AgNO_3$		1:100
$MnSO_4$		<1:10
$ZnCl_2$	1:8	
ZnI_2	1:30	1:10
$Zn(C_2H_3O_2)_2 \cdot 2H_2O$	1:3	
$ZnSO_4 \cdot 7H_2O$	1:3	
$NiCl_2 \cdot 6H_2O$	1:5	1:3
$Ni(NO_3)_2 \cdot 6H_2O$	1:3	1:3
$CoCl_2 \cdot 6H_2O$	1:20	1:6
$Co(NO_3)_2$	1:5	1:5
$CoCl_2 \cdot 2H_2O$	1:300	1:7
$Co(NO_3)_2 \cdot 6H_2O$	1:50	1:6
$Pb(NO_3)_2$	1:3	
$Pb(C_2H_3O_2)_2 \cdot 3H_2O$	1:3	
$SnCl_2 \cdot 2H_2O$	1:350	1:70
$SnCl_4$		1:50
$FeCl_3 \cdot 6H_2O$	1:200	1:10
$FeCl_2$		<1:10
$Fe_2(SO_4)_3$	1:500	1:30
$FeSO_4$		<1:10
	Nonmetals	
phenol		1:60
cresol		1:300
hexylresorcinol		1:6,000
iodine		1:20,000

Gaseous Sterilants

Ethylene Oxide. Ethylene oxide [*75-21-8*] (qv), C_2H_4O, is the gas most used in hospitals to sterilize items that cannot be sterilized at high temperatures. Formaldehyde, used more in Europe than in the U.S., is second to ethylene oxide in importance as a gaseous sterilant. Items sterilized include most of the plastics, medical and biologic preparations, surgical implants, medical instrumentation, hospital bedding, and oxygen tents. Ethylene oxide is also used to sterilize dry powders and food spices. It is a colorless, highly reactive gas that reacts by alkylation with alcohols, thiols, amines, organic acids, and amides in microbial cells and spores. It also reacts with nucleic acids, which is thought to be the primary cause of its biocidal activity. Alkylation of guanine and adenine components of DNA was shown to cause bacterial spores to lose the power of reproduction (278).

$$\text{protein} \begin{cases} -\text{COOH} \\ -\text{NH}_2 \\ -\text{C}_6\text{H}_4\text{OH} \\ -\text{SH} \end{cases} + \text{H}_2\text{C}\underset{\diagdown \text{O} \diagup}{-}\text{CH}_2 \longrightarrow \text{protein} \begin{cases} -\text{COO(CH}_2)_2\text{OH} \\ -\text{NH(CH}_2)_2\text{OH} \\ -\text{C}_6\text{H}_4\text{O(CH}_2)_2\text{OH} \\ -\text{S(CH}_2)_2\text{OH} \end{cases}$$

Ethylene oxide boils at 10.4°C, is completely soluble in water, and is flammable and explosive. To prevent flammability, it is diluted with carbon dioxide or fluorocarbon mixtures. The latter are preferred over carbon dioxide; however, the effect of fluorocarbons on the ozone layer is expected to restrict usage (279,280). Ethylene oxide has been classified as a mutagen and carcinogen, resulting in strict regulations on use. A special ethylene oxide sterilizer (autoclave) (Fig. 8) with safety devices is used. The sterilizer is a pressure vessel having the capacity to evacuate air, warm and moisturize the product uniformly, add the gas at a specific temperature, remove the gas after the specified interval, and finally replace it with fresh, filtered air. Because some materials, like plastics, absorb ethylene oxide and do not release it quickly, products containing such materials are placed in special aerated cabinets after sterilization, until the residual ethylene oxide is removed.

The primary factors affecting inactivation of microorganisms using ethylene oxide are contact with the gas, gas concentration, temperature, relative humidity, and exposure time (281). The inactivation rate increases with gas concentration from 50–500 ppm, and there is no significant increase in spore inactivation over 500 ppm (282). In practice, however, a concentration of 800–1200 ppm is used to ensure adequate penetration and allow for the gas that is absorbed by the packaging material. The gas provides sterilization at room temperature, but the long time factor makes it impractical. The rate of sporicidal action doubles for every 10°C rise in temperature (283); the time for sterilization is reduced 8-fold with a temperature increase from 20 to 50°C. A temperature range of 45–60°C, which is nondestructive to most materials, is generally used in ethylene oxide sterilization.

Humidity can be a problem. Whereas it was shown (284) that 33% RH was best for spore inactivation, and that at least 30% RH was needed for effective sterilization (285), dried spores are difficult to kill, and the spore substrate material and wrappings compete with the spore for the available moisture (286). Therefore, the relative humidity is adjusted to 50–70% to provide sufficient moisture for the spores to equilibrate. The exposure time depends upon the gas mixture, the concentration of ethylene oxide, the load to be sterilized, the level of contamination, and the spore reduction assurance required. It may be anywhere

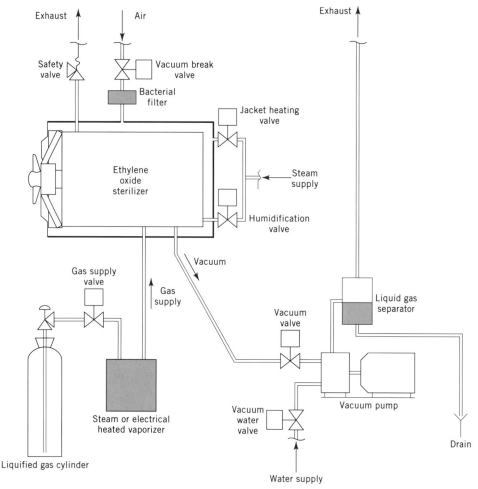

Fig. 8. An ethylene oxide sterilization vessel (autoclave) and supportive system.

from 4–24 hours. In a run, cycles of pre-conditioning and humidification, gassing, exposure, evacuation, and air washing (Fig. 9) are automatically controlled.

Ethylene oxide is able to inactivate all microorganisms. Bacterial spores are more resistant than vegetative cells, yeasts, and molds (287). Spores are 5 to 10 times more resistant than the vegetative cells (288). *Bacillus subtilis* spores were the most resistant of those tested (289). Ethylene oxide was also shown to be virucidal (290).

Propylene Oxide.   Propylene oxide [75-56-9] (qv), C_3H_6O, is a higher homologue of ethylene oxide that boils at 35°C. Propylene oxide is not as germicidally active as ethylene oxide (291), but has one distinct advantage: it hydrolyzes to produce nontoxic propylene glycol (292), allowing use for treating foods. Three hours at 37°C reduced the microbial count of cocoa powder by 50–70% and molds by 90–99% (293). Powdered cosmetics and toiletries are treated with 1–2% of

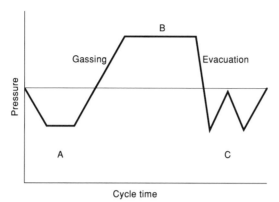

Fig. 9. Ethylene oxide sterilization 4-h cycle, 12/88; 12 wt % ethylene oxide/88 wt % chlorofluorocarbon (Freon 12), where the horizontal line represents standard barometric pressure; A, preconditioning and humidification at 87.8–94.55 kPa, 30–40 min; B, exposure for ¾ h at 55.17–68.95 kPa above standard barometric pressure; C, air washes at 81.04–87.8 kPa, 20 min. To convert kPa to psi, multiply by 0.145.

liquid propylene oxide in sealed containers, and the temperature is raised to cause vaporization and increased activity (294).

Beta-Propiolactone. β-Propiolactone [57-57-8], $C_3H_4O_2$, was reported to be 4000 times more active than ethylene oxide, and 25 times more effective than formaldehyde (295). It is not recommended as a substitute for ethylene oxide because of its lack of penetrating power, and is also carcinogenic in mice. As with ethylene oxide sterilization, the control of humidity is critical (296). β-Propiolactone is sporicidal (297) and virucidal (298). Because of its carcinogenicity, it has been banned in interstate shipping by EPA for use as a pesticide.

Ozone. Ozone [10028-15-6] (qv), O_3, is an allotropic form of oxygen having outstanding properties as an oxidant. It is also a powerful germicide, being lethal to all forms of microorganisms. Because bacterial spores are 10–15 times more resistant than vegetative forms (299), the instability of the gas limits its usefulness as a sporicide (300). Although it is more stable in dry air, it is more active in moist air and in water (301). The reported threshold relative humidity value for disinfection was 50%, and the optimum range was 60–80%. Its temperature coefficient is opposite to that of most disinfectants in that it is more effective at low rather than high temperatures. Bacteria were found to be killed by 150 ppm on agar at 0°C, whereas 210–480 ppm were necessary at 20°C. For molds, the respective values were 5–10 ppm at 0°C, and 700–9000 ppm at 25°C (302). The odor of ozone can be detected with as little as 0.02–0.04 ppm in air, and 20–30 ppm causes irritation to the eyes, nose, and throat. Prolonged exposure to a concentration of 1000 ppm can cause death (303). The toxicity, instability, and corrosiveness have limited the practical applications of ozone. In addition, it is necessary to generate it electrically as it is used. Being very reactive, organic matter greatly reduces the germicidal effectiveness of ozone. Its main use over the years has been for the disinfection of drinking water and process water for manufacturing pharmaceuticals, cosmetics, and household products. However, it has been suggested (304) that because of its high oxidative properties, ozone could be used in hospitals for resterilizing instruments composed of materials such as noble metals, titanium, stainless steel, silicone rubber, ceramics, polyvinyl chloride, and polyurethane.

Chlorine Dioxide.   Like ozone, chlorine dioxide [10049-04-4] is a powerful oxidant. It is usually generated as used. It has been used for disinfecting drinking water and bleaching paper pulp. Its effectiveness in killing microorganisms is well documented (305,306), and it has received recent study as a gas to sterilize medical devices. It requires 50% rh or higher to be effective. Bacterial cells had a D-value of 2.6 min and spores of 24 min (307).

Analytical Test Methods

Disinfectants.   Analytical test methods have been improved over the years to make them applicable to the various disinfectants and their uses. Conditions under which disinfectants must perform vary greatly. Factors that influence activity include: concentration of the disinfectant; temperature; pH and osmotic pressure of the medium; nature and number of the microorganisms employed; presence of chemicals that react with the disinfectant; and the duration of contact between the disinfectant and the microorganisms. Scientific organizations that have developed and published procedures to standardize conditions of tests are as follows: AOAC International in the United States; the German Society for Hygiene and Microbiology (DGHM) in Germany; the British Standards Institute (BSI) in the United Kingdom; the French Association of Normalization (AFNOR) in France; and the Dutch Committee of Phytopharmacy in the Netherlands, etc. A comparison of the methods in different countries is given (308).

Disinfection tests can be classified according to the test organism, ie, whether the test employs certain species of bacteria, fungi, or viruses; classified as to whether it is a static test or a cidal test, as in a bactericidal vs bacteriostatic test or sporicidal vs sporistatic test; or classified as to whether it is a microbial reduction test or an end-point test where all the organisms in the test are apparently killed. Procedures may be distinguished by *in vitro* or *in vivo* testing. Another way to consider tests is whether they are screening tests, practical type laboratory tests, or field tests.

Phenol Coefficient Test.   The first important attempt at standardizing testing methods was known as the phenol coefficient test (96). It has been modified several times, and is an official AOAC screening test recognized by EPA and FDA. The phenol coefficient test compares the activity of disinfectants to that of phenol, under specific conditions, to give a number that measures the activity of the chemical tested with respect to that of phenol, ie, the phenol coefficient. The AOAC method employs visual examination of bacterial growth in a nutrient medium. The Kelsey-Sykes test (1969) is a modified method popular in British circles.

Some tests of the AOAC (309,310) are listed in Table 10. The phenol coefficient test employs three test organisms and a standard concentration of phenol for each. In the test, a suspension of bacteria is added to the phenol dilutions and several dilutions of the disinfectant to be tested. After contact for 5, 10, and 15 min, samples are transferred to a nutrient-rich recovery medium that contains a neutralizer that nullifies inhibition of bacterial growth by the disinfectant and

Table 10. Approved Tests of the AOAC[a]

Name	Type	Test organisms	Resistance standard
phenol coefficient method	suspension	*S. typhi* ATCC 6539 *S. aureus* ATCC 6538 *P. aeruginosa* ATCC 15442	phenol, 1:90, 1:100 phenol, 1:60, 1:70 phenol, 1:80, 1:90
use dilution method	surface (stainless steel carriers)	*S. choleraesuis* ATCC 10708 *S. aureus* ATCC 6538 *P. aeruginosa* ATCC 15442	as above
hard surface carrier test	surface (fire-polished glass carriers)	*S. choleaesuis* ATCC 10708 *S. aureus* ATCC 6558 *P. aeruginosa* ATCC 15442	none
chlorine (available) germicidal equivalent concentration	suspension (capacity test)	*S. typhi* ATCC 6539 and/or *S. aureus* ATCC 6538	sodium hypochlorite 200, 100 & 50 ppm available chlorine
sporicidal activity	surface (porcelain cylinders, silk suture loops)	*B. subtilis* ATCC 19659 *C. sporogenes* ATCC 3584	hydrochloric acid, 2.5 N
tuberculocidal activity	surface (porcelain cylinders)	*M. smegmatis* PRD No. 1 (presumptive test) *M. bovis* (BCG) (confirmative test)	phenol, 1:50, 1:75
fungicidal activity	suspension	*T. mentagrophytes* (eg, ATCC 9533)	phenol, 1:60, 1:70

[a]Refs. 309, 310.

permits surviving organisms to multiply. The phenol coefficient is determined by dividing the greatest dilution of the disinfectant being tested that kills the bacteria in 10 min but not in 5 min, by the dilution of phenol that gives the same result.

Use Dilution Test and Hard Surface Carrier Test. While the phenol coefficient is useful as a screening test, the use dilution and hard surface carrier tests are examples of practical testing for a disinfectant's activity on pre-cleaned hard surfaces against residual dried bacteria, as on surgical blades. To simulate such a practical situation, stainless steel carriers or fire-polished glass carriers in a solution of asparagine are sterilized and immersed in a bacterial culture. Thus inoculated, the glass carriers are dried and put into 10 tubes containing the disinfectant solution. After 10 min immersion, the cylinders are transferred to tubes of recovery medium and incubated to determine whether the bacteria survived the disinfectant treatment. If there is growth in any of the 10 tubes, the test is repeated using a higher concentration of the disinfectant; killing in 59 of 60 replicates is necessary for a confidence level of 95%. The hard surface carrier test and the use dilution test are similar in nature in that they test the same organism on carriers for 10 min contact. The hard surface carrier test is an improvement over the use dilution test in that it is accurate and reproducible. The improvement is due to several important changes, ie, the carrier is changed from stainless steel to glass; a standardized inoculum is employed; critical control points are identified (310).

Available Chlorine Test. The chlorine germicidal equivalent concentration test is a practical-type test. It is called a capacity test. Under practical conditions of use, a container of disinfectant might receive many soiled, contaminated instruments or other items to be disinfected. Eventually, the capacity of the disinfectant to serve its function would be overloaded due to reaction with the accumulated organic matter and organisms. The chlorine germicidal equivalent concentration test compares the load of a culture of bacteria that a concentration of a disinfectant will absorb and still kill bacteria, as compared to standard concentrations of sodium hypochlorite tested similarly. In the test, 10 successive additions of the test culture are added to each of 3 concentrations of the hypochlorite. One min after each addition a sample is transferred to the subculture medium and the next addition is made 1.5 min after the previous one. The disinfectant is then evaluated in a manner similar to the phenol coefficient test. For equivalence, the disinfectant must yield the same number of negative tubes as one of the chlorine standards.

Sporicidal Test. The sporicidal test, like the use dilution test, measures the ability of a disinfectant to kill organisms in a practical type situation. In this case, spores are dried on the surface of solid substances, namely porcelain cylinders and silk suture loops. The organisms used are spores of aerobic and anaerobic bacteria. The spores are dried on 60 suture loops and 60 porcelain cylinders for each organism, to make a total of 240 carriers (Table 10). The spores must be able to survive 2 min in 2.5 N HCl, and are tested for resistance up to 20 minutes. The inoculated carriers are added in 12 groups of 5 to separate tubes of the disinfectant being tested, and transferred individually after a selected contact time to thioglycollate subculture medium. They are then incubated at 37°C for 21 days. If no growth has occurred, the tubes are heated for 20 min at 80°C to heat shock the spores, and are reincubated for 3 more days. If no more than 1 out of 60 carriers for each kind of spore is positive, the disinfectant can be claimed to be sporicidal. If all 240 cultures are negative, the product may be considered to be a sterilant. To further substantiate efficacy of a sporicide or sterilant for official registration with U.S. government agencies, testing on three samples of the product, including a 60 day aging test, is required.

Tuberculocidal Test. The tubercle bacillus is resistant to disinfectants because the cells are protected with a waxy coating that is not readily penetrated. The tuberculocidal test is a use dilution practical type test that employs porcelain cylinders. The bacteria are different from those in the use dilution method (Table 10), the incubation time is longer, and the details of the procedure are different. For example, in the tuberculocidal test the test is divided into two parts, a presumptive test and a confirmatory test. The former employs *Mycobacterium smegmatis* and the latter employs *Mycobacterium bovis* (BCG). For the presumptive test the incubation time is 12 days, as against 48 hours for other bacteria used in the use-dilution method. For the confirmatory test the incubation time is 60 days, with an additional 30 days in case there is no growth. As shown in Table 10, the concentrations of the phenol standard are higher than used with other bacteria.

Fungicidal Testing. The AOAC test for fungicidal activity is a screening test, as is the phenol coefficient test. It employs a suspension of fungus spores, which are generally less resistant to disinfectants than bacterial spores. The fungi selected for testing are usually dermatophytes that produce skin infections in man and animals.

Virucidal Testing. Virucidal tests have not been approved for official recognition by the AOAC because of the many problems in standardizing these tests (311). However, there are tests that compare the action of several disinfectants (311), and procedures have been developed for determining germicidal effectiveness in eliminating specific viruses in environmental disinfection programs (312). Viruses are found in a multitude of places, in dust and water, and on inanimate objects where they may survive but do not multiply until they infect the host. A surface test was developed (313–315) that uses herpes simplex virus and poliovirus. A comparison of surface tests and suspension tests of virucides has been made (316,317). Of the disinfectants tested in the suspension test, only 70% ethanol, 70% isopropanol, and 2% glutaraldehyde were effective against rotavirus. In the surface test only the 2% glutaraldehyde was effective. A comprehensive report on the effectiveness of numerous chemicals tested as virucides is available (318), and has been reviewed (319).

In the procedure for the surface test (313), the virus is grown in a monolayer of baby hamster kidney cells and incubated in Eagles medium supplemented with tryptose phosphate broth and calf serum. After separation of the virus from the cells by sonification and centrifugation, amounts of the suspension containing 3×10^9 plaque-forming units are dried on coverslips. The inoculated coverslips are placed in 5 ml of the disinfectant for 1, 5, or 10 min, then rinsed, sonicated, and assayed.

Field Testing. Field tests cannot be readily standardized because the fields of use are different from each other, as are the sources of contamination and the many other factors. Each investigator, having performed the screening and practical type tests, will then be able to select suitable disinfectants to evaluate under the specific conditions that pertain to the local environment. If, following disinfection, the rate of contamination is lowered and continues to remain so after careful testing over a period of time, one may attribute the change to the disinfectant, when all other factors that could be responsible for the difference have been taken into account. As an example, a controlled field test was conducted against rotavirus (63). A disinfectant spray registered with EPA (following laboratory tests) against rotavirus was found to prevent infection in human volunteers after they ingested treated virus. Of the volunteers ingesting untreated virus, 93% were infected.

In field testing, there are many commercial products and instruments available that are helpful in rapidly detecting, identifying, and quantifying microorganisms.

It should be recognized that disinfectant tests are less than perfect. The United States General Accounting Office (GAO), an agency of the U.S. Congress, issued a report in 1990 that questioned the validity of test methods for disinfectants employed by the EPA (320). In this report, GAO states that methods used by EPA have been widely criticized by industry and academia for producing highly variable results, and that laboratory tests may not accurately predict how a disinfectant will perform in actual use; attention focused principally on the AOAC use dilution test and tuberculocidal test. Some scientists claim that the carrier methods can be refined using more stringent procedures and different carrier materials; other scientists question whether carrier materials, like glass, approximate real surfaces, and even if they do, if the variability will be reduced to an acceptable level. Disinfection is a new science and the problems presented are not simple. It should be stated that careful studies of the AOAC use dilution test (321) have resulted in the modification of that test (310), and studies are being made on the sporicidal and tuberculocidal tests to improve them as well.

Antiseptics.   Because antiseptics are applied to the human body, they are considered to be drugs, and come under the control of the U.S. Food and Drug Administration (FDA). The FDA issues guidelines on the testing of antiseptics. Antiseptics must have special properties not required of disinfectants. Antiseptics must not be toxic or irritating to skin or tissue, and must not be inactivated in the presence of skin, blood, serum, or mucous. These requirements greatly restrict the candidates for use as antiseptics. Screening tests like the phenol coefficient method and the Kelsey-Sykes test are accepted by FDA as preliminary data on the germ killing power of a chemical. However, greater weight is put on *in vivo* methods and clinical trials that relate more closely to the use of the antiseptic. Methods of testing for these applications have been developed, but because of the inherent difficulties involved in *in vivo* testing as of the early 1990s, they have not yet been given approved official status. However, the FDA, through its OTC Antimicrobial Panel and its report on testing methods, has issued specific protocols related to the methods (322). The FDA recognizes and has defined certain antiseptic products as follows.

A *skin antiseptic* is a nonirritating antimicrobial-containing preparation that prevents overt skin infection.

A *health-care personnel handwash* is a nonirritating antimicrobial-containing preparation designed for frequent use; it reduces the number of transient microorganisms on intact skin to an initial baseline level after adequate washing, rinsing, and drying; and it is broad-spectrum, fast-acting, and, if possible, persistent.

A *patient preoperative skin preparation* is a fast-acting broad-spectrum antimicrobial-containing preparation that significantly reduces the number of microorganisms on intact skin.

A *surgical hand scrub* is a nonirritating antimicrobial-containing preparation that significantly reduces the number of microorganisms on the intact skin. A surgical hand scrub should be broad-spectrum, fast-acting, and persistent.

A *skin wound protectant* is a nonirritating antimicrobial-containing preparation applied to small cleansed wounds; it provides a protective physical barrier and a chemical (antimicrobial) barrier that neither delays healing nor favors the growth of microorganisms.

A *skin wound cleanser* is a nonirritating, liquid preparation (or product to be used with water) that assists in the removal of foreign material from small superficial wounds, does not delay wound healing, and may contain an antimicrobial ingredient.

An *antimicrobial soap* is a soap containing an active ingredient with both *in vitro* and *in vivo* activity against skin microorganisms.

In all antiseptic testing, it is recognized that skin and mucous membranes to which products are applied cannot be disinfected or sterilized; but it is possible to significantly reduce the population of transient and resident pathogenic bacterial flora. All *in vivo* test methods require a determination of the bacteria on the skin before and after treatment. Because of the normal variation in bacterial population of the skin of different people, a number of people must be tested in order to make a statistical analysis of the results. Different parts of the body are used for different tests. In all of the tests the details of the protocol are extremely important and must be strictly adhered to in order to obtain reproducible results.

Early tests for skin degerming that involve measuring organisms residing on the skin have been developed (323–325). These methods, which employ hand-washing techniques, have numerous applications, such as the evaluation of antimicrobial soaps. More recent versions (326–328) involve treating the hands with known bacterial cultures and the antiseptic to be tested. The cup scrubbing technique (329), another method for measuring organisms on the skin, uses a small area of the skin delineated by a round glass cup and employs buffer solution with scrubbing to remove the organisms. A simple, qualitative method is the skin stripping technique (322) which strips off a sampling from the skin with cellophane tape. Methods have been developed (330,331) in which organisms on the hands are transferred to gloves and sampled. These methods assess both immediate and persistent effects of antiseptics. In all tests with antiseptics, suitable neutralizing chemicals for the antiseptics employed are necessary so that surviving bacteria may be recovered and counted. Some of these are lecithin, Lubrol W, Polysorbate 80, and sodium thiosulfate (322).

Tests have been developed that test different products for their effectiveness as a healthcare personnel handwash (327); evaluate hand disinfectants for use in surgery (333); determine the effectiveness of a surgical hand scrub, ie, the glove juice test (311,329); evaluate antiseptics for the oral cavity to be used in mouthwashes (334,335); and test antiseptics for the periurethral area and application to catheters (336,337). A method used for a test comparing four antiseptic products was adopted as recommended practice by the Association of Practitioners of Infection Control (338).

Mice are utilized for testing antiseptics for application to cuts, wounds, and incisions (339). The test bacteria, type 1 pneumococcus and hemolytic streptococcus, are applied to the tails of anaesthetized mice. The tip of the tail is then dipped into the antiseptic for 2 min, after which one-half inch of the tail is removed and inserted into the peritoneal cavity and the incision is closed. If after 10 days the animals survive, the product is considered satisfactory for use as a skin antiseptic. The blood of dead animals is sampled and streaked on blood agar for confirmation of infection from the test bacteria as the cause of death. Since lack of toxicity is another requirement of a product to be applied to wounds, this test has been combined with a toxicity test (340).

Reviews of antiseptics for skin and wound cleansers (341), a review of methods for testing antiseptics (329), and an overall review of antiseptics and their testing methods (342) are available.

BIBLIOGRAPHY

"Antiseptics, Disinfectants, and Fungicides" in *ECT* 1st ed., Vol. 2, "Survey" pp. 77–91, by W. C. Tobie, American Cyanamid Co.; "Methods of Testing" pp. 91–105, by G. F. Reddish, Lambert Pharmaceutical Co.; "Antiseptics and Disinfectants" in *ECT* 2nd ed., Vol. 2, pp. 604–608, by Emil G. Klarmann, Lehn & Fink, Inc.; "Disinfectants and Antiseptics" in *ECT* 3rd ed., Vol. 7, pp. 793–832, by William Gump, Consultant.

1. S. S. Block in S. S. Block, ed., *Disinfection, Sterilization, and Preservation*, 4th ed., Lea & Febiger, Philadelphia, 1991, pp. 3–17.
2. G. Fracastoro, *Contagion, Contagious Diseases and their Treatment*, trans. W. C. Wright, The Putnam Publishing Group, Inc., New York, 1930.
3. A. van Leeuwenhoek, *Phil. Trans. Roy. Soc. London* **14,** 568–574 (1684).
4. I. P. Semmelweis, *The Etiology, Conception and Prophylaxis of Childbed Fever*, Hartleben. Pest, Vienna & Leipsig, 1861.
5. L. Pasteur, *Compt. Rend. Acad. Sci.* **52,** 344–347 (1861).
6. J. Lister, *Lancet* **1,** 326–329, 352–359, 387, 507–509, **2,** 95–96, 353–356, 668–669 (1867).
7. R. Koch, *Mittheilungen aus dem Kaiserlichen Gesunheitsamte* **1,** 234–282 (1881).
8. H. Chick and C. J. Martin, *J. Hyg.* **8** 654–697 (1908).
9. C. M. Beck-Sague and W. B. Jarvis in S. S. Block, ed., *Disinfection, Sterilization, and Preservation*, 4th ed., Lea & Febiger, Philadelphia, 1991, pp. 663–675.
10. R. W. Haley and co-workers, *Am. J. Epidemiol.* **111,** 472–485 (1980).
11. R. E. Dixon, *Ann. Intern. Med.* **89,** 749–753 (1978).
12. *Gainesville (Florida) Sun*, quoting Pierre Payment, Institute Armand-Frappier (Montreal), Judy Lew and Christine Moe at Centers for Disease Control, Atlanta, and Phillip Berger, EPA, 5A, Dec. 8, 1990.
13. S. S. Block in S. S. Block, ed., *Disinfection, Sterilization, and Preservation*, 4th ed., Lea & Febiger, Philadelphia, 1991, pp. 1107–1119.
14. U.S. General Accounting Office, *GAO/RCED-90-139*, Aug. 1990, pp. 11–12.
15. A. J. Marisca in S. S. Block, ed., *Disinfection, Sterilization, and Preservation*, 4th ed., Lea & Febiger, Philadelphia, 1991, pp. 999–1008.
16. K. R. Fulton, *Food Technol.* **35,** 80 (1981).
17. H. W. Rossmoore in S. S. Block, ed., *Disinfection, Sterilization, and Preservation*, 3rd ed., Lea & Febiger, Philadelphia, 1983, pp. 289–290.
18. G. C. White, *The Handbook of Chlorination*, 2nd ed., Van Nostrand Reinhold Co., Inc., New York, 1986.
19. R. Burgess, *J. Soc. Dyers Colourists* **50,** 138–142 (1934).
20. G. M. Hunt and G. A. Garrett, *Wood Preservation*, McGraw-Hill Book Co., New York, 1953.
21. G. Dychdala in S. S. Block, ed., *Disinfection, Sterilization, Preservation*, 4th ed., Lea & Febiger, Philadelphia, 1991, pp. 131–151.
22. J. N. Mbithi and co-workers, *Appl. Env. Microbiol.* **56,** 3601–3604 (1990).
23. N. Lloyd-Evans, *J. Hyg. Camb.* **97,** 163–173 (1986).
24. D. Rupp and co-workers, *Ann. Meet. ASM*, Dallas, Tex., 1991.
25. A. D. Russell, *The Destruction of Bacterial Spores*, Academic Press, Inc., London, 1982, pp. 202–205.
26. T. Kristoffersen and I. A. Gould, *J. Dairy Sci.* **41,** 940–955 (1958).
27. H. Farkas-Himsley, *Appl. Microbiol.* **12,** 1–6 (1964).
28. I. C. Warren and J. Ridgway, *Technical Report TR90*, Water Research Center, Medmenham, UK, 1978.
29. S. L. Chang, *J. Am. Water Works Assoc.* **36,** 1192–1206 (1944).
30. F. C. Schmelkes and E. S. Horning, *J. Bacteriol.* **29,** 323 (1935).

31. S. Bloomfield, *ASM Int. Symp. on Chem. Germicides*, Atlanta, Ga. 1990.
32. S. L. Chang and G. Berg, *U.S. Armed Forces Med. J.* **10,** 33 (1959).
33. B. Carroll, *J. Bacteriol.* **69,** 413 (1955).
34. O. Wyss and F. B. Strandskov, *Arch. Biochem.* **6,** 261 (1945).
35. P. B. Price, *Drug Stand.* **19,** 161 (1951).
36. L. Sykes, *Disinfection and Sterilization*, 2nd ed., Lippincott, Philadelphia, 1965, pp. 400–409.
37. L. Gershenfeld in S. S. Block, ed., *Disinfection, Sterilization, and Preservation*, 2nd ed., Lea & Febiger, Philadelphia, 1977, pp. 196–218.
38. W. Gottardi, *Zentralbl. Bakt. Kyg. 1. Abt. Orig. B,* **162,** 384–388 (1976).
39. R. J. Connolly and J. J. Shepherd, *Austral. N.X. J. Surg.* **42,** 94–97 (1972).
40. H. Voherr and co-workers, *J.A.M.A. Med. Assoc.* **244,** 25–28 (1980).
41. R. L. Berkelman and co-workers, *J. Clin. Microbiol.* **15,** 635–639 (1982).
42. R. L. Anderson and co-workers, *ASM Int. Symp. on Chem. Germicides*, Atlanta, Ga., 1990, Abstract 12, p. 17.
43. F. W. Tilley and J. M. Schaffer, *J. Bacteriol.* **12,** 303 (1926).
44. G. Weitzel and E. Schraufstätter, *Z. Physiol. Chem.* **285,** 172 (1950).
45. R. D. Fletcher and co-workers, *Antimicrob. Agents Chemother.* **19,** 917–921 (1981).
46. F. W. Tanner and F. L. Wilson, *Proc. Soc. Exp. Biol. Med.* **52,** 138–140 (1943).
47. M. Rotter, *ASM Int. Symp. on Chem. Germicides,* Atlanta, Ga., 1990, p. 23.
48. C. Harrington and H. Walker, *Boston Med. Surg. J.* **148,** 548–552 (1903).
49. E. G. Klarmann, L. W. Gates, and V. A. Shternov, *J. Am. Chem. Soc.* **54,** 3315 (1932).
50. H. E. Morton, *Ann. N.Y. Acad. Sci.* **53,** 191 (1950).
51. P. B. Price, *Arch. Surg.* **38,** 528 (1939).
52. P. B. Price, *J. Infect. Dis.* **63,** 301 (1938).
53. P. B. Price, *Arch. Surg.* **38,** 528 (1929); **60,** 492 (1950).
54. C. R. Smith, *Public Health Rep. (U.S.)* **62,** 1285 (1947).
55. H. Neves, *Arch Dermatol.* **84,** 132 (1961).
56. E. C. McCulloch, *Disinfection and Sterilization*, Lea & Febiger, Philadelphia, 1945, p. 319.
57. C. H. Cunningham, *Am. J. Vet. Res.* **9,** 195 (1948).
58. M. Klein and A. Deforest, *Soap. Chem. Spec.* **39,** 70–72, 95–97 (1963).
59. S. Stockman and F. C. Minett, *J. Comp. Pathol. Therap.* **39,** 1 (1926).
60. M. Klein and A. Deforest in S. S. Block, ed., *Disinfection, Sterilization, and Preservation*, 3rd ed., Lea & Febiger, Philadelphia, 1983, p. 425.
61. L. Resnick and co-workers, *JAMA Med. Assoc.* **255,** 1887 (1986).
62. H. Kobayashi and co-workers, *J. Clin. Microbiol.* **20,** 214–216 (1984).
63. R. L. Ward and co-workers, *J. Clin. Microbiol.* **29,** 1991–1996 (1991).
64. C. E. Coulthard and G. Sykes, *Pharm. J.* **137,** 79–81 (1936).
65. P. B. Price in W. Modell, ed., *Drugs of Choice*, C. V. Mosby, 1966 pp. 133–144.
66. E. L. Larson and H. E. Morton in S. S. Block, ed., *Disinfection, Sterilization, and Preservation*, 4th ed., Lea & Febiger, Philadelphia, 1991, pp. 191–203.
67. W. A. Altemeier in S. S. Block, ed. *Disinfection, Sterilization, and Preservation*, 3rd ed., Lea & Febiger, Philadelphia, 1983, pp. 493–504.
68. P. B. Price, *Ann. Surg.* **134,** 476 (1951).
69. E. H. Spaulding in S. S. Block, ed. *Disinfection, Sterilization, and Preservation*, Lea & Febiger, Philadelphia, 1968, p. 527.
70. D. O. O'Connor and J. R. Rubino in S. S. Block, ed., *Disinfection, Sterilization, and Preservation*, 4th ed., Lea & Febiger, Philadelphia, 1991, pp. 204–224.
71. J. F. Gardner and M. M. Peel, *Introduction to Sterilization, Disinfection, and Infection Control*, 2nd ed., Churchill Livingstone, Melbourne, 1991, p. 157.
72. J. E. Death and D. Coates, *J. Clin. Pathol.* **32,** 148–153 (1979).

73. E. G. Klarmann, V. A. Shternov, and L. W. Gates, *J. Lab. Clin. Med.* **19,** 835; **20,** 40 (1934).
74. R. P. Christiansen and co-workers, *J. Am. Dent. Assoc.* **119,** 493–505 (1989).
75. K. H. Wallhausser, *Seifen, Oele, Fette Wocheschr.* **106,** 107–111 (1980).
76. W. S. Gump, *Soap Sanit. Chem.* **21,** 36 (1945).
77. U.S. Pat. 2,535,077 (Dec. 26, 1950), E. C. Kunz and W. S. Gump (to Sindar Corp.).
78. C. V. Seastone, *Surg. Gynecol. Obstet.* **84,** 355 (1947).
79. P. B. Price and A. Bonnett, *Surgery* **24,** 542 (1948).
80. I. H. Blank and M. H. Coolidge, *J. Invest. Dermatol.* **15,** 257 (1950).
81. P. B. Price, *Ann. Surg.* **134,** 476 (1951).
82. W. S. Gump, *J. Soc. Cosmet. Chem.* **20,** 173 (1969).
83. B. P. Vaterlaus and J. J. Hostynek, *J. Soc. Cosmet. Chem.* **24,** 291 (1973).
84. R. D. Kimbrough and T. B. Gaines, *Arch. Environ. Health* **23,** 114 (1971).
85. T. B. Gaines, R. D. Kimbrough, and R. E. Linder, *Toxicol. Appl. Pharmacol.* **25,** 332 (1973).
86. J. A. Santolucito, *Toxicol. Appl. Pharmacol.* **22,** 276 (1972).
87. H. M. Powell, *J. Indiana State Med. Assoc.* **38,** 303 (1945).
88. *Fed. Reg.* **37,** 160, 219 (Jan. 7, 1972).
89. E. S. Barghoorn, *Office of Scientific Research and Development Report No. 4807,* 1945.
90. A. D. Russell, W. B. Hugo, and G. A. J. Ayliffe, eds., *Principles and Practice of Disinfection, Preservation, and Sterilization,* Blackwell Scientific Publications, London, 1982, p. 27.
91. R. Pfleger and co-workers, *Naturforsch.* **46,** 344 (1949).
92. O. F. Jillson and R. D. Baughma, *Arch. Dermatol.* **88,** 409 (1963).
93. W. S. Gump in S. S. Block, ed., *Disinfection, Sterilization, and Preservation,* 2nd ed., Lea & Febiger, Philadelphia, 1977, pp. 252–281.
94. E. L. Richardson, *Cosmetics and Toiletries* **92,** 85–86 (1977).
95. R. Zinkernagel and M. Koenig, *Seifen-Öle Fette-Wachse* **93,** 670 (1967).
96. S. Rideal and J. T. A. Walker, *J. Royal Sanit. Inst.* **24,** 424–441 (1903).
97. W. E. Finch, *Pharm. J.* **170,** 59–60 (1953).
98. T. R. Aalto, M. C. Firman, and N. E. Rigler, *J. Am. Pharm. Assoc. Sci. Ed.* **42,** 449 (1953).
99. H. Sokol, *Drug. Stand.* **20,** 89 (1952).
100. H. F. Cremer, *Z. Untersuch. Lebensm.* **70,** 136 (1935).
101. K. Schubel and I. Manger, Jr., *Münch. Med. Wochschr.* **77,** 13 (1929).
102. L. Gershenfeld and D. Perlstein, *Am. J. Pharm.* **111,** 227 (1939).
103. N. S. Gottfried, *Am. J. Hosp. Pharm.* **19,** 310 (1962).
104. C. Mathews and co-workers, *J. Am. Pharm. Assoc. Sci. Ed.* **45,** 260 (1956).
105. M. Huppert, *Antibiot. Chemother.* **7,** 29 (1957).
106. W. E. Rosen and co-workers, *J. Soc. Cosmetic Chem.* **28,** 83–87 (1977).
107. T. A. Bell and co-workers, *J. Bact.* **77,** 573–580 (1959).
108. P. A. Wolf, *Food Technol.* **4,** 294–297 (1950).
109. H. N. Prince, *J. Bact.* **78,** 788 (1959).
110. M. F. Kuhrt and co-workers, *Antimicrob. Agents and Chemotherapy* **26,** 924–927 (1984).
111. R. E. Pepper and V. L. Chandler, *Appl. Microbiol.* **11,** 384–388 (1963).
112. S. D. Rubbo and co-workers, *J. Appl. Bacteriol.* **30,** 78–87 (1967).
113. J. B. Kelsey, *J. Appl. Bacteriol.* **30,** 92–100 (1067).
114. V. G. Alder and R. A. Simpson in A. D. Russell, W. B. Hugo, and G. A. J. Ayliffe, eds., *Principles and Practice of Disinfection, Preservation, and Sterilization,* Blackwell, London, 1982, pp. 446–451.

115. E. A. Christensen and H. Kristensen in A. D. Russell, W. B. Hugo, and G. A. J. Ayliffe, eds., *Principles and Practice of Disinfection, Preservation, and Sterilization*, Blackwell, London, 1982, pp. 562–566.

116. M. S. Favero and W. W. Bond in S. S. Block, ed., *Disinfection, Sterilization, and Preservation*, 4th ed., Lea & Febiger, Philadelphia, 1991, pp. 623–635.

117. H. W. Rossmoore in S. S. Block, ed., *Disinfection, Sterilization, and Preservation*, 4th ed., Lea & Febiger, Philadelphia, 1991, pp. 290–304.

118. P. M. Borick, *Adv. Appl. Microbiol.* **10**, 291–312 (1968).

119. P. Gelinas and J. Goulet, *J. Appl. Bacteriol.* **54**, 243–247 (1983).

120. P. Gelinas and co-workers, *J. Food Protection* **47**, 841–847, 852 (1984).

121. S. P. Gorman and E. M. Scott, *Int. J. Pharm.* **4**, 57–65 (1979).

122. ASM *Manual of Clinical Microbiol.*, 5th ed., 1991, Chapt. 24.

123. G. A. J. Ayliffe and co-workers, *J. Hosp. Infect.* **7**, 295–309 (1986).

124. F. D. Loffer, *Reprod. Med.* **25**, 263–266 (1980).

125. S. J. Jachuck and co-workers, *J. Soc. Occup. Med.* **39**, 69–71 (1989).

126. Y. Yoshe-Purer and E. Eylan, *Health Lab. Sci.* **5**, 233–238 (1968).

127. A. L. Rosensweig, *Lancet* **8070**, 944 (1978).

128. K. Naguib and L. Hussein, *Milchwessenschaft* **27**, 758–762 (1972).

129. M. D. Wardle and G. M. Renninger, *Appl. Microbiol.* **30**, 710–711 (1975).

130. F. J. Turner in S. S. Block, ed., *Disinfection, Sterilization, and Preservation*, 3rd ed., Lea & Febiger, Philadelphia, 1983, pp. 240–250.

131. L. D. Sabath and co-workers, in R. L. Simmons and R. J. Howard, ed., *Surgical Infectious Disease*, Appleton-Century-Croft, New York, 1982, pp. 409–416.

132. B. von Bockelmann in S. S. Block, ed., *Disinfection, Sterilization, and Preservation*, 4th ed., Lea & Febiger, Philadelphia, 1991, pp. 833–845.

133. N. A. Klapes, *ASM Int. Symp. Chem. Germicides*, Atlanta, Ga., 1990, Abstract 20, pp. 14–15.

134. J. Suen and co-workers, *ASM Int. Symp. Chem. Germicides*, Atlanta, Ga., 1990, Abstract 33, p. 21.

135. U.S. Pat. 4,643,876 (1987), P. T. Jacobs and S. M. Lin.

136. H. R. Dittmar, J. L. Baldwin, and S. B. Miller, *J. Bacteriol.* **19**, 203 (1930).

137. C. E. Bayliss and W. M. Waites, *J. Gen. Microbiol.* **96**, 401–407 (1976).

138. C. E. Bayliss and W. M. Waites, *J. Food Technol.* **17**, 467–470 (1982).

139. F. I. K. Ahmed and C. Russell, *J. Appl. Bacteriol.* **39**, 31–40 (1975).

140. H. Eggensberger, *Zentralbl. Bakteriol. Mikrobiol. Hyg.* [B] **168**, 517–524 (1979).

141. M. G. C. Baldry, *J. Appl. Bact.* **54**, 417–423 (1983).

142. H. Krzywicka and co-workers in W. B. Kedzia, ed., *Resistance of Microorganisms to Disinfectants*, Polish Acad. of Sciences, Warsaw, 1975.

143. M. Sprossig in Ref. 142.

144. I. J. Hutchins and H. Xezones, *Proc. 49th Ann. Mtg. Soc. Am. Bacteriol*, Abstract 50–51, 1949.

145. S. S. Block in S. S. Block, ed., *Disinfection, Sterilization, and Preservation*, 4th ed., Lea & Febiger, Philadelphia, 1991, pp. 167–181.

146. R. Orth and H. Mrozeck, *Fleischwirtsch* **69**, 1575–1576 (1989).

147. D. M. Portner and R. K. Hoffman, *Appl. Microbiol.* **16**, 1782 (1968).

148. J. A. Reyniers, *Lobund Rept. No. 1*, University of Notre Dame Press, 1946, pp. 87–120.

149. G. R. Dychdala, *Proc. 4th Conf. Prog. Chem. Disinfection*, Binghamton, N.Y., 1988, pp. 315–342.

150. F. G. Bock and co-workers, *J. Nat. Cancer Inst.* **55**, 1359–1361 (1975).

151. M. G. C. Baldry, *J. Appl. Bacteriol.* **57**, 499–503 (1984).

152. G. Domagk, *Dtsch. Med. Wochschr.* **61**, 829 (1935).

153. J. J. Merianos in S. S. Block, ed., *Disinfection, Sterilization, and Preservation*, 4th ed., Lea & Febiger, Philadelphia, 1991, pp. 225–255.
154. W. S. Mueller and D. B. Seeley, *Soap* **27,** 131 (1951).
155. I. M. Maurer, *Hospital Hygiene*, 3rd ed., Edward Arnold, London, 1985.
156. C. A. Lawrence, *Surface-Active Quaternary Ammonium Germicides*, Academic Press, Inc., New York, 1950.
157. P. F. D'Arcy and E. P. Taylor, *J. Pharm. and Pharmacol.* **14,** 129–146, 193–216 (1962).
158. A. Rembaum, *App. Polymer Symp.* **22,** 299–317 (1973).
159. T. Ikeda and co-workers, *Antimicrobial Agents Chemother.* **26,** 139–144 (1984).
160. M. Ghosh, *Polymer News* **13,** 71–77 (1988).
161. O. W. May in S. S. Block, ed., *Disinfection, Sterilization, and Preservation*, 4th ed., Lea & Febiger, Philadelphia, 1991, pp. 322–333.
162. U.S. Pat. 4,525,346 (1985), R. L. Stark.
163. F. L. Rose and G. Swain, *J. Chem. Soc.* **1956,** 4422–4425 (1956).
164. N. Bodor and co-workers, *J. Med. Chem.* **23,** 469–474 (1980).
165. C. A. Lawrence and A. J. Maffia, *Bull. Am. Soc. Hosp. Pharm.* **14,** 164 (1957).
166. R. A. Benson and co-workers, *J. Pediat.* **31,** 369 (1947); **34,** 49 (1949).
167. A. N. Petrocci and P. Clark, *J. Assoc. Off. Anal. Chem.* **52,** 836–842 (1969).
168. R. L. Gettings and B. L. Triplett, *Book of Papers, Am. Soc. Text. Chemists Colorists Natl. Tech. Conf.* 1978, pp. 259–261.
169. T. Lewandowski, G. R. Dychdala, and J. A. Lopes in S. S. Block, ed., *Disinfection, Sterilization, and Preservation*, 4th ed., Lea & Febiger, Philadelphia, 1991, pp. 256–262.
170. F. Bartnik and K. Kuenstler in J. Falbe, ed., *Surfactants in Consumer Products: Theory, Technology, and Application*, Springer-Verlag, New York, 1987.
171. K. Siebert and D. Boltersdorf, *Second World Surfactant Congress*, Paris, 1988, pp. 646–661.
172. P. Gelinas and co-workers, *J. Food Protection* **47,** 841–847 (1984).
173. H. Hays and P. R. Elliker, *J. Milk Food Technol.* **2,** 109–111 (1959).
174. U.S. Pat. 3,969,258 (July 13, 1976), C. M. Carrandang and G. R. Dychdala (to Penn Walt Corp.).
175. U.S. Pat. 4,404,040 (Sept. 13, 1983), Y. Wang (to Economics Laboratory, Inc.).
176. G. Sykes, *Disinfection and Sterilization*, 2nd ed., Lippincott, Philadelphia, 1965, p. 378.
177. A. Schmitz, *Milchwissenschaft* **7,** 250–257 (1952).
178. P. Gelinas and J. Goulet, *Can. J. Microbiol.* **39,** 1715–1730 (1983).
179. B. Sorenson and co-workers, *Biologico* **35,** 3–7 (1969).
180. M. Sainclivier and L. Kerhare, *Ind. Aliment Agric.* **83,** 127–136 (1966).
181. T. Nagai, *Naika, Hokan* **26,** 155–160 (1979).
182. T. Nada and co-workers, *Esei Kensa* **29,** 929–937 (1980).
183. M. Ichikawa and Y. Miyoshi, *Bokin Bobai* **8,** 143–147 (1980).
184. P. Gelinas and co-workers, *J. Food Protection* **47,** 841–847 (1984).
185. K. Wagener, *Technical data*, Germany Institute of Hygiene, College of Veterinary Science, Hanover, July 17, 1954.
186. B. R. Frisby, *Lancet* **2,** 829 (1961).
187. S. G. Hesselgren, *SV Tondlaek Tidskr.* **66,** 181–196.
188. A. Albert and co-workers, *Br. J. Exp. Pathol.* **28,** 69–87 (1947).
189. A. Albert, M. I. Gibson, and S. D. Rubbo, *Br. J. Exp. Pathol.* **34,** 119 (1953).
190. S. D. Rubbo, A. Albert, and M. I. Gibson, *Br. J. Exp. Path.* **31,** 425 (1950).
191. W. Liese, *Zentr. Bakteriol.* **105(I),** 137 (1927).
192. K. A. Oster and M. J. Golden, *J. Am. Pharm. Assoc. Sci. Ed.* **37,** 283 (1947).

193. K. A. Oster and M. J. Golden, *J. Am. Pharm. Assoc.* **37**, 283–288 (1947).

194. K. A. Oster and M. J. Golden, *Exp. Med. Surg.* **7**, 37–45 (1949).

195. V. D. Warner, *J. Periodont* **47**, 664 (1976).

196. W. Jadassohn and co-workers, *Schweiz. Med. Wochschr.* **74**, 168 (1944).

197. *Ibid.* **77**, 987 (1947).

198. K. Sigg, *Schweiz, Med. Wochschr.* **77**, 123 (1947).

199. S. S. Block, *Agr. & Food Chem.* **3**, 222–234 (1955).

200. A. Albert, *Selective Toxicity*, 5th ed., Chapman and Hall, London, 1968.

201. N. Orentreich, *J. Soc. Cosmet. Chem.* **23**, 189–194 (1972).

202. H. W. Rossmoore, *Dev. Ind. Microbiol.* **20**, 41–71 (1979).

203. A. M. Kligman and W. Rosenzweig, *J. Invest. Dermatol.* **10**, 59 (1947).

204. C. R. Miller and W. O. Elson, *J. Bacteriol.* **57**, 47 (1949).

205. *Ibid.*

206. A. D. Russell in W. B. Hugo, ed., *Inhibition and Destruction of the Microbial Cell*, Academic Press, Inc., 1979, pp. 209–225.

207. D. Jaconia, *Preservatives in Pharmaceutical Products in Quality Control in the Pharmaceutical Industry*, Vol. 1. Academic Press, Inc., 1972.

208. A. D. Russell and J. R. Furr, *J. Appl. Bact.* **43**, 253–260 (1977).

209. G. W. Denton in S. S. Block, ed., *Disinfection, Sterilization, and Preservation*, 4th ed., Lea & Febiger, Philadelphia, 1991, pp. 274–289.

210. R. A. Simpson and co-workers, *Program abstract, 29th ICAAC*, Sept. 17–20, 1989, Houston, pp. 212, 663.

211. L. A. Shaker and co-workers, *Int. J. Pharm.* **34**, 51–56 (1986).

212. G. A. J. Ayliffe and co-workers, *Public Health Laboratory Service*, London, 1984.

213. D. Taplin in H. I. Maibach and R. Aly, eds., *Skin Microbiology*, Springer-Verlag, New York, 1981, pp. 113–124.

214. S. Bygdeman, *Infection Control* **5**, 275–278 (1984).

215. H. G. Smylie and co-workers, *Br. Med. J.* **4**, 586–589 (1973).

216. M. A. K. LaRocca and co-workers, *Adv. Therap.* **2**, 269–274 (1985).

217. A. J. Ball and co-workers, *J. Urol.* **138**, 491–494 (1987).

218. K. K. Christiansen and co-workers, *Europ. J. Obstet. Gynecol. Reproduct. Biol.* **19**, 231–236 (1985).

219. A. A. Colombo, *Minerva Chir.* **42**, 23–24 (1987).

220. K. S. Kornman, *J. Periodont. Res.* **21**, (Suppl.), 5–22 (1986).

221. M. Okano and co-workers, *Arch Dermatol.* **125**, 50–52 (1989).

222. P. Woodcock in K. R. Payne, ed., *Industrial Biocides*, John Wiley & Sons, Inc., New York, 1988, pp. 19–36.

223. H. W. Rossmoore in S. S. Block, ed., *Disinfection, Sterilization, and Preservation*, 4th ed. Lea & Febiger, Philadelphia, 1991, pp. 290–321.

224. S. S. Block in S. S. Block, ed., *Disinfection, Sterilization, and Preservation*, 4th ed. Lea & Febiger, Philadelphia, 1991, pp. 901–947.

225. H. C. Stecker in S. S. Block, ed., *Disinfection, Sterilization, and Preservation*, 2nd ed. Lea & Febiger, Philadelphia, 1977, pp. 282–300.

226. P. S. Herman and W. M. Sams, *Soap Photodermatitis and Photosensitivity to Halogenated Salicylanilides*, Charles C Thomas, Springfield, Ill., 1972.

227. L. J. Vinson and R. S. Flatt, *J. Invest. Dermatol.* **38**, 327 (1962).

228. D. S. Wilkinson, *Br. J. Dermatol.* **73**, 213 (1961); **74**, 302 (1962).

229. A. Kraushaar, *Arzneim. Forsch.* **4**, 548 (1954).

230. U.S. Pat. 2,745,874 (May 15, 1956), G. Schetty, W. Stammbach, and R. Zinkernagel (to J. R. Geigy A.G.).

231. U.S. Pat. 2,906,711 (Sept. 29, 1959), H. C. Stecker.

232. U.S. Pat. 3,041,236 (June 26, 1962), H. C. Stecker.

233. U.S. Food and Drug Administration, *Federal Register 39, (179), 33102-33104,* Washington, D.C., 1974.
234. D. J. Beaver and co-workers, *J. Am. Chem. Soc.* **79,** 1236–1245 (1957).
235. M. B. Finkey and co-workers, *J. Soc. Cosmetic Chem.* **35,** 351–355 (1984).
236. W. B. Hugo and A. D. Russell, eds, *Pharmaceutical Microbiology,* 3rd ed., Blackwell Scientific Publications, London, 1983, pp. 103–105.
237. R. J. Lukens in S. S. Block, ed., *Disinfection, Sterilization, and Preservation,* 4th ed., Lea & Febiger, Philadelphia, 1991, pp. 763–764.
238. R. J. Schanno and co-workers, *J. Soc. Cosmetic Chem.* **31,** 85–96 (1980).
239. P. A. Berk and W. E. Rosen, *Am. Perfumes and Cosmetics* **85,** 55–60 (1970).
240. K. H. Wallhauser, *Parfumerie Kosmetik* **62,** 387–479 (1981).
241. H. W. Rossmoore and M. Sondossi, *Adv. Appl. Microbiol.* **33,** 223–277 (1988).
242. H. W. Rossmoore, *Dev. Ind. Microbiol.* **20,** 41–71 (1979).
243. R. J. Lukens in D. Torgeson, ed., *Fungicides,* Vol. 2, Academic Press, Inc., New York, 1969, pp. 395–445.
244. H. H. Shepard, *The Chemistry and Toxicity of Insecticides,* Burgess Publishing Co., Minneapolis, 1939.
245. W. Forsyth, *A Treatise on the Culture and Management of Fruit Trees,* London, 1803.
246. J. G. Horsfall, *Principles of Fungicidal Action,* Chronica Botanica Co., Waltham, Mass., 1956, pp. 2–6.
247. E. C. McCulloch, *Disinfection and Sterilization,* 2nd ed., Lea & Febiger, Philadelphia, 1945, p. 369.
248. L. B. Kingery and A. Adkinson, *Arch. Dermatol. and Syphilol.* **17,** 449 (1928).
249. L. Weinstein and co-workers, *N. Engl. J. Med.* **263,** 793–800, 842–849, 900–907 (1960).
250. S. S. Block and J. P. Weidner, *Dev. Ind. Microbiol.* **4,** 213–222 (1963).
251. S. S. Block and J. P. Weidner, *Nature* **214,** 478–479 (1967).
252. G. A. Woods, *Chem. Eng.* **80,** 81–84 (1973).
253. O. K. Stark and M. Montgomery, *J. Bact.* **29,** 6 (1935).
254. K. E. Birkhaug, *J. Inf. Dis.* **53,** 250 (1933).
255. J. H. Brewer, *J. Am. Med. Assoc.* **112,** 2009 (1939).
256. R. Koch, *Mitteilungen ans dem Kaiserlichen Gesundheitsamte* **1,** 234–282 (1881).
257. J. Geppert, *Berlin Klin. Woch. Schr.* **26,** 789–794 (1889).
258. A. E. Elkhouly and R. T. Yousef, *J. Pharm. Sci.* **63,** 681 (1974).
259. N. Grier in S. S. Block, ed., *Disinfection, Sterilization, and Preservation,* 3rd ed. Lea & Febiger, Philadelphia, 1983, pp. 346–374.
260. J. Raulin, *Sci. Nat.* **11,** 93 (1869); R. G. Berk, *Proj. WS768 Engineer Board,* Corps of Engineers, U.S. Army, Ft. Belvoir, Va., Abstr. 1, 1947.
261. C. E. Hartford and S. E. Ziffren, *J. Trauma* **12,** 682 (1972).
262. H. S. Carr, T. J. Wodkowski, and H. S. Rosenkranz, *Antimicrob. Agents Chemotherap.* **4,** 585 (1973).
263. C. L. Fox, *Arch. Surg.* **96,** 184 (1968).
264. C. H. Brandes, *Ind. Eng. Chem.* **26,** 962 (1934).
265. R. K. Hoffman and co-workers, *Ind. Eng. Chem.* **45,** 287 (1953).
266. N. Grier in S. S. Block, ed., *Disinfection, Sterilization, and Preservation,* 3rd ed., Lea & Febiger, Philadelphia, 1983, pp. 375–389.
267. I. B. Romans in C. A. Lawrence and S. S. Block, eds., *Disinfection, Sterilization, and Preservation,* Lea & Febiger, Philadelphia, 1968, pp. 469–475.
268. S. E. A. McCallan, in D. C. Torgeson, ed., *History of Fungicides,* Academic Press, Inc., New York, 1967, p. 4.
269. G. M. Hunt and G. A. Garratt, *Wood Preservation,* McGraw-Hill Book Co., Inc., New York, 1953.

270. S. S. Block in S. S. Block, ed., *Disinfection, Sterilization, and Preservation*, 4th ed. Lea & Febiger, Philadelphia, 1991, pp. 911–912.

271. C. Yeager in S. S. Block, ed., *Disinfection, Sterilization, and Preservation*, 4th ed. Lea & Febiger, Philadelphia, 1991, pp. 358–361.

272. P. A. Wolf and W. H. Riley, *Appl. Microbiol.* **13,** 28–33 (1965).

273. R. T. Ross, *Am. Paint J.* **55,** 23–37 (1971).

274. D. H. Stormont, *Oil Gas J.* **60,** 133–134 (1962).

275. M. H. Gitlitz and C. B. Beiter in S. S. Block, ed., *Disinfection, Sterilization, and Preservation*, 4th ed. Lea & Febiger, Philadelphia, 1991, pp. 344–357.

276. J. B. Sprowls and C. F. Poe, *J. Am. Pharm. Assoc.* **32,** 41 (1943).

277. A. J. Salle in S. S. Block, ed., *Disinfection, Sterilization, and Preservation*, 2nd ed., Lea & Febiger, Philadelphia, 1973, pp. 408–416.

278. F. G. Winaro and C. R. Stumbo, *J. Food Sci.* **36,** 892–895 (1971).

279. U.S. Environmental Protection Agency, *Federal Register 53, 30566,* Aug. 12, 1988, Washington, D.C.

280. U.S. Environmental Protection Agency, *Federal Register 54, 13502,* Apr. 3, 1989, Washington, D.C.

281. C. R. Phillips, *Amer. J. Hyg.* **49,** 280 (1949).

282. R. A. Caputo and K. J. Rohn, *Med. Device Diagn. Ind.* **4,** 37–41 (1982).

283. T. S. Liu and co-workers, *Food Technol.* **22,** 86–89 (1968).

284. G. L. Gilbert and co-workers, *Appl. Microbiol.* **12,** 496–503 (1964).

285. S. Kaye and C. R. Phillips, *Am. J. Hyg.* **50,** 296–300 (1949).

286. C. W. Bruch and M. K. Bruch in M. A. Benarde, ed., *Disinfection*, Marcel Dekker, New York, 1970.

287. C. W. Bruch, *Ann. Rev. Microbiol.* **15,** 245–262 (1961).

288. C. R. Phillips in S. S. Block, ed., *Disinfection, Sterilization, and Preservation*, 2nd ed., Lea & Febiger, Philadelphia, 1977, pp. 592–610.

289. K. Kereluk and co-workers, *Appl. Microbiol.* **19,** 152–156 (1970).

290. A. Klarenbeek and H. A. E. van Tongeren, *J. Hyg.* **52,** 525–528 (1954).

291. S. Kaye, *Am. J. Hyg.* **50,** 289 (1949).

292. K. Kereluk in D. J. D. Hockenfull, ed., *Progress in Industrial Microbiology*, Vol. 10, Churchill Livingstone, Edinburgh, UK, 1971, pp. 105–128.

293. C. W. Bruch and M. G. Koesterer, *J. Food Sci.* **26,** 428–435 (1961).

294. U.S. Pat. 2,809,879 (Oct. 15, 1957), J. N. Masci (to Johnson & Johnson Inc.).

295. R. K. Hoffman and B. Warshowsky, *Appl. Microbiol.* **6,** 358–362 (1958).

296. R. K. Hoffman, *Appl. Microbiol.* **16,** 641–644 (1968).

297. H. R. Curran and F. R. Evans, *J. Inf. Dis.* **99,** 212–218 (1956).

298. A. Yaprov and co-workers, *Can. J. Microbiol.* **24,** 72–74 (1978).

299. J. R. Rickloff, *Appl. and Environ. Microbiol.* **53,** 683–686 (1985).

300. P. M. Foegeding and M. L. Fulp. *App. Bacteriol.* **65,** 249–259 (1988).

301. W. J. Elford and J. van den Ende, *J. Hyg. Comb.* **42,** 240 (1942).

302. M. Ingram and R. B. Haines, *J. Hyg. Camb.* **47,** 146 (1949).

303. R. K. Hoffman in W. B. Hugo, ed., *Inhibition and Destruction of the Microbial Cell*, Academic Press, Inc., New York, 1971, pp. 251–253.

304. A. N. Parisi and W. E. Young in S. S. Block, ed., *Disinfection, Sterilization, and Preservation*, 4th ed., Lea & Febiger, Philadelphia, 1991, p. 585.

305. M. A. Benarde and co-workers, *Appl. Microbiol.* **13,** 776–780 (1965).

306. V. P. Olivieri and co-workers, *5th Chem. Environ. Impact Health Eff. Proc. Conf.*, 1985, pp. 619–634.

307. A. A. Rosenblatt and J. E. Knapp, *Health Industry Manufacturer's Association (HIMA) Conference Proceedings*, 1988, Washington, D.C., pp. 47–50.

308. A. Cremieux and J. Fleurette in S. S. Block, ed., *Disinfection, Sterilization, and Preservation*, 4th ed. Lea & Febiger, Philadelphia, 1991, pp. 1009–1027.

309. Association of Official Analytical Chemists, in K. Helrich, ed., *Official Methods of Analysis,* Vol. 1, 15th ed., AOAC, Arlington, Va., 1990, p. 133.

310. J. R. Rubino and co-workers, *J. Assoc. Offic. Anal. Chem.,* July (1992).

311. J. H. S. Chen in S. S. Block, ed., *Disinfection, Sterilization, and Preservation*, 4th ed., Lea & Febiger, Philadelphia, 1991, pp. 1076–1093.

312. M. Klein and A. Deforest, *Proc. Chem. Spec. Mfg. Assoc. 49th Mid-year Mtg.,* 1963, pp. 116–118.

313. R. Tyler and G. A. J. Ayliffe, *J. Hosp. Infec.,* **9**, 22–29 (1987).

314. R. Tyler, G. A. J. Ayliffe, and C. Bradley, *J. Hosp. Infect.* **15**, 339–345 (1990).

315. P. J. V. Hanson and co-workers, *Brit. Med. J.* **298**, 862–864 (1989).

316. V. S. Springthorpe and co-workers, *J. Hyg.* **97**, 139–161 (1986).

317. N. Lloyd-Evans, V. S. Springthorpe, and S. A. Sattar, *J. Hyg.* **97**, 163–173 (1986).

318. H. N. Prince and co-workers, *J. Pharm. Sci.* **67**, 1629–1631 (1978).

319. H. N. Prince and co-workers in S. S. Block, ed., *Disinfection, Sterilization, and Preservation*, 4th ed., Lea & Febiger, Philadelphia, 1991, pp. 411–444.

320. U.S. General Accounting Office, *GAO/RCED-90-139,* Aug. 1990.

321. J. R. Rubino and co-workers, *ASM Int. Symp. on Chem. Germicides,* Atlanta, Ga., 1990, abstract 11.

322. U.S. Food and Drug Administration, *OTC Topical Antimicrobial Products, Federal Register,* Jan. 6, 1978, pp. 1210–1249.

323. P. B. Price, *J. Infect. Dis.* **63**, 301–318 (1938).

324. C. W. Walter, *Publ. Health Nurs.* **28**, 825–826 (1936).

325. A. R. Cade, *J. Soc. Cosmet. Chem.* **2**, 281–290 (1951).

326. G. A. J. Ayliffe, J. R. Babb, and A. H. Quoraishi, *J. Clin. Pathol.* **31**, 923–928 (1978).

327. G. A. J. Ayliffe and co-workers, *J. Hosp. Inf.* **11**, 226–243 (1988).

328. M. L. Rotter and co-workers, *J. Hyg.* **96**, 27–37 (1986).

329. M. K. Bruch in S. S. Block, ed., *Disinfection, Sterilization, and Preservation*, 4th ed., Lea & Febiger, Philadelphia, 1991, pp. 1028–1046.

330. E. J. L. Lowbury, H. A. Lilly, and J. P. Bull, *Brit. Med. J.* **2**, 531–536 (1964).

331. A. Rosenberg, S. D. Alatary, and A. F. Peterson, *Surg., Gynecol., & Obstet.* **143**, 789–792 (1976).

332. A. D. Russell in C. H. Collins and co-workers, eds., *Disinfectants: Their Use and Evaluation of Effectiveness,* Academic Press, Inc., London, 1981, pp. 45–49.

333. J. W. A. Bendig, *J. Hosp. Infect.* **15**, 143–148 (1990).

334. P. D. Fine, *J. Hosp. Infect.* **6** (Suppl.), 189–193 (1985).

335. P. Axelsson and J. Lindhe, *J. Clin. Peridontol.* **14**, 205–212 (1987).

336. A. Cohen, *J. Hosp. Infect.* **6**, (Suppl.), 155–161 (1985).

337. D. J. Stickler and J. C. Chawla, *J. Hosp. Infect.* **10**, 219–228 (1987).

338. E. L. Larson and B. E. Laughton, *Antimicrob. Agents Chemother.* **31**, 1572–1574 (1987).

339. W. J. Nungester and A. H. Kempf, *J. Inf. Dis.,* **71**, 174–178 (1942).

340. E. H. Spaulding and J. A. Bondi, *J. Infect. Dis.* **80**, 194–200 (1947).

341. H. Laufman, *Am. J. Surg.* **157**, 359–365 (1989).

342. J. F. Gardner and M. M. Peel, *Introduction to Sterilization, Disinfection, and Infection Control,* Vol. 12, 2nd ed., Churchill Livingstone, Melbourne, 1991, pp. 193–216.

SEYMOUR S. BLOCK
University of Florida, Gainesville

DISPERSANTS

Dispersants are materials that help maintain fine solid particles in a state of suspension, and inhibit their agglomeration or settling in a fluid medium. With the help of mechanical agitation, dispersants can also break up agglomerates of particles to form particle suspensions. Another use of dispersants is to inhibit the growth of crystallites in a supersaturated solution. This characteristic is also known as precipitation inhibition, threshold inhibition, or antinucleation. Overall, dispersants are useful in preventing settling, deposition, precipitation, agglomeration, flocculation, coagulation, adherence, or caking of solid particles in a fluid medium.

Specific terms used in this article to compare various states of particles that are undispersed are

Aggregation: a general term to describe an association of particles.

Agglomeration: a process where precipitation particles grow by collision with other particles. Pigment agglomerates can be broken into smaller primary particles with the aid of mechanical shear.

Flocculation: a relatively reversible aggregation often associated with the secondary minimum of a potential energy diagram. Particles are held together loosely with considerable surface separations.

Coagulation: a relatively irreversible aggregation often associated with the primary minimum of a potential energy diagram of two approaching particles. Particles are held together closely.

Physical Chemistry of Dispersants

A convenient way to understand particle dispersion is to consider the process in four successive parts: the nature of particles and surfaces, adsorption onto particles, interface properties, and forces of attraction and repulsion.

Particles and Surfaces.   Dispersants are primarily used to increase stability (prevent settling) of solid particles in liquid media, whereas surfactants are used more frequently to stabilize liquid (including polymer latex) surfaces within liquids. When the surface of a liquid is increased (stressed), molecules of the liquid flow to the surface to lower its energy, "healing" it. In contrast, solids exhibit no significant flow to the surface. Any stresses applied therefore remain in the form of a higher energy surface. Thus the history of a particle is important to its surface properties. Treatments that alter particle surface properties include freshly cleaving a surface along lowest energy crystal faces, adsorption of molecules and ions, heating or cooling, friction, corrosion, and grinding or polishing. The process by which a particle is formed also affects its surface properties. Examples of this include screw and spiral dislocations, missing layers, and other defects due to contamination or stress during formation (1).

Crystals.   A crystal in equilibrium with a solution contains a point whose distance to a crystal face is proportional to the specific surface free energy of that face (Wulff theorem). If equilibrium is disturbed, the system shifts toward the condition of minimum energy by precipitating more substance on, or dissolving from, the crystal faces where the greatest amount of energy will be released. Thus crystal faces with smaller initial linear growth rates become larger, and those with greater growth rates become smaller. The relationships among the growth rates of the individual faces are affected by temperature, pressure, and the presence of substances such as dispersants or antinucleation agents. Crystal growth inhibition was first described in the early 1930s (2). According to this description, threshold inhibitors attach to growing crystal nuclei, inhibiting their further growth. Maintaining small crystal size favors their redissolution due to high surface to volume ratio. The crystal growth inhibitor or antinucleation agent may then be free to interact with other embryonic nuclei (3). An example of the effect of a dispersant on crystal habit modification is shown in the scanning electron micrograph (Fig. 1).

Surface Charge. Inorganic particles have a surface charge in water that is a function of both the particle's character and the pH of the water. Each particle has an isoelectric pH value where the negative and positive charges on the surface just neutralize each other. Isoelectric points for some common inorganic particles are shown in Table 1.

As the pH is increased or decreased from the isoelectric point, the particles acquire a charge (surface potential) that can enhance repulsion. Surface charge on the particle can be approximated by measuring zeta potential, which is the

(a)   (b)

(c)

Fig. 1. Scanning electron micrograph of dried crystals of 5200 ppm calcium sulfate solution containing (**a**) zero, (**b**) 10, and (**c**) 25 mg/L of added poly(acrylic acid) (500 × magnification). Photograph courtesy of Rohm and Haas Co.

Table 1. Point of Zero Charge (Isoelectric Point) of Selected Inorganic Particles

Particle	CAS Registry Number	Isoelectric point pH
hydroxyapatite, $3[Ca_3(PO_4)_2]\cdot Ca(OH)_2$	[1306-06-5]	7
calcite, $CaCO_3$	[471-34-1]	9.5
hematite, Fe_2O_3 (dehydrated)	[1309-37-1]	5.2
hematite (freshly formed in water)		8.0–8.7
rutile, TiO_2	[13463-67-7]	4.7
quartz, SiO_2	[14808-60-7]	2.2

electrostatic potential at the Stern layer surrounding a particle. The Stern layer is the thickness of the rigid or nondiffuse layer of counterions at a distance (δ) from the particle surface, which corresponds to the electrostatic potential at the surface divided by e (2.718...).

As shown in Figure 2, adsorption of dispersants on particle surfaces can increase zeta potential further, enhancing electrostatic repulsion. Increased repulsion between particles is evidenced by lower viscosity in concentrated slurries, or decreased settling rates in dilute suspensions. The effect of added dispersants on settling of (anhydrous) iron oxide particles is shown in Figure 3.

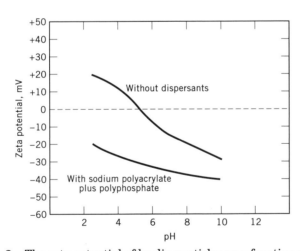

Fig. 2. The zeta potential of kaolin particles as a function of pH.

Particle Motion. All suspended micrometer-size particles are in motion due to the thermal energy they possess. At any given temperature, the average kinetic energy due to thermal motion of an individual particle is equal to kT, where k is the Boltzmann constant (k = the gas constant, R, divided by Avogadro's number):

$$mv^2 = kT$$

In other words, the lower the mass of the particle, m, the higher its velocity, v, because the average energy of any particle at a given temperature is constant,

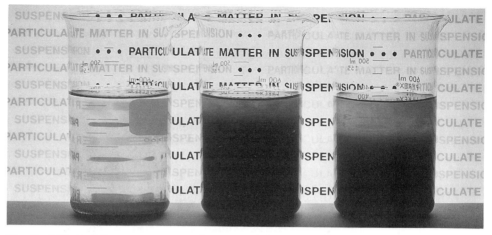

Fig. 3. Effect of dispersants on settling rate of 700 mg/L dehydrated iron oxide in water. Left, no dispersant. Right, 3 mg/L styrenesulfonate–maleic acid copolymer. Center, 3 mg/L acrylic acid–sulfonate–nonionic terpolymer. Photograph courtesy of Rohm and Haas Co.

kT. A dispersed particle is always in random thermal motion (Brownian motion) due to collisions with other particles and with the walls of the container (4). If the particles collide with enough energy and are not well dispersed, they will coagulate or flocculate.

Adsorption onto Particles.   The Gibbs Adsorption Law relates how adsorption (qv) onto surfaces affects interfacial tension,

$$d\gamma = -RT\ \Gamma\ d\ ln\ c$$

where γ = interfacial or surface tension, in N/m (1 N/m = 1000 dyn/cm); R = gas constant; T = absolute temperature; Γ = interfacial or surface concentration, in mol/unit area (ie, adsorption); and c = dimensionless concentration ($d\ ln\ c = dc/c$, thus units cancel).

If adsorption occurs ($\Gamma > 0$), then increasing the concentration of dispersant in the bulk water reduces interfacial tension (5).

Most adsorption processes are exothermic (ΔH is negative). Adsorption processes involving nonspecific interactions are referred to as physical adsorption, a relatively weak, reversible interaction. Processes with stronger interactions (electron transfer) are termed chemisorption. Chemisorption is often irreversible and has higher heat of adsorption than physical adsorption. Most dispersants function by chemisorption, in contrast to surfactants, which tend to physically adsorb. With polymeric adsorbates, not all the monomeric units are adsorbed at any instant, even though the dispersant may be irreversibly bound. Even though each polymer segment may be adsorbed reversibly, the probability is small that all adsorbed segments will desorb at the same time. A good dispersant needs some segments of the chain extending into the solvent (water) to increase repulsion and prevent collision with other particles. The exact configuration of the polymer chain at the interface at any instant results from the balance of the enthalpy of adsorption, the entropy decrease of the chain due to adsorption, and the entropy gain by freeing solvent molecules when the segments adsorb. Computer modeling of polymeric dispersants has been used as a tool to study the effects of configuration and counterion binding on end use properties such as detergent builder assists (6) and water-treatment applications (7).

Adsorption of dispersants at the solid–liquid interface from solution is normally measured by changes in the concentration of the dispersant after adsorption has occurred, and plotted as an adsorption isotherm. A classification system of adsorption isotherms has been developed to identify the mechanisms that may be operating, such as monolayer vs multilayer adsorption, and chemisorption vs physical adsorption (8). For moderate to high mol wt polymeric dispersants, the low energy (equilibrium) configurations of the adsorbed layer are typically about 3–30 nm thick. Normally, the adsorption is monolayer, since the thickness of the first layer significantly reduces attraction for a second layer, unless the polymer is very low mol wt or adsorbs by being nearly immiscible with the solvent.

Interface Properties   A polymeric dispersant may have segments extended into the solution, or the segments may be coiled, depending on whether the solvent is good (polymer–solvent interactions energetically favored) or poor (polymer–polymer and solvent–solvent contacts favored). Between these two solvent–polymer interactions is a θ (theta) solvent, in which neither condition is favored. If the polymer–solvent interaction is better than θ conditions, the extending chains or segments will repel chains adsorbed on other particles, as well as making the distance between two particles greater and enhancing steric repulsion. If the interaction is worse than θ conditions, the particles may flocculate due to mutual attraction of the polymer layers (9). On the other hand, if the polymer–solvent interaction is too strong, the polymeric dispersant may be adsorbed only weakly or not at all. This can lead to depletion flocculation, which is due to desorbed chains that are squeezed out from between two approaching particles. The desorbed chains can cause solvent to flow from between the particles by osmotic forces, leaving a bare area so that attraction between the particles is increased.

Attractive and Repulsive Forces.   The force that causes small particles to stick together after colliding is van der Waals attraction. There are three van der Waals forces: (*1*) Keesom-van der Waals, due to dipole–dipole interactions that have higher probability of attractive orientations than nonattractive; (*2*) Debye-van der Waals, due to dipole-induced dipole interactions (ie, uneven charge distribution is induced in a nonpolar material); and (*3*) London dispersion forces, which occur between two nonpolar substances.

London forces are caused by the electrons in nonpolar molecules circulating with extremely high frequency. This causes nonpolar molecules to be polar at any given instant, even though the polarity changes with high frequency. As with true dipoles, attractive alignments are more favored because the molecules are free to rotate. Two nonpolar molecules begin to oscillate as they approach each other, and the attraction between them becomes stronger as the distance decreases. The first oscillator relays its orientation to the second at the speed of electromagnetic waves. Thus, when the molecules are far apart, there is a delay in the response of the second oscillator to the first, leading to weaker attraction at larger distances. The term dispersion forces comes from their effect on the index of refraction, which is frequency dependent. Hamaker (10) applied London's expression for the attraction between two isolated molecules to calculate the energy of attraction for all the molecules in two separate particles. The result is the Hamaker constant, which is used for calculating the attractive force between particles. A further refinement is to include terms for the effect of solvent such as water. Hamaker constants for representative particles are shown in Table 2 (higher constants indicate greater attraction).

Table 2. Hamaker Constants of Selected Materials

Material	Hamaker constant, J $\times$ 10^{20}	
	In space	In water
gold	45.3	33.5
magnesium oxide	10.5	1.6
polystyrene	~8	0.93
C_8 alkane	5.02	0.41
C_5 alkane	3.94	0.336

Electrostatic Repulsive Forces.   As the distance between two approaching particles decreases, their electrical double layers begin to overlap. As a first approximation, the potential energy of the two overlapping double layers is additive, which is a repulsive term since the process increases total energy. Electrostatic repulsion can also be considered as an osmotic force, due to the compression of ions between particles and the tendency of water to flow in to counteract the increased ion concentration.

DLVO Theory.   The overall stability of a particle dispersion depends on the sum of the attractive and repulsive forces as a function of the distance separating the particles. DLVO theory, named for Derjaguin and Landau (11) and Verwey and Overbeek (12), encompasses van der Waals attraction and electrostatic repulsion between particles, but does not consider steric stabilization. The net energy, ΔG_T, between two particles at a given distance is the sum of the repulsive and attractive forces:

$$\Delta G_T = \text{(electrostatic repulsive forces)} - \text{(van der Waals attractive forces)}$$

The electrostatic repulsive forces are a function of particle kinetic energy (kT), ionic strength, zeta potential, and separation distance. The van der Waals attractive forces are a function of the Hamaker constant and separation distance.

Figure 4 is a potential energy curve calculated for a titanium dioxide pigment mixed with different concentrations of sodium polymethacrylate dispersant, which changes the zeta potential. Vt is the net potential energy calculated from DLVO theory, and kT is the particle energy. Normally, Vt should be about 5 or 6 times higher than kT to promote stability in a dispersion. The fluidity point is the point at which enough dispersant has been added to a pigment paste to create an initial fluid dispersion.

A summary of the effects of basic colloid variables predicted by DLVO theory is presented in Table 3.

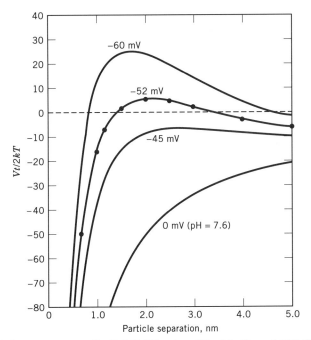

Fig. 4. Potential energy curves for R-900 Titanium Dioxide/Tamol 850 (T-850) dispersant mixtures. Tamol is a trademark of Rohm and Haas Co. Numbers on curves indicate the zeta potential in mV. The fluidity point = 0.21% T-850 = -52 mV. Figure courtesy of P. R. Sperry, Rohm and Haas Co.

Table 3. Effects of Basic Colloid Variables

Increasing	Coagulation resistance	Flocculation resistance
particle size	increase	decrease
surface potential	increase	increase
electrolyte	decrease	decrease
Hamaker constant	decrease	decrease

Steric Repulsion. Although some progress has been made in calculating steric repulsive forces (13), the theory concerning them is not as completely developed as DLVO theory. The adsorbed polymer layers of two particles (in a good solvent) begin to interpenetrate as the particles approach each other. The interaction between these polymer layers can have an osmotic effect due to an increase in the local concentration of the adsorbed polymer layers, and can have an entropic or volume restriction effect due to crowding of the interacting chains. In both cases, entropy decreases, which is unfavorable. Moreover, the osmotic effect can create unfavorable enthalpic changes due to desolvation of closely packed chains. To regain lost entropy, the particles must separate to allow the chains more freedom of movement, while the solvent moves in to resolvate the polymer layer. As with electrostatic repulsion, an energy barrier is created. A common approximation used is that the strength of the energy barrier rises steeply at slightly less than the adsorbed layer thickness. Some of the practical differences between sterically and electrostatically stabilized dispersions may be summarized as follows (14):

Steric stabilization	Electrostatic stabilization
insensitive to electrolyte	coagulation occurs with increased electrolyte
effective in aqueous and nonaqueous media	more effective in aqueous media
effective at high and low concentrations	more effective at low concentrations
reversible flocculation common	coagulation often irreversible
good freeze–thaw stability	freezing often induces irreversible coagulation

Advances have been made in directly measuring the forces between two surfaces using freshly cleaved mica surfaces mounted on supports (15), and silica spheres in place of the sharp tip of an atomic force microscopy probe (16). These measurements can be directly related to theoretical models of surface forces.

Comparisons with Other Materials

Surfactants vs Dispersants. Surface-active agents or surfactants (qv) are functionally related to dispersants. Surfactants are used primarily in systems where a second, nonaqueous liquid phase is present. As a result, surfactants are used to stabilize oil–water emulsions such as lubricating oils, lotions, resins, and latices. On solid particles surfactants adsorb and wet the surface, but do not yield stable dispersions due to small barrier thickness. Although the terms dispersant and surfactant are frequently confused, there are several important differences between the two classes of materials. For example, surfactants, which are small molecules containing both a hydrophilic and a hydrophobic portion, are defined by the nature of the molecule. Surfactants tend to orient at the air–water interface, oil–water interface, or sometimes at a liquid–solid interface depending on the length of the hydrophobic portion and the nature of the hydrophilic part (anionic, cationic, or nonionic). In contrast, dispersants, which tend to be larger polymeric molecules, are defined more by their use, which is to disperse a solid in a liquid. Surfactants adsorb at surfaces, preferring to be out of the water phase. Dispersants adsorb by means of chemisorption or electron transfer, using specific anchoring groups. As an example, most dispersants will not adsorb at latex surfaces, in contrast to surfactants, which do. Although there are some dispersants that have significant surfactant-like properties, a quantitative way to distinguish the two classes is by measuring their surface tension in water. A 0.02% by weight solution of a surfactant produces a surface tension of <40 mN/m (=dyn/cm), whereas in a dispersant solution it will generally be above that number.

Chelants and Precipitation Inhibitors vs Dispersants. Dispersants can inhibit crystal growth, but chelants, such as ethylenediaminetetraacetic acid [60-00-4] (EDTA), and pure precipitation inhibitors such as nitrilotris(methylene)trisphosphonic acid [6419-19-8], commonly known as amino trismethylene phosphonic acid (ATMP), can be more effective under certain circumstances. Chelants can prevent scale by forming stoichiometric ring structures with polyvalent cations (such as calcium) to prevent interaction with anions (such as carbonate). Chelants interact stoichiometrically with polyvalent cations, in preference to adsorbing on surfaces. Pure crystal growth inhibitors such as ATMP adsorb nearly completely to surfaces, and thus do not provide enough residual negative charge for coulombic particle repulsion and dispersion.

Flocculants vs Dispersants. In direct contrast to dispersants, flocculants or coagulants are used to aggregate fine particles or liquid droplets in aqueous media to improve separation of the two components. Flocculants function by charge neutralization, double-layer compression, particle bridging, or by forming large nets that engulf masses of particles (sweep flocculation). Dispersants normally function by charge repulsion, steric repulsion, or both. Some polymeric flocculants can be chemically similar to dispersants, differing only in molecular weight and dosage level used. However, dispersants are normally anionic materials of lower molecular weight and higher charge density than polymeric flocculants (see FLOCCULATING AGENTS).

In summary, dispersants are effective for particle dispersion and crystal growth inhibition, but do not normally have surface-active properties such as oil emulsification. Chelants and antiprecipitants frequently inhibit crystal growth better than dispersants, but are ineffective for particle dispersion. Flocculants are effective for aggregating particles, the opposite function of a dispersant.

An illustration of the key performance differences among these different classes is given in Table 4.

Table 4. Comparison of Performance of Materials by Functional Class

Material	Function	Typical dosage, ppm	Ratio of cation to material	Oil emulsification	Kaolin dispersing	$CaCO_3$ growth inhibition
EDTA	chelant	20–100	1:1	poor	poor	100%
ATMP	crystal growth inhibitor	5–10	10^2–10^4:1	poor	poor	90–100%
polycarboxylic acid[a]	dispersant	5–1000+	10–10^6+:1	poor–fair	very good	80–90%
anionic surfactant	surface-active agent	100	na	very good	fair–poor	<30%
anionic polyacrylamide[b]	flocculant	5–100	na	poor	very poor	<10%

[a] 2000 mol wt.
[b] 5,000,000 mol wt.

Dispersant Materials

Condensed Phosphates. The term condensed phosphate refers to any dehydrated, condensed orthophosphate for which the $M_2O:P_2O_5$ ratio is less than 3:1 (17). (**1**) is disodium orthophosphate [7558-79-4] and (**2**) tetrasodium pyrophosphate [7722-88-5].

$$2 \ Na_2HPO_4 \longrightarrow Na_4P_2O_7 + H_2O$$

(**1**)   (**2**)

The most active dispersants for many applications (18) are amorphous polyphosphate salt glasses which contain long phosphate chains. Most condensed phosphates are prepared by calcining the appropriate ratio of $Na_2O:P_2O_5:H_2O$ under controlled conditions to form the desired compositions directly. Despite the widespread applicability of the condensed phosphates, ease of manufacture, and low cost, the total demand for them throughout the 1980s and 1990s has declined. Probably the single biggest blow to phosphates was the discovery of their role in the eutrophication of enclosed bodies of water (19). In addition, polyphosphates suffer reversion or degradation to orthophosphates through hydrolysis in neutral or acidic solution. Orthophosphate is not an active dispersant, and orthophosphate anions may actually favor flocculation or deposition under conditions of high calcium levels. As conditions of use become more severe, reversion of polyphosphates becomes a more significant problem. Thus better products have frequently been developed for particular applications.

Organic Polymeric Dispersants. Table 5 lists dispersant materials by types and trademarked names for each class of materials.

Polyacrylates (**3**), where R = H, CH_3; $n < 100,000$; and Y = OH, OCH_3, O^-Na^+, etc, or copolymers with compatible monomers, are probably the most flexible dispersant products, because they are produced in a variety of molecular weights and degrees of anionic charge. Moreover, reaction of acrylic acid with other monomers confers additional properties that make them more adaptable for niche applications.

R
|
$+CH_2-C+_n$
|
C=O
|
Y

(**3**)

$+CH-CH+_n$
| |
C=O   C=O
| |
Y   Y

(**4**)

Polymaleates (**4**), where $n < 100,000$ and Y = OH, O^-Na^+, or copolymers with compatible monomers such as styrene, acrylic acid, etc, generally show properties similar to those of polyacrylates, but maleic acid is not as easily copolymerized with other functional monomers to allow tailoring to specific applications. Naphthalenesulfonic acid–formaldehyde condensates and melamine–formaldehyde condensates are used in cost-sensitive applications such as cement.

Natural product-derived dispersants, such as tannins, lignins, and alginates, are still widely used as drilling mud thinners or in specialty applications where

Table 5. Examples of Dispersants

Chemical name	Manufacturers	Trademarks
poly(meth)acrylates	Alco Chemical[a]	Alcosperse, Aquatreat
	Allied Colloid, Ltd.	Antiprex, Alcomer
	American Cyanamid Co.	Cyanamer
	FMC Corp.	Belsperse
	Goodrich Chemical Co.	Goodrite
	Rohm and Haas Co.	Acusol, Acumer, Tamol
	W.R. Grace & Co.	Daxad
polymaleates	BASF Corp.	Sokolan
	FMC Corp.	Belgard, Belasol
	NorsoHaas S.A.	Norasol
condensed phosphates	Calgon Corp.	[b]
	FMC Corp.	[b]
	Monsanto Chemical Co.	[b]
polysulfonates[c]	National Starch & Chemical Corp.	Versa TL
	Westvaco Corp.	Polyfon, Reax, Indulin
	Borregaard Industries, Ltd.	Vanisperse, Borresperse, Ultrazine
sulfonated polycondensates[d]	Diamond-Shamrock Corp.	Lomar
	Rohm and Haas Co.	Tamol
tannins, lignins, glucosides, alginates[e]	Georgia-Pacific Corp.	[b]
	Kimberly-Clark Corp.	[b]
	Marathon Paper Co.	Marasperse

[a]Division of National Starch and Chemical Corp.
[b]Material sold either generically or by code number only.
[c]Polymeric structures having pendent SO_3 groups, lignosulfonates, and polystyrene sulfonates.
[d]Including naphthalene sulfonate–formaldehyde condensates.
[e]Polymeric materials derived from natural products.

their low toxicity is a crucial property, eg, in boilers producing steam for food applications.

Uses

Recirculating Cooling Water. Water used to cool plant processes and buildings contains contaminants that can accelerate corrosion of metal surfaces and leave scale or particle deposits on pipes and heat-transfer surfaces. To prevent corrosion and scale, cooling water treatment formulations contain corrosion inhibitors, biocides, phosphonates, and dispersants. Dispersants aid in the prevention of inorganic fouling (silt, iron oxide), scaling (calcium carbonate, calcium sulfate), and corrosion. The dispersant minimizes settling of inorganic foulants by adsorbing on particles to increase their mutual repulsion (20). Dispersants prevent scale by adsorbing on active sites on growing crystals, ideally minimizing their particle size to less than the wavelength of visible light, making them invisible to the unaided eye. Small particles settle out on pipes and surfaces less easily, following Stokes law. In addition, the relatively large surface area of these particles favors their redissolution. Dispersants also restrict the particle size of corrosion-inhibiting agents such as calcium phosphonate and zinc. In alkaline cooling water treatment, corrosion inhibitors form films at high pH cathodic areas on a corroding metal surface. By helping to minimize particle size of these corrosion inhibitors in the bulk water, dispersants enable the inhibitors to precipitate preferentially at cathodic surfaces as a thin film of closely packed particles in the form of a hydroxide salt or complex (21). Specialty dispersants have been developed to prevent specific inorganic scales and particles from fouling cooling water surfaces. For example, calcium phosphate and iron are controlled with copolymers of acrylic acid–nonionic monomers (22,23), acrylic acid–sulfonate monomers (24), or acrylic acid–sulfonate–nonionic terpolymers (25). Calcium carbonate is controlled with poly(maleic acid) (26) and organic phosphonates. Dispersants designed to control silica and magnesium silicate (7,27,28) have been introduced.

Boiler Water. Dispersants are used to prevent scale buildup on the boiler tubes and drums of boilers that operate at pressures less than about 10.3 MPa (1500 psig). Above that pressure, the temperature is too high for most dispersants to withstand without breaking down into less effective fragments (29). Boiler water treatments using dispersants generally fall into three categories: precipitation programs, where water hardness ions are preferentially precipitated as calcium carbonate or hydroxyapatite (calcium phosphate hydroxide) and are dispersed in the bulk water rather than at the metal surface; dispersant and chelant combinations (30), where the dispersant is used to control precipitates formed from excess hardness leakage from water pretreatment; and dispersant only (31–33) or dispersant plus sequestrant (34), where dispersants bind hardness ions to prevent precipitation and scaling with polyvalent anions. A significant fouling problem in boilers is iron, which can be controlled with specialty dispersants (35) that are not inactivated by ferrous(ic) ions.

Geothermal Fluids. Geothermal fluids are used to provide energy for power generation and home heating. When heat or steam is extracted from geothermal fluids, the fluid (water) has a greater tendency to deposit scales. Scaling causes pipes, wells, and drains to become blocked, reinjection wells to become less receptive to waste fluid, loss of heat energy, and environmental problems (36–38). In most cases, the scale is comprised of either silica or calcium carbonate. Dispersants are used to change the surface of the precipitates formed so that they no longer adhere to the other surfaces. It is likely that the dispersants minimize scaling from geothermal fluids by modifying crystals and by increasing coulombic repulsion of the particles formed.

Seawater Distillation. The principal thermal processes used to recover drinking water from seawater include multistage flash distillation, multi-effect distillation, and vapor compression distillation. In these processes, seawater is heated, and the relatively pure distillate is collected. Scale deposits, usually calcium carbonate, magnesium hydroxide, or calcium sulfate, lessen efficiency of these units. Dispersants such as poly(maleic acid) (39,40) inhibit scale formation, or at least modify it to form an easily removed powder, thus maintaining cleaner, more efficient heat-transfer surfaces.

Reverse Osmosis. In contrast to distillation, reverse osmosis (qv) (RO) uses hydraulic pressure as its energy source to purify water. In RO, a fraction of the water content of seawater or brackish water is driven under pressure through a semipermeable membrane. The impure water, flowing past the membrane in several stages, becomes progressively more concentrated. The membrane, which is usually cellulose acetate or polyamide, can become fouled with silt, calcium carbonate, iron, or silica. Dispersants are used to minimize fouling and decrease the frequency of flushing or cleaning (41,42). Dispersants are also used in proprietary membrane cleaning agents (42). Since the water quality, measured by the amount of salt passage through the membrane, from RO treatment can be traded off for increased flux (volume), this method is often combined with multistage flash distillation to produce drinking water of acceptable quality and high volume (43). To conform to regulations for the production of drinking water by RO, specially produced grades of poly(acrylic acid) (44) or other dispersants must be used.

Sugar Processing. Dispersants are used in the production of cane and beet sugar to increase the time between evaporator clean outs. Typical scales encountered include calcium sulfate, calcium oxalate, calcium carbonate, and silica. Dispersants are fed at various points in the process to prevent scale buildup, which would interfere with efficient heating of the vessels. Only certain dispersants, conforming to food additive regulations, can be used, since a small amount of the dispersant may be adsorbed on the sugar crystals.

Oilfield. Scales can plug a producing well, requiring expensive remediation or even requiring a new well to be drilled. Scale also forms on topside equipment and piping, which is usually less difficult to handle. One primary cause of scale formation is the mixing of incompatible waters from either two different aquifers or, in the case of off-shore drilling, from the mixing of seawater with formation water (45,46). The problem is aggravated by the release of gas pressure, eg, carbon dioxide, at the wellbore, which raises pH and increases the risk of calcium carbonate formation. The most frequently encountered scales are calcium carbonate, and barium, strontium, and calcium sulfate. Polymeric dispersants (47) and organic phosphonates are most often used to prevent oilfield scaling by delaying precipitation and preventing scale adherence on pipes and surface equipment. Scale inhibitors are normally injected (squeezed) into the producing well periodically at high dosages to force adsorption onto formation surfaces near the wellhead (48,49). The inhibitor then slowly desorbs and is produced with the oil and formation water at low levels over time, preventing scale formation. This process is repeated when the phosphonate or polymeric dispersant falls below effective levels.

Drilling Muds. Aqueous drilling muds normally consist of bentonite clay, weighting agents such as barite, dispersants such as lignite, lignosulfonate, and various polymers, and fluid loss agents. Bentonite clay is used to modify mud rheology so that drilled cuttings can be carried to the surface, and to help seal off the drilled hole so that fluid does not easily penetrate into the surrounding formation causing wash-out or hole collapse. Bentonite clay consists of flat plate particles with negatively charged faces and positively charged edges, which attract each other to form an open card-house structure that does not have good wall-sealing and rheological properties. A dispersant neutralizes the positive edge charges of bentonite particles, allowing them to lay flat against the sides of the drilled hole to minimize water intrusion into the formation. Weighting agents, for increasing the density of the mud, must also be dispersed. Seawater muds (offshore drilling) and gypsum muds require specialty dispersants that have higher divalent cation tolerance than dispersants used in freshwater muds. Chrome lignosulfonate has been the preferred dispersant for these wells, but is now being replaced by polymeric acrylic dispersants (50–52) because of the harmful effects of chromium on the environment.

Cement. Although water is needed in the hydration reactions of cement (qv), excess water added for workability of the concrete and morter creates voids that decrease strength and increase water permeability. Dispersants are used as plasticizers in cements to cut water demand by 2–4% and decrease void volume. More recently, superplasticizers have been used that can reduce water demand by about 5–15%. This results in much higher early strength and less water permeability in the concrete (53–55). Superplasticizers function by adsorbing on the surface of alite (tricalcium silicate, the principal constituent of Portland cement). This adsorption increases the charge repulsion of the particles, allowing less water to be used while maintaining low slurry viscosity (increased workability). Reportedly, superplasticizers retard the hydration of alite by limiting its interaction with water molecules and minimizing the dissolution of calcium ions that normally further react as the cement cures (56). The two main classes of superplasticizers are melamine–formaldehyde condensates and naphthalenesulfonate condensates. Newer superplasticizers derived from acrylic acid-based polymers require only about 30% of the dosages of earlier types (57–60).

Paints and Pigments. Typically, a paint (qv) contains pigments (qv) to provide hiding or coverage, polymeric binders to hold the coating onto the surface, dispersants, thickeners, flow and leveling aids, defoamers, and biocides. As they are customarily supplied, pigments are powders consisting of agglomerates of individual pigment particles. Pigments must be dispersed in the liquid medium and stabilized at their primary particle size to provide maximum hiding and film properties (61,62). The dispersion process involves three steps. The first step involves wetting the pigment particles. This is done by the dispersant or with an auxiliary surfactant, if the dispersant is not an efficient wetting agent. The second step is to break down agglomerates, requiring energy supplied by a high speed disperser. The resulting dispersion must then be stabilized, which is the primary function of the dispersant. This requires maintaining the pigment particles in their dispersed state during the manufacture, storage, application, and drying of the paint. Normally, polymeric acrylic dispersants are used to obtain a stable pigment dispersion, although polyphosphates and other low cost dispersants are sometimes used in grinding the initial dispersion. There are three principal types of polymeric dispersants, each having a balance of advantages and disadvantages (63). Dispersants based on poly(carboxylic acid) are usually the cheapest and the most efficient in producing the initial dispersion, but ordinarily do not produce high gloss in latex paints. Copolymers of an acid and a hydrophilic comonomer are less efficient than the polyacids, but frequently provide much higher gloss, and have good compatibility with various thickeners used in paints. Copolymers of an acid and a hydrophobic comonomer are also less efficient than the polyacids, and can also provide high gloss. The hydrophobic comonomer, which contributes surfactant-like properties, can provide better color acceptance with organic pigments (64–67).

Mineral Processing. Dispersants are used as mineral processing aids for grinding and improving slurry stability. Key mineral processes using dispersants include calcium carbonate and kaolin manufacture, and gold beneficiation. As a grinding aid, the dispersant reduces the amount of mechanical energy required to break down the ore to smaller particle size. For grinding, polyphosphates are frequently used because of their low cost. If the slurry is spray-dried to a powder, then little or no additional dispersant is required. If the mineral is kept in slurry form, then an auxiliary polymeric dispersant such as sodium polyacrylate is added to improve storage stability.

Caulks, Sealants, and Roof Coatings. These have similar technology to paints, except that they are formulated at higher solids to produce a thicker coating. As with paints, two types of dispersant are used; a primary dispersant such as KTPP (potassium tripolyphosphate) to disperse pigment agglomerates, and a secondary dispersant, such as sodium polymethacrylate, to provide storage stability of the formulation. The polyphosphate is a good dispersant for pigment, but hydrolyzes during storage to orthophosphate, which is not effective. Caulks, sealants, and roof coatings differ from paints in that they must minimize water leakage (permeance), but still allow water-vapor transport. Although water permeability of these films lessens with time due to leaching of water-sensitive materials such as dispersants, a variety of less sensitive dispersants have been developed to overcome this problem. Less water-sensitive dispersants include zinc sodium hexametaphosphate (primary dispersant), and hydrophobic polymeric dispersants (secondary dispersant) (68).

Agricultural Uses. Dispersants are used to formulate pesticides into aqueous dispersions (flowables), wettable powders, water-dispersible granules, and dry flowables. Aqueous pesticide dispersions or flowables are liquid suspensions of 20–50% active ingredient (A.I.), and usually contain wetting agent, thickeners such as polysaccharide gums, and dispersants. When the formulation is mixed in water, the A.I. particles can settle in spray tanks unless a dispersant is used to prevent settling and agglomeration of the pesticide before application. Powders, granules, and dry flowables contain A.I., a carrier or diluent (eg, clay or silica), wetting agent, and dispersing agent. These formulations are diluted with water before spraying on crops. The dispersant keeps the A.I. uniformly suspended during application to provide uniform coverage and prevent nozzle clogging. Dispersants used include lignosulfonates, naphthalenesulfonate condensates, and sodium polymethacrylate (69). Another agricultural use of dispersants is in the production of animal feed, where lignosulfonate acts as a binder to enhance pelletizing.

Detergents and Cleaners. Dispersants function as builder assists (70,71) in cleaning formulations to increase particulate soil removal, prevent redeposition of soils to maintain whiteness or eliminate residues (spots) on hard surfaces, prevent precipitation of inorganic salts (carbonates, phosphates, and silicates), increase water-wettability of soiled surfaces, and promote physical stability of slurried formulations. They improve spray-drying of powders by improving slurry (crutcher) homogeneity, increasing solids of crutcher mix (time and energy savings), reducing dusty fines, and increasing bead strength. Dispersants increase the rate of solution for powdered detergents and buffer the water to maintain optimum cleaning pH (see DETERGENCY).

A typical cleaning or detergent formulation contains surfactants (eg, linear alkyl sulfonic acid), builders (eg, sodium metasilicate, soda ash, polyphosphates), and dispersants, eg, polyphosphates, poly(acrylic acid). In fabrics, soil redeposition during cleaning leads to graying, and on glasses it leads to spotting and filming. Dispersants increase the coulombic repulsive barrier around particles to enhance their removal from surfaces and inhibit their readherence (72,73). Historically, polyphosphates have been used as dispersants in detergents and cleaners. However, detergent manufacturers have been eliminating polyphosphate from some formulations due to its impact on eutrophication of lakes and streams. Revised cleaning formulations now contain a combination of inorganic builder and synthetic polymeric dispersant to replace polyphosphates (70,72,74). Nonphosphate detergents and cleaners that use soda ash as a builder can form calcium carbonate with the hardness in the water. The addition of polymeric dispersants to the formulation inhibits calcium carbonate precipitation, thus preventing the encrustation of the carbonate salt on the fabric or surface.

Environmental Considerations

Biodegradability of Dispersants. Because dispersants are water soluble, their environmental fate is less obvious to the general public than, for example, packaging plastic. Most reviews on biodegradable polymers suggest that, with the exception of poly(vinyl alcohol) and poly(ethylene glycol)s, most synthetic organic dispersants are recalcitrant in the environment (75). More recently developed dispersants displaying biodegradability are polymers containing ester linkages (76) and ether linkages (77). There is currently a great deal of research activity to develop dispersants that are both effective and biodegradable. Consequently, developments in this area should be rapid and changing into the year 2000.

Future of Dispersants. The dispersant market has been changing continuously since the early 1970s. The primary driving forces in this market are environmental; reduction or elimination of phosphate from detergents (replacement of polyphosphate builders with polymeric dispersants), elimination of chromate in cooling water (requiring high pH, dispersant-dependent formulations), removal of ferrochrome lignosulfonate from drilling muds, and elimination of organic solvents (requiring aqueous dispersants) from paints, inks, coatings, and agricultural formulations to minimize air pollution.

Continuing the current trend, there will be an even greater need to provide highly effective dispersants with minimum environmental impact. Future development efforts will focus on improved performance at low dosage to further reduce the environmental load, accountability (ability to detect dispersants at ppm levels and minimize dose level), and, ultimately, biodegradability. Improved performance will mean ever more specialized dispersants to fill specialized technical requirements. The trend to eliminate organic solvents from paints, coatings, and agricultural formulations will continue.

BIBLIOGRAPHY

"Dispersants" in *ECT* 3rd ed., Vol. 7, pp. 833–848, by R. M. Goodman, American Cyanamid Co.

1. D. Myers, *Surfaces, Interfaces and Colloids*, VCH Publishers, Inc., New York, (1991).
2. G. B. Hatch and A. Rice, *Ind. Eng. Chem.* **31,** 51 (1939).
3. G. R. McCartney and A. G. Alexander, *J. Colloid Sci* **13,** 383 (1958).
4. T. C. Patton, *Paint Flow and Pigment Dispersion*, 2nd ed., John Wiley & Sons, New York, 1979.
5. J. Lyklema, *Fundamentals of Interface and Colloid Science*, Vol. I, Academic Press, Inc., San Diego, 1991.
6. S. J. Fitzwater and M. B. Freeman, paper presented at *82nd Am. Oil Chem. Soc. Annual Meeting*, Chicago, 1991.
7. W. M. Hann and S. T. Robertson, *Industrial Water Treatment* **23**(6), 12–21 (1991).
8. C. H. Giles and co-workers, *J. Chem. Soc.*, 3973 (1960).
9. W. B. Russel, D. A. Saville, and W. R. Schowalter, *Colloidal Dispersions*, Cambridge University Press, New York, 1989.
10. H. C. Hamaker, *Physica* **4,** 1058 (1937).
11. B. V. Derjaguin and L. D. Landau, *Acta Physiochim (USSR)* **14,** 633 (1941).
12. E. J. W. Verwey and J. Th. G. Overbeek, *Theory of Stability of Lyophobic Colloids*, Elsevier, Amsterdam, 1948.
13. Th. F. Tadros, *The Effect of Polymers on Dispersion Properties*, Academic Press, Inc., London, 1982.
14. R. J. Hunter, *Foundations of Colloid Science*, Vol I, Oxford Science Publications, New York, 1988.
15. J. Klein, *Molecular Conformation and Dynamics of Macromolecules in Condensed Systems*, Elsevier, Amsterdam, 1988, p. 333.
16. W. A. Ducker, T. J. Senden, and R. M. Pashley, *Nature* **353** (Sept. 19, 1991).
17. J. R. Van Wazer, *Phosphorous and Its Compounds*, Interscience Publishers, New York, 1958.
18. U.S. Pat. 3,130,152 (Apr. 21, 1964), R. J. Fuchs (to FMC Corp.).
19. D. W. Schindler, *Science* **184,** 897 (May 24, 1974).
20. J. W. McCoy, *The Chemical Treatment of Cooling Water*, 2nd ed., Chemical Publishing Co., New York, 1983.
21. W. M. Hann, S. T. Robertson, and P. Zini, *Proceedings of 11th Int. Corrosion Congress*, Associazione Italiana di Metallurgica, Milan, 1990, pp. 115–122.
22. U.S. Pat. 4,029,577 (June 14, 1977), I. T. Godlewski, J. J. Schuck, and B. L. Libutti (to Betz Laboratories).
23. U.S. Pat. 4,209,398 (June 24, 1980), M. Ii and co-workers (to Kurita Water Industries and Sanyo Chemical Industries, Ltd.).
24. U.S. Pat. 3,928,196 (Dec. 23, 1975), L. J. Persinski, P. H. Raiston, and C. G. Gordon (to Calgon Corp.).
25. U.S. Pat. 4,711,725 (Dec. 8, 1987), D. R. Amick, W. M. Hann, and J. Natoli (to Rohm and Haas Co.).
26. U.S. Pat. 3,810,834 (May 14, 1975), T. I. Jones, N. Richardson, and A. Harris (to FMC Corp.).
27. U.S. Pat. 5,100,558 (Mar. 31, 1992), J. M. Brown and co-workers (to Betz Laboratories).
28. U.S. Pat. 5,158,685 (Oct. 27, 1992), D. T. Freese (to Betz Laboratories).
29. J. W. McCoy, *The Chemical Treatment of Boiler Water*, Chemical Publishing Co., New York, 1981.
30. U.S. Pat. 4,566,972 (Jan. 28, 1986), J. J. Bennison, S. W. Longwerth, and J. G. Baker (to Dearborn Chemical Co.).

31. U.S. Pat. 4,545,920 (Oct. 18, 1985), W. F. Lorenc, J. A. Kelly, and F. S. Mandel (to Nalco Chemical Co.).

32. U.S. Pat. 4,457,847 (June 3, 1984), W. F. Lorenc, J. A. Kelly, and F. S. Mandel (to Nalco Chemical Co.).

33. U.S. Pat. 4,680,124 (July 14, 1987), P. R. Young, M. E. Koutek, and J. A. Kelly (to Nalco Chemical Co.).

34. U.S. Pat. 4,576,722 (Mar. 18, 1986), L. W. Gaylor and J. W. Beard (to Mogul Corp.).

35. W. M. Hann, J. H. Bardsley, and S. T. Robertson, paper no. 428, presented at *Corrosion/89*, New Orleans, La., 1989.

36. K. L. Brown, *Scaling and Geothermal Development*, course taught at U.S. Dept. of the Interior Geological Survey, Denver, Colo., 1991.

37. D. M. Thomas and J. S. Gudmundson, *Geothermics* **18**(1,2), 5–15 (1989).

38. *Ibid.*, 337–341 (1989).

39. A. Harris, M. A. Finan, and M. N. Elliot, *Desalination* **14**, 325–340 (1974).

40. S. W. Walinsky, B. J. Morton, and J. J. O'Neill, *Proceedings Int. Congress on Desalination and Water Re-Use* **1** (1981).

41. D. Comstock, *Industrial Water Treatment* **23**(4), 39–42 (1991).

42. Z. Amjad, *Ultrapure Water* (4), 57–60 (1989).

43. D. C Brandt, *Desalination* **52**, 177–186 (1985).

44. *Acumer 4000 Scale Inhibitor*, Rohm and Haas Co., Philadelphia, Pa., 1992.

45. J. C. Cowan and D. J. Weintritt, *Water-Formed Scale Deposits*, Gulf Publishing Co., Houston, Tex., 1976.

46. C. C. Patton, *Applied Water Technology*, Campbell Petroleum Series, Norman, Okla., 1986, pp. 77–87.

47. S. R. Jakobsen, P. A. Read, and T. Schmidt, paper presented at *UK Corrosion/89*, Blackpool, UK, 1989.

48. M. C. van der Leeden and co-workers, "Role of Polyelectrolytes in Barium Sulphate Precipitation," *Innovators Digest, Report No. D221H*, Technical University of Delft, The Netherlands, 1991.

49. A. T. Kan and co-workers, paper No. 33, presented at *Corrosion/92, NACE*, Houston, Tex., 1992.

50. U.S. Pat. 4,544,719 (Mar. 18, 1986), D. M. Giddings, D. G. Ries, and A. R. Syrinek (to Nalco Chemical Co.).

51. U.S. Pat. 4,476,029 (Oct. 9, 1984), A. O. Sy and D. G. Cuisia (to W. R. Grace & Co.).

52. U.S. Pat. 4,507,422 (Mar. 26, 1985), D. Farrar and M. Hawe (to Allied Colloids, Inc.).

53. L. Miljkovic and co-workers, *Cement and Concrete Research* **1**, 864–870 (1986).

54. P. J. Anderson, D. M. Roy, and J. M Gaidis, *Cement Concrete Res.* **17**, 805–813 (1987).

55. P. C. Hewlett, *Superplasticizers in Concrete*, American Concrete Institute Publication SP-62, 1979, Detroit, Mich., pp. 1–20.

56. N. B. Singh and S. P. Singh, *J. Materials Science* **22**, 2751–2758 (1987).

57. U.S. Pat. 4,906,298 (Mar. 6, 1990), T. Natsuume, H. Kadono, and Y. Miki (to Nippon Zeon Co., Ltd.).

58. U.S. Pat. 4,460,720 (July 17, 1984), J. M. Gaidis and A. M. Rosenberg (to W. R. Grace & Co.).

59. U.S. Pat. 4,870,120 (Sept. 26, 1989), T. Tsubakimoto and co-workers (to Nippon Shokubai Kagaku Co., Ltd.).

60. U.S. Pat. 4,888,059 (Dec. 19, 1989), K. Yamaguchi and T. Goto (to Dainippon Ink and Chemicals, Inc.).

61. H. Jakubuskas, *Use of Pigment Dispersions in Paints*, E. I. Du Pont De Nemours & Co., Athens, Ga.,1985.

62. See Ref. 4.

63. E. A. Johnson and E. J. Schaller, *Additives for Coatings*, Center for Professional Advancement, E. Brunswick, N.J., 1991.

64. E. A. Johnson and co-workers, *Resin Review* **41**(3), 3–19 (1991).

65. S. A. Ellis and co-workers, *Resin Review* **27**(2), 18–26 (1977).

66. J. J. Gambino and E. J. Schaller, *Resin Review*, **34**(1), 14–19 (1984).

67. M. P. Daut and J. Ross, *Resin Review*, **36**(2), 27–31 (1986).

68. W. A. Kirn, *Mod. Paint Coatings*, 164–170 (Oct. 1984).

69. *Tamol Dispersants*, Rohm and Haas Co., Philadelphia, Pa., 1991.

70. G. T. McGrew, *Soap/Cosm./Chem Spec.*, 53–79 (Apr., 1986).

71. D. Witiak, paper presented at *Chemical Specialties Manufacturers Association*, Chicago, Ill., May 7–9, 1986.

72. A. P. Hudson, F. E. Woodward, and G. T. McGrew, *J. Am. Oil Chem. Soc.* **65**(8), 1353–1356 (1988).

73. M. B. Freeman, paper presented at *the 80th Amer. Oil Chem. Soc. National Meeting*, Cincinnati, Ohio, 1989.

74. M. K. Nagarajan, *J. Am. Oil Chem. Soc.* **62**(5), 949–956 (1985).

75. G. Swift in E. J. Glass and G. Swift, eds., *Agricultural and Synthetic Polymers*, ACS Symposium Series 433, Washington, D.C., 1990.

76. S. Matsumura, Y. Yamanaka, and K. Nomoto, *Yukagako* **36**(1), 32–37 (1987).

77. S. Matsumura, K. Hashimoto, and S. Yoshikawa, *Yukagako* **36**(110), 874–881 (1987).

WILLIAM M. HANN
Rohm and Haas Company

DISPERSION OF POWDERS IN LIQUIDS. See POWDER

HANDLING.

DISTILLATION

Distillation is a method of separation that is based on the difference in composition between a liquid mixture and the vapor formed from it. This composition difference arises from the dissimilar effective vapor pressures, or volatilities, of the components of the liquid mixture. When such dissimilarity does not exist, as at an azeotropic point, separation by simple distillation is not possible. Distillation as normally practiced involves condensation of the vaporized material, usually in multiple vaporization/condensation operations, and thus differs from evaporation (qv), which is usually applied to separation of a liquid from a solid but which can be applied to simple liquid concentration operations.

Distillation is the most widely used industrial method of separating liquid mixtures and is at the heart of the separation processes in many chemical and petroleum plants (see SEPARATION SYSTEMS SYNTHESIS). The most elementary form of the method is simple distillation in which the liquid is brought to boiling and the vapor formed is separated and condensed to form a product. If the process is continuous with respect to feed and product flows, it is called flash distillation. If the feed mixture is available as an isolated batch of material the process is a form of batch distillation and the compositions of the collected vapor and residual liquid are thus time dependent. The term fractional distillation, which may be contracted to fractionation, was originally applied to the collection of separate fractions of condensed vapor, each fraction being segregated. In modern practice the term is applied to distillation processes in general, where an effort is made to separate an original mixture into several components by means of distillation. When the vapors are enriched by contact with counterflowing liquid reflux, the process is often called rectification. When fractional distillation is accomplished with a continuous feed of material and continuous removal of product fractions, the process is called continuous distillation. When steam (qv) is added to the vapors to reduce the partial pressures of the components to be separated, the term steam distillation is used.

Most distillations conducted commercially operate continuously, with a more volatile fraction recovered as distillate and a less volatile fraction recovered as bottoms or residue. If a portion of the distillate is condensed and returned to the process to enrich the vapors, the liquid is called reflux. The apparatus in which the enrichment occurs is usually a vertical, cylindrical vessel called a still or distillation column. This apparatus normally contains internal devices for effecting vapor–liquid contact; the devices may be categorized as plates or packings.

Distillation has been practiced in one form or another for centuries. It was of fundamental importance to the alchemists and was in use well before the time of Christ. The historical development of distillation has been published (1) as has the history of vapor–liquid contacting devices (2).

Vapor–Liquid Equilibria

The equilibrium distributions of mixture component compositions in the vapor and liquid phases must be different if separation is to be made by distillation. It is important, therefore, that these distributions be known. The compositions at thermodynamic equilibrium are termed vapor–liquid equilibria (VLE) and may be correlated or predicted with the aid of thermodynamic relationships. The driving force for any distillation is a favorable vapor–liquid equilibrium, which provides the needed composition differences. Reliable VLE are essential for distillation column design and for most other operations involving liquid–vapor phase contacting. Many VLE have been measured and reported in the literature, and compilations of such data are available (3,4). Also, bibliographic guides have been published, providing source references for thousands of publications presenting VLE (5–7). If data are not to be found, they may be measured, or estimated by generalized methods (8–10), with some sacrifice in reliability. Even if carefully measured data are available, thermodynamic models are usually required to extrapolate or interpolate the data for conditions not represented by the experiments. Whatever the source and extent of the VLE, some evaluation should be made with regard to accuracy.

The VLE for the system at hand may be simple and easily represented by an equation or, in some systems, may be so complex that they cannot be adequately measured or represented. Excellent treatises are available for selection and implementation of vapor–liquid equilibrium studies (11–14). Typical VLE for binary systems are shown graphically in Figure 1. Figure 1a is a representative boiling point diagram showing equilibrium compositions as functions of temperature at a constant pressure. The lower line is the liquid bubble point line, the locus of points at which a liquid on heating forms the first bubble of vapor. The upper line is the vapor dew point line, representing points at which a vapor on cooling forms the first drop of condensed liquid. The liquid and vapor compositions are conventionally plotted in terms of the low boiling (more volatile) substance, L, in the mixture. The system point A has a vapor composition of $y_L{}^A$ in equilibrium with a liquid composition of $x_L{}^A$ at a temperature of T^A. Figure 1b is a typical isobaric phase or $y–x$ diagram. For further discussion, several textbooks are available (15,16).

Thermodynamic Relationships. A closed container with vapor and liquid phases at thermodynamic equilibrium may be depicted as in Figure 2, where at least two mixture components are present in each phase. The components distribute themselves between the phases according to their relative volatilities. A distribution ratio for mixture component i may be defined using mole fractions:

$$K_i = y_i{}^*/x_i \qquad (1)$$

where the asterisk is used to denote an equilibrium condition. This K term, known as the vapor–liquid equilibrium ratio, or often the K value, is widely used, especially in the petroleum (qv) and petrochemical industries. For any two mixture components i and j, their relative volatility, often called the alpha value, is defined as

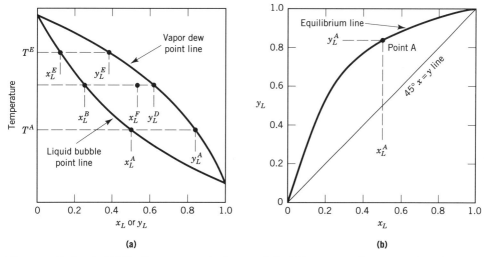

Fig. 1. Isobaric VLE diagrams: (**a**) dew and bubble point; (**b**) vapor–liquid ($y–x$) equilibrium.

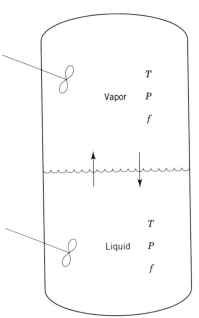

Fig. 2. Equilibrium between vapor and liquid. The conditions for equilibrium are $T^V = T^L$ and $P^V = P^L$. For a given T and P, phase fugacities are equal, ie, $f^V = f^L$ and $f_i^V = f_i^L$.

$$\alpha_{ij} = \frac{K_i}{K_j} = \frac{y_i\, x_j}{x_i\, y_j} = \frac{y_i\, (1 - x_i)}{x_i\, (1 - y_i)} \qquad (2)$$

Equation 2 may be rearranged to form an expression for the equilibrium curve in Figure 1**b**.

$$y_i = \frac{\alpha_{ij}\, x_i}{1 + (\alpha_{ij} - 1)x_i} \qquad (2a)$$

The relative volatility, α, is a direct measure of the ease of separation by distillation. If $\alpha = 1$, then component separation is impossible, because the liquid-and vapor-phase compositions are identical. Separation by distillation becomes easier as the value of the relative volatility becomes increasingly greater than unity. Distillation separations having α values less than 1.2 are relatively difficult; those which have values above 2 are relatively easy.

When both phases form ideal thermodynamic solutions, ie, no heat of mixing, no volume change on mixing, etc, Raoult's law applies:

$$p_i^V = x_i P_i^0 \qquad (3)$$

where P_i^0 is the vapor pressure of i at the equilibrium temperature. Combining this expression with Dalton's law of partial pressures, K values and relative volatilities may be obtained:

$$K_i = P_i^0/P \tag{4}$$

$$\alpha_{ij} = P_i^0/P_j^0 \tag{5}$$

Examples of ideal binary systems are benzene–toluene and ethylbenzene–styrene; the molecules are similar and within the same chemical families. Thermodynamics texts should be consulted before making the assumption that a chosen binary or multicomponent system is ideal. When pressures are low and temperatures are at ambient or above, but the solutions are not ideal, ie, there are dissimilar molecules, corrections to equations 4 and 5 may be made:

$$K_i = \gamma_i^L P_i^0/P \tag{6}$$

$$\alpha_{ij} = \gamma_i^L P_i^0/(\gamma_j^L P_j^0) \tag{7}$$

where the Raoult's law correction factor, γ^L, is a thermodynamically important liquid-phase activity coefficient.

The development and thermodynamic significance of activity coefficients is discussed in most chemical engineering thermodynamics texts. The liquid-phase coefficients are strong functions of liquid composition and temperature and, to a lesser degree, of pressure. A system with positive deviation, ie, the two components having activity coefficients greater than one such that the logarithm of the coefficient is positive, is shown in Figure 3a; a system with negative deviation, the coefficients less than unity and logarithms negative, is shown in Figure 3b. In a few cases one component of a binary mixture has a positive deviation and the other a negative deviation. Most commonly, however, both coefficients have positive deviations.

Terminal activity coefficients, γ_i^∞, are noted in Figure 3. These are often

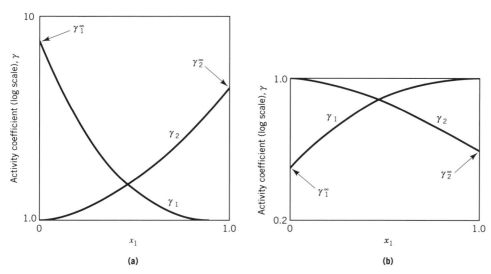

Fig. 3. Binary activity coefficients for two component systems having (**a**) positive and (**b**) negative deviations from Raoult's law. Conditions are either constant pressure or constant temperature and terminal coefficients, γ_i^∞, are noted.

called infinite dilution coefficients and for some systems are given in Table 1. The hexane–heptane mixture is included as an example of an ideal system. As the molecular species become more dissimilar they are prone to repel each other, tend toward liquid immiscibility, and have large positive activity coefficients, as in the case of hexane–water.

If the molecular species in the liquid tend to form complexes, the system will have negative deviations and activity coefficients less than unity, eg, the system chloroform–ethyl acetate. In azeotropic and extractive distillation (see DISTIL-LATION, AZEOTROPIC AND EXTRACTIVE) and in liquid–liquid extraction, nonideal liquid behavior is used to enhance component separation (see EXTRACTION, LIQUID–LIQUID). An extensive discussion on the selection of nonideal addition agents is available (17).

A great deal of study and research has gone into the development of working equations that can represent the curves of Figure 3. These equations are based on solutions of the Gibbs-Duhem equation:

$$x_i \left(\frac{\alpha \ln \gamma_i}{\alpha x_i} \right)_{T,P} + \ldots x_n \left(\frac{\alpha \ln \gamma_n}{\alpha x_i} \right)_{T,P} = 0 \tag{8}$$

One of the simplest and often used equations, or models, is that of Van Laar (18). For a binary system of components 1 and 2, these equations are

$$\ln \gamma = \frac{A_{12}}{\left(1 - \frac{A_{12}x_1}{A_{21}x_2} \right)^2} \tag{9}$$

$$\ln \gamma_2 = \frac{A_{21}}{\left(1 - \frac{A_{21}x_2}{A_{12}x_1} \right)^2} \tag{10}$$

Table 1. Terminal Activity Coefficients at Atmospheric Pressure[a]

Component 1	Component 2	γ_1^∞	γ_2^∞
chloroform	ethyl acetate	0.3	0.3
chloroform	benzene	0.9	0.7
n-hexane	n-heptane	1.0	1.0
ethyl acetate	ethanol	2.5	2.5
ethanol	toluene	6.0	6.0
benzene	methanol	9.0	9.0
ethanol	isooctane	11.0	8.0
methyl acetate	water	20.0	7.0
ethyl acetate	water	100.0	15.0
water	water	>100.0	>100.0

[a]Values are approximate.

It should be noted that only two parameters are involved. They are directly related to the terminal activity coefficients:

$$\ln \gamma_1^\infty = A_{12} \tag{11}$$

$$\ln \gamma_2^\infty = A_{21} \tag{12}$$

A useful and quite popular model is given (19):

$$\ln \gamma_1 = -\ln(x_1 + \Lambda_{12}x_2) + \left(\frac{\Lambda_{12}}{x_1 + \Lambda_{12}x_2} - \frac{\Lambda_{21}}{\Lambda_{21}x_1 + x_2} \right) \tag{13}$$

$$\ln \gamma_2 = -\ln(x_2 + \Lambda_{21}x_1) - \left(\frac{\Lambda_{12}}{x_1 + \Lambda_{12}x_2} - \frac{\Lambda_{21}}{\Lambda_{21}x_1 + x_2} \right) \tag{14}$$

This, the Wilson model, is more complex than the Van Laar model, but it does retain the two-parameter feature. The terminal activity coefficients are related to the parameters:

$$\ln \gamma_1^\infty = 1 - \ln \Lambda_{12} - \Lambda_{21} \tag{15}$$

$$\ln \gamma_1^\infty = 1 - \ln \Lambda_{21} - \Lambda_{12} \tag{16}$$

Whereas the Wilson model has been found to represent a wide variety of nonideal VLE, it cannot handle the case of partial immiscibility of the liquid phase; for this purpose a three-parameter relationship, the nonrandom, two-liquid (NRTL) model was developed (20).

The most recently developed model is called UNIQUAC (21). Comparisons of measured VLE and predicted values from the Van Laar, Wilson, NRTL, and UNIQUAC models, as well as an older model, are available (3,22). Thousands of comparisons have been made, and Reference 3, which covers the Dortmund Data Base, available for purchase and use with standard computers, should be consulted by anyone considering the measurement or prediction of VLE. The predictive VLE models can be accommodated to multicomponent systems through the use of certain combining rules. These rules require the determination of parameters for all possible binary pairs in the multicomponent mixture. It is possible to use more than one model in determining binary pair data for a given mixture (23).

To estimate VLE when no experimental data or model parameters are available and the cost of special measurements cannot be justified, a group contribution method based on the molecular structures involved called UNIFAC (8) has been developed. Not all possible groups have been evaluated, but regular progress reports are published (24,25). The UNIFAC method, as well as the other models, are critically important in extending limited data to conditions in distillation columns that can cover wide ranges of temperatures, pressures, and compositions. Handling of all these models by computer solution has been described in some detail (26) (see also ENGINEERING, CHEMICAL DATA CORRELATIONS).

The vapor–liquid equilibria of dilute solutions are frequently expressed in terms of Henry's law:

$$p_i^V = H_i^* x_i \tag{17}$$

where, from equation 6, the Henry's law coefficient is

$$H_i^* = \gamma_i^L P_i^0 \tag{18}$$

Henry's law is useful for handling equilibria associated with gas absorption (qv) and stripping problems. Henry's law coefficients are useful for estimating terminal activity coefficients and have been tabulated for many compounds in dilute aqueous solutions (27).

The foregoing discussion has dealt with nonidealities in the liquid phase under conditions where the vapor phase mixes ideally and where pressure–temperature effects do not result in deviations from the ideal gas law. Such conditions are by far the most common in commercial distillation practice. However, it is appropriate here to set forth the completely rigorous thermodynamic expression for the K value:

$$K_i = \frac{\gamma_i^L \phi_i^0 p_i^0 \exp\left(\dfrac{1}{R'T} \displaystyle\int_{p_i^0}^{p} v_i^L dP\right)}{\hat{\phi}_i P} \tag{19}$$

For nonideal vapor-phase behavior, the fugacity coefficient for component i in the mixture must be determined:

$$\ln \hat{\phi}_1 = \frac{1}{R'T} \int_0^P \left(v_i^v - \frac{R'T}{P}\right) dP \tag{20}$$

If the vapor forms an ideal solution,

$$v_i^v = v^v = zR'T/P \tag{21}$$

where z is the compressibility factor for the mixture. The right side of the numerator in equation 19 is called the Poynting correction, PC:

$$PC = \exp\left(\frac{1}{R'T} \int_{p_i^0}^{p} v_i^L dP\right) \tag{22}$$

when the liquid is incompressible,

$$PC = \exp\left(\frac{v_i^L (P - p_i^0)}{R'T}\right) \tag{23}$$

At pressures less than 2 MPa (20 bar) and temperatures greater than 273 K, PC $\sim$ 1.0. When the vapor obeys the ideal gas law, $z = 1.0$; then for ideal vapor solutions and for conditions such that $PC = 1.0$, equation 19 reduces to equation 6.

The fugacity coefficient departure from nonideality in the vapor phase can be evaluated from equations of state or, for approximate work, from fugacity/compressibility estimation charts. References 11, 14, and 27 provide valuable insights into this matter.

Journals for the publication of VLE data are available as is a comprehensive tabulation of azeotropic data (28); if the composition and temperature of the azeotrope are known (at a given pressure), then such information may be used to calculate activity coefficients. At the azeotropic point, by definition, $y_i = x_i$; from equation 6,

$$\gamma_i^L = P/P_i^0 \qquad (24)$$

The vapor pressure P_i^0 can be obtained from any of many sources such as handbooks.

The measurement of VLE can be carried out in several ways. A common procedure is to use a recycle still which is designed to ensure equilibrium between the phases. Samples are then taken and analyzed by suitable methods. It is possible in some cases to extract equilibrium data from chromatographic procedures. Discussions of experimental methods are available (5,11). For the more challenging measurements, eg, conditions where one or more components in the mixture can decompose or polymerize, commercial laboratories can be used.

Azeotropic Systems. An azeotropic mixture is one that vaporizes without any change in composition. Figures 4 and 5 represent homogeneous azeotropic systems. Figure 4 depicts a minimum boiling azeotropic system such as ethanol–water; Figure 5 describes a maximum boiling azeotropic system such as acetone–chloroform. The point Z defines the azeotropic composition; this azeotropic point

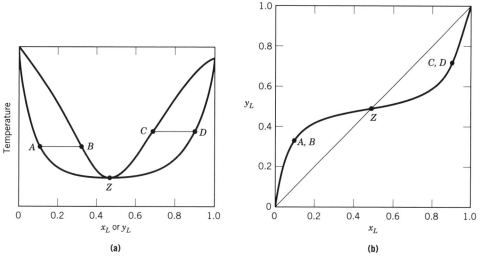

(a)   (b)

Fig. 4. Boiling point (**a**) and phase diagram (**b**) for a minimum boiling binary azeotropic system at constant pressure. A, B and C, D are representative equilibrium points; Z is the azeotropic point.

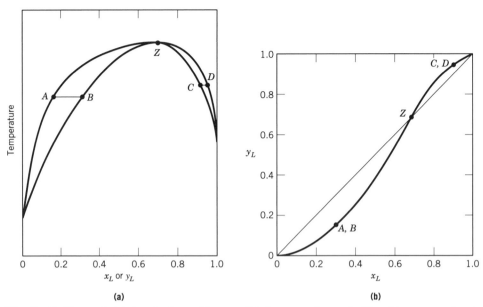

Fig. 5. Boiling point (**a**) and phase diagram (**b**) for a maximum boiling binary azeotropic system at constant pressure. A, B and C, D are representative equilibrium points; Z is the azeotropic point.

is also called the constant boiling mixture (CBM). Positive activity coefficients tend to produce minimum boiling azeotropes, and negative coefficients tend to produce maximum boiling azeotropes.

Heterogeneous azeotropes are formed when the positive activity coefficients are sufficiently large to produce two liquid phases which exist at the boiling point, and a constant boiling mixture which is formed at some composition, generally within the liquid immiscibility composition range. An example of a heterogeneous azeotropic system is the water/1-butanol system shown in Figure 6. Within the immiscible range, M–N, the equilibrium vapor is the heterogeneous azeotrope, Z, of constant composition and the equilibrium temperature is constant. At liquid compositions lower in water than in the azeotrope, the relative volatility of water/1-butanol is greater than one; at liquid compositions higher in water than in the azeotrope, the relative volatility of water/1-butanol is less than one.

Distillation Processes

Basic distillation involves application of heat to a liquid mixture, vaporization of part of the mixture, and removal of the heat from the vaporized portion. The resultant condensed liquid, the distillate, is richer in the more volatile components and the residual unvaporized bottoms are richer in the less volatile components. Most commercial distillations involve some form of multiple staging in order to obtain a greater enrichment than is possible by a single vaporization and condensation.

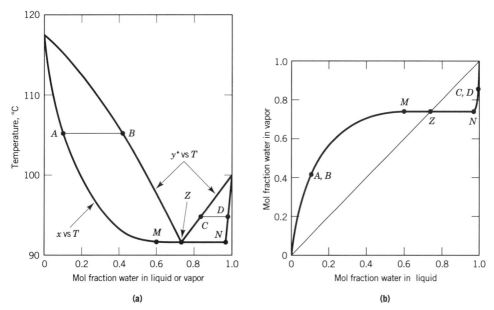

Fig. 6. Boiling point (**a**) and phase diagram (**b**) for the heterogeneous azeotropic system, water/1-butanol at atmospheric pressure. A, B and C, D are representative equilibrium points; Z is the azeotropic point; M and N are liquid miscibility limits.

For ease of presentation and understanding, the initial discussion of distillation processes involves binary systems. Examining the binary boiling point (Fig. 1a) and phase (Fig. 1b) diagrams, the enrichment from liquid composition x_L to vapor composition y_L represents a theoretical step, or equilibrium stage.

Simple Distillations. Simple distillations utilize a single equilibrium stage to obtain separation. Simple distillation, also called differential distillation, may be either batch or continuous, and may be represented on boiling point or phase diagrams. In Figure 1a, if the batch distillation begins with a liquid of composition x_L^A the initial distillate vapor composition is y_L^A. As the distillate is removed, the remaining liquid becomes less rich in the low boiler, L, and the boiling liquid composition moves to the left along the bubble point line. If the distillation is continued until the liquid has a composition of x_L^E, the last vapor distillate has a composition of y_L^E. Simple batch distillation is not widely used in industry, except for the processing of high valued chemicals in small production quantities, or for distillations requiring regular sanitization. Calculation methods are found in most standard distillation texts, and computer programs, such as BATCHFRAC of Aspen Technology (29), are available for handling the more complex, multicomponent batch distillations.

Simple continuous distillation, also called flash distillation, has a continuous feed to a single equilibrium stage; the liquid and vapor leaving the stage are considered to be in phase equilibrium. On the boiling point diagram (Fig. 1a), the feed is represented by x_L^F, the bottoms liquid by x_L^B, and the equilibrium vapor distillate by y_L^D. The overall mass balance is

$$F = D + B \tag{25}$$

the component L balance is

$$x_L^F F = y_L^D D + x_L^B B \tag{26}$$

Flash distillations are widely used where a crude separation is adequate. Examples of flash multicomponent calculations are given in standard distillation texts (30).

Multiple Equilibrium Staging.   The component separation in simple distillation is limited to the composition difference between liquid and vapor in phase equilibrium. To overcome this limitation, multiple equilibrium staging is used to increase the component separation. Figure 7 schematically represents a continuous distillation that employs multiple equilibrium stages stacked one upon another. The feed, F, enters the column at equilibrium stage f. The heat $\bar{q}^s$ required for vaporization is added at the base of the column in a reboiler or calandria. The vapors V^T from the top of the column flow to a condenser from which heat $\bar{q}^c$ is removed. The liquid condensate from the condenser is divided into two streams:

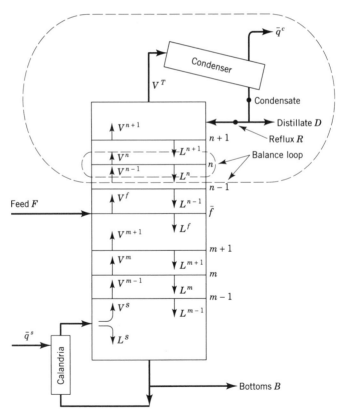

Fig. 7.   Distillation column with stacked multiple equilibrium stages. Terms are defined in text.

the first, a distillate D, which is the overhead product (sometimes called heads or make), is withdrawn from the system, and the second, a reflux R, which is returned to the top of the column. A bottoms stream B is withdrawn from the reboiler. The overall separation is represented by feed F separating into a distillate D and a bottoms B.

Above the feed a typical equilibrium stage is designated as n; the stage above n is $n + 1$ and the stage below n is $n - 1$. The section of column above the feed is called the *rectification section* and the section below the feed is referred to as the *stripping section*.

The mass balance across stage n is (*1*) vapor (V^{n-1}) from the stage below $(n - 1)$ flows up to stage n; (*2*) liquid (L^{n+1}) from the stage above $(n + 1)$ flows down to stage n; (*3*) on stage n the vapors leaving V^n are in equilibrium with the liquid leaving L^n. The vapors moving up the column from equilibrium stage to equilibrium stage are increasingly enriched in the more volatile components. Similarly, the liquid streams moving down the column are increasingly diminished in the more volatile components.

The overall column mass balances are

$$F = D + B \qquad (27)$$

and for any component i,

$$Fx_i^F = Dx_i^D + Bx_i^B \qquad (28)$$

The overall enthalpy balance is

$$H^F F + H^S = H^D D + H^B B + H^C \qquad (29)$$

A mass balance around plate n and the top of the column gives:

$$V^{n-1} = L^n + D \qquad (30)$$

And for any component:

$$V^{n-1} y_i^{n-1} = L^n x_i^n + D\, x_i^D \qquad (31)$$

$$y_i^{n-1} = \left(\frac{L^n}{V^{n-1}}\right) x_i^n + \left(\frac{D}{V^{n-1}}\right) x_i^D \qquad (32)$$

Below the feed, a similar balance around plate m and the bottom of the column results in:

$$y_i^{m-1} = \left(\frac{L^m}{V^{m-1}}\right) x_i^m + \left(\frac{B}{V^{m-1}}\right) x_i^B \qquad (33)$$

Equation 32 represents the upper (or rectifying) operating line equation and equation 33 represents the lower (or stripping) operating line equation. The slopes L^n/V^{n-1} and L^m/V^{m-1} can vary, depending on heat effects.

Graphical Method. The graphical McCabe-Thiele (31) design method facilitates a visualization of distillation principles while providing a solution to the material balance and equilibrium relationships. Here, the subscripts L and H are not used and x and y refer to the lower boiler, ie, more volatile component, in the binary system. A McCabe-Thiele diagram is given in Figure 8 where P, Q, and S are the x^B, x^F, and x^D compositions on the $y = x$, 45° construction line, respectively. Line OP is the stripping operating line and line OS is the rectifying operating line.

The McCabe-Thiele method employs the simplifying assumption that the molal overflows in the stripping and the rectification sections are constant. This assumption reduces the rectifying and stripping operating line equations to:

$$y^{n-1} = \left(\frac{\overline{L}}{\overline{V}}\right)_R x^n + \left(\frac{D}{\overline{V}_R}\right) x^D \tag{34}$$

$$y^{m-1} = \left(\frac{\overline{L}}{\overline{V}}\right)_S x^m + \left(\frac{B}{\overline{V}_S}\right) x^B \tag{35}$$

The constant molal flows in each section are designated by L and V. The McCabe-Thiele assumption of constant molal overflow implies that the molal latent heats of the two components are identical, the sensible heat effects are negligible, and the heat of mixing and the heat losses are zero. This simplified situ-

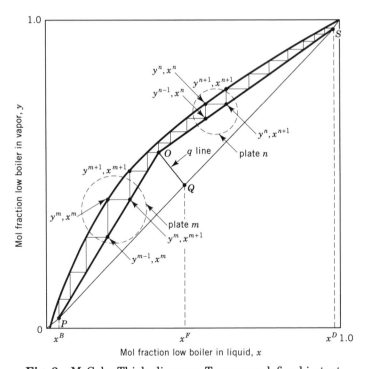

Fig. 8. McCabe-Thiele diagram. Terms are defined in text.

ation is closely approximated for many distillations. Equation 34 now represents the straight upper operating line OS and equation 35 represents the straight lower operating line OP. The upper operating line has the slope $(L/V)_R$ and the intercept at x^D ($=y^D$) on the $x = y$ line. Note that this operating line slope is less than one. Similarly, the lower operating line has a slope of $(L/V)_S$ and the intercept is at x^B on the $y = x$ line. This operating line has a slope greater than one. The line QO from the feed intercept Q to the intersection of the operating lines at O is called the q line.

The equilibrium curve gives the vapor–liquid relationships of y^n and x^n above the feed and of y^m and x^m below the feed. The upper operating line gives the relationship between y^{n-1} and x^n and the lower operating line gives the relationship y^{m-1} and x^m, ie, the streams passing each other. The graphical representation of theoretical equilibrium stages n and m is shown. The y^{m-1}, x^m to y^m, x^{L+1} represent the mass balance and phase equilibrium for theoretical stage m. Similarly, y^{n-1}, x^n to y^n, x^{n+1} represent theoretical stage n. The total number of theoretical stages in the column can now be stepped off starting either at the composition x^B and stepping upward or starting at x^D and stepping downward.

Condition of Feed (q Line).   The q line, which marks the transition from rectifying to stripping operating lines, is determined by mass and enthalpy balances around the feed plate. These balances are detailed in distillation texts (15).

The slope of the q line is $q/(q - 1)$ where:

$$q = \frac{\text{heat needed to vaporize one mole of feed}}{\text{molal latent heat of feed}} \tag{36}$$

The q line, therefore, depends on the enthalpy condition of the feed. Types of q lines are shown in Figure 9 and are listed below.

Feed enthalpy condition	q	Slope of q line	q Line coordinates
cold liquid	>1	+	$Q–E$
saturated liquid	1	∞	$Q–D$
partially vaporized	0–1	–	$Q–C$
saturated vapor	0	0	$Q–B$
superheated vapor	<0	+	$Q–A$

Reflux and Reflux Ratio.   The liquid returned to the top of the column is called reflux. The molar ratio R/D is the external reflux ratio. The ratio $(L/V)_R$, which is the slope of the rectifying operating line, is the rectifying internal reflux ratio. Similarly, the ratio $(L/V)_S$, which is the slope of the stripping operating line, is the stripping internal reflux ratio. As the ratio R/D increases, the rectifying internal reflux ratio increases and numerically approaches unity; similarly, the stripping internal reflux ratio decreases and numerically approaches unity. In the McCabe-Thiele plot the two operating lines move away from the equilibrium line toward the $y = x$ diagonal as the reflux ratio increases, and the individual theoretical stage steps become larger; accordingly, fewer theoretical stages are required to make a given separation.

McCabe-Thiele Example.   Assume a binary system $L–H$ that has ideal vapor–liquid equilibria and a relative volatility of 2.0. The feed is 100 mol of $x^F =$

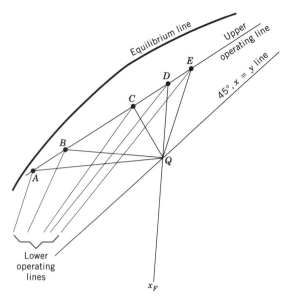

Fig. 9. McCabe-Thiele q lines for various feed enthalpy conditions. Terms are defined in text.

0.6; the required distillate is $x^D = 0.95$, and the bottoms $x^B = 0.05$, with the compositions identified and the lighter component L. The feed is at the boiling point. To calculate the minimum reflux ratio, the minimum number of theoretical stages, the operating reflux ratio, and the number of theoretical stages, assume the operating reflux ratio is 1.5 times the minimum reflux ratio and there is no subcooling of the reflux stream, then:

(*1*) Calculate the vapor composition in equilibrium with the liquid feed. From equation 2a and for $x = 0.60$ mol fraction;

$$y^* = \frac{2(0.6)}{1 + (2.0 - 1)0.6} = 0.75 \text{ mol fraction} \tag{37}$$

(*2*) Similarly, the entire equilibrium curve is calculated and is plotted in Figure 10. The feed is at the boiling point so the q line is drawn vertically with an infinite slope.

(*3*) Calculate mass balances on the basis of 100 mol of feed:

$F = D + B$ (overall column balance)
$0.60F = 0.95D + 0.05B = 60$ (component L balance)
$D = 61.11$ mol distillate
$B = 38.89$ mol bottoms

(*4*) Calculate reflux ratios. The minimum internal reflux ratio is a line from the intercept of the q line with the equilibrium curve to the x^D point on the 45° line:

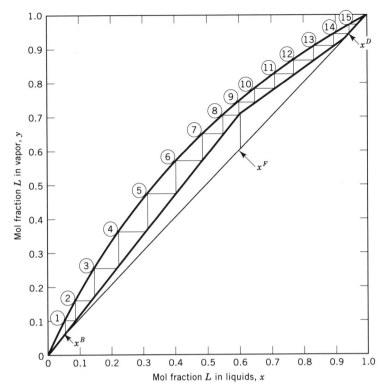

Fig. 10. McCabe-Thiele example. See text.

slope $= (L/V)_R = (0.95 - 0.75)/(0.95 - 0.60) = 0.5714$ minimum internal reflux ratio

$V = L + D = 0.5714V + 61.11$.

$V = 142.58$ mol (at minimum reflux)

$L = 81.47$ mol $= R$ (at minimum reflux)

$(R/D)_{min} = 81.47/61.11 = 1.333$ (minimum reflux ratio)

(5) Operating reflux ratio $= (R/D)_{operating} = 1.5 \times 1.333 = 2.0$.

(6) Reflux flow $= R = 2.0\,(61.11) = 122.22$ mol $= L$ (at operating reflux ratio).

(7) Rectifying section vapor flow $= V = L + D = 122.22 + 61.11 = 183.33$ mol.

(8) Upper operating line (eq. 34):

$$y^{n-1} = (122.22/183.33)\,x^n + (61.11/183.33)0.95 = 0.667\,x^n + 0.317$$

(9) Stripping section liquid and vapor flows, because the feed is at the boiling point,

$$L_S = L_R + F = 122.22 + 100 = 222.22 \text{ mol}$$
$$V_S = V_R = 183.33 \text{ mol}$$

(*10*) Lower operating line (eq. 35):

$$y^{m-1} = (222.22/183.33) \, x^m + (38.89/183.33)(0.05) = 1.212 \, x^m - 0.0106$$

(*11*) Theoretical stages: the complete construction is shown in Figure 10. Stages were stepped off starting at the base. Approximately 14.2 theoretical stages are required. This includes the reboiler, which normally functions as an equilibrium stage. Therefore, a capability of 13.2 theoretical stages in the column is needed. If the condenser were to condense only reflux, with the distillate product leaving the process as a vapor, it could be counted also as an equilibrium stage, making 12.2 stages needed for the column.

Unequal Molal Overflow.   The McCabe-Thiele method is based on the simplifying assumption that the molal overflow is constant in both the rectifying and stripping sections. For many problems this assumption is not valid and more precise calculations are necessary. For the more general case, detailed enthalpy balances are made around individual stages or groups of stages. Standard distillation texts discuss the internal enthalpy calculations by algebraic balances or by graphical procedures; eg, Reference 15 details the stage-to-stage mass and enthalpy balances with equilibrium calculations and also by means of the graphical Ponchon-Savarit procedure (32,33). Hand algebraic and graphical methods requiring internal enthalpy calculations have been largely superseded by simulations performed on modern computing devices, including personal computers (see COMPUTER TECHNOLOGY).

Minimum Number of Theoretical Stages and Minimum Reflux Ratio.   There are infinite combinations of reflux ratios and numbers of theoretical stages for any given distillation separation. The larger the reflux ratio, the fewer the theoretical stages required. For any distillation system with its given feed and its required distillate and bottoms compositions, there are two constraints within which the variables of reflux ratio and number of theoretical stages must lie: the minimum number of theoretical stages and the minimum reflux ratio. The minimum reflux ratio occurs when the reflux ratio is reduced so that the upper and lower operating lines and the *q* line are coincident at a single point on the equilibrium line as shown in Figure 11**a**. When this condition exists, an infinite number of theoretical stages would be required to make the separation. The minimum number of theoretical stages occurs when the system is at total reflux: no feed, distillate, or bottoms. This is illustrated in Figure 11**b**, where the operating lines are coincident with the 45°, $y = x$ line. For the McCabe-Thiele example presented above, the graphical procedure would give slightly less than nine minimum theoretical stages including the reboiler.

Simple analytical methods are available for determining minimum stages and minimum reflux ratio. Although developed for binary mixtures, they can often be applied to multicomponent mixtures if the two key components are used. These are the components between which the specification separation must be made;

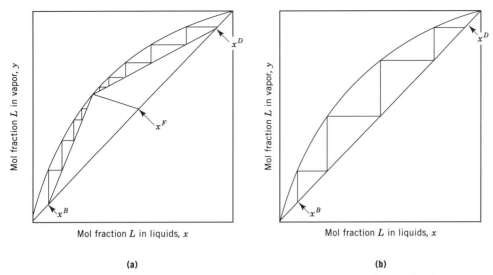

Fig. 11. Limiting conditions in binary distillation. (**a**) Minimum reflux and infinite number of theoretical stages; (**b**) total reflux and minimum number of theoretical stages.

frequently the heavy key is the component with a maximum allowable composition in the distillate and the light key is the component with a maximum allowable specification in the bottoms. On this basis, minimum stages may be calculated by means of the Fenske relationship (34):

$$N_{\min} = \frac{\ln\,[(y_i/y_j)D(x_j/x_i)_B]}{\ln\,\alpha_{ij,\mathrm{avg}}} \tag{38}$$

where i and j are the light and heavy components of a binary mixture, or the light key and heavy key in a multicomponent mixture. The average relative volatility is often taken as the geometric average of the relative volatilities at the top and bottom of the column. For the McCabe-Thiele example,

$$N_{\min} = \frac{\ln\,[(0.95/0.05)(0.95/0.05)]}{\ln\,2.0} = 8.50\text{ stages}$$

For minimum reflux ratio, the following equations (35) may be used:

$$\sum_i \frac{\alpha_i x_{if}}{\alpha_i - \phi} = 1 - q \tag{39}$$

$$\sum_i \frac{\alpha_i (x_{id})}{\alpha_i - \phi} = R_{\min} + 1 \tag{40}$$

where the value of q is determined as in the McCabe-Thiele procedure. Equation 39 is solved for root ϕ, the value of which must lie between 1.0 and the light key volatility. The root value so determined is then used in equation 40 to obtain the value of R_{min}. Although a trivial example, the McCabe-Thiele problem would yield

$$\frac{2.0(0.6)}{2.0 - \phi} + \frac{1.0(0.4)}{1.0 - \phi} = 1 - q = 0 \text{ (because } q = 1)$$

solving $\phi = 1.25$. Substituting in equation 40, for the given distillate compositions,

$$\frac{2.0(0.95)}{2.0 - 1.25} + \frac{1.0(0.05)}{1.0 - 1.25} = R_{min} + 1$$

from which $R_{min} = 1.333$.

Both of these limits, the minimum number of stages and the minimum reflux ratio, are impractical for useful operation, but they are valuable guidelines within which the practical distillation must lie. As the reflux ratio decreases toward the minimum reflux, the required number of stages increases rapidly. Similarly, as the minimum number of stages is approached, the required reflux ratio increases rapidly. A representative plot of the number of theoretical stages vs reflux ratio for some distillation separation is shown in Figure 12. Both minimum limits may be calculated for any distillation, thereby bracketing the practical design. Actual operating reflux ratios for most commercial columns are in the range of 1.1 to 1.5 times the minimum reflux ratio.

The operating, fixed, and total costs of a distillation system are functions of the relation of operating reflux ratio to minimum reflux ratio. Figure 13 shows a typical plot of costs; as the operating to minimum reflux ratio increases, the oper-

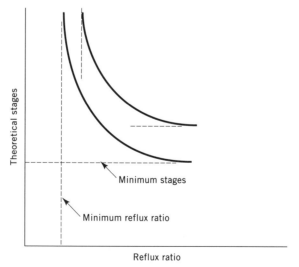

Fig. 12. Representative plot of theoretical stages vs reflux ratio for a given separation. Each curve is the locus of points for a given separation. Note the limiting conditions of minimum reflux and minimum stages.

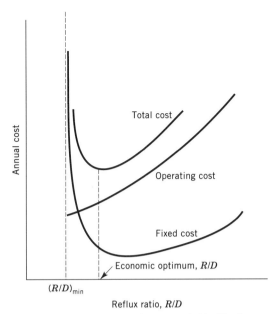

Fig. 13. Fixed, operating, and total costs of a typical distillation, as a function of reflux ratio.

ating cost (principally energy cost for the boil-up) increases almost linearly. Similarly, the fixed costs at first decrease from the infinite number of stages, pass through a minimum, and then increase again as the diameter of column increases with increased reflux ratio. These costs for typical distillations have been calculated (36); the ratio of the economic optimum reflux to the minimum reflux is often 1.2 or less.

Minimum Reflux with Pinch Zone. There are some distillations where the minimum reflux does not occur at the intersection of the upper and lower operating lines and the q line. These cases arise when the equilibrium is skewed from positive activity coefficients and when the operating line intersects the equilibrium line in a zone of constant composition, a pinch zone, which is not at the q line intersection. Figure 14 illustrates such a case. An example of such a pinch zone in an ethanol–water column is available (37).

Multicomponent Calculations. The calculations that determine the reflux and stage requirements are more difficult to make for multicomponent systems than for binary systems. When the concentration of a component in the distillate and in the bottoms is specified for the overall solution of a binary distillation, the component balance around the column also is completely specified. In the multicomponent case, only a single high boiling key component can be specified in the distillate and a single low boiling key component in the bottoms; the split of other components can be determined only by detailed calculations. These require a series of trial and error computations to obtain the solution at any given reflux ratio and number of stages. As the number of components and number of stages become large, the mathematical problem becomes formidable. Two approaches may be followed: use of approximate, ie, shortcut, methods, or use of a suitable computer

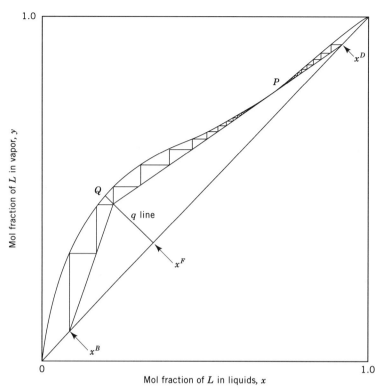

Fig. 14. False minimum reflux for system of skewed equilibria. Minimum reflux occurs at intersection P of operating line and equilibrium line, not at intersection of q line and equilibrium line. Terms are defined in text.

program that provides rigorous solutions. The former are used when approximate solutions are adequate or when a computer is not available. For the latter, numerous commercial programs are available and may be used with personal computers.

Most shortcut methods involve: (*1*) calculating the minimum number of stages; (*2*) calculating the minimum reflux ratio; and (*3*) estimating, from empirical correlations, the actual number of stages at an operating reflux. For minimum stages the Fenske relationship (eq. 38) is used, whereas for minimum reflux ratio the Underwood relationships (eqs. 39 and 40) are used. The relationship of operating to minimum reflux ratio and of operating to minimum number of plates is then estimated from the Gilliland correlation (38), or from a more recent correlation such as that of Reference 39.

The Gilliland correlation is in graphical form and the curve has been fitted by several workers (40):

$$\frac{N_t - N_{\min}}{N_t + 1} = 0.75 - 0.75 \left(\frac{R - R_{\min}}{R + 1}\right)^{0.5668} \tag{41}$$

For the McCabe-Thiele example, and using equation 41

$$\frac{N_t - 8.50}{N_t + 1} = 0.75 - 0.75 \left(\frac{2.0 - 1.333}{2.0 + 1}\right)^{0.5668}$$

from which $N_t = 15.7$ stages. The original plot of Gilliland would give $N_t = 14.8$, closer to the McCabe-Thiele value of 14.2 stages.

Discussions of shortcut methods have appeared many times in the literature (16,36), accompanied by the usual admonition to use such methods only for approximate designs or analyses. For multicomponent systems having significant nonidealities, the shortcut methods can be grossly in error.

Rigorous computer solutions are used for complex distillations involving multiple stages, multiple components, nonideal phase equilibria, multiple feeds and drawoffs, and heat addition or removal at intermediate stages. Most calculations are made by computer and the algorithms are generally based on the Thiele-Geddes model (41), which rates a given number of stages and reflux ratio for separation capability. A detailed discussion of computer solutions, including the handling of convergence problems, is available (42).

Computer solutions entail setting up component equilibrium and component mass and enthalpy balances around each theoretical stage and specifying the required design variables as well as solving the large number of simultaneous equations required. The explicit solution to these equations remains too complex for present methods. Studies to solve the mathematical problem by algorithm or iterative methods have been successful and, with a few exceptions, the most complex distillation problems can be solved.

Multiple Products.   If each component of a multicomponent distillation is to be essentially pure when recovered, the number of columns required for the distillation system is $N^* - 1$, where N^* is the number of components. Thus, in a five-component system, recovery of all five components as essentially pure products requires four separate columns. However, those four columns can be arranged in 14 different ways (43).

The number of columns in a multicomponent train can be reduced from the $N^* - 1$ relationship if side-stream draw-offs are used for some of the component cuts. The feasibility of multicomponent separation by such draw-offs depends on side-stream purity requirements, feed compositions, and equilibrium relationships. In most cases, side-stream draw-off distillations are economically feasible only if component specifications for the side-stream are not tight. If a single component is to be recovered in an essentially pure state from a mixture containing both lower and higher boiling components, a minimum of two columns is required, one column to separate the lower boilers from the desired component and another column to separate the component from the higher boilers.

The economics of the various methods that are employed to sequence multicomponent columns have been studied. For example, the separation of three-, four-, and five-component mixtures has been considered (44) where the heuristics (rules of thumb) developed by earlier investigators were examined and an economic analysis of various methods of sequencing the columns was made. The study of sequencing of multicomponent columns is part of a broader field, process synthesis, which attempts to formalize and develop strategies for the optimum overall process (45) (see SEPARATION SYSTEMS SYNTHESIS).

Distillation Columns

Distillation columns are vertical, cylindrical vessels containing devices that provide intimate contacting of the rising vapor with the descending liquid. This contacting provides the opportunity for the two streams to achieve some approach to thermodynamic equilibrium. Depending on the type of internal devices used, the contacting may occur in discrete steps, called plates or trays, or in a continuous differential manner on the surface of a packing material. The fundamental requirement of the column is to provide efficient and economic contacting at a required mass-transfer rate. Individual column requirements vary from high vacuum to high pressure, from low to high liquid rates, from clean to dirty systems, and so on. As a result, a large variety of internal devices has been developed to fill these needs. The column devices discussed herein are used for absorption (qv) and stripping as well as distillation. The principal operational difference is that in absorption or stripping, the gas flowing up the column is primarily a noncondensable phase at column conditions, whereas in distillation the gas phase is a condensable vapor.

Plate Columns. There are two general types of plates in use: crossflow and counterflow. These names refer to the direction of the liquid flow relative to the rising vapor flow. On the cross-flow plate the liquid flows across the plate and from plate to plate via downcomers. On the counterflow plate liquid flows downward through the same orifices used by the rising vapor.

Crossflow Plates. As indicated in Figure 15, liquid enters a crossflow plate from the bottom of the downcomer of the plate above and flows across the active or bubbling area where it is aerated by the vapors flowing through orifices from the plate below. It is in this aerated zone where most of the vapor–liquid mass transfer occurs. The aerated mixture flows over the exit weir into a downcomer. A vapor–liquid disengagement takes place in the downcomer and most of the trapped vapor escapes from the liquid and flows back to the interplate vapor space. The liquid, essentially free of entrapped vapor, leaves the plate by flowing under the downcomer to the inlet side of the next lower plate. The vapor, disengaging from the aerated mass on the plate, rises to the next plate above.

The pressure drop incurred by the vapor as it passes through the orifices of the plate is fundamental to plate operation. In most plate designs, the pressure drop prevents the crossflowing liquid from falling through the plate. The pressure drop also results from the energy consumed to disperse the vapor–liquid mixture, eg, to atomize a portion of the liquid to provide increased interfacial area for mass transfer. Diameters of commercial crossflow plate columns range from 0.3 to 15 m and plate spacings range from 0.15 to 1.2 m. The total pressure drop per plate is often in the range of 0.25 to 1.6 kPa (2–12 mm Hg).

Three principal vapor–liquid contacting devices are used in current crossflow plate design: the sieve plate, the valve plate, and the bubble cap plate. These devices provide the needed intimate contacting of vapor and liquid, requisite to maximizing transfer of mass across the interfacial boundary.

Sieve Plates. The conventional sieve or perforated plate is inexpensive and the simplest of the devices normally used. The contacting orifices in the conventional sieve plate are holes that measure 1 to 12 mm diameter and exhibit ratios

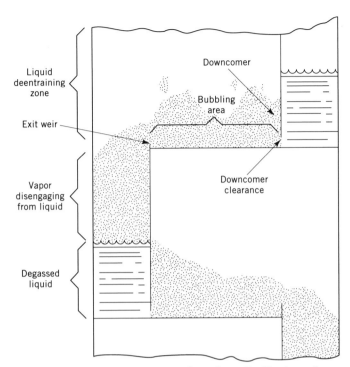

Fig. 15. Flow pattern in a crossflow plate distillation column.

of open area to active area ranging from 1:20 to 1:7. If the open area is too small, the pressure drop across the plate is excessive; if the open area is too large, the liquid weeps or dumps through the holes.

Valve Plates. Valve plates are categorized as proprietary and details of design vary from one vendor to another. These represent a variation of the sieve plate in which the holes are large and are fitted with liftable valve units such as those shown in Figure 16. The principal advantage over sieve plates is the ability to maintain efficient operation over a wider operating range through the use of variable orifices (valves) which open or close depending on vapor rate. The most common valve units consist of flat disks having attached legs that allow the valve to open or close. Sometimes two weights of valves are used on a single plate to

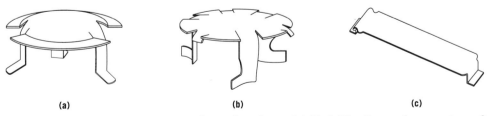

Fig. 16. Individual valve units used in valve plates: (**a**) Koch Flexitray valve, courtesy of Koch Engineering Co.; (**b**) Glitsch Ballast valve, courtesy of Glitsch, Inc.; and (**c**) Nutter Float Valve, courtesy of Nutter Engineering Co.

extend operating range and improve vapor distribution. The valve units usually have a tab or indentation that provides a minimum open area of vapor flow, even when the valve is closed, and also prevents the valve from sticking under corrosive or fouling conditions. Details on valve plate geometry, along with methods for valve plate design, are available (46–48).

Bubble Cap Plates. Until the early 1950s, bubble caps were the standard design in the chemical industry. Usage in newer installations is limited to low liquid flow rate applications, or to those cases where the widest possible operating range is desired. A typical bubble cap is shown in Figure 17. The vapor flows through a hole in the plate floor, through the riser, reverses direction in the dome of the cap, flows downward in the annular area between the riser and the cap, and exits through the slots in the cap. Commercial caps range from 50 to 150 mm in diameter and many slot design variations have been used. Bubble cap trays are more expensive and have lower capacity than sieve or valve plates; therefore, use has dropped to a very small percentage of newer column designs.

Multiple Liquid-Path Plates. As the liquid flow rate increases in large diameter crossflow plates (ca 4 m or larger), the crest heads on the overflow weirs and the hydraulic gradient of the liquid flowing across the plate become excessive. To obtain improved overall plate performance, multiple liquid-flow-path plates

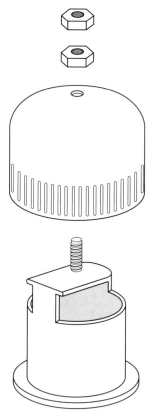

Fig. 17. Expanded view of a bubble cap. Courtesy of Vulcan Manufacturing Co.

may be used, with multiple downcomers. These designs are illustrated and discussed in detail in the literature (49).

Counterflow Plates. Counterflow plates are used less frequently than crossflow plates. The liquid flows downward and the vapor upward through the same orifices in a counterflow plate and the plate does not have downcomers. The openings are round holes (dualflow tray) or slots (Turbogrid tray). A variation of the dualflow tray is the Ripple tray in which the tray floor is shaped in a corrugated fashion (50). Counterflow plates are used advantageously in fouling services because for each hole vapor and liquid flow alternately, providing a self-cleaning action that is quite effective. The dualflow and Turbogrid plates have similar operating characteristics, and typical operating data have been published (51).

Another important plate which has characteristics similar to a counterflow plate is the Multiple Downcomer (MD) plate (52). This is a plate where the active area occupies the full column cross section but with a plurality of small downcomers interspersed among the perforations. The downcomers are specially sealed to prevent upflow of vapor through them. The plate has been used successfully in many high liquid flow cases.

Vapor Capacity Parameters. The diameter of a distillation column is determined by the capacity of the column to handle the required flows of vapor and liquid. The vapor capacity parameter is

$$C_{sb} = V^* \left(\frac{\rho_g}{\rho_L - \rho_g} \right)^{0.5} \tag{42}$$

and its simplification

$$F^* = V^*(\rho_g)^{0.5} \tag{43}$$

The term C_{sb} in equation 42 is called a Souders-Brown capacity parameter and is based on the tendency of the upflowing vapor to entrain liquid with it to the plate above. The term F^* in equation 43 is called an F-factor. For C_{sb} and F^* to be meaningful the cross-sectional area to which they apply must be specified. The capacity parameter is usually based on the total column cross section minus the area blocked for vapor flow by the downcomer(s). For the F-factor, typical operating ranges for sieve plate columns are

	Area basis	$(kg/(m \cdot s^2))^{0.5}$	$(lb/(ft \cdot s^2))^{0.5}$
F_S^*	total cross section	0.6–3.0	0.5–2.5
F_A^*	active area	0.85–4.3	0.7–3.5
F_H^*	hole area	8.5–30	7–25

Entrainment Flooding. The vapor capacity of a column is limited by excessive entrainment, usually called flooding. A flooding condition can be observed when the holdup of liquid becomes excessive, the pressure drop increases dramatically, and the mass-transfer efficiency falls precipitously. Estimates of the vapor velocity for a flooding condition may be made from the chart in Figure 18 (53). The abscissa term $L/G(\rho_g/\rho_L)^{0.5}$ is called a flow parameter and its value can indicate several things about the character of the aerated mass on the plate. For example, a very low value can indicate a phase inversion in which the vapor flow is continuous (spray flow) whereas a high value can indicate a bubbly mass (emulsion flow). The value of the flow parameter is easily determined from the stage calculations (reflux and boilup ratios) and densities of the phases. The ordinate value in Figure 18 leads to a value of the flooding velocity, and prudent design calls for limiting actual flows to 70–80% of this velocity.

Downcomer Flooding. For cases of very high liquid-to-vapor flow ratios the limiting capacity of the column is based on the ability of the downcomers to move the de-aerated liquid from a plate to the next plate below. It is clear that there can be constrictions in the downcomer design or that even with no constrictions there is simply not enough flow area to accommodate the high volume of liquid. Thus, the downcomer can flood, or choke, when it becomes completely filled with liquid or aerated mass. Typical design heuristics include limiting the downcomer velocity (clear liquid basis) to no more than 0.12 m/s. Also, to allow for complete disengagement of vapor from liquid in the downcomer, a minimum residence time of 4 s is often used. The actual limiting values of these parameters varies somewhat with the properties of the fluids and the exact dimensions of the plate components.

Stable Operating Range. All plates have a stable operating envelope bound by a range of liquid and vapor flow rates as shown in Figure 19. The size and shape of the stable area depends on the plate design and on the system properties. The line *AD* represents the minimum operable vapor flow rate at various liquid flow rates. Below *AD*, the vapor rate is too low to maintain the liquid on the plate and, as a result, the liquid weeps excessively or dumps through the plate orifices. Above line *BC* the column floods by entrainment. To the right of *CD* the high liquid rate causes downcomer flooding. The area to the left of *AB* represents high

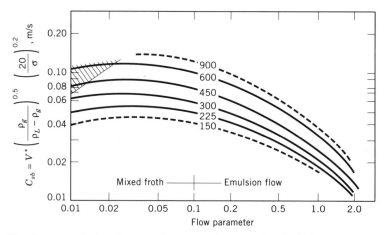

Fig. 18. Flooding correlation for crossflow trays (sieve, valve, bubble-cap) where the numbers represent tray spacing in mm. Also shown are approximate boundaries of the spray zone, ▧ , and mixed froth and emulsion flow regimes.

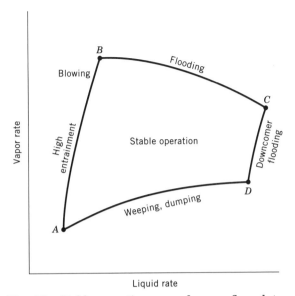

Fig. 19. Stable operating range for crossflow plates.

entrainment at low liquid flow rates, with vapor jets at the orifices. Design procedures for bubble caps (49) and sieve trays (53,54) have been published. Additionally, vendors of valve trays make available their design methods.

Plate Efficiencies. Column requirements are calculated in terms of theoretical stages or plates. Actual plates must, however, be specified in the design. Thus the effectiveness of the plate in approaching the equilibrium condition must be predicted. This approach is called the plate efficiency, which is a measure of the rate of mass transfer on the actual plate. This efficiency, expressed either as a fraction or as a percentage, depends on three principal factors: the geometry of the plate (hole arrangement, valve design, etc); the loading of vapor and liquid traffic on the plate; and the diffusional properties of the fluids.

The simplest efficiency is the overall column efficiency which is the number of theoretical plates in a column divided by the number of actual plates:

$$E_o = N_t/N_a \tag{44}$$

Thus the overall efficiency is an averaged efficiency of all the individual plates.

A more useful plate efficiency for theoretical prediction is the Murphree plate efficiency (55):

$$E_{mv} = \frac{y^n - y^{n-1}}{y^{n*} - y^{n-1}} \tag{45}$$

where y^n and y^{n-1} are the vapor compositions from plate n and $n-1$ (the plate below n), and y^{n*} is the vapor composition that would be in equilibrium with the liquid composition leaving plate n. Thus, for a given plate, E_{mv} is a ratio of the actual vapor composition change to the change that would occur if the plate were effective enough to bring the vapor and liquid to thermodynamic equilibrium. This definition is based on the outlet liquid composition, and says nothing about the average liquid composition on the plate. In cases where a significant concentration gradient exists in the liquid composition across the plate, it is possible for E_{mv} to have a value greater than 1.0 (100%). Equation 45 is written in terms of vapor composition. A similar equation can be written in terms of the liquid compositions and is denoted as E_{mL}.

Of still more theoretical importance is the efficiency at some point on the plate:

$$E_{og} = \left(\frac{y^n - y^{n-1}}{y^{n*} - y^{n-1}}\right)_{point} \tag{46}$$

This parameter is called the point efficiency (or local efficiency). It cannot have a value greater than 1.0, and it has a counterpart term for liquid compositions.

Prediction of Plate Efficiency.   As of this writing, the most comprehensive study of plate efficiency known was made in the mid-1950s, based on the then still popular bubble cap plates (56). Unfortunately, the predictive model developed has been shown to be inadequate for many industrial distillations. There has been continuing research effort directed toward a better understanding of the mechanisms that occur in the rather complex aerated mass on the typical plate (57,58). A complicating factor is the lack of uniform liquid flow across the plate, and situations have been found where the liquid actually stagnates in certain zones of larger diameter plates. For larger columns it is possible for the observed Murphree efficiency to exceed 100%. A satisfactory method for predicting plate efficiency does not exist. Most recently there have been studies of the various types of flow regimes that occur on operating plates and of the effect of these regimes on tray performance, including plate efficiency. Pursuit of the flow regime studies (59–62) may lead to improved plate efficiency prediction methods. For example, a newer model (63) takes into account the regime as well as the vapor bubble (froth flow) or liquid drop (spray flow) characteristics in determining mass-transfer coefficients in the aerated zone on the plate.

Empirical Efficiency Prediction Methods.   Numerous empirical methods for predicting plate efficiency have been proposed. Probably the most widely used method correlates overall column efficiency as a function of feed viscosity and relative volatility (64). A statistical correlation of efficiency and system variables has been developed from numerous plate efficiency data (65).

General Comments on Plate Efficiency.   The plate efficiencies of well-designed commercial bubble cap, sieve, and valve plates are approximately the same when the plates are operated within their normal design range. The plate efficiency decreases both at the low end of the plate's operating range, where the liquid tends to leak through the plate, and at the high end of the operating range, where liquid entrainment becomes substantial.

Most distillation systems in commercial columns have Murphree plate efficiencies of 70% or higher. Lower efficiencies are found under system conditions of a high slope of the equilibrium curve (Fig. 1b), of high liquid viscosity, and of large molecules having characteristically low diffusion coefficients. Finally, most experimental efficiencies have been for binary systems where by definition the efficiency of one component is equal to that of the other component. For multicomponent systems it is possible for each component to have a different efficiency. Practice has been to use a pseudo-binary approach involving the two key components. However, a theory for multicomponent efficiency prediction has been developed (66,67) and is amenable to computational analysis.

Packed Columns. In packed columns, the vapor–liquid contacting takes place in continuous beds of solid packing elements rather than in discrete individual plates. The contacting can be visualized as occurring in differential increments across the height of the packing; thus packings are known as counterflow devices rather than stagewise devices. Mechanically, the packed column is a relatively simple structure. In its simplest form the packed column comprises a vertical shell having dumped or carefully arranged packing elements on an open-type support, together with a suitable liquid distribution device above the packed bed. A packed column having two packed beds and a midcolumn feed is shown in Figure 20. The vapor enters the column below the bottom bed and flows upward through the column. The liquid (reflux or other liquid stream) enters at the top

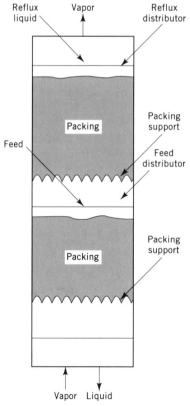

Fig. 20. Packed column shell and internals. Column shown has single packed beds above and below the feed. For separations requiring a large number of stages, additional beds, separated by redistribution devices, are likely to be needed.

through the liquid distributor and flows downward through the packing counter-currently to the rising vapor. The height of the individual packed beds is limited to 2–9 m by the mechanical strength of the packing or by the need to redistribute the liquid so that good mass-transfer efficiency can be maintained.

Packings. For many years packed columns consisted of randomly dumped packings almost exclusively, with occasional applications of regularly stacked packings or pads of woven or knitted wire. In the late 1960s a partial trend away from random packings began when a special structured packing made of wire gauze was introduced by Sulzer Brothers in Switzerland (68). The indicated advantages of the structured packings were high mass-transfer efficiency and very low pressure drop. These devices appeared to be ideal for high vacuum distillations. However, cost of fabrication was very high and they were considered mainly for the vacuum distillation of specialty chemicals. In 1977, a lower cost sheet metal version was introduced (69), and since that time a large business in structured sheet metal packings has arisen. At the same time, improved random packings have been developed and a comprehensive discussion of their characteristics has been published (70). Some of the common random packings are shown in Figure 21. The Raschig ring, one of the oldest of packings, is an open cylinder of equal height and diameter. The Berl saddle and the ceramic Intalox saddle (Norton Co.) have a higher capacity and efficiency than the Raschig ring. The Pall ring is a modification of the Raschig ring which allows through-flow of liquid and vapor, with consequent lower pressure drop and better efficiency. The newer Intalox metal saddle (IMTP) is an example of a random packing having a very high void fraction and low resistance to the flowing phases. Other newer random packings, not shown in Figure 21, include the CMR ring (Glitsch, Inc.) and the Nutter ring (Nutter Engineering Co). The random-type packings can generally be made from metal, plastic, or ceramic materials; the approximate nominal size range for the individual elements is 12–75 mm.

Common structured packing geometries are shown in Figure 22. Flat plates of gauze or sheet metal are perforated, or embossed or lanced, and corrugated. Corrugated sheets are then stacked together such that adjacent sheets have opposite corrugation directions. The corrugations have angles with the horizontal of 45 to 60 degrees. Vapor and liquid contact each other in wetted-wall fashion, and the perforations plus other surface enhancements, eg, texturing, serve to promote liquid spreading into thin films. Dimensions, performance character-

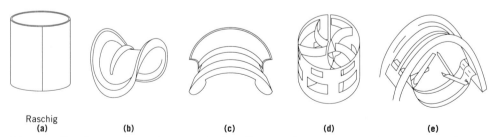

Raschig
(a)   **(b)**   **(c)**   **(d)**   **(e)**

Fig. 21. Random packing elements for distillation columns: (**a**), Raschig ring (metal); (**b**), Berl saddle (ceramic); (**c**), Intalox saddle (ceramic); (**d**), Pall ring (metal); and (**e**), Intalox saddle (metal).

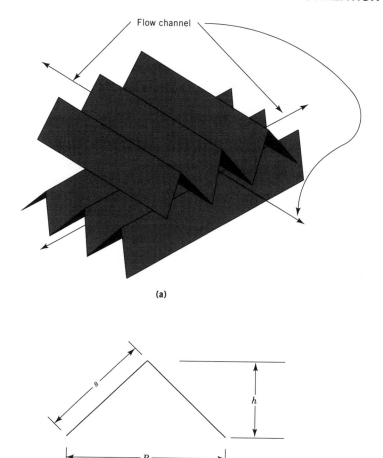

Fig. 22. (a) Flow channel arrangement; (b) flow channel triangular cross section where for angles of 90°, $D_{EQ} = 4\,R_H = 4\left(\dfrac{S \cdot S}{2}\right)\dfrac{1}{2S} = S.$

istics, and design procedures for the structured packings are summarized in Reference 71.

Packed Column Internals. In order to ensure good packed column mass-transfer efficiency, the liquid must be distributed uniformly over the surface of the packing. As a general rule there should be at least 100 pour points per square meter (10 points/ft²), although fewer points may be used for random packings of the bluff-body type such as Raschig rings and Berl saddles. Although they have capacity and pressure drop limitations, the bluff-body packing elements are able to divide the downflowing liquid and thus improve on an initially marginal distribution. On the other hand, the through-flow type random packings, eg, Pall rings and Intalox metal tower packings, as well as the structured packings, are not able to correct the initial distribution and in fact may allow some deterioration of distribution if the bed heights are greater than about five meters.

Considerable research is in progress on methods for ensuring good liquid distribution in large diameter columns, and the packing manufacturers maintain large test stands where a particular design of distributor can be tested using water before being installed in the column. The distributor design problem becomes more severe at low, ie, <700 cm^3/(s·m^2) (1 gal/(min·ft^2)) liquid rates or in large (>3 m) diameter towers. An example of a more fundamental study of liquid distribution is available (72) as are typical liquid distributor designs and typical packing supports (54).

Packed Column Operation. In the packed column, liquid flows downward in opposition to the upward flow of vapor; both phases flow through the same open space or interstices between the packing elements. At low liquid and gas flow rates, the descending liquid occupies only a small fraction of the interstices and, therefore, offers little hindrance to the rising vapor flow. Figure 23 shows a schematic plot of pressure drop per unit of height as a function of the gas rate at low and high liquid flow rates. At a low rate of gas flow, the log slope of each curve is approximately 2. As the gas flow rate increases, there is an increasing tendency for the liquid to be held up in the void space, thereby decreasing the space available for the gas flow. As the gas flow rate increases further, more liquid is held up until at some high gas rate the packing floods. At this point, the liquid is essentially filling the interstices and can no longer flow downward. At flooding, the log slope is practically infinite. The pressure drop at the inception of flooding ranges from 1.6 to 3.3 kPa/m (2 to 4 in. water/ft) of packing. More comprehensive discussions of packed column hydraulics may be found in distillation texts (15,17,73), monographs (70,74), or handbooks (75,76).

Capacity of Packed Columns. Packed columns are usually designed to operate at some percentage approach to flooding, eg, 60–70%, or at some specified pressure drop per unit height of packing, eg, 0.8 kPa/m (1 in. water/ft) of packing. Flooding correlations have been proposed (77), one revision introducing constant pressure drop lines (78). The most recent revision in these correlations is shown

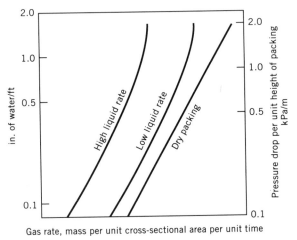

Fig. 23. Log–log plot of pressure drop per unit height of typical packing as a function of gas rate at two liquid rates and for the unirrigated packing.

in Figure 24 (70). The idea of flooding has been eliminated from the chart with the stipulation that the topmost curve represents the maximum capacity. Experimentally determined packing factors F_p, presented in Table 2, should be used in the ordinate group. These factors distinguish between the various shapes and sizes of the available packings. The curves are for constant pressure drop and thus the chart enables estimation of both capacity and pressure drop.

Packing Mass-Transfer Characteristics. The contacting for mass transfer (qv) in a packed column occurs differentially along the length of the column. The separation calculations can thus be made on a differential basis along this length, using mass-transfer coefficients or heights of transfer units. The calculations are somewhat imprecise because of the uncertainty in the fundamental mass-transfer mechanisms in larger-scale columns. Useful models for predicting the mass-transfer efficiency of randomly packed columns have been published (79,80), using the same database of commercial-scale performance data. These models cover the better known packings, eg, metal and ceramic Raschig rings, ceramic Berl saddles, and metal Pall rings, in nominal sizes in the range of 12 to 50 mm. It has been found that to avoid excessive maldistribution of liquid near the wall, a ratio of column diameter to packing element size of at least

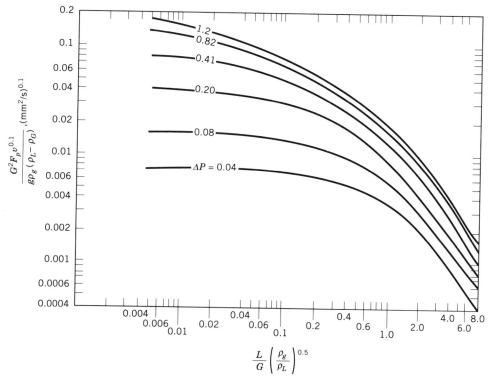

Fig. 24. Generalized method using log scales for estimating packed column flooding and pressure drop, ΔP, in kPa/m; g = gravitational constant, 9.81 m/s^2; v = kinematic viscosity in mm^2/s ($=$cSt); L, G have units of kg/(m^2·s); ρ_L, ρ_g are in kg/m^3; and the packing factor, F_p, in m^{-1} can be found in Table 2. To convert kPa/m to mm Hg/m, multiply by 7.5 (77).

Table 2. Characteristics of Packing[a]

Packing	Nominal size, mm[b]	Surface area, m²/m³	Void fraction	Packing factor, F_p, m⁻¹
		Dumped (random) packing		
Intalox saddles				
ceramic	13	625	0.78	660
	25	255	0.77	300
	50	118	0.79	130
metal (IMTP)	25		0.97	135
	40		0.97	79
	50		0.98	59
plastic	25	206	0.91	130
	50	108	0.93	92
	75	88	0.94	59
Berl saddles, ceramic	13	465	0.62	790
	25	250	0.68	360
	50	105	0.72	150
Pall rings				
metal	16		0.92	265
	25	205	0.94	183
	50	115	0.96	88
plastic	16	341	0.87	310
	25	207	0.90	180
	50	100	0.92	85
Raschig rings, ceramic	13	370	0.64	1902
	25	190	0.74	587
	50	92	0.74	215
		Structured packing		
Flexipac				
1	6	558	0.91	108
2	12	246	0.93	72
3	37	134	0.96	52
4	50	69	0.98	30
Sulzer-BX	6	490	>0.90	66

[a]Ref. 70.
[b]For structured packings, values correspond to crimp height.

eight should be maintained. Thus if one wishes to conduct pilot-scale packed column tests, a minimum column diameter of about 100 mm would be used together with 12-mm packing elements. The models would then permit scale-up to large columns containing 50-mm size elements of the same type, eg, Pall rings.

These models provide values of the height of a transfer unit for the liquid-phase H_L and the vapor-phase H_V. These values are combined to form the height of an overall transfer unit, H_{ov}:

$$H_{ov} = H_v + (m'V/L)H_L \qquad (47)$$

where V and L are molar flow rates of vapor and liquid and m' is the slope of the $y-x$ equilibrium curve (Fig. 1b) in the concentration range of interest. The required total height of the packed section is then obtained from the simple relationship,

$$Z_p = (N_{ov})(H_{ov}) \qquad (48)$$

In order to determine the packed height Z_p it is necessary to obtain a value of the overall number of transfer units N_{ov}; methods for doing this are available for binary systems in any standard text covering distillation (73) and, in a more complex way, for multicomponent systems (81). However, it is simpler to calculate the number of required theoretical stages and make the conversion:

$$N_{ov} = N_t (\ln m'V/L)(m'V/L - 1) \qquad (49)$$

An alternative to determining packed height is through the use of an empirical term, height equivalent to a theoretical plate (HETP). This term can be measured in a fashion similar to that used for the overall plate efficiency of a column (eq. 44):

$$\text{HETP} = \frac{\text{total packed height}}{\text{no. of theoretical plates}} = \frac{Z_p}{N_t} \qquad (50)$$

Typical experimental values of HETP for a random packing such as 50-mm Pall rings, and a structured packing, such as Intalox 2T of Norton Co., under the same system conditions, are shown in Figure 25. Many designers of packed columns prefer the use of HETP instead of H_{ov}, but the latter is more fundamental and discriminates between liquid- and vapor-phase resistances. It should be noted that terms such as H_{ov} and N_{ov} are based on vapor-phase concentrations; equivalent terms based on liquid concentrations could be used.

For structured packings, methods for predicting H_v and H_L are somewhat more reliable than those for random packings. Perhaps the most used efficiency correlation for these packings is that in Reference 82; a slightly different model that covers a broader packing size range is found in Reference 83. Methods for predicting pressure drop and flooding in beds of structured packings have been reviewed (71).

Packed vs Plate Columns. Relative to plate towers, packed towers are more useful for multipurpose distillations, usually in small (under 0.5 m) towers or for the following specific applications: severe corrosion environment where some corrosion-resistant materials, such as plastics, ceramics, and certain metallics, can easily be fabricated into packing but may be difficult to fabricate into plates; vacuum operation where a low pressure drop per theoretical plate is a critical requirement; high (eg, above 49,000 kg/(h·m²) (~10,000 lb/(h·ft²)) liquid rates; foaming systems; or debottlenecking plate towers having plate spacings that are relatively close, under 0.3 m.

Plate columns have the advantage of lower fabrication cost, less dependence on good liquid and gas distribution, and protection against vapor bypassing the

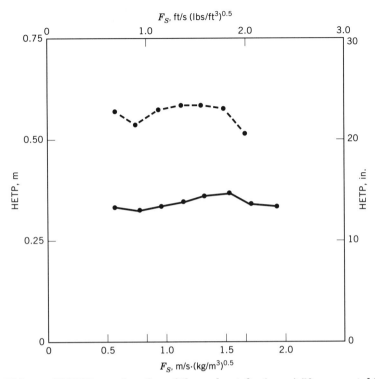

Fig. 25. Values of HETP as a function of throughput for (– – –) 50-mm metal Pall rings and (—) No. 2 structured packing at 12-mm crimp height. Conditions are cyclohexane/ n-heptane system, 165 kPa (24 psia) operating pressure, total reflux, 0.43-m diameter column. Courtesy of The University of Texas at Austin.

liquid in critical zones, eg, regions of extremely low impurities. Further, methods for the design on plate columns are somewhat more reliable than those for many of the packings, especially those packings of a proprietary nature.

There are notable cases where plate columns have been converted to packed columns to gain advantage of the low pressure drop exacted from the vapor stream. More recently the packings have been largely of the structured type. Illustrative of this is the trend toward the use of structured packing in ethylbenzene–styrene fractionators, some of which have diameters of 10 m or higher.

Steam Distillation

Steam distillation is used to lower the distillation temperatures of high boiling organic compounds that are essentially immiscible with water. If an organic compound is immiscible with water, both liquids exert full vapor pressure upon vaporization from the immiscible two-component liquid. At a system pressure of P, the partial pressures would be:

$$P = p_{\text{water}} + p_{\text{organic}} \tag{51}$$

and because the water and organic compound are immiscible:

$$P = P^0_{water} + P^0_{organic} \qquad (52)$$

The steam distillation of N-ethylaniline at atmospheric pressure (73) gives the following: the vapor pressures at 99.15°C of water and N-ethylaniline are 98.27 and 3.04 kPa (737 and 22.8 mm Hg), respectively. Thus, according to equation 52,

$$P = 98.27 + 3.04 = 101.3 \text{ kPa} \qquad (53)$$

and the concentration of the N-ethylaniline in the vapor is

$$y = 3.04/101.3 = 0.030 \text{ mol fraction} \qquad (54)$$

The normal boiling point of N-ethylaniline is 204°C. Therefore, steam distillation makes possible the distillation of N-ethylaniline at atmospheric pressure at a temperature of 99.15°C instead of its normal boiling point of 204°C. Commercial applications of steam distillation include the fractionation of crude tall oil (qv) (84), the distilling of turpentine (see TERPENOIDS), and certain essential oils (see OILS, ESSENTIAL). A detailed calculation of steam distillation of turpentine has been reported (85).

Molecular Distillation

Molecular distillation occurs where the vapor path is unobstructed and the condenser is separated from the evaporator by a distance less than the mean-free path of the evaporating molecules (86). This specialized branch of distillation is carried out at extremely low pressures ranging from 13–130 mPa (0.1–1.0 μm Hg) (see VACUUM TECHNOLOGY). Molecular distillation is confined to applications where it is necessary to minimize component degradation by distilling at the lowest possible temperatures. Commercial usage includes the distillation of vitamins (qv) and fatty acid dimers (see DIMER ACIDS).

Distillation as a Separation Method

Distillation is the most important industrial method of separation and purification of liquid components. Liquid separation methods in less common use include liquid–liquid extraction (see EXTRACTION, LIQUID–LIQUID), membrane diffusion (see DIALYSIS; MEMBRANE TECHNOLOGY), ion exchange (qv), and adsorption (qv). However, distillation does not require a mass separating agent such as a solvent, adsorbent, or membrane, and distillation utilizes energy in a convenient heating medium (often steam). Also, a wealth of experience with design and operations makes distillation column performance prediction more reliable than equivalent predictions for other methods. At times distillation also competes indirectly with methods involving solid–liquid separations such as crystallization (qv). An extensive discussion of the selection of alternative separation methods is available (30) (see SEPARATION SYSTEMS SYNTHESIS).

The suitability and economics of a distillation separation depend on such factors as favorable vapor–liquid equilibria, feed composition, number of components to be separated, product purity requirements, the absolute pressure of the distillation, heat sensitivity, corrosivity, and continuous vs batch requirements. Distillation is somewhat energy-inefficient because in the usual case heat added at the base of the column is largely rejected overhead to an ambient sink. However, the source of energy for distillations is often low pressure steam which characteristically is in long supply and thus relatively inexpensive. Also, schemes have been devised for lowering the energy requirements of distillation and are described in many publications (87).

Favorable Vapor–Liquid Equilibria. The suitability of distillation as a separation method is strongly dependent on favorable vapor–liquid equilibria. The absolute value of the key relative volatilities directly determines the ease and economics of a distillation. The energy requirements and the number of plates required for any given separation increase rapidly as the relative volatility becomes lower and approaches unity. For example: given an ideal binary mixture having a 50 mol % feed and a distillate and bottoms requirement of 99.8% purity each, the minimum reflux and minimum number of theoretical plates for assumed relative volatilities of 1.1, 1.5, and 4, are

Relative volatility	Minimum reflux ratio	Minimum no. of theoretical plates
1.1	20	130
1.5	4.0	31
4	0.66	9

In the example, the minimum reflux ratio and minimum number of theoretical plates decreased 14- to 33-fold, respectively, when the relative volatility increased from 1.1 to 4. Other distillation systems would have different specific reflux ratios and numbers of theoretical plates, but the trend would be the same. As the relative volatility approaches unity, distillation separations rapidly become more costly in terms of both capital and operating costs. The relative volatility can sometimes be improved through the use of an extraneous solvent that modifies the VLE. Binary azeotropic systems are impossible to separate into pure components in a single column, but the azeotrope can often be broken by an extraneous entrainer (see DISTILLATION, AZEOTROPIC AND EXTRACTIVE).

Feed Composition. Feed composition has a substantial effect on the economics of a distillation. Distillations tend to become uneconomical as the feed becomes dilute. There are two types of dilute feed cases, one in which the valuable recovered component is a low boiler and the second when it is a high boiler. When the recovered component is the low boiler, the absolute distillate rate is low but the reflux ratio and the number of plates is high. An example is the recovery of methanol from a dilute solution in water. When the valuable recovered component is a high boiler, the distillate rate, the reflux relative to the high boiler, and the number of plates all are high. An example for this case is the recovery of acetic acid from a dilute solution in water. For the general case of dilute feeds, alternative recovery methods are usually more economical than distillation.

Product Purity. Product purity requirements influence choice of separation methods. For favorable equilibria, distillation energy requirements do not increase significantly as purity specifications become tighter. For example, in an ideal binary distillation of 60 mol % of A in the feed, the minimum and operating reflux ratios would be essentially the same whether the required purity of A was 99 or 99.9999%. The number of plates would increase substantially, however, as the purity requirements became more stringent. The shortcut methods of calculating minimum reflux ratio, minimum number of plates, operating reflux ratio, and number of operating plates allow a rapid evaluation of the effect of changes in purity requirements on the key economic factors in distillation.

Operating Pressure. The absolute pressure of the distillation may have substantial economic impact. The temperatures at which heat is supplied to the reboiler and removed from the condenser determines the unit cost of the energy. The cost of removing heat in the condenser increases rapidly as the condensing temperature drops below the range of air or water cooling capability, eg, the cost of removing a unit quantity of heat at $-25°C$ may be one hundred times as high as removing it at 100°C. Similarly, the cost of the energy required for the reboiler increases rapidly as the boiler temperature increases above some level determined by local conditions. For example, at a particular site low pressure waste steam at 110°C may be essentially without cost, but if a temperature level of 200°C is required, the unit cost of the heat is much higher. The relative cost of the heat being removed and supplied is the controlling factor determining the design of some distillations. The use of multiple interstage reboilers and condensers at different energy levels as well as the use of other operational modes used to optimize the overall economics, has been discussed (88).

The absolute pressure may have a significant effect on the vapor–liquid equilibrium. Generally, the lower the absolute pressure the more favorable the equilibrium. This effect has been discussed for the styrene–ethylbenzene system (30). In a given column, increasing the pressure can increase the column capacity by increasing the capacity parameter (see eqs. 42 and 43). Selection of the economic pressure can be facilitated by guidelines (89) that take into consideration the pressure effects on capacity and relative volatility. Low pressures are required for distillation involving heat-sensitive material.

Heat Sensitivity. The heat sensitivity or polymerization tendencies of the materials being distilled influence the economics of distillation. Many materials cannot be distilled at their atmospheric boiling points because of high thermal degradation, polymerization, or other unfavorable reaction effects that are functions of temperature. These systems are distilled under vacuum in order to lower operating temperatures. For such systems, the pressure drop per theoretical stage is frequently the controlling factor in contactor selection. An excellent discussion of equipment requirements and characteristics of vacuum distillation may be found in Reference 90.

Corrosivity. Corrosivity is an important factor in the economics of distillation. Corrosion rates increase rapidly with temperature, and in distillation the separation is made at boiling temperatures. The boiling temperatures may require distillation equipment of expensive materials of construction; however, some of these corrosion-resistant materials are difficult to fabricate. For some materials, eg, ceramics (qv), random packings may be specified, and this has been a classical application of packings for highly corrosive services. On the other hand, the extensive surface areas of metal packings may make these more susceptible to corrosion than plates. Again, cost may be the final arbiter (see CORROSION AND CORROSION CONTROL).

Batch vs Continuous Distillation. The mode of operation also influences the economics of distillation. Batch distillation is generally limited to small-scale operations where the equipment serves several different distillations.

Research. Much of the research on commercial-size distillation equipment is being done by Fractionation Research, Inc. (FRI), a nonprofit, industry-sponsored, research corporation. The industrial sponsors are fabricators, designers, and constructors, or users of distillation equipment. Publications include liquid mixing on sieve plates (91), bubble cap plate efficiency (92), and sieve plate efficiency (93,94). A motion picture of downcomer performance is also available (95). References 96 and 97 cover the literature from 1967 to 1990.

Equipment Costs

A compilation of costs of distillation and related equipment is available (98). Some of the commercial computer-aided process design packages contain equipment cost information (see COMPUTER-AIDED DESIGN AND MANUFACTURING). For specialized internals, such as distributors, support plates, packings, crossflow plates, and so on, it is usually necessary to obtain cost information directly from the equipment vendors. It is important to recognize that the cost of a distillation system includes many components in addition to the column itself. For example, an expensive packing may be justified on the basis that it can reduce the cost of the column shell, foundations, piping, and so on. Discussions of economics of distillation systems are available (99,100).

Column Control

Distillation columns are controlled by hand or automatically. The parameters that must be controlled are (1) the overall mass balance, (2) the overall enthalpy balance, and (3) the column operating pressure. Modern control systems are designed to control both the static and dynamic column and system variables. For an in-depth discussion, see References 101–104.

NOMENCLATURE

Symbol	Definition	Units
A_{12}, A_{21}	constants in the Van Laar activity coefficient equation	
B	bottoms from column	mol/s

C_{sb}	vapor capacity parameter	m/s
D	distillate from column	mol/s
E_o	overall column plate efficiency (eq. 44)	fractional
E_{mv}	Murphree plate efficiency (eq. 45)	fractional
E_{og}	local, or point, efficiency based on vapor concentrations	fractional
f	fugacity	kPa
F	feed	mol/s
F^*	F-factor (eq. 43)	m/s·(kg/m³)^0.5
F_A^*	F-factor based on active (bubbling) area	m/s·(kg/m³)^0.5
F_H^*	F-factor based on hole area	m/s·(kg/m³)^0.5
F_p	packing factor from Table 2	1/m
G	gas mass rate	kg/s
$\overline{H}$	enthalpy per mole	
H	enthalpy per unit time	
H^*	Henry's law constant (eq. 17)	kPa/mol fraction
HETP	height equivalent of theoretical plate	m
H_L	height of a liquid-phase transfer unit	m
H_{ov}	height of an overall transfer unit, vapor concentrations	m
H_v	height of a vapor-phase transfer unit	m
K	y^*/x, vapor–liquid equilibrium ratio (eq. 1)	
L	liquid rate	mol/s
L	average liquid rate for section	mol/s
$\overline{L}$	liquid mass rate	kg/s
m	an equilibrium stage below the feed	
m'	slope of equilibrium line	
n	an equilibrium stage above the feed	
N	number of stages	
N^*	number of components	
N_a	number of actual stages	
N_{ov}	number of transfer units, vapor concentration basis	
N_t	number of theoretical stages	
P	total pressure of system	kPa
p	partial pressure	kPa
P^0	vapor pressure	kPa
q	heat to vaporize 1 mol feed divided by molal latent heat of feed (eq. 36)	
$\overline{q}$	heat removed or added at column auxiliaries	
R	reflux	mol/s
R'	gas law constant	
T	temperature	K
v	vapor molar volume	m³/mol
V	vapor molar rate	mol/s
$\overline{V}$	average molar vapor rate for section	mol/s
V^*	vapor velocity	m/s
x	mole fraction in liquid	
y	mole fraction in vapor	
y^*	mole fraction vapor in equilibrium with x	
z	compressibility factor in gas law	
Z_p	height of packed bed	m

Symbol	Definition	Units
α	relative volatility (eq. 2)	
γ^L	liquid-phase activity coefficient (eq. 6)	
γ^∞	terminal activity coefficient, at infinite dilution	
$\Lambda_{12}, \Lambda_{21}$	constant in Wilson activity coefficient model (eq. 13)	
ρ	fluid-phase density	kg/m^3
ϕ	fugacity coefficient (eq. 20)	

Superscripts

B	bottoms	
C	condenser	
D	distillate	
E	end	
F	feed	
f	feed stage	
L	liquid	
m	stage number m	
$m-1$	stage below m	
$m+1$	stage above m	
n	stage number n	
$n-1$	stage below n	
$n+1$	stage above n	
N	Nth component of components i to n	
P	pressure	
S	reboiler	
T	top column	
V	vapor	

Subscripts

$1,2,3 \ldots n$	component numbers	
B	bottoms	
D	distillate	
F	feed	
g	gas	
H	component H of binary system L–H, H is the high boiler	
i,j	components of mixture $1 \ldots i, j, \ldots n$	
L	component L of binary system L–H, L is the low boiler	
L	liquid	
min	minimum	
P	pressure	
R	rectifying section	
S	stripping section	
T	temperature	

BIBLIOGRAPHY

"Distillation" in *ECT* 1st ed., Vol. 5, pp. 156–187, by E. G. Scheibel, Hoffmann-LaRoche, Inc.; in *ECT* 2nd ed., Vol. 7, pp. 204–248, by C. D. Holland and J. D. Lindsey, Texas A & M University; in *ECT* 3rd ed., Vol. 7, pp. 849–891, by E. R. Hafslund, E. I. du Pont de Nemours & Co., Inc.

1. A. J. V. Underwood, *Trans. I. Chem. E.* **13,** 34 (1935).
2. J. R. Fair, *AIChE Symp. Ser. No. 235* **79,** 1 (1984).
3. J. Gmehling, U. Onken, and W. Arlt, *Vapor–Liquid Equilibrium Collection* (continuing series), DECHEMA, Frankfurt, Germany, 1979.
4. M. Hirata, S. Ohe, and K. Nagahama, *Computer Aided Data Book of Vapor–Liquid Equilibria*, Elsevier, Amsterdam, The Netherlands, 1975.
5. E. Hala, J. Pick, V. Fried, and O. Vilim, *Vapor–Liquid Equilibrium*, 2nd ed., Pergamon Press, Oxford, UK, 1967.
6. E. Hala, I. Wichterle, J. Polak, and T. Boublik, *Vapor–Liquid Equilibrium at Normal Pressures*, Pergamon Press, Oxford, UK, 1968.
7. I. Wichterle, J. Linek, and E. Hala, *Vapor–Liquid Equilibrium Data Bibliography*, Elsevier, Amsterdam, The Netherlands, 1975.
8. A. Fredenslund, J. Gmehling, and P. Rasmussen, *Vapor–Liquid Equilibria Using UNIFAC*, Elsevier, Amsterdam, The Netherlands, 1977.
9. J. H. Hildebrand, J. M. Prausnitz, and R. L. Scott, *Regular and Related Solutions*, Van Nostrand Reinhold Co., Inc., New York, 1970.
10. E. L. Derr and C. H. Deal, *I. Chem. E. Symp. Ser. No. 32* **3**(40), (1969).
11. D. A. Palmer, *Handbook of Applied Thermodynamics*, CRC Press, Inc., Boca Raton, Fla., 1987.
12. J. M. Prausnitz, R. N. Lichtenthaler, and E. G. Azeredo, *Molecular Thermodynamics of Fluid-Phase Equilibria*, 2nd ed., Prentice-Hall, Inc., Englewood Cliffs, N.J., 1986.
13. R. C. Reid, J. M. Prausnitz, and B. Pohling, *The Properties of Gases and Liquids*, 4th ed., McGraw-Hill Book Co., Inc., New York, 1987.
14. S. M. Walas, *Phase Equilibria in Chemical Engineering*, Butterworths, Reading, Mass., 1985.
15. E. J. Henley and J. D. Seader, *Equilibrium-Stage Separation Operations in Chemical Engineering*, John Wiley & Sons, Inc., New York, 1981.
16. P. Wankat, *Equilibrium-Staged Separations*, Elsevier Science Publishing Co., Inc., New York, 1988.
17. M. Van Winkle, *Distillation*, McGraw-Hill Book Co., Inc., New York, 1967.
18. J. J. Van Laar, *Z. Physik. Chem.* **72,** 723 (1910); **83,** 599 (1913).
19. G. M. Wilson, *J. Am. Chem. Soc.* **86,** 127 (1964).
20. H. Renon and J. M. Prausnitz, *AIChE J.* **14,** 135 (1968).
21. D. S. Abrams and J. M. Prausnitz, *AIChE J.* **21,** 116 (1975).
22. Margules, *Sitzber. Math.-Naturw. Kl. Kaiserlichen Akad. Wiss. (Vienna)* **104,** 1243 (1895).
23. H. H. Chien and H. R. Null, *AIChE J.* **18,** 1177 (1972).
24. D. Tiegs, J. Gmehling, P. Rasmussen, and A. Fredenslund, *Ind. Eng. Chem. Res.* **26,** 159 (1987).
25. H. K. Hansen, P. Rasmussen, A. Fredenslund, M. Schiller, and J. Gmehling, *Ind. Eng. Chem. Res.* **30,** 2352 (1991).
26. J. M. Prausnitz and co-workers, *Computer Calculations for Multicomponent Vapor–Liquid and Liquid–Liquid Equilibria*, Prentice-Hall, Inc., Englewood Cliffs, N.J., 1980.
27. *Technical Data Book, Petroleum Refining*, 3rd ed., Vols. I and II, American Petroleum Institute, New York, 1976.

28. L. Horsley, *Azeotropic Data—III*, Advances in Chemistry Series No. 116, American Chemical Society, Washington, D.C., 1973.
29. J. F. Boston, H. I. Britt, S. Jiraphongphan, and V. B. Shah, "An Advanced System for the Simulation of Batch Distillation Operations," in *Foundations of Computer-Aided Chemical Process Design*, Vol. 2, American Institute of Chemical Engineers, New York, 1981.
30. C. J. King, *Separation Processes*, 2nd ed., McGraw-Hill Book Co., Inc., New York, 1980.
31. W. L. McCabe and E. W. Thiele, *Ind. Eng. Chem.* **17,** 605 (1925).
32. M. Ponchon, *Tech. Mod.* **13,** 20, 55 (1921).
33. R. Savarit, *Arts Metiers* **65,** 145, 178, 266, 307 (1922).
34. M. R. Fenske, *Ind. Eng. Chem.* **24,** 482 (1932).
35. A. J. V. Underwood, *Chem. Eng. Progr.* **44,** 603 (1948).
36. J. R. Fair and W. L. Bolles, *Chem. Eng.* **75**(9), 156 (Apr. 22, 1968).
37. G. G. Brown and co-workers, *Unit Operations*, John Wiley & Sons, Inc., New York, 1950.
38. E. R. Gilliland, *Ind. Eng. Chem.* **32,** 918 (1940).
39. J. H. Erbar and R. N. Maddox, *Petrol. Ref.* **40**(5), 183 (1961).
40. H. E. Eduljee, *Hydrocarbon Proc.* **54**(9), 120 (1975).
41. E. W. Thiele and R. L. Geddes, *Ind. Eng. Chem.* **25,** 290 (1933).
42. C. D. Holland, *Fundamentals of Multicomponent Distillation*, McGraw-Hill Book Co., Inc., New York, 1981.
43. R. N. S. Rathore, K. A. Van Wormer, and G. J. Powers, *AIChE J.* **20,** 491 (1974).
44. D. C. Freshwater and B. D. Henry, *Chem. Eng. (London)* (301), 533 (1975).
45. J. E. Hendry, D. F. Rudd, and J. D. Seader, *AIChE J.* **19,** 1 (1973).
46. *Ballast Tray Design Manual*, Bulletin 4900, Glitsch, Inc., Dallas, Tex., 1974.
47. *Flexitray Design Manual*, Bulletin 960, Koch Engineering Co., Wichita, Kans., 1960.
48. *Float Valve Tray Design Manual*, Nutter Engineering Co., Tulsa, Okla., 1976.
49. W. L. Bolles, in B. D. Smith, ed., *Design of Equilibrium Stage Processes*, McGraw-Hill Book Co., Inc., New York, 1963, Chapt. 14.
50. M. H. Hutchinson and R. F. Baddour, *Chem. Eng. Progr.* **52**(12), 503 (1956).
51. F. Kastanek, M. V. Huml, and V. Braun, *I. Chem. E. Symp. Ser. No. 32*, 5(100), (1969).
52. W. V. Delnicki and J. L. Wagner, *Chem. Eng. Progr.* **52**(1), 28 (1956).
53. J. R. Fair, in Ref. 49, Chapt. 15.
54. J. R. Fair, in R. H. Perry and D. Green, eds., *Perry's Chemical Engineers' Handbook*, 6th ed., McGraw-Hill Book Co., Inc., New York, 1984, section 18.
55. E. V. Murphree, *Ind. Eng. Chem.* **17,** 747, 960 (1925).
56. *Bubble-Tray Design Manual*, American Institute of Chemical Engineers (AIChE), New York, 1958.
57. M. J. Lockett, *Distillation Tray Fundamentals*, Cambridge University Press, Cambridge, Mass., 1986.
58. M. M. Dribika and M. W. Biddulph, *Trans. I. Chem. E.* **70,** Part A, 142 (1992).
59. K. E. Porter, M. J. Lockett, and C. T. Lim, *Trans. I. Chem. E.* **50,** 91 (1972).
60. W. V. Pincezewski, N. D. Benke, and C. J. D. Fell, *AIChE J.* **21,** 1210 (1975).
61. K. E. Porter, A. Safekouri, and M. J. Lockett, *Trans. I. Chem. E.* **51,** 265 (1973).
62. M. Prado, K. L. Johnson, and J. R. Fair, *Chem. Eng. Progr.* **83**(3), 32 (1987).
63. M. Prado and J. R. Fair, *Ind. Eng. Chem. Res.* **29,** 1031 (1990).
64. H. E. O'Connell, *Trans. AIChE* **42,** 741 (1946).
65. G. E. English and M. Van Winkle, *Chem. Eng.* **70**(23), 241 (1963).
66. H. L. Toor and J. K. Burchard, *AIChE J.* **6,** 202 (1960).
67. R. Krishna, H. F. Martinez, R. Sreedhar, and G. L. Standart *Trans. I. Chem. E.* **55,** 178 (1977).

68. A. Sperandio, M. Richard, and M. Huber, *Chem.-Ing.-Tech.* **37,** 22 (1965).
69. W. D. Stoecker and B. Weinstein, *Chem. Eng. Progr.* **73**(11), 71 (1977).
70. R. F. Strigle, *Random Packings and Packed Tower Design*, Gulf Publishing, Houston, Tex., 1987.
71. J. R. Fair and J. L. Bravo, *Chem. Eng. Progr.* **86**(1), 19 (1990).
72. P. J. Hoek, J. A. Wesselingh, and F. J. Zuiderweg, *Chem. Eng. Res. Des.* **64,** 431 (1986).
73. R. E. Treybal, *Mass Transfer Operations*, 3rd ed., McGraw-Hill Book Co., Inc., New York, 1980.
74. W. S. Norman, *Absorption, Distillation and Cooling Towers*, John Wiley & Sons, Inc., New York, 1961.
75. P. A. Schweitzer, ed., *Handbook of Separation Techniques for Chemical Engineers*, 2nd. ed., McGraw-Hill Book Co., Inc., New York, 1988, Chapts. 1.1–1.8.
76. R. W. Rousseau, ed., *Handbook of Separation Process Technology*, John Wiley & Sons, Inc., New York, 1987, Chapt. 5.
77. T. K. Sherwood, G. H. Shipley, and F. A. L. Holloway, *Ind. Eng. Chem.* **30,** 765 (1938).
78. M. Leva, *Chem. Eng. Progr. Symp. Ser. No. 10* **50,** 51 (1954).
79. W. L. Bolles and J. R. Fair, *Chem. Eng.* **89**(14), 109 (July 12, 1982).
80. J. L. Bravo and J. R. Fair, *Ind. Eng. Chem. Proc. Des. Dev.* **21,** 162 (1982).
81. R. Krishnamurthy and R. Taylor, *AIChE J.* **31,** 449, 456 (1985).
82. J. L. Bravo, J. A. Rocha, and J. R. Fair, *Hydrocarbon Proc.* **64**(1), 91 (1985).
83. L. Spiegel and W. Meier, *I. Chem. E. Symp. Ser. No. 104*, A203 (1987).
84. J. Drew and M. Propst, eds., *Tall Oil*, Pulp Chemicals Association, New York, 1981.
85. W. L. McCabe and J. C. Smith, *Unit Operations of Chemical Engineering*, 3rd ed., McGraw-Hill Book Co., Inc., New York, 1976, Chapt. 19.
86. K. C. D. Hickman, in R. H. Perry and C. H. Chilton, eds., *Chemical Engineers' Handbook*, 5th ed., McGraw-Hill Book Co., Inc., New York, 1973, section 13.
87. J. R. Fair, in Y. A. Liu, H. A. McGee, and W. R. Epperly, eds., *Recent Developments in Chemical Process and Plant Design*, John Wiley & Sons, Inc., New York, 1987, Chapt. 3.
88. W. C. Petterson and T. A. Wells, *Chem. Eng.* **84**(20), 79 (1977).
89. H. Z. Kister and I. D. Doig, *Hydrocarbon Proc.* **56**(7), 132 (1977).
90. P. G. Nygren and G. K. S. Connolly, *Chem. Eng. Progr.* **67**(3), 49 (1971).
91. T. Yanagi and B. D. Scott, *Chem. Eng. Progr.* **69**(10), 75 (1973).
92. B. D. Scott and H. S. Myers, in Ref. 91, p. 73.
93. M. Sakata and T. Yanagi, *I. Chem. E. Symp. Ser. No. 56*, **3.2**(21), (1979).
94. T. Yanagi and M. Sakata, *Ind. Eng. Chem. Proc. Des. Dev.* **21,** 712 (1982).
95. T. Yanagi, *Performance of Downcomers in Distillation Columns* (motion picture), AIChE Meeting, Atlanta, Ga., Feb. 1970. Available from Fractionation Research, Inc., Stillwater, Okla.
96. M. S. Ray, *Chemical Engineering Bibliography, 1967–1988*, Noyes Publications, Park Ridge, N.J., 1990.
97. M. S. Ray, *Sepn. Sci. Technol.* **27,** 105 (1992).
98. M. Peters and K. D. Timmerhaus, *Plant Design and Economics for Chemical Engineers*, 4th ed., McGraw-Hill Book Co., Inc., New York, 1991.
99. H. Z. Kister, *Distillation–Operation*, McGraw-Hill Book Co., Inc., New York, 1990.
100. H. Z. Kister, *Distillation—Design*, McGraw-Hill Book Co., Inc., New York, 1992.
101. A. E. Nisenfeld and R. C. Seeman, *Distillation Columns*, Instrument Society of America, Research Triangle Park, N.C., 1981.
102. F. G. Shinskey, *Distillation Control*, 2nd ed., McGraw-Hill Book Co., Inc., New York, 1984.
103. P. B. Deshpande, *Distillation Dynamics and Control*, Instrument Society of America, Research Triangle Park, N.C., 1984.

104. P. S. Buckley, W. L. Luyben, and J. P. Shunta, *Design of Distillation Control Systems*, Instrument Society of America, Research Triangle Park, N.C., 1985.

General References

Fluid Phase Equilibria.
J. Chem. Eng. Data.

JAMES R. FAIR
The University of Texas at Austin

DISTILLATION, AZEOTROPIC AND EXTRACTIVE

Distillation (qv) is the most widely used separation technique in the chemical and petroleum industries. Not all liquid mixtures are amenable to ordinary fractional distillation, however. Close-boiling and low relative volatility mixtures are difficult and often uneconomical to distill, and azeotropic mixtures are impossible to separate by ordinary distillation. Yet such mixtures are quite common (1) and many industrial processes depend on efficient methods for their separation (see also SEPARATION SYSTEMS SYNTHESIS). This article describes special distillation techniques for economically separating low relative volatility and azeotropic mixtures.

Whereas there is extensive literature on design methods for azeotropic and extractive distillation, much less has been published on operability and control. It is, however, widely recognized that azeotropic distillation columns are difficult to operate and control because these columns exhibit complex dynamic behavior and parametric sensitivity (2–11). In contrast, extractive distillations do not exhibit such complex behavior and even highly optimized columns are no more difficult to control than ordinary distillation columns producing high purity products (12).

At low to moderate pressures the vapor–liquid phase equilibrium (VLE) of many mixtures can be adequately described by:

$$y_i P = x_i \gamma_i P_i^{\text{sat}} \qquad \text{for } i = 1, \ldots, c \qquad (1)$$

where y_i is the mole fraction of component i in the vapor phase; x_i is the mole fraction of component i in the liquid phase; P is the system pressure; P_i^{sat} is the vapor pressure of pure component i; c is the number of components in the mixture; and γ_i is the liquid-phase activity coefficient of component i. The activity coefficient is a measure of the nonideality of a mixture and changes both with temperature and composition. When $\gamma_i = 1$ the mixture is said to be ideal and equation 1 simplifies to Raoult's Law. Nonideal mixtures ($\gamma_i \neq 1$) can exhibit either positive ($\gamma_i > 1$) or negative ($\gamma_i < 1$) deviations from Raoult's Law. In many highly nonideal mixtures these deviations become so large that the pressure-composition (P-x,y) and temperature-composition (T-x,y) phase diagrams exhibit a minimum or maximum point (Fig. 1). At these minima and maxima the liquid phase and its equilibrium vapor phase have the same composition, ie,

$$x_i = y_i \qquad \text{for } i = 1, ..., c \qquad (2)$$

and the dew-point (vapor) and bubble-point (liquid) curves are tangent with zero slope. These are the defining conditions for a homogeneous azeotrope where a single liquid phase is in equilibrium with a vapor phase. Mixtures that form two liquid phases are capable of forming heterogeneous azeotropes where the overall liquid composition is identical to the vapor composition, but the vapor and liquid surfaces are not tangent with zero slope. A maximum boiling azeotrope (Fig. 1b) is equivalent to a minimum pressure azeotrope (Fig. 1a) and a minimum boiling azeotrope (Fig. 1d) is also a maximum pressure azeotrope (Fig. 1c). The majority of the known azeotropes are minimum boiling.

Separation by distillation is dependent on the fact that when a liquid is partially vaporized the vapor and liquid compositions differ. The vapor phase becomes enriched in the more volatile components and depleted in the less volatile components with respect to its equilibrium liquid phase. By segregating the phases and repeating the partial vaporization, it is often possible to achieve the desired degree of separation. One measure of the degree of enrichment or the ease of separation is the relative volatility defined as:

$$\alpha_{ij} = \frac{y_i x_j}{x_i y_j} = \frac{\gamma_i P_i^{\text{sat}}}{\gamma_j P_j^{\text{sat}}} \qquad (3)$$

The relative volatility of most mixtures changes with temperature, pressure, and composition. The larger the value of α_{ij}, the easier it is to separate component i from component j. From equation 2, at a c-component, ie, binary, ternary, etc, homogeneous azeotrope, $x_i = y_i$ for all c components in the mixture. Therefore $\alpha_{ij} = 1$ for all components i and j and it is impossible to further enrich the vapor. Thus homogeneous azeotropes can never be separated into pure components by ordinary fractional distillation. Similarly, any mixture, be it ideal, nonideal, close-boiling, or isomeric, where the relative volatilities are close to unity is difficult to separate by ordinary distillation because little enrichment occurs with each partial vaporization step.

Most methods for distilling azeotropic and low relative volatility mixtures rely on the addition of specially chosen chemicals to facilitate the separation. These separating agents can be divided into distinct classes which define the principal distillation techniques used to separate mixtures containing azeotropes. The five methods for separating azeotropic mixtures are (1) extractive distillation and homogeneous azeotropic distillation where the liquid separating agent is completely miscible. For extractive distillation, separating agents are variously

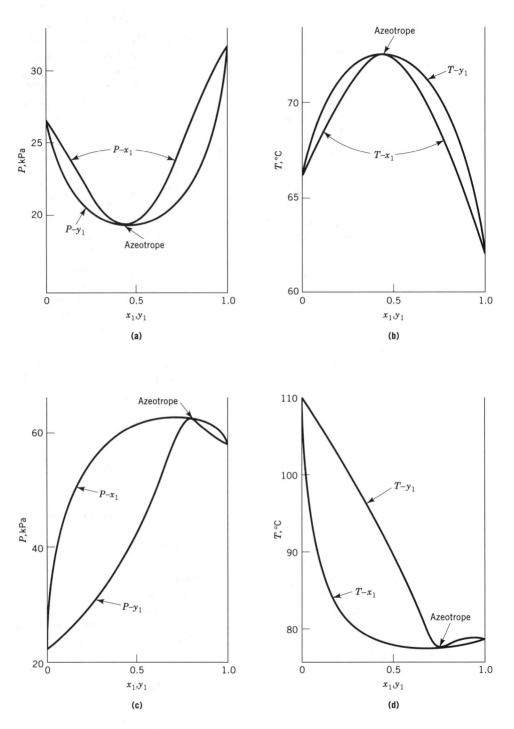

Fig. 1. P-x,y and T-x,y phase diagrams showing maximum and minimum azeotropes. (**a**) Chloroform (1)-tetrahydrofuran (2) at 30°C; (**b**) chloroform (1)-tetrahydrofuran (2) at 101 kPa; (**c**) ethanol (1)-toluene (2) at 65°C; and (**d**) ethanol (1)-toluene (2) at 101 kPa (13). To convert kPa to atm, multiply by 9.87×10^{-3}.

known as solvents, extractive agents, entrainers, or extractants. (2) Heterogeneous azeotropic distillation or, more commonly, azeotropic distillation where the liquid separating agent, called the entrainer, forms one or more azeotropes with the other components in the mixture and causes two liquid phases to exist over a broad range of compositions. This immiscibility is the key to making the distillation sequence work. (3) Distillation in the presence of ionic salts. The salt dissociates in the liquid mixture and alters the relative volatilities sufficiently that the separation becomes possible. (4) Pressure-swing distillation where a series of columns operating at different pressures are used to separate binary azeotropes which change appreciably in composition over a moderate pressure range or where a separating agent which forms a pressure-sensitive azeotrope is added to separate a pressure-insensitive azeotrope. (5) Reactive distillation where the separating agent reacts preferentially and reversibly with one of the azeotropic constituents. The reaction product is then distilled from the nonreacting components and the reaction is reversed to recover the initial component.

Of these five methods all but pressure-swing distillation can also be used to separate low volatility mixtures and all but reactive distillation are discussed herein. It is also possible to combine distillation and other separation techniques such as liquid–liquid extraction (see EXTRACTION, LIQUID–LIQUID), adsorption (qv), melt crystallization (qv), or pervaporation to complete the separation of azeotropic mixtures.

Residue Curve Maps

The most basic form of distillation, called simple distillation, is a process in which a multicomponent liquid mixture is slowly boiled in an open pot and the vapors are continuously removed as they form. At any instant in time the vapor is in equilibrium with the liquid remaining in the still. Because the vapor is always richer in the more volatile components than the liquid, the liquid composition changes continuously with time, becoming more and more concentrated in the least volatile species. A simple distillation residue curve is a graph showing how the composition of the liquid residue in the pot changes over time. A residue curve map is a collection of residue curves originating from different initial compositions. Residue curve maps contain the same information as phase diagrams, but residue curve maps represent this information in a way that is more useful for understanding how to synthesize a distillation sequence to separate a mixture. The liquid composition profiles in a continuous packed or staged distillation column operating at infinite reflux and reboil are closely approximated by simple distillation residue curves.

Residue curves can only originate from, terminate at, or be deflected by the pure components and azeotropes in a mixture. Pure components and azeotropes that residue curves move away from are called unstable nodes (UN), those where residue curves terminate are called stable nodes (SN), and those that deflect residue curves are called saddles (S).

The simplest residue curve map for a ternary mixture is shown in Figure 2. All ternary nonazeotropic mixtures, including ideal and constant volatility mixtures, are represented qualitatively by this map. All of the residue curves origi-

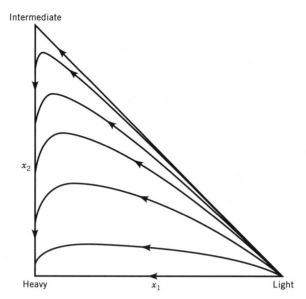

Fig. 2. Residue curve map for a ternary nonazeotropic mixture.

nate at the light (lowest boiling) pure component, move toward the intermediate boiling component, and end at the heavy (highest boiling) pure component. Thus the residue curves point in the direction of increasing temperature. In fact, residue curves must always move in such a way that the boiling temperature of the mixture continuously increases along every curve. From this property and the direction of the arrows in Figure 2, the light component is an unstable node; the intermediate component, which deflects the residue curves, is a saddle; and the heavy component is a stable node. A detailed mathematical treatment of simple distillation residue curve maps can be found in the literature (14,15).

Many different residue curve maps are possible when azeotropes are present. For example, for ternary mixtures containing only one azeotrope there are six possible residue curve maps that differ by the binary pair forming the azeotrope and by whether the azeotrope is minimum or maximum boiling. Figure 3 represents the case where the intermediate and heaviest components form a minimum boiling binary azeotrope. Pure component D is an unstable node, pure components A and B are stable nodes, the minimum boiling binary azeotrope C is a saddle, and the boiling point order from low to high is $D \longrightarrow C \longrightarrow A$ or B. The residue curve connecting component D to the azeotrope C has the special property that it divides the composition triangle into two separate distillation regions. Any initial still-pot composition lying to the left of the curve $D–C$ results in the last drop of liquid being pure B; any initial still-pot composition lying to the right of the curve $D–C$ yields pure A as the last drop of liquid. Residue curves like $D–C$ that divide the composition space into different distillation regions are called simple distillation boundaries or infinite reflux boundaries, often referred to simply as distillation boundaries. There must be a saddle on at least one end of every distillation boundary and each distillation region must contain a stable node, an unstable node, and at least one saddle.

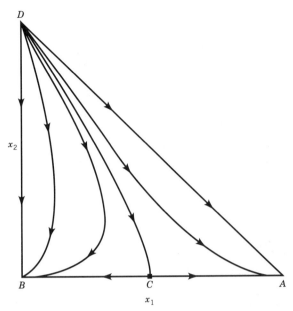

Fig. 3. Residue curve map for a ternary mixture with a distillation boundary running from pure component D to the binary azeotrope C.

Residue curve maps would be of limited usefulness if they could only be generated experimentally. Fortunately that is not the case. The simple distillation process can be described (14) by the set of equations:

$$\frac{dx_i}{d\xi} = x_i - y_i \qquad \text{for } i = 1, ..., c \qquad (4)$$

where x_i and y_i are the liquid and vapor mole fractions of component i, respectively, ξ is a nonlinear time scale, and c is the number of components in the mixture. Given a method for calculating the vapor–liquid phase equilibrium (VLE) for the mixture of interest (information that is required before a process simulator can be used to design a distillation column), equation 4 can be numerically integrated forward and backward in time from a number of initial conditions to generate a residue curve map. An alternative method for sketching residue curve maps that requires only knowledge of the boiling points of the pure components and azeotropes in the mixture has also been published (16–18). This latter method is particularly well suited to mixtures for which detailed VLE data is lacking and thus can be useful in entrainer/solvent screening.

Even though the simple distillation process has no practical use as a method for separating mixtures, simple distillation residue curve maps have extremely useful applications. These maps can be used to test the consistency of experimental azeotropic data (16,17,19); to predict the order and content of the cuts in batch distillation (20–22); and, in continuous distillation, to determine whether a given mixture is separable by distillation, identify feasible entrainers/solvents, predict the attainable product compositions, qualitatively predict the composition profile shape, and synthesize the corresponding distillation sequences (16,23–30). By identifying the limited separations achievable by distillation, residue curve maps are also useful in synthesizing separation sequences combining distillation with other methods.

Residue curve maps exist for mixtures having more than three components but cannot be visualized when there are more than four components. However, many mixtures of industrial importance contain only three or four key components and can thus be treated as pseudo-ternary or quaternary mixtures. Quaternary residue curve maps are more complicated than their ternary counterparts but it is still possible to understand these maps using the boiling point temperatures of the pure components and azeotropes (31).

Homogeneous Azeotropic Distillation

The most general definition of homogeneous azeotropic distillation is the separation of any single liquid-phase mixture containing one or more azeotropes into the desired pure component or azeotropic products by continuous distillation. Thus, in addition to azeotropic mixtures which require the addition of a miscible separating agent in order to be separated, homogeneous azeotropic distillation also includes self-entrained mixtures that can be separated without the addition of a separating agent.

The first step in the synthesis of a homogeneous azeotropic distillation sequence is to determine the separation objective. For example, sometimes it is desirable to recover all of the constituents in the mixture as pure components, other times it is sufficient to recover only some of the pure components as products. In other cases an azeotrope may be the desired product. Not every objective is attainable and those that are feasible may require different distillation sequences.

The second step is to sketch the residue curve map for the mixture to be separated. The residue curve map allows one to determine whether the goal can be reached and if so how to reach it, or whether the goal needs to be redefined. The addition of a separating agent to meet a separation objective carries with it the additional responsibility of finding an effective method for its recovery for reuse.

Distillation boundaries for continuous distillation are approximated by simple distillation boundaries. This is a very good approximation for mixtures with nearly linear simple distillation boundaries. Although curved simple distillation boundaries can be crossed to some degree (16,25–30,32,33), the resulting distillation sequences are not normally economical. Mixtures such as nitric acid–water–sulfuric acid, that have extremely curved boundaries, are exceptions. Therefore, a good working assumption is that simple distillation boundaries should not be crossed by continuous distillation. In other words, for a separation to be feasible by distillation it is sufficient that the distillate and bottoms compositions lie in the same distillation region.

An overall material balance for a continuously operated distillation column requires that the feed, distillate, and bottoms compositions lie on a straight line in the composition space (composition triangle for ternary mixtures). Thus feasible distillation sequences for separating homogeneous azeotropic mixtures can first be identified by noting whether the desired products lie in the same distillation region and then can be synthesized by superimposing material balance lines onto simple distillation residue curve maps. When determining the column products for the sequence: (1) the distillate composition must have a lower boiling temperature than the bottoms composition, however the component with the lowest (highest) boiling point is not necessarily removed as the distillate (bottoms); (2) pure components and azeotropes which are nodes on the residue curve map are easier to obtain as pure products than saddles; and (3) double-feed columns are almost always required to obtain saddles as the product, eg, extractive distillations.

As an example, consider the residue curve map for the nonazeotropic mixture shown in Figure 2. It has no distillation boundary so the mixture can be separated into pure components by either the direct or indirect sequence (Fig. 4). In the direct sequence the unstable node (light component, L) is taken overhead in the first column and the bottom stream is essentially a binary mixture of the intermediate, I, and heavy, H, components. In the binary I–H mixture, I has the lowest boiling temperature (an unstable node) so it is recovered as the distillate in the second column and the stable node, H, is the corresponding bottoms stream. The indirect sequence removes the stable node (heavy component) from the bottom of the first column and the overhead stream is an essentially binary L–I mixture. Then in the second column the unstable node, L, is taken overhead and I is recovered in the bottoms.

A second example is the azeotropic mixture of acetone, water, and 2-propanol that arises in the production of acetone by 2-propanol dehydrogenation. The goal is to recover the acetone product, recycle the unreacted 2-propanol to the reactor, and discard the wash water. From the residue curve map (Fig. 5a), it is clear that it is impractical to obtain the three pure components as in the previous example because of the distillation boundary running from the acetone vertex to the 2-propanol–water azeotrope which divides the triangle into two distillation regions. However, by modifying the objective slightly and recycling the 2-propanol–water azeotrope to the reactor instead of pure 2-propanol, a feasible separation sequence is possible (Figs. 5b and c). The acetone (unstable node) is taken overhead in the first column leaving an essentially binary 2-propanol–water mixture to be distilled in the second column. In this binary mixture the azeotrope is an unstable node so it becomes the distillate and the stable node, water, is removed from the bottom of the second column.

The overwhelming majority of all ternary mixtures that can potentially exist are represented by only 113 different residue curve maps (35). Reference 24 contains sketches of 87 of these maps. For each type of separation objective, these 113 maps can be subdivided into those that can potentially meet the objective, ie, residue curve maps where the desired pure component and/or azeotropic products lie in the same distillation region, and those that cannot. Thus knowing the residue curve for the mixture to be separated is sufficient to determine if a given separation objective is feasible, but not whether the objective can be achieved economically.

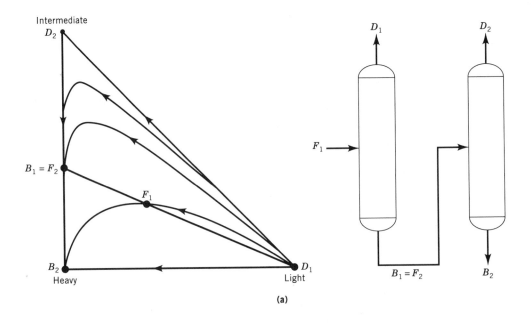

(a)

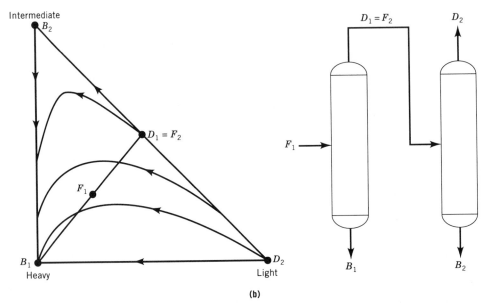

(b)

Fig. 4. Column sequences and material balance lines for the (**a**) direct and (**b**) indirect sequences for separating nonazeotropic mixtures.

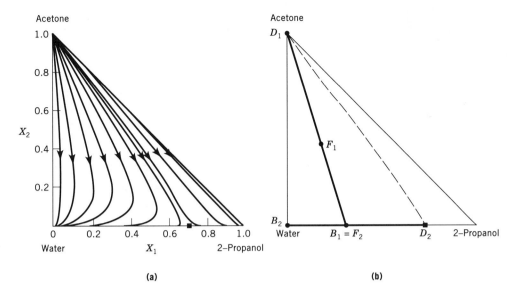

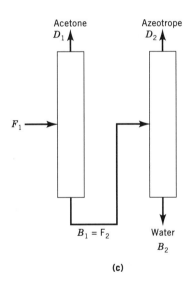

Fig. 5. The acetone–2-propanol–water system where ■ represents the 2-propanol–water azeotrope. (**a**) Residue curve map (34); (**b**) material balance lines showing the (– – –) distillation boundary; and (**c**) column sequence.

In the most common situation, a separating agent is added to separate a minimum boiling binary azeotrope into its two constituent pure components by homogeneous azeotropic distillation. The seven most favorable residue curve maps for this task are shown in Figure 6. Other maps have the potential to meet the stated objective (25–30), however they must be studied much more carefully because the design and operation of the resulting sequences are sensitive to the

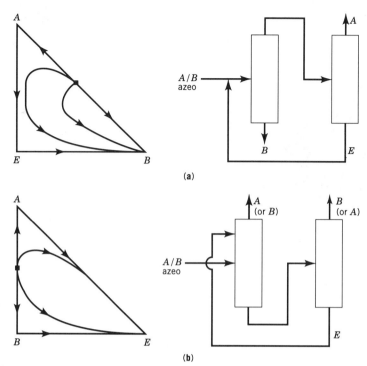

Fig. 6. The seven most favorable residue curve maps and corresponding column sequences for homogeneous azeotropic distillation of a minimum boiling azeotrope where ■ represents an azeotrope (16). (**a**) Case I, where the separating agent is intermediate boiling and does not introduce a new azeotrope; (**b**) Case II, extractive distillation with a heavy solvent which introduces no new azeotrope. In some cases, B can come off the top of the first column; (**c**) Case III, where the separating agent is intermediate boiling and forms a maximum boiling azeotrope with the lighter of the two pure components (ie, A). The agent may or may not form a minimum boiling azeotrope with B, with or without a minimum boiling ternary azeotrope lean in A; and (**d**) the same column configuration as case III, but the separating agent is lower boiling.

detailed VLE behavior. Thus, for initial screening purposes, only those separating agents that result in one of these seven maps need be considered for meeting the stated objective. Of these seven, the map representing extractive distillation (Fig. 6**b**) is by far the most common and the most important.

Extractive Distillation

Extractive distillation is defined as distillation in the presence of a miscible, high boiling, relatively nonvolatile component, the solvent, that forms no azeotropes with the other components in the mixture (23). It is widely used in the chemical

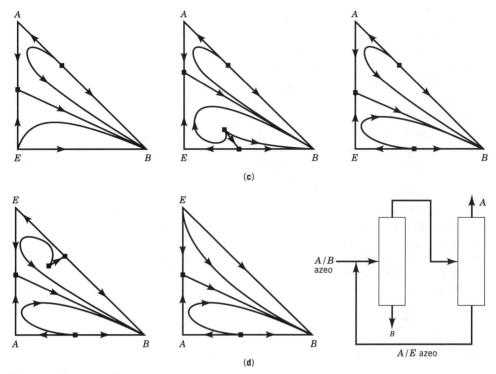

Fig. 6. (*Continued*)

and petrochemical industries for separating azeotropic, close-boiling, and other low relative volatility mixtures, including those forming severe tangent pinches.

Extractive distillation works because the solvent is specially chosen to interact differently with the components of the original mixture, thereby altering their relative volatilities. Because these interactions occur predominantly in the liquid phase, the solvent is continuously added near the top of the extractive distillation column so that an appreciable amount is present in the liquid phase on all of the trays below. The mixture to be separated is added through a second feed point further down the column (see Fig. 6**b**). In the extractive column the component having the greater volatility, not necessarily the component having the lowest boiling point, is taken overhead as a relatively pure distillate. The other component leaves with the solvent via the column bottoms. The solvent is separated from the remaining components in a second distillation column and then recycled back to the first column.

A selection of industrial applications of extractive distillation includes: (*1*) the separation of the *n*-butane–butadiene azeotrope in mixed C_4-hydrocarbon streams using furfural [*98-01-1*], $C_5H_4O_2$, as the solvent (*36*); (*2*) the dehydration of ethanol using ethylene glycol [*107-21-1*] (*37*–*39*); (*3*) the separation of acetone and methanol using water (*40*); (*4*) the Ryan-Holmes process for separating the ethane–carbon dioxide azeotrope that arises when carbon dioxide is used for enhanced oil-field recovery (41); (5) the separation of the pyridine–water azeotrope using bisphenol (42); (6) the dehydration of tetrahydrofuran using monopropylene glycol (43); (7) the separation of the cumene–phenol azeotrope using trisubstituted phosphates (44); (8) the separation of the azeotropes formed by alcohols and their esters using aromatic hydrocarbons (45); (9) the separation of the methanol–methylene bromide azeotrope using ethylene bromide (46); (10) the separation of the phenol–cyclohexanone azeotrope using adipic acid diester (47); (11) the removal of close-boiling heptane isomers from cyclohexane, an important raw material for the manufacture of nylon precursors, using a mixture of solvents (48); (12) the separation of the low relative volatility mixture of propylene and propane using acrylonitrile (49); and (13) separating toluene from nonaromatics with phenol (50).

All extractive distillations correspond to one of three possible residue curve maps; one for mixtures containing minimum boiling azeotropes, one for mixtures containing maximum boiling azeotropes, and one for nonazeotropic mixtures. Thus extractive distillations can be divided into these three categories.

Minimum Boiling Azeotropes. All extractive distillations of binary minimum boiling azeotropic mixtures are represented by the residue curve map and column sequence shown in Figure 6**b**. Typical tray-by-tray composition profiles are shown in Figure 7.

In the distillation of ideal mixtures, the component having the lowest boiling point is always the one recovered as the distillate. This is not true for extractive distillations. Neither can the design engineer freely pick which azeotropic component to recover overhead despite the apparent symmetry of the residue curve map (Fig. 6**b**). For a given solvent, one and only one component can be recovered in the extractive column and it need not be the pure component having the lowest

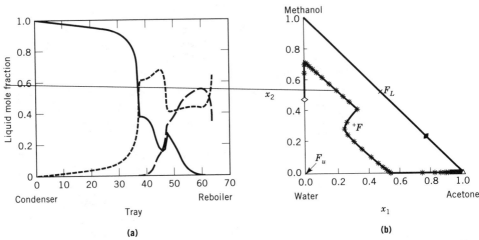

Fig. 7. Extractive distillation column profiles for the acetone–methanol–water separation (40). (**a**) Liquid composition versus theoretical tray location where (——) represents acetone, (– – –) methanol, and (·····) water; and (**b**) liquid composition profiles in mole fraction coordinates where ■ represents the azeotrope; △, the distillate; ×, feed; ◇, bottoms; and ∗, tray composition (51).

boiling point. For example, the extractive distillation of ethanol and water using gasoline (38), some phenols (53), cyclic ketones, or cyclic alcohols (54) causes water to be removed as the overhead product of the extractive column and the lower boiling ethanol to leave in the bottom stream with the solvent. The higher ketones distill methanol overhead from methanol–acetone mixtures (55,56) and furfural reverses the natural volatility of butane–butadiene mixtures (36). This phenomenon of having the intermediate boiling pure component distill overhead in the extractive column results entirely from the way in which the solvent modifies the volatilities of the other components in the mixture.

Drawing pseudo-binary $y–x$ phase diagrams for the mixture to be separated is the easiest way to identify the distillate product component. A pseudo-binary phase diagram is one in which the VLE data for the azeotropic constituents (components 1 and 2) are plotted on a solvent-free basis. When no solvent is present, the pseudo-binary $y–x$ diagram is the true binary $y–x$ diagram (Fig. 8**a**). At the azeotrope, where the VLE curve crosses the 45° line, $\alpha_{12} = 1.0$. To determine which component is the distillate, a series of pseudo-binary $y–x$ plots must be drawn at increasing solvent compositions until the pseudo-azeotrope, the point where the solvent-free VLE curve crosses the 45° line, ie, where $\alpha_{12} = 1.0$, disappears into one of the pure component corners. The resulting pseudo-binary phase diagram is either Figure 8**b** where the solvent increases the volatility of component 1 relative to component 2, making component 1 the distillate, or Figure 8**c**, where the solvent has the opposite effect, making component 2 the distillate. The expressions to "break" or "negate" an azeotrope originated from the use of these psdudo-binary phase diagrams because that is what appears to happen as the solvent composition increases. For example, aniline increases the volatility of cyclohexane (bp = 80.8°C) relative to benzene (bp = 80.1°C) (Fig. 8**d**). Thus the higher boiling cyclohexane would be recovered as the distillate in an extractive distillation. Pseudo-binary $y–x$ phase diagrams can also be used to determine the product component for an extractive distillation of a nonazeotropic mixture. For example, phenol enhances the volatility of the nonaromatics relative to toluene so the nonaromatics can be distilled overhead (50).

Because binary homogeneous azeotropic mixtures cannot be separated into pure component products by isobaric distillation without a solvent, but can be separated in the presence of a sufficient amount of solvent, there is clearly some minimum amount of solvent that just makes the separation possible. The minimum solvent flow depends on the solvent used and offers one method for discriminating between solvents. A solvent having a small minimum solvent flow is a better solvent and should result in a lower cost design than a solvent having a large minimum solvent flow. The solvent composition required to just make the pseudo-azeotrope disappear on a pseudo-binary $y–x$ phase diagram, ie, to "break" the azeotrope, can be taken as a qualitative measure of the minimum solvent flow necessary to make a separation feasible. A simple method for estimating minimum solvent flows has been published (57) as has an exact quantitative method (51,58).

Optimization. Optimization of the design variables is an important yet often neglected step in the design of extractive distillation sequences. The cost of the solvent recovery (qv) step affects the optimization and thus must also be included. Optimization not only yields the most efficient extractive distillation de-

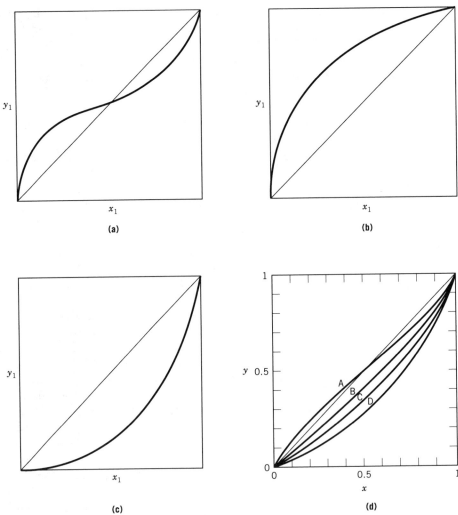

Fig. 8. Pseudo-binary (solvent-free) y–x phase diagrams for determining which component is to be the distillate where (——) is the 45° line. (**a**) No solvent; (**b**) and (**c**) sufficient solvent to eliminate the pseudo-azeotrope where the distillate is component 1 and component 2, respectively (51); and (**d**) experimental VLE data for cyclohexane–benzene where A, B, C, and D represent 0, 30, 50, and 90 mol % aniline, respectively (52).

sign, it is also a prerequisite for valid comparisons with other separation sequences and methods.

When several simple heuristics are used the optimization procedure usually reduces to a single-variable optimization of the feed ratio, ie, the molar ratio of the solvent to process feed flow rates, which has the greatest effect on the sequence cost (39). The simple heuristics are for calculating the minimum purity of the recycled solvent (see eqs. 3–6 of Ref. 39) and for setting the reflux ratio in all columns to 1.2–1.5 times the minimum. Minimum reflux ratios for extractive distillations can be calculated using the methods presented in References 59 and 60 to solve the equations given by References 61 and 62. Figure 9 shows how the

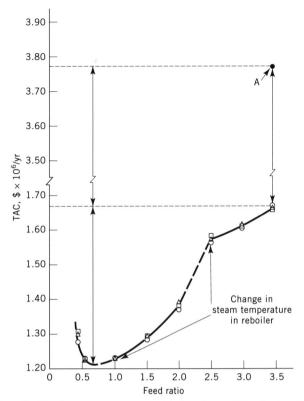

Fig. 9. Extractive distillation sequence cost as a function of the feed ratio for the production of anhydrous ethanol from azeotropic ethanol using ethylene glycol at reflux ratios of △, 1.15 r_{min}; ○, 1.2 r_{min}; and □ 1.3 r_{min} (39). Point A represents a previously published design for the same mixture (37).

feed ratio influences the total annualized cost (TAC), defined by Reference 63, for dehydrating ethanol using ethylene glycol. Cost diagrams for all extractive distillations have the same distinctive shape: a very high cost near the minimum feed ratio, ie, the minimum solvent flow, which rapidly decreases to the minimum cost and then slowly increases at higher feed ratios. As a rough rule-of-thumb, the economically optimal feed ratio is often between two and four times the minimum feed ratio (51,58). Many industrial extractive distillations operate at feed ratios between 1 and 4. Figure 9 also shows the significant cost reduction in an extractive distillation sequence published in 1972 (37) that is made possible merely by applying the reflux ratio heuristic.

Maximum Boiling Azeotropes. Maximum boiling azeotropes are far less common than minimum boiling azeotropes. Successful extractive distillations of maximum boiling azeotropes using high boiling solvents are even more rare because the combination of a high boiling solvent and a maximum boiling azeotrope gives rise to a distillation boundary running from the maximum boiling azeotrope to the heavy solvent (Fig. 10) that divides the desired pure components into different distillation regions. Consequently, other methods are often used for sepa-

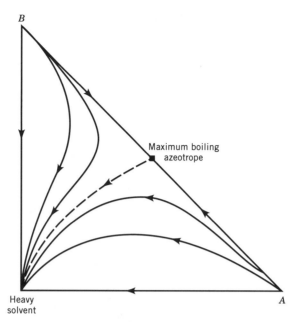

Fig. 10. Residue curve map for separating a maximum boiling azeotrope using a high boiling solvent where (– – – –) represents the distillation boundary and ■, the azeotrope.

rating maximum boiling azeotropes. The only way a high boiling solvent can yield an economically viable means for separating a maximum boiling azeotrope is if the resulting distillation boundary is extremely curved. The classic example is the concentration of aqueous nitric acid mixtures using sulfuric acid as the solvent (64–66). Figure 11 shows that the distillation boundary in this mixture is indeed highly curved. Applying the lever arm rule to the material balance lines superimposed on the residue curve map indicates that this process has a relatively high internal recycle rate relative to the product rates (D_1 and D_2).

Nonazeotropic Mixtures. The residue curve map representing the extractive distillation of any close-boiling, isomeric, or other low relative volatility mixtures using a high boiling solvent is represented by Figure 2. Although this map is different from that for extractive distillation of minimum boiling azeotropes, the distillation sequence is identical (see Fig. 6**b**). This process also works because the solvent alters the relative volatilities of the components to be separated via liquid-phase interactions. Depending on the nature of these interactions, the intermediate boiling pure component can become the distillate.

Because there is no azeotrope, these mixtures could be separated without adding a solvent. This, however, would be a difficult and expensive separation. Thus there is no minimum feed ratio (minimum solvent flow) and the only way to determine the optimal solvent-to-process feed ratio is by determining the sequence cost over a range of feed ratios. The best reflux ratios are again 1.2–1.5 times the minimum.

Solvent Selection. One of the most important steps in developing a successful (economical) extractive distillation sequence is selecting a good solvent. A

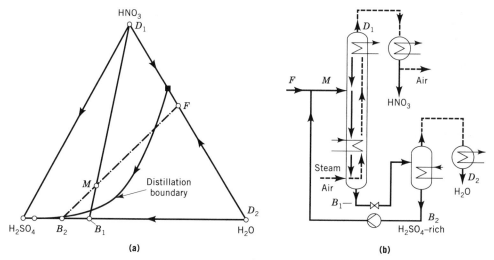

Fig. 11. Separation of nitric acid, HNO_3, and water, H_2O, using sulfuric acid, H_2SO_4, as the solvent. (**a**) Residue curve map and material balance lines where ■ represents the maximum boiling azeotrope and (**b**), column sequence (67).

number of approaches have been proposed. The solvent selection procedure can be thought of as a two-step process. The first step requires techniques for rapidly identifying a group of feasible solvents from among all compounds that could potentially be used and winnowing this group down to a small number of the most promising. In the second step, detailed vapor–liquid equilibrium measurements are made for the top two or three candidates and the final selection is made, preferably by an economic comparison of the optimal sequence for each of the solvents. Residue curve maps are of limited usefulness at the preliminary screening stage because there is usually insufficient information available to sketch them, but they are valuable and should be sketched or calculated as part of the second stage of solvent selection, ie, after the VLE data are available because the screening methods discussed herein are not sufficient. The remaining discussion focuses on the first stage.

As a starting point for identifying candidate solvents, all compounds having boiling points below that of any component in the mixture to be separated should be eliminated. This is necessary to yield the correct residue curve map for extractive distillation, but this process implicitly rules out other forms of homogeneous azeotropic distillation. In fact, compounds which boil as much as 50°C or more above the mixture have been recommended (68) in order to minimize the likelihood of azeotrope formation. On the other hand, the solvent should not boil so high that excessive temperatures are required in the solvent recovery column.

Because extractive solvents work by altering the relative volatility between the components to be distilled, a good solvent causes a substantial change in relative volatility when present at moderate compositions. As seen from equation 3, the solvent modifies the relative volatilities by affecting the ratio of the liquid-phase activity coefficients (γ_i/γ_j). Whereas it is possible to find solvents which increase or decrease this ratio, it is usually preferable to select a solvent which accentuates the natural difference in vapor pressures between the components to be separated; that is, a solvent that increases γ_i relative to γ_j when $P_i^{sat} > P_j^{sat}$ is favored, over one that increases γ_j relative to γ_i. In the latter case, adding small amounts of the solvent actually makes the separation more difficult, and relatively large quantities would be required to completely overcome the natural volatility difference and enhance the separability of the original mixture (65). This second situation corresponds to the intermediate boiling component becoming the distillate. Thus a heuristic is to favor solvents that cause the naturally more volatile component to distill overhead.

To force the naturally more volatile component i overhead, the solvent should either behave essentially ideally with component j and cause positive deviations from Raoult's Law for component i ($\gamma_j \sim 1$ and $\gamma_i > 1$), or behave essentially ideally with component i and cause negative deviations from Raoult's Law for component j ($\gamma_i \sim 1$ and $\gamma_j < 1$). Compounds of similar type and size, eg, pentane–hexane or methanol–ethanol, tend to behave ideally in the liquid phase and thus have activity coefficients close to unity. Dissimilar molecules tend to repel each other causing positive deviations from Raoult's Law and, in the extreme, resulting in liquid-phase immiscibilities. Compounds that tend to associate in the liquid phase exhibit negative deviations from Raoult's Law. Because systems showing positive deviations are more common, the usual approach is to force the lower boiling component overhead by selecting a solvent which is chemically similar to the higher boiling species and dissimilar to the lower boiling species.

Deviations from ideality are often attributed to hydrogen bonding or polarity. General guidelines for predicting the type of deviation that occurs in a mixture based on the hydrogen-bonding tendency of each component are available (69,70). Successful extractive solvents are typically highly hydrogen-bonded liquids such as phenols, aromatic amines, alcohols, glycols (qv), etc (see AMINES, AROMATIC AMINES; ALCOHOLS, HIGHER ALIPHATIC; ALCOHOLS, POLYHYDRIC; PHENOL) (70). Homologues of the component having the smaller natural volatility have also been advocated (54). For example, a higher alcohol can be used to force acetone overhead from the acetone–methanol azeotrope. Polarity arguments suggest using a highly polar solvent to increase the relative volatility of the less polar component in the mixture and a nonpolar solvent to have the opposite effect (65), for example, using water as the solvent to distill the less polar acetone overhead from the more polar methanol. Dipole moment interactions in hydrocarbon systems are discussed in Reference 71.

These qualitative methods are often useful for identifying general classes of compounds that may make good solvents. There are, however, two experimental methods that provide a more quantitative means for screening potential solvents. The first method entails measuring the relative volatility of a fixed composition mixture of the components to be separated (often 50% each) at a constant solvent-to-feed ratio that is typically one to three times the binary mixture on a molar basis (68). The objective is to find the candidate solvent(s) that cause the largest increase in the relative volatility of the components to be separated. Favorable solvents selected by this technique can, however, actually be infeasible solvents. For example, methyl ethyl ketone (MEK) has been identified as a solvent for separating acetone and methanol (55,56). Yet in reality methanol and MEK form a minimum boiling azeotrope (1) which results in a distillation boundary that puts acetone and methanol into different distillation regions, making the separation impossible.

The most common method for screening potential extractive solvents is to use gas–liquid chromatography (qv) to determine the infinite-dilution selectivity of the components to be separated in the presence of the various solvent candidates (71,72). The selectivity or separation factor is the relative volatility of the components to be separated (see eq. 3) in the presence of a solvent divided by the relative volatility of the same components at the same composition without the solvent present. A potential solvent can be examined in as little as 1–2 hours using this method. The tested solvents are then ranked in order of infinite-dilution selectivities, the larger values signify the better solvents. Favorable solvents selected by this method may in fact form azeotropes that render the desired separation infeasible.

In addition to its ability to make the separation feasible or easier, the ideal solvent is inexpensive, readily available, nontoxic, noncorrosive, thermally stable, and nonreactive with and easily separated from the other components in the mixture. In reality some compromise in solvent properties is almost always required.

Distillation Using Ionic Salts

Distillation using ionic salts (salt-effect distillation) is analogous to extractive distillation, but, rather than using a high boiling liquid solvent to alter the relative volatility of the mixture to be separated, a nonvolatile and soluble ionic salt is added to modify the volatility. Examples include adding calcium chloride to separate ethanol (qv) and water (73), using magnesium nitrate to dehydrate nitric acid (qv) (see NITRIC ACID), and using ferrous, lithium, or calcium chloride to separate acrylic acid and water (74) (see ACRYLIC ACID AND DERIVATIVES). Note that like liquid solvents, salts which have a significant effect on the volatility are preferred and some salts reverse the natural volatility difference, causing the intermediate boiling pure component to be distilled overhead. See Reference 74 for an example.

For a more complete discussion of salt-effect distilllation see References 75–77.

Pressure-Swing Distillation

It is well known that changing the system pressure can affect the azeotropes in a mixture. This effect can be exploited to separate a binary mixture containing either a minimum or maximum boiling azeotrope which appreciably changes composition over a moderate pressure range in a sequence of two columns operated at different pressures. This process, called pressure-swing distillation, is often used industrially to separate tetrahydrofuran (THF) and water (78–81) (see FURAN DERIVATIVES). Pressure-swing distillation has also been proposed for separating ethanol and water (38,82), a variety of alcohol–ketone azeotropes (83), and the maximum boiling hydrogen chloride–water azeotrope (84). For a binary mixture forming a pressure-sensitive minimum boiling azeotrope, the separation sequence works as shown in Figure 12. The fresh feed, F, is mixed with the re-cycled stream from the second column to form the feed stream, F_1, to the first column, which operates at pressure P_1. Because F_1 lies to the right of the azeotrope at pressure P_1 (Fig. 12a), pure A is removed as the bottom product, B_1, and a mixture near the azeotropic composition at pressure P_1 is the distillate, D_1. Stream D_1 is changed to pressure P_2 and fed to the next column as stream F_2. Because F_2 now lies to the left of the azeotropic composition at pressure P_2 (Fig. 12a), the other pure component, B, can be recovered in the bottom stream, B_2, and a near azeotropic mixture becomes the distillate, D_2, for recycling to the first column. An analogous procedure is used for binary maximum boiling azeotropes (86).

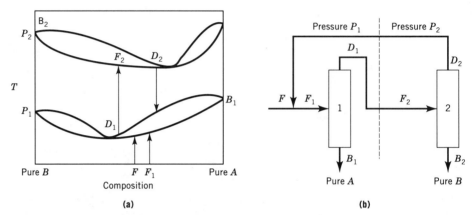

Fig. 12. Pressure-swing distillation of a minimum boiling binary azeotrope. (**a**) Temper-ature–composition phase diagram showing the effect of pressure on the azeotropic com-position; (**b**) column sequence (85).

Only a fraction of the known azeotropes are sufficiently pressure-sensitive for the conventional pressure-swing distillation process to work. However, the concept can be extended to pressure-insensitive azeotropes by adding a separating agent which forms a pressure-sensitive azeotrope and distillation boundary. Then the pressure is varied to shift the location of the distillation boundary (85).

Heterogeneous Azeotropic Distillation

Heterogeneous azeotropic distillation, or simply azeotropic distillation, is widely used for separating nonideal mixtures. The technique uses minimum boiling azeo-tropes and liquid–liquid immiscibilities in combination to overcome the effect of other azeotropes or tangent pinches in the mixture that would otherwise prevent the desired separation. The azeotropes and liquid heterogeneities that are used to make the desired separation feasible may either be induced by the addition of a separating agent, usually called the entrainer, or they may be intrinsically present, in which case the mixture is sometimes called self-entrained. The most common case is the former; it includes such classic separations as ethanol dehy-dration using either benzene, heptane, ethyl ether, etc, as the entrainer, and acetic acid recovery from water using either ethyl acetate, 1-propyl acetate, or 1-butyl acetate as the entrainer. In ethanol dehydration the entrainer is used to break the homogeneous minimum boiling azeotrope between ethanol and water; in the acetic acid recovery process the entrainer is used to overcome the tangent pinch between acetic acid and water.

The first successful application of heterogeneous azeotropic distillation was in 1902 (87) and involved using benzene to produce absolute alcohol from a binary mixture of ethanol and water. This batch process was patented in 1903 (88) and later converted to a continuous process (89). Good reviews of the early development and widespread application of continuous azeotropic distillation in the pre-war chemical industry are available (90).

Historically azeotropic distillation processes were developed on an individual basis using experimentation to guide the design. The use of residue curve maps as a vehicle to explain the behavior of entire sequences of heterogeneous azeotropic distillation columns as well as the individual columns that make up the sequence provides a unifying framework for design. This process can be applied rapidly, and produces an excellent starting point for detailed simulations and experiments.

Phase Diagrams. For binary mixtures, it is well known that when a liquid–liquid envelope merges with a minimum boiling vapor–liquid-phase envelope the resulting azeotropic phase diagram has the form shown in Figure 13. When the liquid composition $x_1 = x_1^{AZ}$, as in Figure 13**a**, then the vapor composition, y_1, is also equal to x_1^{AZ} and the mixture boils at constant temperature and at constant (and equal) composition in each phase. Thus a homogeneous azeotrope is formed. When the overall liquid composition $x_1^o = (x_1^o)^{AZ}$, as in Figure 13**b**, then y_1 is also equal to $(x_1^o)^{AZ}$ and again the mixture boils at constant temperature and at constant composition in each phase. However, the liquid of composition $(x_1^o)^{AZ}$ splits into two liquid phases so that there are three coexisting equilibrium phases which have different compositions. That is, a heterogeneous azeotrope is formed. Ho-

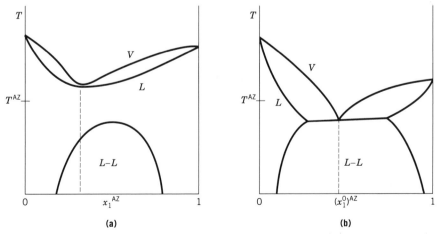

Fig. 13. Schematic isobaric phase diagrams for binary azeotropic mixtures (AZ). (**a**) Homogeneous azeotrope; (**b**) heterogeneous azeotrope.

mogeneous and heterogeneous azeotropes share the common property that the overall liquid composition is equal to the vapor composition providing a means for identifying azeotropes experimentally and computationally.

The properties of ternary heterogeneous vapor–liquid–liquid equilibrium (VLLE) phase diagrams are not well documented, however, these diagrams are important for understanding azeotropic distillation. The simplest VLLE phase diagram, where a liquid–liquid envelope merges with a vapor–liquid equilibrium (VLE) surface containing a single minimum boiling binary azeotrope, is shown in Figure 14. The characteristic feature is the existence of a heterogeneous liquid boiling surface (Fig. 14**b**). When the overall liquid composition lies inside the heterogeneous boiling envelope, and the temperature lies on the heterogeneous boiling surface, then the liquid boils and splits into two equilibrium liquid phases and one coexisting vapor phase. When the overall liquid composition lies on the boiling surface, the Gibbs phase rule requires that the locus of all the corresponding equilibrium vapor compositions forms a curve in T–y space and *not* a surface, as happens in the homogeneous region.

A convenient way of representing the T–x–y phase diagram (Fig. 14**b**) is by projection onto the composition triangle at the base of the figure. It is understood that the temperature varies from point to point on the projected vapor line and on the projected boiling envelope. The latter looks like an isothermal liquid–liquid binodal envelope, but is not. Each tie line across the boiling envelope is associated with a different boiling temperature (Fig. 15).

Little experimental vapor–liquid–liquid phase equilibrium data for ternary mixtures has been published and thus good model parameters are not available.

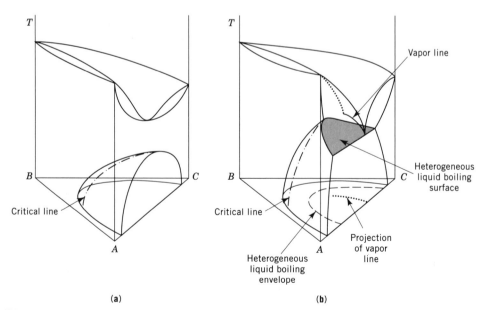

(a)   (b)

Fig. 14. Schematic isobaric phase diagrams for ternary (A,B,C) azeotropic mixtures. (**a**) Homogeneous liquid phase at all boiling points; (**b**) heterogeneous liquid phase for some boiling points.

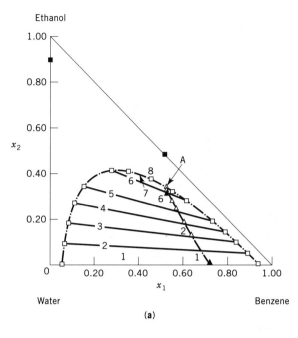

(a)

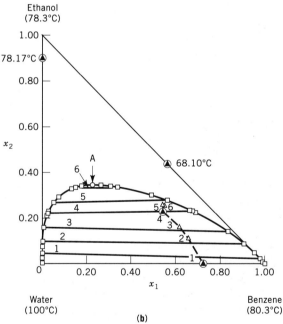

(b)

Fig. 15. Isobaric vapor–liquid–liquid (VLLE) phase diagrams for the ethanol–water–benzene system at 101.3 kPa; (□–□) represent liquid–liquid tie-lines; (△–△), the vapor line; ■, homogeneous azeotropes; ▲, heterogeneous azeotropes; ◬, Horsley's azeotropes. (**a**) Calculated, where A is the end point of the vapor line and the numbers correspond to boiling temperatures in °C of 1, 70.50; 2, 68.55; 3, 67.46; 4, 66.88; 5, 66.59; 6, 66.46; 7, 66.47, and 8, the critical point, 66.48. (**b**) Experimental, where A is the critical point at 64.90°C and the numbers correspond to boiling temperatures in °C of 1, 67; 2, 65.5; 3, 65.0; 4, 64.85; 5, 64.9; and 6, 64.90.

Nor is it possible to test the reliability of the modeling predictions. Figure 15**a** shows a model-predicted phase diagram for the ethanol–water–benzene mixture at 101.3 kPa (1 atm) pressure (91). In this diagram, the liquids and vapors in equilibrium with each other are signified by a common number. For example, the coexisting liquids on tie-line number 2 are in equilibrium with vapor number 2 at a boiling temperature of 68.55°C. The ethanol–water–benzene mixture has a minimum boiling heterogeneous ternary azeotrope. Experimental VLLE data for this mixture is shown in Figure 15**b** (91). Heterogeneous saddle azeotropes are also possible, eg, formic acid–water–*m*-xylene (92), water–acetone–chloroform (93), however, maximum boiling heterogeneous azeotropes cannot exist (94). This differs from homogeneous azeotropes where all three types are found in nature.

The typical phase equilibrium problem encountered in distillation is to calculate the boiling temperature and the vapor composition in equilibrium with a liquid phase of specified composition at a given pressure. If the liquid phase separates, then the problem is to calculate the boiling temperature and the compositions of the two equilibrium liquid phases plus the coexisting vapor phase at the specified overall liquid composition. Robust and practical numerical methods have been devised for solving this problem (95–97) and have become the recommended techniques (98,99).

Thus, using these techniques and a nonideal solution model that is capable of predicting multiple liquid phases, it is possible to produce phase diagrams comparable to those of Figure 15. These predictions are not, however, always quantitatively accurate (2,6,8,91,100).

Residue Curve Maps.   Residue curve maps are useful for representing the infinite reflux behavior of continuous distillation columns and for getting quick estimates of the feasibility of carrying out a desired separation. In a heterogeneous simple distillation process, a multicomponent partially miscible liquid mixture is vaporized in a still and the vapor that is boiled off is treated as being in phase equilibrium with all the coexisting liquid phases. The vapor is then withdrawn from the still as distillate. The changing liquid composition is most conveniently described by following the trajectory (or residue curve) of the overall composition of all the coexisting liquid phases. An extensive amount of valuable experimental data for the water–acetone–chloroform mixture, including binary and ternary LLE, VLE, and VLLE data, and both simple distillation and batch distillation residue curves are available (93,101). Experimentally determined simple distillation residue curves have also been reported for the heterogeneous system water–formic acid–1,2-dichloroethane (102).

Using the new generation of VLLE techniques (96), it is possible to calculate residue curve maps for heterogeneous liquid systems (Fig. 16). In this map the heterogeneous liquid boiling envelope has been superimposed on the residue curves in order to distinguish the homogeneous and heterogeneous regions of the triangular diagram.

Systems Having a Ternary Heterogeneous Azeotrope.   Binary or ternary heterogeneous azeotropes are restricted to being either unstable nodes or saddles in the residue curve map (94). An example is the computed residue curve map for the ethanol–water–benzene mixture which exhibits a ternary minimum boiling heteroazeotrope (see Fig. 16). This map exhibits three distillation regions. The three distillation boundaries all begin at the minimum boiling ternary hetero-

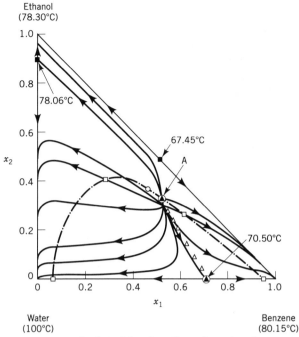

Fig. 16. Residue curve map calculated for the ethanol–water–benzene mixture where A is the end point of the vapor line; ■ represents a homogeneous azeotrope; (▲), heterogeneous azeotropes; (△—△), the vapor line; (–·–), the heterogeneous liquid boiling envelope; ○, the critical point; and □, the end points of tie-lines (103).

azeotrope and end at each of the binary azeotropes. If ethanol is the desired product of the separation, the initial condition for the simple distillation must lie in the upper region of Figure 16. The computed phase diagram is shown in Figure 15a. The acetone–chloroform–water system exhibits a ternary heterogeneous saddle azeotrope (93).

    Column Sequences. The analysis of residue curve maps and distillation boundaries for homogeneous azeotropic mixtures provides a simple and useful technique for distinguishing between feasible and infeasible sequences of distillation columns. Heterogeneous mixtures also exhibit distillation boundaries. Residue curves cross continuously through the liquid boiling envelope from one side to the other. Thus a simple distillation boundary inside the heterogeneous region does not stop abruptly or exhibit a discontinuity at the liquid boiling envelope but passes continuously through it, becoming a homogeneous distillation boundary thereafter (see Fig. 16). As for homogeneous systems, residue curves cannot cross heterogeneous distillation boundaries. However, if the two individual equilibrium liquid phases resulting from a point x^o on a residue curve inside the heterogeneous region lie in two different distillation regions, then a liquid–liquid phase separation can be exploited to jump across heterogeneous distillation boundaries in a way that is not possible for homogeneous systems. This is the key to devising feasible sequences of columns for separating heterogeneous mixtures.

Binary Mixtures. A binary mixture containing a homogeneous azeotrope can be distilled up to, but not beyond, the azeotropic composition. If the binary azeotrope is heterogeneous, however, the situation is more favorable and a simple sequence of columns, shown in Figure 17, is capable of isolating each pure component. The process feed, x_F, is fed as a saturated liquid to a decanter where it phase-separates into an A-rich phase and a B-rich phase. The A-rich phase, x_{F1},

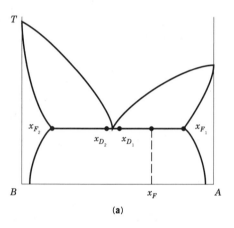

(a)

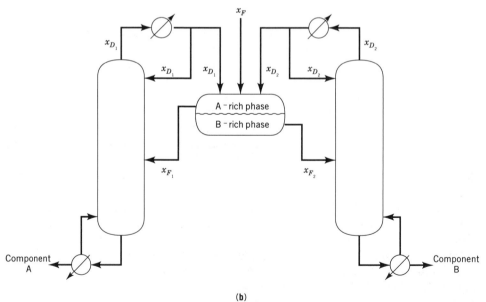

(b)

Fig. 17. Column sequence for separating a binary heterogeneous azeotropic mixture, A and B, where x_F represents the process feed mole fraction. (**a**) Phase diagram; (**b**) column sequence.

is then fed to column 1; the B-rich phase, x_{F2}, is fed to column 2. As can be seen in Figure 17**a**, because of the liquid–liquid-phase split the compositions of these two feed streams lie on either side of the azeotrope. Therefore, column 1 produces pure A as a bottoms product and the azeotrope as distillate, whereas column 2 produces pure B as a bottoms product and the azeotrope as distillate. The two distillate streams are fed to the decanter along with the process feed to give an overall decanter composition partway between the azeotropic composition and the process feed composition acording to the lever rule. This arrangement is well suited to purifying water–hydrocarbon mixtures, such as a C_4–C_{10} hydrocarbon, benzene, toluene, xylene, etc; water–alcohol mixtures, such as butanol, pentanol, etc; as well as other immiscible systems.

If the process feed does not lie in the liquid–liquid region it can be made to do so by deliberately feeding either pure A or pure B to the decanter, as required. This may only be necessary during start-up or for control purposes because the recycled azeotrope has the beneficial effect of dragging the decanter composition further into the liquid–liquid region.

Ternary Mixtures. When the binary mixture containing the minimum boiling azeotrope is completely homogeneous, ie, the liquid is homogeneous for all compositions, the method given in Figure 17 requires modification. In this case a third component, called the entrainer, is added which induces a liquid–liquid-phase separation over a limited portion of the ternary composition diagram. Many options for sequencing ternary heterogeneous azeotropic distillation systems exist (90). These sequences generally consist of two, three, or four columns using various techniques for handling the entrainer recycle stream. The feasibility of such sequences rests on the use of a liquid–liquid-phase split to provide each column with a feed composition in a different distillation region. In this regard the sequences for ternary mixtures resemble the sequences for binary mixtures. In all cases the heart of the process is the azeotropic column and its decanter.

The classical example is the separation of ethanol from water using benzene as the entrainer. Many other separations fit this mold, eg, ethanol–water–carbon tetrachloride, and 2-propanol–water–benzene. The task is to separate a homogeneous binary mixture of ethanol and water, which contains a minimum boiling binary azeotrope, into its pure components. An entrainer, which has limited miscibility with one of the components, in this case water, is used. In addition, entrainers in this class cause two more minimum boiling binary azeotropes (one homogeneous, the other heterogeneous) to form together with a ternary minimum boiling heterogeneous azeotrope. The resulting residue curve map is similar to that shown in Figure 16, in which there are three distillation regions. Figure 18 shows the column sequence and material balance lines for this system. The map indicates that the components to be separated lie in two different distillation regions, I and II.

The azeo-column must be designed to meet the following target compositions: (*1*) the bottoms composition is specified to be almost pure ethanol, ie, x_B is specified to lie close to the vertex of distillation region II; and (*2*) the overhead vapor composition leaving the last vapor–liquid equilibrium stage of the column, y_N, is specified to lie in the wedge-shaped portion of region II inside the heterogeneous region near the ternary azeotrope. Typical target values for x_B and y_N are shown in Figure 18. The final values of x_B and y_N are subject to optimization

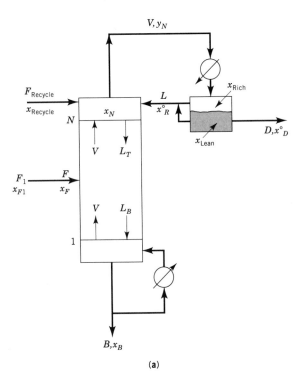

(a)

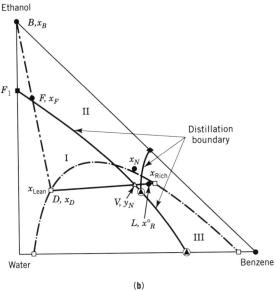

(b)

Fig. 18. Separation of ethanol from an ethanol–water–benzene mixture using benzene as the entrainer. (**a**) Schematic representation of the azeo-column; (**b**) material balance lines where ■ denotes the homogeneous and ◉ the heterogeneous azeotropes; □, the end points of the liquid tie-line and △, the overhead vapor leaving the top of the column. The distillate regions, I, II, and III, and the boundaries are marked. Other terms are defined in text.

and may differ slightly from those shown. All values of y_N inside the wedge-shaped region, except those special values which lie on the vapor line, are in equilibrium with a homogeneous liquid. Therefore, the liquid composition leaving the top tray, x_N, lies in the homogeneous portion of region II as shown (Fig. 18). The azeo-column must therefore be designed so that the steady-state liquid composition profile runs from x_B to x_N, and this can normally be done in such a way that every stage inside the column is homogeneous, ie, has only one liquid phase. A method for doing this is given in Reference 104. Not all azeo-columns have actually been designed this way. Neither do columns that have been designed in this manner necessarily remain homogeneous under the action of disturbances (2–10,99,105).

The overhead vapor of composition y_N is totally condensed into two equilibrium liquid phases, an entrainer-rich phase of composition x_{rich} and an entrainer-lean phase of composition x_{lean}. The relative proportion of these two liquid phases in the condenser, ϕ, is given by the lever rule, where ϕ represents the molar ratio of the entrainer-rich phase to the entrainer-lean phase in the condensate.

The two condensate liquids must be used to provide reflux and distillate streams. Normally, the reflux ratio, r, is chosen so that $r = L/D \geq \phi$. This requires that the reflux rate be greater than the condensation rate of entrainer-rich phase and that the distillate rate be correspondingly less than the condensation rate of entrainer-lean phase. This means that the distillate stream consists of pure entrainer-lean phase, ie, $x_D = x_{lean}$, and the reflux stream consists of all the entrainer-rich phase plus the balance of the entrainer-lean phase. Thus, the overall composition of the reflux stream, x^o_R, lies on the tie-line between the points x_{rich} and y_N, as shown in Figure 18.

Completing the Separation Sequence. In the remainder of the separation sequence the distillate stream leaving the azeotropic column, column 2 in Fig. 19**a**, must be separated into a product stream and a recycle stream so that the entire sequence is closed with respect to the entrainer.

Kubierschky Three-Column Sequence. If only simple columns are used, ie, no side-streams, side-rectifiers/strippers etc, then the separation sequence can be completed by adding an entrainer recovery column, column 3 in Figure 19**a**, to recycle the entrainer, and a preconcentrator column (column 1) to bring the feed to the azeotropic column up to the composition of the binary azeotrope.

The entrainer recovery column takes the distillate stream, D_2, from the azeo-column and separates it into a bottoms stream of pure water, B_3, and a ternary distillate stream for recycle to column 2. The overall material balance line for column 3 is shown in Figure 19**b**. This sequence was one of two original continuous processes disclosed in 1915 (106). More recently, it has been applied to other azeotropic separations (38,107,108).

Extensive design and optimization studies have been carried out for this sequence (108). The principal optimization variables, ie, the design variables that have the largest impact on the economics of the process, are the reflux ratio in the azeo-column; the position of the tie-line for the mixture in the decanter, dertermined by the temperature and overall composition of the mixture in the decanter; the position of the decanter composition on the decanter tie-line (see Reference 104 for a discussion of the importance of these variables); and the distillate composition from the entrainer recovery column.

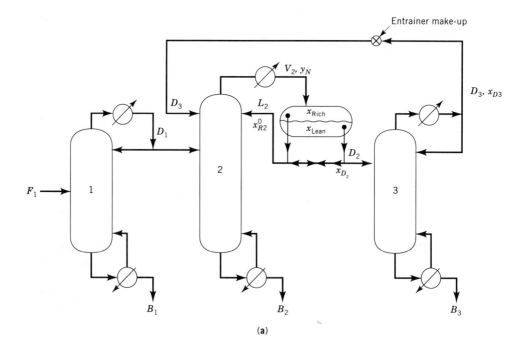

(a)

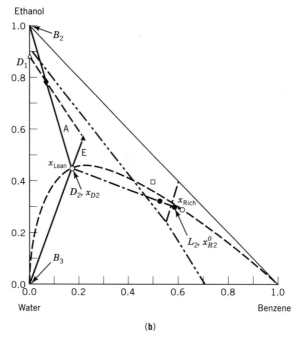

(b)

Fig. 19. Separation of ethanol and water from an ethanol–water–benzene mixture. Bottoms B_1 and B_3 are water, B_2 is ethanol. (**a**) Kubierschky three-column sequence where columns 1, 2, and 3 represent the preconcentration, azeotropic, and entrainer recovery columns, respectively. (**b**) Material balance lines from the azeotropic and the entrainer recovery columns, A and E, respectively, where ● represents the overall vapor composition from the azeo-column, V_2, y_N; □, the liquid in equilibrium with overhead vapor composition from the azeo-column, x_N; ▲, distillate composition from entrainer recovery column x_{D3}; and ◆, overall feed composition to the azeo-column, $D_1 + D_3$.

388

Figure 20 shows material balance lines for three different decanter tie-lines. The process feed to the preconcentrator in each sequence is a binary mixture of 4.2 mol % ethanol and 95.8 mol % water. The product purity from the azeo-column is set at 99.9 mol % ethanol, and the water purity leaving the entrainer recovery column is set at 99.5 mol % water. These specifications are essentially identical to those used for studying the optimal extractive distillation sequences (39,40). For design 1, the decanter tie-line is set at the bubble-point of the mixture leaving the top of the azeo-column, having composition y_N. This temperature is 337.57 K. For design 2, the decanter composition is the same as for design 1 but subcooled to 298.0 K. For design 3 the decanter composition is placed nearer to the ternary azeotrope than for designs 1 and 2, and the decanter temperature is set at the bubble-point of the mixture. In each design, the distillate composition from the entrainer recovery column is placed close to the distillation boundary. The position of the tie-line has a significant influence on the distillate composition from the azeo-column and this influences the position of the material balance line for the entrainer recovery column. The position of the distillate composition from the entrainer recovery column also influences the overall feed composition to the azeo-column.

Optimization studies indicate that the distillate composition from the entrainer recovery column has a strong influence on the process economics and the optimum position is always close to the distillation boundary. This decreases the amount of water being recycled, or equivalently, makes the overall feed to the

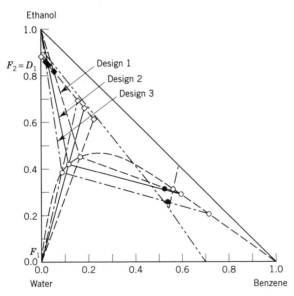

Fig. 20. Three sets of material balance lines for the Kubierschky three-column sequence where design 1 corresponds to the upper tie-line having $r_{min} = 8.78$; design 2, to the subcooled upper tie-line having $r_{min} = 12.23$; and design 3, to the lower tie-line having $r_{min} = 17.31$; ● represents overall decanter composition; ◆, the overall feed composition to the azeo-column; ◇, the distillate composition from the entrainer recovery column; and ○, the end points of the liquid–liquid tie-lines.

azeo-column richer in ethanol. For each tie-line, the optimal position of the decanter liquid composition is found by the method proposed in Reference 104. The minimum reflux ratio for a given tie-line can be reduced by as much as 50% by making small changes in the decanter composition. Calculations indicate that the optimal reflux ratio for the azeo-column is normally in the range 1.1–1.5 r_{min} and that the cost of the sequence is insensitive to this factor, leaving the position of the decanter tie-line as the sole remaining optimization variable.

The intrasequence flows, compositions, and reflux ratios are quite sensitive to relatively small changes in the position of the decanter tie-line as evidenced by the r_{min} value for the azeo-column: 8.78 for design 1, 12.23 for design 2, and 17.3 for design 3. Based on knowledge of homogeneous distillations, the vapor rate and total annualized cost for sequence 1 would appear to be the lowest. For homogeneous distillations at the design stage the feed and product flow rates, as well as their compositions, can be held constant as the reflux ratio is changed from one design to another. Thus there is a direct relationship between increased minimum reflux ratio and increased costs. However, such a relationship does not occur for heterogeneous distillations.

A good approximation for the vapor rate leaving the reboiler, V, for any type of distillation is

$$V = (r + 1)D \qquad (5)$$

where r is the reflux ratio and D is the distillate flow rate. For homogenous distillations, D is constant so that V increases as r increases. For azeotropic distillation, however, both r and D change from one tie-line to another. These effects may tend either to reinforce or to cancel each other depending on the mixture. There is no general rule and each mixture must be treated separately. In the ethanol–benzene–water system the reflux ratio increases from one design to another, but the distillate flow rate decreases, as can be seen from the material balance lines for the azeo-column in Figure 20. The net effect is that the vapor rate in the azeo-column hardly changes from one design to another. All the sequences shown in Figure 20 have approximately the same cost. This fortuitous cancellation of effects does not occur in general. It is always worthwhile exploring the economic impact of variations in the position of the decanter tie-line.

Other Sequences.   The Kubierschky sequence is not the only way to perform the separation. Alternatives include: (*1*) If the process feed already has a composition at or near the composition of the binary azeotrope then the preconcentrator is not needed. (*2*) Recycling the distillate stream from the entrainer recovery column directly to the decanter analogous to the binary process shown in Figure 17. This recycle alternative causes the reflux ratio in the azeotropic column to be much larger than necessary (109) and should be avoided even though it has been studied extensively in the literature (2,3,110–114). (*3*) Use of the Kubierschky two-column sequence (106). For the ethanol–water–benzene system, this alternative has lower capital costs but higher operating costs than the Kubierschky three-column sequence so that the total annualized cost is about the same for both sequences (108). (*4*) Use of the Steffen three-column sequence, the basic layout of which is the same as the Kubierschky three-column sequence. Reflux to the azeo-column is provided by condensing the overhead vapor and returning part of the ternary liquid mixture before it goes to the decanters. The entrainer-rich phases from the decanters then get returned to the azeo-column as a second feed stream. (5) Use of the Ricard-Allenet four-column sequence (90,106, 108,109,115,116). Ricard-Allenet proposed a variation on this sequence (106) in which the overhead vapors from the entrainer recovery column are provided with a separate condenser–decanter system, similar to the one provided for the over-head vapors from the azeo-column. (6) Use of the Ricard-Allenet three-column sequence which has good economic possibilities, but there are no available studies of this alternative.

In summary, for systems of the ethanol–water–benzene type, the three most attractive sequences for carrying out azeotropic distillation are the Kubierschky three-column sequence, the Kubierschky two-column sequence, and the Ricard-Allenet three-column sequence. For each of these there is the added possibility of putting a liquid–liquid extraction step after the azeo-column.

Other Classes of Entrainers.   Not all azeotropic mixtures are of the ethanol–water–benzene type. The number of azeotropes in the mixture may vary from system to system as may the character, ie, maximum or minimum boiling, het-ereogeneous or homogeneous. In addition, the size and shape of the liquid–liquid region varies greatly from system to system. The feasibility and sequencing strategy for each new system is most conveniently established using residue curve maps such as those shown in Figure 21.

Any entrainer that induces a liquid-phase heterogeneity over a portion of the composition triangle and which does not divide the components to be sepa-rated into different distillation regions is always a feasible entrainer. Examples are shown in Figures 21**a** and **b** and it is possible to construct many more such maps by strategically placing heterogeneous regions on the feasible residue curve maps given in Figure 6. Sequences based on these maps normally have multiple liquid phases on some of the stages in the column, which may lower the mass-transfer efficiency on those stages. In practice, mass-transfer efficiency and hy-drodynamic performance of three-phase columns does not present a problem. As early as 1938 it was stated that "The efficiency of the plates was apparently un-diminished by the heterogeneity of the boiling liquid and that was undoubtedly due to the violent agitation produced on the plates by the rapid bubbling of the vapours through the liquid . . ." (90). A more recent contribution and assessment of the literature from a practitioner's viewpoint is available (117). However, mass-transfer efficiency is of secondary importance to feasibility.

The map shown in Figure 21**b** is relevant to the separation of acetic acid and water which is of commercial significance. Although this binary mixture does not form an azeotrope, it does have a severe tangent pinch at high aqueous compo-sitions preventing the distillate from being acid free. It is not economical to sepa-rate this mixture into pure product streams without the aid of an entrainer. All known commercial entrainers for this separation are heterogeneous and produce residue curve maps similar to the one shown in Figure 21**b**.

Clarke-Othmer Process for Acetic Acid–Water Separation.   Large amounts of dilute acetic acid are purified industrially (see ACETIC ACID AND ITS DERIVA-TIVES). The entrainer, eg, ethylene dichloride, ethyl acetate, 1-propyl acetate, or 1-butyl acetate, is charged to the azeo-column, which has a process feed consisting of acetic acid and water. The bottom stream from the column is pure acetic acid,

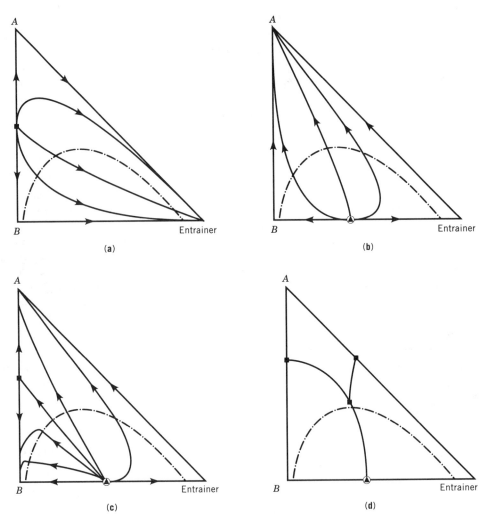

Fig. 21. A selection of feasible residue curve maps for ternary heterogeneous mixtures where ■ represents homogeneous and Ⓐ heterogeneous azeotropes. See text.

and the overhead vapor is close to the composition of the minimum boiling heterogeneous binary azeotrope formed by the entrainer and water (see Fig. 21**b**). The azeotropic vapors are condensed and decanted into a water stream that leaves as distillate, and an entrainer-rich stream that is returned to the column as reflux. Additional reflux, if needed, is achieved by returning some of the water stream. It is typical for these systems to have multiple liquid phases present on many stages in the rectifying section. This process was invented by Clarke and Othmer (118); more detailed operating information and entrainer comparison for this separation is available (119).

Wentworth Process for Ethanol–Water Separation. In the Wentworth process ethyl ether is used as the entrainer for ethanol–water separation, producing a residue curve map similar to the one shown in Figure 21**c**. Ethyl ether and water form a minimum boiling heterogeneous azeotrope at 34.15°C containing 98.75 wt % ether and 1.25 wt % water at atmospheric pressure. There is no azeotrope between ethyl ether and ethanol and no ternary azeotrope. The dilute ethanol process feed is first preconcentrated up to the composition of the ethanol–water azeotrope. This stream is fed to the azeo-column which produces pure ethanol as bottoms product and an overhead vapor close to the composition of the ethyl ether–water azeotrope. The overhead vapors are condensed and decanted into an ether-rich layer which is returned to the column as reflux, and a water-rich layer which leaves as distillate. The distillate contains no alcohol and very little ether. Pure water may be obtained from this stream by sending it to a stripping column where water is the bottom product and the overhead vapor has a composition near the ethyl ether–water azeotrope. These vapors are condensed and recycled to the decanter. The azeo-column is normally operated at about 700 kPa (7 atm) pressure because this increases the amount of water in the ether–water azeotrope, thereby reducing the amount of ether needed in the system.

This process is described in the literature in more detail (120–122). Its main advantage was a lower energy consumption (about 6,000 kJ/L (20,000 Btu/gal) ethanol product) relative to the benzene process (12,000 kJ/L (43,000 Btu/gal) ethanol) (121). Since the 1940s the gap has been narrowed by better designs for the benzene process which is capable of producing 99.8 mol % ethanol at an energy consumption of 8,400 kJ/L (30,000 Btu/gal) ethanol product (40,108), or using thermally integrated columns for 5,000 kJ/L (18,000 Btu/gal) ethanol product (40,123). In recent years homogeneous separating agents have shown great promise for the ethanol–water separation, and extractive distillation processes using ethylene glycol as the solvent, have been designed having energy consumptions of ca 6,000 kJ/L (22,000 Btu/gal) ethanol product (40), or using thermally integrated columns for 2,200–3,300 kJ/L (8,000–12,000 Btu/gal) ethanol product (40,124).

Rodebush Sequence for Ethanol–Water Separation. When ethyl acetate is used as the entrainer to break the ethanol–water azeotrope the residue curve map is similar to the one shown in Figure 21**d**, ie, the ternary azeotrope is homogeneous. Otherwise the map is the same as for ethanol–water–benzene. In such cases the liquid leaving the condenser from the azeo-column does not separate into two liquid phases, and the sequence is infeasible unless special tricks are employed. In the Rodebush sequence water is continuously added to the decanter in order to shift the overall composition into the two-liquid phase region. Each of the liquid phases from the decanter is fed to a separate distillation column which produces pure water and pure ethyl acetate, respectively. Some of the water is recycled to the decanter and all of the ethyl acetate is recycled to the azeo-column (106). More recently, a clever variation on this sequence was patented (125) for separating a ternary feed consisting of ethanol, water, and diethoxymethane. This variation also has a residue curve map similar to the one shown in Figure 21**d**.

More Complex Mixtures. All the sequences discussed are type I liquid systems, ie, mixtures in which only one of the binary pairs shows liquid–liquid behavior. Many mixtures of commercial interest display liquid–liquid behavior in two of the binary pairs (type II systems), eg, secondary butyl alcohol–water–di-secondary butyl ether (SBA–water–DSBE), and water–formic acid–*meta*-xylene (92). Sequences for these separations can be devised on the basis of residue curve maps. The SBA–water–DSBE separation is practiced by ARCO and is considered in detail in the literature (4,5,105,126).

NOMENCLATURE

B	bottom stream or bottoms flow rate
c	number of components in a mixture
D	distillate stream or distillate flow rate
F	feed stream or feed flow rate
P	pressure
P_i^{sat}	vapor pressure of component i
r	reflux ratio
T	temperature
TAC	total annualized cost (63)
V	vapor flow rate
x_i	liquid-phase composition of component i (mol fraction)
y_i	vapor-phase composition of component i (mol fraction)
α_{ij}	volatility of component i relative to component j (defined by eq. 3)
γ_i	liquid-phase activity coefficient of component i
ξ	nonlinear time scale (see eq. 4)

Subscripts

B	bottom
D	distillate
F	feed
lean	entrainer-lean
N	stage N
R	reflux
rich	entrainer-rich

Superscripts

o	overall (composition)

BIBLIOGRAPHY

"Azeotropes" in *ECT* 1st ed., under "Distillation," Vol. 5, pp. 176–179, by E. G. Scheibel, Hoffmann-LaRoche Inc.; "Azeotropy and Azeotropic Distillation" in *ECT* 2nd ed., Vol. 2, pp. 839–858, by D. F. Othmer, Polytechnic Institute of Brooklyn;' "Azeotropic and Extractive Distillation" in *ECT* 3rd ed., Vol. 3, pp. 352–377, by D. F. Othmer, Polytechnic Institute of New York; in *ECT* 3rd ed., Supl. Vol., pp. 145–158, by G. Prokopakis, Columbia University.

1. L. H. Horsley, *Azeotropic Data III*, Advances in Chemistry Series No. 116, American Chemical Society, Washington, D.C., 1973.
2. G. J. Prokopakis and W. D. Seider, *AIChE J.*, **29**, 49–60 (1983).
3. *Ibid.*, pp. 1017–1029.
4. J. W. Kovach III and W. D. Seider, *AIChE J.* **33**, 1300–1314 (1987).

5. J. W. Kovach III and W. D. Seider, *Comput. Chem. Eng.* **11,** 593–605 (1987).
6. M. Rovaglio and M. F. Doherty, *AIChE J.,* **36,** 39–52 (1990).
7. B. P. Cairns and I. A. Furzer, *Ind. Eng. Chem. Res.* **29,** 1349–1363 (1990).
8. *Ibid.*, pp. 1383–1395.
9. D. S. H. Wong, S. S. Jang, and C. F. Chang, *Comput. Chem. Eng.* **15,** 325–335 (1991).
10. S. Widagdo, W. D. Seider, and D. H. Sebastian, *AIChE J.* **38,** 1229–1242 (1992).
11. M. Rovaglio, T. Faravelli, G. Biardi, P. Gaffuri, and S. Soccol, *Comput. Chem. Eng.* **17,** in press (1993).
12. E. W. Jacobsen, L. Laroche, M. Morari, S. Skogestad, and H. W. Andersen, *AIChE J.* **37**(12), 1810–1824 (1991).
13. J. M. Smith and H. C. Van Ness, *Introduction to Chemical Engineering Thermodynamics*, 3rd ed., McGraw-Hill Publishing Co., New York, 1975.
14. M. F. Doherty and J. D. Perkins, *Chem. Eng. Sci.* **33,** 281–301 (1978).
15. *Ibid.*, pp. 569–578.
16. E. R. Foucher, M. F. Doherty, and M. F. Malone, *Ind. Eng. Chem. Res.* **30,** 760–772 (1991).
17. M. F. Doherty, *Chem. Eng. Sci.* **40,** 1885–1889 (1985).
18. M. F. Doherty and J. D. Perkins, *Chem. Eng. Sci.* **34,** 1401–1414 (1979).
19. Y. Yamakita, J. Shiozaki, and H. Matsuyama, *J. Chem. Eng. Jpn.* **16,** 145–146 (1983).
20. D. B. Van Dongen and M. F. Doherty, *Chem. Eng. Sci.* **40,** 2087–2093 (1985).
21. C. Bernot, M. F. Doherty, and M. F. Malone, *Chem. Eng. Sci.* **45,** 1207–1221 (1990).
22. C. Bernot, M. F. Doherty, and M. F. Malone, *Chem. Eng. Sci.* **46,** 1311–1326 (1991).
23. M. Benedict and L. C. Rubin, *Trans. AIChE* **41,** 353–370 (1945).
24. M. F. Doherty and G. A. Caldarola, *Ind. Eng. Chem. Fund.* **24,** 474—485 (1985).
25. E. Rev, *Ind. Eng. Chem. Res.* **31,** 893–901 (1992).
26. L. Laroche, N. Bekiaris, H. W. Andersen, and M. Morari, *Ind. Eng. Chem. Res.* **31,** 2190–2209 (1992).
27. L. Laroche, N. Bekiaris, H. W. Andersen, and M. Morari, *AIChE J.* **38,** 1309–1329 (1992).
28. O. M. Wahnschafft, J. W. Koehler, E. Blass, and A. W. Westerberg, *Ind. Eng. Chem. Res.* **31,** 2345–2362 (1992).
29. J. G. Stichlmair and J-R. Herguijuela, *AIChE J.* **38,** 1523–1535 (1992).
30. Z. T. Fidkowski, M. F. Doherty, and M. F. Malone, *AIChE J.* **39,** in press (1993).
31. V. Julka, *A Geometric Theory of Multicomponent Distillation*, Ph.D. dissertation, University of Massachusetts, Amherst, 1992.
32. D. B. Van Dongen, *Distillation of Azeotropic Mixtures: The Application of Simple-Distillation Theory to Design of Continuous Processes*, Ph.D. dissertation, University of Massachusetts, Amherst, 1983.
33. S. G. Levy, *Design of Homogeneous Azeotropic Distillations*, Ph.D. dissertation, University of Massachusetts, Amherst, 1985.
34. S. G. Levy, D. B. Van Dongen, and M. F. Doherty, *Ind. Eng. Chem. Fund.* **24,** 463–474 (1985).
35. H. Matsuyama and H. Nishimura, *J. Chem. Eng. Jpn.* **10,** 181–187 (1977).
36. C. K. Buell and R. G. Boatright, *Ind. Eng. Chem.* **39**(6), 695–705 (1947).
37. C. Black and D. E. Ditsler, in R. F. Gould, ed., *Azeotropic and Extractive Distillation*, Advances in Chemistry Series No. 115, American Chemical Society, Washington, D.C. 1972, pp. 1–15.
38. C. Black, *Chem. Eng. Prog.* **76**(9), 78–85 (1980).
39. J. R. Knight and M. F. Doherty, *Ind. Eng. Chem. Fund.* **28,** 564–572 (1989).
40. J. P. Knapp and M. F. Doherty, *AIChE J.* **36,** 969–984 (1990).
41. R. L. Schendel, *Chem. Eng., Prog.* **80**(5), 39–43 (1984).
42. U.S. Pat. 3,804,722 (Apr. 16, 1974), E. D. Oliver (to Montecatini Edison S. p. A.).

43. U.S. Pat. 4,918,204 (Apr. 17, 1990), T. T. Shih and T. Chang, (to Arco Chemical Technology, Inc.).
44. U.S. Pat. 4,166,772 (Sept. 4, 1979), T. P. Murta (to Phillips Petroleum Co.).
45. U.S. Pat. 4,473,444 (Sept. 25, 1984), J. Feldman and J. M. Hoyt (to National Distillers and Chemical Corp.).
46. U.S. Pat. 3,794,568 (Feb. 26, 1974), G. A. Daniels and J. A. Wingate (to Ethyl Corp.).
47. U.S. Pat. 4,016,049 (Apr. 5, 1977), G. B. Fozzard and R. A. Paul (to Phillips Petroleum Co.).
48. R. E. Brown and F.-M. Lee, *Hydroc. Proc.* **70**(5) 83–86 (1991).
49. E. R. Hafslund, *Chem. Eng. Prog.* **65**(9) 58–64 (1969).
50. H. G. Drickamer and H. H. Hummel, *Trans. AIChE* **41,** 631–644 (1945).
51. J. P. Knapp, *Exploiting Pressure Effects in the Distillation of Homogeneous Azeotropic Mixtures,* Ph.D. dissertation, University of Massachusetts, Amherst, 1991.
52. B. Kolbe, J. Gmehling, and U. Onken, *I. Chem. E. Symp. Ser.* (56), 1.3/23–1.3/40 (1979).
53. U.S. Pat. 4,428,798 (Jan. 31, 1984), D. Zudkevitch, S. E. Belsky, and P. D. Krautheim (to Allied Corp.).
54. U.S. Pat. 4,455,198 (June 19, 1984), D. Zudkevitch, D. K. Preston, and S. E. Belsky (to Allied Corp.).
55. E. G. Scheibel, *Chem. Eng. Prog.* **44**(12), 927–931 (1948); U.S. Pat. 4,501,645 (Feb. 26, 1985), L. Berg, and An-I. Yeh.
56. An-I. Yeh, L. Berg, and K. J. Warren, *Chem. Eng. Comm.* **68,** 69–79 (1988).
57. L. Laroche, N. Bekiaris, H. W. Andersen, and M. Morari, *Can. J. Chem. Eng.* **69,** 1309–1319 (1991).
58. J. P. Knapp and M. F. Doherty, "Minimum Entrainer Flows for Extractive Distillation. A Bifurcation Theoretic Approach," submitted to *AIChE J.* (1993).
59. V. Julka and M. F. Doherty, *Chem. Eng. Sci.* **48,** 1367–1391 (1993).
60. Z. T. Fidkowski, M. F. Malone, and M. F. Doherty, *AIChE J.* **37,** 1761–1779 (1991).
61. S. G. Levy and M. F. Doherty, *Ind. Eng. Chem. Fund.* **25,** 269–279 (1986).
62. J. R. Knight, *Synthesis and Design of Homogeneous Azeotropic Distillation Sequences,* Ph.D. dissertation, University of Masschusetts, Amherst, 1986.
63. J. M. Douglas, *Conceptual Design of Chemical Processes,* McGraw-Hill Publishing Co., New York, 1988.
64. U.S. Pat. 1,074,287 (Sept. 30, 1913), H. Pauling.
65. C. S. Robinson and E. R. Gilliland, *Elements of Fractional Distillation,* 4th ed., McGraw-Hill Publishing Co., New York, 1950.
66. U.S. Pat. 4,966,276 (Oct. 30, 1990), A. Guenkel.
67. J. G. Stichlmair, J. R. Fair, and J. L. Bravo, *Chem. Eng. Prog.* **85**(1), 63–69 (1989).
68. C. S. Carlson and J. Stewart, in E. S. Perry and A. Weissberger, eds., *Techniques of Organic Chemistry,* Vol. IV, *Distillation,* Wiley-Interscience, New York, 1965.
69. R. H. Ewell, J. M. Harrison, and L. Berg. *Ind. Eng. Chem.* **36**(10), 871–875 (1944).
70. L. Berg, *Chem. Eng. Prog.* **65**(9), 52–57 (1969).
71. D. P. Tassios, in Ref. 37, pp. 46–63; D. P. Tassios, *Hydroc. Proc.* **49**(7), 114–118 (1970).
72. D. P. Tassios, *Ind. Eng. Chem. Proc. Des. Dev.* **11**(1), 43–46 (1972).
73. D. Barba, V. Brandini, and G. DiGiacomo, *Chem. Eng. Sci.* **40,** 2287–2292 (1985).
74. U.S. Pat. 4,269,666 (May 26, 1981), C. G. Wysocki (to Standard Oil Co.).
75. W. F. Furter, *Can. J. Chem. Eng.* **55**(6), 229 (1977).
76. W. F. Furter and R. A. Cook, *Int. J. Heat Mass Trans.* **10,** 23–26 (1967).
77. H. R. Galindez and A. Fredenslund, *I. Chem. E. Symp. Ser.* (104), A397–A403 (1987).
78. S. I. Abu-Eishah and W. L. Luyben, *Ind. Eng. Chem. Proc. Des. Dev.* **24,** 132–140 (1985).

79. U.S. Pat. 4,257,961 (May 24, 1981), J. S. Coates (to E. I. du Pont de Nemours & Co., Inc.).
80. U.S. Pat. 4,348,262 (Sept. 7, 1982), A. M. Stock and W. S. Tse (to E. I. du Pont de Nemours & Co., Inc.).
81. U.S. Pat. 4,093,633 (June 6, 1978), Y. Tanabe, J. Toriya, M. Sato, and K. Shiraga (to Mitsubishi Chemical Industries Co., Ltd.).
82. U.S. Pat. 1,676,700 (July 10, 1928), W. K. Lewis.
83. U.S. Pat. 2,324,255 (July 13, 1943), E. C. Britton, H. S. Nutting, and L. H. Horsley, (to The Dow Chemical Company).
84. U.S. Pat. 3,394,056 (July 23, 1968), M. Nadler and co-workers (to Esso Research and Engineering).
85. J. P. Knapp and M. F. Doherty, *Ind. Eng. Chem. Res.* **31,** 346–357 (1992).
86. M. Van Winkle, *Distillation*, McGraw-Hill Publishing Co, New York, 1967.
87. S. Young, *J. Chem. Soc.* **81,** 707–717 (1902).
88. Ger. Pat. 142,502 (June 25, 1903), S. Young.
89. Ger. Pat. 287,897 (Oct. 11, 1915), Kubierschky.
90. H. M. Guinot and F. W. Clark, *Trans. Inst. Chem. Eng.* **16,** 189–199 (1938).
91. H. N. Pham and M. F. Doherty, *Chem. Eng. Sci.* **45,** 1823–1836 (1990).
92. W. Reinders and C. H. De Minjer, *Recl. Trav. Chim.* **66,** 564–572 (1947).
93. *Ibid*, pp. 573–604.
94. H. Matsuyama, *J. Chem. Eng. Jpn.* **11,** 427–431 (1978).
95. L. E. Baker, A. C. Pierce, and K. D. Luks, *Soc. Petrol. Engrs. J.* **22,** 731–742 (1982).
96. M. L. Michelsen, *Fluid Phase Equil.* **9,** 1–19 (1982).
97. *Ibid.*, pp. 21–40.
98. D. J. Swank and J. C. Mullins, *Fluid Phase Equil.* **30,** 101–110 (1986).
99. Ref. 7, pp. 1364–1382.
100. J. M. Prausnitz and co-workers, *Computer Calculations for Multicomponent Vapor–Liquid and Liquid–Liquid Equilibria*, Prentice Hall, Englewood Cliffs, N.J., 1980, Chapt. 4.
101. W. Reinders and C. H. De Minjer, *Recl. Trav. Chim.* **59,** 207–230 (1940).
102. I. N. Bushmakin and P. Ya. Molodenko, *Russ. J. Phys. Chem.* **37,** 2618–2624 (1964).
103. Ref. 91, pp. 1837–1843.
104. H. N. Pham, P. J. Ryan, and M. F. Doherty, *AIChE J.* **35,** 1585–1591 (1989).
105. S. Widagdo, W. D. Seider, and D. H. Sebastian, *AIChE J.* **35,** 1457–1464 (1989).
106. D. B. Keyes, *Ind. Eng. Chem.* **21,** 998–1001 (1929).
107. D. W. Townsend, private communication, 1982.
108. P. J. Ryan and M. F. Doherty, *AIChE J.* **35,** 1592–1601 (1989).
109. Ref. 91, pp. 1845–1854.
110. W. S. Norman, *Trans. Inst. Chem. Eng.* **23,** 66–75 (1945).
111. Ref. 65, pp. 312–324.
112. Zh. A. Bril', A. S. Mozzhukhin, F. B. Petlyuk, and L. A. Serafimov, *Theor. Found. Chem. Eng.* **9,** 761–770 (1975).
113. Zh. A. Bril', A. S. Mozzhukhin, F. B. Petlyuk, and L. A. Serafimov, *Russ. J. Phys. Chem.* **11,** 675–681 (1977).
114. G. J. Prokopakis, W. D. Seider, and B. A. Ross, in R. S. Mah and W. D. Seider, eds., *Foundations of Computer-Aided Chemical Process Design*, Engineering Foundation, New York, 1981.
115. C. D. Holland, S. E. Gallun, and M. J. Lockett, *Chem. Eng.* **88,** 185–200 (1981).
116. C. J. King, *Separation Processes*, 2nd ed., McGraw-Hill Publishing Co., New York, 1980.
117. M. E. Harrison, *Chem. Eng. Prog.* **86**(11), 80–85 (1990).
118. U.S. Pat. 1,804,745 (1931), H. T. Clarke and D. F. Othmer.

119. D. F. Othmer, *Chem. Metall. Eng.* **40,** 91–95 (1941).

120. D. F. Othmer and T. O. Wentworth, *Ind. Eng. Chem.* **32,** 1588–1593 (1940).

121. T. O. Wentworth and D. F. Othmer, *Trans. Am. Inst. Chem. Eng.* **36,** 785–799 (1940).

122. T. O. Wentworth, D. F. Othmer, and G. M. Pohler, *Trans. Am. Inst. Chem. Eng.* **39,** 565–578 (1943).

123. U.S. Pat. 4,217,178 (Aug. 12, 1980), R. Katzen, G. D. Moon, Jr., and J. D. Kumans (to Raphael Katzen Associates).

124. S. Lynn and D. N. Hanson, *Ind. Eng. Chem. Proc. Des. Dev.* **25,** 936–941 (1986).

125. U.S. Pat. 4,740,273 (Apr. 26, 1988), D. L. Martin and P. W. Raynolds (to Eastman Kodak Co.).

126. J. W. Kovach III and W. D. Seider, *J. Chem. Eng. Data* **32,** 16–20 (1988).

General Reference

C. L. Dunn, R. W. Miller, G. J. Pierotti, R. N. Shiras, and M. Souders, Jr., *AIChE J.* **41,** 631–644 (1945).

MICHAEL F. DOHERTY
University of Massachusetts, Amherst

JEFFREY P. KNAPP
E. I. du Pont de Nemours & Co., Inc.

DIURETIC AGENTS

Diuresis is defined as the excretion of urine, and is derived from the Greek word *diourein* which means to urinate. It more commonly denotes production of unusually large volumes of urine. Diuretics are agents that increase urine output or flow. The term is generally used to describe all drugs that act on the kidney to increase the production of urine. More specifically, the terms saliuretic or natriuretic are used to describe those agents that exert diuretic effects by primarily increasing the excretion of sodium chloride. Aquaretics are agents that increase urine output by producing a water diuresis but do not promote the urinary excretion of electrolytes. Both the use of diuretics in the treatment of hypertension, and the effects of diuretics on the kidney to promote the excretion of urine to normalize derangements in body fluid distribution leading to edematous states are discussed herein. The use of diuretics in the treatment of congestive heart failure and hypertension is discussed elsewhere (see CARDIOVASCULAR AGENTS).

Disturbances in body fluid distribution may occur at three principal sites: (*1*) within the interstitial space, as occurs in peritonitis or cirrhosis with ascites; (*2*) between the interstitial space and the vascular tree, as in the nephrotic syndrome; or (*3*) within the vascular tree, as in congestive heart failure. Thus, the

underlying disease may be of cardiac, hepatic, or renal origin. These derangements provide what amounts to a low blood volume signal to the kidneys that
activates renal mechanisms to retain salt and water. If the retained salt and water
do not terminate the low volume signal, the kidneys continuously retain salt and
water, resulting in edema. The clinical outcome of excessive accumulation of salt
and water depends on the particular sector of the extracellular space to which the
retained fluid is relegated. Clinical signs of edema appear when the volume of the
extracellular space is exceeded by several liters. Diuretics are used in the treatment of edematous states because, in most cases, they produce satisfactory mobilization and subsequent prevention of fluid accumulation in the interstitial
space, the abdominal cavity, the lungs, and/or thoracic cavity. However, diuretic
therapy is symptomatic in nature, and unless the underlying pathology is corrected, the kidneys continue to retain salt and water and the retained fluid and
electrolytes are redistributed to the various compartments described. The principal indications of diuretics in the treatment of edema are in congestive heart
failure (1); renal disease (2); hepatic cirrhosis with ascites (3); obesity, where salt
and water retention are prominent (see also ANTIOBESITY DRUGS) (4); premen-
strual tension (5); edema of pregnancy, including toxemia (6); and steroid administration (7). Edema may also be associated with other clinical conditions such as
inflammation or hypersensitivity reactions.

Renal Function

In addition to being involved in the formation of urine, the kidney acts as an
endocrine organ secreting renin, erythropoietin, prostaglandins (qv), and kinins;
it is also capable of synthesizing substances such as $1\alpha,25$-dihydroxycholecalci-
ferol [32222-06-3], $C_{27}H_{44}O_3$. One of the principal functions of the kidney is to
maintain the body's extracellular fluid composition in relatively narrow concentration ranges by the simultaneous regulation of water and multiple solute excretion. This is of particular interest in regard to the effects of diuretic drugs. The
kidney also eliminates unwanted metabolic products, drugs, and their metabolites
from the body via the urine. The latter effect also may be modified by some diuretic
drugs.

Anatomically, the kidneys are bilaterally paired organs located against the
posterior abdominal wall in retroperitoneal pockets below the diaphragm. The
kidney is divided into an outer cortical region and an inner medullary region.
Each kidney contains approximately 1 million distinct functional units called
nephrons. The individual nephron consists of a glomerulus, proximal convoluted
tubular segment, loop of Henle, distal convoluted tubular segment, and multi-
branched collecting duct common for several nephrons. Depending on the anatom-
ical location in the kidney, two types of nephrons can be identified (Fig. 1). The
cortical nephron originates in the outer regions of the renal cortex, and the principal portion of the unit is contained in the cortex. The juxtamedullary nephron,
which has a longer loop of Henle than the cortical nephron, originates close to the
corticomedullary junction, and part of its tubule descends deep into the inner
medulla along its osmolar gradient. Although the blood supply to these two types
of nephrons may change as a result of a regulatory mechanism, in principle they
operate similarly (8,9).

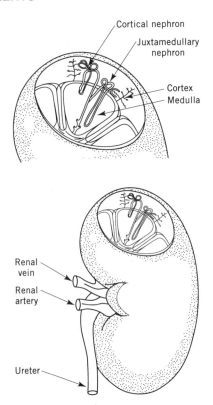

Fig. 1. Position of the two types of nephrons (not drawn to scale) in the kidney.

Under normal conditions, ca 25% of the resting cardiac output passes through the kidney. Blood flowing through the renal artery and the afferent arterioles of all glomeruli is filtered through the glomerular capillary plexus, resulting in ca 120 mL of total glomerular filtrate (ultrafiltrate) per minute. This is also called the glomerular filtration rate (GFR). Owing to the nature of the glomerular membrane, which is said to be as much as 300 times more permeable than other systemic beds, the filtrate contains all of the plasma constituents except lipids, proteins, and protein-bound substances. The driving force is the blood pressure within the glomerular capillaries. On its way through the nephron, where it ends up as urine, the ultrafiltrate rapidly loses its identity by both passive and active transport processes based on physical forces or involving consumption of energy from cellular metabolism, respectively. The final urine volume is ca 1% of the ultrafiltrate. Thus, of the ca 180 L of glomerular filtrate formed daily, only 1–1.5 L reaches the urinary bladder. It is virtually cleared of filtered glucose and amino acids, and contains, with respect to the main plasma electrolytes, only ca 1% of the filtered sodium and chloride ions, together with traces of bicarbonate ion. Substantial fractions of the remaining solutes are reabsorbed, and exogenous organic acids and bases are added to the tubular filtrate by secretion (8,9).

The transport processes for relevant electrolytes along the nephron are schematically shown in Figure 2. Under normal circumstances, ca 60% of the ultra-

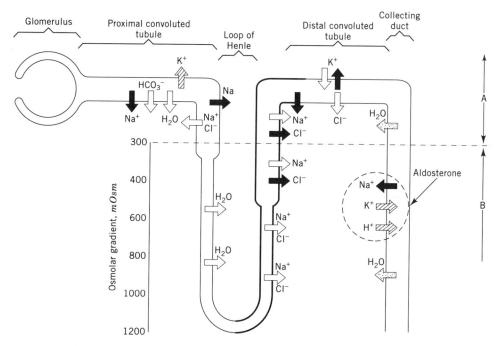

Fig. 2. Schematic representation of relevant electrolyte transport through the renal tub-ule, depicting the osmolar gradient in medullary interstitial fluid in $mOsm$ where ➡ represents active transport, ⇨ passive trasport, ⬆ both active and passive transport, and ⇨ passive transport of H_2O in the presence of ADH, in A, the cortex, and B, the medulla. An osmole equals a mole of solute divided by the number of ions formed per molecule of the solute. Thus one mole of sodium chloride is equivalent to two osmoles, ie, $1\ M\mathrm{NaCl} = 2\ Osm\ \mathrm{NaCl}$. ADH = antidiuretic hormone. $mOsm$ = milliosmolar = milli-osmoles per liter.

filtrate from the glomerulus is isotonically reabsorbed in the proximal tubule and the thick descending segment of the loop of Henle, where both active and passive transport of sodium ion, Na^+, and passive reabsorption of chloride ion, Cl^-, and water occur. Bicarbonate is also reabsorbed, but the HCO_3^- is derived from the filtered HCO_3^-. The filtered HCO_3^- combines with H^+ secreted by the proximal tubular cells into the tubular lumen to form H_2CO_3, which in turn dissociates to form CO_2 and water. The CO_2 moves passively into the proximal tubular cell where, depending on cellular carbonic anhydrase activity, it is rehydrated to form H_2CO_3. This H_2CO_3 dissociates into H^+ and HCO_3^-; it is this HCO_3^- that is reab-sorbed with actively transported Na^+. Substantial parts of the filtered load of K^+ are also reabsorbed by active and passive transport processes.

    Owing to a high water and low NaCl permeability of the thin descending limb of the loop of Henle, the tubular fluid, when it enters the ascending limb, has become hypertonic by osmotic equilibration along the osmolar gradient in the medulla. This effect is not simply reversed when the fluid passes into the ascend-ing limb, as the latter is impermeable to water. On the contrary, as the tubular fluid moves up the ascending limb, it decreases in osmolarity owing to Na^+ and Cl^- reabsorption. Active Cl^- reabsorption, and passive Na^+ and K^+ reabsorption, without an osmotic equivalent of water, provides a hypotonic fluid that moves into the distal convoluted tubule. The reabsorbed NaCl is part of the solute supply to the interstitium, and is necessary for the osmolar gradient in the renal medulla.

The principle of fluid concentration and dilution in the loop of Henle is known as countercurrent multiplication. The ascending limb of the loop of Henle is also referred to as the diluting segment of the nephron, and is the basis of the kidney's ability to produce concentrated or dilute urine. It is here that 20–35% of the filtered sodium ion is reabsorbed. In the distal convoluted tubules and collecting ducts, an additional 5% of the tubular filtrate is reabsorbed by active Na^+ and passive Cl^- reabsorption, whereas K^+ might be secreted or absorbed, depending on the potassium balance of the organism. In the collecting duct, Na^+ is reabsorbed under the influence of aldosterone [52-39-1]. At this part of the nephron, the rate of Na^+ reasborption is intimately connected to K^+ and carbonic anhydrase-dependent H^+ secretion.

The collecting ducts are distinguished from the other segments of the nephron by changing their water permeability as a result of the action of anti-diuretic hormone (ADH), also called vasopressin [9034-50-8], an octapeptide released from the posterior pituitary gland into the systemic circulation when the body is in the hydropenic state. In the presence of ADH, the distal convoluted tubules and collecting ducts are freely permeable to water, which allows back-diffusion of water into the hypertonic interstitial space of the medulla, resulting in a urine of high osmolarity. On the other hand, in the absence of ADH the impermeability to water during hydration leads to water diuresis because the flow of water from these segments into the osmolar medullary interstitium is blocked (8,9).

Pharmacology and Mechanism of Action

Low Ceiling Diuretics. The designation of low ceiling diuretics denotes that the total excretion of the filtered sodium ion load is less than 10% compared to about 30% for the high ceiling (loop) diuretics (2). There are many chemical classes in this category, ie, thiazides, quinazoline sulfonamides, chlorthalidone, indapam-ide, etc, but their site of action in the kidney is similar, and they are grouped as thiazide-type diuretics for general discussion (Table 1).

The most popular diuretics in this class are hydrochlorothiazide and chlor-thalidone; there are more potent low ceiling diuretics available (10). The long duration of action of chlorthalidone, 24 to 72 h, makes once a day dosing possible, and achieves good patient compliance. Cyclothiazide, polythiazide, and trichlor-methiazide are about 15 to 30 times more potent than hydrochlorothiazide, and about 500 to 1000 times more potent than chlorothiazide, the first member of the thiazide family marketed (2).

The low ceiling diuretics increase urinary sodium excretion by acting directly on the $Na^+ - Cl^-$ transport mechanism in the convoluted distal tubules of the kidney (7,11–15). The thiazides cause a maximal saluretic response of about 10% of the filtered sodium ion load, or ca 300 mEq/min, increase above the predrug control; about 90% of the sodium ions in the filtrate have been reabsorbed in the proximal tubule and the loop of Henle (12). Diuretic dose-response curves of low ceiling diuretics are rather flat, indicating that increased dosage only causes a small increment of sodium excretion (14). The lowest effective dose should be used to minimize side effects, particularly hypokalemia, ie, potassium-losing effect.

Indapamide has been shown to possess diuretic and independent vasodilatory effects (16). It lowers the elevated blood pressure and reduces total peripheral resistance without an increase in heart rate. Indapamide antagonizes the vasoconstricting effects of the catecholamines and angiotensin II (16), a property not shared by other thiazide-type diuretics. Tripamide is also reported to have direct vasodilatory effects (13).

Weight loss is a good indicator of fluid loss and excretion. The first wave of fluid mobilized is from the periphery. The excretion of chloride and water is considered passive, and the excretion of potassium and magnesium is increased. In long-term use, the excretion of calcium is decreased.

The thiazides are actively secreted into the proximal tubules, where they exert their action on the luminal side of the tubules. The diuretic effect occurs within 1 h after oral administration, and the duration of action varies from 4 to 24 h depending on which thiazide-type diuretic is used. The diuretic effects of the thiazides are not influenced by acid–base conditions of the blood or urine. Probenecid, which also is secreted into the proximal tubules, may block the diuretic effects of the thiazides.

Hydrochlorothiazide and chlorthalidone are absorbed at about 60–70%, and the plasma half-life is about 2.5 and 44 h, respectively (12). About 90% of an oral dose of indapamide is absorbed, and its plasma half-life is about 18 h (18). In the quinazoline–sulfonamide family, quinethazone has the longest experience, and is about 10 to 20 times less potent than metolazone. A newer member, fenquizone (19,20), is claimed to have less hyperglycemic and hyperuricemic effects as compared to other thiazides.

In long-term treatment, the thiazides may produce hypokalemia, hyperglycemia, hyperuricemia, and a 5% increase in plasma cholesterol; indapamide has been shown not to increase plasma cholesterol or lipids at therapeutic doses (21–23). The decrease of plasma potassium, ie, hypokalemic effect, is dose-dependent, and can be avoided if high doses are avoided (24,25). Thiazides can cause hyponatremia in patients with large water intake while on the drug (26,27); hyponatremia may be associated with nausea, vomiting, and headaches.

Paradoxically, the thiazides are efficacious, especially if combined with a prostaglandin synthetase inhibitor such as indomethacin or aspirin, in the treatment of nephrogenic diabetes insipidus, in which the patient's renal tubules fail to reabsorb water despite the excessive production of ADH (28). Thiazides can decrease the urine volume up to 50% in these patients.

High Ceiling (Loop) Diuretics. The principal action of the loop diuretics (Table 2) is inhibition of sodium and chloride reabsorption in the thick ascending limb of the loop of Henle (7,11–13,29–31). They can produce an excretion of 20 to 30% of the filtered sodium ion load. The increase of sodium excretion can be greater than 1000 mEq/min above the predrug control. This pronounced magnitude of sodium excretion cannot be achieved by other classes of diuretics available as of this writing. Most high ceiling diuretics have a rapid onset of action, steep diuretic dose-response curves, and usually are short acting (32,33).

Table 1. Low Ceiling Diuretics (Thiazide-like)

Generic name	CAS Registry Number	Molecular formula	Trade name	Structure
		Thiazides		
bendroflumethazide	[73-48-3]	$C_{15}H_{14}F_3N_3O_4S_2$	Benuron Bristuron	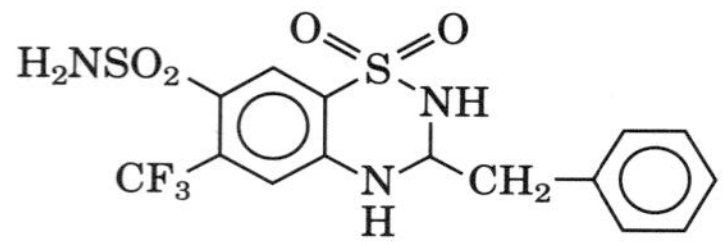
benzthiazide	[91-33-8]	$C_{15}H_{14}ClN_3O_4S_3$	Aquatag Exna Proaqua	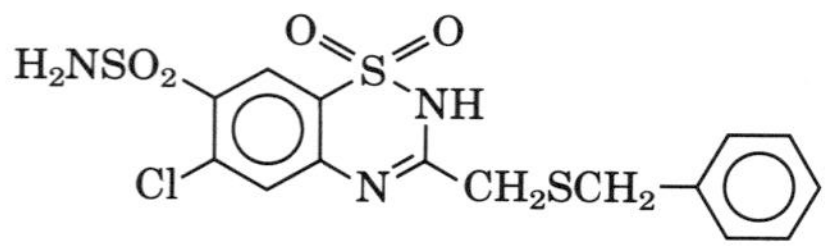
buthiazide	[2043-38-1]	$C_{11}H_{16}ClN_3O_4S_2$		
chlorothiazide chlorothiazide sodium	[58-94-6] [7085-44-1]	$C_7H_6ClN_3O_4S_2$ $C_7H_5ClN_3NaO_4S_2$	Diuril Diuril Sodium	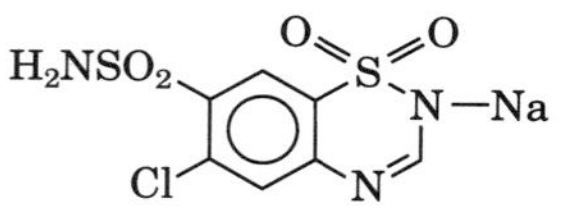
cyclothiazide	[2259-96-3]	$C_{14}H_{16}ClN_3O_4S_2$	Anhydron	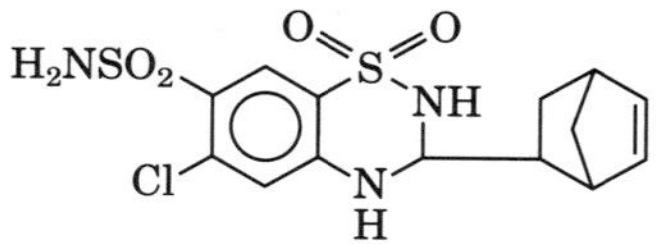
epithiazide[a]	[1764-85-8]	$C_{10}H_{11}ClF_3N_3O_4S_3$		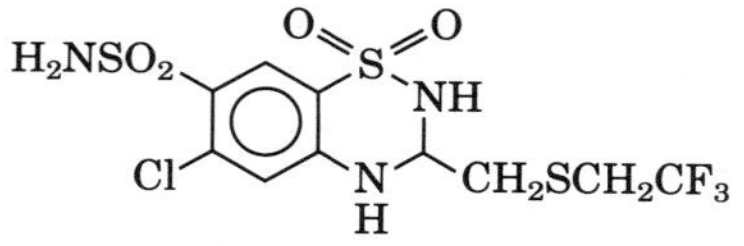

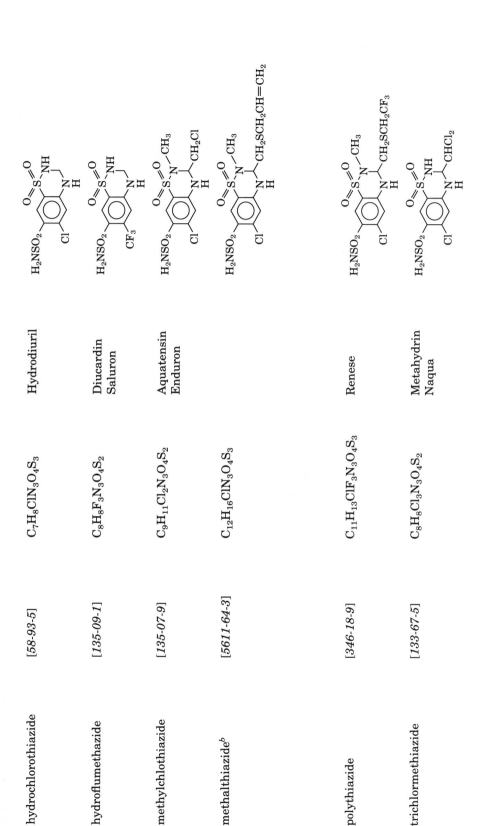

hydrochlorothiazide	[58-93-5]	$C_7H_8ClN_3O_4S_3$	Hydrodiuril	
hydroflumethazide	[135-09-1]	$C_8H_8F_3N_3O_4S_2$	Diucardin Saluron	
methylchlothiazide	[135-07-9]	$C_9H_{11}Cl_2N_3O_4S_2$	Aquatensin Enduron	
methalthiazide[b]	[5611-64-3]	$C_{12}H_{16}ClN_3O_4S_3$		
polythiazide	[346-18-9]	$C_{11}H_{13}ClF_3N_3O_4S_3$	Renese	
trichlormethiazide	[133-67-5]	$C_8H_8Cl_3N_3O_4S_2$	Metahydrin Naqua	

405

Table 1. (*Continued*)

Generic name	CAS Registry Number	Molecular formula	Trade name	Structure
Quinazoline sulfonamides				
fenquizone	[20287-37-0]	$C_{14}H_{12}ClN_3O_3S$	Idrolone	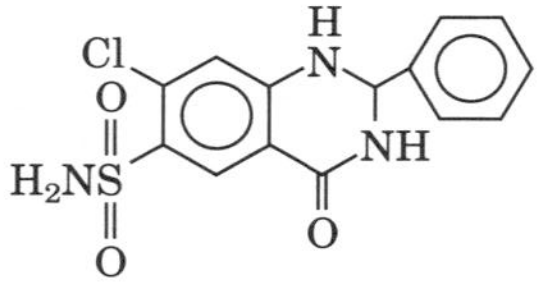
metolazone	[17560-51-9]	$C_{16}H_{16}ClN_3O_3S$	Diulo Mycrox Zaroxolyn	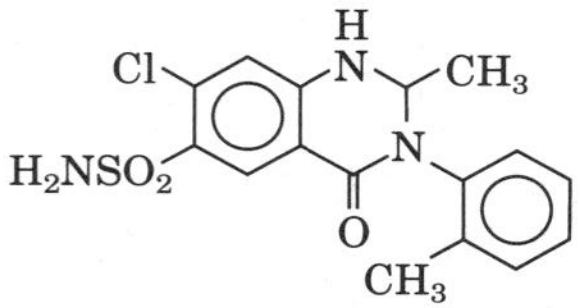
quinethazone	[73-49-4]	$C_{10}H_{12}ClN_3O_3S$	Hydromox	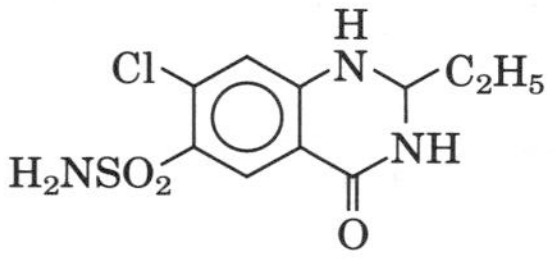
Other low ceiling diuretics				
alipamide[c]	[3184-59-6]	$C_9H_{12}ClN_3O_3S$		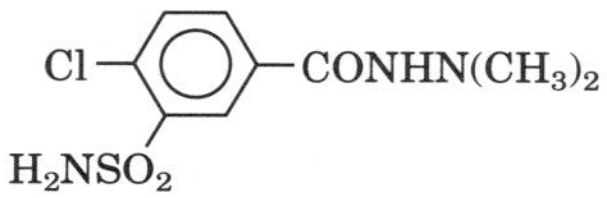

chlorthalidone	[77-36-1]	$C_{14}H_{11}ClN_2O_4S$	Hygroton Thalitone 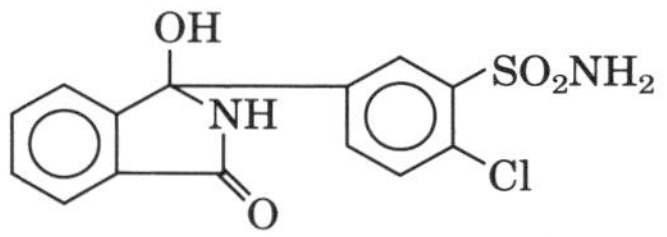
indapamide	[26807-65-8]	$C_{16}H_{16}ClN_3O_3S$	Lozol 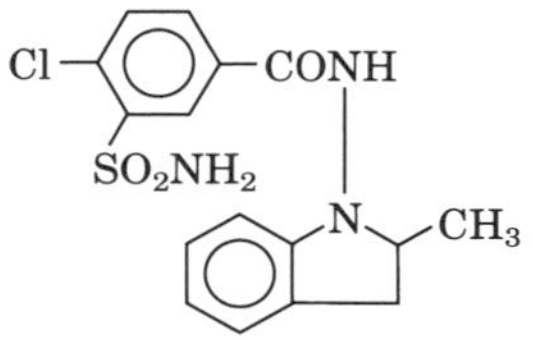
tripamide[d]	[73803-48-2]	$C_{16}H_{20}ClN_3O_3S$	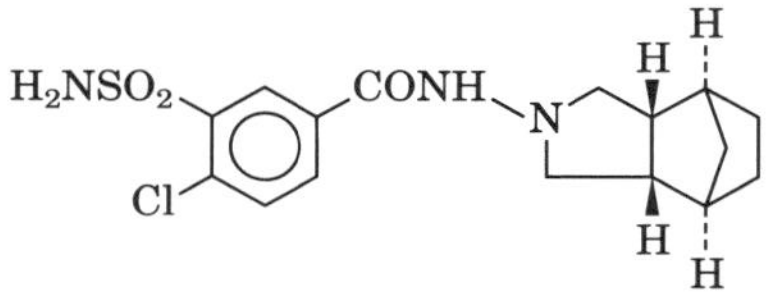
xipamide[e]	[14293-44-8]	$C_{15}H_{15}ClN_2O_4S$	Aquaphor 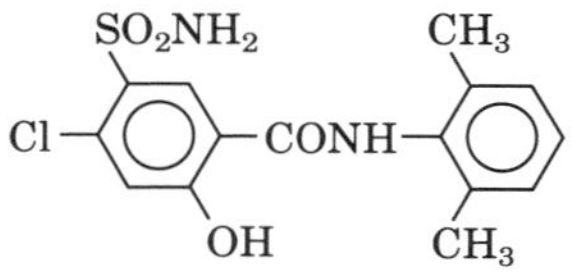

[a]Also known as Epitizide-INN [1764-85-8], P-2015, NSC-108164.
[b]Also known as P-2530.
[c]Also known as CI-546; CN-38,474; D-1721.
[d]Also known as ADR-033, E-614.
[e]Also known as MIF 10,938; Be-1293.

407

Table 2. High Ceiling (Loop) Diuretics

Generic name	CAS Registry Number	Molecular formula	Trade name	Structure
bumetanide	[28395-03-1]	$C_{17}H_{20}N_2O_5S$	Bumex	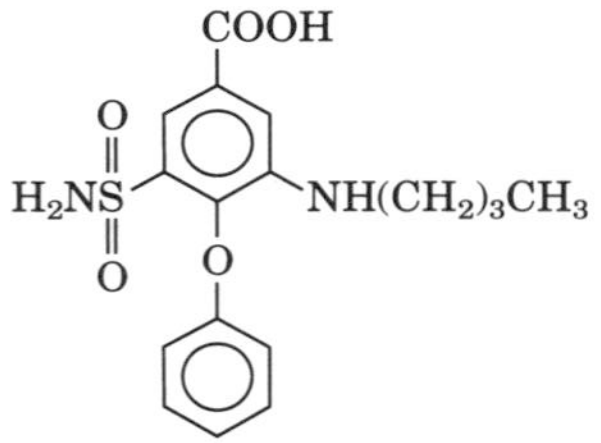
ethacrynic acid[a]	[58-54-8]	$C_{13}H_{12}Cl_2O_4$	Edecrin	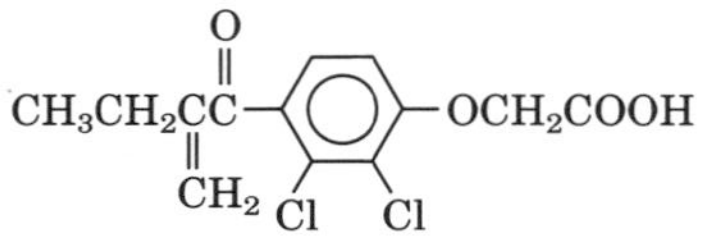
ethacrynic acid sodium	[6500-81-8]	$C_{13}H_{11}Cl_2NaO_4$	Edecrin Sodium	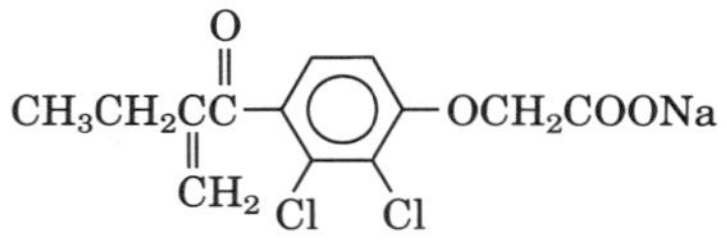
furosemide	[54-31-9]	$C_{12}H_{11}ClN_2O_5S$	Lasix Disal	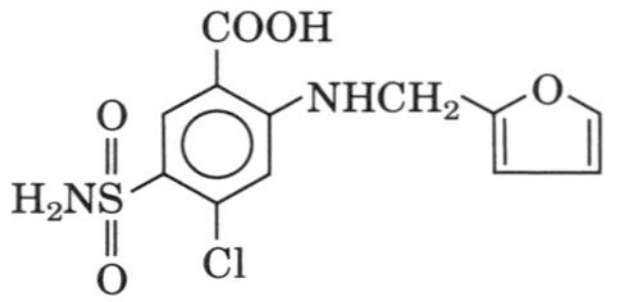
piretanide	[55837-27-9]	$C_{17}H_{18}N_2O_5S$	Arelix	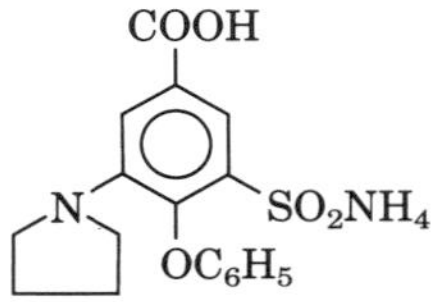

torasemide[b]

[56211-40-6]

$C_{16}H_{20}N_4O_3S$

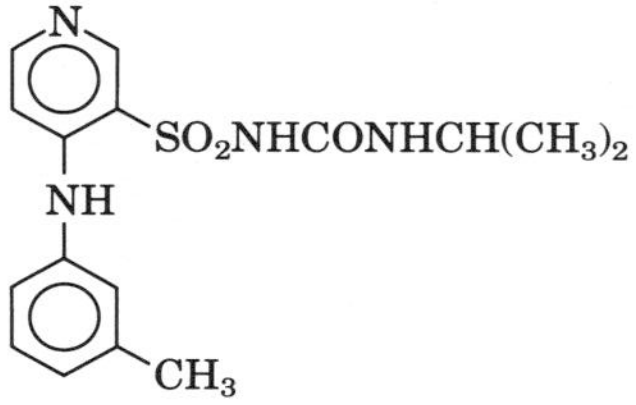

AY-31906

[124788-41-1]

$C_{14}H_{21}N_6O_3S$

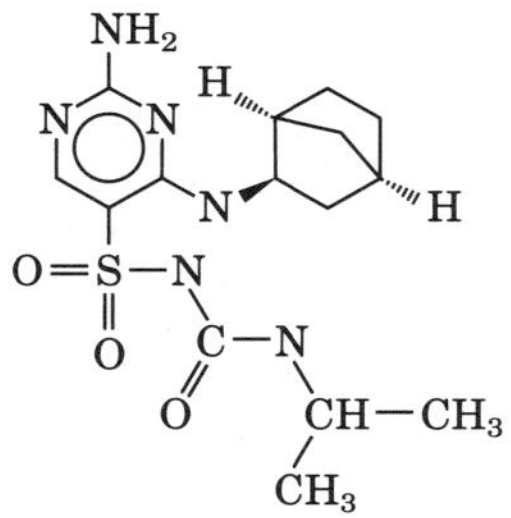

M-17055

[114417-20-8]

$C_{17}H_{15}ClN_2O_5SK$

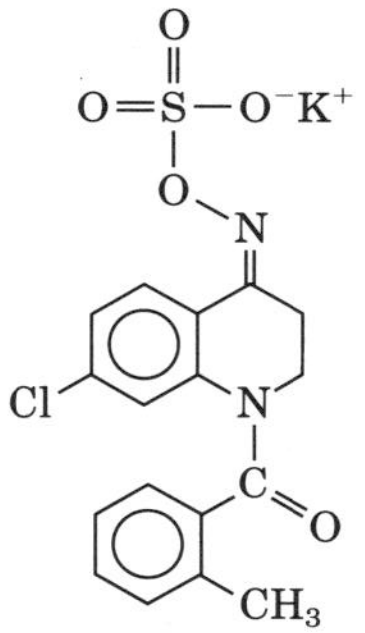

409

[a]Also known as Mk-595.
[b]Also known as BM 02 015, or AC 4464.

The most commonly used high ceiling (loop) diuretics are furosemide, bumetanide, and ethacrynic acid. Newer agents available in some countries include torasemide and piretanide. The potency ratio for furosemide:ethacrynic acid:torasemide:piretanide:bumetanide is 1:1:2:4:40. The onset of action is rapid following oral administration, and the duration of action of furosemide, bumetanide, and piretanide is about 2 to 3 h (34,35). The duration of action of torasemide, an analogue of furosemide, is reported to be much longer and single daily dosing has been reported to be efficacious (35,36).

After long-term use of a high ceiling diuretic, the extracellular fluid volume contracts and water reabsorption in the proximal and distal tubules increases, thus overriding and diminishing the diuretic's effects. A second high ceiling diuretic may sometimes induce diuresis again. This may be due to additional mechanisms. The high ceiling (loop) diuretics increase urinary excretion of potassium and magnesium, as do the thiazides, but high ceiling (loop) diuretics also increase urinary excretion of calcium; the excretion of chloride ion is greater than that of sodium ion, suggesting the inhibition of active chloride transport by the high ceiling diuretics in the loop of Henle (37,38). Alkalosis may develop in patients treated with the high ceiling diuretics, and plasma renin activity (PRA) is markedly elevated. The high ceiling diuretics inhibit the ability of the kidney to concentrate urine even in the presence of high concentrations of vasopressin. Like the thiazides, these diuretics act at the luminal side of the tubule so the secretion process of the proximal tubules of the nephron is critical in delivering the drug to the site of action. Probenecid competes for the secretion process, and decreases the efficacy of the loop diuretics. The loop diuretics increase renal blood flow, eg, bumetanide is reported to increase renal blood flow by as much as 40% (39,40).

Furosemide works mainly by inhibiting $Na^+-2Cl^--K^+$ co-transport across the luminal membrane and diminishes free-water clearance and the ability of the nephrons to concentrate urine (30). Both the effects on blood flow and the diuretic effects of furosemide are blocked by cyclooxygenase inhibitors, such as aspirin and indomethacin (40). These observations suggest that the diuretic effects of furosemide, and perhaps other high ceiling diuretics, may be via endogenous prostaglandin formation (41). Prostaglandins and their analogues have been shown to produce diuretic effects (42). Therefore, the concomitant use of nonsteroidal antiinflammatory drugs with the loop diuretics is contraindicated.

Ototoxicity, as evidenced by transient or permanent hearing loss, is a serious side effect of ethacrynic acid, and occurs less frequently with furosemide. Bumetanide is claimed to have only 20% of the ototoxic potential of furosemide (43). It has been reported that patients treated with torasemide at high doses for four weeks did not suffer hearing loss (36).

Potassium-Sparing Diuretics. Potassium-sparing diuretics act on the aldosterone-sensitive portion of cortical collecting tubules, and partially in the distal convoluted tubules of the nephron. The commonly used potassium-sparing diuretics are triamterene, amiloride, and spironolactone (Table 3). Spironolactone is a competitive aldosterone receptor antagonist, whereas triamterene and amiloride are not (44,45).

Triamterene and amiloride have been shown to act in the same portion of the nephron by interrupting the Na^+ transport process, and by inhibiting the potassium—sodium exchange mechanism, resulting in a decrease of the electrical potential across the membrane. They exert their effects in the presence or absence of aldosterone. The elimination of the potential gradient may inhibit the secretion of potassium and therefore produce the potassium-sparing or antikaliuretic effects (46–48). Amiloride is far more soluble than triamterene, and is the most widely studied potassium-sparing diuretic. Its natriuretic effect is minimal because only 2 to 3% of the filtered sodium ion load reaches the collecting tubules of the nephron (44). Etozolin (49) is a newer, long-lasting agent that has a gradual onset of action.

Spironolactone antagonizes the effects of aldosterone by binding at the aldosterone receptor in the cytosol of the late distal tubules and renal collecting ducts. Side effects of spironolactone are gynecomastia, decreased libido, and impotency.

Potassium-sparing by diuretic agents, particularly spironolactone, enhances the effectiveness of other diuretics because the secondary hyperaldosteronism is blocked. This class of diuretics decreases magnesium excretion, eg, amiloride can decrease renal excretion of potassium up to 80%. The most important and dangerous adverse effect of all potassium-sparing diuretics is hyperkalemia, which can be potentially fatal; the incidence is about 0.5% (50). Therefore, blood potassium concentrations should be monitored carefully.

Natriuretic Peptide Diuretics. Atrial natriuretic peptide (ANP), an endogenous diuretic, natriuretic, and vasodilator, is a peptide hormone primarily synthesized and stored by atrial cardiocytes, and secreted by the atria in response to mechanical stretch of the atria. It was discovered in the crude extracts of atria in 1981 (51). ANP is also known as anaritide [95896-08-5], $C_{112}H_{175}N_{39}O_{35}S_3$; atrial natriuretic factor [104595-79-1] (ANF); auriculin; cardionatrin; and atriopeptide. Its primary action is in the kidney and the vascular system. Human ANF is a 28 amino acid residue stored in atrial granules as pro-ANF, a 126 amino acid peptide. It has a very short half-life, and is mainly degraded by neutral endopeptidases, ie, atriopeptidase.

H—Arg—Ser—Ser—Cys—Phe—Gly—Gly—Arg—Met—Asp—Arg—Ile—Gly—
   1   2 |3   4   5   6   7   8   9   10   11   12

   S————————S
            |
Ala—Gln—Ser—Gly—Leu—Gly—Cys—Asn—Ser—Phe—Arg—Tyr—OH·xCH$_3$COOH
13   14   15   16   17   18   19   20   21   21a   21b   21c

ANP

It has been suggested that both the increased glomerular filtration rate (GFR) caused by ANP and the direct epithelial action in the collecting ducts by ANP are necessary to explain the diuretic effects of ANP. It appears that ANP may increase GFR by relaxing the glomerular mesangial cells resulting in increased surface area for filtration. Due to the augmented GFR, ANP increases the delivery of sodium and water to the renal tubules beyond the distal convoluted tubule. In the collecting ducts, ANP reduces sodium and free-water reabsorption by antagonizing the action of vasopressin. Therefore, the increased loads of so-

Table 3. Potassium-Sparing Diuretics

Generic name	CAS Registry Number	Molecular formula	Trade name	Structure
amiloride	[17440-83-4]	$C_6H_8ClN_7O \cdot 2H_2O \cdot HCl$	Midamor	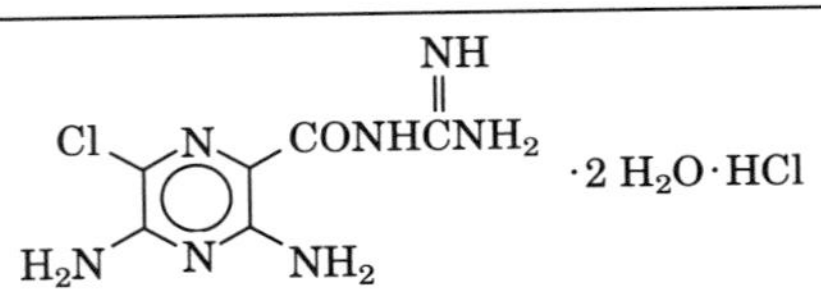
triamterene	[396-01-0]	$C_{12}H_{11}N_7$	Dyrenium	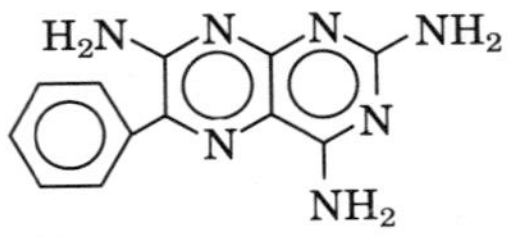
etozolin	[73-09-6]	$C_{13}H_{20}N_2O_3S$	Elkapin	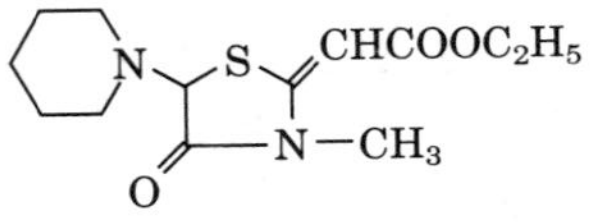
		Aldosterone antagonists		
spironolactone	[52-01-7]	$C_{24}H_{32}O_4S$	Aldactone	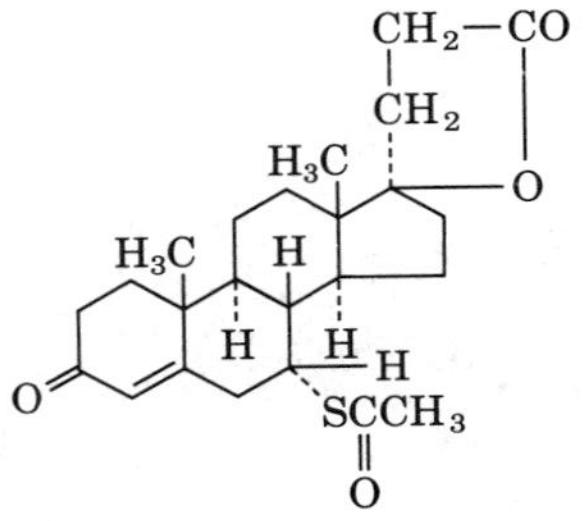

412

prorenoate potassium   [49847-97-4] $C_{23}H_{31}O_4K$

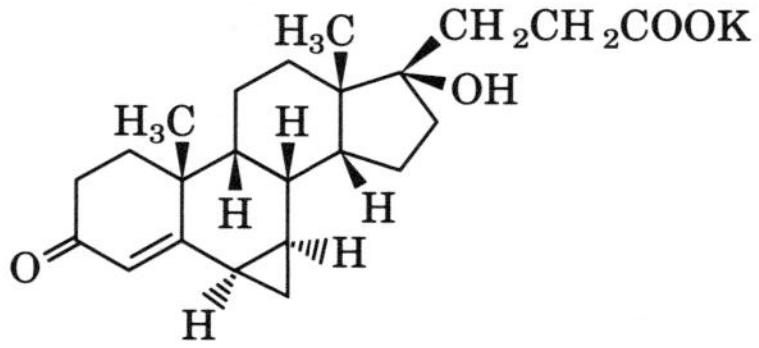

canrenone   [976-71-6] $C_{22}H_{28}O_3$

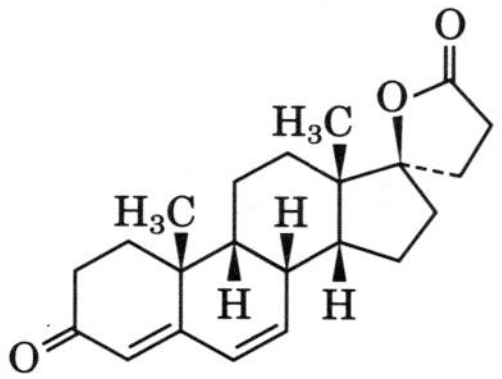

dium and water passing through the collecting ducts without the increased compensatory reabsorption result in profound diuresis and natriuresis (52).

In normal human subjects, ANP infusion for one hour causes increased absolute and fractional sodium excretion, urine flow, GFR, and water clearance (53–55). As shown in many *in vitro* and *in vivo* animal studies, ANP achieves this by direct effect on the sodium reabsorption in the inner medullary collecting duct, ie, by reducing vasopressin-dependent free-water and sodium reabsorption leading to diuresis; and by indirect effect through increased hemodynamic force upon the kidney. ANP inhibits the release of renin and aldosterone resulting in the decreased plasma renin activity and aldosterone concentration (56,57).

Urodilantin, a 32-amino acid natriuretic peptide synthesized by the kidney and found in urine but not in plasma, is believed to complicate the interpretation of the natriuretic effects of ANP (58). There have been reports of natriuretic peptides related to ANP that have been isolated from several sources, including the brain, eg, brain natriuretic peptide (BNP) [*114471-18-0*] (59–61).

Atrial Natriuretic Peptide Potentiator Diuretics. Neural endopeptidase (EC 3.4.24.11) inhibitors or atrial peptidase inhibitors (Table 4) are compounds that inhibit the enzyme that degrades ANP, resulting in higher plasma concentrations, and longer duration of action, of ANP; eg, neural endopeptidase cleaves ANP at the Cys_{105}–Phe_{106} and Ser_{123}–Phe_{124} bonds resulting in the loss of activity of ANP. The diuretic effects of this class of compounds resemble those due to administration of ANP. Compounds such as thiorphan, candoxatril, SCH-34826, and SCH-39370 have been studied in hypertension and congestive heart failure in humans with only limited success (62–67).

Osmotic Diuretics. An effective osmotic diuretic is a nonionic compound, freely filterable at the glomerulus, not reabsorbed by the tubules of the nephron, and biologically inert except for its osmotic properties. One of the best examples of an osmotic diuretic is mannitol, used to prevent acute renal failure in many major surgeries and traumatic injuries (Table 5) (12,68). An osmotic diuretic increases urine flow, rather than the excretion of sodium, by maintaining a high osmotic gradient resulting from the presence of large amounts of nonreabsorbable solutes in the luminal side of the proximal tubules. Under such conditions, the reabsorption of water is impaired along the descending loop of Henle as well as the collecting ducts, and urine flow increases. At high concentrations of an osmotic diuretic, the urinary excretion of sodium is also increased. This is due to the reduced reabsorption of sodium in the proximal tubules.

Carbonic Anhydrase Inhibitor Diuretics. Carbonic anhydrase [*9001-03-0*] accelerates the hydration of carbon dioxide to carbonic acid in aqueous solution, up to 7500-fold as compared to the nonenzymatic reaction. The hydrogen ions liberated from carbonic acid in the epithelial cells of the proximal tubules of the nephron are exchanged for sodium ions in the renal tubular lumen. When the generation of hydrogen ions is inhibited by a carbonic anhydrase inhibitor, the exchange of hydrogen for sodium ions is greatly diminished and the diuretic effect ensues. The site of action of carbonic anhydrase inhibitors (Table 5) is in the proximal tubules. In addition to sodium, the excretion of bicarbonate and potassium is also increased; the maximal excretion is about 2 to 4% of the filtered sodium load. Due to the increased urinary bicarbonate excretion, the urine be-

Table 4. Neural Endopeptidase (Atrial Peptidase) Inhibitors

Generic name	CAS Registry Number	Molecular formula	Structure
SCH-34826	[105262-04-2]	$C_{27}H_{34}N_2O_7$	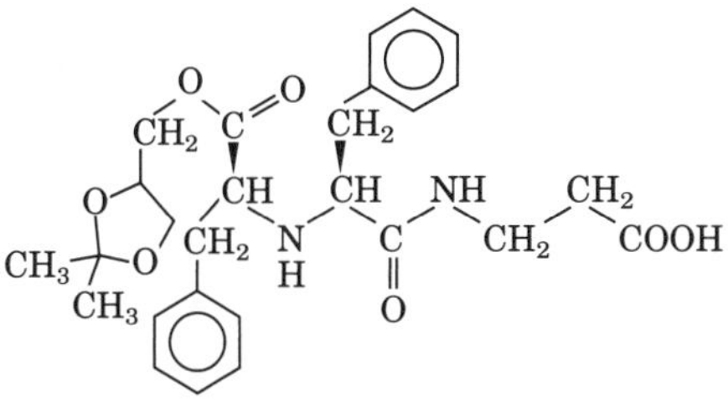
SCH-39370		$C_{22}H_{26}N_2O_6$	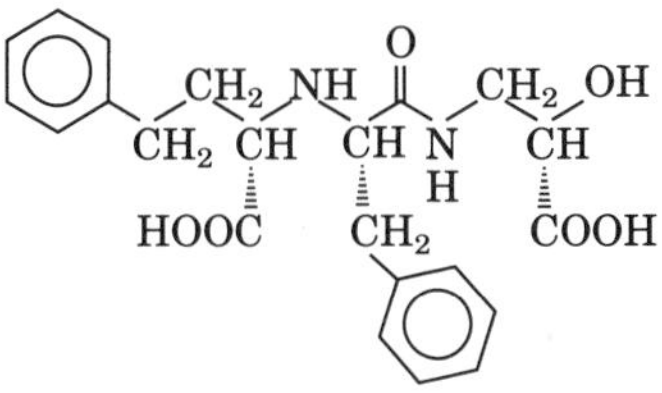
SQ-29,072		$C_{17}H_{24}NO_3S$	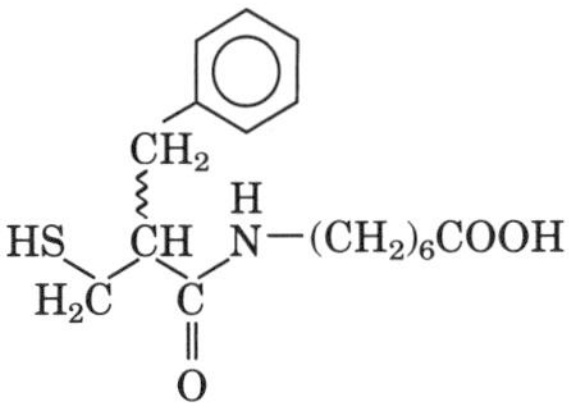
UK-69578	[123898-42-0]	$C_{20}H_{33}NO_7$	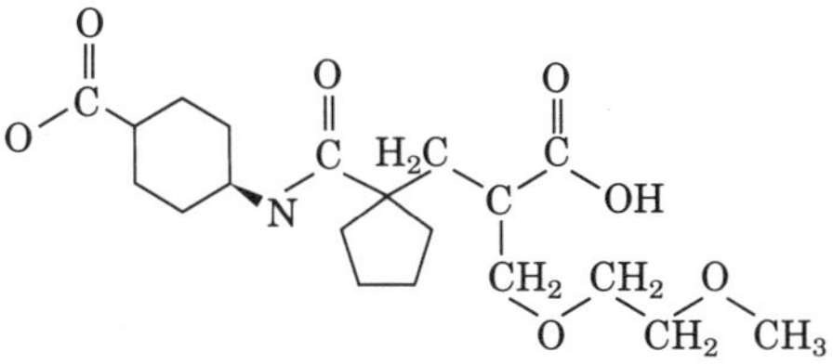
candroxatril[a]	[123122-55-4]	$C_{20}H_{33}NO_7$	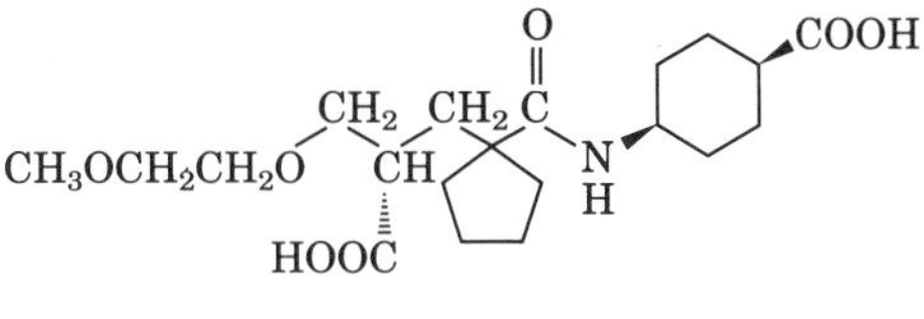
candro-xatrilet[b]	[123122-54-3]	$C_{29}H_{33}NO_7$	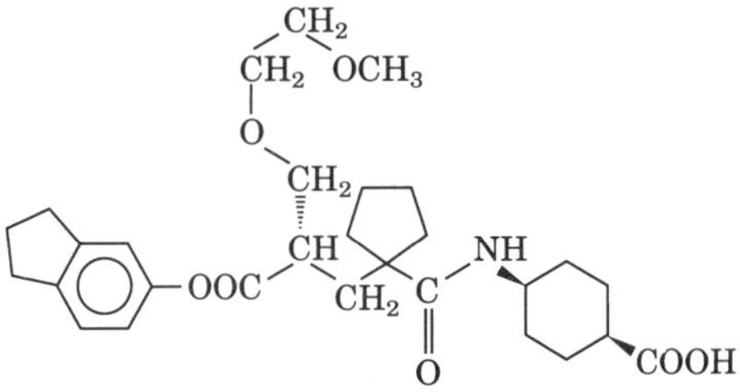

[a]Also known as UK-79300.
[b]Also known as UK-73967.

Table 5. Miscellaneous Diuretics

Generic name	CAS Registry Number	Molecular formula	Trade name	Structure
		Osmotic diuretics		
mannitol[a]	[69-65-8]	$C_6H_{14}O_6$	Osmitrol Resectisol	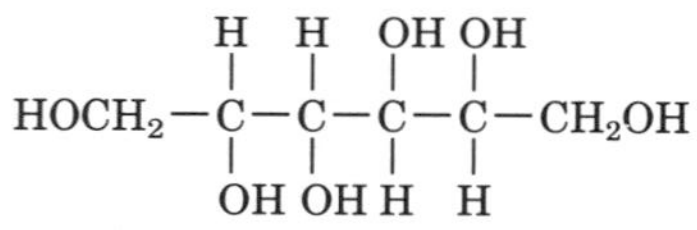 $HOCH_2-C-C-C-C-C-CH_2OH$ (with H, H, OH, OH and OH, OH, H, H substituents)
		Carbonic acid anhydrase inhibitors		
acetazolamide acetazolamide sodium	[59-66-5] [1424-27-7]	$C_4H_6N_4O_3S_2$ $C_4H_5N_4NaO_3S_2$	Diamox Diamox Sodium	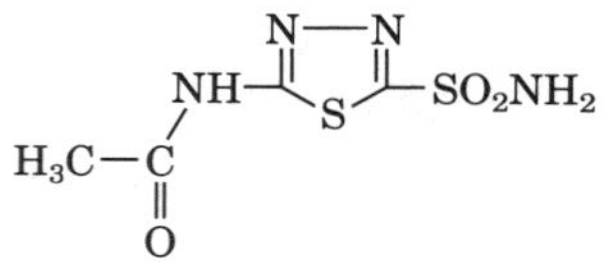 $H_3C-C(=O)-NH-$ (thiadiazole) $-SO_2NH_2$
dichlorphenamide	[120-97-8]	$C_6H_6Cl_2N_2O_4S_2$	Daranide	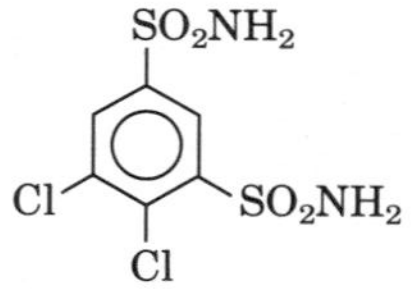 (benzene ring with SO_2NH_2, SO_2NH_2, Cl, Cl)
methazolamide	[554-57-4]	$C_5H_8N_4O_3S_2$	Neptazane	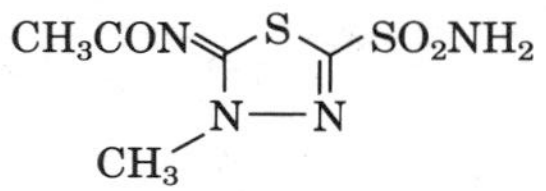 $CH_3CON=$ (thiadiazole with $N-CH_3$) $-SO_2NH_2$

Methylxanthines

theophylline
monohydrate

$C_7H_8N_4O_2$ [58-55-9]
$C_7H_8N_4O_2H_2O$ [5967-84-0]

Elixophyllin,
Constant-T
Theo-Dur
Theovent
Somophyllin
Respbid
Theochron
Duraphyl
Liniphyl

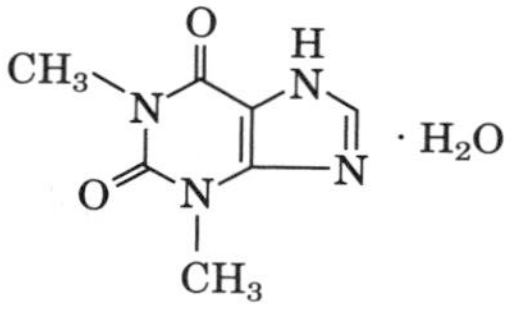

Organomercurials

mersalyl $C_{13}H_{16}HgNNaO_6$ [492-18-2] Salyrgan

mercaptomerin $C_{16}H_{27}HgNO_6S$ [20223-84-1]

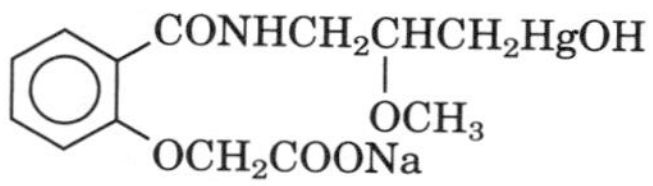

mercaptomerin
sodium $C_{16}H_{25}HgNNa_2O_6S$ [21259-76-7]

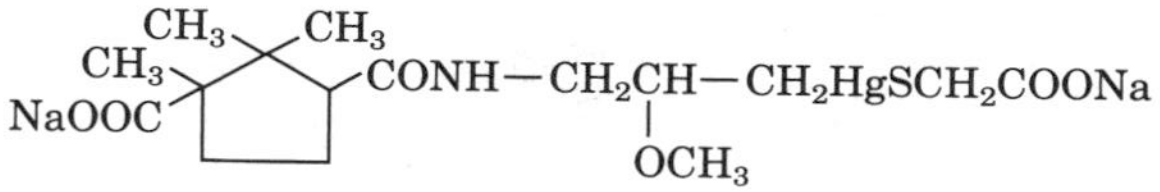

^{a}Also known as D-Mannitol.

417

comes alkaline and the blood becomes acidotic. The diuretic effect ceases once metabolic acidosis occurs.

Acetazolamide, the best example of this class of diuretics (69,70), is rarely used as a diuretic since the introduction of the thiazides. Its main use is for the treatment of glaucoma and some minor uses, eg, for the alkalinization of the urine to accelerate the renal excretion of some weak acidic drugs, and for the prevention of acute high altitude mountain sickness.

Methylxanthine Diuretics.   The mild diuretic effect of drinking coffee, from caffeine, and tea, mainly from theophylline, has been recognized for a long time. But the methylxanthines (Table 5) are of very limited efficacy when used as diuretics. The excretion of sodium and chloride ions are increased, but the potassium excretion is normal. Methylxanthines do not alter the urinary pH. Even though the methylxanthines have been demonstrated to have minor direct effects in the renal tubules, it is believed that they exert their diuretic effects through increased renal blood flow and GFR (71).

Organomercurial Diuretics.   Before the advent of the thiazide diuretics, mercurial and organomercurial diuretics were the mainstay therapy for the treatment of edema (72). They have become obsolete and are of historical value only. It is generally accepted that the main site of action of the organomercurial diuretics is in the loop of Henle, although they have also been shown to inhibit sodium and chloride ion reabsorption in the proximal and distal tubules. The organomercurial diuretics, which do not cause the excretion of an amount of bicarbonate equivalent to chloride, produce hypochloremic alkalosis; when this happens the diuretic effect disappears. Metabolic acidosis potentiates their diuretic effects. The organomercurials have to be given parenterally to be effective, and they produce toxic effects on the heart, kidney, and liver. Sudden cardiac death due to ventricular fibrillation has occurred with the use of this class of diuretic agents.

Water Diuretics (Aquaretics).   A water diuretic, ie, aquaretic, decreases urinary osmolality by influencing the kidney to excrete water selectively without a concomitant proportionally increased excretion of sodium ions (73,74) (Table 6). A water diuretic should be efficacious for the treatment of hyponatremia, ie, low plasma sodium concentration, and the syndrome of inappropriate antidiuretic hormone secretion (SIADH). In many diseases and conditions, when water is retained to a greater extent as related to sodium ions, hyponatremia results. This is seen in many edema cases arising from congestive heart failure (CHF), hepatic cirrhosis, renal failure, and nephrotic syndrome. In the treatment of these conditions, the conventional diuretics will lose their effectiveness once hyponatremia occurs. Diseases of the brain and the lung, certain surgeries, and some tumors also will cause hyponatremia; in these conditions, with any given plasma osmolality, the plasma antidiuretic hormone (ADH) concentrations are inappropriately high. When this occurs, the patient is inferred to have SIADH.

Logically, ADH receptor antagonists, and ADH synthesis and release inhibitors can be effective aquaretics. ADH, 8-arginine vasopressin [113-79-1], is synthesized in the hypothalamus of the brain, and is transported through the supraopticohypophyseal tract to the posterior pituitary where it is stored. Upon sensing an increase of plasma osmolality by brain osmoreceptors or a decrease of blood volume or blood pressure detected by the baroreceptors and volume recep-

Table 6. Water Diuretics (Aquaretics)[a]

Generic name	CAS Registry Number	Molecular formula	Trade name	Structure
OPC-31260	[137975-06-5]	$C_{27}H_{31}N_3O_2$		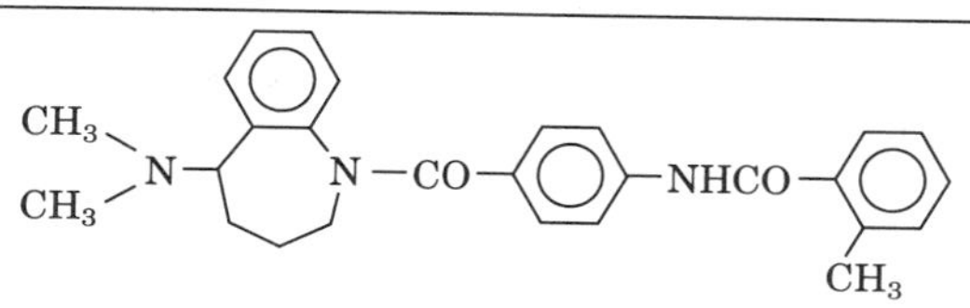
SK&F-101926	[90332-82-4]	$C_{51}H_{74}N_{12}O_{10}S_2$		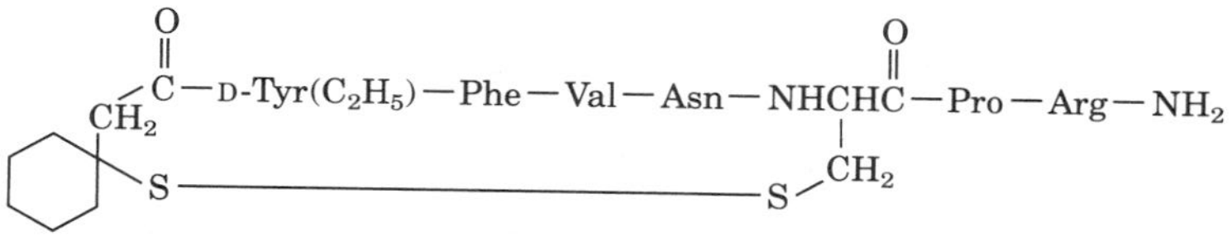
SK&F-105494	[114923-99-8]	$C_{54}H_{83}N_{15}O_{10}$		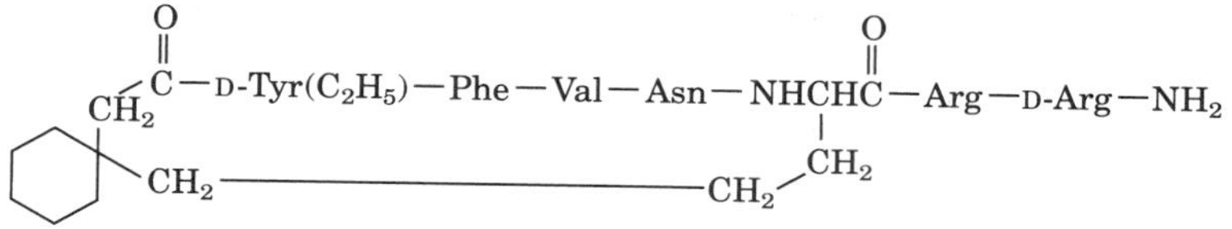
spiradoline mesylate	[87173-97-5]	$C_{22}H_{30}Cl_2N_2O_2 \cdot CH_4O_3S$		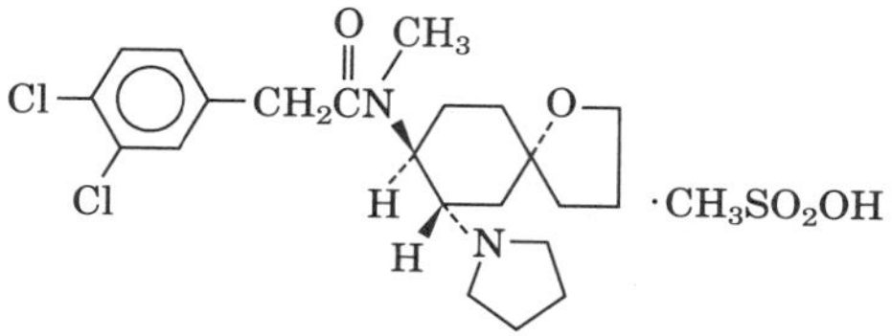
demeclocycline	[64-73-3]	$C_{21}H_{21}ClN_2O_8 \cdot HCl$	Declomycin	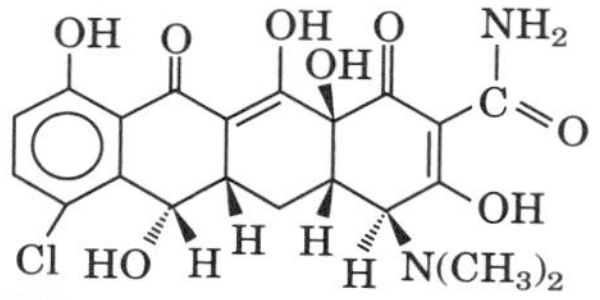

[a]Antidiuretic hormone (ADH) antagonists.

419

tors, ADH is released into the blood circulation; it activates vasopressin V_1 receptors in blood vessels to raise blood pressure, and vasopressin V_2 receptors of the nephrons of the kidney to retain water and electrolytes to expand the blood volume.

There is no specific water diuretic marketed as of this writing. Demeclocycline has been used clinically with only limited success; in preclinical pharmacology experiments, it has been shown to antagonize the effects of vasopressin in conscious rats (75). Studies in human volunteers have shown a diuretic profile for spiradoline resembling that of an aquaretic by selectively increasing diuresis without increasing excretion of electrolytes (76). Many peptides, including $d(CH_2)_5[D\text{-}Ile^2, Ile^4]AVP$, SK&F 101926, and SK&F 105494, have been reported to have vasopressin receptor (V_2-receptor type) antagonistic effects. SK&F 101926 and 105494 have been shown to be *in vitro* and *in vivo* vasopressin V_2-receptor antagonists in rat, dog, and monkey, and in humans, *in vitro*. However, both compounds were found to exert vasopressin V_2-agonist properties in humans *in vivo* (77,78); the reason for the discrepancy between *in vitro* and *in vivo* results in humans is obscure. OPC-31260 (79,80), a nonpeptide, has been shown to be 100 times more potent in displacing arginine vasopressin from V_2 than V_1 vasopressin receptors, and has been reported to potently antagonize the antidiuretic effects, ie, V_2-receptor function, of vasopressin in rats. However, OPC-31260 also antagonized the vasoconstricting effects of vasopressin in isolated perfused dog femoral arteries; this indicates that OPC-31260 may not be a highly selective vasopressin V_2-receptor antagonist (79). It has been proposed that the dog femoral artery contains vasopressin V_2 receptors, and these receptors also cause vasoconstriction like the V_1 receptors if activated *in vitro* (79).

Health and Safety

Diuretic agents have long been considered relatively safe drugs (81); however, they do produce prominent side effects and toxic effects. The side effects result from the pharmacology of natriuresis, leading to hyponatremia and diuresis, resulting in decreased plasma volume. The decrease in plasma volume reduces cardiac output, which may produce postural hypotension. The decrease in plasma volume may also decrease renal blood flow and glomerular filtration rate producing pre-renal azotemia, or increased proximal reabsorption of uric acid [69-93-2] and distal tubular reabsorption of calcium, leading to hyperuricemia and hypercalcemia, respectively; hypercalcemia may be a prominent feature of thiazide diuretic therapy only.

In an attempt to conserve sodium, the kidney secretes renin; increased plasma renin activity increases the release of aldosterone, which regulates the absorption of potassium and leads to kaliuresis and hypokalemia. Hypokalemia is responsible in part for decreased glucose intolerance (82). Hyponatremia, postural hypotension, and pre-renal azotemia are considered of little consequence. Hyperuricemia and hypercalcemia are not unusual, but are not considered harmful. However, hypokalemia, progressive decreased glucose tolerance, and increased serum cholesterol [57-88-5] levels are considered serious side effects or even toxic effects of diuretics.

Hypokalemia. Hypokalemia associated with thiazide diuretic therapy has been implicated in the increased incidence of cardiac arrhythmias and sudden death (82). Several large clinical trials have been conducted in which the effects of antihypertensive drug therapy on the incidence of cardiovascular complications were studied. The antihypertensive regimen included diuretic therapy as the first drug in a stepped care (SC) approach to lowering the blood pressure of hypertensive patients.

One study (83) indicated that in mildly hypertensive male patients treated with an antihypertensive drug, those below the age of 50 or having no clinical evidence of cardiovascular disease had no significant improvement from cardiovascular diseases within 3.3 years; those over the age of 50 or having pre-existing cardiovascular disease benefited significantly.

Another study (84), which enrolled men and women between the ages of 21–55 who had mild hypertension and no recognizable cardiovascular risk factors, showed no significant differences in mortality between drug- and placebo-treated patients. Significant reductions in hypertensive complications were noted, but atherosclerotic complications were not reduced.

A third study (85) enrolled 7825 hypertensive patients (55% males and 45% females) having diastolic blood pressures (DBP) of 99–104 mm Hg (13–14 Pa); there were no placebo controls. Forty-six percent of the patients were assigned to SC antihypertensive drug therapy, ie, step 1, chlorthalidone; step 2, reserpine [50-55-5] or methyldopa [555-30-6]; and step 3, hydralazine [86-54-4]. Fifty-four percent of the patients were assigned to the usual care (UC) sources in the community. Significant reductions in DBP and in cardiovascular and noncardiovascular deaths were noted in both groups. In the SC group, deaths from ischemic heart disease increased 9%, and deaths from coronary heart disease (CHD) and acute myocardial infarctions were reduced 20 and 46%, respectively.

In a study with 3427 male and female patients having DBP of 95–109 mm Hg (12–15 Pa), and no clinical evidence of cardiovascular diseases, half of the patients were placebo-treated and half were SC antihypertensive drug-treated, ie, step 1, chlorothiazide; step 2, methyldopa, propranolol [525-66-6], or pindolol [13523-86-9]; and step 3, hydralazine, or clonidine [4205-90-7] (86). Overall, when the DBP was reduced below 100 mm Hg (13 Pa), there were more deaths in the drug-treated group than in the placebo group. The data suggest reduction of blood pressure by antihypertensive drug treatment that includes a diuretic is accompanied by increased cardiovascular risks.

The Oslo Trial (87) enrolled 785 male patients <50 years of age with DBP <110 mm Hg (15 Pa) and free of clinical evidence of cardiovascular disease. If the initial DBP was <100 mm Hg (13 Pa), there were no differences in mortality or cardiovascular events in the placebo- or drug-treated groups. If the initial DBP was >100 mm Hg, then the incidence of cardiovascular disease was greater in the drug-treated than in the placebo-treated group.

The multiple risk factor intervention trial (MRFIT) (88,89) examined control of the three primary risk factors, ie, smoking, hypercholesterolemia, and hypertension, in 2338 males ages 40–59 who were assigned either to special intervention care (SIC) or UC antihypertensive drug-treatment groups. The patients had initial DBP >90 mm Hg, and included those with normal and abnormal electrocardiograms (ECG). There were no significant differences in mortality in those patients assigned to the SIC or UC groups if their initial DBP were <100 mm Hg. All patients with initial DBP >100 mm Hg or those with ECG abnormalities in the SIC group had significantly higher mortality. The data were interpreted as suggesting that patients having mild hypertension and ECG abnormalities were at increased risk of mortality from ischemic heart disease.

There is no overwhelming evidence to suggest that patients on thiazide diuretic therapy are at greater risk of CHD. The Oslo Study (87) and MRFIT Study using patients who had baseline ECG abnormalities (88,89) indicate that patients on diuretic therapy are at greater risk of CHD, whereas the remainder of the studies (83–86) and the MRFIT study using patients having baseline exercise ECG abnormalities (88,89) indicate that patients on diuretic therapy are not at greater risk of CHD.

Diuretics, Arrhythmias, and Sudden Death.   Diuretic-induced hypokalemia may increase the frequency and severity of arrhythmias in patients having pre-existing cardiac problems, eg, enlarged hearts, ECG abnormalities, and frequent ventricular ectopic beats. However, the evidence that patients without cardiac problems are at greater risk of sudden death or developing cardiac arrhythmias is not overwhelming. Three studies reported the same number of cases of sudden death in the control or placebo groups and in the thiazide-treated groups (83,84,87), and the MRFIT (88,89) had a statistically insignificant higher incidence of sudden death in the SIC group who had minor ECG abnormalities than in the UC group; in the group having baseline ECG abnormalities, the incidence of sudden death was approximately four times higher in the usual care group than in the special intervention care group.

Diuretics and Lipid Metabolism.   It is well recognized that thiazide diuretics elevate plasma total cholesterol, very low density lipoproteins (VLDL), low density lipoproteins (LDL), triglycerides, and phospholipid concentrations. The levels of high density lipoprotein (HDL), ie, the scavenger lipoprotein, are decreased (83,90,91). The changes in lipid profile are characterized as having the potential to increase the risk for coronary heart disease. The mechanism(s) by which the thiazides induce the changes in blood lipids are not well understood (91). However, based on indirect evidence, increased production or decreased clearance of the aforementioned lipids may be involved. The changes are small, ca 5%, but if sustained for several years, these may potentially increase the risk for coronary heart disease. Several studies have re-examined the effects of thiazides on blood lipids. Short-term high dose thiazide diuretic therapy increases plasma cholesterol, but the levels are reduced to or below pretreatment values after several months to a year of daily therapy. The data suggest that long-term thiazide diuretic therapy does not increase the risk for development of atherosclerosis (92).

Diuretics and Glucose Tolerance.   The thiazide diuretics have been known for many years to decrease glucose tolerance in patients with established diabetes receiving oral hypoglycemic agents (46,93). *De novo* glucose intolerance occurs at a rate of ca 9 cases per 1000 persons annually, and appears to be reversible when diuretic therapy is stopped (94,95). The mechanism by which glucose intolerance occurs may result in part from the hypokalemia induced by the thiazide diuretics (96) because strict control of potassium balance results in minimal disturbances of glucose metabolism (97). Hypokalemia decreases secretion of insulin from the β-islet cells of the pancreas, and decreases insulin-like activity in the blood, probably by inhibiting conversion of proinsulin to insulin (98).

When administered long-term for the treatment of hypertension, diuretics fulfill the goals of preventing cardiovascular disease and increasing longevity. However, diuretic therapy may produce both side and toxic effects that are significant in certain patient subgroups, eg, diabetics and cardiac patients.

Therapeutic Uses of Diuretics

Diuretics are one of the drug categories most frequently prescribed. The principal uses of diuretics are for the treatment of hypertension, congestive heart failure, and mobilization of edema fluid in renal failure, liver cirrhosis, and ascites. Other applications include the treatment of glaucoma and hypercalcemia, as well as the alkalinization of urine to prevent cystine and uric acid kidney stones.

Hypertension. Diuretics were the first drugs used in the stepped care (SC) approach to the treatment of hypertension popular before 1988. They became the cornerstone of all antihypertensive therapies. As of this writing, an individualized or patient-oriented approach, rather than the rigid SC approach, is emphasized, but diuretics, particularly the thiazide-type, still remain one of the first four drugs to be used in antihypertensive therapy (99–102). The importance of the diuretics, especially hydrochlorothiazide (HCTZ), in the treatment of hypertension is evidenced by the large number of combination antihypertensive drug preparations containing HCTZ. HCTZ can be found in combination with almost every class of antihypertensive agents regardless of their mechanism of action, including angiotensin-converting enzyme inhibitors and calcium channel blocking agents. The diuretic in the combination therapy will counteract the fluid retention side effects of other antihypertensive agents. Therefore, an additive or sometimes a potentiative effect is observed. Diuretics as monotherapy can decrease blood pressure up to 15 mm Hg (2 Pa). They can normalize the blood pressure of about 50% of the hypertensives treated, and are most effective in patients with low plasma renin activity (PRA); they are also very effective in elderly and black patients (103,104). Indapamide, because of an additional vasodilatory property, can normalize the blood pressure in up to 60 to 80% of the patients; it can decrease the systolic blood pressure up to 20 to 35 mm Hg, and the diastolic up to 10 to 20 mm Hg (105). The main mechanism for lowering the blood pressure of hypertensive patients in long-term diuretic treatment is the contraction of blood volume (106). However, the reduction of excessive sodium and water in the vascular wall induced by the long-term use of a diuretic, rendering the vessel wall less sensitive to neurohormonal stimuli, may play a minor role (106,107). The possibility of alteration of various ion distribution in the vascular smooth muscle leading to decrease in vascular tone and reactivity also has been proposed. In meta analysis of numerous clinical studies, a reduction of blood volume of 5% is needed before the antihypertensive efficacy of a diuretic is evident. It is emphasized, in the early 1990s, that antihypertensive treatment should not be merely to lower the elevated blood pressure, but also to improve quality of life and to lower cardiovascular morbidity and mortality. Diuretics have been shown to significantly protect against stroke but not myocardial infarction; it has been speculated that the undesirable effects of diuretics on blood lipids and potassium may have contributed to the unexpected negative outcome on myocardial infarction. All factors considered, use of diuretics for the treatment of hypertension is considered to be most cost-effective (108). It is particularly suitable for use in patients at high risk of cerebrovascular disease and in hypertensive patients with renal failure.

Furosemide is more efficacious than the thiazides in hypertensive patients with reduced renal functions. Piretanide has been shown to be more efficacious than hydrochlorothiazide in patients with uncomplicated essential hypertension. However, diuretic monotherapy should be avoided in patients with left ventricular hypertrophy or with irregular ventricular heartbeat since diuretics will not reverse cardiac hypertrophy, and can cause additional electrical instability of the heart due to possible serum potassium and magnesium depletion. The lowest effective dose of a diuretic should be used, since this will decrease side effects. Diuretics, particularly potassium-sparing diuretics, should not be used in hypertensive patients with diabetes. It has been shown that low doses of ANP infusions produce a prolonged lowering of blood pressure and increases in heart rate of patients with hypertension (53,54,109,110). Despite some natriuresis, no changes in urine volume or cardiac output were observed. Side effects in some patients included severe postural hypotension and bradycardia (109). Nearly all known antihypertensive agents are more efficacious when combined with a diuretic. Withdrawal of diuretics from patients who have been treated for more than six months may result in low blood pressure for many months.

Congestive Heart Failure.   Congestive heart failure (CHF) occurs when the heart fails as a pump, leading to a decrease in the cardiac output and venous blood pooling. The amount of blood perfusing the vital organs decreases, and receptors on the arterial side of the systemic circulation sense the decrease of the effective volume perfusion and initiate various compensatory mechanisms attempting to correct the deficiency and to restore normal blood supply. The adrenergic nervous system, the renin–angiotensin–aldosterone system, and the vasopressin system are activated. In turn the kidney excretory functions are changed to favor water and salt retention. Capillary pressures finally are increased sufficiently to cause fluid accumulation in the tissue space, resulting in edema. Therefore, the symptoms of CHF are mainly contributed by hemodynamic and neurohormonal forces. Edema through salt retention is one of the factors in increasing ventricular wall stress. Under such conditions, the diuretic can increase renal water and salt excretion so as to facilitate the movement of excess fluid out of the tissues and organs.

Diuretics have become the cornerstone of all treatment regimens of CHF (111–113). They can relieve symptoms of pulmonary and peripheral edema. In mild CHF, the thiazide-type diuretics are adequate unless the GFR falls below 30 mL/min, as compared to 120 mL/min in normal subjects. Diuretics improve left ventricular function in CHF due in part to decrease of preload. Indapamide has been shown to cause reduction of pulmonary arterial pressure and pulmonary wedge pressure.

The high ceiling (loop) diuretics, such as furosemide, bumetanide, and ethacrynic acid, are preferred in moderate CHF because they have higher efficacy and can retain their efficacy until the GFR falls below 5 mL/min. Furosemide has been shown to decrease afterload, as demonstrated by improving left ventricular function. It also causes a reduction of left ventricular filling pressure, improvement of pulmonary compliance, and a decrease of pulmonary wedge pressure. There is evidence to show that the loop diuretics, such as piretanide, will cause a decrease of preload, as demonstrated by decreasing pulmonary wedge pressure and right atrial pressure.

As mentioned earlier, the long-term employment of diuretics can cause low serum levels of potassium and magnesium, and can predispose the patients to lethal ventricular fibrillation and sudden cardiac death (112). Therefore, it is highly useful if the regimen includes the use of a thiazide-type diuretic or a loop diuretic in combination with a potassium-sparing diuretic. Attention must be paid to monitor the serum potassium and magnesium levels. Even though diuretics alone can relieve the symptoms of patients with CHF, their disease continues to deteriorate (112). This may be due, in part, to the activation of the renin–angiotensin–system (RAS) by the diuretics, ie, increased plasma renin activity is a common adverse effect of the diuretics. The combined use of a diuretic and an angiotensin-converting enzyme (ACE) inhibitor, such as captopril, enalapril, or lisinopril, will have additive efficacy. Similarly, the combined use of a diuretic with ANP and/or a neutral endopeptidase (atriopeptidase) inhibitor will have beneficial effects, since ANP will suppress the RAS. Diuretics will also enhance the efficacy of vasodilators by counteracting their fluid retention side effects. Furosemide is effective for the treatment of hyponatremia if used concomitantly with hypertonic sodium chloride solution, as it will antagonize the ability of the nephrons to generate free water (114).

Renal Failure. High ceiling (loop) diuretics, such as furosemide, bumetanide, ethacrynic acid, piretanide, and torasemide, are the drugs of choice for the treatment of acute and chronic renal failure, with or without hypertension. Acute renal failure is characterised by a rapid loss of renal function, leading to a rapid develoment of azotemia. In acute renal failure that occurs after certain surgical procedures, both mannitol, an osmotic diuretic, and furosemide have been shown to be successful in increasing urine flow and creatinine clearance, but they do not always decrease mortality (115,116). Chronic renal failure results from destruction of nephrons and much reduced GFR, leading to uremia. In these patients, high ceiling diuretics can increase sodium excretion to >50% of the filtered load. The greater the renal impairment, the higher the dose of diuretic is needed. The half-life of the diuretics can increase two- to fourfold in these patients. Mannitol has been found to be useful in maintaining urine flow even at low GFRs, such as in hypotension and dehydration (116).

Ascites. Patients with cirrhosis, especially liver cirrhosis, very often develop ascites, ie, accumulation of fluid in the peritoneal cavity. This is the final event resulting from the hemodynamic disturbances in the systemic and splanchnic circulations that lead to sodium and water retention. When therapy with a low sodium diet fails, the drug of choice for the treatment of ascites is furosemide, a high ceiling (loop) diuretic, or spironolactone, an aldosterone receptor antagonist/potassium-sparing diuretic.

Since nonsteroidal antiinflammatory drugs, such as aspirin, an inhibitor of the synthesis of prostaglandins, will block the diuretic effects of furosemide, their concomitant use should be avoided (13,117).

In severe ascites, therapeutic paracentesis, the nonpharmacological treatment, should be used first, followed by a diuretic to prevent the reaccumulation of fluid in the abdominal cavity (117). Furosemide is effective in only ca 50% of these patients. One of the factors for patients not responding to a high ceiling diuretic may be that they have hyperaldosteronism; in this instance spironolactone has been shown to be more efficacious than furosemide in many patients. The combination of a high ceiling diuretic and spironolactone offers a greater natriuretic effect and a lower incidence of hyperkalemia (117). Patients with ascites may be refractory to diuretic treatment due to the coexistence of functional renal failure; this results in impairment of the delivery of the diuretics to the sites of action in the renal tubules. It has been suggested that the longer-acting loop diuretic, torasemide, is more effective in the treatment of ascites in cirrhosis (118).

Economic Impact

Worldwide sales of diuretics in 1991 were $1.602 billion (119). U.S. sales of diuretics in 1992 U.S. dollars were over $650 million (120). Worldwide sales of the leading diuretics Dyazide, ie, triamterene plus hydrochlorothiazide (HCTZ); Moduretic, ie, amiloride plus HCTZ; and Aldactone, ie, spironolactone, were $280, $145, and $206 million, respectively, in 1989–1990 (121,122). Lozol and Maxzide, a lower dosage form of Dyazide, commanded sales of $74 and $90 million, respectively (123,124). In the United States, the market shares for potassium-sparing diuretics, the high ceiling (loop) diuretics, and the low ceiling diuretics were 72, 14, and 14%, respectively (125). Between 1984 and 1991, the sales of diuretic agents in constant dollars, adjusted to 1991 prices, declined 40%. In 1992 dollars not adjusted for price increases, sales increased ca 10% (120).

The sales of oral diuretics are declining, and are forecast to continue their decline in constant dollars during the 1990s (119,120). Several possible explanations can be offered for these trends. The patents of market leaders are expiring, leading to the introduction of generic brands at ca 40% below the cost of the branded market leaders; physicians are switching to newer treatments for hypertension, eg, calcium channel blockers and angiotension-converting enzyme inhibitors; and concerns are growing about the possible adverse effects of diuretics, eg, hypokalemia, the progression of atherosclerosis, and the increase in mortality, serum cholesterol, glucose tolerance, and diabetes (120,121).

NOMENCLATURE

acidosis	pH of blood or plasma below normal; the normal range in the adult male is 7.33–7.45
active transport	the movement of materials across cell membranes and epithelial layers resulting directly from the expenditure of metabolic energy

afterload	the load against which cardiac muscle exerts its contractile force; the arterial pressure against which the ventricle must contract; the higher the pressure, the greater the afterload
alkalosis	pH of blood or plasma above normal
angiotensin II	a potent octapeptide vasopressor and stimulant of aldosterone secretion from the adrenal cortex
ascites	accumulation of fluid in the peritoneal cavity
azotemia	an excess of urea or other nitrogenous bodies in the blood
baroreceptor	a sensory nerve ending that is sensitive to changes in pressure, as those in the walls of blood vessels
cardiac output	the quantity of blood pumped into the aorta per minute by the heart
catecholamines	one of a group of similar compounds possessing a catechol pharmacophore and having sympathomimetic action, eg, epinephrine, norepinephrine, or dopamine
clearance (renal)	a calculated volume (mL) of blood (plasma, serum) which is cleared of a compound per minute by renal elimination
collecting duct	part of the nephron (Fig. 2); in the original definition of the nephron the collecting duct is excluded
cortex	outer part of the kidney (Fig. 1)
diastole	period of dilation of the heart, especially of the ventricles
diuretic profile	the pattern of urinary volume and electrolyte excretion
glomerulus	the filtering unit of the nephron consisting of Bowman's Capsule and the glomerular capillary network (Fig. 2)
glomerular filtration rate (GFR)	the quantity of glomerular filtrate formed each min in all nephrons of both kidneys
gynecomastia	excessive development of the male mammary glands, even to the functional state
hydropenia	water deficiency
hypercalcemia	serum calcium levels above normal; normal range in humans is 4.5–5.5 mEq/L
hyperkalemia	serum potassium levels above normal; normal range in humans is 3.5–5.0 mEq/L
hypernatremia	serum sodium levels above normal; normal range in humans is 136–145 mEq/L
hypertonic	tonicity (osmolarity) more than isotonic; normal range in humans is 285–295 $mOsm$/kg serum water
hyperuricemia	serum uric acid levels above normal; normal range is 2.5–8.0 mg/dL for men and 1.5–7.0 mg/dL for women
hypocalcemia	serum calcium levels below normal

hypoglycemia	an abnormally diminished concentration of glucose in the blood; normal range in humans is 70–115 mg/dL
hypokalemia	serum potassium levels below normal
hyponatremia	serum sodium levels below normal
hypotonic	tonicity (osmolarity) less than isotonic
interstitium	the interspace of a tissue, eg, of the kidney
isotonic	having the same tonicity (osmolarity); usually in comparison with plasma or other body fluids
juxtamedullary	close to the medulla of the kidney
kaliuresis	excretion of potassium in the urine
loop of Henle	part of the nephron between the proximal convoluted tubule and the distal convoluted tubule; divided into a descending limb and an ascending limb (Fig. 2)
medulla	inner part of the kidney (Fig. 1)
nephrogenic diabetes insipidus	a rare congenital and familial form of diabetes insipidus, resulting from the failure of the renal tubules to reabsorb water; there is excessive production of ADH, but the renal tubules fail to respond to it
nephron	the functioning unit of the kidney
osmolarity	the concentration of a solution expressed as osmols (mol wt of a solute in g, divided by the number of ions into which it dissociates in solution) of solute per L of solution
paracentesis	the passage into a cavity of a hollow instrument for the purpose of removing fluid
parenteral	introduced by any other route than by way of the digestive tract
passive transport	the movement of materials into and out of cells and across epithelial layers that is dependent on concentration gradients and not on expenditure of energy
plasma volume	total blood volume minus blood cells
pleural effusion	the presence of fluid in the pleural space
preload	the amount of tension on the cardiac muscle when it begins to contract; the volume of blood in the ventricle at the time of diastole; also known as central venous filling pressure
pulmonary compliance	the extent to which the lungs expand for each unit increase in transpulmonary pressure
systemic	relating to the body as a whole
tubule, distal convoluted	part of the nephron between the ascending limb of the loop of Henle and the collecting ducts (Fig. 2)
tubule, proximal convoluted	part of the nephron between the glomerulus and the descending limb of the loop of Henle (Fig. 2)
uremia	the retention of excessive by-products of protein metabolism in the blood, and the toxic condition produced thereby
uricosuric	increasing the urinary excretion of uric acid

BIBLIOGRAPHY

"Diuretic and Antidiuretics" in *ECT* 1st ed., Vol. 5, pp. 188–194, by E. Di Cyan, Di Cyan and Brown; "Diuretics" in *ECT* 2nd ed., Vol. 7, pp. 248–271 by G. deStevens, Ciba Pharmaceutical Co.; in *ECT* 3rd ed., Vol. 8, pp. 1–33, by P. W. Feit, Leo Pharmaceutical Products.

1. H. L. Cohn, Jr. and O. Horwitz, *Cardiac and Vascular Disease*, Vol. 1, Lea and Febiger, Philadelphia, 1972, p. 486.
2. C. K. Friedberg, *J. Am. Med. Assoc.* **174,** 2129 (1960).
3. R. V. Ford and J. Bush, *J. Conn. Med.* **24,** 704 (1960).
4. I. S. Eskwith and co-workers, *Am. J. Cardiol.* **9,** 194 (1962).
5. A.M.A. Council on Drugs, *Drug Evaluations*, 1st ed., American Medical Association, Chicago, 1971, p. 43.
6. J. J. Sanders and co-workers, *N. Y. J. Med.* **65,** 762 (1965).
7. I. M. Weiner in A. G. Gilman and co-workers, eds., *The Pharmacological Basis of Therapeutics*, Pergamon Press, Inc., Elmsford, N.Y., 1990, p. 713.
8. A. C. Guyton, *Textbook of Medical Physiology*, 8th ed., W. B. Saunders Co., Philadelphia, 1991, p. 273.
9. N. D. Larkin and D. D. Fanestil in J. B. West, ed., *Physiological Basis of Medical Practice*, 12th ed., Williams & Wilkins Co., Baltimore, Md., 1991, p. 406.
10. E. J. Cragoe, Jr., *Diuretics, Chemistry, Pharmacology, and Medicine*, John Wiley & Sons, Inc., New York, 1983.
11. E. H. Blaine in Ref. 10, p. 19.
12. J. B. Hook and R. Z. Gussin in M. Antonaccio, ed., *Cardiovascular Pharmacology*, 2nd ed., Raven Press, New York, 1984, p. 35.
13. D. E. Hutcheon and J. C. Martinez, *J. Clin. Pharmacol.* **26,** 567 (1986).
14. H. Velázquez, *Renal Physiol.* **10,** 184 (1987).
15. B. J. Materson and M. Epstein in F. H. Messerli, ed., *Cardiovascular Drug Therapy*, W. B. Saunders Co., Philadelphia, Pa., 1990, p. 338.
16. P. R. Wilson and D. Kem in Ref. 15, p. 348.
17. L. Z. Benet and R. L. Williams in Ref. 7, pp. 1669 and 1684.
18. F. S. Caruso and co-workers, *Am. Heart J.* **106,** 212 (1983).
19. N. Glorioso and co-workers *Curr. Ther. Res.* **35,** 483 (1984).
20. F. V. Costa and co-workers, *Curr. Ther. Res.* **32,** 359 (1982).
21. P. Weidman and co-workers, *Curr. Med. Res. Opin. (Suppl. 3)* **8,** 123 (1983).
22. A. Gerber and co-workers, *Hypertension (Suppl. 2)* **7,** II-164 (1985).
23. A. Scalabraino and co-workers, *Curr. Ther. Res.* **35,** 17 (1984).
24. J. Parijs and co-workers, *Am. Heart J.* **85,** 22 (1973).
25. B. J. Materson and co-workers, *J. Hypertension (Suppl. 4)* **6,** S-751 (1988).
26. N. Ashraf and co-workers, *Am. J. Med.* **70,** 1163 (1981).
27. P. A. Gross and co-workers, *Kidney Int. (Suppl. 21)* **32,** S-67 (1987).
28. J. R. Seckl and D. B. Dunger, *Drugs* **44,** 216 (1992).
29. R. Greger and P. Wangemann, *Renal Physiol.* **10,** 174 (1987).
30. M. Wittner and co-workers, *Drugs (Suppl. 3)* **41,** 1 (1991).
31. R. Greger, *Physiol. Rev.* **65,** 760 (1985).
32. M. Epstein and B. J. Materson in Ref. 15, p. 318.
33. A. Whelton and P. K. Whelton in Ref. 15, p. 328.
34. J. M. Kitzen in A. Scriabine, ed., *New Drugs Annual: Cardiovascular Drugs*, Vol. 3, Raven Press, New York, 1985, p. 21.
35. D. C. Brater, *Drugs (Suppl. 3)* **41,** 14 (1991).
36. H. A. Friedel and M. M.-T. Buckley, *Drugs* **41,** 81 (1991).
37. M. Burg and L. Stoner, *Ann. Rev. Physiol.* **38,** 37 (1976).
38. M. Imai, *Eur. J. Pharmacol.* **41,** 409 (1977).

39. T. Higashio and co-workers, *J. Pharmacol. Exp. Ther.* **207,** 212 (1978).
40. D. C. Brater and co-workers, *J. Clin. Pharmacol.* **21,** 647 (1981).
41. T. W. Wilson and co-workers, *Hypertension* **4,** 634 (1982).
42. G. J. Quirk and co-workers, *Prostaglandin Leuko. Med.* **13,** 219–226 (1984).
43. W. Flamenbaum and R. Friedman, *Pharmacotherapy* **2,** 213 (1982).
44. J. D. Horisberger and G. Giebisch, *Renal Physiol.* **10,** 198 (1987).
45. P. Corvol and co-workers, *Kidney Int.* **20,** 1 (1981).
46. R. P. Ames, *Drugs* **32,** 260 (1986).
47. W. A. Baba and co-workers, *Clin. Sci.* **27,** 181 (1964).
48. G. Eknoyan in Ref. 15, Chapt. 28, p. 368.
49. *Pharma-projects* **C37,** 106 (May, 1992).
50. V. Papademetriou and co-workers, *Am. J. Cardiol.* **54,** 1015 (1984).
51. A. J. DeBold and co-workers, *Life Sci.* **28,** 89 (1981).
52. M. L. Zeidel, *Ann. Rev. Physiol.* **52,** 747 (1990).
53. A. M. Richards and co-workers, *Lancet* **i,** 545 (1985).
54. M. G. Nicholls and co-workers, *Endocrinol. Metab. Clin. North Am.* **16,** 199 (1987).
55. R. C. Cuneo and co-workers, *J. Clin. Endocrinol. Metab.* **63,** 943 (1986).
56. K. Atarashi and co-workers, *J. Clin. Invest.* **76,** 1807 (1985).
57. A. S. Hollister and T. Inagami, *Am. J. Hypertension* **4,** 850 (1991).
58. K. L. Goetz, *Am. J. Physiol.* **261,** F-921 (1991).
59. T. Sudoh and co-workers, *Nature* **332,** 78 (1988).
60. K. Nakao and co-workers, *Hypertension* **15,** 774 (1990).
61. T. Sudoh and co-workers, *Biochem. Biophys. Res. Commun.* **168,** 863 (1990).
62. G. Achilihu and co-workers, *J. Clin. Pharmacol.* **31,** 758 (1991).
63. A. A. Seymour and co-workers, *Hypertension* **14,** 87 (1989).
64. A. J. Trapani and co-workers, *J. Cardiovasc. Pharmacol.* **14,** 419 (1989).
65. E. J. Sybertz, *Clin. Nephrology* **36,** 187 (1991).
66. J. E. O'Connell and co-workers, *J. Hypertension* **10,** 271 (1992).
67. E. G. Bevan and co-workers, *J. Hypertension* **10,** 607 (1992).
68. F. Lang, *Renal Physiol.* **10,** 160 (1987).
69. T. H. Maren, *Physiol. Rev.* **47,** 595 (1967).
70. P. A. Preisig and co-workers, *Renal Physiol.* **10,** 136 (1987).
71. B. B. Fredholm in G. A. Spiller, ed., *Methylxanthine Beverages and Foods: Chemistry, Consumption and Health Effects,* Alan R. Liss, New York, 1984, p. 303.
72. A. Vogl, *Am. Heart J.* **39,** 881 (1950).
73. R. M. Hays in Ref. 7, Chapt. 29, p. 732.
74. F. A. László and co-workers, *Pharmacol. Rev.* **43,** 73 (1991); M. Manning and W. H. Sawyer, *J. Lab. Clin. Med.* **114,** 617 (1989).
75. P. S. Chan, *Fed. Proc.* **38,** 749 (1979).
76. *Pharma-projects* **C3Z,** 359 (May, 1992).
77. R. R. Ruffolo and co-workers, *Drug News Perspectives* **4**(4), 217 (1991).
78. N. Allison and co-workers in A. W. Cowley, Jr., and co-workers, eds., *Vasopressin: Cellular and Integrative Functions,* Raven Press, New York, 1988, p. 207; B. E. Ilson and co-workers, *Kidney Intl.* **37,** 583 (1990).
79. S. Chiba and M. Tsukada, *Jpn. J. Pharmacol.* **59,** 133 (1992).
80. Y. Yamamura and co-workers, *Br. J. Pharmacol.* **105,** 787 (1992).
81. W. M. Bennet, in J. H. Dirks and R. A. L. Sutton, eds., *Diuretics: Physiology, Pharmacology and Clinical Use,* W. B. Saunders Co., Philadelphia, Pa., 1986, p. 370.
82. B. J. Materson and M. Epstein in F. H. Messerli, ed., *Cardiovascular Drug Therapy,* W. B. Saunders Co., Philadelphia, Pa., 1990, p. 343.
83. Veterans Administration Cooperative Study Group on Antihypertensive Agents, *J. Am. Med. Assoc.* **248,** 2004 (1982).

84. U.S. Public Health Service Cooperative Study Group, *Circ. Res. (Suppl. I)* **40,** 1098 (1977).
85. The Hypertension Detection and Follow-up Program Cooperative Research Group, *Circulation* **70,** 996 (1984).
86. The Management Group, *Lancet* **i,** 1261 (1980).
87. A. Helgelund, *Am. J. Med.* **69,** 725 (1980).
88. The Multiple Risk Factor Intervention Trial Research Group, *Am. J. Med.* **35,** 1 (1985).
89. The Multiple Risk Factor Intervention Trial Research Group, *Am. J. Cardiol.* **35,** 16 (1985).
90. Veterans Administration Study Group on Antihypertensive Agents, *J. Am. Med. Assoc.* **248,** 2004 (1982).
91. J. Alcazar and co-workers, *Proc. Ninth Int. Soc. Hypertension* Abstr. No. 8, Mexico City, 1982.
92. R. W. Williams and co-workers, *J. Am. Coll. Cardiol.* **1,** 623 (1983).
93. A. Lant, *Drugs* **29,** 162 (1985).
94. C. Bengtssohn and co-workers, *Brit. Med. J.* **289,** 1495 (1984).
95. Medical Research Council Working Party on Mild to Moderate Hypertension, *Lancet,* **ii,** 539 (1981).
96. M. B. Murphy and co-workers, *Lancet,* **ii,** 1293 (1982).
97. J. B. Wyngarden and L. H. Smith, Jr., *Cecil Textbook of Medicine,* 17th ed., W. B. Saunders Co., Philadelphia, Pa., 1985; E. Perez-Stable and P. V. Corvalis, *Am. Heart J.* **106,** 245 (1983).
98. J. E. Caldwell, *Sports Med.* **4,** 290 (1987).
99. K. A. Conrad in G. A. Ewy and R. Bressler eds., *Cardiovascular Drugs and the Management of Heart Disease,* 2nd ed., Raven Press, New York, 1992, p. 89.
100. N. M. Kaplan in E. Braunwald, ed., *Heart Disease: A Textbook of Cardiovascular Medicine,* 4th ed., W. B. Saunders Co., Philadelphia, Pa., 1992, p. 852.
101. P. S. Chan and P. Cervoni in P. B. Goldberg and J. Roberts, eds., *CRC Handbook on Pharmacology of Aging,* CRC Press, Boca Raton, Fla., 1983, p. 51.
102. M. S. Pecker in J. H. Laragh and B. M. Brennen, eds., *Hypertension: Pathophysiology, Diagnosis, and Management,* Raven Press, New York, 1990, p. 2143.
103. R. W. Gifford in Ref. 15, p. 298.
104. F. B. Müller and J. H. Laragh in Ref. 102, p. 2107.
105. P. R. Wilson and D. C. Kem in Ref. 15, p. 348.
106. J. Conway in F. Gross, ed., *Antihypertensive Agents,* Springer-Verlag, Berlin, 1977, p. 477.
107. W. H. Birkenhäger, *J. Hypertension (Suppl. 2)* **8,** S3 (1990).
108. J. T. Edelson and co-workers, *J. Am. Med. Assoc.* **263,** 408 (1990).
109. R. Franco-Saenz and co-workers, *Am. J. Hypertension* **5,** 266 (1992).
110. A. M. Richards and co-workers, *Hypertension* **7,** 812 (1985).
111. T. W. Smith and co-workers in Ref. 100, Chapt. 17, p. 464.
112. M. Packer, *Lancet* **340,** 92 (1992).
113. N. K. Hollenberg in Ref. 15, Chapt. 22, p. 310.
114. D. Hantman and co-workers, *Ann. Intern. Med.* **78,** 870 (1973).
115. R. A. Kelly and W. E. Mitch in Ref. 15, Chapt. 20, p. 284.
116. T. Risler and co-workers, *Drugs (Suppl. 3)* **41,** 69 (1991).
117. P. Ginès and co-workers, *Drugs* **43,** 316 (1992).
118. Y. Laffi and co-workers, *Hepatology* **13,** 1101 (1991).
119. J. Moran, *Scrip: Hypertension Therapy, Research and Market Opportunities,* 2nd ed., PJB Publication, Richmond, UK, 1991, p. 139.

120. *Prospects: The Pharmaceutical Industry*, Vol. IV, no. 34, Health Forecasting, Inc., Glastonbury, Conn., 1992, pp. S20–S33.
121. *Medical Advertising News*, May, 1992, pp. S3 and S9.
122. *International Pharmaceutical Service*, CountyNatWestWoodMac, Data Base, Section 1, Monsanto U.S., Dec. 13, 1991.
123. *Investext*, Feb. 22, 1991, pp. 1–14.
124. J. Moran in Ref. 119, p. 123.
125. *Medical Advertising News*, May 15, 1990, p. 3.

PETER CERVONI
PETER S. CHAN
American Cyanamid Company

DNA.   See BIOPOLYMERS; GENETIC ENGINEERING; NUCLEIC ACIDS.

DOPAMINE.   See EPINEPHRINE AND NOREPINEPHRINE; NEUROREGULATORS; PYSCHOPHARMACOLOGICAL AGENTS.

DRIERS AND METALLIC SOAPS

Metal soaps as a class of compounds have been defined as the reaction products of alkaline, alkaline-earth, or transition metals with monobasic carboxylic acids containing 6–30 carbons. Commercially important metal soaps include those of aluminum, barium, cadmium, calcium, cobalt, copper, iron, lead, lithium, magnesium, manganese, potassium, nickel, zinc, and zirconium. Their solubility or solvation in a variety of organic solvents accounts for their many and varied uses. Significant application areas for metal soaps include lubricants and heat stabilizers in plastics as well as driers in paint (qv), varnishes, and printing inks. Other uses are as processing aids in rubber, fuel and lubricant additives, catalysts, gel thickeners, emulsifiers, water repellents, and fungicides.

The first use of metallic soaps as drying promoters was not actually recorded, but it is clear that improved drying characteristics in vegetable oil coatings became associated with some of the natural earth colors used as pigments and hence small quantities of these pigments were added, not for coloring purposes, but for improving the drying properties. Deliberate use was practiced during the early Egyptian civilization, or at least as early as 2000 BC. The useful materials were various compounds of lead, iron, and manganese, which had sufficient reactivity to form soaps with the fatty acids in vegetable oils. Deposits taken from the axle of a chariot dating back to 1400 BC contained quartz, iron, and a sufficient quantity of fat and lime to indicate the use of a lime soap as a lubricant. There is also evidence of the use of lead in combination with oils in mummification (1).

Early efforts to prepare metal soaps involved attempts to dissolve the natural materials in oils. By the latter part of the nineteenth century, substantial progress had been made in the preparation of fused resinates and linoleates of lead and manganese. The utility of cobalt as a drying catalyst was discovered close to the turn of the century, but the factors that led to its ultimate discovery are not recorded.

Whereas the addition of early metal soaps to a coating for the specific purpose of improving the drying performance did so, the compounds lacked uniformity of composition and therefore did not give predictable results. Even if all of the metal reacted with the acid to give an expected metal ion concentration, which seldom happened, the ions were subject to oxidation, which resulted in loss of solubility in the vehicle and therefore a loss of activity.

A significant advance in metal soap technology occurred in the 1920s with the preparation of the metal naphthenates. Naphthenic acids (qv) are not of precise composition, but rather are mixtures of acids isolated from petroleum. Because the mixture varies, so does acid number, or the combining equivalent of the acid, so that the metal content of the drier would not always be the same from lot to lot. The preparation of solvent solutions of these metal naphthenates gave materials that were easy to handle and allowed the metal content to be standardized. Naphthenates soon became the standard for the industry.

Octoates were the next drier development. Because these driers are produced from synthetic 2-ethylhexanoic acid, the chemical composition can be controlled and uniformity assured. Also, other synthetic acids, eg, isononanoic and neodecanoic, became available and are used for metal soap production. Compared to naphthenic acid, these synthetic acids have high acid values, are more uniform, lighter in color, and do not have its characteristic odor. It is also possible to produce metal soaps with much higher metal content by using synthetic acids.

More recently, so-called overbased driers with even higher metal contents have become available. These driers are made with combinations of monocarboxylic acids and carbon dioxide or polyfunctional acids.

Composition and Properties

Metal soaps are composed of a metal and acid portion supplied as solutions in solvent or oil. The general formula for a metal soap is $(RCOO)_x M$. In the case of neutral soaps, x equals the valence of the metal M. Acid soaps contain free acid (positive acid number) whereas neutral (normal) soaps contain no free acid (zero acid number); that is, the ratio of acid equivalents to metal equivalents is greater than one in the acid soap and equal to one in the neutral soap. Basic soap is characterized by a higher metal-to-acid equivalent ratio than the normal metal soap. Particular properties are obtained by adjusting the basicity.

Properties are furthermore determined by the nature of the organic acid, the type of metal and its concentration, the presence of solvent and additives, and the method of manufacture. Higher melting points are characteristics of soaps made of high molecular-weight, straight-chain, saturated fatty acids. Branched-chain unsaturated fatty acids form soaps with lower melting points. Table 1 lists the properties of some solid metal soaps.

The anion used to prepare the metal soap determines to a large extent whether it will meet fundamental requirements, which can be summed up as follows: solubility and stability in various kinds of vehicles (this excludes the use of short-chain acids); good storage stability; low viscosity, making handling the material easier; optimal catalytic effect; and best cost/performance ratio.

Table 1. Properties of Solid Metal Soaps[a]

Compound	CAS Registry Number	Total ash	Free fatty acid	Sp gr	Mp, °C	Color	Fineness, μm[b]
			Stearate				
Al	[637-12-7]	5.5–16.0	3.0–3.5	1.01	110–150	white	95–98% <74 (−200)
Ba	[6865-35-6]	19–28	0.5–1.0	1.23	dec	white	<44 (−325)
Ca	[1592-23-0]	8.8–10.6	0.5	1.12	145–160	white	<44 (−325)
Cd	[2223-93-0]	19.0	0.5	1.21	104	white	99% <74 (−200)
Co	[13586-84-0]	8.2	2.0	1.13	140	violet	99% <74 (−200)
Cu	[660-60-6]	14	1.0	1.10	112	blue-green	99% <44 (−325)
Fe	[2980-59-8], [555-36-2]	13	4.0	1.12	100	red-brown	99% <74 (−200)
Pb	[7428-48-0]	30.2–57.0	0.1–0.6	1.34–2.0	103–dec	white	<74 (−200)
Li	[4485-12-5]	2.5	nil	1.01	212	white	99% <149 (−100)
Mg	[557-04-0]	8.0	0.5	1.03	145	white	<44 (−325)
Mn	[10476-84-3]	12.5	1.0	1.22	110	light brown	99% <74 (−200)
Ni	[2223-95-2]	9.4	5.2	1.13	180	green	<44 (−325)
Sr	[10196-69-7]	17.5	0.5	1.03	155	white	<44 (−325)
Zn	[557-05-1]	13.5–17.7	0.5–0.9	1.09–1.11	120	white	<44 (−325)
			Octanoate				
Al	[6028-57-5]	15.7	3.0	1.03	dec	white	98% <74 (−200)
Li	[16577-52-0]	4.7	nil	1.01	dec	white	<149 (−100)
			Oleate				
Al	[688-37-9]	10.0	8.5	1.01	134	cream	85% <74 (−200)
			Palmitate				
Al	[555-35-1]	8.3	12.5	1.01	120	white	97% <74(−200)
Zn	[4991-47-3]	14.7	1.4	1.12	123	white	<44 (−325)

[a]Ref. 2.

[b]U.S. standard sieve in parentheses.

Manufacture

Metallic soaps are manufactured by one of three processes: a fusion process, a double decomposition or precipitate process, or a direct metal reaction (DMR). The choices of process and solvent depend on the metal, the desired form of the product, the desired purity, raw material availability, and cost.

Fusion Process.   In the fusion process, a metal oxide, carbonate, or hydroxide reacts with a carboxylic acid at temperatures up to 230°C. Water is split out and the resulting metal soap is solubilized in a hydrocarbon solvent because the metal soaps themselves are generally hard, glassy, and difficult to grind.

$$MO + 2\ RCOOH \longrightarrow M(OOCR)_2 + H_2O$$

or

$$M(OH)_2 + 2\ RCOOH \longrightarrow M(OOCR)_2 + 2\ H_2O$$

Double Decomposition.   In the double decomposition reaction, an inorganic metal salt such as a sulfate, chloride, acetate, or nitrate reacts with the sodium salt of the carboxylic acid in a hot aqueous solution. The metal soap precipitate is filtered, washed, dried, and milled.

$$NaOH + RCOOH \longrightarrow RCOONa + H_2O$$

and

$$2\ RCOONa + MCl_2 \longrightarrow M(OOCR)_2 + 2\ NaCl$$

Direct Metal Reaction.   The DMR process is carried out over a catalyst with fatty acids in a melted state or dissolved in hydrocarbons. The acid reacts directly with the metal, supplied in a finely divided state, producing the metal soap and in some cases hydrogen. Catalysts include water, aliphatic alcohols, and low molecular-weight organic acids.

$$2\ RCOOH + M \longrightarrow M(OOCR)_2 + H_2$$

The DMR process has no aqueous effluent, gives high purity products, and is less expensive. However, if hydrogen is produced, it has to be removed carefully and should not reach explosive limits. Not all metals are sufficiently reactive to be suitable for the DMR process.

Basic metallic soaps are prepared by overbasing. That is, a normal metallic soap is treated in the presence of excess metal, metal oxide, hydroxide, or various salts, with reactive species, such as CO_2 or SO_2, capable of forming covalent bonds with the metal. The resulting moiety generally contains metal–oxygen bonds free of carboxy groups. Overbasing can be carried out simultaneously with the general manufacturing techniques although the fusion and DMR processes are preferred; it can also be used as a post treatment.

Economic Aspects

Production figures for metal soaps are given in Tables 2–5.

Analysis

Metal Content. Two common analytical methods for determining metal content are by titration and by atomic absorption spectrophotometry (aas). The titra-

Table 2. U.S. Production[a,b] of Stearates as Metal Soaps, 10^3 t

Year	Al	Ba	Ca	Mg	Zn	Other	Total
1976[c]	1.10	0.45	20.64	2.51	10.09	1.64	36.59
1977	1.50	0.36	22.91	2.27	12.09	2.37	41.52
1978	1.50	0.32	24.66	2.94	12.11	1.25	42.78
1979	1.65	0.27	24.08	2.97	10.78	1.42	41.17
1980	0.55	0.30	19.94	2.48	7.80	2.53	33.60
1981	0.93	0.47	20.78	6.77	10.54	5.69	45.18
1982	0.83	0.61	26.14	7.71	8.13	1.33	44.75
1983	1.81	0.52	29.46	9.50	10.10	1.09	52.46
1984[d]	1.79	0.28	32.31	10.79	12.84	0.70	59.01
1985[e]	1.46	0.40	31.35	10.26	12.66	0.87	57.27
1986	0.63	0.92	26.62	2.99	11.82		
1987	0.83	0.76	34.97	11.60	15.15		63.40
1988	2.17	1.12	39.86	9.57	16.55	1.80	71.07

[a]Courtesy of Synthetic Organic Chemicals, U.S. Production and Sales, U.S. International Trade Commission.
[b]Less than 100 t of cadmium stearates produced in all years except 1986 for which production was 350 t.
[c]Also 160 t cobalt stearates.
[d]Also 250 t cobalt stearates.
[e]Also 230 t cobalt stearates.

Table 3. U.S. Production of Naphthenates as Metal Soaps, 10^3 t

Year	Ca	Co	Pb	Mg	Zn	Cu	Other	Total
1976	0.38	1.31	2.10	0.45	0.44	0.41	0.44	5.53
1977	0.51	1.70	2.22	0.70	0.60	0.58	1.29	7.60
1978	0.30	1.68	1.84	0.36	0.84	0.54	0.85	6.41
1979	0.37	1.41	2.92	0.39	0.87	0.73	1.26	7.95
1980	0.34	1.07	2.33		0.81	1.01	1.30	6.86
1981	0.22	0.97	1.80		0.61	0.16	1.25	5.01
1982	0.22	0.94		0.14			2.27	3.57
1983	0.31	1.23		0.27		1.45	2.93	6.19
1984	0.21	1.50		0.26		1.36		na
1985	0.22	1.64		0.28		1.60	2.54	6.28
1986	0.23	1.71		0.21		na	2.34	4.49
1987	na	1.41	na	na	na	1.67	2.07	5.15
1988	na	1.26	na	na	na	1.55	2.15	4.96

Table 4. U.S. Production[a] of 2-Ethylhexanoates as Metal Soaps, 10^3 t

Year	Ca	Co	Pb	Mn	Zn	Zr	Other	Total
1976	1.14	2.00	1.18	0.45	0.68	1.18	0.82	7.45
1977	0.90	2.00	0.86	0.63	0.71	1.21	1.07	7.38
1978[b]	1.00	2.22	0.97	0.71	0.44	1.26	1.10	7.80
1979	1.07	1.70	0.83	0.58	0.42	1.36	1.55	7.51
1980	0.86	1.33	0.84	0.40	0.36	1.22	1.67	6.68
1981	0.81	1.01	0.54	0.39	0.49	1.12	1.94	6.30
1982	0.73	1.26	0.40	0.37	0.30	1.07	1.57	5.74
1983[c]	1.01	2.46	0.42	0.50	0.28	1.69	2.05	8.75
1984[d]	0.92	2.29	0.49	0.49	0.48	1.49	2.84	9.44
1985[e]	0.97	2.00	0.43	0.49	0.44	1.60	2.71	9.47
1986	1.08	2.38	0.37	0.61	0.42	1.80	2.82	9.48
1987	1.14	2.28	0.38	0.44	0.66	1.95	0.46	7.31
1988	1.26	2.39		0.52	0.60	2.11	4.84	11.72

[a]Courtesy of Synthetic Organic Chemicals, U.S. Production and Sales, U.S. International Trade Commission.
[b]Also 100 t of the Ni soap.
[c]Also 340 t of the Ni soap.
[d]Also 440 t of the Ni soap.
[e]Also 830 t of the Ni soap.

Table 5. U.S. Production[a] of Metal Soaps, 10^3 t

Year	2-Ethyl hexanoates	Stearates	Naphthenates[b]	Tallates[c]	Oleates	Total[d]
1965	1.99	17.02	11.22	3.93	0.13	34.29
1970	na[d]	20.93	12.24	3.75	0.26	39.23
1973	5.16	34.34	9.87	3.38	0.46	53.21
1975	5.39	26.38	5.24	na	0.24	37.25
1977	7.38	41.51	7.57	1.31		57.78
1978	7.80	42.78	6.27	3.67		60.68
1980	6.68	33.61	6.85	2.77		49.91
1981	6.41	45.17	5.01			56.58
1982	5.73	44.75	3.56	1.00		55.04
1983	8.74	52.48	6.19	1.67	0.07	69.13
1984	9.44	59.01	na	1.06	0.17	73.01
1985	9.46	57.26	6.27			72.98
1986	9.25	na	4.49	1.48		na
1987	7.31	63.42	6.16	1.33	0.11	78.33
1988	11.72	71.09	4.96	0.85	0.10	88.72

[a]Courtesy of Synthetic Organic Chemicals, U.S. Production and Sales, U.S. International Trade Commission.
[b]Data include copper naphthenate in all years except 1986.
[c]Data include production of calcium, cobalt, lead, manganese, iron, copper, and other salts.
[d]Certain years (1955–1969, 1973) may include small quantities of palmitates, linoleates, and resinates in the reported total.

tion method is a complexiometric procedure utilizing the disodium salts of ethylenediaminetetraacetic acid (EDTA). The solvent, indicator, and titrating solution depend on the specific metal being analyzed. In aas the sample is heated to a high temperature in a flame. The flame dissociates the chemical bonds in the metal soap forming a cloud of individual atoms floating in the sample area. In this condition the atoms can absorb radiation. The absorption wavelength differs for each specific element.

Oxalate Acid Number. A metal soap solution is treated with a measured excess of organic acid. Potassium oxalate solution is added to precipitate the metal and the total sample is back-titrated with alkali to determine its acidity. Acidity is expressed in acid number units, equivalent to mg KOH per g. A neutral soap gives a zero acid number, an acidic soap solution a positive acid number, and a basic soap solution a negative acid number.

Water Determination. The sample is refluxed with toluene and the resultant toluene–water azeotrope is distilled into a gradual water-trap receiver (Dean and Stark apparatus). Here the water and toluene separate into two distinct layers, permitting the volume of water to be read and its percentage calculated.

Ash. After the sample is heated in a crucible over a hot plate to drive off volatile solvents and moisture, it is charred over a Bunsen burner and then transferred to a muffle furnace where final ignition is completed. The weight of the ash is determined and reported as a percentage of the weight of the original sample.

Nonvolatile. A weighed amount of sample is placed in a drying oven maintained at a temperature of 105°C. After three hours, the sample is removed from the oven, placed in a desiccator to cool, and reweighed. The weight of the residue, consisting of solids and nonvolatiles, is calculated as a percentage of the total sample weight taken for analysis.

Gardner Color. Color measurement is obtained by comparing the sample with 18 separately numbered Gardner color standards. These are conveniently mounted as glass disks on wheels. The entire apparatus, consisting of two wheels containing nine disks each, a case to enclose the wheels, and a slot in the case for the sample, is commercially known as the Hellige Comparator (3).

Health and Safety Factors

The hazards encountered in the manufacture, processing, handling, and use of metal soaps are largely associated with the inherent toxicity of the metals and solvents. In general, the acid portion of the metal soap is low in toxicity. Material Safety Data Sheets (MSDS) are available from the commercial suppliers of these metal soaps specifying the inherent hazards. The Hazardous Material Identification System (HMIS) rating for liquid metal soaps may be summarized as follows, where the hazard rating index rates 0–4 as minimal to extreme.

Factor		Rating
health	2	moderate toxicity, may be harmful if inhaled or absorbed
flammability	2	combustible, requires moderate heating to ignite flash point 38 to 93°C
reactivity	0	normally stable, does not react with water

Metal soaps may cause skin irritation or sensitization. They are harmful if swallowed or ingested, which could result in gastrointestinal irritation and vomiting. Inhalation of concentrated vapors can lead to headaches and incoordination.

Solid metal soaps, when finely divided, may present an explosion hazard and are capable of spontaneous combustion. Inhalation of the dust can cause eye and/or respiratory irritation, so they require adequate ventilation.

Uses

The principal applications for metal soaps are as heat stabilizers and lubricants (both internal and external) in plastics, and as driers in paint and printing inks.

METAL STEARATES

More than half the metal stearates produced in the United States are applied as lubricants and heat stabilizers (qv) in plastics, particularly in the processing of poly(vinyl chloride) (PVC) resins.

The versatility of PVC allows it to be processed by a number of techniques such as calendering, extrusion, injection molding, blow molding, vacuum, and press forming. However, PVC suffers from a disadvantage in that it has a tendency to degrade on heating or exposure to uv light. Since it is necessary to heat the polymer to relatively high temperatures in order to soften it during the processing operation, this could contribute a limitation to its uses. Heat degradation leads to embrittlement of the plastic and a loss in tensile strength. Incorporation of metallic stabilizers like barium cadmium soaps or calcium zinc soaps into the PVC matrix stabilizes the polymer against the degradation process.

Paper coating applications represent 20% of total demand for metal stearates. Calcium is preferred to aluminum, sodium, ammonium, and zinc stearates (4). Smaller volumes of metal stearates are used in cements (aluminum, calcium, ammonium); drugs, food, and cosmetics (aluminum, calcium, sodium, magnesium, zinc); rubber (zinc, magnesium); grease (aluminum, lithium, zinc); petroleum products (aluminum, lithium, lead, cobalt, manganese); textiles (aluminum, calcium, zinc); and waxes (aluminum, calcium). Their functions in these applications include water repellency, thickening, pigment suspension, gelation, anticaking, flatting, lubrication, and plasticizing (4) (Table 6).

METAL 2-ETHYLHEXANOATE (OCTOATES)

The principal applications of metal 2-ethylhexanoates are as paint driers. The drying process of coatings, which contains oxidatively drying vehicles such as air-drying alkyd resins (qv) or drying oils (qv), is characterized by solvent evaporation followed by chemical reactions. This leads to the transition of a liquid to a solid film. The chemical reactions result in oxidative cross-linking and polymerization and are of greatest significance for film formation. Oxidative cross-linking is accelerated and modified through the addition of driers.

Drier Mechanism.   Oxidative cross-linking may also be described as an autoxidation proceeding through four basic steps: induction, peroxide formation, peroxide decomposition, and polymerization (5). The metals used as driers are categorized as active or auxiliary. However, these categories are arbitrary and a

Table 6. Applications of Stearates

Function	Cosmetics and toiletries	Food and food packaging	Lacquers and varnishes	Metal working	Paints	Paper	Pharmaceuticals	Plastics
anticaking[a]	Ca, Zn, Mg	Ca, Mg						
antifoaming	Mg	Ca, Mg					Mg	
binding						Zn, Al		
corrosion resisting								Ca, Zn, Ba
dusting	Zn, Mg							
emulsifying	Ca, Zn, Al, Mg	Ca, Mg					Ca	
flatting			Ca, Zn, Al, Mg		Ca, Zn, Al, Mg			
gelling[b]	Al, Mg		Al		Al	Al	Al	Ca, Mg
hardening						Al		
lubrication[a] dry				Ca, Zn, Al				Ca, Zn, Mg, Ba
int/ext	Ca, Zn, Mg	Ca, Mg		Ca, Zn, Al, Mg, Ba		Ca	Ca, Zn, Mg	Ca, Zn, Ba
pigment suspension[b]	Mg	Ca, Zn, Mg	Ca, Zn, Al		Ca, Zn, Al			Ca, Zn, Al, Mg
release[a]	Zn						Mg	
sanding sealing		Ca, Zn	Zn		Zn			
vinyl stabilizing								Ca, Zn, Ba
viscosity[b]	Ca, Mg		Ca		Ca, Al	Ca, Al		
water repellent[b]	Ca, Zn							
wetting[b]					Al			

[a]Also, Ca and Zn function in powdered metals.
[b]Also, Al functions in printing inks.

considerable amount of overlap exists between them. Drier systems generally contain two or three metals but can contain as many as five or more metals to obtain the desired drying performance.

Active driers promote oxygen uptake, peroxide formation, and peroxide decomposition. At an elevated temperature several other metals display this catalytic activity but are ineffective at ambient temperature. Active driers include cobalt, manganese, iron, cerium, vanadium, and lead.

Auxiliary driers do not show catalytic activity themselves, but appear to enhance the activity of the active drier metals. It has been suggested that the auxiliary metals improve the solubility of the active drier metal, can alter the redox potential of the metal, or function through the formation of complexes with the primary drier. Auxiliary driers include barium, zirconium, calcium, bismuth, zinc, potassium, strontium, and lithium.

Cobalt. Without a doubt cobalt 2-ethylhexanoate [136-52-7] is the most important and most widely used drying metal soap. Cobalt is primarily an oxidation catalyst and as such acts as a surface or top drier. Cobalt is a transition metal which can exist in two valence states. Although it has a red-violet color, when used at the proper concentration it contributes very little color to clear varnishes or white pigmented systems. Used alone, it may have a tendency to cause surface wrinkling; therefore, to provide uniform drying, cobalt is generally used in combination with other metals, such as manganese, zirconium, lead, calcium, and combinations of these metals.

Manganese. Although generally less active than cobalt, manganese 2-ethylhexanoate [15956-58-8] is also an active drier. As an accelerator of the polymerization in baking finishes, manganese is usually more effective than cobalt. Manganese gives better low temperature drying performance than cobalt, and films containing manganese do not suffer from wrinkling under high humidity conditions, as do films with cobalt alone. It also has the advantage that it does not cause baked film to embrittle. As a result of its slower activity, systems prone to skinning can be improved with manganese. It is seldom used alone.

Iron. This is a specialty drier that is considered active at temperatures of about 130°C. For this reason iron 2-ethylhexanoate [19583-54-1] is used in bake coatings that require maximum hardness. The principal drawback of using iron driers is that iron contributes a characteristic brownish red color to the coating and should only be used in dark pigmented systems. It has been reported that iron aids the dispersion of carbon black pigment and reduces the tendency for orange peel film defects (6).

Cerium/Rare Earth. Cerium 2-ethylhexanoate [56797-01-4] and rare-earth driers promote polymerization and through dry. Like iron they are active at elevated temperature and, since they do not contribute to film discoloration, are recommended for white bake finishes and overprint varnishes where color is critical. Rare earths also find use at the other end of the temperature spectrum in coatings dried at low temperature and high humidity.

Lead. Lead 2-ethylhexanoate [16996-40-0] has traditionally been the most commonly used auxiliary drier because it gives excellent through dry in most oleoresinous coatings. It improves flexibility, toughness, and durability of the film. Better water and salt spray resistance is also noted. It also serves as a pigment dispersing and wetting agent, thereby assisting in the dispersion phase of formulating. Use of lead as a grinding aid reduces the tendency for loss of dry.

Because of the toxicity associated with lead compounds, governmental rulings have severely limited the use of lead drier in coatings. From a performance viewpoint the use of lead in aluminum paint will destroy the leafing characteristics of the film. Coatings containing lead that are exposed to sulfur fumes will discolor.

Calcium.   Calcium 2-ethylhexanoate [136-51-6] has little drying action itself but is very useful in combination with other metals. Calcium driers are also good as pigment wetting and dispersing agents. They help improve hardness and gloss and have been reported to prevent blooming and silking (1).

Zirconium.   Zirconium 2-ethylhexanoate [22464-99-9] is classified as an auxiliary drier and is the most widely used replacement for lead. Zirconium improves through dry mainly by the formation of coordination bonds. It has excellent color, a low tendency to yellow, and better durability compared to other auxiliary metals.

Zinc.   The primary function of zinc 2-ethylhexanoate [136-53-8] is to keep the film "open," thus permitting hardening throughout and preventing surface wrinkling. It is an excellent pigment wetting agent and therefore improves pigment dispersion and reduces loss of dry when included in the grind phase of manufacture.

Bismuth.   Bismuth 2-ethylhexanoate [72877-97-5] is an auxiliary drier that has been promoted for drying under adverse conditions. Like rare earths, in some coatings it is reported to give better results than zirconium at low temperature and high humidity.

Vanadium.   Vanadium differs from the other drier metals because its greatest stability is at the higher valence state. A considerable disadvantage is its propensity to stain the final film. Vanadium also seems to be particularly prone to loss of dry problems, which again limits its use.

Barium.   This drier has been used to some extent in Europe as a substitute for lead. Barium 2-ethylhexanoate [2457-01-4] has been under governmental scrutiny because of its fairly acute toxicity.

Potassium.   Potassium 2-ethylhexanoate [3164-85-0] functions best in conjunction with cobalt. Potassium strongly activates cobalt in aqueous coatings and in high solids paints based on low molecular-weight vehicles.

Strontium.   Strontium 2-ethylhexanoate [2457-02-5] is a candidate to replace lead in lead-free paints. It functions well under adverse weather conditions and promotes through drying. Outdoor performance of low pigmented coatings like stains may be effected negatively by using strontium instead of zirconium.

Lithium.   Lithium functions best with cobalt. Lithium 2-ethylhexanoate [15590-62-2] is mainly used as a replacement for lead and in low molecular-weight vehicle, high solids coatings. Lithium promotes through drying and hardness and reduces the sensitivity of high solids coatings to wrinkling.

Water-Borne Coatings.   Water-borne air-drying coatings have found a wide acceptance as a replacement for organic solvent coatings. Essentially the same mechanism for drying applies to both water-borne and solvent air-drying vehicles but the drying performance is quite different. Besides the solvent composition, the vehicle system is responsible for various drying deficiencies associated with water-borne coatings, such as slow initial dry time, loss of dry, poor through drying, and hardness (see COATINGS).

Water may hydrolyze the vehicle or the metal drier resulting in loss of dry. Water can also slow the oxygen uptake of the vehicle thereby slowing the autoxidation process. Water is considered a strong ligand and complexes metal ions such as cobalt. The resulting complex has a weaker oxidizing potential resulting in a reduction of the performance of cobalt as an autoxidation catalyst.

The vehicle system of water-borne coatings can be divided into two categories: water-emulsifiable vehicles and water-soluble vehicles. Water-emulsifiable vehicles contain emulsifiers that may act as plasticizers after film formation, affecting the hardness. Water-soluble vehicles usually contain a neutralizing amine, the primary purpose of which is to solubilize the resin. These amines can influence the drying properties as they tend to complex the metal drier, thus affecting the catalytic activity. Acceptable results are usually obtained with trialkylamines such as dimethylethanolamine, trimethylamine, and aminomethylpropanol (7).

The differences in composition between water-borne and solvent-borne airdrying paints necessitate change in driers and drier combinations. Since traditional driers are dissolved in mineral spirits, xylene, or other aliphatic/aromatic solvents, they are not readily dispersed in an aqueous system. If traditional driers are used, they must be dissolved in the vehicle before neutralization, which may result in a severe viscosity increase and processing problems.

Traditional driers tend to destabilize emulsified vehicles. The compatibility of these driers with such vehicles is improved by adding suitable surfactants and diluting the drier with cosolvent.

High Solids Coatings.　In the coatings industry, many states have or will be placing severe restrictions on the Volatile Organic Content (VOC) of solventbased enamels. To meet these requirements, a new generation of longer oil and lower molecular-weight alkyd resins are being offered. Although they dry by the same oxidative mechanism, they behave differently from conventional vehicles. The structural variation and lower molecular weight result in slower through dry and/or faster set times with the consequence being soft films and/or possible wrinkling. Other film irregularities such as discoloration or gloss reduction can also occur. Proper selection of the standard metal soaps plus the right balance between active and auxiliary driers is critical in obtaining an acceptable final film.

Loss of Dry.　When the initial dry time of a solvent-based coating becomes substantially longer after aging, it is said to lose dry. The primary cause of this problem has been identified as adsorption of the drier on the pigment surface. Pigments with large surface areas are the worst offenders. There are two common approaches to preventing the phenomenon. The first method is to include a sacrificial drier such as zinc or calcium in the grind as a dispersing agent. The idea is that the sacrificial drier will be preferentially adsorbed on the pigment surface leaving the active driers to function. Another way is to use a "feeder" drier, such as cobalt hydroxy naphthenate, which is not completely soluble but will disperse. Therefore it functions by reacting with the acidity in the vehicle, becoming more soluble as the paint ages, and thereby feeding soluble drier into the paint.

The main reason for loss of dry in water-borne paints is the hydrolysis of the metal soap. In the presence of water the drier is first hydrated. These hydrates are unstable and result in hydrolysis of the metal soap and subsequently the insolubility of the basic metal soap.

Other Octoate Uses. Metal octoates are also used as driers in printing inks. Another application of octoates includes the use of the aluminum salt to gel paint. Stannous, dibutyltin, and bismuth carboxylates find application as catalysts in polyurethane foam applications in order to obtain a reaction efficiency suitable for industrial production. In polyurethane foam manufacture the relative rate of polymerization and gas foaming reactions must be controlled so that the setting of the polymer coincides with the maximum expansion of the foam.

METAL NAPHTHENATES

Naphthenates of cobalt, manganese, calcium, copper, iron, zinc, and zirconium are used as driers in printing inks. Their use in coatings is declining as a result of the use of higher metal content synthetic driers and the overall trend to latex paint in architectural coatings.

Copper and zinc napthenates are wood preservatives and are of more interest now that there are governmental rules regarding the application of material containing creosote and use of pentachlorophenol. Copper soaps have a good history for protecting wood, textiles, and cordage from decay and mildew. For above ground use, brushing or spraying is adequate. Effective below ground use requires pressure treatment or hot and cold soak treatment. Copper soaps impart a characteristic green color to the treated substrate necessitating the use of zinc soaps when color must be avoided. Zinc soaps are not as effective as copper and must be used in higher amounts.

OTHER METAL SOAPS

Metal Linoleates and Resinates. The calcium and cobalt salts of linoleates and resinates are used chiefly as components of metallic driers used in printing inks. Copper linoleate [7721-15-5] is used in antifouling paints for marine use (4).

Metal Oleates, Palmitates, and Tallates. Oleate applications include use as driers, fungicides, lubricants, and waterproofing agents. Palmitates are consumed in pigment suspending, waterproofing, and rubber and plastics compounding. Tallate salts are used almost exclusively as paint and printing ink driers (8).

OTHER USES

In the cosmetic, toilet, and pharmaceutical industries, alkali and heavy-metal soaps are employed as ingredients of face powders, talcum powders, tablet formulations, and creams. When added to waxes in candle and crayon manufacture, metallic soaps increase the melting point and prevent softening and sagging at high temperatures. Metal soaps promote adhesion of rubber to steel as required in steel-belted radial tires and similar applications. Because of their hydrophobic nature, metallic soaps are good waterproofing agents.

With the increasing emphasis on energy conservation and environmental considerations, additives for fuels that can correct combustion-related problems have aroused considerable interest. Many commercial fuel additives are combinations of organometallics, dispersants, emulsifiers, and carrier solvents. The organometallic, often a metal soap, acts as a combustion catalyst, increasing efficiency with reduction of smoke, deposits, and corrosion.

BIBLIOGRAPHY

"Driers and Metallic Soaps," in *ECT* 1st ed., Vol. 5, pp. 195–206, by S. B. Elliott, Ferro Chemical Corp.; in *ECT* 2nd ed., Vol. 7, pp. 272–287, by G. C. Whitaker, Harshaw Chemical Co.; in *ECT* 3rd ed., Vol. 8, pp. 34–49, by F. J. Buono and M. L. Feldman, Tenneco Inc.

1. W. S. Stewart, in W. H. Madison, ed., *Paint Driers and Additives, Federation Series of Coatings Technology*, Federation of Societies for Paint Technology, Philadelphia, Pa., 1969, Unit 11, pp. 1–26.
2. *Stearate Product Specifications*, Tenneco Chemicals, Inc., Piscataway, N.J.; *Witco Metallic Stearates, Bulletin 55-4R-5-63*, Witco Chemical Corp., New York, May 1963.
3. ASTM D1544, American Society for Testing and Materials, Philadelphia, Pa., 1992.
4. A. Bujold, *Metal Soaps—United States Chemical Economics Handbook*, SRI International, Menlo Park, Calif., Jan. 1990, p. 674.400 A-P.
5. R. G. Middlemiss, *J. Water-borne Coat.*, (Nov. 1985).
6. C. Gardner, *J. Am. Oil Chem. Soc. XXXVI* **11,** 568 (1959).
7. J. H. Bieleman, *Farbe und Lack* **94,** 434 (June 1988).
8. Ref. 4, p. 647.400 A-P.

MARVIN LANDAU
Huls America Inc.

DRILLING FLUIDS. See PETROLEUM.

DRILLING MUDS. See PETROLEUM.

DRUG DELIVERY SYSTEMS

The realization of sensitive bioanalytical methods for measuring drug and metabolite concentrations in plasma and other biological fluids (see AUTOMATIC INSTRUMENTATION; BIOSENSORS) and the development of biocompatible polymers that can be tailor made with a wide range of predictable physical properties (see PROSTHETIC AND BIOMEDICAL DEVICES) have revolutionized the development of pharmaceuticals (qv). Such bioanalytical techniques permit the characterization of pharmacokinetics, ie, the fate of a drug in the plasma and body as a function of time. The pharmacokinetics of a drug encompass absorption from the physiological site, distribution to the various compartments of the body, metabolism (if any), and excretion from the body (ADME). Clearance is the rate of removal of a drug from the body and is the sum of all rates of clearance including metabolism, elimination, and excretion.

Biological responses that can reflect either drug efficacy or an adverse reaction may be related to a concentration of bound drug at the appropriate biological receptors. The relationship of this biological response to time is known as pharmacodynamics (qv). Any dependence of this biological response on the plasma level of a drug is a pharmacokinetic/pharmacodynamic (PK/PD) relationship. Upon constant input of drug, steady-state conditions may be achieved. At steady state there is an equilibrium between the bound drug on the receptor sites and the free drug in plasma, and the PK/PD relationships are greatly simplified. Under this simplified steady-state condition, plasma levels of drug, side effects, and efficacy in healthy subjects or in the patient population may be characterized and should resemble *in vitro* enzyme binding curves. From this characterization of steady-state PK/PD, the desired drug input profile may be specified to design an efficacious therapy with fewer side effects in the patient population. The range of plasma levels that provides efficacy and avoids side effects in most of the patient population is known as the therapeutic window or range. Design of a dosage form that inputs drug into the physiological system at a specified rate profile for the longest convenient dosing interval is the goal and definition of controlled release drug delivery (see also CONTROLLED RELEASE TECHNOLOGY, PHARMACEUTICALS).

Controlled release drug delivery applies polymer chemistry to design a well-characterized, reproducible dosage form that produces the desired drug delivery profile. This profile includes not only the dosage form itself, but absorption across the relevant biological membranes into the systemic circulation. Controlled release dosage form design centers around understanding the relationship between tailoring the physical or polymer chemistry, the release of drug from the dosage form, and membrane permeation; membrane permeation is often the central tenet. Controlled release drug delivery involves control of the time course or location of drug delivery. Much research focuses on either the localized delivery of drugs, or the site-specific delivery to specific organs, classes of cells, and physiological compartments.

The two most common temporal input profiles for drug delivery are zero order (constant release), and half order, ie, release that decreases with the square root of time. These two profiles correspond to diffusion through a membrane and desorption from a matrix, respectively (1,2). In practice, membrane systems have a period of constant release, ie, steady-state permeation, preceded by a period of either an increasing (time lag) or decreasing (burst) flux. This initial period may affect the time of appearance of a drug in plasma on the first dose, but may become insignificant upon multiple dosing.

Design of a controlled release dosage form requires sufficient knowledge of both the desired therapy to specify a target plasma level and the pharmacokinetics. The desired drug input rate from a zero order system may be calculated by:

$$\text{drug input rate} = (\text{target plasma level}) \times (\text{clearance}) \qquad (1)$$

where the clearance, ie, the rate of removal of drug from the body, is preferably obtained from intravenous pharmacokinetic data by (3,4):

$$\text{clearance} = \text{dose}/\text{AUC} \tag{2}$$

AUC is the area under the curve or the integral of the plasma levels from zero to infinite time. Conversely, equation 1 may be used to calculate input rates of drug that would produce steady-state plasma levels that correspond to the occurrence of minor or major side effects of the drug.

Permeation and Desorption Studies

To design a drug delivery system, release from the device and its component materials should be investigated, and absorption across the relevant biological membrane for the selected route of access into the body must be measured. The two most common experimental paradigms for evaluating drug absorption or materials for drug delivery devices are membrane permeation and desorption (or dissolution) (see MEMBRANE TECHNOLOGY). A gradient in chemical potential or activity is the driving force for diffusion (5). In both experiments, the temperature is controlled, the receiver solution is well-stirred, and the chemical activity in the receiver solution is maintained near zero concentration by either very large volumes, frequent replacement, or flow through diffusion cells.

For membrane transport experiments, the relevant membrane is sandwiched between two solutions: a donor typically at constant drug concentration, $C = C_0$, and a receiver at zero concentration, $C = 0$. The drug concentration in the receiver is monitored as a function of time and Q, the cumulative amount transported, has a linear asymptote with time where A is the area, J_S is the steady-state flux, t is the time, and t_L is the time lag.

$$Q = AJ_S(t - t_L) \tag{3}$$

This linear regime is achieved to within 90% of J_S by $2t_L$. For an ideal homogeneous membrane

$$J_S = \frac{DS}{l} \tag{4}$$

where D is the diffusion constant, S is the solubility in the membrane, and l is the membrane thickness. Moreover, for an ideal membrane, t_L is a time lag when $C = 0$ initially ($t = 0$) in the membrane:

$$t_L = l^2/6D \tag{5}$$

On the other hand, if the membrane is initially full of drug, ie, at $t = 0$, $C = C_0$ in the membrane, there is a burst time, t_B, that may be calculated from

$$t_B = -l^2/3D \tag{6}$$

The flux is related both to an equilibrium quantity, ie, the chemical activity, through the solubility; and to a nonequilibrium coefficient D. In contrast, for an ideal membrane t_L emphasizes the path length and the diffusion constant. Binding and heterogeneity of the membrane may complicate these simple relationships.

In the other common paradigm, desorption from a system or a material is studied. At $t = 0$, the system is loaded with a known concentration or activity of drug and is immersed in an infinite, well-stirred receiver solution maintained at concentration $C = 0$. The amount released per unit area Q is then measured over time. For a uniform slab of thickness l where the release of drug is from both sides, the fraction released Q/Q_0 is linear with the square root of time and is given by:

$$Q/Q_0 = \left(\frac{Dt}{\pi l^2}\right)^{1/2} \tag{7}$$

where $Q/Q_0 \leq 0.6$. A non-zero intercept reflects either surface excess or depletion of drug. Deviations from this behavior may reflect polymer swelling or non-Fickian behavior. Semiempirical expressions have been developed for these anomalous cases. The appearance of non-Fickian behavior, as for non-Newtonian behavior in studies of viscoelasticity, may most easily be interpreted in terms of a Deborah number De defined as (6):

$$De = \frac{\lambda D}{l^2} \tag{8}$$

where λ is the material characteristic time. For $De \ll 1$, diffusion of drug is the slower process, and classical Fickian behavior is observed. If $De \gg 1$, polymer relaxation and swelling predominate and once again there is classical Fickian behavior.

If $De \simeq 1$, non-Fickian behavior may appear, ie, a dependence other than linearity with the square root of time may be observed.

Physiological Routes for Drug Delivery

Design of a drug delivery device is dictated by the properties of the physiological barrier, the effective plasma levels, and the total dosage.

Oral. The oral route for drug delivery includes the gastrointestinal (GI) tract and the oral cavity including the buccal mucosa. The buccal mucosa is considered separately because of differences in the approach to drug delivery via this route.

The primary function of the GI tract is the digestion and absorption of food. Thus, drugs entering the GI tract are exposed to a wide range of pH values, from 1–2 in the stomach to 5.0–6.5 in the small intestine, as well as high levels of various enzymes involved in the digestion of proteins, fats, and carbohydrates. The absorptive surface area of the small intestine is greatly increased over that of a simple tube by the presence of mucosal folds, villi (finger-like infoldings of the intestinal wall), and microvilli (found on the luminal surface of the enterocyte), resulting in a total surface area estimated to be between 250 and 1000 m^2. The colon has a large predominantly anaerobic bacterial population, 10^{11} to 10^{12} per gram on a dry wt basis (7,8), that constitute 40 to 55% of fecal solids in subjects consuming an average Western diet (9,10). Bacteria may contribute to the metabolism of xenobiotics, including drugs, and have a wide variety of enzymatic activities.

The transit of a dosage form through the GI tract can have a profound influence on its performance. Total GI transit time is between 24 and 48 h on average. Recovery of oral osmotic dosage forms gave a median transit time of 27.4 h with a wide range of 5.1 to 58.3 hours (11). Gastric emptying time is affected by the physical state of the drug (liquid vs solid), the size of the dosage form, the presence of food, the emotional state of the patient, the presence of disease, and certain medications. Gastric emptying time in the fasted state in humans varies between 0.5 and 2 h, and drugs given as small pellets or single units appear to leave the stomach en masse (12). After feeding, drugs that are dispersed or given as small pellets (2 to 5 mm) tend to empty from the stomach with the meal, whereas larger dosage forms are retained by the stomach until the meal has emptied (13,14). The delay in gastric emptying appears to be a function of the caloric content of the meal (15). Gastric emptying is delayed during pregnancy, severe exercise (moderate exercise accelerates it), stress, and in the elderly. Gastric stasis is associated with certain diseases, including diabetes and migraine. Emptying is accelerated in duodenal ulcer disease and slowed in gastric ulcer disease. Several drugs and other agents also affect gastric emptying. Anticholinergics, antihistamines, tricyclic antidepressants, phenothiazines, and narcotic analgesics cause delay. Metoclopramide [*364-62-5*], domperidone [*57808-66-9*], cisapride [*81098-60-4*], anticholinesterase, sodium bicarbonate, and cigarette smoking accelerate emptying.

The transit time in the small intestine is 3–4 h, and is the least variable part of total transit (16). It is independent of the size and physical state of the drug dosage form or the presence of food, and is unaffected by certain disease states such as constipation, diarrhea, and ulcerative colitis that might reasonably be expected to influence transit (12,17). Pooling at the ileo-cecal junction for 4 to 12 h has been observed (18), as well as pooling at the hepatic flexure despite the absence of a sphincter. Colonic residence time is about 80% of total GI transit time, and on average is 10 to 20 h, although it is extremely variable. In humans, the colon is 1 to 1.5 m long and has a surface area of about 1300 cm^2. It lacks the villi present in the small intestine, but the decrease in surface area may be compensated for by an increase in residence time. The main physiological function of the colon is the reabsorption of water and ions, reducing the volume of ileal effluent entering the colon (1 to 1.5 L) to an average stool output of 100 to 150 g per day. Colonic transit is characterized by periods of quiescence interspersed with bursts of activity. Solutions and small particles pass through the colon more slowly than large capsules (19).

Absorption of drugs across the wall of the GI tract is primarily the result of passive diffusion. Absorption is believed to take place by partitioning of the drug from the aqueous GI environment into the lipoidal membrane, diffusion through the membrane, and partitioning into the blood and body fluids. Most drugs are weak acids or weak bases that exist in an equilibrium between the ionized and unionized forms. The neutral species exhibit greater oil solubility than the ionic forms, and thus absorption of the unionized form of the drug predominates. Drug absorption depends on the pH at the absorptive site, the pK_a of the drug, and its oil/water partition coefficient. Although the pH partition hypothesis (20) may provide a useful approximation for drug absorption, some drugs are extremely well absorbed even though they are ionized throughout the GI tract. Active transport systems exist for the absorption of hexoses, amino acids (qv), and di- and tripeptides (21,22); a few drugs are taken up by these systems, eg, baclofen [1134-47-0], some β-lactam antibiotics (see ANTIBIOTICS, -LACTAMS), and angiotensin converting enzyme (ACE) inhibitors. After passage across the enterocyte, drugs enter the mesenteric circulation and are transported to the liver via the portal vein.

Drugs, such as opiates, may undergo metabolism both in the intestinal wall and in the liver (first-pass metabolism). The metabolism may be extensive and considerably reduce the amount of drug reaching the systemic circulation. Alternatively, the metabolite may be metabolically active and contribute significantly to the action of the parent drug. Some compounds undergo enterohepatic circulation in which they are secreted into the GI tract in the bile and are subsequently reabsorbed. Enterohepatic circulation prolongs the half-life of a drug.

For those compounds absorbed from a small part of the intestine, the amount of drug absorbed can be increased by extending the residence time of the dosage form in the GI tract. The two basic approaches used are gastric flotation or retention devices, and bioadhesive delivery systems. Flotation devices may be designed that float on the surface of the stomach contents in a way similar to that of lipids in the meal. These devices are sufficiently large to be retained by the stomach for prolonged periods. After deflation, the device is small enough to pass safely through the pylorus. Bioadhesive polymers have thus far shown little success in humans, although some increase in transit time has been shown in animal studies (23,24).

Drug absorption from the colon has become the subject of much attention. The development of dosage forms that release drug for 16 to 24 h depends on the drug being absorbed from the colon, because the bulk of the delivery period may be spent there. Only those compounds that exhibit good colonic absorption are suitable for extended delivery dosage forms, eg, metoprolol [37350-58-6] (25,26). Protein and peptide drugs are more readily available since the advent of recombinant DNA technology, and the oral delivery of these compounds has become the holy grail of drug delivery. However, proteins (qv) present several challenges because of size, hydrophilicity, and susceptibility to hydrolysis and degradation by proteases. Compared to the upper GI tract, proteolytic activity is lower in the colon (27) and the residence time is longer, which has led to interest in the development of dosage forms targeted to the colon. However, it appears unlikely that significant absorption of proteins occurs from the colon in the absence of either protease inhibitors, absorption enhancers, or both.

Targeting drugs to the colon has followed two basic approaches, ie, delayed release and exploitation of the colonic flora. Delayed release generally relies upon enteric coating to ensure safe passage through the stomach, and a delay of 4 to 6 h before drug release. Enteric coating is a special polymeric coating, such as cellulose acetate phthalate [9004-38-0], that is resistant to gastric fluids at low (1 to 3) pH but dissolves upon exposure to the higher pH of the intestinal contents (pH 5 to 7). The delay capitalizes on the fairly reproducible small intestinal transit time to release drug in the desired region of the GI tract. The time delay may be modified to release the drug in different regions of the GI tract. The possible disadvantage of this approach is that some patients have transit times outside the normal range.

Exploitation of the colonic flora relies on bacterial enzymes to cleave specific bonds. These enzymes include azoreductases, which cleave sulfasalazine [599-79-1] to 5-acetylsalicylic acid [13110-96-8] and sulfapyridine [144-83-2] (28); glycosidases, for the cleavage of dexamethasone glucoside (29); and glucuronidases, which cleave narcotic antagonist glucuronides (30). In all these examples, the active drugs are administered as prodrugs, which are then cleaved to yield the active moiety. Another approach has been to design polymeric drug delivery systems which contain diazo linkages (31,32) that are cleaved in the colon. The drug may be loaded into the polymer or the polymer may be used to coat a more conventional dosage form. There are concerns that must be addressed with regard to this approach, ie, treatment with antibiotics may decrease both the total number of bacteria and the balance between them; some patient groups, particularly the elderly, may have bacterial overgrowth in the ileum and also in the stomach, the latter being related to achlorhydria (33); and the toxicology of polymeric dosage forms containing azo bonds must be evaluated very carefully.

Rectal. The rectal route for drug delivery is an extremely unpopular one in the United States, but may present advantages in certain situations. Enemas containing either steroids or 5-acetylsalicylic acid for the treatment of proctitis, ie, inflammation of the rectum, offer good therapy in inflammatory bowel disease. The rectal route may be used when gastric stasis or vomiting is present, making the oral route of drug delivery untenable, eg, ergotamine [113-15-5] for the treatment of migraine. The vascular drainage of the rectum may partially avoid first-pass metabolism which offers definite advantages for those drugs undergoing extensive metabolism.

Transdermal. The skin offers a formidable barrier to the entry of foreign compounds, including drugs, into the body, both in terms of a physical barrier and an immunological one. The principal barrier to drug diffusion lies in the outer few layers of the epidermis, the stratum corneum, which is 10–20 μm thick in humans and consists of sheets of keratinized epithelial cells joined by tight junctions. The remainder of the epidermis, which is about 100 μm thick in humans, consists of living cells that are metabolically active. A drug applied to the skin must therefore diffuse through the epidermis to reach the blood capillaries in the dermis for distribution to the systemic circulation. Blood supply to the skin can vary tremendously from 200 to 4000 mL/(m²·min) (34) as a result of its role in the control of body temperature. A fall in body temperature results in vasoconstriction and a rise in temperature in vasodilation. Drug delivery by the transdermal route avoids presystemic metabolism in the gastrointestinal tract or first-pass metabolism in the liver. The permeability of skin is low, which limits the usefulness of this route to highly permeable, potent compounds. Permeability varies somewhat with regions of the body. The greatest permeability is in the scrotum (35).

Generally, permeation is higher for low (<400) mol wt compounds that have adequate oil and water solubility. Highly lipophilic compounds penetrate easily through the stratum corneum, but a degree of hydrophilicity is necessary for penetration through more aqueous regions. Lipid–water partition coefficients have been correlated with permeation of compounds through skin. The use of absorption enhancers for transdermal delivery may be necessary as a result of the low permeability of a drug through skin. As of this writing ethanol is the only enhancer in use in a commercially available system, and the flux of estradiol and nitroglycerin is linearly correlated with the flux of ethanol (36,37). Other absorption enhancers such as 1-dodecylazacycloheptan-2-one [59227-89-3] (Lauroca-pram) (38,39), terpenes (40), oleic acid [112-80-1] (41), pyrrolidones (42), n-alkan-ols (43,44), and alkyl esters (45) are under evaluation.

Dermal irritation and sensitization are issues specific to the transdermal route of drug delivery and can result in the cessation of therapy. Irritation may be defined as a local, reversible inflammatory response of the skin to the application of an agent without the involvement of an immunological mechanism. Acute, primary irritation occurs in response to a single application of an agent; cumulative irritation occurs following repeated applications of an agent that does not induce primary irritation. Irritation is manifested as erythema and edema, and is assessed using a standardized scoring system (46). The degree of irritation has been correlated with pK_a for a series of acids and bases (47,48). Sensitization results from an immune response to an antigen, which may lead to an exaggerated response upon repeated exposure to the antigen. A well-known example of contact sensitization is the response to poison ivy. Irritation or sensitization may occur in response to either the drug or a component of the transdermal system. Careful testing of both active and placebo patches is needed.

Transdermal drug delivery is associated with a relatively long time lag before the onset of efficacy, and removal of the system is followed by a correspondingly extended fall in plasma concentration, which probably results from formation of a drug depot in the skin that dissipates slowly. The time lag is approximately 3 to 5 h for many drugs that have low binding in the skin (49–51), but may be considerably longer. In contrast, plasma drug levels may be obtained between 2 and 5 min by the oral, buccal, or nasal routes.

Despite the limitations imposed by the physiology of the skin, several marketed controlled release transdermal drug delivery systems are available in the United States; for example, scopolamine [51-34-3] for the treatment of motion sickness, nitroglycerin [55-63-0] for angina, estradiol [50-28-2] for the relief of postmenopausal symptoms and osteoporosis, clonidine [4205-90-7] for the treatment of hypertension, fentanyl [437-38-7] as an analgesic, and nicotine [54-11-5] as an aid to smoking cessation. These systems are designed to deliver drug for periods of one to seven days.

Buccal. The oral mucosa consists of several different types of mucosa. In humans, the gingiva and hard palate are keratinized squamous epithelium; the buccal and sublingual mucosa are nonkeratinized. The total oral surface area is about 100 cm^2, about 30 cm^2 of which is made up of the buccal mucosa. Buccal mucosa has a high blood flow of 20–30 mL/min for each 100 g of tissue (52,53) and good lymphatic drainage. Vascular drainage is directly into the systemic circulation, and thus first-pass metabolism is avoided. Buccal epithelium has an average thickness of 0.58 mm (54) and is penetrated by connective tissue papillae that may reach to within 0.1 mm of the surface (55,56). The epithelial thickness varies from 20 cell rows over the papillae to 40 to 50 cell rows throughout the rest of the tissue. The buccal mucosa is readily accessible to the patient for self-administration of drugs, as well as rapid removal of the dosage form should it be necessary.

The use of a bioadhesive, polymeric dosage form for sustained delivery raises questions about swallowing or aspirating the device. The surface area is small, and patient comfort should be addressed by designing a small (less than 2 cm^2), thin (less than 0.1 mm (4 mil) thick) device that conforms to the mucosal surface. The buccal route may prove useful for peptide or protein delivery because of the absence of protease activity in the saliva. However, the epithelium is relatively tight, based on its electrophysiological properties. An average conductance in the dog is 1 mS/cm^2 (57) as compared to conductances of about 27 and 10 mS/cm^2 in the small intestine and nasal mucosa, respectively (58,59); these may be classified as leaky epithelia. Absorption of proteins and peptides, which has been reviewed (60), is generally low and somewhat erratic. The judicious use of absorption enhancers may be necessary and can be accomplished in a very controlled manner in this area. The mouth is routinely exposed to a wide variety of agents of different pH and osmolarity and appears to be more robust than many other epithelia. Exposure to a wide range of pH values produced damage only at the extremes of pH 1, 2, and 14 (61).

Commercially available buccal or sublingual dosage forms include nitroglycerin for angina, buprenorphine [52485-79-7] for pain relief, ergotamine for the treatment of migraine, methyltestosterone [58-18-4] for hypogonadism, captopril [62571-86-2] for hypertensive emergencies, and nifedipine [21829-25-4] for hypertensive emergencies and acute angina. Nicotine gum is available as a smoking cessation aid. Absorption is predominantly from the oral cavity, with a minor contribution from intestinal absorption of swallowed drug (62). These dosage forms are essentially tablets that dissolve rapidly over a few minutes. An alternative approach is the use of a bioadhesive, polymeric system that would provide sustained drug delivery over an extended period of time. The use of a backing material that is impermeable to the drug and saliva directs the drug toward the mucosa and prevents drug loss because of swallowing. The feasibility of this approach has been demonstrated in clinical trials (63,64).

Nasal. The nasal passages serve several physiological functions, eg, filtration of particulates from inspired air, warming and humidification of air, and olfaction. The surface area of about 180 cm^2 in the adult (65) is lined with pseudostratified columnar epithelium, which forms the primary barrier to drug absorption. The nose has good vascular drainage and an estimated blood supply of 40 mL/min for each 100 g of tissue (66). The nasal cavity is obviously accessible, absorption is very rapid, and first-pass metabolism in the liver is avoided. A potential disadvantage is the rapid mucociliary clearance rate for removal of trapped particles from the nose. The estimated turnover rate is 15 minutes (67). Both the common cold and conditions such as allergic rhinitis can affect clearance as well as the extent of absorption (68,69). The nasal route is used primarily for topical delivery of drugs, generally in aerosol form, for the treatment of allergic rhinitis and cold/flu symptoms. This route may have utility for rapid delivery of proteins or peptides, ie, compounds which may require pulsatile rather than sustained delivery.

The permeability of the nasal mucosa is similar to that of the ileum, and it is therefore a leaky epithelium. The structural requirements for drug absorption from the nasal cavity have been analyzed (70). Examination of data for 24 compounds has shown that the nasal route is suitable for the efficient, rapid delivery of many drugs having mol wts <1000. Mean bioavailability is 70%, without the use of adjuvants. This limit may be extendable to compounds of at least 6000 mol wt using adjuvants. Many adjuvants, however, enhance absorption by disruption of the cells, eg, the degree of enhancement of nasal insulin absorption was positively correlated with the membrane lytic activity of a series of nonionic surfactants (71). Another approach is to prolong the residence time using bioadhesive agents such as methylcellulose [9004-67-5], carboxymethylcellulose [9004-42-6], hydroxypropyl cellulose [9005-18-9], and polyacrylic acid [9003-01-4] (72–74), or bioadhesive microspheres that also protect proteins from degradation (75).

The effects of drugs and adjuvants must be assessed, both in short-term administration and during chronic treatment. Local effects include changes in mucociliary clearance, cell damage, and irritation. Chronic erosion of the mucous membrane may lead to inflammation, hyperplasia, metaplasia, and deterioration of normal nasal function (76).

Pulmonary. The trachea and bronchi are lined with pseudostratified ciliated columnar epithelium, similar to that found in the nasal passages. The bronchi divide to give rise to bronchioles, the larger ones being lined with simple columnar ciliated epithelium and the smaller ones with simple cuboidal nonciliated epithelium. The goblet cell population also decreases, and clearance of mucus by the cilia of the respiratory tract is of lesser importance in the deep lung. The barrier to drug absorption in the alveolae is the thin alveolar–capillary barrier that consists of squamous epithelial cells. Drug delivery to the lung is primarily for local therapy, but pulmonary delivery may offer opportunities for systemic delivery of compounds, including vaccines (see VACCINE TECHNOLOGY)(77).

Two primary factors involved in pulmonary drug delivery are delivery of the drug to the desired region of the respiratory tract, and permeation through the epithelial barrier. Delivery of the drug is highly dependent on the particle size; larger (5–10 µm) particles are lost by impaction in the upper airways and small (<0.5 µm) particles are expired. The ideal particle size for reaching the lung appears to be 1–2 µm (77,78). The barrier properties of the epithelium are not well-defined and, as in many other epithelia, are not solely dependent on the mol wt of the compound. Difficulties may be encountered in calculating bioavailability by the pulmonary route as a result of failure to deliver drug to the appropriate region of the respiratory tract.

Ocular. Drug delivery to the eye presents several challenges based on anatomy and physiology. The eye is isolated from the rest of the body by blood–eye barriers that include the retinal pigment layer, the ciliary epithelium which provides a barrier to proteins and antibiotics, and the thick walls of the blood vessels in the iris. The sclera and cornea provide physical protection to the eye. The cornea has an outer epithelial layer about five cells thick, an aqueous layer, and an inner endothelium. Drugs therefore have to cross two lipid layers and an aqueous layer to enter the eye, and compounds such as acetazolamide [59-66-5] that are readily absorbed elsewhere cannot effectively cross the corneal barrier. The epithelium is rate-limiting for most drugs; the aqueous region, ie, the stroma, is rate-limiting for very lipophilic drugs.

The eye is highly innervated, and patient comfort is of paramount importance in order to achieve good compliance. The eye is designed to keep the surface free of foreign bodies by blinking, tear production, and rapid drainage into the nasolacrimal duct. The average volume of tears is 7 μL, which is replaced at the rate of 16%/min except when sleeping or under anesthesia. The total precorneal volume is 20 μL, and excess solution applied to the eye is lost by spillage. Two approaches used to increase the residence time of drugs in the eye, and consequently the amount of drug absorbed, are increasing the viscosity of the solution and the use of an implant, such as Ocusert, or hydrogel contact lenses (qv) loaded with drug. Polymers that undergo a phase change from a liquid to a gel in response to temperature, pH, or ionic strength also show promise in this field.

Vaginal. The vaginal mucosa consists of stratified squamous epithelium, thrown into numerous transverse folds or rugae. The area is well supplied with both blood and lymphatic drainage. Changes in the human vaginal epithelium are considerably less pronounced during the estrous cycle than those observed in subhuman primates and many other animals. Although vaginal drug delivery is topical as of this writing, eg, for local infections and contraceptive purposes, this route may offer opportunities for systemic delivery for the treatment of diseases, such as osteoporosis, in which the patient population is predominantly female. In a study in the rat, the ovulation-inducing activity of leuprolide [53714-56-0] was compared after intravenous, subcutaneous, oral, rectal, nasal, and vaginal administration (79). Vaginal administration exhibited the greatest potency of all the nonparenteral routes studied.

Classes of Controlled Drug Delivery Systems

Controlled release drug delivery systems include those that are diffusion controlled, chemically controlled, swelling controlled, and externally controlled. Osmotically controlled systems are a subset of diffusion controlled systems and are often classified separately.

Drug Delivery Rates. Membrane devices, a class of diffusion-controlled systems, such as the form-fill-and-seal transdermal nitroglycerin systems (80) or osmotic pumps (81), present zero-order delivery of drug (eqs. 3 and 4) after an initial period (eqs. 5 and 6) and before depletion of the drug reservoir is in evidence, typically within 10–20% of the initial driving force. That is, a large excess of drug is typically present in a membrane system, and this may be impractical for expensive drugs. Dispersions of drug or control of pH may be used to maintain a constant chemical potential in an adequately large drug reservoir. Prior to onset of steady-state, a burst of drug (eq. 6) often results from migration of drug into the membrane during storage.

For monolithic systems that do not swell or erode, release of drug is dependent on diffusion of drug and may be classified by the drug loading relative to the solubility as dissolved or dispersed drug. For polymeric systems, this classification may be artificial, because metastable states can endure as a result of slow relaxation. Thus a nonequilibrium solution state may be maintained for a long time even at a drug loading above saturation. However, the stability of such metastable systems is difficult to control, and specifications on drug release are difficult to establish. Release from a planar dissolved monolith is linear with the square root of time (eq. 7). In a dispersed monolith, the drug is released from that portion of the matrix where the concentration of drug is below saturation, ie, there is a moving front where the concentration of dissolved drug is the solubility limit and meets the region with dispersed drug. If steady-state transport is assumed across the region of the matrix containing only dissolved drug, then it may be shown that for a slab where L is the drug loading (82):

$$Q = [S(2L - S)Dt]^{1/2} \tag{9}$$

A more accurate approximation may be obtained by (83):

$$Q = S(1 + H)(Dt/3H)^{1/2} \tag{10}$$

where

$$H = \frac{5L}{S} + \left[\left(\frac{L}{S} \right)^2 - 1 \right]^{1/2} - 4 \tag{11}$$

Both treatments assume that the dissolution of drug is rapid compared to the diffusion of drug, and both predict release of drug that is linear with $t^{1/2}$.

In swelling-controlled systems, glassy hydrogels in particular, the release process is a combination of the diffusion of water into the system and drug from the system. Empirically, release from these systems may be expressed as (83):

$$Q/Q_o = \alpha t^n \tag{12}$$

where α is a constant and n characterizes the mechanism of transport. As stated earlier, except where the Deborah number is near unity, $n = 0.5$. The special case of n = 1, or zero-order delivery, from the swelling monolith is known as Case 2 diffusion, and there is a large body of literature interpreting this process (84,85).

Erodible devices may be designed to exhibit either physical or chemical erodibility. Physical erosion may be either heterogeneous, ie, only at the surface, or homogeneous, ie, throughout the bulk. In cases of physical erosion, the system maintains its shape, is reduced in size, and results in zero-order release for planar geometry. For more hydrophilic polymers, water permeates into the polymer and homogeneous erosion occurs. Whereas release of drug from systems undergoing bulk erosion is more difficult to predict, the release intuitively represents a product of both release from a dispersed matrix that is proportional to $t^{1/2}$ and bulk erosion that typically increases exponentially. Loss of integrity from bulk erosion may be a safety limitation if, for example, an implant must be removed.

Chemical erosion has been classified as typically involving three approaches to degradation (86), ie, cross-linked water soluble macromolecules; insoluble polymers with labile bonds such as poly(lactic acid), poly(glycolic acid), and poly (orthoesters); and conversion of side groups in insoluble polymers to become water soluble through protonation or hydrolysis. Drugs may also be covalently bound to polymers, ie, polymer–drug conjugates, and liberated by either hydrolysis or enzymatic cleavage from the polymeric backbone.

Diffusion Controlled. When two materials are in contact and contain components at different activities, the components diffuse in such a way as to equalize the activities. This driving force provides a method of controlling drug delivery systems. The flux of the drug mass is defined by Fick's First Law (87) which states that the drug flux is equivalent to the concentration gradient in the system multiplied by the diffusivity of the drug in the medium. This expression incorporates the driving force of concentration difference and the drug's diffusivity, ie, mobility in the medium. A reservoir system and a monolithic system are the two basic types of drug delivery systems designed from this principle. The reservoir system separates the drug compartment from the polymer membrane which presents a diffusional barrier to control the drug flux. The diffusion rate is controlled by the polymer properties, as well as by the solubility and diffusivity of the drug in the polymer. It is possible to achieve constant release of the drug through the polymer membrane if the driving force is kept constant, ie, keeping the difference in drug activity at the reservoir side of the polymer and the drug activity at the external side of the polymer constant.

A monolithic system is comprised of a polymer membrane with drug dissolved or dispersed in it. The drug diffuses toward the region of lower activity causing the release of the drug. It is difficult to achieve constant release from a system like this because the activity of the drug in the polymer is constantly decreasing as the drug is gradually released. The cumulative amount of drug released is proportional to the square root of time (88). Thus, the rate of drug release constantly decreases with time. Again, the rate of drug release is governed by the physical properties of the polymer, the physical properties of the drug, the geometry of the device (89), and the total drug loaded into the device.

Norplant System. The Norplant System is a long-term system for birth control (see CONTRACEPTIVES). Investigation of this type of system began in the early 1960s, and a patent on the technology was issued in 1966 (90).

The commercial system contains levonorgestrel implants, a set of six flexible closed capsules, each containing 36 mg of the progestin, levonorgestrel [797-63-7], mol wt 312.46. The rate-controlling membrane is made of Silastic (dimethylsiloxane/methylvinylsiloxane copolymer) which surrounds the drug reservoir. Each capsule is 2.4 mm in diameter and 34 mm in length. The capsules are inserted by a physician beneath the skin of the upper arm and are effective for a period of up to 5 yrs for continuous contraceptive protection. The capsules release the drug initially at 85 μg/d followed by a decline to about 50 μg/d by 9 months and to about 35 μg/d by 18 months followed by a further decline thereafter to about 30 μg/day. Clinical studies have shown plasma concentrations of 0.30 ng/mL over 5 yrs, but are highly variable as a function of individual metabolism and body weight (11). The typical failure rate with this device in the first year is 0.2% compared to 3% with oral contraceptives. The Norplant System in the first year of use is second only to male sterilization (0.15% failure rate) in minimizing typical failure rate (81). The capsules are removed by a physician at the end of the therapy or the five year period. This procedure requires an aseptic technique and a local anesthetic (81).

Implantable controlled release drug delivery systems, including Norplant and other contraceptive steroids in silicone-based devices, have been reviewed (91,92).

Ocusert Pilo-20 and Ocusert Pilo-40. The Ocusert ocular therapeutic system is an elliptically shaped device designed for the continuous release of pilocarpine [*92-13-7*], mol wt 208.25, after placement in the conjunctival cul-de-sac of the eye (93). The system is effective for one week and two strengths are available, Pilo-20 and Pilo-40. The Ocusert system is indicated for control of elevated intraocular pressure in pilocarpine-responsive patients (81). The rate controlling membrane is comprised of a hydrophobic copolymer of ethylene–vinyl acetate (EVA). The inner core reservoir consists of pilocarpine and alginic acid [*9005-52-1*], mol wt ~240,000, a thickening agent to aid in manufacture of the drug delivery system. The Pilo-40 also contains di(2-ethyl-hexyl)phthalate [*117-81-7*], mol wt 390.56, to

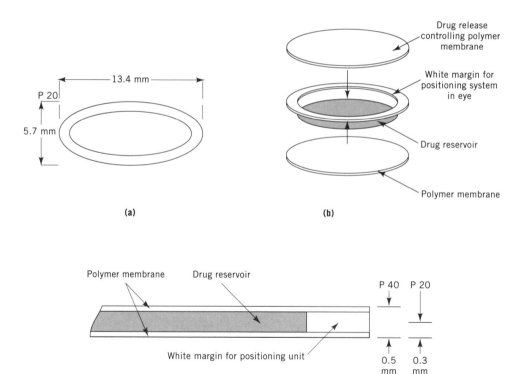

Fig. 1. Ocusert Pilo-20 and Ocusert Pilo-40 therapeutic systems. (**a**) Dimensions of the P20 system; P40 systems are 5.5 mm and 13.0 mm, respectively; (**b**) components of the system; and (**c**) cross-sectional view of the system (94). Courtesy of Georg Thieme Verlag, Stuttgart.

increase the rate of drug delivery across the EVA membrane. Figure 1 shows the dimensions of the systems. The indicated white margin of the system contains titanium dioxide [13463-67-7], TiO$_2$, for visibility and aids in positioning and retrieval of the system. Further information is available (81).

Figures 2 and 3 illustrate the constant release of pilocarpine over the seven day treatment period. An initial burst of drug into the eye is seen in the first few hours. This is temporary and the system drops to the rated value in approximately six hours. The total amount of drug released in this transitory period is less than that normally given in pilocarpine ophthalmic solutions. The ocular hypotensive effect of these devices is fully developed within 2 hours of placement in the conjunctival sac, and the hypotensive response is maintained throughout the therapy. This system replaces the need for eyedrops applied four times per day to control intraocular pressure.

Nitroglycerin Delivery Systems. Transderm-Nitro, Nitro-dur, and Minitran are all transdermal therapeutic systems that deliver nitroglycerin [55-63-0], mol wt 227.09, at a continuous, controlled rate through intact skin for treatment of angina (95).

Transderm-Nitro, marketed by Summit Pharmaceuticals, Division of Ciba-Geigy Corp., is a reservoir system having a rate-controlling membrane to regulate the diffusion of drug to the skin surface. Figure 4 shows a schematic of the device. There is a protective peel strip to prevent drug release prior to application, an adhesive to keep the system on the skin for the dosing period, a semipermeable rate-controlling ethylene–vinyl acetate membrane permeable to nitroglycerin, a reservoir that contains the drug, and a backing material impermeable to nitroglycerin to prevent loss of the drug to the environment.

The Nitro-dur system, marketed by Key Pharmaceuticals, Inc., contains nitroglycerin in acrylic-based polymeric adhesives with a cross-linking agent for polymer stability. This depot of drug provides a continuous source of active ingredient. An impermeable backing prevents release of nitroglycerin away from

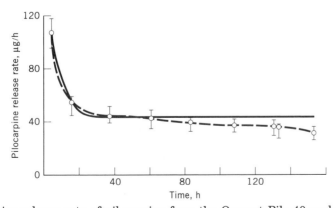

Fig. 2. *In vivo* release rate of pilocarpine from the Ocusert Pilo-40 ocular therapeutic system where (—) represents the calculated rate, and (– –), the experimental. Data points are shown with standard deviation error bars (94). Courtesy of Georg Thieme Verlag, Stuttgart.

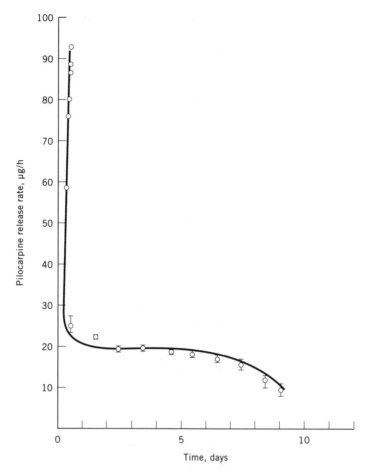

Fig. 3. Release rate of pilocarpine from Ocusert Pilo-20. Data points shown with standard deviation error bars (94). Courtesy of Georg Thieme Verlag, Stuttgart.

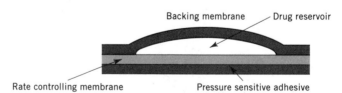

Fig. 4. Cross-section of a reservoir type of transdermal drug delivery system. The backing material and heat seals at the periphery of the system prevent loss of drug from the reservoir.

the skin. The systems are individually sealed in a paper polyethylene-foil pouch (81).

The rate of release of nitroglycerin in these two systems is linearly proportional to the surface area of the system, ie, 0.02 mg nitroglycerin are delivered per h per cm^2 area. The systems are available in doses of 0.1, 0.2, 0.4, and 0.6 mg nitroglycerin per h and the corresponding sizes are 5, 10, 20, and 30 cm^2 systems.

The Minitran system, by 3M Health Care, is a monolithic transdermal system that delivers nitroglycerin at a continuous rate of 0.03 mg/(cm^2·h) (81). The drug flux through the skin is higher than the previous two systems; thus the Minitran system is a smaller size for equivalent dosing. For example, the 0.1 mg/h dose is achieved with a 3.3 cm^2 system rather than the 5 cm^2 systems of Transderm-Nitro or Nitro-dur. Because the skin is rate-controlling in a monolithic system and the Minitran flux is higher than the similar monolithic Nitro-dur system flux, it appears that 3M Health Care has included an additive to increase the skin flux to 0.03 mg/(cm^2·h). Whereas this information is not apparent in Reference 81, patent information supports the hypothesis (96).

Each of the transdermal nitroglycerin systems is effective in treating angina pectoris when worn for 12–16 h followed by an off period. The FDA recommends the period without drug (8–12 h) to mitigate the possibility of the patient acquiring a tolerance to the antianginal effects of nitrate therapy. Thus, a noncontinuous dosing schedule of nitrates is recommended even with the transdermal delivery system (81).

Duragesic. Duragesic, developed by Alza and marketed by Janssen Pharmaceutical, is a form-fill-and-seal drug reservoir and an ethylene–vinyl acetate membrane designed to deliver transdermally the opioid painkiller, fentanyl [*437-38-7*], for up to 72 h (81) (see ANALGESICS, ANTIPYRITICS, AND ANTI-INFLAMMATORY AGENTS). The reservoir contains an aqueous ethanolic solution of fentanyl to increase its skin permeation yet limit its solubility to minimize the drug contents of this abusable drug (97). When a Duragesic patch is applied for a 24 h period, the plasma levels increase until the patch is removed and then remain elevated because of a skin depot through 72 h (98,99). Duragesic is available at mean delivery rates of 25, 50, 75, and 100 μg/h corresponding to surface areas of 10, 20, 30, and 40 cm^2, respectively (81). Whereas it is only approved for use in chronic pain (81), it has also been studied and found efficacious in treatment of post-operative pain (81,99–101).

Nicotine Delivery Systems. For all transdermal nicotine products, the hypothesis is that continuous delivery of nicotine [*54-11-5*] near trough levels during smoking should alleviate physical nicotine withdrawal symptoms and allow the smoker to concentrate on eliminating the behavioral aspects of addiction.

Habitrol (102), co-developed by Lohmann Therapie Systeme and Ciba-Geigy Ltd., and marketed by Ciba-Geigy, consists of an impermeable backing laminate with a layer of adhesive and a nonwoven pad to which a nicotine solution is applied (103). Multiple layers of adhesive on a release liner are then laminated on the patch. The systems come in 10, 20, and 30 cm^2 sizes corresponding to 7, 14, and 21 mg/day, respectively, delivered over 24 hours. 10 cm^2 systems are used only as a weaning dose (104).

Nicoderm (105), developed by Alza and marketed by Marion Merrell Dow Incorporated, is a multilaminate containing an impermeable backing layer, a nicotine-containing ethylene–vinyl acetate drug layer, a polyethylene membrane, and an adhesive layer. Because the system delivers nicotine for one day, delivery is controlled for a substantial time by the nicotine partitioned into the adhesive layer during storage as well as a period during which it is controlled by delivery through the membrane (106). It is available with delivery rates of 7, 14, and 21 mg/day over 24 hours.

ProStep, developed by Elan and marketed by Lederle, delivers nicotine over 24 hours (107) and is available in a 22 mg dose (108). The system consists of a nicotine-containing carrageenan gel embedded and heat-sealed into a foil–polyethylene laminate to isolate the nicotine from a peripheral acrylic adhesive during storage.

Nicotrol, developed by Cygnus Corp. and to be marketed by Warner-Lambert, is the fourth nicotine transdermal system to be approved in the United States. It delivers nicotine only during waking hours to mimic better the habits of smokers (109). In one controlled study, no statistical differences in efficacy and good tolerability were observed for the waking and continuous regimen (112).

Chemically Controlled.   These systems are classified together because of the hydrolysis or enzymatic cleavage of a chemical bond that allows delivery of the drug. There are two main types of systems, ie, pendent chain systems and bioerodible systems.

In the pendent chain systems, the drug is chemically bound to a polymer backbone and is released by hydrolytic or enzymatic cleavage of the chemical bond. The drug may be attached directly to the polymer or may be linked via a spacer group. The spacer group may be used to affect the rate of drug release and the hydrophilicity of the system. These systems allow very high drug loadings (over 80 wt %) (89) which decrease the cost of the polymeric materials used in the systems. These systems have been examined by many investigators (111,112).

In bioerodible systems the drug is distributed uniformly throughout a polymer in a manner similar to the matrix systems. However, the polymer phase changes with time through cleavage of bonds in the polymer. The cleavage of polymer bonds degrades the polymer, and consequently the polymer surrounding the drug erodes allowing the drug to diffuse from the polymer matrix. There are significant advantages in this type of system over nonerodible systems in many therapeutic areas because the biodegradable polymer is literally absorbed by the body and obviates the need for surgical removal of an implanted system. However, degradation must be uniform so that mechanical integrity is retained and the system may be removed for safety reasons.

The length of time for these products to come to the market is related to the ability to erode. Extensive study must be devoted to verify that each byproduct of the degradation of the polymer is nontoxic, nonimmunogenic, and noncarcinogenic. More specific information is available (86,113).

There are two commercially available bioerodible/biodegradable systems used in the palliative treatment of advanced prostate cancer. Both systems are implants effective for 28 days and provide an alternative treatment of prostate cancer when orchiectomy or estrogen administration are either not indicated or unacceptable to the patient. Each of these drug delivery systems uses D,L-lactic acid-glycolic acid copolymer [*34346-01-5*] to control the rate of drug release. The drugs differ in each system. Studies of D,L-lactic acid and glycolic acid copolymer have indicated that it is completely biodegradable and has no demonstrable antigenic effect. This copolymer family is also used in biodegradable suture material that has been on the market for some time (114).

Zoladex. Zoladex (goserelin acetate, mol wt 1269 for the free base form, implant) contains a synthetic analogue of luteinizing hormone-releasing hormone (LHRH) [*9034-40-6*]. Each dosage unit of Zoladex contains 3.6 mg of drug in a sterile biodegradable product designed for subcutaneous injection with continuous release over a 28-day period. The drug is dispersed in a matrix of D,L-lactic acid and glycolic acid copolymer (13.3–14.3 mg/dose) containing less than 2.5% acetic acid. The formulation is supplied in a special preloaded, single use syringe with a 16 gauge needle.

In clinical trials using the 3.6 mg formulation of Zoladex, peak concentrations in serum were achieved 12 to 15 days after subcutaneous administration (81). Goserelin [*65807-02-5*] was absorbed at a slow rate initially for the first 8 days, followed by more rapid and continuous absorption for the remainder of the 28 day dosing period.

Lupron Depot. Lupron depot contains leuprolide acetate [*74381-53-6*], mol wt 1182.33, which is a synthetic analogue of gonadotropin releasing hormone (115). Each dosage unit is available in a vial containing sterile lyophilized microspheres which, when mixed with diluent, becomes a suspension that is administered as a monthly intramuscular injection. The single dose vial contains 7.5 mg leuprolide acetate, 1.3 mg purified gelatin, 66.2 mg D,L-lactic and glycolic acids copolymer, and 13.2 mg D-mannitol. The accompanying diluent contains 7.5 mg sodium carboxymethylcellulose [*9004-32-4*], 75 mg D-mannitol [*69-65-8*], 1.5 mg polysorbate 80 [*9005-32-4*], and water for injection. One mL of diluent is added to the vial of microspheres for administration.

Studies of the pharmacokinetics of this delivery system in two animal models have been reported in the literature. After injection of these microspheres at three doses, leuprolide concentrations were sustained for over four weeks following an initial burst (116). The results indicated that linear pharmacokinetic profiles in absorption, distribution, metabolism, and excretion were achieved at doses of 3 to 15 mg/kg using the drug loaded microspheres in once-a-month repeated injections.

Swelling Controlled. The mechanism of swelling-controlled release of drugs is derived from the glassy/rubbery nature of polymers. A penetrant may lower the glass transition temperature of the polymer below the ambient conditions, thus changing it from the glassy state to its rubbery state. Some polymers when dry are in a glassy state, and as water migrates into the polymer, the polymer relaxes and swells. The swollen polymer is in a rubbery state and allows the drug to diffuse toward areas of lower activity. The rate of drug release from these systems has been investigated (88,117,118). The polymer must be chosen very carefully to achieve zero order release as this occurs only when the drug diffuses more rapidly than the swelling front, and the swelling front is moving with constant velocity (Case 2 diffusion) (89).

There are a few systems in industrial development that utilize the swelling mechanism. Ciba-Geigy has worked extensively with cross-linked poly (hydroxyethyl methacrylate) gels (119,120) in clinical trials. Many researchers are exploring the use of various hydrogels for swelling-controlled drug release. Polymers being investigated include: poly(2-hydroxyethyl methacrylate) [*25249-16-5*] (118, 121–124), poly(vinyl alcohol) [*9002-89-5*] (125–131), poly(ethylene–vinyl alcohol) (131), poly(ethylene oxide) [*25322-68-3*] (131,132), and cellulosic polymers [*9004-62-0*], [*9004-65-3*], [*9004-64-2*] (125,134,135).

Osmotic Control.   Several oral osmotic systems (OROS) have been developed by the Alza Corporation to allow controlled delivery of highly water-soluble drugs. The elementary osmotic pump (94) consists of an osmotic core containing drug surrounded by a semi-permeable membrane having a laser-drilled delivery orifice. The system looks like a conventional tablet, yet the outer layer allows only the diffusion of water into the core of the unit. The rate of water diffusion into the system is controlled by the membrane's permeability to water and by the osmotic activity of the core. Because the membrane does not expand as water is absorbed, the drug solution must leave the interior of the tablet through the small orifice at the same rate that water enters by osmosis. The osmotic driving force is constant until all of the drug is dissolved; thus, the osmotic system maintains a constant delivery rate of drug until the time of complete dissolution of the drug.

If the drug itself cannot provide the osmotic driving force, then a push-pull design of the osmotic system is available with other salts as the osmotic force. This system is schematized in Figure 5. The outer surface is a rigid semi-permeable membrane that surrounds the osmotic layer of salt (propellant). Inside the osmotic layer is a compressible membrane that surrounds the drug solution. As the salt layer swells with water, the inner membrane compresses and pushes out the drug solution.

The advantages of this type of system are that the release rates are independent of the drug properties, macromolecules and ionic species may be delivered, fluxes may be high, and release rates are not dependent upon environmental conditions such as pH. The disadvantages are that the system is subject to dose-dumping if it is chewed. It is also more expensive to formulate than coating tablets, and there is a possibility of hole plugging.

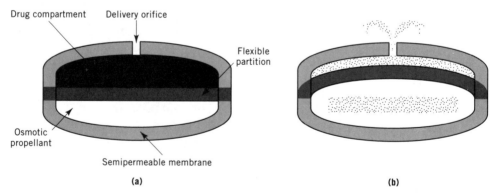

Fig. 5. (**a**) Cross-section of the push-pull OROS, which has an inner flexible partition to segregate the osmotic propellant from the drug compartment. (**b**) Push-pull OROS in operation with the propellant imbibing water, increasing in volume, and pushing the drug out of the device through the delivery orifice (94). Courtesy of Georg Thieme Verlag, Stuttgart.

Acutrim 16 Hour Steady Control Tablets. Acutrim is an appetite suppressant diet aid available without a prescription and marketed by CIBA Consumer Pharmaceuticals. The active ingredient is phenylpropanolamine hydrochloride [*154-41-6*], a sympathomimetic amine (see ANTIOBESITY DRUGS). Acutrim delivers its dosage at a precisely controlled rate for up to 16 hours. This is achieved through the OROS technology.

Using this dosage form design, the tablets contain a total of 75 mg active ingredient. The target plasma level for appetite suppression is 60 ng/mL. Based upon this information and pharmacokinetic data for the clearance of phenylpropanolamine, the OROS was designed to deliver 3.5 mg/h in addition to a 20 mg bolus dose of drug coated onto the exterior of the OROS system. The drug itself provides the osmotic force necessary to deliver the drug through the open orifice of the system. Thus, it is an elementary OROS design with an immediate-release bolus of drug. The bolus provides immediate release of the drug and allows the plasma levels to rise quickly and then be maintained with the constant release rate of drug from the system. Figure 6 shows a schematic of the OROS system with the exterior coating of drug. The plasma concentration of drug following Acutrim administration remains at or above 60 ng/mL for 16 hours. This differs greatly from the concentration profile of a normal capsule drug delivery system which has a high peak in plasma concentration followed by a rapid decline in the plasma concentration of the drug (81).

Procardia XL. Procardia XL extended-release capsules, marketed by Pfizer Labs Division of Pfizer, Inc., contain nifedipine [*21829-25-4*], a calcium channel blocker of mol wt 346.3. The extended release tablet is formulated as a once-a-day controlled release capsule for oral administration delivering either 30, 60, or 90 mg nifedipine. Procardia XL is indicated for use in the management of vasospastic angina, chronic stable angina, and hypertension (see CARDIOVASCULAR AGENTS).

The tablet consists of an oral osmotic system based upon the push-pull design (81). The osmotically active drug core is surrounded by a semipermeable membrane, possibly cellulose acetate [*9004-40-6*]. The core itself is divided into two regions, ie, a region containing drug and a push region containing pharmacologically inert but osmotically active components (Fig. 5). As water from the

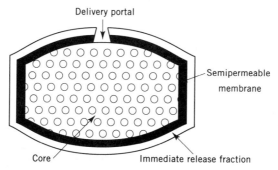

Fig. 6. Schematic representation of the elementary osmotic pump containing a reservoir of drug for controlled release and an immediately available overcoat of drug as designed for Acutrim.

gastrointestinal tract enters the tablet, pressure increases in the osmotic region and pushes against the drug region, releasing drug through the precision laser-drilled tablet orifice in the active region. Upon swallowing, the biologically inert components of the tablet remain intact during GI transit, and are eliminated in the feces as an insoluble shell.

Plasma drug concentrations rise at a gradual, controlled rate after dosing, and reach a plateau at approximately 6 h after the first dose with minimal fluctuations over the 24 h dosing interval. Subsequent doses maintain the plasma concentration at this plateau. The extended release tablets taken once daily have reduced by fourfold the fluctuations (ratio of peak to trough plasma concentration) observed with the conventional immediate release Procardia tablets taken three times daily (81).

External Control. The use of external control to govern the release of drugs from delivery systems has largely been experimental. A number of mechanisms have been explored, and include external sources such as electrical currents, magnetism, ultrasound, temperature changes, and irradiation. Each of these systems relies upon an external trigger to activate drug release, eg, iontophoretic devices (136,137) that depend on release of drug by an electric field or polymer-drug conjugates that are light-activated (138). The area of responsive polymeric delivery systems has been addressed and a general review of both externally regulated and self-regulated drug delivery systems is available (139). External systems must adhere to specific guidelines. The trigger should produce the desired drug delivery rate and should not produce any harmful effects including a permanent alteration to membrane permeability. There should be a quantitative relationship between the drug flux and the applied trigger for flux. Finally, the triggering device must be capable of maintaining the desired output trigger to control the drug flux.

An example of external control for drug delivery in clinical trials is iontophoresis, electrically controlled or enhanced transdermal drug delivery. Since the 1940s, electric stimulation has been applied for localized drug delivery to muscles and joints. Current and voltage conditions are used to trigger the drug release and the actual permeation rate of charged drugs across skin is enhanced by the electric current. Care must be taken to prevent charge build up, irritation, pain, or damage to the skin. The iontophoretic transdermal devices are similar to the current transdermal devices yet a small battery pack must be designed to enable commercialization.

Infusion Systems. There are many different infusion pumps, both external and implantable, available (140–143). Infusion devices deliver drug to a specific site in the body at constant or variable rates, depending on the needs of the patient, on a prolonged basis. Drugs have been delivered intravenously, intra-arterially, subcutaneously, and intrathecally/epidurally. Ambulatory infusion devices can reduce the length of a hospital stay and improve the patients' quality of life. The devices range from simple pumps capable of infusing drugs at a constant, nonadjustable rate to programmable pumps that can deliver constant infusions at variable rates with stimuli as needed, eg, in pain control.

The mechanisms that control drug delivery from pumps may be classified as vapor-pressure, electromechanical, or elastomeric. The vapor-pressure controlled implantable system depends on the principle that at a given temperature, a liquid in equilibrium with its vapor phase produces a constant pressure that is independent of the enclosing volume. The two-chamber system contains infusate in a flexible bellows-type reservoir and the liquid power source in a separate chamber (142). The vapor pressure compresses the drug reservoir causing drug release at a constant rate. Drug may be added to the reservoir percutaneously via a septum, compressing the fluid vapor into the liquid state.

Electromechanical pumps may be either the syringe or peristaltic variety. In the former, a prefilled syringe is depressed by a battery-powered electric motor and the pump may range in complexity from a simple to a programmable system. Peristaltic pumps may be linear or rotary and rely on the movement of an external force to propel drug along the tubing from the drug reservoir to the delivery site. The elastomeric device produces a constant flow rate based on the kinetic elastomeric properties of the balloon reservoir and does not require an external power source (141). The reservoir is coupled to a glass capillary to produce the constant flow rate. Dosage alterations are made by changing drug concentration, and the duration of the infusion depends on the fill volume.

Infusion devices have been used for diabetes, cancer chemotherapy, pain control (patient-controlled analgesia, ie, PCA), infection, Alzheimer's disease, Parkinson's, nausea, thalassemia, thromboembolism, and to treat severe spasms resulting from spinal cord injury (140–143).

System Characterization and Manufacturing

A drug delivery system is a vehicle that provides a stable environment to store an active ingredient prior to usage, and controls the release of the drug during usage. Typically, it is desirable for the system to be stable for a period of at least two years from the date of manufacturing. The type of release profile depends on the strategy to optimize the therapeutic effect, eg, sustained, constant, or specific temporal patterns.

Material Characterization. Drug delivery systems are made from polymeric materials. Chemical and physical characterization needs to be performed at each stage of development and manufacturing to prove the safety and efficacy of the final product. Chemical analyses are used to identify and quantify the compositions of the polymer, and the minor constituents. The material compositions can be determined by elemental analysis, nmr, and various spectroscopic methods such as Fourier transform ir (ftir) (see INFRARED AND RAMAN SPECTROSCOPY, INFRARED TECHNOLOGY; MAGNETIC SPIN RESONANCE). The size of the polymer is characterized by its average mol wt (MW), or by its mol wt distribution (MWD) to reflect the distribution of chain lengths formed during polymerization (144). The number-average mol wt (M_n) emphasizes the low mol wt species, and can be measured from the colligative properties of the polymer by osmometry, ebulliometry (boiling-point elevation), and cryoscopy (melting-point depression). The weight-average (M_w) and z-average (M_z) mol wts depict the population of higher molecular species. Equation 13 shows the relative magnitude of these average mol wts, in a polydisperse polymer where N_i is the number of polymer chains, each having a mol wt M_i (14).

$$\left[M_n = \frac{\Sigma N_i M_i}{\Sigma N_i} \right] < \left[M_w = \frac{\Sigma N_i M_i^2}{\Sigma N_i M_i} \right] < \left[M_z = \frac{\Sigma N_i M_i^3}{\Sigma N_i M_i^2} \right] \qquad (13)$$

Light scattering and gel-permeation chromatography (qv) can be used to measure the weight-average mol wt, whereas ultracentrifugation (qv) can provide a measure of the z-average mol wt.

The macroscopic properties of a polymer are strongly influenced by the intermolecular forces of the polymeric chains. The melting temperature and glass transition temperature are respectively first and second order transitions that depict the mobility of these chains. Polymers are amorphous above their melting temperature. However, highly crystalline polymers usually degrade at high temperature instead of melting because intermolecular forces can exceed covalent bond strengths. The segmental motions of the polymer chains decrease with decreasing temperature, and these motions are virtually frozen below the glass transition temperature. The melting and glass transition temperatures can be measured by differential scanning calorimetry (dsc) (see THERMAL, GRAVIMETRIC, AND VOLUMETRIC ANALYSES) (145). In addition, chemical reactions and degradation can be detected from a dsc. The chemical composition of the byproducts can be further analyzed by mass spectrometry (qv).

The solubility parameter of a polymer is a measure of its intermolecular forces, and provides an estimate of the compatibility of a polymer with another polymer or a polymer with a solvent. Two components are compatible if they have similar solubility parameters. The solubility parameter can be determined by various methods, such as intrinsic viscosity and swelling measurements. The solubility parameters of various polymers and solvents are tabulated in reference handbooks (146,147). It also can be estimated from the structure of the polymer (148).

The resistance of a polymer to deformation is described by its rheological properties. As such, the material characteristic time t_M and its relative magnitude to the time scale of deformation t_D are the key parameters that need to be evaluated. This dimensionless ratio of time scales (148) is the Deborah number (De $= t_M/t_D$), and is defined analogously to that for diffusion (eq. 8). When the Deborah number is small (De $<<1$), the polymer has ample time to react to an imposed deformation. Consequently, the response appears instantaneous, and this type of response is termed viscous. The rheological model can be greatly simplified to a Newtonian fluid model. At the other extreme of a very large Deborah number (De $>>1$), the opposite behavior is observed, and this response is classified as elastic. The corresponding rheological model is also simple, and can be described by a linear Hookean spring model. When the Deborah number is near unity (De ~ 1), viscous and elastic contributions coexist and produce unexpected behaviors, such as polymers climbing the shaft of a mixing rod. The rheological models become very complicated for describing these nonlinear responses (see RHEOLOGICAL MEASUREMENTS). For polymers in the solid state, the elongational and shear moduli can be determined in a tensile tester (149), which is operated at a constant rate or a constant force. Dynamic moduli can be measured in a dynamic mechanical analyzer that measures the material response to an oscillatory input. The material characteristic time can be calculated from these results.

Material Processing.   There are many types of drug delivery systems and the manufacturing of each system consists of fabricating a device as well as loading the drug and other excipients into this device. The drug can be introduced into various components of the system by physical and chemical means, and at various stages of system fabrication (150).

The first processing step is the handling of raw materials. These materials can be in the form of solids such as pellets, granules, and powders; fluids such as solvents and liquids of various viscosities, and polymeric solutions; and suspensions. All these materials need to be transported from storage units to the next processing station. Solid particulates are transported by gravitational flow from bins and hoppers (see CONVEYING), by fluidization using a gaseous carrier, or by means of a liquid slurry. Liquid media are commonly pumped through closed conduits.

Next, these materials are converted into a deformable state by melting, plasticizing, or forming a solution. Melting is the easiest and most convenient approach because it does not require the addition and subsequent removal of a second component. However, this approach is limited because of potential degradation of the polymer and of the active ingredients. Plasticization involves adding a low mol wt component to reduce the glass transition temperature of a polymer (see PLASTICIZERS). The net effect is to render the polymer more deformable, particularly at room temperature. The plasticizer may not have to be removed, depending on its interaction with the active ingredients, and the mechanical property requirement of the drug delivery system. The addition of a solvent to form a polymeric solution overcomes the problems of thermal degradation. The selection of the solvent can be facilitated by matching the solubility parameter of the solvent to that of the polymer. However, additional consideration should be given to the toxicity of the solvent, the interaction of the solvent with the active ingredients, and the removal of the solvent by volatilization or extraction.

When a material is in a deformable state, it can be shaped into desirable configurations such as films, discs, rods, or microspheres. The primary shaping processes are extrusion and molding. Extrusion is the positive displacement of a material through a die that is designed to produce a desired geometry in the extrudate. The extruder can be a simple piston that is driven by a hydraulic system, or it can be a set of twin screws used to deform, mix, and extrude. The design of the die is critical because the geometry of the extrudate can be significantly different. The morphological and mechanical properties of the product depend on the rate of solidification. For instance, structural gradients can result when the outer hardened shell impedes the heat or mass transfer of the inner layers. Nonequilibrium solidification, such as spinodal decomposition, has not been fully exploited to produce interesting transport properties in polymeric materials. When properly designed, extrusion can be used to make sophisticated drug delivery systems, such as multicomponent thin film composites.

Molding is a shaping operation where a mold is used to provide the desired geometry for the product. An injection molding machine is a versatile piece of equipment that can be used to deform and mix a polymer with other ingredients prior to pushing the mixture into a mold. Another molding operation is to cast a polymeric solution into a mold, or on a flat surface to form a film. The solvent can be removed by heat or by extraction with a nonsolvent. As an example, many monolithic drug delivery systems are made by solvent casting of an adhesive.

Drug loading can be accomplished by dispersion or adsorption. In dispersed systems, a drug is blended into a polymer by mechanical means, such as a kneader. The viscosity of the polymer, and the size and concentration of the drug, need to be optimized to minimize aggregates. Drugs can also be absorbed by equilibrating a polymer in a drug solution. The absorption rate can be accelerated by introducing an appropriate solvent to swell the polymer. All solvents would then have to be removed.

System Characterization. After all the components are assembled to produce a drug delivery system, extensive testing is performed to ensure that the performance and stability of the drug delivery system meet specifications.

The extent of chemical and physical interactions among the components of a drug delivery system are characterized. Changes in chemical composition can be detected by analytical methodologies. The drug formulation and the occurrence of byproducts need to be identified. Physical changes, such as swelling and delamination, also need to be identified so that corrective actions can be taken.

The performance of the drug delivery system needs to be characterized. The rate of drug release and the total amount of drug loaded into a drug delivery system can be determined in a dissolution apparatus or in a diffusion cell. Typically, the drug is released from the drug delivery system into a large volume of solvent, such as water or a buffer solution, that is maintained at constant temperature. The receiver solution is well stirred to provide sink conditions. Samples from the dissolution bath are assayed periodically. The cumulative amount released is then plotted vs time. The release rate is the slope of this curve. The total drug released is the value of the cumulative amount released that no longer changes with time.

In addition to transport properties, the adhesive properties are characterized by tensile measurements. For instance, the peel strength is determined by measuring the force required to pull the adhesive from a substrate at a constant speed in a controlled temperature and humidity environment.

BIBLIOGRAPHY

1. J. Crank, *The Mathematics of Diffusion*, 2nd ed., Clarendon Press, Oxford, 1975, p. 44.
2. H. S. Carslaw and J. C. Jaeger, *Conduction of Heat in Solids*, 2nd ed., Clarendon Press, Oxford, 1959, p. 99.
3. M. Gibaldi, *Biopharmaceutics and Clinical Pharmacokinetics*, 3rd ed., Lea & Febiger, Philadelphia, 1984, pp. 131–155.
4. J. G. Wagner, *Fundamentals of Clinical Pharmacokinetics*, Drug Intelligence Publications, Hamilton, Ill., 1975, p. 342.
5. A. Katchalsky and P. F. Curran, *Nonequilibrium Thermodynamics in Biophysics*, Harvard University Press, Cambridge, Mass., 1965, pp. 113–132.
6. J. S. Vrentas, C. M. Jarzebski, and J. L. Duda, *AIChE J.* **21**, 894–901 (1975).
7. W. E. C. Moore and L. V. Holdeman, *Appl. Microbiol.* **27**, 961–979 (1974).
8. S. M. Finegold and co-workers, *Am. J. Clin. Nutr.* **30**, 1781–1792 (1977).
9. A. M. Stephen and J. H. Cummings, *J. Med. Microbiol.* **13**, 45–56 (1980).

10. L. M. Cabotaje and co-workers, *Appl. Env. Microbiol.* **56,** 1786–1792 (1990).
11. V. A. John and co-workers, *Br. J. Clin. Pharmac.* **19,** 203S–206S (1985).
12. S. S. Davis, J. G. Hardy, and J. W. Fara, *Gut.* **27,** 886–892 (1986).
13. M. Feldman, H. J. Smith, and T. R. Simon, *Gastroenterology* **87,** 895–902 (1984).
14. J. H. Meyer and co-workers, *Gastroenterology* **88,** 1502 (1985).
15. S. S. Davis and co-workers, *Int. J. Pharm.* **21,** 331–340 (1984).
16. N. W. Read and co-workers, *Gut.* **27,** 300–308 (1986).
17. J. G. Hardy and co-workers, *Int. J. Pharm.* **48,** 79–82 (1988).
18. W. Fischer and co-workers, *Pharm. Res.* **4,** 480–485 (1987).
19. J. G. Hardy, C. G. Wilson, and E. Wood, *J. Pharm. Pharmacol.* **37,** 874–877 (1985).
20. C. A. M. Hogben and co-workers, *J. Pharmacol. Exp. Ther.* **120,** 540–545 (1957).
21. U. Hopfer, *Physiology of the Gastrointestinal Tract,* 2nd ed., Raven Press, New York, 1987, pp. 1499–1526.
22. D. M. Matthews, *Physiol. Rev.* **55,** 537–608 (1975).
23. D. Harris and co-workers, *J. Contr. Rel.* **12,** 45–53 (1990).
24. D. Harris and co-workers, *J. Contr. Rel.* **12,** 55–65 (1990).
25. J. Godbillon and co-workers, *Br. J. Clin. Pharmac.* **19,** 113S–118S (1985).
26. J. W. Fara, R. E. Myrback, and D. R. Swanson, *Br. J. Clin. Pharmac.* **19,** 91S–95S (1985).
27. S. A. W. Gibson and co-workers, *Appl. Env. Microbiol.* **55,** 679–683 (1989).
28. K. M. Das, *Gastroent. Clin. N. Am.* **18,** 1–20 (1989).
29. D. R. Friend and co-workers, *J. Pharm. Pharmacol.* **43,** 353–355 (1991).
30. J. W. Simpkins and co-workers, *J. Pharm. Exp. Ther.* **244,** 195–205 (1988).
31. M. Saffran and co-workers, *Science* **233,** 1081–1084 (1986).
32. P. Kopeckova and J. Kopecek, *Makromol. Chem.* **191,** 2037–2045 (1990).
33. S. H. Roberts, O. James, and E. H. Jarvis, *Lancet,* **2,** 1193–1195 (1977).
34. L. B. Rovell, *Physiology and Biophysics,* W. B. Saunders, Philadelphia, 1974, pp. 185–199.
35. R. J. Scheuplein and I. H. Blank, *Physiol. Rev.* **51,** 702–747 (1971).
36. U.S. Pat. 4,379,454 (1983), P. S. Campbell and S. K. Chandrasekaran (to ALZA Corp.).
37. B. Berner and co-workers, *J. Pharm. Sci.* **78,** 402–407 (1989).
38. R. B. Stoughton, *Arch. Dermatol.* **118,** 474–477 (1982).
39. R. B. Stoughton and W. O. McClure, *Drug Dev. Ind. Pharm.* **9,** 725–744 (1983).
40. A. C. Williams and B. W. Barry, *Int. J. Pharm.* **74,** 157–168 (1991).
41. M. Walker and J. Hadgraft, *Int. J. Pharm.* **71,** R1–R4 (1991).
42. H. Sasaki and co-workers, *J. Pharm. Pharmacol.* **42,** 196–199 (1990).
43. D. Friend and co-workers, *J. Contr. Rel.* **7,** 243–250 (1988).
44. T. Kai and co-workers, *J. Contr. Rel.* **12,** 103–112 (1990).
45. D. Friend, P. Catz, and J. Heller, *J. Contr. Rel.* **9,** 33–41 (1989).
46. J. H. Draize, G. Woodward, and H. O. Calvary, *J. Pharmacol. Exp. Ther.* **82,** 377–390 (1944).
47. B. Berner and co-workers, *Pharm. Res.* **5,** 660–663 (1988).
48. B. Berner and co-workers, *Fund. Appl. Tox.* **15,** 760–766 (1990).
49. K. Tojo, C. C. Chiang, and Y. W. Chien, *J. Pharm. Sci.* **76,** 123–126 (1987).
50. H. Okamoto, M. Hashida, and H. Sezaki, *J. Pharm. Sci.* **77,** 418–424 (1988).
51. E. Cooper, *J. Pharm. Sci.* **65,** 1396–1397 (1976).
52. D. Nanny and C. A. Squier, *J. Dent. Res., Abs.* 465 (1982).
53. J. Hoke and co-workers, *J. Dent. Res.* **68,** 237 (1989).
54. J. Meyer and S. J. Gerson, *Periodontics* **2,** 284–291 (1964).
55. A. J. Klein-Szanto and H. E. Schroeder, *J. Anat.* **123,** 93–103 (1977).
56. H. E. Schroeder, *Differentiation of Human Oral Stratified Epithelia,* Karger, Basel, 1981.

57. E. Quadros and co-workers, *J. Contr. Rel.* **19,** 77–86 (1991).
58. D. Fromm, *Am. J. Physiol.* **224,** 110–116 (1973).
59. M. A. Wheatley and co-workers, *Proceed. Intern. Symp. Control. Rel. Bioact. Mater.* **14,** 25–26 (1987).
60. H. P. Merkle and coworkers, *Peptide and Protein Drug Delivery*, Marcel Dekker, Inc., New York, 1990, pp. 545–578.
61. V. Place and co-workers, *Clin. Pharmacol. Ther.* **43,** 233–241 (1988).
62. N. L. Benowitz, P. Jacob, and C. Savanapridi, *Clin. Pharmacol. Ther.* **41,** 467–473 (1987).
63. J. Cassidy and co-workers, *Proceed. Intern. Symp. Control. Rel. Bioact. Mater.* **16,** 91–92 (1989).
64. H. P. Merkle and co-workers, *Delivery Systems for Peptide Drugs*, Plenum Publishing Corp., New York, 1986, pp. 159–175.
65. J. P. Schreider, *Toxicology of the Nasal Passages*, Hemisphere Publishing Corp., Washington, D.C., 1986, pp. 1–23.
66. M. Bende and co-workers, *Acta Otolaryngol.* **96,** 277–285 (1983).
67. G. S. M. J. E. Duchateau and co-workers, *Laryngoscope* **95,** 854–859 (1985).
68. S. W. Bond, J. G. Hardy, and C. G. Wilson, in J. M. Aiacha and J. Hirtz, eds., *Proceedings of the Second European Congress of Biopharmaceutics and Pharmacokinetics*, Salamanca, Lavoisier, Paris, France, 1984, pp. 93–98.
69. A. Agosti and G. Bertaccini, *Lancet* **1,** 580–581 (1969).
70. C. McMartin and co-workers, *J. Pharm. Sci.* **76,** 535–540 (1987).
71. S. Hirai, T. Yashiki, and H. Mima, *Int. J. Pharm.* **9,** 173–184 (1983).
72. L. S. Olanoff and R. E. Gibson, *Controlled-Release Technology, Pharmaceutical Applications*, American Chemical Society, Washington, D.C., 1987, pp. 301–309.
73. T. Nagai and co-workers, *J. Contr. Rel.* **1,** 15–22 (1984).
74. K. Morimoto, K. Morisaka, and A. Kamada, *J. Pharm. Pharmacol.* **37,** 134–136 (1985).
75. L. Illum and co-workers, *Int. J. Pharm.* **46,** 261–265 (1988).
76. W. A. Lee and J. P. Longnecker, *Biopharm.* **1,** 30–37 (1988).
77. D. T. O'Hagan and L. Illum, *Crit. Rev. Ther. Drug Carrier Systems* **7,** 35–97 (1990).
78. I. Gonda, *Crit. Rev. Ther. Drug Carrier Systems* **6,** 273–313 (1990).
79. H. Okada and co-workers, *J. Pharm. Sci.* **71,** 1367–1371 (1982).
80. W. R. Good, *Drug Dev. Indust. Pharm.* **9,** 647–670 (1983).
81. Edward R. Barnhart, *Physician's Desk Reference*, 45th ed., Medical Economics Co., Inc., Oradell, N.J.
82. T. Higuchi, *J. Pharm. Sci.* **50,** 874–875 (1961).
83. P. I. Lee, *J. Membrane Sci.* **7,** 255–275 (1980).
84. N. L. Thomas and A. H. Windle, *Polymer* **23,** 529–542 (1983).
85. H. B. Hopfenberg in D. R. Paul and F. W. Harris, eds., *Controlled Release Polymeric Formulations*, ACS Symp. Ser. No. 33, ACS, Washington, D.C., pp. 26–32.
86. J. Heller, *CRC Crit. Rev. Ther. Drug Carrier Sys.* **1,** 39–90 (1984).
87. R. B. Bird, W. E. Stewart, and E. N. Lightfoot, *Transport Phenomena*, John Wiley & Sons, Inc., 1964, p. 502.
88. N. Peppas in V. F. Smolen and L. A. Ball, eds., *Controlled Drug Bioavailability*, John Wiley & Sons, Inc., New York, 1984, pp. 203–237.
89. R. S. Langer and N. A. Peppas, *Biomaterials* **2,** 201–214 (1981).
90. U.S. Pat. 3,279,996 (1966), D. M. Long and M. J. Folkman.
91. H. A. Nash in R. S. Langer and D. L. Wise, eds., *Medical Applications of Controlled Release*, Vol. 2, CRC Press, Inc., Boca Raton, Fla., 1984, pp. 35–64.
92. Y. W. Chien, *Novel Drug Delivery Systems, Drugs and the Pharmaceutical Sciences*, Vol. 14, Marcel Dekker, Inc., New York, 1982, pp. 311–412.

93. K. Heilmann, *Therapeutic Systems: Rate-Controlled Drug Delivery, Concept and Development*, 2nd rev. ed., Thieme-Stratton, Inc., New York, 1984, pp. 66–82.

94. K. Heilmann, *Therapeutic Systems: Rate-Controlled Drug Delivery, Concept and Development*, 2nd rev. ed., Thieme-Stratton, Inc., New York, 1984.

95. Y. W. Chien, *Drug Del. Ind. Pharm.* **13,** 589–651 (1987).

96. U.S. Pat. 4751-087-A (1988) (Riker Industries, Inc.).

97. Belg. Pat. 0,905,568 (1987), R. M. Gale and co-workers.

98. P. M. Plezia and co-workers, *Pharmacotherapy* **9,** 2–9 (1989).

99. D. J. R. Duthie and co-workers, *Br. J. Anaesth.* **60,** 614–618 (1988).

100. F. O. Holley and C. Van Stennis, *Br. J. Anaesth.* **60,** 608–613 (1988).

101. R. A. Caplan and co-workers, *J. Am. Med. Assoc.* **261,** 1036–1039 (1990).

102. T. Abelin and co-workers, *Lancet*, 7–9 (1989).

103. Ger. Pat. 3,629,304 (1988), A. Hoffmann.

104. Package insert, Ciba-Geigy Corp., 1992.

105. Transdermal Nicotine Study Group, *J. Am. Med. Assoc.* **266,** 3133–3138 (1991).

106. Package insert, Marion Merrell Dow, 1992.

107. S. C. Mulligan and co-workers, *Clin. Pharmacol. Ther.* **47,** 331–337 (1990).

108. Package insert, Lederle, 1992.

109. *Scrip* (*World Pharmaceutical News*), Vol. 1677, PJB Publications Ltd., London, 1992, p. 23.

110. D. M. Daughton and co-workers, *Arch. Intern. Med.* **151,** 749–752 (1991).

111. R. V. Petersen and co-workers, *Polym. Prepr.* **20,** 20–33 (1979).

112. S. W. Kim, R. V. Petersen, and J. Feijen, in E. J. Ariens, ed., *Drug Design*, Vol. 10, Academic Press, New York, 1980, pp. 193–250.

113. R. Langer, *Science* **249,** 1527–1533 (1990).

114. E. J. Frazza and E. E. Schmitt, *Biomed. Mater. Symp.* **1,** 43–58 (1971).

115. M. Windholz and S. Budavari, eds., *The Merck Index: An Encyclopedia of Chemicals, Drugs, and Biologicals*, 10th ed., Merck & Co., Inc., Rahway, N.J., 1983.

116. H. Okada and co-workers, *Pharm. Res.* **8,** 787–791 (1991).

117. R. W. Korsmeyer and N. A. Peppas in T. J. Roseman and S. J. Mansdorf, eds., *Controlled Release Delivery Systems*, Marcel Dekker, Inc., New York, 1983.

118. S. S. Shah, M. G. Kulkarni, and R. A. Mashelkar, *J. Control. Rel.* **15,** 121–132 (1991).

119. W. R. Good in R. J. Kostelnik, ed., *Polymeric Delivery Systems*, Gordon and Breach Science Publishers, New York, 1978, pp. 139–153.

120. W. R. Good and K. F. Mueller, *Controlled Release of Bioactive Materials*, Academic Press, Inc., New York, 1980, pp. 155–175.

121. J. H. Kou, D. Fleisher, and G. L. Amidon, *J. Contr. Rel.* **12,** 241–250 (1990).

122. W. E. Roorda and co-workers, *Pharm. Weekbl. [Sci]* **8,** 165–189 (1986).

123. W. E. Roorda and co-workers, *Thermochimica Acta* **112,** 111–116 (1987).

124. S. W. Kim and co-workers, *ACS Symp. Ser.* **127,** 347–359 (1980).

125. U. Conte and co-workers, *Biomaterials* **9,** 489–493 (1988).

126. R. W. Korsmeyer and co-workers, *Intern. J. Pharmaceut.* **15,** 25–35 (1983).

127. C. T. Reinhart, R. W. Korsmeyer, and N. A. Peppas, *Int. J. Pharm. Tech. & Prod. Mfr.* **2,** 9–16 (1981).

128. F. Urushizaki and co-workers, *Intern. J. Pharm.* **58,** 135–142 (1990).

129. B. Gander and co-workers, *Int. J. Pharm.* **58,** 63–71 (1990).

130. P. Colombo and co-workers, *Acta Pharm. Technol.* **33,** 15–20 (1987).

131. H. B. Hopfenberg, A. Apicella, and D. E. Salesby, *J. Membr. Sci.* **8,** 273–282 (1981).

132. N. B. Graham in E. Piskin and A. S. Hoffman, eds., *Polymeric Biomaterials NATO ASi Series E: Applied Sci.*, Vol. 106, 1986, pp. 170–194.

133. N. B. Graham in N. A. Peppas, ed., *Hydrogels in Medicine and Pharmacy*, vol. 2, 1987, pp. 95–113.

134. K. V. R. Rao and K. P. Devi, *Int. J. Pharm.* **48,** 1–13 (1988).
135. S. K. Bajeva and K. V. R. Rao, *Int. J. Pharm.* **39,** 39–45 (1987).
136. A. K. Banga and Y. W. Chien, *J. Controlled Rel.* **7,** 1–24 (1988).
137. J. DeNuzzio and B. Berner, *J. Controlled Rel.* **11,** 105–112 (1990).
138. N. L. Krinick and co-workers, *Proc. Int. Symp. Controlled Rel. Bioact. Mater.* **16,** 138–139 (1989).
139. J. Kost and R. Langer, *Adv. Drug Del. Rev.* **6,** 19–30 (1991).
140. P. Lukacsko and G. S. May in P. Tyle, ed., *Drug Delivery Devices, Fundamentals and Applications*, Marcel Dekker, Inc., New York, 1988, pp. 177–211.
141. D. A. Winchell and J. A. Tune in Ref. 140, pp. 213–234.
142. T. D. Ronde, H. Buchwald, and P. J. Blackshear, in Ref. 140, pp. 235–260.
143. R. E. Fischell in Ref. 140, pp. 261–284.
144. P. J. Flory, *Principles of Polymer Chemistry*, Cornell University Press, Ithaca, N.Y., 1986, p. 266.
145. J. F. Rabek, *Experimental Methods in Polymer Chemistry*, John Wiley & Sons, Inc., New York, 1980.
146. A. F. M. Barton, *CRC Handbook of Polymer-Liquid Interaction Parameters and Solubility Parameters*, CRC Press, Boca Raton, Fla., 1990.
147. H. Burrell, in J. Brandrup, E. H. Immergut, and W. McDowell, eds., *Polymer Handbook*, 2nd ed., John Wiley & Sons, Inc., New York, 1975.
148. M. Reiner, *Physics Today*, 62 (1964).
149. J. L. Throne, *Plastics Process Engineering*, Marcel Dekker, Inc., New York, 1979.
150. Z. Tadmor and C. G. Gogos, *Principles of Polymer Processing*, John Wiley & Sons, Inc., New York, 1979.

General References

Y. W. Chien in J. Swarbrick, ed., *Drugs and the Pharmaceutical Sciences*, Vol. 14, Marcel Dekker, Inc., New York, 1982.
V. Lenaerts and R. Gurny, eds., *Bioadhesive Drug Delivery Systems*, CRC Press, Inc., Boca Raton, Fla., 1990.
M. Rosoff, ed., *Controlled Release of Drugs, Polymers and Aggregate Systems*, VCH Publishers, Inc., New York, 1989.
A. Kydonieus, *Treatise on Controlled Drug Delivery*, Marcel Dekker, Inc., New York, 1991.
K. Heilmann, *Therapeutic Systems: Rate-Controlled Drug Delivery, Concept and Development*, 2nd rev. ed., Thieme-Stratton, Inc., New York, 1984.
R. W. Duncan and L. W. Seymour, *Controlled-Release Technologies*, Elsevier Advanced Technology, Oxford, N.Y., 1989.

ELIZABETH QUADROS
ANN R. COMFORT
STEVEN M. DINH
BRET BERNER
Ciba-Geigy Corporation

DRUG DESIGN. See MOLECULAR MODELING.

DRYCLEANING AND LAUNDERING. See DETERGENCY;
SURFACTANTS.

DRYING

Drying is an operation in which volatile liquids are separated by vaporization from solids, slurries, and solutions to yield solid products. In dehydration, vegetable and animal materials are dried to less than their natural moisture contents, or water of crystallization is removed from hydrates. In freeze drying (lyophilization), wet material is cooled to freeze the liquid; vaporization occurs by sublimation. Gas drying is the separation of condensable vapors from noncondensable gases by cooling, adsorption (qv), or absorption (qv) (see also ADSORPTION, GAS SEPARATION). Evaporation (qv) differs from drying in that feed and product are both pumpable fluids.

Reasons for drying include user convenience, shipping cost reduction, product stabilization, removal of noxious or toxic volatiles, and waste recycling (qv) and disposal. Environmental factors, such as emission control and energy efficiency, increasingly influence equipment choices. Drying operations involving toxic, noxious, or flammable vapors employ gas-tight equipment combined with recirculating inert gas systems having integral dust collectors, vapor condensers, and gas reheaters (see EXHAUST CONTROL, INDUSTRIAL).

Drying is an applied science; ie, drying theory is based on the laws of physics, physical chemistry, and the principles underlying the transfer processes of chemical and mechanical engineering: heat (see HEAT EXCHANGE TECHNOLOGY, HEAT TRANSFER), mass and momentum transfer (see MASS TRANSFER), vaporization, sublimation, crystallization (qv), fluid mechanics, mixing (see MIXING AND BLENDING), and material handling. Drying is one of several unit operations involving simultaneous heat and mass transfer. However, drying is complicated by the presence of solids that interfere with heat, liquid, and vapor flow and retard the transfer processes, at least during the final drying stages or when a solids phase is continuous.

Because all drying operations involve processing of solids, equipment material handling capability is of primary importance. In fact, most industrial dryers are derived from material handling equipment designed to accommodate specific forms of solids. If possible, liquid separation from solids as liquid, by dewatering (qv) in a mechanical separation operation, should precede drying. Solids handling is made easier, and liquid separation without vaporization is less costly (see FILTRATION; SEDIMENTATION; SEPARATION, CENTRIFUGAL). Evaporators, which have lower investment and operating costs than dryers, are also used to minimize dryer loads.

Several methods are employed to classify commercial dryers by process application (1–3). Mode of heat transfer is the conventional choice. The principal heat-transfer mechanisms in drying are (1) convection from a hot gas that contacts the material, used in direct-heat or convection dryers; (2) conduction from a hot surface that contacts the material, used in indirect-heat or contact dryers; (3) radiation from a hot gas or hot surface that contacts or is within sight of the material, used in radiant-heat dryers; and (4) dielectric and microwave heating in high frequency electric fields that generate heat inside the wet material by molecular friction, used in dielectric, or radio frequency, and microwave dryers. In the last group, high internal vapor pressures develop and the temperature inside the material may be higher than at the surface (see MICROWAVE TECHNOLOGY). Many dryers effect more than one heat-transfer mechanism, but most dryers can be identified by the one that predominates.

In order of priority, the factors that govern the selection of industrial dryers are (1) personnel and environmental safety; (2) product moisture and quality attainment; (3) material handling capability; (4) versatility for accommodating process upsets; (5) heat- and mass-transfer efficiency; and (6) capital, labor, and energy costs.

Costs are determined by energy, labor, capacity, and equipment materials of construction. Continuous dryers are less expensive than batch dryers and drying costs rise significantly if plant size is less than 500 t/yr. Vacuum batch dryers are four times as expensive as atmospheric-pressure batch dryers and freeze dryers are five times as costly as vacuum batch dryers. Once-through air dryers are half as costly as recirculating inert-gas dryers. Per unit of liquid vaporization, freeze and microwave dryers are the most expensive. The cost difference between direct- and indirect-heat dryers is minimal because of the former's large dust recovery requirement. Drying costs for particulate solids at rates of $1 \times 10^3 - 50 \times 10^3$ t/yr are about the same for rotary, fluid-bed, and pneumatic conveyor dryers, although few applications are equally suitable for all three (4,5).

Terminology

Bound moisture is liquid held by a material that exerts a vapor pressure less than that of the pure liquid at the same temperature. Liquid can be bound by solution in cell or fiber walls, by homogeneous solution throughout the material, and by chemical or physical adsorption on solid surfaces.

Capillary flow is liquid flow through the pores, interstices, and over the surfaces of solids which is caused by liquid–solid molecular attraction and liquid surface tension.

Constant rate period is the drying period during which the liquid vaporization rate remains constant per unit of drying surface.

Critical moisture content is that obtained when the constant rate period ends and the falling rate periods begin. Second critical moisture content specifies that remaining in a porous material when capillary flow dominance is replaced by vapor diffusion.

Dry basis describes material moisture content as weight of moisture per unit weight of dry material.

Dryer efficiency is the fraction of total energy consumed which is used to heat and vaporize the liquid.

Equilibrium moisture content is that which a material retains after prolonged exposure to a specific ambient temperature and humidity.

Evaporative efficiency in a direct-heat dryer compares vaporization obtained to that which would be obtained if the drying gas were saturated adiabatically.

Falling rate period is a drying period during which the liquid vaporization rate per unit surface or weight of dry material continuously decreases.

Fiber saturation point is the bound moisture content of cellular materials such as wood.

Free moisture content is the liquid content that is removable at a specific temperature and humidity. Free moisture may include bound and unbound moisture, and is equal to the total average moisture content minus the equilibrium moisture content for the specific drying conditions.

Humidity denotes the amount of condensable vapor present in a gas, expressed as weight of vapor per unit weight of dry gas; ie, dry basis weight.

Internal diffusion occurs during drying when liquid or vapor flow obeys the fundamental diffusion laws.

Moisture is a word used commonly to describe any volatile liquid or vapor involved in drying; ie, it is not used selectively to mean only water.

Moisture gradient is the moisture profile in a material at a specific moment during drying, which usually reveals the mechanisms of moisture movement in the material up to the moment of measurement.

Percent saturation is the ratio of the partial pressure of a condensable vapor in a gas to the vapor pressure of the liquid at the same temperature, expressed as a percentage. For water vapor in air this is called percent relative humidity.

Unaccomplished moisture change is the ratio of free moisture present in a material at any moment during drying to that present initially.

Unbound moisture in a hygroscopic material is moisture that exerts the same vapor pressure as the pure liquid at the same temperature. Unbound moisture behaves as if the material were not present. All moisture in a nonhygroscopic material is unbound.

Wet basis is a material's moisture content expressed as a percentage of the weight of wet material. Although commonly employed, this basis is less satisfactory for drying calculations than the dry basis for which the percentage change of moisture per unit weight of dry material is constant at all moisture contents.

Figure 1 shows the relationship between dry and wet bases. When the wet basis is used to state moisture content, a 2–3% change at moisture contents above 50% may represent a 10–30% change in evaporative load per unit weight of dry material.

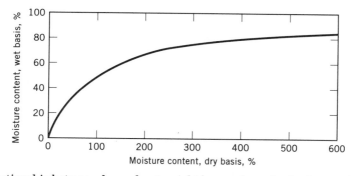

Fig. 1. Relationship between dry and wet weight bases where dry basis = weight moisture per weight of dry material; wet basis = weight moisture per weight of moisture + dry material.

Psychrometry

Before drying can begin, a wet material must be heated to such a temperature that the vapor pressure of the contained liquid exceeds the partial pressure of vapor already present in the surrounding atmosphere. The effect of a dryer's atmospheric vapor content and temperature on performance can be studied by construction of a psychrometric chart for the particular gas and vapor. Figure 2 is a standard chart for water vapor in air (6).

The wet bulb or saturation temperature curve indicates the maximum weight of vapor that can be carried by a unit weight of dry gas. For any temperature on the abscissa, saturation humidity is found by reading up to the saturation temperature curve, then across to the ordinate, kg/kg dry air. At saturation, the partial pressure of vapor in the gas is the vapor pressure of the liquid at the specific temperature:

$$H_s = \frac{p_s}{(P - p_s)} \cdot \frac{M_v}{M_g} \tag{1}$$

where H_s = saturation humidity, the weight ratio of moisture/kg dry gas; P = system total pressure; p_s = liquid vapor pressure at the gas temperature; and M_v/M_g = molecular weight ratio of vapor to dry gas. Pressure is in units of kPa. At any condition less than saturation, humidity, H, is expressed similarly:

$$H = \frac{p}{(P - p)} \cdot \frac{M_v}{M_g} \tag{2}$$

where p = partial pressure of the vapor in the gas in kPa. Percent relative humidity, % rh, curves indicate percent saturation and are related to vapor pressure:

$$\%(\text{rh}) = 100 \frac{p}{p_s} \tag{3}$$

In Figure 2 the lines, volume, m³/kg dry air, indicate humid volume, which includes the volume of 1.0 kg of dry gas plus the volume of vapor it carries. Enthalpy at saturation data are accurate only at the saturation temperature and humidity; however, for air–water vapor mixtures, the diagonal wet bulb temperature lines are approximately the same as constant-enthalpy adiabatic cooling lines. The latter are based on the relationship:

$$(H - H_s) = -\frac{C_s}{L_s}(t - t_s) \tag{4}$$

where H_s and t_s are adiabatic saturation humidity, kg/kg, and temperature, K, respectively, corresponding to gas conditions represented by H and t; C_s = humid heat for humidity H in units of kJ/(kg·K); and L_s = latent heat of vaporization

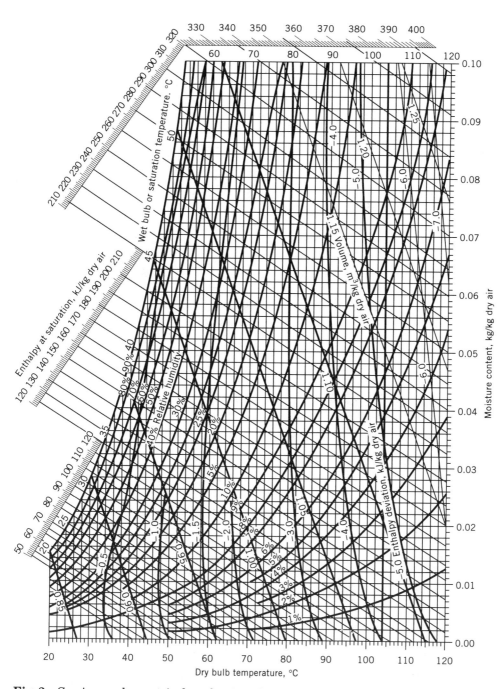

Fig. 2. Carrier psychrometric chart for air and water vapor at 101.325 kPa (1 atmosphere) total pressure. Courtesy of Carrier Corp. (6).

479

at t_s in kJ/kg. The slope of the constant-enthalpy adiabatic cooling line is $-C_s/L_s$, which is the relationship between temperature and humidity of gas passing through a totally adiabatic direct-heat dryer. The humid heat of a gas–vapor mixture per unit weight of dry gas includes the specific heat of the vapor

$$C_s = c_{pg} + c_{pv}H \tag{5}$$

where, c_{pg} = specific heat of the dry gas (air) and c_{pv} = specific heat of the vapor (water); both are in units of kJ/(kg·K). The wet bulb temperature is established by a steady-state, nonequilibrium relationship between heat transfer and mass transfer when liquid evaporates from a small mass; eg, the wet bulb of a thermometer, into a sufficiently large mass of flowing gas, so that the latter undergoes no temperature or humidity change. Provided radiant heat transfer is insignificant, steady-state conditions are expressed by the relationship:

$$h_c\,(t - t_w) = -\,k'L_w\,(H - H_w) \tag{6}$$

where h_c = heat transfer by convection only in units of kW/(m²·K); k' = mass-transfer coefficient in kg/(s·m²)(kg/kg); t = gas dry bulb temperature, K; t_w = gas wet bulb temperature, K; H = humidity at t; H_w = saturation humidity at t_w; and L_w = latent heat of vaporization at t_w, kJ/kg. For air–water vapor mixtures, it happens that $h_c/k' \simeq C_s$. Therefore, because the ratio $(H - H_w)/(t - t_w)$ $= -\,h_c/k'L_w$, the slope of the wet bulb temperature line in equation 6 is also approximately equal to $-C_s/L_s$, the slope of the constant enthalpy adiabatic cooling line in equation 4, and t_w is approximately the same as t_s. Enthalpy deviation curves in Figure 2 permit enthalpy corrections for humidities less than saturation and reveal the extent to which the wet bulb temperature lines do not coincide with constant enthalpy adiabatic cooling lines. For thorough treatment of wet bulb thermometry, see References 7–9. If system pressure is different from 101.3 kPa (1 atm), the humidity at measured dry bulb and wet bulb temperatures can be corrected (4). A separate chart is preferably constructed for the pressure of interest.

It is a coincidence that for water vapor in air, $h_c/k'C_s$ has a value of approximately one. For most organic vapors encountered in drying, wet bulb temperatures are considerably higher than the adiabatic saturation temperatures. This is because of the higher molecular weights. Larger molecules do not diffuse as easily through air or through most inert gases. For example, values of $h_c/k'C_s$ at 101.3 kPa and 0°C for carbon tetrachloride, benzene, and toluene are 2.17, 1.87, and 1.98, respectively. Psychrometric charts for these vapors in air have been prepared (10). It is necessary to employ such charts for evaluating humidities and particularly material temperatures when vaporizing organic liquids in direct-heat dryers.

To illustrate changes in temperatures and humidities as gas passes through a direct-heat (convection) dryer, such as a pneumatic conveyor (see CONVEYING), rotary, or fluid-bed, several temperature profiles are drawn on Figure 3. Line AB is an adiabatic saturation line. Gas that enters the dryer at H_1 and t_1 cools and humidifies along line a. Assuming ideal adiabatic operation, the gas could leave at H_2 and t_2. The maximum humidity gain, were the gas cooled to saturation

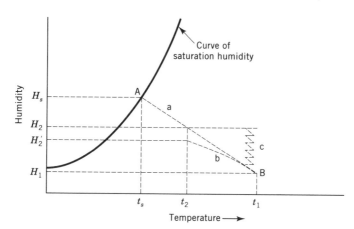

Fig. 3. Humidity chart illustrating changes in air temperature and humidity in adiabatic direct-heat (convection) dryers. AB is an adiabatic saturation line. Terms are defined in text.

adiabatically, would be $(H_s - H_1)$ at t_s. The ratio, $(H_2 - H_1)/(H_s - H_1)$, is the ideal evaporative efficiency of this direct-heat dryer; however, as there are usually heat losses through the dryer enclosure and sensible heat absorbed by the solids, the gas temperature change is rarely accounted for completely by humidity gain; the outlet humidity, H_2', is less than the adiabatic saturation humidity at t_2, and the drying profile traces the path of line b. In dryers that employ internal steam coil gas reheaters, gas may pass several times through the material and heaters. Line c is a characteristic temperature–humidity profile.

Drying Mechanisms

Drying Periods. The goal of most drying operations is not only to separate a volatile liquid, but also to produce a dry solid of a desirable size, shape, porosity, density, texture, color, or flavor. An understanding of liquid and vapor mass-transfer mechanisms is essential for quality control. Mass-transfer mechanisms are best understood by measuring drying behavior under controlled conditions in a prototypic, pilot-plant dryer. No two materials behave alike and a change in material handling method or any operating variable, such as temperature or gas humidity, also affects mass transfer. For example, a layer of sand on a belt conveyor exhibits a different drying profile than sand dried on a vibrating (fluid-bed) conveyor.

Figure **4a** shows drying time profiles for one material dried under three conditions. Corresponding rate profiles are in Figure **4b**. Three products having uniquely different characteristics were produced by three different kinds of agitation. Other controllable drying conditions were constant. These profiles show that during drying several distinct periods may occur, which depend on how the material is handled. These are (*1*) an induction period during which wet material is heated to drying temperature; (*2*) a constant rate drying period indicated by

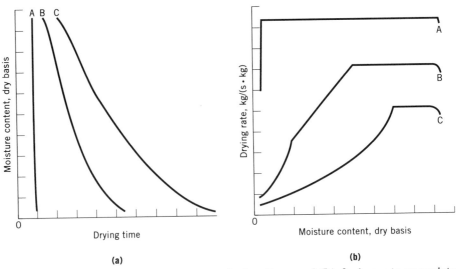

Fig. 4. (**a**) Profiles of moisture content vs drying time; and (**b**) drying rate vs moisture content for a slightly soluble, water-wet organic powder centrifuge cake at 4.0 kPa (0.58 psi) absolute pressure using 120°C indirect heat. Profile A was produced in a continuous, high speed agitator dryer provided with scrapers to maintain a clean heating surface. Drying time was 45 s and at an almost entirely constant rate because of high solids surface exposure to the heating surface; the product particle size was 100% less than 150 μm. Profile B was produced in a paddle-agitated batch dryer also having scrapers. Drying time was 70 min, including periods of constant rate, capillary, and diffusion drying. Because of the much slower agitator, the product was a porous, 100–500 μm powder having some dust. Profile C was produced in a double-cone batch dryer using some dry recycle. Drying time was 120 min, and was almost entirely liquid and vapor diffusion-controlled because turning of the double-cone pelletized the wet material early in the cycle; the product was composed of rather dense (200–800 μm) spheres, having negligible dust.

the horizontal portions of the profiles in Figure 4**b**; (*3*) a period of decreasing rate shown by the sloping portions of two rate profiles during which the drying rate appears proportional to moisture content; and (*4*) a period of decreasing rate shown by the curved portions of two rate profiles during which the drying rate is evidently a more complex function of moisture content than simple proportionality.

The moisture content at the end of constant rate drying is the critical moisture content. Drying periods following are falling rate periods. The curved portion of profile B in Figure 4**b** is a second falling rate period; moisture content at the second break is the second critical moisture content. Profile C shows that drying may occur almost entirely in a falling rate period; a slight change in specified product moisture content can have a significant effect on drying time.

Constant Rate Drying. During constant rate drying, vaporization occurs from a liquid surface of constant composition and vapor pressure. Material structure has no influence except moisture movement from within the material must be fast enough to maintain the wet surface. The vaporization rate is controlled by the heat-transfer rate to the surface. The mass-transfer rate adjusts to the heat-transfer rate and the wet surface reaches a steady-state temperature. The drying rate remains constant, therefore, as long as external conditions are constant. If heat is supplied solely by convection, the steady-state temperature is the gas wet bulb temperature. When conduction and radiation contribute, eg, the material contacts and/or receives radiation from a warm surface, a liquid surface temperature between the wet bulb temperature and the liquid's boiling point is obtained. In indirect-heat and radiant-heat dryers, where conduction and radiation predominate, surface liquid may boil regardless of ambient humidity and temperature. During constant rate drying, material temperature is controlled more easily in a direct-heat dryer than in an indirect-heat dryer because in the former the material temperature does not exceed the gas wet bulb temperature as long as all surfaces are wet. For convection, all principles relating to simultaneous heat and mass transfer between gases and liquids apply. The steady-state relationship between heat and mass transfer at the liquid surface is

$$-\frac{dw}{d\theta} = \frac{h_t A}{L'_s}(t - t'_s) = k'_a A(p'_s - p) \tag{7}$$

where $dw/d\theta$ = moisture loss in kg/s; h_t = sum of all convection, conduction, and radiation components of heat transfer in kW/(m^2·K); A = effective surface area for heat and mass transfer in m^2; L'_s = latent heat of vaporization at t'_s in kJ/kg; k'_a = mass-transfer coefficient in kg/(s·m^2·kPa); t = mean source temperature for all components of heat transfer in K; t'_s = liquid surface temperature in K; p'_s = liquid vapor pressure at t'_s in kPa; p = partial pressure of vapor in the gas environment in kPa. It is often useful to express this relationship in terms of dry basis moisture change. For vaporization from a layer of material:

$$-\frac{dW}{d\theta} = \frac{h_t}{\rho_m d_m L'_s}(t - t'_s) \tag{8}$$

where $dW/d\theta$ = dry basis drying rate in units of kg/(s·kg); ρ_m = dry material bulk density in kg/m^3; and d_m = layer depth in m. A similar equation describes through-circulation drying in which gas flows through a bed of particles:

$$-\frac{dW}{d\theta} = \frac{h_t a}{\rho_m L'_s}(t - t'_s) \tag{9}$$

where a = the effective heat-transfer area per unit of bed volume in units of m^2/m^3. For a static bed, ρ_m can be measured, but the effective area of a particle bed is difficult to estimate except for uniform shapes such as cylinders and spheres (4). The practice is to conduct drying tests from which a value of the quantity $h_t a$ can be calculated by inserting property and drying data into equation 9. A modification of equation 9 is used to describe dryers in which gas flows among dispersed particles; eg, direct-heat rotary dryers. In dispersed-particle dryers, particle concentration per unit volume of dryer changes continuously and varies from place to place. For these, tests are conducted in a prototype of the commercial dryer and scale-up is based on average dispersion. The designer's concern is that the quality of particle dispersion in the gas in the scaled-up dryer duplicates that in the prototype or that a proper allowance is made for differences. From this procedure comes the concept of the volumetric heat-transfer coefficient:

$$U_a = h_t a \tag{10}$$

where U_a = an average volumetric heat-transfer coefficient having units of kW/(m³·K). For constant rate drying in dispersed-particle dryers, the general relationship is

$$-\frac{dw}{d\theta} = \frac{U_a V \, \Delta t_m}{L'_s} \tag{11}$$

where V = effective dryer volume in units of m³; Δt_m = log-mean temperature difference between all convection, conduction, and radiation heat sources and the material in K; and L'_s = latent heat of vaporization at the material surface temperature in kJ/kg. For estimating effective areas for various heat-transfer components, methods have been developed for tray dryers that may serve as a guide for other arrangements (11,12).

Convection heat transfer is dependent largely on the relative velocity between the warm gas and the drying surface. Interest in pulse combustion heat sources anticipates that high frequency reversals of gas flow direction relative to wet material in dispersed-particle dryers can maintain higher gas velocities around the particles for longer periods than possible in simple cocurrent dryers. This technique is thus expected to enhance heat- and mass-transfer performance. This is apart from the concept that mechanical stresses induced in material by rapid directional reversals of gas flow promote particle deagglomeration, dispersion, and liquid stream breakup into fine droplets. Commercial applications are needed to confirm the economic value of pulse combustion for drying.

Gas impingement from slots, orifices, and nozzles at 10–100 m/s velocities is used for drying sheets, films, coatings (qv), and thin slabs, and as a secondary heat source on drum dryers and paper (qv) machine cans. The general relationship for convection heat transfer is (13,14):

$$h_c = \alpha G^{0.78} \tag{12}$$

where α is a factor dependent on orifice-plate open area, hole, or slot size, and spacing between the plate, slots, or nozzles, and the heat-transfer surface; and G = hole, slot, or nozzle gas velocity in mass flow terms having units of kg/(s·m²). Convection heat- and mass-transfer performance is enhanced by thinning of the laminar gas film immediately above the wet surface caused by direct gas impact on the surface. In a float dryer, cloth, sheet, or film is supported and conveyed on layers of gas which impinge on both sides of the material. This noncontact dryer is an impingement dryer modification.

Contact Drying. Contact drying occurs when wet material contacts a warm surface in an indirect-heat dryer (15–18). A sphere resting on a flat heated surface is a simple model. The heat-transfer mechanisms across the gap between the surface and the sphere are conduction and radiation. Conduction heat transfer is calculated, approximately, by recognizing that the effective conductivity of a gas approaches 0, as the gap width approaches 0. The gas is no longer a continuum and the rarified gas effect is accounted for in a formula that also defines the conduction heat-transfer coefficient:

$$\gamma = \frac{\gamma_g}{[1 + (p/s)]} \quad \text{or} \quad \frac{\gamma}{s} = h_d = \frac{\gamma_g}{(s + p)} \tag{13}$$

where γ_g = the continuum gas heat conductivity; γ = the rarified gas heat conductivity; s = local width of the gas gap; p = the product of the mean-free path of the gas molecules and a function of the accommodation coefficient (19); and h_d = the conduction heat-transfer coefficient based on heating surface. At the contact point, the gap goes to 0, and the coefficient reaches its maximum:

$$h_d(0) = \frac{\gamma_g}{p} \tag{14}$$

Figure 5 shows conduction heat transfer as a function of the projected radius of a 6-mm diameter sphere. Assuming an accommodation coefficient of 0.8, $h_d(0)$ = 3370 W/(m²·K); the average coefficient for the entire sphere is 72 W/(m²·K). This variation in heat transfer over the spherical surface causes extreme non-uniformities in local vaporization rates and if contact time is too long, wet spherical surface near the contact point dries. The temperature profile penetrates the sphere and it becomes a continuum to which Fourier's law of nonsteady-state conduction applies.

If the sphere is one of a mass of wet spherical particles, fastest drying occurs when the specific contact time of each particle approaches 0 in an ideally mixed bed. In general, gas thermal conductivity is independent of pressure between 150 Pa (0.022 psi) and 10^6 Pa (145 psi); below 15 Pa (0.11 mm Hg), conductivity is almost proportional to absolute pressure (20). The mean-free paths of gas mole-

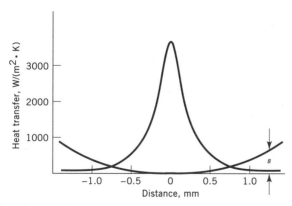

Fig. 5. Profile of conduction heat transfer across the gap between a sphere and a flat plate vs projected radius, R = 3 mm, of the sphere at 40°C and 2.1 kPa (0.30 psi); s is the width of the gas gap (17).

cules are inversely proportional to pressure. Equations 13 and 14 state that conduction heat transfer must decrease as pressure decreases. At very low pressures, conduction heat transfer approaches 0 and radiation alone is effective. Conduction heat transfer is also influenced by particle shape, surface roughness, and probably specific gravity. In agitator-stirred dryers, agitator speed, mixing efficiency, and heating surface clearance are important variables.

Between 1 s and 1 min specific contact time, conduction heat-transfer performance decreases theoretically as the 0.29 power of contact time. This is consistent with empirical data from several forms of indirect-heat dryers which show performance variation as the 0.4 power of rotational speed (21). In agitator-stirred and rotating indirect-heat dryers, specific contact time can be related to rotational speed provided that speed does not affect the physical properties of the material. To describe the mixing efficiency of various devices, the concept of a mixing parameter is employed. An ideal mixer has a parameter of 1.

$$N_{\text{mech}} = \theta_s N \tag{15}$$

where N_{mech} is the mixing parameter, independent of thermal conditions; $\theta_s =$ specific particle contact time in seconds; $N =$ agitator or dryer vessel rotational speed in units of 1/s. Values of N_{mech} reported for various dryers lie between 2 and 25. The principle applies to materials other than particulate solids. In the case of drum dryers used for solutions and slurries, capacity often can be increased by increasing drum speed, reducing specific contact time, and laying a thinner film on the drum surface. In the situations of cloth, paper, film, and fiber tow dried on heated cans, an additional resistance to drying is the inability of vapors and entrained gas to escape from between the moving material and the can surface, which also encourages shorter specific contact times.

Critical Moisture Content. Critical moisture content, which is the average material moisture content at the end of constant rate drying, is a function of material properties, the constant drying rate, particle size, or bed depth. Critical moisture content cannot be determined except by a prototypic drying test. For example, while a bed of material is drying during the constant rate period, assume that the drying rate is increased by increasing gas velocity over the material surface. The moisture gradient below the surface which causes liquid flow to replenish the surface becomes steeper and as the surface approaches dryness, the internal moisture content of the material is greater than it would have been had the gas velocity and drying rate not been increased. Critical moisture data are misleading, therefore, unless the exact conditions of drying are known (22,23).

Particle size distribution determines surface-to-mass ratios and the distance internal moisture must travel to reach the surface. Large pieces thus have higher critical moisture contents than fine particles of the same material dried under the same conditions. Pneumatic-conveyor flash dryers work because very fine particles are produced during initial dispersion and these have low critical moisture contents.

Case hardening refers to a circumstance in which a mass of nonporous, soluble, or colloidal material is dried at such a high rate during initial constant rate drying that the surface overheats and shrinks. Because liquid diffusivity decreases with moisture content, the barrier formed by the overdried surface prevents moisture flow from the interior of the mass to the surface. Case hardening of nonporous materials can be minimized by initially maintaining a high relative humidity environment and consequently a high surface equilibrium moisture content until internal moisture has time to escape.

Equilibrium Moisture Content. Equilibrium moisture content is the steady-state equilibrium reached by the gain or loss of moisture when material is exposed to an environment of specific temperature and humidity for a sufficient time. The equilibrium state is independent of drying method or rate. It is a material property. Only hygroscopic materials have equilibrium moisture contents. Clean beach sand is nonhygroscopic and has an equilibrium moisture content of 0. The same rules apply to organic vapors. Hygroscopic material retains a constant fraction of moisture under specific ambient humidity and temperature conditions. At constant temperature, if ambient humidity increases or decreases, an increase or decrease in moisture content follows. This is called equilibrium moisture because it is held in vapor pressure equilibrium with the partial pressure of vapor in the atmosphere. The reason it is retained even when the atmosphere is quite dry is that the retention mechanism reduces effective liquid vapor pressure. It is bound moisture because it is bound to material in solution or by adsorption and bound moisture behaves as if the atmosphere were saturated even when the atmosphere is not saturated relative to the unbound liquid's normal vapor pressure. Chemically combined liquid may behave like bound moisture depending on the nature of the chemical bond. Because equilibrium is influenced by partial vapor pressure in the atmosphere and the effective vapor pressure of the bound liquid, temperature and humidity are both important. For many materials in the 15–50°C temperature range, equilibrium moisture content can be plotted vs relative humidity as an essentially straight line. Equilibrium moisture content appears independent of temperature and relates to Henry's law:

$$p = H(x) \tag{16}$$

where p = partial vapor pressure in the atmosphere in kPa; H = Henry's constant; and x = dry basis moisture content. Henry's constant is a function of the unbound liquid's vapor pressure:

$$H = f(p_s) \tag{17}$$

where p_s = the unbound liquid's vapor pressure; therefore, $p = f(p_s)(x)$, and since relative humidity = $100(p/p_s)$:

$$100(p/p_s) = 100f(x) \tag{18}$$

At any given relative humidity, x is constant. In a typical silica gel–air–water vapor system at 5–50°C, $p/p_s = 1.79(x)$, where p = partial pressure of vapor in the air; p_s = water-vapor pressure at the adsorption temperature; x = gel moisture content, kg/kg (24). For many materials as the temperature increases above 50°C, equilibrium moisture content decreases at constant relative humidity. Above 100°C at atmospheric pressure, saturation humidity for water vapor goes to infinity and the concept of relative humidity becomes meaningless; in fact, hygroscopic materials often can be dried in 100% superheated vapor atmospheres.

A profile of equilibrium moisture content vs percent relative humidity (Fig. 6) often is not a perfectly straight line because at high humidities a porous material may retain condensed capillary moisture, whereas at low humidities moisture may be adsorbed in a single-molecular layer on capillary surfaces. The maximum bound moisture a material can hold is identified by the intersection of the equilibrium profile with the 100% relative humidity ordinate. A difference between adsorption and desorption profiles may have several causes. When material initially dries, shrinkage often closes many small capillaries which do not reabsorb moisture when the material is rewet. Also, some capillaries may be cul-de-sacs that resist vapor reentry once filled with gas. One reason for freeze drying solid foods is to minimize shrinkage and capillary closing so the dry residue can be reconstituted to its original moisture content more easily. On the other hand, for laboratory and pilot-plant studies, a once-dried material must never be rewet and reused for a drying test. In fact, once-dried material can never be returned, physically, to its original wet condition.

Falling Rate Drying.   Heat transfer is limited by material conductivity, but the drying rate usually is controlled by internal liquid and vapor mass transfer. The principal mass-transfer mechanisms are (1) liquid diffusion in continuous, homogeneous materials; (2) vapor diffusion in porous and granular materials; (3) capillarity in porous and fine granular materials; (4) gravity flow in granular materials; (5) flow caused by shrinkage-induced pressure gradients; and (6) pressure flow of liquid and vapor when porous material is heated on one side, but vapor must escape from the other.

Liquid flows by diffusion through materials in which the liquid is soluble, eg, single-phase systems like soap and gelatin. Movement of other bound moisture by liquid diffusion may occur, but the mechanisms probably are more complicated. Vapor flows by diffusion through the gas phase when liquid vaporizes below the

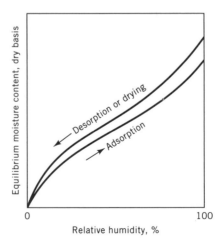

Fig. 6.   Equilibrium moisture content profiles vs relative humidity for a hygroscopic material.

surfaces of porous and granular materials. Wood and other cellular materials at moisture contents less than fiber saturation and the final drying stages of paper, textiles, and hydrophilic solids are examples. Vapor diffusion also occurs in the laminar sublayer of the gas film adjacent to the material surface during all drying periods. Diffusion-controlled mass transfer (qv) is assumed in drying when liquid or vapor flow conforms to Fick's second law of diffusion, which applies to non-steady-state systems (see also DIFFUSION SEPARATION METHODS):

$$\frac{\delta c_A}{\delta \theta} = D_{AB} \frac{\delta^2 c_A}{\delta z^2} \tag{19}$$

where c_A = concentration of one component in a two-component phase of A and B; θ = diffusion time; z = distance in the direction of diffusion; and D_{AB} = binary diffusivity of the two components. This equation applies to diffusion in solids, liquids, and gases. The analogous Fourier laws apply to heat conduction. The units of the diffusion coefficient, D_{AB} = (mol/(s·m^2))(m^3/mol)(m) = m^2/s, is the abbreviation usually employed both for this coefficient and the quantity thermal diffusivity in Fourier's heat conduction equations.

In porous and granular materials, liquid movement occurs by capillarity and gravity, provided passages are continuous. Capillary flow depends on the liquid material's wetting property and surface tension. Capillarity applies to liquids that are not adsorbed on capillary walls, moisture content greater than fiber saturation in cellular materials, saturated liquids in soluble materials, and all moisture in nonhygroscopic materials.

When clay or similar material is dried, often a pressure gradient is developed by the forces of repulsion between particles as shrinkage brings the particles close together (25). This gradient forces liquid toward the surface and the resulting moisture profile resembles that characteristic of liquid diffusion.

When a layer of material pervious to gas flow is dried by through-circulation, a drying front usually moves through the layer in the direction of gas flow. In circumstances when material moisture content is sufficiently high, incoming gas is sufficiently hot and the material is of sufficient depth, the gas may cool adiabatically to saturation before it passes fully through the layer. The gas may then be subcooled by material contact with consequent liquid condensation within the layer. Condensed liquid fills the flow passages and is forced through the layer as liquid by the pressure of the blocked gas stream. This phenomenon is known as a vaporization-condensation sequence. It is an ingenious drying process usable most commonly on stationary or moving combination filter-dryers. Thermal efficiency exceeds 100%.

Usually, only one mass-transfer mechanism predominates at any given time during drying, although several may occur together. In most materials, the mechanisms of internal liquid and vapor flow during falling rate drying are complex. Simultaneous heat transfer is a factor and falling rate drying rarely can be described with mathematical precision. Computer models for some materials are published (26), but most employ data from actual drying tests. In the absence of tests, the falling rate drying periods usually are studied on the assumption that internal mass transfer is controlled either by diffusion or capillarity depending on whether the material is porous or nonporous, soluble or not.

Diffusion. Characteristic drying time and rate profiles for liquid diffusion appear as profile C in Figure 4. Figure 7 contains a diffusion drying curve for corn kernels and introduces the complicating factor in solids drying that liquid diffusivity is affected by material moisture content. The diffusion coefficient is not constant, but decreases as material moisture content decreases. Nonetheless, for evaluation of falling rate drying by liquid diffusion an integration of equation 19 is employed using four simplifying assumptions: (1) liquid diffusivity is independent of moisture content; (2) initial moisture distribution is uniform; (3) material size, shape, and density are unchanging; and (4) the material's equilibrium moisture content is constant. For material in the form of a slab:

$$\frac{W_\theta - W_e}{W_c - W_e} = \frac{8}{\pi^2} \left[\sum_{n=0}^{n=\infty} \frac{1}{(2n+1)^2} \exp[-(2n+1)^2 D\theta(\pi/2d)^2] \right] \qquad (20)$$

where W_θ, W_c, W_e are the moisture content at time θ, the first critical moisture content, and the equilibrium moisture content, respectively, so that the term on the left-hand side is the unaccomplished moisture change as defined; D = liquid diffusivity in units of m^2/s; θ = drying time in s; d = one-half the slab thickness for drying from both sides or total thickness for one-side drying in m. For long drying times, when $D\theta/d^2$ exceeds about 0.1,

$$\frac{W_\theta - W_e}{W_c - W_e} = \frac{8}{\pi^2} \exp[-D\theta(\pi/2d)^2] \qquad (21)$$

from which a drying rate expression can be derived

$$-\frac{dW}{d\theta} = \frac{\pi^2 D}{4d^2} (W_\theta - W_e) \qquad (22)$$

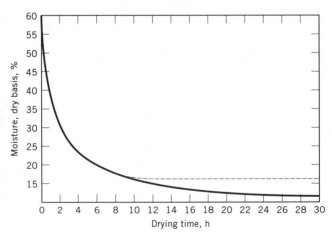

Fig. 7. Drying of corn kernels by liquid diffusion. The dashed line is that predicted by theory based on constant diffusivity. The solid curve shows actual performance at 38°C in air.

where $dW/d\theta$ = dry basis, drying rate in units of kg/(s·kg). When internal diffusion controls the drying rate, the rate is proportional to the free moisture content and diffusivity, and inversely proportional to the square of material thickness. When equation 20 is plotted semilogarithmically with unaccomplished moisture change as the ordinate and the quantity, $D\theta/d^2$ as the abscissa, a straight line results for $(W_\theta - W_e)/(W_c - W_e)$ values less than 0.6. Equation 21 describes the straight line portion of this plot. An approximate relationship for falling rate drying by liquid diffusion is

$$\theta_f = \frac{4d^2}{\pi^2 D} \ln\left[\frac{8(W_c - W_e)}{\pi^2(W_\theta - W_e)}\right] \tag{23}$$

where θ_f = falling rate drying time in s. For nonhygroscopic materials, where $W_e = 0$, the unaccomplished moisture change becomes simply W_θ/W_c. Equations 20–23 apply to materials in layers or slabs. Equation 19 is solved for other shapes as well (27), and these equations also may be used to study vapor diffusion in porous and granular materials (28).

An analogy exists between mass transfer by diffusion and heat transfer by conduction. Each involves collisions between molecules and a gradient as the driving force which causes flow. For diffusion, this is a concentration gradient; for conduction, the driving force is an energy gradient. Fourier's nonsteady-state conduction equation is analogous to equation 19, Fick's second law of diffusion,

$$\frac{\delta T}{\delta \theta} = \left(\frac{k}{\rho c_p}\right) \frac{\delta^2 T}{\delta z^2} \tag{24}$$

where T = material temperature in K; θ = conduction time in s; z = distance in the direction of conduction in m; k = material thermal conductivity in units of kW·m/(m²·K); ρ = material density in kg/m³; and c_p = material specific heat in kJ/(kg·K). The quantity, $k/\rho c_p$ is the thermal diffusivity of the material. This term is needed because temperature is a scale and not a quantity. The units of thermal diffusivity are m²/s, the same as for the diffusion coefficient. Solutions for equation 24 are available for many material shapes (29–32). During falling rate drying, material conductivity is important because liquid in porous materials is vaporized below the surface and heat must be conducted to this liquid by the dry material.

Capillarity. The outer surface of porous material has pore entrances of various sizes. As surface liquid is evaporated during constant rate drying, a meniscus forms across each pore entrance and interfacial forces are set up between the liquid and material. These forces may draw liquid from the interior to the surface. The tendency of liquid to rise in porous material is caused partly by liquid surface tension. Surface tension is defined as the work needed to increase a liquid's surface area by one square meter and has the units J/m². The pressure increase caused by surface tension is related to pore size:

$$\Delta p = s_t/r \tag{25}$$

where Δp = the meniscus pressure increment resulting from surface tension in Pa; s_t = surface tension in J/m²; and r = pore radius in m. The excess pressure resulting from surface tension is always directed from the concave toward the convex surface of the meniscus.

A second property important to capillarity is surface wetting ability which depends on properties of both the liquid and material. Wetting ability is indicated by the contact angle formed at the liquid–material interface; eg, water and clean glass have a contact angle of 0° and water rises in glass capillaries; mercury has a contact angle with glass of 132° and does not rise. When the contact angle is less than 90°, the force of surface adhesion exceeds liquid cohesive strength. Liquid molecules climb the capillary wall and surface tension causes a liquid column to follow. Liquids rise higher in fine capillaries because adhesive force is a wall effect. In very fine capillaries, meniscus radii become so small that the pressure increase caused by surface tension suppresses liquid vapor pressure. The effect is described in the Kelvin equation

$$\ln(p/p_s) = -2s_t M_1 \cos\phi / (r\rho RT) \qquad (26)$$

where p = effective vapor pressure of the capillary liquid in kPa; p_s = normal liquid vapor pressure in kPa, at temperature T in K; M_1 = liquid molecular weight in kg/mol; ρ = liquid density in kg/m³; ϕ = liquid contact angle; R = 8314 J/(mol·K), the ideal gas constant; and r = pore radius in m. Calculated effects for water at 50°C and an s_t of 0.06791 J/m² give the following:

p/p_s	$r(\mu m)$
0.999	0.910
0.990	0.091
0.960	0.022
0.920	0.011
0.900	0.009
0.800	0.004
0.700	0.003
0.500	0.001

This is the capillary condensation phenomenon, which partly accounts for the hysteresis observed in adsorption profiles of porous materials.

At the critical moisture content, at the end of constant rate drying, dry areas begin to appear on the material surface. Menisci in the larger pores begin to withdraw below the surface. As drying continues, the surface becomes completely dry and liquid withdraws in even the smallest pores. This completes the first falling rate drying period which is represented by the straight line, decreasing rate portions of profiles A and B in Figure 4b. Final drying during the second falling rate period is accomplished by heat conduction to the liquid pockets and vapor diffusion through the pores to the material surface. In most porous materials during this period, the drying rate profile has the concave upward shape of diffusion control as appears in profile B, Figure 4b. In granular materials where pores are large and capillary forces are weak, gravity contributes to retreat of the liquid surface and both falling rate profiles may be straight lines (33). In other capillary porous materials dried from two sides, or in thin layers from one side, complete drying also may occur at a drying rate proportional to residual moisture content. Figure 8 is an example and falling rate drying often may be approximated by assuming the rate proceeds in this manner

$$-\frac{dW}{d\theta} = K(W_\theta - W_e) \tag{27}$$

where K is a function of the constant rate, drying rate at the critical moisture content,

$$-\frac{dW}{d\theta_c} = K(W_c - W_e) \tag{28}$$

For a layer of wet material, using equation 8,

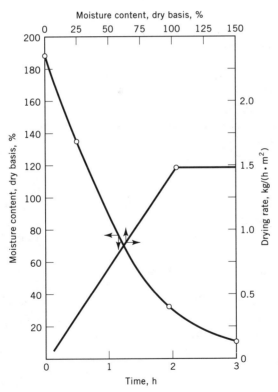

Fig. 8. Drying time and rate profiles for leather pasted on glass plates and dried in two temperature stages. Gas velocity = 5 m/s in parallel flow, 71°C in the first stage, 57°C in the second. The falling rate, drying rate is proportional to residual moisture content.

$$K = \frac{h_t}{\rho_m d_m L_s'} \left[\frac{(t - t_s')}{(W_c - W_e)} \right] \tag{29}$$

and

$$-\frac{dW}{d\theta} = \frac{h_t(t - t_s')}{\rho_m d_m L_s'} \left[\frac{(W_\theta - W_e)}{(W_c - W_e)} \right] \tag{30}$$

For materials that follow equation 30 the drying rate is inversely proportional to material thickness; falling rate drying time is estimated,

$$\theta_f = \frac{\rho_m d_m L_s'(W_c - W_e)}{h_t(t - t_s')} \ln \left[\frac{(W_c - W_e)}{(W_\theta - W_e)} \right] \tag{31}$$

A relationship for through-circulation drying analogous to equation 30 employs equation 9.

$$-\frac{dW}{d\theta} = \frac{h_t a(t - t_s')}{\rho_m L_s'} \left[\frac{(W_\theta - W_e)}{(W_c - W_e)} \right] \tag{32}$$

Drying Profiles. An application of diffusion principles to falling rate drying is exemplified in Figure 9 (34). Single drops of whole milk were dried by suspension in a warm air stream. Because of the rapid formation of surface films, drying was mostly by vapor diffusion. Drying times to an unaccomplished moisture change of 0.1 were 190 s and 300 s, respectively, for drops B and A. Based on equations 22 and 23 and employing the square of the initial drop diameter for the d^2 dimension, drying time for the larger drop should have been 317 s. Neglecting the initially 8.0% greater moisture content of the larger drop, drying time and rate varied with the 1.8 power of initial drop diameter. The error in predicting drying time for the larger drop is roughly 6%, which is probably attributable to material shrinkage during drying.

Figure 10 depicts freeze drying data for two milk products (35). Heat and mass transfer involved radiation to the material surface, conduction through the material to the retreating ice phase, and vapor diffusion through ice-free capillaries above the ice phase to the surface. Because both conduction heat transfer and vapor diffusion are distance dependent, a continuously falling rate drying profile would be expected. In freeze drying, however, it is necessary to control heat input to prevent temperature rise and melting of the ice. Also, it is necessary to limit vapor transport to condenser capacity so as to prevent drying chamber pressure rise. In this situation, dryer operation is controlled by external factors. The drying profile of necessity is mostly constant rate.

Dryers

Industrial dryers may be broadly classified by heat-transfer method as being either direct or indirect heat. Dryers evolved from material handling equipment

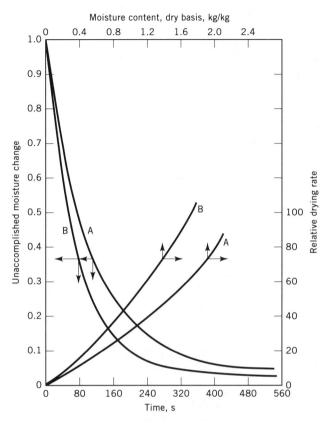

Fig. 9. Drying profiles for single drops of whole milk at 94°C and 0.6 m/s relative air flow. The initial diameter of drop A = 1900 μm, initial moisture content = 2.6 kg/kg, dry basis; drop B = 1470 μm initial diameter and 2.4 kg/kg moisture.

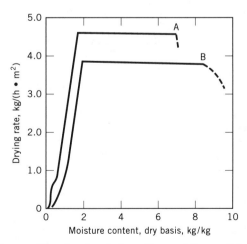

Fig. 10. Freeze drying profiles for A, whole milk, and B, nonfat milk. Heat was transmitted by radiation from heated wires above the frozen milk which rested in a transparent plastic tray. (– – –) is the induction period. Total pressure was 33 Pa (0.25 mm Hg).

and thus most types of industrial dryers are specially suited for certain forms of material. Dryers are also classified as being batch or continuous.

A batch dryer is best suited for small lots and for use in single-product plants. This dryer is one into which a charge is placed; the dryer runs through its cycle, and the charge is removed. In contrast, continuous dryers operate best under steady-state conditions drying continuous feed and product streams. Optimum operation of most continuous dryers is at design rate and steady-state. Periods of low rate operation are energy inefficient; shutdowns and start-ups waste fuel and frequently include periods of offgrade production. Continuous dryers are unsuitable for short operating runs in multiproduct plants.

The material suitability of industrial dryers may be summarized as

Dryer	Material form
spray dryer	pumpable, heat-sensitive pastes, slurries, and solutions; all pumpables at high capacities
indirect-heat drum dryer	pumpable, heat-insensitive pastes, slurries, and solutions
pneumatic conveyor dryer	materials instantly dispersible into discrete particles in the drying gas
fluid-bed dryer	fluidizable particulate materials
spouted-bed dryer	particulate materials too coarse or uniform in size to fluidize adequately
hopper dryer	preheated coarse or uniform materials and beds pervious to gas throughflow
direct-heat rotary dryer	particulate materials too coarse, sticky, or unpredictable to be fluidized or spouted
indirect-heat rotary dryer	fine, dusty materials
double-cone (vacuum) dryer	particulate materials that do not stick together, ball, or pelletize during drying
agitator (vacuum) dryer	particulate materials that may stick together, ball, or pelletize until almost dry
through-circulation or band dryer	materials that can be formed into static beds pervious to gas throughflow
continuous conveyors	continuous webs, paper, fabric, film, and fiber tow
batch, cabinet, and tray dryers	small lots, batch identification, and single-product plants less than 500 t/yr

Direct-Heat Dryers.   In direct-heat dryers, steam-heated, extended-surface coils are used for gas heating up to about 200°C. Electric and hot oil or vapor heaters are added for higher temperatures. Diluted combustion products are used for all temperatures. An increasingly popular technique for producing inert gas is to recycle the dryer exit gas and vapor as secondary dilution gas for incoming combustion products. Thereby the oxygen level in the dryer gas stream is reduced to safe levels for organic materials. This is usually less than 10% oxygen, but always material-dependent. These are called self-inerting heaters.

If material must be protected from combustion product contact, gas may be heated indirectly by passing it through tubes in a furnace. A clean, high temperature gas is obtained, but fuel efficiency is 50–70% of direct combustion gas heaters. Unless metal surfaces are protected by insulating or refractory lining (see REFRACTORY COATINGS), maximum usable gas temperature is about 1000°C. For low temperature operations, gas may be dehumidified. It usually is more economical to recirculate gas back through the dehumidifier after each dryer pass than to continuously dehumidify fresh gas. Polymers dried to very low moisture contents for extrusion or solid-state polymerization require very dry gas regardless of drying temperature. It is more economical to predry these materials as low as possible using ambient humidity gas before final drying in very dry or inert gas.

In most direct-heat dryers, more gas is needed to transport heat than to purge vapor. Larger dust recovery installations are needed than for indirect-heat dryers handling the same vapor load. Strict environmental regulations have eliminated the capital cost advantage of direct-heat dryers, eg, dust recovery investment for a modern spray dryer often exceeds the dryer investment. The greater the gas velocity over, through, or impinging upon a material, the greater the convection heat-transfer coefficient. The more completely material is dispersed, ie, greater surface-to-mass exposure, the faster the drying rate. Gas and material flowing in the same direction in a continuous dryer is called cocurrent flow; gas flow opposing material flow is countercurrent flow. Gas flow across a material is parallel flow or crossflow. Gas flow normal to material flow is impingement flow or through-circulation.

Batch Compartment Dryers. Direct-heat batch compartment dryers are often called tray dryers because of frequent use for drying materials loaded in trays on trucks or shelves. Figure 11 illustrates a two-truck tray dryer. The compartment enclosure comprises insulated panels designed to limit exterior surface temperatures to less than 50°C. Slurries, filter cakes, and particulate solids are placed in stacks of trays; large objects are placed on shelves or stacked in piles. An important design requirement is to assure gas flow uniformity, top-to-bottom of the compartment and back-to-front. This is essential but consumes fan power. Unless the material is dusty, gas is recirculated through an internal heater as shown. Only enough purge is exchanged so as to maintain needed internal humidity. For

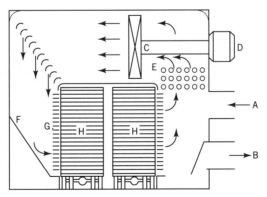

Fig. 11. Two-truck tray dryer. A, air inlet duct; B, air-exhaust duct with damper; C, axial flow fan; D, fan motor, 2–15 kW; E, air heaters; F, air-distribution plenum; G, distribution slots; and H, wheeled trucks and trays. The arrows indicate air and vapor flow pattern.

inert gas operation, purge gas is sent through an external dehumidifier and returned.

These dryers are economical only for a single-product rate less than 500 t/yr, large objects on drying cycles greater than 8 h, multiproduct operations, and batch identification. Tray loading depth and spacing must be uniform throughout the compartment. Tray loading is usually 2–10 cm. Parallel flow gas velocity is 1–10 m/s. Two-speed or variable speed fans are employed to provide higher gas velocity over the material during early drying stages. To minimize dusting, the fans reduce velocity after constant rate drying when heat transfer at the material surface is no longer the limiting drying mechanism. Deep tray loading reduces labor, but reduces overall capacity because falling rate drying time usually varies with the square of the loading depth. Shallow loading yields faster drying, but care is needed to ensure depth uniformity and labor is increased. Metal trays enhance heat flow through the tray walls and bottom. Screen-bottom trays permit vapor escape through the bottom; ie, two-side drying. Dryer efficiency should be 50–70%. Based on exposed material surface, vaporization rates are 0.2–2.0 kg/(h·m^2).

Through-circulation compartments employ perforated or screen bottom trays and suitable flow baffles so gas is forced through the material. If material is not inherently pervious to gas flow, it may be mechanically shaped into noodles, pellets, or briquettes. These dryers are used in small-scale operations to dry explosives, foods, and pigments. Dryer efficiency is 50–70%. Based on tray area, water vaporization rates are 1–10 kg/(h·m^2).

Continuous Conveyors.   Continuous conveyors are characterized by continuous material flow without mixing. Dryer residence time is uniform for all material increments. Tray and tunnel conveyors comprise long insulated compartments through which material is moved on trucks or trays fastened together. Gas flow is usually parallel to material surface and may be cocurrent, countercurrent, or crossflow through recirculation fans and reheaters installed on each side of the compartment. Conveyor movement may be continuous, but usually must be interrupted periodically for introduction to new work at the wet end. Performance is otherwise comparable to batch compartments. Figure 8 data were obtained from a continuous conveyor dryer in which the glass plates were suspended vertically from an overhead chain conveyor.

Turbotray Dryers.   The turbotray dryer is a continuous tray dryer comprising a stack of circular trays rotating slowly inside a vertical, insulated, cylindrical housing. Each rotating tray has uniformly spaced radial slots through which material is discharged to the tray below by a stationary plough once per revolution. Material falling through a slot is leveled to a uniform depth on the tray below by a stationary rake. After another revolution the process repeats. Wet material fed onto the uppermost tray moves down the stack in this manner and exits at the bottom. Circulating fans are carried on the central rotating shaft. Gas reheaters are mounted on the housing walls and gas flows across the trays parallel to material surface. Free flowing, nonsticky, and nondusty materials are dried more rapidly than in static beds. Having a stationary housing, the dryer finds employment in inert gas, solvent recovery (qv), and sublimation processes.

Foam-mat drying is a process in which a suspension, slurry, or solution is transformed into a stable foam by inert gas injection. The foam structure provides porosity and the mat is dried in trays or on a belt in a tunnel compartment, either under vacuum or with circulating gas. A free-flowing powder capable of rapid rehydration results. Fruit juices (qv) are dried successfully in this manner.

Continuous Web Dryers.   Web dryers are used for polymer films, paper (qv), cloth, nonwoven fabrics (qv), printed and coated films, and printed fabrics. Gas impinges on or flows parallel to the moving material, called a web, that is supported by various methods. Electric and gas-fired radiant heaters also are usable on some dryer types. On a festoon conveyor the web is draped over sticks that are carried on chains through a heated enclosure. The web is unrestrained and free to shrink or stretch. Gas flow must be comparatively gentle to avoid excessive material movement. On single or multipass roll conveyors web is conveyed either vertically or horizontally over a series of driven rolls while web tension is controlled by differential roll speeds. The rolls are crowned slightly to hold axial alignment, but there is no restriction to lateral shrinkage. Because the web is restrained axially, however, high velocity gas impingement slots or nozzles may be employed on one or both faces. Radiant heaters are used in these dryers. For one-side drying of printed or coated webs, roll conveyors often are installed in ceiling-hung housings to conserve floor space. Tenter frames restrain a web in two directions and are employed to control shrinkage or to stretch a web during drying. It is an ideal setup for two-side gas impingement heating because the nozzles or orifice plates can be mounted close to the stretched web. The closer the spacing between the nozzles and web, the more effective are the impinging jets for heat and mass transfer. Tenter frames are also ideal for electric and gas-fired radiant heaters. Float dryers have closely mounted nozzles both for heat and mass transfer and to support and convey with minimum web tension. For long drying times, festoon or multipass roll conveyors are economical because a long length of web can be contained in a tall enclosure that occupies little floor space.

Through-Circulation Dryers.   In through-circulation dryers, permeable materials are conveyed through enclosures on perforated plate or screen conveyors. The enclosures comprise a series of independent compartments, each having its own fans and recirculating gas heaters. Humid gas is removed at the material feed end of the enclosure; fresh dry gas is introduced at the dry end. Conveyor widths are 0.5–5.0 m; length may be 50 m in single or multiple conveyor tiers. Textile fibers, elastomer crumbs, plastic pellets, vegetables, and centrifuge and filter cakes are dried in 1–20-cm deep layers. On a conveyor area basis, drying rates up to 50 kg/(h·m^2) of water are obtained at gas temperatures up to 400°C. Dryer efficiency is 50–70%. Centrifuge and filter cakes are preformed by extrusion into small noodles or by granulation in knife mills. Thin pastes and slurries are predried on indirect-heat drum dryers to form short sticks. Shear-thinning materials are scored and cut in small pieces. Powders are briquetted or pelleted (see POWDERS, HANDLING). Pin elevator feeders open fibrous clumps and lay a uniform bed on the conveyor. Fiber tow is distributed by oscillating chutes. Materials that shrink during drying may be redistributed on a second conveyor to prevent gas bypassing by ensuring full conveyor coverage.

Perforated-drum dryers are through-circulation conveyors specially suited for fiber staple, tow, and nonwoven fabrics. Material is continuously supported and conveyed on a series of perforated screen-covered suction drums installed in compartments similar in form to horizontal conveyor compartments. Compared to the former, drying is more uniform because the material is turned over as it passes from one drum to the next. Drying rates are greater than on perforated plate and screen conveyors because greater pressure drop and higher gas velocities can be taken through the drums. Because material is retained by drum suction, edge sealing is less of a problem as well.

Dispersed-Particle Dryers. Through-circulation conveyors realize relatively high drying rates because gas flows through the material and contacts more material surface than do parallel flow, crossflow, and impingement arrangements. Nonetheless, if layer depth or material porosity is not uniform, gas channels through thin areas or larger passages and drying is not uniform. For drying particulate material, a better arrangement is one in which particles are separated completely, so that gas can flow freely among them. The drying rate for all particles of a given size then should be uniform. This is the purpose of all dispersed-particle dryers. Each is intended to provide optimum conditions of particle separation and surface exposure for materials having specific material handling requirements. A disadvantage, compared to continuous conveyors, is the loss of material plug-flow. Particle residence time in dispersed-particle dryers varies around an average and only the average can be calculated from feed rate, dryer fillage, and material density. Variations may be narrowed by various devices, but never eliminated completely. Figure 12 depicts relative particle residence time distributions among four dryers.

Rotary Dryers. A direct-heat rotary dryer is a horizontal rotating cylinder through which gas is blown to dry material that is showered inside. Shell diameters are 0.5–6 m. Batch dryers are usually one or two diameters long. Continuous dryers are at least four and sometimes ten diameters long. At each end, a stationary hood is joined to the cylinder by a rotating seal. These hoods carry the inlet and exit gas connections and the feed and product conveyors. One hood also attaches to the inlet gas heater. For continuous drying, the cylinder may be

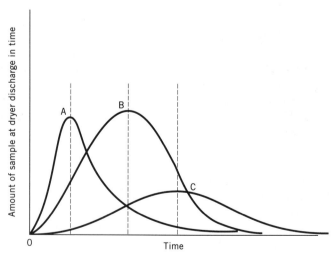

Fig. 12. Material residence time profiles in A, a pneumatic conveyor, B, a spray, and C, a rotary or fluid-bed dryer.

slightly inclined to the horizontal to control material flow. An array of material showering flights of various shapes is attached to the inside of the cylinder, as shown in Figure 13. Knockers are use to dislodge wet material that sticks to metal surfaces. A flight circle is usually 0.5 diameters long and adjacent circles are offset to minimize gas bypassing. Dry product may be recycled for feed conditioning if material is too fluid or sticky initially for adequate showering. Slurries also may be sprayed into the shell in a manner that the feed strikes and mixes with a moving bed of dry particles. Material fillage in a continuous dryer is 10–18% of cylinder volume. Greater fillage is not showered properly and tends to flush toward the discharge end.

Gas flow in these rotary dryers may be cocurrent or countercurrent. Cocurrent operation is preferred for heat-sensitive materials because gas and product leave at the same temperature. Countercurrent operation allows a product temperature higher than the exit gas temperature and dryer efficiency may be as high as 70%. Some dryers have enlarged cylinder sections at the material exit end to increase material holdup, reduce gas velocity, and minimize dusting. Indirectly heated tubes are installed in some dryers for additional heating capacity. To prevent dust and vapor escape at the cylinder seals, most rotary dryers operate at a negative internal pressure of 50–100 Pa (0.5–1.0 cm of water).

Direct-heat rotary dryers are the workhorses of industry. Most particulate materials can somehow be processed through them. These dryers provide reasonably good gas contacting, positive material conveying without serious backmixing, good thermal efficiency in either cocurrent or countercurrent use, and good flexibility for control of gas velocity and material residence time. Usual drying rates are 10–50 kg water/(h·m^3) of cylinder volume.

Fig. 13. Partial view through a direct-heat rotary dryer.

Fluid and Spouted Beds.   A fluid bed of particulate material is produced by introducing gas through a perforated plate, bubble caps, nozzles, or a ceramic grid beneath a static bed of material in such a manner that the solids are lifted uniformly and the particulate material and gas behave together like a boiling fluid (see FLUIDIZATION). For drying, the upward gas velocity is less than the terminal settling velocity of the particles, so few particles are conveyed out of the bed. At the same time, gas bubbles rise fast enough to lift particles directly above them. Particle motion is violent and a fluid bed exhibits intensive splashing at its surface. A substantial freeboard is included above the bed in the fluidizing vessel to allow particle disentrainment and full-back into the bed. Possibly because of a gas cushion that surrounds each particle as it circulates in the bed, particle attrition is usually moderate. For proper fluidization, it is essential that sufficient pressure drop be taken through the gas distributor so that the gas is distributed uniformly across the entire bed area independently of bed depth or bed behavior.

Spouted beds are used for coarse particles that do not fluidize well. A single, high velocity gas jet is introduced under the center of a static particulate bed. This jet entrains and conveys a stream of particles up through the bed into the vessel freeboard where the jet expands, loses velocity, and allows the particles to be disentrained. The particles fall back into the bed and gradually move downward with the peripheral mass until reentrained. Particle-gas mixing is less uniform than in a fluid bed.

Hopper Dryers.   Gas-purged hopper dryers are used for granular materials that need holdup times measured in hours. Gas flow may be upflow or crossflow. Drying of pelleted and extruded animal feeds at less than 100°C are typical operations. Applications also include the continuous final drying of polymer pellets, such as nylon and polyester at temperatures of 150–200°C, prior to melt extrusion or solid-state polymerization. Gas flow is countercurrent to material flow and rarely exceeds 0.25 m/s. Because flow is insufficient to provide needed sensible heat, it is necessary to preheat the polymers to drying temperature before introduction into the hoppers. Drying usually starts at 0.1–0.5% moisture, so evaporative thermal loads are small. The hoppers serve essentially as holding vessels, at temperature, to permit release by diffusion of minute quantities of moisture. This prevents polymer degradation during later processing. A free-flowing character is the principal material requirement. Uniform material holdup is the principal hopper requirement, so all particles are retained the minimum required time. Average holdup usually is 2–3 times the minimum required.

Fluid and spouted beds offer ideal conditions for drying provided the feed material is consistently suitable for fluidization or spouting; however, if the drying operation is preceded by mechanical liquid separation, eg, centrifugation, use of these dryers should be considered with caution. Fluid and spouted beds do not tolerate sticky materials and oversize lumps. Successful applications are particulate and pelleted polymers, grain, sand, coal, and mineral ores, applications wherein the physical size and character of the feed material is known and controllable 100% of the time. Fluid and spouted beds are attractive for inert gas and organic liquid drying because the vessels are stationary. Superheated steam drying is carried out in fluid beds. This is an attractive alternative environmentally, but a process which was stalled for many years because of the lack of a suitable process vessel. The volumetric drying capacity of a fluid bed is many times that of a rotary dryer. The reason is that gas flowing through the latter moves between a series of parallel particle curtains in which the gas must be entrained and mixed to contact particle surfaces. In the former, small bubbles of gas enter through the distributor and immediately penetrate and mix with a cloud of particles. Figure 14 shows that whereas the dryer efficiency of a cocurrent rotary dryer and fluid bed may be comparable, because both are single-stage vessels, the vessel size requirements are quite different. To approach the dryer efficiency of a countercurrent rotary dryer, two or more fluid beds with countercurrent gas flow must be operated in series. Figure 15 shows one form of a two-stage fluid bed.

A vibrating conveyor fluid bed dries while conveying particulate material on a screen-covered perforated deck. Gas is blown up through the material as it is conveyed mechanically and a particle dispersion much like that in a shallow fluid bed may be produced. Both mechanical and fluid energy contribute to fluidization. To minimize dusting, a lower fluidizing velocity is used than in a stationary fluid bed. Bed depth rarely exceeds 50 mm because mechanical energy is not transmitted through deeper beds effectively. Mechanical conveying encourages plug flow and several temperature stages may be incorporated in a single conveyor,

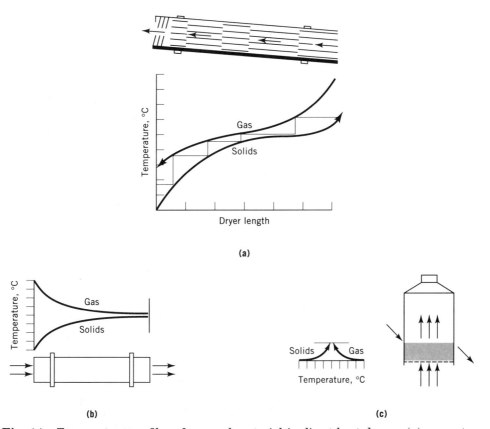

Fig. 14. Temperature profiles of gas and material in direct-heat dryers: (**a**) a countercurrent rotary dryer; (**b**) a cocurrent rotary dryer; and (**c**) a single-stage fluid bed.

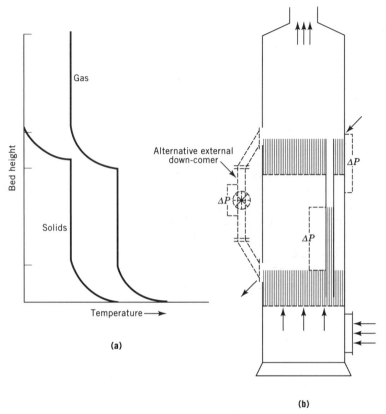

Fig. 15. A two-stage fluid bed dryer: (**a**) gas and material temperature profiles; (**b**) bed arrangements. ΔP = pressure drop through the upper stage distributor and bed.

but maximum material residence time is about 5 min. As in stationary fluid beds, all feed material must be nonsticky and free-flowing.

Pneumatic Conveyors. Conveyors are adapted for drying by heating the conveying gas, although for drying, gas-to-material ratios must be greater than those sufficient for conveying (qv). Particle residence time is only a few seconds; in fact, most drying takes place near the feed point where the velocity difference between gas and material is the greatest. Conveying tubes rarely need to be over 10 diameters long. For drying accompanied by deagglomeration, two or three pneumatic conveyors may be used in series and dry product may be recycled to the first stage for feed conditioning. Figure 16 shows a simple dryer consisting of a venturi feeder, tube, and product collector. Knife, hammer, and roller mills are alternative feeding devices. A paddle conveyor, its paddles inclined to retard material flow, may be installed in place of the venturi to increase residence time and enhance dispersion. Conveying tube gas velocity is usually 25–35 m/s; venturi throat velocity is 100–140 m/s. The conveyor is a low power fluid energy mill and particle attrition may be severe. This dryer is single-stage and cocurrent, like the cocurrent rotary, but has much lower residence time. Indirect heat may be combined with direct heat by jacketing the conveying tube.

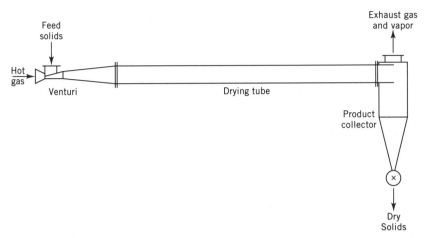

Fig. 16. Single-stage pneumatic conveyor dryer using a venturi for material acceleration and gas–solids mixing.

The principal applications of pneumatic conveyors are for materials that are nonsticky and readily dispersible in the gas stream as drying must be entirely constant rate. Many are employed as predryers ahead of longer residence time fluid-bed and rotary dryers in polymer drying operations.

Spray Dryers. A spray dryer is a large, usually vertical chamber through which hot gas is blown and into which a solution, slurry, or pumpable paste is sprayed by a suitable atomizer. Three atomizers are commonly used: (*1*) two-fluid pneumatic nozzles for very fine particles, between 10 μm and 100 μm, at rates less than 2 t/h; (*2*) single-fluid pressure nozzles for large particles, 125–500 μm, for dust-free products; and (*3*) centrifugal atomizers for various particle sizes at rates up to 150 t/h. The largest spray-dried particle is about 1000 μm; the smallest is about 5 μm. For large capacity dryers, the first two atomizers require multiple-nozzle setups, which may result in spray interference and particle agglomeration. In two-fluid nozzles, particle size is controlled by atomizing fluid pressure, usually compressed air or steam, and atomizing fluid-to-liquid ratio; therefore, very fine particles are obtainable with high fluid-to-liquid ratios and these nozzles are preferred for small, low capacity dryers. Pressure nozzles are limited by the fact that a change in feed rate or atomizing pressure causes a significant change in particle size distribution unless the nozzle orifice size is changed; high pressure feed pumps are required and, with abrasive materials, orifice wear may be rapid with a consequent increase in particle size distribution. Centrifugal atomizers are characterized by large capacity ranges, simple feed systems, narrow particle size distributions, and ease of particle size control by disk speed changes.

Because all drops must reach a nonsticky state before striking a chamber wall, the largest drop produced determines the size of the drying chamber. Chamber shape is determined by nozzle or disk spray pattern. Nozzle chambers are tall towers, usually having height/diameter ratios of 4–5. Disk chambers are large diameter and short, height being fixed by the fact that the discharge cone slope must be at least 60°, preferably 70° to discourage dry product accumulation on the sloping wall. For any evaporative load, chamber volume may be estimated by assuming a usable inlet gas temperature, exit gas and product at 100°C, and an average gas residence time of 15 s based on total chamber volume and exit gas humid volume, ie, from a simple dryer heat balance. This calculation does not obviate pilot-plant demonstrations, of course.

A spray dryer may be cocurrent, countercurrent, or mixed flow. Cocurrent dryers are used for heat-sensitive materials because relatively high inlet gas temperatures, up to 800°C, may be used while holding the exit gas and product near 100°C. Material temperature usually does not exceed the exit gas temperature provided chamber wall sticking is avoided. At any rate, a maximum dryer inlet gas temperature is about 1100°C because of limits for materials of construction. Countercurrent spray dryers yield higher bulk density products and minimize hollow particle production. Figure 17 shows an open-cycle, cocurrent, disk atomizer chamber with a pneumatic conveyor following for product cooling. Alternative cocurrent chamber discharge arrangements, two of which accomplish particle classification, are shown in Figure 18. The scheme of a closed-cycle, gas recirculation dryer, the self-inerting system, is shown in Figure 19. Spray dryers are often followed by fluid beds for second-stage drying or fines agglomeration. Some two-stage dryers make the fluid bed the bottom of the spray chamber, in effect copying the method of slurry and solution drying obtained by spraying directly onto a fluid bed of dry particles (36). Overall, two-stage efficiency is improved.

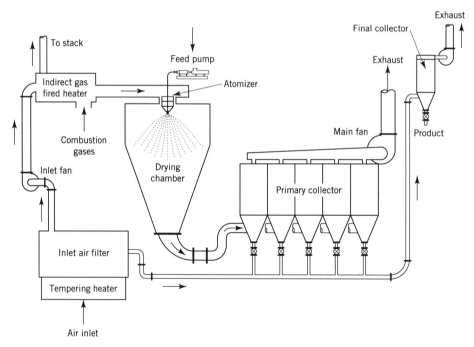

Fig. 17. Open-cycle, cocurrent, disk atomizer spray dryer. Courtesy of Niro Atomizer, Inc., Columbia, Md.

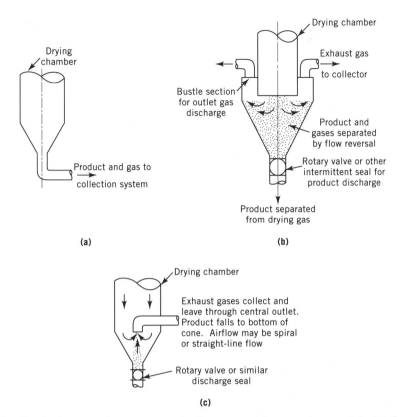

Fig. 18. Product removal arrangements for cocurrent spray dryers: (**a**) simple outlet; (**b**) product separation in an agglomeration chamber; and (**c**) classifying cone to reduce collector load.

Spray dryer applications include coffee and milk powders, detergents, instant foods, pigments, dyes, and chemical reactions, eg, flue gas desulfurization.

Indirect-Heat Dryers. In indirect-heat dryers, heat is transferred mostly by conduction, but heat transfer by radiation is significant when conducting surface temperatures exceed 150°C. For jacketed vessels, steam is the common heating medium because the condensing-side film resistance is insignificant compared to material-side resistance. Hot water is circulated for low temperature heating. Heat-transfer oils or condensing organic vapors are used for high temperatures. Liquid film resistance to heat transfer is much greater than that of condensing vapor; therefore, liquids are better suited for simple heating jobs rather than for drying operations having high evaporative thermal loads. Indirect-heat rotary dryers and calciners operating at temperatures exceeding 200°C usually are furnace-enclosed. The cylinders are heated externally by electric or gas-fired radiant heaters and circulating combustion products. Regardless of heating medium or method, the primary heat-transfer resistance in indirect-heat drying is on the material side. The material-side heat-transfer coefficient is affected by the rapidity of material agitation, particle size, shape, porosity, density, and degree of wetness.

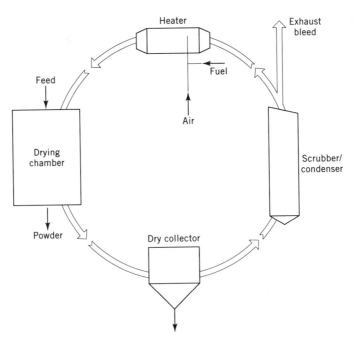

Fig. 19. Self-inerting spray dryer. Courtesy of Niro Atomizer, Inc., Columbia, Md.

Based on dryer cost alone, indirect-heat dryers are more expensive to build and install than direct-heat dryers designed for the same duty. As environmental concerns and resulting restrictions on process emissions increase, however, indirect-heat dryers are more attractive because they employ purge gas only to remove vapor and not to transport heat as well. Dust and vapor recovery systems for indirect-heat dryers are smaller and less costly: to supply heat for drying, gas throughput in direct-heat dryers is 3–10 kg/kg of water evaporated; indirect-heat dryers require only 1–1.5 kg/kg of vapor removed. System costs vary directly with size, so whereas more money may be spent for the dryer, much more is saved in recovery costs. Wet scrubbers are employed for dust recovery on indirect-heat dryers because dryer exit gas usually is close to saturation. Where dry systems are employed, all external surfaces must be insulated and traced to prevent vapor condensation inside.

Atmospheric Dryers. The rotary steam-tube dryer is a horizontal rotating cylinder in which are installed one or more circumferential rows of steam-heated tubes. These tubes extend axially the length of the cylinder and are connected to a steam and condensate manifold at one end. Figure 20 is a dryer installation that incorporates a dry product recycle system for feed conditioning. An essential component of this system is the recycle storage hopper which must be sufficient for one dryer fillage and kept full at all times, so that an empty dryer can be started without immediately fouling tube surfaces with nonconditioned feed. Cylinder diameters are 0.5–4 m; length may be 40 m. A large dryer carrying 1500 m² of tube surface in three circumferential rows may evaporate 8 t/h water. A nominal water evaporation rate in a dryer operating with 1.0 MPa (150 psig)

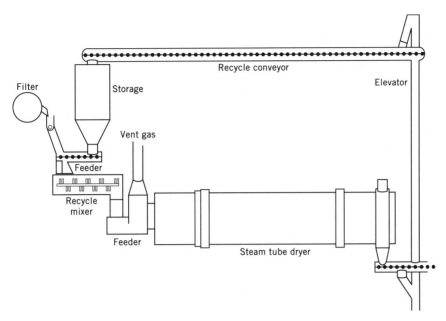

Fig. 20. Steam-tube rotary dryer using a dry product recycle and feed conditioning system.

saturated steam is 5 kg/(h·m²). Steam-tube dryers are built for steam-tube pressures up to 3.5 MPa (500 psig). Steam is introduced and condensate is removed through a rotary joint attached to the manifold at the product discharge end. The need for a rotary joint to introduce and remove the heating medium discourages the use of media other than steam and hot water. Feed is introduced and purge gas usually is removed through a stationary throat piece attached to the rotating cylinder by a sliding seal at the end opposite the manifold. The cylinder is slightly inclined to the horizontal to direct material flow and aid condensate drainage from the tubes. The diameter of the throat seal fixes maximum cylinder fillage, which rarely exceeds 20% of cylinder volume, but must be sufficient to ensure that the inner tube row is submerged in material.

To prevent dust and vapor escape at the cylinder seals, a negative internal pressure of 50–100 Pa (0.5–1.0 cm of water column) is maintained. Steam-tube dryers are suitable for any particulate material that can be conditioned so as not to stick to metal when dry. Because of relatively inexpensive heating surface and large capacities, these dryers are probably the most commonly used of the indirect-heat dryers. Gas- and vapor-tight seals sometimes are built for operations involving dangerous vapors and inert gas circulation, but these seals are expensive and high maintenance. Small installations excepted, stationary vessels are preferable.

Calciners. The indirect-heat rotary dryer or calciner resembles the direct-heat type except that the cylinder is enclosed in a furnace and showering flights are replaced by short turning bars attached to the inner cylinder wall, so that material rolls on itself but remains in contact with the wall. Heat is transferred by convection and radiation from the furnace heaters to the cylinder, then by conduction and radiation to the material. Cylinder volume fillage is limited to 5–8% to avoid formation of a core of unheated material in the center of the rolling bed. By choosing suitable metal alloys, cylinders can be fabricated for temperatures up to 1150°C, for drying operations that are too dusty for direct-heat dryers, at temperatures that are too high for steam-tube dryers. Evaporation rates for water are 5–15 kg/(h·m²), based on total cylinder surface. Cylinders less than 1 m in diameter are employed commonly for calcining and heat-treating operations which require special atmospheres, including hydrogen. Gas-tight rotary seals work successfully on these vessels because cylinder and seal diameters are relatively small.

Fluid-Bed Dryers. Indirect-heat fluid-bed dryers are usually rectangular vessels in which are installed vertical pipe or plate coils. Figure 21 is a diagram of a two-stage, indirect-heat fluid bed incorporating plate coil heaters. This dryer comprises a fully back-mixed section with a centrifugal feed distributor above, followed by a plug-flow section for final drying or cooling. The general design is used for drying several particulate polymers. Vertical plate coils provide heating surface up to 8 m²/m³ of bed volume. Excellent heat transfer is obtained in an environment of intense particle agitation and mixing. Water evaporation rates, based on panel surface, are 10–25 kg/(h·m²). The fluidizing gas is also the vapor purge gas but contributes little heat to the operation, all of which is transferred indirectly from the coils. Because of its favorable heat- and mass-transfer capabilities, flexibility for staging both temperatures and fluidizing gas velocities and the fact that the vessel is stationary, so rotary seals are not needed except for the

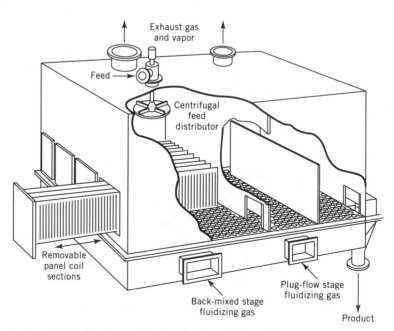

Fig. 21. Indirect-heat, two-stage, back mixed, and plug flow fluid-bed dryer. Courtesy of Niro Atomizer, Inc., Columbia, Md.

feed distributor shaft, an indirect-heat fluid bed is an ideal vessel for vapor recovery and drying in special atmospheres.

Screw Conveyors.   Indirect-heat screw conveyor dryers carry hollow, double-wall screws heated with saturated steam or hot oil. The conveyor trough may be jacketed as well, but the trough is only a small fraction of the total heating surface, and because there is a fairly wide clearance between the screw tips and the jacket, transfer coefficients are low. One trough may carry up to four parallel heated screws. A popular arrangement used for both batch and continuous drying consists of two screws that convey in opposite directions in a single trough. This internal recycle arrangement is used for drying slurries and solutions. A bed of dry particles is loaded, circulated, and heated. Feed material is sprayed onto the hot moving bed on one side of the dryer, mixed, and dried by contact with the circulating hot solids. Product is discharged through an overflow weir on the opposite side. Recycle rates as high as 2000 can be obtained. The feed rate is controlled so that the circulating dry solids contain enough sensible heat to fully dry the wet feed falling on and mixing with them before wet feed can contact heating surface. Purge gas is circulated through a vapor hood covering the screws.

Agitator Dryers.   Increasing interest in indirect-heat drying has brought out a variety of relatively slow speed, batch, and continuous agitator dryers. These combine the advantages of low purge gas flow, characteristic of all indirect-heat dryers, with a material handling versatility lacking in fluid beds and an ease of gas-tight operation lacking in rotary dryers. Material holdup may be varied from a few minutes to several hours. Figure 22 is an example of this dryer type, called a porcupine dryer. It may have one or two parallel shafts and, as shown, can be

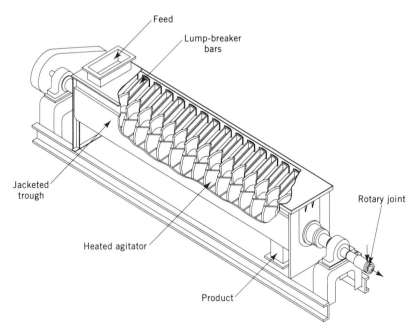

Fig. 22.   Indirect-heat, paddle-type agitator dryer. Courtesy of the Bethlehem Corp., Easton, Pa.

provided with stationary, lump-breaker bars that intermesh with the moving paddles. Other types may provide scrapers for both jacket and agitator heating surfaces. Most can be operated under vacuum or well above atmospheric pressure and at temperatures up to 400°C using hot oil, steam, and water. The vessels are stationary and shaft seals are small to minimize leakage.

Paddle dryers are used for drying lumpy materials. Disk dryers contain a series of closely-spaced, parallel, internally-heated plates mounted on the rotating shaft. Disk dryers are suitable for granular, essentially free-flowing materials; these often include adjustable scrapers for continuous disk doctoring to maintain a clean heating surface. Water evaporation capacity of these dryers when operated using 1.0 MPa (145 psig) saturated steam is about 15 kg/(h·m^2). Dryer capacities vary with type, eg, a single, quad-screw conveyor dryer can provide 150 m^2 of heating surface, whereas one disk dryer can carry 400 m^2 of heating surface on a single shaft. Agitator dryers are ideally suited for vapor recovery and drying in special atmospheres.

Both theory and empirical data demonstrate that the faster the movement of particulate material in contact with heated surface, the greater the heat-transfer rate. Nonetheless, agitator speed of the agitator dryers discussed herein rarely exceeds 10 min^{-1}. Because these dryers usually run between 50% and 90% full, mechanical stresses and power demand would become intolerable at higher agitator speeds, especially when drying sticky or sluggish materials. The tradeoff is between heat-transfer efficiency and fillage, power, and mechanical construction being dependent variables. For high fillage, long holdup, and the need to accommodate pasty, sticky, and sluggish materials, the best choice is usually a low speed and as much heating surface as possible built into the agitator.

High speed agitator dryers operate at about 10% fillage and are used for continuous evaporation of easily removable surface moisture. The example in Figure 23 consists of a stationary, horizontal, jacketed cylinder inside of which is carried an array of paddles mounted on a central shaft. Only the cylinder is heated. Paddle tip speed is about 15 m/s. The largest dryer has 100 m^2 of jacket surface. Water evaporation rate when operating using 1.0 MPa (145 psig) saturated steam is 20–25 kg/(h·m^2). To handle fluidlike or sticky feed materials, some high speed agitators include jacket scrapers. These dryers are suitable, however, only for materials that do not stick to metal when dry.

Drum Dryers.   Indirect-heat drum dryers, like spray dryers, are usable only for materials that are fluid initially and pumpable. Drying is effected by applying a thin film of material onto the outer surface of a rotating heated drum using applicator rolls, spray nozzles, or by dipping the drum into a reservoir. Usually the drum is cast iron or steel and chrome-plated to provide a smooth surface for ease of product release by doctoring. Drum rotational speed is such that drying occurs in a few seconds. Little thermal damage is experienced and acceptable milk powders have been produced on drum dryers. Nonetheless, material surface directly in contact with the drum reaches drum temperature, so heat effects are less favorable than in spray dryers. Single-, twin-, and double-drum arrangements are used, depending on material properties and feed method.

Drum dryers are operated at atmospheric pressure and enclosed in vacuum housings for heat-sensitive materials. Twin drums are merely two single drums using a common feed system. On double-drum dryers, feed is retained and par-

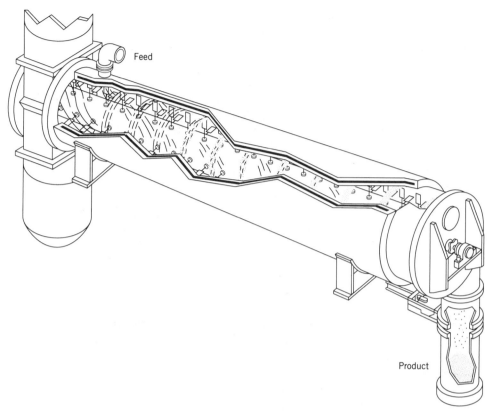

Fig. 23. Indirect-heat, high speed agitator dryer. Courtesy of Bēpex Corp., Minneapolis, Minn.

tially concentrated in a reservoir formed by the nip between two drums. Drum clearance is adjusted to fix film thickness. Material in the nip is drawn through the nip clearance, dried, and released from the back side of each drum by spring-loaded doctor knives. Single and twin drums may be provided with nip applicators by installing auxiliary feed rolls (see COATING PROCESSES). Water evaporation rates on drum dryers, based on total drum surface, are $50-80 \ kg/(h \cdot m^2)$. Greater capacities usually follow higher rotational speeds and thinner films.

The fin-drum dryer is a preforming device for fluid and pastelike materials intended for through-circulation drying. Slurry or paste is forced by a feed roll into circumferential grooves machined in the outer drum surface, partially dried and released in the form of short sticks by finger scrapers. Total drying rarely is attempted. Thin films of solutions, slurries, and pastes also are dried on horizontal belt dryers which are heated by radiant heaters mounted above and below the belt and operate at atmospheric pressure or under vacuum. Thicker films are handled than on drums because residence time can be longer. Temperature staging is feasible.

Can dryers, also called cylinder dryers, are similar in construction to drum dryers and used to dry paper, fiber tow, cloth, and other continuous webs that are insufficiently self-supporting for accommodation by festoon, roll, or tenter-frame equipment. A can dryer may be one 3–5-m diameter can, eg, the Yankee dryer, or it may comprise a number of cans arranged so that material passes over them in series, eg, a paper machine. To enhance conduction heat transfer initially, the web may be forced against the cans by an endless fabric belt, or felt, that also absorbs liquid and is dried separately. Further along, gas impingement nozzles may be mounted close to the web surface to add convection heat transfer. Humid air is removed through hoods above the cans. Paper (qv) drying is the largest application.

Vacuum Dryers. The indirect-heat form of batch compartment dryer usually operates under vacuum and is called a vacuum shelf dryer. Wet material is spread on trays that rest on heated shelves in an insulated vacuum chamber. The shelves are heated by steam, hot oil, or water and vacuum is produced by steam jets or pumps (see VACUUM TECHNOLOGY). Heat is transferred by conduction and radiation from the supporting shelf and by radiation from the shelf above the material. Conduction heat-transfer rates are low, however, because contact between the tray bottom and its supporting shelf rarely is continuous or uniform. In chambers maintained at 1.5 kPa (0.22 psi) pressure, with shelves heated by 200 kPa (29 psig) saturated steam, water evaporation based on exposed tray area is 1–2 (kg/(h·m^2). Because of low drying rates, dust losses are negligible, and these dryers are suitable for small lot drying of valuable products. Batch identification is maintained, but if there are alternative choices, shelf dryers rarely are economical for production rates exceeding 200 t/yr.

A rotating vacuum dryer is formed by equipping a double-cone mixer with a jacket and an internal vapor exit tube passing through a rotary joint in one trunnion. Volume capacity is 0.1–30 m^3. Fillage is 50–70% of total volume. Internal operating pressures are 1–10 kPa (0.15–1.5 psi). In dryers operated at 1.5 kPa, having jackets heated by 200 kPa saturated steam, water evaporation based on total heated surface is 4–5 kg/(h·m^2). Rotating vacuum dryers are suitable for materials that do not stick to metal when wet or dry and do not pelletize during drying. Feed conditioning is an option. The ratio of jacket surface to operating volume decreases as dryer size increases, so large dryers often include internal plate or pipe coils to compensate. These internal elements partially destroy a principal attraction of the dryers, however, which is ease of complete emptying and cleaning between batches.

The rotary vacuum dryer is a horizontal stationary jacketed cylinder having an internal rotating ribbon or paddle agitator. If material does not stick to metal when wet or dry, the rotating shaft, ribbon arms, and paddles also may be heated. For materials that are sticky, jacket scrapers can be included. Feed conditioning by wet/dry blending externally and inside the dryer are options. Volume capacity is 0.1–30 m^3. Fillage is 50–90% of total volume. In dryers operated at 1.5 kPa, using jackets and internals heated by 200 kPa saturated steam, water evaporation based on total heated surface is 5–7 kg/(h·m^2). Agitator speed is 2–8 min^{-1}, but dust carryover may be severe during initial drying. Vacuum bag-type dust collectors usually are provided to recover dust. Rotary dryers are more versatile than the rotating type, but are difficult to empty completely and are less attractive for multiproduct operations and batch identification.

The vacuum pan dryer, the workhorse of this vacuum group of batch dryers, is a vertical stationary jacketed cylinder having a jacketed dished or flat bottom and a vertical top-driven plough-type agitator to overcome torque loads presented by heavy, sticky, and doughlike materials that would overload or break the ribbons and paddles in rotary vacuum dryers. The agitator stirs these heavy materials at $1-4$ min^{-1} until they are dry enough to break down into particulate form. Power usually peaks just before the material breaks apart. The largest pan is about 4 m diameter. Maximum fillage is about 10 m^3. In pans operated at 1.5 kPa pressure, using jackets heated by 200 kPa saturated steam, water evaporation based on total heated surface is $2-4$ kg/(h·m^2). These dryers also operate at atmospheric pressure; a purge gas is employed for vapor removal.

Freeze Dryers. The original freeze dryer was a vacuum shelf dryer, operated at much lower pressure: 100 Pa (0.8 mm Hg) for seafood, meat, and vegetables, 50 Pa (0.4 mm Hg) for fruits, 20 Pa (150 μm Hg) for concentrated beverages, and 10 Pa (80 μm Hg) for pharmaceuticals (qv). The material is first frozen to effect separation of solutes and solvents by crystallization. Frozen material is placed in trays in a closed compartment that is evacuated and ice is caused to sublime by careful introduction of heat. The purpose is to protect heat-sensitive materials from thermal damage and prevent shrinkage of porous materials, so they can be instantly and fully rehydrated (see FOOD PROCESSING). The sublimation driving force is the difference between the vapor pressure of the ice and the condenser pressure. The drying rate is controlled by heat input and the conductivity of the material.

Most rapid drying and uniform product quality is obtained when all material surfaces are heated uniformly. Use of metal rib trays is helpful because metal conducts heat better than most organic materials and these trays distribute heat more effectively bottom-to-top. Channels for vapor escape are also opened; however, bottom material may still overheat. A suspended rib tray depends on heat transfer entirely by radiation. Higher shelf temperatures may be used without a danger of local material overheating and both top and bottom are heated uniformly. Typical food dryer shelf temperatures of the suspended tray type are $50-150$°C, using a refrigeration system at -50°C. Many biological dryers contain movable shelves, so product vials can be stoppered after drying and before compartment venting. The vials are assembled in trays that rest directly on heated shelves. Typical shelf temperatures are -20°C to 50°C, the condenser temperature is -60°C. Based on exposed material surface, sublimation capacity of shelf-type food dryers is $0.2-2.0$ kg/(h·m^2).

Freeze drying has also been carried out at atmospheric pressure in fluid beds using circulating refrigerated gas. Vacuum-type vibrating conveyors, rotating multishelf dryers and vacuum pans can be used as can dielectric and microwave heating.

Radiant-Heat Dryers. Heat transfer by radiation occurs in all dryers to some degree and is controlled by the temperature and emissivity of the source and the temperature and absorptivity of the receiver. For drying, sources may consist of a number of incandescent lamps, reflector-mounted quartz tubes, electrically heated ceramic surfaces, and ceramic-enclosed gas burners. Usual source temperatures are $800-2500$ K. Radiant energy does not penetrate most material surfaces. Heat penetration below the surface is dependent on material conductivity. In situations wherein radiant-heat flow to a surface is high while material thermal conductivity is relatively low, the surface temperature may rise above the liquid boiling point at dryer operating pressure. When drying printed cloth, film, and coatings in continuous conveyor dryers, adjustments to source spacing and temperature often can prevent boiling, skin formation on films, and bubble formation in coatings. For thicker materials, radiant-heat sources are installed alternately with gas impingement and parallel flow zones of more moderate temperature to allow time for liquid diffusion to the material surface, evaporation, and vapor dispersion.

Radiant heaters are most suitable for the drying of thin films, eg, paint films. They are not suitable for large objects and deep material layers in which drying rates are controlled by material internal heat- and mass-transfer mechanisms. When drying heat-sensitive materials, low temperature sources should be used. In continuous dryers, banks of radiant sources are placed above, below, and on both sides of the material in an enclosed tunnel designed to minimize direct and reflection losses to the outside. When materials are dried that may degrade or burn if exposed too long, means are provided to shut off and shutter all radiating surfaces instantly when material flow is interrupted. Purge gas must be provided to remove vapors from atmospheric radiant-heat dryers. A common practice is to pass the incoming purge gas behind the source enclosures to cool the enclosures and preheat the gas. On roll conveyor and tenter frame dryers, water evaporation based on exposed material surface is $10-100$ kg/(h·m^2).

Dielectric and Microwave Dryers. Dielectric, also called radio frequency, dryers operate in the frequency range of $1-100$ MHz. Microwave dryers in the United States operate at 915 MHz and 2450 MHz (see MICROWAVE TECHNOLOGY). As depicted in Figure 24, a dielectric dryer may consist of two flat metal plates between which material is placed or conveyed. The arrangement forms a capacitor, the plates of which are connected to a high frequency generator. During one-half of a cycle, one plate has a positive charge, the other a negative charge. One-half cycle later, the charges are reversed. Flat plate or platen electrodes are used for bulky objects. Parallel rods of alternating charge, called stray field electrodes, are employed for thin webs and are installed directly in line below or above the moving web. Parallel rods, called staggered-type electrodes, are used for thick webs and boards and are installed alternately above and below the material. The web moves between them.

Microwave applicators are single, like a microwave oven, or multimode cavities in which material is placed or through which it is conveyed, or rectangular waveguides which in effect surround material as it is conveyed. Rapid reversal of

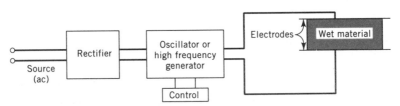

Fig. 24. Diagram of a dielectric (radio frequency) dryer.

electrode polarity generates heat in the material. In a mechanism called dipole rotation, dipoles, which normally are in random orientation, become ordered in the electrical field. As the field dies, they return to random orientation; as the field reverses, they again become ordered but in the opposite direction. Electrical energy is converted to potential energy, to random kinetic energy, and to heat. In ionic conduction, ions are accelerated by the electrical field. They collide with nonionized molecules in random billiard ball fashion. Electrical energy is converted to kinetic energy and to heat. These are two primary mechanisms of energy conversion.

Industrial applications of dielectric and microwave energy for drying are many; however, response to high frequency electromagnetic radiation depends on a material's dielectric constant and dissipation factor, the product of which is its loss factor. A material having a loss factor greater than 0.05 is a potential drying candidate. Air, glass, some ceramics, and plastics transmit high frequency radiation. Metals may reflect radiation. Water, alcohols, aldehydes, ketones, unsymmetrical halogenated hydrocarbons, and ionic solutions absorb radiation. Hydrocarbons and symmetrical halogenated hydrocarbons do not. Depth of material penetration at which half the energy is absorbed is called a half power depth and is proportional to wave length. For dielectric heating, this may be measured in meters; for microwaves, in centimeters. If the material is large or wide, dielectric heating is preferred. If watt density is high because of a low loss factor, microwaves are preferred; however, if power requirement exceeds 50 kW, economics favor dielectric equipment.

The cost of microwave equipment per kilowatt output is about twice that of the dielectric. For irregular shapes, microwaves are preferable because to avoid hot spots during heating, dielectric electrodes are needed that conform to the material shape. Industrial dielectric dryers are employed for lumber drying, plywood bonding and drying, furniture parts drying, textile skeins and package drying, paper moisture leveling, tire cord drying, and many food products. Dielectric heating frequently is combined with radiant heat and hot air for print and coating drying. Microwave dryers are employed for drying cloth, lumber, and foods. Microwaves are used as an energy source in vacuum and freeze dryers.

Dielectric and microwave heating are generally more costly than alternative methods. Thus many applications involve material preheating and second-stage drying where energy demand is low and cycle times can be reduced significantly. Dielectric and microwave heating are chosen mostly when other methods will not work or are impractical. Material behavior in high frequency electromagnetic fields is frequency-dependent and varies with moisture content, salt concentrations, and other factors. Laboratory testing is necessary. Water evaporation is about 1.0 kg/kWh, and overall power efficiency is about 60% (see FURNACES, ELECTRIC).

BIBLIOGRAPHY

"Drying" in *ECT* 1st ed., Vol. 5, pp. 232–265, by W. R. Marshall, Jr., University of Wisconsin; in *ECT* 2nd ed., Vol. 7, pp. 326–378, by W. R. Marshall, Jr., University of Wisconsin;

in *ECT* 3rd ed., Vol. 8, pp. 75–113, by P. Y. McCormick, E. I. du Pont de Nemours & Co., Inc.

1. D. W. Green, ed., *Perry's Chemical Engineers' Handbook*, 6th ed., McGraw-Hill, Inc., New York, 1984, pp. 20-14.
2. F. W. Dittman, *Chem. Eng.* **84**(2), 106 (1977).
3. J. H. Perry, ed., *Chemical Engineers' Handbook*, 3rd ed., McGraw-Hill, Inc., New York, 1950, pp. 813–817.
4. Ref. 1, Sect. 20.
5. S. F. Sapakie, D. R. Mihalik, and C. H. Hallstrom, *Chem. Eng. Progr.* **75**(4), 44 (1979).
6. *Carrier Psychrometric Chart, Catalog No. 794-005*, copyrighted by Carrier Corp., Syracuse, N.Y., 1975.
7. Ref. 1, Sect. 12.
8. W. L. McCabe, J. C. Smith, and P. Harriott, *Unit Operations of Chemical Engineering*, 4th ed., McGraw-Hill, Inc., New York, 1985.
9. T. K. Sherwood, R. L. Pigford, and C. R. Wilke, *Mass Transfer*, McGraw-Hill, Inc., New York, 1975.
10. Ref. 1, pp. 20-7 and 20-8.
11. Ref. 1, pp. 20-20 and 20-21.
12. Ref. 3, p. 804.
13. Ref. 1, p. 20-20.
14. H. Martin, in J. P. Hartnett and T. F. Irvine, Jr., eds., *Advances in Heat Transfer*, Vol. 13, Academic Press, Inc., New York, 1977.
15. E. U. Schlünder, *Heat Exchangers*, McGraw-Hill, Inc., New York, 1974, pp. 1–19.
16. E. U. Schlünder, *Heat Exchangers*, Hemisphere Publishing Corp., New York, 1981, pp. 177–208.
17. E. U. Schlünder, *Drying '80*, Vol. 1, Hemisphere Publishing Corp., New York, 1980, pp. 184–193.
18. N. Mollekopf and E. U. Schlünder, *Proceedings of the Third International Drying Symposium*, Vol. 2, Drying Research Ltd., Wolverhampton, UK, 1982, pp. 502–513.
19. M. L. Wiedmann and P. R. Trumpler, *Trans. Am. Soc. Mech. Eng.* **68,** 57–64 (1946).
20. Ref. 1, p. 3-282.
21. V. W. Uhl and W. L. Root, *Chem. Eng. Progr.* **58**(6), 37–44 (1962).
22. Ref. 1, p. 20-12.
23. Ref. 3, p. 808.
24. Ref. 3, p. 883.
25. H. H. Macey, *Trans. Br. Ceram. Soc.* **41,** 73 (1942).
26. C. W. Hall and A. S. Mujumdar, eds., *Drying Technology*, Vols. 1–11, Marcel Dekker, Inc., New York, 1983–1993.
27. R. E. Treybal, *Mass Transfer Operations*, 3rd ed., McGraw-Hill, Inc., New York, 1980, p. 91.
28. T. K. Sherwood and R. L. Pigford, *Absorption and Extraction*, McGraw-Hill, New York, 1952, pp. 1–28.
29. Ref. 1, p. 10–10.
30. Ref. 3, p. 462.
31. W. H. McAdams, *Heat Transmission*, 3rd ed., McGraw-Hill, Inc., New York, 1954, pp. 31–54.
32. Ref. 8, pp. 278–285.
33. Ref. 8, p. 796.
34. D. H. Charlesworth and W. R. Marshall, *AIChE J.* **6**(1), 9 (1960).
35. J. Lambert and W. R. Marshall, *Conference on Freeze-Drying of Foods*, National Academy of Sciences, National Research Council, 1962.
36. D. E. Metheny and S. W. Vance, in Ref. 21, pp. 45–48.

General References

D. W. Green, ed., *Perry's Chemical Engineers' Handbook*, 6th ed., McGraw-Hill, Inc., New York, 1984, Sect. 20, pp. 1–74.

J. H. Perry, ed., *Chemical Engineers' Handbook*, 3rd ed., McGraw-Hill, Inc., New York, 1950, Sect. 13, pp. 800–884. This remains the best edition on Drying as a unit operation.

C. M. van'tLand, *Industrial Drying Equipment*, Marcel Dekker, Inc., New York, 1991.

E. M. Cook and H. D. DuMont, *Process Drying Practice*, McGraw-Hill, Inc., New York, 1991.

A. S. Mujumdar, ed., *Handbook of Industrial Drying*, Marcel Dekker, Inc., New York, 1987.

J. L. Ryans and D. L. Roper, *Process Vacuum System Design and Operation*, McGraw-Hill, Inc., New York, 1986.

K. Masters, *Spray Drying Handbook*, 4th ed., Halstead Press, Inc., New York, 1985.

W. R. Marshall, *Chem. Eng. Progr. Monogr. Ser.* **50,** 2(1954). This monograph is still extremely useful.

R. B. Keey, *Drying of Loose and Particulate Materials*, Hemisphere Publishing Corp., New York, 1991.

R. B. Keey, *Introduction to Industrial Drying Operations*, Pergamon press, Elmsford, N.Y., 1978.

R. B. Keey, *Drying, Principles and Practice*, Pergamon Press, New York, 1972.

G. Nonhebel and A. A. H. Moss, *Drying of Solids in the Chemical Industry*, CRC Press, Cleveland, Ohio, 1971.

A. Williams-Gardner, *Industrial Drying*, CRC Press, Cleveland, Ohio, 1971.

PAUL Y. MCCORMICK
Drying Unincorporated

DRYING AGENTS.   See DESICCANTS.

DRYING OILS

Drying oils oxidize upon exposure to air from a liquid film to a solid, dry film. Linseed oil was used in making paints at least as early as the Roman Empire. By the nineteenth century, drying oils, and varnishes made from them, had become the primary vehicles for paints, artists' colors, printing inks, putty, oil cloth, and linoleum. Since the 1920s, drying oils increasingly have been replaced by other film-forming materials (see also COATINGS). In paints and coatings, replacements include alkyds, acrylic copolymers, polyesters, epoxy resins, vinyl acetate and acrylic copolymer latexes, and nitrocellulose, among others. In artists' colors, acrylic latexes have been the principal replacement. Hydrocarbon and phenolic resins, alkyds, nitrocellulose, and polyamides are among the replacements in printing inks. Drying oil putties have been replaced in large measure by glazing compounds made with polybutenes and polysulfides. Oil cloth has been replaced by supported and unsupported vinyl films, and linoleum has been supplanted by a variety of polymer-based tile materials.

Consumption of drying oils in the United States peaked in the late 1940s or early 1950s; in the early 1990s much smaller, but still significant, amounts are used. Resins, such as oxidizing alkyds, epoxy esters, and urethane oils (uralkyds), are synthetic drying oils made from drying and semidrying oils. The use of synthetic drying oil-based resins has exceeded the use of natural drying oils; however, their consumption is declining due to discoloration and embrittlement caused by continued oxidation after film formation.

Natural Oils

Occurrence and Isolation. Most drying oils are derived from plant seeds. The largest volume drying oil, linseed oil, is obtained from flaxseed, *Linum usitatissimum*. In the United States flax is grown in North Dakota, Minnesota, and South Dakota. Flax for oil is also raised in Canada, Argentina, India, and parts of the former USSR. Linseed oil is isolated by continuous pressing, ie, expelling, from the seed; further yield is obtained by solvent extraction. Soybean oil, the second most important oil, is obtained from soybeans, the seed of *Glycine Max* (L) Merrill. It is produced in the United States, Brazil, Argentina, and China. Without modification, it is a semidrying oil, not a drying oil. Perilla, safflower, sunflower, and walnut oils have limited uses as drying oils. Tung oil, also called wood oil or chinawood oil, is obtained from the seed kernels of the tung tree, *Aleurites fordii*. The principal source of the oil is China. Tung trees were grown in the Gulf states until the early 1970s, but a combination of freeze and hurricane damage and reduced demand resulted in cessation of U.S. production. Limited quantities of another conjugated oil, oiticica oil from the oiticica tree, *Licania rigida*, also are used. The only animal oils used as drying oils on a significant scale are marine fish oils, primarily from anchovy, menhaden, pilchard, and sardines. Fish oil is isolated by steam treatment of the fish. For use as drying oil, the fraction of esters of saturated fatty acids is reduced by winterizing, ie, cooling followed by filtering off the solid, primarily saturated, triglycerides that freeze out. Residual saturated triglycerides act as plasticizers (qv).

Castor oil, derived from the beans of *Ricinus communis*, is converted to a drying oil by heating with catalysts to yield dehydrated castor oil.

Trees, especially conifers, contain tall oils. Tall oil is not isolated directly; tall oil fatty acids are isolated from the soaps generated as a by-product of the sulfate pulping process for making paper. Refined tall oil fatty acids are obtained by acidification of the soaps, followed by fractional distillation to separate the fatty acids from the rosin acids and terpene hydrocarbons that also are present in the crude tall oil fatty acids (see CARBOXYLIC ACIDS; FATTY ACIDS FROM TALL OIL).

A wide variety of other oils have been investigated and, in many cases, used commercially over the years. More complete listings are available (1,2).

Composition and Analysis. Naturally occurring drying oils are triglycerides, ie, triesters of glycerol [56-81-5] (1,2,3-propanetriol) with mixtures of fatty acids (3). The reactivity of the oils results from the presence of esters of fatty acids with two or more nonconjugated double bonds separated by single methylene groups, —CH=CHCH$_2$CH=CH—, or those with two or more conjugated double bonds. The most common nonconjugated unsaturated fatty acids are linoleic acid [60-33-3] ((Z,Z)-9,12-octadecadienoic acid) and linolenic acid [463-40-1] ((Z,Z,Z,)-9,12,15-octadecatrienoic acid). Fish oils contain esters of eicosanoic and docosanoic acids with four to six double bonds separated by single methylene groups, eg, (all-Z)-4,7,10,13,16,19-docosahexaenoic acid [6217-54-5] in menhaden oil (4). The fatty acid with conjugated double bonds in tung oil is α-eleostearic acid [506-23-0] ((E,Z,E)-9,11,13-octadecatrienoic acid); the predominant fatty acid in oiticica oil is licanic acid [17699-20-6] (4-oxo-(E,Z,E)-9,11,13-octadecatrienoic acid). In all oils, other fatty acids including palmitic acid [57-10-3] (hexadecanoic acid), stearic acid [57-11-4] (octadecanoic acid), and oleic acid [112-80-1] ((Z)-9-octadecenoic acid) also are present.

Typical compositions of some of the more important oils are listed in Table 1. There can be significant variations in the compositions of natural oils with variations in plant strain, climate, soil, and other growth conditions. In general, oils derived from seeds grown in colder climates contain larger fractions of more highly unsaturated fatty acids as esters (6). Fish oils are not included in Table 1 because of the wide variation depending on the variety of fish and the degree of removal of the relatively saturated triglycerides (4). Tall oil fatty acids also have a wide range of compositions (7); eg, saturated acids, 2.5%; oleic acid, 51%; linoleic acid, 37%; geometric isomers of linoleic acid, 6%; other, 1.5%.

Oils are mixtures of mixed esters with different fatty acids distributed among the ester molecules. Generally, identification of specific esters is not attempted; instead the oils are characterized by analysis of the fatty acid composition (8,9). The principal methods have been gas–liquid and high performance liquid chromatographic separation of the methyl esters of the fatty acids obtained by transesterification of the oils. Mass spectrometry and nmr are used to identify the individual esters. It has been reported that the free fatty acids obtained by hydrolysis can be separated with equal accuracy by high performance liquid chro-

Table 1. Typical Fatty Acid Composition of Drying Oils From Seeds,[a] %

Oil	Saturated[b]	Oleic	Linoleic	Linolenic
linseed	10	22	16	52
perilla	7	14	16	63
safflower	10	13	77	
soybean	16	24	51	9
sunflower[c]	14	14	72	
sunflower[c]	9	72	19	
tung[d]	6	4	8	
walnut	8	16	72	

[a]Adapted from Ref. 5; actual compositions can vary greatly.

[b]Palmitic and stearic acids.

[c]Examples of the especially large variations in composition of available sunflower oils.

[d]Also 82% α-eleostearic acid.

matography (10). A review of the identification and determination of the various mixed triglycerides is available (11).

Autoxidation. Oils are classified as drying oils, which form solid films on exposure to air; semidrying oils, which form tacky, sticky films; and nondrying oils, which do not undergo marked increase in viscosity on exposure to air. Drying oils are further classified as nonconjugated and conjugated oils, depending on whether the double bonds in the predominant fatty acids are separated by one methylene group or are conjugated.

Nonconjugated oils having a drying index larger than 70 are drying oils (12). Drying index is calculated as the sum of the percentage of linoleic acid plus twice the percentage of linolenic acid in the oil. A more general statement, useful in considering synthetic drying oils as well as natural oils, is that if the average number of methylene groups between two double bonds per molecule is greater than 2.2, the oil is a drying oil; if less than 2.2, the oil is a semidrying oil. There is no sharp dividing line between semidrying and nondrying oils. Drying, semidrying, and nondrying oils also are defined based on their iodine values, ie, the number of grams of iodine required to saturate the double bonds of 100 grams of an oil. The types of oils have been defined as follows: drying oils, iodine value >140; semidrying oils, iodine value 125 to 140; and nondrying oils, iodine value <125 (13). Although iodine values can serve as satisfactory quality control specifications, they are not useful, and can be misleading, as a means of defining a drying oil or predicting reactivity.

The reactivity of nonconjugated drying oils is related to the average number of methylene groups between double bonds per molecule. Such methylene groups are allylic to two double bonds, and show much greater reactivity than methylene groups allylic to only one double bond. This fact is demonstrated by the relative rates of autoxidation of triolein [537-39-3] (glyceryl trioleate), trilinolein [537-40-6], and trilinolenein [14465-68-0], which are 1:120:330 (14). The number of methylene groups between double bonds for the three esters are 0, 3, and 6, respectively; the theoretical iodine values are 86, 173, and 262. The autoxidation rates are more closely related to the number of methylene groups between double bonds; the iodine values are proportional to the average number of double bonds per molecule. Use of the average number of double bonds per molecule to represent average functionality (15) is as unsatisfactory as is the use of iodine values.

Based on the data in Table 1, the average number of methylene groups between double bonds, ie, the functionality, for the typical linseed oil is 3.6; it is a drying oil. The corresponding number for soybean oil is 2.07; it is a semidrying oil. The higher the average functionality is above 2.2, the more rapidly a solvent-resistant, cross-linked film forms on exposure to air.

The reactions taking place during drying are complex, with many side reactions. Many of the studies of the chemistry of drying were done before modern analytical instrumentation was available (16,17). More recent studies have applied high performance liquid chromatography, nmr, and time-lapse Fourier transform infrared spectroscopy (ftir) to investigate autoxidation of unsaturated fatty acids (18) and drying of films (19,20).

Films form from a drying oil, such as linseed oil, in the following steps: an induction period during which naturally present antioxidants, mainly tocopherols, are consumed; a period of rapid oxygen uptake with a weight gain of about

10% (ftir shows an increase in hydroperoxides and appearance of conjugated dienes during this stage); a complex sequence of autocatalytic reactions in which hydroperoxides are consumed and the cross-linked film is formed; and cleavage reactions to form low mol wt by-products. In one study (19), when linseed oil was catalyzed by a drier, the first, second, and third steps were evident at 4, 10, and 50 h, respectively. After film formation, slow continuing reactions lead to further cross-linking, embrittlement, formation of volatile by-products, and discoloration. Films from drying oils with significant amounts of esters of fatty acids with three or more double bonds are particularly subject to discoloration.

The following scheme illustrates some of the many reactions that occur during cross-linking. Naturally present hydroperoxides decompose to form free radicals:

$$ROOH \longrightarrow RO\cdot + HO\cdot$$

At first, these highly reactive free radicals react with the antioxidant, but as the antioxidant is consumed, the free radicals react with other compounds. Hydrogens on methylene groups between double bonds are particularly susceptible to abstraction to yield the resonance stabilized free radical (**1**).

$$RO\cdot \text{ (or } HO\cdot) + -CH=CHCH_2CH=CH- \longrightarrow -CH=CH\overset{H}{\underset{\cdot}{C}}CH=CH- + ROH \text{ (or } H_2O)$$

(1)

This free radical (**1**) exists as three resonance contributors. It reacts with oxygen to yield a conjugated peroxy free radical such as (**2**). In a molecule, like trilinolein with its multiple functionality, there are many possible hydroperoxy products that are formed (21); radical (**2**) illustrates the principal type of structure formed.

$$-\overset{OO\cdot}{\underset{H}{C}}-CH=CH-CH=CH-$$

(2)

The peroxy free radicals can abstract hydrogens from other activated methylene groups between double bonds to form additional hydroperoxides and generate additional free radicals like (**1**). Thus a chain reaction is established resulting in autoxidation. The free radicals participate in these reactions, and also react with each other resulting in cross-linking by combination.

$$R\cdot + R\cdot \longrightarrow R-R$$
$$RO\cdot + R\cdot \longrightarrow R-O-R$$
$$RO\cdot + RO\cdot \longrightarrow R-O-O-R$$

Free radicals also add to conjugated double bonds, resulting in cross-links.

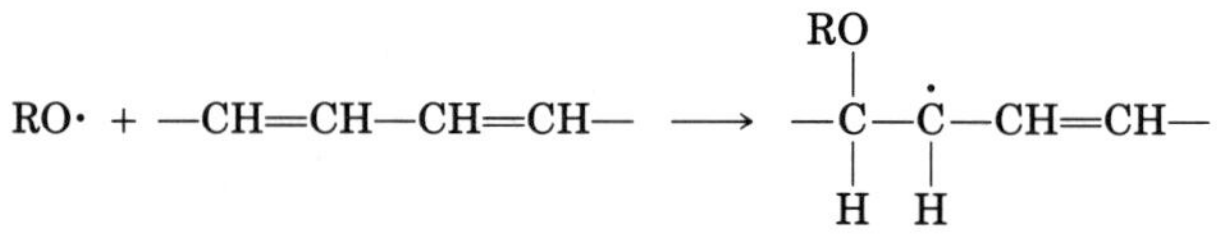

$$RO\cdot\ +\ -CH=CH-CH=CH-\ \longrightarrow\ \begin{array}{c} RO \\ | \\ -C-\overset{\displaystyle \cdot}{C}-CH=CH- \\ |\ \ | \\ H\ \ H \end{array}$$

(**3**)

Free radical (**3**) can rearrange, add oxygen to form a peroxy free radical, abstract a hydrogen from a methylene group between double bonds, combine with another free radical, or add to a conjugated double-bond system.

Studies by ^{1}H and ^{13}C nmr of the reactions of ethyl linoleate with oxygen in the presence of cobalt driers indicate that the cross-linking reactions were only those which formed ether and peroxy cross-links (20). It seems probable, however, that carbon to carbon bonds are formed in films of drying oils as well as the ether and peroxy cross-links.

Rearrangement and cleavage of hydroperoxides leads to a wide range of low mol wt, volatile by-products. The characteristic odor of oil and alkyd paints during drying is attributable to such volatile by-products as well as to the odor of the organic solvents used in the paints. This undesirable odor has encouraged the replacement in paints of oils and alkyds with latex vehicles, particularly for interior applications. The reactions leading to odors have been studied extensively in the analogous problem of flavor changes of vegetable cooking oils (23). Aldehydes have been shown to be significant by-products of the catalyzed autoxidation of drying oil-modified alkyd resins and of methyl esters of oleic, linoleic, and linolenic acids (20,23).

The rates at which nonconjugated drying oils dry are slow. Metal salts (driers) are known to catalyze the drying rate. The most widely used are the oil-soluble cobalt, manganese, lead, zirconium, and calcium salts of 2-ethylhexanoic acid [149-57-5] or naphthenic acids (see DRIERS AND METALLIC SOAPS). Cobalt and manganese salts, so-called top driers or surface driers, primarily catalyze drying at the film surface where oxygen concentration is highest. This results from their catalysis of hydroperoxide decomposition (24,25).

$$Co^{2+}\ +\ ROOH\ \longrightarrow RO\cdot\ +\ OH^-\ +\ Co^{3+}$$

$$Co^{3+}\ +\ ROOH\ \longrightarrow ROO\cdot\ +\ H^+\ +\ Co^{2+}$$

The net result is formation of water and a high concentration of free radicals. The cobalt cycles between the two oxidation states. Lead and zirconium salts catalyze drying throughout the film and are called through driers. Calcium salts show little, if any, activity alone, but may reduce the amount of other driers needed.

Combinations of metal salts are almost always used. Although mixtures of lead with cobalt and/or manganese are particularly effective, toxicity regulations ban the use of lead driers in consumer paints sold in interstate commerce in the United States. Combinations of cobalt and/or manganese with zirconium, and frequently also with calcium, are commonly used. The amounts of driers needed are very system specific. Their use should be kept to the minimum possible level since they not only catalyze drying but also catalyze the post-drying embrittlement and discoloration reactions.

Oils containing conjugated double bonds, such as tung oil, dry more rapidly than nonconjugated drying oils. Free-radical polymerization of the conjugated diene systems can lead to chain-growth polymerization rather than just combination of free radicals to form cross-links. High degrees of polymerization are unlikely because of the high concentration of readily abstractable hydrogens acting as chain-transfer agents. However, the free radicals formed by chain transfer also yield cross-links. In general, the water and alkali resistance of films formed using conjugated oils are superior, presumably because more of the cross-links are stable carbon to carbon bonds. However, since α-eleostearic acid in tung oil has three double bonds, discoloration on baking or aging is severe.

Both nonconjugated and conjugated drying oils can be polymerized by heating under an inert atmosphere to form so-called bodied oils. Bodied oils have higher viscosities and are often used in oil paints to improve application and performance characteristics. Process temperatures may be as high as 300 to 320°C for nonconjugated oils, and 225 to 240°C for conjugated oils, although the reactions occur at an appreciable rate at somewhat lower temperatures. At least in part, bodying may result from thermal decomposition of hydroperoxides, always present in natural oils, to yield free radicals resulting in a limited degree of cross-linking. It has also been shown that thermal rearrangement to conjugated systems occurs with subsequent Diels-Alder reactions leading to formation of dimers (26). Since tung oil has a high concentration of conjugated double bonds, it underegoes thermal polymerization much more rapidly than nonconjugated oils such as linseed oil; heating of tung oil must be carefully controlled or the polymerization will lead to gelation.

Viscosity of drying oils also can be increased by passing air through the oil at relatively moderate temperatures, 140 to 150°C, to produce blown oils. Presumably, reactions similar to those involved in cross-linking cause autoxidative oligomerization of the oil.

Synthetic and Modified Drying Oils

Varnishes. The drying rate of drying oils can be increased by dissolving a solid resin in the oil and diluting with a hydrocarbon solvent. Such a solution is called a varnish. The solid resin serves to increase the glass-transition temperature, T_g, of the solvent-free film so that film hardness is achieved more rapidly. There is no increase in the rate of cross-linking, so the time required for the film to become solvent resistant is not shortened. Essentially any high melting thermoplastic resin that is soluble in drying oil will serve the purpose; the higher the melting point of the resin and the ratio of resin to oil, the greater the effect on drying time. Both naturally occurring resins, such as congo, copal, damar, and kauri resins, and synthetic resins, such as ester gum (glyceryl esters of rosin), phenolic resins, and coumarone-indene resins, have been used.

In varnish manufacture, the drying oil, ie, linseed oil, tung oil, or mixtures of the two, and the resin are cooked together to high temperatures to yield a homogeneous solution of the proper viscosity. The varnish is then thinned with hydrocarbon solvents to application viscosity. During cooking some dimerization or oligomerization of the drying oil occurs. Reactions between the oil and the resin have been demonstrated.

Varnishes were widely used in the nineteenth and early twentieth centuries. They have been replaced almost completely by a wide variety of other products, especially alkyds, epoxy esters, and urethane oils. The term varnish has come to be used for transparent coatings, such as trade sales varnishes, even though few of them are varnishes in the original meaning of the word.

Synthetic Conjugated Oils. Tung oil dries rapidly, but is expensive, and its films discolor rapidly due to the presence of three double bonds. These defects led to efforts to synthesize conjugated oils, especially those containing esters of fatty acids with two conjugated double bonds.

Castor oil contains 89% ricinoleic acid [141-22-0] ((R)-12-hydroxy-(Z)-9-oc-tadecenoic acid), as the glyceryl ester (27). Heating with an acid catalyst yields dehydrated castor oil; a mixture of geometric isomers of the 9,11-conjugated and the 9,12-nonconjugated fatty acid esters are formed (2). Dehydrated castor oil dries rapidly at room temperature, but on further exposure to air the surface becomes tacky. This aftertack has been attributed to the presence of various geometric isomers formed during the dehydration (28–30). Dehydrated castor oil and its fatty acids are used in preparing alkyds and epoxy esters for baking coatings where aftertack does not occur.

Nonconjugated oils can be partially isomerized to conjugated oils by heating with a variety of catalysts, mostly alkaline hydroxides (31,32). These modified oils also contain a mixture of geometric isomers and exhibit aftertack. A similar process can be used to partially conjugate the double bonds of tall oil fatty acids. Synthesis of conjugated fatty acids by treatment of oils at high temperature under pressure with aqueous alkali hydroxides accomplishes isomerization and saponification simultaneously (32). The principal use of such conjugated oils and fatty acids has been in making alkyds and epoxy esters for baking coatings.

Esters of Higher Functionality Polyols. The time required for nonconjugated oils to form solvent-resistant cross-linked films decreases as the average number of methylene groups between double bonds per molecule increases. When oil-derived fatty acids react with polyols having more than three hydroxyl groups per molecule, the average number of cross-linking sites per molecule increases proportionally to the functionality of the polyol. Since the number of reactive sites in soybean oil with the composition listed in Table 1 is 2.07, soybean oil is a semidrying oil. However, the pentaerythritol (2,2-bis(hydroxymethyl)-1,3-propanediol [115-77-5]) ester of soybean fatty acids with 2.76 reactive sites per molecule is a drying oil. The pentaerythritol ester of linseed oil fatty acids has about 4.8 methylene groups between double bonds per average molecule and gives dry, solvent-resistant films more rapidly than linseed oil. Still faster drying rates can be achieved with still higher functionality polyols such as dipentaerythritol [126-58-9] (2,2'-[oxybis(methylene)]-bis[2-(hydroxymethyl)-1,3-propanediol]) and tripentaerythritol [78-24-0] (2,2-bis{[3-hydroxy-2,2-bis(hydroxymethyl)propoxy]methyl}-1,3-propanediol).

Oxidizing alkyds can be considered as still higher functionality drying oils. Their drying speed is faster owing to both the higher functionality and the higher T_g of the rigid aromatic rings in the phthalate esters. The time required to provide sufficient cross-linking to achieve solvent-resistant films with a series of alkyds from a particular drying oil reaches a minimum at an oil length of about 60 (see ALKYD RESINS). At still shorter oil lengths, average functionality decreases, and longer times are required to achieve solvent resistance. However, since the phthalic content is higher, such alkyds are converted more rapidly to hard films. An oil length of about 50 is commonly considered optimum at room temperature. Similarly, drying oil fatty acid esters of bisphenol A epoxy resins dry more rapidly than the corresponding triglycerides because of a higher functionality and the effect of the rigid aromatic rings. Since the backbone does not contain ester groups, epoxy esters are superior to alkyds in applications requiring high saponification resistance, such as metal primers. Similarly, urethane oils dry faster than the drying oil from which they were made, and offer superior abrasion and water resistance. Another example of commercial synthetic drying oils is the fatty acid esters of low mol wt styrene–allyl alcohol copolymers, eg, resins RJ-100 and 101 produced by Monsanto Resins and Plastics Co. Again, these oils dry faster than the corresponding triglycerides and have better saponification resistance.

Oils Modified with Maleic Anhydride (Maleated Oils). Oils, with either conjugated or nonconjugated double bonds, react with maleic anhydride [108-31-6] (2,5-furandione) to form adducts. Conjugated oils, such as dehydrated castor oil, react at moderate temperatures by a Diels-Alder reaction. Nonconjugated oils, such as soybean and linseed oils, react at temperatures above 100°C to form a variety of adduct structures. Model compound studies with methyl linoleate indicate maleic anhydride reacts to give succinyl anhydride adducts at the 8 and 14 positions as well as at the 9 and 13 positions; in the latter case, rearrangement of the double bonds to conjugated positions occurs (33). The conjugated bonds can undergo a Diels-Alder reaction with further maleic anhydride to form a dianhydride.

The products of these reactions with maleic anhydride, termed maleated oils, react with polyols to give moderate mol wt derivatives that dry faster than the unmodified oils. For example, maleated, esterified soybean oil is a drying oil with a drying rate comparable to that of a bodied linseed oil with a similar viscosity. Maleated linseed oil can be converted to a water-dilutable form by hydrolysis with aqueous ammonium hydroxide to convert the anhydride groups to ammonium salts of the diacid. Such products have not found significant commercial use, but similar reactions with alkyds and epoxy esters are used on a large scale to make water-dilutable derivatives.

Vinyl-Modified Oils. Both conjugated and nonconjugated drying oils react in the presence of free-radical initiators with such vinyl monomers as styrene, vinyltoluene, acrylic esters, and cyclopentadiene. High degrees of chain transfer cause the formation of wide varieties of products, including low mol wt homopolymers of the vinyl monomer, short-chain graft copolymers, and dimerized drying oil molecules. The reaction products with drying oils, except cyclopentadiene, are not commercially important, but the same principle is widely used in making modified alkyds. Linseed oil modified with cyclopentadiene has found fairly sizable commercial use. This product is made by heating a mixture of linseed oil and dicyclopentadiene above 170°C (34). At this temperature a reverse Diels-Alder reaction liberates monomeric cyclopentadiene at an appreciable rate; it in turn reacts with the linseed oil. The product is inexpensive and dries faster than linseed oil, but a residual odor and dark color limit applications.

Economic Aspects

Use of an oil as a drying oil, or as an intermediate for making synthetic drying oils such as alkyds, may be only one of many possible uses; frequently the principal use is as edible oil. Therefore, production data for oils are of limited value in assessing their importance as drying oils or as raw materials for synthetic drying oils. Extensive data on United States and world production and some data on consumption of all fats and oils are available (35). Since linseed oil is used primarily (but not exclusively) as a drying oil, data on it may be somewhat more significant than, for example, soybean oil, which is used primarily as an edible oil. World production of linseed oil was reported as 918,000 t in 1978 and 565,000 t in 1989 (35); United States production was 260,000 t in 1978 and 170,000 t in 1988. United States use of oils as drying oils was reported to be 909,000 t in 1965 and 479,000 t in 1979. The last year in which the Agricultural Statistics reports give this breakdown of consumption is 1979. These numbers do not include the usage of fatty acids such as tall oil fatty acids, which are widely used in alkyds, epoxy esters, and urethane oils. More data on consumption of oils is available in Table 2.

Soybean oil and tall oil fatty acids are not used in paints without modification. These products, listed as used in paints, first must be converted to alkyds or other synthetic drying oils. Presumably significant amounts of the linseed oil listed under paints are also converted to alkyds or other derivatives before use. In addition to the numbers given in Table 2, relatively large amounts of the oils

Table 2. Consumption of Drying Oils, 10^3 t[a]

Oil	Paints[b]		Resins[c]	
	1972	1989	1972	1989
fish	4.2	0.05		
linseed	73.7	46.2	10.8	1.1
soybean	35.4	15.9	25	56
tall oil	9.6	14.5	8.2	19.1
tung oil	8.2	1.6	1.7	0.8
Total[d]		82		92

[a]Ref. 36.
[b]Including varnish, enamel, and similar products.
[c]Plastics, alkyds, and plasticizers.
[d]Total includes small amounts of other drying oils, such as safflower oil, consumed by individual consumers, and nondrying oils, such as castor oil and coconut oil, probably used in making nondrying alkyds; presumably does not include dehydrated castor oil.

are reported to have been consumed by conversion into fatty acids. Some indeterminate fraction of the fatty acids, especially tall oil fatty acids, are presumably converted into derivatives that are used like drying oils.

Prices for oils, like prices of many agricultural products, vary substantially over time. Average prices for a series of years are provided in Table 3.

Table 3. Average Price of Oils, $/kg[a]

Oil	1967	1979	1987	1989
linseed, raw, tanks, Minneapolis	0.284	0.631	0.546	0.889
soybean, crude, tanks, midwest	0.211	0.609	0.339	0.464
tung, drums, imported	0.363	1.412	0.902	1.075

[a]Ref. 36.

Uses

Although some drying oils continue to be used as drying oils, the largest use of drying and semidrying oils in the early 1990s is as raw materials, either directly or as the fatty acids obtained by saponification, in the manufacture of oxidizing alkyds, epoxy esters, urethane oils, and synthetic drying oils. Tall oil fatty acids also are used to manufacture dimer acids (qv) (see CARBOXYLIC ACIDS, FATTY ACIDS FROM TALL OIL). U.S. government and ASTM specifications for drying oils are listed in Table 4.

Table 4. Drying Oil Specifications

Oil	Specification
U.S. government	
linseed oil, alkali-refined	TT-L-1155 (1967)
	A-A-714 (1980)
linseed oil, boiled	A-A-371A (1987)
linseed oil, heat polymerized	A-A-379A (1980)
ASTM	
castor oil, dehydrated	D961-86
linseed oil, boiled	D260-86
linseed oil, raw	D234-82 (1987)
oiticica oil, permanently liquid	D601-87
safflower oil	D1392-87
soybean oil, degummed	D124-88
soybean oil, refined	D1462-87
sunflower oil, once-refined	D3169-89
tall oil fatty acids, distilled	D1984-69 (1988)
tung oil, raw	D12-88
tung oil, quality	D1964-85 (1989)

Since drying oils are considered a renewable resource, they may again become important in paints and printing inks, depending on the cost of drying oils compared with petroleum-derived raw materials. However, in many cases, the properties that can be obtained with synthetic binders, especially retention of flexibility, gloss, and color, are superior to those that can be obtained with a drying oil or drying oil-derived binder.

Paints. Although most drying oils have been replaced as paint vehicles by latexes and other synthetic resins, oils are still being used to a degree in paint and allied products. In exterior house paints, linseed oil or oxidizing alkyds are used when paint must be applied at temperatures as low as 4 to 5°C, ie, temperatures at which latexes do not coalesce satisfactorily. They also are used in primers over chalky surfaces where latex paints do not provide adequate adhesion. Drying oils and synthetic drying oil esters, such as those derived from low mol wt styrene–allyl alcohol copolymers, improve the adhesion of latex paints to chalky surfaces. On a solids basis, about 15% of the latex polymer is replaced by the drying oil in emulsion form. After application the emulsion breaks, permitting the drying oil to penetrate between the loose pigment particles of the chalky surface where the latex particles cannot penetrate. The oil reaches the continuous substrate of the old paint, thus serving to cement the chalky particles together and provide adhesion.

Most stains used for finishing shingles and other natural wood exterior products are made with pigmented linseed oil, diluted with hydrocarbon solvents for penetration. The oil penetrates into the porous wood surface, sealing it against water, and the stain gives a desired uniform color. The use of latex stains is increasing; however, the latex stain wood finishes are less transparent and conceal more of the grain pattern of the wood than do oil stains. Oil stains based on linseed oil also are used in nonprofessional finishing of furniture; color uniformity is more easily controlled than with the dye solutions used in commercial furniture finishing, but the transparency is not as good. Most varnish sold for nonprofessional furniture finishing is based on urethane oils rather than the older resin/oil varnishes. Linseed oil and tung oil as penetrating finishes also are used to protect the surfaces of wood furniture against staining without leaving an apparent film on the surface.

Red lead-in-oil primers have been used for many years for corrosion protection of steel when it is not practical to remove oily rust particles from the surface of the steel; concern about lead content has resulted in use reduction. The drying oil and the hydrocarbon solvents in the red lead-in-oil primer dissolve the contaminating oil from the surface of the rust particles, permitting wetting. The low viscosity and slow drying of the oil permit penetration of the primer vehicle through the rust deposits down to the surface of the steel, thus binding the rust particles together in the film and providing adhesion to the steel substrate. The adhesion of these primers to steel, especially in the presence of water, and their saponification resistance is inferior to other types of steel primers, such as epoxy–amine systems; hence, red lead-in-oil primers are used only when the steel surface cannot be thoroughly cleaned (see PAINT).

Printing Inks. The use of drying oils in printing inks has decreased. Some inks based on drying oils are still used for sheet-fed letterpress printing. The principal remaining use is in lithographic printing, particularly sheet-fed lithographic printing. Because of their low surface tension, oil-based lithographic inks do not wet the hydrophilic portions of the plates. Since the viscosity of these inks must be high, bodied linseed oils or esterified maleic-modified soybean oils are used. Drying oil-modified alkyds are also being used in lithographic inks.

Other Uses. Linseed oil and modified linseed oil have been used to cure concrete and treat surfaces of concrete highways and bridge decks to reduce scaling and spalling. Some linseed oil is used as core oil to bind sand cores in metal casting. High viscosity bodied linseed oil is used as a binder in replacement brake linings. Drying oils also are used as binders in making hardboard. Epoxidized oils are stabilizing plasticizers for poly(vinyl chloride) and, by reaction with acrylic acid, are used in uv-curable oligomers. Dimer acids derived from tall oil fatty acids are used in making hydroxy-terminated polyesters for use with melamine–formaldehyde resins in baking enamels. Dimer acids react with polyamines to form amine-terminated polyamides, which are used with epoxy resins (qv) in making air-dry high performance steel primers. Alcohol-soluble polyamides derived from dimer acids are used as vehicles for flexographic printing inks.

BIBLIOGRAPHY

"Drying Oils" in *ECT* 1st ed., Vol. 4, pp. 277–299, by O. Grummitt, Western Reserve University, H. J. Lanson, Mastercraft Paint Manufacturing Co., and A. E. Rheineck, Hercules Powder Co.; in *ECT* 2nd ed., Vol. 7, pp. 398–428, by O. Grummitt, Western Reserve University, J. Mehaffy, Sherwin-Williams Co., A. E. Rheineck, North Dakota State University, and H. J. Lanson, Lanson Chemicals Corp.; in *ECT* 3rd ed., Vol. 8, pp. 130–150, by J. C. Cowan, Bradley University.

1. A. E. Rheineck and R. O. Austin in R. R. Myers and J. S. Long, eds., *Treatise on Coatings*, Vol. 1, No. 2, Marcel Dekker, Inc., New York, 1968, pp. 181–248.
2. D. H. Solomon, *The Chemistry of Organic Film Formers*, 2nd ed., Robert E. Krieger Publishing Co., Huntington, N.Y., 1977, pp. 35–74.
3. R. J. Harwood, *Chem. Rev.* **62**, 99 (1962).
4. F. D. Gunstone, *Chemistry and Biochemistry of Fatty Acids and Their Glycerides*, 2nd ed., Chapman and Hall, Ltd., London, 1967, p. 151.
5. *Ibid.*, p. 156.
6. A. C. Dillman and T. H. Hopper, *U.S. Dept. Agric. Tech. Bull.*, 844 (1943).
7. L. V. Berman and M. L. Loeb in E. C. Leonard, ed., *The Dimer Acids*, Humko Sheffield Chemical Operation of Kraftco Corp., Memphis, Tenn., 1975, p. 4.
8. R. G. Ackman, *Prog. Chem. Fats Other Lipids* **12**, 165 (1972).
9. G. R. Khan and F. Scheinmann, *Prog. Chem. Fats Other Lipids* **17**, 343 (1977).
10. J. W. King, E. C. Adams, and B. A. Bidlingmeyer, *J. Liq. Chromatog.* **5**, 275 (1982).
11. A. Kuksis, *Prog. Chem. Fats Other Lipids* **12**, 1 (1972).
12. J. H. Greaves, *Oil Colour Trades J.* **113**, 949 (1948).
13. Ref. 4, p. 107.
14. J. R. Chipault, E. E. Nickell, and W. O. Lundberg, *Off. Dig. Fed. Paint Varn. Prod. Clubs* **23**, 740 (1951).
15. Ref. 1, p. 238.
16. E. H. Farmer and D. A. Sutton, *J. Chem. Soc.*, 10 (1946).
17. H. Wexler, *Chem. Rev.* **64**, 591 (1964).
18. N. A. Porter and co-workers, *J. Am. Chem. Soc.* **103**, 6447 (1981).
19. J. H. Hartshorn, *J. Coat. Technol.* **54**(687), 53 (1982).

20. W. J. Muizebelt, J. W. van Velde, and F. G. H. van Wijk, *Proc. XVth Intl. Conf. Org. Coatings Sci. Technol.*, 299 (1989).

21. E. N. Frankel, W. E. Neff, and K. Miyashita, *Lipids* **25,** 40 (1990).

22. E. N. Frankel, *Prog. Lipid Res.* **19,** 1 (1980).

23. R. A. Hancock, N. J. Leeves, and P. F. Nicks, *Prog. Org. Coat.* **17,** 321, 337 (1989).

24. C. E. H. Bawn, *Discuss. Faraday Soc.* **38,** 356 (1942).

25. G. A. Russel, *J. Chem. Ed.* **36,** 111 (1959).

26. D. H. Wheeler and J. White, *J. Am. Oil Chem. Soc.* **44,** 298 (1967).

27. Ref. 4, p. 157.

28. J. C. Cowan, *Ind. Eng. Chem.* **41,** 294 (1949).

29. J. C. Cowan, *J. Am. Oil Chem. Soc.* **27,** 492 (1950).

30. A. E. Rheineck and D. D. Zimmermann, *Fette Seifen Anstrichm.* **71,** 869 (1969).

31. J. P. Kass and G. D. Burr, *J. Am. Chem. Soc.* **61,** 3292 (1939).

32. T. F. Bradley and G. H. Richardson, *Ind. Eng. Chem.* **34,** 237 (1942).

33. A. E. Rheineck and T. H. Khoe, *Fette Seifen Anstrichm.* **71,** 644 (1969).

34. L. I. Hansen, J. C. Konen, and M. W. Formo, *Prepr. Book, Div. Paint, Varn., Plast. Chem. Am., Chem. Soc.*, 57 (Sept. 1949).

35. U.S. Dept. of Agriculture, *Agriculture Statistics 1990*, U.S. Government Printing Office, Washington, D.C., 1990.

36. *Current Industrial Reports, Fats and Oils Production, Consumption, and Stocks*, U.S. Dept. of Commerce, Bureau of Standards M20K-13 (89), Mar. 1990.

General References

R. R. Myers and J. S. Long, eds., *Treatise on Coatings*, Vol. 1, No. 2, Marcel Dekker, Inc., New York, 1986.

D. H. Solomon, *The Chemistry of Organic Film Formers*, 2nd ed., Robert E. Krieger Publishing Co., Huntington, N.Y., 1977.

M. W. Formo in D. Swern, ed., *Bailey's Industrial Oil and Fat Products*, Vol. 1, 4th ed., John Wiley & Sons, Inc., New York, 1979, Chapt. 10; Vol. 2, Chapt. 5.

ZENO W. WICKS, JR.
Consultant

DUST, ENGINEERING ASPECTS. See AIR POLLUTION CONTROL

METHODS; POWDER HANDLING.

DUST, HYGIENE ASPECTS. See AIR POLLUTION; INDUSTRIAL

HYGIENE AND PLANT SAFETY.

DYE CARRIERS

Dye carriers are needed for complete dye penetration of polyester fibers. Carriers cause the glass-transition temperature, T_g, of the polyester polymer to become lower and allow the penetration of water-insoluble dyes into the fiber.

It is difficult for dye solutions in water to penetrate synthetic fibers such as polyester, cellulose triacetate, polyamides, and polyacrylics which are somewhat hydrophobic. The rate of water imbibition differs with each fiber as shown in Table 1 as compared to viscose (see FIBERS, REGENERATED CELLULOSICS), which imbibes water at the rate of 100% (1). The low imbibition rate is attributed to the high T_g obtained when the polymeric fibers are drawn. During this drawing operation the polymer chains become highly oriented and tightly packed, forming a structure practically free of voids.

Disperse dyes commonly used to dye polyester are nonionic, and dye the polyester fiber through a diffusion mechanism. Prolonged boiling of the dyebath loosens the forces binding the polymer chains to each other, causing the fiber to swell. This allows a limited penetration of the fiber surface by the dye. The rate of absorption or diffusion of disperse dyes in polyester is much lower than that on nylon (see FIBERS, POLYAMIDE) or cellulose triacetate fibers (see FIBERS, CELLULOSE ESTERS). This low dyeing rate is too costly to meet the economic requirements of industrial processing. In addition, deep shades are difficult to achieve, and the final dyeing does not meet the minimum fastness required by commercial standards (see DYES, APPLICATION AND EVALUATION).

Deep shades and full fastness properties on polyester can be achieved using disperse dyes and carriers, or temperatures over 100°C with or without carriers.

Dye carriers, occasionally called dyeing accelerants, are used on cellulose triacetate fibers, but have found their greatest use in the dyeing of polyester. Many theories have been advanced to explain the mechanism of carrier dyeing. One of these is based on the ability of carriers to solubilize disperse dyes. However, this theory seems untenable because many organic solvents do this and yet do not exhibit carrier properties. Another theory suggests that the carrier coats the individual fibers, forming a layer through which the dye transfers to the fiber. Still another theory suggests that the carrier loosens the binding forces holding

Table 1. Rate of Water Imbibition of Fibers
Compared to Viscose[a]

Substrate, fiber	Water imbibition, %
viscose (rayon)	100
cellulose acetate	25
cellulose triacetate	10
polyamide (nylon)	11–13
polyacrylic	8–10
polyester	3
polypropylene	0

[a]Ref. 1.

the polymer chains together, thus providing suitable spaces for the dye molecules; this is reflected in a lower T_g. An excellent discussion of these theories has been given by a research committee of the American Association of Textile Chemists and Colorists (2). No universal agreement exists on the mechanism of carrier dyeing.

Carrier Properties

Many substances show carrier behavior, and some have found more acceptance than others for various reasons, eg, availability, cost, environmental concerns, ease of handling, odor, etc. Most carriers are aromatic compounds, and have similar solubility parameters to the poly(ethylene terephthalate) fibers and to some disperse dyes (3).

There are many chemicals, by lowering T_g, suitable as carriers. Their bp is one of the principal criteria in selection. If bp is too low, the compound will evaporate from the dyebath at dyeing temperatures, and will be lost before it is effective in its role as a carrier. It may also steam distill (condense on the cooler parts of the equipment) and cause drips that will spot the fabric. On the other hand, if the bp is too high, the compound cannot be removed from the fabric under normal plant drying conditions and will affect lightfastness of finished goods, leave residual odor, and possibly cause skin irritation to the wearer.

o-Phenylphenol was one of the earliest carrier-active compounds used industrially. Originally it was used as its water-soluble sodium salt (4). By lowering the pH of the dyebath, the free phenol was precipitated in fine form and made available to the fiber. However, proprietary liquid preparations containing the free phenol are available that afford a greater ease of handling.

Table 2 lists the four main groups of compounds most commonly used as dye carriers. In order for these compounds to act effectively as carriers, they must be homogeneously dispersed in the dyebath. Because the carrier-active compounds have little or no solubility in water, emulsifiers are needed to disperse these compounds in the dyebath (see EMULSIONS).

Emulsification

Proper emulsification is essential to the satisfactory performance of a carrier. A well-formulated carrier readily disperses when poured into water, and forms a milky emulsion upon agitation or steaming. It should not cause oil separation upon heating or crystallization and sedimentation upon cooling.

Many proprietary carriers are available as solids (flakes or pellets) or in preemulsified form. These present some difficulties in the dyehouse. The former require dispersion in water through steam injection and addition to a preheated dyebath. The latter suffer from short storage life owing to separation of the emultion. Currently the industry prefers clear products easily emulsified by premixing with water at the time of use.

Manufacturing of the flake and pellet forms requires melting of the various components. The mass solidifies upon cooling and can be flaked or pelletized according to need.

Table 2. Compounds Most Commonly Used as Dye Carriers

Compounds	CAS Registry Number	Mol wt	Bp, °C
phenolics			
o-phenylphenol	[90-43-7]	170.2	280–284
p-phenylphenol	[92-69-3]	170.2	305–308
methyl cresotinate	[23287-26-5]	166.0	240
chlorinated aromatics			
o-dichlorobenzene	[95-50-1]	147.0	172–178
1,3,5-trichlorobenzene	[108-70-3]	181.45	214–219
aromatic hydrocarbons and ethers			
biphenyl	[92-52-4]	154.2	255.9
methylbiphenyl	[28652-72-4]	168.24	255.3
diphenyl oxide	[101-84-8]	170.0	259.0
1-methylnaphthalene	[90-12-0]	142.2	244.6
2-methylnaphthalene	[91-57-6]	142.2	241
aromatic esters			
methyl benzoate	[93-58-3]	136.14	198–200
butyl benzoate	[136-60-7]	178.22	250
benzyl benzoate	[120-51-4]	212.24	323–324
phthalates			
dimethyl phthalate	[131-11-3]	194.18	298
diethyl phthalate	[84-66-2]	212.18	298
diallyl phthalate	[131-17-9]	246.25	290
dimethyl terephthalate	[120-61-6]	194.18	284

The preemulsified carriers contain water. These products usually require homogenization through colloidal mills or similar equipment to reduce the particle size and ultimately stabilize the product. The preemulsified as well as the clear self-emulsifying products require the use of a solvent when the carrier-active material is a solid.

Carrier Formulation

The formulation of a carrier depends on four considerations: (1) the carrier-active chemical compound; (2) the emulsifier; (3) special additives; and (4) environmental concerns. Additional parameters to be considered in the formulation of a carrier product with satisfactory and repeatable performance arise from the equipment in which the dyeing operation is to be carried out. The choice of equipment is usually dictated by the form in which the fiber substrate is to be processed, eg, loose fiber, staple, continuous or texturized filament, woven or knot fabric, yarn on packages or in skeins (see TEXTILES).

The carrier-active chemical is selected according to its effectiveness at various temperatures. Members of the phenolic group (Table 2), considered to be stronger carriers, are employed for formulations to be used in open equipment at the boil. Weaker carriers, such as the members of the aromatic ester group, are utilized generally for high temperature dyeing.

The emulsifier is seldom a single surfactant but a blend selected according to the hydrophilic–lipophilic balance (HLB) required by the chemicals employed. The performance of the final carrier preparation is limited to the effectiveness of its emulsification system. It provides the initial fine dispersion throughout the dyebath, and prevents the carrier from separating and causing spotting or uneven dyeing throughout the dyeing cycle. The system must supply effective reemulsification of the steam-distilled chemicals that drop back into the dyebath and onto the fabric. Finally, the emulsifier produces the foaming characteristics of the product that determine the suitability of the carrier for use with different types of equipment. For example, a carrier intended for use on jet dyeing equipment is formulated with low foaming nonionic surfactants (see SURFACTANTS).

Special additives are often included in a carrier formulation to provide specific properties such as foam control, stability, and fiber lubrication during dyeing. Most important are the solvents used to solubilize the solid carrier-active chemicals. These often contribute to the general carrier activity of the finished product. For example, chlorinated benzenes and aromatic esters are good solvents for biphenyls and phenylphenols. Flammable compounds (flash point below 60°C) should be avoided.

Carrier Selection

A carrier is selected by the dyer according to various criteria. The type of equipment and conditions under which it is to be used have already been mentioned. Other considerations include color yield, dye migration, and product and emulsion stability.

Color Yield. The amount of dye successfully removed from the dyebath by the fiber results in a specific depth of shade or intensity of coloration (see COLOR). This is commonly called color yield. The color-building properties of a carrier are influenced by other factors, such as the fiber to liquor ratio (its concentration in the dyebath), the maximum temperature that can be reached on the chosen machine, and other dyebath additives. Most carriers give optimum dye utilization in atmospheric dyeing (95–100°C) at 8–12% calculated on the weight of the fiber (owf). In low pressure dyeing equipment (106–110°C), 4–6% owf is usually required. In high pressure equipment (126–132°C), the amount of carrier used can be reduced to 1–3% owf. Above these limits of optimum carrier concentrations, the equilibrium between fiber, dye, and water is switched in favor of the dyebath, and higher amounts of carriers act as shade-reducing agents, as shown in Figure 1.

Dye Migration. Dye carriers promote dye migration and transfer, thus producing level and satisfactory dyeings.

The dyestuff is exhausted out of the dyebath and into the fiber with the assistance of carriers and heat. This process can be reversed by the use of an excessive amount of carrier, the depth of shade can be greatly reduced, and the dyestuff returned to the dyebath (stripping effect). This phenomenon is utilized to level an unevenly dyed fabric. Some carriers are more effective than others in

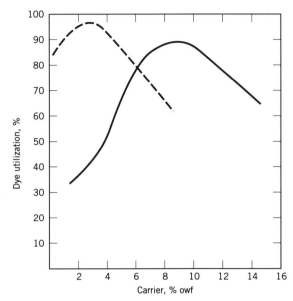

Fig. 1. Dyestuff utilization as a function of carrier concentration. (——), 98°C; (– – –), 120°C.

promoting dye migration and equalizing the dye distribution through the fiber substrate. Defectively dyed fabrics showing streaks or unlevelness (shade variations from side to center or from one end of a length of fabric to the other) are usually repaired by treating them with two or three times the amount of carrier normally required in dyeing.

Barré is caused by uneven tension in knitting, defective yarn, improper needle action, or other similar factors that are not recognized until the fabric is dyed. It appears as a repetitive characteristic pattern of varying intensity that is easily recognizable.

Its causes may be mechanical or chemical: variations in knitting tension cause lack of uniformity of fabric density that in dyeing produce the barré effect, and different degrees of crystallinity of the polymeric fiber owing to uneven drawing or subsequent heat treatments of the filament are also responsible for barré. Dye carriers in combination with selected dyes are instrumental in overcoming this condition.

Product Stability and Emulsion Stability. These properties are not necessarily related, but are both highly prized in the selection of a carrier. The first refers to the storage or shelf stability of the product. Many carrier preparations are not properly balanced, or unsuitable emulsifiers have been used. Upon storing, these products separate in layers, particularly when exposed to temperature changes.

Emulsion stability refers to the stability of the emulsion in water. It must withstand various dyebath conditions.

The Dilution Factor of the Dyebath. The amount of water in a dyebath may vary from 5–25 times the weight of fiber to be dyed, according to the capacity and type of equipment employed. The larger water-to-fiber ratio dilutes the emulsifier, thus reducing its effect.

Elevated Temperatures of the Dyebath.   Emulsions that are stable in cold or warm water lose their stability at higher temperatures. The carrier-emulsifier equilibrium undergoes stress, particularly when the time at high temperature is prolonged.

Agitation and Shear Action of the Dyebath.   In some equipment fabrics are rotated through the dyebath, thus keeping it in constant agitation. In other equipment the dyebath is circulated through the fabric or yarn. In jet equipment the high speed circulation of the liquid is responsible for the rotation of the fabric. Although agitation contributes to the stability of certain emulsions, the high shear of the pumps used in the equipment with forced circulation is often detrimental.

pH and Electrolyte Content of the Dyebath.   A pH of 4–5 provides the best conditions for dyeing polyester with disperse dyes. A carrier emulsion must be stable under this condition. In addition, when cellulose is present in a fiber blend with polyester and it is to be dyed in the same dyebath with direct or fiber-reactive dyes, the carrier emulsion needs to have considerable stability to large amounts of inorganic salts. Sodium chloride or anhydrous sodium sulfate are employed for this purpose in 10–50% owf.

Other Considerations.   Some carrier-active products, especially o-phenylphenol and methylnaphthalenes, have an adverse effect on the lightfastness of the finished dyeing. The reason for this is not clear, but the effect is readily established. This problem is overcome by submitting the dyed material to temperatures higher than those normally required in drying. Under the conditions (150–175°C) that are usually required to heat-set dyed fabrics or to cure resins applied in finishing operations, the residual carrier is volatilized.

The cost of a carrier, in addition to its satisfactory performance in dyeing, is often a considerable factor in selection. The rising cost of petroleum-derived chemicals is a factor in the price structure of carrier-active chemicals and most carriers, unfortunately, fall in this category.

Dyeing Procedures

Dyeing procedures vary according to the fiber content of the textile material and the equipment to be used. Examples of basic carrier dyeing procedures are as follows.

100% Polyester Fabric Dyed Atmospherically.   The dyebath is prepared with water conditioning chemicals as needed to control the hardness of the water being used. Dispersion of the dyes is done by pasting them with cold water and diluting them with warm water (70°C). The dye dispersion is added to the bath and the equipment run for 10 minutes. Addition of 5–10 g/L of carrier is done following the dilution procedures recommended by the manufacturer. Acetic acid is used to adjust pH to ca 5. The system is brought to a boil in 30–45 min, and held at the boil for at least 1 h before checking the shade. After slow cooling and rinsing completely, the bath is dropped and an afterscour is done as required to remove residual carrier and unfixed dye.

100% Polyester (Textured or Filament) Dyed Under Pressure. The dyebath (50°C) is set with water conditioning chemicals as required, acetic acid to ca 5 pH, properly prepared disperse dyes, and 1–3 g carrier/L. The bath is run for 10 minutes, then the temperature is raised at 2°C/min to 88°C and the equipment is sealed. Temperature is raised at 1°C/min to 130°C, and the maximum temperature held for ½–1 h according to the fabric and depth of shade required. Cooling to 82°C is done at 1–2°C/min, the machine is depressurized, and the color sampled. The shade is corrected if needed. Slow cooling avoids shocking and setting creases into the fabric. Afterscour is done as needed.

Economic Aspects

Manufacturing capacities and sales volumes of dye carriers are difficult to assess because the materials used are made by many companies for use in a variety of applications. Companies manufacturing dye carriers include:

Apollo Chemical Corp.
Applied Textile Technologies Inc.
BASF, Fibers Division, Dispersions & Textile Chemicals Group
Burlington Chemical Co.
Chemonic Industries Inc.
Ciba-Geigy Corp., Dyestuff & Chemicals Division
CNC International LP
Crompton & Knowles Corp., Dyes & Chemicals Division
Dexter Chemical Corp.
Eastern Color & Chemical Co., Pigment Division
Emkay Chemical Co.
Finetex Inc.
Glo-Tex Chemicals Inc.
Grant Industries Inc.
Gresco Mfg. Co.
A. Harrison & Co.
High Point Chemical Corp.
Hoechst-Celanese Corp., Colorants & Surfactants Division
Hydro Labs Inc.
ICI Americans Inc.

IVAX Industries Inc., Textile Products Division
Leatex Chemical Co.
Lenmar Chemical Corp.
Marlowe-Van Loan Corp.
MFG's Chemical & Supply Co.
Mobay Corp., Dyes, Pigments & Organics Division
Novachem Corp.
Organic Dyestuffs Corp.
Piedmont Chemical Industries Inc.
Reilly-Whiteman Inc.
Rhône-Poulenc Inc.
RPM American Emulsions Co.
Sandoz Chemicals Corp.
Specialty Chemicals Co.
Stockhausen Inc.
Surpass Chemical Co.
Sybron Chemicals Inc.
Troy Chemical Corp.
Union Carbide Chemicals & Plastics Co.
Unitex Chemical Corp.
The Virkler Co.
Witco Corp., Organics Division

Health and Safety Factors

Most carrier-active compounds are based on aromatic chemicals with characteristic odor. An exception is the phthalate esters, which are often preferred when ambient odor is objectionable or residual odor on the fabric cannot be tolerated. The toxicity of carrier-active compounds and of their ultimate compositions varies with the chemical or chemicals involved. The environment surrounding the dyeing equipment where carriers are used should always be well-ventilated, and operators should wear protective clothing (eg, rubber gloves, aprons, and safety glasses or face shields, and possibly an appropriate respirator). Specific handling information can be obtained from the supplier or manufacturer.

OSHA and EPA have established exposure limits that must be carefully considered in relation to the waste disposal method available and the environment in which dye carriers are to be used. Some are only skin irritants, whereas others can contribute to air and water pollution. Atmospheric dyeing equipment should be exhausted to scrubbers that wash the vapors into the general plant effluent for chemical or biological degradation. The same system applies to drying and heat-setting equipment where residual carriers are completely volatilized out of the dyed fibers. A schematic illustration for the control of air and water pollution is shown in Figure 2 (see AIR POLLUTION CONTROL METHODS).

Sara title III, section 313, Clean Air Act 1990, threshold limit values, and LD_{50}s are given in Table 3 for the substances for which data are available. Additional information is continuously being developed to provide guidelines for the safe handling of dye carriers and carrier-active chemicals.

The increasingly stringent government regulations and the introduction of carrierless–dyeable polyester have not substantially affected the use of carriers. The factor with the greatest impact on their use has been provided by the spectacular technological advances in dyeing equipment that have taken place since the early 1980s. High temperature dyeing has greatly reduced the time element (dyeing cycle) and the need for carriers.

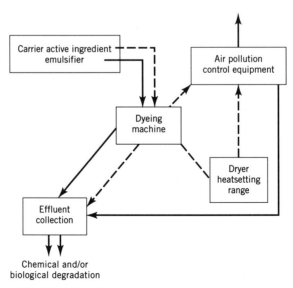

Fig. 2. Air and water pollution controls for dye carriers.

Table 3. LD$_{50}$ and Threshold Limit Values for Compounds Most Commonly Used as Dye Carriers[a]

Compound	TLV (in air)[b]		LD$_{50}$ mg/kg[c]	Title III Sara[d]	Clean air[d]
phenolics					
o-phenylphenol			2700	yes	
p-phenylphenol					
methyl cresotinate					
chlorinated aromatics					
o-dichlorobenzene	50	ppm	500	yes	
trichlorobenzene	5	ppm	756	yes	yes
aromatic hydrocarbons and ether					
biphenyl	0.2	ppm	3280	yes	yes
methylbiphenyl					
diphenyl oxide	1	ppm	3370		
methylnaphthalene			4360		
aromatic esters					
methyl benzoate			2170		
butyl benzoate			5140		
benzyl benzoate			1700		
phthalates					
dimethyl phthalate	5 mg/m^3		6900	yes	yes
diethyl phthalate	5 mg/m^3		5058	yes	
diallyl phthalate			770		
dimethyl terephthalate			4390		

[a]Data extracted from Ref. 5.
[b]TLV indicates the upper limit to which workers can be exposed without adverse effect in an 8-h day.
[c]LD$_{50}$ values are for rat-oral (ingestion) except for diethyl phthalate which is rat-intraperitoneal.
[d]Yes indicates that compound is listed in act.

BIBLIOGRAPHY

"Dye Carriers" in *ECT* 3rd ed., Vol. 8, pp. 151–158, by R. Wannemacher and A. DeMaria, Tanatex Chemical Co.

1. K. Tandy, Jr., "Characteristics of Polyester Homopolymer Fiber Which Affect Dyeing Properties," paper presented at the *14th AATCC New England Regional Technical Conference*, New Hampshire, May 19–21, 1977.
2. S. Salvin and co-workers, *Am. Dyest. Rep* (22) (Nov. 2, 1959).
3. C. M. Hansen, *J. Paint Technol* **39**, 104 (1967).
4. M. C. Keen and R. J. Thomas, "Absorption Properties of Latyl Disperse Dyes on Application to Dacron Polyester Fibers," *Dyes and Chemicals Technical Bulletin*, E. I. du Pont de Nemours & Co., Inc., Organic Chemicals Dept., Wilmington, Del., 1992.
5. *Registry of Toxic Effect of Chemical Substances*, NIOSH, U.S. Dept. of Health, Education, and Welfare, Washington, D.C., 1992.

ERNESTO DE GUZMAN
BOYCE SUTTON, JR.
Sybron Chemicals Inc.

DYES AND DYE INTERMEDIATES

The first synthetic dye, Mauveine, was discovered by Perkin in 1856. Hence the dyestuffs industry can rightly be described as mature. However, it remains a vibrant, challenging industry requiring a continuous stream of new products because of the quickly changing world in which we live. The early dyes industry saw the discovery of the principal dye chromogens (the basic arrangement of atoms responsible for the color of a dye). Indeed apart from one or two notable exceptions, all the dye types used today were discovered in the nineteenth century (1). The introduction of the synthetic fibers, nylon, polyester, and polyacrylonitrile during the period 1930–1950, produced the next significant challenge. The discovery of reactive dyes in 1956 heralded a big breakthrough in the dyeing of cotton; intensive research into reactive dyes followed over the next two decades and, indeed, is still continuing today (1) (see DYES, REACTIVE). The oil crisis in the early 1970s, which resulted in a steep increase in the prices of raw materials for dyes, created a drive for more cost-effective dyes, both by improving the efficiency of the manufacturing processes and by replacing tinctorially weak chromogens, such as anthraquinone, with tinctorially stronger chromogens, such as azo and benzodifuranone. These themes are still important and ongoing, as are the current themes of product safety, quality, and protection of the environment. There is also considerable activity in dyes for high technology applications, especially for the electronics and reprographics industries (see DYES, ANTHRAQUINONE).

The scale and growth of the dyes industry is inextricably linked to that of the textile industry. World textile production has grown steadily to an estimated 35 million tons in 1990 (2,3). The two most important textile fibers are cotton, the largest, and polyester. Consequently, dye manufacturers tend to concentrate their efforts on producing dyes for these two fibers. The estimated world production of dyes in 1990 was 1 million tons (2,3). This figure is significantly smaller than that for textile fiber because a little dye goes a long way. For example, 1 t of dye is sufficient to color 16,650 cars or 42,000 suits (3).

Perkin, an Englishman, working under a German professor, Hoffman, discovered the first synthetic dye, and even today the geographical focus of dye production lies in Germany (BASF, Bayer, Hoechst), England (Zeneca), and Switzerland (CIBA-GEIGY and Sandoz). Far Eastern countries, such as Japan, Korea, and Taiwan, and Third World countries, such as India, Brazil, and Mexico, also produce dyes.

Classification Systems for Dyes

Dyes may be classified according to chemical structure or by their usage or application method. The former approach is adopted by practicing dye chemists who use terms such as azo dyes, anthraquinone dyes, and phthalocyanine dyes. The latter approach is used predominantly by the dye user, the dye technologist, who speaks of reactive dyes for cotton and disperse dyes for polyester. Very often, both terminologies are used, for example, an azo disperse dye for polyester and a phthalocyanine reactive dye for cotton.

Chemical Classification. The most appropriate system for the classification of dyes is by chemical structure, which has many advantages. First, it readily identifies dyes as belonging to a group that has characteristic properties, for example, azo dyes (strong and cost-effective) and anthraquinone dyes (weak and expensive). Secondly, there are a manageable number of chemical groups (about a dozen). Most importantly, it is the classification used most widely by both the synthetic dye chemist and the dye technologist. Thus, both chemists and technologists can readily identify with phrases such as an azo yellow, an anthraquinone red, and a phthalocyanine blue.

The classification given in this article maintains the backbone of the Colour Index classification, but attempts to simplify and update it. This is done by showing the structural interrelationships of dyes that are given separate classes by the Colour Index, and the classification is chosen to highlight some of the more recent discoveries in dye chemistry (4).

Usage Classification. It is advantageous to consider the classification of dyes by use or method of application before considering chemical structures in detail because of dye nomenclature and jargon that arises from this system.

Classification by usage or application is the principal system adopted by the Colour Index (5). Because the most important textile fibers are cotton (qv) and polyester, the most important dye types are those used for dyeing these two fibers, including polyester–cotton blends (see FIBERS, POLYESTER). Other textile fibers include nylon, polyacrylonitrile, and cellulose acetate (see FIBERS, ACRYLIC; FIBERS, CELLULOSE ESTERS; FIBERS, POLYAMIDE).

Classification of Dyes by Use or Application Method

The classification of dyes according to their usage is summarized in Table 1, which is arranged according to the CI application classification. It shows the principal substrates, the methods of application, and the representative chemical types for each application class.

Although not shown in Table 1, dyes are also being used in high technology applications, such as in the medical, electronics, and especially the reprographics industries. For example, they are used in electrophotography (qv) (photocopying and laser printing) in both the toner and the organic photoconductor, in ink jet printing, and in direct and thermal transfer printing (6). As in traditional applications, azo dyes predominate; phthalocyanine, anthraquinone, xanthene, and triphenylmethane dyes are also used. These applications are low volume (tens of kg up to several hundred t per annum) and high added value (hundreds of dollars to several thousand dollars per kg), with high growth rates (up to 60%).

Reactive Dyes. These dyes form a covalent bond with the fiber, usually cotton, although they are used to a small extent with wool and nylon. This class of dyes, first introduced commercially in 1956 by ICI, made it possible to achieve extremely high washfastness properties by relatively simple dyeing methods. A marked advantage of reactive dyes over direct dyes is that their chemical structures are much simpler, their absorption spectra show narrower absorption bands, and the dyeings are brighter. The principal chemical classes of reactive

Table 1. Usage Classification of Dyes

Class	Principal substrates	Method of application	Chemical types[a]
acid	nylon, wool, silk, paper, inks, and leather	usually from neutral to acidic dyebaths	azo, including premetallized anthraquinone, triphenylmethane, azine, xanthene, nitro, and nitroso
azoic components and compositions	cotton, rayon, cellulose acetate, and polyester	fiber impregnated with coupling component and treated with a solution of stabilized diazonium salt	azo
basic	paper, polyacrylonitrile-modified nylon, polyester, and inks	applied from acidic dyebaths	diazacarbocyanine cyanine, hemicyanine, diazahemicyanine, diphenylmethane, triarylmethane, azo, azine, xanthene, acridine, oxazine, and anthraquinone
direct	cotton, rayon, paper, leather, and nylon	applied from neutral or slightly alkaline baths containing additional electrolyte	azo, phthalocyanine, stilbene, and oxazine
disperse	polyester, polyamide, acetate, acrylic, and plastics	fine aqueous dispersions often applied by high temperature–pressure or lower temperature carrier methods; dye may be padded on cloth and baked on or thermofixed	azo, anthraquinone, styryl, nitro, and benzodifuranone
fluorescent brighteners[b]	soaps and detergents, all fibers, oils, paints, and plastics[c]	from solution, dispersion, or suspension in a mass	stilbene, pyrazoles, coumarin, and naphthalimides
food, drug, and cosmetic[d]	foods, drugs, and cosmetics		azo, anthraquinone, carotenoid, and triarylmethane
mordant[e]	wool, leather, and anodized aluminum	applied in conjunction with chelating Cr salts	azo and anthraquinone
natural[f]	food	applied as mordant, vat, solvent, or direct and acid dyes	anthraquinone, flavonols, flavones, indigoids, chroman
oxidation bases	hair, fur, and cotton	aromatic amines and phenols oxidized on the substrate	aniline black and indeterminate structures

Table 1. (*Continued*)

Class	Principal substrates	Method of application	Chemical types[a]
pigments[g]	paints, inks, plastics, and textiles	printing on the fiber with resin binder or dispersion in the mass	azo, basic, phthalocyanine, quinacridone, and indigoid
reactive[h]	cotton, wool, silk, and nylon	reactive site on dye reacts with functional group on fiber to bind dye covalently under influence of heat and pH (alkaline)	azo, anthraquinone, phthalocyanine, formazan, oxazine, and basic
solvent	plastics, gasoline, varnish, lacquer, stains, inks, fats, oils, and waxes	dissolution in the substrate	azo, triphenylmethane, anthraquinone, and phthalocyanine
sulfur	cotton and rayon	aromatic substrate vatted with sodium sulfide and reoxidized to insoluble sulfur-containing products on fiber	indeterminate structures
vat	cotton, rayon, and wool	water-insoluble dyes solubilized by reducing with sodium hydrosulfite, then exhausted on fiber and reoxidized	anthraquinone (including polycyclic quinones) and indigoids

[a]*Encyclopedia* articles on specific chemical types of dyes are AZINE DYES; AZO DYES; CYANINE DYES; DYES, ANTHRA-QUINONE; PHTHALOCYANINE COMPOUNDS; POLYMETHINE DYES; STILBENE DYES; SULFUR DYES; THIAZOLE DYES; TRIPHENYLMETHANE AND RELATED DYES; XANTHENE DYES.
[b]See FLUORESCENT WHITENING AGENTS.
[c]See COLORANTS FOR PLASTICS.
[d]See COLORANTS FOR FOODS, DRUGS, COSMETICS, AND MEDICAL DEVICES.
[e]See DYES, APPLICATIONS AND EVALUATION.
[f]See DYES, NATURAL.
[g]See PAINT; PIGMENTS; INKS.
[h]See DYES, REACTIVE.

dyes are azo, triphendioxazine, phthalocyanine, formazan, and anthraquinone (see DYES, REACTIVE).

Direct Dyes. These water-soluble anionic dyes, when dyed from aqueous solution in the presence of electrolytes, are substantive to, ie, have high affinity for, cellulosic fibers. The principal use is the dyeing of cotton and regenerated cellulose, paper, leather, and, to a lesser extent, nylon. Most of the dyes in this class are azo compounds with some stilbenes, phthalocyanines, and oxazines. After-treatments, frequently given to the dyed material to improve washfastness properties, include chelation with salts of metals (usually copper or chromium), and treatment with formaldehyde or a cationic dye-complexing resin.

Vat Dyes. These water-insoluble dyes are applied mainly to cellulosic fibers as soluble leuco-salts after reduction in an alkaline bath, usually with sodium hydrosulfite. Following exhaustion onto the fiber, the leuco forms are reoxidized to the insoluble keto forms and aftertreated, usually by soaping, to redevelop the crystal structure. The principal chemical classes of vat dyes are anthraquinone and indigoid.

Sulfur Dyes.   These dyes are applied to cotton from an alkaline-reducing bath with sodium sulfide as the reducing agent. Numerically this is a relatively small group. However, the low cost and good washfastness properties of the dyeings make this class important from an economic standpoint (see SULFUR DYES).

Disperse Dyes.   These are substantially water-insoluble nonionic dyes for application to hydrophobic fibers from aqueous dispersion. They are used predominantly on polyester and to a lesser extent on nylon, cellulose, cellulose acetate, and acrylic fibers. Thermal transfer printing, in which disperse dyes are printed onto paper and subsequently transferred to the fiber by a dry-heat process, represents a niche market for selected members of this class. They are also used in the Dye Diffusion Thermal Transfer (D2T2) process for electronic photography (6) (see ELECTROPHOTOGRAPHY).

Basic Dyes.   These water-soluble cationic dyes are applied to paper, polyacrylonitrile (eg, Dralon), modified nylons, and modified polyesters. Their original use was for silk, wool, and tannin-mordanted cotton when brightness of shade was more important than fastness to light and washing. Basic dyes are watersoluble, and yield colored cations in solution. For this reason they are frequently referred to as cationic dyes. The principal chemical classes are diazahemicyanine, triarylmethane, cyanine, hemicyanine, thiazine, oxazine, and acridine. Some basic dyes show biological activity and are used in medicine as antiseptics (see DISINFECTANTS AND ANTISEPTICS).

Solvent Dyes.   These water-insoluble dyes are devoid of polar solubilizing groups such as sulfonic acid, carboxylic acid, or quaternary ammonium. They are used for coloring plastics, gasoline, oils, and waxes. The dyes are predominantly azo and anthraquinone, but phthalocyanines and triarylmethane dyes are also used.

Acid Dyes.   These water-soluble anionic dyes are applied to nylon, wool, silk, and modified acrylics. They are also used to some extent for paper, leather, food, and cosmetics. The original members of this class all had one or more sulfonic or carboxylic acid groups in their molecules. This characteristic probably gave the class its name. Chemically, the acid dyes consist of azo (including preformed metal complexes), anthraquinone, and triarylmethane compounds with a few azine, xanthene, ketone imine, nitro, nitroso, and quinophthalone compounds.

Nomenclature of Dyes

Dyes are named either by their commercial trade name or by their Colour Index (CI) name. In the Colour Index (5) these are cross-referenced.

The commercial names of dyes are usually made up of three parts. The first is a trademark used by the particular manufacturer to designate both the manufacturer and the class of dye, the second is the color, and the third is a series of letters and numbers used as a code by the manufacturer to define more precisely the hue, and also to indicate important properties the dye possesses. The code letters used by different manufacturers are not standardized. The most common letters used to designate hue are R for reddish, B for bluish, and G for greenish shades. Some of the more important letters used to denote the dyeings and fast-ness properties of dyes are W for washfastness and E for exhaust dyes. For solvent and disperse dyes, the heatfastness of the dye is denoted by letters A, B, C, or D, A being the lowest level of heatfastness and D being the highest. In reactive dyes for cotton, M denotes a warm (ca 40°C) dyeing dye and H a hot (ca 80°C) dyeing dye. Examples that follow illustrate the use of these letters.

Consider Dispersol Yellow B-6G. Dispersol is the Zeneca trade name for its range of disperse dyes for polyester. Therefore, it reveals the manufacturer and the usage. Yellow denotes the main color of the dye. "B" denotes its heatfastness, ie, rather low, and 6G denotes that it is six steps of green away from a neutral yellow, so it is a very greenish yellow, ie, a lemon yellow.

In the name Procion Red H-E 7B, Procion is the Zeneca trade name for its range of reactive dyes for cotton. Red denotes the main color of the dye. H-E denotes the dye to be hot dyeing and an exhaust dye (high fixation), and 7B de-notes it to be a very bluish red dye, ie, a magenta.

There are instances in which one manufacturer may designate a bluish red dye as Red 4B and another manufacturer uses Violet 2R for the same dye. To resolve such a problem the manufacturers' pattern leaflets should be consulted. These show actual dyed pieces of cloth so the colors of the dyes in question can be compared directly in the actual application. Alternatively, colors can be speci-fied in terms of color space coordinates. The Cielab system is becoming the stand-ard; in this system the color of a dye is defined by three numbers, the L, a, and b coordinates (see COLOR).

The CI name for a dye is derived from the application class to which the dye belongs, the color or hue of the dye, and a sequential number, eg, CI Acid Yellow 3, CI Acid Red 266, CI Basic Blue 41, and CI Vat Black 7. A five digit CI number is assigned to a dye when its chemical structure has been made known by the manufacturer. The following example illustrates these points, where CA indicates *Chemical Abstracts* and CI *Colour Index*.

chemical structure:

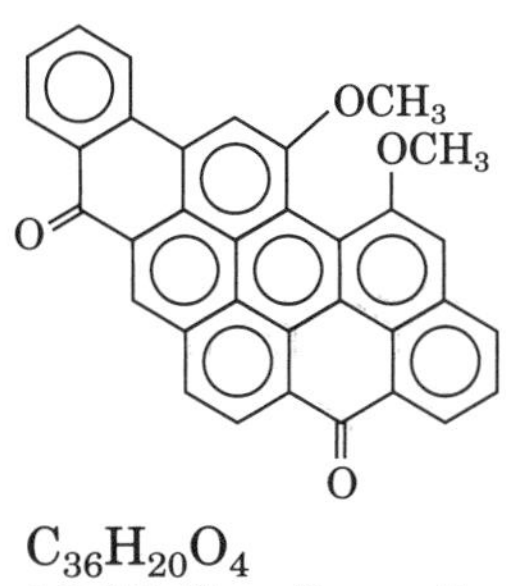

molecular formula: $C_{36}H_{20}O_4$

CA name: 16,17-dimethoxydinaphtho[1,2,3−*cd*:3′,2′,1′,−*lm*] perylene-5,10-dione

(*Continued on next page*)

trivial name:	jade green
CI name:	CI Vat Green 1
CI number:	CI 59825
application class:	vat
chemical class:	anthraquinone
CAS Registry Number:	[*128-58-5*]
commercial names:	Solanthrene Green XBN, Zeneca Specialties
	Cibanone Brilliant Green, BF, 2BF, BFD, CIBA-GEIGY SA
	Indanthrene Brilliant, Green, B, FB, Badische Anilin-und Soda-Fabric AG (BASF)

Classification of Dyes by Chemical Structure

The two overriding trends in dyestuffs research for many years have been improved cost-effectiveness and increased technical excellence. Improved cost-effectiveness usually means replacing tinctorially weak dyes such as anthraquinone, the second largest class after the azo dyes, with tinctorially stronger dyes such as heterocyclic azos, triphendioxazines, and benzodifuranones. This theme will be pursued throughout this section discussing dyes by chemical structure.

Azo Dyes. These dyes are by far the most important class, accounting for over 50% of all commercial dyes, and having been studied more than any other class (see AZO DYES). Azo dyes contain at least one azo group (—N=N—) but can contain two (disazo), three (trisazo), or, more rarely, four or more (polyazo) azo groups. The azo group is attached to two radicals of which at least one, but, more usually, both are aromatic. They exist in the trans form where the bond angle is ca 120° and the nitrogen atoms are sp^2 hybridized and may be represented as follows. The designation of A and E groups is consistent with CI usage (5).

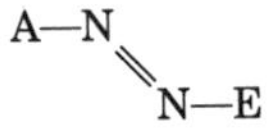

$$A—N \atop \diagdown N—E$$

In monoazo dyes, the most important type, the A radical often contains electron-accepting groups, and the E radical contains electron-donating groups, particularly hydroxy and amino groups. If the dyes contain only aromatic radicals such as benzene and naphthalene, they are known as carbocyclic azo dyes. If they contain one or more heterocyclic radicals, the dyes are known as heterocyclic azo dyes. Examples of various azo dyes are shown in Figure 1. These illustrate the enormous structural variety possible in azo dyes, particularly with polyazo dyes.

Synthesis. Almost without exception, azo dyes are made by diazotization of a primary aromatic amine followed by coupling of the resultant diazonium salt with an electron-rich nucleophile. The diazotization reaction is carried out by treating the primary aromatic amine with nitrous acid, normally generated *in situ* with hydrochloric acid and sodium nitrite. The nitrous acid nitrosates the

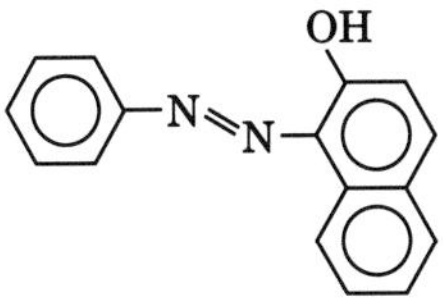

(**1**)

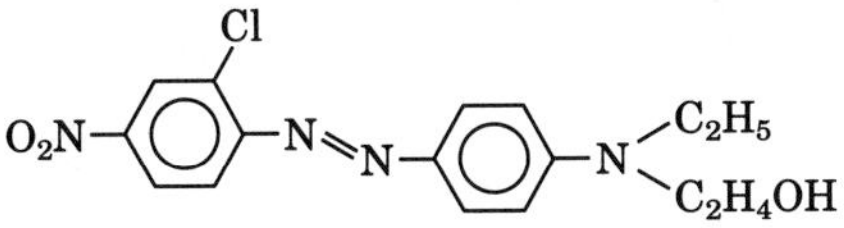

(**2**)

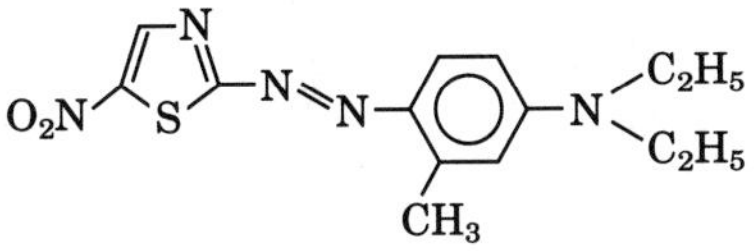

(**3**)

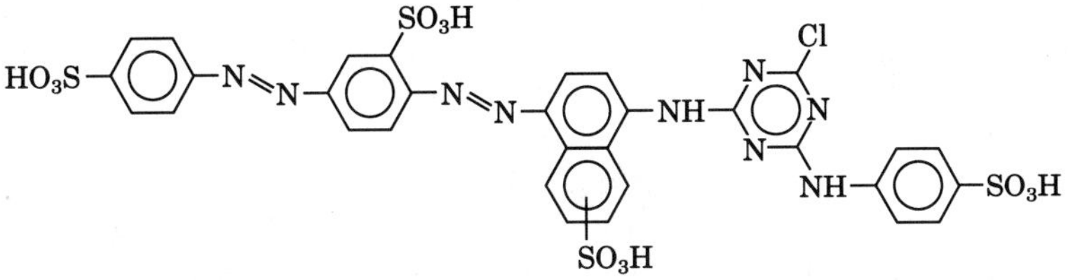

(**4**)

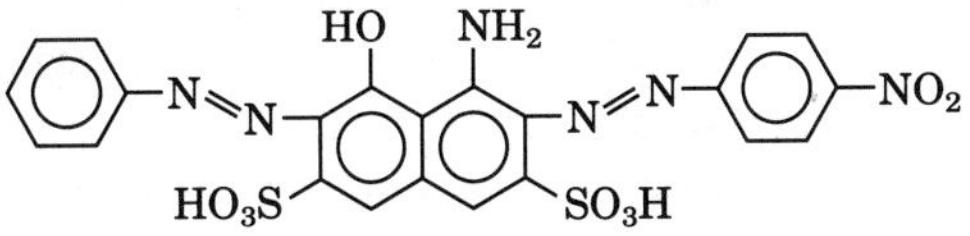

(**5**)

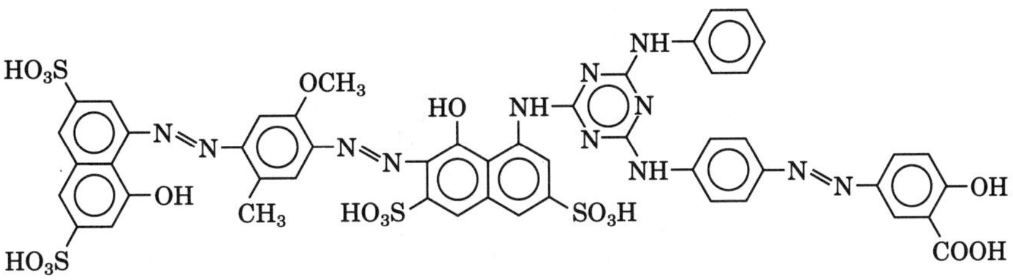

(**6**)

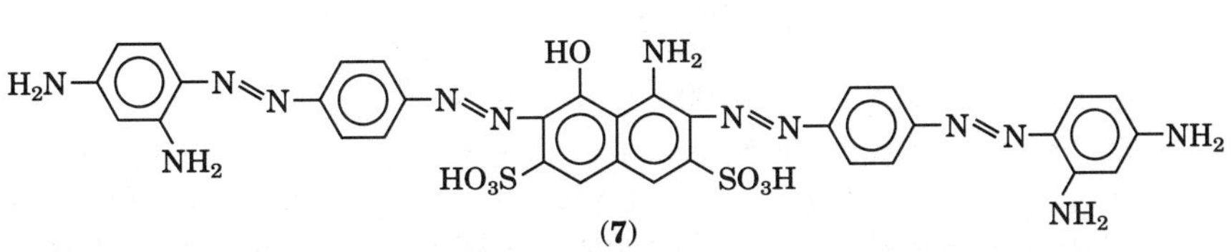

(**7**)

Fig. 1. Azo dyes. (**1**), CI Solvent Yellow 14 [*842-07-9*] (CI 12055); (**2**), CI Disperse Red 13 [*3180-81-2*] (CI 11115); (**3**), CI Disperse Blue [*70693-64-0*]; (**4**), CI Reactive Brown 1 [*12238-04-9*] (CI 26440); (**5**) CI Acid Black 1 [*1064-48-8*] (CI 20470) (**6**), CI Direct Green 26 [*25780-48-7*] (CI 34045); (**7**) CI Direct Black 19 [*6428-31-5*] (CI 35255).

amine to generate the *N*-nitroso compound, which tautomerizes to the diazo hydroxide.

$$Ar—NH—N{=}O \rightleftharpoons Ar—N{=}N—OH$$

Protonation of the hydroxy group followed by the elimination of water generates the resonance-stabilized diazonium salt.

$$Ar—N{=}N^{+} \longleftrightarrow Ar—\overset{+}{N}{\equiv}N$$

For weakly basic amines, ie, those containing several electron-withdrawing groups, nitrosyl sulfuric acid ($NO^{+}HSO_{4}^{-}$) is used as the nitrosating species in sulfuric acid, which may be diluted with phosphoric, acetic, or propionic acid.

A diazonium salt is a weak electrophile, and thus reacts only with highly electron-rich species such as amino and hydroxy compounds. Even hydroxy compounds must be ionized for reaction to occur. Consequently, hydroxy compounds such as phenols and naphthols are coupled in an alkaline medium (pH $\geq$ pK_a of phenol or naphthol; typically pH 7–11), whereas aromatic amines such as *N*,*N*-dialkylamines are coupled in a slightly acid medium, typically pH 1–5. This provides optimum stability for the diazonium salt (stable in acid) without deactivating the nucleophile (protonation of the amine).

Coupling components containing both amino and hydroxy groups, such as H-acid (1-amino-8-naphthol-3,6-disulfonic acid) (**8**) [*90-20-0*], can be coupled stepwise. Coupling is first carried out under acid conditions to effect azo formation in the amino-containing ring. The pH is then raised to ionize the hydroxy group (usually to pH $\geq$ 7) to effect coupling in the naphtholate ring, with either the same or a different diazonium salt. Doing it in the reverse order fails because the nucleophilicity of the amino group is insufficient to facilitate the second coupling step.

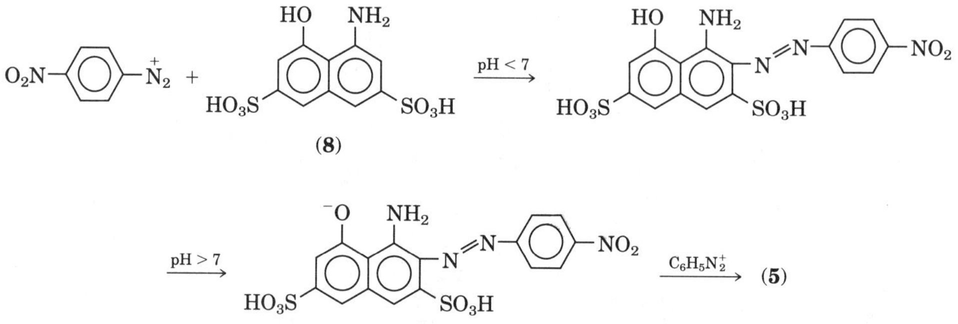

The unusual conditions needed to produce an azo dye, namely, strong acid plus nitrous acid for diazotization, the low temperatures necessary for the unstable diazonium salt to exist, and the presence of electron-rich amino or hydroxy compounds to effect coupling, means that azo dyes have no natural counterparts.

Tautomerism. In theory, azo dyes can undergo tautomerism: azo/hydrazone for hydroxyazo dyes; azo/imino for aminoazo dyes, and azonium/ammonium for protonated azo dyes. A more detailed account of azo dye tautomerism can be found elsewhere (7).

Azo/hydrazone tautomerism was discovered in 1884 (8). The same orange dye was obtained by coupling benzene diazonium chloride with 1-naphthol and by condensing phenylhydrazine with 1,4-naphthoquinone. The expected products were the azo dye (**9**) (R = H) [3651-02-3] and the hydrazone (**10**) (R = H) [19059-71-3]. It was correctly assumed that there was a mobile equilibrium between the two forms, ie, tautomerism.

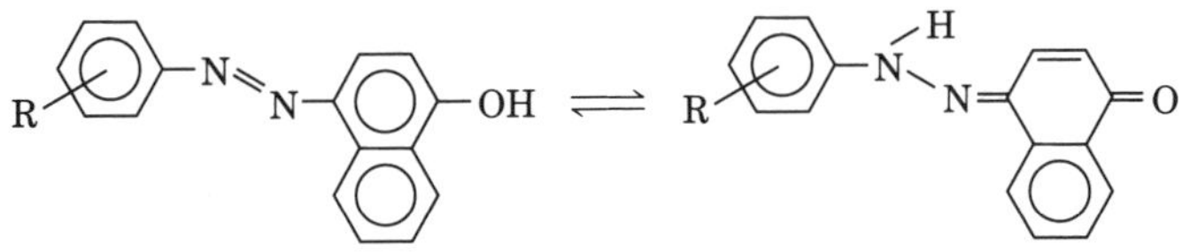

(**9**)   (**10**)

The discovery prompted extensive research into azo/hydrazone tautomerism, a phenomenon which is not only interesting but also extremely important as far as commercial azo dyes are concerned because the tautomers have different colors, different properties, eg, lightfastness, different toxicological properties, and, most importantly, different tinctorial strengths. Since the tinctorial strength of a dye primarily determines its cost-effectiveness, it is desirable that commercial azo dyes should exist in the strongest tautomeric form. This is the hydrazone form.

Hydroxyazo dyes vary in the proportion of tautomers present from pure azo tautomer to mixtures of azo and hydrazone tautomers, to pure hydrazone tautomer. Almost all azophenol dyes (**11**) exist totally in the azo form, except for a few special cases (9).

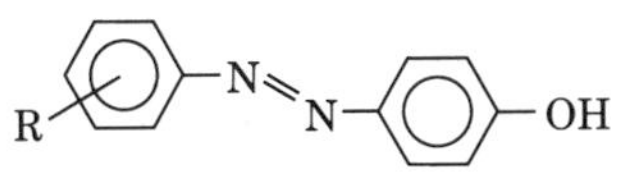

(**11**)

The energies of the azo and hydrazone forms of 4-phenylazo-1-naphthol dyes are similar, so both forms are present. The azo tautomers (**9**) are yellow ($\lambda_{max} \sim$ 410 nm, $\epsilon_{max} \sim$ 20,000) and the hydrazone tautomers (**10**) are orange ($\lambda_{max} \sim$ 480 nm, $\epsilon_{max} \sim$ 40,000). The relative proportions of the tautomers are influenced by both solvent (Fig. 2) and substituents (Fig. 3).

The isomeric 2-phenylazo-1-naphthols (**12**) [1602-36-4] and 1-phenylazo-2-naphthols (**13**) [1602-30-8] exist more in the hydrazone form than the azo form as shown by their uv spectra. Their λ_{max} values are each about 500 nm.

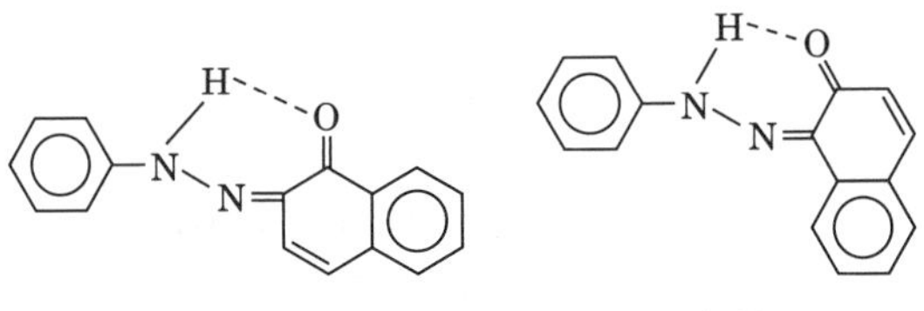

(**12**)   (**13**)

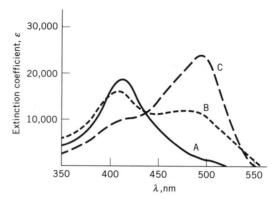

Fig. 2. Effect of solvent on 4-phenylazo-1-naphthol absorption. A, pyridine; B, methanol; C, acetic acid.

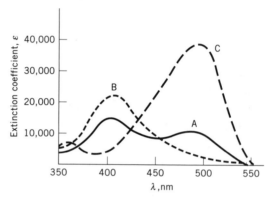

Fig. 3. Electronic effect of substituents in 4-phenylazo-1-naphthol (**9**). A, R = H; B, R = p-OCH$_3$; C, R = p-NO$_2$.

Important classes of dyes that exist totally in the hydrazone form are azopyrazolones (**14**), azopyridones (**15**), and azoacetoacetanilides (**16**).

(**14**)   (**15**)   (**16**)

All aminoazo dyes exist exclusively as the azo form; there is no evidence for the imino form. Presumably, a key factor is the relative instability of the imino grouping.

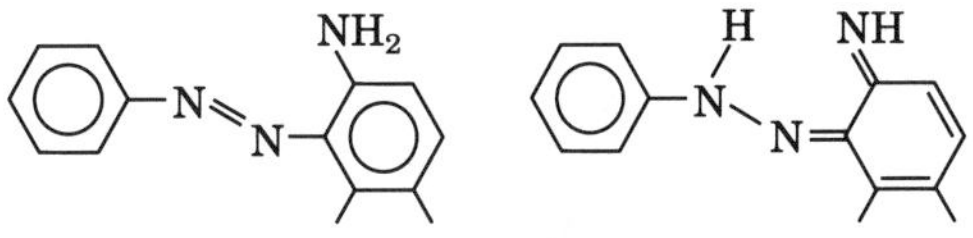

azo   imino

In disazo dyes from aminonaphthols, one group exists as a true azo group and one as a hydrazo group (**17**).

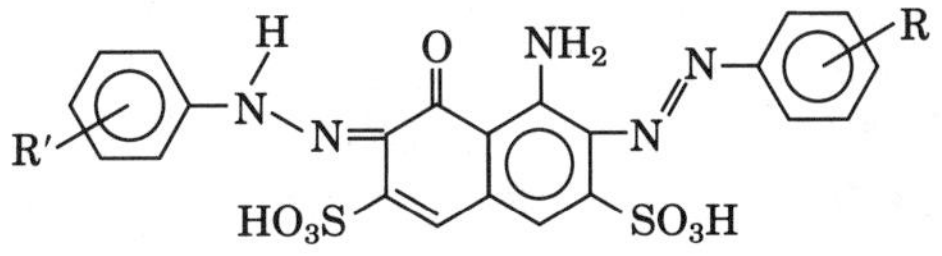

(**17**)

Aminoazo dyes undergo protonation at either the terminal nitrogen atom to give the essentially colorless ammonium tautomer (**18**) ($\lambda_{max} \sim 325$ nm),

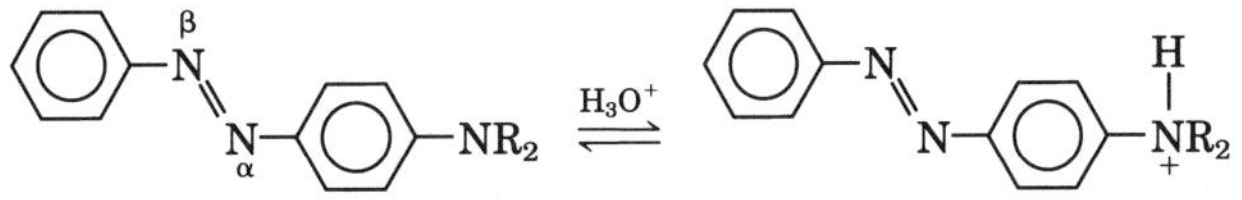

(**18**)

or at the β-nitrogen atom of the azo group to give a resonance-stabilized azonium tautomer (**19**) as shown for methyl orange. The azonium tautomer is brighter, generally more bathochromic and stronger ($\epsilon_{max} \sim 70,000$) than the neutral azo dye. The azonium tautomers are related to diazahemicyanine dyes used for coloring polyacrylonitrile. The most familiar use of the protonation of azo dyes is in indicator dyes such as methyl orange (**20**) [*502-02-3*] and methyl red [*493-52-7*].

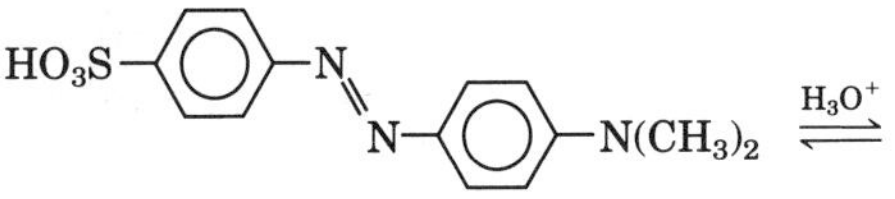

(**20**)
Yellow

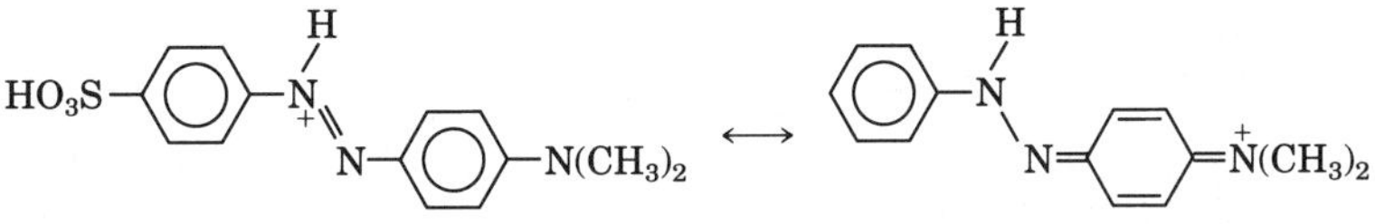

(**19**)
Red

Metallized Azo Dyes.   The three metals of importance in azo dyes are copper, chromium, and cobalt. The most important copper dyes are the 1:1 copper(II): azo dye complexes of formula (**21**); they have a planar structure.

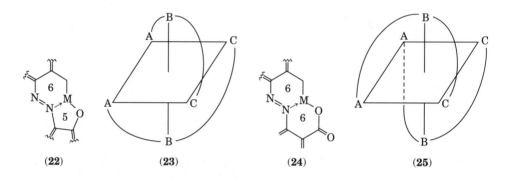

(21)

In contrast, chromium(III) and cobalt(III) form 2:1 dye:metal complexes that have nonplanar structures. Geometrical isomerism exists. The *o,o'*-dihydroxyazo dyes (**22**) form the Drew-Pfitzner or *mer* type (**23**) (A = C = O) whereas *o*-hydroxy–*o'*-carboxyazo dyes (**24**) form the Pfeiffer-Schetty or *fac* type (**25**), where A = CO_2 and C = O.

(22)   (23)   (24)   (25)

Metallization of dyes was originally carried out during the mordanting process to help fix the dye to the substrate. Premetallized dyes are now used widely in various outlets to improve the properties of the dye, particuarly lightfastness. However, this is at the expense of brightness, since metallized azo dyes are duller than nonmetallized dyes.

Carbocyclic Azo Dyes.   These dyes are the backbone of most commercial dye ranges. Based totally on benzene and naphthalene derivatives, they provide yellow, red, blue, and green colors for all the major substrates such as polyester, cellulose, nylon, polyacrylonitrile, and leather. Typical structures (**26–30**) are shown in Figure 4.

Most azoic dyes belong to the carbocyclic azo class, but these dyes are formed in the fiber pores during the application process.

The carbocyclic azo dye class provides dyes having high cost-effectiveness combined with good all-around fastness properties. However, they lack brightness, and consequently, they cannot compete with anthraquinone dyes for brightness. This shortcoming of carbocyclic azo dyes is overcome by heterocyclic azo dyes.

Heterocyclic Azo Dyes.   One long-term aim of dyestuffs research has been to combine the brightness and high fastness properties of anthraquinone dyes with the strength and economy of azo dyes. This aim is now being realized with heterocyclic azo dyes, which fall into two main groups: those derived from het-

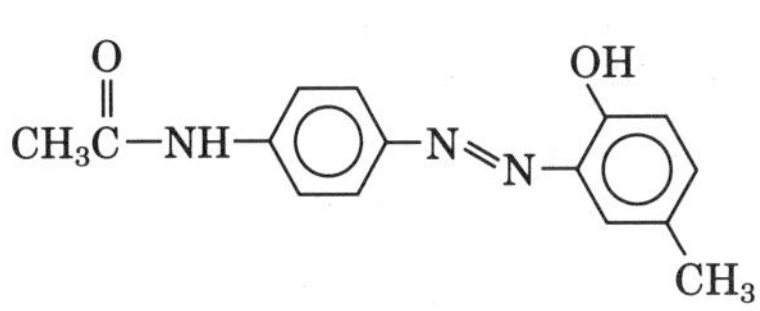

Fig. 4. Carbocyclic azo dyes. Disperse Yellow 3 [*2832-40-8*] (CI 11855) (**26**) is used to dye polyester; Reactive Orange 1 [*6522-74-3*] (CI 17907) (**27**) is a cotton dye; Direct Orange 26 [*25188-23-2*] (CI 29150) (**28**) is a dye for paper; Synacril Fast Red 2G [*48222-26-0*] (CI 11085) (**29**) dyes acrylic fibers; Acid Red 138 [*93762-37-9*] (CI 18073) (**30**) dyes nylon and wool.

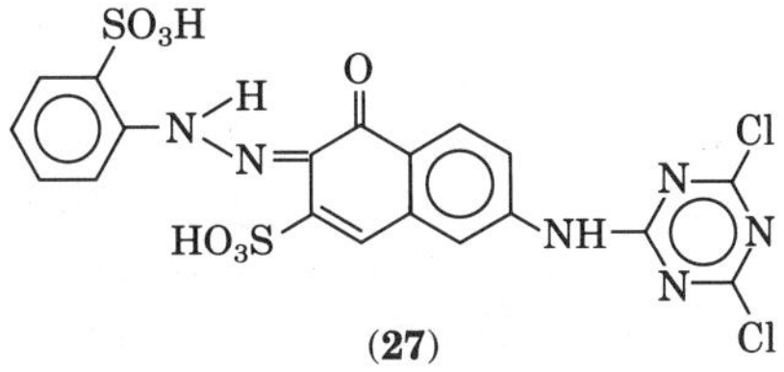

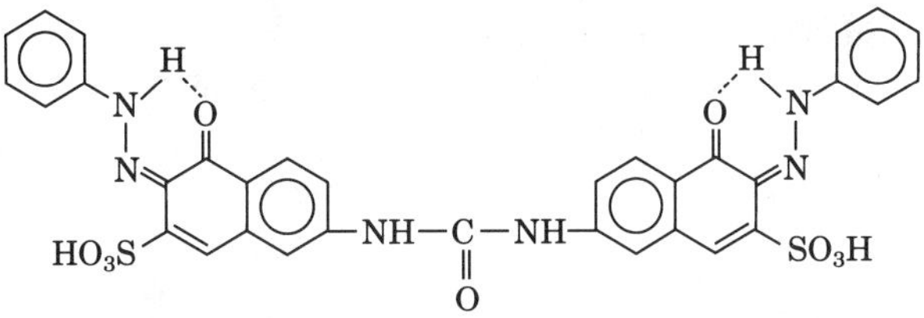

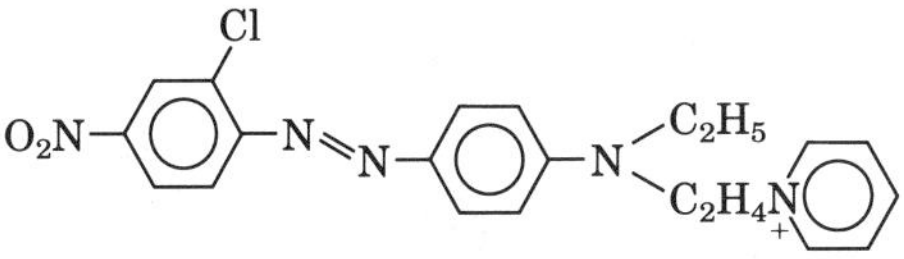

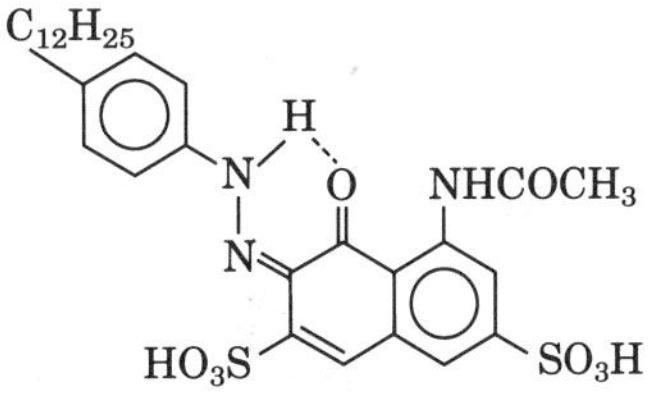

erocyclic coupling components, and those derived from heterocyclic diazo components.

All the heterocyclic coupling components that provide commercially important azo dyes contain only nitrogen as the hetero atom. They are indoles (**31**), pyrazolones (**32**), and especially pyridones (**33**); they provide yellow to orange dyes for various substrates.

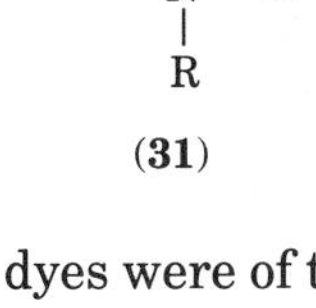

Many yellow dyes were of the azopyrazolone type, but nowadays, these have been largely superseded by azopyridone dyes. Azopyridone yellow dyes are brighter, stronger, and generally have better fastness properties than azopyrazolone dyes. They exist in the hydrazone tautomeric form.

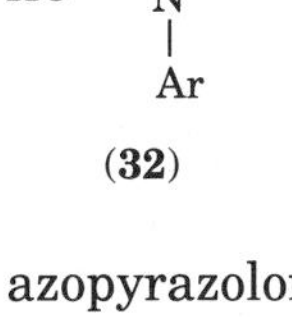

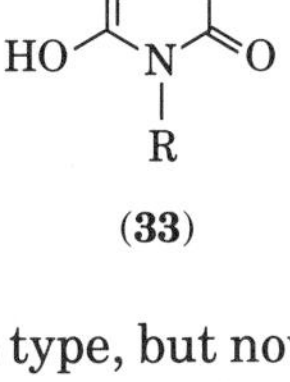

In contrast to the heterocyclic coupling components, virtually all the heterocyclic diazo components that provide commercially important azo dyes contain sulfur, either alone or in combination with nitrogen (the one notable exception is the triazole system). These S or S/N heterocylic azo dyes provide bright, strong shades that range from red through blue to green, and therefore complement the yellow–orange colors of the nitrogen heterocyclic azo dyes in providing a complete coverage of the entire shade gamut. Two representative dyes are the thiadiazole red [88779-75-3] (**34**) and a thiophene greenish blue (**35**). Both are disperse dyes for polyester, the second most important substrate after cellulose.

(**34**)   (**35**)

Anthraquinone Dyes.   This second most important class of dyes also includes some of the oldest dyes; they have been found in the wrappings of mummies dating back over 4000 years. In contrast to the azo dyes, which have no natural counterparts, all the important natural red dyes were anthraquinones (see DYES, NATURAL). However, the importance of anthraquinone dyes is declining due to their low cost-effectiveness.

Anthraquinone dyes are based on 9,10-anthraquinone (**36**) [84-65-1] which is essentially colorless. To produce commercially useful dyes, powerful electron-donor groups such as amino or hydroxy are introduced into one or more of the four alpha positions (1,4,5, and 8). The most common substitution patterns are 1,4–, 1,2,4–, and 1,4,5,8–. To maximize the properties, primary and secondary amino groups (not tertiary) and hydroxy groups are employed. These ensure the maximum degree of π-orbital overlap, enhanced by intramolecular hydrogen bonding, with minimum steric hindrance. These features are illustrated in CI Disperse Red 60 [17418-58-5] (CI 60756) (**37**).

(**36**)   (**37**)

The strength of electron-donor groups increase in the order: $OH < NH_2 < NHR < HNAr$. Tetra-substituted anthraquinones (1,4,5,8–) are more bathochromic than di- (1,4–) or trisubstituted (1,2,4–) anthraquinones. Thus, by an appropriate selection of donor groups and substitution patterns, a wide variety of colors can be achieved (see DYES, ANTHRAQUINONE).

Synthesis. Anthraquinone dyes are prepared by the stepwise introduction of substituents onto the preformed anthraquinone skeleton (**36**) or ring closure of appropriately substituted precursors. The degree of freedom for producing a variety of different structures is restricted, and the availability of only eight substitution centers imposes a further restriction on synthetic flexibility. Therefore, there is significantly less synthetic flexibility than in the case of azo dyes, and consequently less variety; this is a drawback of anthraquinone dyes.

Tautomerism. Although tautomerism is theoretically possible in amino and hydroxy anthraquinone dyes, none has been observed. Studies by ^{13}C nmr have shown convincingly that amino and hydroxy dyes of 9,10-anthraquinone exist as such.

Metal Complexes. The main attributes of anthraquinone dyes are brightness and good fastness properties, including lightfastness. Metallization would detract from the former, and there is no need to improve the latter. Consequently, metallized anthraquinone dyes are of no importance.

Properties. The principal advantages of anthraquinone dyes are brightness and good fastness properties, but they are both expensive and tinctorially weak. However, they are still used extensively, particularly in the red and blue shade areas, because other dyes cannot provide the combination of properties offered by anthraquinone dyes, albeit at a price.

Benzodifuranone Dyes. These (BDF) dyes are challenging anthraquinone dyes. The BDF chromogen is one of the very few novel chromogens to have been discovered in this century. As with many other discoveries (10) the BDF chromogen was detected by accident. The pentacenequinone structure [*100734-53-0*] (**38**) assigned to the intensely red-colored product obtained from the reaction of *p*-benzoquinone with cyanoacetic acid was questioned (11). A compound such as (**38**) should not be intensely colored owing to the lack of conjugation. Instead, the red compound was correctly identified as the 3,7-diphenylbenzodifuranone [*64501-49-1*] (**39**) (λ_{max} = 466 nm; ϵ = 51,000 in chloroform).

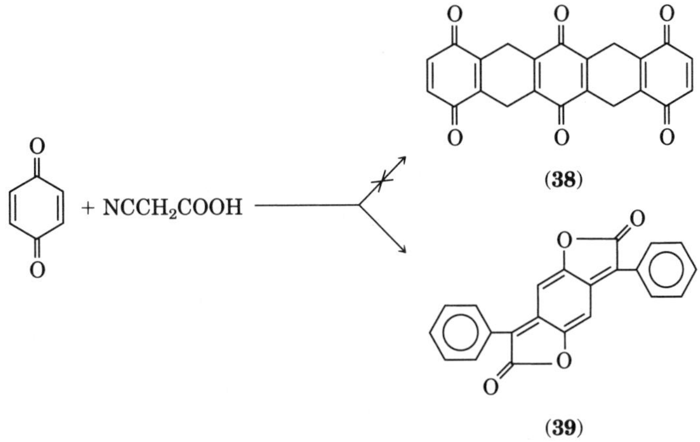

Improved syntheses from arylacetic acids and hydroquinone [*123-31-9*] or substituted quinones have been devised for BDF dyes (12).

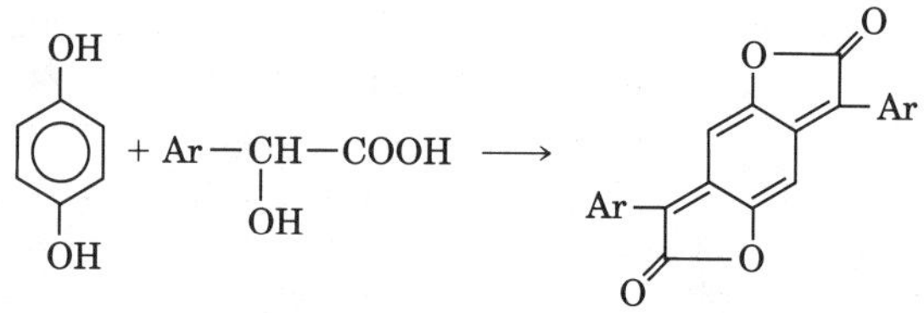

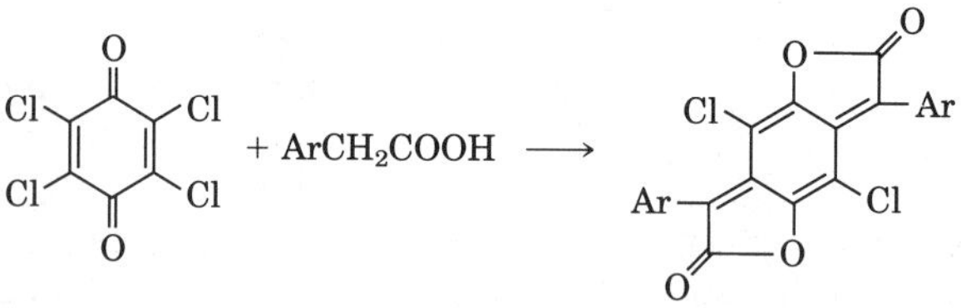

BDFs are unusual in that they span the whole color spectrum from yellow through red to blue, depending on the electron-donating power of the R group on the phenyl ring of the aryl acetic acid, ie, Ar = —C_6H_4R (R = H, yellow–orange; R = alkoxy, red; R = amino, blue). The first commercial BDF, Dispersol Red C-BN, a red disperse dye for polyester, is already making a tremendous impact. Its brightness even surpasses that of the anthraquinone reds, while its high tinctorial strength (ca 3–4 times that of anthraquinones) makes it cost-effective.

Polycyclic Aromatic Carbonyl Dyes. Structurally, these dyes contain one or more carbonyl groups linked by a quinonoid system. They tend to be relatively large molecules built up from smaller units, typically anthraquinones. Since they are applied to the substrate (usually cellulose) by a vatting process, the polycyclic aromatic carbonyl dyes are often called the anthraquinonoid vat dyes.

Although the colors of the polycyclic aromatic carbonyl dyes cover the entire shade gamut, only the blue dyes and the tertiary shade dyes, namely, browns, greens, and blacks, are important commercially. Typical dyes are the blue indanthrone [81-77-6] (**40**), the brown CI Vat Brown 3 [131-92-0] (CI 69012), (**41**), the black CI Vat Black 27 [2379-81-9] (**42**), and the green CI Vat Green 1 [128-58-5] (CI 59825) (**43**), probably the most famous of all the polycyclic aromatic carbonyl dyes.

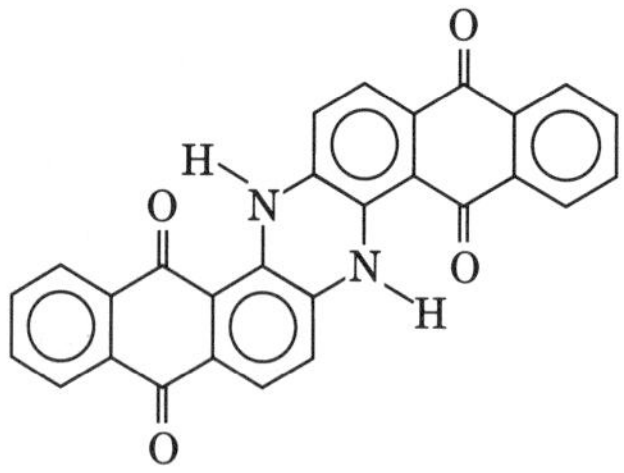

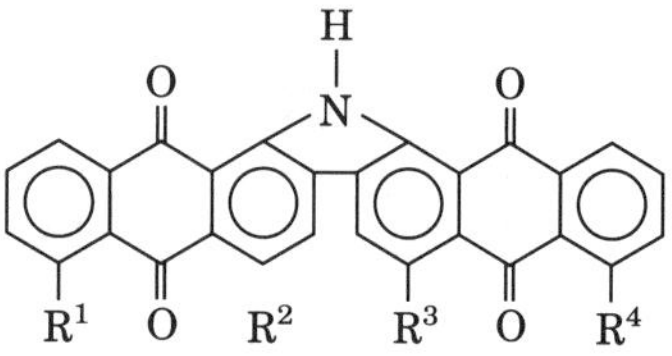

(**41**) $R^1 = R^3 = C_6H_5CONH$; $R^2 = R^4 = H$
(**42**) $R^1 = R^4 = H$; $R^2 = R^3 = C_6H_5CONH$

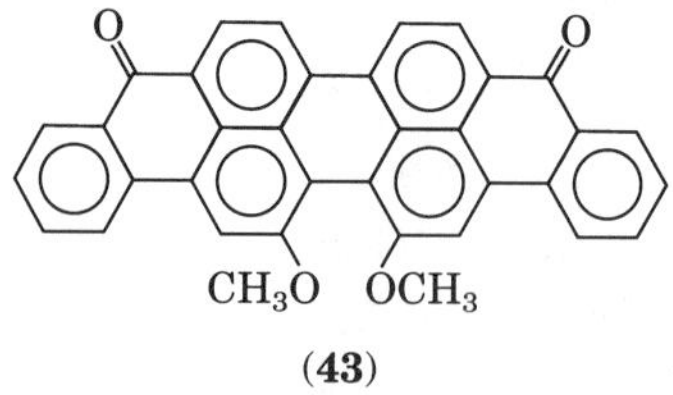

(**40**)

(**43**)

As a class, the polycyclic aromatic carbonyl dyes exhibit the highest order of lightfastness and wetfastness. The high lightfastness is undoubtedly associated with the absence of electron-donating and any further electron-withdrawing groups, other than carbonyl, thus restricting the number of photochemical sites in the molecule. The high wetfastness is a direct manifestation of the application process of the dyes.

Indigoid Dyes. Like the anthraquinone, benzodifuranone, and polycyclic aromatic carbonyl dyes, the indigoid dyes also contain carbonyl groups. They are also vat dyes.

Indigoid dyes represent one of the oldest known classes of dyes. For example, 6,6'-dibromoindigo [*19201-53-7*] (**44**) is Tyrian Purple, the dye made famous by the Romans. Tyrian Purple was so expensive that only the very wealthy were able to afford garments dyed with it. Indeed, the phrase "born to the purple" is still used today to denote wealth.

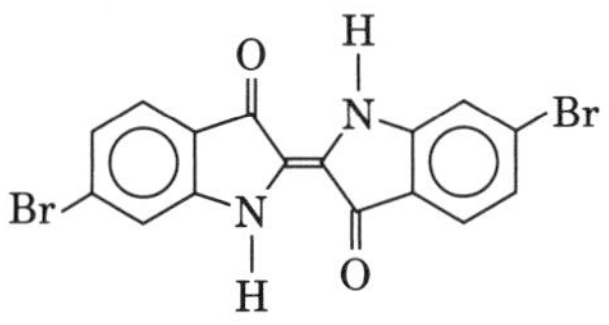

(**44**) 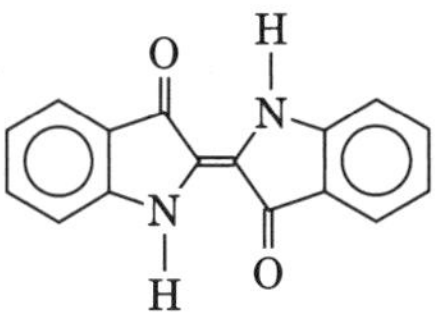   (**45**)

Although many indigoid dyes have been synthesized, only indigo [*482-89-3*] itself (**45**) is of any importance today. Indigo is the blue used almost exclusively for dyeing denim jeans and jackets and is held in high esteem because it fades in tone to give progressively paler blue shades.

One of the main fascinations of indigo is that such a small molecule should be blue. Normally, extensive conjugation, eg, phthalocyanines and/or several powerful donor and acceptor groups, eg, azo and anthraquinone dyes, are required to produce blue dyes. After much controversy it was proven that the chromogen, ie, that molecular species responsible for the color of indigo, is the crossed conjugated system (13–15). This surprising discovery prompted a new term for this kind of chromogen: it was called an H-chromogen because the basic shape resembled a capital H (16).

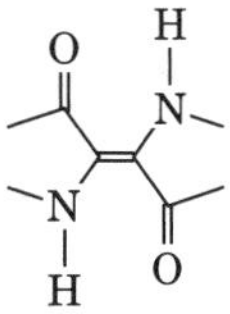

H-chromogen

Polymethine and Related Dyes. Cyanine dyes (qv) (**46**) are the best known polymethine dyes. Nowadays, their commercial use is limited to sensitizing dyes for silver halide photography. However, derivatives of cyanine dyes provide important dyes for polyacrylonitrile.

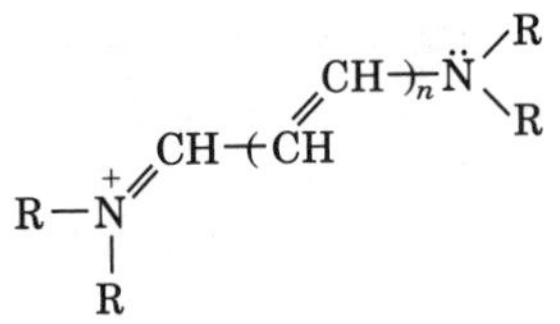

where n = 0, 1, 2, ... and R groups are typically part of ring systems

(46)

Azacarbocyanines. A cyanine containing three carbon atoms between heterocyclic nuclei is called a carbocyanine (n = 1 in (**46**)). Replacing these carbon atoms by one, two, and three nitrogen atoms produces azacarbocyanines, diazacarbocyanines, and triazacarbocyanines, respectively. Dyes of these three classes are important yellow dyes for polyacrylonitrile, eg, CI Basic Yellow 28 [*52757-89-8*] (CI 48054) (**47**).

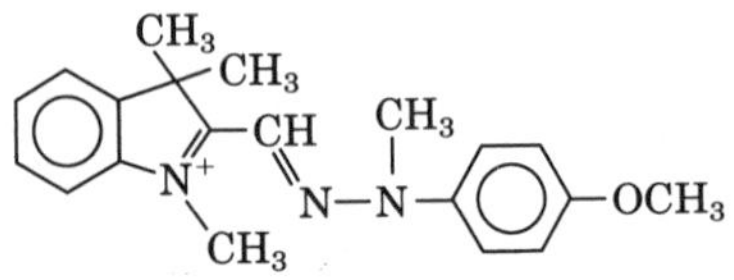

(47)

Hemicyanines. These half-cyanine dyes may be represented by structure (**48**). They may be considered as cyanines in which a benzene ring has been inserted into the conjugated chain. Hemicyanines provide some bright fluorescent red dyes for polyacrylonitrile.

Diazahemicyanines. Diazahemicyanine dyes are arguably the most important class of polymethine dyes. They have the general structure (**49**)

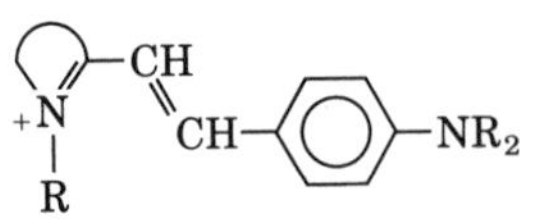

(48) 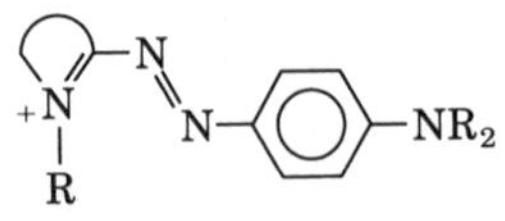   **(49)**

The heterocyclic ring is normally composed of one (eg, pyridinium), two (eg, pyrazolium and imidazolium), or three (eg, triazolium) nitrogen atoms, or sulfur and nitrogen atoms, eg, (benzo)thiazolium and thiadiazolium. Triazolium dyes (**50**) provide the market-leading red dyes for polyacrylonitrile, and a benzothiazolium dye (**51**) is the market-leading blue dye.

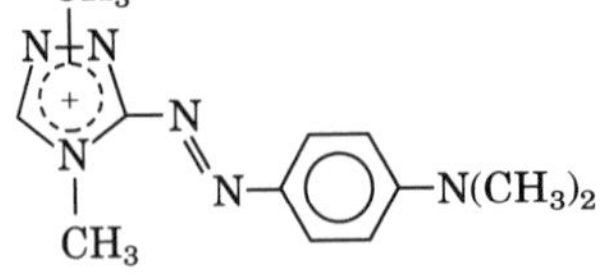

(50) 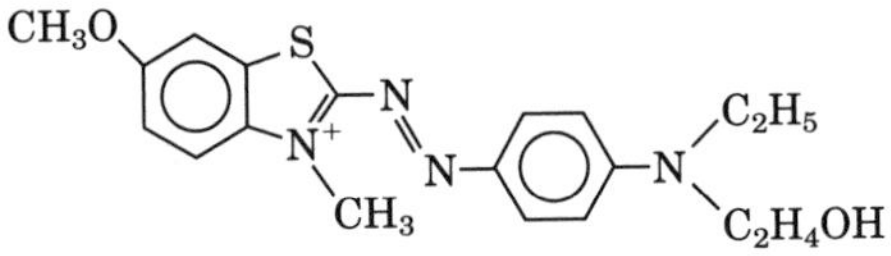   **(51)**

Styryl Dyes. The styryl dyes are uncharged molecules containing a styryl group C_6H_5—CH=C usually in conjugation with an *N,N*-dialkylaminoaryl group. Styryl dyes were once a fairly important group of yellow dyes for a variety of substrates. They are synthesized by condensation of an active methylene compound, especially malononitrile [*109-77-3*] (**52**); X = Y = CN with a carbonyl group, especially an aldehyde. As such, styryl dyes have small molecular structures and are ideal for dyeing densely packed hydrophobic substrates such as polyester. CI Disperse Yellow 31 [*4361-84-6*] (**53**), R = C_4H_9, R^1 = C_2H_4Cl, R^2 = H, X = CN, Y = $COOC_2H_5$, is a typical dye.

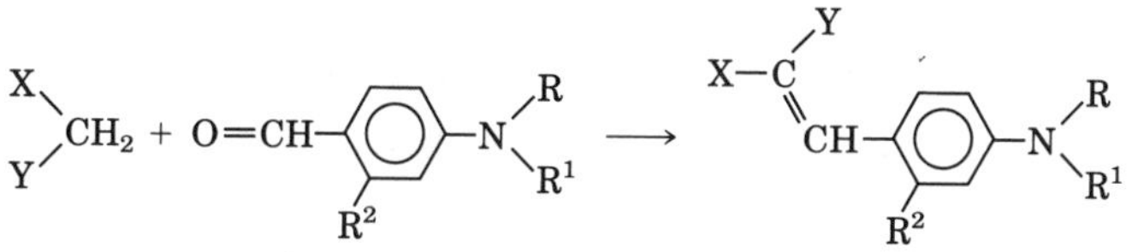

(**52**)   (**53**)

Yellow styryl dyes have now been largely superseded by superior dyes such as azopyridones, but there has been a resurgence of interest in red and blue styryl dyes. The addition of a third cyano group to produce a tricyanovinyl group causes a large bathochromic shift: the resulting dyes, eg, (**54**), are bright red rather than the greenish yellow color of the dicyanovinyl dyes. These tricyanovinyl dyes have been patented by Mitsubishi for the transfer printing of polyester substrates. Two synthetic routes to the dyes are shown: one is by the replacement of a cyano group in tetracyanoethylene, and the second is by oxidative cyanation of a dicyanovinyl dye (**55**) with cyanide. The use of such toxic reagents could hinder the commercialization of the tricyanovinyl dyes (see CYANOCARBONS).

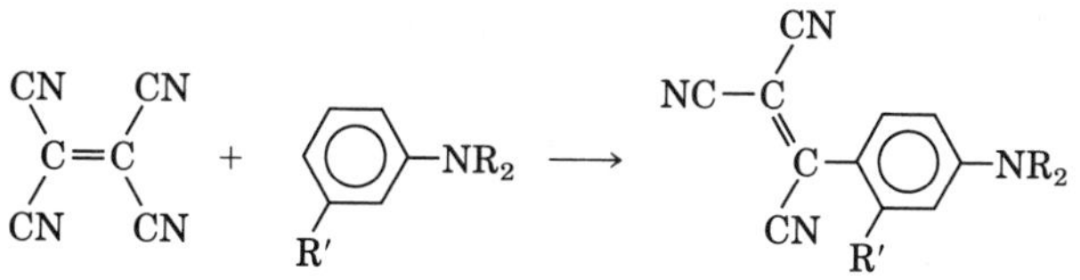

(**54**)

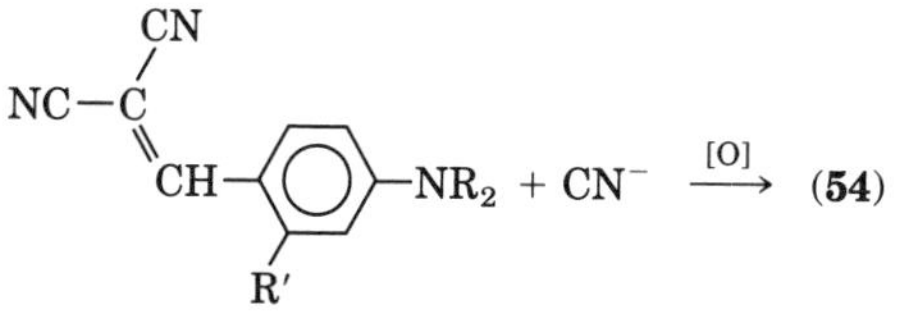

(**55**)

Blue styryl dyes are produced when an even more powerful electron-withdrawing group than tricyanovinyl is used. Thus, Sandoz discovered that the condensation of the sulfone (**56**) [*74228-25-4*] with an aldehyde gives the bright blue dye (**57**) for polyester. In addition to exceptional brightness, this dye also possesses high tinctorial strength (ϵ_{max} ~ 70,000). However, its lightfastness is only moderate.

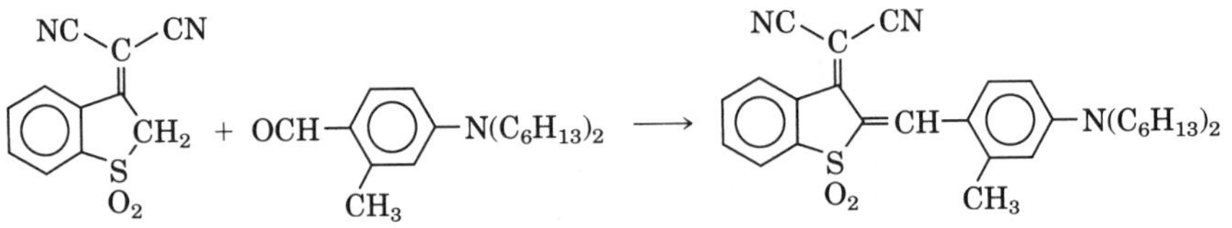

(56)   (57)

Di- and Triaryl Carbonium and Related Dyes. The structural interrelationships of the diarylcarbonium dyes (**58**), triarylcarbonium dyes (**59**), and their heterocyclic derivatives are shown in Figure 5. As a class, the dyes are bright and strong, but are generally deficient in lightfastness. Consequently, they are used in outlets where brightness and cost-effectiveness, rather than permanence, are paramount, for example, the coloration of paper. Many dyes of this class, especially derivatives of pyronines (xanthenes), are among the most fluorescent dyes known.

Typical dyes are the diphenylmethane, Auramine O ((**60**); R = CH$_3$, X = C—NH$_2$; CI Basic Yellow 2) [2465-27-2]; the triphenylmethane, Malachite Green ((**59**) R = CH$_3$; R' = H; CI Basic Green 4) [569-64-2]; the thiazine dye Methylene Blue ((**62**); R = CH$_3$; CI Basic Blue 9) [61-73-1] (CI 52015) used extensively in the Gram staining test for bacteria; the oxazine dye, CI Basic Blue 3 ((**63**); R = C$_2$H$_5$) [47367-75-9]; and the xanthene dye, CI Acid Red 52 (**64**) [121313-93-7].

Notable advances have been made in recent years in triphendioxazine dyes (see AZINE DYES). Triphendioxazine direct dyes have been known for many years; CI Direct Blue 106 (**65**) is a typical dye. Resurgence of interest in triphendioxazine dyes arose through successful modification of the intrinsically strong and bright triphendioxazine chromogen to produce blue reactive dyes for cotton (17). These blue reactive dyes combine the advantages of azo dyes and anthraquinone dyes. Thus they are bright, strong dyes with good fastness properties. Structure (**66**) is typical of these reactive dyes. R represents the reactive group.

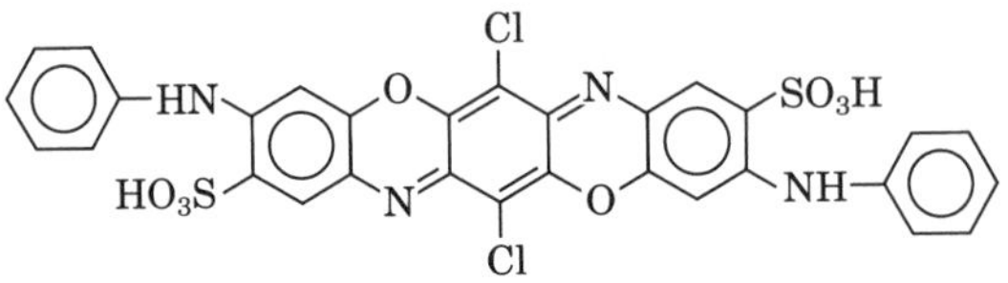

(65)

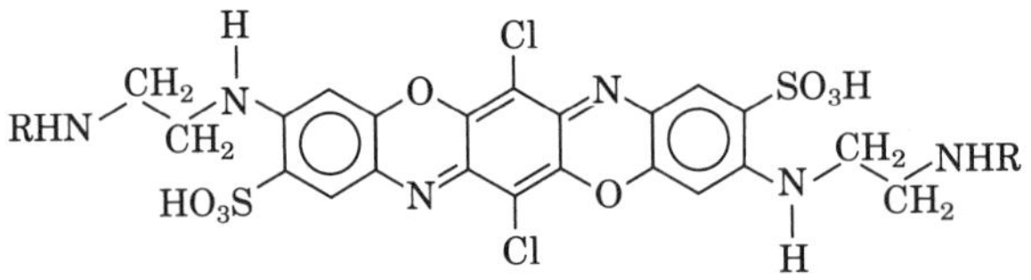

(66)

Like phthalocyanine dyes, triphendioxazine dyes are large molecules, and therefore their use is restricted to coloring the more open-structured substrates such as paper and cotton.

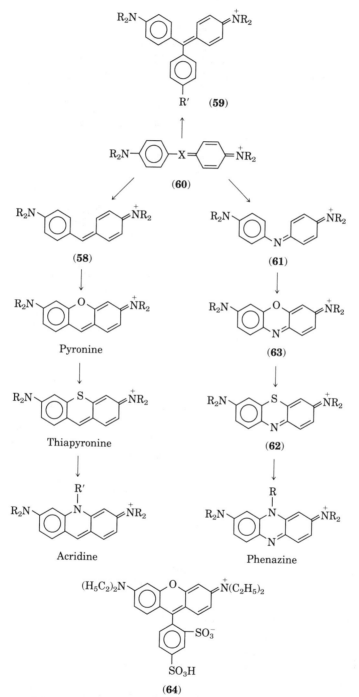

Fig. 5. Structural interrelationships among diaryl and triaryl carbonium dyes. Pyronine = xanthene; (**61**), R = CH₃ is Bindschedler's Green.

563

Phthalocyanines. Phthalocyanine is the only novel chromogen of commercial importance discovered since the nineteenth century. It was discovered accidently in 1928, when during the routine manufacture of phthalimide from phthalic anhydride and ammonia it was found that the product contained a blue contaminant. Chemists of Scottish Dyes Ltd, now part of Zeneca, carried out an independent synthesis of the blue material by passing ammonia gas into molten phthalic anhydride containing iron filings. The importance of the colorant was realized (it was intensely colored and very stable), and a patent application was filed in the same year.

The structure of the blue material was not elucidated until 1934, when it was shown to be the iron complex of (**67**). The new material was christened phthalocyanine [574-93-6], reflecting both its origin from phthalic anhydride and its beautiful blue color (like cyanine dyes). A year later the structure was confirmed by one of the first uses of x-ray crystallography.

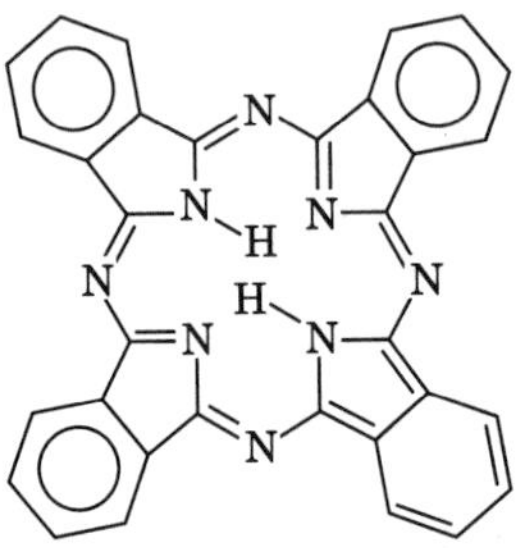

(**67**) 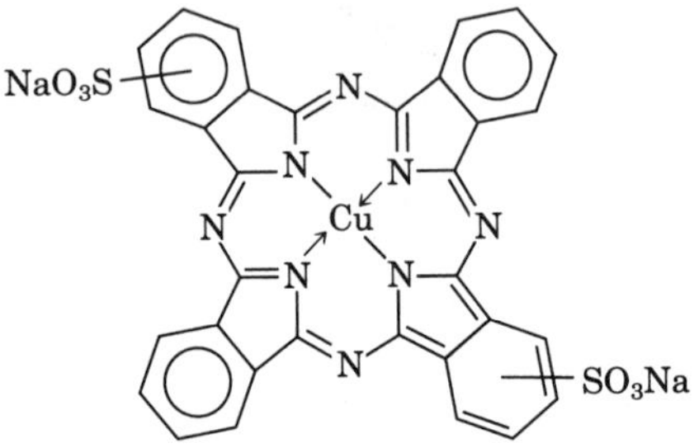   (**68**)

Phthalocyanines are analogues of the natural pigments chlorophyll and heme. However, unlike these natural pigments, which have extremely poor stability, phthalocyanines are probably the most stable of all the colorants in use today. Substituents can extend the absorption to longer wavelengths, into the near infrared, but not to shorter wavelengths, and so their hues are restricted to blue and green.

Of all the metal complexes evaluated, copper phthalocyanines give the best combination of color and properties and consequently the majority of phthalocyanine dyes are based on copper phthalocyanine; CI Direct Blue 86 [1330-38-7] (CI 74180) (**68**) is a typical dye.

As well as being extremely stable, phthalocyanines are bright and tinctorially strong ($\epsilon_{max} \sim 100,000$); this renders them cost-effective. Consequently, phthalocyanines are used extensively in printing inks and paints. The preponderance of blue and green labeling on products is testament to the popularity of phthalocyanine-based printing inks, and most blue and green cars are the product of phthalocyanine-based paints (see PHTHALOCYANINE COMPOUNDS).

Quinophthalones. Like the hydroxy azo dyes, quinophthalone dyes can, in theory, exhibit tautomerism. Because the dyes are synthesized by the condensation of quinaldine derivatives (**69**) with phthalic anhydride, they are often depicted as structure (**70**), but this is incorrect, since the two single bonds prevent any conjugation between the two halves of the molecule.

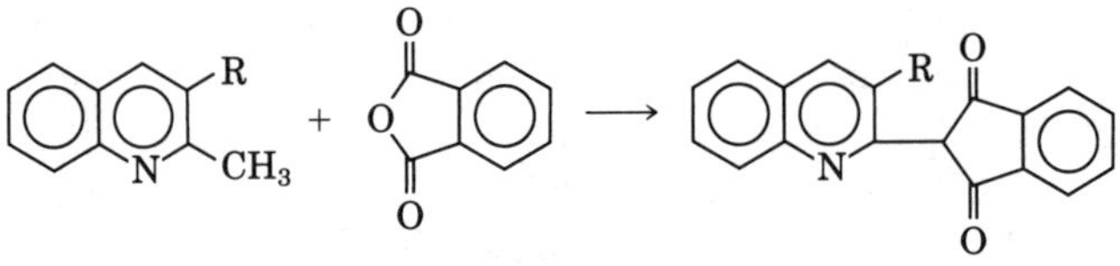

(69)   (70)

The dyes exist as structure (71),

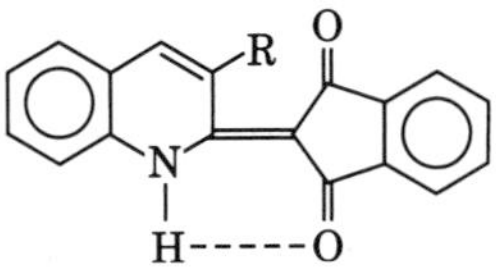

(71)

in which the donor pyrrole-type nitrogen atom is conjugated to the two acceptor carbonyl groups via an ethylenic bridge. In addition to the increased conjugation, structure (71) is stabilized further by the six-membered intramolecular hydrogen bond between the imino hydrogen atom and the carbonyl oxygen atom.

Quinophthalones provide important dyes for the coloration of plastics (eg, CI Solvent Yellow 33 (71), R = H [5662-03-3]) and for the coloration of polyester. For example, CI Disperse Yellow 54 (71) R = OH, is the leading yellow dye for the transfer printing of polyester.

Sulfur Dyes. These dyes are synthesized by heating aromatic amines, phenols, or nitro compounds with sulfur or, more usually, alkali polysulfides. Unlike most other dye types, it is not easy to define a chromogen for the sulfur dyes (qv). It is likely that they consist of macromolecular structures of the phenothiazone-thianthrone type (72), in which the sulfur is present as (sulfide) bridging links and thiazine groups (1).

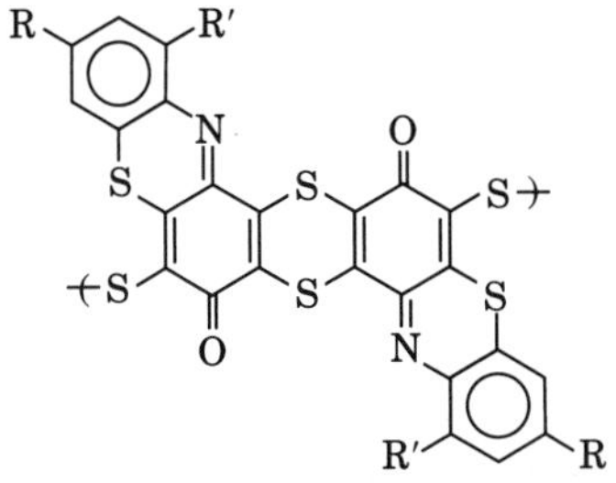

(72)

Sulfur dyes are used for dyeing cellulosic fibers. They are insoluble in water and are reduced to the water-soluble leuco form for application to the substrate by using sodium sulfide solution. The sulfur dye proper is then formed within the fiber pores by atmospheric oxidation (5). Sulfur dyes constitute an important class of dye for producing cost-effective tertiary shades, especially black, on cellulosic fibers. One of the most important dyes is CI Sulfur Black 1 [1326-82-5] (CI 53185), prepared by heating 2,4-dinitrophenol with sodium polysulfide.

Nitro and Nitroso Dyes.   These dyes are now of only minor commercial importance, but are of interest for their small molecular structures. The early nitro dyes were acid dyes used for dyeing the natural animal fibers such as wool and silk. They were nitro derivatives of phenols, eg, picric acid [88-89-1] (**73**) (CI 10305), or naphthols, eg, CI Acid Yellow 1 [846-70-8] (**74**) (CI 10316).

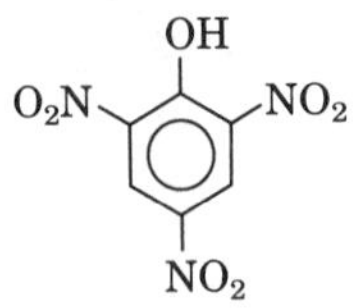

(**73**) 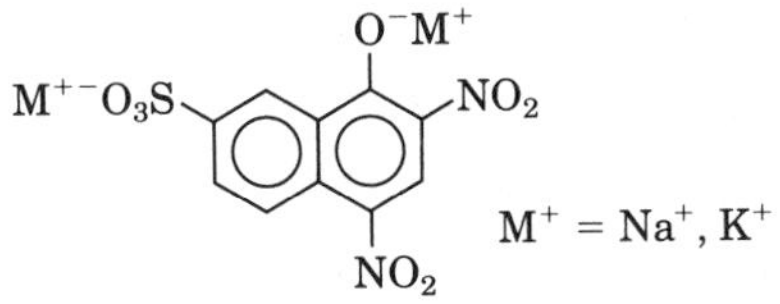   (**74**)

The most important nitro dyes are the nitrodiphenylamines of general structure (**75**).

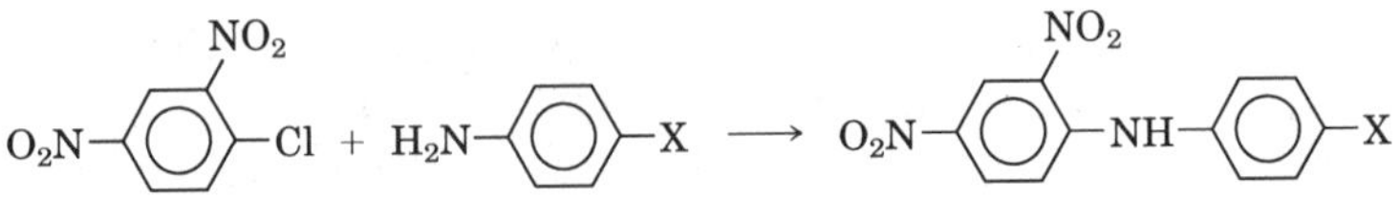

(**75**)

These small molecules are ideal for penetrating dense fibers such as polyester, and are therefore used as disperse dyes for polyester. All the important dyes are yellow: in (**75**) X = H is CI Disperse Yellow 14 [961-68-2]; X = OH is CI Disperse Yellow 1 [119-15-3]; and X = NH$_2$ is CI Disperse Yellow 9 [6373-73-5]. Although the dyes are not terribly strong ($\epsilon_{max} \sim 20{,}000$), they are cost-effective because of their easy synthesis from inexpensive intermediates. CI Disperse Yellow 42 [5124-25-4] and CI Disperse Yellow 86 are important lightfast dyes for automotive-grade polyester.

Nitroso dyes are metal-complex derivatives of o-nitrosophenols or naphthols. Tautomerism is possible in the metal-free precursor between the nitrosohydroxy tautomer (**76**) and the quinoneoxime tautomer (**77**).

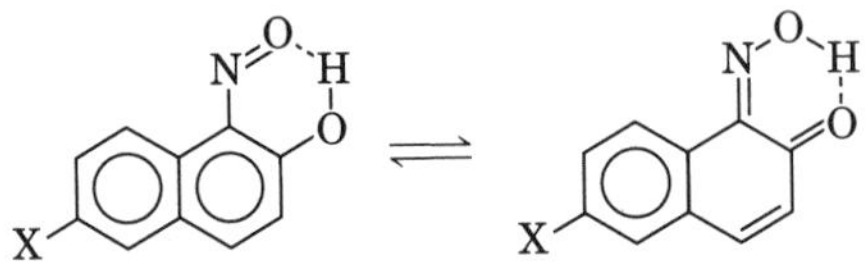

(**76**)   (**77**)

The only nitroso dyes important commercially are the iron complexes of sulfonated 1-nitroso-2-naphthol, eg, CI Acid Green 1 [57813-94-2] (**78**) (CI 10020); these inexpensive colorants are used mainly for coloring paper.

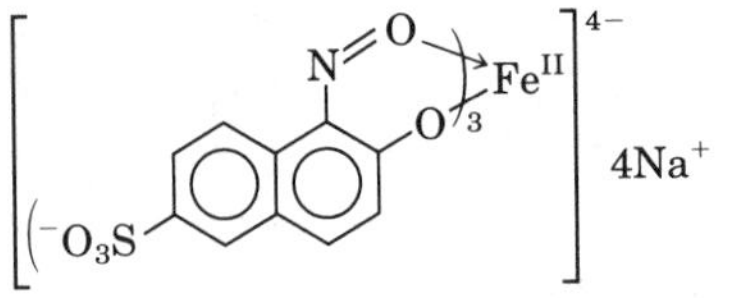

(**78**)

Miscellaneous Dyes.   Other classes of dyes that still have some importance are the stilbene dyes and the formazan dyes. *Stilbene dyes* are in most cases mixtures of dyes of indeterminate constitution that are formed from the condensation of sulfonated nitroaromatic compounds in aqueous caustic alkali either alone or with other aromatic compounds, typically arylamines (5). The sulfonated nitrostilbene [128-42-7] (**79**) is the most important nitroaromatic, and the aminoazobenzenes are the most important arylamines. CI Direct Orange 34 [2222-37-6] (CI 40215-40220), the condensation product(s) of (**79**) and the aminoazobenzene [104-23-4] (**80**), is a typical stilbene dye.

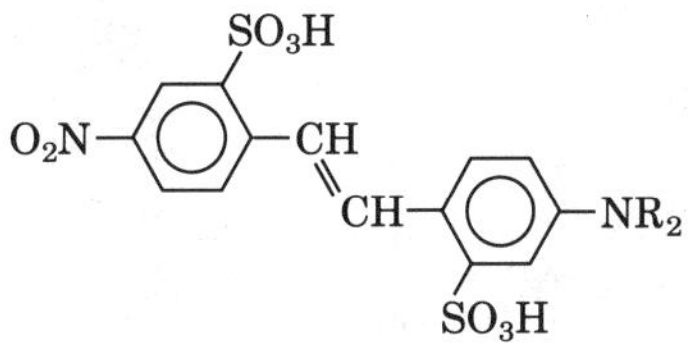

(**79**) 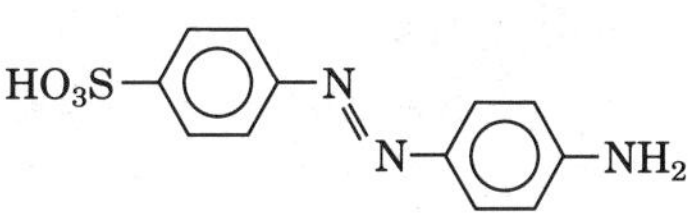   (**80**)

Formazan dyes bear a formal resemblance to azo dyes, since they contain an azo group but have sufficient structural dissimilarities to be considered as a separate class of dyes. The most important formazan dyes are the metal complexes, particularly copper complexes, of tetradentate formazans. They are used as reactive dyes for cotton; (**81**) is a representative example.

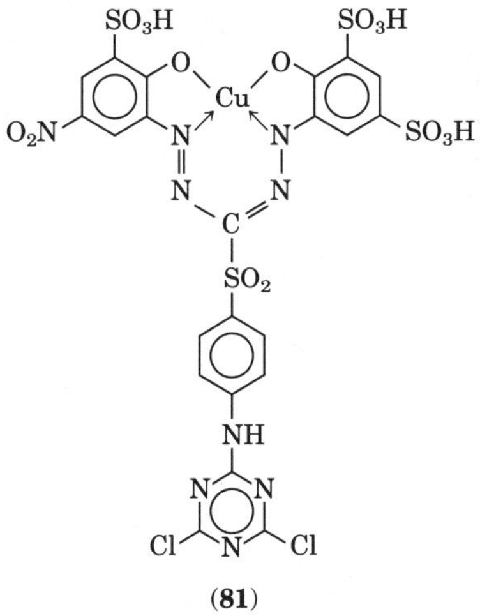

(**81**)

Dye Intermediates

The precursors of dyes are called dye intermediates. They are obtained from simple raw materials, such as benzene and naphthalene, by a variety of chemical reactions. Usually, the raw materials are cyclic aromatic compounds, but acyclic precursors are used to synthesize heterocyclic intermediates. The intermediates are derived from two principal sources, coal tar and petroleum (qv).

Sources of Raw Materials. Coal tar results from the pyrolysis of coal (qv) and is obtained chiefly as a by-product in the manufacture of coke for the steel industry (see COAL, CARBONIZATION). Products recovered from the fractional distillation of coal tar have been the traditional organic raw material for the dye industry. Among the most important are benzene (qv), toluene (qv), xylene, naphthalene (qv), anthracene, acenaphthene, pyrene, pyridine (qv), carbazole, phenol (qv), and cresol (see also ALKYLPHENOLS; ANTHRAQUINONE; XYLENES AND ETHYLBENZENES).

The petroleum industry is now the principal supplier of benzene, toluene, the xylenes, and naphthalene (see BTX PROCESSING; FEEDSTOCKS). Petroleum displaced coal tar as the primary source for these aromatic compounds after World War II because it was relatively cheap and abundantly available. However, the re-emergence of king coal is predicted for the twenty-first century, when oil supplies are expected to dwindle and the cost of producing chemicals from coal (including new processes based on synthesis gas) will gradually become more competitive (3).

Intermediates Classification. Intermediates may be conveniently divided into primary intermediates (primaries) and dye intermediates. Large amounts of inorganic materials are consumed in both intermediates and dyes manufacture.

Inorganic Materials. These include acids (sulfuric, nitric, hydrochloric, and phosphoric), bases (caustic soda, caustic potash, soda ash, sodium carbonate, ammonia, and lime), salts (sodium chloride, sodium nitrite, and sodium sulfide) and other substances such as chlorine, bromine, phosphorus chlorides, and sulfur chlorides. The important point is that there is a significant usage of at least one inorganic material in all processes, and the overall tonnage used by, and therefore the cost to, the dye industry is high.

Primary Intermediates. Primary intermediates are characterized by one or more of the following descriptions, which associate them with raw materials rather than with intermediates.

(1) Manufactured in a dedicated plant, ie, one devoted to a single product or at most two or three closely related products.

(2) At least 1000 t/yr capacity from a single plant and may be up to 100,000 t/yr, eg, aniline.

(3) Manufacturing process and/or operation is continuous or semicontinuous, ie, at least one stage is in a continuous, as distinct from batch, mode.

(4) A primary intermediate has established usage in basic industries such as rubber, polymers, or agrochemicals in addition to dyes.

Primary intermediates were originally manufactured within the dyes industry. All the significant primaries, about 30 different products, are derived from benzene, toluene, or naphthalene. Actual production figures for primaries are not readily available, and in any event the amounts used within the dyes industry are variable. The primaries are listed here with a reference to the *Encyclopedia* article that covers them in detail including production and consumption figures.

The following amines are covered under the title AMINES, AROMATIC:

aniline	dimethylaniline
p-nitroaniline	*m*-phenylenediamine
o-toluidine	*p*-phenylenediamine
p-toluidine	

The article NITROBENZENES AND NITROTOLUENES covers the primaries:

nitrobenzene	*o*-chloronitrotoluene
p-chloronitrobenzene	*p*-nitrotoluene

Some primaries have articles devoted to them and their derivatives, ie, BENZOIC ACID, PHENOL, SALICYLIC ACID, and PHTHALIC ANHYDRIDE as a derivative of phthalic acid. The primary β-naphthol is discussed in NAPHTHALENE DERIVATIVES.

Dye Intermediates. Dye intermediates are defined as those precursors to colorants that are manufactured within the dyes industry, and they are nearly always colorless. Colored precursors are conveniently termed color bases. As distinct from primaries they are only rarely manufactured in single-product units because of the comparatively low tonnages required. Fluorescent brightening agents (FBAs) are neither intermediates nor true colorants. Basic manufacturing processes for FBAs are described in Reference 18 (see FLUORESCENT WHITENING AGENTS).

There are at least 3000 different intermediates in current manufacture (over half that number are specifically mentioned in the *Colour Index*), and in addition there is a comparatively small number of products manufactured by individual companies for their own specialties. Only a selection of intermediates can be discussed here, but since 300 of the products probably account for 90% of the quantity of intermediates used, most of the important aspects can be covered. No meaningful quantification of world tonnage requirements of primaries and intermediates for dyes can be made.

Intermediates vary in complexity, usually related to the number of chemical and operational stages in their manufacture, and therefore cost. Prices may be classed as cheap (less than $1500/t, as with primaries), average ($1500 to $5000/t) or expensive (more than $5000/t).

The Chemistry of Dye Intermediates

The chemistry of dye intermediates may be conveniently divided into the chemistry of carbocycles, such as benzene and naphthalene, and the chemistry of heterocycles, such as pyridones and thiophenes.

CHEMISTRY OF AROMATIC CARBOCYCLES

Benzene and naphthalene are by far the most important aromatic carbocycles used in the dyes industry. The hundreds of benzene and naphthalene intermediates used can be prepared from these parent compounds by the sequential introduction of a variety of substituents eg, NO_2, NR^1R^2, Cl, SO_3H, etc. Introduction of these groups are known as unit processes. The substituents are introduced into the aromatic ring by either electrophilic or nucleophilic substitution. In general, aromatic rings, because of their inherently high electron density, are much more susceptible to electrophilic attack than to nucleophilic attack. Nucleophilic attack only occurs under forcing conditions unless the aromatic ring already contains a powerful electron-withdrawing group, eg, NO_2. In this case, nucleophilic attack is greatly facilitated because of the reduced electron density at the ring carbon atoms.

Electrophilic Substitution.   The most common mechanism for electrophilic attack at an aromatic system involves the initial attack of an electrophile E^+ to give an intermediate containing a tetrahedral carbon atom; loss of Y^+, usually a proton, from the intermediate, then gives the product:

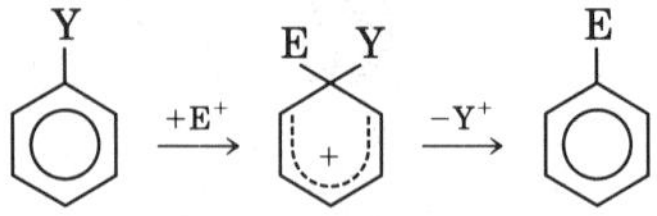

Attack of an unsubstituted benzene ring can lead to only one monosubstitution product. However, when electrophilic attack occurs at a benzene ring already containing a group, there are three possible sites of attack.

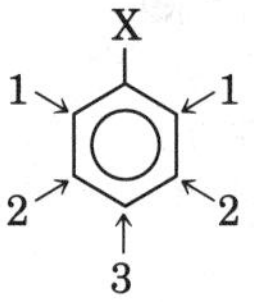

Fortunately, the position of attack may be predicted with a fair degree of accuracy because the various X groups are known to fall into two main categories: those directing the incoming electrophile to the ortho and para positions, and those directing it to the meta position (19) (see BENZENE). The directing effect of the substituent is not exclusive, but is usually sufficient to give one type of product. Thus nitration of nitrobenzene gives approximately 93% of meta, 6% of ortho, and 1% of para dinitrobenzene.

The majority of ortho-para directing groups are also activating substituents (with respect to hydrogen) whereas meta directing substituents are deactivating. The halogens, F, Cl, Br, and I, are peculiar in that they deactivate the benzene ring to electrophilic attack, yet direct the incoming substituent to the ortho and para positions.

When the benzene ring contains more than one group it is usually harder to predict where the incoming substituent will enter. However, a few simple rules have been formulated for disubstituted benzene rings: (1) if the two groups favor attack at one position, then the electrophile will attack there; (2) if a strongly activating group competes with a group that deactivates or only weakly activates the benzene ring, then the position of attack is controlled by the strongly activating group; (3) for steric reasons, an electrophile is least likely to attack the position between two groups in a meta aspect to each other; (4) in the situation shown by the following, attack will take place at position A rather than B.

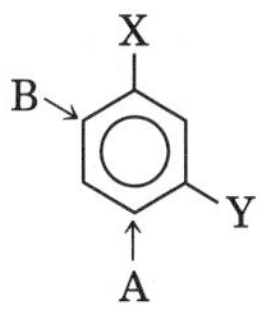

X electron donor (o, p-director)
Y electron acceptor (m-director)

The other important carbocyclic ring system used in dyes is naphthalene [91-20-3]. Here the preferred position of attack is the 1-position. However, 2-substituted naphthalenes are thermodynamically more stable, and under equilibrating conditions the 2-isomer is formed in preference to the 1-isomer.

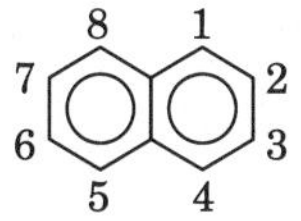

Activating substituents in one of the rings of naphthalene promote substitution in that ring. Deactivating substituents deactivate the ring to which they are attached and electrophilic substitution occurs in the other ring. These effects may be explained by similar arguments to those for benzene substitution patterns and many analogies may be drawn between the two systems.

Nucleophilic Substitution. The unsubstituted benzene ring is not susceptible to nucleophilic attack. However, if the benzene ring contains a good leaving group, eg, Cl, and a strong electron-withdrawing substituent, eg, NO_2, in an ortho and para position, then nucleophilic substitution is greatly facilitated.

The most important mechanism for nucleophilic aromatic substitution is the S_NAr mechanism. The first step is usually rate determining since this is the step in which the aromaticity is lost.

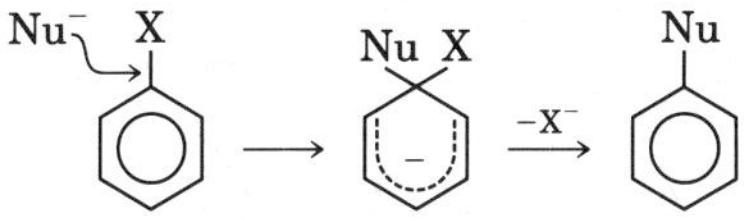

The S_N1 mechanism, in which the substituent leaves before the incoming nucleophile attacks, is less frequently encountered, although it is known to occur in the substitution of the diazonium group ($-\overset{+}{N}\equiv N$).

Orientation in nucleophilic aromatic substitution is not problematic since the nucleophile usually replaces the leaving group, although there are exceptions, eg, those involving benzyne intermediates. As in nucleophilic aliphatic substitution, the common leaving groups are the halides, sulfonates, and ammonium salts. In addition, nitro, alkoxy, sulfone, and sulfonic acid groups are frequently encountered as leaving groups in aromatic substitution (20).

Unit Processes. The unit processes encountered in intermediate and dye chemistry are summarized in Table 2.

Nitration. The unit process of nitration (qv) is concerned with the introduction of one or more nitro (NO_2) groups into an aromatic nucleus by the replace-

Table 2. Unit Processes in Dyes Manufacture

Process	Primaries[a]	Intermediates (common usage)	Colorants (common usage)
nitration	6	✓	
reduction	8	✓	
sulfonation	4	✓[b]	✓
oxidation	5	✓	
fusion/hydroxylation	3	✓	
amination	3	✓[c]	
alkylation	2	✓	✓
halogenation	2	✓	✓
hydrolysis	2	✓	
condensation	1	✓	✓
alkoxylation	1	✓	
esterification	1	✓	
carboxylation	1	✓	
acylation	1	✓	✓
phosgenation	1	✓	✓
diazotization	1	✓	✓
coupling (azo)	1	✓	✓

[a]Number of occurrences within 30 identified product manufactures.
[b]Includes chlorosulfonation.
[c]Includes the Bucherer reaction.

ment of a hydrogen atom. For the more important intermediates, the reaction is achieved by the addition of nitric acid, usually in combination with sulfuric acid (mixed acid), to a sulfuric acid solution of the reactant. The nitronium ion (NO_2^+) is the active nitrating species, and the key process variables that affect yield and quality of product are concentration, temperature, and mixed acid composition. The nitro product is obtained by drowning out (ie, adding to excess water with internal or external cooling as required) followed by separation or extraction.

The largest scale nitrations are typically those in which the nitro compounds are precursors of amine primaries. When nitrating benzene, toluene, chlorobenzene, and naphthalene, continuous operation in custom-built plants can normally be justified. These primary nitro products are also typical of the isomer patterns and separation processes that are encountered in more highly substituted molecules. Apart from the normal economic advantages of continuous operation, continuous nitration offers better controlled conditions, which together with a lower in-process inventory, lead to safer operation. The remaining smaller scale nitrations are typically carried out in multiproduct nitration units. Occasionally nitration is combined with sulfonation to give a telescoped two-stage reaction in a single unit.

The reaction vessel (nitrator) is constructed of cast iron, mild carbon steel, stainless steel, or glass-lined steel depending on the reaction environment. It is designed to maintain the required operating temperature with heat-removal capability to cope with this strongly exothermic and potentially hazardous reaction.

Secondary problems are the containment of nitric oxide fumes and disposal or reuse of the dilute spent acid. Examples of important intermediates resulting from nitration are summarized in Table 3.

Reduction. The most important reduction process is the conversion of an aromatic nitro or dinitro compound into an arylamine or arylene diamine. The six hydrogen atoms per nitro group that are required ($RNO_2 + 6H \rightarrow RNH_2 + 2H_2O$)

Table 3. Dye Intermediates Obtained by Nitration

Product	CAS Registry Number	Starting material	Nitration process
1,2-dichloro-4-nitrobenzene	[99-54-7]	o-dichlorobenzene	mixed acid/30°C and purification from isomers
1,4-dichloro-2-nitrobenzene	[89-61-2]	p-dichlorobenzene	mixed acid/30°C
1,8-dinitronaphthalene	[602-38-0]	naphthalene	mixed acid/80°C
1,5-dinitronaphthalene	[605-71-0]		(50% yield 1,8 + 25% 1,5 after separation)
8-nitronaphthalene-1-sulfonic acid	[112-41-9]	naphthalene-1-sulfonic acid	HNO₃/30°C
5-nitronaphthalene-1-sulfonic acid	[17521-00-5]		
3-nitronaphthalene-1,5-disulfonic acid	[117-86-2]	naphthalene-1,5-disulfonic acid	HNO₃/oleum to minimize 4-nitro
1-nitronaphthalene-3,6,8-trisulfonic acid	[38267-31-1]	naphthalene-1,3,6-trisulfonic acid	mixed acid/40°C through process
6-nitrodiazo-1,2,4-acid		diazo-1,2,4-acid	mixed acid/5°C (95% yield)
1-nitroanthraquinone	[82-34-8]	anthraquinone	HNO₃/85% sulfuric acid
1,8-dinitroanthraquinone	[129-39-5]	1-nitroanthra-quinone	HNO₃/monohydrate/ 100°C separation by fractional crystallization
1,5-dinitroanthraquinone	[82-35-9]		
4,8-dinitroanthrarufin-2,6-disulfonic acid	[6449-09-8]	anthrarufin (1,5-dihydroxy-anthraquinone)	oleum sulfonation followed by mixed acid/30°C
4,5-dinitrochrysazin (plus an isomeric byproduct)	[81-55-0]	chrysazin (1,8-dihydroxy-anthraquinone)	HNO₃/oleum/0°C
2-amino-5-nitrophenol	[121-88-0]	2-methylbenz-oxazolone (protected form of 2-aminophenol)	mixed acid/0–10°C followed by hydrolysis and separation from 4-nitro isomer
2-amino-5-nitrothiazole	[121-66-4]	2-aminothiazole nitrate	low temperature rearrangement

are provided either directly, by catalytic hydrogenation, or indirectly, using a wide range of reagents and operating conditions.

The industrial processes used for reduction are catalytic hydrogenation (qv), iron reduction (aqueous neutral or acidic, or solvent), and sulfide reduction. Sulfide is used for the selective reduction of dinitro compounds to nitroarylamines, and for metal-sensitive systems, such as certain substituted 2-aminophenols. It is also used to selectively reduce nitro groups, and not the azo groups, in nitroazo dyes. These are illustrated in Table 4.

Benzidine chemistry involves reduction. The nitro precursor is reduced either by zinc and alkali or electrolytically to the hydrazo intermediate, which is then transformed to the benzidine by treatment with acid.

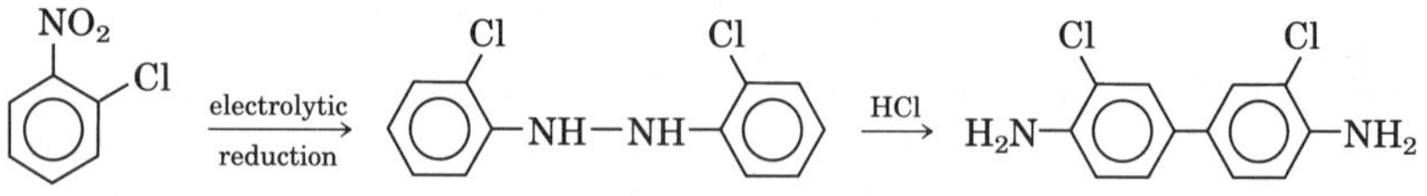

Although manufacture of benzidine itself has virtually ceased in Europe and the United States, similar rearrangement processes are operated for 3,3'-dichloro-benzidine [91-94-1], o-dianisidine (3,3'-dimethoxybenzidine), and benzidine-2,2'-disulfonic acid.

Another important reduction process is that of aryldiazonium salts with sulfite/bisulfite at controlled pH to produce arylhydrazines. Arylhydrazines are important intermediates for the preparation of pyrazolones and indoles.

In the benzene and naphthalene series there are few examples of quinone reductions other than that of hydroquinone itself. There are, however, many intermediate reaction sequences in the anthraquinone series that depend on the generation, usually by employing aqueous "hydros" (sodium dithionite) of the so-called leuco compound. The reaction with leuco quinizarin [122308-59-2] is shown because this provides the key route to the important 1,4-diaminoanthraquinones.

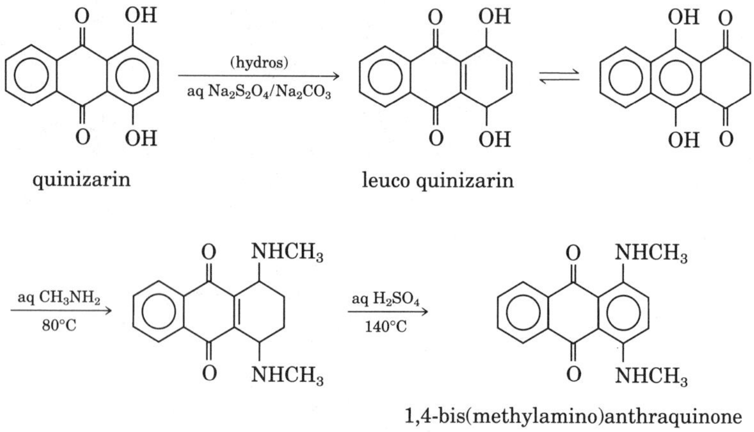

quinizarin   leuco quinizarin

1,4-bis(methylamino)anthraquinone

Arylalkylsulfones are important intermediates obtained by alkylation of arylsulfinic acids. The latter are obtained by reduction of the corresponding sul-

Table 4. Arylamine Dye Intermediates Obtained by Reduction

Product	CAS Registry Number	Starting material	Reduction process
aniline	[62-53-3]	nitrobenzene	catalytic hydrogenation
m-phenylenediamine	[108-45-7]	m-dinitrobenzene	catalytic hydrogenation
m-nitroaniline		m-dinitrobenzene	sulfide
p-chloroaniline	[95-51-2]	p-chloronitro-benzene	iron
2,5-dichloroaniline	[95-82-9]	2-nitro-1,4-dichlorobenzene	iron
m-toluidine	[108-44-1]	m-nitrotoluene	iron or catalytic hydrogenation
p-toluidine	[106-49-0]	p-nitrotoluene	catalytic hydrogenation
1-naphthylamine	[134-32-7]	1-nitronaphthalene	catalytic hydrogenation
o-aminophenol	[95-55-6]	o-nitrophenol	sulfide
2-amino-4-nitrophenol	[99-57-0]	2,4-dinitrophenol	sulfide/ammonia
2-amino-4-chlorophenol	[95-85-2]	4-chloro-2-nitrophenol	iron/H$_2$SO$_4$
2-amino-6-chloro-4-nitrophenol	[6358-09-4]	6-chloro-2,4-dinitrophenol	sulfide
o-anisidine	[90-04-0]	o-nitroanisole	iron/formic acid
orthanilic acid	[88-21-1]	o-nitrobenzene-sulfonic acid	iron
metanilic acid	[121-47-1]	m-nitrobenzene-sulfonic acid	iron or catalytic hydrogenation
Laurent's acid	[84-89-9]	1-nitronaphthalene-5-sulfonic acid	iron; may be separated after reduction of mixed nitro isomers
peri acid	[82-75-7]	1-nitronaphthalene-8-sulfonic acid	
2-naphthylamine-4,8-disulfonic acid	[131-27-1]	3-nitronaphthalene-1,5-disulfonic acid	iron
Koch acid	[117-42-0]	8-nitronaphthalene-1,3,6-trisulfonic acid	iron
1,6-Cleves acid	[119-79-9]	5/8-nitronaph-thalene-2-sulfonic acid	iron
1,7-Cleves acid	[119-28-8]		
2-aminophenol-4-sulfonic acid	[98-37-3]	2-nitrophenol-4-sulfonic acid	sulfide
6-nitro-2-aminophenol-4-sulfonic acid	[96-93-5]	2,6-dinitrophenol-4-sulfonic acid	sulfide

Table 4. (*Continued*)

Product	CAS Registry Number	Starting material	Reduction process
p-phenylenediamine	[*106-50-3*]	*p*-nitroaniline	iron (90% yield) or catalytic hydrogenation
p-aminoacetanilide	[*122-80-5*]	*p*-nitroacetanilide	iron
p-phenylenediamine-sulfonic acid	[*88-45-9*]	4-nitroaniline-2-sulfonic acid	iron
4,4′-diaminostilbene-2,2′-disulfonic acid	[*81-11-8*]	4,4′-dinitrostilbene-2,2′-disulfonic acid	iron
1-aminoanthraquinone		1-nitroanthraquinone	sulfide
1,5-diaminoanthraquinone	[*129-44-2*]	1,5-dinitroanthraquinone	sulfide or NH_3 amination
1,8-diaminoanthraquinone	[*129-42-0*]	1,8-dinitroanthraquinone	
4,5-diaminochrysazin (crude)	[*128-94-9*]	4,5-dinitrochrysazin	sulfide
4,8-diaminoanthrarufin-2,6-disulfonic acid	[*128-86-9*]	4,8-dinitroanthrarufin-2,6-disulfonic acid	sulfide

fonyl chloride. This reduction process is simple and of general application involving the addition of the isolated sulfonyl chloride paste to excess aqueous sodium sulfite followed by salting-out the product and isolation. With more rigorous reduction conditions, such as zinc/acid, sulfonyl chlorides are reduced through to arylmercaptans, eg, 2-mercaptonaphthalene is manufactured from naphthalene-2-sulfonyl chloride.

Sulfonation. The sulfonic acid group is used extensively in the dyes industry for its water-solubilizing properties, and for its ability to act as a good leaving group in nucleophilic substitutions. It is used almost exclusively for these purposes since it has only a minor effect on the color of a dye.

The sulfonic acid group can be introduced into the aromatic ring by a variety of reagents, eg, H_2SO_4, oleum, SO_3, and $ClSO_3H$, and under a variety of conditions. All these reagents are merely SO_3 carriers, and, not surprisingly, free SO_3 is found to be the most active sulfonating agent. However, the latter has to be used in aprotic solvents. As a result, sulfuric acid, oleum, and chlorosulfonic acid are the preferred sulfonating agents on the manufacturing scale.

The sulfonic acid group can be introduced into molecules containing a variety of other groups such as halogen, hydroxy, acylamino, etc, without interfering with this functionality. Sulfonation is retarded by the presence of deactivating substituents, especially nitro groups. For instance, under normal conditions 2,4-dinitrobenzene cannot be sulfonated. Sulfonation of benzene gives initially the monosulfonated benzene: this is then converted to the *m*-disulfonic acid under more severe conditions or longer reaction times. Sulfonation of toluene leads to a mixture of ortho and para isomers, which can be sulfonated further to toluene-2,4-disulfonic acid.

As in the nitration of naphthalene, sulfonation gives the 1-substituted naphthalene. However, because the reverse reaction (desulfonation) is appreciably fast at higher temperatures, the thermodynamically controlled product, naphthalene-2-sulfonic acid, can also be obtained. Thus it is possible to obtain either of the two possible isomers of naphthalene sulfonic acid. Under kinetically controlled conditions naphthalene-1-sulfonic acid [85-47-2] (**82**) is obtained; thermodynamic control gives naphthalene-2-sulfonic acid [120-18-3] (**83**).

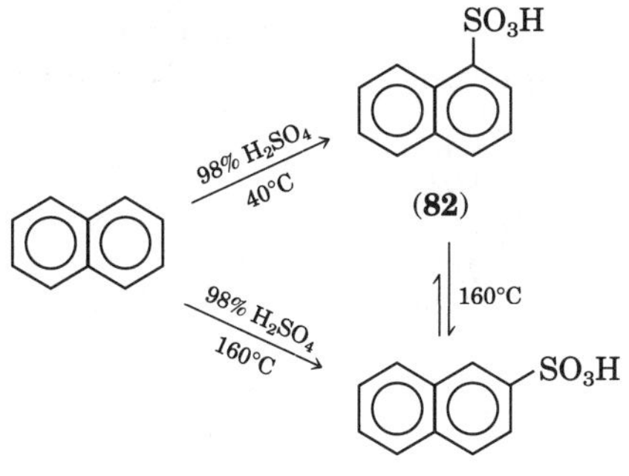

Prolonged heating and stronger reagents lead to the introduction of further sulfonic acid groups into naphthalene. In this way disulfonated and trisulfonated derivatives can be obtained. By careful manipulation of the reaction conditions a high yield of just one isomer is possible, eg, in the preparation of 1,3,6-naphthalenetrisulfonic acid [86-66-8] (**84**). When sulfonic acid groups are introduced into a naphthalene ring, they always enter at a vacant position that is not ortho, para, or peri to another sulfonic acid group. The result of this interesting experimental observation is that only one tetrasulfonated naphthalene [6654-67-7] (**85**) is obtained, and pentasulfonated naphthalenes cannot be obtained by sulfonation.

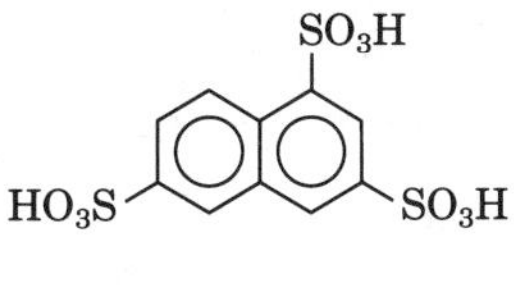

Bake sulfonation is an important variant of the normal sulfonation procedure. The reaction is restricted to aromatic amines, the sulfate salts of which are prepared and heated (dry) at a temperature of approximately 200°C *in vacuo*. The sulfonic acid group migrates to the ortho or para positions of the amine to give a mixture of orthanilic acid [88-21-1] and sulfanilic acid [121-57-3], respectively. This tendency is also apparent in polynuclear systems so that 1-naphthylamine gives 1-naphthylamine-4-sulfonic acid.

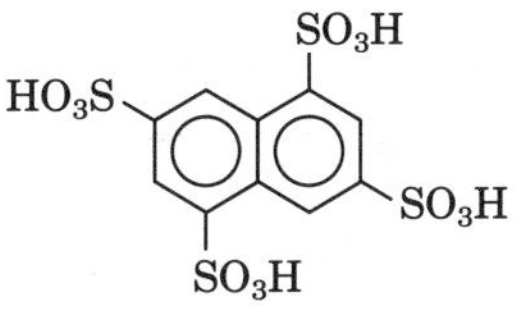

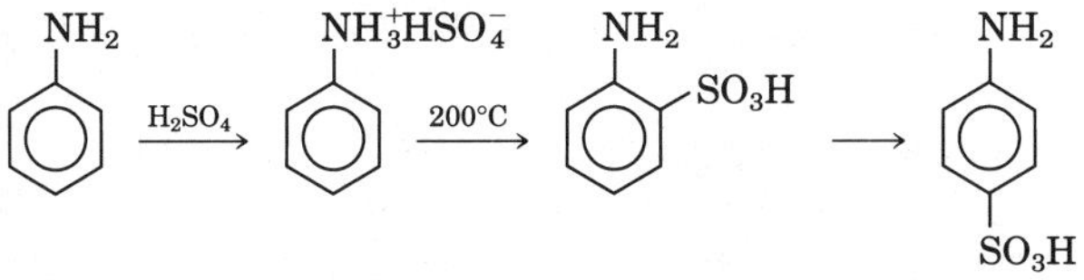

The reaction is therefore useful for the introduction of a sulfonic acid group into a specific position. An example where bake sulfonation complements conventional sulfonating procedures is given by the sulfonation of 2,4-dimethylaniline. Conventional sulfonating conditions result in the sulfonic acid group entering the 5-position (directed ortho and para to the methyl groups and meta to the NH_3^+ group) to give the 4,6-dimethylaniline-3-sulfonic acid [6370-23-6] (**86**), whereas bake sulfonation gives the 4,6-dimethylaniline-2-sulfonic acid [88-22-2] (**87**).

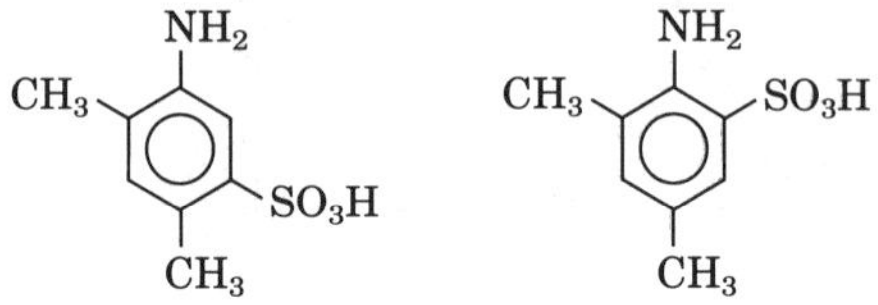

(**86**)   (**87**)

In cases where a large excess of acid is undesirable, chlorosulfonic acid is employed. An excess of chlorosulfonic acid leads to the introduction of a chlorosulfonyl group which is a useful synthon for the preparation of sulfonamides and sulfonate esters.

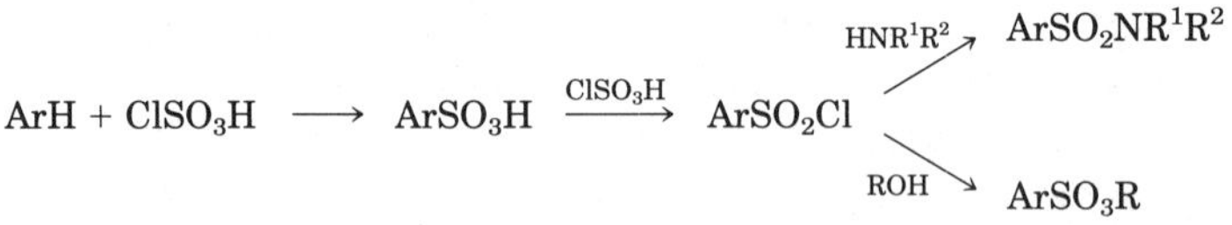

$$ArH + ClSO_3H \longrightarrow ArSO_3H \xrightarrow{ClSO_3H} ArSO_2Cl \begin{array}{c} \xrightarrow{HNR^1R^2} ArSO_2NR^1R^2 \\ \searrow_{ROH} ArSO_3R \end{array}$$

It is possible to introduce sulfonic acid groups by alternative methods, but these are little used in the dyes industry. However, one worth mentioning is sulfitation, because it provides an example of the introduction of a sulfonic acid group by nucleophilic substitution. The process involves treating an active halogen compound with sodium sulfite. This reaction is used in the purification of *m*-dinitrobenzene.

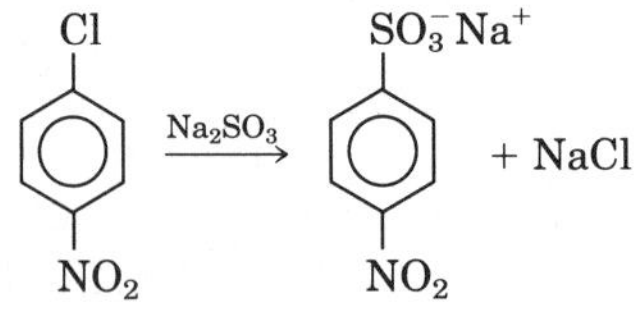

Table 5 lists some intermediates obtained by sulfonation and the sulfonating conditions employed.

Oxidation. This process, ie, the introduction of oxygen into or the removal of hydrogen from a molecule, mainly occurs at an early stage in the syntheses;

Table 5. Intermediates Obtained by Sulfonation

Product	CAS Registry Number	Starting material	Sulfonation process
benzene-1,3-disulfonic acid	[98-48-6]	benzene	30% oleum/180°C
p-toluenesulfonic acid	[104-15-4]	toluene	monohydrate 100°C (100% H_2SO_4)
1-naphthalenesulfonic acid	[85-47-2]	naphthalene	98% H_2SO_4/40°C (96% yield)
2-naphthalenesulfonic acid	[20-18-3]	naphthalene	96% H_2SO_4/165°C (85% yield)
m-nitrobenzenesulfonic acid	[98-47-5]	nitrobenzene	65% oleum/100°C
4-chloro-3-nitrobenzene-sulfonic acid	[121-18-6]	o-chloronitro benzene	SO_3 or $ClSO_3H$/solvent
naphthalene-2,7 disulfonic acid	[92-41-1]	naphthalene-2-sulfonic acid	monohydrate/165°C (65% yield + 25% 2,6)
2-chloro-5-nitrobenzene-sulfonic acid	[96-73-1]	p-chloronitro benzene	65% oleum/100°C
2-hydroxy-1-naphthalene-sulfonic acid	[567-47-5]	2-naphthol	SO_3 or $ClSO_3H$/solvent
naphthalene-1,3,6-trisulfonic acid	[86-66-8]	naphthalene-2,7-disulfonic acid	20% oleum/180°C through process
7-hydroxy-1,3-naphthalene disulfonic acid[a]	[118-32-1]	2-naphthol-6-sulfonic acid	98% H_2SO_4/120°C and separation by fractional salting
3-hydroxy-2,7-naphthalene disulfonic acid[b]	[148-75-4]	2-naphthol-6-sulfonic acid	
7-hydroxy 1,3,6-naphthalene trisulfonic acid	[6259-66-1]	2-naphthol-6,8-disulfonic acid	40% oleum/120°C through process
sulfanilic acid	[21-57-3]	aniline	dry bake/200°C
aniline-2,5-disulfonic acid	[98-44-2]	metanilic acid	65% oleum/160°C
m-phenylenediamine-4-sulfonic acid	[88-63-1]	m-phenylene-diamine	
p-phenylenediamine-2,5-disulfonic acid	[7139-89-1]	p-phenylenediamine sulfonic acid	
naphthionic acid	[84-86-6]	1-naphthylamine	dry bake/170°C
1,5-naphthalenedisulfonic acid	[117-62-4]	2-naphthylamine-1-sulfonic acid (Tobias acid)	65% oleum/80°C
6-amino-1,3,5-naphthalenetrisulfonic acid	[55524-84-0]	2-naphthylamine-1-sulfonic acid	65% oleum/80°C
anthraquinone-1-sulfonic acid	[82-49-5]	anthraquinone	20% oleum/HgO catalyst/150°C (75% yield)[c]

Table 5. (*Continued*)

Product	CAS Registry Number	Starting material	Sulfonation process
anthraquinone-2-sulfonic acid	[*84-48-0*]	anthraquinone	20% oleum/150°C (90% yield after 50% conversion)[d]
anthraquinone-1,5-disulfonic acid	[*117-14-6*]	anthraquinone	65% oleum/HgO catalyst/120°C[e]
p-toluenesulfonyl chloride	[*98-59-9*]	toluene	ClSO₃H/20°C isomer separation ClSO₃H/100°C

[a]G-acid.
[b]R-acid.
[c]Plus 22% anthraquinone for recycle.
[d]Plus anthraquinone for recycle.
[e]After separation 40% yield of 1,5 plus 25% 1,8.

the more highly substituted molecules are less amenable to oxidation. The processes for manufacturing anthraquinone, phthalic anhydride, phenol, hydroquinone, benzaldehyde, and benzoic acid are virtually all catalytic and sited in dedicated plants. The Hofmann process for making anthranilic acid should also be mentioned as a special example of hypochlorite oxidation.

Fusion/Hydroxylation. The conversion of arylsulfonic acids to the corresponding hydroxy compound is normally effected by heating with caustic soda (caustic fusion). The primary examples are β-naphthol in the naphthalene series and resorcinol in the benzene series; further examples are *m*-aminophenol from metanilic acid and *N,N*-diethyl-*m*-aminophenol from *N,N*-diethylmetanilic acid. In the naphthalene series the hydroxy group is much more commonly introduced at a later stage in the synthesis. Some examples are listed in Table 6, including the most important of the letter acids, ie, acids known by a single letter. In most cases selective hydroxylation takes place at the α-position in α,β-disulfonated naphthalenes, eg, G-acid.

Amination. Amination describes the introduction of amino groups into aromatic molecules by reaction of ammonia or an amine with suitably substituted halogeno, hydroxy, or sulfonated derivatives by nucleophilic displacement. Although reaction and operational conditions vary, the process always involves the heating of the appropriate precursor with excess aqueous ammonia or amine under pressure.

For the amination of aryl halides the reaction proceeds smoothly only when the halogen is activated by electron-withdrawing groups, particularly nitro groups. Examples of suitable precursors are 2-chloronitrobenzene, 4-chloronitrobenzene, 2,4-dinitrochlorobenzene, 1,2-dichloro-4-nitrobenzene, and 1,4-dichloro-2-nitrobenzene, for the respective manufacture of 2-nitroaniline, 4-nitroaniline, 2,4-dinitroaniline, 2-chloro-4-nitroaniline, and 4-chloro-2-nitroaniline. Manufacture of this family of products requires an autoclave operating at temperatures of around 175°C and pressures of 3.5–4.0 MPa (500–600 psi). Sulfonated nitro-

Table 6. Substituted Naphthols Obtained by Fusion

Product	CAS Registry Number	Starting material
H-acid (1-amino-8-naphthol-3,6-disulfonic acid)	[90-20-0]	1-naphthylamine-3,6,8-trisulfonic acid
J-acid (6-amino-1-naphthol-3-sulfonic acid)	[87-02-5]	2-naphthylamine-5,7-disulfonic acid
gamma acid (7-amino-1-naphthol-3-sulfonic acid)	[90-51-7]	2-naphthylamine-6,8-disulfonic acid
4,6-dihydroxynaphthalene-2-sulfonic acid	[6357-93-3]	7-hydroxynaphthalene-1,3-disulfonic acid (G-acid)
K-acid (8-amino-1-naphthol-3,5-disulfonic acid)	[130-23-4]	1-naphthylamine-4,6,8-trisulfonic acid
Chicago acid (1,8-dihydroxy-naphthalene-2,4-disulfonic acid)	[82-47-3]	naphthasultam-2,4-disulfonic acid
chromotropic acid (1,8-dihydroxy naphthalene-3,6-disulfonic acid)	[148-25-4]	1-naphthol-3,6,8-trisulfonic acid

chlorobenzenes react under less vigorous conditions, and the important 4-nitro-aniline-2-sulfonic acid is manufactured in high yield by amination of 4-chloroni-trobenzene-3-sulfonic acid with aqueous ammonia at 120°C and 600 kPa (6 atm).

Catalysts (eg, copper salts) may help with aryl halides of lower activity, but amination of chlorobenzene and 1,4-dichlorobenzene is generally not feasible. However, the 4-chloro atom of 1,2,4-trichlorobenzene is sufficiently reactive for high temperature amination to yield 3,4-dichloroaniline.

Amination of phenolic derivatives is limited to specially developed catalytic processes for aniline and *m*-toluidine (3). More general conditions apply to amination of naphthols by the Bucherer reaction. Important intermediates made by a Bucherer reaction include Tobias acid and gamma acid.

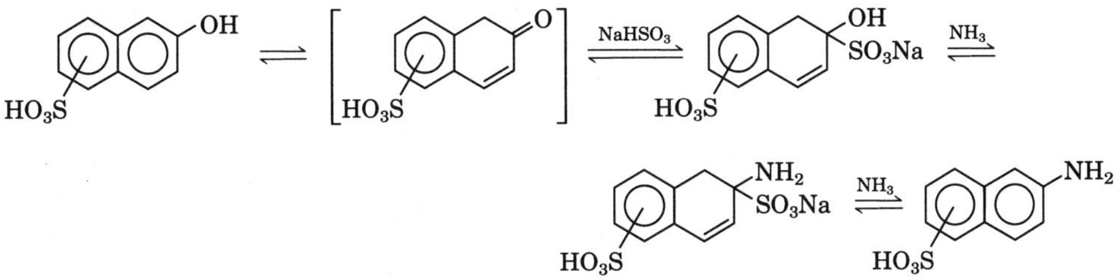

In the anthraquinone series, apart from the special case of the amination of leuco-quinizarin, sulfonic acid and nitro are the preferred leaving groups. 1-Aminoan-thraquinone is manufactured from anthraquinone-1-sulfonic acid or 1-nitroan-thraquinone, and 2-aminoanthraquinone (betamine) from anthraquinone-2-sulfonic acid.

Alkylation. The substitution of a hydrogen atom by an alkyl group can take place at a carbon, nitrogen, or oxygen atom.

N-Alkylation is important in producing a wide range of substituted anilines for use in many dye classes. The products made in the highest quantities, ie, mono- and di-*N*-methyl- and *N*-ethylanilines, are manufactured by continuous catalytic processes, but other products require a wide variety of batch processes. These include alkyl halides, such as ethyl chloride and benzyl chloride, ethylene oxide, which introduces a 2-hydroxyethyl group, acrylonitrile, which introduces a 2-cyanoethyl group, and acrylic acid, which introduces a 2-carboxyethyl group.

N-Phenylglycine [103-01-5], the key intermediate for indigo, may be manufactured by alkylation of aniline with chloroacetic acid, but it is much more economical, even though three *in situ* stages are required, to use formaldehyde as the alkylating agent.

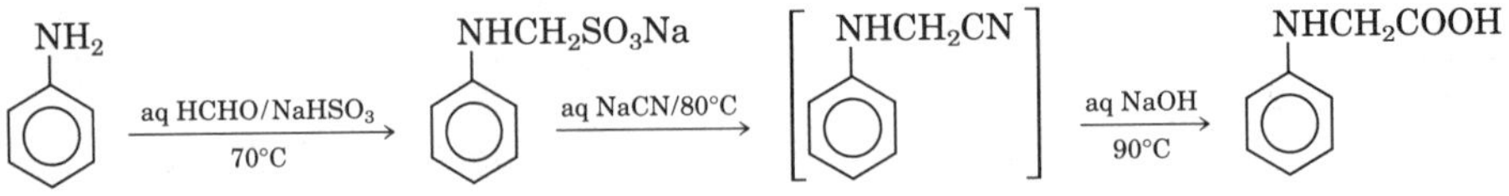

N-Alkyl and *N*-aryl substituted naphthylamines are also important, eg, letter acid derivatives, but are usually manufactured by the Bucherer reaction.

O-Alkylation is comparable to *N*-alkylation, but since the sodium salts are water-soluble it is most convenient to treat the phenol or naphthol in aqueous caustic solution with dimethyl sulfate or diethyl sulfate. These are comparatively expensive reagents, and therefore, alkoxy groups are introduced at a prior stage by a nucleophilic displacement reaction whenever possible.

Manufacture of alkylsulfones, important intermediates for metal-complex dyes and for reactive dyes, also depends on *O*-alkylation. An arylsulphinic acid in an aqueous alkaline medium is treated with an alkylating agent, eg, alkyl halide or sulfate, by a procedure similar to that used for phenols. In the special case of β-hydroxyethylsulfones (precursors to vinylsulfone reactive dyes) the alkylating agent is ethylene oxide or ethylene chlorohydrin.

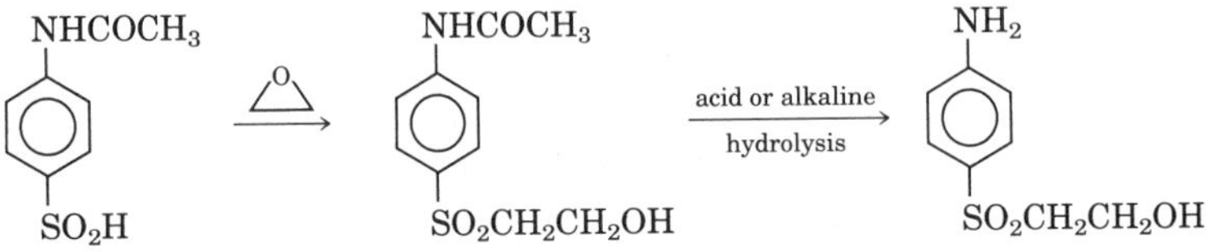

C-Alkylations may be discussed under the headings of alkene reactions and *N*-alkyl rearrangements. The isopropylation of benzene and naphthalene are two important examples of alkylation with alkenes (see ALKYLATION). Manufacture of *p*-butylaniline, by heating *N*-butylaniline with zinc chloride, typifies the rearrangement reaction appropriate to C_4 and higher alkyl derivatives.

Halogenation. Halogenation is the process of introducing one or more halogen atoms (F, Cl, or Br) into an organic molecule. The two most important halogens, chlorine and bromine, are usually introduced into benzenoid aromatics by direct halogenation. In contrast, they are frequently introduced into the naphthalene nucleus using the Sandmeyer or Gattermann reactions. When halogens are introduced directly in their elemental forms, ie, chlorine gas or bromine liquid, a catalyst is required, usually the corresponding iron III halide ($FeCl_3$ or $FeBr_3$). The iron catalyst is added either directly or is formed *in situ* by the addition of iron to the reaction mixture ($2\ Fe + 3\ X_2 \longrightarrow 2FeX_3$). In the presence of catalyst, halogenation proceeds by electrophilic attack of the halonium ion ($FeX_3 + X_2 \longrightarrow X^+ + FeX_4^-$) at the aromatic nucleus to give the corresponding halobenzene. Thus benzene is chlorinated by passing a stream of chlorine gas into benzene containing a catalytic amount of $FeCl_3$ at approximately 30°C. Similarly, toluene is chlorinated to give a mixture of *o*- and *p*-chlorotoluenes that is separated by fractional distillation. Predictably, deactivated benzenes, eg, nitrobenzene, require higher temperatures (>50°C) for reaction to occur. In contrast, activated systems such as phenols and anilines require no catalyst, and the reaction proceeds readily at room temperature. In fact, phenols are so active that dilute sodium hypochlorite (NaOCl) solutions must be used to obtain the monochlorinated derivative, and aromatic amines have to be deactivated by acetylation to avoid over-chlorination.

If the aromatic hydrocarbon is substituted by alkyl groups, then, under the right conditions, substitution can occur in the side chain. The mechanism involves radicals, and therefore the conditions for ionic nuclear halogenation must be avoided, eg, no Lewis acid catalysts (FeX_3) must be present. As a result these reactions are carried out in iron-free reaction vessels (enamel or glass-lined). Higher temperatures, which facilitate homolytic cleavage of the halogen, are generally required than for the ionic reaction and the presence of radical initiators, eg, uv light or benzoyl peroxide, is beneficial. Benzyl chloride, a useful alkylating agent, may thus be obtained from toluene. Further chlorination of benzyl chloride gives a mixture of the dichloro derivative and the trichloro derivative.

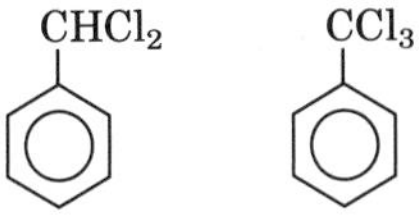

The latter two compounds are important sources of benzaldehyde and benzoyl chloride/benzoic acid respectively (see CHLOROCARBONS AND CHLOROHYDRO-CARBONS, BENZYL CHLORIDE, BENZAL CHLORIDE, AND BENZOTRICHLORIDE).

Fluorination and iodination reactions are used relatively little in dye synthesis. Fluorinated species include the trifluoromethyl group, which can be obtained from the trichloromethyl group by the action of hydrogen fluoride or antimony pentafluoride, and various fluorotriazinyl and pyrimidyl reactive systems for reactive dyes, eg, Cibacron F dyes.

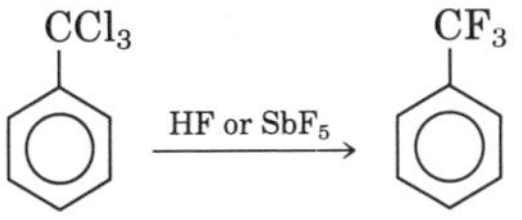

Hydrolysis. The general process definition for hydrolysis embraces all double-decomposition reactions between water (usually in the form of acid or alkali solutions of a wide range of strengths) and an organic molecule.

The main type of hydrolysis reaction is that of halogenoaryl compounds to hydroxyaryl compounds, eg, the aqueous caustic hydrolysis of *o*- and *p*-chloronitrobenzene derivatives to nitrophenols. Another important reaction is the hydrolysis of *N*-acyl derivatives back to the parent arylamine, where the acyl group is frequently used to protect the amine.

Condensation. This term covers all processes, not previously included in other process definitions, where water or hydrogen chloride is eliminated in a reaction involving the combination of two or more molecules. The important condensation reactions are nitrogen and sulfur heterocycle formation, amide formation from acid chlorides, formation of substituted diphenylamines, and miscellaneous cyclizations.

A significant group of carbonamides are the Naphtol products formed by condensation of 2-hydroxy-3-naphthoic acid (via its acid chloride) with a wide range of arylamines. The simplest example, Naphtol AS from aniline, is typical, and manufacture is accomplished by suspending the acid in a solvent such as toluene, preforming the acid chloride by addition of phosphorus trichloride, and then adding the aniline.

Sulfonamides, as a class, are simple to manufacture once the isolation conditions for the moderately stable sulfonyl chloride have been established. Basically all processes involve the addition of the sulfonyl chloride paste to excess ammonia or amine in aqueous solution. The product can usually be filtered off in a reasonably pure form with only the hydrolysis product remaining in the liquor.

The reaction of substituted chloronitrobenzenes with arylamines to form substituted diphenylamines is typified by 4-nitrodiphenylamine-2-sulfonic acid where 4-chloronitrobenzene-3-sulfonic acid (PN salt) is condensed with aniline in an aqueous medium at 120°C and 200 kPa (2 atm) in the presence of alkaline buffer at low pH to avoid the competing hydrolysis of the PN salt.

Alkoxylation. The nucleophilic replacement of an aromatic halogen atom by an alkoxy group is an important process, especially for production of methoxy-containing intermediates. Alkoxylation is preferred to alkylation of the phenol wherever possible, and typically involves the interaction of a chloro compound, activated by a nitro group, with the appropriate alcohol in the presence of alkali. Careful control of alkali concentration and temperature are essential, and formation of by-product azoxy compounds is avoided by passing air through the reaction mixture (21).

Carboxylation. This is the process of introducing a carboxylic acid group into a phenol or naphthol by reaction with carbon dioxide under appropriate conditions of heat and pressure. Important examples are the carboxylation of phenol and 2-naphthol to give salicylic acid and 2-hydroxy-3-naphthoic acid, respectively.

Acylation. Somewhat analogous to alkylation, the substitution of a hydrogen atom by an acyl group can take place at a carbon, nitrogen, or oxygen atom. The most common method of *C*-acylation is the Friedel-Craft reaction (qv) between an acyl chloride or diacid anhydride and an aromatic residue. Typically, *o*-carboxybenzophenones are formed by heating benzenoid substrates, phthalic anhydride, and aluminum chloride. These are normally cyclized to anthraquinone using concentrated sulfuric acid.

N-Acetyl and *N*-benzoyl derivatives are formed by treating the arylamine in water with acetic anhydride or benzoyl chloride. Apart from the occasional requirement of an *N*-acyl group in the final product, the *N*-acetyl group is frequently used in the manufacturing chain to temporarily protect the amine group during nitration, chlorination or chlorosulfonation; the acetyl group is readily removed by hydrolysis. Protection of o-aminophenols in this way is a slight variation since the N-acetyl derivative ring closes to form the 2-methylbenzoxazolone. Selective acetylation of diamines is utilized in the manufacture of 4-acetylamino-2-amino-benzenesulfonic acid from m-phenylenediamine sulfonic acid, and 5-acetylamino-2-aminobenzenesulfonic acid from p-phenylenediamine sulfonic acid. In these cases use is made of the increased difficulty of acylating the amino group sterically hindered by a bulky ortho substituent. The corresponding dichlorotriazinyl derivatives, using cyanuric chloride as the acylating agent, are important reactive dye intermediates.

O-Acylation is used to modify side-chain properties, and is typically achieved by heating an N-β-hydroxyethylaniline with acetic anhydride to form the O-acetoxy derivative.

Phosgenation. Reaction of phosgene with arylamines to form ureas, and with reactive aryl species to form substituted benzophenones, are special cases of acylation. They are dealt with separately since a more specialized plant is required than for other acylations. Urea formation takes place readily with water-soluble arylamines by simply passing phosgene through a slightly alkaline solution. An important example is carbonyl-J-acid from J-acid.

Esterification. The formation of an ester from an acid (or its derivative) and an alcohol is of limited application since carboxylic esters are comparatively rare substituents in dyes. Esters of N-β-hydroxyethylanilines are important intermediates for azo disperse dyes for polyester. Another example is methyl anthranilate, formed by the classical esterification of anthranilic acid using methanol and sulfuric acid.

Figure 6 summarizes the preparation of key intermediates from benzene, and Figure 7 shows a dye synthesis from benzene. Figures 8 and 9 show the preparation of key intermediates from naphthalene and β-naphthol.

CHEMISTRY OF AROMATIC HETEROCYCLES

In contrast to the benzenoid intermediates, it is unusual to find a heterocyclic intermediate that is synthesized via the parent heterocycle. They are synthesized from acyclic precursors.

The most important heterocycles are those with five- or six-membered rings; these rings may be fused to other rings, especially a benzene ring. Nitrogen, sulfur, and to a lesser extent oxygen, are the most frequently encountered heteroatoms. They are often considered in two groups: those containing only nitrogen, such as pyrazolones, indoles, pyridones, and triazoles which, except for triazoles, are used as coupling components in azo dyes, and those containing sulfur (and also optionally nitrogen), such as thiazoles, thiophenes, and isothiazoles, that are used as diazo components in azo dyes. Triazines are treated separately since they are used as the reactive system in many reactive dyes.

***N*-Heterocycles.** *Pyrazolones.* Pyrazolones are used as coupling components since they couple readily at the 4-position under alkaline conditions to give important azo dyes in the yellow-orange shade area.

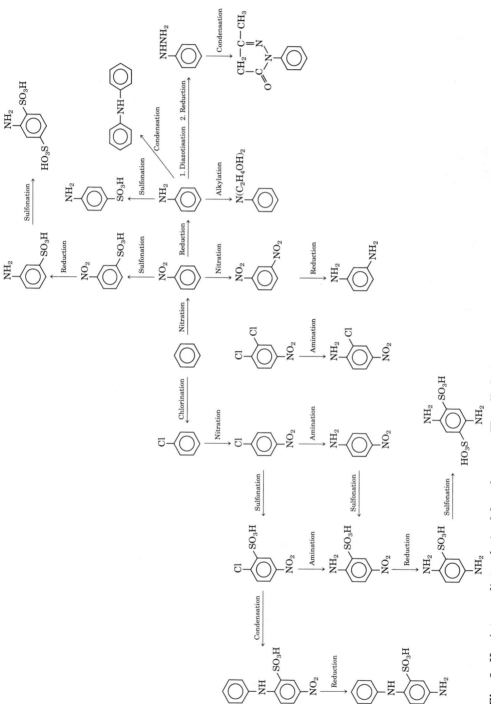

Fig. 6. Key intermediates derived from benzene. The alkylation reaction shown employs ethylene oxide. Hydrazine condenses with acetoacetic acid to form the heterocyclic ring shown.

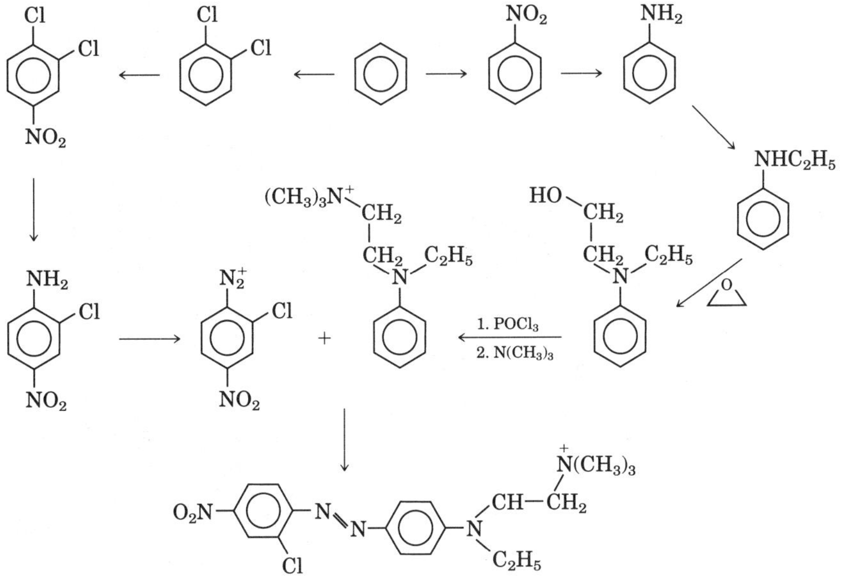

Fig. 7. Synthesis of a dye, CI Basic Red 18, from benzene.

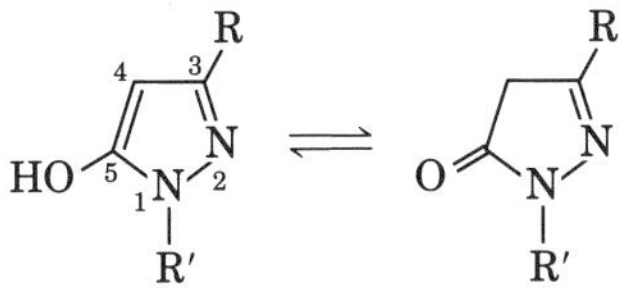

The most important synthesis of pyrazolones involves the condensation of a hydrazine with a β-ketoester such as ethyl acetoacetate. Commercially important pyrazolones carry an aryl substituent at the 1-position, mainly because the hydrazine precursors are prepared from readily available and comparatively inexpensive diazonium salts by reduction. In the first step of the synthesis the hydrazine is condensed with the β-ketoester to give a hydrazone: heating with sodium carbonate then effects cyclization to the pyrazolone. In practice the condensation and cyclization reactions are usually done in one pot without isolating the hydrazone intermediate.

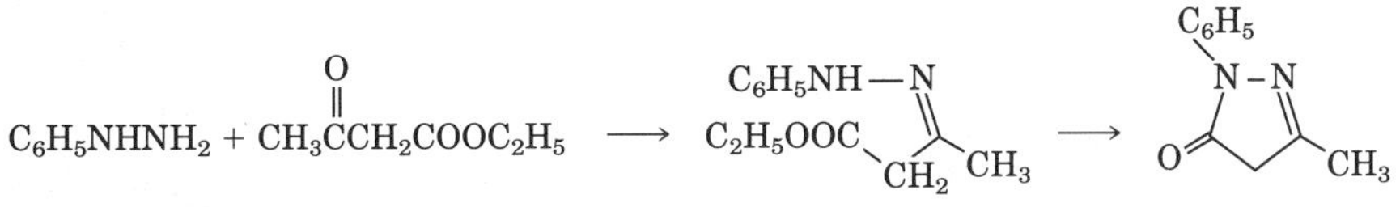

Pyridones. Pyridine itself has little importance as a dyestuff intermediate. However, its 2,6-dihydroxy derivatives have achieved prominence in recent years as coupling components for azo dyes, particularly in the yellow shade area.

The most convenient synthesis of 6-hydroxy-2-pyridones is by the condensation of a β-ketoester, eg, ethyl acetoacetate, with an active methylene com-

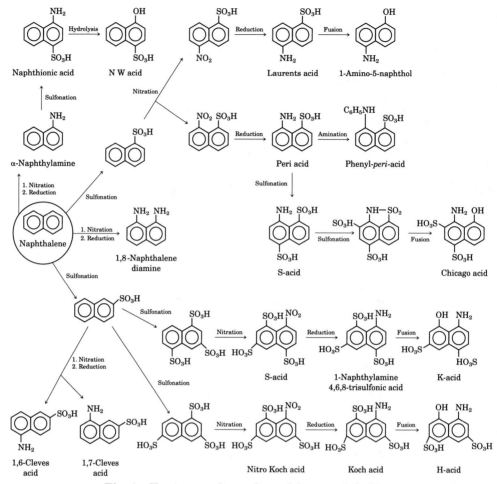

Fig. 8. Key intermediates derived from naphthalene.

pound, eg, malonic ester, cyanoacetic ester, and an amine. The amine can be omitted if an acetamide is used and in some cases this modification results in a higher yield.

$$X = CN, COOR, NR_3^+$$

R is usually alkyl or less frequently aryl; R^1 is H, and the R^2 group can be alkyl, aryl, and even an amino or a hydroxy group. By virtue of the synthetic method,

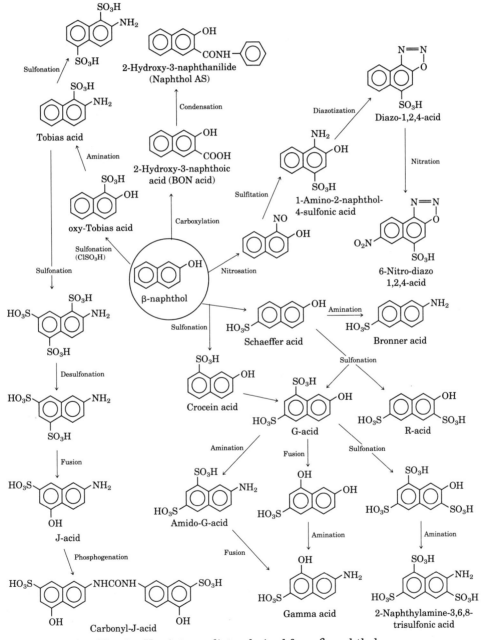

Fig. 9. Key intermediates derived from β-naphthol.

589

X is an electron-withdrawing group, eg, CN, CO_2R, although simple derivatives, eg, X = $CONH_2$, H, are readily obtained by hydrolysis. Dyes from pyridones, in which X is an ammonium group (from $R_3N^+CH_2CONH_2$), are used for polyacrylonitrile fibers. An interesting situation is encountered in the synthesis of the pyridone (**88**) using malononitrile as the active methylene compound. Here an amine is not required since the nitrogen is provided by the nitrile which undergoes a facile cyclization from the nitrile ester intermediate.

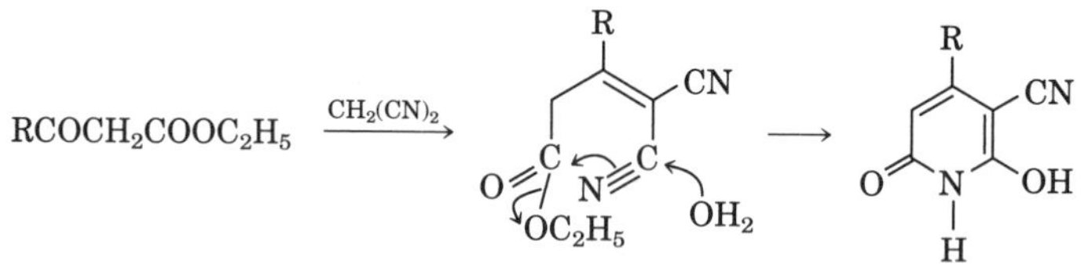

(**88**)

Indoles. Indoles (**89**) are used as coupling components to give azo dyes; the structurally related indolenines (**90**) are also employed as intermediates in cyanine dyes. The arrows indicate coupling positions. In (**89**) R = H, aryl, alkyl; R^1 = aryl, alkyl; in (**90**) R, R^1 = aryl, alkyl; and R^2 = alkyl.

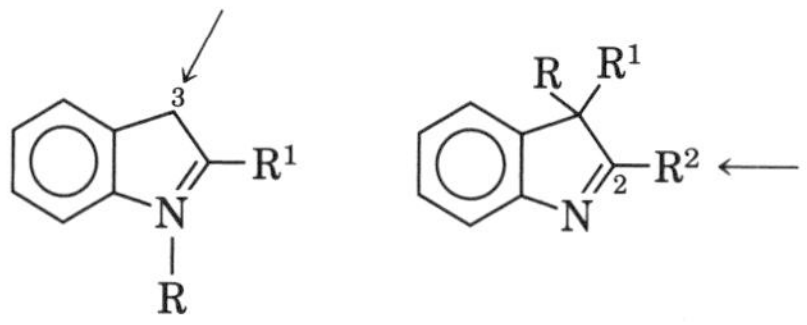

(**89**)   (**90**)

The majority of commercial indoles are synthesized by way of the Fischer-Indole synthesis, which involves the rearrangement of phenylhydrazones to indoles and their derivatives. Such hydrazones must contain at least one α-hydrogen for rearrangement to occur and acid catalysts such as $ZnCl_2$, HCl, polyphosphoric acid, and H_2SO_4 are used to effect the rearrangement. Fischer's Base [*118-12-7*] (**91**) is one of the most important indole derivatives.

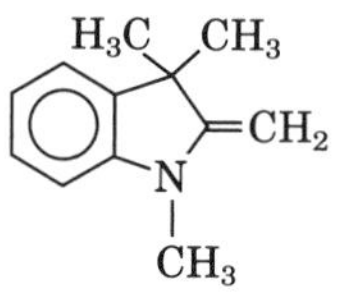

(**91**)

Triazoles. The triazoles are fairly representative of five-membered heterocycles containing three heteroatoms. The only other commercially important heterocycles containing three heteroatoms are 5-amino-1,2,4-thiadiazoles, and the isomeric 2-amino-1,3,4-thiadiazoles. Heterocycles containing more heteroatoms are not generally found in dyestuffs, and even triazoles are not of widespread importance, although they do provide some useful red dyes, eg, (**92**) for poly-acrylonitrile fibers.

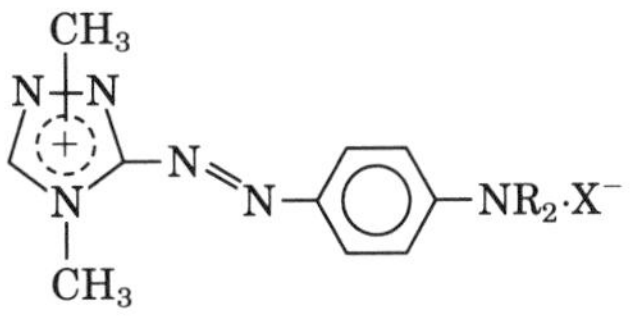

(**92**)

The most important triazole is 3-amino-1,2,4-triazole itself, used in the synthesis of diazahemicyanine dyes. The preparation of 3-amino-1,2,4-triazole [*61-82-5*] (**93**) is simple, using readily available and quite inexpensive starting materials. Thus cyanamide [*420-04-2*] reacts with hydrazine, to give aminoguanidine [*54852-84-5*] which then condenses with formic acid. Substituents in the 5-position are introduced merely by altering the carboxylic acid used.

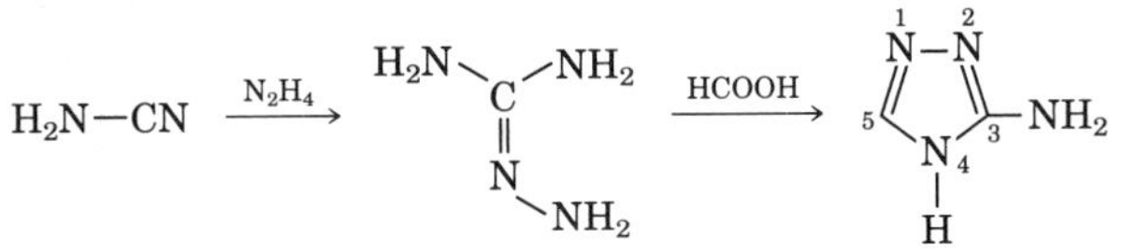

(**93**)

Sulfur and Sulfur–Nitrogen Heterocycles. *Aminothiazoles.* In constrast to the pyrazolones, pyridones, and indoles just described, aminothiazoles are used as diazo components. As such they provide dyes that are more bathochromic than their benzene analogues. Thus aminothiazoles are used chiefly to provide dyes in the red-blue shade areas. The most convenient synthesis of 2-aminothiazoles is by the condensation of thiourea with an α-chlorocarbonyl compound; for example, 2-aminothiazole [*96-50-4*] (**94**) is prepared by condensing thiourea [*62-56-6*] with α-chloroacetaldehyde [*107-20-0*] both readily available intermediates.

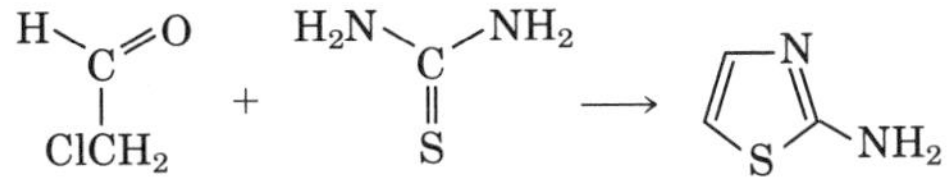

(**94**)

Substituents can be introduced into the thiazole ring either by using suitably substituted precursors or by direct electrophilic attack on the ring. An interesting example of the latter method is the preparation of 2-amino-5-nitrothiazole [*121-66-4*] from the nitrate salt of 2-aminothiazole.

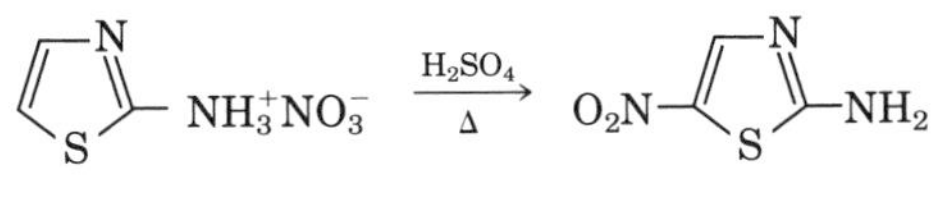

Aminobenzothiazoles. These compounds are prepared somewhat differently than the thiazoles. The thiazole ring is annelated onto a benzene ring, usually via an aniline derivative. Thus 2-amino-6-methoxybenzothiazole [*1747-60-0*] (**95**) is obtained from *p*-anisidine [*104-94-9*] and thiocyanogen. The thiocyanogen, which is formed *in situ* from bromine and potassium thiocyanate (the Kaufman reaction), reacts immediately with *p*-anisidine to give the thiocyanate derivative, which spontaneously ring closes to the aminobenzothiazole.

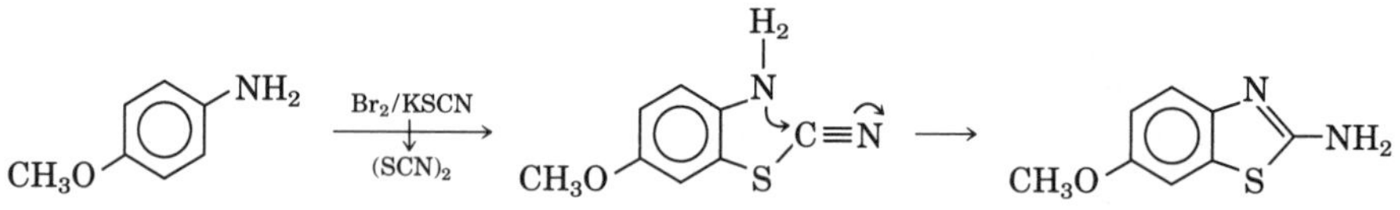

(**95**)

Benzoisothiazoles. 5-Aminoisothiazoles are relatively difficult to prepare cheaply, but they are used as diazo components in magenta dyes for Dye Diffusion Thermal Transfer, D2T2. However, the benzo-homologues are far easier to prepare and are important diazo components. Aminobenzoisothiazoles give dyes that are even more bathochromic than the corresponding dyes from aminobenzothiazoles. Aminobenzoisothiazoles [*2400-12-6*] can be prepared from *o*-nitroanilines through the intermediacy of *o*-cyanoanilines.

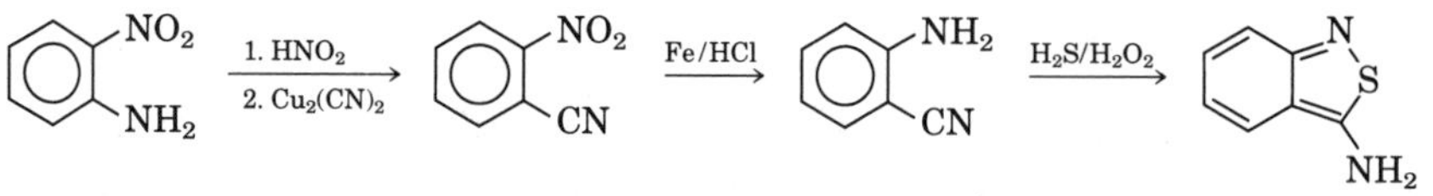

The most important commercial benzoisothiazole is the 3-amino-5-nitrobenzoisothiazole [*14346-19-1*] (**96**) prepared by an analogous route but starting from 2-cyano-4-nitroaniline [*17420-30-3*].

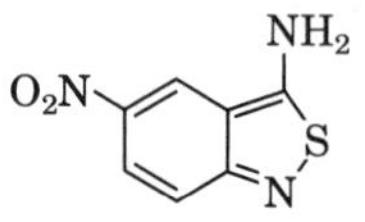

(**96**)

Thiophenes. The most important thiophenes, ie, 2-aminothiophenes, are used as diazo components for azo dyes and are capable of producing very bathochromic dyes, eg, greens, having excellent properties. Part of the reason for their late arrival on the commercial scene has been the difficulty encountered in attaining a good synthetic route. However, this problem has been overcome by ICI (22). The synthetic route is very flexible so that a variety of protected 2-aminothiophenes can be obtained merely by altering the reaction conditions or alternatively by stopping the reaction at an intermediate stage. Thus the initially formed thiophene [*51419-38-4*] (**97**) can be converted to the useful 2-amino-3,5-dinitrothiophene [*2045-70-7*] (**98**) by either of two routes.

COOH

S NHCOCH$_3$

(**97**)

NO$_2$

O$_2$N S NH$_2$

(**98**)

Triazines. The most commercially important triazine is 2,4,6-trichloro-*s*-triazine [*108-77-0*] (cyanuric chloride, (**99**)). Cyanuric chloride has not achieved prominence because of its value as part of a chromogen but because of its use for attaching dyestuffs to cellulose, ie, as a reactive group (see DYES, REACTIVE). This innovation was first introduced by ICI in 1956, and since then other active halogen compounds have been introduced.

Cl

N N

Cl N Cl

(**99**)

$\longrightarrow$

R

N N

dye—N N O—cellulose

H

On the large scale, cyanuric chloride is produced by the trimerization of cyanogen chloride. The cyanogen chloride is produced by chlorination of hydrogen cyanide and is trimerized by passing it over charcoal impregnated with an alkaline-earth metal chloride at a high temperature (250–480°C).

$$3\,\text{H}-\text{C}\equiv\text{N} \xrightarrow{\text{Cl}_2} 3\,\text{Cl}-\text{C}\equiv\text{N} \longrightarrow (\mathbf{99})$$

Equipment and Manufacture

The basic steps of dye (and intermediate) manufacture are shown in Figure 10. There are usually several reaction steps or unit processes.

The reactor itself, in which the unit processes to produce the intermediates and dyes are carried out, is usually the focal point of the plant, but this does not mean that it is the most important part of the total manufacture, nor that it absorbs most of the capital or operational costs. Operations subsequent to reaction are often referred to as work-up stages. These vary from product to product with intermediates (used without drying wherever practicable) needing less finishing operations than colorants.

The reactions for the production of intermediates and dyes are carried out in bomb-shaped reaction vessels made from cast iron, stainless steel, or steel lined

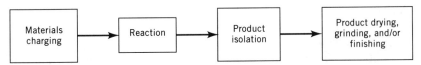

Fig. 10. Operation sequence in dye and intermediate manufacture.

with rubber, glass (enamel), brick, or carbon blocks. Wooden vats are also still used in some countries, eg, India. These vessels have capacities of 2–40 m^3 (ca 500–10,000 gal) and are equipped with mechanical agitators, thermometers or temperature recorders, condensers, pH-probes, etc, depending on the nature of the operation. Jackets or coils are used for heating and cooling by circulating through them high boiling fluids (eg, hot oil, or Dowtherm), steam, or hot water to raise the temperature, and air, cold water, or chilled brine to lower it. Unjacketed vessels are often used for aqueous reactions, where heating is affected by direct introduction of steam, and cooling by addition of ice or by heat exchangers. The reaction vessels normally span two or more floors in a plant to facilitate ease of operation (see REACTOR TECHNOLOGY).

Products are transferred from one piece of equipment to another by gravity flow, pumping, or by blowing with air or inert gas. Solid products are separated from liquids in centrifuges, on filter boxes, on continuous belt filters, and perhaps most frequently, in various designs of plate-and-frame or recessed plate filter presses. The presses are dressed with cloths of cotton, Dynel, polypropylene, etc. Some provide separate channels for efficient washing, others have membranes for increasing the solids content of the presscake by pneumatic or hydraulic squeezing.

The plates and frames are made of wood, cast iron, or now usually hard rubber, polyethylene, and polyester.

When possible, the intermediates are taken for the subsequent manufacture of other intermediates or dyes without drying because of savings in energy costs and handling losses. There are, however, many cases where products, usually in the form of pastes discharged from a filter, must be dried. Even with optimization of physical form, the water content of pastes varies from product to product in the range of 20 to 80%. Where drying is required, air or vacuum ovens (in which the product is spread on trays), rotary dryers, spray dryers, or less frequently drum dryers (flakers) are used. Spray dryers have become increasingly important. They need little labor and accomplish rapid drying by blowing concentrated slurries, eg, reaction masses, through a small orifice into a large volume of hot air. Dyes, especially disperse dyes, that require wet-grinding as the penultimate step, are now often dried this way. In this case their final standardization, ie, addition of desired amounts of auxiliary agents and solid diluents, is achieved in the same operation (see DRYING).

The final stage in dye manufacture is grinding or milling. Dry grinding is usually carried out in impact mills (Atritor, KEK, or ST); considerable amounts of dust are generated, and well-established methods are available to control this problem. Dry grinding is an inevitable consequence of oven drying, but more modern methods of drying, especially continuous drying, allow the production of materials that do not require a final comminution stage. The ball mill, consisting of a rotating or vibrating hollow cylinder partially filled with hard balls, has always been of limited use for dry grinding because of product removal problems. Although historically significant for wet milling, the ball mill has been superseded by sand or bead mills. Wet milling has become increasingly important for pigments and disperse dyes. Many recently patented designs, particularly from Draiswerke GmbH and Gebruder Netzsch Maschinfabrik, consist of vertical or horizontal cylinders equipped with high speed agitators of various configurations with appropriate continuous feed and discharge arrangements. The advantages and disadvantages of vertical and horizontal types have been discussed (23); these are so finely balanced as to lead to consideration of tilting versions to combine the advantages of both.

In the past the successful operation of batch processes depended mainly on the skill and accumulated experience of the operator. This operating experience was difficult to codify in a form that enabled full use to be made of it in developing new designs. The gradual evolution of better instrumentation, followed by the installation of sequence control systems, has enabled much more process data to be recorded, permitting maintenance of process variations within the minimum possible limits.

Full computerization of multiproduct batch plants is much more difficult than with single-product continuous units because the control parameters vary fundamentally with respect to time. The first computerized azo (24) and intermediates (25) plants were brought on stream by ICI Organics Division (now Zeneca Specialties) in the early 1970s, and have now been followed by many others. The additional cost (ca 10%) of computerization has been estimated to give a saving of 30 to 45% in labor costs (26). However, highly trained process operators and instrument engineers are required. Figure 11 shows the layout of a typical azo dye manufacturing plant.

Economic Aspects

Early this century, about 85% of world dyes requirements were manufactured in Germany, with other European countries (Switzerland, UK, and France) accounting for a further 10%. Seventy years and two world wars have seen dramatic changes in this pattern. Table 7 shows that, in weight terms, the Western European share of world production had moved to 50% by 1938 and 40% by 1974. However, since a large part of U.S. manufacture and some of "others" is based on Western European subsidiaries, their overall share remains at over 50%. Since 1974 many of the national figures have not been published, but those that are available indicate that 1974 was the peak production year of the 1970s, with 1975 being the nadir. World recession caused a 20% slump in production which has now been more than recovered.

Consequently, the figures of Table 7 are still applicable to the present day. More recent independent reviews (27,28) indicate a slightly lower world output of 750,000 to 800,000 t/yr. With the present state of the world economy in 1993 the growth rate for the industry is likely to be as low as 2–3%, but even this low figure represents something like an additional 20,000 t/yr. The major dye producing countries and companies in Western Europe are shown in Table 8.

Divergent post-war developments within the three principal German companies, Bayer, BASF, and Hoechst (including Cassella which is 75% owned by Hoechst) have led to contrasting profiles. In common with ICI, development of their dyes businesses, representing less than 5% of a total chemicals business, must be seen in the wider context of both heavy chemicals (including plastics and fibers) and fine/specialty chemicals (including dyes). Bayer's strength in intermediates manifested itself in a DM 500 million joint investment with CIBA-

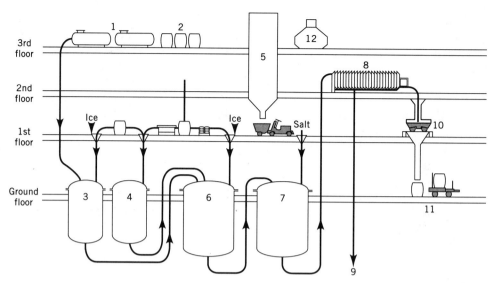

Fig. 11. Layout of azo dye manufacturing plant. 1, storage tanks for liquid starting materials; 2, storage drums for solid starting materials; 3, diazotization vessel; 4, coupling component vessel; 5, ice machine; 6, coupling vessel; 7, isolation vessel; 8, filter presses; 9, filtrate to waste liquor treatment plant; 10, dryers; 11, emptying of dyestuffs to go to the mill; 12, outgoing air purification plant.

Table 7. World Dyestuff Production Trends, 10^3 t

Year	W. Europe	U.S.	Eurasia[a]	Japan	Others	Total
1938	110	37	35		28	210
1948		110				
1958	112	80	127		27	346
1966	191	130		49		
1974	300	138	200 +	68	44	750 +

[a]Eastern Europe, former USSR, and China.

GEIGY for new capacity. This Schelde-Chemie plant at Brunsbuttel is by far the largest single investment in new dye/intermediate capacity that the Western world has seen since its main manufacturing bases were established in the 1930s and 1940s. Establishment of an important new works on the Elbe estuary will significantly lessen the effluent load on the Rhine, close to which most of the Swiss and the German industries have been based. Unlike the Germans and the British, the Swiss have confined themselves to their original fine chemicals base, so that dyes represent 20–30% of their chemical industry. This must, however, be continually decreasing in the face of more rapidly expanding pharmaceuticals and agrochemicals sectors. They export 80% of their dye production, compared with 65% for UK and 50% for Germany.

Dyes production in the United States has remained much more fragmented than that of European countries. Manufacture is shared between at least 10 man-

Table 8. West European Dyestuff Production Estimates[a]

Country	Company	Production, 10^3 t	Value, 10^6 $
West Germany		148	976
	Bayer		
	BASF		
	Hoechst		
UK		54	268
	ICI		
Switzerland		30	370
	CIBA-GEIGY		
	Sandoz		
France		31	147
Italy		15	62
Spain		15	42
Belgium		6	23
rest of Western Europe		1	4
Total		*300*	*1892*

[a]1974 Data extrapolated to 1993 as explained in the text.

ufacturers rather than three or four for each European country. All European majors have a significant manufacturing presence in the United States, gradually replacing exports from Europe. The demise of the long-standing Du Pont dye business in 1979, the withdrawal by American Cyanamid in 1980, the liquidation of the American Color and Chemical business in 1982, and the takeover of the Sodyeco Division of the Martin-Marietta Corporation by Sandoz in 1983, have left manufacturers with wide product ranges. Eastman Chemical Products, a significant producer of disperse dyes, with a restricted approach stemming from its association with a dominant photographic colors manufacturer (Eastman Kodak), has sold its disperse dye line to CIBA-GEIGY. Atlantic has been acquired by Crompton and Knowles.

In the United States during the 1970s and 1980s, reactive dyes (mostly azo) increased, whereas anthraquinone vat dyes decreased. The most notable change is the increase in indigo dyes, which may be as high as 10 million kilograms owing to the popularity of blue jeans.

A historically fragmented Japanese industry, with rapid growth to meet internal demand, changed in 1979 with an agreement by its five principal manufacturers (Mitsubishi, Sumitomo, Nippon Kayaku, Mitsui Toatsu, and Hodogaya) to the formation of a cartel to coordinate research as well as manufacture. Successful operation of this cartel could have profound effects on the international situation by the end of the twentieth century.

Comparatively little is known of manufacture in China and the former Eastern block countries, although their aim must be to satisfy all their own requirements. The signs are that they need to import technology to hasten the process toward self-sufficiency. They are currently significant importers of dyes, although Czechoslovakia in particular, more recently joined by China, does export a limited number of intermediates. These countries were the only ones to announce large-scale investment in new dye-manufacturing capacity for the 1980s. These are a

4000 t/yr disperse dye plant and a 2000 t/yr leather dye plant in the former USSR, both using Montedison/ACNA technology. The Italian group is also reported to be providing know-how for a dyestuff finishing plant to be built in northeast China. Manufacture in India, South America, and Mexico is mainly through subsidiaries, or partly owned companies associated with the main European companies. Within Far Eastern countries, other than Japan, a large number of small manufacturing facilities are evolving, especially in Taiwan and South Korea.

Health and Safety Aspects

Toxicology and Registration.   The toxic nature of some dyes and intermediates has long been recognized. Acute, or short-term, effects are generally well known. They are controlled by keeping the concentration of the chemicals in the workplace atmosphere below prescribed limits and avoiding physical contact with the material. Chronic effects, on the other hand, frequently do not become apparent until after many years of exposure. Statistically higher incidences of benign and malignant tumors, especially in the bladders of workers exposed to certain intermediates and dyes, were recorded in dye-producing countries during the period 1930–1960. The specific compounds involved were 2-naphthylamine [*91-59-8*], 4-aminobiphenyl [*92-67-1*], benzidine (4,4'-diaminobiphenyl) [*92-87-5*], fuchsine [*632-99-5*] (CI Basic Violet 14), auramine [*2465-27-2*] (CI Solvent Yellow 2). There is considerable evidence that metabolites of these compounds are the actual carcinogenic agents (29,30). Strict regulations concerning the handling of known carcinogens have been imposed in most industrial nations. In the United States the regulations (31) caused virtually all the dye companies to discontinue use of the compounds. Other actual or suspected carcinogens, such as the nitrosamines, or *N*-nitroso compounds (30,32) (see *N*-NITROSAMINES), polycyclic hydrocarbons, alkylating agents (30,33), and other individual compounds, such as the dichromates, should be considered in the wider context of industrial chemistry rather than as dye intermediates (see COLORANTS FOR FOODS, DRUGS, COSMETICS, AND MEDICAL DEVICES; INDUSTRIAL HYGIENE AND TOXICOLOGY).

The positive links between benzidine derivatives and 2-naphthylamine with bladder cancer prompted the introduction of stringent government regulations to minimize such occurrences in the future. Currently, the three principal regulatory agencies worldwide are European Core Inventory (ECOIN) and European Inventory of Existing Commercial Substances (EINECS) in Europe, Toxic Substances Control Act (TOSCA) in the United States, and Ministry of Technology and Industry (MITI) in Japan. Each of these has its own set of data and testing protocols for registration of a new chemical substance. For registration in the European Community (EEC) the following items are required: (*1*) identity of the substance; (*2*) information on the substance; (*3*) physicochemical properties of the substance; (*4*) toxicological studies; (*5*) ecotoxicological studies; and (*6*) possibility of rendering the substance harmless. Items (*1*) and (*2*) refer to the chemical structure of the substance (or its method of preparation if the structure is unknown) and information on its appearance.

Physicochemical properties required include melting/boiling point, vapor pressure, solubility, and flammability/explosion characteristics. The toxicological studies include acute toxicity tests, oral, inhalation, and dermal; skin and eye irritation; skin sensitization; subacute toxicity, oral, inhalation, and dermal; and mutagenicity tests. *In vitro*: reverse mutation assay (Ames test) on *Salmonella typhimurium* and/or *Escherichia coli* and mammalian cytogenic test. *In vivo*: mouse micronucleus test.

Finally, the ecotoxicological studies, designed to assess the impact of the substance on the environment, embrace acute toxicity tests to fish and Daphnia, and a battery of tests for the biodegradability of the substance and its biological oxygen demand characteristics.

Registration of a new chemical substance in the United States and Japan requires similar comprehensive sets of data, although there are some differences. Obtaining all the data for a full registration can be time-consuming and costly. In 1989 it cost approximately $150,000 and took about a year to register a new substance in Europe.

In order to expedite the launch of a new chemical and allow further time to complete the toxicological package for full registration, a "limited announcement" is normally used. This requires only parts of the full toxicological packages, usually acute toxicity and Ames test. Consequently, it is less expensive ($20,000) and quicker (90 days) than full registration. However, only 1 t or less of the chemical per year is allowed to be sold in the EEC.

The outcome of these toxicological tests determines the fate of the chemical. If the chemical is a potential human carcinogen it is abandoned. If it is nontoxic in all the tests then it is free to be sold as a commercial product. If the chemical gives inconclusive results at the first stage of screening, for example, an Ames positive response, then one of two courses of action is taken: either further *in vivo* testing is authorized or the chemical is abandoned. Which course of action is taken depends on the likely economic viability of the chemical. For a technically excellent product with a high profit margin aiming at a large market, further expensive animal testing would be justified. However, a chemical with a borderline technical profile and/or aimed at a smaller, more uncertain market would probably be abandoned.

Environmental Concerns. Dyes, because they are intensely colored, present special problems in effluent discharge; even a very small amount is noticeable. However, the effect is more aesthetically displeasing rather than hazardous, eg, red dyes discharged into rivers and oceans. Of more concern is the discharge of toxic heavy metals such as mercury and chromium.

Effluents from both dye works and dyehouses are treated both before leaving the plant, eg, neutralization of acidic and alkaline liquors and heavy metal removal, and in municipal sewage works. Various treatments are used (34).

Biological treatment is the most common and most widespread technique used in effluent treatment, having been employed for over 140 years. There are two types of treatment, aerobic and anaerobic. The aerobic system needs air (oxygen) in order for the bacteria to perform the degradation process on the activated sludge, whereas anaerobic bacteria operate in the absence of air. Activated sludge usually removes only a moderate amount (10–20%) of the color. Red reactive dyes are a problem because they are very visible, and losses in manufacturing are especially high.

Removal of color by adsorption using activated carbon is also employed. Activated carbon is very good at removing low levels of soluble chemicals, including dyes. Its main drawback is its limited capacity. Consequently, activated carbon is best for removing color from dilute effluent (see CARBON, ACTIVATED CARBON).

Chemical treatment of the effluent with a flocculating agent is the most robust and generally most efficient way to remove color. The process involves adding a flocculating agent, such as ferric ion (Fe^{3+}) or aluminium (Al^{3+}), to the effluent. This induces flocculation. A coagulant may also be added to assist the process. The final product is a concentrated sludge that is easy to dispose of.

Chemical oxidation is a more recent method of effluent treatment, especially chemical effluent. This procedure uses strong oxidizing agents like ozone, hydrogen peroxide, chlorine, and potassium permanganate in order to force degradation of even some of the more resilient organic molecules. It has even been demonstrated that ozone, for example, is quite capable of decolorizing most textile effluents. The use of ozone in combination with uv light has also shown some added potential in its ability to neutralize many common pesticides. For the moment, these treatments remain very expensive and of limited size, though they may have some promise in the future.

Further strategies being implemented to minimize dye and related chemical effluent include designing more environmentally friendly chemicals, more efficient (higher yielding) manufacturing processes, and more effective dyes, eg, reactive dyes having higher fixation (see DYES, ENVIRONMENTAL CHEMISTRY).

BIBLIOGRAPHY

"Dyes and Dye Intermediates" in *ECT* 1st ed., Vol. 5, pp. 327–354, by G. E. Goheen and J. Werner, General Aniline & Film Corp., and A. Merz, American Cyanamid Co.; in *ECT* 2nd ed., Vol. 7, pp. 462–505, by D. W. Bannister and A. D. Olin, Toms River Chemical Corp.; in *ECT* 3rd ed., Vol. 8, pp. 152–212, by D. W. Banister, A. D. Olin, and H. A. Stingl, Toms River Chemical Corp.

1. P. F. Gordon and P. Gregory, *Organic Chemistry in Color,* Springer-Verlag, Berlin, 1983.
2. A. Calder, *Dyes in Non-Impact Printing, IS and T's Seventh International Congress on Advances in Non-Impact Printing Technologies,* Portland, Oreg., Oct., 1991.
3. G. Booth, *The Manufacture of Organic Colorants and Intermediates,* Society of Dyers and Colorists, Bradford, UK, 1988.
4. P. Gregory in D. R. Waring and G. Hallas, eds., *The Chemistry and Application of Dyes,* Plenum Publishing Corp., New York, 1990, pp. 17–47.
5. *Colour Index,* Vol. 4, 3rd ed., The Society of Dyers and Colorists, Bradford, UK, 1971.
6. P. Gregory, *High Technology Applications of Organic Colorants,* Plenum Publishing Corp., New York, 1991.
7. P. F. Gordon and P. Gregory, *Organic Chemistry in Color,* Springer-Verlag, Berlin, 1983, pp. 96–115.
8. T. Zincke and H. Binderwald, *Chem. Ber.* **17,** 3026 (1884).
9. Ref. 7, pp. 104–108.
10. Ref. 7, pp. 5–21.
11. C. W. Greenhalgh, J. L. Carey, and D. F. Newton, *Dyes and Pigments* **1,** 103 (1980).
12. Brit. Pat. 2,151,611 (1985), J. S. Hunter, R. J. Lindsay, R. W. Kenyon, and D. Thorp (to ICI).

13. W. Luttke and M. Klessinger, *Chem. Ber.* **97,** 2342 (1964).
14. W. Luttke, H. Hermann, and M. Klessinger, *Angew. Chem. Int. Ed. Engl.* **5,** 598 (1966).
15. M. Klessinger, *Tetrahedron* **19,** 3355 (1966).
16. M. Klessinger and W. Luttke, *Tetrahedron* **19** (Suppl. 2), 315 (1963).
17. Brit. Pat. 1,349,513 (1970), B. Parton (to ICI).
18. Zahradnik, *Production and Application of Fluorescent Brightening Agents*, John Wiley & Sons, Inc., New York, 1982.
19. J. March, *Advanced Organic Chemistry*, 4th ed., John Wiley & Sons, Inc., New York, 1992, pp. 507–511.
20. *Ibid.*, p. 652.
21. N. Donaldson, *The Chemistry and Technology of Naphthalene Compounds*, Edward Arnold, London, 1958.
22. Brit. Pats. 1,419,074 and 1,419,075 (1975), D. B. Baird, R. Baker, R. R. Fishwick, and R. D. McClelland (to ICI).
23. von K. Engels, *Farbe Lack* **85,** 267 (1979).
24. H. Pinkerton and P. R. Robinson, *Institute of Electrical Engineers Conference Publication No. 104*, Oct. 1973.
25. P. R. Robinson and J. M. Trappe, *Melliand Textilber.*, 557, (1975).
26. S. Baruffaldi, *Chim. Ind.* **60,** 213 (1978).
27. O'Sullivan, *Chem. Eng. News* **16** (Feb. 26, 1979).
28. Schundehutte, *Defazet* **3,** 202 (1979).
29. W. C. Hueper, *Occupational and Environmental Cancers of the Urinary System*, Yale University Press, New Haven, Conn., 1969, p. 216.
30. Ref. 6, Chapt. 12.
31. *Fed. Reg.* **38,** 10929 (1973).
32. C. E. Searles, *ACS Monograph*, Vol. 173, American Chemical Society, Washington, D.C., 1977, p. 491.
33. *Ibid.*, p. 83.
34. K. Socha, *Textile Month*, 52 (Dec. 1992).

General References

R. Allen, *Color Chemistry*, Appleton-Century-Crofts, New York, 1971.
E. Abrahart, *Dyes and Their Intermediates*, 2nd ed., Edward Arnold Ltd., London, 1977; 1st American ed., Chemical Publishing Co., Inc., New York, 1977.
W. Beech, *Fibre-Reactive Dyes*, SAF International, New York, 1970.
G. Booth, *The Manufacture of Organic Colorants and Intermediates*, Society of Dyers and Colorists, Bradford, UK, 1988.
W. Bradley, *Recent Progress in the Chemistry of Dyes and Pigments*, The Royal Institute of Chemistry, London, 1958.
E. Clayton, *Identification of Dyes on Textile Fibres*, 2nd ed., Society of Dyers and Colorists, Bradford, UK, 1963.
N. Donaldson, *The Chemistry and Technology of Naphthalene Compounds*, Edward Arnold Ltd., London, 1958.
J. Fabian and H. Hartmann, *Light Absorption of Organic Colorants: Theoretical Treatment and Empirical Rules*, Springer-Verlag, Heidelberg, 1980.
H. E. Fierz-David and L. Blangey, *Fundamental Processes of Dye Chemistry*, trans. P. W. Vittum, Interscience Publishers, New York, 1949.
T. S. Gore and co-workers, *Recent Progress in the Chemistry of Natural and Synthetic Coloring Matters and Related Fields*, Academic Press, Inc., New York, 1962.
P. F. Gordon and P. Gregory, *Organic Chemistry in Color*, Springer-Verlag, Berlin, 1983.
P. Gregory, *High Technology Applications of Organic Colorants*, Plenum Publishing Corp., New York, 1991.

J. Griffiths, *Color and Constitution of Organic Molecules*, Academic Press, Inc., London, 1976.

E. Gurr, *Synthetic Dyes in Biology, Medicine and Chemistry*, Academic Press, Inc., London, 1971.

H. A. Lubs, ed., *The Chemistry of Synthetic Dyes and Pigments*, ACS Monograph, American Chemistry Society, Washington, D.C., 1955, p. 127.

F. H. Moser and A. L. Thomas, *Phthalocyanine Compounds*, Reinhold Publishing Corp., London, 1963.

A. T. Peters and H. S. Freeman, *Color Chemistry: The Design and Synthesis of Organic Dyes and Pigments*, Elsevier, London, 1991.

P. Rys and H. Zollinger, *Fundamentals of the Chemistry and Application of Dyes*, Wiley-Interscience, New York, 1972.

K. Venkataraman, *The Chemistry of Synthetic Dyes*, Vols. I–VIII, Academic Press, Inc., New York, 1952–1974.

D. R. Waring and G. Hallas, eds., *Practical Dye Chemistry*, Plenum Publishing Corp., New York, 1990.

H. Zollinger, *Diazo and Azo Chemistry*, Interscience Publishers, New York, 1961.

H. Zollinger, *Color Chemistry*, VCH, 1987.

H. Zollinger, *Color Chemistry: Synthesis, Properties and Applications of Organic Dyes and Pigments*, 2nd ed., VCH, 1991.

Patents

These contain most of the dyes and intermediates on chemistry and technology, but in general are more difficult to access than books and journals. Dyes are classified under Section E2 of Derwents World Patents Index Classification.

PETER GREGORY
Zeneca Specialties

DYES, ANTHRAQUINONE

Anthraquinone chemistry began in 1868 with the elucidation of the structure of the naturally occurring compound alizarin (**1**) (1,2-dihydroxyanthraquinone) [*72-48-0*] by C. Graebe and C. Liebermann.

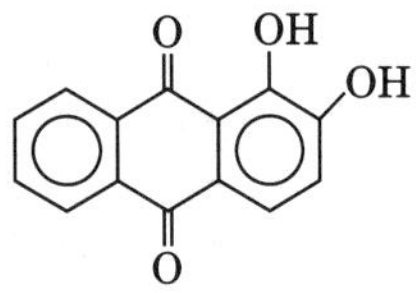

(**1**)

Subsequently, H. Caro and W. H. Perkin independently developed the commercial manufacturing process of alizarin from anthraquinone (qv) through anthraquinone-2-sulfonic acid. Taking advantage of these inventions, many manufacturers came to produce various kinds of hydroxyanthraquinones, which were used as mordant dyes for dyeing cotton and wool.

Mordant dyes have excellent lightfastness. However, their colors are not so brilliant, and they need treatment of fibers with metal salts such as those of Cr, Al, Fe, or Ni before dyeing, which makes the dyeing process complicated and leveling properties unsatisfactory.

In 1894 the first two anthraquinone acid dyes, CI Acid Violet 43 [4430-18-6] (2) (CI 60730) and CI Acid Green 25 [4403-90-1] (3) (CI 61570) were invented. This encouraged the subsequent development of various kinds of anthraquinone acid dyes, which were used to dye wool in fast, brilliant shades without need for pretreatment.

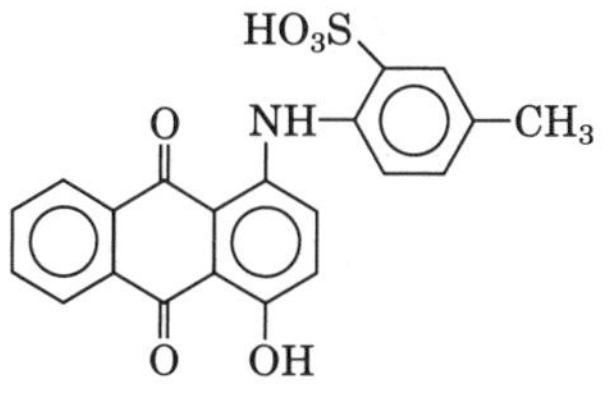

(2) 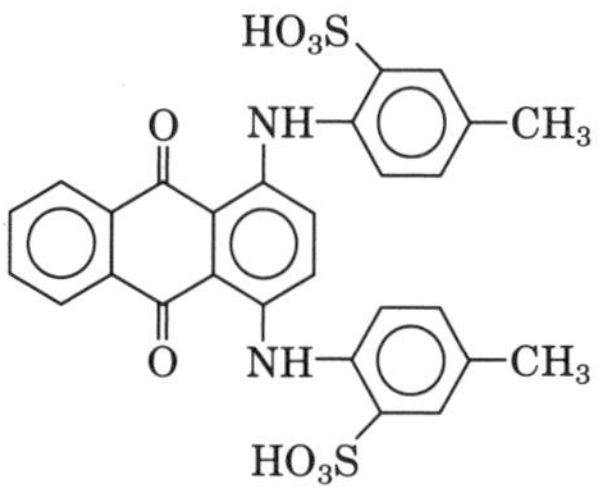   (3)

In 1901, mercury catalyzed α-sulfonation of anthraquinone was discovered, and this led to the development of the chemistry of α-substituted anthraquinone derivatives (α-amino, α-chloro, α-hydroxy, and α,α′-dihydroxyanthraquinones). In the same year R. Bohn discovered indanthrone. Afterward flavanthrone, pyranthrone, and benzanthrone, etc, were synthesized, and anthraquinone vat dyes such as benzoylaminoanthraquinone, anthrimides, and anthrimidocarbazoles were also invented. These anthraquinone derivatives were widely used to dye cotton with excellent fastness, and formed the basis of the anthraquinone vat dye industry.

The appearance of synthetic fibers in the 1920s accelerated the further development of anthraquinone dyes. Soon after British Celanese succeeded in commercializing cellulose acetate fiber in 1921, anthraquinone disperse dyes for this fiber were invented by Stepherdson (British Dyestuffs Corp.) and Celatenes (Scottish Dyes) independently. Anthraquinone disperse dyes for polyester fiber were developed after the introduction of this fiber by ICI and Du Pont in 1952. These dyes were improved products of the disperse dyes that had been developed for cellulose acetate fiber 30 years before.

In the 1950s acid dyes were successively developed to dye nylon carpet with excellent fastness and uniform leveling. Development of polyacrylonitrile fiber stimulated the invention of anthraquinone basic dyes, modified disperse dyes in which quaternary ammonium groups are introduced.

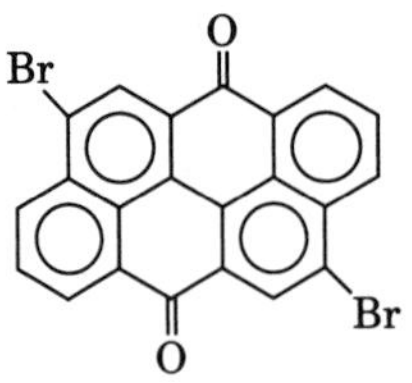

Fig. 1 Anthraquinone dyes used as organic pigments: (**4**) = dibromoanthanthrone [*4378-61-4*] (CI Pigment Red 168; CI Vat Orange 3; CI 59300); (**5**) = an anthrapyrimidine [*4216-01-7*] (CI Pigment Yellow 108; CI Vat Yellow 20; CI 68420); (**6**) = indanthrone blue [*81-77-6*] (CI Pigment Blue 60; CI Vat Blue 4; CI 69800); (**7**) = a bisanthraquinonyl [*4051-63-2*] (CI Pigment Red 177, CI 65300).

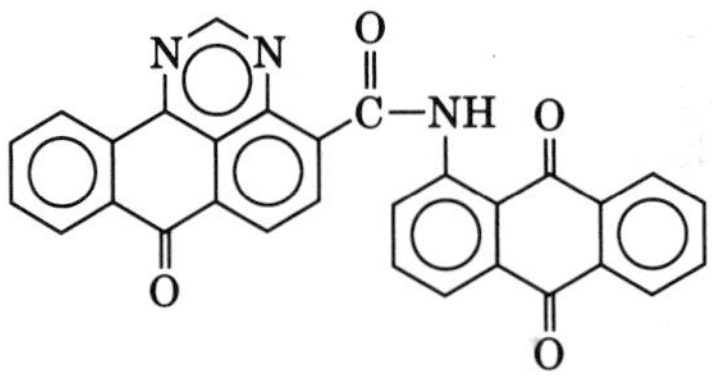

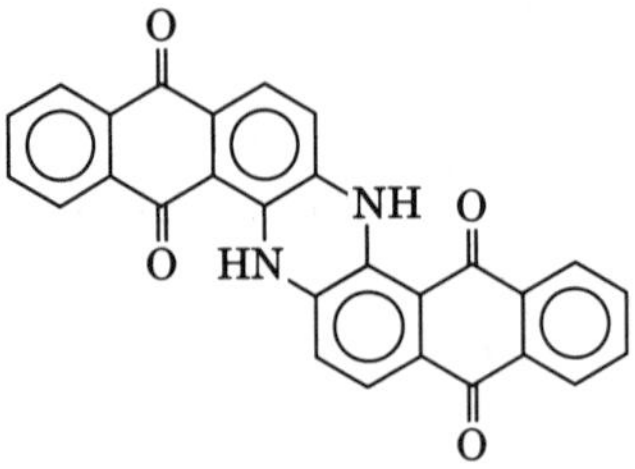

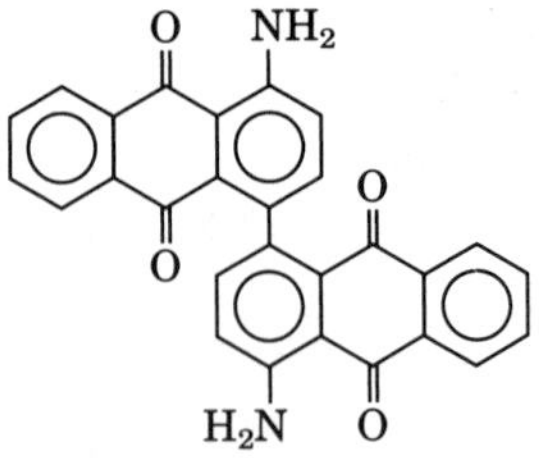

Some anthraquinone dyes are employed as organic pigments (see PIGMENTS, ORGANIC). Examples appear in Figure 1. Indanthrone blue (**6**) is an important automotive paint pigment as is CI Pigment Red 177 (**7**), a bisanthraquinonyl.

Dyes for cellulose fiber include the direct, sulfur, vat, azoic, and reactive dyes. R&D activities of world dye manufacturers have been focused on the area of reactive dyes, because reactive dyes offer brighter shades and excellent wet-fastness and have been increasingly used for dyeing cotton.

Production of anthraquinone reactive dyes based on derivatives of broma-mine acid (**8**) was first commercialized in 1956. Some improvements have been made, and now they are predominantly used among the reactive blue dyes. CI Reactive Blue 19 [*2580-78-1*] (**9**) (CI 61200) (developed by Hoechst in 1957) has the greatest share among them including dye chromophores other than anthra-quinones.

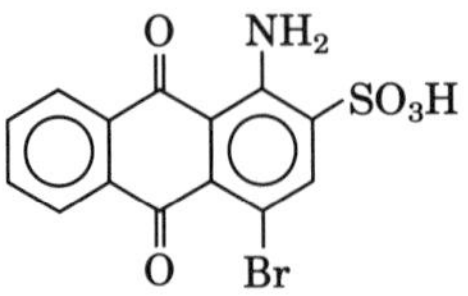

(**8**) 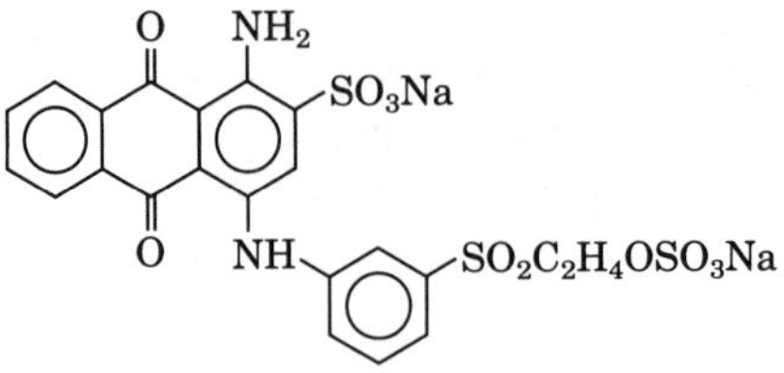   (**9**)

MANUFACTURING OVERVIEW

The synthesis of an anthraquinone dye generally involves a large number of steps. For example, CI Disperse Red 60 [*17418-58-5*] (**10**) (CI 60756) (a typical disperse red dye) requires five steps starting from anthraquinone, and CI Disperse Blue 56 [*31810-89-6*] (**11**) (CI 63285) requires six steps.

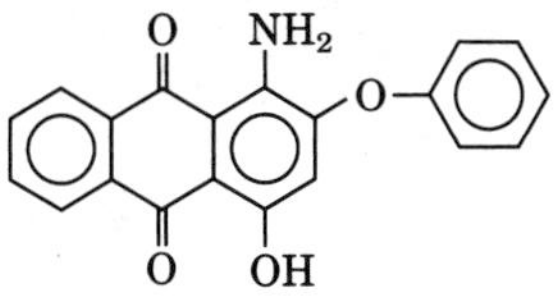

(**10**) 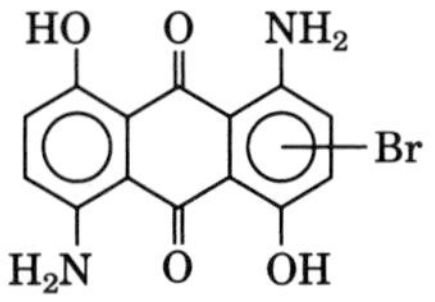   (**11**)

The manufacturing process of anthraquinone vat dyes is more complicated, and, in the extreme case of CI Vat Blue 64 [*15935-52-1*] (**12**) (CI 66730), requires 11 steps starting from phthalic anhydride.

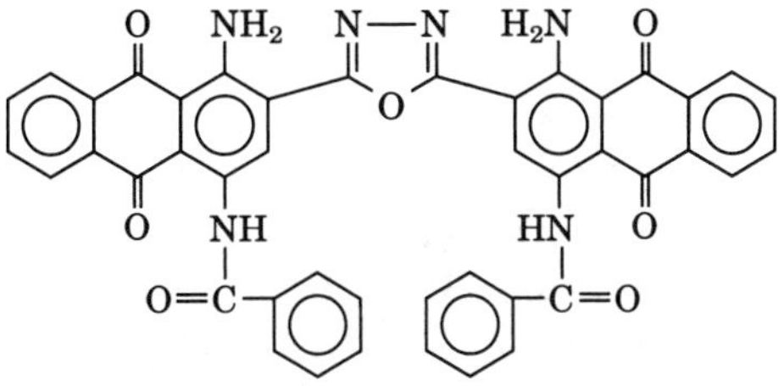

(**12**)

Highly toxic metals such as mercury or chromium(VI) are sometimes required. Some processes need to employ a large amount of organic solvent, and others involve a great quantity of waste acids. With the increasing demand for environmental protection, the regulation of pollutant effluents has become more stringent year after year, which has caused a sharp increase in the costs for wastewater treatment. This situation has led to intensive improvement of conventional methods and the development of new synthetic routes as well. A typical example is the development of nonpolluting processes for the production of 1-aminoanthraquinone and CI Disperse Blue 56. These compounds have been produced conventionally via anthraquinone-α-sulfonic acid or anthraquinone-α,α'-disulfonic acid prepared by mercury-catalyzed sulfonation of anthraquinone. In 1980 Sumitomo Chemical and Mitsubishi Chemical developed a mercury-free production process for 1-aminoanthraquinone and CI Disperse Blue 56 (**11**), respectively. These processes involve α-nitration of anthraquinone instead of α-sulfonation.

Efforts have also been made to overcome complicated processes. Methods to reduce the number of steps or to use new starting materials have been studied extensively. 1-Amino-2-chloro-4-hydroxyanthraquinone (the intermediate for disperse red dyes) conventionally requires four steps from anthraquinone and four separation (filtration and drying) operations. In recent years an improved process has been proposed that involves three reactions and only two separation operations starting from chlorobenzene (Fig. 2).

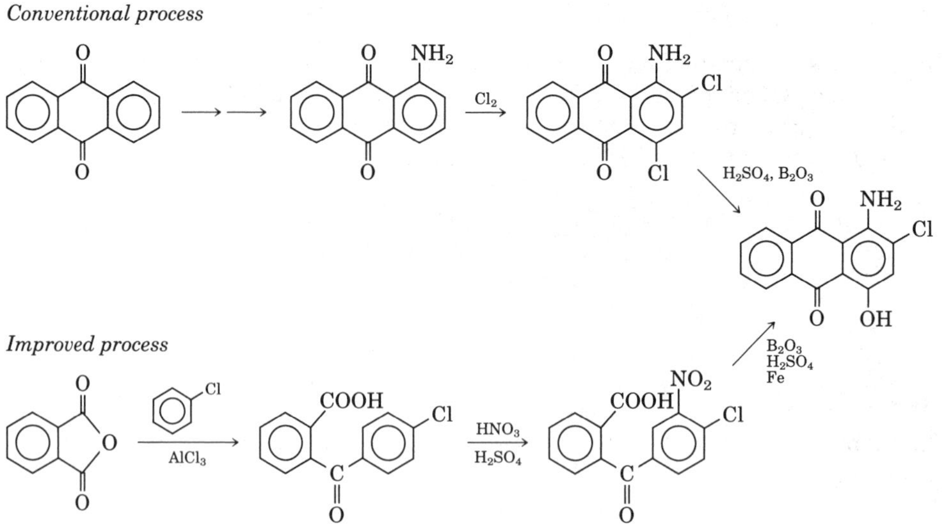

Conventional process

Improved process

Fig. 2. Manufacturing processes for the dye intermediate 1-amino-2-chloro-4-hydroxyanthraquinone [*2478-67-3*].

Because of their small extinction coefficients anthraquinone dyes have less tinctorial strength than azo dyes; that is the intrinsic disadvantage of anthraquinones. This fact and the complexity of preparation have made their production costs higher than those of azo dyes (qv). However, the anthraquinone dyes have excellent properties that are not attainable by azo dyes, such as brilliancy of color, fastness, and excellent dyeing properties (leveling and dye bath stability). Thus the anthraquinone dyes have been widely used in the areas where these properties are required. Cotton or polyester–cotton blend fibers for military wear and working wear that require extreme fastness are dyed mainly with anthraquinone vat dyes. Most polyester fabrics for automobile seats are dyed with anthraquinone disperse dyes, since the requirement for lightfastness is extremely high and, simultaneously, bright shades are needed.

World dye manufacturers have already begun to develop new types of dyes that can replace the anthraquinones technically and economically (1). Some successful examples can be found in azo disperse red and blue dyes. Examples are brilliant red [*68353-96-6*] and CI Disperse Blue 165 [*41642-51-7*] (CI 11077). They have come close to the level of anthraquinone reds and blues, respectively, in terms of brightness. In the reactive dye area intensive studies have continued to develop triphenodioxazine compounds, eg, (**13**), which are called new blues, to replace anthraquinone blues. In this representation R designates the substituents having reactive groups (see DYES, REACTIVE).

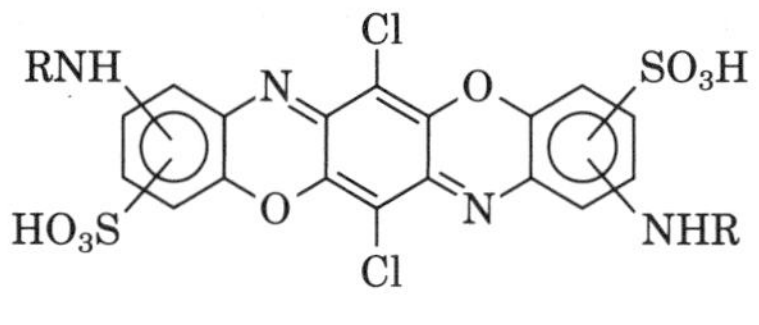

(**13**)

COLOR AND STRUCTURE

The uv–vis spectrum of anthraquinone shows an absorption maximum at 323 nm (ϵ = 4500) due to a π-π^* transition and very weak absorption in the visible range, 405 nm (ϵ = 60) due to a n-π^* transition. Thus anthraquinone is almost colorless. Introduction of electron-donating substituents causes a bathochromic shift. This is due to the charge-transfer band from the lone pair of amino or hydroxyl groups to the oxygen atom of the carbonyl group. By increasing the electron-donating ability of substituents, the bathochromic shifts are enhanced (Table 1). In the case of the same substituent, the bathochromic shift is larger when the substituent is in the 1-position rather than in the 2-position. The introduction of an electron-withdrawing group has little effect on the absorption maximum of the spectrum.

Table 1. Spectral Data for Some Monosubstituted Anthraquinones^a in Methanol

| Substituent | 1-position | | 2-position | |
	λ_{max}, nm	ϵ	$_{max}$, nm	ϵ
	Electron-donating groups			
OCH_3	378	5200	363	3950
OH	402	5500	368	3900
$NHCOCH_3$	400	5600	367	4200
NH_2	475	6300	440	4500
$NHCH_3$	503	7100	462	5700
$N(CH_3)_2$	503	4900	472	5900
	Electron-withdrawing groups			
NO_2	325	4300	323	5200
Cl	333	5000	325	3900

^{a}Unsubstituted anthraquinone λ_{max} = 323 nm; ϵ = 4500.

A methylamino group is more effective than a dimethylamino group as an electron donor. This is interpreted in terms of hydrogen-bonding of the substituent with the adjacent carbonyl group that promotes the conjugation of the lone pair of electrons of the donor. Also, a sterically hindered dimethylamino group in the 1-position is unable to conjugate, which decreases the extinction coefficient as well.

The absorption maximum of a disubstituted anthraquinone greatly depends on the substituents and their positions (Table 2). The 1,4-disubstituted compound shows a remarkable bathochromic shift. The effects of β-substituents on 1,4-diaminoanthraquinones (**14**) are shown in Table 3. Larger bathochromic shifts are observed with increasing electron-withdrawing ability of β-substituents.

1,4,5,8-Tetrasubstituted anthraquinones give a slightly reddish blue tint to greenish blue color depending on the substituents and their positions, eg, 1,4,5,8-tetraaminoanthraquinone is blue green.

R = OH, slightly reddish blue R_1 = NH_2; R_2 = OH; slightly greenish blue
R = NH_2, neutral blue R_1 = OH; R_2 = NH_2, greenish blue

Table 2. Spectral Data for Some Disubstituted Anthraquinones in Methanol

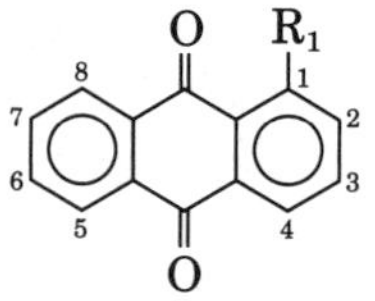

R_1	R_2^a	λ_{max}, nm	ϵ
NH_2	H	475	6,500
NH_2	5-NH_2	487	12,600
NH_2	8-NH_2	507	10,000
NH_2	4-NH_2	550, 590	15,850, 15,850
NH_2	4-OH	528, 563	11,670, 9,540
OH	H	402	5,500
OH	5-OH	425	10,000
OH	8-OH	430	10,960
OH	4-OH	470	17,000

^{a}Numerical locant indicates substituent position.

Table 3. Spectral Data for 2- and 3-Substituted 1,4-Diaminoanthraquinones^a in *N,N*-Dimethylformamide

R	Monosubstitution (2-) λ_{max}, nm	Disubstitution (2-,3-) λ_{max}, nm
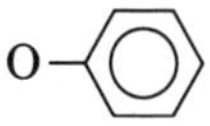 O–⟨phenyl⟩		550, 588
H	553, 594	553, 594
Cl	559, 598	563, 601
SO_3H	562, 603	596, 637
COOH	603	
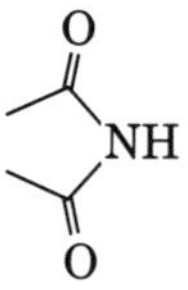 (imide group)		666

^{a}Structure (**14**).

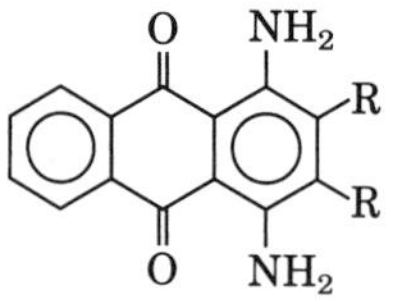

(**14**)

In addition to the color and the tinctorial strength, which are very important factors for the molecular design of anthraquinone dyes, affinity for fibers, various kinds of fastness (light, wet, sublimation, nitrogen oxides (NO_x) gas, washing, etc), and application properties (sensitivity for dyeing temperature, pH, etc) must be considered thoroughly as well.

METHOD OF SYNTHESIS

Anthraquinone dyes are derived from several key compounds called dye intermediates, and the methods for preparing these key intermediates can be divided into two types: (*1*) introduction of substituent(s) onto the anthraquinone nucleus, and (*2*) synthesis of an anthraquinone nucleus having the desired substituents, starting from benzene or naphthalene derivatives (nucleus synthesis). The principal reactions are nitration and sulfonation, which are very important in preparing α-substituted anthraquinones by electrophilic substitution. Nucleus synthesis is important for the production of β-substituted anthraquinones such as 2-methylanthraquinone and 2-chloroanthraquinone. Friedel-Crafts acylation using aluminum chloride is applied for this purpose. Synthesis of quinizarin (1,4-dihydroxyanthraquinone) is also important.

Key Intermediates

1-Aminoanthraquinone and Related Compounds. 1-Aminoanthraquinone [*82-45-1*] (**17**) is the most important intermediate for manufacturing acid, reactive, disperse, and vat dyes. It has been manufactured from anthraquinone-1-sulfonic acid [*82-49-5*] (**16**) by ammonolysis of the sulfo group with aqueous ammonia in the presence of an oxidizing agent such as nitrobenzene-3-sulfonic acid.

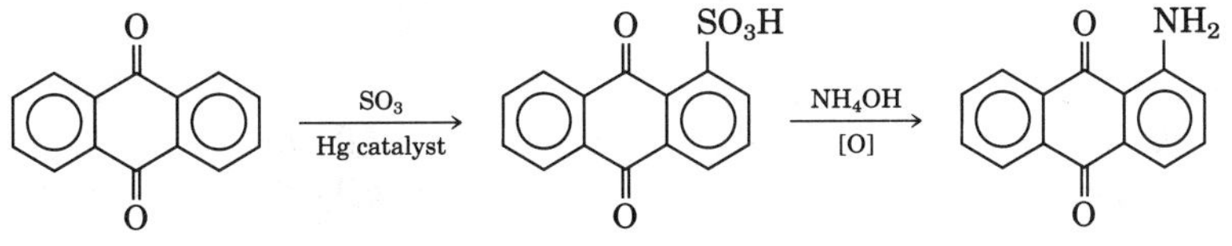

In this process the starting material can only be obtained by mercury-catalyzed sulfonation of anthraquinone [*84-65-1*] (**15**) with oleum. For improved ecology, the alternative route based on 1-nitroanthraquinone [*82-34-8*] (**18**) was established. 1-Nitroanthraquinone is prepared from anthraquinone by nitration in sulfuric acid or organic solvent. 1-Aminoanthraquinone can be prepared from 1-nitroanthraquinone by reduction with sodium sulfide, sodium hydrogen sulfide in water (2), in organic solvent (3), with hydrazine hydrate (4), or by catalytic hydrogenation (5).

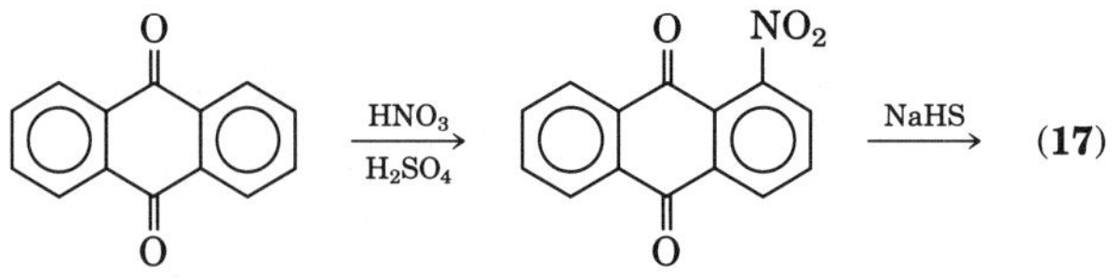

(18)

Purification is carried out by recrystallization from organic solvent (6) or from sulfuric acid (7). Highly purified product is manufactured by continuous vacuum distillation (8).

One purification method applies the difference in oxidation rate between the leuco forms of the mono- and diaminoanthraquinones. Thus a mixture of 1-aminoanthraquinone and diaminoanthraquinones are first reduced by treating with dithionite in aqueous alkaline solution or by catalytic hydrogenation to convert to the leuco form, ie, corresponding anthrahydroquinones. Then, by subsequent partial oxidation by air, diaminoanthrahydroquinones are oxidized selectively to regenerate the quinoide form, and are separated from water-soluble 1-aminoanthrahydroquinone (9,10).

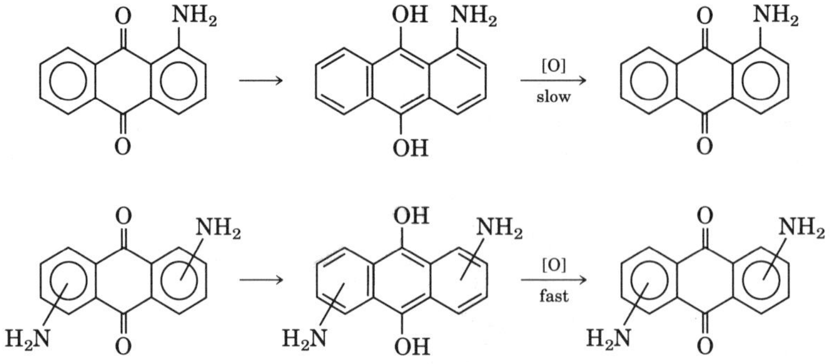

leuco (hydroquinone) forms

1-Nitroanthraquinone (18) is now the key intermediate for 1-aminoanthraquinone. The classical route from anthraquinone-1-sulfonic acid has become less competitive, because perfect recovery of mercury catalyst is demanded, and this requires a large investment.

1-Nitroanthraquinone is prepared from anthraquinone by nitration in sulfuric acid (11), or in organic solvent (12). Nitration in nitric acid is dangerous. The mixture of anthraquinone and nitric acid forms a Sprengel mixture (13,14) which may detonate. However, detonation can be prevented by adding an inert third component such as sulfuric acid. Experimental results of the steel-tube detonation tests for the anthraquinone–HNO_3–H_2SO_4 system have been published (13).

The nitration route shows insufficient alpha-selectivity in addition to producing considerable dinitro product. Although a large amount of work has been done to maximize the yield of 1-nitro compound, the best result is less than 80%. Several methods have been developed to remove 2-nitroanthraquinone and dinitroanthraquinones from crude 1-nitroanthraquinone. Purification is carried out, for example, by recrystallization from nitric acid, or from organic solvents (15).

The oxidation of 1-nitronaphthalene by ceric ammonium nitrate has been reported (16). The resulting 1-nitronaphthoquinone condenses with 1,3-butadiene followed by air oxidation under alkaline conditions to form 1-nitroanthraquinone, or 1-aminoanthraquinone is formed directly by an intramolecular redox reaction.

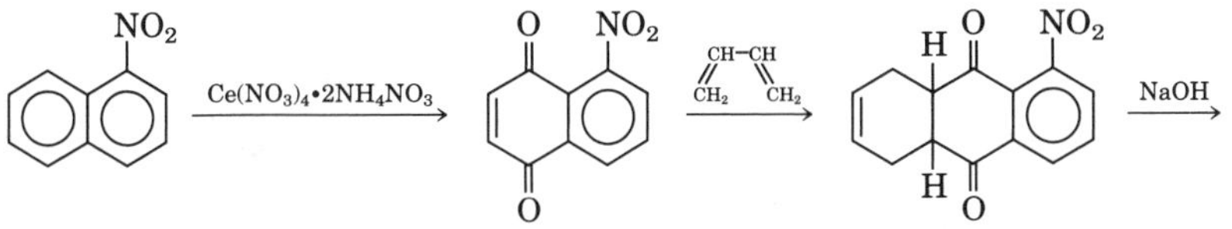

Efforts to raise the alpha-selectivity have been made. Thus nitration of anthraquinone using nitrogen dioxide and ozone has been reported (17).

1-Amino-4-bromoanthraquinone-2-sulfonic acid (bromamine acid) [116-81-4] (8) is the most important intermediate for manufacturing reactive and acid dyes. Bromamine acid is manufactured from 1-aminoanthraquinone-2-sulfonic acid [83-62-5] (19) by bromination in aqueous medium (18–20), or in concentrated sulfuric acid (21). 1-Aminoanthraquinone-2-sulfonic acid is prepared from 1-aminoanthraquinone by sulfonation in an inert, high boiling point organic solvent (22), or in oleum with sodium sulfate (23).

In the first case (22), almost stoichiometric amounts of sulfuric acid or chlorosulfonic acid are used. The amine sulfate or the amine chlorosulfate is, first, formed and heated to about 180 or 130°C, respectively, to rearrange the salt. The introduction of the sulfonic acid group occurs only in the ortho position, and an almost quantitative amount of 1-aminoanthraquinone-2-sulfonic acid is obtained. On the other hand, the use of oleum (23) requires a large excess of SO_3 to complete the reaction, and inevitably produces over-sulfonated compound such as 1-aminoanthraquinone-2,4-disulfonic acid. Addition of sodium sulfate reduces the by-product to a certain extent. Improved processes have been proposed to make the isolation of the intermediate (19) unnecessary (24,25).

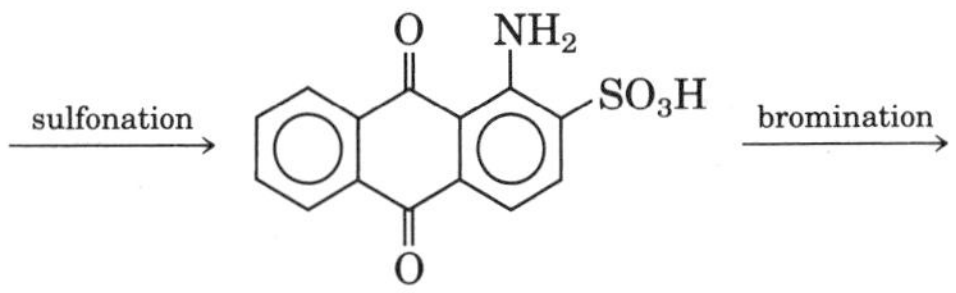

Contamination by water-insoluble reaction by-products such as 1-amino-2,4-dibromoanthraquinone affects the quality of dyestuff significantly. Therefore, several methods for purification have been reported. Examples are extraction of impurities with organic solvent (18), or precipitation of bromamine acid from concentrated (60–85%) sulfuric acid (26).

Many anthraquinone reactive and acid dyes are derived from bromamine acid. The bromine atom is replaced with appropriate amines in the presence of

copper catalyst in water or water–alcohol mixtures in the presence of acid binding agents such as alkali metal carbonate, bicarbonate, hydroxide, or acetate
(Ullmann condensation reaction).

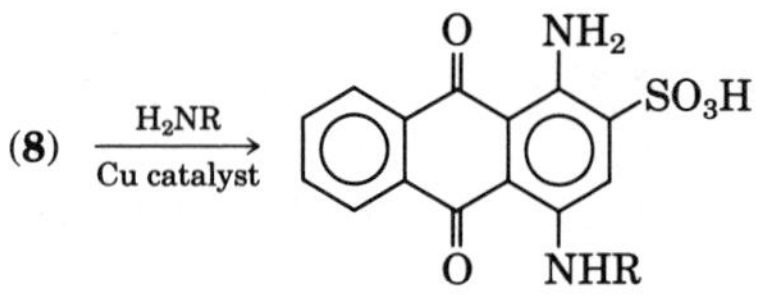

$$(8) \xrightarrow[\text{Cu catalyst}]{\text{H}_2\text{NR}}$$

(20)

Yields depend on the reactivity of the amines and the choice of reaction conditions,
including the choice of copper catalyst. Generally, the reactivity increases with
increasing amine basicity. Thus, *para*-toluidine ($pK_a = 5.1$) reacts four times
faster than aniline ($pK_a = 4.7$) (27). Sterically hindered amines such as 3,5-di-
amino-2,4,6-trimethylbenzenesulfonic acid react very slowly.

The main by-products of the Ullmann condensation are 1-aminoanthraqui-
none-2-sulfonic acid and 1-amino-4-hydroxyanthraquinone-2-sulfonic acid. The
choice of copper catalyst affects the selectivity of these by-products. Generally,
metal copper powder or copper(I) salt catalyst has a greater reactivity than cop-
per(II) salts. However, they are likely to yield the reduced product (1-aminoan-
thraquinone-2-sulfonic acid). The reaction mechanism has not been established.
It is very difficult to clarify which oxidation state of copper functions as catalyst,
since this reaction involves fast redox equilibria where anthraquinone derivatives
and copper compounds are concerned. Some evidence indicates that the catalyst
is probably a copper(I) compound (28,29).

1-Amino-2-bromo-4-hydroxyanthraquinone (bromo pink) [*116-82-5*] (**22**) is
one of the most important intermediates for manufacturing red disperse dyes. It
is prepared by dibrominating 1-aminoanthraquinone (**17**) in concentrated sulfuric
acid and subsequent hydrolysis in the presence of boric acid. These two reactions
are carried out in one pot without isolation of 1-amino-2,4-dibromoanthraquinone
[*81-49-2*] (**21**) (30–32).

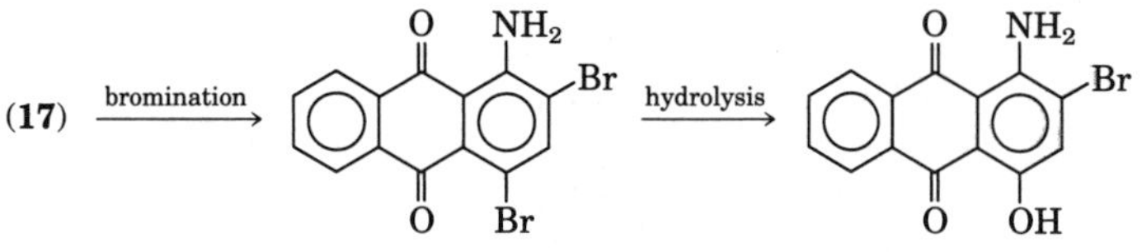

$$(17) \xrightarrow{\text{bromination}} \quad (21) \xrightarrow{\text{hydrolysis}} \quad (22)$$

(21)   (22)

Furthermore, a method using formaldehyde in the second step (hydrolysis) instead of boric acid has been reported recently (33). 1-Amino-4-hydroxyanthraqui-
none [*116-85-8*] (**26**) is also brominated to form 1-amino-2-bromo-4-hydroxyan-
thraquinone [*116-82-5*]. Bromination is carried out in an inert organic solvent
such as nitrobenzene (34).

1-Amino-2-chloro-4-hydroxyanthraquinone (chloro pink) [*2478-67-3*] (**23**) is
another important intermediate in red disperse dye manufacture. 1-Amino-2-

chloro-4-hydroxyanthraquinone is prepared via a route from chlorobenzene and phthalic anhydride as the raw materials (35) (see Fig. 2). 2-(4′-Chlorobenzoyl)-benzoic acid is nitrated in concentrated sulfuric acid, then reduction of the nitro group, ring closure, and hydrolysis occur simultaneously in concentrated sulfuric acid in the presence of a reducing agent and boric acid. Thus obtained crude chloro pink is purified by selective precipitation from sulfuric acid in order to separate it from by-produced 2-amino-3-chloro-1-hydroxyanthraquinone (**24**) (36).

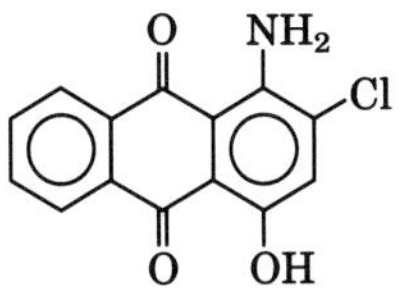

(23) 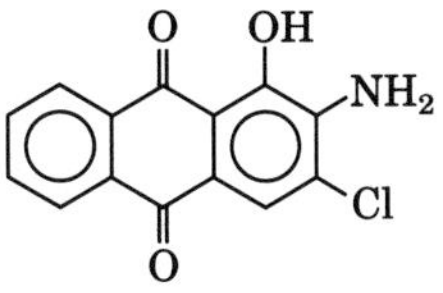   (24)

1-Amino-2-chloro-4-hydroxyanthraquinone can be prepared from 1-amino-2,4-dichloroanthraquinone by selective hydrolysis of the chlorine atom in the 4-position in concentrated sulfuric acid in the presence of boric acid in the same manner as bromo pink. However, 1-amino-2,4-dichloroanthraquinone [*13432-32-1*] (**25**) cannot be practically obtained by chlorination of 1-aminoanthraquinone in concentrated sulfuric acid. Rather, 1-amino-2,4-dichloroanthraquinone must be prepared by chlorinating 1-aminoanthraquinone in an organic solvent (37), or in thionyl chloride (38), and isolating it.

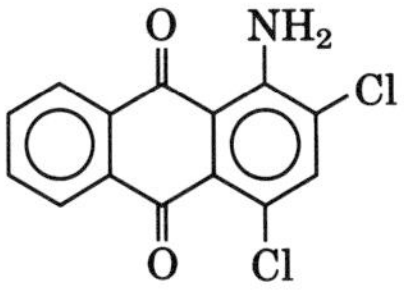

(25)

1-Amino-4-hydroxyanthraquinone (**26**) can be used as the starting material for 1-amino-2-chloro-4-hydroxyanthraquinone (**23**). Chlorination is carried out in concentrated sulfuric acid with chlorine gas in the presence of a catalytic amount of FeCl$_3$ (39). 1-Amino-4-hydroxyanthraquinone is prepared by oxidizing 1,4-di-aminoanthraquinone in concentrated sulfuric acid with a catalytic amount of manganese dioxide (39).

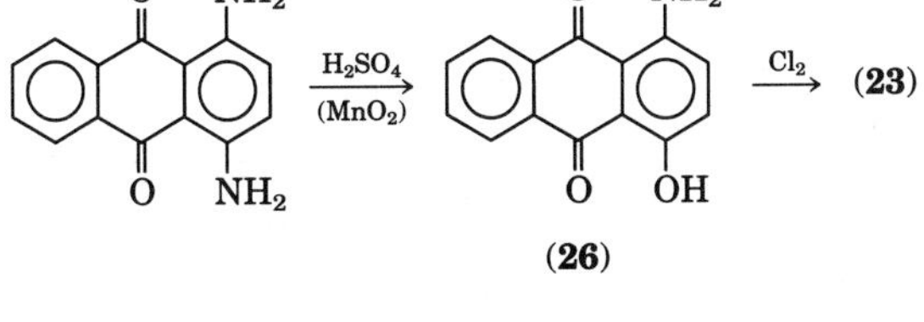

(26)

1,4-Dihydroxyanthraquinone. This anthraquinone, also known as quini-zarin [*81-64-1*] (**29**), is of great importance in manufacturing disperse, acid, and vat dyes. It is manufactured by condensation of phthalic anhydride (**27**) with 4-chlorophenol [*106-48-9*] (**28**) in oleum in the presence of boric acid or boron trifluoride (40,41). Improved processes for reducing waste acid have been reported (42), and yield is around 80% on the basis of 4-chlorophenol.

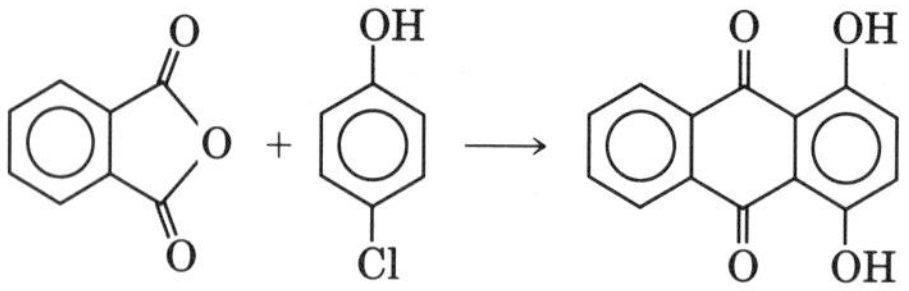

(**27**)   (**28**)   (**29**)

In this reaction, three steps, ie, acylation, cyclization, and replacement of the chlorine atom by the hydroxyl group, take place simultaneously in concentrated sulfuric acid. In the course of cyclization 2,7-dichlorofluoran (**31**) may be formed as a by-product presumably through the carbonium ion (**30**) illustrated as follows. The addition of boric acid suppresses this pathway and promotes the regular cyclization to form the anthraquinone structure. The stable boric acid ester formed also enables the complete replacement of chlorine atoms by the hydroxyl group. Hydrolysis of the boric acid ester of quinizarin is carried out by heating in dilute sulfuric acid. The purity of quinizarin thus obtained is around 90%. Highly pure product can be obtained by sublimation.

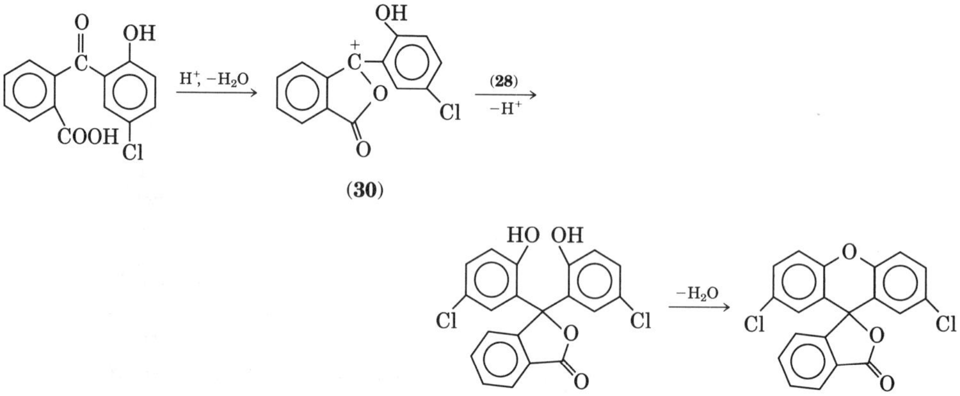

(**30**)

(**31**)

Hydroquinone may also be used in place of 4-chlorophenol. In this case an aluminum chloride–sodium chloride melt is usually employed. However, the yield is not satisfactory (43). It has also been reported that the reaction of hydroquinone with substantially stoichiometric phthalic acid dichloride in the presence of anhydrous aluminum chloride in moderately polar solvents, such as nitrobenzene at around 100°C gives quinizarin (44). The reported yield is 65% after purification by crystallization from toluene.

1,4-Diaminoanthraquinone and Related Compounds. Leuco-1,4-diaminoanthraquinone [*81-63-0*] (leucamine) (**32**) is an important precursor for 1,4-diaminoanthraquinone [*128-95-0*] (**33**) and is prepared by heating 1,4-dihydroxy-anthraquinone (**29**) with sodium dithionite in aqueous ammonia under pressure.

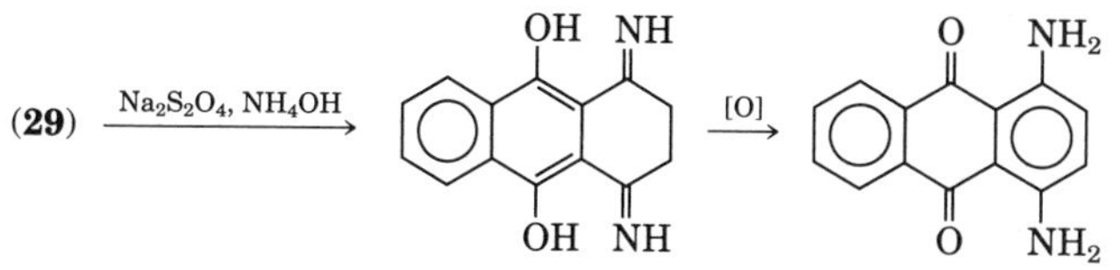

1,4-Diaminoanthraquinone is an important intermediate for vat dyes and disperse dyes, and is prepared by oxidizing leuco-1,4-diaminoanthraquinone with nitrobenzene in the presence of piperidine. An improved process has been reported (45).

1,4-Diaminoanthraquinone-2-sulfonic acid [*4095-85-6*] (**34**) is a possible precursor of 1,4-diamino-2,3-dicyanoanthraquinone, and is prepared from 1-amino-4-bromoanthraquinone-2-sulfonic acid (**8**) by reaction with liquid ammonia in the presence of copper catalyst (46,47).

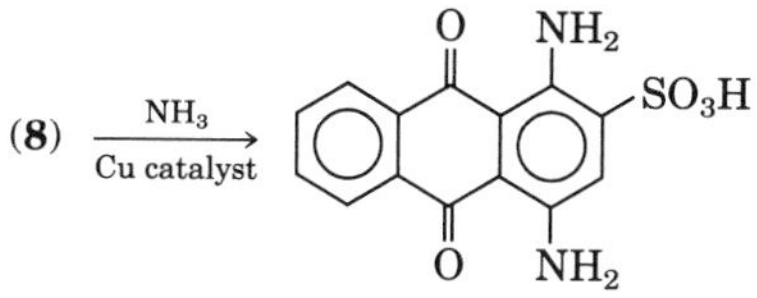

Instead of liquid ammonia, aqueous ammonia is also used together with a polar aprotic solvent such as formamide (48). It is also prepared by sulfonating 1,4-diaminoanthraquinone (**33**) with chlorosulfonic acid (49), sulfuric acid, or oleum (50).

1,4-Diamino-2,3-dichloroanthraquinone [*81-42-5*] (**35**) (CI Disperse Violet 28) is an important compound as an intermediate for CI Disperse Blue 60 and CI Disperse Violet 26 (**116**), and is prepared by chlorination of leuco-1,4-diaminoanthraquinone (**32**) with chlorine gas or sulfuryl chloride in an inert organic solvent such as nitrobenzene (51,52).

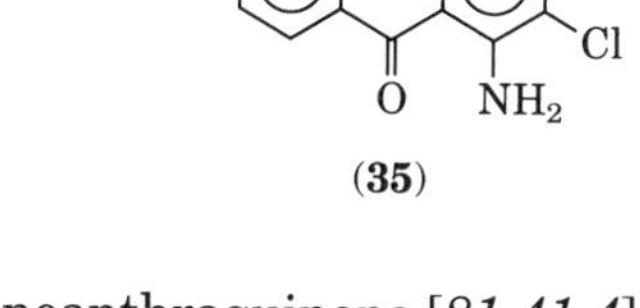

1,4-Diamino-2,3-dicyanoanthraquinone [*81-41-4*] (**37**) is the key intermediate for manufacturing CI Disperse Blue 60. 1,4-Diamino-2,3-dicyanoanthraquinone is manufactured by reaction of 1,4-diaminoanthraquinone-2,3-disulfonic acid (**36**) with alkali metal cyanide (53). A one-pot process from 1,4-diamino-2,

3-dichloroanthraquinone, ie, sulfonation with alkali metal sulfite in the presence of quaternary ammonium compound and subsequent cyanation without isolation of the intermediate, has been proposed (54).

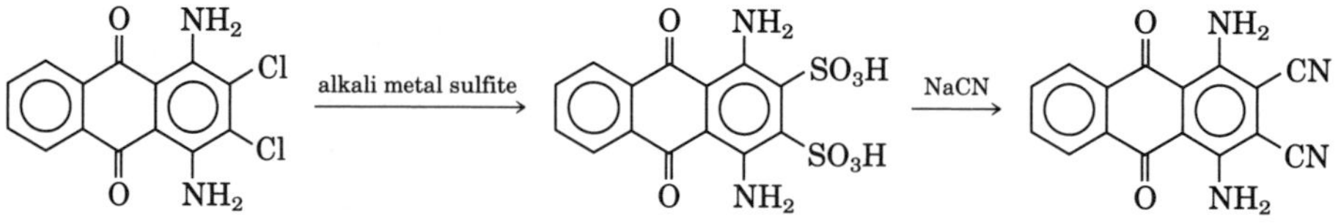

(**35**)   (**36**)   (**37**)

It is also prepared by direct cyanation of 1,4-diamino-2,3-dichloroanthraquinone (**35**) or 1,4-diaminoanthraquinone (**33**) in an aprotic organic solvent. In the latter case, the presence of an ammonium compound and a dehydrogenating agent is necessary (55).

$$(\textbf{33}), (\textbf{35}) \xrightarrow{\text{NaCN, [O]}} (\textbf{37})$$

A process from 1,4-diaminoanthraquinone-2-sulfonic acid (**34**) has also been proposed. In this case, cyanation is preferably carried out in an aqueous medium in the presence of a dehydrogenating agent such as nitrobenzene-3-sulfonic acid, and a quarternary ammonium compound (56). Cyanation in an aprotic solvent such as formamide or 1-methoxypropan-2-ol has also been proposed (57,58). Cyanation of the derivative (**38**) of 1,4-diaminoanthraquinone in 2-pyrrolidinone has also been proposed. The starting material is prepared by reaction of 1,4-diaminoanthraquinone with boron trifluoride–diethyl ether complex (59).

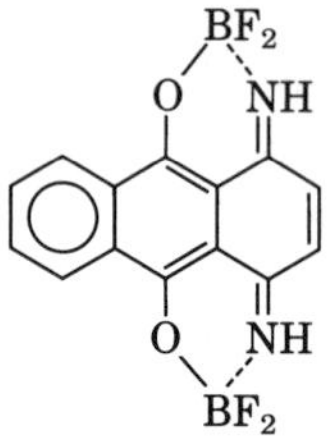

(**38**)

1,4-Diaminoanthraquinone-2,3-dicarboxyimide [*128-81-4*] (**39**) is the intermediate for CI Disperse Blue 60 (**40**), in which the imide H is replaced by the R group —$CH_2CH_2CH_2OCH_3$. (**39**) is prepared by hydrolysis of 1,4-diamino-2,3-dicyanoanthraquinone in concentrated sulfuric acid.

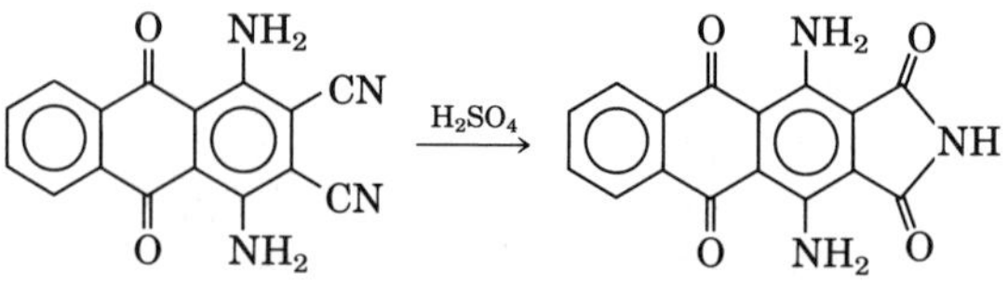

(**37**)   (**39**)

Anthraquinone-1-sulfonic acid and Its Derivatives. Anthraquinone-1-sulfonic acid [82-49-5] (**16**) has become less competitive than 1-nitroanthraquinone as the intermediate for 1-aminoanthraquinone. However, it still has a great importance as an intermediate for manufacturing vat dyes via 1-chloro-anthraquinone.

Anthraquinone-1-sulfonic acid is prepared from anthraquinone by sulfonation with 20% oleum in the presence of mercury catalyst, a Hg(II) salt such as $HgSO_4$ or HgO, at 120°C. Direct sulfonation with the oleum in the absence of mercury catalyst at the same temperature produces the 2-sulfonic acid exclusively. Although the 1-position in anthraquinone is much more reactive than is the 2-position, as predicted by calculations of the localization energy of π-electrons, the formation of 2-isomer, which is not sterically hindered by an adjacent carbonyl group, is thermodynamically favored. Anthraquinone-1-sulfonic acid is isolated as the potassium salt. Thus, after completion of sulfonation, the reaction mixture is charged into a large excess of water, heated, and unreacted anthraquinone is separated by filtration. The potassium salt is precipitated from the filtrate by adding potassium chloride solution, and isolated by filtration. Demercuration of the sulfonation mixture using sulfur compounds such as CH_3CSNH_2 etc, has been proposed (60).

1-Chloroanthraquinone [82-44-0] (**41**) is an intermediate for manufacturing vat dyes such as CI Vat Brown 1. 1-Chloroanthraquinone is prepared by chlorination of anthraquinone-1-sulfonic acid with sodium chlorate in hydrochloric acid at elevated temperature (61). An alternative route from 1-nitroanthraquinone (**18**) using elemental chlorine at high temperature has been reported (62).

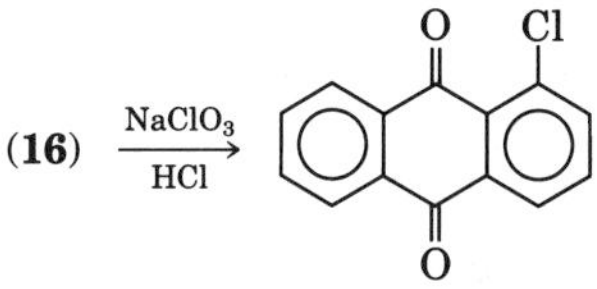

$$(16) \xrightarrow[\text{HCl}]{\text{NaClO}_3}$$

(**41**)

1-Methylaminoanthraquinone [82-38-2] (**42**) is an important intermediate for manufacturing solvent dyes and acid dyes, and is prepared from anthraquinone-1-sulfonic acid (**16**) by replacing the SO$_3$H group with methylamine. An oxidizing agent such as *m*-nitrobenzenesulfonic acid is usually added to oxidize the liberated sulfite. 4-Bromo-1-methylaminoanthraquinone [128-93-8] (**43**) is a precursor of *N*-methylanthrapyridone and is prepared from 1-methylaminoanthraquinone by bromination (63).

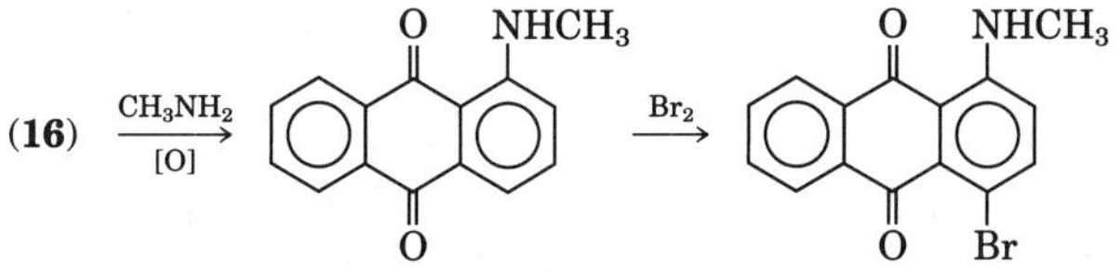

$$(16) \xrightarrow[\text{[O]}]{\text{CH}_3\text{NH}_2} \qquad \xrightarrow{\text{Br}_2}$$

(**42**)   (**43**)

Anthraquinone-α,α'-disulfonic acids and Related Compounds. Anthraquinone-α,α'-disulfonic acids and their derivatives are important intermediates for manufacturing disperse blue dyes (via 1,5-, or 1,8-dihydroxyanthraquinone, or 1,5-dichloroanthraquinone) and vat dyes (via 1,5-dichloroanthraquinone).

Anthraquinone-1,5-disulfonic acid [117-14-6] (**44**), and anthraquinone-1,8-disulfonic acid [82-48-4] (**45**) are produced from anthraquinone by disulfonation in oleum; a higher concentration of SO_3 than that used for 1-sulfonic acid is employed in the presence of mercury catalyst (64,65). After completion of sulfonation, 1,5-disulfonic acid is precipitated by addition of dilute sulfuric acid and separated. After clarification with charcoal, 1,5-disulfonic acid is precipitated as the sodium salt by addition of sodium chloride. The 1,8-disulfonic acid is isolated as the potassium salt from the sulfuric acid mother liquor by addition of potassium chloride solution.

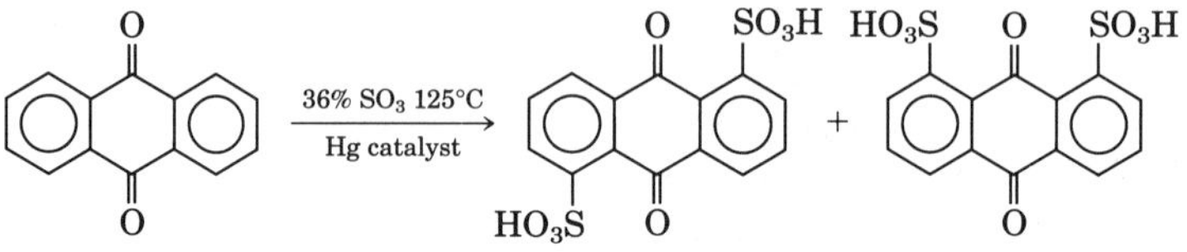

1,5-Dichloroanthraquinone [82-46-2] (**46**) is an important intermediate for vat dyes and disperse blue dyes. Examples are CI Vat Violet 13 [4424-87-7] (**170**), CI Vat Orange 15 [2379-78-4] (**154**), and CI Disperse Blue 56 [31810-89-6] (**11**). 1,5-Dichloroanthraquinone is prepared by the reaction of anthraquinone-1,5-disulfonic acid with $NaClO_3$ in hot hydrochloric acid solution. Alternative methods from 1,5-dinitroanthraquinone (**49**) by reaction of chlorine at high temperature in the presence of phthalic anhydride have been proposed (66).

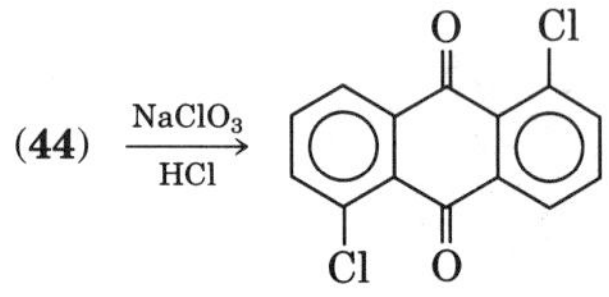

1,5-Dihydroxyanthraquinone (anthrarufin) [117-12-4] (**47**) is an important intermediate for manufacturing disperse blue dyes, eg, CI Disperse Blue 73 (**113**), and is prepared from anthraquinone-1,5-disulfonic acid by heating with an aqueous suspension of calcium oxide and magnesium chloride under pressure at 200–250°C (67). Alternative methods have been proposed, ie, direct replacement of the NO_2 groups of 1,5-dinitroanthraquinone (**49**) (68) or the route via 1,5-dimethoxyanthraquinone [6448-90-4] (**48**) and subsequent hydrolysis (69).

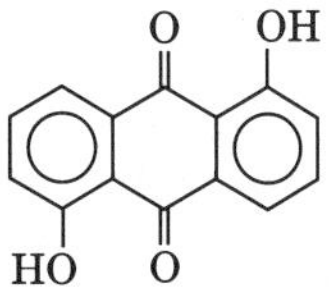

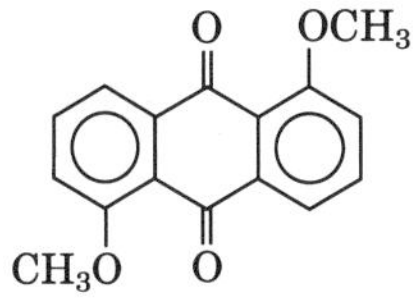

(47)   (48)

1,8-Dihydroxyanthraquinone (chrysazin) [117-10-2] is prepared in a similar manner to that for anthrarufin.

α,α'-Dinitroanthraquinones and Related Compounds. 1,5- and 1,8-Dinitroanthraquinone are the key intermediates for manufacturing disperse blue dyes via dinitrodihydroxyanthraquinone and vat dyes via diaminoanthraquinones. 1,5-Dinitroanthraquinone [82-35-9] (49) and 1,8-dinitroanthraquinone [129-39-5] (50) are prepared by nitration of anthraquinone with nitric acid in sulfuric acid. α,β'-Dinitroanthraquinones are also formed in the reaction.

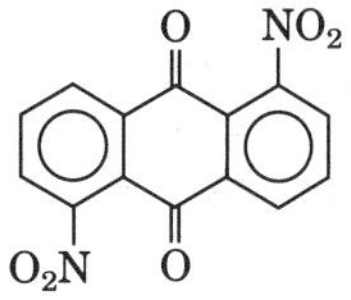

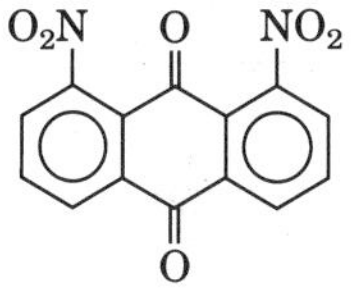

(49)   (50)

1,5-Dinitroanthraquinone and 1,8-dinitroanthraquinone can also be prepared by nitration of anthraquinone in concentrated nitric acid (70). The 1,5-isomer can then be easily separated from the reaction mixture by filtration, since 1,8- or other isomers than 1,5-dinitroanthraquinone are completely dissolved in concentrated nitric acid. However, this process is unsuitable for industrial production for safety reasons; the mixture of dinitroanthraquinone and concentrated nitric acid forms a detonation mixture (71). Addition of sulfuric acid makes it possible to work outside the detonation area.

Dinitroanthraquinones are industrially prepared by nitration of anthraquinone in mixed nitric–sulfuric acid at 0–50°C. The reaction mixture is then heated to a temperature slightly higher than the nitration reaction temperature to enrich the content of 1,5-dinitroanthraquinone in solid phase, and then cooled and filtered to obtain the 1,5-dinitroanthraquinone wet cake. Mother liquor is concentrated by distillation of nitric acid and crystallized 1,8-isomer is separated. The filtrate is again distilled, and precipitated β-isomers are filtered off and filtrate is recycled to the nitration step (72–74).

1,5-Diaminoanthraquinone [129-44-2] (51) is prepared from 1,5-dinitroanthraquinone (49) by ammonolysis in organic solvents (75), in aqueous ammonia (76), by catalytic hydrogenation in an organic solvent (77), or by reduction with sodium sulfide. 1,5-Diaminoanthraquinone is also prepared from anthraquinone 1,5-disulfonic acid (44) by ammonolysis in the presence of an oxidizing agent such as m-nitrobenzenesulfonic acid (78,79). 1,5-Diaminoanthraquinone is an important intermediate for manufacturing vat dyes such as CI Vat Brown 3 [131-92-0] (156) (CI 69015).

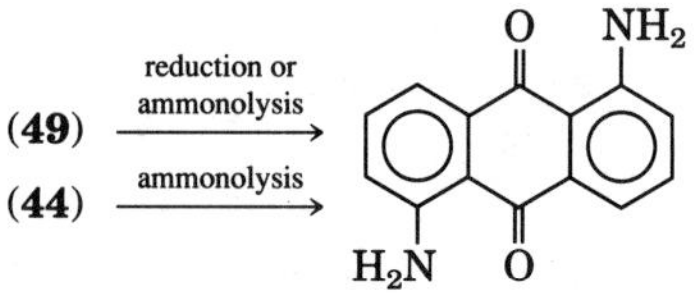

(49) $\xrightarrow[\text{ammonolysis}]{\text{reduction or}}$

(44) $\xrightarrow{\text{ammonolysis}}$

(51)

1,5-Diphenoxyanthraquinone [*82-21-3*] (**52**) is a precursor of 1,5-dihydroxy-4,8-dinitroanthraquinone, and is prepared from 1,5-dinitroanthraquinone and alkali metal phenoxide in phenol (80), or in an inert organic solvent (81). 1,5-Dimethoxyanthraquinone (**48**) is also a precursor for 1,5-dihydroxy-4,8-dinitroanthraquinone and is prepared from 1,5-dinitroanthraquinone with methanolic alkali metal hydroxide (82,83).

High purity of 1,5-dimethoxyanthraquinone is required for manufacturing disperse blue dyes (CI Disperse Blue 56 (**11**)). A small amount of unreacted 1,5-dinitroanthraquinone in 1,5-dimethoxyanthraquinone affects the brightness of the dye and makes it much duller. Improved processes have been reported (84,85).

1,5-Dihydroxy-4,8-dinitroanthraquinone [*128-91-6*] (**54**) is an important dye precursor for CI Disperse Blue 56, and is prepared from 1,5-diphenoxyanthraquinone by hexanitration in sulfuric acid and subsequent hydrolysis with aqueous alkali.

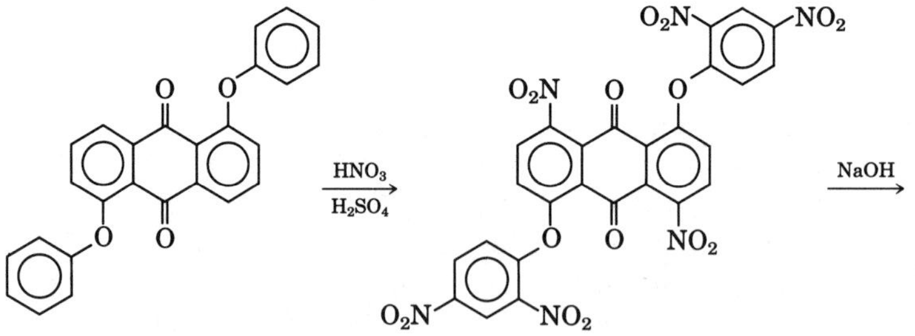

(52) $\xrightarrow[\text{H}_2\text{SO}_4]{\text{HNO}_3}$ (53) $\xrightarrow{\text{NaOH}}$

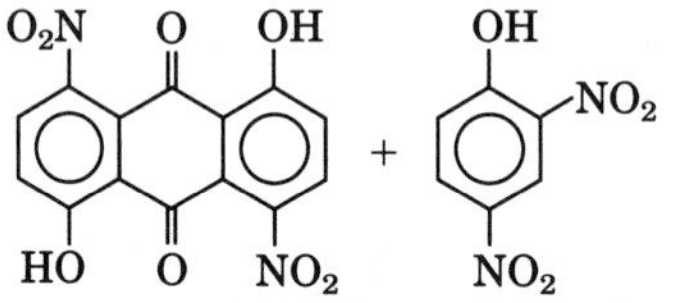

(54)

This compound can be converted to 1,5-diamino-4,8-dihydroxyanthraquinone by reduction of nitro groups with sodium sulfide.

1,5-Dinitro-4,8-dihydroxyanthraquinone is also prepared from 1,5-dimethoxyanthraquinone (**48**) as illustrated in the following.

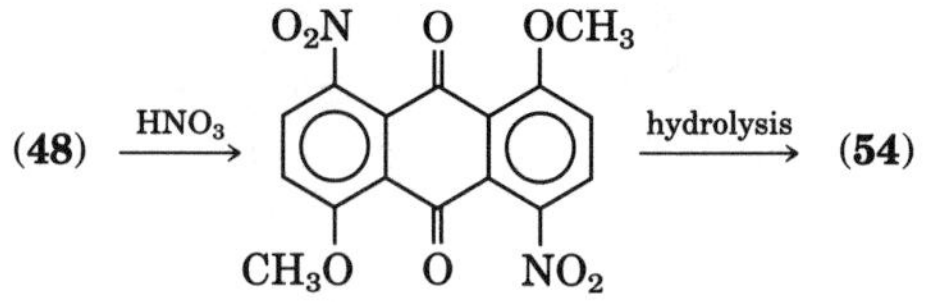

(48) $\xrightarrow{\text{HNO}_3}$ [structure] $\xrightarrow{\text{hydrolysis}}$ (54)

2-Methylanthraquinone and Related Compounds. 2-Methylanthraquinone and its derivatives are important as intermediates for manufacturing various kinds of vat dyes and brilliant blue (turqoise blue) disperse dyes. 2-Methylanthraquinone [84-54-8] (56) is prepared from phthalic anhydride and toluene via a benzoylbenzoic acid (55) (86).

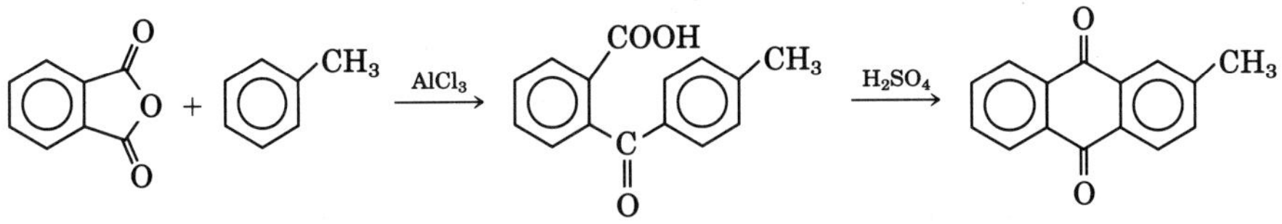

(55)   (56)

2-Methyl-1-nitroanthraquinone [129-15-7] (57) is an important precursor for 1-nitroanthraquinone-2-carboxylic acid (58) and is prepared by nitration of 2-methylanthraquinone (56) (87). This compound is probably carcinogenic (88).

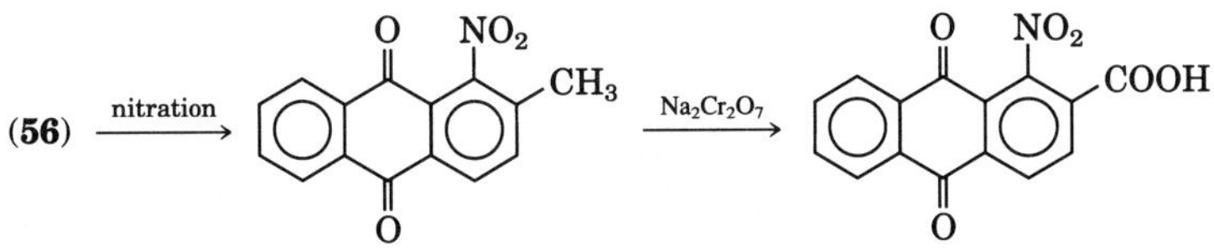

(57)   (58)

1-Nitroanthraquinone-2-carboxylic acid [128-67-6] is of great importance as an intermediate for manufacture of vat dyes as well as disperse dyes. Examples are CI Vat Yellow 20 [4216-01-7] (172), CI Vat Orange 13 [6417-38-5] (169), CI Vat Red 10 [2379-79-5] (166), CI Vat Red 21 [4430-70-0] (164), Red (59), CI Vat Blue 64 [15935-52-1] (12), and CI Disperse Blue 87 [12222-85-4] (107).

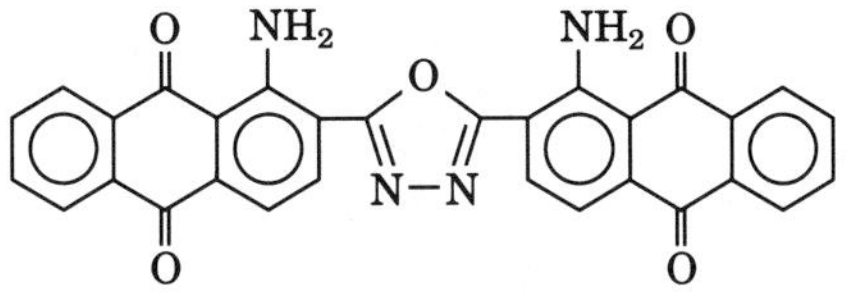

(59)

1-Nitroanthraquinone-2-carboxylic acid (58) is conventionally prepared from 2-methyl-1-nitroanthraquinone (57), by oxidation in sulfuric acid with sodium dichromate. In recent years this process has been faced with the problem of treatment of wastewater containing Cr^{6+}, which cannot be reclaimed, together with a large amount of waste acid. The following improved process has been proposed to avoid the use of Cr^{6+} (89).

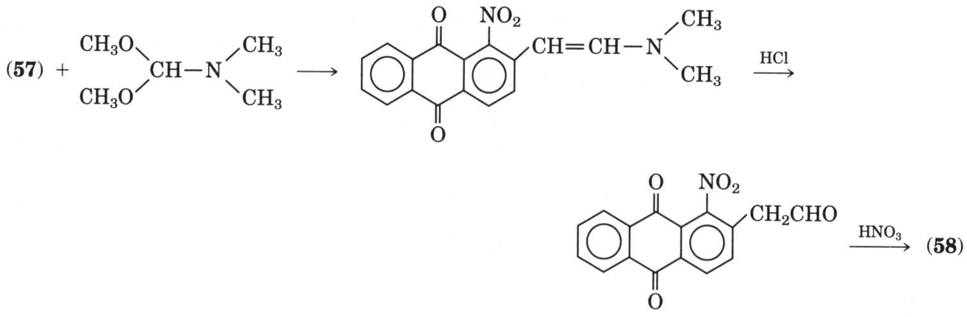

1-Aminoanthraquinone-2-carboxylic acid [*82-24-6*] (**60**) is also an important intermediate for vat dyes and disperse dyes and is prepared from 1-nitroanthraquinone-2-carboxylic acid by reaction with ammonia.

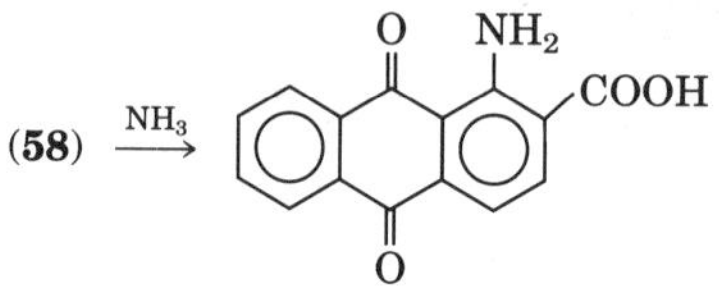

(**60**)

However, the preparation of 1-nitroanthraquinone-2-carboxylic acid has the difficulties mentioned previously. Therefore, new processes for preparing this compound not from the 1-nitro compound but from other precursors have been intensively studied. 1-Aminoanthraquinone derivatives have been proposed for this purpose (90).

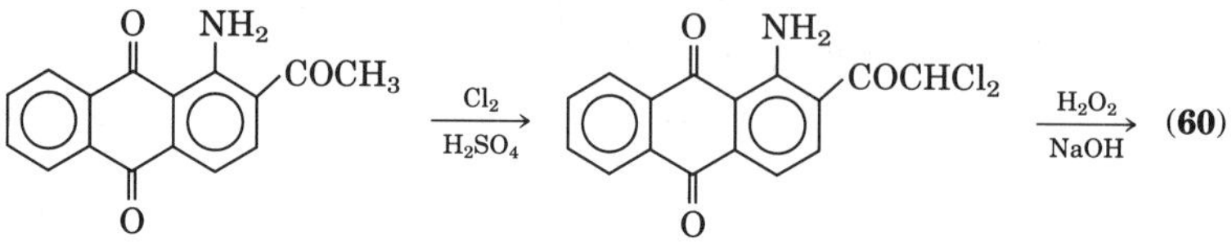

2-Chloroanthraquinone and Its Derivatives.   2-Chloroanthraquinone and its derivatives are the most important intermediates for vat dyes and high performance organic pigments. Examples are CI Vat Blue 4 [*81-77-6*] (**6**), CI Vat Blue 6 [*130-20-1*] (**147**), CI Vat Blue 11 [*130-19-8*] (**61**) (CI 69815), CI Vat Blue 12 [*1324-28-3*] (**63**) (CI 69840), CI Vat Blue 13 [*6871-71-2*] (**62**) (CI 69845), CI Vat Yellow 1 [*475-71-8*] (**177**), CI Vat Red 10 [*2379-79-5*] (**166**), CI Vat Orange 16 [*10142-57-1*] (**64**) (CI 69540), and CI Vat Blue 30 [*6492-78-0*] (**167**).

CI Vat Blues 11, 12, and 13 all have the 6,15-dihydroanthrazinetetrone structure (**6**) as follows. CI Vat Blue 11 (**61**) has Br at positions 7 and 16. CI Vat Blue 13 (**62**) has OH at C-8 and C-17. CI Vat Blue 12 (**63**) has one OH substituent in an undesignated position. In CI Vat Orange 16 (**64**), position 15 is a carbonyl rather than NH.

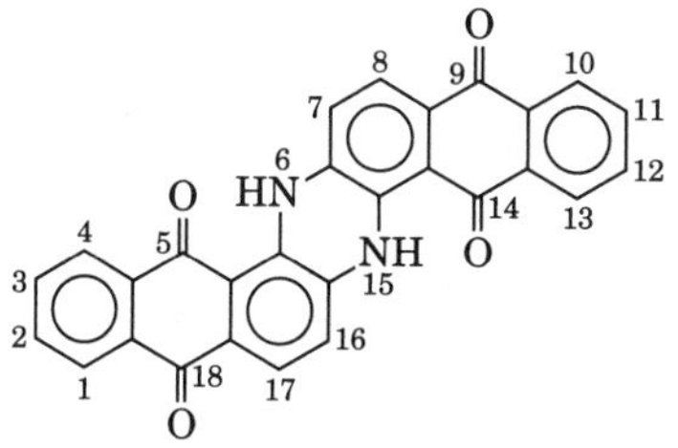

(**6**)

2-Chloroanthraquinone [*131-09-9*] (**65**) is prepared by Friedel-Crafts reaction of chlorobenzene and phthalic anhydride in the presence of aluminum chloride followed by ring closure in concentrated sulfuric acid (91).

2-Aminoanthraquinone [*117-79-3*] (**66**) is prepared by replacement of the chlorine atom in 2-chloroanthraquinone by ammonia (92). This compound has been found to have probable carcinogenicity (93).

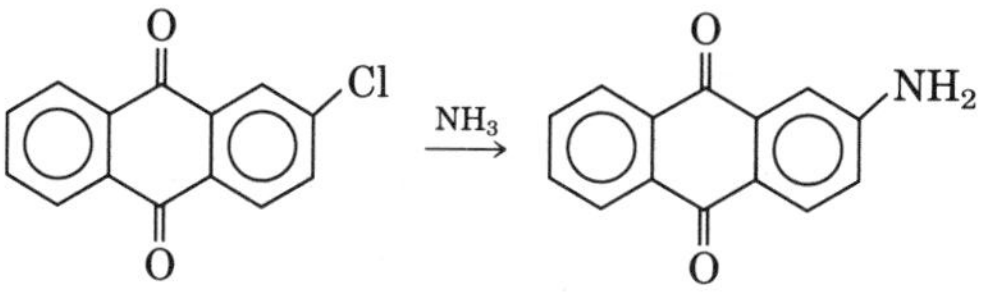

(**65**)   (**66**)

2-Amino-3-chloroanthraquinone [*84-46-8*] (**68**) is prepared from 2,3-dichloroanthraquinone by partial chlorine replacement by a NH_2 group. 2,3-Dichloroanthraquinone [*84-45-7*] (**67**) is prepared by Friedel-Crafts reaction of phthalic anhydride and 1,2-dichlorobenzene followed by ring closure of the resultant benzoylbenzoic acid in sulfuric acid (94).

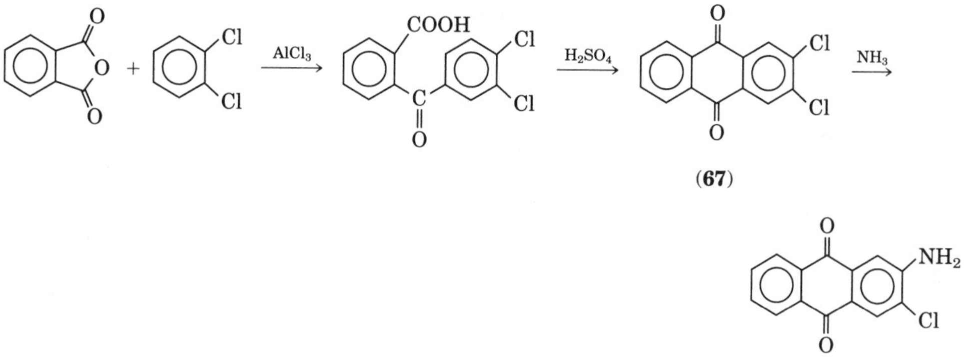

(**67**)

(**68**)

2-Amino-3-bromoanthraquinone [*6337-00-4*] (**69**) is prepared from 2-aminoanthraquinone by bromination in an organic solvent or in sulfuric acid (95).

2-Amino-3-hydroxyanthraquinone [*117-77-1*] (**70**) is prepared by heating 5-benzoylbenzoxazolone-2'-carboxylic acid in sulfuric acid (96). This compound is an intermediate for CI Vat Red 10 (**166**).

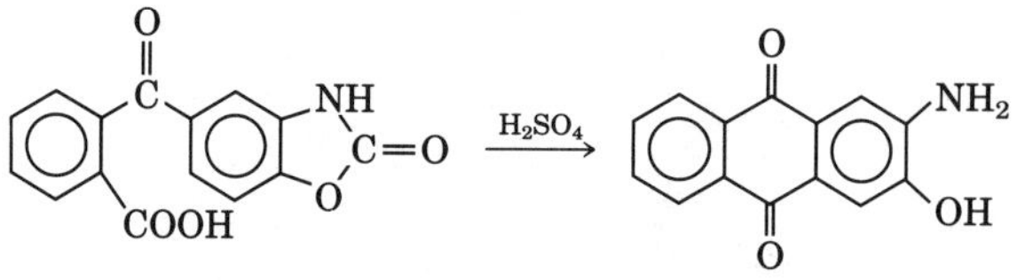

(**70**)

Benzanthrone and Related Compounds.   Benzanthrone [*82-05-3*] (**71**) is prepared by the reaction of anthraquinone (**15**) with glycerol, sulfuric acid, and a reducing agent such as iron.

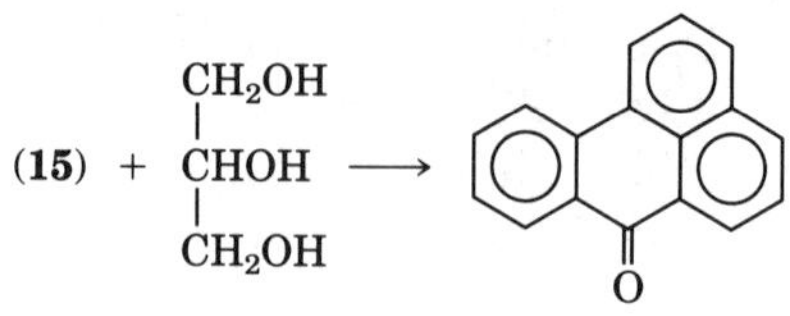

(**71**)

Benzanthrone is an important intermediate for manufacturing vat dyes. Examples are CI Vat Green 1 [*128-58-5*] (**138**) via 4,4′-dibenzanthronyl (**72**), CI Vat Blue 20 [*116-71-2*] (**135**) via direct KOH–NaOH fusion of benzanthrone, CI Vat Green 3 [*3271-76-9*] (**142**), CI Vat Green 5 [*1328-37-6*] (**143**), and CI Vat Black 25 [*4395-53-3*] (**144**) via 3,9-dibromobenzanthrone [*81-98-1*] (**74**).

4,4′-Dibenzanthronyl [*116-90-5*] (**72**) is a precursor of violanthrone dyes and is prepared from an alcoholic alkali melt of benzanthrone.

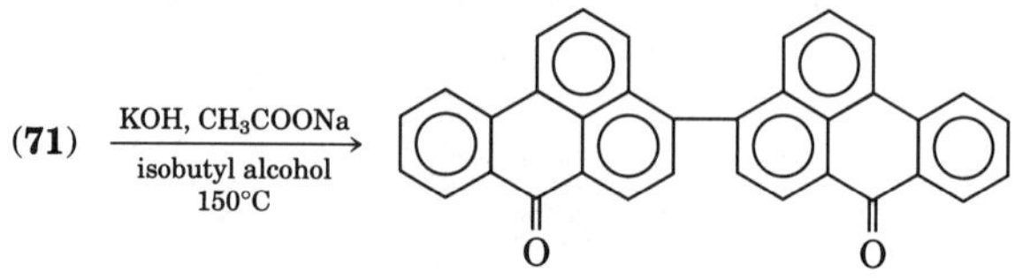

(**72**)

3-Bromobenzanthrone [*81-96-9*] (**73**) is prepared from benzanthrone by bromination in hydrochloric acid or in sulfuric acid or in an organic solvent such as nitrobenzene. An improved process has been cited (97). 3,9-Dibromobenzanthrone [*81-98-1*] (**74**) is prepared from benzanthrone by bromination in chlorosulfonic acid, concentrated sulfuric acid, or an organic solvent such as nitrobenzene.

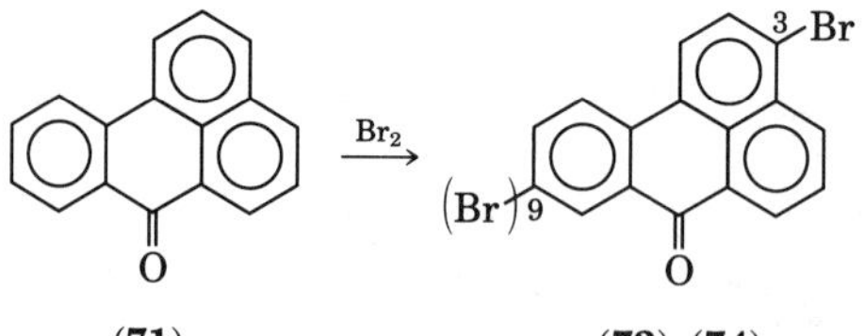

(**71**)   (**73**), (**74**)

N-Methylanthrapyridone and Its Derivatives.   6-Bromo-3-methylanthra-pyridone [81-85-6] (**75**) is an important intermediate for manufacturing dyes soluble in organic solvents. These solvent dyes are prepared by replacing the bromine atom with various kinds of aromatic amines. 6-Bromo-3-methylanthrapyridone is prepared from 1-methylamino-4-bromoanthraquinone (**43**) by acetylation with acetic anhydride followed by ring closure in alkali. The starting material of this route is anthraquinone-1-sulfonic acid (**16**).

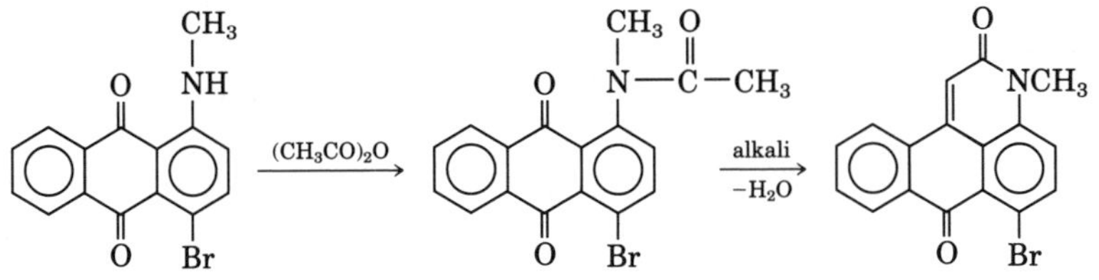

An alternative route from 1-aminoanthraquinone (**17**) has been proposed. Methylation is preferably carried out using dimethyl sulfate or methyl iodide in an organic solvent in the presence of alkali metal hydroxide and a catalytic amount of quaternary ammonium compound (98).

Reactive Dyes

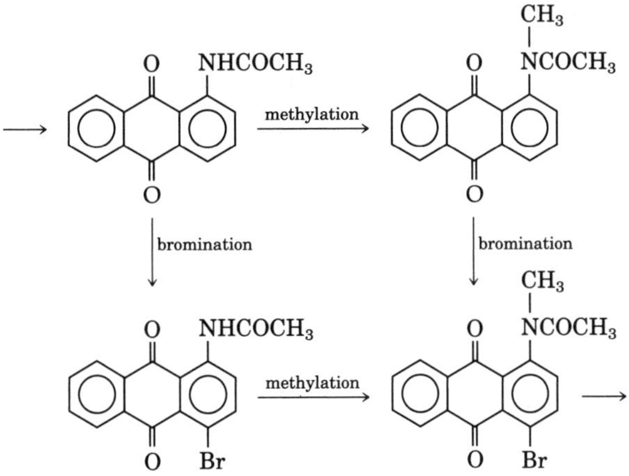

Most of the anthraquinone reactive dyes are derived from bromamine acid. These dyes give a bright blue shade and excellent lightfastness. A great number of reactive groups have been proposed; typical examples include sulfatoethylsulfone, dichlorotriazine, monochlorotriazine, monofluorotriazine, and other heterocyclic groups (see DYES, REACTIVE).

Reactive Sulfatoethylsulfonyl Groups.   CI Reactive Blue 19 [2580-78-1] (**9**) (CI 61200) is most widely used for dyeing cellulose fibers in exhaustion dyeing, and it gives a brilliant reddish blue shade and excellent lightfastness.

CI Reactive Blue 19 (**9**) is prepared by the reaction of bromamine acid (**8**) with *m*-aminophenyl-*β*-hydroxyethylsulfone [*5246-57-1*] (**76**) in water in the presence of an acid-binding agent such as sodium bicarbonate and a copper catalyst (Ullmann condensation reaction) and subsequent esterification to form the sulfuric ester.

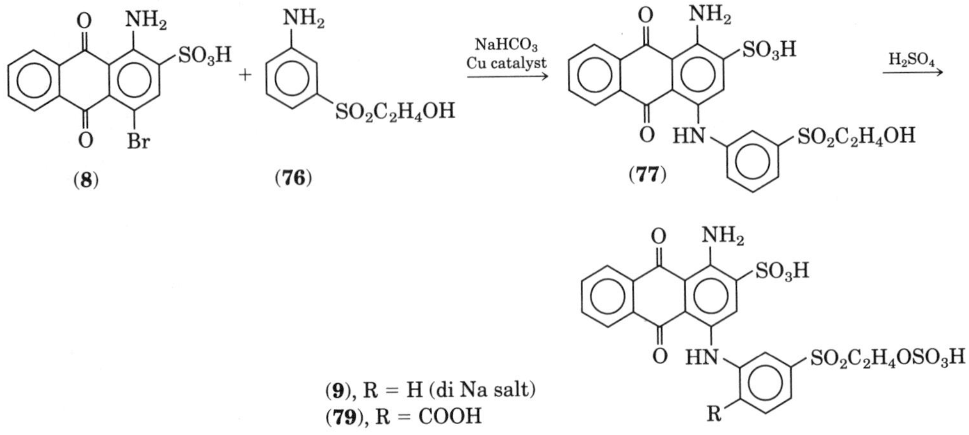

(**8**)   (**76**)   (**77**)

(**9**), R = H (di Na salt)
(**79**), R = COOH

The conventional esterification process requires a large excess of sulfuric acid or oleum; improved processes which minimize the acid wastewater or inorganic salt have been proposed (99–101). One example is esterification of 1-amino-4-[3-(*β*-hydroxyethylsulfonyl)phenylamino] anthraquinone-2-sulfonic acid [*39582-26-8*] (**77**) in an organic solvent containing tertiary amines with a stoichiometric amount of sulfuric acid or sulfamic acid (99). A method in a machine operating with a kneading action using a small excess amount of sulfuric acid or oleum has also been proposed (100). In order to improve the solubility in water or aqueous alkaline solution, a dye composition of CI Reactive Blue 19 (**9**) with its para-isomer, ie, 1-amino-4-[4-(*β*-sulfatoethylsulfonyl)phenylamino]-anthraquinone-2-sulfonic acid, disodium salt [*16102-99-1*] (**78**) has been proposed (102). Another example of a dye with a sulfatoethylsulfonyl group is CI Reactive Blue 27 [*20640-71-5*] (**79**) (103).

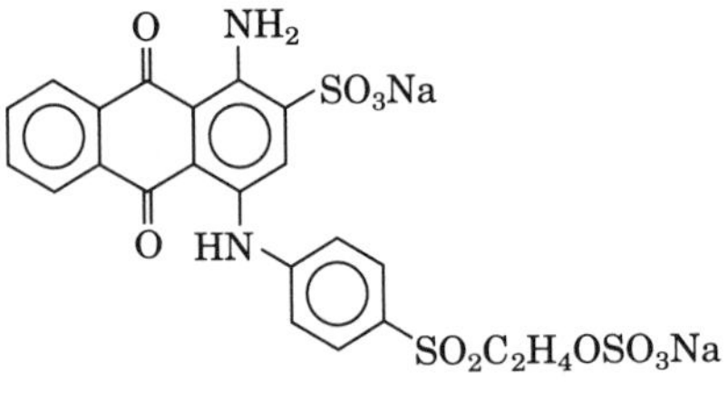

(**78**)

Monochloro- or Dichlorotriazine Groups.   Examples of commercial importance are CI Reactive Blue 2 [*12236-82-7*] (**80**) (CI 61211), CI Reactive Blue 5 [*16823-51-1*] (**81**) (CI 61205:1), reddish brilliant blue [*72927-99-2*] (**82**) and CI Reactive Blue 4 [*13324-20-4*] (**83**) (CI 61205). As was the case for (**9**), (**78**), and

(**79**), the substituent having the reactive group replaces the Br of (**8**) and the dyes can be represented as follows. The R groups for dyes (**80–83**) appear in Figure 3.

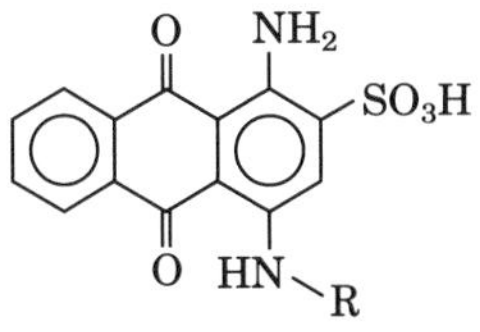

(**80–86**)

The use of dyes having dichlorotriazine groups is rather limited because the reactivity of this group is so high that the stability of the dyed fiber as well as that of dyestuff itself is not satisfactory.

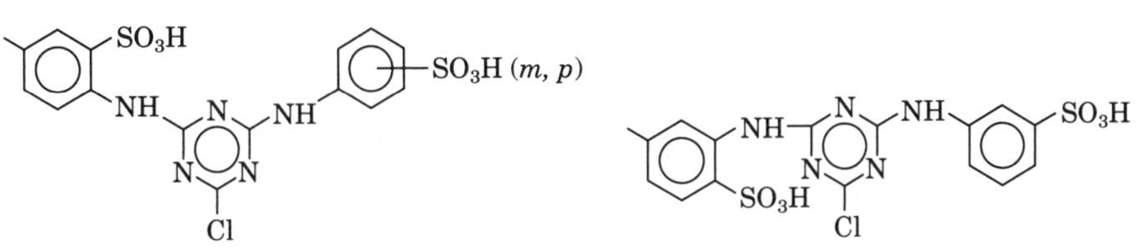

(**80**)

(**81**)

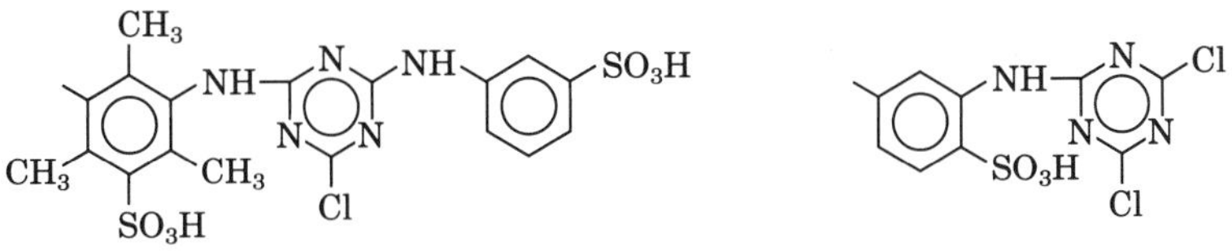

(**82**)

(**83**)

Fig. 3. R groups having reactive mono- and dichlorotriazine groups attached to (**8**) through NH.

Dyes with Other Heterocyclic Reactive Groups. Some heterocyclic reactive components have been developed; eg, R in Figure 3. Other examples of the dyes of commercial importance are brilliant blue [88318-06-3] (**84**) (monofluorotriazine), blue [83399-87-5] [104601-66-3], (**85**) (dichloroquinoxaline), and brilliant blue [64387-69-5] (**86**) (difluorochloropyrimidine). The R components are shown in Figure 4 for (**84–86**).

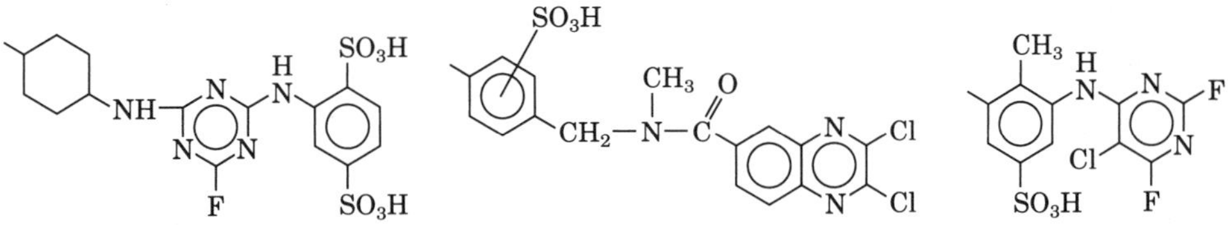

(**84**)

(**85**)

(**86**)

Fig. 4. R groups having reactive heterocyclic groups attached to (**8**) through NH.

Reactive green dyes are obtained by combination of a blue chromophore (a bromamine acid derivative) and a yellow chromophore with a triazinyl group. Green [70210-47-8] (**87**) (104) is an example. The yellow chromophore of this dye was invented by ICI for dichlorotriazine dyes and exhibits good lightfastness and chlorine resistance.

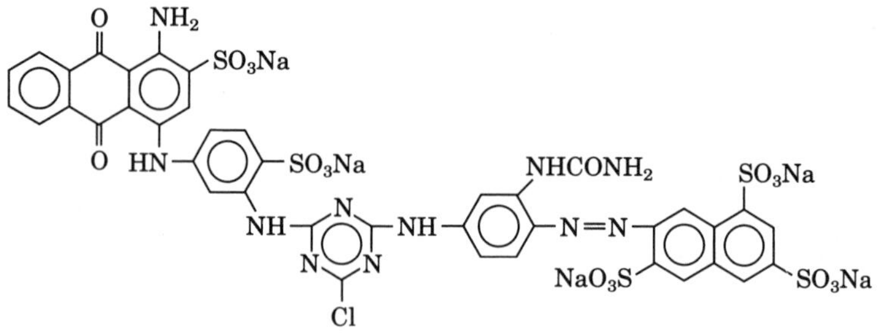

(**87**)

In recent years attempts to replace anthraquinone blue reactive dyes by derivatives of triphenodioxazine chromophores have been successful to a certain extent (105). The triphenodioxazine chromophore has an intrinsically brighter shade and much greater tinctorial strength than anthraquinones. Examples are CI Reactive Blue 198 (**88**) (105) and CI Reactive Blue 204 (**89**) (105) in which R is a substituted alkylamino group having a reactive group. In (**88**) the reactive group is chlorotriazine, in (**89**) it is fluorotriazine.

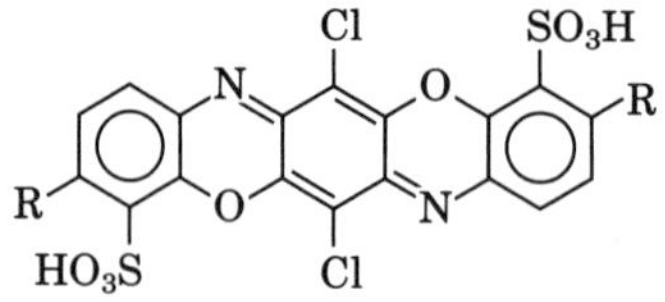

(**88**), (**89**)

Disperse Dyes

Disperse dyes are water-insoluble, aqueous dispersed materials that are used for dyeing hydrophobic synthetic fibers, including polyester, acetate, and polyamide.

In 1923, the first disperse dye was developed for dyeing cellulose acetate fibers. However, in recent years the most important application of disperse dyes has been to dye polyester fibers. Accompanied by the rapid growth of polyester fibers after World War II, disperse dyes have currently achieved the largest production among all dye classes in terms of quantity (106).

By introducing amino, hydroxy, or methyl groups onto the anthraquinone moiety as the principal auxochromes, dyes that have yellow through greenish blue shades are obtained. Among these dyes many that have brilliant red, violet, blue, and greenish blue shades have great industrial importance in view of their affinity for polyester or cellulose acetate fibers and lightfastness and sublimation resistance. On the contrary, yellow or orange dyes are not satisfactory because of the rather simple molecular structure. Therefore these shades are obtained from other chromophores.

On the basis of the kind and the position of their substituents and their color range, the anthraquinoid disperse dyes may be classified as follows:

Color range	Chemical description
red	1-amino-4-hydroxyanthraquinones
blue, greenish blue	1,4,5,8-substituted anthraquinones
greenish blue	1,4-diaminoanthraquinone-2,3-dicarboxyimides
violet, blue	1,4-diaminoanthraquinone derivatives
violet, blue	N-substituted 1-amino-4-hydroxyanthraquinones

1-Amino-4-hydroxyanthraquinone Derivatives.   These compounds in general have bright red shades, good lightfastness, and good affinity for polyester fibers. CI Disperse Red 60 [*17418-58-5*] (**90**) (CI 60756) is the most typical red dye for polyester fibers. It is widely used for mainly pale and medium shades and for exhaustion dyeing, and exhibits good lightfastness and affinity. It is manufactured by reaction of 1-amino-2-halo(Cl,Br)-4-hydroxyanthraquinone (**22,23**) with potassium phenoxide in phenol as the solvent (Fig. 5). Improved methods for reducing the amount of phenol by employing an inert organic solvent such as dimethyl sulfoxide (107), sulfolane (108), or water in the presence of a phase-transfer catalyst (109) have been reported.

1-Amino-2-alkoxy-4-hydroxyanthraquinones have generally brighter and yellower shades than the 2-phenoxy type, and have better lightfastness. Examples are (**91–96**) in Table 4. These dyes have good sublimation fastness as well.

1-Amino-2-alkoxy-4-hydroxyanthraquinones are prepared by reaction of 1-amino-2-phenoxy-4-hydroxyanthraquinone with the corresponding alcohols

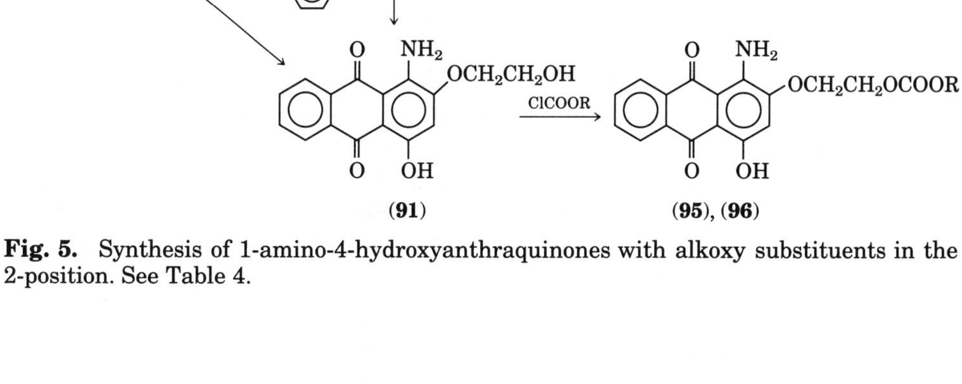

(23) X = Cl
(22) X = Br

(90)

(91)

(95), (96)

Fig. 5.   Synthesis of 1-amino-4-hydroxyanthraquinones with alkoxy substituents in the 2-position. See Table 4.

Table 4. Disperse Red Dyes, 2-Substituted 1-Amino-4-hydroxyanthraquinones

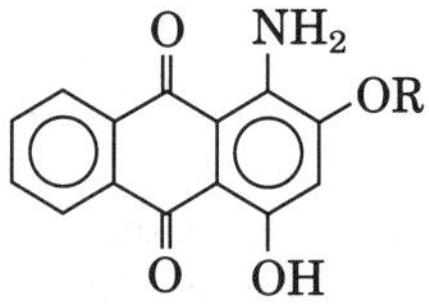

R	Structure number	CI Name	CAS Registry Number
	Alkoxy		
—C$_2$H$_4$OH	(91)	Disperse Red 55[a]	[17869-07-7]
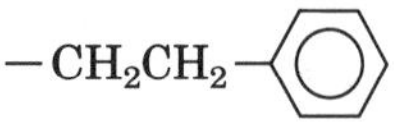 —CH$_2$CH$_2$—⬡	(92)		[23753-49-3]
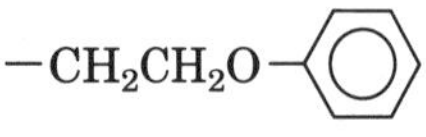 —CH$_2$CH$_2$O—⬡	(93)		[17418-59-6]
—C$_6$H$_{12}$OH	(94)[b]	Disperse Red 91	[12236-10-1]
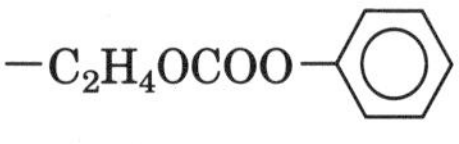 —C$_2$H$_4$OCOO—⬡	(95)		[28173-59-3]
—C$_2$H$_4$OCOOC$_2$H$_5$	(96)		[40530-60-7]
	Aryloxy		
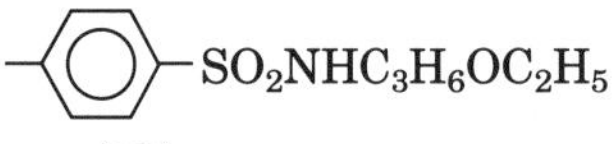 ⬡—SO$_2$NHC$_3$H$_6$OC$_2$H$_5$	(97)[b]	Disperse Red 92	[12236-11-2]
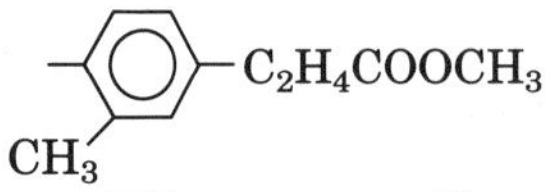 ⬡—C$_2$H$_4$COOCH$_3$, CH$_3$	(98)[b]	Disperse Red 127	[16472-09-6]
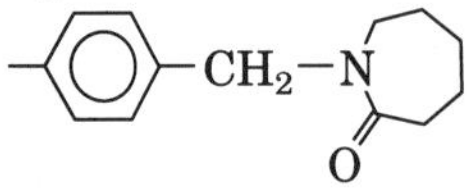 ⬡—CH$_2$—N(ring)O	(99)		[19014-53-0]
 ⬡—OCH$_2$CH$_2$—⬡	(100)		[55154-34-2]
 ⬡—OCH$_2$CH$_2$O—⬡	(101)		[55154-36-4]

[a] CI number 60757.
[b] Ref. 110.

(Fig. 5). Improved methods, using 1-amino-2-halo-4-hydroxyanthraquinone as a starting material, have been reported. Thus, 1-amino-2-chloro-4-hydroxyanthraquinone (23) reacts with the alcohol in an inert organic solvent such as chlorobenzene in the presence of an acid binding agent, phenol, and a phase-transfer catalyst (111). For the dyes (95) and (96) 1-amino-2-(2-hydroxyethoxy)-4-hydroxyanthraquinone is first prepared by the method just described, and then reacts with phenyl or ethyl carbonate (112).

1-Amino-2-(substituted phenoxy)-4-hydroxyanthraquinones, that is (**90**) with substituents on the phenyl group, have also been developed to improve sublimation fastness. Examples are (**97–101**) in Table 4.

1,4-Diaminoanthraquinone-2,3-dicarboxyimide Derivatives. These dyes have a bright turquoise blue shade and excellent lightfastness and good sublimation fastness. Commercially important examples are CI Disperse Blue 60 [*12217-80-0*] (**40**) (CI 61104) (113), and CI Disperse Blue 87 [*12222-85-4*] (**107**) (114,115).

CI Disperse Blue 60 is prepared by alkylating 1,4-diaminoanthraquinone-2,3-dicarboxyimide [*128-81-4*] (**39**) with methoxypropylamine in water (116–118) with or without organic solvent.

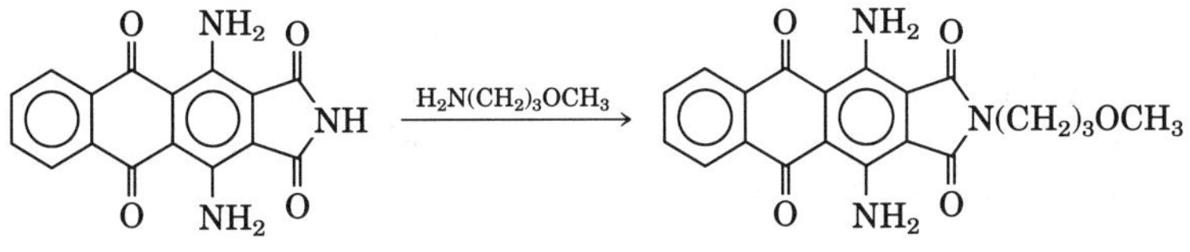

(**39**)   (**40**)

CI Disperse Blue 60 is often used as a mixture with slightly different derivatives (116), that is, the methyl group of (**40**) may be replaced by ethyl (**102**), propyl (**103**), or $CH_2CH_2OCH_3$ (**104**).

CI Disperse Blue 87 (**107**) and related dyestuffs (**105,106**) are illustrated as follows:

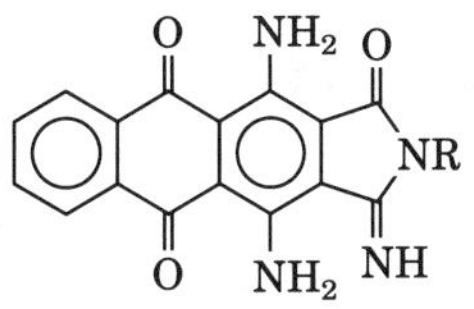

(**105**) R = —$C_2H_4OC_2H_5$
(**106**) R = —$C_2H_4OC_4H_9$
(**107**) R = —$C_3H_6OCH_3$

These dyes are prepared by the reaction of 1-amino-4-nitroanthraquinone-2-carboxylic acid amide (**108**) with cyanide in water (119).

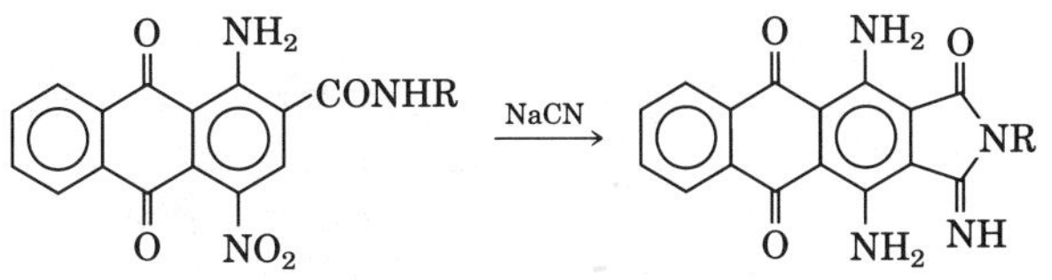

(**108**)   (**109**)

1-Amino-4-nitroanthraquinone-2-carboxylic acid amide (**108**) is prepared from 1-nitroanthraquinone-2-carboxylic acid (**58**). An improved process has been proposed (120).

1,4-Diaminoanthraquinone-2-carboxylic acid alkylamide (**110**) is also used as a starting material. In this case, a dehydrogenating agent such as air is necessary (121).

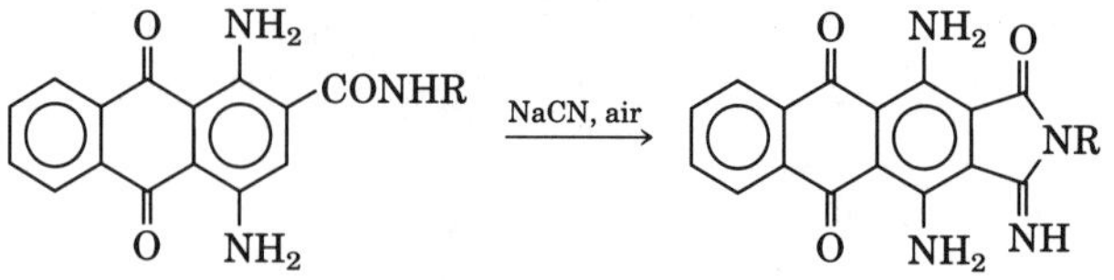

CI Disperse Blue 87 (**107**) and related dyestuffs are also prepared from 1-oxo-3-imino-4,7-diamino-5,6-phthaloylisoindoline [*13418-50-3*] (**111**) by alkylation with corresponding alkyl halides (122), sulfonic esters (123), or alkyl amines (124), ie, X of RX = halogen, *p*-toluenesulfonyloxy, or NH_2.

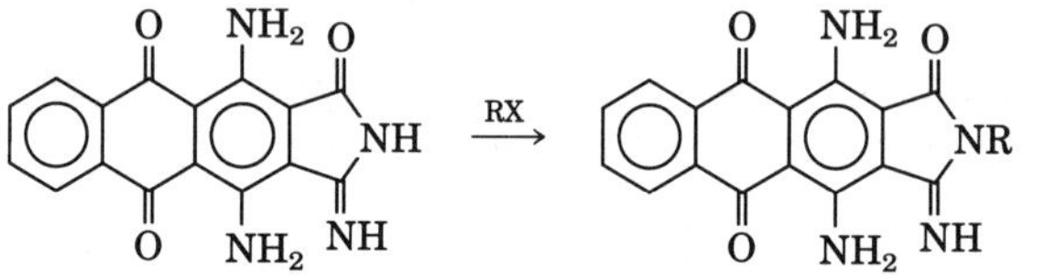

1-Oxo-3-imino-4,7-diamino-5,6-phthaloylisoindoline is prepared by hydrolysis of 1,4-diamino-2,3-dicyanoanthraquinone (**37**) in alcoholic alkaline conditions (125).

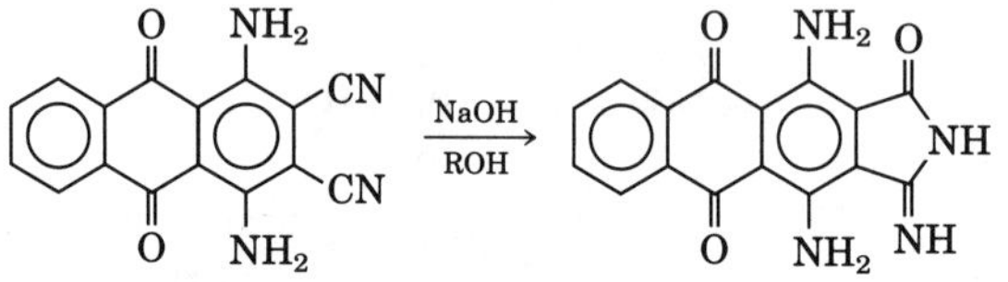

1,4,5,8-Substituted Anthraquinones. Commercially important blue disperse dyes are derived from 1,4,5,8-substituted anthraquinones. Among them, diaminodihydroxyanthraquinone derivatives are most important in view of their shades and affinity. Representative examples are CI Disperse Blue 56 [*31810-89-6*] (**11**) (CI 63285) (126), and CI Disperse Blue 73 (**113**) (115). Introduction of a halogen atom ortho to the amino group improves affinity and lightfastness.

CI Disperse Blue 56 is the most important blue dye for polyester fibers because it has a brilliant shade, excellent lightfastness, and good leveling properties.

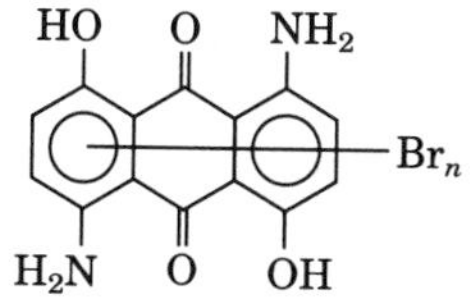

(112)

Mixtures of material (**112**) with $n = 0$, 1, and 2 show greater affinity than that of one pure component. CI Disperse Blue 56 (**11**) or (**112**) $n = 1$, is manufactured from (**54**):

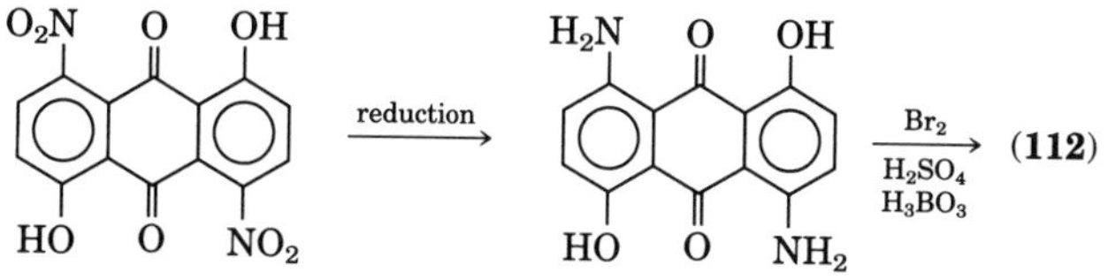

(54)

Conventional 1,5-dichloroanthraquinone has also been used as a starting material. A route via 1,5-diaminoanthraquinone has also been proposed (127); treatment of this compound with Br_2 in H_2SO_4 and H_3BO_3 gives (**11**).

CI Disperse Blue 73 [*12222-75-2*] (CI 63265) is an example of a dye that was developed to improve sublimation fastness for special use, eg, thermosol dyeing or printing. This dye also has a bright shade, excellent lightfastness and good leveling properties. CI Disperse Blue 73 (**113**) is prepared as follows, where R = H or CH_3.

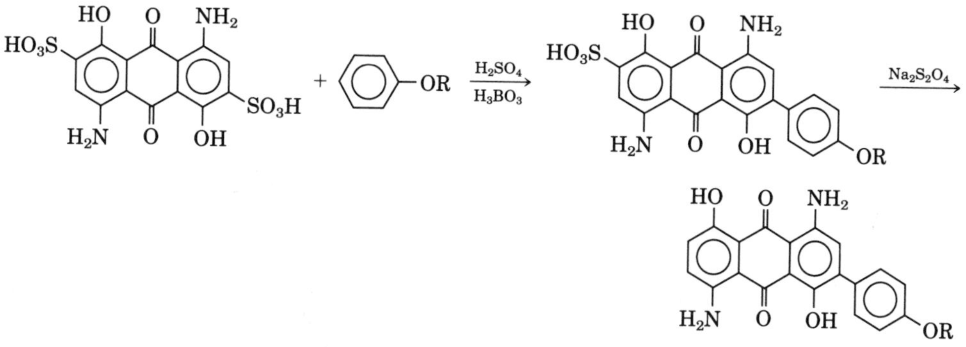

(113)

The starting material is an acid dye, ie, CI Acid Blue 45 which is prepared from 1,5-dihydroxyanthraquinone by sulfonation followed by nitration and then reduction.

CI Disperse Blue 81 [*12222-79-6*] (**114**) (CI 63603) is a 1,8-diamino-4,5-dihydroxyanthraquinone derivative. Its shade is greenish blue and much duller than CI Disperse Blue 56. Bromo-derivatives of pure 1,8-diamino-4,5-dihydroxyanthraquinone have a bright greenish blue shade, but CI Disperse Blue 81 generally contains other isomers derived mainly from the starting material 1,8-dinitroanthraquinone, which affect the shade significantly. The manufacturing process of CI Disperse Blue 81 is as follows.

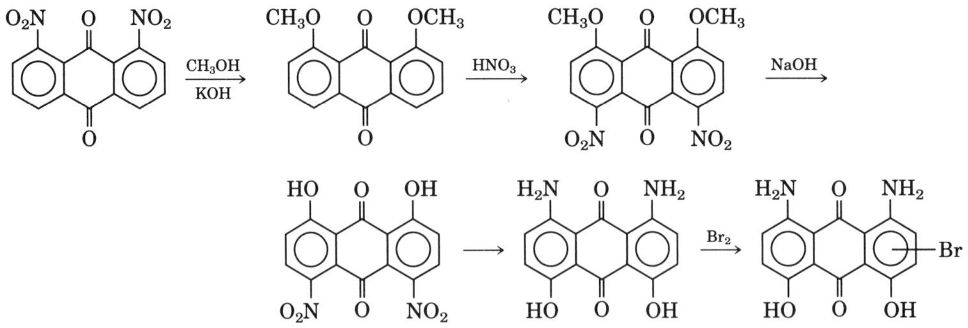

(114)

1-Arylamino-8-nitro-4,5-dihydroxyanthraquinones are of importance because of their greenish blue shade and good fastness to light and sublimation. A representative example is 1-anilino-4,5-dihydroxy-8-nitroanthraquinone [20241-76-3] (115) (128). It is prepared by the reaction of 1,8-dinitro-4,5-dihydroxyan-thraquinone with aniline.

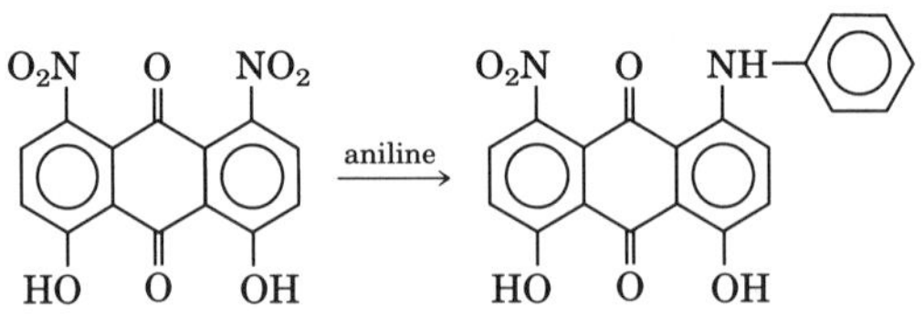

(115)

1,4-Diaminoanthraquinones (Except Turquoise Blue Dyes). This dye class generally gives reddish violet, violet, and reddish blue shades. Dyes with such neutral tints are relatively less important today, since dyeing of polyester fibers with the blend of three primary colors has become popular. However, some dyes that have excellent affinity, lightfastness, and sublimation resistance remain commercially important. Examples are CI Disperse Violet 28 [81-42-5] (CI 61102) (35), and CI Disperse Violet 26 [6408-72-6] (CI 62025) (116). CI Disperse Violet 28 is obtained by introducing chlorine atoms in the 2,3-position of 1,4-diaminoan-thraquinone. The introduction of chlorine leads to a bathrochomic shift and improves lightfastness considerably.

CI Disperse Violet 26 is prepared by the reaction of 1,4-diamino-2,3-dichloroanthraquinone (CI Disperse Violet 28 (35)) with potassium phenoxide in phenol as a solvent at high temperature. Introduction of phenoxy groups into the 2,3-position shifts the shade to bright, reddish violet and improves the lightfastness and sublimation resistance.

(35)   (116)

N-Substituted-1-Amino-4-hydroxyanthraquinones. These dyes show good affinity and lightfastness and give violet to blue shades. However, the sublimation fastness is in general not satisfactory. An example is CI Disperse Blue 72 [81-48-1] (117) (CI 60725), prepared from leucoquinizarin and *p*-toluidine.

(117)

Acid Dyes

Acid dyes are used for dyeing wool, synthetic polyamides, and silk in aqueous media. Anthraquinone acid dyes give brilliant reds, violets, blues, and greens and exhibit excellent lightfastness. Because of their relatively high cost, they are used to dye high grade textiles in pale and moderate shades. Various kinds of anthraquinone acid dyes have been developed so far mainly by IG-Farbenindustrie in Germany applying chemical reactions that were studied in developing vat dyes. However, the number of commercial products has declined because of poor properties or unavailable raw materials. Anthraquinone acid dyes may be classified into two groups: bromamine acid derivatives and quinizarin derivatives.

Bromamine Acid Derivatives. Acid dyes derived from bromamine acid (8) are important because they give bright blue shades with excellent lightfastness that are not obtainable with azo dyes. Among the bromamine acid derivatives CI Acid Blue 25 [2786-71-2] (118) (CI 62055) (129) and CI Acid Blue 40 [6424-85-7] (119) (CI 62125) (129) are the first acid dyes, invented in 1913. These dyes are obtained from bromamine acid by reaction with aniline and *p*-aminoacetoanilide respectively. They show good leveling properties in acidic media. However, the wetfastness is not so good.

(118), R = H
(119), R = NHCOCH₃

Dyes with better wetfastness and better affinity in neutral or weakly acid bath have been developed by introducing more hydrophobic amines. Examples are CI Acid Blue 129 [6397-02-0] (**120**) (CI 62058), CI Acid Blue 126 [72152-61-5] (**121**) (130), and CI Acid Blue 230 [12269-82-8] (**122**) (CI 62073). However, uniform leveling cannot be obtained with these dyes.

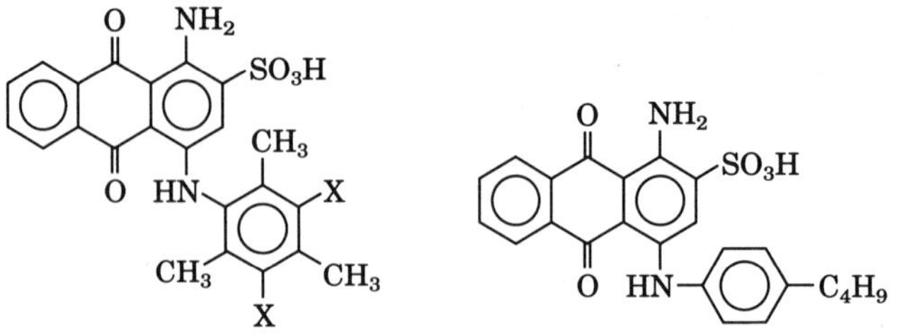

(**120**), X = H
(**121**), X = Br

(**122**)

Linking of two dye molecules is another method for improving wetfastness. An example is CI Acid Blue 127:1 [12237-86-4] (**123**) (130), which is obtained by condensing two molecules of CI Acid Blue 25 (**118**) with formaldehyde. CI Acid Blue 127 is also exemplified by [6471-01-8] (**124**) (CI 61135).

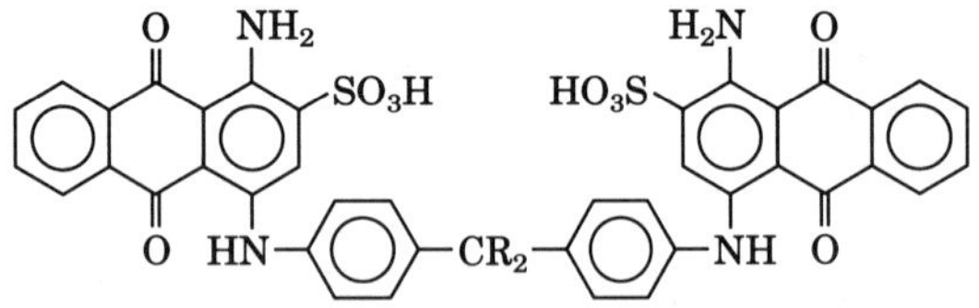

(**123**), R = H
(**124**), R = CH$_3$

The shade may be varied by choosing amines. For aromatic amines, the steric effect of substituents in the ortho position reduces the conjugation of the anilino group with the anthraquinone moiety, and the result is a hypsochromic shift and brighter shade. Thus CI Acid Blue 129 (**120**) has a more reddish and brighter shade than CI Acid Blue 25 (**118**). Cycloalkylamines have a similar effect on the shade. CI Acid Blue 62 [5617-28-7] (**125**) (CI 62045) is an example.

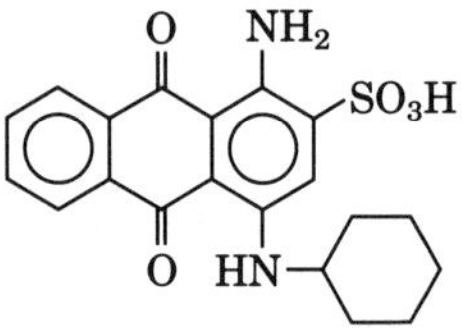

(**125**)

CI Acid Blue 40 (**119**) has a greener and somewhat duller shade than the parent dye (ie, CI Acid Blue 25) (**118**), which is considered to be due to the electronic effect of the *para*-acetylamino group.

In recent years excellent lightfastness and leveling properties have been required for application to nylon carpet. The dye [*66736-54-7*] (**126**) is an example invented for this application (131).

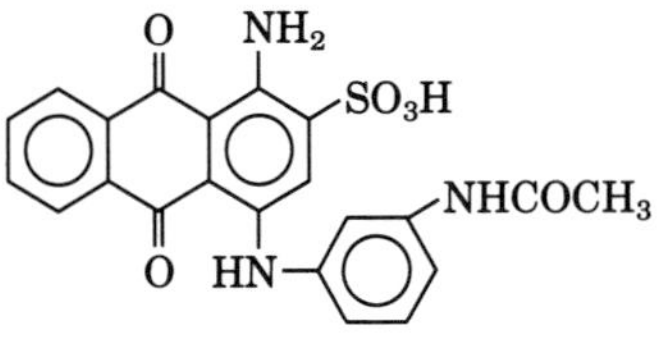

(**126**)

Quinizarin Derivatives.   Acid dyes derived from the reaction products of quinizarin with aromatic or aliphatic amines are important commercially and predominant in number. They are prepared from leucoquinizarin (**127**), the reduced form of quinizarin (**29**). The dominant structure of leucoquinizarin is the 2,3-dihydrotautomer (**128**) (132).

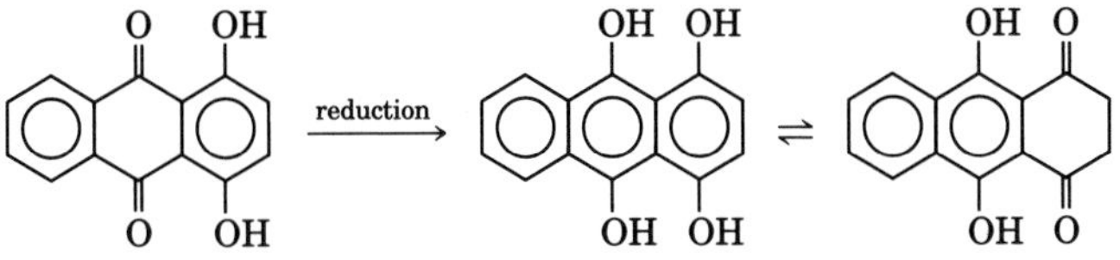

(**29**)   (**127**)   (**128**)

Alkylamines react with leucoquinizarin in a stepwise manner to give 1-alkylamino-4-hydroxyanthraquinone, and 1,4-dialkylamino derivatives after air oxidation. Aromatic amines react similarly in the presence of boric acid as a catalyst. The complex formed (**129**) causes the less nucleophilic aromatic amines to attack at the 1-, and 4-positions.

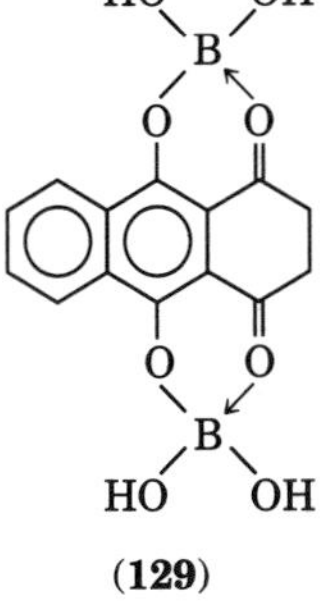

(**129**)

CI Acid Violet 43 [*4430-18-6*] (**2**) (CI 60730) is one of the first acid dyes, invented in 1894. This dye exhibits good leveling and is available from leucoquinizarin by reaction with *p*-toluidine in the presence of boric acid, followed by oxidation and subsequent sulfonation (133).

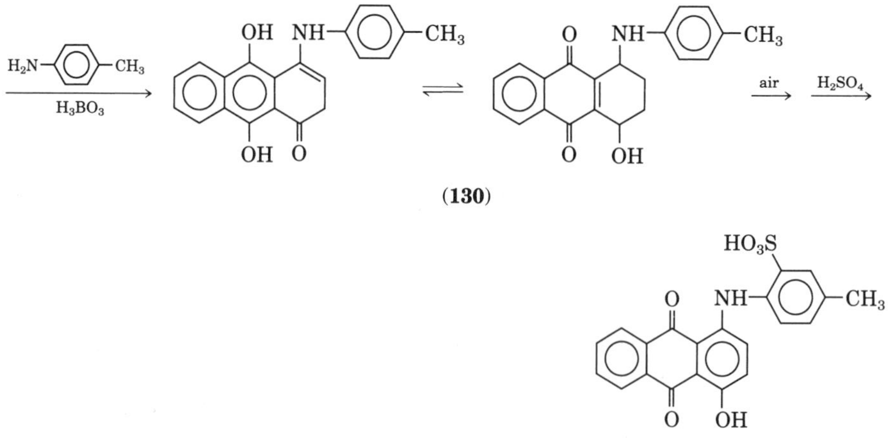

(**130**)

(**2**)

CI Acid Green 25 [*4403-90-1*] (**3**) (CI 61570) was also invented in 1894. This dye shows improved wetfastness, and is prepared from leucoquinizarin by reaction with 2 moles of *p*-toluidine in a similar manner to the preparation of CI Acid Violet 43 (134). Wetfastness and leveling properties may be altered by choosing the substituents of arylamines. The introduction of alkyl groups into aromatic amines improves the wetfastness and affinity in neutral or weekly acid baths. Examples are CI Acid Blue 80 [*4474-24-27*] (**131**) (CI 61585) and CI Acid Green 27 [*6408-57-7*] (**132**) (CI 61580).

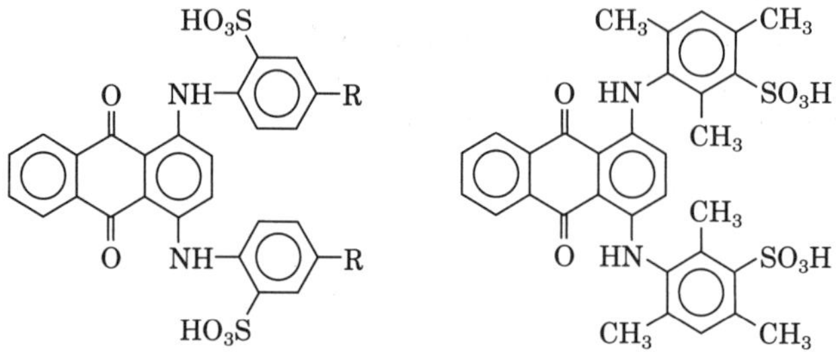

(**3**), R = CH₃
(**132**), R = C₄H₉

(**131**)

1,4-Diamino-2,3-disubstituted anthraquinones are also used as acid dyes. Examples are CI Acid Violet 41 [*6408-71-5*] (**133**) (CI 62020), and CI Acid Violet 42 [*6408-73-7*] (**134**) (CI 62026). They are prepared from quinizarin (**29**) through the intermediacy of leucamine (**32**) and 1,4-diamino-2,3-dichloroanthraquinone (**35**) (133).

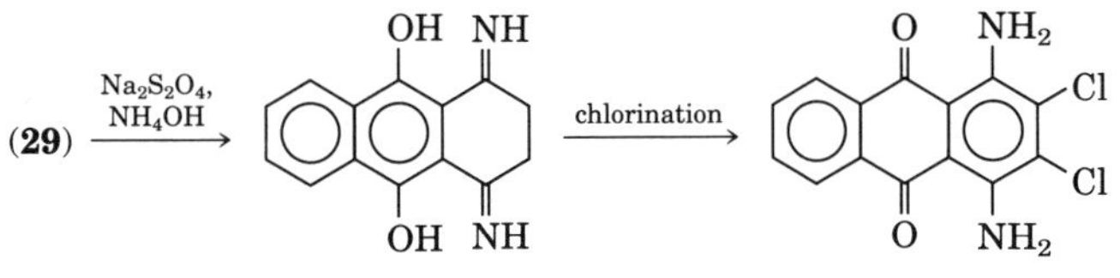

(29) $\xrightarrow[\text{NH}_4\text{OH}]{\text{Na}_2\text{S}_2\text{O}_4,}$ (32) $\xrightarrow{\text{chlorination}}$ (35)

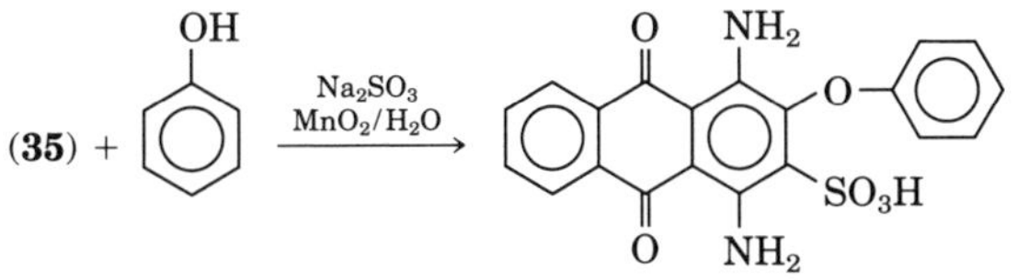

(35) + [phenol] $\xrightarrow[\text{MnO}_2/\text{H}_2\text{O}]{\text{Na}_2\text{SO}_3}$ (133)

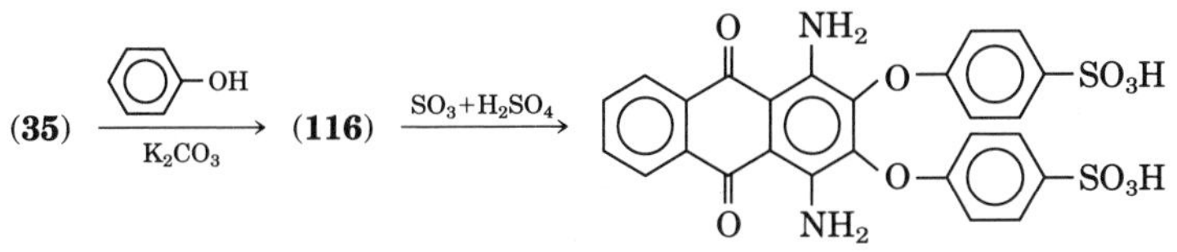

(35) $\xrightarrow[\text{K}_2\text{CO}_3]{}$ (116) $\xrightarrow{\text{SO}_3 + \text{H}_2\text{SO}_4}$ (134)

Vat Dyes

Anthraquinone vat dyes have been used to dye cotton and other cellulose fibers for many decades. Despite their high cost, relatively muted colors, and difficulty in application, anthraquinone vat dyes still form one of the most important dye classes of synthetic dyes because of their all-around superior fastness.

Anthraquinone vat dyes are water-insoluble dyes. They are converted to leuco compounds (anthrahydroquinones) by reducing agents such as sodium hydrosulfite in alkaline conditions. These water-soluble leuco compounds have an affinity to cellulose fibers and penetrate them. After reoxidation by means of air or other oxidizing agents, the dye becomes water-insoluble again and fixes firmly on the fiber.

The anthraquinone vat dyes can be classified into several groups on the basis of their chemical structures: (1) benzanthrone dyes, (2) indanthrones, (3) anthrimides, (4) anthrimidocarbazoles, (5) acylaminoanthraquinones, (6) anthraquinoneazoles, (7) anthraquinone acridones, (8) anthrapyrimidines, and (9) highly condensed ring systems. Most currently (1993) available dyes have been known for many decades, and very few new dyes have been commercialized since the 1970s. Recently, research and development efforts have focused on improved manufacturing of traditional vat dyes.

Benzanthrone Dyes. Vat dyes derived from benzanthrone may be divided into two groups: violanthrones and isoviolanthrones, dyes that have the perylene ring in their molecular structure; and benzanthrone pyrazolanthrones and benzanthrone acridones, the peri ring closure products of 3-anthraquinonylaminobenzanthrone.

Violanthrone and Isoviolanthrone Dyes. Violanthrone dyes cover a wide range of shades from reddish blue to green and grey. Examples of dyes that have industrial importance appear in Table 5.

Table 5. Commercially Important Violanthrone Dyes

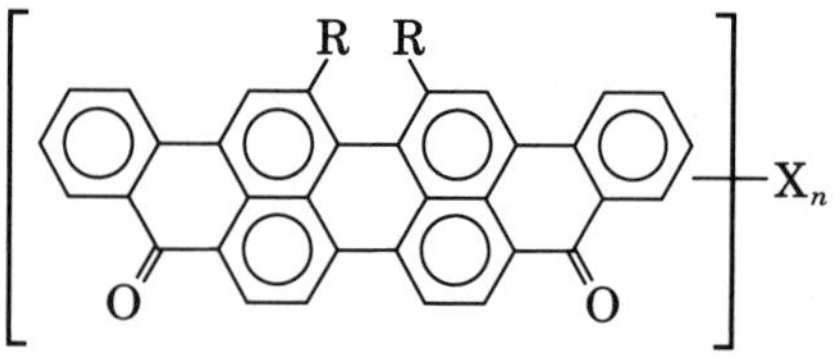

CI Vat name	CAS Registry Number	CI number	Structure number	R	X_n
Blue 20	[116-71-2]	59800	(135)	H	H
Blue 18	[1324-54-5]	59815	(136)	H	Cl_3
Blue 19	[1328-18-3]	59805	(137)	H	Br
Green 1	[128-58-5]	59825	(138)	OCH_3	H
Green 2	[25704-81-8]	59830	(139)	OCH_3	Br_2

CI Vat Blue 20 (135) is prepared from benzanthrone as follows:

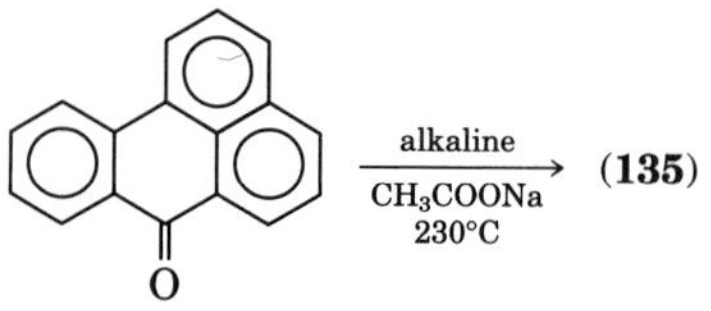

(71)

CI Vat Green 1 (138) is prepared from benzanthrone as follows:

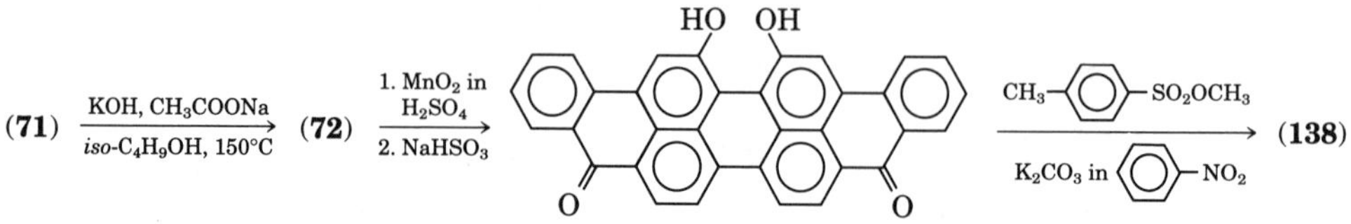

Isoviolanthrones that are currently produced include: CI Vat Violet 1 [1324-55-6] (CI 60010) (140) X = Cl, CI Vat Violet 9 [1324-17-0] (CI 60005) (140) X = Br, and CI Vat Blue 26 [4430-55-1] (CI 60015) (140) X = OCH_3.

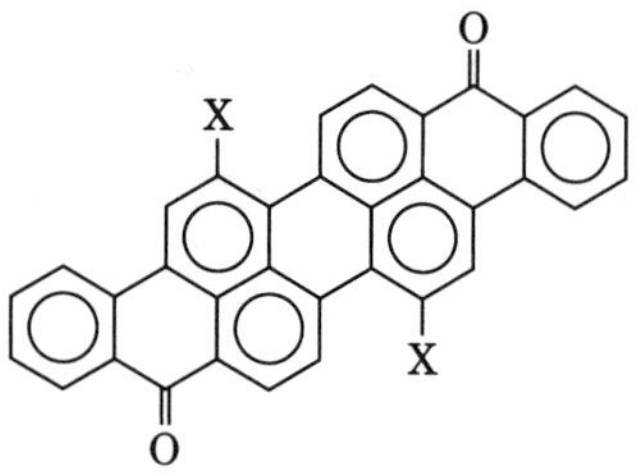

(140)

Benzanthrone Pyrazolanthrones and Benzanthrone Acridones. Benzanthrone pyrazolanthrones give from navy blue to gray shades and have good fastness. However, the only example of industrial use is CI Vat Black 8 [*2278-50-4*] (**141**) (CI 71000). CI Vat Blue 25 [*6247-39-8*] (CI 70500) has the basic structure of this dye class, but it is not produced today.

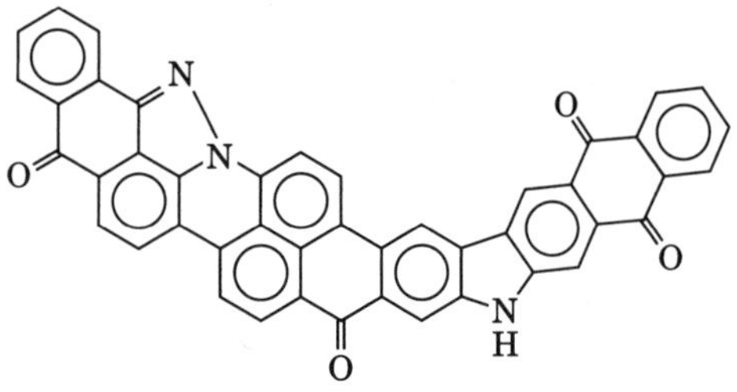

(**141**)

Benzanthrone acridones play an important role in vat dyes, since they give from dark green to olive green shade and have a high color value and excellent fastness. Examples are given in Table 6.

Table 6. Benzanthrone Acridone Vat Dyes

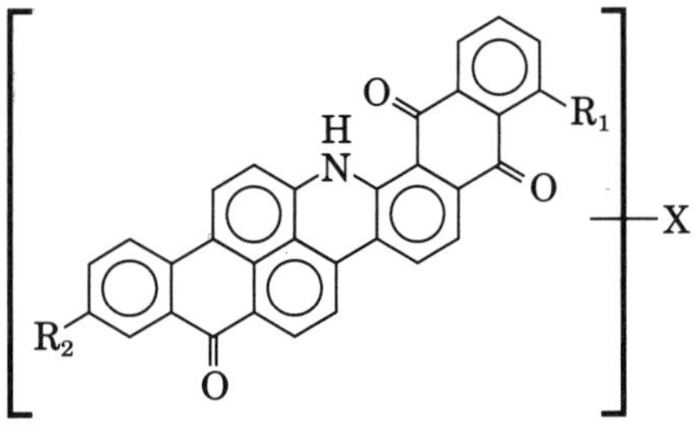

CI Vat name	CAS Registry Number	CI number	Structure number	R_1	R_2	X
Green 3	[*3271-76-9*]	69500	(**142**)	H	H	H
Green 5	[*1328-37-6*]	69520	(**143**)	—NHCO— 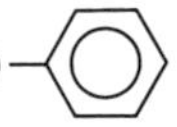	H	Cl
Black 25	[*4395-53-3*]	69525	(**144**)	H	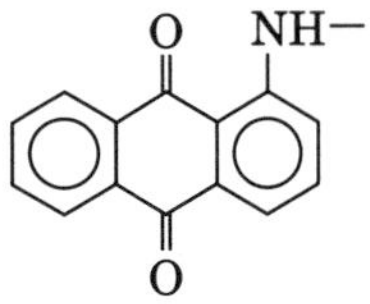	H

CI Vat Green 3 is prepared from 3-bromobenzanthrone (**73**) and 1-aminoan-thraquinone (**17**) as follows:

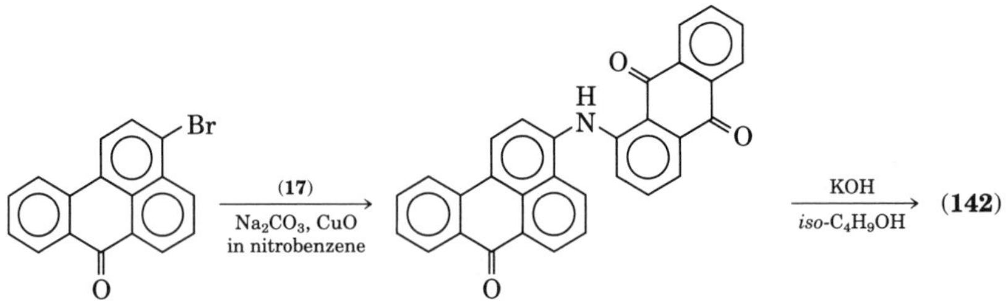

(**73**)

CI Vat Black 25 is prepared from dibromobenzanthrone (**74**) and (**17**) by a similar route. This dye is known as Indanthrene Olive T.

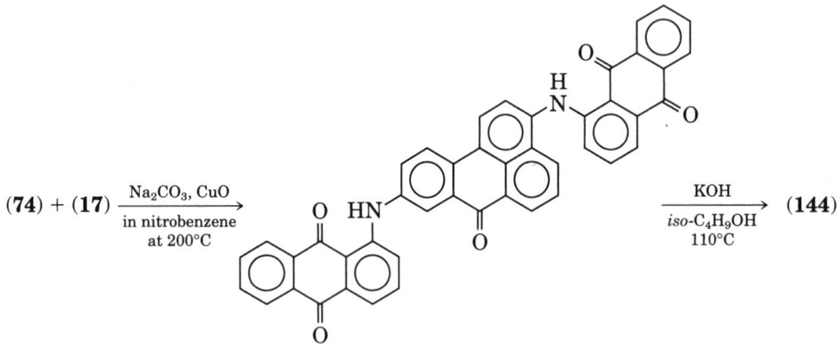

Indanthrones. Indanthrone blue (CI Vat Blue 4) [*81-77-6*] (**6**) (CI 69800) is the first invented anthraquinone vat dye, and has been extensively used as the most important vat dye for many decades because of its bright color as well as excellent affinity and fastness. These advantages are considered to be due to the stable structure attained by the intramolecular hydrogen bonding (**145**).

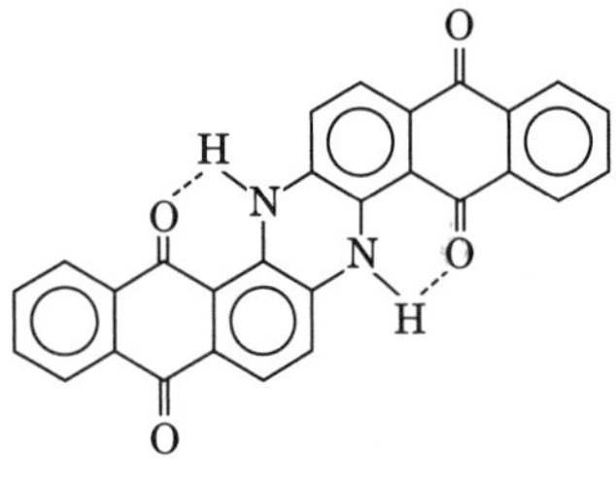

(**145**)

The only drawback of this dye is poor chlorine-fastness, which is considered to be due to the oxidation by chlorine to form a yellowish green azine (**146**).

(145) $\underset{\text{reduction}}{\overset{\text{oxidation}}{\rightleftharpoons}}$

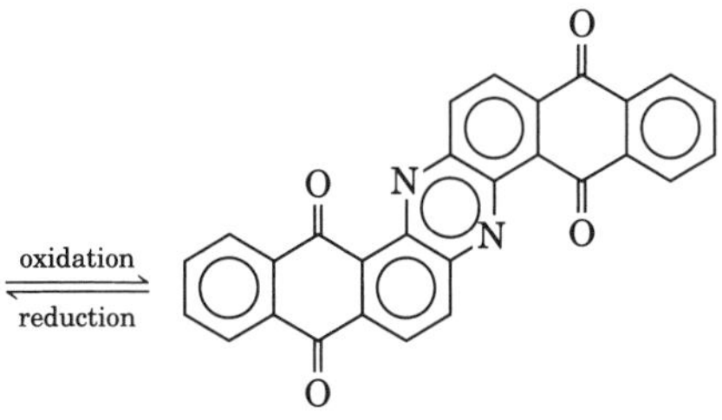

(146)

The post-halogenation of indanthrone improves its chlorine resistance to some extent and gives dyes of slightly greener shade. Examples are CI Vat Blue 6 [*130-20-1*] (**147**) (CI 69825) and CI Vat Blue 14 [*1324-27-2*] (**148**) (CI 69810) in which the position of the Cl is not indicated. (**148**) is a monochloro derivative of CI Vat Blue 4 (**6**).

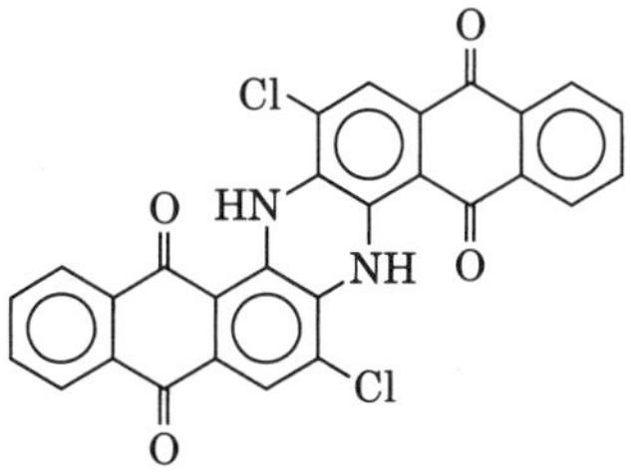

(147)

CI Vat Blue 4 is prepared from 2-aminoanthraquinone (**66**) by potash fusion in the presence of an oxidizing agent such as sodium nitrite or air. An alternative method by dimerization of 1-aminoanthraquinone (**17**) by using such solvents as dimethyl sulfoxide or tetramethylurea has been reported, and improved methods for this reaction have been cited (135–138). These methods are considered to be advantageous in terms of the yield as well as the availability of starting compounds.

Chlorination of indanthrone in sulfuric acid by passing through chlorine gas with addition of a small amount of manganese dioxide affords CI Vat Blue 14 as well as CI Vat Blue 6.

Anthrimides. Despite the facts that anthrimides have excellent leveling properties and cover a wide range of colors from yellow to black, they have little commercial importance because of their duller shades and lower tinctorial strength. However, the anthrimides are important intermediates for manufacturing anthrimidocarbazoles. Some examples of anthrimide dyes are CI Vat Orange 20 [*6370-69-0*] (**149**) (CI 65025), CI Vat Violet 16 [*4003-36-5*] (**150**) (CI 65020), CI Vat Black 28 [*128-79-0*] (**151**) (CI 65010), and CI Vat Red 48 [*4478-06-2*] (**152**) (CI 65205) (Fig. 6).

Anthrimidocarbazoles. Anthrimidocarbazoles cover a wide range of colors (yellow, orange, reddish brown, brown, and green) and have excellent leveling properties. Examples of commercial importance are CI Vat Orange 11 [*2172-33-0*] (**153**) (CI 70805), CI Vat Orange 15 [*2379-78-4*] (**154**) (CI 69025), CI Vat

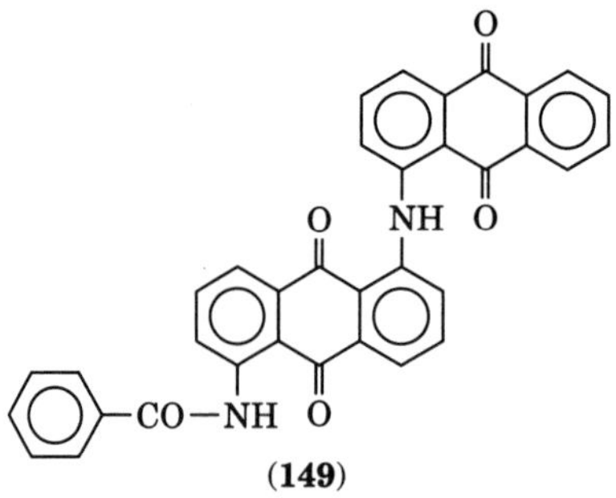

Fig. 6. Anthrimide vat dyes.

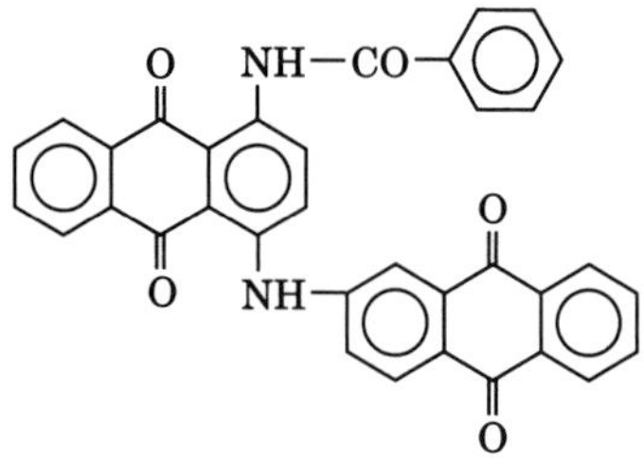

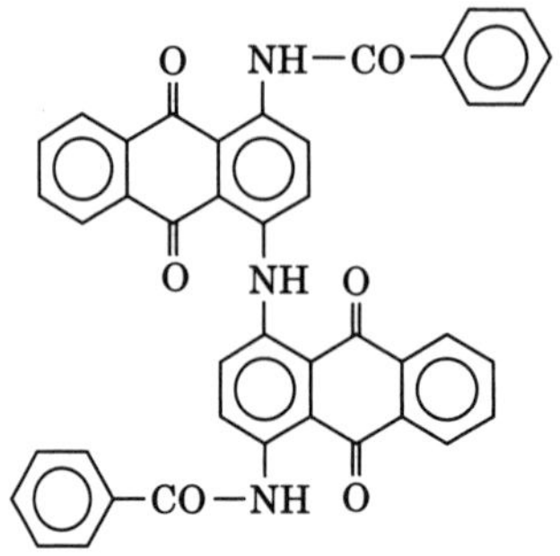

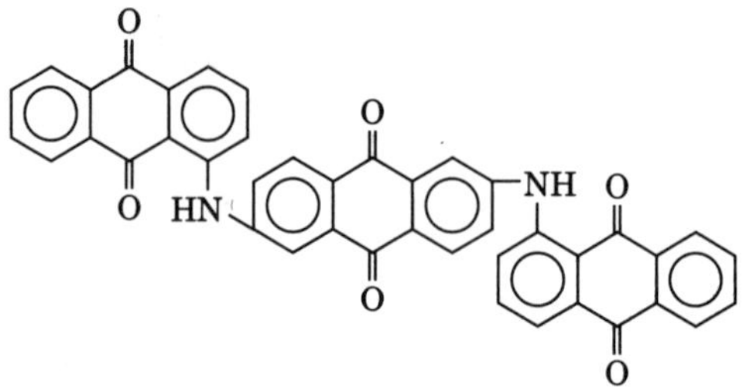

Brown 1 [*2475-33-4*] (**155**) (CI 70800), CI Vat Brown 3 [*131-92-0*] (**156**) (CI 69105), and CI Vat Brown 55 [*4465-47-8*] (**157**) (CI 70905). CI Vat Green 8 [*14999-97-4*] (**158**) (CI 71050) is generally called Indanthrene Khaki GG and has great significance (Fig. 7).

CI Vat Orange 15 is prepared by the reaction of 1-amino-5-benzoylamino-anthraquinone and 1-benzoylamino-5-chloroanthraquinone [*117-05-5*] (**159**) followed by ring closure. (**159**) is prepared by benzoylation of 1-amino-5-chloroanthraquinone [*117-11-3*] with benzoyl chloride.

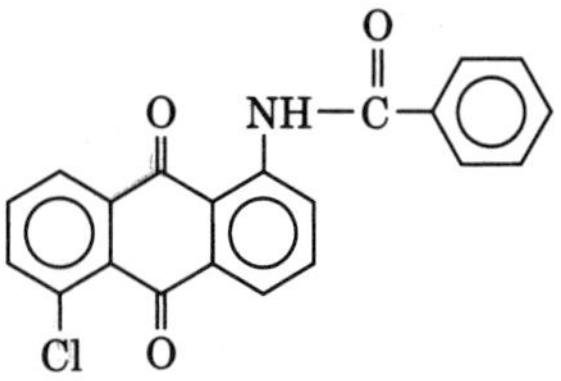

(**159**)

CI Vat Brown 1 is prepared from 1,1′:4,1″-trianthrimide(= 1,4-bis-(1-anthra-quinonylamino)anthraquinone [*116-76-7*] (**160**) in aluminum chloride–pyridine melts, followed by work up (by adding to sodium hydroxide solution) and oxidative cleanup (with sodium hypochlorite). 1,1′:4,1″-Trianthrimide is prepared by the condensation of 1,4-diaminoanthraquinone (**33**) with two moles of 1-chloroan-

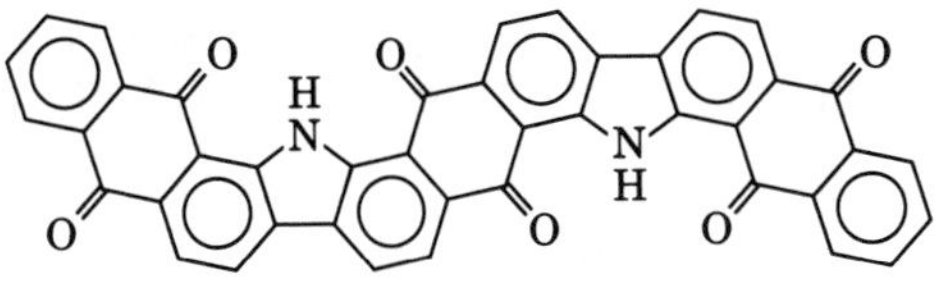

(153)

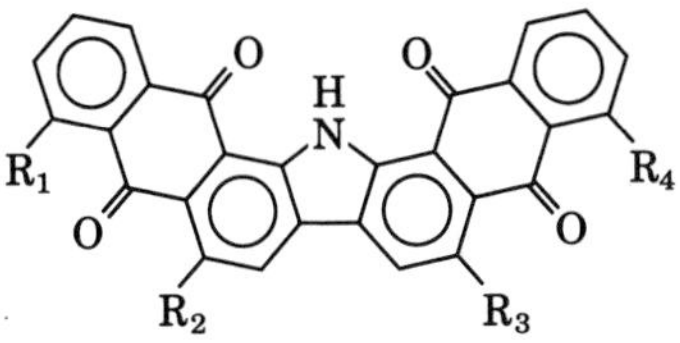

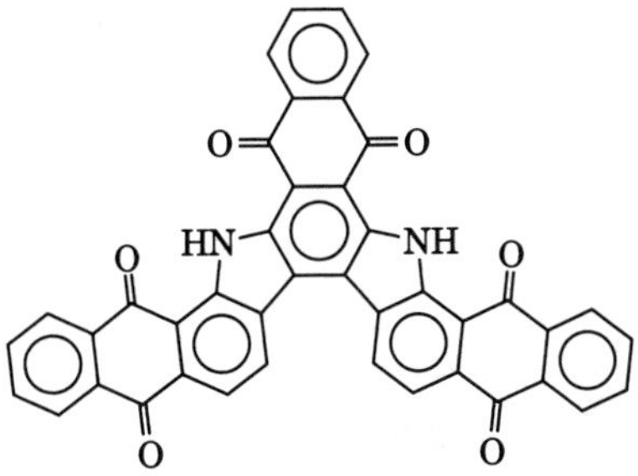

(154) R_2, R_3 = H; R_1, R_4 = $-NH-\overset{\overset{\displaystyle O}{\|}}{C}-$⟨○⟩

(155)

(156) R_2, R_4 = $-NH-\overset{\overset{\displaystyle O}{\|}}{C}-$⟨○⟩ ; R_1, R_3 = H

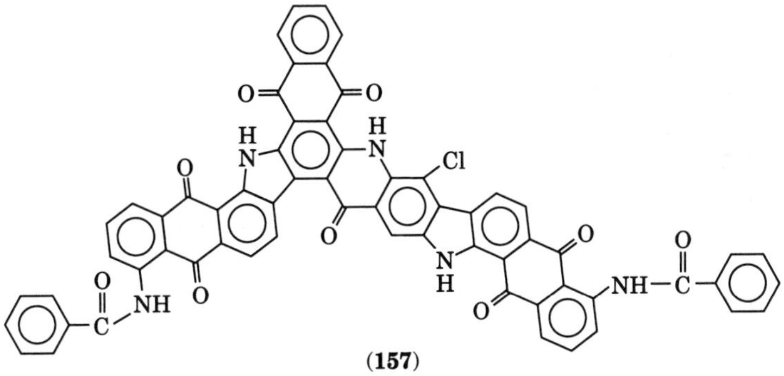

(157)

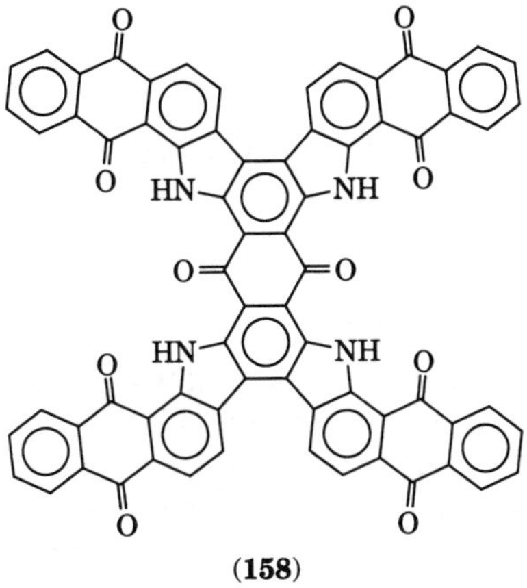

(158)

Fig. 7. Anthrimidocarbazole vat dyes.

645

thraquinone at elevated temperature in a high boiling point organic solvent in the presence of copper or copper(I) compound and acid-binding agent (139).

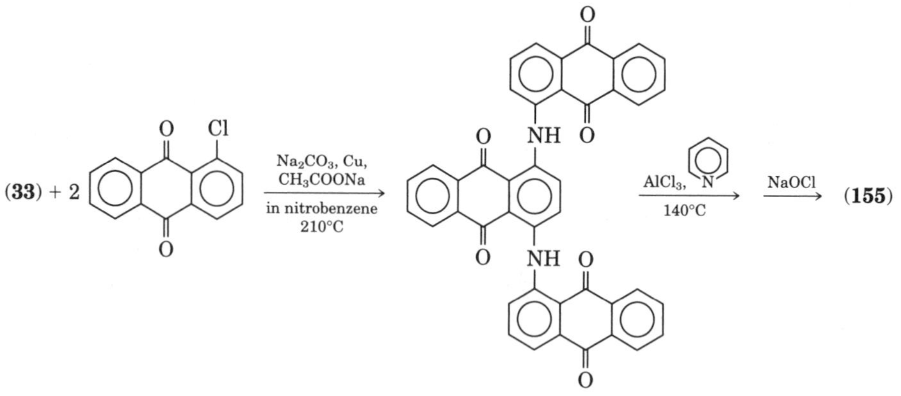

CI Vat Brown 3 is prepared from 4,5'-bisbenzoylamino-1,1'-dianthrimide by ring closure in concentrated sulfuric acid. 4,5'-Bisbenzoylamino-1,1'-dianthrimide is prepared from 1-amino-5-benzoylaminoanthraquinone [*117-06-6*] (**161**) and 1-benzoylamino-4-chloroanthraquinone [*81-45-8*] (**162**) at elevated temperature in an organic solvent in the presence of copper and an acid-binding agent. 1-Amino-5-benzoylaminoanthraquinone is prepared from 1,5-diaminoanthraquinone with benzoyl chloride.

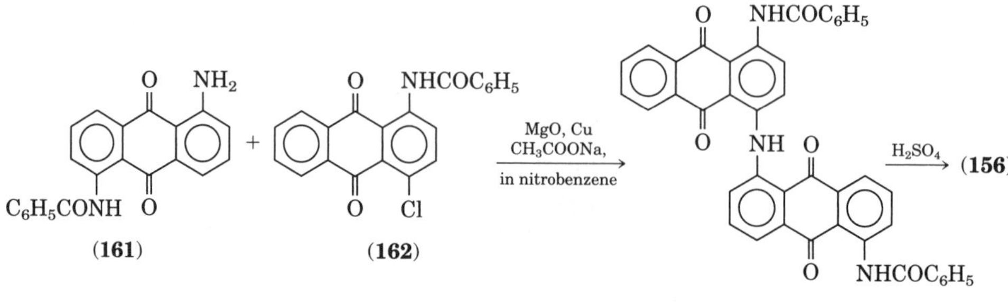

Indanthrene Khaki GG (**158**) is prepared from the corresponding pentanthrimide with aluminum chloride or aluminum chloride–sodium chloride melts.

Acylaminoanthraquinones. This dye class consists mainly of benzoyl derivatives of aminoanthraquinones. Due to the relatively low molecular weight, this dye class is applied in dyeing at low temperature. Yellow, orange, red, and even violet colors are covered by acylaminoanthraquinones.

Examples of industrial importance are CI Vat Yellow 12 [*6370-75-8*] (**163**) (CI 65405), CI Vat Red 21 [*4430-70-0*] (**164**) (CI 61670), and CI Vat Yellow 33 [*12227-50-8*] (**165**) (CI 65429) (Fig. 8).

CI Vat Yellow 12 is prepared by condensing 1-amino-5-benzoylaminoanthraquinone [*117-06-6*] (**161**) with oxalyl chloride in nitrobenzene. CI Vat Yellow

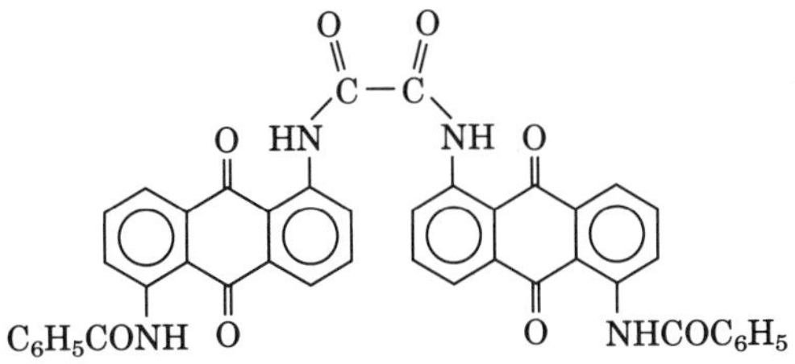

Fig. 8. Acylamino anthraquinone vat dyes.

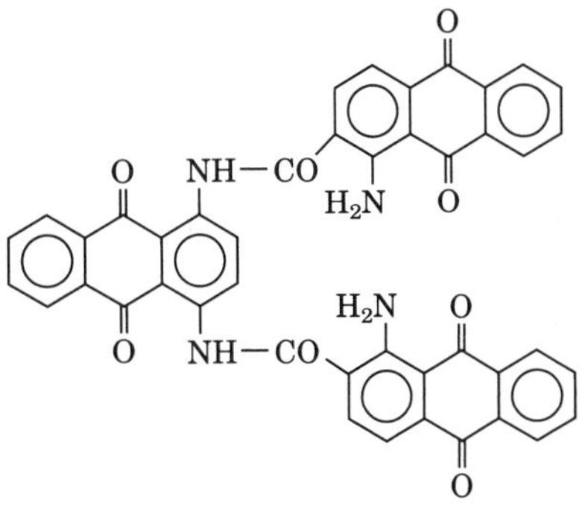

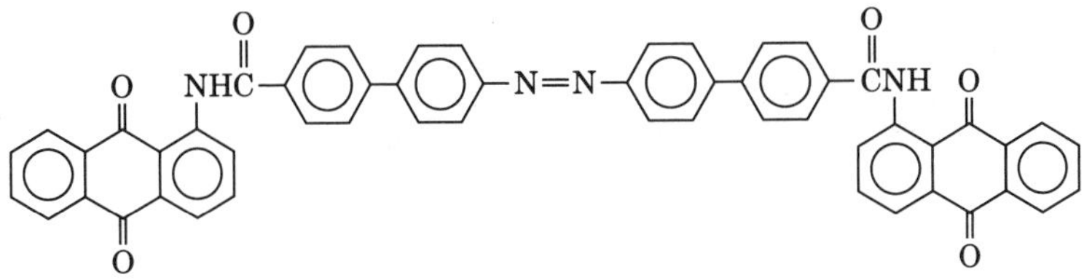

33 is prepared by condensation of two moles of 1-aminoanthraquinone with one mole of 4′,4‴-azobis(4-biphenylcarbonyl chloride). CI Vat Red 21 is prepared from 1,4-diaminoanthraquinone (**33**) and 1-nitroanthraquinone-2-carboxylic acid (**58**) by the following process:

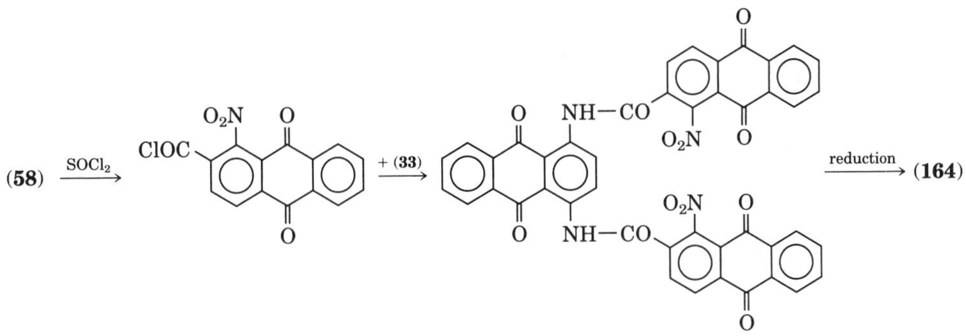

Anthraquinoneazoles. The representative anthraquinone oxazole vat dye is CI Vat Red 10 [2379-79-5] (**166**) (CI 67000). This dye is extensively used as a typical red vat dye because of its excellent dyeing properties and fastness. This fastness is considered to be due to hydrogen bonding of the amino hydrogen with an oxygen of the carbonyl group or the oxazole ring.

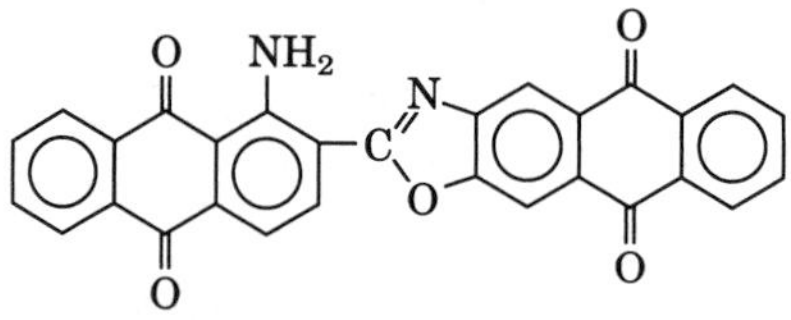

(166)

CI Vat Red 10 is prepared by condensation of 1-nitroanthraquinone-2-carboxylic acid chloride with 2-amino-3-hydroxyanthraquinone followed by ring closure in sulfuric acid and subsequent replacement of the nitro group with aqueous ammonia.

In recent years extensive studies have been carried out to develop a method of synthesis of CI Vat Red 10 starting from 1-aminoanthraquinone-2-carboxylic acid (**60**) rather than 1-nitroanthraquinone-2-carboxylic acid because of problems with Cr^{6+} in the production of the nitro compound (140,141).

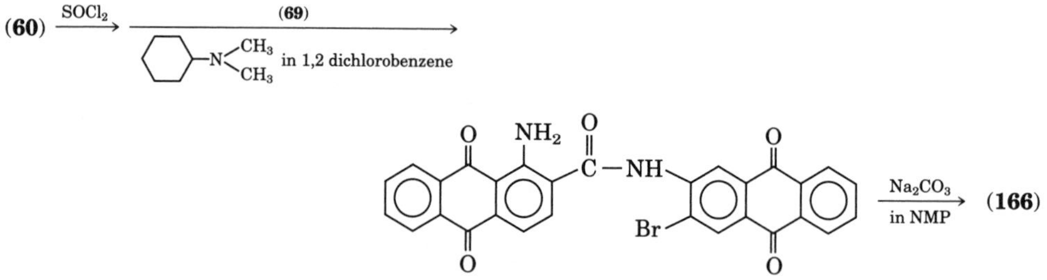

An example of an anthraquinone thiazole that has commercial importance is CI Vat Blue 30 [6492-78-0] (**167**) (CI 67110). This dye exhibits good fastness to light and chlorine.

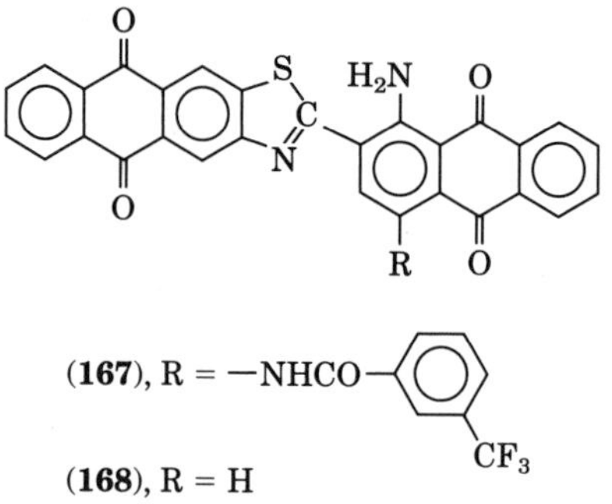

(**167**), R = —NHCO—⟨ ⟩
                              CF$_3$
(**168**), R = H

CI Vat Red 20 [6371-49-9] (**168**) (CI 67100) has also been known as an anthraquinone thiazole dye with good fastness. However, this dye is no longer produced commercially because of its duller shade.

Anthraquinone Acridones. Anthraquinone acridones give a wide range of shades including orange, red, violet, blue, green, and brown with good fastness especially to light. Representatives are CI Vat Orange 13 [6417-38-5] (**169**)

(CI 67820), and CI Vat Violet 13 [*4424-87-7*] (**170**) (CI 68700). CI Vat Orange 13 is prepared from 1-nitroanthraquinone-2-carboxylic acid as shown in Figure 9a. CI Vat Violet 13 is prepared from 1,5-dichloroanthraquinone as shown in Figure 9b.

The deeper shades are obtained by addition of other heterocyclic systems, such as quinazoline, to the molecular structure. CI Vat Green 12 [*6661-46-7*] (**171**) (CI 70700) is an example.

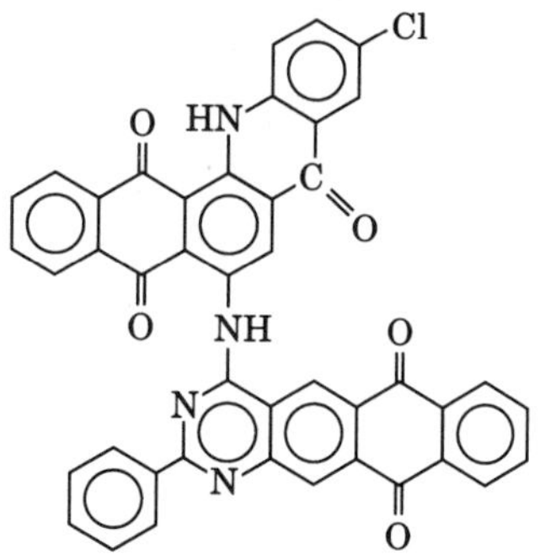

(**171**)

Anthrapyrimidines. Anthrapyrimidines are yellow vat dyes and exhibit good fastness. CI Vat Yellow 20 [*4216-01-7*] (**5**) (CI 68420) is the only example that is currently produced. The production method is shown in the following scheme. This compound is also used as a high performance organic pigment (CI Pigment Yellow 108).

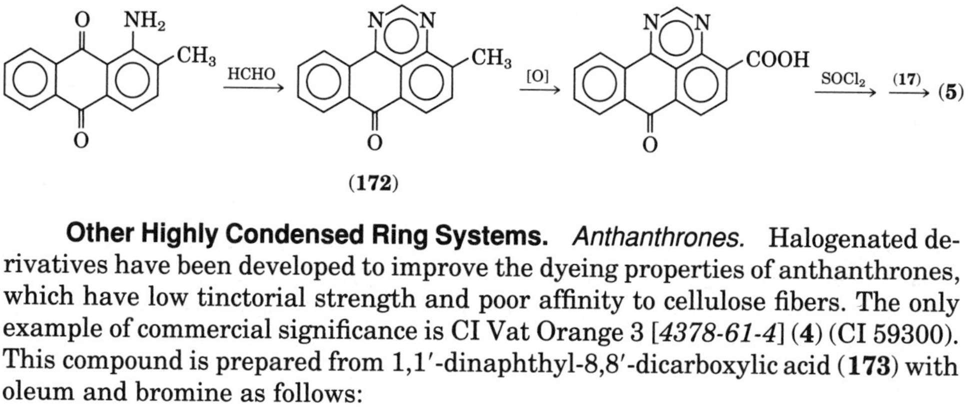

(**172**)

Other Highly Condensed Ring Systems. *Anthanthrones.* Halogenated derivatives have been developed to improve the dyeing properties of anthanthrones, which have low tinctorial strength and poor affinity to cellulose fibers. The only example of commercial significance is CI Vat Orange 3 [*4378-61-4*] (**4**) (CI 59300). This compound is prepared from 1,1'-dinaphthyl-8,8'-dicarboxylic acid (**173**) with oleum and bromine as follows:

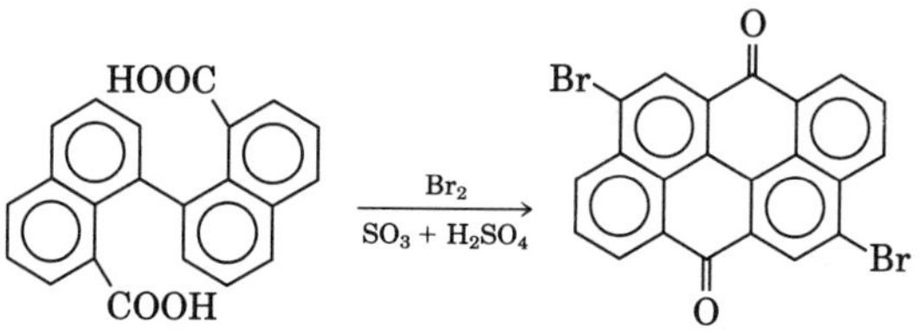

(**173**)   (**4**)

Dibenzopyrenquinones. These compounds give brilliant shades from yellow to orange and have high tinctorial power and good dyeing properties. Representa-

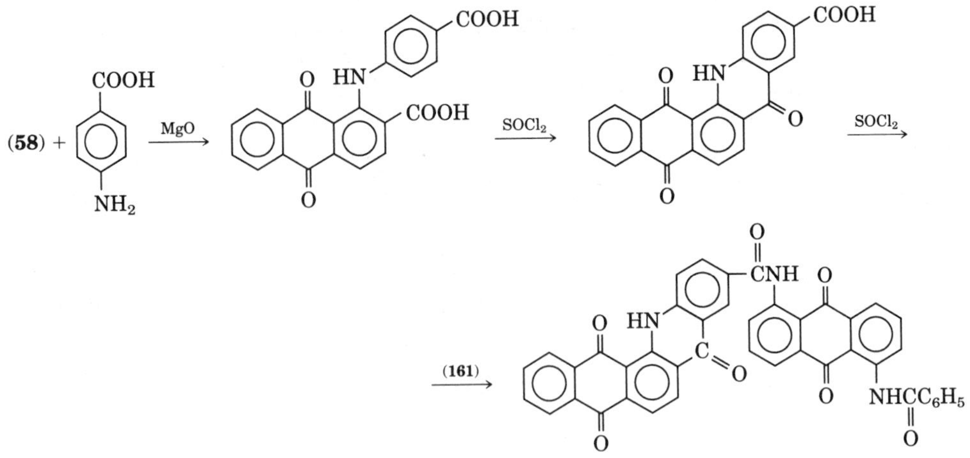

(169)

(a)

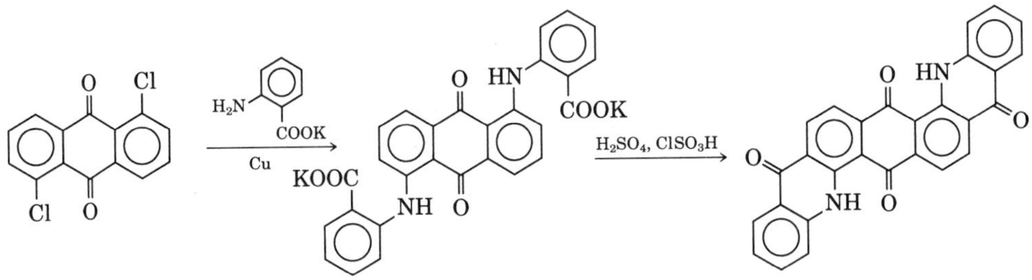

(170)

(b)

Fig. 9. Synthesis of anthraquinone acridones. (**a**) CI Vat Orange 13 (**169**); (**b**) CI Vat Violet 13 (**170**).

tives are CI Vat Yellow 4 [*128-66-5*] (**174**) (CI 59100), and its dibrominated derivative CI Vat Orange 1 [*1324-11-4*] (**175**) (CI 59105). These dyes are prepared as shown from 1,5-dibenzoylnaphthalene, which can be made by Friedel-Crafts acylation (benzoylation) of naphthalene. An improved method for the oxidative cyclization step using a catalytic amount of ferric halides has been given (142).

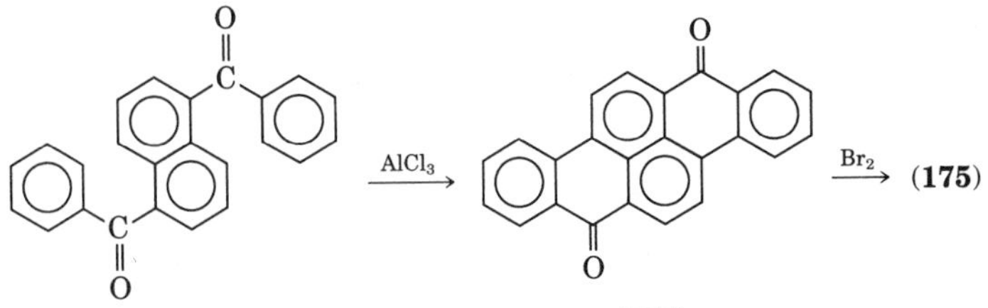

(174)

The dibromo derivative exhibits improved fastness to light, washing, and bleach.

Pyranthrones. Pyranthrones give orange shades and halogenated compounds improve lightfastness. A representative that is still widely used industrially is CI Vat Orange 2 [*1324-35-3*] (**176**) (CI 59705). Its synthesis begins with 1-chloro-2-methylanthraquinone, which is coupled by heating with copper powder in dichlorobenzene–pyridine. Ring closure is effected by KOH in isobutyl alcohol, and the rings are brominated.

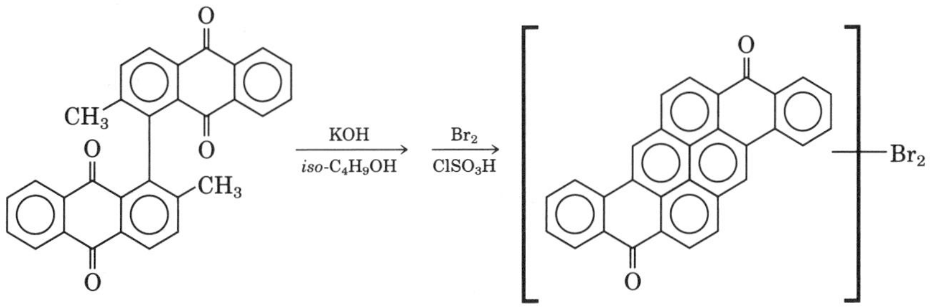

(**176**)

Flavanthrone. Flavanthrone [*475-71-8*] (**177**) (CI 70600) has excellent dyeing properties, which are due to the stability of the leuco form, but its fastness is not satisfactory. Only the unsubstituted flavanthrone is used as a vat dye, ie, CI Vat Yellow 1 (**177**). It is mainly used as a pigment, ie, CI Pigment Yellow 24. Synthetic routes are illustrated in Figure 10.

Mordant Dyes

Mordant dyes have hydroxy groups in their molecular structure that are capable of forming complexes with metals. Although a variety of metals such as iron, copper, aluminum, and cobalt have been used, chromium is most preferable as a mordant. Alizarin or CI Mordant Red 11 [*72-48-0*] (**1**) (CI 58000), the principal component of the natural dye obtained from madder root, is the most typical mordant dye (see DYES, NATURAL). The aluminum mordant of alizarin is a well-known dye by the name of Turkey Red and was used to dye cotton and wool with excellent fastness. However, as is the case with many other mordant dyes, it gave way to the vat or the azoic dyes, which are applied by much simpler dyeing procedures.

Alizarin is prepared from anthraquinone-2-sulfonic acid by heating with aqueous sodium hydroxide and sodium nitrate at 200°C (143,144).

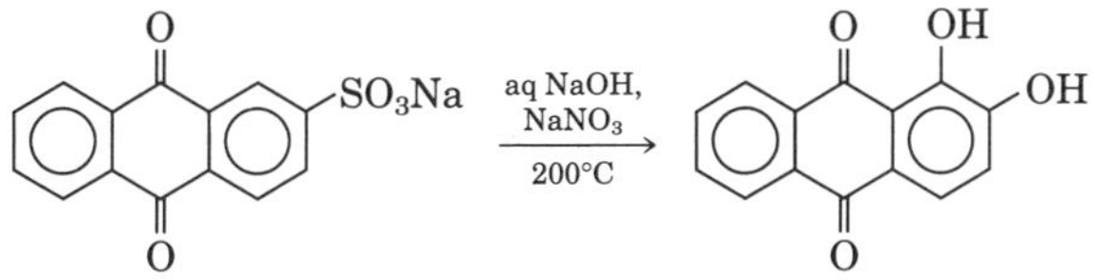

(**1**)

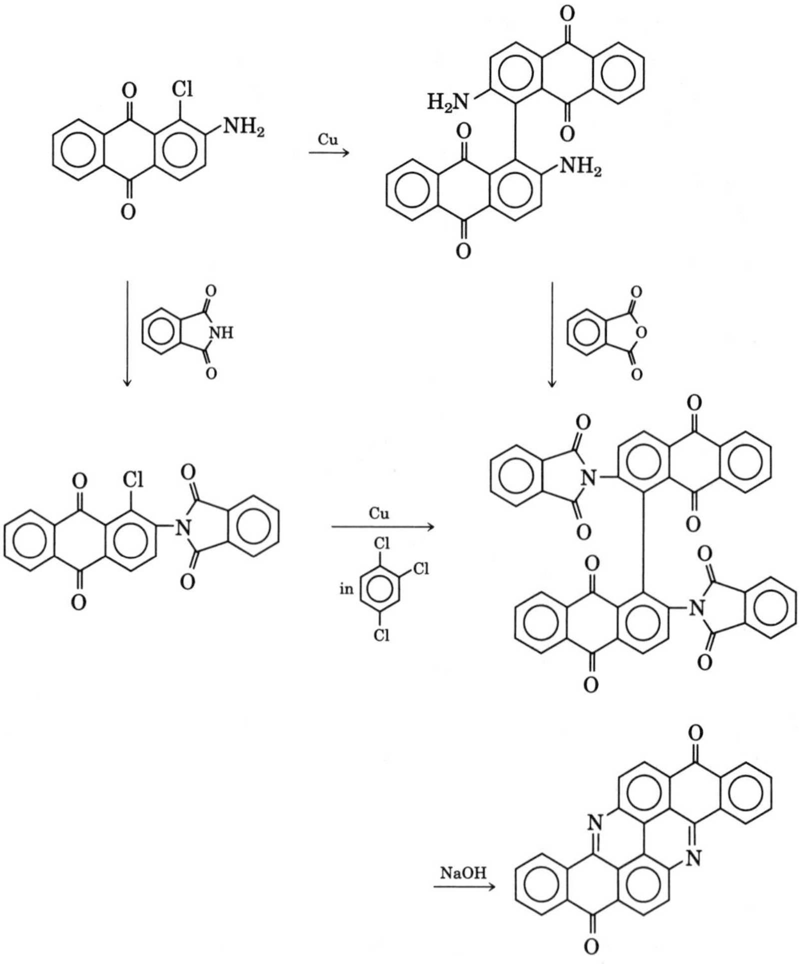

Fig. 10. Synthetic routes to CI Vat Yellow 1 [475-71-8] (**177**).

A variety of derivatives have been produced from alizarin and used as mordant dyes for cotton and wool. Examples are given in Table 7.

Purpurin [81-54-9] (**179**) is a useful intermediate for preparing acid–mordant dyes, and is prepared by oxidation of alizarin with manganese dioxide and sulfuric acid (145).

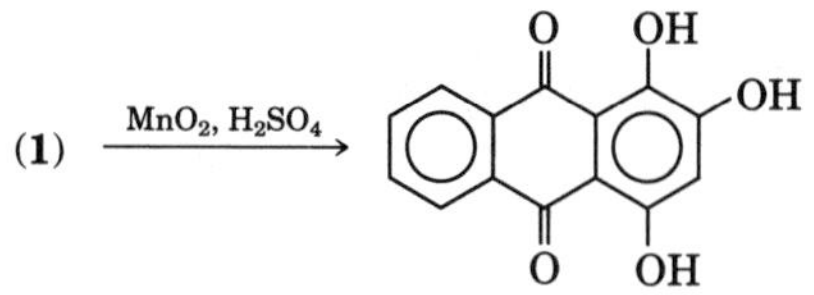

Table 7. Alizarin Derivatives as Mordant Dyes

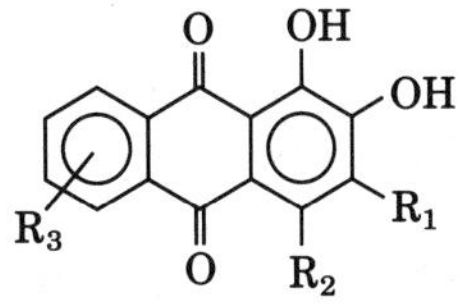

Structure number	CI Mordant name or number	R_1	R_2	R_3	Color/Mordant	CAS Registry Number	Common name
(**1**)	Red 11	H	H	H	rose red/Al	[72-48-0]	alizarin
(**178**)	Brown 42	OH	H	H	brown/Cr	[602-64-2]	anthragallol
(**179**)	58205	H	OH	H	scarlet/Al	[81-54-9]	purpurin
(**180**)	58215	NO_2	OH	H	red/Al	[6486-91-5]	3-nitropurpurin
(**181**)	Red 45	H	H	5-OH	bordeaux/Al	[6486-93-7]	5-hydroxyalizarin
(**182**)	Red 4	H	H	6-OH	dull red/Cr	[82-29-1]	flavopurpurin

Acid–mordant dyes have characteristics similar to those of acid dyes which have a relatively low molecular weight, anionic substituents, and an affinity to polyamide fibers and mordant dyes. In general, brilliant shades cannot be obtained by acid–mordant dyes because they are used as their chromium mordant by treatment with dichromate in the course of the dyeing procedure. However, because of their excellent fastness for light and wet treatment, they are predominantly used to dye wool in heavy shades (navy blue, brown, and black). In terms of chemical constitution, most of the acid–mordant dyes are azo dyes; some are triphenylmethane dyes; and very few anthraquinone dyes are used in this area. CI Mordant Black 13 [1324-21-6] (**183**) (CI 63615) is one of the few examples of currently produced anthraquinone acid–mordant dyes. It is prepared by condensation of purpurin with aniline in the presence of boric acid, followed by sulfonation and finally by conversion to the sodium salt (146,147).

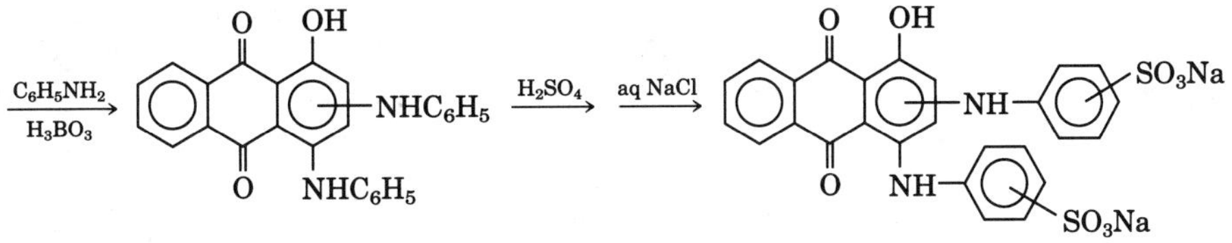

(**183**)

Functional Dyes

The investigation of new dyes has always been focused on the development of fast, brilliant, inexpensive, and easy applicable dyes. Because a great emphasis has been placed especially on fastness, the dyes with poorer fastness have been ignored in the past. However, in recent years new needs for dyes that change color in response to low energy stimuli including light, electricity, or heat have arisen in the electronics industry. This new application includes information recording, information display, and energy conversion. The term functional dye has been applied to dyes that are used in advanced technologies based on optoelectronics since 1981 when the book entitled *The Chemistry of Functional Dyes* was published in Japan (148).

In order to develop the dyes for these fields, characteristics of known dyes have been re-examined, and some anthraquinone dyes have been found usable. One example of use is in thermal-transfer recording where the sublimation properties of disperse dyes are applied. Anthraquinone compounds have also been found to be useful dichroic dyes for guest-host liquid crystal displays when the substituents are properly selected to have high order parameters. These dichroic dyes can be used for polarizer films of LCD systems as well. Anthraquinone derivatives that absorb in the near-infrared region have also been discovered, which may be applicable in semiconductor laser recording.

SUBLIMATION THERMAL-TRANSFER PRINTING

The need to obtain color hard copies from electronic systems such as TV and video sets, or from personal computers, has been increasing. Several methods have been proposed to obtain hard copies of full color images. Among them, sublimation thermal-transfer printing has the following characteristics: the quality of the printed color picture is extremely high, and the equipment is compact, quiet, and easy to manipulate as well as to maintain.

This system consists of a thermal-transfer sheet (ink sheet) containing sublimation dyes, an acceptor sheet containing resin that accepts the sublimated dyes, and a thermal head. The dye in the ink sheet sublimes and is transfer-printed on an acceptor sheet with the thermal energy given by the thermal head. Since the amount of dye transferred can be controlled according to the thermal energy given, good, continuous-gradation color images can be obtained, and full color images can be formed by using ink sheet containing dyes of three primary colors, ie, yellow, magenta, and cyan (see COLOR PHOTOGRAPHY).

The dyes used in the ink sheet must satisfy various requirements: (*1*) optimum color characteristics of the three primary colors (hue, color density, shape of absorption spectrum); (*2*) sensitivity, ie, sublimability from ink sheet to acceptor sheet; (*3*) fastness for light and migration; and (*4*) compatibility with the resin in the ink sheet. With respect to these characteristics, a large number of anthraquinone dyes have been proposed particularly for magenta and cyan colors. Typical examples are given in Table 8 and Table 9.

DYES FOR LIQUID CRYSTAL DISPLAY SYSTEMS

Liquid crystal display systems have been increasingly used in electro-optical devices such as digital watches, calculators, televisions, instrument panels, and displays of various kinds of electronic equipment, ie, lap-top computers and word processors. The dominant reason for their success is their extremely low power consumption. Furthermore, the liquid crystal display systems have been remarkably improved in recent years, and today they have high resolution (more than 300,000 pixels) and full color capability almost equivalent to those of a cathode ray tube.

Table 8. Examples of Magenta Dyes for Sublimation Thermal-Transfer Printing

General structure	Examples	References
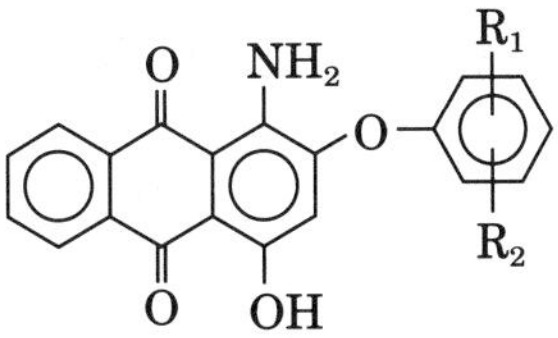	$R_1 = C_5–C_{12}$ alkyl $R_2 = H, C_1–C_{12}$ alkyl	149
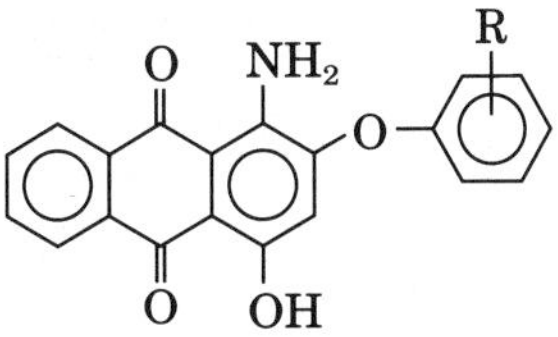	R = alkyl	150
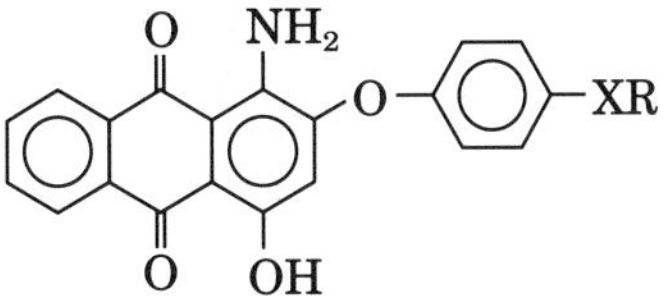	$X = SO_2NH, CONH, NHCO,$ $SO_2O, COO, NH,$ $NHSO_2$ R = alkyl	151
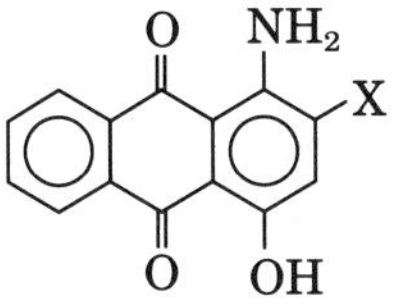	X = H, Cl, Br	152 153
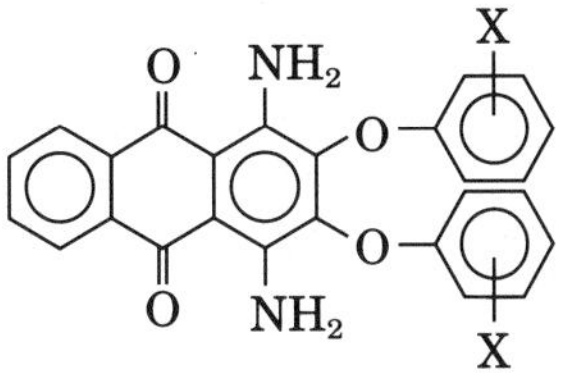	X = H, Cl, Br, CF_3, alkyl, alkoxy	154

Guest-Host Mode LCD Systems.   Guest-host liquid crystal display systems consisting of dichroic dyes (guest) and liquid crystal media (host) have a wider angle of vision and better color reproducibility than systems that consist of colored polarizers and twisted nematic liquid crystal cells. These dyes are required to satisfy the following criteria: (1) brilliant colors with panchromaticity and high extinction coefficient; (2) solubility in liquid crystal medium; (3) high order parameters; (4) stability to light, heat, humidity, and electronic current; (5) high purity; and (6) high resistivity.

Several basic chromophore structures have been proposed for this purpose. Anthraquinone dyes appear to be predominant since they have a wider color range, excellent photostability, good solubility in liquid crystal media, and very high order parameters. Typical basic structures of the three primary colors are illustrated in Figure 11. Some examples are given in Table 10. The appropriate combination of three primary colors gives a black display.

Table 9. Examples of Cyan Dyes for Sublimation Thermal-Transfer Printing

General structure	Examples	References
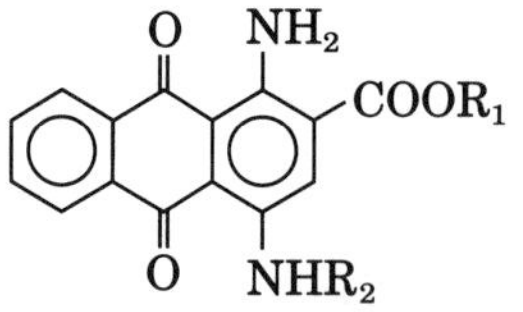	R_1 = alkoxyalkyl, alkenyl, hydroxyalkyl, etc R_2 = alkyl, alkoxyalkyl, hydroxyalkyl, etc	155
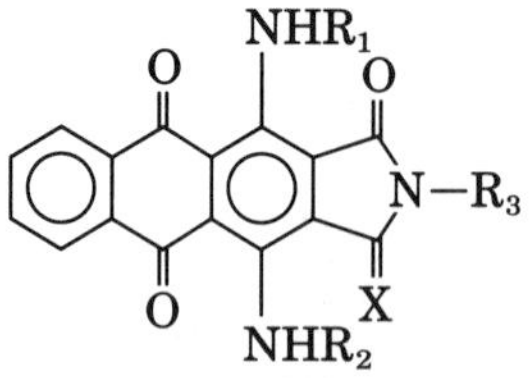	R_1, R_2 = H,C_1–C_4 alkyl R_3 = (substituted)C_1–C_4 alkyl X = O, NH	156
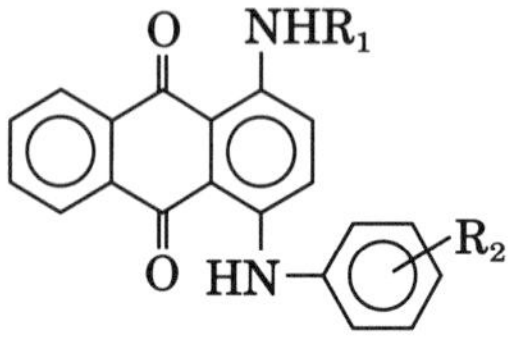	R_1 = C_1–C_{20}alkyl R_2 = alkyl, alkoxy, alkoxyalkyl	157 158
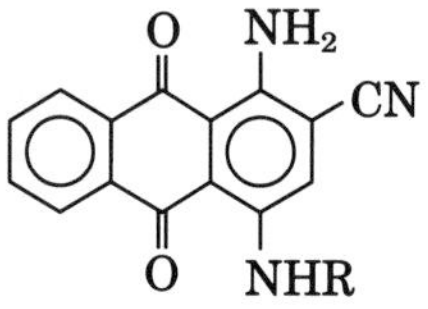	R = H,C_1–C_4 alkyl	159
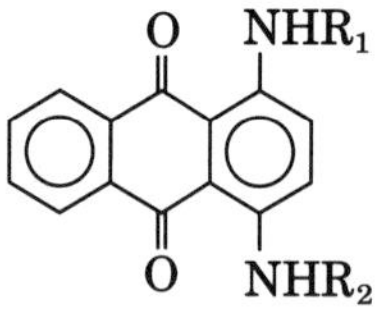	R_1, R_2 = (substituted)alkyl, alkenyl, etc	160

Dichroic Dyes for Polarizer Filters.   Super twisted nematic liquid crystal display (STN–LCD), now predominant in the market of black and white liquid crystal display systems, consists of a STN cell placed between two polarizers. Most polarizers are poly(vinyl alcohol) films containing iodine molecules aligned in one direction. They have enough quality for the displays of wrist watches, calculators, and other electronic appliances used mainly indoors. However, they have problems due to sublimable iodine, and lack of durability to high temperature, high humidity, and light; they cannot be applied to displays used outdoors or in severe conditions. Recently, polarizers containing dichroic dyes have been intensively studied for such needs as displays of electronic dashboard components for automobiles. Poly(vinyl alcohol) films with dichroic direct dyes and polyester (PET) or polyamide (nylon-6) films containing dichroic disperse dyes have been developed for this purpose. Some anthraquinone disperse dyes have been proposed in the latter system. Examples are shown in Table 11. Dyes having excellent durability and moisture resistance with high polarizing degree have been found (166).

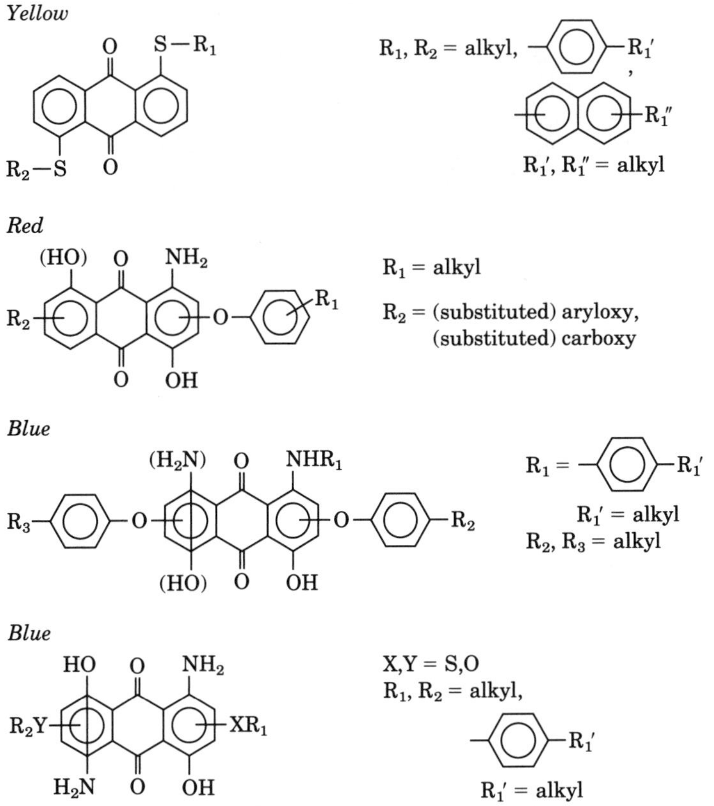

Fig. 11. Basic structural concepts for the three primary color anthraquinone liquid crystal dyes.

Dyes for Color Filters. Color liquid crystal display systems consist of LSI drivers, glass plates, polarizers, electrodes (indium–tin oxide), and microcolor filters. The independent microcolor filter containing dyes is placed on each liquid crystal pixel addressed electrically and acts as an individual light switch. All colors can be expressed by the light transmitted through each filter layer of the three primary colors, ie, red, green, and blue (Fig. 12).

Four significant fabrication processes have been proposed and commercialized so far: a dyeing method using mainly acid or reactive dyes in a gelatin or casein base; a printing method using ink containing pigments and organic vehicles and resins; a pigment dispersion method using photosensitive resins in which pigments are dispersed; and an electrodeposition method using polymer resins with pigments dispersed in an aqueous medium and resins electrochemically deposited on electrode (ITO).

The principle of each process is briefly described in the literature (170). Some anthraquinone dyes and pigments appear to be used in combination with other dye or pigment classes such as phthalocyanines and carbazole violets, etc. Two examples described in patents are the red pigment and blue dye that follow:

Table 10. Some Examples of Liquid Crystal Dyes

Structure	Order parameter[a]	References
Yellow		

| 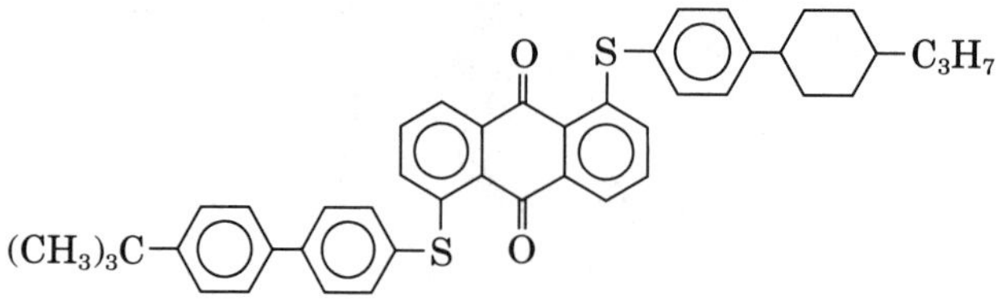 | 0.79 | 161 |

| 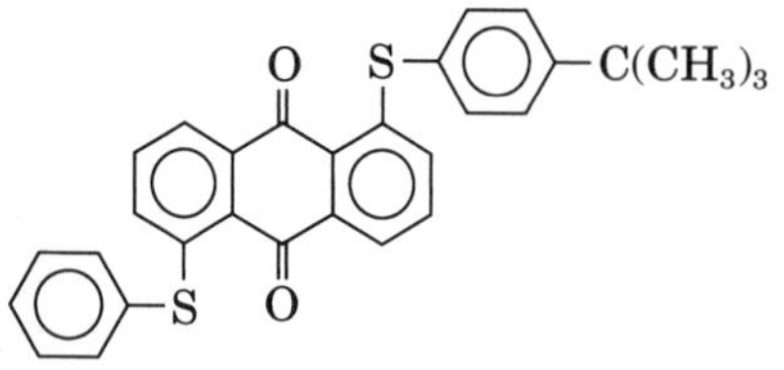 | | 162 |

Red

| 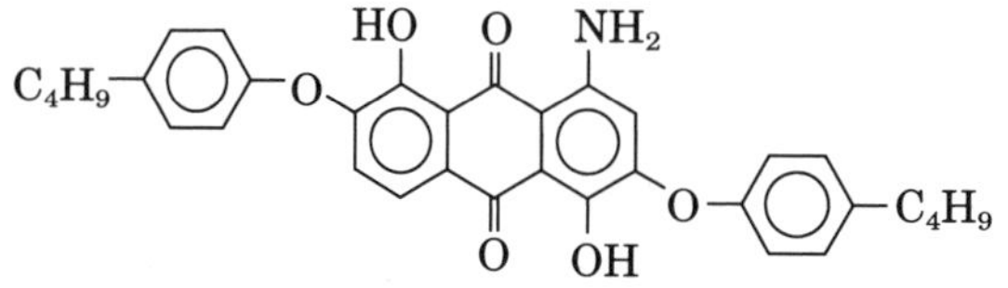 | 0.76 | 163 |

| 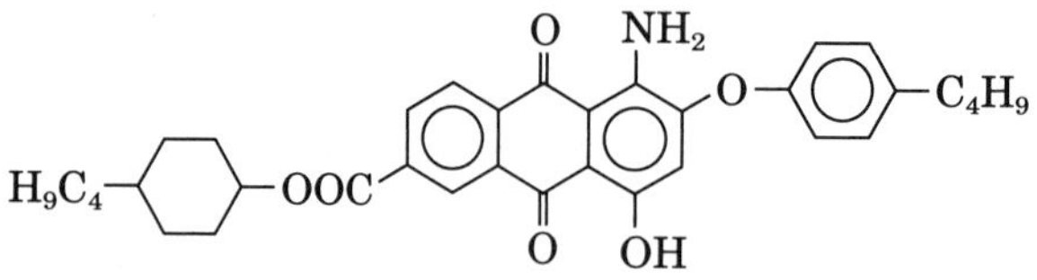 | | 162 |

Blue

| 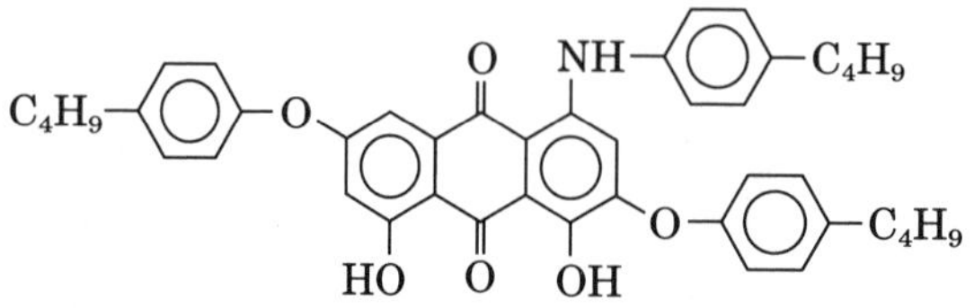 | 0.76 | 164 |

| 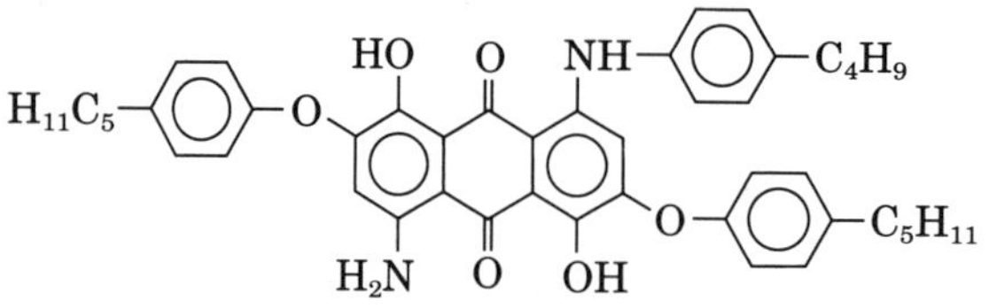 | 0.77 | 164 |

| 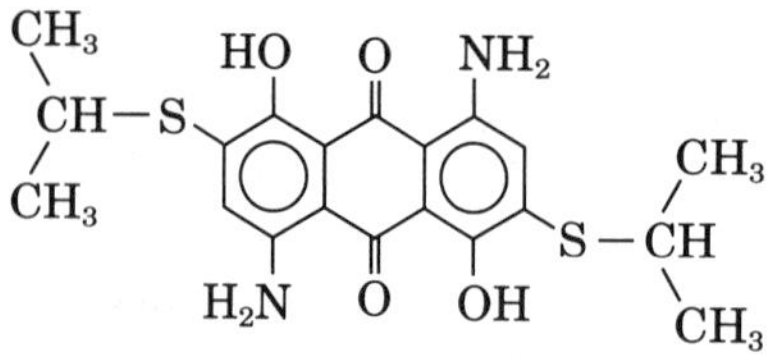 | 0.77[b] | 165 |

[a] In ZLI-1565[c] unless otherwise noted.
[b] In ZLI-1840[c].
[c] Liquid crystal mixtures of E. Merck.

Table 11. Some Examples of Anthraquinone Dyes for Polarizer Films[a]

Chemical structure	Hue	λ_{max}, nm	Polarization degree, %	Reference
R—⬡—⬡—R				
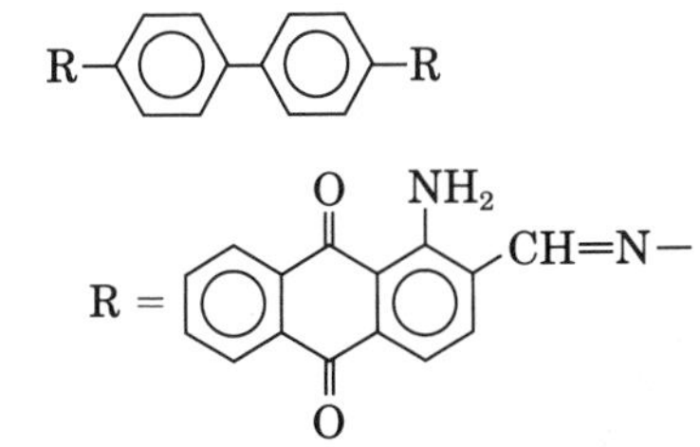 R = (structure)	red	540	99.9	166
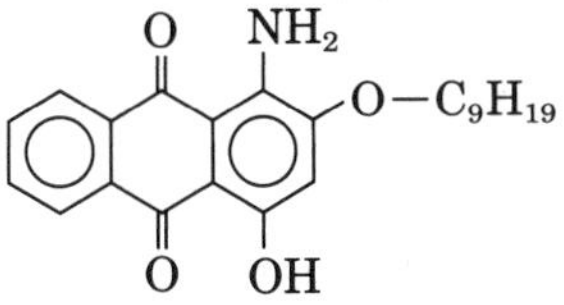 (structure, $O-C_9H_{19}$)	pink	520	91	167
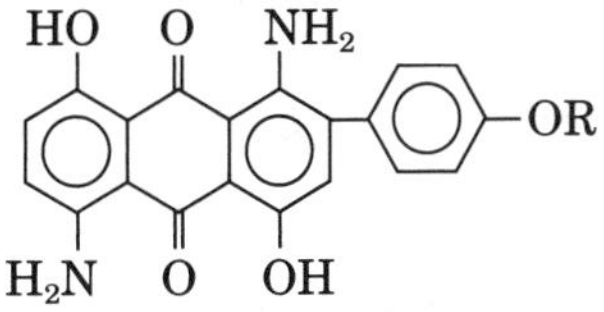 (structure) $R = C_8H_{17}$	blue	640	93	167
$R = C_7H_{15}$	blue	640 / 648	96^b / 91^b	168
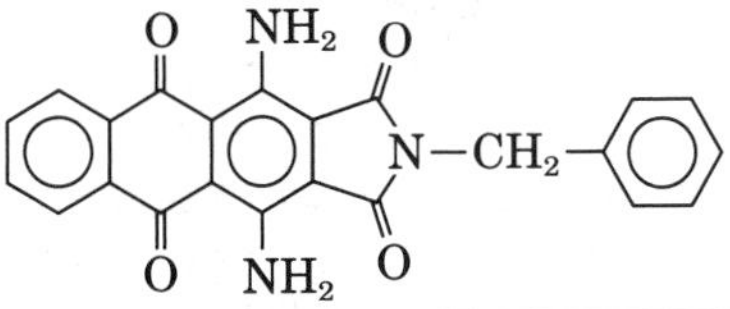 (structure, $N-CH_2-⬡$)	brilliant blue	685	88	169

[a]The film is poly(ethylene terephthalate) PET unless otherwise stated.
[b]Film is nylon-6.

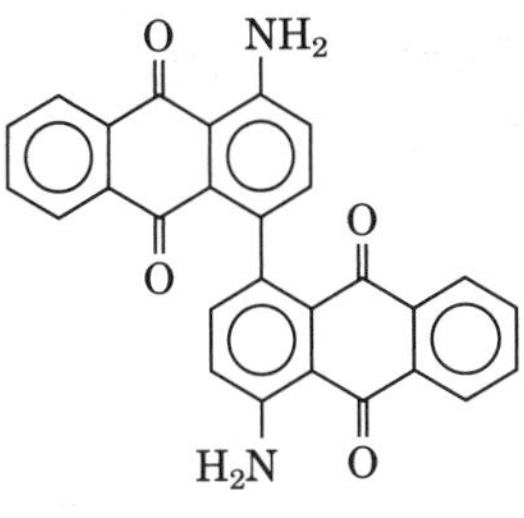

red

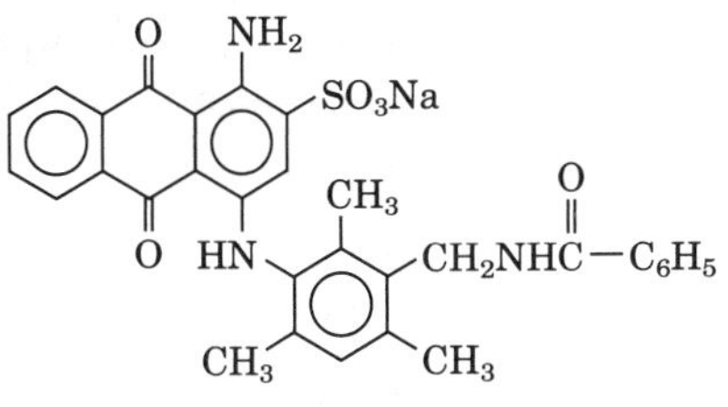

blue

The red pigment has been proposed to be usable in both the pigment dispersion method and the electrodeposition method (171,172). The blue dye may be used in the dyeing method fabrication process (173).

NEAR-INFRARED ABSORBING DYES

Optical disk data storage systems have been increasingly used in offices for document storage (see INFORMATION STORAGE MATERIALS). The disk consists of a

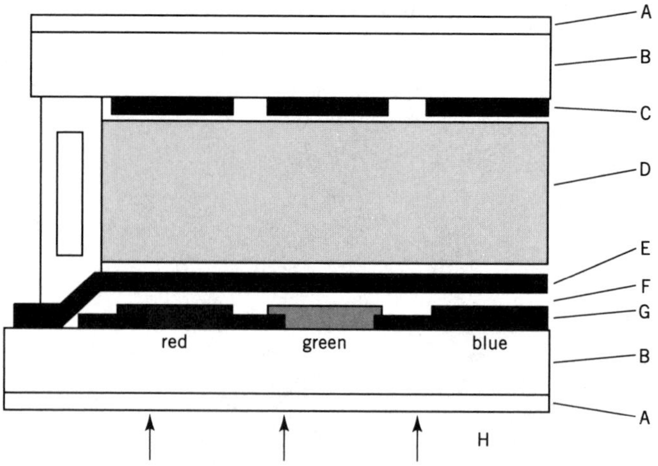

Fig. 12. General structure of LCD. A, polarizer plate; B, glass plate; C, electrodes (indium–tin oxide); D, liquid crystal; E, common electrode (ITO); F, overcoated layer; G, colored pixel; H, back light. In an improved color LCD system today, retardation films are placed between A and B.

polycarbonate base overlaid with a polymer layer containing photosensitive material. Localized thermal changes in this layer can be induced by a semiconductor diode laser, which can be read afterward by optical means. In this system laser light energy is absorbed by the dyes contained in the polymer layer to melt locally and form micropits there. Semiconductor diode lasers emit in the wavelength range around 800 nm, and generally the wavelength range of the AlGaAs laser (780–830 nm) is used. Accordingly, the dyes are required to absorb near-infrared light at this wavelength with a large extinction coefficient. It is also required that the dye has stability to light, resistance to heat and humidity, and solubility in organic solvents. The last is related to coating characteristics. Some anthraquinone dyes have been proposed for this purpose together with other dye or pigment classes such as polymethine dyes, phthalocyanines, and metal complex dyes. Examples are given in Table 12.

Economic Aspects

Anthraquinone dyes are derived from several key compounds, ie, dye intermediates. Production of these dye intermediates often requires sophisticated production processes and a large amount of investment in plant construction. The competitiveness of final products, dyestuffs, depends on that of the intermediates, ie, quality, cost, and availability.

Production Capacity and Demand. The production capacity for each dye or dye intermediate has rarely been announced officially by the individual manufacturers. However, the world demand of anthraquinone colorants can be roughly estimated as in Table 13 and, more specifically, in Figure 13. Principal manufacturers of anthraquinone dyes and their intermediates are as follows:

Table 12. Examples of Near-Infrared Absorbing Dyes

Chemical structure	Optical characteristics	Reference
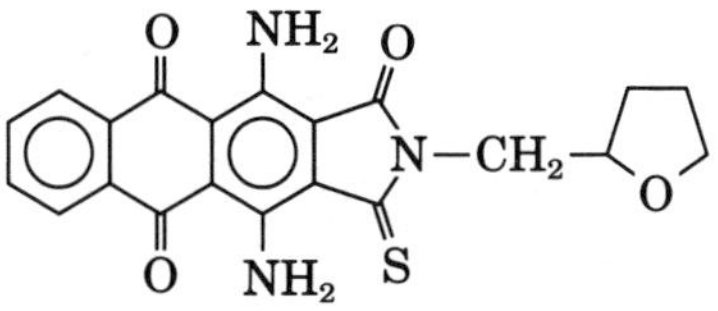	λ_{max} = 753 nm	174
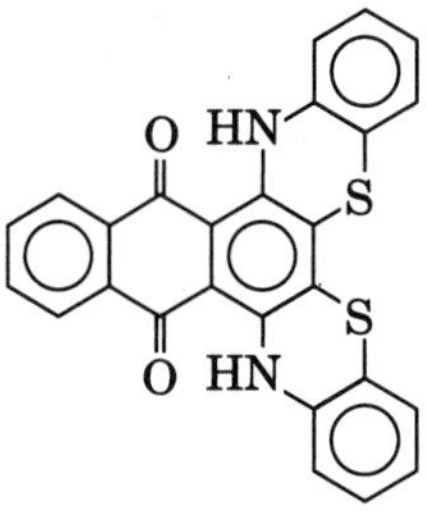	λ_{max} = 712 nm	175
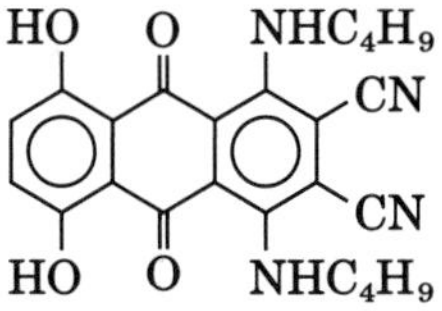	Reflectance at 830 nm = 21%[a]	176

[a]Thickness of recording dye layer = 110 nm when spincoated from a $CHCl_3$ solution on an acrylate resin plate.

Germany
  Bayer
  BASF
  Hoechst
Japan
  Hoechst Mitsubishi Kasei
  Sumitomo Chemical
  Mitsui Badische Dyes
  Nippon Kayaku

Switzerland
  Ciba-Geigy
  Sandoz
UK
  Zeneca
  Holliday
U.S.
  Crompton & Knowles
India
  IDI

In addition to these, some anthraquinone dyes and their intermediates are also produced in Eastern Europe, Russia, China, and Korea. As the result of the history of anthraquinone chemistry, most manufacturers are still located in Western Europe. Most former manufacturers in the United States abandoned the dyestuff business or were acquired by European companies by the middle of the 1980s.

Anthraquinone dyes have been produced for many decades and have covered a wide range of dye classes. In spite of the complexity of production and relatively high costs, they have played an important role in the areas where excellent properties are required, because they have excellent lightfastness and leveling prop-

Table 13. World Demand of Anthraquinone Colorants[a]

Dye classes	Demand, t/yr
vat dyes	24,000
disperse dyes	21,000
reactive dyes	7,000
acid dyes, etc[b]	8,500
solvent dyes	2,000
pigments	1,000

[a]1991 estimates.
[b]Includes mordant dyes.

erties with brilliant shades that are not attainable with other chromophores. However, recent increases in environmental costs have become a serious problem, and future prospects for the anthraquinone dye industry are not optimistic. Some traditional manufacturers have stopped the production of a certain dye class or dye intermediates that were especially burdened by environmental costs, eg, vat dyes and their intermediates derived from anthraquinone-1-sulfonic acid and 1,5-disulfonic acid. However, several manufacturers have succeeded in process improvement and continue production, even expanding their capacity. In the forthcoming century the worldwide framework of production will change drastically.

Consumption. Anthraquinone dyes are the most important dye class after azo dyes. World textile production is estimated in Table 14. Estimates of the consumption of dyes for textiles are given in Figure 14, together with the figures for fiber consumption. This shows that the consumption of each dye class or classes is approximately parallel to the consumption of fibers to which they are applied.

Among these dye classes, anthraquinone dyes are in an important position in reactive dyes and vat dyes for cellulose fibers, disperse dyes for polyester, and acid dyes for polyamide. Application for high performance organic pigments for plastics and paints are also important areas.

Table 14. World Textile Production[a]

Textile	Production, 10^3 t
cotton	18,500
viscose/acetate	3,200
wool/silk	1,950
polyester	8,500
polyamide	3,800
acrylic	2,300
others	1,100

[a]1990 Estimates.

Health and Safety Information

In general, anthraquinone dyes and their intermediates have not been reported as strongly toxic substances, but for many compounds safety data have not been

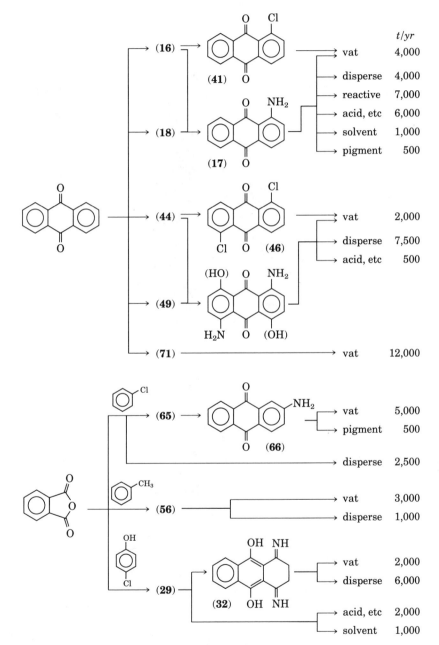

Fig. 13. 1991 estimates in t/yr of anthraquinone dye intermediates and their use. (**16**) is anthraquinone-1-sulfonic acid; (**18**), 1-nitroanthraquinone; (**29**), 1,4-dihydroxyanthraquinone; (**44**), anthraquinone-1,5-disulfonic acid; (**49**), 1,5-dinitroanthraquinone; (**56**), 2-methylanthraquinone; (**65**), 2-chloroanthraquinone; and (**71**) benzanthrone.

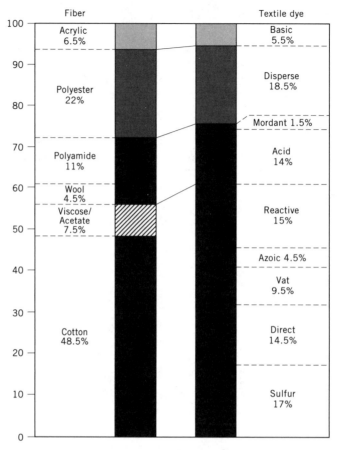

Fig. 14. Comparison of fiber and textile dye consumption (1990 estimates).

evaluated. Acute toxicity data (LD_{50}) of some anthraquinone compounds are given in Table 15 (177). 1-Nitroanthraquinone, 1-chloroanthraquinone, and benzanthrone are reported to cause mild skin irritation in a test with rabbits, 500 mg/24 h (177). Some eye irritation data have been reported (177). Most of the compounds cause mild irritation of eyes in a test with rabbits, 500 mg/24 h. These include:

1,2-dihydroxyanthraquinone	1,5-diphenoxyanthraquinone
1-amino-2,4-dibromoanthraquinone	3-bromobenzanthrone
1-amino-2-bromo-4-hydroxyanthra-	3,9-dibromobenzanthrone
quinone	1-amino-4-hydroxy-2-phenoxy-
1,4-dihydroxyanthraquinone	anthraquinone
1,4-diaminoanthraquinone	1-benzoylamino-5-chloroanthra-
1-methylamino-4-bromoanthra-	quinone
quinone	1-amino-5-chloroanthraquinone
1,5-dichloroanthraquinone	1,1':4,1''-trianthrimide
1,5-dihydroxyanthraquinone	CI Disperse Blue 56
1,5-dimethoxyanthraquinone	CI Vat Brown 1
1,5-dinitroanthraquinone	CI Vat Green 1
1,5-diaminoanthraquinone	CI Vat Green 2

Table 15. Acute Toxicity Data for Some Anthraquinone Compounds[a]

Compound	Structure number	LD_{50}[b] mg/kg
1-nitroanthraquinone	(18)	1,540[c]
		1,050
1,5-dinitroanthraquinone	(49)	4,750[c]
		3,130
1-nitroanthraquinone-2-carboxylic acid	(58)	2,000[d]
leuco-1,4-diaminoanthraquinone	(32)	454[d]
1,4-diaminoanthraquinone	(33)	5,790[d]
		250
1,5-dimethoxyanthraquinone	(48)	16,600[d]
CI Acid Blue 62	(125)	983[e]
benzanthrone	(71)	290[e]
		1,500
3-bromobenzanthrone	(73)	300[e]
		2,400
4,4′-dibenzanthronyl	(72)	1,100[e]
		2,400
3,9-dibromobenzanthrone	(74)	1,410[e]
		4,900
CI Vat Blue 20	(135)	2,600[e]
		5,000
1-nitro-2-methylanthraquinone	(57)	1,100
1,5-diaminoanthraquinone	(51)	1,300
1-aminoanthraquinone	(17)	>10,000[c]
		6,026[e]
		1,500
2-aminoanthraquinone	(66)	1,500
1-benzoylamino-4-chloroanthraquinone	(162)	2,000
1,4-dihydroxyanthraquinone	(29)	2,100
2-amino-3-chloroanthraquinone	(68)	2,400
CI Vat Green 1	(138)	2,600
1-amino-4-hydroxyanthraquinone	(27)	2,700
1-benzoylamino-5-chloroanthraquinone	(159)	3,000
2-chloroanthraquinone	(65)	4,310

[a]Ref. 177.
[b]Intraperitonial, rat, unless otherwise noted.
[c]Oral, mouse.
[d]Oral, rat.
[e]Intraperitonial, mouse.

A few have been found to be moderate eye irritants in rabbits at 100 mg/24 h, including 1-aminoanthraquinone, 1-nitroanthraquinone, leuco-1,4-diaminoanthraquinone, 1-chloroanthraquinone, and benzanthrone. 4,4′-Dibenzanthronyl is reported to cause severe eye irritation (rabbit, 500 mg/24 h).

There are some tumorigenic data for anthraquinone dyes and intermediates which have been evaluated thoroughly. Data for 2-aminoanthraquinone and 2-methyl-1-nitroanthraquinone are available (88,93). 2-Aminoanthraquinone has been assessed by the United Nations International Agency for Research on Cancer (IARC) from studies in animals, and is judged to fall into the *Animal: Limited*

Evidence group. 2-Aminoanthraquinone has been evaluated by EPA (Genetic Toxicology program) and a positive carcinogenic effect for rat and mouse is designated. 2-Methyl-1-nitroanthraquinone has been assessed by IARC and judged as belonging in the *Animal: Sufficient Evidence* group. 2-Methyl-1-nitroanthraquinone has been evaluated by the National Cancer Institute (NCI) and clear evidence of carcinogenicity for rat and mouse is demonstrated.

Most anthraquinone dyes and their intermediates are handled in a powder form. Their dust poses the threat of contact to eyes and skin or contamination of surroundings. Attention must be paid to avoid these hazards. Special attention should be paid to avoid contact with compounds that are recognized to have probable carcinogenicity.

In the case of handling in relatively small quantities, ie, for laboratory use, normal personal equipment, ie, dust masks, safety glasses, and gloves, and hoods with local exhaust ventilation should be used. In plant operations, special technical handling measures should be taken because the possibility of contact is extremely high, especially when charging the raw materials or isolating or packaging the intermediates or final products (see INDUSTRIAL HYGIENE AND PLANT SAFETY).

BIBLIOGRAPHY

"Anthraquinone and Related Quinonoid Dyes" in *ECT* 1st ed., Vol. 1, pp. 959–982, by J. Werner, General Aniline Works, Div. General Aniline & Film Corporation; "Anthraquinone Dyes" in *ECT* 2nd ed., Vol. 2, pp. 501–533, by A.J. Cofrancesco, General Aniline & Film Corporation; "Dyes, Anthraquinone" in *ECT* 3rd ed., Vol. 8, pp. 212–279, by R. H. Chung, General Electric Co. and R. E. Farris, Sandoz Colors & Chemicals; "Anthraquinone Derivatives" in *ECT* 3rd ed., Vol. 2, pp. 708–759, by R. H. Chung, GAF Corp.

1. O. Annen and co-workers, *Rev.Prog.Coloration* **17**, 72 (1987).
2. Brit. Pat. 2,021,622 (Dec. 5, 1979), A. Fukasawa and co-workers (to Sumitomo Chemical).
3. Jpn. Kokai 53 44,550 (Apr. 12, 1976), A. Fukasawa, Y. Maekawa, and M. Yoshimura (to Sumitomo Chemical).
4. Ger. Pat. 2,452,413 (May 13, 1976), G. Epple (to BASF).
5. Ger. Pat. 2,715,072 (Oct. 27, 1977), M. Yoshimura and co-workers (to Sumitomo Chemical).
6. Jpn. Kokai 50 64,257 (May 31, 1975), Y. Hirai and K. Miyata (to Mitsui Toatsu).
7. Jpn. Kokai 50 134,026 (Oct. 23, 1975), K. Yoshiura and co-workers (to Mitsui Toatsu).
8. Ger. Pat. 3,029,302 (Mar. 26, 1981), M. Takahashi and co-workers (to Sumitomo Chemical).
9. Jpn. Kokai 51 109,029 (Sept. 27, 1976), H. Mori, K. Mukai, and K. Yoshiura (to Mitsui Toatsu).
10. Jpn. Kokai 51 92,828 (Aug. 14, 1976), H. Mori and co-workers (to Mitsui Toatsu).
11. Ger. Pat. 2,751,666 (July 20, 1978), A. Fukasawa, S. Masaki, and N. Serizawa (to Sumitomo Chemical).
12. Jpn. Kokai 54 19,958 (Feb. 15, 1979), T. Sakai and co-workers (to Mitsui Toatsu).
13. K. Obata and co-workers, *Kogyo Kayaku* **46** (2), 100 (1985).
14. T. Urbanski, *Chemistry and Technology of Explosives*, Vol. 3, Pergamon Press, Oxford, UK, 1967, p. 288; H. Sprengel, *J. Chem. Soc.* **26**, 796 (1873).

15. Ger. Pat. 2,343,978 (Apr. 3, 1975), K. W. Thiene, R. Neeff, and W. Auge (to Bayer).
16. Jpn. Kokai 2 311,445 (Dec. 27, 1990), N. Sugishima and co-workers (to Nippon Shokubai).
17. H. Suzuki and co-workers, *Chemistry & Industry* **547** (15), (1991).
18. Ger. Offen. 2,205,300 (Sept. 21, 1972), P. Wirth, V. Schmied-Kowarzik, and D. Gudel (to Sandoz).
19. Jpn. Kokai 60 42,363 (Mar. 6, 1985), R. Schmitz and co-workers (to Bayer).
20. Czech. Pat. 169,459 (May 15, 1977), K. Brezina and H. Taeublova; *Chem. Abstr.* **88,** 89440f.
21. Czech. Pat. 122,317 (Mar. 15, 1967), L. Hruska and Rybitui; *Chem. Abstr.* **69,** 96343n.
22. U.S. Pat. 2,581,016 (Jan. 1, 1952), P. Grossmann (to Ciba).
23. *Fridlaender*, Vol. 16, I.G. Farbenind., 1927, p. 1248.
24. U.S. Pat. 4,213,910 (July 22, 1980), R. Muders and co-workers (to Bayer).
25. Jpn. Kokai 58 131,964 (Aug. 6, 1983), M. Hattori and co-workers (to Sumitomo Chemical).
26. U.S. Pat. 4,246,180 (Jan. 20, 1981), H. Leister, H. Dittmer, and H. Schoenhagen (to Bayer).
27. T. Fujita and co-workers, *Sumitomo Kagaku*, **1989**(2), 43 (Nov. 27, 1989).
28. S. Tuong and M. Hida, *J. Chem. Soc.*, 676, (1974).
29. Z. Vrba, *Coll. Czech. Chem. Comm.* **46,** 92 (1981).
30. Ger. Pat. 2,713,575 (Oct. 12, 1978), J. Redekar, H. Hiller, and E. Spohler (to BASF).
31. U.S. Pat. 4,292,247 (Sept. 29, 1981), M. Nishikuri, A. Takeshita, and H. Kenmochi (to Sumitomo Chemical).
32. Jpn. Kokai 53 135,962 (Nov. 28, 1978), K. Kato and co-workers (to Mitsui Toatsu).
33. Eur. Pat. 190,994 (Aug. 13, 1986), R. Muellers, H. Niederers, and O. Hahn (to Ciba-Geigy).
34. Jpn. Kokai 2 204,473 (Aug. 14, 1990), G. Epple (to BASF).
35. U.S. Pat. 3,880,892 (Apr. 29, 1975), H. Hiller and A. Schuhmacher (to BASF).
36. Ger. Offen. 3,843,790 (May 10, 1990), R. Kohlhaupt (to BASF).
37. Ger. Pat. 2,428,337 (Dec. 18, 1975), A. Schuhmacher (to BASF).
38. Ger. Pat. 2,353,532 (May 7, 1975), H. Jaeger (to Bayer).
39. Jpn. Kokai 2 129,157 (May 17, 1990), G. Epple (to BASF).
40. Ger. Pat. 2,830,554 (Jan. 24, 1980), R. Schmitz (to Bayer).
41. Ger. Pat. 2,014,566 (Oct. 21, 1971), A. Schuhmacher (to BASF).
42. Ger. Pat. 2,758,397 (July 13, 1978), M. Grelat (to Ciba-Geigy).
43. Eur. Pat. 423,583 (Apr. 24, 1991), J. Pfister, M. Schiessl, and R. Nachtrab (to BASF).
44. U.S. Pat. 4,739,062 (Apr. 19, 1988), F. Bigi and co-workers (to Consiglio Nazionalle Dolle Ricerche).
45. Jpn. Kokai 61 246,155 (Nov. 1, 1986), R. Blattner (to Ciba-Geigy).
46. Ger. Pat. 1,155,786 (Oct. 17, 1963), K. Maier (to BASF).
47. U.S. Pat. 4,510,087 (Apr. 9, 1985), M. Hattori, A. Taguma, and A. Takeshita (to Sumitomo Chemical).
48. Jpn. Kokai 57 108,059 (July 5, 1982), F. W. Kroeck and R. Neeff (to Bayer).
49. Jpn. Kokai 52 8,032 (Jan. 21, 1977), S. Fujii (to Mitsui Toatsu).
50. Eur. Pat. 189,376 (Jan. 20, 1986), M. Grelat (to Ciba-Geigy).
51. Brit. Pat. 271,023 (Jan. 11, 1926), A. Shepherdson and co-workers (to British Dyestuffs Corp.).
52. Ger. Pat. 488,684 (Feb. 9, 1927), S. Gassner and F. Baumann (to I. G. Farbenind.).
53. U.S. Pat. 3,203,751 (Aug. 31, 1965), J. D. Hildreth (to Toms River Chemical Co.).
54. U.S. Pat. 4,519,947 (May 28, 1985), M. Hattori and co-workers (to Sumitomo Chemical).

55. U.S. Pat. 4,402,605 (Aug. 16, 1977), E. Hartwig (to BASF).
56. U.S. Pat. 4,299,771 (Nov. 10, 1981), A. Takeshita, K. Yokoyama, and M. Hattori (to Sumitomo Chemical).
57. Ger. Offen. 2,931,981 (Feb. 26, 1981), F. W. Kroeck, R. Neeff, and H. Scheiter (to Bayer).
58. Ger. Offen. 3,644,824 (July 14, 1988), P. Miederer and E. Michaelis (to BASF).
59. Jpn. Kokai 62 234,057 (Oct. 14, 1987), J. M. Adam (to Ciba-Geigy).
60. Ger. Pat. 2,163,674 (June 28, 1973), R. Schmitz and C. Wittig (to Bayer).
61. *B.I.O.S. Final Report* **1493,** 29 (1946).
62. Ger. Pat. 2,654,650 (June 8, 1978), H. Seidler, N. Majer, and H. Judat (to Bayer).
63. Belg. Pat. 626,554 (June 27, 1963), (to Ciba).
64. *B.I.O.S. Final Report* **1484,** 10 (1946).
65. *F.I.A.T. Report 1313* **2,** 54 (1948).
66. Ger. Pat. 2,720,965 (Nov. 16, 1978), H. Herzog, W. Hohmann, and H. Seidler (to Bayer).
67. *B.I.O.S. Final Report* **1484,** 16 (1946).
68. Jpn. Kokai 54 22,357 (Feb. 20, 1979), H. Aiga and M. Kawakami (to Mitsui Toatsu).
69. Jpn. Kokai 49 5,429 (Jan. 18, 1974), E. Yamada and co-workers (to Sumitomo Chemical).
70. Ger. Offen. 2,351,590 (Apr. 17, 1975), W. Hohmann (to Bayer).
71. Y. Takeda, H. Imafuku, and H. Kawamura, *Nikkakyo Geppo*, **38**(11), 20 (1985).
72. Jpn. Kokai 53 34,765 (Mar. 31, 1978), Y. Takeda, S. Koga, and F. Fukuda (to Mitsubishi Chemical).
73. Jpn. Kokai 54 30,145 (Mar. 6, 1979), Y. Takeda and co-workers (to Mitsubishi Chemical).
74. Jpn. Kokai 53 132,552 (Nov. 18, 1978), Y. Takeda, S. Koga, and Y. Fukuda (to Mitsubishi Chemical).
75. Ger. Pat. 2,827,197 (Jan. 10, 1980), R. Braden and co-workers (to Bayer).
76. Swiss Pat. 586,188 (Mar. 31, 1977), F. Krenmuller and co-workers (to Sandoz).
77. Jpn. Kokai 51 92,829 (Aug. 14, 1976), E. Yamada and co-workers (to Sumitomo Chemical).
78. *B.I.O.S. Final Report* **1484,** 14 (1946).
79. *F.I.A.T. Report 1313* **2,** 43 (1948).
80. Jpn. Kokai 54 3,051 (Jan. 11, 1977), Y. Takeda and co-workers (to Mitsubishi Chemical).
81. Ger. Pat. 2,855,939 (July 10, 1980), H. Herzog and G. Gehrke (to Bayer).
82. Ger. Pat. 2,607,036 (Sept. 9, 1976), Y. Kimura and co-workers (to Mitsubishi Chemical).
83. Jpn. Kokai 50 131,962 (Oct. 18, 1975), Y. Shimizu and co-workers (to Sumitomo Chemical).
84. Jpn. Kokai 53 101,355 (Sept. 4, 1978), Y. Kimura and co-workers (to Mitsubishi Chemical).
85. Jpn. Kokai 51 98,262 (Aug. 30, 1976), Y. Kimura and co-workers (to Mitsubishi Chemical).
86. *F.I.A.T. Final Report 1313* **2,** 39 (1948).
87. *B.I.O.S. Final Report* **987,** 13 (1946).
88. *RTECS Microfisch Quarterly* no. 10853, (Jan. 1992).
89. Ger. Offen. 3,840,341 (June 7, 1990), J. Henkelmann and co-workers (to BASF).
90. Eur. Pat. 410,315 (Jan. 30, 1990), R. Helwig and H. Hoch (to BASF).
91. *B.I.O.S. Final Report* **987,** 8,11 (1946).
92. *B.I.O.S. Final Report* **987,** 20 (1946).
93. *RTECS Microfisch Quarterly* no. 10750 (Jan. 1992).
94. *B.I.O.S. Final Report* **987,** 10 (1946).

95. Ger. Offen. 2,705,106 (Aug. 10, 1978), W. Sweckendiek, H. Hiller, and R. Kohlhaupt (to BASF).
96. *B.I.O.S. Final Report* **987,** 14 (1946).
97. Ger. Pat. 2,631,833 (Jan. 19, 1978), A. Schuhmacher and E. Kling (to BASF).
98. Jpn. Kokai 2 225,569 (Sept. 7, 1990), T. Kondo and co-workers (to Daiei Chemical and Sumitomo Chemical).
99. U.S. Pat. 4,351,765 (Sept. 28, 1982), T. Fujita and co-workers (to Sumitomo Chemical).
100. U.S. Pat. 4,315,865 (Feb. 16, 1982), E. Hoyer, H. H. Steuernagel, and D. Wagner (to Hoechst).
101. H. Kenmochi, Y. Tezuka, and A. Takeshita, *Sumitomo Kagaku*, **1980**(2), 25(1980).
102. Jpn. Kokai 60 108,472 (June 13, 1985), S. Fujino, Y. Yamada, and N. Miyahara (to Kasei Hoechst).
103. *Registry Handbook, Registry Number Update, (1965–1990)*, 11119-73-6, American Chemical Society, Washington, D.C.
104. Jpn. Pat. 39 24,655 (Nov. 2, 1964), H. Riet and co-workers (to Ciba-Geigy).
105. A. H. M. Renfrew, *Rev. Prog. Coloration* **15,** 15 (1985).
106. *Chemical Week*, **147**(13), 26 (Oct. 3, 1990).
107. Jpn. Kokai 48 37,432 (June 2, 1973), T. Mori and M. Saito (to Nippon Kayaku).
108. Ger. Pat. 2,910,716 (Oct. 2, 1980), H. Hiller, E. Eilingsfeld, and H. Reinicke (to BASF).
109. U.S. Pat. 4,710,320 (Dec. 1, 1987), M. Hattori, M. Nishikuri, and Y. Ueda (to Sumitomo Chemical).
110. S. Abeta and K. Imada, *Kaisetsu Senryo Kagaku (Comprehensive Dyestuff Chemistry)*, Shikisensha, Osaka, 1989, p. 310.
111. Jpn. Kokai 58 27,752 (Feb. 18, 1983), P. Kniel (to Ciba-Geigy).
112. Jpn. Kokai 63 43,961 (Feb. 25, 1988), (to Ciba-Geigy).
113. U.S. Pat. 2,628,963 (Feb. 17, 1951), J. F. Laucius and S. B. Speck (to E. I. duPont de Nemours & Co.).
114. Ger. Pat. 1,918,696 (Dec. 17, 1969), E. Hartwig (to BASF).
115. Ref. 110, p. 312.
116. Eur. Pat. 119,465 (Sept. 26, 1984), U. Karlen, R. Putzar, and R. Schaulin (to Ciba-Geigy).
117. Jpn. Kokai 49 18,918 (Feb. 19, 1974), M. Kurosawa and co-workers (to Nippon Kayaku).
118. Jpn. Kokai 51 116,830 (Oct. 14, 1976), M. Yamada, and M. Nishikuri (to Sumitomo Chemical).
119. Ger. Offen. 3,705,386 (Sept. 1, 1988), E. Michaelis and H. Hoch (to BASF).
120. Ger. Pats. 2,436,459 and 2,436,460 (Feb. 20, 1975), M. Grelat (to Ciba-Geigy).
121. Ger. Pat. 2,536,051 (Feb. 17, 1977), K. Maier and E. Hartwig (to BASF).
122. Jpn. Kokai 51 132,219 (Nov. 17, 1976), M. Nishikuri and H. Korenaga (to Sumitomo Chemical).
123. Jpn. Kokai 59 89,360 (May 23, 1984), M. Nishikuri and H. Korenaga (to Sumitomo Chemical).
124. Ger. Offen. 3,109,951 (Sept. 23, 1982), R. Niess and K. Schaeffer (to BASF).
125. Jpn. Kokai 47 26,413 (July 17, 1972), Y. Sato and co-workers (to Mitsubishi Chemical).
126. G. Hallas in J. Shore, ed., *Colorants and Auxiliaries*, Society of Dyers and Colourists, Bradford, UK, 1990, pp. 243–244.
127. Swiss Pat. 460,212 (Dec. 21, 1961), O. Fuchs and F. Ische (to Hoechst).
128. Brit. Pat. 770,570 (Mar. 20, 1957), P. Grossmann, W. Jenny, and W. Kern (to Ciba).
129. *B.I.O.S. Final Report* **1484,** 41 (1946); *B.I.O.S. Final Report* **987,** 135 (1946).
130. Ref. 110, p. 66.

131. Ger. Pat. 2,710,152 (Mar. 9, 1978), W. Harms and H. G. Otten (to Bayer).
132. M. Kikuchi, T. Yamagishi, and M. Hida, *Dyes and Pigments* **2,** 143 (1981).
133. *B.I.O.S. Final Report* **1484,** 48 (1946).
134. *F.I.A.T. Final Report 1313* **2,** 215 (1948).
135. Jpn. Kokai 54 93,021 (July 23, 1979), Y. Torisu and co-workers (to Mitsui Toatsu).
136. Jpn. Kokai 55 142,053 (Nov. 6, 1980), Y. Torisu and co-workers (to Mitsui Toatsu).
137. Jpn. Kokai 60 1,169 (Jan. 7, 1985), K. Kato and co-workers (to Mitsui Toatsu).
138. Jpn. Kokai 62 25,167 (Feb. 3, 1987), A. Hirayama, K. Kato, and T. Sakai (to Mitsui Toatsu).
139. Eur. Pat. 199,670 (Oct. 29, 1986), R. Blattner (to Ciba-Geigy).
140. Ger. Offen. 3,903,623 (Aug. 9, 1990), H. Hoch and G. Kilpper (to BASF).
141. Ger. Pat. 2,259,329 (June 13, 1974), A. Schuhmacher (to BASF).
142. Ger. Offen. 3,910,596 (Oct. 4, 1990), H. Steuernagel (to Hoechst).
143. *B.I.O.S. Final Report* **1484,** 39 (1946).
144. *F.I.A.T. Final Report 1313* **2,** 44 (1948).
145. *F.I.A.T. Final Report 1313* **2,** 57 (1948).
146. *B.I.O.S. Final Report* **1484,** 26 (1946).
147. *F.I.A.T. Final Report 1313* **2,** 210 (1948).
148. M. Okawara, N. Kuroki, and T. Kitao, *Kinousei-Sikiso no Kagaku* (*The Chemistry of Functional Dyes*), CMC, Tokyo, 1981.
149. Jpn. Kokai 62 25,092 (Feb. 3, 1987), W. Black, R. Bladbally, and P. Gregoly (to ICI).
150. Jpn. Kokai 62 138,559 (June 27, 1987), M. Kutsukake and J. Kanto (to Dai Nippon Printing).
151. Jpn. Kokai 62 156,371 (July 1, 1987), M. Kutsukake and J. Kanto (to Dai Nippon Printing).
152. Jpn. Kokai 60 253,595 (Dec. 14, 1985), K. Hashimoto and A. Takeshita (to Sumitomo Chemical).
153. Jpn. Kokai 61 262,190 (Nov. 20, 1986), K. Hashimoto, A. Takashita, and M. Nishikuri (to Sumitomo Chemical).
154. Eur. Pat. 365,392 (Apr. 25, 1990), K. Hashimoto and Y. Suzuki (to Sumitomo Chemical).
155. Jpn. Kokai 60 131,292 (June 12, 1985), T. Niwa, Y. Murata, and S. Maeda (to Mitsubishi Chemical).
156. Jpn. Kokai 61 284,489 (Dec. 15, 1986), T. Suzuki, J. Kanto, and M. Akada (to Dai Nippon Printing).
157. Jpn. Kokai 62 124,152 (June 5, 1987), M. Kutsukake and J. Kanto (to Dai Nippon Printing).
158. Jpn. Kokai 62 292,858 (Dec. 19, 1987), M. Kutsukake and J. Kanto (to Dai Nippon Printing).
159. Jpn. Kokai 60 53,563 (Mar. 27, 1985), T. Niwa, Y. Murata, and S. Maeda (to Mitsubishi Chemical).
160. Jpn. Kokai 60 151,097 (Aug. 8, 1985), T. Niwa, Y. Murata, and S. Maeda (to Mitsubishi Chemical).
161. Jpn. Kokai 63 278,969 (May 11, 1987), K. Miura and co-workers (to Mitsubishi Chemical).
162. Jpn. Kokai 2 173,090 (July 4, 1990), M. Kaneko and N. Nakajima (to Mitsubishi Chemical).
163. Jpn. Kokai 61 51,062 (Mar. 13, 1986), Y. Morishita and co-workers (to Nippon Kayaku).
164. Jpn. Kokai 62 5,941 (Jan. 12, 1987), Y. Morishita and co-workers (to Nippon Kayaku).
165. Jpn. Kokai 62 127,352 (June 9, 1987), Y. Shimizu (to Sumitomo Chemical).
166. Eur. Pat. 300,770 (Jan. 25, 1989), T. Misawa and co-workers (to Mitsui Toatsu).

167. Jpn. Kokai 58 124,621 (July 25, 1983), J. Fujii and co-workers (to Mitsui Toatsu).
168. Jpn. Kokai 58 68,008 (Apr. 22, 1983), Y. Yamada and co-workers (to Mitsui Toatsu).
169. Jpn. Kokai 61 87,757 (May 6, 1986), K. Nakamura and co-workers (to Mitsui Toatsu).
170. T. Uchida, *Nikkei New Materials* **93,** 48 (Feb. 25, 1991).
171. Jpn. Kokai 63 314,501 (Dec. 22, 1988), T. Shimizu and co-workers (to Matsushita Electric).
172. Jpn. Kokai 64 35,417 (Feb. 6, 1989), T. Shimizu and M. Soga (to Matsushita Electric).
173. Jpn. Kokai 3 36,502 (Feb. 18, 1991), Y. Hirasawa and co-workers (to Nippon Kayaku).
174. Jpn. Kokai 60 255,853 (Dec. 17, 1985), K. Miura, T. Ozawa, and J. Iwanami (to Mitsubishi Chemical).
175. Jpn. Kokai 60 250,065 (Dec. 10, 1985), T. Kitao and M. Matsuoka (to Mitsubishi Chemical).
176. Jpn. Kokai 62 21,584 (Jan. 29, 1987), N. Ito and H. Aiga (to Mitsui Toatsu).
177. *Registry of Toxic Effects of Chemical Substances, 1985–1986; RTECS Microfisch Quarterly* (Jan. 1992).

General References

H. Zollinger, *Color Chemistry*, VCH Verlagesgesellshaft mbH, Weinheim, Germany, 1987.

H. Zollinger, *Color Chemistry*, 2nd rev. ed., VCH Verlagesgesellshaft mbH, Weinheim, Germany, 1991.

G. Hallas in J. Shore, ed., *Colorants and Auxiliaries*, Society of Dyers and Colourists, Bradford, UK, 1990, pp. 230–267.

G. Booth, *The Manufacture of Organic Colorants and Intermediates*, Society of Dyers and Colourists, Bradford, UK, 1988.

S. Abeta and K. Imada, *Kaisetsu Senryo Kagaku (Comprehensive Dyestuff Chemistry)*, Shikisensha, Osaka, 1989.

K. Venkataraman, *The Chemistry of Synthetic Dyes*, Vol. 2, Academic Press, Inc., New York, 1952.

F. B. Stilmar and co-workers in H. A. Lubs, ed., *The Chemistry of Synthetic Dyes and Pigments*, Reinhold Publishing Corp., New York, 1955, pp. 335–550.

H. R. Schweizer, *Künstriche Organische Farbstoffe und ihre Zwischenproducte*, Springer-Verlag, Berlin, 1964, pp. 301–320.

MAKOTO HATTORI
Sumitomo Chemical Company

DYES, APPLICATION AND EVALUATION

The global consumption of textiles is estimated at around 30 million t, and this is expected to grow at 3% per year. The coloration of this amount needs ca 700,000 t of dye, with a value of $4400 million (1,2). The principal reasons for coloring textiles are for aesthetic appearance and decoration or for utilitarian purposes, and unless there is an unpredicted change in human behavior the majority of textiles will continue to be dyed to produce colored apparel, home furnishings, carpets, etc. Among the aesthetic uses are fashion garments and household articles such as drapes, towels, and carpets. In the utilitarian group are uniforms (military and civil), and work wear.

In order for a colored substance to be regarded as a dyestuff, a number of requirements must be satisfied. A dyestuff must be substantive for a textile and exhaust from an aqueous solution into the fiber; have a high exhaustion; exhaust at a rate allowing economic processing; give a uniform level dyeing; and have satisfactory fastness for the particular end use the textile is intended for. The process of dyeing is therefore a combination of chemistry, application technology, economics, and customer needs.

Dyes can be classified according to chemistry, shade, application conditions, fastness, etc (see DYES AND DYE INTERMEDIATES). In this article the traditional classification by application method is used.

Initially all coloring matters were of natural origin obtained from plants and even animals. Woad and indigo (violet), madder (red), and weld (yellow) were typical of coloring matters extracted from plants. Coloring matters obtained from animals included Tyrian Purple from a species of whelk and the reds cochineal and kermes from insects. The actual colors used would depend on geographical availability. In one area the coloring matters would not necessarily be the same as other areas. As the number of available coloring matters increased, either by discovery, improved communications, or trading, it became possible to be selective. Compounds with poorer properties were no longer used. For example the use of kermes in Europe, which was available locally, declined in favor of first lac (from India), and then cochineal from India and the Americas as a result of the improved brightness and color strength of the latter compounds. By the time of the Renaissance, coloring matters derived from safflower and Brazilwood were available. Gradually the range of dyes available increased but always from natural sources (see DYES, NATURAL).

The first synthetic dyestuff, mauveine, was discovered by Perkin in 1856 in the UK and led to many investigations of the derivatives of coal tar as potential coloring matters. The first diazonium salt derived from picramic acid was prepared in 1858 and is still the basic chemistry behind countless commercial products. Despite these inventions, in 1900 the vast majority of dyestuffs were still of natural origin. Now at the end of the century the situation has changed dramatically; synthetic dyes dominate.

The Case For Synthetic Dyes

With the increased awareness of environmental and green issues there has been an increased interest in natural dyes accompanied by a lobby for natural fibers in favor of synthetic fibers. However, studies have shown (3) that if 1990's volume of cotton were colored with natural dyes, at least 31% of the available world's agricultural land would be needed to cultivate the requisite plants. The estimate is based on it taking, on average, 440 g of fresh dye plant to achieve the same tinctorial effect as 1 g of synthetic dye.

Contrary to proponents' assumptions, natural dyes are not necessarily environmentally friendly. First, large amounts of plant waste would be produced because of the low dye content in plants, eg, 170 million t of waste to color the cotton volume. Also, in order to dye fabric with natural dyes, "mordants," which are usually based on heavy metals, have to be used. For example, madder is applied with tin (Sn) or aluminum (Al). Weld is applied with tin salts. Salts based on copper and iron can also be used.

Also, the cost of dyeing cotton with natural vegetable dyes is $31–77/kg compared to a synthetic approach costing 35 cents/kg. In short, the use of natural vegetable dyes to color the world's textiles is both environmentally and logistically impossible (4).

Classification of Dyestuffs According to Application

The *Colour Index* (5), categorizes all coloring matters according to application characteristics. The following are the main types currently of interest as dyes.

Acid Dyes. These are anionic dyes, usually containing sulfonic acid groups, that are substantive to wool, other protein fibers, and polyamides when dyed from an acidic dyebath. The lower the pH the more rapid the dyeing, and exhaustion efficiency is enhanced by increased acidity.

Mordant Dyes. This group includes many natural as well as synthetic dyes. They have no or low substantivity for textile fibers and are therefore applied to cellulosic or protein fibers that have been treated (mordanted) with metallic oxides to give points of attraction for the dye. The dye forms a complex with the metal and depending on the metal and fiber can simply form a large macromolecule incapable of desorbing, or a dye molecule bound to the fiber resulting from chelation with the metal. An important subgroup is chrome dyes where wool is treated with Cr^{3+} with which it reacts; dye is then applied which in turn complexes with the chromium.

Metal Complex Dyes.   Metals such as chromium and cobalt can be introduced into dye molecules to give larger molecules. They can be regarded as being a special form of mordant dye. The complexes can be formed by chelating one or two molecules of dye with metal. They are applied in a similar manner to acid dyes.

Direct Dyes.   These are defined as anionic dyes, again containing sulfonic acid groups, with substantivity for cellulosic fibers. They are usually azo dyes (qv) and can be mono-, dis-, or polyazo, and are in general planar structures. They are applied to cellulosic fibers from neutral dyebaths, ie, they have direct substantivity without the need of other agents. Salt is used to enhance dyebath exhaustion. Some direct dyes can be applied to wool and polyamides under acidic conditions, but these are the exception.

Fiber-Reactive Dyes.   These dyes can enter into chemical reaction with the fiber and form a covalent bond to become an integral part of the fiber polymer. They therefore have exceptional wetfastness. Their main use is on cellulosic fibers where they are applied neutral and then chemical reaction is initiated by the addition of alkali. Reaction with the cellulose can be by either nucleophilic substitution, using, for example, dyes containing activated halogen substituents, or by addition to the double bond in, for example, vinyl sulfone, $-SO_2CH{=}CH_2$, groups.

Basic Dyes.   These are usually the salts of organic bases where the colored portion of the molecule is the cation. They are therefore sometimes referred to as cationic dyes. They are applied from mild acid, to induce solubility, and applied to fibers containing anionic groups. Their main outlet is for dyeing fibers based on polyacrylonitrile (see FIBERS, ACRYLIC).

Vat Dyes.   The basic mechanism of vat dye application is the conversion of an insoluble complex polycylic molecule based on the quinone structure into a soluble leuco form by treatment with alkaline-reducing agents. This leuco form is then absorbed onto cellulose. Once the dye has been exhausted into the cellulose it is reconverted *in situ* to the insoluble pigment form which is trapped within the fiber. These dyes have high wet- and lightfastness. A subgroup of vat dyes is the solubilized vat dyes which are temporarily solubilized to allow easy application without reducing agents followed by regeneration of the insoluble dye after dyeing. These dyes are no longer of commercial importance.

Sulfur Dyes.   These are complex molecules containing sulfur obtained from the reaction between selected organic intermediates such as 4-aminophenol, or *p*-phenylenediamine and molten sulfur or polysulfide. The actual structures of sulfur dyes are largely unknown although it is considered that they possess sulfur-containing heterocyclic rings. They are applied like vat dyes with the leuco form being generated by using alkaline sodium sulfide as a reducing agent.

Disperse Dyes.   These are substantially water-insoluble dyes applied from aqueous dyebath in a finely dispersed form. They are the most important class of dye for dyeing hydrophobic synthetic fibers such as polyester and acetates.

Ingrain Dyes/Azoic Dyes.   These are dyes that are formed in the fiber by applying precursors. An example of this class are the azoic dyes. With these dyes a coupling component is applied to the fabric followed by a diazonium compound to form the insoluble dyes. Alternatively, a stabilized mix of the two can be applied and the insoluble azoic dye created on the fiber in a separate treatment, eg, acid steaming. This is a traditional method for obtaining bright heavy shades cheaply but has lost some popularity as a result of poor rubbing fastness and the decreasing availability of the amines and diazonium salts.

Other dyes in this group are phthalocyanine compounds which still have commercial importance, particularly in textile printing.

Other Dyes.   Other dye classes listed in the *Colour Index* (5) include dyes for leather, solvents, paper, and food. Leather dyes are those acid, direct, mordant, and basic dyes that show substantivity for leather, good diffusion into it, and acceptable fastness. They are essentially applied in an analogous manner to acid or basic dyes. Paper is colored by both inorganic pigments and natural and synthetic organic colorants. The main dyes used are basic, acid, and direct dyes. Solvent dyes can be regarded as similar to disperse dyes. They are small, unsulfonated molecules, plus a few basic dyes that show high solubility in solvents. Finally food dyes, nontoxic colored substances that can be added to food, are included (see COLORANTS FOR FOOD, DRUGS, COSMETICS, AND MEDICAL DEVICES).

Fluorescent Whitening Agents.   These are fluorescent substances that transform invisible ultraviolet (uv) light into visible blue light. Fluorescent whitening agents (qv) change the appearance of substrates in two ways: by emitting light and therefore increasing the luminosity (brightness); and by changing the shade from yellowish white to bluish white. They are unfavorably influenced by factors such as low uv content light and high uv light absorption of the substrate or other chemicals present on the substrate. Like dyes, fluorescent whitening agents are available in classes analogous to acid, basic, direct, and disperse dyes for application to all substrates, including paper.

Physical and Organic Chemistry of Dyes and the Dyeing Process

The practical characteristic of a dyestuff is that when a textile is immersed in a solution containing a dye, the dye preferentially adsorbs onto and diffuses into the textile. The thermodynamic equations defining this process have been reviewed in detail (6–9). The driving force for this adsorption process is the difference in chemical potential between the dye in the solution phase and the dye in the fiber phase. The chemical potential μ is defined as the increase in the molar free energy of the system resulting from the addition at constant temperature and pressure of 1 mole of component to such a large quantity of the system that its composition remains virtually unchanged. In practice it is only necessary to consider changes in chemical potential and to understand that the driving force is the reduction in free energy associated with the dye molecule moving from one phase to the other, as the molecule always moves to the state of lowest chemical potential. If the chemical potential is lower in the fiber phase than the aqueous phase (the dyebath), the dye molecules will adsorb onto the fiber. The greater the drop in chemical potential the stronger the driving force for dye molecules to move from the dyebath onto the fiber.

Thus the necessary chemical properties of dyes are a chemical structure that imparts colors and chemical characteristics that result in the molecule having a lower chemical potential in a fiber than in the dyebath.

The chemical potential in the dyebath solution, μ_s, is defined in equation 1 where μ°_s is the standard chemical potential in the solution, R is the gas constant, T is temperature in K, and a is activity.

$$\mu_s = \mu^\circ_s + RT \log_e a_s \tag{1}$$

For the dye in the fiber:

$$\mu_f = \mu^\circ_f + RT \log_e a_f \tag{2}$$

In the dyeing process absorption from the dyebath solution to the fiber eventually stops when an equilibrium exists between the dye in the fiber phase and the dye in the solution phase. At this point $\mu_s = \mu_f$ by definition (no movement of dye molecules), therefore

$$\mu^\circ_f + RT \log_e a_f = \mu^\circ_s + RT \log_e a_s \tag{3}$$

or rearranging

$$-\Delta\mu^\circ = -(\mu^\circ_f - \mu^\circ_s) = RT \log_e a_f/a_s \tag{4}$$

The standard affinity of the dyeing process, $-\Delta\mu^\circ$, is a measure of the desire of the dye to move from its standard state in the dyebath to its standard state in the fiber and it can be regarded as the driving force of the dyeing process. It is often referred to simply as affinity. The negative sign is used in order that positive values can be quoted to describe the decrease in chemical potential. A high affinity describes a dye that has a strong tendency to move from the dyebath to the fiber and stay there. A dye of low affinity has a low tendency to move to the fiber phase, and also is more easily desorbed than a high affinity dye.

Affinity values are obtained by substituting concentration for activity in equation 4 for the dye and, where appropriate, other ions in the system. A number of equations are used depending on the dye–fiber combination (6). An alternative term used is the substantivity ratio which is simply the partition between the concentration of dye in the fiber and dyebath phases. The values obtained are specific to a particular dye–fiber combination, are insensitive to liquor ratios, but sensitive to all other dyebath variables. If these limitations are understood, substantivity ratios are a useful measure of dyeing characteristics under specific application conditions.

Influence of the Fiber.   In order for a dye to move from the aqueous dyebath to the fiber phase the combination of dye and fiber must be at a lower energy level than dye and water. This in turn implies that there is a more efficient, lower energy sharing of electrons or intramolecular energy forces, and there are a number of mechanisms that allow this to happen.

Fibers exist as natural, or synthetic, hydrophilic, hydrophobic, nonionic, and ionic. Natural fibers have complex chemical structures with a multitude of possible points of attraction for a dyestuff and are difficult to characterize because of the structure being strongly influenced by regional, climatic variations and the species of plant or animal. Dyeing of natural fibers is therefore much more complex than dyeing synthetic fibers where structures can be characterized and the availability of points of attraction can be deliberately engineered into the fiber's molecular chain. The various types of fiber are summarized in Table 1. The fiber type dictates the type of dye needed. Hydrophilic fibers are hydrated and need water-soluble dyestuffs. This water solubility is brought about by introducing ionic groups into the molecule (groups capable of being hydrated) which can be either cationic or anionic.

Hydrophobic fibers are difficult to dye with ionic (hydrophilic) dyes. The dyes prefer to remain in the dyebath where they have a lower chemical potential.

Table 1. Fiber–Dye Property Requirements

Fiber name[a]	Type/general classification	Chemical constitution	Ionic nature in dyebath
cotton, linen, and other vegetable fibers	natural, hydrophilic	cellulose	anionic
viscose rayon	synthetic,[b] hydrophilic	regenerated cellulose	anionic
wool, silk, hair	natural, hydrophilic	complex proteins	cationic
nylon	synthetic, somewhat hydrophobic	polyamide	usually cationic; can be anionic depending on pH and/or chemical modification
acrylics	synthetic, hydrophobic	modified polyacrylonitriles	anionic
acetate	synthetic,[b] hydrophobic	acetylated cellulose	nonionic
triacetate	synthetic,[b] hydrophobic	acetylated cellulose	nonionic
polyester	synthetic, hydrophobic	polyester	usually nonionic; can be made dyeable with basic dyes by chemical modification
polypropylene	synthetic, hydrophobic	polyolefin	nonionic; not dyeable unless modified

[a]See FIBERS, SURVEY.
[b]Some references distinguish between synthetic fibers made from synthetic polymers and those made by modification of cellulose (man-made fibers).

Therefore nonionic, hydrophobic dyes are used for these fibers. The exceptions to the rule are polyamide and modified polyacrylonitriles and modified polyester where the presence of a limited number of ionic groups in the polymer, or at the end of polymer chains, makes these fibers capable of being dyed by water-soluble dyes.

MODES OF ATTRACTION

The force of attraction between a dye and fiber results from the usual electronic interactions.

Ionic Forces (Coulombic Attraction). Water-soluble dyes and hydrophilic fibers containing salt groups dissociate to varying degrees in water depending on pH. Thus, in the dyebath, ionic fibers are charged either negatively or positively and so is the chromogen, the colored part of the dyestuff. Soluble dyestuffs are salts of complex chromogens and a simple anion, eg, Cl^-, or cation Na^+.

There is a strong attraction between dye and fiber when oppositely charged. The mobile charged dyestuff chromogen "moves" from the external water phase to the stationary internal fiber phase. For like-charged entities repulsion occurs, and no movement of dyestuff from the water to the fiber takes place.

Ion-Dipole Forces. Ion-dipole forces bring about solubility resulting from the interaction of the dye ion with polar water molecules. The ions, in both dye and fiber, are therefore surrounded by bound water molecules that behave differently from the rest of the water molecules. If when the dye and fiber come together some of these bound water molecules are released, there is an increase in the entropy of the system. This lowers the free energy and chemical potential and thus acts as a driving force to dye absorption.

Hydrogen Bonds. Originally the formation of hydrogen bonds between the dye and fiber chain was regarded as being a primary force of attraction, but this is not the case for hydrophilic fibers because it was realized that all hydrogen bonding possibilities are accounted for by the water of hydration. Hydrogen bonding may provide attraction for hydrophobic fibers where water molecules have not satisfied their hydrogen bonding potential.

Charge-Transfer Forces. An electron-rich atom, or orbital, can form a bond with an electron-deficient atom. Typical examples are lone pairs of electrons, eg, in nitrogen atoms regularly found in dyes and protein and polyamide fibers, or π-orbitals as found in the complex planar dye molecules, forming a bond with an electron-deficient hydrogen or similar atom, eg, $-O^{\delta-}H^{\delta+}$. These forces play a significant role in dye attraction.

van der Waals Forces. The key feature of van der Waals forces is that they are exceptionally strong at close range, but weak at long range. Although there is long-range coulombic repulsion between negatively charged dyes and negatively charged fibers, if the dye and fiber molecules can be brought close together, then the stronger van der Waals attractive forces dominate and cause the dye to be strongly attracted to the fiber. This is the key phenomenon present in dyeing cellulosic fibers.

Hydrophobic Interaction. This is the tendency of hydrophobic groups, especially alkyl chains such as those present in synthetic fibers, and disperse dyes to associate together and escape from the aqueous environment. Hydrophobic bonding is considered (7) to be a combination of van der Waals forces and hydrogen bonding taking place simultaneously rather than being a completely new type of bond or intermolecular force.

Covalent Bonds. Fiber-reactive dyes, ie, dyestuff molecules containing reactive groups, are adsorbed onto the fiber and react with specific sites (chemical groups) in the fiber polymer to form covalent bonds. The reaction is irreversible, so active dye is removed from the equilibrium system (it becomes part of the fiber) and this causes more dye to adsorb onto the fiber to re-establish the equilibrium of active dye between fiber and aqueous dyebath phases (see DYES, REACTIVE).

DYESTUFF ORGANIC CHEMISTRY

Dyestuffs impart color to textiles because of their ability to absorb electromagnetic radiation in the wavelengths visible to the human eye (400–650 nm). When white light strikes a dyestuff molecule certain wavelengths are absorbed, depending on the molecular construction, and others are reflected. The wavelengths of the reflected light give the specific color of the dyestuff. A black or dark shade absorbs across all wavelengths reflecting only a small quantity of light. A white or pale shade absorbs very little light.

Dyestuff organic chemistry is concerned with designing molecules that can selectively absorb visible electromagnetic radiation and have affinity for the specified fiber, and balancing these requirements to achieve optimum performance. To be colored the dyestuff molecule must contain unsaturated chromophore groups, such as azo, nitro, nitroso, carbonyl, etc. In addition, the molecule can contain auxochromes, groups that supplement the chromophore. Typical auxochromes are amino, substituted amino, hydroxyl, sulfonic, and carboxyl groups.

There is little correlation between classifications according to chemical type and application properties. Application classifications are of most practical usefulness to the dyer, and therefore the chemical constitutions of dyes are described here only briefly. Further detailed information on dye types (10) and their chemical manufacture (11) can be found elsewhere, and in many other *Encyclopedia* articles to which references are made.

Azo Dyes. Azo dyes (qv) are the most numerous chemical group and cover all shades from yellow to navy. They have relatively high extinction coefficients, ϵ, a measure of their efficiency in absorbing specific wavelengths and reflecting others. Lesser amounts, as a % by weight on the fiber, are required of a dye with a high extinction coefficient than dyes with low extinction coefficients to give the same depth of shade, ie, high ϵ means high tinctorial strength.

Azo dyes are obtained by diazotizing aromatic amines and coupling the diazonium salt to organic coupling components. The products may be monoazo $R—N=N—R_1$, diazo $R—N=N—R_1—N=N—R_2$, or polyazo.

The disadvantage of azo dyes is that with bathochromic (dark) shades the ability to preferentially absorb light at only specific wavelengths is reduced with a resulting dulling of the shade. Further, if the azo chromophore is destroyed, the molecule becomes colorless or at best pale yellow and this can happen by reductive and oxidative photochemical reactions causing azo dyes to slowly fade on exposure to strong light. Nevertheless they are used to color the vast majority of apparel and household textiles.

By careful selection of the diazonium and coupling components and the other substituents in the molecule that are either electron withdrawing or donating, fine tuning of the absorption characteristics to give the desired shade is possible. Sulfonic acid groups are incorporated to give solubility, and the bigger the molecule the more are needed to achieve the desired solubility. The selection of the diazonium compound and the substituents also influence the forces of attraction between dye and fiber and hence chemical potentials in fiber and dyebath. For example, NH_2 and OH groups introduce lone electron pairs and the ability to hydrogen bond, and sulfonic acid groups alter the charge on the dyestuff molecule and its coulombic attraction or repulsion with various fiber types. With molecules containing hydroxyl groups ortho to the azo linkage it is possible to obtain complexes with copper or chromium to give violets, olives, browns, and blacks. The dyestuff chemist has to therefore strike a balance between desired shade and desired dyeing properties.

In order to obtain more bathochromic shades the dyestuff molecule must become increasingly large and complex. This in turn increases the need for sulfonic acid groups and hence bigger molecules have a much stronger negative charge than smaller ones. Further, these larger molecules have greater potential for van der Waals charge-transfer and ion-dipole attraction. When applying two dyes in admixture, eg, a small yellow molecule and large navy molecule to give an olive, these differences in forces of attraction can lead to different rates of exhaustion and incompatibility, unless the dyeing process is carefully controlled.

A similar trend exists with azo dyes for nonionic hydrophobic fibers (12). These are the so-called disperse dyes which are applied as dispersions and contain no solubilizing or ionic groups. With these dyes the molecular size has to be relatively small in order to obtain good penetration of the fiber polymer, and the use of naphthalene-based coupling components is not possible. Greater use is made of polar groups such as NO_2, OH, Cl, tertiary amines, etc to give both forces of attraction and desired shades. Again the more bathochromic the dye the more complex the molecule. The greater use of strongly polar groups such as CN, Cl, Br, or NO_2 increase van der Waals, ion-dipole hydrogen bonding and as a result of the use of long hydrophobic chains, hydrophobic attractions. In order to extend the shade range, heterocyclic diazonium compounds containing nitrogen and sulfur can also be used.

Azo dyes can also be positively charged to form basic dyes by quaternizing the nitrogen in a heterocyclic diazonium component or by incorporating a quaternized group in a side chain.

Anthraquinone Dyes.   Simple anthraquinone dyes are used mainly to obtain bright reds, pinks, blues, greenish blues, turquoises, and bluish greens. They give a purer and brighter shade than found with most commonly available azo dyes (see DYES, ANTHRAQUINONE).

The compact and similar size of anthraquinone disperse dyes in all shades makes them excellent in obtaining level dyeings in pale shades on difficult to dye fabrics, eg, difficult to penetrate or fabrics with physical variations in the yarns. A further advantage is that these dyes have good to excellent lightfastness. It is possible to produce basic dyes but the use of simple anthraquinone dyes is generally restricted to acid, fiber-reactive, and disperse dyes. The principal disadvantage of simple anthraquinone dyes is that they are tinctorially weak and based on complex and expensive chemicals and chemical processes. They are therefore generally an expensive way of obtaining heavy shades. Because of this more recent research has been directed to producing new dyes of similar brightness based on either complex heterocyclic azo dyes or new chromophores (13).

The use of simple anthraquinone dyes may be declining, but the complex anthraquinone dyes remain important. These are the vat dyes that contain a number of anthraquinone building bricks fused together. There are three basic types: anthrimides, benzanthrones, and indanthrones. The anthrimide dyes cover all shades yellow, orange, brown, olive, khaki, and black; the benzanthrone dyes offer green and navy; and the indanthrones are blue. Vat dyes have excellent light- and wetfastness and are widely used in uniforms, outerwear, furnishings, etc. Because of their polycyclic nature they are highly conjugated unsaturated molecules absorbing electromagnetic radiation over considerable wavelengths. They are therefore relatively dull shades. As a result of the nature of the dyes they have a high potential for charge-transfer attraction and van der Waals forces, and these strong forces overcome the coulombic repulsion of the dyes in their leuco form and cotton. These dyes exhibit high affinity for cellulose.

Phthalocyanine Dyes. The phthalocyanine molecule is much too big to be used on hydrophobic fibers and therefore is only used in its sulfonated form as the basis for direct and reactive dyes (see PHTHALOCYANINE COMPOUNDS). Its forces of attraction are different from a small linear yellow azo dye with which it is used to form bright greens. Compatibility between the two is likely to be a problem in practice and to overcome this, green dyestuffs containing a phthalocyanine dye linked via a saturated chromophore blocker (—x—) have been made, eg,

azo yellow chromogen —x— phthalocyanine turquoise

Other Chemical Groups. A new chemical class, benzodifuranones, has been introduced (13) for disperse dyes. The claimed advantages of these dyes are high extinction values and excellent wetfastness. Indigo and sulfur dyes are special cases and will be discussed later. Other chemical types are possible but the majority of dyes are covered by the foregoing types (10). Some other chemical groups used to synthesize dyes include nitroso, stilbene, diphenylmethane, triarylmethane, triphendioxine, xanthene, acridine, quinoline, methine, thioindigoid, thiazole, indamine, indophenol, azine, oxazine, thiazine, lactone, aminoquinones, and hydroxyketones (see AZINE DYES; CYANINE DYES; POLYMETHINE DYES; STILBENE DYES; and XANTHENE DYES).

THE DYEING PROCESS

The physical chemistry associated with dyeing has been described elsewhere both in detail (6,8) and in summary (14). The purpose of this treatment is to outline those basic concepts that have a direct impact on dyeing in order to appreciate the fundamental processes taking place.

Zeta Potential. When a textile is immersed in water a negative charge is developed on its surface. This is called the zeta potential. This happens even with ionic fibers in neutral dyebaths. Negatively charged dyes therefore are coulombically repelled.

Internal and External Phases. When dyeing hydrated fibers, for example, hydrophilic fibers in aqueous dyebaths, two distinct solvent phases exist, the external and the internal. The external solvent phase consists of the mobile molecules that are in the external dyebath so far away from the fiber that they are not influenced by it. The internal phase comprises the water that is within the fiber infrastructure in a bound or static state and is an integral part of the internal structure in terms of defining the physical chemistry and thermodynamics of the system. Thus dye molecules have different chemical potentials when in the internal solvent phase than when in the external phase. Further, the effects of hydrogen ions (H^+) or hydroxyl ions (OH^-) have a different impact. In the external phase acids or bases are completely dissociated and give an external or dyebath pH. In the internal phase these ions can interact with the fiber polymer chain and cause ionization of functional groups. This results in the pH of the internal phase being different from the external phase and the theoretical concept of internal pH (6).

Isotherms. When a fiber is immersed in a dyebath, dye moves from the external phase into the fiber. Initially the rate is quick but with time this slows and eventually an equilibrium is reached between the concentration of dye in the fiber and the concentration of dye in the dyebath. For a given initial dyebath concentration of a dye under given dyebath conditions, eg, temperature, pH, and conductivity, there is an equilibrium concentration of dye in fiber, D_f, and dye in the dyebath external solution, D_s. Three models describe this relationship: simple partition isotherm, Freundlich isotherm, and Langmuir isotherm.

With simple partition the situation is comparable to the partition of a solute between two solvents. The bonding forces involved between uncharged dye and uncharged fiber, and uncharged dye and uncharged solvent are considered to be the same. The dye is sometimes referred to as in solid solution in the fiber. This type of isotherm is found in practice with disperse dyes on cellulose acetate and polyester. It represents the dyeing situation with the minimum restrictions for the dye to enter the fiber; the only restriction is when the fiber solution becomes saturated.

Below saturation, D_f is directly proportional to D_s and a plot of D_f against D_s is linear. With more complex systems having bigger charged molecules and charged fibers, there are a multiplicity of bonding sites, and forces of attraction and repulsion. In these situations the overall net charge over an area of the fiber and the charge in the dyestuff together with all the other bonding forces are vital. This puts a greater constraint on dye absorption than for simple partition, although saturation values, ie, when all the potential bonding possibilities have been exhausted, can be high. This mechanism for absorption is best described by the Freundlich isotherm where the dye in fiber D_f is directly proportional to $(D_s)^x$ and a plot of $\log D_f$ against $\log D_s$ gives a straight line. Freundlich isotherms are generally found with cellulosic and other ionic hydrophobic fibers.

In synthetic fibers the number of ionic groups or dye sites is relatively small, and may have been introduced deliberately to make the base polymer dyeable. The restrictions on dye absorption are therefore very great; the dye molecule must find an available specific site from among the limited number of sites in the fiber. This situation follows a Langmuir isotherm, where the reciprocal of dye in fiber

$1/DF_f$ is directly proportional to the reciprocal of dye in the dyebath $1/DF_s$. A plot of $1/D_f$ against $1/D_s$ therefore gives a straight line.

As can be expected when the sites available for dye are limited, saturation values for the fiber are low, much lower than Freundlich or simple partition, and within the range of normal depth of shades dyed. This saturation value can be derived from the plot of $1/D_f$ against $1/D_s$, where the point the line intersects the $1/D_f$ axis represents the value $1/S_f$, where S_f is the saturation value.

Langmuir isotherms are typically found with ionic synthetic fibers and ionic dyes, eg, dyeing polyacrylonitrile with modified basic dyes, and on hydrophilic fibers in situations when the number of sites becomes very low. This may arise when the internal pH is such that only a small number of sites ionize.

Aggregation, Activity, and Solution.   Theoretical treatments use the term activity which assumes that the dyestuff is present in a monomolecular disso- ciated state in solution. Dyes are not generally in this state except at very dilute concentrations; some molecular interaction is more likely. Thus practical situa- tions are not fully described or characterized by theoretical treatments assuming monomolecularity, but the errors involved are usually of no practical consequence. However, when gross aggregation takes place there is significant interference as the dyestuff available for absorption is removed and previous theoretical consid- erations become invalid. Where aggregation takes place less dye is absorbed than predicted by theory. In general, aggregation is therefore to be avoided, except in exceptional situations where it is introduced to deliberately slow down the rate of dyeing. Dyebath conditions are usually adopted to minimize the potential for aggregation.

Rate of Diffusion.   Diffusion is the process by which molecules are trans- ported from one part of a system to another as a result of random molecular motion. This eventually leads to an equalization of chemical potential and con- centration throughout the system, and in the case of dyeing an equilibrium be- tween dye in the fiber and dye in the dyebath. In dyeing there are three stages to diffusion: diffusion of dye through the bulk solution of the dyebath to the fiber surface, diffusion through this surface, and diffusion of dye from the surface into the body of the fiber to allow for more dye to diffuse through the surface layer. These processes have been summarized elsewhere (9).

The mechanisms by which molecules diffuse from one part of a system to another is complex. The rate of diffusion is influenced by the relative concentra- tions in the parts of the system, the size of the molecules, the amount of molecular motion, the forces of attraction between the diffusing molecules, and any static molecules such as in an insoluble fiber polymer. In the early stages of dyeing, the dye concentration in the dyebath is much higher than in the fiber and this high concentration gradient makes diffusion rapid. As the dyeing continues the con- centration gradient between external and internal phases decreases and the rate of diffusion slows. The movement of dye molecules depends on their relative mo- bility, chemical and physical means of altering the mobility of the fiber chains or water molecules, and the presence of nondiffusing points of attraction within the fiber structure. Finally the size and shape of the dyestuff molecule also plays a significant part, and for each dye–fiber combination there is a diffusion coefficient that characterizes the dyestuff. Big molecules in general have lower diffusion coefficients than small molecules.

Diffusion is important in defining the rate of dyeing. Diffusion through the dyebath solution to the fiber surface can be the rate-determining process when there is no or little mobility in the dyebath. Temperature increases mobility. Dyebath mobility can also be increased by physical means, by increasing the physical movement of the fiber in relation to the dyebath solution. This is the basis of dyeing machine engineering. When there is little or no relative movement between dye liquor and fiber either static or laminar flow is found and diffusion of dye through the dyebath to the fiber surface is rate controlling. If by engineering, turbulent flow is produced, then diffusion through the dyebath to the fiber surface is no longer rate controlling. All dyeing machines introduce turbulent flow by either the simple physical movement of fiber through the liquid or vice versa. With the correctly engineered dyeing machine, diffusion through the fiber is the rate-controlling process.

If diffusion through the fiber is not carried out efficiently then not only will the rate of dyeing be slow, with a chance that equilibrium between dye and fiber is not reached, but also the fibers will be dyed unevenly and possibly be ring dyed leading to poor fastness properties. Diffusion through the fiber is dependent on the actual dye and fiber chain molecular structure and configuration, and also, especially with hydrophobic fibers, the mobility of the chemical chain (7).

Hydrophobic fibers can change from a glasslike solid where there is little motion in the chain, other than vibrational motion, to one where the chains are free to flex and move, thus increasing the ease by which dyestuff molecules can move through the voids left in the polymer structure. The temperature at which this happens is the glass-transition temperature, T_g. As the temperature of a glasslike polymer is increased, vibrational motions increase until they are so strong that they are able to break the bonds of attraction between the polymer chains. This causes a dramatic change in fiber structure. Hydrophobic fibers are always dyed above their glass-transition temperature.

Level Dyeing. The concept of obtaining a level dyeing in reasonable time is fundamental in practical processes. In general, dyes with high affinity show a strong tendency to move quickly from the dyebath to the fiber. Once at the fiber surface they diffuse slowly as the result of their strong forces of attraction to groups in the fiber polymer. If these dye molecules are absorbed unevenly, chances are they will remain unevenly distributed at the end of the dyeing. They are said to exhibit a high rate of strike and poor migration and are likely to produce unlevel dyeings. There is an inverse relationship between dyestuff affinity and diffusion coefficient. Thus dyes that have a low affinity are likely to diffuse to the fiber surface slowly (a low rate of strike), and once in the fiber quickly diffuse through the fiber to give a uniform distribution of dye molecules. Because of the low affinity of these dyes there are no strong forces of attraction restricting movement. These dyes exhibit good migration and are likely to give level dyeings.

Level dyeings can be obtained by applying dyes of low affinity that dye quickly if unevenly, and allow them to migrate to give uniform level dyeings by extending the time of dyeing. This simple approach has the disadvantage that such low affinity dyes produce dyeings of relatively low wetfastness. Good wetfastness requires high affinity dyes. Because these exhibit low migration it is necessary to ensure that a level dyeing is obtained from the start and maintained throughout the dyeing process.

In order to ensure level dyeings from high affinity dyes it is necessary to make them behave as if they were of low affinity. This is done by reducing the difference between their chemical potential in solution and fiber. The techniques used include: higher temperature for more diffusion; change in pH to control the number of sites available; addition of an electrolyte to either compete with the dye for sites, or neutralize the sites preventing ionic attraction; and addition of auxiliary agents that either compete with the dye for the fiber by lowering the chemical potential in the dyebath phase thus making it a more attractive environment for the dye, or by removing dye from the equilibrium by forming a temporary complex or aggregate in the dyebath. In essence, in order to obtain level dyeings with higher affinity dyes the objective is to slow down the rate of dyeing.

The heavier the depth of shade and closer to the saturation value for the fiber under its dyebath conditions (pH, electrolyte, etc), the more likely it is to obtain a level dyeing. Thus for heavy shades relatively little control is needed in the dyeing process.

The most difficult shades to produce level are pale shades, and the most difficult time in the process to obtain level dyeing is during the early stages of dyeing when the number of potential sites of attraction greatly exceeds the number of dyestuff molecules. Reducing the sites available is achieved by altering pH, electrolyte, etc. In practical dyeing situations the early stages of dyeing are always carried out under conditions that promote level dyeing, ie, slow dyeing rates and working close to relative saturation. Once there is enough dye absorbed in a level manner to dictate that the remainder will dye in a level fashion then it is possible to change the conditions to increase the rate of dyeing. This is done in order to obtain commercially acceptable rates of dyebath exhaustion. The definition of application processes and classification of dyes is determined by the need for this control and how it can be achieved.

Compatibility. To produce a desired shade more than one dye is usually needed. Often combinations of three dyes are used, eg, yellow, red, and blue, in order to obtain the maximum number of shades available from the minimum number of dyestuffs. In order to give uniform coloration it is necessary to apply such mixtures of dyes from the same dyebath. As the dyeing proceeds the textile takes up more of the dyes. If the hue of the fabric is the same throughout all stages of dyeings, simply becoming stronger with time, and the hue of the final textile is reproducibly uniform both on its surface and within its interior, eg, for a wound package of yarn, then the dyes are said to be compatible under the dyeing conditions used.

Compatibility is mainly a function of exhaustion rate, but can also be influenced by migration. For dyes with high migration potential under the specific dyebath conditions, the relative rates of exhaustion of individual dyes do not need to be identical. For dyes of high affinity and low migration potential it is necessary to ensure that any mixture of dyes exhausts as if it were a single dye, ie, that all dyes are exhausting at the same rate under the dyebath conditions used. If dyes in combination are compatible then the overall dyeing process can be shorter, reproducibility improved, and the chances of obtaining level dyeings improved (15).

Dyeing of Cellulosic Fibers

OVERVIEW

Preparation for Dyeing.   Cotton fibers are coated with natural waxes and pectins. These can be removed by aqueous alkalies at 80°C or above or by solvent treatment to improve the absorbancy and dyeability of the fibers. Cotton (qv) may be made suitable for dyeing in a variety of forms, such as raw stock, yarn, or piece goods. Raw stock is normally dyed without thorough dewaxing, since the natural waxes aid in subsequent spinning operations. Surfactants are employed to aid the penetration of dyestuffs through the protective waxes. Flaws in the dye levelness are overcome by subsequent carding. Hard-twisted ply yarns are frequently given a kier-boiling prior to dyeing to improve levelness. Careful preparation of cotton piece goods is essential to achieve suitable dye penetration, fastness, and general appearance. Fabric construction dictates whether the fabrics will be processed in rope or open-width forms. Heavy piece goods, and those which are subject to rubs and crease marks, are handled in open width.

The first step in open-width preparation of wovens is singeing which removes lint and fuzz from the fabric surface. The goods are then usually impregnated with an enzyme solution to effect solubilization of the sizing. Desizing may be accomplished by pad-batch, jig, or pad-steam techniques. Material that must be handled in open width requires hot scouring on jigs or on open-width boil-off machines in caustic soda solution (sp gr 1.05–1.06) containing a surfactant. Materials that may be roped are scoured continuously with caustic or soda ash. The goods are then rinsed, scoured with acetic acid, rinsed, and dried.

Before dyeing in light or bright shades, the goods should be bleached with hydrogen peroxide and caustic soda to bleach the motes. This operation also helps in the removal of trace impurities that remain after boil-off. Although kiers are still used for boil-off and bleaching, they have been replaced largely by steam-heated J-boxes or steamers which allow continuous processing and reduction of processing time.

Mercerizing is accomplished by passing the cotton fabric through 15–30% caustic soda. Improved luster and increased dye affinity results. The fabric is normally held under tension during processing. In order to obtain cotton stretch fabrics, mercerization is carried out on fabric in a slack condition. Mercerization techniques using liquid ammonia have gained a limited success.

With knitted fabrics it is necessary to remove the knitting oils by either alkali treatment or solvents. Where water-immiscible oils have been used and the fabric is to be hot dyed (80°C or above), a minimum scour to remove dirt and stains can be sufficient, the rest of the oil being removed during the dyeing process.

Viscose rayon, because of its low wet strength, must be processed under minimum tension at all stages of preparation. Skeins contain few impurities and require only light scouring. Piece goods may contain starch or sizing compounds, which were applied prior to weaving, and may require an open-width enzyme desizing followed by scouring, either in open-width or in rope form, depending on fabric construction and weight. In general, fabrics may be prepared for dyeing by scouring in mildly alkaline synthetic–surfactant solutions.

The Dyeing Process.   When cotton fiber is immersed in water it develops a negative charge. In order for dyes to show good buildup on cotton the dyes must be soluble, planar, aromatic structures. Solubility is obtained by incorporating negatively charged sulfonic acid groups, ie, the anions. Therefore the dyes show long-range, coulombic (ionic) repulsion, but very strong short-range van der Waals forces of attraction. Thus there is a potential barrier that the dyestuff molecule has to overcome. In a static state, assuming no thermal agitation, the ionic concentration varies as in Figure 1a as the negatively charged molecule approaches the fiber.

Figure 1b shows the effects of natural and introduced thermal agitation which tend to equalize the distribution of ions. The differences in profile between dye^- and Cl^- are due only to the dye exhibiting strong close-range forces of attraction.

The use of salt or similar electrolyte is critical in the dyeing of cellulose. When sodium chloride is added to the dyebath, sodium ions (Na^+) diffuse to the negative charges on the cellulose and neutralize them. In the total system electrical neutrality exists, the net negative charge equals the net positive charge. The influence of the Na^+ ions is, however, of critical importance because in the local environment of the fiber surface they neutralize the nondiffusing negative charges on the insoluble fiber. All other ions distribute through the two phases to give electrical neutrality. With the introduction of the sodium ions (Na^+) the coulombic repulsion force between fiber and negatively charged dye is removed and only the strong attraction forces exist. It is then possible for negatively charged dye to diffuse unhindered to the fiber surface.

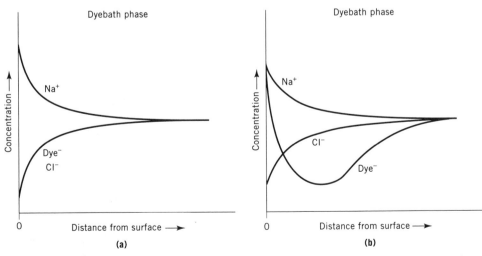

Fig. 1. (**a**) Ionic concentrations in static state. The y-axis represents the cellulose surface. (**b**) Distribution of ions in practice.

DIRECT DYES

The simplest way of coloring cellulosic fibers is with direct dyes. The dyeing mechanism follows exactly the outline just described where the addition of salt is used to allow dyestuff to be absorbed on the fiber. This is done carefully to ensure that level dyeing is achieved, especially during the early stages of dyeing. If a high affinity dye is being applied then the dyeing process can start without added salt. Some dye is absorbed as a result of thermal agitation, but if dyeing was started cold this is minimal. The addition of a small amount of salt neutralizes the negative surface charge and allows dyeing to continue. However, each dye has negative charges from the sulfonic acid groups in the molecule, and the larger molecules contain five or more. As each molecule of dye is absorbed onto the fiber it increases the fiber's net negative charge which is then neutralized by Na^+. Eventually there is insufficient Na^+ to neutralize this charge, the fiber again develops a negative charge at the surface, and dyeing stops. Further addition of salt allows dyeing to start again by the same process. It is therefore possible to control the rate of dyeing by salt addition during the early stages. Once the initial period is over the temperature can be increased to speed up diffusion, increase the rate of dyeing, and promote migration.

Leveling Power.   Direct dyes are classified according to their leveling characteristics (16). Class A direct dyes migrate well and have high leveling power, ie, they have low affinity and high diffusion. They may dye unevenly at first but prolonged treatment at temperatures at or near the boil brings about even distribution. The dyes are applied by adding to the dyebath all the dye and all the salt needed, depending on the depth of shade. The amount of salt added is usually 5% on weight of fabric for pale shades increasing to 20% for heavy shades. The dyebath is set at 40–50°C, the cellulose added, and the temperature raised to the boil over 30–40 min, and kept there until exhaustion is complete or leveling achieved.

Class B direct dyes have poor leveling power and exhaustion must be brought about by controlled salt addition. If these dyes are not taken up uniformly in the initial stages it is extremely difficult to correct the unlevelness. They are dyes that have medium–high affinity and poor diffusion. In their application the cellulose is entered into a dyebath containing only dye. The salt is added gradually and portionwise as the temperature is increased and possibly the final additions made after the dyebath has come to the boil.

Class C direct dyes are dyes of poor leveling power which exhaust well in the absence of salt and the only way of controlling the rate of exhaustion is by temperature control. These dyes have high neutral affinity where, resulting from the complexity of the molecules, the nonionic forces of attraction dominate. When dyeing with these dyes it is essential to start at a low temperature with no added electrolyte, and to bring the temperature up to the boil very slowly without any addition of electrolyte. Once at the boil the dyeing is continued for 45–60 min with portionwise addition of salt to complete exhaustion.

Although the rates of migration vary considerably from dye to dye and with different dyebath conditions, the generalized relationships between Classes A, B, and C hold (17).

In all these application methods the same procedure is adopted. At the beginning the rate of dyeing must be as slow as needed to give levelness; once this has been achieved the rate of dyeing is systematically increased to give complete exhaustion. When dyeing a shade needing two or more dyes, only dyes from the same Class can be mixed to give compatible dyeing behavior. If dyes from different classes must be used, eg, to obtain specific shades not possible any other way, the dyeing procedure appropriate for the poorest leveling dye must be adopted.

Wetfastness. Class A direct dyes offer the most trouble-free process for dyeing cellulose. However, they do not always provide sufficient wetfastness.

When a dyed textile is placed in a hot solution of water, as is the case in domestic washing, the dye desorbs from the fabric to re-establish the equilibrium between dye in fiber and dye in solution. This happens every time the textile is washed. The isotherm in operation, usually Freundlich with direct dyes, dictates that there will always be a given amount of dye in the solution for a given amount of dye in the fiber at equilibrium. For Class A direct dyes their low affinity and high diffusion means that this equilibrium is established relatively quickly and certainly within the times associated with domestic washing. Further, Class A dyes show a high propensity to migrate from dyed fabric to undyed, or different shade, fabrics in the same wash liquors. Left long enough all would become the same shade.

The Class B and C dyes show better resistance to desorption, ie, they show higher wetfastness, but they do not overcome it fully, and even the Class C direct dyes show inadequate wetfastness and poor staining of adjacents in fastness tests as a result of the reversible nature of the dyeing process. Attempts to overcome this problem have concentrated on chemical treatment of the direct dyes after they were applied. These treatments essentially make the direct dye molecules already on the fiber much bigger and thereby increase the nonionic forces of attraction or reduce solubility in order to reduce desorption and give good wetfastness. Methods used include applying dyes containing free amino groups, diazotizing them and coupling with a base; aftertreatment with formaldehyde; applying dyes containing hydroxyl groups ortho to the azo group and then aftertreating with metallic salts, eg, Cu; and applying 1–4% of a cationic surface-active agent (over 15–30 min at 25–60°C) to form a sparingly soluble complex with the dye.

Today only the latter two processes are used; cationic treatments are the most popular and although they give good wetfastness their use is limited because of the complexity of the process, shade changes because of the treatment, often adverse effect on lightfastness, environmental considerations, etc.

FIBER-REACTIVE DYES

Because of the limitations of direct dyes and the ability to use simple acid dye chromophores to give bright washfast dyeings, fiber-reactive dyes have become a well-established, popular way of dyeing cellulose. A market of 56,000 t of reactive dyes was forecast for cellulose fibers in 1989 (18), and the growth rate of reactive dye consumption of 3.9% per annum is four times the growth rate of other dyes for cellulosic fibers (19).

A reactive dye for cellulose contains a chemical group that reacts with ionized hydroxyl ions in the cellulose to form a covalent bond. When alkali is added to a dyebath containing cellulose and a reactive dye, ionization of cellulose and the reaction between dye and fiber is initiated. As this destroys the equilibrium more dye is then absorbed by the fiber in order to re-establish the equilibrium between active dye in the dyebath and fiber phases. At the same time the addition of extra cations, eg, Na^+ from using Na_2CO_3 as alkali, has the same effect as adding extra salt to a direct dye. Thus the addition of alkali produces a secondary exhaustion.

Fiber-reactive dye is also hydrolyzed by reaction with free OH^- ions in the aqueous phase. This is a nonreversible reaction and so active dye is lost from the system. Hydrolysis of active dye can take place both in the dyebath and on the fiber, although in the latter case there is a competition between the reactions with free hydroxyl ions and those with ionized cellulose sites. The hydrolyzed dye establishes its own equilibrium between dyebath and fiber which could be different from the active dye because the hydrolyzed dye has different chemical potentials in the two phases. The various reactions taking place can be summarized as in Figure 2.

At the end of the dyeing process there is fixed dye on the fiber, and hydrolyzed dye in both the dyebath and fiber. All active dye disappears. In order to take advantage of the fastness offered by covalently bonding the dye to the cellulose it is necessary to remove all the hydrolyzed dye from the fiber. Even very small amounts of residual direct dye (1–2%) give heavy stains on the adjacent fabrics used in perspiration tests. Unfortunately, the hydrolyzed dye does exhibit some affinity for the cellulose and removing it is a desorption process rather than a rapid physical removal. Thus the removal is a relatively difficult procedure that is a critical part of the total dyeing process.

The chemistry and application processes for reactive dyes concentrate on maximizing exhaustion and fixation without a risk of unlevel dyeing, minimizing the amount of hydrolyzed dye present, and maximizing its rate of diffusion. Once a dye is fixed it cannot migrate and therefore dyeings achieved using fiber-reactive dyes must be level before they are fixed. All this must also be done against a background of commercial considerations of cost and time (productivity), and new reactive types and processes are still being introduced in attempts to improve the overall efficiency of the process (20).

Basic Theory of Fiber-Reactive Dye Application. The previously described mechanisms of dyeing for direct dyes apply to the application of reactive dyes in neutral dyebaths. In alkaline solutions important differences are found. The de-

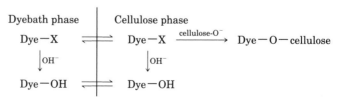

Fig. 2. Reactions taking place in a fiber-reactive dyebath. X represents the reactive group.

tailed theoretical treatments are described elsewhere (6) but it is important to consider some of the parameters and understand how they influence the application of fiber-reactive dyes.

The equilibrium deprotonation of cellulose in alkali can be represented as follows, where $[H_f^+]$ is the concentration of hydrogen ions in the internal fiber phase.

$$\text{cell—OH} \rightleftharpoons \text{cell—O}^- + H_f^+$$

Thus the concentration of cell–O^- is dependent on the internal pH ($-\log[H_f^+]$) of the fiber. This is essentially a theoretical concept because $[H_f^+]$ cannot be measured directly.

When values for the internal pH are calculated (6) it is found that the relationship between internal and external pH is strongly influenced by the presence of electrolyte. With no added electrolyte the internal pH is always lower than the external pH, and for pH values below 12 considerably lower. With the addition of increasing amounts of electrolyte this variance decreases and an approximate linear relationship between internal and external pH exists in a 1 M electrolyte solution. The cell–O^- concentration is dependent on the internal pH, and the rate of reaction of a fiber-reactive dye is a function of cell–O^- (6,16). Thus the higher the concentration of cell–O^- the more rapid the reaction and the greater the number of potential dye fixation sites.

An alkaline pH is needed to hydrolyze cellulose resulting in the cellulose becoming strongly negatively charged, and relative large amounts of electrolyte are needed in the dyebath to prevent desorption of dye at that stage and to enable the economic advantages of secondary exhaustion to take place. It is unwise to use too high an external dyebath pH to generate cell–O^- because this promotes rapid hydrolysis of unfixed dye in the dyebath that is lost from the system. The addition of electrolyte results in a relative increase in internal pH, and a considerable increase in the concentration of cell–O^- leading to efficient dye–fiber reaction.

Electrolyte therefore plays three important roles: increasing absorption in the neutral state, preventing desorption/promoting secondary exhaustion, and increasing the amount of ionized cellulose. Thus the amounts of salt used in the application of fiber-reactive dyes are larger than for direct dyes.

Removal of unfixed hydrolyzed dye after the fixation process is essential to obtaining good wetfastness. Efficient removal of this dye is obtained by having a high degree of fixation and hence only small amounts of dye to remove; using dyes that have low affinity and high diffusion characteristics; working at high temperatures to reduce the affinity and increase diffusion out of the fiber; using wash liquors that are electrolyte-free and removing the salt from the fiber as quickly as possible; using rapid circulation, agitation of the wash liquors, to create turbulent flow at the fiber surface; and replacing the wash bath before equilibrium is established between dye and electrolyte in the wash liquor and fiber phases.

Given all the above conditions the washing-off process efficiency is controlled by the efficiency of salt removal, which is achieved by progressively diluting its concentration in the fiber phase. Providing the dye–fiber bond is stable to hot alkali, a slight alkalinity helps washing off in that it keeps the fiber strongly negatively charged and hence ionic repulsion forces exist between hydrolyzed dye and fiber.

The Ideal Fiber-Reactive Dye Profile. Figure 3 shows the general profile for the application of a reactive dye. In addition to showing the rate profile of fixation between dye and fiber, three other practical parameters (A–C) are noted.

The overall objective is to make the fixation (C) as high as possible for economic reasons. The closer the fixation (C) is to the total dye on the fiber (B) then the smaller the concentration of [dye–OH] and the less that needs to be removed in the washing-off process after dyeing.

If the exhaustion in neutral conditions (A) is low as a result of using dyes with low affinity, then as with direct dyes it can be expected that level dyeings will be obtained at this stage. In order for the dye to be economically viable the secondary exhaustion from the addition of alkali (B–A) must be high. Migration, and hence leveling, is impossible once the dye has formed a covalent bond with the fiber, thus during the secondary exhaustion phase little migration is possible. Dyes that need a large secondary exhaustion therefore have an inherent risk of unlevelness. This may not be a problem in heavy shades but does become of increasing importance in pale shades. Unlike direct dyes, extending the dyeing time will not improve the situation. Methods of overcoming this problem are to dye at low temperatures when the affinity is at its highest and so increase the neutral exhaustion, and to add the alkali slowly during the secondary exhaustion period to maximize control over and reproducibility of the process.

The ability to dye cold (20°C) or warm (60°C) is often regarded as an advantage because of the energy savings. In order to obtain economic levels of fixation in a practical time period at these temperatures the rate of reaction between dye and fiber has to be high. These dyes are described as being of high reactivity. A further advantage of these dyes is that because of their low affinity all hydrolyzed

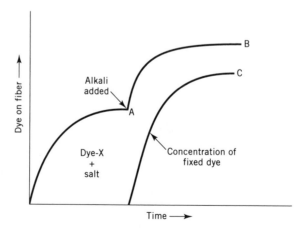

Fig. 3. Amounts and forms of fiber-reactive dye on the fiber as a function of time for a low affinity dye, where X represents the reactive group. Point A represents the amount of dye exhausted in neutral conditions; B is the total amount of dye exhausted at the end of the dyeing process, ie, [dye–OH] + [dye–X] + [dye–O–cell]; and C is the amount of dye fixed [dye–O–cell].

dye is rapidly removed from the fiber during the washing-off process. However, even with these dyes it is necessary to wash off at elevated temperatures and ideally at or close to the boil.

The extreme alternative is where the initial exhaustion in neutral conditions is high which is achieved by having high affinity dyes. Although level dyeing is more difficult than with low affinity dyes, the advantage is that, resulting from the low secondary exhaustion, little can go wrong after the addition of alkali. The dyeing procedures concentrate on obtaining a level dyeing during the time period before alkali addition by changing dyebath variables, eg, by portionwise addition of salt and dyeing at temperatures in the region of 80°C. Because the fixation is carried out at 80°C dyes of relatively low reactivity can be used. Because these dyes are of high affinity it is more difficult to remove hydrolyzed dye and it is essential therefore that the fixation be as high as possible. As a result of their low reactivity these dyes are not sensitive to hot alkali and therefore it is possible to wash off hot without intermediate cold rinses, offsetting the problems associated with the high affinity of these dyes.

Alkali is usually added in a second stage. However, with low reactivity high affinity dyes it is possible to add the alkali at the beginning of the dyeing process and control the rate of uptake and chemical reaction by temperature control. With high affinity dyes the exhaustion takes place at low temperature rapidly before the chemical reaction becomes significant. If dyes are carefully selected or synthesized to have identical dye uptake it is possible to include all the electrolyte from the beginning and operate an "all-in" technique.

All reactive dye chemistry has been developed to either improve the efficiencies within a profile or to modify the profile to give better performance. For example, bifunctional reactive dyes, those containing two reactive groups, were developed to increase the fixation levels (C) and minimize wash off.

Application Methods. There are many detailed application methods used for applying reactive dyes, and all have been described in detail (16). Examples of the main methods may be summarized as follows.

Cold Exhaust Dyeing Fiber-Reactive Dyes. Start at 25–30°C optionally with a sequestrant and maintain. The dye is added over 5 min, then there is portionwise addition of salt every 10–15 min, increasing the size of the addition each time over 1 h. The amount of salt used (10–100 g/L) depends on the depth of shade. After the final addition of salt, wait 15 min, portionwise add soda ash (10–20 g/L) over 15 min, and continue dyeing for 30–45 min. Drop dyebath, cold water rinse, and use a sequence of hot washes to remove all loose "unfixed" dye.

Warm, Hot Exhaust Dyeing Dyes. This is an isothermal technique for depths of shade above 0.5% based on weight of fiber. Start at 40°C with salt and sequestrant at pH 7.0, run fabric, and bring temperature rapidly up to the dyeing temperature (60–80°C depending on the dye). Add the dye over 10–15 min and run for 30 min before adding alkali portionwise over 10–20 min. Run for 45–60 min before dropping the dyebath and starting the wash-off sequence.

Migration Exhaust Technique for Less than 0.5% Depth of Shade. Start at 50°C with sequestrant at pH 7.0. Add dye over 20 min and raise temperature to highest safe level depending on the dye (95°C with monochlorotriazinyl reactive system) at a rate of 1.5°C/min. Hold for 20 min and cool back to dyeing temperature. Alkali is portionwise added, or mechanically dosed, over 20 min and after a further 30 min the dyebath is dropped and the washing sequence begun.

All-in Method. Start at 25°C with sequestrant, salt, and alkali (mixed soda ash and caustic), run for 15 min, and add dyes over 10 min. Run for a further 15 min and bring the temperature up to 80°C over 45 min. Maintain this temperature for 30–60 min before dropping the dyebath and starting the washing process.

Continuous Dyeing. In this case the fabric is passed through a trough containing dye solution and then squeezed through a padding nip at controlled pressure. The dyestuff is therefore applied to the fabric by physical means. The alkali can be added to the system by either applying with the dye or in a second padding treatment. The reactivity of the dye and ambient temperatures dictate whether dye and alkali are applied together in either simple admixture (low reactivity) or by sophisticated dosing techniques (high reactivity), or by a second application process (both). Urea is sometimes added to promote dye yield.

Dyes, whether reactive or others, with low to medium affinity are needed in this continuous application. If the affinity is too high then exhaustion of the dye can take place during the few seconds the cloth is immersed in it causing a gradual depletion of the dye concentration, and the subsequently dyed fabric changing shade with time during the run. This is referred to as tailing. If dyes have very low affinity as the fabric is dried after the application of either dye and/or alkali, the water moves to the areas of greatest rate of evaporation, and if the dye has a low affinity it will migrate with the water giving an uneven dyeing.

After dyeing, the fabric is dried quickly to a moisture content below that at which migration can occur and then either steamed or baked (thermofixed), followed by a washing-off process. All the processes run continuously and in tandem, the fabric passing from one process and machine to the next.

Cold Pad-Batch Dyeing. Dye and alkali (eg, 5–30 g/L soda ash or sodium silicate, depending on shade and dyestuff) are padded, and the whole batched for 4–24 h at ambient temperature, the time depending on the dye. As in all other methods a final washing-off process is needed to remove hydrolyzed dye. This application method is of growing popularity especially for woven fabrics because of the increased demand for short yardages to a particular shade which are not economically suited for coloration by continuous methods, the simplicity and low capital cost of the equipment, and the low energy usage. The technique is less important on knitted fabrics where the traditional batch exhaust methods are suited to short yardages.

In the United States tubular knitted fabric is dyed by the pad-batch method for outlets where the edge mark formed as the double thickness tube passes through the padder is accepted. In Europe it is usual to slit tubular knitted fabric before padding in order to avoid any edge marks.

Pad-batch dyeing shows its primary benefits over other methods where it is possible to apply the dye on grey (unbleached) minimally prepared fabric, eg, T-shirts, followed by washing on the same perforated beam used to batch the fabric after dyeing. Where a bleached fabric base is needed, the bleaching is best carried out using peroxide by pad-batch technique in order to ensure an even effect and pick up when padding the dye liquor. Similarly all other preparation is best carried out in open-width.

Chemical Types. A wide range of reactive groups have been investigated, with 20–30 used commercially and over 200 patented. These have been described in detail elsewhere (10,20). Because these reactive groups differ chemically the activation of the reactive systems is different as are the rates of reaction with cellulose, from one reactive system to another. This rate of reaction with cellulose, or reactivity, dictates the temperature and pH needed for dyeing.

The most important reactive groups are those based on halotriazine or halopyrimidine systems, where an activated halogen substituent undergoes a nucleophilic substitution reaction with ionized cellulose, or dyes based on sulfatoethylsulfonyl groups (Table 2). With dyes containing sulfatoethylsulfonyl groups the group converts under alkaline dyebath conditions to a vinyl sulfone, $SO_2CH{=}CH_2$, group that reacts with cellulose by an addition reaction at the double bond.

For any reactive group a range of reactivity levels is possible, although some general observations can be made (Table 2) (10). Variation occurs because the chromogen as an integral part of the molecule can activate or deactivate the reactive group. For example, in general azo dyes based on H-acid coupling components (usually reds) have higher reactivity, those based on anthraquinone (usually blues), lower reactivity. The deliberate introduction of electron-donating and withdrawing groups either into the heterocyclic reactive group or adjacent to it in the dye molecule alters reactivity.

Because of this range of reactivities, and the further possibility of modifying affinity and diffusion by molecular design of the dye chromogen, it is difficult to classify reactive dyes into groups. Usually the dyes are referred to by trade name rather than any other classification. Interchanging one dye for another manufacturer's in a dyebath formula is risky, and mixing dyes containing different reactive groups requires care if reproducible dyeings are wanted.

Bifunctional fiber-reactive dyes have been developed. The first ones introduced were the Procion Supra Dyes (ICI) and Procion H-E Dyes (ICI) in 1967, based on two monochlorotriazine groups. More recently the Sumifix Supra dyes (Sumitomo, early 1980s) and Cibacron C Dyes (CIBA-GEIGY, 1988) have been introduced which makes use of different types of reactive groups. Other manufacturers have also been active in researching these mixed bifunctional approaches (20).

The concept behind bifunctional dyes is that if two distinct reactive groups are used, the probability of obtaining dye covalently bonded to the cellulose at the end of dyeing instead of being hydrolyzed is increased because each molecule must react twice with OH to be fully hydrolyzed. Dyestuff containing one hydrolyzed reactive group is still capable of covalently bonding to cellulose through its second reactive group. When different reactive groups are introduced, usually a combination of a halogen heterocyclic (eg, monochlorotriazinyl) reactive group and a sulfatoethylsulfonyl (vinyl sulfone) reactive group, it is possible to have groups of differing reactivity in the same molecule. The claimed benefits (20) of bifunctional reactive dyes are a generally higher level of fixation, and in the case of mixed reactive groups, suitability for application over a range of temperatures and methods.

Correlation of Application, Affinity, and Reactivity. Figure 4 correlates fiber-reactive dye application suitability to reactivity and affinity.

Table 2. Commonly Used Reactive Groups and Their Relative Reactivity

Reactive group	Structure	General reactivity	Commercial examples[a]
triazine	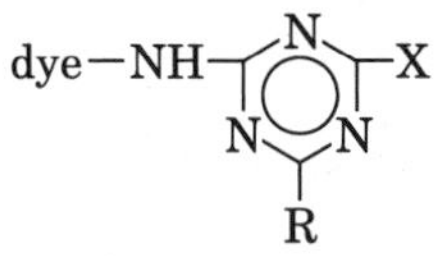 X = Cl, R = nonreactive substituent	low–med	Procion H Procion P Cibacron E Cibacron P
	X = F, R = nonreactive substituent	med	Cibacron F
	X = Cl, R = Cl	high	Procion MX
pyrimidine	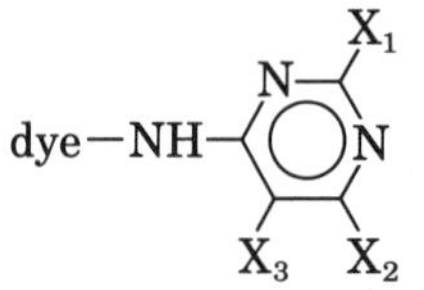 X_1 = F and either X_2 = F, X_3 = Cl or X_2 = Cl, X_3 = F	med	Levafix EA Drimarene K[b]
	X_1, X_2, X_3 = Cl	low–med	Drimarene X[c]
quinoxaline	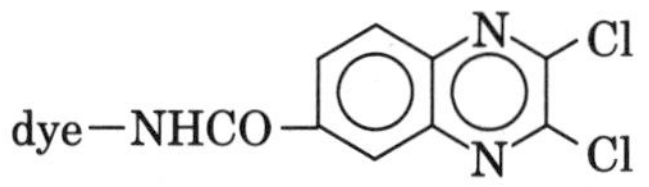	low–med	Levafix E
vinyl sulfone	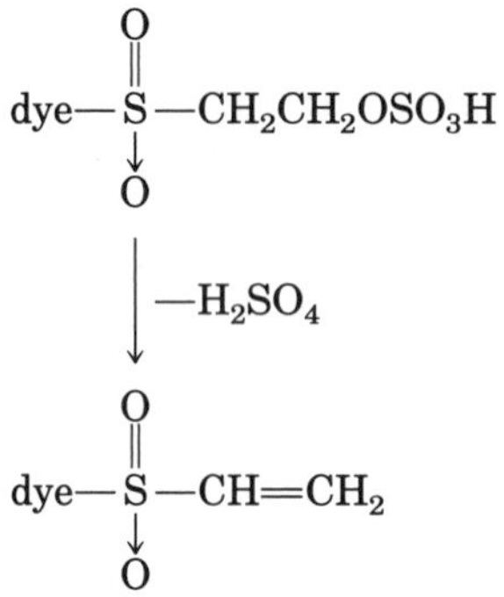	med–high	Remazol[d]

[a]Procion dyes are made by ICI; Cibacron dyes by CIBA-GEIGY; and Levafix by Bayer.
[b]Sandoz.
[c]Bayer.
[d]Hoechst.

VAT DYES ON CELLULOSE

Most vat dyes are based on the quinone structure and are solubilized by reduction with alkaline reducing agents such as sodium dithionite. Conversion back to the insoluble pigment is achieved by oxidation. The dyes are applied by either exhaust or continuous dyeing techniques. In both cases the process is comprised of five stages: preparation of the dispersion, reduction, dye exhaustion, oxidation, and soaping.

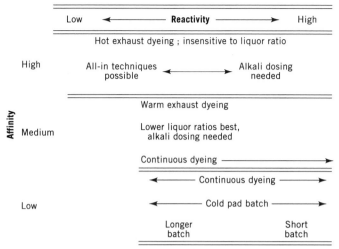

Fig. 4.   Summary of dyeing techniques related to dye reactivity and affinity characteristics.

The Process.   *Preparation of Dispersion.*   The reduction process is a two-phase reaction between soluble reducing agent and insoluble dye particles, and therefore the rate of reduction is influenced by the particle size distribution of the dye dispersion. The smaller the particle size the greater the surface area and hence the more rapid the reduction process. However, if the particles are too small, migration will occur in continuous dyeing. It is therefore extremely important to control the size and range of particle size and this is a closely guarded piece of dyestuff manufacturers' know-how.

Reduction.   This has been studied in great detail (21). The reduction is reversible and an equilibrium between oxidized and reduced forms can be obtained in the presence of a reducing agent. This leads to the definition of a redox potential, the electrical potential in volts obtained at a platinum electrode when there are equal concentrations of oxidized and reduced forms at unit activity of hydrogen ions at 25°C. Products with a positive redox potential are oxidizing agents, those with a negative potential reduce compounds whose redox potential is higher. As vat dyes are applied in their reduced leuco-soluble form it is essential to have a system with a redox potential considerably lower than that of the vat dyes being used to ensure an excess of the reduced form. This is achieved by using a solution of alkaline sodium dithionite (hydrosulfite) which is a strong reducing agent with a redox potential well below that of vat dyes.

Dyeing.   The exhaustion of the negatively charged leuco form of the dye is covered by the general theory already described. Dyeing rates are high, with the main mode of attraction due to charge-transfer and van der Waals forces. In the batchwise dyeing process it takes around 30 min to reduce the dye, and 45 min for dyeing. In continuous applications where the dye is physically applied to the fiber the combined time needed for reduction and dyeing is seconds.

In the batchwise process the temperature can be raised to 80°C to promote levelness providing dyes not sensitive to reductive breakdown are used. In the continuous application method the vat dye is padded onto fabric and dried under conditions that avoid migration, passed through a solution of sodium hydrosulfite and caustic, through a pad mangle, and then steamed in saturated steam for up to 60 s.

Oxidation. In view of the low redox potentials of vat dyes this presents little problem. Air oxidation, or oxidation by rinsing off the reducing agent, and subsequent washing in water are possible but because most vat dyes oxidize too slowly for the required processing speeds it is usual to use a solution of hydrogen peroxide or sodium perborate as an oxidizing agent.

Soaping. During the oxidation process the vat dye pigment is reformed inside the cellulose. Deposition of the pigment onto the surface of the fiber also takes place which reduces the fastness to rubbing. This surface pigment is removed by soaping at or near the boil. The soaping process also causes a shade change that for some dyes can be quite marked, and the shade of the fabric needs to be stabilized by this soaping treatment. Two schools of thought exist as to the causes of this: that soaping causes crystal growth and produces crystals orientated in such a manner that the main crystal axis is along the fiber axis and the resonating system arranged perpendicular to this axis (22), or that the soaping process preferentially stabilizes a particular crystal form of the vat dye (23). Soaping is not a problem in batchwise dyeing, but in continuous dyeing multibath large soaping ranges are needed to allow for sufficient immersion time to develop the final shade.

Uses of Vat Dyes. The main characteristic of vat dyes is their excellent fastness to light, water, and other agents, eg, chlorine. Vat dyes are therefore widely used in outlets demanding high lightfastness such as outerwear, furnishings, drapes, etc; high wetfastness and fastness to repeated washing such as workwear; high chlorine fastness such as institutional articles; or where general excellent fastness is required as in the case of sewing threads where it is impossible to know the use of final garment they will be used to construct.

The majority of vat dyes used worldwide are applied by continuous dyeing; polyester–cotton blends are the most important substrate. The fabric is padded with vat dye dispersion, dried, padded with sodium hydrosulfite, caustic soda, and salt, steamed for 30–60 s at 102°C, rinsed, and dried.

Sulfur Dyes. These are a special case of vat dyes and behave in an analogous manner except that the reducing agent used is sodium sulfide. In order to obtain rapid oxidation chemical oxidizing agents are used. The main outlet for these dyes is in the economic production of navy and black shades on woven fabrics by continuous dyeing, often applying the pre-reduced form of the sulfur dye.

Indigo. Indigo is similar to vat dyes in its application, however, it is not based on a quinone structure but on indigotin. In the presence of alkaline-reducing agents the C=O group is reduced to CH—OH and the dye rendered soluble. As with vat dyes the reaction is reversible, oxidation being achieved with atmospheric oxygen.

The application procedures of reduction with caustic and reducing agents such as sodium dithionite (hydrosulfite) and subsequent oxidation are identical to those for vat dyes. The principal difference is that indigo is applied by a "dipping" method. When cellulose is dipped into reduced indigo the fiber quickly (within seconds) absorbs indigo and reaches an equilibrium. At this stage the fiber is removed from the indigo vat and atmospheric oxygen reforms the insoluble indigotin in the fiber. The fiber is then re-immersed in the indigo vat, more leuco is absorbed, and the process repeated until the desired depth of shade is obtained. Each time the fabric is dipped there is more or less the same incremental increase in depth of shade.

The principal use for indigo is in denim where indigo is dyed on cotton warp yarns which are subsequently woven with white cotton weft yarns to give the typical denim look. Special effects such as "stone wash," "ice wash," etc, are obtained by chemical treatments of the dyed fabric, for example, with potassium permanganate and pumice. The "wash-down" behavior, where the shade reduces during each wash, is achieved by deliberately ring dyeing the yarn by altering the conditions of reduction and oxidation. The more penetration of indigo into the yarn the less wash-down, the less the penetration the greater the wash-down.

Dyeing of Wool

OVERVIEW

Preparation for Dyeing. Raw wool must be cleaned before it can be efficiently carded, combed, otherwise processed, or dyed. Loose wool, as obtained from the sheep, contains 30–80% impurities consisting of wool grease, dried perspiration (suint), dirt, seeds, and burrs. Methods of scouring the wool vary widely, depending on the type of wool and the amount and types of soil. The equipment for washing wool by the countercurrent method usually consists of four to five bowls in sequence, each approximately a meter deep and fitted with perforated trays that support the wool. Forks keep the wool moving along toward heavy squeeze rollers that are located at the end of each bowl and are designed to squeeze out dirty liquor before the wool is passed on to the next bowl. Soap and soda ash, historically used for wool scouring, largely have been replaced by nonionic surfactants with or without soda ash. In order to prevent harshening of the wool, soda ash concentrations are kept at a minimum. Scouring temperatures are normally 45–60°C, depending on the system and type of wool to be cleaned.

Mechanism of Dyeing. Wool (qv) is a complex protein polymer based on amino acid building blocks, and the polymer chains are cross-linked by disulfide groups. Amino acids (qv) contain both amino and carboxylic acid groups and therefore the wool polymer contains both $-NH_2$ and $-COOH$ side chains and can react with both acid, to give protonated $-NH_3^+$ groups, and alkali to give $-COO^-$ groups. These groups are not equal in their ability to react with agents. For example, for the equilibrium $-NH_2 \rightleftarrows -NH_3^+$ for arginine, $pK = 13.20$, whereas for lysine, $pK = 8.95$.

Thus wool is a fiber whose net charge is a balance of the numbers of $-NH_3^+$ and $-COO^-$ depending on the individual amino acids in the polymer and the pH. At strong alkali pH the charge is negative. As the pH moves toward neutral the net negative charge is reduced as individual groups, and depending on their position in the molecule, these are protonated. The fiber eventually becomes neutral at a pH known as the isoelectric point. In the case of wool this occurs within a wide range between pH 5.0 to 12.0. Strict neutrality is not found;

the true isoelectric point is around pH 5.0, but between pH 5.0 and 12.0 (the isoelectric region) only a gradual increase in net negative charge is found.

Below pH 5.0, as the pH becomes more acid, increasing amounts of amino groups are progressively protonated resulting in an ever increasing net positive charge. At pH 2.0 all the amino groups are protonated and the maximum net positive charge is obtained.

In theory wool could be dyed in its positive, neutral, or negatively charged form but this is not possible in practice because under alkaline conditions the disulfide cross-linking groups are broken causing the polymer structure to break down, ie, the wool dissolves in the alkali. Wool is either dyed in its neutral or net positively charged form. As with cellulose, wool, being a hydrophilic fiber, is dyed with water-soluble dyes that contain sulfonic acid groups to impart solubility.

The dyeing of wool is carried out by applying a negatively charged dye to a neutral (or slightly negatively charged substrate) or to a strongly positively charged fiber depending on pH. There are therefore no problems of ionic repulsion as with cellulose. Instead strong ionic attraction exists that results in high affinity and rapid uptake of dye, so rapid that it is essential to control this rate of uptake if level dyeing is to be achieved. The wool dyeing processes are therefore designed around methods of obtaining level dyeings under practical application conditions.

ACID DYES

Classes.   There are three classes of acid dyes: acid leveling, acid milling, and super milling.

Acid Leveling Dyes.   These are molecular dispersions at low temperatures (true solutions) and are simple molecules. They have low affinity at neutral pH and exhibit ionic attraction at acidic pH when the wool becomes charged. They exhibit good leveling and migration behavior. Their low affinity however also results in low fastness.

Acid Milling Dyes.   These are colloidal dispersions at low temperatures and true solutions at high temperatures. They are generally larger molecules than acid leveling dyes but although they have a higher nonionic attraction the main mode of attraction is electrostatic. These dyes have medium fastness and medium leveling, migration, and fastness properties.

Super Milling Dyes.   These are colloidal dispersions at both low and high temperatures and are complex molecules containing low alkyl hydrophobic chains to enhance fastness. These dyes exhibit high affinity at neutral pH resulting from attraction through van der Waals forces, charge-transfer, and hydrophobic interaction. They also exhibit ionic attraction at acidic pH values. They exhibit high affinity and high fastness but poor leveling and migration.

Controlling Dyeing Behavior.   As with all other dyes the dyeing process concentrates on obtaining level dyeings within an economic time period, and once again slower dyeing means better control and level dyeing is enhanced. When dyeing wool with acid dyes four factors control the dyeing behavior: pH of the dyebath, presence and concentration of electrolyte, temperature of the dyebath, and choice of dyestuff class.

Effect of pH.   The simple acid leveling dyes exhibit little affinity for wool with a net neutral charge. When wool is charged the dye is absorbed by ionic attraction, until the negative charges taken onto the wool by the charged dye neutralize all the charges, and then dyeing stops. The lower the pH the more positive sites on the wool and an excess is needed to obtain full exhaustion in deep shades. This is achieved at around pH 3.0 by using dilute sulfuric acid.

Acid milling dyes have medium neutral affinity and dyeing can commence on neutrally charged wool. If excess positive charges exist then rapid absorption and unlevel dyeing result. Usually acetic acid is sufficient to generate the requisite number of sites.

Super milling dyes have a high affinity for neutrally charged wool and the rate of dyeing is sufficiently high without any acid in the system to require control to give level dyeing behavior. An acidic pH is needed to achieve full exhaustion and this is obtained by using acid generators, eg, ammonium acetate, that cause the pH to drift increasingly acidic during the dyeing process. The aim is to have a minimum rate of dyeing initially.

Electrolyte. The usual electrolyte used is sodium sulfate. The sulfate ion, SO_4^{2-}, has a low affinity for the wool, but because of its smaller size exhausts onto the sites before the dye. As the dyeing process progresses the higher affinity dye molecule displaces the sulfate ions. The sulfate ions therefore act as a leveling agent as they slow down the overall rate of dye uptake, and act as "retarders." The degree of retardation of the dyes depends on the importance of ionic attractions and the affinity of the dye for neutrally charged wool. With acid leveling dyes the retardation effect is maximum, whereas with the acid milling dyes having less dependence on ionic attraction the effect is less. In the case of the super milling dyes the effect is minimal. Further, because the presence of electrolyte is likely to increase the risk of aggregation with what are already colloidal dispersions, it is usually avoided with super milling dyes.

Temperature. No dye transfer takes place below 40°C, above which, because the fiber swells, the rate of dyeing increases.

Practical Processes. With acid leveling dyes no real problems exist because the dyes show good migration, electrolyte is added from the beginning, and rather like Class A direct dyes level dyeing is achieved by prolonging the times at the boil.

The other extreme is found with super milling dyes when at the start ammonium acetate, sulfate, or an organic ester is present without any electrolyte. Dyeing is carried out more slowly taking some 60 min to reach the boil, and often the dye is applied with a cationic leveling agent. The action of this leveling agent is to form a soluble complex with the dye in the dyebath which then breaks down as the dyeing temperature is increased. The rate of breakdown of the complex is slow and is the rate-controlling process in the early stages of dyeing, so the possibility of obtaining controlled level dyeing conditions is increased.

Acid milling dyes are intermediate in behavior being applied with acetic or formic acid in the presence of sodium sulfate.

A disadvantage of acid dyes is that their wetfastness depends on the formation and maintenance of a salt linkage between the charged wool and dye. This requires an acidic internal pH to be maintained in the wool. When there are excess positive sites in the wool good fastness results. Once these are neutralized, either in processing or usage, dye can desorb very readily with acid leveling dyes where there are no other forces of attraction to prevent desorption in, for example, alkaline wash liquors.

MORDANT DYES/METAL COMPLEX DYES

Certain acid dyes can have their fastness properties improved by combining the dye with a metal atom (chelation). The most common metal is chromium, although cobalt is sometimes used, and this can be introduced in a number of ways. The basic mechanism is donation of electron pairs by groups in the dye (ligands) to a metal ion. For example, Cr^{3+} has a coordination number of 6, and therefore will

accept six lone pairs of electrons. Typical ligand groups are $\diagdown C = O$, $-N = N-$,

$-COOH$, and $C_6H_5O^-$. At least two coordinate bonds must be formed between positive metal and negative dye sites, namely $-OH$, or $-COOH$ groups ortho to an azo bond (24).

Methods of Introducing Metal. *Prechroming.* Sodium dichromate is applied to wool with which it reacts because the wool reduces the chromate to give chromous ions which are able to chelate with dye in subsequent dyeing methods. The advantage is that good fastness is obtained. The disadvantage is that the dye–metal chelate coordinated bond formation is irreversible, like the covalent bonds of reactive dyes, and migration is impossible. It is therefore difficult to obtain level dyings in pale shades, so this method is best used with heavy shades where leveling is not a problem.

Chroming and Dyeing Together. In this method a high affinity dye and chromate are applied together. It is difficult to control and is therefore not a popular method.

Afterchroming. Acid leveling dyes are applied and once full exhaustion and levelness have been obtained the dyed wool is treated with chromate salts to give a dyeing of very high fastness. This is the most popular of the methods so far described. The only practical problem is that the afterchroming process causes a severe shade change which makes shade matching difficult. The process is therefore usually reserved for standard repeating shades. Chromium is toxic to the environment and effluent from the dyeing process needs treatment before discharging. The environmental considerations and high cost of chromium salts are likely to make this process of declining interest.

Metal Introduced Into the Dyestuff Molecule in Manufacture. These dyes fall into two categories according to the ratio of metal cation to dye molecule anion: 1:1 metal dye complexes comprise one metal cation and one chromophoric ligand and 1:2 metal dye complexes comprise one metal cation and two chromophoric ligands.

An example of a 1:1 metal dye complex is shown in Figure 5a. The net charge on the molecule is derived from the plus three charge on chromium, Cr^{3+}, and the negative charges on two phenoxy groups. Solubility is induced by introducing sulfonic acid groups into the structure. These are ionized in solution, SO_3^-, and therefore each sulfonic acid group introduces a negative charge. The number of sulfonic acid groups has a marked influence on the net charge: the overall net charge is positive if there are no SO_3^-, overall neutrality is the case for one SO_3^-, and two or more SO_3^- provide an overall net negative charge.

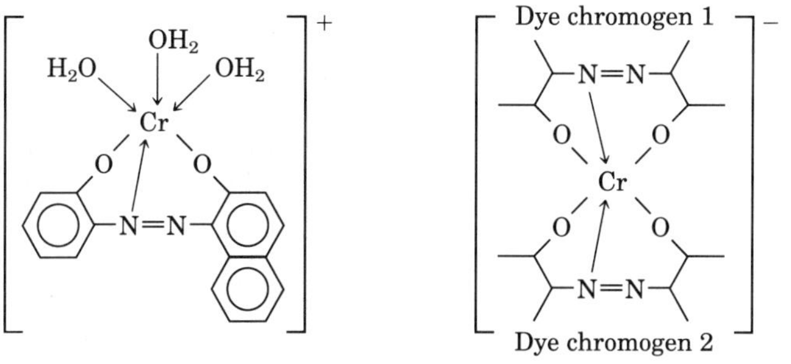

Fig. 5.   Chemical formulas for metal complex dyes: (**a**), 1:1 complex; (**b**), 1:2 complex.

The dyes with no sulfonic acid groups are applied as dispersions, although sometimes groups such as —SO_2NH_2 are introduced to impart some solubility without ionic charge. The positively charged dyes show typical cationic dye behavior and form salt linkages with the COO^- groups in wool. Dyes containing two or more sulfonic groups behave as typical acid dyes forming salt linkages with the NH_3^+ groups in wool and exhibiting a high affinity for neutrally charged wool.

The positively charged 1:1 complex dyes are no longer popular. The greatest interest is in the monosulfonated dyes. With these dyes the mode of attraction is by the lone pair of electrons on the nitrogen groups in wool acting as a ligand and coordinating with the metal in the dye. These neutral 1:1 metal complex dyes therefore act as reactive dyes with all the accompanying excellent fastness and poor level dyeing properties and migration. In order to control levelness it is necessary to control the availability of the lone pair of electrons on the secondary amine groups. This is done by protonation, but because the secondary amines are weak bases, a pH of 2.0 (8% H_2SO_4) is needed. Dyeing is carried out at the boil to enhance levelness, and once finished the act of rinsing increases the pH and reforms the secondary amine groups allowing fixation to take place *in situ*. At these high temperatures and low pH, damage to the wool surface can occur, but the resulting dyeing is of excellent fastness.

With 1:2 metal–dye complexes the coordination number of the metal is fully satisfied so no reaction can take place with the amino groups in the wool. A representative structure is shown in Figure 5**b**.

The net charge resulting from this complex is calculated as before, namely three positive charges from the Cr^{3+} cation and four single negative charges from the four phenoxy groups. The overall net change is negative, and the addition of ionic solubilizing groups increases the net negativity. The higher the net negativity the greater the ionic attraction and because these dyes are already macromolecules the rate of dyeing is very high with very poor migration.

It is therefore necessary to minimize the net negative charge by minimizing the number of ionic solubilizing groups. In practice there are three types of dye: (*1*) dyes containing SO_2NH_2 groups only to give net negative charge of −1, with the best leveling and migration properties, but lowest fastness of the group; (*2*) unsymmetrical dyes where the two dye chromogens used are different, one containing a sulfonic acid group, one without. These dyes have a net −2 charge so are not quite as good in terms of migration, but have better fastness; and (3) symmetrical dyes where both dye chromogens are the same and have the same ionic solubilizing group or groups. These dyes have poor migration behavior but excellent fastness, and in view of their ease of manufacture compared to the other two types, are the most economical. They are widely used for heavy shades such as navy, black, brown, bordeaux, etc.

The 1:2 metal complex dyes are dyed either at neutral pH or with ammonium acetate, and the exhaustion achieved by the effect of van der Waals forces. The pH is then allowed to go slightly acidic to form salt linkages between the dye anion and the protonated primary amine groups in the wool (NH_3^+). All the dyes have similar dyeing properties and the conditions of application do not damage the wool.

DYEING WOOL WITH FIBER-REACTIVE DYES

Fiber-reactive dyes are by no means as popular for dyeing wool as they are for cotton because the fastness of fiber-reactive dyes on wool is not that much better than other dyes. They are difficult to apply and the need for bright high fast shades on wool is not as high as on cotton because severe washing treatments damage the wool itself. They are used on specially treated wools that have been made suitable for washing in automatic washing machines by treating with a polymer. The dye reacts with cations in the polymer.

These dyes are not very commercially important, and the dyeing mechanism has been described in detail elsewhere (15,25). The difficulty in applying fiber-reactive dyes to wool is the result of the same reactions already described. They are negatively charged and the wool is positively charged so ionic attraction exists. The fiber-reactive dyes are essentially acid leveling or milling dyes and so this attraction can be controlled by pH. Once the dye is fixed no migration can take place, and in this case the reaction is with primary amine sites. Assuming a level dyeing is obtained, as with cellulose, all hydrolyzed dye must be removed from the wool in order to have good fastness. This is extremely difficult because of the dangers of damaging the wool during this process.

These constraints cause different reactive groups to be developed for dyeing wool. They fall into two groups: novel reactive groups with more than one type of group in each molecule in order to increase the amount of dye fixed and minimize the wash-off problem, eg, Lanasol Dyes by CIBA-GEIGY, and those based on cellulosic reactive dyes, but where the dye is applied to the wool in a nonreactive form. On boiling, this form slowly converts to the reactive form that fixes rapidly *in situ*, eg, the Hostalan Dyes by Hoechst, introduced in 1971 were precursors of vinyl sulfone dyes.

SILK

Because it is also a protein, silk can be dyed as wool, but in practice the dyes used are generally acid dyes in view of the fiber not being treated to any severe washing in its life. The main difference between wool and silk is in the preparation of the fiber for dyeing.

Silk in its raw state is coated with sericin. It is necessary to remove this gum in order to develop the silk luster and dyeability. Historically, 25–30% soap on weight of fiber (owf) was used for degumming. Synthetic detergent systems, such as higher alcohol sulfates, and soda ash and boric acid have replaced soap to a large extent. Buffered alkalies, especially polyphosphates, are frequently used with synthetic detergent systems. Strongly alkaline systems must be avoided to prevent attack on the proteinaceous silk fibers.

Dyeing of Synthetic Polyamides

Polyamides (nylons) are thermoplastic fibers that retain their form produced by heat treatment. They are usually given an alkaline scour and then heat-set. The heat-setting treatment is conducted at ca 10°C above the subsequent wet processing steps; this ensures good form retention after processing. Woven fabrics are usually heat-set on a contact heat-setting machine and nylon tricot is generally heat-set on a tenter frame or in steam chambers.

When bleaching is required, sodium chlorite is preferred as the bleaching agent. However, for economy of operation, bleaching is often accomplished with peracetic acid in the alkaline scouring operation. Cotton–nylon blends are frequently bleached first with hydrogen peroxide to remove motes, and sometimes are bleached further with sodium chlorite. Bleaching is followed with an antichlor treatment, involving chemicals such as 0.1% sodium bisulfite and 0.15% tetrasodium pyrophosphate.

Scouring may be conducted on jigs, boil-off machines, or kettles, depending on fabric weight, construction, and crease tendency in the rope form. A combination of a synthetic detergent and soda ash is usually used and scouring is conducted at 85–100°C. Certain nylon blends may require less stringent conditions and the use of less alkaline builders, such as tetrasodium pyrophosphate.

DYEING MECHANISM

Nylon is similar in its general chemical structure to the natural fiber wool, and therefore all the previously described processes for wool are applicable to dyeing nylon with acid, metallized, and other dyes. There are, however, significant differences. Nylon is synthetic, it has defined chemical structure depending on the manufacturing process, and it is hydrophobic (see FIBERS, POLYAMIDES).

Chemically there are important differences. There are no side chains and unlike wool the number of amino and carboxylic groups differs; there is an excess of carboxylic groups. The numbers of amino groups can be changed by chemical modification, eg, in deep dyeing nylon, but for the most part nylon fibers can be considered to have a limited number of sites, which can differ from one chemical type to another. For example, a given acid dye may build up more readily and show superior leveling performance on nylon-6 than on nylon-6,6, but have a better wetfastness on nylon-6,6 than on nylon-6. The polymer is not as complex as wool, reducing the scope for attraction other than by ionic forces.

Physically there are differences. Like all polymeric fibers nylon contains crystalline and noncrystalline areas. Only amino groups in the noncrystalline regions are accessible.

Finally, because the fiber is synthetic, polymer formation followed by drawing into a yarn presents the likelihood of chemical and physical variations in the yarn. It is usual to stabilize the fibers by a heat-setting process before dyeing, and again further physical variation can be introduced at this stage (7). The manufacturing history of the polymer therefore plays a role in determining the dyeing performance. In order to obtain level dyeings it is necessary to consider not only the basic chemical reactions taking place but also the relative sensitivities of the dyes to physical and chemical variation in the fiber.

ACID DYES

The majority of acid dyes is applied to nylon rather than to wool. There are three groups of dyes: Group 1 includes dyes with little affinity at neutral or acidic pH but which exhaust under strongly acidic conditions; Group 2, the largest group of dyes which exhaust onto nylon in the pH range 3.0–5.0; and Group 3, dyes with a high affinity for nylon under neutral or weakly acidic pH. Only dyes within one group should be used together, and dyestuff manufacturers assist in this by having different nomenclature for dyes in each group.

Group 1. With dyes in Group 1 the mode of absorption is best considered by the Langmuir isotherm where the negatively charged dye is attracted to specific NH_3^+ sites. The fiber can therefore be saturated because of the limited number of sites depending on the pH. With pH control it is possible to ensure controlled level dyeing by making the rate-controlling process the generation of NH_3^+ groups, and always dyeing at saturation for the given pH. Drawing an analogy to a theatre, this is rather like issuing blocks of tickets over a time period where each block of tickets contains seats uniformly spread across the theatre. As each block of uniformly distributed seat tickets is released (in practice by dropping the pH another step) more people can enter, but if the demand for seats is always greater than the availability, a uniform distribution is ensured.

Once the majority of sites is uniformly occupied the remaining ones will be uniformly available by definition, and so the pH can be dropped quickly to give full exhaustion. The advantage of the Group 1 dyes is that they are small molecules having high rates of diffusion with little hindrance to their movement through the hydrophobic nylon structure. Thus high levels of exhaustion approaching 100% are possible with all available sites occupied, and the dyes are excellent in covering physical variations in the polymer and reasonable in covering chemical variations. Their main problem is that they give poor wetfastness in neutral or alkaline washing. The main outlet for these dyes is for dyeing carpets.

Group 2. These dyes show some affinity under neutral conditions and therefore it is preferable to use acid-generating salts (or mixtures of these with acetic acid) which decompose on heating to lower the pH. This gives low exhaustion rates and level dyeing. These dyes have larger molecules than Group 1, and hence are slower diffusing. This means they are less good at covering physical variations. Because their affinity is to a degree based on attraction by forces other than ionic forces they are also not quite as good as Group 1 in covering chemical variations. They do have higher wetfastness. Dyeing is usually carried out as for super milling dyes.

Group 3. These dyes have high affinity under neutral conditions and are large complex molecules. From the previous considerations it is clear that it is difficult to obtain level dyeings with these dyes, and they are sensitive to physical and chemical variations in the nylon. They do have excellent fastness and therefore it is often worthwhile overcoming these application problems. This is done by using specially developed auxiliary agents that are added to the dyebath. The detailed mechanism has been described in detail elsewhere (26).

The principle of the technique is that dyeing is carried out in the presence of both anionic and cationic auxiliary agents. The anionic agent competes with the dye for the fiber sites and the cationic agent provides an alternative site of attraction for the dye. The net results of these competing reactions is that during the initial stages of dyeing the NH_3^+ are not freely available to the dye anion, but as the process continues more and more sites and dye anions are slowly made available. The attraction between different dyes and cation depends on the relative charge on the dye. The effect is therefore to equalize the forces of attraction and diffusion enhancing dyestuff compatibility and the ability to overcome physical and chemical variances in the fiber.

Tanning Agents. It is possible to improve the wetfastness of acid dyes by aftertreatment. The original method was to apply tannic acid, tartar emetic, and formic acid. The mode of action is not clear but the qualitative explanation is that this compound permanently fills the "pores" in the hydrophobic fiber through which dye can desorb. This method is long and time consuming and dulls the shade of the dye which suggests that some interaction with the dye cannot be ruled out. Today synthetic tanning agents (syntans) are used. These are anionic phenolic condensates that have a similar effect, although they do not improve the fastness quite as much as the classic treatment. Being anionic they are applied from a treatment bath at pH 3–5.

Metal Complex Dyes. The 1:1 metal–dye complexes cannot be used on nylon because the low pH values needed cause fiber degradation. The 1:2 metal–dye complexes are of commercial interest because of their excellent lightfastness in pale shades. These macromolecules are difficult to apply level and are sensitive to both chemical and physical variations. In their application they are treated as the Group 3 acid dyes.

OTHER SOLUBLE HYDROPHILIC DYES

Although nylon is a hydrophobic fiber, water-soluble dyes diffuse into it to a greater or lesser degree. Nylon also contains both positive and negative sites, and sites that can form the basis for coordinate bonds. Some direct dyes have profiles on nylon very similar to Group 3 dyes and therefore, to supplement the range of shades available, they are sometimes applied with Group 3 dyes. Some dyestuff manufacturers give these selected direct dyes the same nomenclature as their Group 3 acid dyes. For basic dyeable nylon used in special effects, basic dyes are absorbed onto COO^- groups.

DISPERSE DYES

The insoluble, hydrophobic disperse dyes readily dye nylon, and because their mode of attraction is completely nonionic they are completely insensitive to chemical variations and pH. Small molecular-sized disperse dyes (ca mol wt 400) show very high rates of diffusion and excellent migration properties and they are insensitive to physical variations in the nylon. As the molecular size of disperse dyes increases they show increasing sensitivity to physical variation.

Although when using disperse dyes on nylon they are readily absorbed at temperatures up to the boil, they are also readily desorbed when the dyed fabric is immersed in wash liquors. There are no known methods of overcoming this problem although at one time disperse dyes containing fiber-reactive groups were unsuccessfully promoted. Conventional disperse dyes give such low fastness because of the open structure of nylon. The dye can freely move within the non-crystalline regions. The glass-transition temperature of the fiber is exceeded at normal temperatures, and because of the relative low hydrophobicity desorption into an external aqueous phase is readily possible.

The main use for disperse dyes is where excellent coverage of fibers likely to have physical and chemical variations is needed, and where wetfastness is not critical. The small molecular-weight dyes are therefore widely used for pale shades on continuous filament yarns used in hosiery. There is also some use made in exhaust dyeing of carpets made from continuous bulk filament nylon to give good coverage. The higher molecular-weight disperse dyes are not of importance on nylon because of their poor fastness and poor coverage of physical variations.

Dyeing is relatively simple. The disperse dye is added to a dyebath containing a nonionic dispersing agent, sodium hexametaphosphate, and sometimes acetic acid is added to give pH 5.5 to prevent decomposition of some disperse dyes. Dyeing is carried out by bringing the dyebath to the boil, and continuing until exhaustion is completed.

CARPET COLORATION

Some 50% of all nylon is in the form of carpets almost exlusively colored with acid dyes, and around 50% of the carpet manufacturing industry is located in the United States. The acid dyes from Group 1 are those most widely used because they exhibit the rapid diffusion needed to penetrate the bulky yarns used in carpets, especially bulk continuous filament yarn used in tufted constructions, with high exhaustion. Their wetfastness properties are generally adequate for most outlets.

The most popular coloration method is to apply the dyes continuously, usually by padding, but printing, spray jet, and droplet applications are used. In order to obtain patterned effects, chemical or physical resisting agents can be applied first, and deep and normal dyeing nylon and basic dyeable nylon blends can be used. In the latter case the basic dyeable nylon is dyed with cationic dyes. Carpets can be printed in an analogous method to other textiles and this process is more popular in Europe than the continuous application techniques used in the United States.

The acid dyes are applied with strong acids, eg, sulfamic acid, to give the low pH needed to give an adequate number of NH_3^+ groups for rapid and high exhaustion. After physically applying the dyestuff the carpet is usually steamed wet, its bulk making drying difficult and expensive, for 5–8 min at 100–105°C. After steaming, the carpet is immediately washed to remove unfixed dye and auxiliaries. Alternatively the carpets can be dyed on the winch, starting neutral and slowly adding acid over a period of time to promote exhaustion. This is known as the "pH-swing method" and is relatively popular in Europe.

Where high wetfastness is needed, for example in hotel lobbies and bars where liquid spillages are likely, the higher fastness acid dyes (Groups 2 and 3) and even metal complex dyes are used.

Dyeing of Acrylic Fibers

Pure polymeric acrylonitrile is not an interesting fiber and it is virtually undyeable. In order to make fibers of commercial interest acrylonitrile is copolymerized with other monomers such as methacrylic acid, methyl methacrylate, vinyl compounds, etc, to improve mechanical, structural, and dyeing properties. Fibers based on at least 85% of acrylonitrile monomer are termed acrylic fibers; those containing between 35–85% acrylonitrile monomer, modacrylic fibers. The two types are in general dyed the same, although the type and number of dye sites generated by the fiber manufacturing process have an influence (see FIBERS, ACRYLIC).

Because of polymerization and spinning procedures there is no such thing as a standard acrylic fiber. Acrylic fibers differ in the relative amounts of strong anionic groups, SO_3^-, and weak anionic groups, COO^-, and their total number, and whether the fiber structure is open or not depending on the manufacturing history. Because the potential dye sites are introduced deliberately their number is limited. The Langmuir isotherm best describes the absorption behavior of dyes, and saturation values differ from fiber to fiber. The saturation value for a fiber has been defined (27) as the percentage of dye on weight of fabric of a basic dye of molecular weight 400 that occupies all the available (those in the amorphous regions) sites in the fiber. A factor of almost three has been found (28) between different commercial fibers, the lowest values ~1.2% and the highest 3.5%. Basic dyes are the most popular class applied to acrylic fibers. Like nylon, acrylic can be dyed with disperse dyes, but with the same reservations of fastness. Disperse dyes are therefore only used for pale shades where excellent levelness is needed or difficult to obtain by any other method due to variations in the fiber.

Preparation for Dyeing.   Fabrics are scoured with a synthetic detergent at 45–65°C and are rinsed before further processing to remove tints, size, wax, grease, spinning oils, or other impurities that were applied or picked up during the manufacturing operation. Bleaching, when required, is usually accomplished by means of a sodium chlorite bleach, a selected optical brightener, or a suitable combination of the two. Acrylic-blend fabrics may require other bleaching agents if chlorine-sensitive fibers are present. Most acrylic fibers require a presetting in open-width in boiling water to avoid dimensional stability problems during subsequent wet-processing steps.

Dyeing Mechanism.   The original basic dyes were characterized by a delocalized charge in the molecule. As the importance of acrylic fibers grew basic dyes were developed with localized charge in one specific part of the molecule allowing stronger salt links to be formed than with the delocalized type.

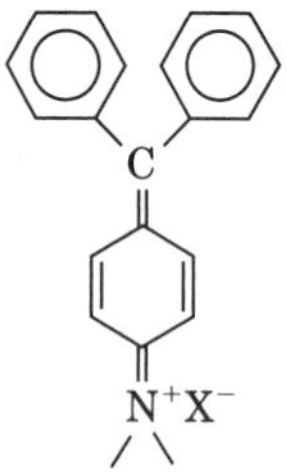

These newer dyes are often referred to as modified basic dyes. Essentially their structure is that of a disperse dye that has been protonated. These dyes therefore have high rates of diffusion into the fiber, and their mode of attraction is almost entirely ionic.

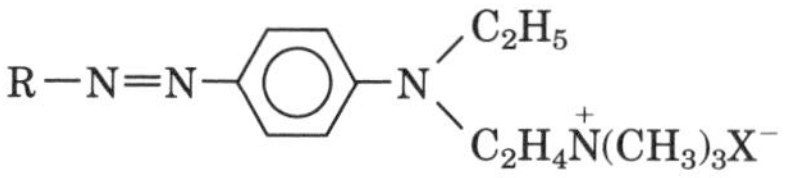

As with wool and nylon when applying dyes to the fiber where dye and site are oppositely charged the need is to control the rate of exhaustion to promote level dyeing. With acrylic this need is made all the more important by two additional factors: first, the modified basic dyes show poor migration because they form very strong salt bonds especially between dyes with delocalized charges and fibers with strongly negative SO_3^- sites; secondly acrylic fibers do not readily dye below their glass-transition temperature, T_g, which is usually around 80°C. Above this temperature the rapid transition within the fiber structure is accompanied by a rapid uptake of dyestuff. Up to 80°C little dye is absorbed, but immediately as T_g is reached the dye exhausts very quickly, uptake of dye is virtually uncontrollable, and level dyeing very difficult to achieve. The dyeing methods proposed have been reviewed in detail (29,30).

The effect of pH depends on the fiber type. The SO_3^- groups on fibers are so strong that they are deprotonated even in neutral dyebaths. The dyebath pH therefore has no influence on the availability of these sites in the fiber and therefore pH cannot be used to control the uptake and level dyeing behavior of the dye. For carboxylic acid groups the pK_a is about 5.5. At lower pH values there are considerably fewer sites available for the dye and at higher pH values considerably more COO^-. For example, with Courtelle (Courtaulds PLC) which contains only carboxylic acid groups the saturation value has been measured (15) as 2.5 at pH 3.6 and 3.5 at pH 4.5. With such fibers pH can be used to control the rate of dyeing and hence levelness.

Compatibility Values.   The need to apply dyes in admixture to give more shades necessitates a way of measuring the compatibility of dyes. Depending on charge, the degree of localization, and molecular shape and size, dyes have different affinities and behavior and hence different dyeing rates. A qualitative testing procedure has been defined (27) where dyes are applied with a range of known dyes and the unknown dye is ascribed the same compatibility value as that already given to the known dye with which it dyes compatibly under all practical exhaust-dyeing conditions except in the presence of anionic dyes or auxiliaries. There are five values (1–5) and in combinations the dye with the lowest value exhausts most rapidly. For best results dyes should be mixed with dyes having the same compatibility value, or at least no more than one value different.

The compatibility value is mainly related to the affinity of the dye for the particular fiber because for basic dyes on modified acrylic fibers there is little possibility for migration and therefore this does not play a significant part in determining compatibility. The rate of dyeing of a specific mixture of dyes of the same compatibility value is not determined by the value itself. The adsorption of cationic dyes is influenced by the presence of others in the dyebath; the presence of cationic retarding agents and electrolytes also influences the rate of exhaustion. It is therefore possible to have a combination of dyes with a compatibility value 5 that under specific dyebath conditions exhausts more rapidly than a combination based on dyes of compatibility value 3.

Level Dyeing Techniques. It is exceptionally difficult to obtain level dyeings on acrylic, and temperature and pH control depend on fiber type and are not always adequate. Sodium sulfate in limited amounts can be used to some effect. The sulfate ions compete for the dye with the fiber SO_3^- sites and so retard the rate of dyeing by forming a dye complex with the SO_4^{2-} ions. The effect of sodium sulfate is best with dyes having the lowest compatibility values.

Other anionic retarding agents can be used, where a dye–agent complex is formed. The breakdown of this takes place slowly as a function of temperature, so limiting the availability of the dye chromogen anion and therefore promoting slow rates of dyeing and levelness. The disadvantages of anionic retarding agents are that they behave differently with different dyes and can produce incompatibility, they reduce the final dyebath exhaustion, and they do not affect the availability of sites on the fiber and therefore do not exert total control over the system.

The more popular method to control leveling is to use cationic products that act as colorless dyes competing with the colored cationic dye for the fiber sites. If amounts of colored modified basic dye and colorless modified basic dye equal to the saturation value of the fiber are uniformly dissolved in the dyebath then level dyeing behavior is promoted. In practice the rate of dyeing of the colorless dye or cationic retarder is slightly higher than that of the colored dyes, but must be chosen to be of the same compatibility value. A dyebath is used containing $x\%$ modified basic dye + $y\%$ cationic (colorless basic dye) where $x + y$ is equal to or slightly in excess of the fiber saturation value corrected to allow for differences in molecular weight from the theoretical 400, acetic acid (pH 3.6–4.0), and up to 2.5 g/L sodium sulfate and nonionic surfactant to assist wetting out and dyebath solubility. The temperature is brought rapidly up to just below the glass-transition temperature (80°C) and then slowly raised to 96–106°C depending on the dyes, fiber, and machinery. Temperatures above 100°C are used to promote migration and diffusion after all the dye has exhausted. After exhaustion is complete a further 10–15 min is needed to ensure all the dye is diffused into the fiber.

Gel Dyeing. Continuous methods exist for dyeing wet-spun acrylic yarns while they are still in their swollen nonaligned state. The advantage of these methods is that they are extremely rapid, and the rate of uptake of dye is not dependent on glass-transition temperature so reducing the constraints on dyestuff selection.

Pad-Steam. Acrylic tow or sliver is continuously dyed by padding cationic dye and acetic acid and steaming for 5–90 min depending on depth of shade. Compatibility values are not always valid, and uniformity in fixation is difficult. The process is therefore mainly used for dyeing tow or sliver.

Dyeing of Polyester

Polyester fibers are based on poly(ethylene terephthalate) (PET); some modified versions are formed by copolymerization, eg, basic dyeable polyester. The modified forms dye in analogous manner to other fibers of similar charge.

Preparation for Dyeing. A hot alkaline scour with a synthetic surfactant and with 1% soda ash or caustic soda is used to remove size, lubricants, and oils. Sodium hypochlorite is sometimes included in the alkaline scouring bath when bleaching is required. After bleaching, the polyester fabric is given a bisulfite rinse and, when required, a further scouring in a formulated oxalic acid bath to remove rust stains and mill dirt which is resistant to alkaline scouring.

Other fibers blended with polyesters in numerous blended fabrics require alternative methods of preparation. Generally, the scouring and bleaching procedures used for these blends are those employed for the primary component of the blended fiber or for the component that most influences aesthetic appearance.

In order to remove long fibers, polyester fabrics are sometimes sheared prior to dyeing. Singeing of the polyester fabrics, which prevents pilling, is usually delayed until after dyeing to prevent uneven dye exhaustion in unevenly singed areas.

Polyester fabrics and many other hydrophobic synthetic fabrics require the application of an antistatic agent prior to printing to prevent the buildup of static charges at rapid printing speeds (see ANTISTATIC AGENTS).

Dyeing Mechanism. Unmodified polyester fibers are very hydrophobic and absorb only minimal amounts of water and are therefore only dyeable with hydrophobic disperse dyes. The mechanism of dyeing is by simple partition, the so-called solid solution mechanism. The dyeing process can be described by the general scheme

$$[\text{dye particles}]_{\text{insoluble}} \rightleftharpoons [\text{dye}]_{\text{soluble}} \rightleftharpoons [\text{dye}]_{\text{fiber}}$$

Disperse dyes are only sparingly soluble and therefore high temperatures are needed to increase the amount of soluble dye in the system. At normal temperatures diffusion through the polyester fiber is very slow and rates of dyeing slow. This slow dyeing rate is overcome by dyeing at high temperature and carrier dyeing (see DYE CARRIERS). Polyester fabrics have a glass-transition temperature in the region 110–120°C. Dyeing above this temperature is much more rapid than below, and therefore exhaust dyeing of polyester is carried out under pressure at 125–135°C, or continuously by the use of high temperature steam at 165–180°C, or baking (thermofixation) at 190–220°C. The higher temperatures also increase dyebath solubility. In carrier dyeing, chemicals are added to the dyebath that allow dyeing to be carried out at 98–100°C. The effect of these carriers is to lower the glass-transition temperature of the fiber (7) to temperatures below the boil.

Disperse dyes on polyester generally have good fastness as a result of the fiber being below its glass-transition temperature in wash treatments and the slow rate of desorption. The degree of crystallinity, the drawing, and heat-setting temperatures of polyester all play a role in determining the rate and amount of dye uptake (7). The practical significance is that with slow dyeing fibers, eg, high tenacity filament sewing threads, higher temperatures and times are needed.

Disperse Dyes. There is a general correlation between heat fastness, the propensity to desorb under conditions of dry heat onto a white piece of polyester, and the dyeing properties of disperse dyes. Dyes with low heat fastness tend to be small molecules and are ideal for carrier dyeing. They also have good coverage of any physical differences in the polyester and rapid rates of diffusion and good migration. They are widely used for dyeing pale shades. Their only disadvantage is that because of their high rates of diffusion they show relatively poor wetfastness especially in staining other more rapid dyeing adjacents such as nylon and secondary acetate. They are usually given a special name, or special suffixes by dyestuff manufacturers, to distinguish them, and the overall group is often referred to as *low energy* dyes in view of dyeing being effected with lower energy techniques than the other disperse dyes. Low energy dyes are not usually used in thermofixation as their low heat fastness at the thermofixation temperatures used (200–210°C) results in them subliming from the hot fabric.

Medium energy dyes are based on larger sized molecules than the low energy dyes. They have slower rates of dyeing, better heat fastness, and generally higher wetfastness. They are not suitable for carrier dyeing. Their main application methods are exhaust dyeing at temperatures of 125–135°C, and for continuous dyeing by thermofixation at around 30–60 s at 190–210°C. Because of their medium molecular size these dyes dye rapidly (15–30 min) at 125°C.

High energy dyes are based on large molecules with polar groups. These dyes have excellent heat fastness resulting from extremely low rates of sublimation. Their main use is in dyeing fabrics that are to be given a subsequent high temperature heat treatment, eg, permanent pleating finish, or sewing threads whose future use and treatments are unknown and therefore every possibility must be considered. Although they have very high heat fastness these dyes do not necessarily have any better wetfastness than medium energy dyes because although they desorb slowly, dye that does desorb has a high affinity for hydrophobic adjacent fibers. Dyeing with these dyes requires either longer times or temperatures than with medium energy dyes to achieve full exhaustion, eg, 45–60 min at 125–135°C in exhaust dyeing. They are sensitive to the conditions of thermofixation needing 60–90 s at 210–220°C to give full yield.

Dyeing Processes. Polyester yarns and fabrics are usually dyed by exhaust techniques; continuous dyeing is largely used only for blends with cellulose. The basic dyeing process is relatively simple. The dyebath is set with disperse dye and dispersing agent (a nonionic or anionic surface-active agent) at pH 5.5 obtained with, for example, acetic acid. The temperature is slowly raised up to the dyeing temperature (125–135°C) and kept there to complete exhaustion and promote migration followed by cooling to below the boil for removal of the dyed material.

Dyeing starts at around 80°C, and the rate of dyeing progressively increases as the temperature increases up to and beyond the glass-transition temperature. If dyes are used in mixture shades that are compatible throughout this heating up phase, level on-tone dyeings result. The early stages of dyeing should be slow enough to encourage level dyeing. If these conditions have been met, then at some final temperature all that is needed is to complete the exhaustion. These conditions are met most readily by selecting medium energy dyes. A large amount of these dyes is already exhausted by the time the final dyeing temperature is reached. A further 15–30 min, depending on the fiber, will be sufficient to complete exhaustion. Exhaustions of greater than 95% are obtained with the medium energy dyes, whereas these levels are only obtained with high energy dyes by prolonged dyeing times.

During the cooling process after dyeing, the solubility of the disperse dye remaining in the dyebath decreases rapidly and it can precipitate onto the surface of the polyester fibers. If it is not removed the resulting dyeing will exhibit both poor fastness to rubbing and poor wetfastness. In any subsequent wash fastness test these particles of disperse dye will be able to stain other hydrophobic fibers adjacent to the dyed polyester. This precipitated dye is removed by a combined chemical decomposition and stripping. The dye is destroyed by reduction using hot (70°C) caustic soda and sodium hydrosulfite, optionally in the presence of a detergent.

This process, based on strong reducing agents, can be avoided by the use of disperse dyes that are removed by aqueous alkali alone. Two types of dye are used: dyes containing diesters of carboxylic acid and dyes destroyed by mild alkali. The reaction of diester dyes is shown in equation 5.

$$R-N=N-\langle\bigcirc\rangle-N\Big\langle{}^{C_2H_4COOCH_3}_{C_2H_4COOCH_3} \xrightarrow{OH^-} R-N=N-\langle\bigcirc\rangle-N\Big\langle{}^{C_2H_4COO^-}_{C_2H_4COO^-} \tag{5}$$

The diester dye is applied to the polyester. Then in the presence of 2.0 g/L sodium carbonate at 80°C the ester groups are hydrolyzed to give the soluble sodium salt of a dicarboxylic acid. This ionized form of the dye has no affinity for the hydrophobic polyester and is quickly removed. Dyes that decompose in similar alkaline solution generally give soluble decomposition products exhibiting no affinity for the polyester. Using such a selection of dyes it is possible to significantly reduce the total processing time because washing off is carried out by a simple alkaline washing process.

Thermal Migration.   Although it is possible to obtain excellent fastness properties by either reduction clearing of traditional dyes or alkali clearing of novel alkali-sensitive dyes, this fastness can be short-lived. In any subsequent heat treatment of the polyester such as heat-setting to stabilize the fiber or fabric, or in the application of a finishing agent such as a softener or antistat, the polyester is again taken above its glass-transition temperature and dyestuff molecules again have mobility within the fiber.

Depending on the dyestuff and the conditions, dye molecules diffuse from within the fiber and crystallize on the surface in a similar manner to the situation found after cooling the dyebath. There is no adequate explanation why this takes place to varying degrees with different dyes. Some general observations are that low energy dyes thermally migrate more than medium or high energy dyes, presumably because of their tendency to sublime out of the fiber at high temperature; the behavior of medium and high energy dyes has no correlation to their heat fastness; the more polar the dye the greater the likelihood of thermal migration; the higher the temperature the greater the risk; and the presence of hydrophobic finishing agents increases the likelihood of thermal migration.

Further thermal migration can take place without necessarily causing poor wetfastness. Some high energy dyes thermally migrate to a high degree and crystals of the dye can be observed on the fiber surface. If these crystals are insoluble under aqueous conditions wetfastness does not deteriorate. Fastness to rubbing decreases, however.

In order to obtain dyeings that do not exhibit thermal migration behavior it is necessary to carefully select the dyes and the conditions of subsequent heat treatments. This is done by trial and error or by following dyestuff suppliers' recommendations.

Dyeing of Cellulose Esters

Acetate fibers are dyed usually with disperse dyes specially synthesized for these fibers. They tend to have lower molecular size (low and medium energy dyes) and contain polar groups presumably to enhance the forces of attraction by hydrogen bonding with the numerous potential sites in the cellulose acetate polymer (see FIBERS, CELLULOSE ESTERS). Other dyes can be applied to acetates such as acid dyes with selected solvents, and azoic or ingrain dyes can be applied especially for black colorants. However their use is very limited.

Cellulose Diacetate. When preparing cellulose diacetate for dyeing, strong alkalies must be avoided in the scouring of acetate because the surface of the cellulose acetate would be saponified by such treatment. Many fabrics tend to crease and therefore require open-width handling. Scouring is frequently carried out on a jig or beam using 1.0 g/L of surfactant and 0.5–1.0 g/L tetrasodium pyrophosphate for 30 min at 70–80°C.

Very small quantities of acetate staple are dyed, however, large quantities of acetate filament are found in satin, taffeta, and tricot fabrics; these are usually dyed open-width on a jig owing to their inclination to crease or crack easily. A typical dyeing procedure on the jig involves addition of acetic acid and dispersing agent over two ends at 50°C. The disperse dye is added over two ends and the dyebath temperature is gradually raised to 80°C in 5°C increments with two passes at each temperature. The dyeing is completed after 30–60 min at 80°C.

Cellulose Triacetate. Cellulose acetate having 92% or more of the hydroxyl groups acetylated is referred to as triacetate. This fiber is characteristically more resistant to alkali than the usual acetate and may be scoured, generally, in open-width, with aqueous solutions of a synthetic surfactant and soda ash.

Dyeing Procedure. Triacetate is a hydrophobic fiber, as compared to secondary acetate, and consequently does not dye rapidly. It is necessary to increase the rate of diffusion of the disperse dye into the fiber by increasing the dyeing temperature to 110–130°C or using a dye accelerant or carrier, or both. Trichlorobenzene, butyl benzoate, methyl salicylate, and biphenyl are typical accelerants. Higher amounts (10–15% owf) are used for dyeing at the boil than at 110–130°C where 2–4% carrier (or in some cases, none) suffices. Triacetate, like polyester, achieves dimensional stability by heat-setting which requires disperse dyes with good sublimation fastness.

A typical dyeing procedure for triacetate includes setting a bath containing acetic acid, dispersing agent, and carrier (amount based on dyeing temperature and liquor ratio) to pH 4.5–5.0; circulating 10 min and adding disperse dyestuffs; raising to 100°C or 110–120°C over 30–45 min; and running for 1–2 h, depending on the desired depth of shade. Triacetate may also be dyed continuously by the pad-dry–thermosol-scour method. The disperse dye is padded along with a suitable thickener. The goods are dried and treated in a thermosol oven for 90 s at 190–200°C, rinsed, and scoured.

Dyeing of Fiber Blends

Fiber blends combine the advantageous properties of two or more fibers into one fabric. They are available as blends of natural fibers, synthetic fibers, or natural fibers blended with synthetic. The most important blend in terms of usage and growth is that of polyester with cotton (1). The combination of the esthetics of a natural fiber (eg, hydrophilic properties, hand) and the physical properties of the synthetic fibers (eg, strength, abrasion resistance) is an important factor in the acceptance of these fiber blends. Also the differences in dyeability between the many fibers on the market open a wide field of multicolored yarns and fabrics to the stylist. The multitude of fibers offered have resulted in an almost unlimited number of fiber blends in the field; usually two fibers are used together, but three-fiber blends are also relatively common. There are two different fiber blend types in use. The so-called intimate blend is obtained by mixing the fibers prior to spinning them; blended fabrics can consist of one fiber in the warp and another in the filling or can be comprised of yarns of different fibers woven or knitted in patterns. The dyeing methods for both types of blends are similar but an intimate blend does not show a shade difference between two fibers as clearly.

Fiber blends can be dyed into union shades (tone-on-tone) or multicolor effects can be obtained by coloring the individual components in different shades or by maintaining one fiber in an undyed state (reserving). A complete reserving of a fiber is not possible in all cases.

When dyeing fiber blends it must be decided whether the fibers can be dyed simultaneously from the same dyebath, or separately and in what order from different dyebaths. The benefits of dyeing from separate dyebaths is that the conditions can be chosen to give the maximum dyeing efficiency for each dye–fiber combination. The disadvantage is that it is more time consuming.

When the fiber blend components are capable of being dyed by the same class of dye, eg, acid dyes on wool–nylon blends, then one-bath treatments are the obvious choice. With fiber components that are dyed with completely different dye classes the ability to use single-bath techniques (exhaust and continuous) depends on the interaction between the dyes and the compatibility of their dyeing procedures. Blends of polyester and cellulosic fibers dyed with neutral disperse and anionic dyes can be dyed by single- or two-bath methods. Blends of cotton and acrylic where anionic and cationic dyes are being applied to negatively charged fibers, and wool–acrylic using cationic and anionic dyes but with both positively and negatively charged fibers, are much more difficult to dye from a single bath. When the fibers are dyed separately, the methods used are generally identical to those already described for the individual fibers.

CELLULOSIC FIBER BLENDS

Cellulosic–Polyester Fibers. One of the most important fiber blends on the market is the mix of 35/65 or 50/50 cotton–polyester. High tenacity viscose fibers are sometimes used instead of cotton. Many of the apparel knitgoods consist of this fiber blend as do sheeting, shirting, and work-cloth fabrics. Although the knitgoods are dyed in exhaust dyeing procedures, most of the woven fabrics are dyed according to one of the continuous dyeing processes. The choice of dyes and hence dyeing method is determined by the fastness properties required.

Exhaust Dyeing. The easiest way to dye cellulosic–polyester blends is a one-bath exhaust dyeing process with application of direct and disperse dyes. High temperature dyeing equipment is frequently utilized for the dyeing of cotton-polyester blends. This requires a special selection of direct dyes that are stable to high temperature dyeing conditions. The dyeing is conducted for 30–45 min at 125–130°C. To improve wetfastness properties, the dyeings can be aftertreated or combinations of disperse dyes and fiber-reactive dyes are used. The classical approach is to dye the polyester first with medium or high energy disperse dyes and then clear all loose dye by reduction, before dyeing the cotton (or viscose) using any one of the fiber-reactive dyeing procedures. The intermediate clearing process is vital because the disperse dye also shows some affinity for the cotton during dyeing owing to the cotton being more hydrophobic than the water phase, and can give a heavy stain on this component with very low subsequent wetfastness. The disadvantage of this technique is that total processing times of 6–8 h minimum are needed.

The process time can be reduced by about 1 h by using compatible medium energy alkali clearing disperse dyes where a shorter dyeing process can be used and clearing done with alkali alone, thus reducing chemical costs. This approach gives dyeings that are equal in fastness to the classical approach.

A further refinement that has been adopted is to dye the cotton first, and after rinsing the polyester is dyed at high temperature. With this process it is impossible to reduction clear at the end of the process because the cotton is already dyed with a fiber-reactive dye that would be destroyed by the action of alkaline reducing agents. This results in a risk of lower fastness because of disperse dye precipitated during cooling. If alkali clearing disperse dyes are used then the problem can be avoided. Total dyeing times of around 5 h are typical.

Recently one-bath methods have been successfully introduced based on carefully selecting fiber-reactive dyes that are stable to high temperatures and medium energy disperse dyes that are alkali clearable. The disperse dye and fiber-reactive dye are added together at the start of dyeing in the presence of dispersing agent, acetic acid (pH 5.5), and sodium sulfate as electrolyte. This is used instead of NaCl because of salt's corrosive attack on stainless steel machinery at high temperatures. The dyebath is raised to 130°C for about 30 min depending on shade. The disperse dye exhausts onto the polyester, and the fiber-reactive dye exhausts onto the cotton. The high temperature promotes leveling of the fiber-reactive dye. The dyebath is then cooled to 80°C and alkali added. This alkali both fixes the fiber-reactive dye and clears the alkali clearable disperse dyes. After allowing adequate time to complete the reactive dye fixation the fabric is washed as for cellulose. Total dyeing times of around 4 h are possible with excellent fastness properties.

Continuous Dyeing.   Large quantities of cotton–polyester fabrics are dyed continuously to promote shade consistency and economy. The processes vary with the class of dye applied to the cellulosic fiber: pigments, sulfur, fiber-reactive, or vat dyes depending on the fastness requirements for the finished goods. The simplest way is a pad-dry-cure process using pigments and polymeric binders for pale shades on sheeting and on shirting fabrics. For darker shades and where higher color fastness is needed, combinations of disperse/vat or sulfur and disperse/fiber-reactive dyes are applied. The most common procedures are (*1*) pad–dry–thermofix–chemical, pad–steam–wash; (*2*) pad–dry–thermofix–reduction, clear–dry–pad, fiber-reactive dye–batch–wash; (*3*) pad–dry–thermofix–reduction, clear–dry–pad, fiber-reactive dye–bake–wash; and (*4*) pad–dry–bake–washoff, ie, a single-bath process.

These methods (*2, 3,* and *4*) are only used when applying fiber-reactive dyes. In all methods thermofixation conditions are 60–90 s at 190–220°C depending on the choice of disperse dye. In method (*1*) the chemical pad is caustic and hydrosulfite for vat dyes, and alkali and salt for fiber reactives. This is the most popular method in the United States.

In method (*2*) the fiber-reactive dye is applied with alkali. The choice of alkali and batching times and temperature are dependent on the fiber-reactive dye used.

In method (*3*) the fiber-reactive dye is applied with alkali and urea, the choice of alkali and amounts of alkali and urea being dependent on the depth of shade. Baking conditions are around 60–90 s at 150°C. This method is popular in Asia.

Finally, in method (*4*) the fabric is padded with a mixture of medium energy disperse dyes, carefully selected higher reactivity, and rapid diffusing fiber-reactive dyes, up to 10 g/L sodium bicarbonate depending on depth of shade, and proprietary auxiliary agents.

The trend in continuous dyeing is toward shorter and shorter lengths to be processed. For example, in 1980 the average run length per shade in the United States was 30,000 m, while in Western Europe, Japan, and Southeast Asia it was around 10,000 m (31). By 1990, the average length in the United States had dropped below 10,000 m, and Japan had dropped to 1,000 m. The shorter the runs, the higher the proportion of production time taken in cleaning and restabilizing and equalizing temperatures, humidity, etc, throughout the whole line and the lower the overall processing efficiency is. Further, the more processing steps the longer the line, and hence the greater the amount of fabric that is processed before final inspection. This can be hundreds of meters which becomes a significant factor when only processing 1000 meters. The popularity of methods (*2*) and (*4*) has therefore grown because of the shorter continuous lines used. This trend to shorter runs has also increased interest in dyeing woven fabrics by exhaust dyeing routes and many continuous dyehouses are putting a limited jet dyeing capacity in place to dye the short fabric lengths. An alternative is to switch to dyeing the polyester component by batch methods and to then dye the cellulose with reactive dyes using the pad-batch method.

Cellulosic–Acrylic Fibers.   Commonly this blend is used in knitgoods, woven fabrics for slacks, drapery, and upholstery fabrics. Since anionic direct dyes are used for the cellulosic fiber and cationic dyes for the acrylics, a one-bath dyeing process is only suitable for light to medium shades. Auxiliaries are needed to prevent precipitation of any dye complexes.

In two-bath processes either the cotton or the acrylic can be dyed first. If the cotton is dyed first cationic dye can form a salt linkage with the sulfonic acid groups on the anionic dye on cotton leading to poor fastness. Heavy shades are best dyed by first dyeing the acrylic and then dyeing the cotton under alkaline conditions. In order to prevent desorption of the cationic dye the dyeing temperature for the cotton dyeing must be below the glass-transition temperature for the acrylic of 80°C.

Cotton–acrylic fiber blends are also used for high quality upholstery pile fabrics. Besides the one-bath exhaust dyeing procedure involving a very high ratio of liquor to fabric, a continuous pad-steam process is used to dye these fabrics. After padding, the goods are steamed for 7–15 min at 98–100°C. The material then must be rinsed warm and cold before drying.

Cellulosic Fiber–Nylon Blends. These blends are used in fabrics for apparel, corduroy, and swimwear. If wetfastness requirements are relatively low, the nylon portion can be dyed with disperse dyes and the cellulosic fiber with direct dyes and a one-bath procedure can be employed. For better wetfastness, the nylon portion is dyed with level dyeing acid colors together with the direct dyes in one bath at 95°C using a reserving agent to prevent the direct dyes from dyeing the nylon. An aftertreatment with a cationic fixative improves the wetfastness properties. For swimwear, the cotton portion is dyed with fiber-reactive dyes. After rinsing hot and cold and soaping at the boil, the nylon portion is dyed with a phosphate buffer system. Selected acid and/or acid milling colors are applied. An aftertreatment with a phenolsulfonic acid condensation product results in best wetfastness properties.

WOOL BLENDS

Wool–Cellulosic Fibers. One of the oldest fiber blends in the textile market is the combination of wool and cotton or wool and viscose. Economy was the primary reason for this blend. Selected direct dyes, which dyed both fibers from a neutral bath in a uniform shade, or a combination of neutral-dyeing acid dyes with direct dyes, were used to dye this fiber blend. However, problems with cross-staining, shade reproducibility, and loss of strength of the wool fiber by boiling in a neutral dyebath (pH 6.5–7.0) led to development of the one-bath process. Selected direct and acid dyes are applied at pH 4.5–5.0 at 98–100°C. A phenolsulfonic acid condensation product is added as a reserving agent, to prevent the direct dyes from dyeing the wool under acid conditions. If optimum wetfastness properties are required, fiber-reactive dyes can be applied to both fibers by use of a two-bath process.

Wool–Nylon. Nylon has been blended with wool in order to give additional strength to the yarn or fabric. It is used mainly in the woollen industry for coats and jackets and, to a lesser extent, for socks and carpet yarns. Both fibers are dyed with the same products, however the fibers have different affinity to them. Generally level dyeing acid dyes are applied. Disulfonic acid types are preferred for light to medium shades because they dye both fibers more easily in the same depth as monosulfonic acid types. However, in most instances a reserving agent is needed for light to medium shades to balance the depth between nylon and wool. Without it, the nylon dyes more strongly.

For heavy shades, monosulfonic acid types are preferred to obtain the necessary buildup on the nylon. In case the nylon remains lighter than the wool, a small amount of disperse dye is added to the dyebath to build up the depth on this fiber.

Wool–nylon upholstery fabrics and carpet yarns require higher light- and wetfastness properties. Neutral premetallized dyes are used in these cases. However, they have a much higher affinity to the nylon than the wool. Therefore, stronger retarding agents have to be employed, eg, phenolsulfonic acid condensation products. Higher amounts are required for light shades than for medium depth. Dark shades can be dyed without addition of a retarding agent.

Wool–Acrylic Fibers. This blend is being used for industrial and hand knitting yarns. The acrylic fiber is aesthetically similar to wool, increases the strength of the yarn, and adds bulk to the goods. Special precautions are necessary since the two fibers are colored with dyes of opposite ionic type. Coprecipitation is prevented with the use of an antiprecipitant. Usually, level dyeing acid dyes are used for the wool portion in combination with the cationic dyes for acrylic fiber.

Wool–Polyester Fibers. The 45/55 wool–polyester blend is the most common fiber combination in the worsted industry. Strength and excellent dimensional stability of the polyester fiber enable the creation of lightweight wear fabrics not obtainable before. Economy has modified the fiber ratio and 30/70 and 20/80 wool–polyester blends are as common as the classical 45/55 blend. Disperse dyes for polyester and acid or neutral premetallized dyes for wool are employed in a one-bath process. Should cationic dyes be used for the wool portion, a one-bath procedure can only be employed for light to medium shades, whereas dark shades require a one-bath two-step process. Wool blends should not be dyed above 105°C in order to avoid deterioration of the fiber quality.

BLENDS OF SYNTHETIC FIBERS

Polyester Fiber Blends. Disperse dyeable and cationic dyeable polyester fibers are frequently combined in apparel fabrics for styling purposes. Whereas the disperse dyes dye both fibers, but in different depths, selected cationic dyes reserve the disperse dyeable fiber completely, resulting in color/white effects.

Polyester Fiber–Nylon Blends. This fiber blend is used in apparel fabrics as well as in carpets. Disperse dyes dye both fibers, however they possess only marginal fastness properties on nylon. Therefore it is important to select those disperse dyes that dye nylon least under the given circumstances. The nylon is dyed with acid dyes, selected according to the fastness requirements. The fiber blend is dyed in a one-bath process for one hour at 100°C or at 115–120°C. Cationic dyeable polyester–nylon blends and blends of nylon with both polyester types are also in use. The three-fiber blend is dyed according to a one-bath two-step process. The disperse dyeable portion and nylon are dyed first in one bath, followed by the cationic dyeable polyester.

Polyester Fiber–Acrylic Fiber Blends.   This fiber blend is dyed in a similar fashion to that of the blends of the different polyester fibers. The selection of cationic dyes is substantially larger for the acrylic blend.

Nylon Blends.   Differential dyeing nylon types and cationic dyeable nylon blends are used primarily in the carpet industry. The selection of cationic dyes for nylon is rather limited; most products have very poor fastness to light. These blends are dyed in a one-bath procedure at 95–100°C. Selected acid dyes are used for differential dyeing. Disperse dyes will dye all different types in the same depth.

Elastomeric Fibers.   Elastomeric fibers are polyurethanes combined with other nonelastic fibers to produce fabrics with controlled elasticity (see FIBERS, ELASTOMERIC). Processing chemicals must be carefully selected to protect all fibers present in the blend. Prior to scouring, the fabrics are normally steamed to relax uneven tensions placed on the fibers during weaving. Scouring, which is used to remove lubricants and sizing, is normally conducted with aqueous solutions of synthetic detergents and tetrasodium pyrophosphate, with aqueous emulsions of perchloroethylene or with mineral spirits and sodium pyrophosphate.

When bleaching is required, a reductive bleach with sodium hydrosulfite and sodium metabisulfite is used. Cotton blends may require a hydrogen peroxide bleach at pH 9.0–9.5 prior to or instead of the normal reductive bleach. Chlorine-type bleaches which damage elastomeric fibers are avoided.

Dyeing is carried out by the method best suited to the fiber used as the outer sheath, eg, acid or premetallized dyes for nylon-based, reactive or direct dyes for cotton-based.

Other Application Procedures

Pigment Dyeing.   Many dyers do not look upon this form of coloration as dyeing; nevertheless millions of meters of fabric are dyed by this system each year. A finely dispersed (0.5–5.0 μ diameter) organic pigment is applied by padding together with organic binders and, depending on the binder system, a catalyst. After drying, the fabric is cured at 170–175°C when polymerization and optionally cross-linking of the binder takes place. The typical binder systems used are acrylic and butadiene resins. No wash-off is required as all the pigment is physically bound in the resin systems. However sometimes a washing treatment is carried out to remove chemicals used in the process or to soften the hand of the fabric. The pigments used cover the range of azoics, carbon black, phthalocyanines, triphenylmethanes, and dioxazine derivatives. They give excellent lightfastness and good wash fastness in pale to medium shades.

Of technical concern are abrasions near seams and sharp edges, which tend to become frosty in washing. Further, the crock or rubbing fastness must be carefully observed and controlled by the addition of more binder or by the reduction of the pigment concentration. Fabrics being used range from lightweight poplins and sheetings to corduroy of cellulosic or fabrics of polyester–cellulosic blends.

Speciality Uses.   In addition to polyester–cellulosic and cellulosic fabrics, pigments may also be applied to 100% synthetic fibers of special construction for unique uses. Examples are 100% filament polyester for draperies and glass fabrics for bedspread, curtain, and upholstery uses. This type of fabric is usually run on a finishing frame by going through the padder, with drying and curing in the tenter housing. Quite often these are pastel shades with little or no afterwash required.

Solvent Dyeing.   Solvent dyeing generally refers to dyeing in nonaqueous media. In the early 1970s, solvent dyeing was expected to become the dyeing process of the future and was discussed and researched extensively (32). This interest did not materialize into practical acceptance and the technique has not achieved importance.

Dyeing Machinery

In the application of dyes three techniques are used: the dye liquor is moved as the material is held stationary, the textile material is moved without mechanical movement of the liquor, or both move.

Transportation of Dye Liquor through the Textiles.   Regardless of the form of the textile, raw stock, sliver, yarn, or cloth, the principle is generally the same. A large stainless steel kier, capable of withstanding sufficient pressure to reach a maximum operating temperature of 145°C, has one or more perforated spindles through which the dye liquor is pumped. Around this spindle the textile is packed tightly in the form of a cake in a perforated basket, as yarn, or in a package as a sliver or tow in a can, or as cloth around a beam barrel. The dye liquor is pumped through the textile, then flows to the bottom of the machine and into the return side of the pump. The stock or sliver machine pumps the liquor only in one direction (from the inside spindle to the outer shell) since the material does not have to be dyed as level as would yarn or piece. When dyeing yarn packages on perforated spindles (Fig. 6) or fabric wrapped around a perforated beam (Fig. 7) automatic reversing mechanisms flow from both inside to outside, and outside to inside. This ensures level dyeing in correctly loaded packages and beams.

The kiers are of varying sizes and production units can handle from 225–1400 kg, adjusting proportionally the pump size and speed. Most machines are highly versatile so as to accommodate the kind of textile being dyed; they are able to control rate of temperature rise, volume of liquor flow, and time of dye application.

Significant developments have concentrated on yarn dyeing with the emphasis being on improving process efficiency levels; liquor flow patterns; off-winding by package, package center, and spindle design; and robotization of the loading and unloading. After dyeing, packages are usually centrifuged to remove excess moisture and in a modern operation dried using radio frequency.

Transportation of the Textile Material with No Mechanical Movement of the Liquor.   Probably the oldest example of dyeing in which the textile alone moves is the ancient box-type skein dye machine in which skeins of yarn are hung on wooden pegs with about three-fourths of the skein submerged in the dye liquor. To prevent stick marks and unlevelness, the sticks are turned by hand at frequent intervals.

Chain Warp Dyeing.   This procedure is widely used in the dyeing of indigo on warp yarns which are in the form of ropes or chains. In this procedure, several

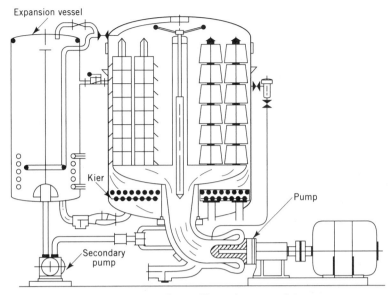

Fig. 6. Vertical-spindle package machine.

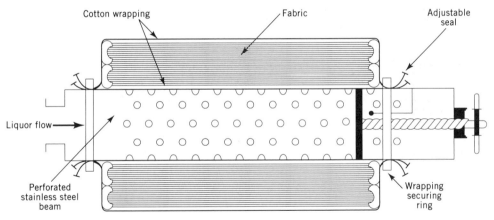

Fig. 7. Beam dyeing.

warp ropes are pulled through a series of tubs containing the dye liquor and gradually are dyed to the desired shade.

Jig Dyeing.   Piece goods are dyed by means of a batch process on a dye jig. Several hundred meters of cloth are wound around a beam roller on one side of a jig. The goods are run off the beam through a small-volume (400 L), V-shaped dyebath and are wound on the opposite beam. There are friction brakes on each beam that control tension. When the entire cloth has been wound on the opposite beam, the clutch is reversed and the cloth travels back in the opposite direction. The process of running the cloth from one beam to another is called an end. The goods are run a sufficient number of ends until the desired shade is achieved. Because contact with the dye liquor is for such a brief time, the cloth temperature is much lower than the dyebath. Any fibers or dyes requiring high temperatures (>80°C) should be dyed on an enclosed jig with open steam jets.

Many improvements have been made to streamline performance and to reduce machine operation labor. Some of these are tensionless jigs using variable speed electric motors with built-in drag for brakes, automatic reversing equipment, and automatic temperature and level controls. These machines are widely used for goods that are easily creased, such as fabrics consisting of filament acetate, heavy filament nylon, or cotton duck. They are also convenient for small dye lots and for sampling purposes.

Winch or Beck Dyeing.   This is one of the oldest forms for dyeing mechanized piece goods (fabric) (Fig. 8). The machine in its basic form consists of a shallow U-shaped box which has a gradual low curvature in the back and a rather high vertical rise in the front. About 2 m above the top of the U-shaped box is a driven elliptical reel. Fabric is passed over the top of this reel and the first end of the fabric sewed to the back end to form an endless loop. The tub is filled with the dye liquor and the fabric is immersed in it. The action of the turning elliptical reel is to lift fabric from the front of the dyebath, over the top of the reel, and redeposit the fabric at the back of the bath. The fabric then slowly moves through the dye liquor from the back to the front of the winch, where it is again lifted by the reel.

Becks and winches are generally constructed so as to run fabrics in either open-width or rope forms. The open-width threading is normally used for heavyweight material, such as carpets, twills, and sateens, which would be damaged if crushed into a rope form. In the second form, the beck has a long bar running over the length of the tub and has pegs every 25–30 cm that keep strands separate and prevent them from tangling. The strands may be threaded as one long spiral rope which runs as a continuous belt, or an alternative method involves threading

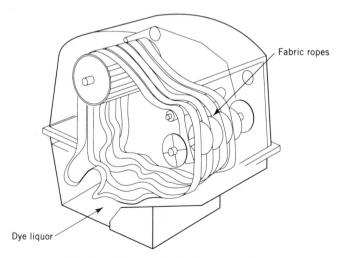

Fig. 8.   Winch or beck dyeing machine.

one piece of cloth per peg division and sewing it together so that there is an endless belt between each set of pegs. All of the pieces in such an arrangement must be approximately the same length to achieve level dyeing.

The modern winch is constructed of stainless steel and may be built for atmospheric temperatures or it may be a heavy-walled, sealed unit suitable for high temperature/high pressure work. Sophisticated controls are added to monitor the entire dye cycle and ensure that dyeings are as consistent as possible.

Continuous Dyeing. Most woven goods, other than stretch fabrics, can be continuously dyed provided each shade consists of sufficient yardage for economical operation. The equipment at hand may be simple padder or a complete dye range. Most dye padders consist of a medium density rubber roller across the width of which pressure can be applied. This roller presses against a stainless steel or a hard rubber roller. These rollers may be mounted either vertically or horizontally (Fig. 9). The cloth is passed through a stainless steel pad box, down under a rod or a roller which is below the dye liquor level, between the squeeze rollers. Wet add-on varies between 50 and 80%, depending on tightness of the squeeze rollers. The tightness of the squeeze is controlled by weight levers, hydraulic pressure, or compressed air. Modern padders are available that allow for different pressures to be introduced across the width of the squeeze roller in order to equalize any variances in fabric adsorption across the piece. If no other fixation equipment exists, the wet dye fabric is simply batched to give full exhaustion, as in the pad-batch dyeing of reactive dyes.

Dye ranges can be of different configuration depending on the composition of the fabrics being processed and dye system used. All contain similar units; differences are mainly in the method of heating. Examples are an electric infrared

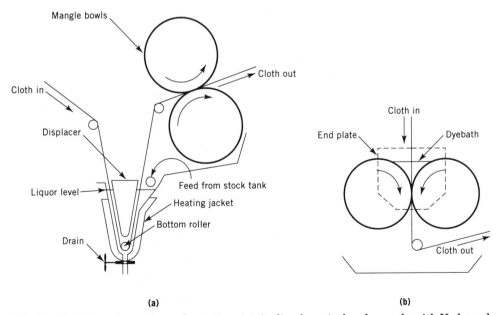

(a)   (b)

Fig. 9. Typical pad mangle configuration: (**a**), inclined vertical pad mangle with V-shaped trough; (**b**), horizontal wedge-nip pad mangle.

(ir) dryer, rather than a gas-fired unit, or a hot-air thermosol unit, rather than an oil-heated contact machine. The following sequence is typical for polyester–cotton blended fabric. The infrared units reduce moisture content 20–30% and greatly minimize uneven migration of the dye on the wet goods when they go onto the drying cylinders. The dried cloth progresses into the thermosol unit where the dyestuff for the synthetic portion of the fabric is fixed. Goods then continue through the chemical pad for immersion into an alkaline (or reducing) solution, depending on whether fiber-reactive or vat dyes are applied to the cellulosic fibers. They then pass through a steamer and finally through 8–10 wash boxes, which contain various chemicals depending on the class of dyestuff applied.

Modifications to continuous dyeing times include reduction of the size of pad troughs, incorporation of rapid cleaning and refill of troughs, and controllers all designed to keep conditions constant throughout the machine and to facilitate rapid changeover of shades. For example, when the line stops, instead of switching off the infrared heater, baffles are brought between the cloth and heater or the heater is revolved to point away from the cloth to both avoid any risk of scorching the fabric or having to heat up the dryer again when starting the next shade.

Continuous dyeing is best suited to dyeing long lengths of fabric to a single shade, eg, 5000 m or more. Demand has increased for shorter lengths to one shade resulting in greater interest in the modifications introduced to reduce changeover times, and also in pad-batch dyeing.

Machines Based on Movement of Both Dye Liquor and Material. One example of a machine in which both yarn and dye liquor are moved is the Klauder-Weldon skein dye machine; not only do the skeins turn, but the liquor is pumped in small streams over the yarn as the threads pass over the spindles. This process assures maximum uniformity and levelness.

Another extremely popular machine of this type is the jet dyeing machine which conserves energy by reducing the cloth-to-liquor ratio to 1:10 or lower as compared to 1:20 for the winch. In this machine, the fabric which is in a rope form is transported by movement of the dye liquor through a Venturi jet. This method provides intimate contact between the dye liquor and each meter of material. The machine operates at 40–135°C.

Although the capital cost of jet machines is higher than winches, several advantages have made them popular. Lengthwise stretching of the fabric is minimal. In many cases, there is shrinkage in the length leading to greater fabric bulk than is obtainable on winches. Shorter dyeing cycles are possible. The high degree of liquor interchange with the fiber and the low liquor ratios enable rapid rates of heating and cooling while maintaining a level dyeing and good fabric physical characteristics. Ability to obtain elevated temperatures and to ensure good uniformity of heating throughout the liquor contribute to good dye exhaustion, penetration of the substrate, and coverage of physical variations in the fiber.

The basic concept of the jet dyeing machine is similar to the winch except that the lifter reel has been replaced by the Venturi jet and its associated pipework and the machine is totally enclosed allowing dyeing above atmospheric pressure. The endless fabric rope has a dwell period in the dye liquor and is then transported by an aqueous jet of dyebath liquid pumped from the actual dyebath through the specially constructed pipework, through the Venturi jet, and like the winch, transported to the back of the machine to recommence the cycle. Each fabric rope needs its own jet. Jet dyeing machines come in a variety of styles although two main types exist: the long horizontal cylindrical machines (Fig. 10) ideally suited for synthetic fabrics where the fabric travels along the length of the cylinder, or (Fig. 11) cylindrical or upright pear or bell-shaped, or J-box autoclaves where the fabric moves around the circumference of the cylinder or equivalent shape. These autoclave types can have more than one fabric rope per autoclave and are suitable for most fabrics.

Since the introduction of the first jet dyeing machines, developments have focused on two main issues: softer flow and lower liquor-to-goods ratios. Soft flow

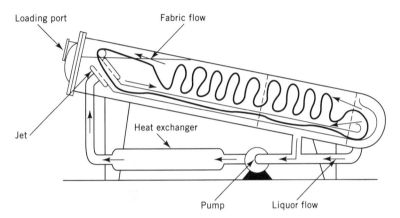

Fig. 10. Long cylindrical jet dyeing machine. The fabric return and jet are shown inside the main tube. In some designs it could be external to the main tube.

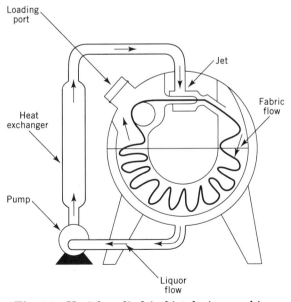

Fig. 11. Upright cylindrical jet dyeing machine.

jets preserve the surface appearance of delicate fabrics and the stitch clarity of knitgoods. Soft flow is achieved by introducing pulleys and driven rollers to assist in transportation of fabric, having large lifting rollers (sometimes with a nip) analogous to the lifting reel in the winch, with the dye jet incorporated at this point to impart minimum energy to the fabric and mechanically moving the fabric through the dyebath liquor at the bottom of autoclaves by slowly revolving cages or moving bars.

Overflow machines (Fig. 12) may be thought of as hybrids of jets and winches. They usually feature a winch reel which, unlike the cloth-guiding roller in a jet, is normally driven and provides motive power to the fabric. There is also some driving force on the cloth from the circulation of the liquor through the overflow tube down which they both pass. Both pressurized and nonpressurized versions exist and are available from a large number of different makers.

The principal advantage that overflow machines possess over jets is a gentler action on the fabric. The degree of fabric-on-fabric abrasion is much less than occurs in jet dyeing, making these machines more suited to the dyeing of fabrics containing spun yarns. The liquor ratios obtainable are more closely comparable to those of jet machines than to conventional winches, consequently, overflow machines retain the low liquor ratio advantages of jets.

Low liquor-to-goods ratio have been obtained by sophisticated plumbing, eg, placing the sump and pumps below the level of the dyebath, dyeing from foam, and incorporating air jets to transport the fabric. On the latest machines air can be used alone, eg, to transport lightweight fabrics, or in combination with water on heavier weights. Using these machines liquor ratios as low as 5:1 and 4:1 (although for cotton 8:1 is around the practical limit) are claimed compared to

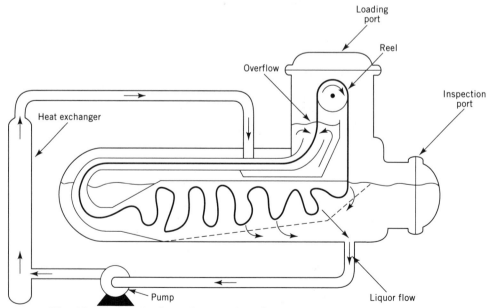

Fig. 12. Pressurized overflow machine having temperatures up to 140°C.

10:1 on the first jet dyeing machines and 20:1 on winches. The lower the liquor-to-goods ratio the less water is used, the less energy used in heating up, the less time taken in filling, emptying, heating, and cooling, and finally the less effluent discharged. The limit to reducing liquor-to-goods ratio is the amount of liquor needed to prime the pump and jet, and that imbibed by the fabric.

Control of Dyeing Equipment. Over the years, the dyer and machinery manufacturer have applied any mechanical or electrical equipment that would enable them, day after day, to produce repeatable dyeings of top quality. First, thermometers were installed in dye lines; these soon evolved into thermocouples with remote recording. Other improvements were soon developed, such as automatic four-way valves with variable-interval controls, flow controls, pressure recorders, hydraulic and air pressure sets on rollers, pH controls, etc.

There are many completely automated computer-controlled exhaust dyehouses. Some firms have a no-add procedure in the dyehouse by which the dyer loads the fabric or yarn, weighs the dye, punches a button, and lets the computer take over the entire process. This procedure ensures a constant dyeing cycle and the only variables are the dye index of the fiber or the quality of the dyestuff.

In continuous dyeing there are many variables and the rapidity of the dyeing process requires many adjustments during the period in which several thousand meters of textile are dyed. Instrumental science has continued to advance rapidly so that continuous ranges are available which are entirely computer-controlled except for the makeup of the dye mix. These units feature computer control and closed-circuit television and continuous color measurement techniques.

Textile Printing

The term textile printing is used to describe the production of colored designs or patterns on textile substrates through a combination of various mechanical and chemical means. In printing on textiles, a localized dyeing process takes place, whereby in general the chemical and physical parameters of dyeing apply.

The process of textile print coloration can be divided into three steps. First, the colorant is applied as pigment dispersion, dye dispersion, or dye solution from a vehicle called print paste or printing ink, containing in addition to the colorant such solutions or dispersions of chemicals as may be required by the colorant or textile substrate to improve and assist in dye solubility, dispersion stability, pH, lubricity, hygroscopicity, rate of dye fixation to the substrate, and colorant-fiber bonding. The required viscosity characteristics of a print paste are achieved by addition of natural or synthetic thickening agents or by use of emulsions.

The second step is the fixation process. During the fixation process the printed textile material is exposed to heat, heat and steam, or chemical solutions at temperatures and lengths of time governed by the type of fiber processed and by the dye class itself. During this treatment, the dyes diffuse from the fiber surface into the fiber structure and undergo physical or chemical bonding to the fiber. In case of pigment prints, latices applied simultaneously with the pigment develop their film-forming properties during a heat treatment and impart to the print the final desired fastness.

During the afterscouring, the third step, the prints are rinsed and scoured in a detergent solution in order to remove auxiliary chemicals, thickening agents, and portions of unfixed dyes remaining on the surface of the printed fibers. Frequently, the scouring operation includes the use of chemicals assisting in removal of dye or preventing redeposition of dye on the white or paler areas of the print. Pigment prints, in general, do not undergo a scouring operation.

Transfer printing employs the intermediate step of printing dye dispersions or dye solutions onto a temporary substrate, usually paper. From the paper, the dye is transferred to the textile by heat and steam, while printed paper and textile are in close contact. The advantages and limitations of the process have been described (33).

COLORANTS FOR TEXTILE PRINTING

Pigments. Pigment-printed textiles represent the highest percentage of all printed textiles, accounting for between 40 and 50% of all cellulose and over 90% of polyester–cotton blend prints. This is primarily the result of the uncomplicated process, low cost of imparting colored patterns to textiles with pigment systems, and their applicability to all fibers.

Water-insoluble pigments have no affinity to textile fibers. To bond the pigments to the textile, an agent, generally of a synthetic latex type, is incorporated in the print paste, which through its film-forming properties holds the embedded pigment firmly on the fiber surface. The binding agent is a dispersion or solution of such polymers as poly(acrylic acid) derivatives and butadiene–styrene copolymers. In most cases a cross-linking agent is applied in the same print paste, consisting of synthetic resin types, eg, melamine–formaldehyde derivatives, for added wash and crock fastness. Softening agents, eg, aluminum stearate, can also be added. Both pigment dispersion and binding agent–cross-linker are incorporated into a clear emulsion or solution containing dispersing agents, emulsifiers, acid donors, lubricants, protective colloids, and synthetic thickening agents of the poly(acrylic acid) and maleic acid anhydride–copolymer type. Emulsion-based thickeners can be used, mostly oil-in-water based on 10–70 wt % of mineral spirits. For economic and environmental reasons emulsion-based systems are declining in favor of printing from synthetic thickeners, especially in the United States and Western Europe.

Pigment print systems have a number of inherent limitations, most notably the effect of the binding agent and cross-linking chemical on the hand and feel of the textile material; also sometimes insufficient fastness to crocking, abrasion, washing, and dry cleaning, noticed especially in heavier shades and on large coverage; and the sometimes objectionable property of pigments to camouflage fiber texture and luster. Pigment prints are fixed by heat treatment at 150–200°C in 1–5 min. Afterscouring is generally not necessary.

Disperse Dyes. Disperse dyes are used in powder or paste form, or ready-to-prepare aqueous dispersions for incorporation into a thickener solution. For disperse dyes that show sensitivity to alkaline hydrolysis or reduction during fixation, an acid donor or acid, and, if necessary, a mild oxidizing agent, are added to the print paste.

For printing on polyester, the fixation conditions are more rigorous than on other disperse dyeable fibers, owing to the slower diffusion of disperse dyes in polyester. For continuous fixation the prints are exposed at atmospheric pressure to superheated steam of 170–180°C for 6–8 min. A carrier may be added to the print paste for accelerated and full fixation. Dry-heat fixation conditions of 170–215°C for 1–8 min are less popular for printed fabrics, but are sometimes employed because of lack of other equipment.

In both these continuous processes medium to high energy disperse dyes should be used to avoid the risk of dye subliming to contaminate the atmosphere of the fixation unit and then staining the print by vapor-phase dyeing, or to produce a loss of definition of the printed mark due to diffusion from the applied thickened paste.

In batch-type fixation, the prints are steamed in a pressurized autoclave with saturated or nearly saturated steam at 14–18 kPa (105–135 mm Hg) and 125–130°C for 45–60 min. In this case, the amount of carrier can be reduced or completely omitted. Generally, a small amount of urea (5 wt % of print paste) is added to the print paste to aid in fixation.

Afterscouring of polyester generally includes a reduction clearing with sodium hydrosulfite and alkali at 60–80°C to remove any dye remaining on the fiber surface unless alkali clearing dyes have been used.

Printing on triacetate follows the same general rules as for polyester. For batch-type pressure steaming, the steam pressure is reduced to 7–10 kPa (50–75 mm Hg) at 115–120°C. Acetate requires a steam pressure of ca 3.5 kPa (25 mm Hg), 108°C for full fixation of disperse dyes. With selected disperse dyes of a higher rate of diffusion in acetate, in combination with a suitable carrier, continuous steam fixation under atmospheric pressure at 100–105°C during 20–30 min is also possible. A light scouring at 40–50°C completes the operation.

On polyamide, disperse dyes have generally low wetfastness properties, making them unsuitable for printed textiles that require even moderate wash or perspiration fastness.

Acid Dyes. These dyes have their greatest importance in printing of polyamide. The dyes are dissolved in hot water, which can, depending on the solubility of a specific dye, have an admixture of a dye–solvent such as thiodiethylene glycol or diethylene monobutyl ether. Acid dyes in liquid form do not require dissolving, but can be added to the thickener solution directly. In case of the print–dry–steam operation, 3–5% of urea or thiourea is also added to the dye solution to act as humectant during fixation. Depending on the processing, additional print-paste additives may be used. This process is usually employed on woven or knitted fabrics for apparel, swimwear, drapery, and upholstery materials. The print paste contains an acid donor, usually 1–3% ammonium sulfate, and in some cases a small amount of sodium chlorate to prevent reduction of the dye during steam fixation. Fixation is by continuous steam treatment at atmospheric pressure of 100–105°C for 30 min or in pressure steam of 3.5 kPa (ca 25 mm Hg) 108°C for one hour.

In afterscouring, most often a synthetic tanning agent is used to prevent redeposition of dye on unprinted areas and to improve wetfastness properties of the print.

Printing of wool or silk with acid dyes is of minor importance. For these fibers the print paste is made with dye solvent, humectant (glycerol and urea), a suitable thickener, and dilute organic acid. An oxidizing agent is also added. Fixation follows the procedure for polyamide with fully saturated steam.

Acid dyes can be printed on acetate, producing prints with very good wet-fastness and exceptional brightness. The print paste contains a solvent, urea, and ammonium thiocyanate, as a fiber swelling agent to aid in diffusion of the dye. Again, fixation and scouring follow the procedures for polyamide.

Premetallized Dyes. This dye group is applied to the same textile fibers and with the same procedures as those with acid dyes. The premetallized dyes offer better fastness properties, but lack brilliancy of shade. Except in printing of carpeting, the neutral dyeing types of premetallized dyes are applied generally without acid or acid donor.

Direct Dyes. A few selected direct dyes are used to complement the acid dyes in printing of polyamide. Printing of cellulosic fibers with direct dyes, as with acid dyes, has lost its importance owing to insufficient wetfastness properties of these dyes on cellulosic fibers, and cumbersome fixation and scouring procedures.

Fiber-Reactive Dyes. This dye class represents, next to pigments, the main dye group for cellulosic fibers, ie, cotton and rayon. Depending on the different levels of reactivity and affinity of the dyes to cellulose, and on the available equipment, a number of different fixation methods can be used. The selection of suitable thickeners is important because of the ability of the fiber-reactive dyes to react with the hydroxyl groups that are found in many traditional thickeners, eg, starch. This prevents efficient removal of the thickener after printing and gives the print a harsh handle. Sodium alginate is therefore the preferred thickening agent, occasionally mixed with an oil-in-water emulsion thickener. The dye powder is dissolved with urea in water or, if suitable, sprinkled as dry powder into the thickener solution, where it dissolves during agitation. Liquid forms of fiber-reactive dyes are also available. For steam and dry-heat fixation, the reactive print pastes contain 5–30 wt % urea and 2–4 wt % soda ash or sodium bicarbonate. Fixation in steam of 100–105°C takes 2–8 min; in superheated steam of 150–180°C, 1–5 min; and in dry heat of 150–200°C, 1–5 min.

In so-called wet fixation methods, the print paste containing 5 wt % urea and no alkali is printed and dried. For fixation, the prints are passed through a solution of strong alkali, such as caustic soda, potassium carbonate, soda ash, or combinations thereof, and salt or sodium silicate, and the fixation takes place within 20–30 s or 6–24 h, depending on temperature and alkali concentration chosen. A combination method is available for flash aging, where a similar alkali-free print is impregnated with a cold alkali solution and immediately passed through steam at 102–115°C for 0.5–2 min.

Basic (Cationic) Dyes. The use of basic dyes is confined mainly to acrylic textile fibers, acetate, and as complementary dyes for acid-modified polyester fibers that accept this class of dyes.

Basic dyes are used in powder or liquid form and are incorporated into the print paste by dissolving powder dyes first in water with the aid of a dye solvent, eg, thiodiethylene glycol, and an organic acid, eg, acetic acid, followed by addition of a suitable thickener. Liquid dyes can be added directly to the thickener solution. Maintaining an acid pH in the print paste, and during the drying and fixation steps, is important. This is accomplished by adding 0.5–1.0% of a nonvolatile organic acid, eg, citric acid, to the print paste. In most cases a fixation accelerator is used in the print paste.

Dye fixation on acrylic fibers and acetate can be done by atmospheric steaming at 100–102°C for 20–30 min. With modified polyester only pressure steam produces full fixation and color yield. Afterscouring is by rinsing and detergent scour. A scouring auxiliary with affinity to the fibers can be used to prevent redeposition of rinsed out dye.

Vat Dyes. Applied to cellulosic fibers, vat dyes yield prints with excellent fastness properties. They are used to print furnishings, drapes, and camouflage where their infrared reflectance resembles natural terrain and foliage. Their application can follow two different procedures.

Flash-Age Fixation. This is analogous to the application of vat dyes in dyeing. The method of flash-age fixation requires printing of a paste consisting of thickener solution and vat dye. The dried print is then impregnated with a solution of alkali, such as caustic soda and potassium carbonate, also containing a reducing agent such as sodium hydrosulfite and sodium formaldehyde sulfoxylate, and immediately passed for 0.5–2 min through air-free steam at 105–125°C. This is followed by oxidation and soaping.

All-In Method. In this alternative process the vat dye, in the form of a dispersion of the water-insoluble dye, is added to a thickener solution that also contains the necessary reducing agent such as sodium formaldehyde sulfoxylate, the alkali such as potassium carbonate, and glycerol as a hygroscopic material. After printing and drying at temperatures not exceeding 80°C the prints are passed through steam at 100–102°C for 5–8 min. It is important that this steam is free of air. During the steam fixation the vat dye is reduced to its water-soluble leuco form and is absorbed by the fiber in this form. Afterscouring consists of reoxidation of the leuco form to the insoluble vat dye, accomplished by addition of hydrogen peroxide or sodium dichromate to the water rinse, followed by boiling detergent solution.

The reducing agents used in the all-in method are less powerful than in the flash-age process, ie, they have a less negative redox potential. Some vat dyes have a more negative redox potential than the all-in reducing agents and therefore cannot be used because they cannot be reduced. Mild reducing agents are needed because they are added to the print paste prior to printing and therefore must be stable to ambient temperature.

Azoic Dyes. These are used to produce cost-effective heavy yellow, orange, red, maroon, navy blue, brown, and black shades and are printed alongside other dye classes to extend the coloristic possibilities for the designer. Two approaches are adopted. The common method in the United States is to use both a naphthol derivative and a stabilized color base, usually in the form of a diazo imino compound in the same print paste. This mixture is soluble in dilute caustic soda and no coupling takes place at this stage. The dried prints are passed through steam at 100–105°C that contains acetic and/or formic acid vapor. As neutralization takes place on the print, the coupling occurs rapidly and the insoluble azoic dye is formed.

The alternative approach is to pad the fabric with the alkaline naphthol and dry, followed by printing directly onto this prepared fabric diazonium salts or stabilized diazonium salts. Coupling is instant and the only further treatment needed is to remove all the uncoupled naphthol and surface azo pigment in a subsequent washing treatment. Because the choice of colors is limited from one naphthol component, other shades are obtained by using other classes of dye alongside the azoic colors, eg, reactives. This approach is widely used in the production of African prints.

Phthalocyanine Dyes.   These dyes are synthesized as the metal complex on the textile fiber from, eg, phthalonitrile and metal salts. A print paste typically contains phthalonitrile dissolved in a suitable solvent and nickel or copper salts. During a heat or steam fixation of 3–5 min, the dye is formed. The color range is restricted to blue and green shades and can be influenced to some extent by the choice of metal salt. A hot acid bath during afterscouring completes the process.

Dye Combinations.   In certain cases it is desirable to print fiber blends with combinations of the appropriate dye classes, rather than with pigments. Only polyester–cellulose blends are of commercial importance and the following dye systems have been developed for them. The dyes of the different classes are contained in the same print paste and, therefore, are applied simultaneously in one print operation.

Disperse–Reactive Combinations.   A careful selection of both disperse and reactive dyes is necessary. Modern systems depend on the use of disperse dyes that can be cleared in alkali without any risk of staining back onto the print during clearing, eg, disperse dyes containing diester groups. Because these disperse dyes are susceptible to hydrolysis strong alkaline fixation conditions must be avoided. The selection of fiber-reactive dyes is therefore based on those dyes that give efficient fixation under weakly alkaline, eg, 1.5% sodium bicarbonate, neutral, and even acidic conditions, eg, reactive dyes containing reactive systems based on phosphoric acids, although this approach has not gained commercial success. Fixation is by high temperature steaming or thermofixation.

The use of disperse–reactive dye combinations is popular for printing polyester–viscose blends for fashion fabrics where the drape, handle, and general aesthetic qualities of such fabrics are adversely affected by pigment printing.

Disperse–Pigment Combinations.   These are applied from a print paste essentially similar in composition to a pigment print paste. However, fixation requires temperatures high enough to accomplish diffusion of the disperse dye into the polyester. The advantage of this system over a straight pigment print is better fastness to abrasion and washing.

Disperse–Vat Combinations.   These require a two-step fixation. The disperse dye is fixed first, usually by dry heat, followed by impregnating of the textile with an alkali and reducing agent solution and short steam fixation for the vat dye. The selected disperse dyes fixed in the polyester fiber are not destroyed by the reducing agent, but disperse dye remaining on the cellulose is destroyed.

STYLES OF PRINTING

Direct Printing.   This is the simplest technique and accounts for the majority of printed fabrics. In this style the print paste is applied directly onto the fabric which can either be white or already dyed with a pale ground shade. Each color requires its own print paste and application and each color is printed immediately after each other without intermediate drying.

Discharge Printing. The limitation of direct printing is that the greater the area of the print to be covered by print paste, the darker and larger the blotch, the more difficult it is to prevent soiling and contamination of the design as a result of the pick up of color by the printing roller or screen from previously printed areas. There is also a limitation on the fineness of the design.

With discharge printing the process is reversed. First the fabric is dyed to give the ground shade and then printed with print pastes containing agents that will destroy the ground shade to give a patterned effect. The ground shade is said to be discharged. Typical approaches are to dye the ground shade with azo dyes and to then use print pastes containing reducing agents to destroy the dye. "White" discharges can be obtained by using only reducing agent, while "colored" discharges are obtained by combining a reducing agent and a dye stable to the chosen reducing agent in the print paste. After printing and dyeing, the fabric is usually steamed when the ground shade is destroyed at the same time the colored discharge dye (also called the illuminating color) is fixed.

On cotton, fiber-reactive dyes give the ground shade and then printing vat dyes and reducing agents, such as sodium formaldehyde sulfoxylate or zinc formaldehyde sulfoxylate, give the discharge print.

With polyester the traditional process was to pad azo disperse dye onto the fabric and after drying print with the same reducing agents as for cotton to obtain white discharge, or with combinations of nonreducing anthraquinone dyes with stannous chloride to obtain colored discharges. This process is rapidly being displaced, for environmental and economic reasons, by one using alkali-sensitive disperse dyes as the ground shade. After padding and drying white discharges are obtained using alkali alone, colored discharges by combining alkali-stable disperse dyes with alkali. On lightweight fabrics it is possible to obtain discharge prints on exhaust dyed polyester with the use of suitable carriers.

Resist Printing. In resist printing, print pastes are used that can inhibit the development or fixation of different dyes that are applied to the textile prior to or after printing. These resists can be of a chemical or mechanical nature, or combine both methods. For example, fiber-reactive dyes, which require alkali for their fixation, can be made resistant by printing a nonvolatile organic acid, such as tartaric acid, on the textile. Colored resists are obtained by printing pigments with a nonvolatile acid.

Wax Printing. This is a special case of resist printing and widely used in African designs. Wax is first printed onto cotton fabric to give a patterned effect, and the fabric is then dyed. The wax resists penetration of the dye and reserves the fabric. The wax is then removed by either alkali washing or solvent when it is also recovered for future use. Other colors can then be printed, if so desired, on the nondyed areas. Azoic dyes or reactive dyes can be used for this.

PRINTING MACHINERY

Textile materials can be printed at different steps of the textile manufacturing process. Woven fabrics comprise the largest percentage of printed goods. In recent years, knitted textile fabrics have considerably increased in importance. However, printing can also be done on yarns in skein form, or on warps being passed from a warp beam to another beam, or as yarn strands. Space printing is a process where a yarn, temporarily knitted into a loose fabric, is printed and then de-knitted. Carpets can be printed in woven or tufted constructions. Vigoureux printing is the printing of woolen slubbing. Regardless of the state of the textile material, any printing process makes use of one of the following methods.

Screen Printing. This printing process essentially consists of the transfer of print paste through a screen to the substrate to be printed. The screen is made of polyamide or polyester (also metal in some cases). The pattern or design is produced in the screen by coating those parts that should not transfer the print paste with an impervious material. The design is transferred to a film and the film placed on the screen, which is coated with light-sensitive material. After light exposure, the unexposed parts, having been protected by the light-impervious areas of the film, are removed by rinsing and thus the pattern is created on the screen. To withstand the mechanical stress of prolonged printing, the light-sensitive coating has to be very strong or has to be reinforced with suitable screen lacquer.

For hand printing, the fabric to be printed is stretched over long tables which are covered with a blanket of layers of felt and back grey or felt- and rubber-coated material. The screen is placed on the textile material and the print paste is scraped across the screen by means of a squeegee, penetrating through the open areas in the screen onto the textile. The screen is then lifted and moved along the table by hand or in a carriage to the next print position. A screen is used to deposit only one color, and for multicolor designs as many screens as there are colors in the design are necessary. It is important that the textile material does not move or shift during the printing process, therefore it is pinned to the back grey or glued to the blanket.

Mechanized screen printing is done on flat-bed screen printing machines. On these machines the screens remain stationary while the textile material is moved underneath the screens in intermittent steps. The textile material is glued to an endless blanket that serves as support for the textile material and which is cleaned with water once during each revolution. When the screens are in a lifted position, the blanket and the textile material with it move forward. When they are in a lowered position the squeegees perform the printing. Flat-bed screen printing is widely used for the more complicated multicolored designs and is popular in Asian countries, eg, Japan.

A further development is the rotary screen printing machine on which the screens have been fashioned into seamless perforated cylinders. In these machines the blanket and fabric move continuously and the cylindrical screens revolve at the same relative speed to allow for continuous printing and hence much higher production rates than the stop–start flat-bed methods. The squeegees remain static applying pressure at the point of contact between blanket, fabric, and screen. Squeegees can be either metal blades where the printing pressure is controlled mechanically or by a pneumatic pressure pad behind the squeegee, or revolving metal rods where the printing pressure is controlled by magnetic force from electromagnets under the printing table. Rotary screen printing is now the most popular method of printing.

Roller Printing. This method of printing has gradually been losing ground to rotary printing as a result of the capital costs associated with storing designs on engraved copper rollers, and the length of time needed to change designs. The roller print machine consists of a copper print roller that has the design to be printed engraved on its surface, and a cast-iron cylinder of a diameter larger than the engraved print roller. The print roller revolves in contact on one side with the cylinder and on the other side with a furnishing roller or furnishing brush, which in turn revolves in a color box (print-paste reservoir) carrying the print paste to the engraved print roller. The fabric to be printed passes between the cylinder and the engraved print roller, absorbing print paste from the engraved areas of the print roller.

To confine the print paste to the engraved parts of the print roller, any excess that may have been applied by the furnishing roller is scraped off the unengraved, smooth surface of the print roller by a metal blade (the doctor blade) prior to making contact with the fabric to be printed.

Paper Coloring

As with textiles the principal reasons for coloring paper are for aesthetic appearance and utilitarian purposes. Aesthetic appearance includes colored background for printed material, colored writing papers, colored household products to harmonize with interior decor, and many other diverse uses dictated by individual tastes. Utilitarian purposes include identification of multicopy forms, identification of manufacturer or marketer of specific materials or products, opaqueness or hiding power of packaged material, or to control consistency of paper manufactured from various colored raw materials.

COLORANTS FOR PAPER

Among the colorants that have been and are being used for the dyeing of paper are natural inorganic pigments (ochre, sienna, etc); synthetic inorganic pigments (chromium oxides, iron oxides, carbon blacks, etc); natural organic colorants (indigo, alizarin, etc); and synthetic organic colorants. The last is the largest and most important group.

Basic Dyestuffs. Basic dyestuffs have a high affinity for mechanical pulps and unbleached pulps that have a large amount of acid groups in the fiber. The cationic dyestuff reacts with these acid groups to produce, by salt formation, very stable lakes that are insoluble in water. Bleached pulps cannot be dyed satisfactorily since they contain few acid groups after bleaching. Although most basic dyestuffs stain ligneous pulps, the addition of alum with or without sizing offers advantages depending on the depth of shade and type of paper required.

Basic dyestuffs usually have brilliant shades with high tinctorial strength, and on ligneous pulps produce good fastness to water, steam, and calendering, and clear backwater. Basic dyestuffs have poor lightfastness, and because of their poor affinity for bleached pulps, have a strong tendency to mottle and/or granite in blended furnishes with ligneous pulps.

Basic dyestuffs are usually used for dyeing of unbleached pulp in mechanical pulp such as wrapping paper, kraft paper, box board, news, and other inexpensive packaging papers. Their strong and brilliant shades also make them suitable for calendar staining and surface coloring where lightfastness is not critical.

Acid Dyestuffs. Acid dyestuffs have no affinity for vegetable fibers, either as bleached or ligneous pulps. This lack of affinity has the advantage that acid dyestuffs do not mottle in blended furnishes and produce level shades and even appearance when properly fixed by use of fixing agents, rosin size, and paper-makers' alum (aluminum sulfate). The acid dyestuffs are precipitated on the fibers by these materials, forming a linkage between the cationic-treated fibers and dye-stuff molecules. Sizing improves the retention of lake particles by formation of aluminum resinate. Because of poor affinity and good solubility, acid dyestuffs have poor bleedfastness and form colored backwater, and are therefore suitable for paper that does not require wetfastness, such as construction grades. Acid dyestuffs are most suitable for calendar staining or surface coloring because of their solubility and brightness of shade.

Direct Dyestuffs. The difference in the affinity of direct dyestuffs for ligneous and bleached pulps is considerably less than basic dystuffs, but direct dyes color bleached pulps more evenly because they are easily wetted and penetrate rapidly.

Direct dyestuffs generally have a high affinity for cellulose fibers and are therefore the most useful dyestuff type for unsized or neutral pH dyeings. Their bonding ability to nonligneous pulps and excellent fastness properties to light and bleeding make them useful for all fine papers. The shades of direct dyestuffs are not as bright as those of acid or basic dyestuffs and in blended furnishes (bleached–ligneous pulps) mottling or graniting may occur.

Pigments. Synthetic organic pigments are replacing the use of some inorganic pigments for ecological reasons. The conditions of dyeing are quite different for pigments than those used for soluble dyestuffs. Pigments do not react chemically with the fiber, but are fixed physically and are dependent on filtration, absorption, occlusion, and flocculation. Paper dyeings with pigments have outstanding fastness properties, but poor affinity, low tinctorial strength, and two-sidedness problems limit their application to paper.

Fluorescent Whitening Agents. These whiteners are fluorescent substances that transform invisible ultraviolet light into visible blue light. Fluorescent whitening agents (qv) change the appearance of paper in two ways: by emitting light and therefore increasing the luminosity (brightness); and by changing the shade from yellowish white to bluish white. They are unfavorably influenced by factors such as low ultraviolet content light and high ultraviolet light absorption of pulps, chemicals, and fillers. They can be used in the wet end, as a surface colorant, or in combination to obtain the greatest efficiency or yield.

DYEING PROCESSES

Paper may be colored by dyeing the fibers in a water suspension by batch or continuous methods. The classic process is by batch dyeing in the beater, pulper, or stock chest. Continuous dyeing of the fibers in a water suspension is adaptive to modern paper machine processes with high production speeds in modern mills. Solutions of dyestuffs can be metered into the high density or low density pulp suspensions in continuous operation.

Paper may also be colored by surface application of dyestuff solutions after the paper has been formed and dried or partially dried by utilizing size-press addition, calendar staining, or coating operations on the paper machine. In addition, paper may be colored in off-machine processes by dip dyeings or absorption of dyestuff solution and subsequent drying, such as for decorative crepe papers.

All dyestuffs and pigments are taken up more readily by the paper fiber in the acid pH range. Aluminum sulfate (papermakers' alum) is more effective than acid in fixing acid and direct dyes or pigments. The aluminum ion forms salts with acid and direct dyes that do not readily dissolve, whereas with pigments the triple positive charge also ensures fixation on the negatively charged fiber. With bleached pulps, only direct dyestuffs may be used to color unsized or alkaline-sized papers. For household tissues of deep shades to be made at neutral pH (no alum), only a selection of direct dyestuffs with particularly high affinity can be used. Blended furnishes of pulp (bleached–ligneous) can be colored successfully with combinations of dyestuff classes such as basic and acid, direct and acid, and direct and basic dyes. Generally, the dyestuff with the greatest affinity should be added first and solutions of the cationic dyes cannot be mixed with solutions of anionic dyes.

Batch Dyeing. Depending on operational conditions, the dyestuffs can be added to the pulper, beater, or stock chest. The advantages of this batch method are its simplicity and the fact that no additional equipment is necessary. Thorough mixing and optimum fixation on the fiber are ensured by adding the dyestuffs at an early stage of processing. However, the relatively long interval from addition to inspection of the dyed paper can lead to prolonged correction and color-change times.

The normal sequence of addition in the dyeing process is pulps, filler, dyestuffs, rosin size, and alum. The dyestuffs are either taken up by the fiber because of their affinity or they must be fixed on the fiber in a finely divided form by suitable fixing agents. Alum, which is required to precipitate the rosin size, has a strong precipitating and fixing effect on dyestuffs.

Continuous Dyeing. Continuous dyeing of the pulp by addition of dyestuff before the head box of the paper machine has been gaining in importance in the 1990s, because this method is well adapted to the high production rates of modern paper machines. Compared to batch dyeing, the advantages of continuous dyeing are faster color changes owing to immediate response and less colored stock in the paper machine system. Continuous dyeing is even more important where the grade assortment includes white paper as well as colored paper. To utilize the advantages of continuous dyeing, the point of addition of the dyestuffs should be as close to the head box as possible. In actual practice this must be balanced against the dyestuff contact time and mixing requirements between dyestuffs and pulps and is dependent on the dyestuff affinity to the pulp. As a rule, only highly substantive dyestuffs can be utilized. To maximize dyestuff yield it is advisable to add the dyestuffs to the stock solution before dilution to facilitate absorption of the dyestuff and to increase contact times and allow adequate mixing.

In the manufacture of colored papers, it is best to add the dyestuffs before addition of rosin size and alum. This is not always possible in continuous dyeing procedures where dyestuffs must be added to stock containing size and/or alum, and this may cause premature laking of the dyestuffs and subsequent loss of tinctorial strength and/or dullness of shade. The proper selection of dyestuffs can help to reduce these disadvantages.

The metering and addition of dyestuff solutions is critical to successful operation of continuous dyeing. It is also just as critical to meter other components of the process including pulp, broke, fillers, size, alum, fixing agents, retention aids, wet-strength resins, and other additions that affect dyeing. Exact and dependable operation, simple maintenance, and accurate control are essential requirements of a successful installation.

Nonimpact Printing.   Interest is growing in the use of nonimpact styles because of the quickness of color changeover and the ability to interface these machines to computer-aided design systems. Two basic types exist: drop on demand and constant drop techniques.

In the drop on demand technique color is "fired" at the substrate when it is needed through fixed valves or bubble jets. Using valves the limit is on the fineness of design owing to the difficulties of packing the number of nozzles and associated hardware into the applicator head needed for such designs. The technique is used to a limited degree on carpets. Bubble jet technology currently is only used on paper. In this dye, solution is boiled in a small cavity and the steam generated propels the rest of the liquid through a fine orifice onto the substrate. Very fine designs result, the limit being in the need to have heat-stable dyes.

Constant flow jets operate by passing a stream of charged particles continuously through an electric field. In order for these to hit the substrate the drops have to be deflected by electrical force. Nondeflected drops fall into a gutter. By changing the electrical field, drops can be deflected different distances and directions, so it is possible to obtain designs using a relatively few number of jets. Both techniques have been used widely for paper.

Dyeing of Leather

Leather (qv) is a less homogeneous product than textiles or paper. It is derived from protein collagen (skin or hide substance) treated with one or more tanning agents. Although the chemical characteristics and properties of collagen have been accurately determined, the variety of materials that convert it into leather leave the chemical composition of leather extremely indefinite. Not only may the compound used to convert hide substance into leather vary chemically over a wide range, but the quantities used, the method of application, and the physical condition of the hide prior to tanning or dyeing may vary, with each factor in turn affecting the dyeing properties of the resultant leather. Also, leather retains many of the properties originally associated with the parent substance, and these affect profoundly and, in many ways, limit the dyeing properties of the final product. Chief among these properties are sensitivity to extremes of pH, thermolability, and the tendency to combine with acidic or basic compounds.

Tanning.   Leather may be produced by use of chrome, vegetable, aldehyde, syntan, oil, and many other tanning agents. Chrome-tanned leather accounts for the greater part of leather production, but chrome is seldom employed alone. Chrome is usually employed first, and then a vegetable, syntan, or resin "retannage" follows. Vegetable tannage is employed to a fairly large degree for specific types of leather. Syntans and aldehydes, eg, glutaraldehyde, are applied combined with chrome or vegetable agents, or as a retan or second tannage over either the chrome or vegetable agents.

Practically all chrome grain leather, used mainly for shoe or garment leathers, is given a retannage with a comparatively small amount of vegetable extracts such as quebracho, wattle, or other wood extracts, or syntans, resin-tanning materials, or possible combinations of two or all three. Such an anionic retannage not only imparts desirable characteristics to the leather, but also promotes more level and uniform dyeing. The use of an anionic retannage also tends to increase the penetration of the dyes into the leather, although full or complete penetration of dye is rarely accomplished in this manner. Larger percentages of dye are therefore required for such retanned leather to produce approximately the same full shades as on untreated leather.

Leather Dyes. The main classes of dyes employed in the coloring of leather are the acid, acid/direct, direct, and basic types. On chrome leather, the direct dyes usually have greater affinity and produce fuller or heavier shades than do acid or chrome dyes. Acid/direct dyes as well as the metallized-type dyestuffs may be classified for the purpose of leather dyeing as the main types in use. Basic dyes color chrome leather weakly and unevenly, unless the leather is first mordanted or retanned with suitable materials, such as vegetable tannin, syntans, or previously applied acid and/or direct dyes. They may be used alone on vegetable-tanned leather to produce full shades or, as is done more frequently, following a preliminary coloring with acid or acid/direct dyes. In the latter case, basic dyes are used to impart fullness of shade with minimum coloring matter and cost.

Acid and direct dyes vary considerably in their penetrating properties on chrome leather. Although it is not possible to make a distinct division, the direct dyes as a class are more surface dyeing than the acid dyes. Chemically, there appears to be a relationship between the number of sulfonic acid groups per dye molecule, the gross molecular weight, and the ability to penetrate into the interior sections of the leather. Thus, dyes with a low molecular weight and a relatively greater proportion of sulfonic acid groups penetrate more deeply than those with either a higher molecular weight or with fewer sulfonic acid groups per molecule. Since the direct dyes fall within the second category, they tend to dye only the surface of chrome leather.

Generally in leather dyeing it is necessary to employ two or more dyestuffs to produce a given shade. Such blending is important, as the selected dyes should have approximately the same rates of diffusion; this is an important factor in the adjustment of shades. The performance of a series of dyes on textiles cannot be used as a guide for their use on leather. For most leather, only superficial dyeing or surface coloring is required. However, some show and garment leathers require a certain degree of penetration, and if this cannot be obtained by simply increasing the degree of neutralization, or the addition of alkali, it may be accomplished by blending surface-coloring dyes with penetrating-type dyes. This procedure is used to obtain enough penetration to resist or minimize the effect of subsequent buffing of the leather surface. If a proper selection of dyes is not made in these combinations, and if one dye tends to diffuse more rapidly than another, or is less firmly bound, the result is a two-layer effect at the expense of shade. The two-layer effect may develop during the drying process and cannot always be observed in the wet state of the dyed leather. Some penetrating dyes also migrate to the interior of the leather during the drying operation, and produce a change in shade on the surface.

In the case of almost all grain leather, except for special leathers, such as glazed calf, glazed kid, and all suede leather, the dye used is applied as "bottom" color and the final shade and finished product are obtained by the application of "top" coatings of pigment finishes. The greater opacity of pigments tends to fill and conceal the many imperfections, eg, those caused by insect bites, barbed-wire scratches, etc, that are normally present on average hides. An exception to this practice is the so-called aniline-finished leathers on which little or no pigment-type finishes are employed. In this type of leather, which is competing against the synthetic leathers, the greater transparency of the dyestuffs reveals the true grain-surface structure of the skin much more strikingly than do pigment finishes. A more rigorous selection of stock is necessary for this type of leather in order to avoid surface defects which are normally obscured by pigments.

Leather Dyeing Methods and Equipment. The methods used in dyeing leather are quite simple and they obtain their names from the equipment employed, such as drum, wheel, paddle, brush, tray, or spray dyeing. Most leather is dyed in drums. Drum dyeing is carried out in revolving cylinders ranging from ca 1–1.5 m in width and 2.4–3.7 m in diameter. Dye and other solutions are fed into the rotating drum through a hollow axle or grudgeon. This method of dyeing is relatively rapid and efficient, as it permits the coloring of large amounts of leather in comparatively small amount of solution, with attendant low utilization of dye and other materials. The amount of solution in drum dyeing ranges from 1–2 times the weight of wet leather or from 4–8 times the weight of dry leather. The usual practice is to employ the necessary quantities of dyestuff, which range from 0.10–5.0 wt % of the blue (ie, wet) stock for grain leather. Dyeing is carried out at 46–52°C. The amounts of 85% formic acid used to exhaust the dyebaths are usually based on the ratio of one-half the amount of formic acid to that of dyestuff employed.

Dyeing can also be carried out in hide processors or automatic stainless steel dyeing machines with a Y-shaped cross section to subdivide the interior into three loading sections, somewhat similar to a large-scale washing machine.

Spray dyeing also accounts for the coloring of large amounts of leather. Usually acid or basic dyes are selected for good solubility and level dyeing properties. At present, proprietary ranges of premetallized or basic liquid dyes are available and used widely. A wetting agent and/or solvent are usually incorporated in the dye solution to prevent droplet formation and to promote levelness and penetration after the solution has been deposited on the grain of the leather by the air-spray gun. This method is extensively used for shoe upper leather where only the grain side is sprayed and the flesh or reverse side is left uncolored with what is termed a natural black. This type of leather is mainly chrome-tanned cowhide that is retanned with relatively small (2–5%) amounts of vegetable-tanning extracts such as quebracho, wattle, or sumac, sometimes combined with small percentages of syntans and/or resin-tanning materials. The leather is then fat-liquored and processed in the normal manner. This method permits the tanning and storage of leather in the dry or natural state, and its rapid conversion into whatever shade is desired by spray-dye application. Spray dyeing is applicable to almost any type of leather where deep penetration is not required, and is the most economical method of applying a dye.

Fastness Tests for Textiles

In an age of consumer activism and increasing international trade, test methods and procedures assume more and more importance. Generally accepted tests, instruments, and standards provide a common basis for the evaluation of quality in dyed textiles. Dyed textiles are evaluated with regard to their fastness to natural destructive agents, such as daylight, weather, and atmospheric gases, as well as to various treatments the material is likely to undergo, such as washing, dry cleaning, ironing, steaming, etc. Test methods should be broad in scope and related to actual processing and use. Such tests are generally accelerated, ie, they reveal the fastness properties exhibited by the material over a lengthy period of time in a test of short duration.

The principal active bodies in the field of colorfastness testing have been the American Association of Textile Chemists and Colorists (AATCC) and the Europaische-Convention für Echtheitprüfung/Groupement d'Etudes des Commissions Européenes pour la Soliditié (ECC). The ISO subcommittee concerned with colorfastness tests is ISO TC 38/SCI. This meets every two or three years to coordinate developments in standard testing methods and to seek international agreement on proposed new tests and modifications to existing tests. The purpose of ISO is solely to produce useful standard test methods. The setting of specifications and levels of acceptance on the basis of such test methods is a matter to be resolved between buyer and seller.

COLORFASTNESS

The principal methods of measuring colorfastness have been described in detail (34). Described below are the general principles for the tests as revised in 1990. For detailed information in English reference should be made to the 1992 AATCC manual equivalent (35).

Colorfastness refers to the resistance of the textile coloration to the different agencies, eg, light or chemicals, to which these materials may be exposed during manufacture or subsequent use. The change of color and staining of undyed adjacent fibers are assessed as fastness ratings on the standard grey scale: grey scale for assessing change in color is ISO 105 A02, and grey scale for assessing staining is ISO 105 A03. The basic scale comprises five pairs of nonglossy grey color chips (or grey fabric) which illustrate the perceived color difference corresponding to fastness ratings 5, 4, 3, 2, 1. Scales illustrating the half steps between the basic levels (ie, 4–5, 3–4, 2–3, 1–2) are also available. The chips used are specified colorimetrically. Fastness is assessed by placing the standard (fabric not exposed to the testing agency), alongside the tested fabric and the difference between them compared to the grey scale and the grey scale pair that has the same perceived magnitude of difference chosen. The fastness of the coloration is then given the value corresponding to the grey scale pair. A value of 5 means no change has occurred to the original sample, or no stain has been found on the adjacent. The lower the value below 5 the poorer the fastness.

Fastness to Light. The ISO test for colorfastness to light is *Daylight ISO 105-B01*. The textile specimen is exposed to daylight under prescribed conditions, including protection from rain, along with a series of blue wool reference samples that fade at defined, prescribed, different rates. The colorfastness is assessed by comparing the change in color of the test specimen with that of the references and is given the rating of the reference which faded most like the test specimen. In Europe eight samples (1 low, 8 high lightfastness) are used: in the United States eight samples are used numbered from 2 (low) to 9 (high). Different results are obtained by the two approaches. The U.S. system is based on blending dyes so that each higher numbered sample is fading at half the rate of the preceding lower numbered sample. The European system is based on selected dyes that give a similar stepwise fading scale.

The test for colorfastness to artificial light *Xenon arc fading lamp test ISO 105-B02*, is identical to B01 except that the sample is exposed to xenon arc light under prescribed conditions in a closed container.

In the test for colorfastness to weathering, *Outdoor exposure ISO 105-B03*, specimens of the textile are exposed in open air without any protection. At the same time and in the same place the eight standard wool samples are exposed to daylight but are covered by glass. The fastness is assessed by comparing the change in color of the specimen with that of the reference. The colorfastness to weathering test, *Xenon arc ISO 105-B04*, is similar to B03 except that the specimen is exposed to light from a xenon arc lamp and to a water spray.

Fastness to Washing and Laundering. In colorfastness to washing, *Test 1 ISO 105-C01*, a sandwich obtained by sewing a specimen of textile in contact with one or two adjacent fabrics is mechanically agitated in a defined closed vessel in 5 g/L of standard soap solution at 40°C and 50:1 liquor-to-goods ratio for 30 min. The adjacent fabric can be a multifiber strip complying to *ISO 105-F10*, or two single-fiber fabrics the choice depending on the final use of the specimen. The change of shade of the specimen and degree of staining of adjacents is carried out after drying by using the grey scale. This is the mildest of the washing tests.

The colorfastness to washing test, *Test 2 ISO 105-C02*, and colorfastness to washing test, *Test 3 ISO 105-C03*, both use the C01 general procedure of 5 g/L soap solution at 50:1 liquor-to-goods ratio, but at different temperatures and times: C02 at 50°C for 45 min and C03 at 60°C for 30 min.

Colorfastness to washing, *Test 4 ISO 105-C04*, is a more severe washing treatment using the same apparatus as C03 but using a solution of 5 g/L soap plus 2 g/L anhydrous sodium carbonate with a liquor-to-goods ratio of 50:1 at 95°C in the presence of 10 noncorrodible stainless steel balls 6 mm in diameter for 30 min. Colorfastness to washing, *Test 5 ISO 105-C05*, is identical in procedure as for C04 but treatment is extended to 4 h. Colorfastness to domestic and commercial laundering, *ISO 105-C06*, covers some 16 possible testing procedures all aimed at assessing the behavior of the test specimen to domestic and commercial laundering. The variations possible include:

temperature	40, 50, 60, 70, or 95°C
liquor-to-goods ratio	from 150:1 for 40, 50°C, 50:1 for others
time	30–45 min
number of steel balls	10, 25, 50, or 100

sodium perborate	0 or 1 g/L
available chlorine	0 or 0.015%
pH	unadjusted or adjusted to pH 10.5
detergent	either 4 g/L AATCC reference detergent or 4 g/L ECC reference detergent

The different detergents are an attempt to copy the typical detergent composition commercially available in the U.S. and European markets. The more severe tests in the C06 series are designed to approximate the effect expected from five repeated launderings. In view of the variety of conditions associated with this test it is vital to ensure good communication and data recording in operating the test.

Dry Cleaning. In colorfastness to dry cleaning, *ISO 105-D01*, a specimen of the textile is placed in a cotton fabric bag together with stainless steel disks and agitated in perchloroethylene (30 min, 20°C) and the effect of the shade and the color of the solvent assessed using the grey scale.

Water Tests. In colorfastness to water, *ISO 105-E01*, the test specimen is placed in contact with the chosen adjacent fabrics, immersed in water, and placed wet between glass plates and left for 4 h at 37°C. After drying, the effect on the test specimen and stain on adjacents are assessed. The test, colorfastness to seawater, *ISO 105-E02*, is the same as E01 but uses 30 g/L anhydrous sodium chloride solution instead of water. To test for colorfastness to chlorinated seawater/swimming baths water, *ISO 105-E03*, the specimen is immersed in sodium hypochlorite solution containing either 100, 50, or 20 mg of active chlorine per liter at pH 7.5 for 1 h at 27°C, rinsed, dried, and assessed.

For colorfastness to perspiration, *ISO 105-E04*, the specimen is immersed in a solution of 0.5 g/L of 1-histidine monohydrochloride monohydrate and 5 g/L sodium chloride buffered to either pH 8.0 (alkali perspiration test) or pH 5.5 (acid perspiration test) in a dish at 50:1 liquor-to-goods ratio, at room temperature for 30 min. The specimen is removed and, as in the water test E01, left for 4 h between plates at 37°C before drying and assessing both test piece and adjacents.

In colorfastness to acid spotting, *ISO 105-E05*, drops of a solution of either acetic acid (300 g of glacial acetic acid per liter of water), sulfuric acid (50 g of concentrated acid per liter), or tartaric acid (100 g of crystalline acid per liter) are spotted onto the test material, which is then dried and assessed. Colorfastness to alkali spotting, *ISO 105-E06*, is like E05 except that a solution of 100 g of anhydrous sodium carbonate per liter of water is used. Colorfastness to water spotting, *ISO 105-E07*, is like E05 but uses drops of water and assessment is made after 2 min wet and after drying. In colorfastness to hot water, *ISO 105-E08*, the textile specimen and adjacents are wound around a glass rod and placed in water adjusted to pH 6 with acetic acid at 70°C for 30 min, dried, and assessed.

The test colorfastness to potting, *ISO 105-E09*, is of importance for dyed wool as potting is one of the processes woven wool fabrics can be given before they are made up into clothing. The procedure is similar to E08 except that the test conditions are 1 h immersion in boiling water.

Colorfastness to decatizing, *ISO 105-E10*, involves a specimen wrapped around a perforated cylinder and through which steam is passed for 15 min at different pressures or temperatures to represent mild or severe conditions. In colorfastness to steaming, *ISO 105-E11*, a specimen and adjacents are rolled into a cylinder and placed in the neck of a flask of boiling water for 30 min.

Colorfastness to alkali milling, *ISO 105-E12*, is another important test for wool because dyed wool fabrics are often given a milling processing treatment. The specimen and adjacents are agitated in three times its mass of a solution of 50 g/L soap and 10 g/L anhydrous sodium carbonate in the presence of 50 steel balls for just over 2 h at 40°C, the time being dependent on testing alongside the specimen a standard material until this shows a change of three on the grey scale.

Colorfastness to Atmospheric Contaminants. The test colorfastness to nitrogen oxides, *ISO 105-G01*, is to assess the fastness of the color to nitrogen oxides that may be present in hot air that has been passed over heated filaments or from the burning of gas, coal, etc. Specimens are exposed to nitrogen oxides in a closed container along with standards until the standards have changed to a predetermined extent.

Colorfastness to burnt gas fumes, *ISO 105-G02*, is similar to G01 except that the specimen and standards are exposed to the fumes from a burning butane gas flame. Colorfastness to ozone in the atmosphere, *ISO 105-G03*, is done as G01 but uses ozone.

Colorfastness to Bleaching. In fastness to hypochlorite bleaching, *ISO 105-N01*, the specimen is agitated in a solution of sodium, calcium, or lithium hypochlorite containng 2 g/L available chlorine buffered to pH 11.0 with sodium carbonate for 1 h at 20°C and 50:1 liquor-to-goods ratio. The specimen is rinsed in water, hydrogen peroxide, or sodium bisulfite solution to remove free chlorine, dried, and assessed.

In fastness to peroxide bleaching, *ISO 105-N02*, the specimen is immersed in a standard bleaching solution containing hydrogen peroxide (or sodium peroxide for viscose) where the composition of the bleaching liquor is dependent on the fibers used in the test specimen as are the pH and time of exposure (1–2 h). The objective of the test is to assess the colorfastness using typical bulk bleaching conditions for the fiber under test.

Fastness to sodium chlorite employs two tests: *ISO 105-N03* (mild test) and *ISO 105-N04* (severe). The specimen and adjacents are immersed without agitation in a solution of 1 g/L sodium chlorite at pH 3.5 for 1 h at 80°C in the mild test. In the severe test 2.5 g/L sodium chlorite is used at pH 3.5, again for 1 h at 80°C.

Colorfastness to Heat Treatment. To test for fastness to dry heat, *ISO 105-P01*, the specimen is sandwiched between adjacent fabrics and placed under slight pressure between heated surfaces where the temperature of the surface is 150, 180, or 210°C for 30 s. The effect on the shade of the pattern and adjacents is then assessed.

In the test for fastness to steam pleating, *ISO 105-P02*, the specimen and adjacents are steamed under pressure for a specified time. The conditions used range from 5 min at 108°C and 135 kPa pressure for the mild test, to 15 min at 130°C and 270 kPa (2.66 atm) pressure for the severe test.

Miscellaneous Fastness Tests. The fastness to hot pressing, *ISO 105-X11*, test is similar to the fastness to dry heat test except that the time of pressing is 15 s (again at 150, 180, and 210°C), and the test can either be carried out dry when a damp cotton fabric containing its own weight of water is placed on top of the dry test fabric, or wet when the test fabric also contains its own weight of water.

To test for fastness to rubbing or crocking, *ISO 105-X12*, specimens are rubbed with a dry rubbing cloth and a wet rubbing cloth using standardized rubbing cloths and specially designed equipment. The equipment passes a standard finger over which the rubbing cloth is placed across the surface of the textiles under standard pressure. The finger is then passed 10 times in 10 s along a 10-cm line. The stain on the rubbing cloths are then assessed.

Other tests include assessing the colorfastness to solvents, felting treatments, stoving, vulcanizing, mercerizing, degumming, etc.

TESTING OF DYES

At the 1989 meeting of ISO/TC38/SC1 in Williamsburg, Va., a new work group (WG11), Characterization of Dyestuffs, was established. The following tests are significant (36), together with alternative techniques currently being considered for introduction as ISO standards.

Evaluation of Dyestuff Migration. A piece of fabric is impregnated with dye and with auxiliaries if the influence of these or a selection of these is under study and a watch glass placed over a portion of the fabric and the whole dried at room temperature. The watch glass is then removed and the degree of migration assessed by assessing the difference between the area of the fabric covered by the watch glass and the body of the fabric either by reflectance measurement or the grey scale.

Thermal Fixation Properties of Disperse Dyes on Polyester–Cotton. This method assesses the fixation properties of disperse dyes as a function of the time, temperature, dyestuff concentration, or presence and amount of auxiliary agents. The polyester–cotton fabric is padded and dried, the cotton dissolved in sulfuric acid and washed out of the blend, and the amount of dye on the polyester component assessed by either reflectance or measuring the optical density of a solution of dye obtained by extracting the dye with boiling chlorobenzene solvent.

Transfer of Disperse Dye on Polyester. A specimen of dyed polyester is placed in a standard dyebath with an equal weight of undyed polyester and the dyeing cycle completed. The rate of transfer from dyed to undyed fabric is compared to that obtained with a range of five standard dyes and the dye under test is given the same number as the dye it most closely resembles.

Transfer of Basic Dyes on Acrylics. This test is identical in concept to the transfer of disperse dye on polyester except that basic dyes, acrylic fiber, and a standard dyebath for dyeing acrylic is used.

Transfer of Acid and Premetallized Dyes in Nylon. A specimen of dyed nylon is placed in a dyebath with undyed nylon and the degree of transfer assessed at pH 4.5, 6.0, and 7.5 at 95°C.

Dispersion Stability of Disperse Dyes at High Temperature. A disperse dye dyebath is treated under the desired test conditions at 130°C in a special apparatus (Gaston County Lab Dye and Chemical Tester) and filtered through cotton and polyester filters. The filter with the heaviest residue is then compared with a series of standard photographs of standard performance and rated equal to the one it most resembles (1 poor, 5 excellent).

Foaming Propensity of Disperse Dyes. A standard weight of disperse dye (5 g of 100% strength) is diluted in distilled water at pH 5.5 and an addition of detergent (0.25 g/L AATCC standard detergent) and then agitated in a kitchen blender (14,000–15,000 rpm 30 s), poured into a cylinder, allowing time to drain the mixer, and after waiting for 150 s the level of foam in the cylinder is assessed. Dye formulations are rated from Class A (very low foam) to Class C (very high foam).

Evaluation of the Dusting Properties of Powder (or Other Solid Dyes). A number of methods exist to test this increasingly important behavior because of the greater awareness of working conditions and good industrial practice present today.

In the 1993 AATCC standard method (re-affirmed in 1988) a standard weight of dye (10 g) is dropped down a funnel into a cylinder of defined size. A wet filter paper is placed as a collar to the funnel 200 mm above the bottom of the cylinder. After 3 min the stain on the filter paper, obtained by the dust created in the funnel dissolving on the paper, is assessed by comparing the standard photographs and the dye given the number of the picture it resembles the most (1 poor, 5 excellent).

There is a new test procedure under review by the ISO for publication as a standard test method (37). This method uses a special apparatus and is again based on dropping a standard weight of dye solid. In this new proposed test the dye is dropped into a container with a defined air-flow through it. The dye powder formed is sucked out of the chamber and through a filter. Dust performance can be assessed visually or qualitatively by eluting the dye from the filter, by measuring the optical density of the solution, or gravimetrically by weighing the filter before and after the test.

Other New Methods. Because the values obtained are dependent on the conditions of measurement, standard test procedures are under review by ISO for: determination of cold-water solubility of water-soluble dyes (38); determination of the solubility and solution stability of water-soluble dyes (39); and determination of the electrolyte stability of reactive dyes (40).

Analysis

Paper and Thin-Layer Chromatography. Both of these techniques are separation methods useful for dye identification. The dyes are extracted from fibers with suitable solvents (41) or dissolved if in powder form and applied to chromatographic paper strips or plates with the help of capillary pipettes. Paper strips or plates are then developed in a tank with a suitable eluent. These vary substantially according to the chemical nature of the dyes to be separated. Not only is it possible in this way to separate dye mixes into their component, but dye identification is feasible based on comparison with known dyes. In such cases, it is generally necessary to compare the dyes in at least two different eluent systems.

For semiquantitative or quantitative analyses, the separated dyes can be measured with a reflectance densitometer or they can be extracted from the plate or paper and measured by transmittance spectrophotometry.

High Pressure Liquid Chromatography. This modern version of the classical column chromatography technique is also used successfully for separation and quantitative analysis of dyes. It is generally faster than thin-layer or paper chromatography; however, it requires considerably more expensive equipment. Visible and uv photometers or spectrophotometers are used to quantify the amounts of substances present.

Solution Spectrophotometry. The relative strength of a dye (compared to a standard) or the dye concentration in solution can be determined accurately for most dyes by this method. Of greatest utility is a computer-interfaced recording visible spectrophotometer or a so-called color-measuring instrument. Solution spectrophotometry is based on Beer's law and the Lambert-Bouger law concerning the relationships between absorption and concentration of the absorbing layer. It is important to prepare dye solutions that are stable and obey Beer's law over the concentration intervals used. A new ISO test method for determining the relative color strength in solutions is being considered (42).

Relative strength determination by solution measurement is difficult in the case of vat and sulfur dyes and leads to unreliable results in cases where some of the dye in the commercial product has no affinity for the fiber such as for certain direct and fiber-reactive dyes.

Reflectance Spectrophotometry. Because of discrepancies that can occur between strength and shade evaluations in solution and on textile substrates, the latter is often the preferred evaluation technique. In the case of dye manufacture, many dyes are standardized in solution but there is always a final control step where dyeings are prepared. Historically, such dyeings have been evaluated visually for the relative strength and the shade of the dye under test on the substrate, compared to the standard. More and more attempts are being made to do such evaluations objectively. Guidelines for the use of this technique have been published (43).

Color Difference Evaluation. Shade evaluation is comparable in importance to relative strength evaluation for dyes. This is of interest to both dye manufacturer and dye user for purposes of quality control. Objective evaluation of color differences is desirable because of the well-known variability of observers. A considerable number of color difference formulas that intend to transform the visually nonuniform International Commission on Illumination (CIE) tristimulus color space into a visually uniform space have been proposed over the years. Although many of them have proven to be of considerable practical value (Hunter Lab formula, Friele-MacAdam-Chickering (FMC) formula, Adams-Nickerson formula, etc), none has been found to be satisfactorily accurate for small color difference evaluation. Correlation coefficients for the correlation between average visually determined color difference values and those based on measurement and calculation with a formula are typically of a magnitude of approximately 0.7 or below. In the interest of uniformity of international usage, the CIE has proposed two color difference formulas (CIELAB and CIELUV) one of which (CIELAB) is particularly suitable for application on textiles (see COLOR).

Safe Handling of Dyes

The Ecological and Toxicological Association of Dyes and Organic Pigments Manufacturers (ETAD), an international body of all primary manufacturers based in Europe but also with standing committees in the United States, Brazil, and Japan, issues clear guidelines for the safe handling of dyes (44). In December 1991 the United States Operating Committee of ETAD joined with the United States Environmental Protection Agency in publishing a pollution prevention guidance manual for the dye manufacturing industry (45) (see also DYES, ENVIRONMENTAL CHEMISTRY).

Colour Index Generic Names

The *Colour Index* (5) assigns CI generic names to commercial dyes. This CI name is defined as "a classification name and serial number which when allocated to a commercial product allows that product to be uniquely identified within any *Colour Index* Application Class." This enables the particular commercial products to be classified along with other products whose essential colorant has the same chemical constitution.

Dyes and pigments are marketed as preparations which include, as well as the essential colorant, diluents, dispersants, or other chemicals. The listing of commercial preparations under the same CI generic name implies only that their essential colorants have the same chemical constitution. The *Colour Index* does not claim that preparations listed under the same CI generic name have identical application fastness or toxicological profiles. It has been demonstrated that products with the same CI generic name do in fact behave differently (46) and that not all preparations are manufactured to the same degree of purity or consistency of strength (47). Thus the application, fastness, and ecotoxicological properties may vary from one preparation to another. It should therefore not be assumed that one commercial brand can replace another in every respect.

For further information regarding this topic, the following organizations can be contacted.

American Association of Textile Chemists and Colorists (AATCC)
P.O. Box 12215
Research Triangle Park, N.C. 27709

Society of Dyers and Colourists (SDC)
P.O. Box 244, Perkin House, 82 Grattan Road
Bradford BD1 2JB, UK

Technological Association of the Pulp and Paper Industry
15 Technology Park
P.O. Box 105113
Atlanta, Ga.

American Leather Chemists Assoc.
Leather Industries Research Lab
University of Cincinnati, ML #14
Cincinnati, Ohio 45221

Ecological and Toxicological Association of Dyes and Organic Pigments
    Manufacturers (ETAD)
P.O. Box CH-4005
Basel Switzerland

U.S. Operating Committee of ETAD
1330 Connecticut Avenue N.W., Suite 300
Washington, D.C. 20036-1702

Japanese Operating Committee of ETAD
Senryo-Kaikan
18-17 Roppongi 5-Chome
Minato-ji
Tokyo 106, Japan

BIBLIOGRAPHY

"Dyes, Application and Evaluation" in *ECT* 1st ed., Vol. 5, pp. 355–445, by P. J. Choquette and co-workers, General Dyestuff Corp. (Application), H. W. Steigler, AATCC, Lowell Textile Institute, (Test Methods for Colorfastness of Textiles, under Evaluation), and F. C. Dexter, Calco Chemical Division, American Cyanamid Division (Spectrophotometric Curves, under Evaluation); in *ECT* 2nd ed., Vol. 7, pp. 505–613, by G. M. Gantz and J. R. Ellis, eds., J. J. Duncan and co-workers, General Aniline and Film Corp.; in *ECT* 3rd ed., Vol. 8, pp. 280–350, R. G. Kuehni and co-workers, Mobay Chemical Corp.

1. *Market Research Data*, ICI Colours, Manchester, UK, 1992.
2. R. Smith and S. Wagner, *Am. Dyestuff Rep.*, 32 (Sept. 1991).
3. B. Glover and J. H. Pierce, *Society of Dyers and Colourists Symposium*, Buxton, UK, Oct. 1992.
4. U. Sewekow, *Melliand Textilberichte* **69,** 271 (1988).
5. *Colour Index*, 3rd ed., Society of Dyers and Colourists, UK (SDC), in collaboration with American Association of Textile Chemists and Colourists, USA (AATCC), 1971.
6. H. H. Sumner, in A. Johnson, ed., *The Theory of Coloration of Textiles*, 2nd ed., Society of Dyers and Colourists, UK, 1989, Chapt. 4.
7. W. C. Ingamells, in Ref. 6, Chapt. 3.
8. B. C. Burdett, in Ref. 6, Chapt. 1.
9. F. Jones, in Ref. 6, Chapt. 5.
10. J. Shore, ed., *Colorants and Auxiliaries, Organic Chemistry and Application Processes*, Vol. 1, *Colorants*, Society of Dyers and Colourists, UK, 1990.
11. G. Booth, *The Manufacture of Organic Colorants and Intermediates*, Society of Dyers and Colourists, UK, 1988.
12. J. F. Dawson, *J. Soc. Dyers Color.* (*JSDC*) **107**(11), 395 (Nov. 1991).
13. A. T. Leaver, *Annual Conference and Exhibition*, American Association of Textile Chemists and Colorists (AATCC), Charlotte, N.C., 1991.
14. J. R. Aspland, *Tex. Chem. Color.* **23,** 14 (Oct. 1991).
15. D. M. Lewis, *Wool Sci. Rev.* **49,** 13 (July 1974).
16. C. Preston, ed., *The Dyeing of Cellulosics Fibres*, Dyers' Company Publications Trust, London, 1986.
17. J. R. Aspland, *Tex. Chem. Color.* **23,** 41 (Nov. 1991).
18. M. Hahnke, *Textilveredlung* **21,** 285 (1986).
19. T. D. Fulmer, *Am. Text. Int.*, 47 (Jan. 1988).
20. A. H. Renfrew and J. A. Taylor, *Rev. Prog. Coloration* **20,** 1 (1990).
21. W. Marshall and R. Peters, *J. Soc. Dyers Colour.* **68,** 289 (1952).
22. H. H. Sumner, T. Vickerstaff, and E. Waters, *J. Soc. Dyers Colour.* **69,** 181 (1953).
23. J. Wegmann, *J. Soc. Dyers Colour.* **76,** 282 (1960).
24. F. R. Hartley, *J. Soc. Dyers Colour.* **85,** 66 (1969).
25. D. M. Lewis, *J. Soc. Dyers Colour.* **91,** 33 (1975).

26. J. A. Hughes, H. H. Sumner, and B. Taylor, *J. Soc. Dyers Colour.* **87,** 463 (1971).

27. *J. Soc. Dyers Colour.* **88,** 354 (1972).

28. W. Beckmann, in D. M. Nunn, ed., *The Dyeing of Synthetic Polymer and Acetate Fibres,* Dyers Company Publications Trust, London, 1979, Chapt. 5.

29. H. Kellet, *J. Soc. Dyers Colour.* **84,** 257 (1968).

30. N. G. Morton and M. G. Kratch, *J. Soc. Dyers Colour.* **85,** 639 (1969).

31. R. F. Hyde and G. Thompson, *J. Soc. Dyers Colour.* **109,** 142 (1993).

32. "Textile Solvent Technology—Update 73", *Proceedings of Symposium American Association of Textile Chemists and Colorists,* Research Triangle Park, N.C., 1973.

33. F. Schlaeppi, *Tex. Res. J.,* 203 (Mar. 1977).

34. *Textiles—Tests for Colour Fastness,* ISO 105, AATCC, Research Triangle Park, N.C., 1990.

35. *British Standard,* BS 1006, British Standards Institute, Milton Keynes, UK, 1992.

36. *AATCC Technical Manual,* Vol. 64, AATCC, Research Triangle Park, N.C., 1989.

37. A. Berger-Schunn and co-workers, *Melliand Textilberichte* **70,** 690 (1989); *J. Soc. Dyers Color.* **107,** 270 (1991).

38. A. Berger-Schunn and co-workers, *Melliand Textilberichte* **67,** 716 (1986); *J. Soc. Dyers Color.* **103,** 140 (1987); *Chem. Color.* **19,** 17 (1987).

39. A. Berger-Schunn and co-workers, *Melliand Textilberichte* **67,** 638 (1986); *J. Soc. Dyers Color.* **103,** 138 (1987); *Chem. Color.* **19,** 33 (1987).

40. A. Berger-Schunn and co-workers, *Melliand Textilberichte,* **67,** 812 (1986); *J. Soc. Dyers Color.* **103,** 272 (1987); *Chem. Color.* **19,** 21 (1987).

41. K. Venkataraman, ed., *The Analytical Chemistry of Synthetic Dyes,* Wiley-Interscience, New York, 1977.

42. R. Brossman and co-workers, *Melliand Textilberichte* **67,** 499 (1986); *J. Soc. Dyers Color.* **103,** 38 (1987); *Chem. Color.* **19,** 25 (1987).

43. W. Baumann and co-workers, *Melliand Textilberichte* **67,** 562 (1986); *J. Soc. Dyers Color.* **103,** 100 (1987); *Chem. Color.* **19,** 32 (1987).

44. *Guidelines for the Safe Handling of Dyestuffs in Colour Storerooms,* Ecological and Toxicological Association of Dyes and Organic Pigment Manufacturers (ETAD), Basel, Switzerland, June 1983.

45. *Pollution Prevention Guidance Manual for the Dye Manufacturing Industry,* U.S. Environmental Protection Agency and U.S. Operating Committee of ETAD, Washington, D.C., Dec. 1991.

46. A. Schmid, *Textilveredlung* **20,** 341 (1985).

47. B. Glover, *J. Soc. Dyers Color.* **107,** 184 (1991).

General References

W. F. Beech, *Fire Reactive Dyes,* Logos Press, London, 1970.

Dyeing of Paper, Bayer Farben Revue, Special Ed. No. 4/1, 1974.

The Application of Vat Dyes, Monograph No. 2, American Association of Textile Chemists and Colorists, Research Triangle Park, N.C., 1953.

R. W. Jones, *Sulfur Dyes, Their Manufacture, Application, Fixation and Removal,* Martin Marietta Chemicals, Sodeyco Division, Rocky Mount, N.C., 1976.

L. Diserens, *The Chemical Technology of Dyeing and Printing,* Vols. I and II, Reinhold Publishing Corp., New York, 1948.

R. Rys and H. Zollinger, *Fundamentals of the Chemistry and Application of Dyes,* John Wiley & Sons, Inc., New York, 1972.

E. R. Trotman, *Dyeing and Chemical Technology of Textile Fibres,* C. Griffin & Co. Ltd., London, 1964.

K. Venkataraman, *The Chemistry of Synthetic Dyes,* Vol. I–VI, Academic Press, Inc., New York, 1951–1978.

T. Vickerstaff, *The Physical Chemistry of Dyeing*, Oliver and Boyd, London, 1950.

F. Jacobs, *Textile Printing–Materials, Methods and Formulae*, Chartwell House Inc., New York, 1952.

H. Knecht and G. C. Fothergill, *The Principles and Practice of Textile Printing*, C. Griffin & Co. Ltd., London, 1952.

K. Johnson, *Dyeing of Synthetic Fibers, Recent Developments*, Noyes Data Corp., Park Ridge, N.J., 1974.

C. Preston, ed., *The Dyeing of Cellulosic Fibres*, Dyers' Company Publications Trust, London, 1986.

D. M. Nunn, ed., *The Dyeing of Synthetic-Polymer and Acetate Fibres*, Dyers' Company Publications Trust, London, 1979.

A. Johnson, ed., *The Theory of Coloration of Textiles*, Society of Dyers and Colourists, UK, 1989.

J. Shore, ed., *Colorants and Auxiliaries*, Vol. 1, *Colorants*, Society of Dyers and Colourists, UK, 1990.

BRIAN GLOVER
Zeneca Colours

DYES, ENVIRONMENTAL CHEMISTRY

Synthetic organic dyes are essential to satisfy the ever growing demands, in terms of quality, variety, fastness, and other technical requirements, for coloration of a growing number of substances. However, these materials present certain hazards and environmental problems. The largest volume use of dyes is for dyeing fibers. Since 1970 the dyestuff and textile industries have become increasingly subject to international, federal, and state regulations designed to improve health, safety, and the environment. This article attempts to provide a perspective of the environmental problems posed by synthetic organic colorants, and the efforts being made by industry, academia, and government to solve these problems.

Effluent Treatment Methods

Methods of effluent treatment for dyes may be classified broadly into three main categories: physical, chemical, and biological (1).

Physical	Chemical	Biological
adsorption	neuralization	stabilization ponds
sedimentation	reduction	aerated lagoons
flotation	oxidation	trickling filters
flocculation	electrolysis	activated sludge
coagulation	ion exchange	anaerobic digestion
foam fractionation	wet-air oxidation	bioaugmentation
polymer flocculation		
reverse osmosis/ultrafiltration		
ionization radiation		
incineration		

There are four stages: preliminary, primary, secondary, and tertiary treatment processes, which differ mainly by the number of operations performed on the waste steams (2,3).

Preliminary treatment processes of dye waste include equalization, neutralization, and possibly disinfection. Primary stages are mainly physical and include screening, sedimentation, flotation, and flocculation. The objective is to remove debris, undissolved chemicals, and particulate matter. Secondary stages are used to reduce the organic load, which essentially is a combination of physical/chemical separation and biological oxidation. Tertiary stages are important because they serve as a polishing of effluent treatment. These methods are adsorption, ion exchange, chemical oxidation, hyperfiltration (reverse osmosis), electrochemical, etc.

PHYSICAL METHODS

Adsorption. Adsorption (qv) is an effective means of lowering the concentration of dissolved organics in effluent. Activated carbon is the most widely used and effective adsorbent for dyes (4) and, it has been extensively studied in the waste treatment of the different classes of dyes, ie, acid, direct, basic, reactive, disperse, etc (5–22). Commercial activated carbon can be prepared from lignite and bituminous coal, wood, pulp mill residue, coconut shell, and blood and have a surface area ranging from 500–1400 m^2/g (23). The feasibility of adsorption on carbon for the removal of dissolved organic pollutants has been demonstrated by adsorption isotherms (24) (see CARBON, ACTIVATED CARBON). Several pilot-plant and commercial-scale systems using activated carbon adsorption columns have been developed (25–27).

Investigations have been undertaken to evaluate cheap alternative materials as potential adsorbents for dyes using activated carbon as a reference. These include peat (2,28), corn stalks (29), chitin (30), carbonized wool (31), sawdust (32), cellulosic graft copolymers (33), fly ash (pulverized fuel ash) (34,35), bagasse pith (36), bentonite (37–39), calcium metasilicate (Wollastonite) (40), organosilicon (41), clays (qv) and Fuller's earth (2,42), activated alumina (43), pig and human hair, meat, bone meal, wheat and rice bran, and turkey feathers (44). Activated sludge which is discussed under biological treatment, has been found to show behavior similar to that of activated carbon in the adsorption of acid, direct reactive, disperse, and basic dyes (45). The adsorption capacity of activated sludge for these classes of dyes can be determined by the Freundlich equation and adsorption isotherms (46–49). The adsorbability of dyes by activated sludge is mainly dependent on dye properties, molecular structure and type, number, and position of the substituents in the dye molecule (46). Adsorption is increased by the presence of hydroxy, nitro, and azo groups. On the other hand, adsorption is decreased by sulfonic acid groups.

Sedimentation. This is the traditional method of treating wastewater in lagoons and uses the force of gravity to remove settable solids (50,51). These solids are separated out as a watery sludge which is removed mechanically.

Flotation. Flotation (qv) is used to remove suspended solids from wastes and for the separation and concentration of sludges (52,53). The waste flow is pressurized in the presence of sufficient air to approach saturation. When the pressurized air–liquid mixture is released to atmospheric pressure in the flotation unit, minute air bubbles are formed. As they rise in the liquor the sludge flocs and suspended solids are floated to the surface where the air–solid mixture can be skimmed off.

Flocculation. Flocculation (qv) uses chemical precipitation to effect separation (54). It is used to increase the rate of sedimentation and flotation. The resulting material known as floc can be removed by filtration, sedimentation, or flotation.

Polymer Flocculation. Polymer flocculation, as referred to in the waste treatment of dyes or dyeing, is the aggregation or coagulation and subsequent removal of dye using a suitable synthetic cationic, nonionic, or anionic polymer. A substantial amount of work has been done by CIBA-GEIGY, Basel, using the cationic polymers, Xantop 885 or Tinofix WS, in decolorizing colored wastewater-containing acid, direct, and reactive dyes. Basic dyes naturally cannot be decolorized by cationic polymer flocculants. Disperse dyes were removed using cationic polymers and alum by Nalco Chemical Co. and Calgon Corp. (55). The Dow Chemical Company has obtained a U.S. patent (56) using polyethylenimines as cationic polymers and carbon adsorptions to treat textile waste streams containing dispersed dyes. Using Armour's cationic polymer, Armeen C, and carbon adsorption reduced the total organic carbon (TOC) of four disperse azo dye dispersion effluents 80.0–96.3% and showed no detectable (0%) dye (20).

Coagulation. Coagulation results from lowering of the zeta potential at the particle surfaces to permit closer approach, followed by association of the particles to form larger flocculated agglomerates (57). Coagulation is an economical and feasible method of treating dye waste, especially in the removal of color. The main inorganic coagulant used is lime (58–62).

Other coagulants used are calcium hydroxide, calcium sulfate, magnesium hydroxide, magnesium sulfate, ferric chloride, ferric sulfate, ferrous sulfate, aluminum sulfate (alum), and a combination of these inorganic salts (63–85). The ferric salts, although commonly used as coagulants, have the disadvantage of being difficult to handle (86,87).

Foam Fractionation. An interesting experimental method that has been performed for wastewater treatment of disperse dyes is foam fractionation (88). This method is based on the phenomenon that surface-active solutes collect at gas–liquid interfaces. The results were 86–96% color removal from a brown disperse dye solution and 75% color removal from a textile mill wastewater. Unfortunately, the necessary chemical costs make this method relatively expensive (see FOAMS).

Reverse Osmosis and Ultrafiltration. Reverse osmosis (qv) (or hyperfiltration) and ultrafiltration (qv) are pressure driven membrane processes that have become well established in pollution control (89–94). There is no sharp distinction between the two: both processes remove solutes from solution. Whereas ultrafiltration usually implies the separation of macromolecules from relatively low molecular-weight solvent, reverse osmosis normally refers to the separation of the solute and solvent molecules within the same order of magnitude in molecular weight (95) (see also MEMBRANE TECHNOLOGY).

Cellulose acetate, the earliest reverse osmosis membrane, is still widely used. Asymmetric polyamide and thin-film composites of polyamide and several other polymers have also made gains in recent years whereas polysulfone is the most practical membrane material in ultrafiltration applications.

The earliest reverse osmosis and ultrafiltration units were based on flat membrane sheets in arrangements similar to that of a plate and frame filter press. Since then, more efficient membrane configurations, ie, tubular, spiral wound, and hollow fiber, have emerged (96–98).

Another type of membrane is the dynamic membrane, formed by dynamically coating a selective membrane layer on a finely porous support. Advantages for these membranes are high water flux, generation and regeneration *in situ*, ability to withstand elevated temperatures and corrosive feeds, and relatively low capital and operating costs. Several membrane materials are available, but most of the work has been done with composites of hydrous zirconium oxide and poly(acrylic acid) on porous stainless steel or ceramic tubes.

Reverse osmosis and ultrafiltration are very effective for the removal of all types of color from dye house discharges; decolorizations on the order of 95–100% are readily obtained (99–119). However, clogging of the membranes with concentrated dyes after prolonged use is a problem. About 10–25% of the original volume of the wastewater treated becomes concentrate which if it cannot be reused has to be disposed of by incineration. Hyperfiltration with dynamically formed dual-layer hydrous zirconium oxide/poly(acrylate) membranes on porous ceramic and carbon tubes has been reported to result in at least 99% color extraction from the effluent of a textile finishing plant processing mostly synthetic fibers (102). TOC removal was greater than 85% without loss in membrane performance. Although the high capital costs and the limited life of membranes are disadvantages of hyperfiltration and ultrafiltration, the continuing increase in the cost of dyes, auxiliary chemicals, water, energy, and the demanding regulations governing wastewater discharge make recovery and reuse by these systems more economically attractive (120).

Ionizing Radiation. Gamma irradiation has been used to decolorize dye-house wastewaters (121–124). Most of the dyes resisting oxidation or reduction can be degraded by ionizing radiation. The rate of reaction is controlled by the radiation dose and the availability of oxygen in solution. For optimum efficiency, the radiation should be totally absorbed by the organic waste. One approach to this is adsorption onto activated charcoal before exposure to the gamma radiation (2).

Incineration. Although incineration or thermal destruction/treatment of waste has met opposition in the United States because of fear of air emission of toxic pollutants and concentration of heavy metals in resulting ash, Europe and Japan readily use this method of waste disposal (125–129). Of special interest is a patent (130) describing dyeing wastewater treated with Fenton's reagent (hydrogen peroxide plus ferrous sulfate) to coagulate the dye. The resulting sludge is then incinerated and iron recovered from ash is then recycled.

CHEMICAL METHODS

Neutralization. The choice of a reagent for pH adjustment depends on cost; ease and safety of storage and handling; effectiveness, eg, for removing heavy metals, buffer characteristics of the pH titration curve as they affect pH control; and availability. The three principal reagents for neutralization of acid wastes are sodium hydroxide, sodium carbonate, and hydrated calcium hydroxide.

The wastewaters of many textile mills are alkaline and the two most commonly employed reagents for pH adjustment are sulfuric and hydrochloric acids. The alkaline wastewater from a dyeing plant can be treated by neutralizing with acid waste combustion gases (131). Another reagent gaining wide acceptance for neutralizing alkaline wastewater in textile production is carbon dioxide (qv). Its manufacturers (132,133) claim that carbon dioxide is safe, effective, and more economical than sulfuric or hydrochloric acids for neutralization. When CO_2 is dissolved in an aqueous system, the familiar H_2CO_3 (carbonic acid), HCO_3^- (bicarbonate), and CO_3^{2-} (carbonate) buffer system forms. A method for reducing the alkalinity of bleaching, mercerizing, and dyeing effluent by using the carbon dioxide in the flue gases from steam power stations has been discussed (134).

Reduction. Many dyes, particularly azo dyestuffs, are susceptible to destructive reduction. The reducing agents that can be used are sodium hydrosulfite, thiourea dioxide, sodium borohydride, zinc sulfoxylate, and ferrous iron.

Sodium hydrosulfite or sodium dithionate, $Na_2S_2O_4$, under alkaline conditions are powerful reducing agents; the oxidation potential is $+1.12$ V. The reduction of p-phenylazobenzenesulfonic acid with sodium hydrosulfite in alkaline solutions is first order with respect to p-phenylazobenzenesulfonate ion concentration and one-half order with respect to dithionate ion concentration (135). The SO_2^- radical ion is a reaction intermediate for the reduction mechanisms. The reaction equation for this reduction is

$$R'-N{=}N-R^2 + 2\,S_2O_4^{2-} + 4\,OH^- \longrightarrow R'NH_2 + R^2NH_2 + 4\,SO_3^{2-}$$

Reduction of dyes, and especially reactive dyes into their respective amines is a well-known analytical technique (136,137).

Although it has been reported (138) that decolorization of wastewater containing reactive azo dyes with sodium hydrosulfite is possible only to a limited extent, others have demonstrated good reduction (decolorization). For example, using zinc hydrosulfite for the decolorization of dyed paper stock (139) resulted in color reduction of 98% for azo direct dyes (139). A Japanese patent (140) describes reducing an azo reactive dye such as Reactive Yellow 3 with sodium hydrosulfite into its respective aromatic amines which are more readily adsorbable on carbon than the dye itself. This report has been confirmed with azo acid, direct, and reactive dyes (22).

Thiourea dioxide as well as sodium hydrosulfite are mainly used in the dyeing industry as reducing agents in vat dyeing, reduction clearing, etc (141,142). One manufacturer of thiourea dioxide working with a dyestuff company is promoting the product as a reducing agent to decolorize a number of direct dyes for dyeing paper (143,144).

Thiourea dioxide, or formamidine sulfinic acid, is an oxygenated thiourea derivative synthesized by the oxidation of thiourea with hydrogen peroxide. It has the chemical formula $(NH_2)NHCSO_2H$ and is tautomeric.

$$H_2N-\underset{\underset{NH}{\|}}{C}-SO_2H \ \rightleftharpoons \ H_2N-\underset{\underset{NH_2}{|}}{C}=SO_2$$

Thiourea dioxide, although stable in acidic solutions, decomposes in alkaline solution to urea and sulfinic acid. Unfortunately, the release of urea can create environmental concerns as it is a fertilizer and causes eutrophication when discharged into bodies of water.

Recent papers by a manufacturer of sodium borohydride, $NaBH_4$ (145,146), have demonstrated that excellent removal of metals and color of acid, direct, and reactive dyes for textiles and paper can be achieved with bisulfite-catalyzed borohydride reduction in combination with polymer flocculation.

Oxidation. Oxidation is one of the main chemical methods to treat and decompose dyes in wastewater. The oxidation agents used are chlorine, bleach, ozone, hydrogen peroxide, Fenton's reagent, and potassium permanganate.

For color removal, ozonization has achieved the greatest practical importance as seen by the plethora of articles and patents on this method (147–163). Ozonization in combination with treatments such as coagulation, flocculation, carbon adsorption, uv irradiation, gamma radiation, and biodegradation significantly and successfully remove dye wastes and reduce costs (156,164–170).

The reaction between ozone and organics does not depend on the ozone concentration, but rather on the concentration of the decomposed products of ozone (171). The free radicals and ions formed by ozone degradation are the chief reacting species with organics.

The identification of the reaction products in the ozonization of dyes have been reported (158–160). The ozonoloysis of 1-phenylazo-2-naphthol proceeds through the intermediacy of ozonides, hydroperoxides, and quinoids. The final products identified were benzene, chlorobenzene, phenol, nitrogen, phthalic anhydride, and phthalide.

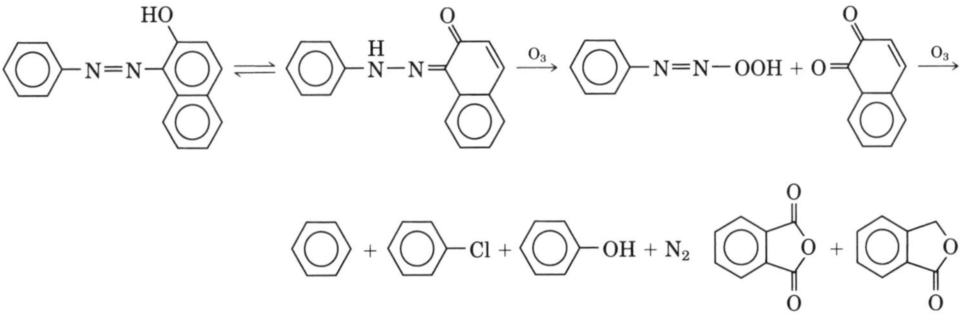

Chlorination was found (172) to be the most suitable and effective method for decolorizing and reducing the COD of waste dyebaths containing azo dyes. These findings have been substantiated for chlorination and biochemical purification (173). A study (174) has been done on the technical and economic feasibility of a chlorination dye wastewater reclamation system for treating effluent that is suitable for reuse in dyeing of polyester/cotton blends with disperse and direct dyes.

In a series of papers (175) comparing chlorination and ozonization, reactive and acid dyes were readily destroyed, but direct and disperse dyes reacted more slowly; ozone was more effective in some instances. Although chlorination is cheaper than ozonization, the possible formation of chlorinated compounds such as dioxin and its environmental impact cannot be overlooked (see CHLORO-CARBONS AND CHLOROHYDROCARBONS, TOXIC AROMATICS).

Hydrogen Peroxide. Uncatalyzed hydrogen peroxide is used commercially to destroy pollutants. Without activation, however, hydrogen peroxide will not destroy the more difficult to oxidize pollutants. Hydrogen peroxide can be activated to form hydroxyl radicals, OH·, which can destroy these organics. This increased ability to destroy organics is related to the substantially higher oxidation potential of the hydroxyl radical (2.80 V) compared to hydrogen peroxide (1.78 V).

One method of generating hydroxyl radicals is by adding a soluble iron salt to an acid solution of hydrogen peroxide (Fenton's reagent) (176–180), ie:

$$Fe^{2+} + H_2O_2 \longrightarrow Fe^{3+} + OH^- + OH·$$

$$\underline{Fe^{3+} + H_2O_2 \longrightarrow Fe^{2+} + H^+ + HOO·}$$

$$2\,H_2O_2 \longrightarrow H_2O + OH· + HOO·$$

Several papers and patents describe treating dye-containing wastewaters with Fenton's reagent with and without other methods such as coagulation, incineration, biodegradation, etc (130,181–187).

Hydrogen peroxide can also be activated by ultraviolet radiation or ozone and ultraviolet radiation (178,188,189). One of the most active fields in waste treatment is ultraviolet-catalyzed oxidation with hydrogen peroxide (190–194). The uv light activates the hydrogen peroxide converting it to hydroxyl radicals (195).

The uv–hydrogen peroxide system has advantages over the iron–hydrogen peroxide (Fenton's reagent) procedures, eg, the reaction is not limited to an acid pH range and the iron catalyst and resulting sludges are eliminated. However, the system to date is not effective for dye wastewaters because of absorption of uv by colored effluent.

Wet Air Oxidation.   An oxidation process of increasing attention and use is wet air oxidation (196–202), which refers to the aqueous phase oxidation of organic and inorganic materials at elevated temperatures and pressures. Oxidation takes place through a family of related oxidation and hydrolysis reactions at temperatures of 175–320°C and a pressure of 2–20 MPa (300–3000 psig). The enhanced solubility of oxygen in aqueous solution at elevated temperature provides a strong driving force for oxidation. The source of oxygen is compressed air or high pressure pure oxygen.

Elevated pressures are required to keep water in the liquid state. Liquid water catalyzes oxidation so that reactions proceed at relatively lower temperatures than would be required if the same materials were oxidized in open flame combustion. At the same time, water moderates oxidation rates by providing a medium for heat transfer and removing excess heat by evaporation.

The wet air oxidation process can treat a wide variety of oxidizable materials. The primary products of oxidation are carbon dioxide and water. Sulfur is oxidized to sulfate which remains in the aqueous phase. Organic nitrogen is converted primarily to ammonia. No sulfur or nitrogen oxides are formed. Metals generally are converted to their highest oxidation state and remain in the aqueous phase as dissolved or suspended solids. Halogens also stay in aqueous phase. The gas discharged from a wet air oxidation unit consists mainly of spent air and carbon dioxide and is essentially free of any air polluting constituents.

Also, wet air oxidation offers an alternative to conventional incineration for the destruction and detoxification of dilute hazardous and toxic waste waters. A 98% removal efficiency of dyehouse effluent has been claimed by wet air oxidation (203).

Electrolysis.   Electrolytic precipitation of concentrated dye wastes by reduction in the cathode space of an electrolytic bath has been reported, but extremely long contact times were required (204). There has been electrochemical treatment of textile effluents containing dyes, auxiliaries, and metals (205–208). Also, two patents (209,210) have been obtained; in one, dissolved sulfonated azo dyes are removed from water by an electrochemical flotation process by feeding the water to an electrolytic cell and electrolyzing it. The dyes are converted to insoluble complexes that are floated to the surface of the water and removed from it. An electrochemical process originally developed for removal of several classes of dyes, including disperse, reactive, acid, and textile auxiliaries such as leveling and stainblocking agents (211–214). In the treatment process, wastewater is circulated between two iron electrodes. An electric current is passed between the electrodes, slowly removing iron from the anode. The reaction releases ferrous and hydroxide ions, thus:

Anode (oxidation) $Fe \longrightarrow Fe^{+2} + 2e^-$

Cathode (reduction) $2\ H_2O + 2e^- \longrightarrow H_2 + 2\ OH^-$

Overall (reaction) $Fe + 2\ H_2O + electrical\ energy \longrightarrow Fe(OH)_2 + H_2$

The ferrous ions that dissolve from the anode combine with the hydroxide ions produced at the cathode to give an iron hydroxide precipitate. The active surface of ferrous hydroxide can absorb a number of organic compounds as well as heavy metals from the wastewater passing through the cell. The iron hydroxide and adsorbed substances are then removed by flocculation and filtration. The separation process was enhanced by the addition of a small quantity of an anionic polymer.

Ion Exchange. Ion-exchange treatment (215,216) entails elution of wastewater through a suitable resin until the available sites for ion exchange become fully occupied and the contaminated ions appear in the outflow. Treatment is then stopped. The bed is backwashed and then regenerated using an appropriate acidic or basic solution. After further washing to remove excess regenerant, the bed is ready for the next treatment cycle. Although highly effective for eliminating dissolved contaminants of known constitution, eg, toxic metal ions, this technique is clearly unsuitable for large volume multicomponent effluents (1) (see ION EXCHANGE).

Biological Methods

There are two methods of biological treatment: aerobic and anaerobic (217). The aerobic systems use free oxygen dissolved in the wastewater to convert wastes in the presence of microorganisms to more microorganisms, energy required for their existence, and carbon dioxide. The anaerobic process occurs in the absence of free oxygen and converts the waste to methane and carbon dioxide, generally in deep tanks or basins, and can produce odor problems when sulfides or sulfates are present in the wastewater.

The four most common aerobic biological treatment processes are stabilization ponds, aerated lagoons, trickling filters, and activated sludge (see AERATION).

Stabilization Ponds. Stabilization ponds have a water depth of 1–2 m and oxygen is supplied by surface entrainment or by algae. The BOD loading must be low and the detention time is 5–25 days (218).

Aerated Lagoons. Aerated lagoons are 2–5 m liquid depth depending on the aeration system and detention times are 2–10 days. They are mainly used because of their efficiency in removing BOD from textile effluents (2).

Trickling Filters. Trickling filters are cylindrical tanks packed either with stone or a synthetic medium (219). The effluent flows onto the filter media by means of a rotating arm that distributes the waste load uniformly over the circular bed. The effluent trickles through the filter and over a slime of bacteria that adheres to the filter media. As bacteria die, they fall off the filter and are removed from the effluent throughout the secondary settling stage. The removal efficiencies depend on the type of media used, the organic loading ratio, the ratio of raw waste to recycle wastewater, and operating temperature. Trickling filters respond well to shock loadings; they offer advantages as to decreasing the loading from highly concentrated organic waste streams (220).

Activated Sludge. Of the aerobic biological treatment methods available, the activated sludge process and its various modifications are the most popular. The activated sludge process relies on microorganisms in suspension to oxidize soluble and colloidal organics with molecular oxygen. Many variables affect the performance of the system including pH, temperature, dissolved oxygen concentration, detention time, nutrients, and the presence of toxic constituents in the raw waste stream. The fundamental principles of aeration and mechanical arrangements needed for maximum efficiency have been outlined (221). The power requirements of aerators, a comparison of aeration and fermentation, and views on related aspects have been given (222).

Although the activated sludge process is capable of providing high BOD, COD, TOC, and TSS removals, it is less effective for color removal. Since synthetic dyes used by the textile industry are specifically formulated to resist breakdown under oxidizing conditions, most dyes are resistant to biological degradation. The color removal in the activated sludge process is primarily by adsorption of the dyes by the sludge to the extent of 40–80% or more, depending on the individual dyestuff and treatment conditions. Complete removal or decolorization can only be achieved by combination with other treatment processes (223–228).

The efficiency of an existing bioaeration plant can be increased by the addition of activated carbon (229). One successful system patented by DuPont and further developed by Zimpro is the powdered activated carbon treatment (PACT) system (230–233). Adding powdered activated carbon to the activated sludge process can be beneficial, ie, more uniform operation and effluent quality; improved BOD, COD, and TOC removals; better removal of phosphorus and nitrogen; less tendency for foaming in the aerator because of the adsorption of detergents; adsorption of refractory solids and thicker sludges; greater treatment flexibility and increased effective plant capacity at little or no added capital investment; and enhanced color removal of dye effluents (95). The mechanism by which pollutant removals are increased with powdered activated carbon addition to activated sludge is not fully understood. The current theory is that the activated carbon aids by direct adsorption of pollutants and by providing a more favorable environment for the microorganisms to propagate. An extensive study (234) was done with the assistance of six dyestuff companies in which laboratory scale processes were conducted to establish the technical feasibility of using ozonization, granular activated carbon adsorption, and the PACT treatment of dye manufacture wastewater. Overall, excellent removals of organic priority pollutants were achieved by the PACT process. In addition, soluble organic carbon (SOC) and color removals were enhanced by the addition of powdered activated carbon to an activated sludge system, generally in direct relation to the steady-state concentration of powdered activated carbon in the reactor. Two dyestuff companies have successfully used the PACT process (235,236).

Anaerobic Digestion. Following biological treatment, the sludge containing adsorbed dyes may be digested under anaerobic conditions. An investigation (237) showed that only 4 out of 42 dyes had an inhibitory effect on sludge bacteria. Of the 29 soluble dyes studied only 4 dyes (Acid Black 1, Acid Blue 45, Acid Green 25, and Basic Blue 3) showed no signs of decolorization (238,239). The rest were either completely decolorized or underwent significant spectral changes. Anaerobic digestion treatment of reactive dye manufacturing wastewater in a multisectional bioreactor removed 88% color and produced an effluent suitable for further biological treatment in conventional systems in one report (240). The addition of excess activated sludge to a wastewater storage tank as a pretreatment stage decreases COD and BOD by 48.2 and 59.0%, respectively.

Bioaugmentation. Bioaugmentation is obtaining maximum biodegradability by addition of nonindigenous microbial cultures in order to have optimum bacterial biomass. This offers flexibility in dealing with problems of bacterial acclimation, toxicity of compounds, and restart of the system (241). Several papers (242–245) have described the biodegradability of azo and triphenylmethane dyes using specific strains of bacteria.

Fate of Dyes

In 1980, approximately 111,000 t of synthetic organic dyestuffs were produced in the United States alone. In addition, another 13,000 t were imported. The largest consumer of these dyes is the textile industry accounting for two-thirds of the market (246). Recent estimates indicate 12% of the synthetic textile dyes used yearly are lost to waste streams during dyestuff manufacturing and textile processing operations. Approximately 20% of these losses enter the environment through effluents from wastewater treatment plants (3).

With few exceptions, the normal use of organic colorants poses few problems in terms of acute ecological effects (247). On the other hand, certain dyestuffs exhibit toxic effects toward microbial populations and can be toxic and/or carcinogenic to animals (248,249). Also, the possible contamination of drinking water supplies is of concern because certain classes of dyes are known to be enzymatically degraded in the human digestive system, producing carcinogenic substances (250–253).

Until recently, few papers appeared on the fate of dyes in the environment. But because of the importance of this subject, work is being done primarily by the U.S. Environmental Protection Agency (USEPA) and the Ecological and Toxicological Association of the Dyestuff Manufacturing Industry (ETAD).

One of the reasons for lack of literature was probably because environmental analysis depends heavily on gas chromatography/mass spectrometry, which is not suitable for most dyes because of their lack of volatility (254). However, significant progress is being made in analyzing nonvolatile dyes by newer mass spectral methods such as fast atom bombardment (FAB), desorption chemical ionization, thermospray ionization, etc.

Fate studies on eighteen dyes were done by the Water Engineering Research Laboratory of the USEPA (49). The objective was to determine the partitioning of water-soluble azo dyes in the activated sludge process (ASP). Specific azo dyes were spiked at 1 and 5 mg/L to pilot-scale treatment systems with both liquid and sludge samples collected. Samples were analyzed by high pressure liquid chromatography and an uv–vis detector. Mass balance calculations were made to determine the amount of the dye compound in the waste activated sludge (WAS) and in the activated sludge effluent (ASE). Of the eighteen dyes studied, eleven compounds passed through the ASP substantially untreated, four were significantly adsorbed into the WAS, and three were apparently biodegraded.

A study of the degradation of two azo disperse dyes, Disperse Orange 5 [6232-56-0] (1) (CI 11100) and Disperse Red 5 [3769-57-1] (CI 11215) showed reduction of the azo linkage into aromatic amines and further dealkylation to p-phenylenediamine [106-50-3] (2) (255).

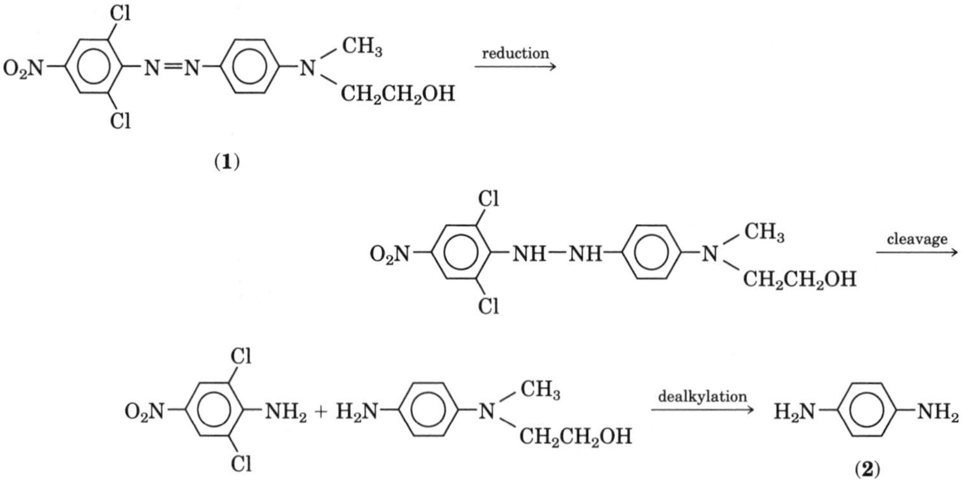

(1)

(2)

A study of Direct Red 28 [573-58-0] (3) (CI 22120) in a sediment–water system indicated that the amount of recovered benzidine [92-87-5] (4) accounted for only 2–5% of lost dye (256).

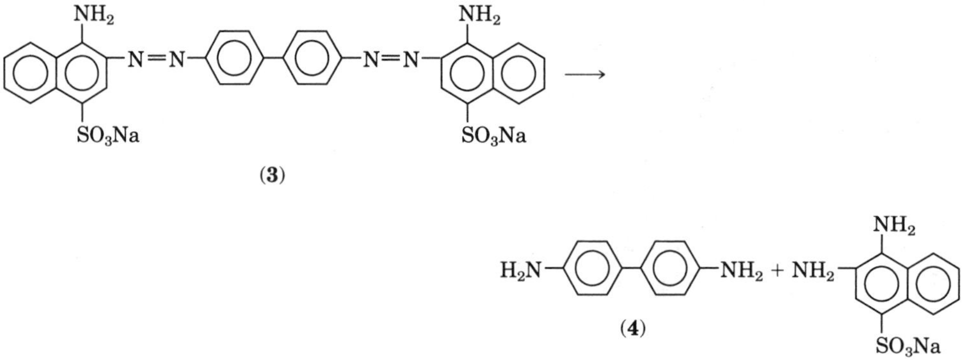

(3)

(4)

For Reactive Blue 19 [2580-78-1] (CI 61200), its reactive form, the vinyl sulfone (**5**), was found in the effluents of a textile mill and a wastewater treatment plant. The hydrolysis product of the vinyl sulfone was detected only in the effluent of the textile mill (257).

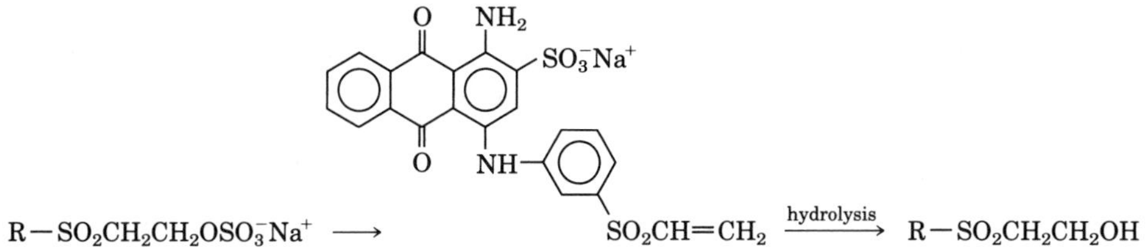

$$R-SO_2CH_2CH_2OSO_3^-Na^+ \longrightarrow \qquad SO_2CH{=}CH_2 \xrightarrow{hydrolysis} R-SO_2CH_2CH_2OH$$

(**5**)

Dyestuffs in general, and azo dyes in particular, are likely to undergo substantial primary biodegradation in an anaerobic environment through reductive cleavage of azo bonds into aromatic amines (258). Lipophilic aromatic primary amines are aerobically degradable, but depending on their precise structure, some sulfonated aromatic amines may not be degradable (258–260).

The fate of one dye that has been thoroughly studied is the azo dye Disperse Blue 79 [12239-34-8] (**6**) (CI 11345) which may be designated 6-bromo-2,4-dinitro-aniline→3-(N,N-diacetoxyethylamino)-4-ethoxyacetanilide (see AZO DYES).

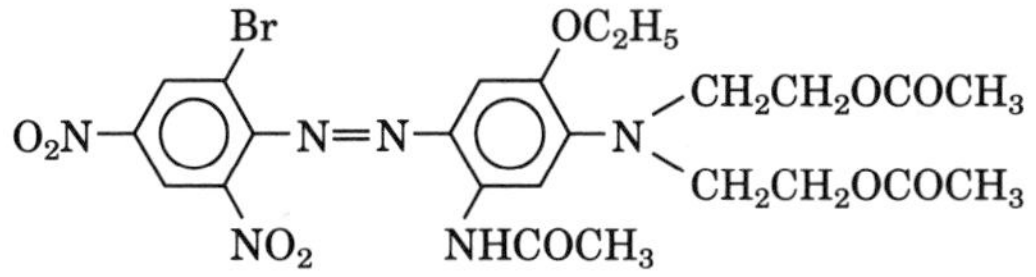

(**6**)

The U.S. International Trade Commission has listed the production of Disperse Blue 79 at 2300–4500 t per year in the United States from 1980 through 1984 (261). Thus Disperse Blue 79 is the largest volume dye on the market today (262); the average annual production in the United States from 1983–1985 was approximately 3200 t.

It has been estimated that during the manufacture of Disperse Blue 79 there would be released the following amounts of dye: 4.5–14 t per year at a total of nine sites with an estimated 3–20 kg per day (261).

Based on one report (250), the U.S. Government was petitioned to look at Disperse Blue 79, and subsequently nine U.S. agencies (EPA, NIOSH, etc) were asked to study the toxicological, epidemiological, chemical, and biological fate assessment of the dye (262). Of particular importance is the degradation, cleavage, or reduction of the dye into aromatic amines, one of which is 6-bromo-2,4-dinitro-aniline. This amine is toxic, mutagenic, and was selected for carcinogenic study (263). The precursor for preparation of 6-bromo-2,4-dinitroaniline [1817-73-8], is 2,4-dinitroaniline [97-02-9] which has been extensively evaluated and found to be highly toxic (264,265), mutagenic (266–269), and was selected for carcinogenic study (263,269).

Small amounts of the following aromatic compounds were found to be present in the effluent after manufacture of Disperse Blue 79: 6-bromo-2,4-dinitroaniline, 6-bromo-2,4-dinitrophenol, 3-diacetoxyethylamino-4-ethoxy acetanilide, 3-dihydroxyethylamino-4-ethoxyacetanilide, 3-[(N-hydroxyethyl, N-acetoxyethyl)amino]-4-ethoxyacetanilide, and 3-acetoxyethylamino-4-ethoxy acetanilide (21). However, after a neutralization and heat stabilization step is added after coupling in manufacture, no diazotizable amine or phenol and only traces of the coupling component and its impurities were found.

The fate study of Disperse Blue 79 (262) in anaerobic sediment–water systems shows the following degradation products:

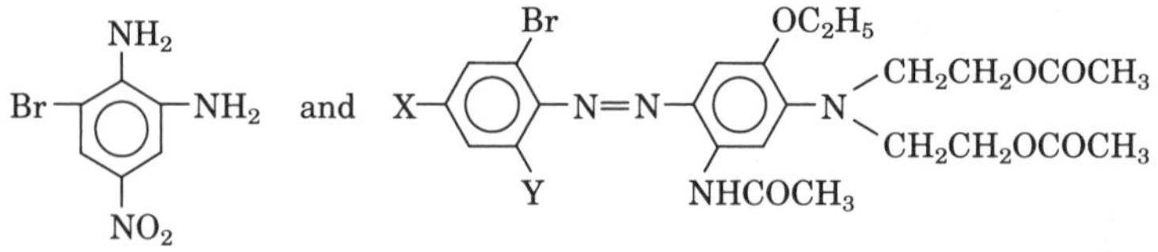

where X = NH_2, Y = NO_2; X = NO_2, Y = NH_2; or X = Y = NH_2

These products suggest that this dye may undergo reduction in bottom sediments in the environment, resulting in the subsequent release of potentially hazardous aromatic amines into water.

A large study (270) was done to determine the fate of Disperse Blue 79 in a conventionally operated activated sludge process and in an anaerobic sludge digestion system. The results showed no degradation in the activated sludge system, but did show degradation in the anaerobic digester of which no positive identification of compounds were made.

Analytical Methods

There is a large amount of literature on the analysis of dyes and the most comprehensive treatment of the analytical chemistry of dyestuffs is found in Reference 271. Earlier papers dealing with the analytical chemistry of dyes in the environment mainly used paper, thin-layer, column, and high pressure liquid chromatography and ultraviolet and visible spectrophotometry (255,272–275).

Volatile dyes and pigments are readily analyzed by gc/ms (276–281), but the majority of dyes causing environmental problems contain solubilizing sulfonic acid groups which make these dyes nonvolatile; in electron impact and chemical ionization mass spectrometry, volatilization of the sample in the mass spectrometer must occur before ionization can take place. With the advent of fast atom bombardment (fab), desorption chemical ionization (dci), thermospray, etc, analyses of these nonvolatile sulfonated dyes by mass spectrometry (282–286) and determination of these dyes in the environment have become possible (287–293). The chemical identification of nonvolatile sulfonated dyes in the environment by these newer techniques will result in more information on the environmental chemistry of dyes.

Pollution Prevention

The United States Environmental Protection Agency (USEPA) has placed a high priority on reducing or minimizing hazardous chemical waste by limiting the amount of pollutants produced, ie, reducing the production of pollutants at their source rather than controlling them at the end of the manufacturing process in water treatment facilities. The USEPA has made official its commitment to waste minimization (294).

A number of states such as New Jersey have passed into law the "Pollution Prevention Act" which requires industry to reduce and eventually eliminate toxic waste in the 1990s (295). Cooperation between industry, government, academia, and private environmental groups to implement these pollution prevention acts have begun in earnest (296–302).

The dye and dyeing industries have also begun to give pollution prevention and its other forms of lessening or eliminating waste generation, such as waste minimization and source reduction, a high priority (303). The USEPA and the Ecological and Toxicological Association of the Dyestuffs Manufacturing Industry (ETAD) have jointly set up a program for pollution prevention in the dyestuff industry (304). The original goals were to develop a pollution prevention guidance manual and conduct a baseline survey of industry prevention practices for dye manufacture. Recently the manual was published and the data for the baseline survey collected by USEPA and ETAD (304).

INFORM, an environmental research group, reported on the waste minimization efforts of two dye manufacturers, Atlantic Industries and CIBA-GEIGY Corp. in New Jersey (305). Two papers have been published describing the reduction of toxic wastewaters in the manufacture of disperse azo dyes (20,21).

There are a number of papers and patents on recycling dye and textile industry wastewater for reuse of dye, textile auxiliaries, and water (99,102, 120,306–311). Recycling is considered a part of pollution prevention.

The North Carolina Department of Environment, Health, and Natural Resources has published a report on the reduction of pollution sources in textile wet processing and a workbook for pollution prevention by source reduction in textile wet processing which includes dyes (312).

Heavy Metals

The heavy metals, copper, chromium, mercury, nickel, and zinc, which are used as catalysts and complexing agents for the synthesis of dyes and dye intermediates, are considered priority pollutants (313).

An example of the use of copper as a catalyst is Acid Blue 25 [6408-78-2] (CI 62055), in which 1-amino-2-sulfonic-4-bromoanthraquinone is condensed with aniline using copper salts (Ullmann reaction) (314). Another example is oxidation to the triazole of Direct Yellow 106 [12222-60-5] (CI 40300) (315,316).

Copper and chromium are used for complexing a number of dyes such as the coppered direct and reactive dyes for cotton and metallized and neutral metal complex acid dyes for nylon, wool, etc. Examples are Direct Blue 218

[*28407-37-6*] (CI 24401) (317), Reactive Violet 2 [*8063-57-8*] (CI 18157) (318), and Acid Black 52 [*5610-64-0*] (CI 15711) (319).

Zinc is used to prepare the double salt for basic dyes. An example is Basic Red 22 (320).

A number of papers have appeared on the removal of heavy metals in the effluents of dyestuff and textile mill plants. The methods used were coagulation (320–324), polymeric adsorption (325), ultrafiltration (326,327), carbon adsorption (328,329), electrochemical (330), and incineration and landfill (331). Of interest is the removal of these heavy metals, especially copper by chelation using trimercaptotriazine (332) and reactive dyed jute or sawdust (333).

Toxicity

The past experience of the dyestuff industry in its use of dye intermediates such as β-naphthylamine and benzidine (4), known human bladder carcinogens (334–343), have led to studies as to whether or not handlers of dyes are exposed to medical hazards such as cancer, dermatitis, and other disorders (344–360).

The National Institute of Occupational Safety and Health (NIOSH) and the Occupational Safety and Health Administration (OSHA) reported in 1978 that the three primary benzidine-based azo dyes, namely Direct Black 38 [*1937-37-7*], Direct Blue 6 [*2602-46-2*], and Direct Brown 95 [*16071-86-6*], were carcinogenic in animals as a result of being converted to benzidine (4) (361). These dyes are characterized by having a biphenyl diazo linkage:

$$\text{\small\leavevmode\raisebox{0pt}{}} N{=}N{-}\bigcirc{-}\bigcirc{-}N{=}N \text{\small}$$

Benzidine-derived azo dyes may be degraded metabolically in the gut or liver in humans to free benzidine or monoacetylbenzidine (362). Although these three benzidine-based dyes produced neoplastic hepatic lesions in rats, no complete bioassay has yet been reported (363). However, this has led to concern about possible carcinogenicity from these dyestuffs, and therefore benzidine-based dyes and pigments are no longer produced by the large dyestuff manufacturers.

Two large studies were done (250,251) for the selection of azo, nitro, and anthraquinone dyes for carcinogen bioassay. Based on previous information or testing, a total of 30 dyes were selected based on chemical structure, potential exposure, and suspicion of carcinogenicity.

Because of the large number of dyestuffs and the fact that most of these colorants have not been tested for carcinogenicity, structure–activity theory may help predict possible candidates for study (250,251,364–366). The underlying concept of structure–activity relationships for carcinogens is the formation of electrophilic species. That is, although most chemical carcinogens act as electrophiles and interact covalently with nucleophilic sites in nucleic acids, proteins, or other cellular macromolecules of the target tissue, the electrophiles must first be activated by metabolism to a reactive electrophilic species carcinogen (364).

In order to minimize the possible toxicity and damage to humans and the environment arising from the production and applications of colorants, an international association, the Ecological and Toxicological Association of the Dyestuff Manufacturing Industry (ETAD), was established in 1974. ETAD coordinates the ecological and toxicological efforts of synthetic organic dyes and pigment manufacturers. To date, ETAD consists of 32 members in Western Europe, North America, Japan, and India (367). The purpose of ETAD's toxicological work is to identify and assess risks caused by colorants and their intermediates with respect to their potential acute toxicity and their chronic effects on human health. This is accomplished by recommended methods to member firms following appraisal and development by appropriate ETAD committees, lectures, and publications (363, 368,369). One of the projects of ETAD was a survey of acute oral toxicity, as measured by LD_{50}, the 50% lethal dose, which showed that of 4461 colorants tested, only 44 had a LD_{50} <250 mg/kg, but 3669 exhibited practically no toxicity (LD_{50} >5 g/kg). The evaluation of these colorants by chemical and coloristic classification showed that the most toxic ones are found among diazo and cationic dyes. Pigments and vat dyes, on the other hand, have a low acute toxicity, presumably because of their low solubility in water and in lipophilic systems.

A number of papers have been published on methods for reducing hazards in the use and handling of dyes and pigments (370–379).

Legislation

There has been a tremendous increase in regulatory activities worldwide aimed at achieving safer manufacture, use, and disposal of chemicals, including colorants (380). Table 1 is a summary of important United States, European, and Japanese Environmental Legislation affecting workers, consumers, and the public and environment (381).

The United States has the most laws regarding environmental safety and health. The National Environmental Policy Act (NEPA) of 1969 has resulted in the following acts: Federal Insecticides, Fungicide and Rodenticide (FIFRA), Resource Conservation and Recovery (RCRA), Superfund (CERCLA), Superfund Amendments and Reauthorization Act (SARA) Plus Title III, Toxic Substance Control Act (TSCA), Clean Water (CWA), Water Quality, Safe Drinking Water (SDWA), and Waste Minimization and Control.

The two most important pieces of chemical control legislation enacted affecting the dye and pigment industries are the United States' Toxic Substance Control Act (TSCA) and EEC's Classification, Packaging, and Labeling of Dangerous Substances and its amendments. Table 2 is a comparison of TSCA and the 6th Amendment of the EEC classifications.

The Toxic Substances Control Act (TSCA) was enacted in 1976 to identify and control toxic chemical hazards to human health and the environment. One of the main provisions of TSCA was to establish and maintain an inventory of all chemicals in commerce in the United States for the purpose of regulating any of the chemicals that might pose an unreasonable risk to human health or the environment. An initial inventory of chemicals was established by requiring companies to report to the United States Environmental Protection Agency (USEPA) all substances that were imported, manufactured, processed, distributed, or disposed of in the United States. Over 50,000 chemical substances were reported.

Table 1. Important Environmental Legislation

Law	Worker	Consumer	Public and environment
United States			
Toxic Substances Control Act (TSCA)	X	X	X
Occupational Safety and Health Act (OSHA)	X		
Food, Drug, and Cosmetic Law		X	
Consumer Product Safety Act		X	
Labeling of Hazardous Materials		X	X
Federal Water Pollution Control Act			X
Clean Air Act			X
Resource Conservation and Recovery Act (RCRA)			X
Superfund Amendments and Reauthorization Act (SARA) Plus Title III	X		X
Superfund (CERCLA)			X
Hazardous Materials Transportation (DOT)			X
CONEG (Heavy Metals)			X
USDA			X
California Proposition 65			X
EEC			
Control of Certain Industrial Activities	X	X	X
Classification, Packaging, and Labeling of Dangerous Substances		X	X
UK			
Health and Safety at Work Act	X		
Carcinogenic Substances Regulations	X		
Toy (Safety) Regulations		X	
Pencil and Graphic Instruments (Safety) Regulations		X	
Poison Act		X	
Clean Air Act			X
Switzerland			
Environment Protection Law		X	X
Poison Law		X	
France			
Control of Chemical Products		X	X
Consumer Protection and Information		X	
Pollution Control Law			X
Germany			
Environmental Chemicals Law	X	X	X
Japan			
Chemical Control Law			X

Table 2. Chemical Control Legislation for Dyes and Pigments

Aspect	6th Amendment[a]	TSCA[b]
general	EEC directive to be incorporated into national legislation of the EEC states	U.S. Federal Law
enforcement	enforcement is the responsibility of individual EEC states	single regulatory authority; Environmental protection Agency (EPA)
scope	notification of new chemicals only; labeling and packaging of new and existing chemicals, premarketing notification; applies to imports also (exports are not regulated); excludes raw materials and intermediates unless they are marketed	applies to new and existing chemicals
concept of new substance	any substance not included in the final European inventory of existing chemical substances (EINECS)	any substances not yet included in TSCA inventory
confidentiality	does not call for disclosure although confidentiality assurance is unclear	calls for extensive public disclosure, but confidentiality of certain data may be retained
testing requirements	established base-set requirements and "Stufen-Plan," which include an escape clause	no mandatory testing
notification	multinotification approach; requires notification updating, 45-day notification review	no second notifier; no updating mechanism unless there is significant new use 90(180)-day notification review

[a]To EEC's Classification, Packaging, and Labeling of Dangerous Substances.
[b]Toxic Substance Control Act.

Following this initial inventory, introduction of all new chemical substances requires a Premanufacturing Notification (PMN) process. To be included in the PMN are the identity of the new chemical, the estimated first year and maximum production volume, manufacture and process information, a description of proposed use, potential release to the environment, possible human exposure to the new substance, and any health or environmental test data available at the time of submission. In the 10 years that TSCA has been in effect, the USEPA has received over 10,000 PMNs and up to 10% of the submissions each year are for dyes (382) (Fig. 1).

The large number of dyes and the potential for toxicity of dyes and dye components have made this class of compounds one of the most extensively reviewed and regulated. Up until 1987, 765 PMNs for dyes have been submitted by 53 different companies and over 50% of these notices by six companies. Figure 2

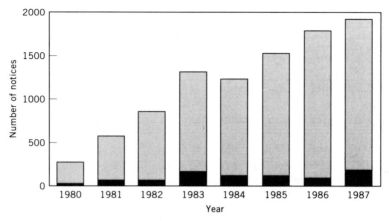

Fig. 1. Number of PMNs received by EPA: (■), dyes; (□), all other.

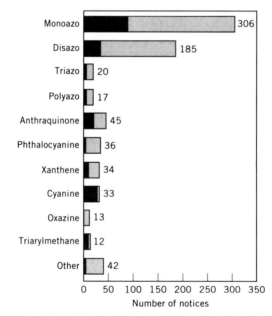

Fig. 2. PMN dyes received by EPA, described by main color forming group: (■), manufacture; (□), import.

shows PMN dyes received by the USEPA, described by principal color forming group (382).

Besides the above federal laws, all 50 U.S. states have also passed environmental laws. As previously mentioned, both the United States and the state of New Jersey have passed Pollution Prevention Acts (295).

There is a difference between the United States, European, and Japanese environmental laws in introducing a new chemical compound. The EEC and Japan require animal and other toxicological and ecological testing, whereas the United States encourages but does not require these tests. This could result in rejection of a U.S. chemical exported product (383). Subsequently, greater harmonization is needed in many federal versus state versus overseas regulatory issues impacting the dye industries.

BIBLIOGRAPHY

1. J. Park and J. Shore, *J. Soc. Dyers Colors* **100,** 383 (1984).
2. G. McKay, *Am. Dyest. Rep.* **68**(4), 29 (1979).
3. E. A. Clarke and R. Anliker, *Environmental Chemistry, Anthropogenic Compounds,* Vol. 3, Part A, Springer-Verlag, New York, 1980, p. 181.
4. J. Porter, *Am. Dyest. Rep.* **60**(8), 17 (1971); **61**(8), 24 (1972).
5. S. G. Hall, *The Adsorption of Disperse Dyes on Powder Activated Carbon,* Ph.D. dissertation, University of North Carolina, Greensboro, 1975, Xerox University Microfilms, Ann Arbor, Mich.
6. A. F. DiGiano, W. H. Frye, and A. S. Natter, *Am. Dyest. Rep.* **64**(8), 15 (1975).
7. A. F. DiGiano, A. S. Natter, and J. Water, *Poll. Cont. Fed.* **49**(2), 235 (1975).
8. G. McKay, M. S. Otterburn, and A. G. Sweeney, *J. Soc. Dyers Colour* **96,** 576 (1980).
9. G. McKay and B. Al-Duri, *Colourage* **36**(13), 15 (1989); **35**(20), 24 (1988).
10. R. J. Posey, *J. Water Poll. Control Fed.* **59**(1), 47 (1987).
11. G. McKay, *J. Chem. Tech. Biotech.* **32,** 759, 773 (1982); **33A,** 196, 205 (1983).
12. G. McKay, *Colourge* **29**(25), 11 (1982).
13. R. H. Horning, *Textile Dyeing Wastewaters—Characterization End Treatment,* EPA-600/2-78-098, U.S. Environmental Protection Agency, Washington, D.C., 1978.
14. O. G. Rhys, *J. Soc. Dyers Colour* **94,** 293 (1978).
15. M. Mitchell, W. R. Ernst, and G. R. Lithsey, *Bull. Environ. Contam. Toxicol.* **19,** 307 (1978).
16. C. A. Rodman, *Tex. Chem. Color.* **3,** 239 (1971).
17. P. B. DeJohn and R. A. Hutchins, *Tex. Chem. Color.* **8,** 69 (1976).
18. E. L. Shunney, A. E. Perrotti, and C. A. Rodman, *Am. Dyest. Rep.* **50**(6), 32 (1971).
19. R. A. Davies, H. J. Kaempf, and M. M. Clemens, *Chem. Ind.* **17,** 827 (1973).
20. A. Reife, *Reduction of Toxic Wastewaters in Disperse Azo Dye Manufacture, Colour Chemistry,* Elsevier Applied Science, London and New York, 1991.
21. A. Reife, "Reduction of Toxic Components and Wastewaters of Disperse Blue 79," *41st Southeast Regional American Chemical Society Meeting,* Raleigh, N.C., Oct. 9–11, 1989.
22. A. Reife, "Waste Treatment of Soluble Azo Acid, Direct and Reactive Dyes Using a Sodium Hydrosulfite Pretreatment Followed by Carbon Absorption," *AATCC Book of Papers, 1990 International Conference and Exhibition,* Philadelphia, Pa., Oct. 1–3, 1990.
23. P. N. Cheremisinoff and F. Ellerbusch, *Carbon Absorption Handbook,* Ann Arbor Science, Ann Arbor, Mich., 1978.
24. W. G. Schwer, "Purification of Industrial Liquids with Granular Activated Carbon: Techniques for Obtaining and Interpreting Data and Selecting the Type of Commercial System," *Carbon Absorption Handbook,* Ann Arbor Science, Ann Arbor, Mich., 1978.
25. E. L. Shunney, A. E. Perotti, and C. A. Rodman, *Am. Dyest. Rep.* **80,** 32 (1971).
26. W. A. Johnston, *Chem. Eng.* **79**(26), 87 (1972).
27. G. M. Lukchis, *Chem. Eng.* **80**(16), 83 (1973).

28. W. Ottemeyer, *Gesundh.-Ing* **53,** 185 (1930).
29. F. Perineau, J. Molinier, and A. Gaset, *Tribune Du Cebedeau* **35**(468), 453 (1982).
30. G. McKay, H. S. Blair, and J. R. Gardner, *J. Appl. Poly. Sci.* **27,** 3043 (1982); **28,** 1767 (1983); **29,** 1499 (1984).
31. G. Malmary, F. Perineau, J. Molinier, and A. Gaset, *J. Chem. Tech. Biotech.* **35A**(8), 431 (1985).
32. H. M. Asfour, O. A. Fadali, M. M. Nassar, and M. S. El-Geundi, *J. Chem. Tech. Biotech.* **35A**(1), 21, 28 (1985).
33. N. Miyata, *Cellulose Chem. Tech.* **21**(5), 551 (1987).
34. M. R. Balasubramanian and I. Muralisankar, *Indian J. Tech.* **29**(10), 471 (1987).
35. S. K. Khare, K. K. Panday, and R. M. Srivastava, *J. Chem. Tech. Biotech.* **38**(2), 99 (1987).
36. G. McKay, M. El-Geundi, and M. M. Nassar, *Water Res.* **21**(12), 1513 (1987).
37. H. Hoppe and H. J. Taglich, *Acta. Hydrochim. Hydrobiol.* **10,** 69 (1982); *Chem. Tech.* **34,** 636 (1982).
38. I. Arvanitoyannis, E. Eleftheriades, and E. Tsotsaroni, *Chemosphere* **18**(9), 1707 (1989).
39. G. McKay, G. Ramprasad, and P. O. Mowli, *Water Air Soil Poll.* **29**(3), 273 (1986).
40. S. K. Khare, R. M. Srivastova, K. K. Panday, and V. N. Singh, *Env. Tech. Lett.* **9**(10), 1163 (1988).
41. T. I. Denisova, S. I. Meleshevich, and I. A. Sheka, *Zur. Prik. Khim.* **62**(5), 1182 (1989).
42. G. McKay, M. S. Otterburn, and J. A. Aga, *Water Air Soil Poll.* **24**(3), 307 (1985); **33**(3/4), 419 (1987); **36**(3/4), 381 (1987).
43. C. A. Rodman, *Am. Dyest. Rep.* **60,** 45 (1971).
44. R. M. Woodby and D. L. Michelsen, *Emerging Technologies for Hazardous Waste Treatment Symposium*, I & EC Division of ACS, Atlantic City, N.J., June 4–7, 1990.
45. M. Nakaoka, S. Tamura, Y. Maeda, and T. Azumi, *Sen-I Gakkaishi* **39**(2), 69 (1983).
46. M. Dohanyas, V. Madera, and M. Sedlacek, *Prog. Water Tech.* **10**(5/6), 559 (1978).
47. G. M. Shaul, R. J. Lieberman, C. R. Dempsey, and K. A. Dostal, "Fate of Azo Dyes in the Activated Sludge Process," presented at the 41st Annual Purdue Industrial Waste Conference, Purdue University, West Lafayette, Ind., May 13–15, 1986.
48. G. M. Shaul, R. J. Lieberman, C. R. Dempsey, and K. A. Dostol, "Treatability of Water Soluble Azo Dyes by the Activated Sludge Process," *Industrial Wastes Symposia Proceedings, 59th Water Pollution Control Federation Annual Conference*, Los Angeles, Calif., Oct. 5–9, 1986.
49. G. M. Shaul, C. R. Dempsey, and K. A. Dostol, *Fate of Water Soluble Azo Dyes in the Activated Sludge Process*, EPA/600/2-88/030, report of the Water Engineering Research Laboratory, Office of Research and Development, USEPA, Cincinnati, Ohio, July, 1988.
50. W. J. Katz and A. Geinopolis, *Chem. Eng.* **75**(10), 78 (1968).
51. P. T. Shannon and E. M. Tory, *Ind. Eng. Chem.* **57**(2), 19 (1965).
52. G. A. Rohlick, *Sewage Ind. Wastes* **26,** 1056 (1954).
53. R. E. Sessler, *Sewage Ind. Wastes* **27,** 1178 (1955).
54. H. E. Hudson, Jr. *J.A.W.W.A.* **57,** 885 (1965).
55. T. Crowe, C. R. O'Melia, and L. Little, *Am. Dyest. Rep.* **67**(12), 52 (1978).
56. U.S. Pat. 3,947,248 (Mar. 30, 1976), J. B. Powers (to The Dow Chemical Company).
57. D. J. Shaw, *Introduction of Colloid and Surface Chemistry*, Butterworths, London, 1968.
58. D. L. Y. Davis, *TAPPI* **52**(11), 2132 (1969).
59. M. Gould, *Chem. Eng.* **78**(2), 55 (1971).
60. Environmental Protection Agency, *12040 ENC 12/71*, Water Pollution Control Research Service, Interstate Paper Corp., 1971.

61. C. Oehme and F. Martinola, *Chem. Ind.* **17,** 823 (1973).

62. Y. H. Oswalt and Y. G. Land, Jr., *EPA-R2-73-086*, technical paper, Environmental Protection Agency, Washington, D.C., 1973.

63. T. A. Sihorwala, *Ind. J. Env. Prot.* **9**(7), 513 (1989).

64. V. I. Ostrovka, V. A. Livke, and N. V. Boldyreva, *Khim. Tekhnol. Vody.* **11**(3), 278 (1989).

65. V. A. Livke, V. I. Ostrovka, and N. P. Gendruseva, *Khim. Tekhnol. Vody.* **11**(2), 185 (1989).

66. G. Gilli, G. Alessi, G. Bellazzi, and M. Pieroni, *Inquinamento* **29**(10), 51 (1987).

67. G. R. Brower and G. D. Reed, *41st Proc. Ind. Waste Conf.*, 612 (1986).

68. G. R. Brower and G. D. Reed, *18th Toxic Hazard Wastes. Proc., Mid-Atl. Waste Conf.*, 340 (1986).

69. G. Song, D. Cao, and Y. Wang, *Huanjing Huaxue* **5**(4), 25 (1986).

70. R. Ya Sid'ko, B. K. Kerzhner, and M. A. Shevchenko, *Khim. Teknol. Vody.* **8**(3), 50 (1986).

71. Z. Inoue, I. Fukunaga, and A. Honda, *Mizu Shori Gijutsu* **27**(7), 457 (1986).

72. B. Yang, X. Qu, and G. Jin, *Huanjing Kexue* **7**(3), 38 (1986).

73. S. N. Suslov and S. S. Timofeeva, *Otkrytiya, Izobret* **2,** 94 (1985).

74. R. J. T. Hung, *Huan Ching Pao Hu*, **7**(2), 11 (1984).

75. Z. Inoue and A. Honda, *Mizie Shori Gijutsu* **25**(3), 169 (1984).

76. N. A. Klimenko, N. F. Lozovskaya, and V. A. Kozhjanov, *Khim. Tekhnol. Vody.* **2**(5), 450 (1980).

77. G. M. Davis, J. H. Koon, and C. E. Adams, *Proc. Ind. Waste Conf.* **32,** 981 (1977).

78. L. A. Kulskii, Z. N. Shkavro, M. I. Medvedev, and O. S. Zul'figarov, *Khim. Tekhnol. Vody.* **6**(3), 268 (1984).

79. S. Kagawa, K. Senba, A. Asano, and S. Shigeyoski, *Ehime-ken Kogyo Shikenjo Kenkyu Hokoku* **18,** 92 (1980).

80. V. A. Kozhanov and co-workers, *Sov. J. Water Chem. Tech.* **7**(3), 74 (1985).

81. R. K. Pandit and M. S. Mayadeo, *Ind. Text. J.* **95**(10), 71 (1985).

82. K. Fytianos, G. Vasilikiotis, T. Agelidis, P. Kofos, and K. Stefanidis, *Chemosphere* **14**(5), 411 (1985).

83. S. Koscielski and O. M. Spivakova, *Przeglad Wlokienniczy* **36**(7), 367 (1982).

84. L. E. Frumin, N. V. Beresneva, G. E. Krichevskii, and R. E. Sorotyuk, *Sov. J. Water Chem. Tech.* **10**(2), 100 (1988).

85. L. A. Kul'skii, A. A. Ivanyuk, and E. I. Potapova, *Sov. J. Water Chem. Tech.* **8**(6), 19 (1986).

86. G. McKay, *Am. Dye Dept.* **68,** 29 (1979).

87. J. F. Kreissl and J. J. Westrick, "Municipal Waste Treatment by Physical–Chemical Methods," *Prog. Water Technol.* **1,** 1 (1972).

88. D. L. Michelsen and T. Fansler, U.S. Clearinghouse of Federal Science Technological Information, PB Report, Blacksburg, Va., 1970, No. 196313.

89. C. H. Gooding, *Chem. Eng.* **92**(1), 56 (Jan. 7, 1985).

90. H. Moes, *Chem. Proc.* **50**(2), 35 (Feb. 1987).

91. J. McCallum, *Chem. Proc.* **50**(13), 23 (Nov. 1987).

92. S. Sourirajan and T. Matsuura, *Env. Sci. Tech.* **13**(8), 946 (1979).

93. U.S. Pat. 4,200,526 (Apr. 29, 1980), E. W. Johnson, M. J. Reider, and R. Anewalt (to Gerber Products).

94. P. Mappelli, M. Santori, A. Chiolle, and G. Gianotti, *Desalination* **24,** 155 (1978).

95. P. J. Halliday and S. Beszedits, *Can. Text. J.* **103**(4), 78 (1986).

96. P. S. Cartwright, *Chem. Proc.* **53**(7), 23 (July 1990).

97. J. Haggin, *C & EN*, (June 6, 1988).

98. H. P. Eder, *Am. Dyest. Rep.* **62**(10), 63 (1973).

99. C. A. Brandon, *Ind. Water Eng.* **12**(6), 14 (1975).
100. C. A. Brandon and J. L. Gladdes, *Tex. World* **128,** 68 (1978); *Desalination* **23,** 19 (1977).
101. H. G. Spencer, C. A. Brandon, G. Goodman, J. J. Porter, and M. Samfield, *Chem. Water Reuse.* **1,** 325 (1981).
102. C. A. Brandon, J. S. Johnson, R. E. Minturn, and J. J. Porter, *Text. Chem. Color.* **5,** 134 (1973).
103. C. A. Brandon and J. J. Porter, *Chemtech,* 402 (June 1976).
104. A. M. El-Nashar, *Desalination* **20,** 267 (1977).
105. M. Jhawar and J. H. Sleigh, *Text. Ind.* **139**(1), 60 (1975).
106. DE 3,116,942 (Nov. 25, 1982), S. Peter, R. Stefan, and T. Peters (to Van Opbergen).
107. DE 3,035,134 (Sept. 19, 1979), C. Linder and M. Perry (to Aligena A.G.).
108. K. Majewska and T. Winnicki, *Desalination* **60**(1), 59 (1986).
109. U.S. Pat. 4,758,347 (July 19, 1988), A. Henz and H. Pfenninger (to CIBA-GEIGY).
110. P. Rivet, *Eau. Indust. Nuisances.* **121,** 48 (1988).
111. D. Lemordant, P. Letellier, M. Rumeau, and C. Soma, *Text. Month.,* 41 (Nov. 1988).
112. D. Kuiper, C. A. Brandon, H. G. Spencer, and G. L. Gaddys, *Textil. Praxis Int.* **40**(10), 1124 (1985).
113. C. A. Brandon, *EPA Report No. 600/2-84/147,* Environmental Protection Agency, Washington, D.C., 1984.
114. G. R. Groves, C. A. Buckley, J. M. Cox, A. Kirk, and M. J. Simpson, Chemsa **10**(4), 350 (1984).
115. A. Fuchs, B. R. Breslan, and A. J. Toompas, *Text. Ind.* **147**(12), 44 (1983).
116. J. L. Short, *Chem. Eng.* **395**(47), 50 (1983).
117. C. A. Brandon, *Text. Ind. Dyegest S. Afr.* **2**(5), 2 (1983).
118. L. Tinghui, T. Matsurra, and S. Sourirajan, *Ind. Eng. Chem. Prod. Res. Dev.* **22**(1), 77 (1983).
119. J. S. Zuk, M. Rucka, and J. Rak, *Chem. Eng. Commun.* **19**(1–3), 67 (1982).
120. J. J. Porter, *Text. Chem. Color.* **22**(6), 21 (1990).
121. T. A. Alspaugh, *Text. Chem. Color.* **5,** 255 (1973).
122. F. N. Case, E. E. Ketchen, and T. A. A. Alspaugh, *Text. Chem. Color.* **5,** 62 (1973).
123. E. E. Ketchen, and F. N. Case, *Proceedings of the AATCC Symposium,* the Textile Industry and the Environment, Research Triangle Park, N.C., 1973, p. 136.
124. I. Perkowski, J. Rouba, and L. Kos, *Przeglad Wlokienniczy* **40**(4), 148 (1986); **36**(5), 250 (1982).
125. H. Tsuruta, "Science for a Better Environment", *Proceedings of the International Congress of Science Hum. Environment,* Pergaman Press, New York, 1975, p. 659.
126. S. Iwai, T. Kitao, and M. Urabe, *Kyoto Daigaku Genshi Enerugi Kenkyusho Iho* **47,** 48 (1975).
127. H. Yamada, Y. Masumoto, M. Sawatari, S. Inoue, and H. Nishikawa *Sen'i Kako* **27**(1), 24 (1975).
128. D.D. 105190, (Apr 12, 1974), H. S. Zuelke, W. Kuhring, and R. Doerfeld.
129. W. Koerbel, *Aufbereit-Tech.* **10**(5), 223 (1969).
130. Jpn. Pat. 540044353 (Apr. 7, 1979), K. Yamaguchi, S. Okataka, Y. Mochizuki, and M. Shimizu (to Gunze).
131. J. David, *Industrie Tex.* **1172,** 1167 (1986).
132. Technical data, *Liquid Carbonic,* Carbon Dioxide Corp., Chicago, Ill., 1992.
133. Technical data, Airco Industrial Gases, Murray Hill, N.J., 1992.
134. I. Torok, *Industria Usoara, Textile, Tricotaje, Confectii Textile* **35**(2), 81 (1984).
135. C. R. Wasmuth, R. L. Donnell, C. E. Harding, and G. E. Shankle, *J. Soc. Dye. Color* **81,** 403 (1965).
136. W. F. Beech, *Fiber Reactive Dyes,* Chapt. X, 1970, pp. 311–334.

137. H. E. Fierz-David and H. E. Matter, *J. Soc. Dyer Color.* **53,** 424 (1937).

138. M. Kolb, P. Korger, and B. Funke, *Meilliand Tex.* **69**(4), 286 (1988).

139. R. W. Barton, *TAPPI* **45**(2), 178A (1962).

140. Jpn. Pat. 52,116,644 (Sept. 30, 1977), Y. Tagashira, H. Takagi, and S. Hayashi (to Asahi Chem.).

141. M. Weiss, *Am. Dyest. Rep.* **67,** 35 (1978); **67,** 72 (1978).

142. L. C. Ellis, *AATCC Nat. Tech. Conf.,* 266 (1981).

143. Technical data, Degussa Corp., Allendale, N.J., 1992.

144. M. Cheek, *Hues & Views,* Vol. 1, Dyestuffs & Chemicals Division, CIBA-GEIGY Corp., Greensboro, N.C., 1991, p. 6.

145. M. M. Cook and co-workers, "Sodium Borohydride Reductions—Novel Approaches to Decolorization and Metals Removal in Dye Manufacturing and Textile Effluent Applications," *203rd National Meeting of the American Chemical Society,* San Francisco, Apr. 5–10, 1992.

146. J. A. Ulman and M. M. Cook, Sodium Borohydride Reductions—"Removal of Color and Metals From Paper Dyes," *204th National Meeting of the American Chemical Society,* Washington, D.C., Aug. 23–28, 1992.

147. M. Matsui, H. Nakabayashi, H. Shibaba, and Y. Takase, *Sen-I Gakkaiski* **39**(3), 1133 (1983).

148. S. V. Zima, A. I. Pivtorak, and L. I. Tkachenko, *Tekstil'Naya Promyshlennost* **46**(6), 49 (1986).

149. J. M. Green and C. Sokol, *Am. Dyest. Rep.* **74**(4), 50 (1985).

150. A. I. El Duwani, G. Hafez, and S. Hawash, *Chemie-Ingenieur-Technik* **59**(8), 653 (1987); **59**(8), 654 (1987).

151. B. G. Nazarov and E. P. Fazullina, *Tekstil'Naya Promyshelennost* **44**(6), 53 (1984); *Sov. J. Water Chem.* **7**(1), 46 (1985).

152. J. Fidrysiak and J. Przbinski, *Przeglad Wlokienniczy* **42**(12), 533 (1988); **38**(7), 224 (1989).

153. J. P. Gould and K. A. Groff, *Ozone Sci. Eng.* **9**(2), 153 (1987).

154. M. Kolb, B. Muller, and B. Funke, *Vom Wasser* **69,** 217 (1987).

155. B. K. Kerzhner, P. N. Taran, and M. A. Shevchenko, *Sov. J. Water Tech.* **7**(4), 18 (1985).

156. M. Matsui, T. Kimura, T. Nambu, and K. Shibata, *J. Soc. Dye Color* **100,** 125 (1984).

157. M. Matsui and T. Kitao, *J. Soc. Dye Color* **101,** 334 (1985).

158. M. Matsui, K. Kobayaski, T. Shibata, and Y. Takase, *J. Soc. Dye Color* **97,** 210 (1981).

159. M. Matsui, H. Hibino, K. Shibata, and Y. Takase, *Am. Dyest. Rep.* **72**(1), 35 (1983); **72**(10), 40 (1983).

160. M. Matsui, K. Shibata, and Y. Takase, *Dyes & Pigm.* **5,** 321 (1984).

161. Y. Onari, *Bull. Chem. Soc. Jpn.* **58**(9), 2526 (1985); *Nippon Kagaku Kaishi,* **1983**(10), 1526; **1978**(11), 1570.

162. J. M. Green and C. Sokol, *Am. Dyest. Rep.* **74.**(4), 50 (1985).

163. A. G. Cox, *Text. Ind.* **147**(7), 26 (1983).

164. J. G. Ginocchio, H. Bischofberger, and A. Gmunder, *Int. Tex. Bull.* **1984**(1), 45.

165. J. Perkowski, L. Kos, and D. Dziag, *Przeglad Wlokienniczy* **41**(8), 311 (1987).

166. Jpn. Kokai 74101264 (Sept. 25, 1974), Y. Nakahira (to Kanebo).

167. Jpn. Kokai 7498055 (Sept. 17, 1974), Y. Kuji, T. Kato, and S. Matsumaru (to Mitsubishi).

168. A. Netzer and H. K. Miyamoto, *Proc. Ind. Waste Conf.* **30,** 804 (1975).

169. L. I. Yakovleva and L. S. Solodova, *Mater. Vses. Nauchn. Simp. Sovren. Probl. Sam. Reg. Kach. Vody. 5th* **6,** 102 (1975).

170. S. Beszedits, *Am. Dyest. Rep.* **69**(8), 37 (1980).

171. C. G. Hewes and R. R. Davison, *A. I. Chem. Eng. J.* **17,** 141 (1971).

172. S. Tong and B. Young, *Water Poll. Cont. Fed.* **73,** 584 (1974).

173. P. N. Endyuskin, *Zhur. Priklad. Khim.* **52,** 815 (1979); *Khim. Prom-ST* **84,** 467, (1979); 9 (1980); 341 (1982).

174. J. R. Mills, *Water Resources Research Institute Bulletin* No. 46, *Water Conservation Technical Text—State of the Art,* Auburn University, Alabama, May 1982, p. 17.

175. W. S. Perkins, J. J. Judkins, and W. D. Perry, *Text. Chem. Col.* **12**(8), 182 (1980); **12**(9), 221 (1980); **12**(10), 262 (1980).

176. H. J. H. Fenton, *J. Chem. Soc.* **65,** 899 (1894).

177. H. J. H. Fenton and H. O. Jones, *J. Chem. Soc.* **77,** 69 (1900).

178. K. Namba and S. Nakayama, *Bull. Chem. Soc. Jpn.* **55,** 3339 (1982).

179. F. Haber and J. Weiss, *Proc. Roy. Soc.* **A147,** 337 (1934).

180. E. J. Keating, *Ind. Water Eng.* **15,** 22 (1978).

181. Jpn. Kokai, 7805853 (Jan. 19, 1978), H. Inoue and M. Takai (to Gosei Chem. Co.).

182. Jpn. Kokai, 54044352 (Apr. 7, 1979), K. Yamada, S. Okataka, and M. Shimizu (to Gunze).

183. T. Kitao and R. Yahashi, *Mizu Shori Gijutsu* **17**(8), 735 (1976).

184. Ger. Offen. 3,032,831 (Mar. 26, 1981), D. Arditti (to Ugine Kuhlmann).

185. V. V. Zenkov, A. I. Rodionov, and I. V. Pestretsova, *Tr.-Mosk Khim-Tekhnol Inst. im. D. I. Mendeleeva* **109,** 78 (1979).

186. M. Hashimoto and Y. Yoshida, *Sen'i Kako* **32**(2), (1980).

187. F. Burzio, *Welt-Tensid-Kongress,* Vol. 147, Munich, May 6, 1984, Tab. 10, Fig. 2, Ref. 6.

188. D. W. Sundstrom, H. E. Klei, T. A. Nalette, D. J. Reidy, and B. A. Wier *Hazard Waste Mat.* **3**(1), 101 (1986).

189. W. H. Glaze, J. W. Kang, and D. H. Chapin, *Ozone Sci. Tech.* **9,** 335 (1987).

190. M. B. Borup, *Nat. Sci. Tech.* **19,** 381 (1987).

191. G. R. Peyton and W. H. Glaze, *Env. Sci. Tech.* **22,** 761 (1988).

192. J. D. Zeff, *AICh.E. Sym. Ser.* **73**(167), 206 (1977).

193. H. W. Prengle and C. E. Mauk, *AICh. E. Sym. Ser.* **74**(178), 228 (1978).

194. G. R. Peyton, F. Y. Huang, J. L. Burleson, and W. H. Glaze, *Env. Sci. Tech.* **16,** 448 (1982).

195. K. A. Roy, *Hazmat World,* 26 (Mar. 1990); 82 (May 1990); 35 (June 1990); 30 (July 1990).

196. M. J. Dietrich, T. L. Randall, and P. J. Canney, *Env. Prog.* **4**(3), 171 (1985).

197. A. R. Wilhehmi and P. V. Knopp, *Chem. Eng. Prog.* 46 (Aug. 1979).

198. F. J. Zimmermann and D. C. Diddams, *TAPPI* **43**(8), 710 (1960).

199. G. H. Teletzki, *Chem. Eng. Prog.* 33 (Jan. 1964).

200. E. Hurwitz and G. H. Teletzki, *Water Sew. Works* **112,** 298 (1965).

201. G. H. Teletzke, W. B. Gitchel, D. G. Diddams, and C. A. Hoffman, *J. Wat. Poll. Cont. Fed.* **39**(6), 994 (1967).

202. T. L. Randal and P. V. Knopp, *J. Water Poll. Cont. Fed.* **52**(8), 2117 (1980).

203. J. N. Foussard, H. Debellefontaine, and J. Besombesvailne, *J. Env. Eng.* **115**(2), 367 (1989).

204. A. G. Zakhorzhevskaya, *Vodosnabzh Kanoliz. Gidrotekh. Sooruzh, Mezhved. Resp. Nauch. Sb.* **1,** 37 (1966).

205. T. A. Kharlamova and N. I. Mitashova, *Khim. Prom-st.* **4,** 206 (1986).

206. I. G. Krasnoborod'ko, *Tekstil'Naya Promyshlennosti* **46**(4), 58 (1986).

207. I. G. Krasnoborod'ko and M. T. Nikiforov, *Tekstil'Naya Promphlennosti* **44** (4), 56 (1984).

208. V. I. Popova, L. I. Safronova, G. E. Krichevskeii, and A. V. Chesnokov, *Tekhnologiya Tekstil' No. I. Promyshlennosti,* **1**(169), 61 (1986); **3**(153), 63 (1983).

209. U.S. Pat. 3,485,729 (Dec. 23, 1969), G. Hertz (to Crompton Knowles).

210. EP 9150413 (Apr. 14, 1982), V. A. Van Hertum (to Morton Thiokol).

211. A. E. Wilcock and S. P. Hay, *Can. Tex. J.* **108**(5), 31 (1991).

212. W. C. Tincher, *Text. Chem. Color.* **21**(12), 33 (1989).

213. W. C. Tincher, M. Weinberg, and S. Stephens, *AATCC Symposium on Safety, Health and Environmental Technology*, Charlotte, N.C., 1988.

214. T. R. Demmin and K. D. Uhrich, *Am. Dyest. Rep.* **77**(6), 13 (1988).

215. R. A. Montanaro and H. A. Noble, *Text. Ind. Env.* **1973**(AATCC), 129.

216. T. V. Arden, *Chem. Brit.* **12**, 285 (1976).

217. E. L. Jones, T. C. Alspaugh, and H. B. Stokes, *J. Water Poll. Cont. Fed.* **34**(5), 495 (1962).

218. D. Evers, *Int. Dyers Text. Print.* **143**(1), 56 (1970).

219. D. B. Purchas, *Chem. Ind.*, 53 (Jan. 1974).

220. M. R. Purvis, *Am. Dyest. Rep.* **63**(8), 19 (1974).

221. G. Rhame, *Am. Dyest. Rep.* **60**(11), 46 (1971).

222. B. Morgeli, *Textilbetrieb* **94**(3), 54 (1976).

223. M. L. Shelley, C. W. Randall, and P. H. King, *J. Water Poll. Cont. Fed.* **48**, 753 (1976).

224. L. E. Shriver and R. R. Daugue, *Am. Dyest. Rep.* **67**(3), 34 (1978).

225. T. A. Alspaugh, *Tect. Chem. Color.* **5**, 255 (1973).

226. J. W. Masselli, N. W. Masselli, and M. G. Burford, *Text. Ind.* **135**(10), 84 (1971).

227. J. J. Porter, and E. H. Snider, *J. Water Poll. Cont. Fed.* **48**, 2198 (1976).

228. D. W. Weeter and A. G. Hodgson, *Am. Dyest. Rep.* **66**(8), 32 (1977).

229. J. J. Porter, *Water Wastes Eng.*, A8 (Jan. 1972).

230. G. Grulich, D. G. Hutton, F. L. Robertaccio, and A. L. Glotzer, *AICHE Symp. Ser.* **129**(69), 127 (1972).

231. D. G. Hutton and F. L. Robertaccio, *Symposium Proceedings: Textile Industry Technology*, Williamsburg, Va., Dec. 5–8, 1978, PB 299132.

232. A. D. Adams, *Am. Dyest. Rep.* **65**(4), 32 (1976).

233. A. B. Scaramelli and F. A. DiGiano, *Water Sew. Works*, 90 (Sept. 1973).

234. T. M. Keinath, *Technology Evaluation for Priority Pollutant Removal from Dyestuff Manufacture Wastewaters*, USEPA report 600/S2-84-055, Washington, D.C., Apr. 1984.

235. B. Dobinsky and C. P. Wickersham, *Chem. Proc.*, 173 (Nov. 1987).

236. Status update, *Improved Wastewater Treatment Facility*, CIBA-GEIGY Corp., Toms River, N.J., 1993.

237. *ADMI Dyes and the Environment*, Vol. 1., American Dye Manufacturers' Institute, New York, 1971.

238. R. H. Horning, *Text. Chem. Color.* **4**, 275 (1972).

239. *Methods for Chemical Analysis of Water and Wastes*, U.S. EPA, Office of Technology Transfer, Washington, D.C., 1974.

240. I. I. Manulyak, V. M. Udod, L. I. Nesynoya, and V. N. Kravchuk, *Khim. Teknol. Vody.* **9**(2), 167 (1987).

241. J. P. Straley, *Am. Dyest. Rep.* **73**(9), 46 (1984).

242. E. Idaka, T. Ogawa, H. Horitsu, and M. Tomoyeda, *J. Soc. Dyer. Color.* **94**(3), 91 (1978).

243. C. Yatome, T. Ogawa, D. Koga, and E. Idaka, *J. Soc. Dyer. Color.* **97**, 167 (1981).

244. T. Ogawa, C. Yatome, and E. Idaka, *J. Soc. Dyer. Color.* **97**, 435 (1981).

245. T. Ogawa, C. Yatome, and E. Idaka, *J. Soc. Dyer. Color.* **102**, 12 (1986).

246. S. V. Kulkarni, C. D. Blackwell, A. L. Blackard, C. W. Stackhouse, and M. W. Alexander, *Textile Dyes and Dyeing Equipment: Classification, Properties, and Environmental Aspects*; EPA 600/2-85/010, U.S. EPA, Research Triangle Park, N.C., 1985.

247. E. A. Clarke and R. Anliker, *Rev. Prog. Color.* **14**, 84 (1984).

248. G. B. Michaels and D. L. Lewis, *Env. Toxicol. Chem.* **4**, 45 (1985).

249. K. T. Chung, G. E. Fulk, and M. Egan, *Appl. Env. Microbiol.* **35**, 558 (1978).
250. C. T. Helmes and co-workers, *J. Env. Sci. Health* **A19**(2), 97 (1984).
251. C. C. Sigman and co-workers, *J. Env. Sci. Health* **A20**(4), 427 (1985).
252. R. A. Levine, W. L. Oller, C. R. Nony, and M. C. Bowman, *J. Anal. Toxicol.* **6**, 157 (1982).
253. C. R. Nony, M. C. Bowman, T. Cairns, L. K. Lowry, and W. P. Tolos, *J. Anal. Toxicol.* **4**, 132 (1980).
254. G. L. Baughman and T. A. Perenich, *Am. Dyest. Rep.* **77**(2), 19 (1988).
255. H. D. Pratt, Jr., Ph.D. dissertation, *A Study of the Degradation of Some Azo Disperse Dyes in Waste Disposal Systems*, Georgia Institute of Technology, Atlanta, Sept. 1968.
256. E. J. Weber, *Studies of Benzidine Based Dyes in Sediment-Water Systems*, U.S. EPA Environmental Research Laboratory, Athens, Ga., Jan. 1990.
257. E. J. Weber, P. E. Sturrock, and S. R. Camp, *Reactive Dyes in the Aquatic Environment: A Case Study of Reactive Blue 19*, Report EPA 600/M-90/009, Washington, D.C., 1990.
258. D. Brown and P. Laboureur, *Chemosphere* **12**(3), 397 (1983).
259. U. Pagga and D. Brown, *Chemosphere* **15**(4), 479 (1986).
260. D. Brown and B. Hamburger, *Chemosphere* **16**(7), 1539 (1987).
261. *Nineteenth Report of the TSCA Interagency Testing Committee to the Administrator*, U.S. EPA, Washington, D.C., Nov. 1986, pp. 70–71.
262. E. J. Weber, *Fate of Textile Dyes in the Aquatic Environment: Degradation of Disperse Blue 79 in Anaerobic Sediment-Water Systems*, Environmental Research Laboratory, U.S. EPA, Athens, Ga., Mar. 1988.
263. C. T. Helmes and co-workers, *J. Env. Sci. Health* **A17**(1), 75 (1982).
264. N. I. Sax, *Dangerous Properties of Industrial Materials*, 6th ed., Van Nostrand Reinhold, New York, p. 1209.
265. *Registry of Toxic Effects of Chemical Substances*, 1983 Suppl., NIOSH, Washington, D.C., 1983, pp. 263–264.
266. *Mut. Res.* **44**, 9 (1977); **67**, 1 (1979).
267. *Proceedings of the 11th Congress of Occupational Health, Chemistry, and Industry*, 1983, pp. 497–405.
268. *J. Nat. Cancer Inst.* **64**(3), 665 (1980).
269. *Chem. Week*, 12, 15 (Sept. 3, 1986).
270. D. A. Gardner, T. J. Holdsworth, G. M. Shaul, K. A. Dostol, and L. D. Betowski, *Aerobic and Anaerobic Treatment of C. I. Disperse Blue 79 Report*, EPA 600/S2-89/051, Washington, D.C., Jan. 1990.
271. K. Venkataraman, ed., *The Analytical Chemistry of Synthetic Dyes, Chemistry of Synthetic Dyes*, John Wiley & Sons, Inc., New York, 1977.
272. W. C. Tincher and J. R. Robertson, *Text. Chem. Color.* **11**(12), 269 (1982).
273. W. C. Tincher, *Survey of the Coosa Basin for Organic Contaminants for Carpet Processing*, No. E-27-630, Environmental Protection Division, Department of Natural Resources, Georgia, Oct. 1978.
274. J. R. Robertson, Jr., Ph.D. dissertation, *The Products of Biodegradation of Selected Carpet Dyes and Dyeing Auxiliaries*, Georgia Institute of Technology, Atlanta, Dec. 1973.
275. D. N. Dickson, Ph.D. dissertation, *The Effectiveness of Biodegradation in the Removal of Acid Dyes and Toxicity from Carpet Dyeing Waste Water*, Georgia Institute of Technology, Atlanta, Mar. 1981.
276. L. M. Games and R. A. Hites, *Anal. Chem.* **49**(80, 1433 (1977).
277. L. D. Betowski and J. M. Bullard, *Anal. Chem.* **56**, 2607 (1984).
278. J. M. Bullard and L. D. Betowski, *Org. Mass Spectrom.* **21**, 575 (1986).

279. L. D. Betowski, S. M. Pyle, J. M. Bullard, and G. M. Shaul, *Biomed. Env. Mass Spectrom.* **14,** 343 (1987).

280. R. D. Voyksner, *Anal. Chem.* **57,** 2600 (1985).

281. J. Yinon, T. L. Jones, and L. D. Betowski, *J. Chromatog.* **482,** 75 (1989).

282. M. Barber, R. S. Bordoli, R. D. Sedgewick, and A. N. Tyler, *Nature* **293** *210 (1981).*

283. J. J. Monaghan, M. Barber, R. S. Bordoli, R. D. Sedgwick, and A. N. Tyler, *Org. Mass Sectrom.* **17,** 569 (1982).

284. J. J. Monaghan, M. Barber, R. S. Bordoli, R. D. Sedgewick, and A. N. Tyler, *Int. J. Mass Spectrom. Ion Phys.* **46,** 447 (1983).

285. H. S. Freeman and co-workers, *Text. Chem. Color.* **22**(5), 23 (1990).

286. C. Shimanskas, K. Ng, and J. Karliner, *Rapid Comm. Mass Spectrom.* **3**(9), 300 (1989).

287. A. P. Bruins, L. O. G. Weidolf, and J. D. Henion, *Anal. Chem.* **59,** 2647 (1987).

288. P. O. Edlund, E. D. Lee, J. D. Henion, and W. L. Budde, *Biomed. Env. Mass Spectrom.* **18,** 233 (1989).

289. M. A. McLean and R. B. Freas, *Anal. Chem.* **61,** 2054 (1989).

290. D. A. Flory, M. W. McLean, M. L. Vestal, and L. D. Betowski, *Rapid Comm. Mass Spectrom.* **1**(3), 48 (1987).

291. J. Yinon, T. L. Jones, and L. D. Betowski, *Biomed. Env. Mass Spectrom.* **18,** 445 (1989).

292. L. D. Betowski, and T. L. Jones, *The Application of High Performance Liquid Chromatography/Mass Spectrometry to Environmental Analysis,* Report EPa 600/4-89/033, Washington, D.C., 1989.

293. J. Yinon and L. D. Betowski, *Analysis of Dyes by LC/MS Using Thermospray Electron Impact and Repeller,* CAD Ionization U.S. EPA Grant No. CR-8158425-01-2, Weizmann Institute of Science, Rehovot, Israel, 1989.

294. "Pollution Prevention Strategy," *Fed. Reg.* (Feb. 26, 1991).

295. *The New Jersey Asbury Park Press,* A12 (June 21, 1991).

296. L. R. Ember, *Strategies for Reducing Pollution at the Source are Gaining Ground,* C&E News, July 8, 1991, p. 7.

297. R. Gager, *Hazmat World,* Sept. 1991, p. 42.

298. H. J. Goldner, *R & D Magazine,* Sept. 1991, p. 48.

299. R. L. Berglund and C. T. Lawson, *Chem. Eng.* Sept. 1991, p. 120.

300. G. V. Cox, *Hazardous Waste: Detection, Control, Treatment,* Elsevier Science Publishers, Amsterdam, The Netherlands, 1988, p. 353.

301. E. Doughterty, *R & D Magazine,* Apr. 1990, p. 62.

302. *Chem. Week,* 22 (Aug. 19, 1987).

303. F. C. Cook, *Text. World* **141,** 84 (May 1991).

304. *Pollution Prevention News,* Office of Pollution Prevention, USEPA, Washington, D.C., Apr. 1990, p. 2; *Pollution Prevention Guidance Manual for the Dye Manufacturing Industry,* USEPA/ETAD, Washington, D.C., 1992.

305. *Cutting Chemical Wastes—What 29 Organic Chemical Plants Are Doing to Reduce Hazardous Wastes,* Inform, Inc., New York, 1985.

306. G. M. Elgal, *Text. Chem. Color.* **18**(5), 15 (1986).

307. U.S. Pat. 4,200,526 (Apr. 29, 1980) R. Anewalt, E. W. Johnson, H. J. King, and M. J. Reider (to Gerber Prod.).

308. B. Lieberherr and W. Beck, *Melliand Testilberichte* **69**(8), 572 (1988).

309. *Chemiefsern/Testilindustrie* **36/88**(11), 888 (1986).

310. V. I. Popova, N. Kravchenko, E. V. Muravena, and G. E. Krichevskii, *Tekh. Texstil Prom.* **4**(166), 67 (1985).

311. D. Fiebig and G. Schulz, *Lenzinger Berichte* **58,** 109 (1985).

312. B. Smith, "Identification and Reduction of Pollution Sources on Textile West Processing," *A Workbook for Pollution Prevention by Source Reduction in Textile Wet Processing*, Pollution Prevention Program, North Carolina Department of Environment, Health, and Natural Resources, Raleigh, N.C., 1986, 1988.

313. E. L. Barnhardt, *Symposium proceedings of the Textile Industry of Technology*, Williamsburg, Va., Dec. 1978, p. 17.

314. *Colour Index*, Vol. 3 CI No. 62055, Society of Dyers and Colorists, Bradford, Yorkshire, UK, 1957, p. 3502.

315. Ref. 314, Suppl., CI No. 40300.

316. U.S. Pat. 2,029,591 (Feb. 4, 1936), H. Schindhelm and C. T. Schults (to General Aniline).

317. Ref. 314, Vol. 4 CI No. 24401, 1971, p. 4209.

318. Ref. 314, Suppl., CI No. 18157, p. 4120.

319. Ref. 314, Vol. 3 CI No. 15711, 1957, p. 3076.

320. A. A. Mamontova, *Khim. Tekhnol. Vody.* **12**(8), 738 (1990).

321. J. He and B. Wang, *Huanjing Wuran Yu Fangzhi* **12**(5), 18 (1990).

322. E. Sindelarova, J. Piskor, J. Vesely, and R. Kocian. CS 196814 (Mar. 31, 1982).

323. J. Urbas, *Przegl. Skorzany* **34**(8), 267 (1979).

324. H. A. Fiegenbaum, *Am. Dyest. Rep.* **67**(3), 43, 46 (1978); *Ind. Wastes* **23**(2), 32 (1977).

325. *Am. Dyest. Rep.* **61**(8), 57 (1972).

326. Fr Demande 2619727 (Mar. 3, 1989), D. Lemordant, P. Letellier, M. Rumeau, and C. Soma (to Universite Pierre et Marie Curie).

327. J. L. Gaddis and H. G. Spenser, report EPA/600/2-79/118, Order No. PB 80-113889, Environmental Protection Agency, Washington, D.C., 1979.

328. U.S. Pat. 4,005,011, (Jan. 25, 1977), C. D. Sweeney.

329. C. A. Pitkat and C. L. Berndt, *Proc. Ind. Waste Conf.* **35**, 178 (1981).

330. K. D. Uhrich and T. R. Demmin, *Book Papers*, AATCC International Conference Exhibition, 97 (1988).

331. F. Petrini and D. Grechi, *Ing. Ambientale* **13**(6), 308 (1984).

332. U.S. Pat. 3,778,368 (Dec. 11, 1973), Y. Nakamura, A. Morioka, and Y. Itsuyo (to Sankyokasel).

333. S. R. Shukla and V. D. Sakhardande, *Am. Dyest. Rep.* **80**(7), 38 (1991).

334. L. Rehn, *Arch Klin. Chir.* **50**, 588 (1895).

335. W. C. Hueper, *Cancer Res.* **21**, 842 (1961); *Occupational and Environmental Cancer of the Urinary System*, Yale University Press, New Haven, Conn., 1969, p. 118.

336. M. W. Goldblatt, *Br. J. Ind. Med.* **6**, 65 (1949).

337. T. S. Scott, *Br. J. Ind. Med.* **9**, 127 (1952); *Br. Med. J.* **2**, 302 (1964).

338. R. A. M. Case, M. E. Hosker, and D. B. McDonald, *Br. J. Ind. Med.* **11**, 75 (1954).

339. G. F. Rubino, G. Scansetti, and G. Piolatto, *Proceedings of the XIX International Congress of Occupational Health*, Vol. 1, *Chemical Hazards*. Institute of Medical Research of Occupational Health, Zagreb, Yugoslovia, 1980, p. 627.

340. R. W. Boyko, R. A. Cartwright, and R. W. Glashan, *J. Occ. Med.* **27**(11), 799 (1985).

341. M. G. Ott and R. R. Langner, *J. Occ. Med.* **25**(10), 763 (1983).

342. G. M. Blackburn and B. Kellard, *Chem. Ind.*, 607 (Sept. 15, 1986).

343. F. B. Stern, L. I. M. Murphy, J. J. Beaumont, P. A. Shulte, and W. E. Halperin, *J. Occ. Med.* **27**(7), 495 (1985).

344. J. Zimnicki and A. Zawadzka, *Tech. Wlok.* **5**, 152 (1990).

345. C. Cavelier and J. Foussereau, *R. Tomb, Cah. Notes Doc.* **133**, 615 (1988); **132**, 421 (1988).

346. Z. W. Myslak and H. W. Bolt, *Zent. Arbeitsmed. Arbeitsschutz, Prophyl. Ergon*, **38**(10), 310 (1988).

347. N. V. Vasilenko and B. A. Kurlyandskii, *Gig. Tr. Prof. Zabol* **9**, 33 (1986).

348. M. Marchisio, *Tinctoria* **83**(6), 99 (1986).
349. A. R. Gregory, *J. Env. Pathol. Toxicol. Oncol.* **5**(4–5), 243 (1984).
350. A. Keil, D. Muller, and K. D. Wozniak, *Textil Technik* **39**(5), 258 (1989).
351. O. P. Hornstein, *Melliand Textilberichte* **70**(3), 222 (1989).
352. V. Foa, *Tinctoria*, **85**(8), 53 (1988).
353. X. Lejeune, *Ind. Tex.* **1191**, 913 (1988).
354. V. N. Valov, L. V. Mikhailova, and N. S. Borisova, *Tekstil Promyshlennost* **47**(12), 46 (1987).
355. V. Baumgarte, *Rev. Prog. Color.* **17,** 29 (1987).
356. J. M. Wattie and E. Marshall, *Int. Dyer* **171**(11), 21 (1986).
357. *J. Soc. Dyer. Color.* **102**(3), 118 (1986).
358. K. L. Hatch, *Text. Res. J.* **54**(10), 664 (1984).
359. C. Frangi, *Tinctoria* **79**(11), 361 (1982).
360. J. S. Gow, *Textilverdelung* **18**(4), 119 (1983).
361. "DHEW/NIOSH/NCI, Joint Intelligence Bulletin No. 24, Direct Black 38, Direct Blue 6, and Direct Brown 95," *Benzidine Derived Dyes*, NIOSH Publication No. 78–148, Dept. of Health, Education, and Welfare, Washington, D.C., 1978.
362. L. Fishbein, "Anthropogenic Compounds," in G. Hutzinger, ed., *Aromatic Amenes, The Handbook of Environmental Chemistry*, Vol. 3, Springer-Verlag, Berlin, Heidelberg, New York, Tokyo, 1984.
363. E. A. Clarke and R. Anliker, "Organic Dyes and Pigments," in Ref. 362, 1980.
364. C. T. Helmes, C. C. Sigman, and P. H. Pappa, *Chem. Toxic. Test*, 15 (1984); *Chem-Tech.*, 48 (Jan. 1985).
365. P. Gregory, *Dyes Pig.* **7**(1), 45 (1986).
366. K. Enslein and H. H. Borgstedt, *Toxic Lett.* **49**(2–3), 107 (1989).
367. *ETAD Annual Report 1989 and General Information 1990/91*, Tokyo, Japan, and Washington, D.C.
368. R. Anliker, *Rev. Prog. Color.* **8,** 60 (1977).
369. R. Anliker, *J. Soc. Dyer. Color.* **95**(9), 317 (1979).
370. R. Anliker, and D. Steinle, *Textilveredlung* **25**(2), 42 (1990); *J. Soc. Dyer. Color.* **104**(10), 377 (1988).
371. L. Arduini and P. A. Porta, *Tinctoria* **85**(8), 56 (1988).
372. *Afr. Text.* **62,** 1 (1987).
373. *Safety in Use; Textile Dyes and Pigments*, Parts I, II, and III, Imperial Chemical Industries PLC, Organics Division, Manchester, UK, 1986, 1987.
374. D. Chambers, *Int. Dyer.* **171**(8), 31 (1986); *Int. Dyer.* **171**(6), 18 (1986).
375. E. A. Clarke, *Can. Text. J.* **101**(12), 47 (1984).
376. E. A. Clarke, and R. Anliker, *Rev. Prog. Color.* **14,** 84 (1984).
377. W. L. Dyson, "Book of Papers," *1982 AATCC National Technical Conference*, Research Triangle Park, N.C., p. 125.
378. R. Anliker, *Textilveredlung* **18**(4), 130 (1983).
379. *ETAD, Guidelines for Safe Handling of Dyes*, U.S. Operating Committee of ETA, Washington, D.C., Nov. 1989.
380. R. Anliker, *Aqua. Ecolog. Chem. (Japan)* **1,** 211 (1979).
381. R. Anliker and E. A. Clarke, *J. Soc. Dyer. Color.* **98,** 42 (1982).
382. J. Houk, M. J. Doa, M. Dezube, and J. M. Rovinski, "Evaluation of Dyes Submitted Under the Toxic Substance Control Act New Chemicals Programme," *Colour Chemistry*, Elsevier Applied Science, London and New York, 1991.
383. H. H. Smith, *Text. Chem. Color.* **23**(10), 29 (1991).

ABRAHAM REIFE
CIBA-GEIGY Corporation

DYES, NATURAL

From the earliest of times humans admired the beautiful natural colors of plants and minerals, and sought to enhance human appearance through color. To do so, they painted their bodies with various natural dyes. The ancient Celtic people used the blue dye from the woad plant; the Aztecs in South America used the red dye from the cochineal bug. As time went on, natural dyes were used more and more as cosmetics: hair was dyed, lips were painted, and cheeks were rouged. Some dyes found questionable use in the treatment of illness. Natural dyes were used to make foods and wines more attractive, although many times they were used to disguise an inferior or spoiled product. But the greatest use for natural dyes occurred when the art of weaving developed. A few synthetic dyes were developed before 1856, but without impact. Then, in 1856, Perkin laid the foundation for the synthetic dyestuff industry by his synthesis and manufacture of the dye Mauve. In quick succession, a number of other synthetic dyes appeared on the market, and the natural dyestuff industry was doomed. Natural dyes were replaced by synthetic dyes, although lately there has been a revival of the use of natural dyes for coloring foods, and some textile manufacturers are using natural dyes for dyeing their products. This article discusses those natural dyes formerly manufactured.

Anthraquinones

The anthraquinone structure occurs in both the plant and animal kingdom. Those natural dyes having this structure surpass all other natural dyes in fastness properties (see DYES, ANTHRAQUINONE).

Alizarin. There is only one significant plant anthraquinone dye, alizarin [72-48-0] (CI Natural Red 6, 8, 9, 10, 11, and 12; CI 75330). In ancient times, alizarin was the preferred red dye. Cloth dyed with it has been found in Egyptian tombs dating 6000 years ago. The dye is found in the madder plant, a member of the Rubiaceae family. In 1944 about 35 species of this plant were known (1), but the use of more sophisticated analytical methods led to the detection of many more species; by 1984 the number had increased to 50 (2). Of these, *Rubia tinctorum* and *R. peregrina* yield the greatest amount of dye, about 1–2% (3). *R. tinctorum* and other species are perennials growing wild in many parts of Asia, Europe, and tropical zones. In Asia Minor madder was known as lizari or alizari, hence the name alizarin for the dye. The madder plant was not only the source of the red dye but was used also as barter. In Tibet, salt was exchanged for madder and taxes were paid with it (4). Madder was not harvested before a minimum of 18 months and a maximum of 28 months in order to obtain the greatest yield of dye. The dye is concentrated, as a glycoside, mainly in the roots of the plant; the roots were dug, washed, dried, and sold as a finely ground powder. The tops of the plant, containing small amounts of dye, were often used as fodder. Sheep that ate such fodder developed a purple wool (5). Also, in 1736, it was observed that the bones of swine fed on madder were tinted red (6). This observation was used later in studying bone growth (7). In France and Holland, farmers were well aware that madder roughage had a tendency to impart a reddish hue to milk and a yellowish hue to butter (6).

Alizarin was sold in various forms depending on how it had been prepared: Garancine, Garanceuz or spent garancine, Flowers of Madder, and Commercial Alizarin (8). The garacine grade was obtained by treating the madder roots with water acidified with sulfuric acid (9). Garancin is the French name for madder.

Cultivation of madder reached its highest degree in France and Holland during the sixteenth century. In North America, growing conditions were quite suitable, but madder was never cultivated there to any great extent (10).

Alizarin is a mordant dye forming various colored coordination complexes with different metallic salts (11,12). Based on analytical results, a structural formula has been proposed for the alizarin complex (13).

The dyeing of cotton with alizarin originated in India and finally reached Turkey. Originally, alizarin was dyed from a hard water bath containing alum, $KAl(SO_4)_2 \cdot 12H_2O$, as a mordant. The resulting red color was quite fast and useful for cotton, wool, linen, and silk. However, its shade was rather dull and lacked the brilliancy of kermisic acid-dyed material. In the hands of the Turks, a rather elaborate, drawn-out, and most secretive dyeing process was developed. The result was a beautiful, fiery red, which became known as Turkey Red. Many attempts were made to shorten the process, but the results were far from satisfactory (14).

Scientific investigation of alizarin began in 1816, at which time it was isolated in a pure form as bright orange needles melting at 289–290°C. Alizarin occurs in the plant as a glycoside (15), which became known as ruberythric acid [152-84-1] (16). Enzymic hydrolysis of this produced alizarin and a sugar, primeverose. Much later, ruberythric acid was synthesized and shown to have the structure (1) (17):

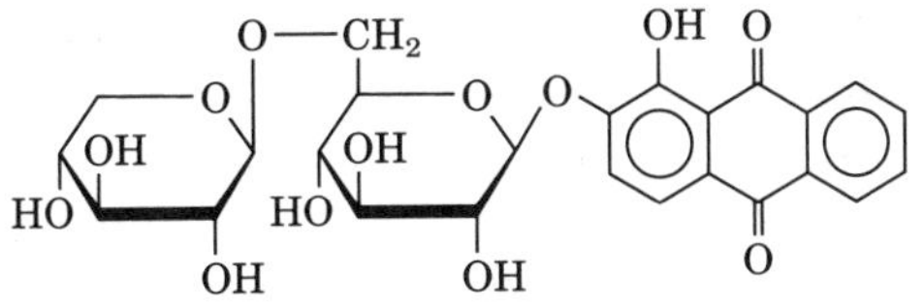

(1)

In 1868, alizarin was thought to be a naphthalene derivative. However, upon subjecting alizarin to a zinc distillation, anthracene was obtained. With this information and other data which indicated that alizarin was 1,2-dihydroxyanthraquinone (2) (18), anthracene was oxidized to anthraquinone, dibrominated, and the dibromo derivative subjected to a caustic fusion. Alizarin was obtained in an impure form and in low yield. This represented the first synthesis of a natural dye.

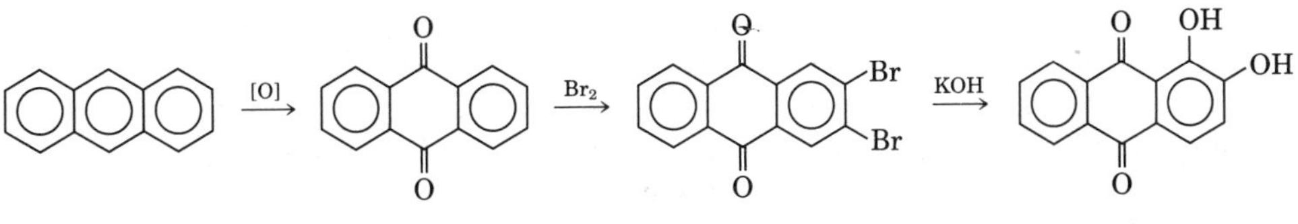

(2)

This process was not acceptable for several reasons: low yields, poor quality, and the high cost of bromine. Later, at BASF, a process was developed for the manufacture of alizarin by the caustic fusion of anthraquinone-2-sulfonic acid (so-called silver salt) which was made by sulfonating anthraquinone with sulfuric acid. This process was patented in England on the 25th of June, 1869. One day later, W. Perkin applied for a patent for the manufacture of alizarin by a process almost identical to the German process except that the "silver salt" was prepared as follows:

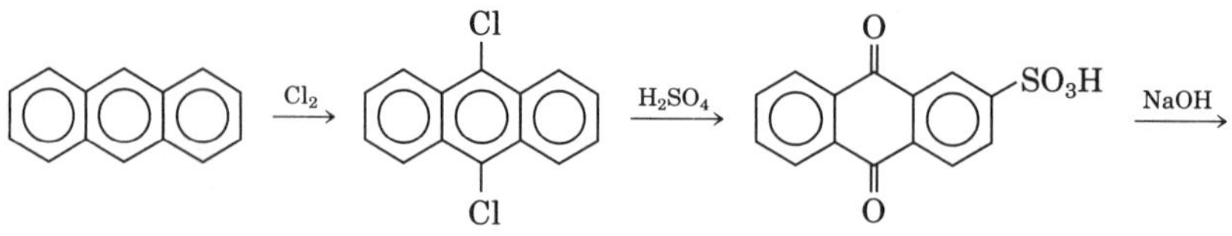

Actually Perkin's process produced a better quality grade of alizarin than that obtained by the German process. To avoid lengthy patent ligations, both parties agreed to a mutual use of the German patent. Later, improvements were made in the process: use of oleum for sulfonating anthraquinone and the addition of an oxidizing agent to the caustic melt.

For years this was the process used to manufacture alizarin (19) although it was claimed that a more economical process would result if 2-chloroanthraquinone was used instead of silver salt (20). In 1870, the market price for 100% synthetic alizarin was 200 German marks, but by 1912 it had fallen to 5–6 marks, thereby sounding the death of natural alizarin (21). Also, dyers welcomed synthetic alizarin since it was 100% 1,2-dihydroxyanthraquinone; natural alizarin always contained varying amounts of other polyhydroxyanthraquinones.

Turkey, for centuries, has been known for the beauty of its handwoven rugs dyed with natural dyes. Nowadays Turkish peasants prefer to dye machine-made rugs with synthetic dyes since these are more readily available and easier to apply. However, there is a growing demand by collectors and connoisseurs for handmade rugs dyed with natural dyes, especially alizarin, and they command premium prices (22). Also, there is a small demand for natural alizarin by artists and home dyers who claim that natural alizarin produces subtle shades not obtainable with synthetic alizarin. Just as synthetic alizarin forced natural alizarin out of the market, synthetic alizarin has been replaced by azoic dyes since they are easier to apply.

Animal Anthraquinone Dyes.   *Kermisic Acid.*   Many accounts claim that kermisic acid [476-35-7] (CI Natural Red 3; CI 75460) is the oldest dyestuff ever recorded (23). The name *kermes* is derived from an Armenian word meaning little worm for which the later Latin equivalent was *vermiculus*, the basis of the English word vermillion. The dye was obtained from an oriental shield louse, *K. ilicis*, which infest the holm oak *Quercus ilex* and the shrub oak *Q. coccifera*. The dye produces a brilliant scarlet color with an alum mordant. Although expensive, it was cheaper than its rival Tyrian Purple. It was in great demand until the sixteenth century when it was displaced by carminic acid.

Although kermisic acid had been obtained pure as early as 1895 (24), no investigation of the dye was undertaken until 1910 when structural elucidation studies by degradation methods began and it was determined to have the structural formula (**3**). Synthetic structural proofs were done much later (25,26) and showed that (**3**) was incorrect. The actual structure of kermisic acid is (**4**), 1,3,4,6-tetrahydroxy-7-carboxy-8-methylanthraquinone.

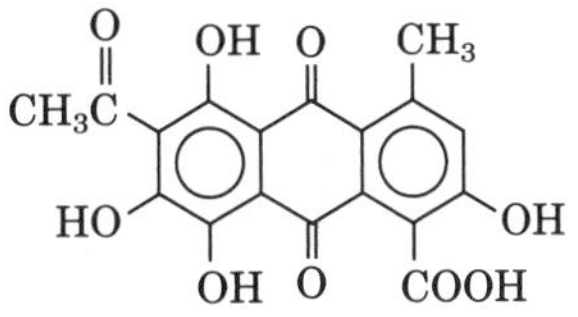

(**3**) 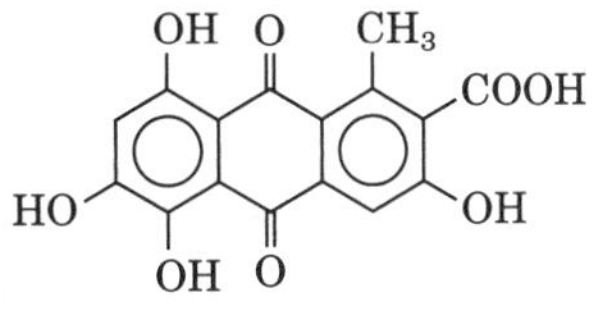   (**4**)

Carminic Acid. Carminic acid [*1260-17-9*] (CI Natural Red 4; CI 75470), is a red dye occurring as a glycoside in the body of the cochineal insect *Dactylopius coccus* of the order Homoptera, family Coccidae. This insect is native to Central and South America. The Aztecs had extracted the dye from the insect centuries before the coming of the Spaniards. For breeding purposes, the insects were collected in the autumn and carefully protected during the winter months. Cochineal was harvested after three months, and then the bugs were killed by immersion in hot water, by placing in hot ovens, or by exposure to the hot sun. The latter method produced the highest quality dye (27). At present, Peru and the Canary Islands are the main source of the dye. Until the advent of synthetic dyes, the principal use for carminic acid was for dyeing tin-mordanted wool or silk. Its aluminum lake, carmine [*1390-65-4*], finds use in the coloring of foods (see COLORANTS FOR FOODS, DRUGS, COSMETICS, AND MEDICAL DEVICES).

Although carminic acid had been known since 1818 (28), it was not obtained in a pure form until 1858 (29). The structure was finally established as (**5**) (30–33).

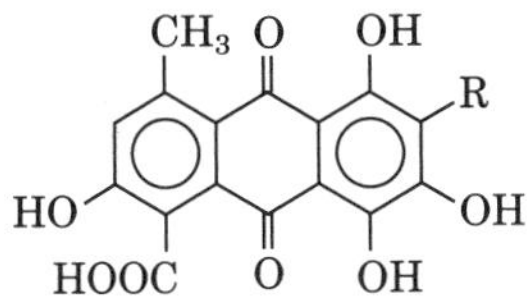

(**5**)

At the time the structure of carminic acid was being studied, the composition of the sugar group R was not known. Much later it was shown to be D-glucopyranosyl. However, carminic acid did not behave like most glycosides. It resisted all attempts to be hydrolyzed into a sugar and the corresponding aglycone. The problem was resolved when it was demonstrated that carminic acid is a *C*-glycoside (34).

Later studies (35–37) showed that the original assignment of (**5**) was not quite correct. The actual structural formula of carminic acid is (**6**) (38).

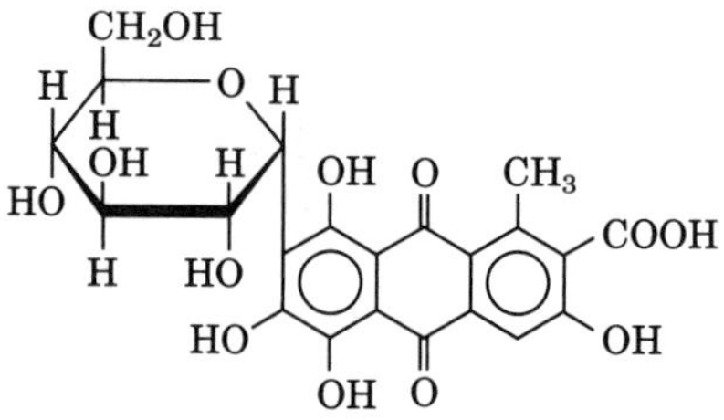

(6)

Laccaic Acid.   This acid has been designated [*6219-66-5*] (CI Natural Red 25; CI 75450). Lac dye ranks as the most ancient of animal dyes. It is found in lac, the resinous secretion of a very small insect, *Coccus laccae,* found growing in India and Southeast Asia. The word lac is derived from the Hindu word *lakh* meaning one hundred thousand or a great number, having reference to the fact that a great number of these insects had to be gathered in order to produce any quantity of dye. The resinous secretion of the insect is found deposited on branches of trees and is called stick-lac. Water extraction of stick-lac produces the lac dye. The remaining water insoluble material contains several other anthraquinone pigments of minor importance. The lac dye was first investigated in 1887 (39) and named laccaic acid. Structures were proposed for this material (40,41), but around 1965 it was discovered, chromographically, that lac dye is actually a mixture of acids (42,43) derived from 2-phenylanthraquinone [*6485-97-8*]. The acids were designated A, B, C, and E (42).

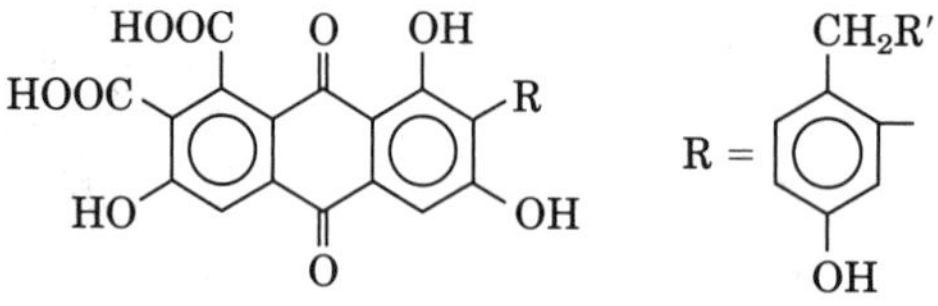

An excellent review of this work has been written (38).

Naphthoquinone Dyes

Although naphthoquinones represent the largest group of naturally occurring quinones, only a small number of these achieved importance as dyestuffs.

Lawsone [*83-72-7*] (CI Natural Orange 6; CI 75420), also known as henna and isojuglone, occurs in the shrub henna (*Lawsone alba*). In England the plant is known as Egyptian privet. The dye was extracted from the leaves of the plant, using sodium bicarbonate, and the extracts used to dye protein fibers an orange shade. Henna is probably the oldest cosmetic known. The ancient Egyptians used it as a hair dye and for staining fingernails. It is said that Mohammed dyed his beard with henna. Lawsone has been identified as 2-hydroxy-1,4-naphthoquinone

(44,45). It has been synthesized by the Thiele acetylation of 1,4-naphthoquinone followed by hydrolysis and oxidation (46):

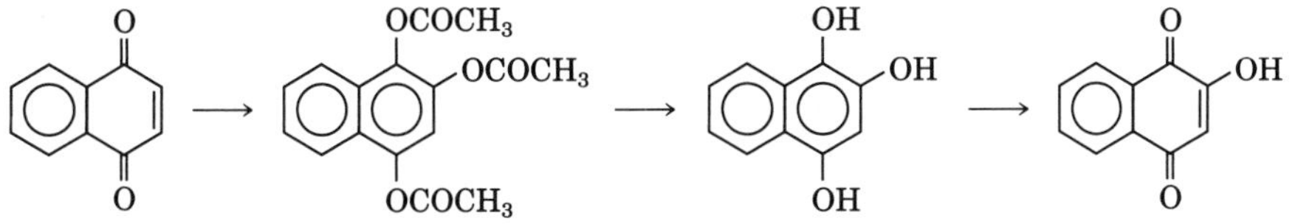

Lapacol [84-79-7] (CI Natural Yellow 16; CI 75490) (lapachic acid, taiguie acid, tecomin) is a yellow pigment occurring in the wood of trees of the genus *Tecoma*, native to the West Indies and tropical South America. The shavings of the wood, treated with lime water, give an extract that dyes cotton yellow.

The pigment was first described in 1866 (47). Based on degradative studies and chemical reactions, it was proposed that lapacol was the alkenylhydroxy-naphthoquinone (7) (48), but it was later determined that its structure is actually (8) (49).

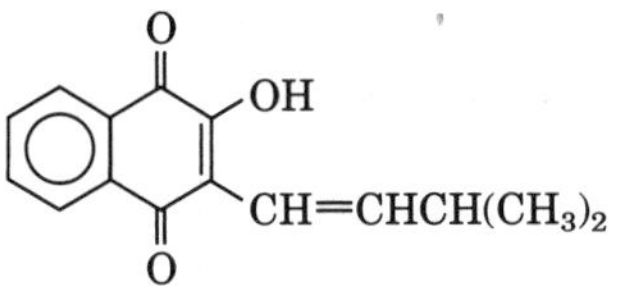

(7) 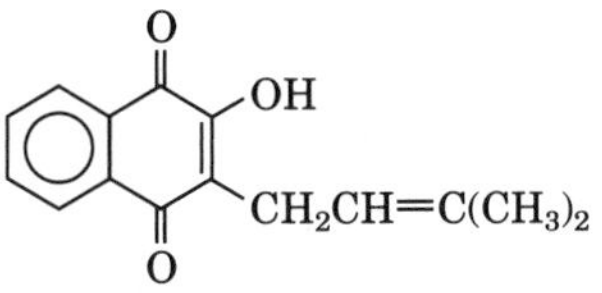   (8)

Juglone [481-39-0] (CI Natural Brown 7; CI 75500) was isolated from the husks of walnuts in 1856 (50). Juglone belongs to the Juglandaceae family of which there are a number of species: *Juglans cinerea* (butter nuts), *J. regia* (Persian walnuts), and *J. nigra* (black walnuts). Persian walnuts were known to the ancient Romans who brought them over from Asia Minor to Europe. As early as 1664, the American colonists knew how to extract the brown dye from the nuts of the black walnut and butternut trees, both native to eastern North America (51).

In 1885, from a detailed study of juglone (52) it was proposed that its structure was 5-hydroxy-1,4-naphthoquinone (9). This structure was confirmed by oxidizing 1,5-dihydroxynaphthalene with potassium dichromate in sulfuric acid (53). Juglone occurs in walnuts as a glycoside of its reduced form, 1,4,5-trihydroxynaphthalene (54). Later it was determined that the sugar is in the 4-position (10) (55).

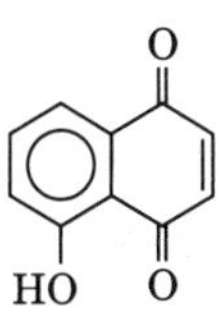

(9) 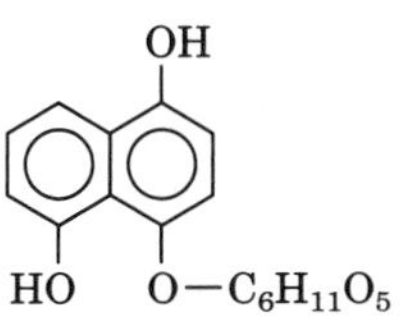   (10)

Juglone is most readily synthesized by Bernthsen's method. However, this method is too drastic and results in low yields (56). Somewhat better yields are obtained by using Fremy's salt (potassium nitroso disulfonate) as the oxidant (57). By using thallium trinitrate to oxidize 1,5-dihydroxynaphthalene, yields as high as 70% of juglone have been reported (58).

In the past, juglone had been used to dye wool and cotton a yellowish brown. Although it no longer has any commercial value as a dye, it is a fungicide and as such finds use in the treatment of skin diseases. Its toxic properties have been made use of in catching fish. Juglone has been used to detect very small amounts of nickel salts since it gives a deep violet color with such salts.

Alkannin, shikonin, and shikalkin are grouped together because the first two are enantiomers and the last one is their racemate. Alkannin [577-88-4] (CI Natural Red 20; CI 75530) (*Anchusa tinctoria* or *alkanna tinctoria*) is a member of the Boraginaceae family. It is found in the roots of alkanet, a perennial shrub native to Southern Europe. This reddish dye was used not only for dyeing cloth but also for coloring olive oil, and as a rouge and lip stain. The generic name Anchuss is derived from the Greek word meaning face paint (59).

In the first century, Dioscorides stated that the roots of the anchusa plant were useful in the treatment of wounds (60); this idea has been verified (61).

Alkannin occurs in the roots of the plant as the alkali-sensitive ester of angelic acid (62). It may be extracted from the roots by using boiling light petroleum ether. Treatment of this extract with dilute sodium hydroxide gives a blue solution from which the dye is precipitated by the addition of acid. The crude product is purified by vacuum sublimation (63). Its structure (**11**) is a hydroxylated naphthoquinone with a long, unsaturated side chain (64,65); it has the (*S*)-configuration.

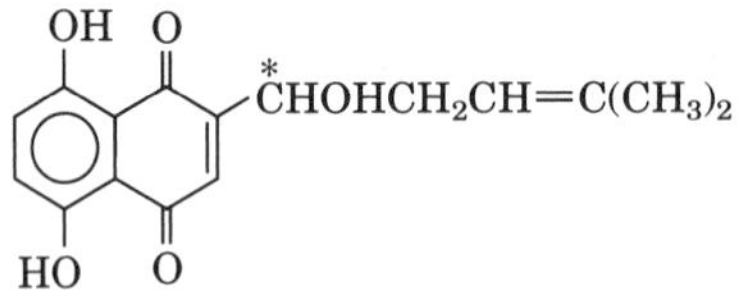

$$\overset{*}{C}HOHCH_2CH=C(CH_3)_2$$

(**11**)

Shikonin [517-89-5] (CI 75535) occurs as an acetyl derivative in the Japanese shikone, *Lithospermum erythrorhizon*, another member of the Boraginaceae family. It is the (*R*)-optical isomer of alkannin (66). Tissue cultures of *L. erythrorhizon* are used in Japan to manufacture shikonin mainly for cosmetic use (67). Both alkannin and shikonin are mordant dyes producing violet to gray colors on fabrics. In Japan, shikonin was used to dye fabrics a color known as Tokyo Violet. Shikalkin [54952-43-1] the racemate (**11**), has been synthesized (68).

Flavones. These compounds are the most widely distributed natural coloring matter formerly used as dyestuffs. The term flavone was first suggested in 1895 (69), and is indicative of their yellow color (*flavus*, Latin for yellow). They have lost their commercial value as dyes since the advent of synthetic dyes in 1856.

Flavone-type dyes occur in all the higher plants: in the leaves, roots, bark, fruits, pollen, and flower petals. None have been found in fungi, mosses, or lichens. The most widespread flavone dyes are quercetin [117-39-5] (**12**) and kaempferol [520-18-3] (**13**):

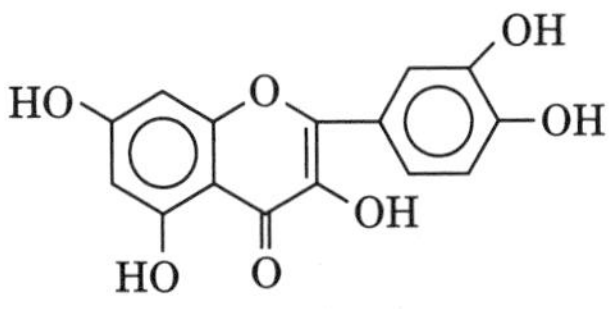

(12) 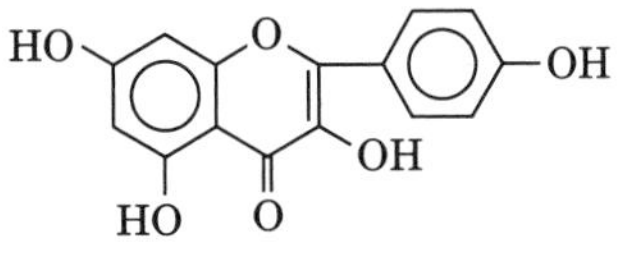   (13)

In general, the dyes occur as glycosides, the most common sugar being glucose. Some flavones contain more than one sugar. Their role as dyes in the plant is not definitely known: a common suggestion is that they protect the plant from harmful uv radiation.

The basic unit of the flavone-type dyes is 2-phenylbenzopyrone (14) which unsubstituted is flavone [525-82-6]; isoflavone [574-12-9] is (15) and flavonol [577-85-5] is (16).

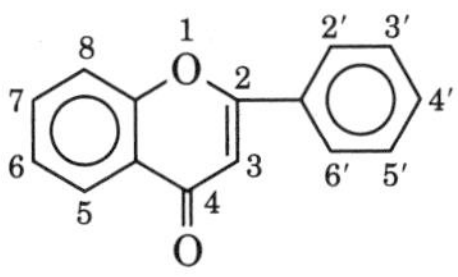

(14) 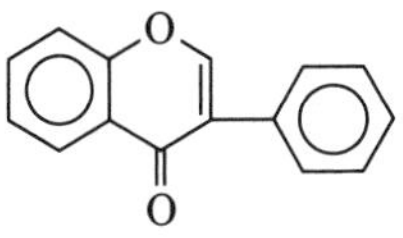   (15) 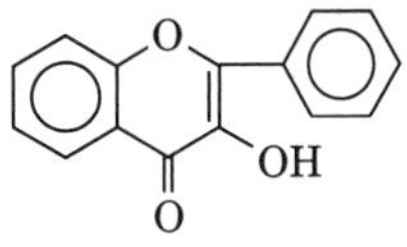   (16)

Flavone dyes having these structures are hydroxylated and methoxylated derivatives. The degree of hydroxylation varies from two in chrysin [480-40-0] (17) to six in gossypetin [489-35-0] (18). Those dyes containing not more than three hydroxyls are generally termed flavones whereas those containing up to and including six are flavonols.

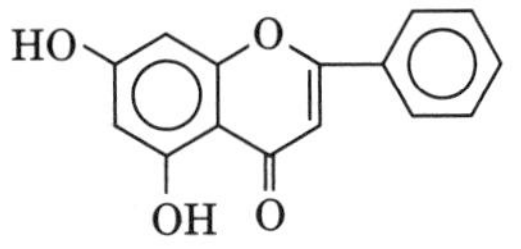

(17) 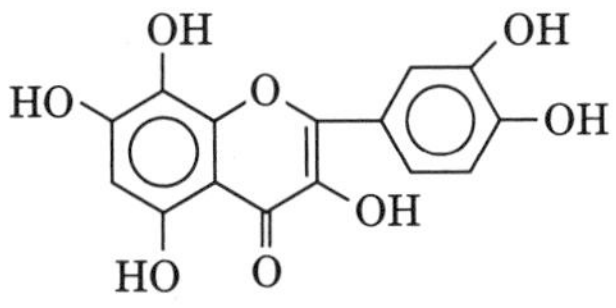   (18)

The flavone, isoflavone, and flavonol-type dyes owe their importance to the presence of an o-hydroxy carbonyl structure within the molecule. Positions 4 and 5 can chelate with different metallic salts to give colored, insoluble complexes. In other words, these dyes require a mordant in order to fix them onto the fiber. Perkin was able to predict the structure of unknown flavones by comparing the color of their complexes with the color of known complexes (70). For example, ferric chloride gives a green color with 5-hydroxyflavones and a brown one with 3-hydroxyflavones (71).

Chrysin (17) was the first flavone to be isolated in a pure form, and its structure was elucidated by identification of its alkaline degradation products (72–74). The structure was confirmed by synthesis (75,76). The same procedures were used to establish the structure of other flavones and in so doing the foundation of flavone chemistry was laid (77).

Of all the flavone dyes luteolin [491-70-3] (5,7,3′,4′-tetrahydroxyflavone) is the oldest known European dye. It is found in weld (*Reseda luteola*) also known as dyer's rocket, dyer's weed, and wild mignonettle. Luteolin was well known to the ancient Romans, who used this very pure yellow pigment to dye the garments and robes of the Vestal Virgins (78). Until the middle of the nineteenth century, luteolin top-dyed with indigotin produced good greens known as Lincoln green and Saxon green (79). The naturally occurring flavone-type dyes used commercially until 1856 are described fully elsewhere (80).

Logwood [8005-32-2] (CI Natural Black 1; CI 75290) is a modified benzopyrone, and was the last of the natural dyes to survive after the appearance of synthetic dyes because of the desirable bluish-black hue it produced on chrome-mordanted fibers. However, for economic reasons and because it requires a mordant, it too is no longer used commercially.

Logwood was discovered originally in Mexico in the Province of Campeche by Spanish settlers who referred to it as *palo de Campeche* (wood of Campeche). The French knew it as *bois de Campeche* and the Germans called it *blauholz* because of its use to give blue dyeings. Sometimes it was referred to as blood wood. Based on this, Linnaeus applied the generic name *Haematoxylon* to the tree from which logwood dyes are obtained. The dye is best known as logwood presumably because it was obtained from logs of wood. These types of dyes were obtained from the wood either by the French or American process. In the American process, the logs are reduced to chips and the dye is extracted with hot water under pressure. The French process did not use pressure. Concentration of the extracts produced dyewood crystals. During the concentration process, the haematein [475-25-2] (**19**) is produced by oxidation of its leuco form haematoxylin [517-28-2] (**20**).

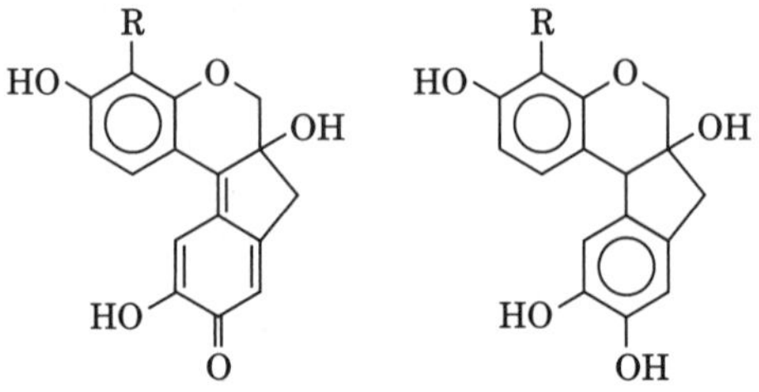

(**19**) R = OH   (**20**) R = OH
(**22**) R = H   (**21**) R = H

Although logwood had been known for several hundred years, it did not achieve technical importance until it was discovered that it combined with metallic salts to give various colored lakes, of which the chrome lake was the most important (81). After an enormous amount of research, the present formulas for the dye components were proposed (82) and later substantiated (83).

Brazilwood [8005-32-1] (CI Natural Red 24; CI 75280) has long disappeared from the market because of the very fugitive character of its dye. The chemistry of the dye parallels that of logwood in that it has a leuco form, brazilin [474-07-7] (**21**) and its oxidation product brazilein [600-76-0] the dye (**22**). It differs from logwood in that it has one less phenolic group. The dye occurs in trees belonging to various species of *Caesalpina*, and is extracted from the wood of these trees by a process similar to that used for logwood. The name brazil has its origin in the Portuguese word *braza* referring to anything having a bright red color. More about the history of brazilwood may be found in the original literature (84).

Anthocyanins. Like the flavones, the anthocyanins are found throughout nature. This class of polyphenolic compounds is responsible for the pink, red, violet, and blue colors found in plants. The term anthocyan is derived from the Greek words *antho* for flower and *kyanos* for blue. It was proposed first to denote the blue color of the cornflower (85). Later, as knowledge about plant pigments increased, the term was extended to include all such pigments and the ending "in" was added. Like many other natural phenolic substances, anthocyanins occur in plants as glycosides; the sugar-free anthocyanins are known as anthocyanidins.

All anthocyanidins have the 2-phenylbenzopyrylium or flavylium cation structure (86,87) (**23**), a resonance hybrid of oxonium forms and carbenium forms [*14051-53-7*]:

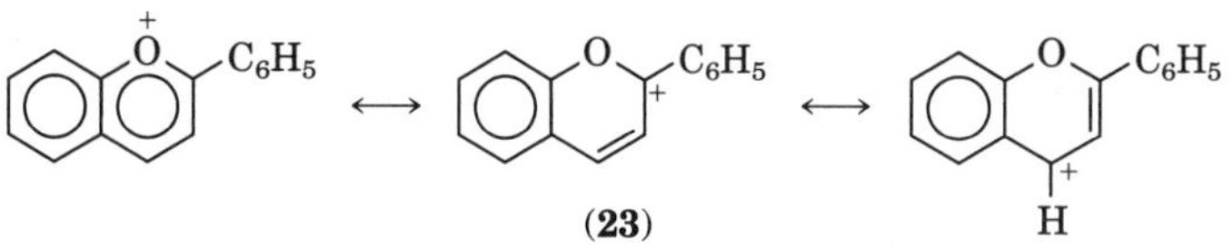

(**23**)

In all these structures, the anions are Cl^-.

There are three fundamental groups of anthocyanidins to which all the other anthocyanidins could be referred (88–90). In the following structure $R_1 = R_2 = H$ designates pelargonidin (**24**); $R_1 = OH$, $R_2 = H$ is cyanidin (**25**); and $R_1 = R_2 = OH$ is delphinidin (**26**).

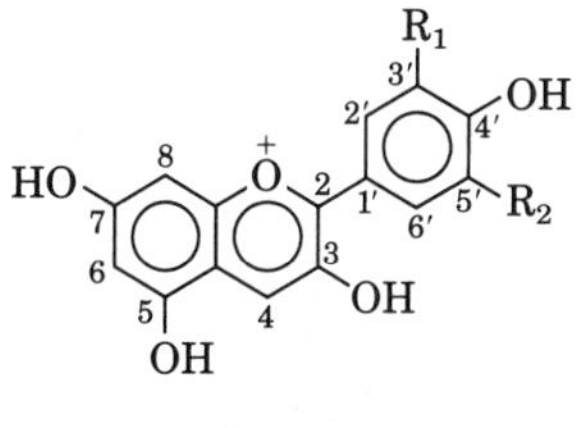

The anthocyanins are pH sensitive. Their color, in part, is determined by the pH of the sap. Cyanin, for example, is red at pH 3, violet at 8, and blue at 11. However, there are other factors that affect the colors of the anthocyanins; metallic salts, notably iron and aluminum, react with those anthocyanins containing vicinal hydroxy groups and produce highly colored complex compounds. Other factors are the colloidal condition of the cell sap and copigmentation (91).

All the anthocyanins, when boiled for a short time with hydrochloric acid, yield one or more sugars and anthocyanidin. The most frequently occurring sugars are glucose, rhamnose, galactose, and gentiobiose. Of these, glucose is the most prevalent. In most cases the sugars are found in the 3- and occasionally in the 5-position. In some cases, the hydroxyl group of the anthocyanin or that of the sugar may be esterified with an organic acid such as the *p*-hydroxy derivative of benzoic, cinnamic, coumaric, or malonic acid (92).

Alkaline degradation was used to determine the structure of pelargonidin [*134-04-3*], cyanidin [*528-58-5*], and delphenidin [*528-53-0*] by treating each of them with concentrated potassium hydroxide at 140–150°C (89). Each of them produced phloroglucinol (27) and the corresponding phenolic benzoic acid:

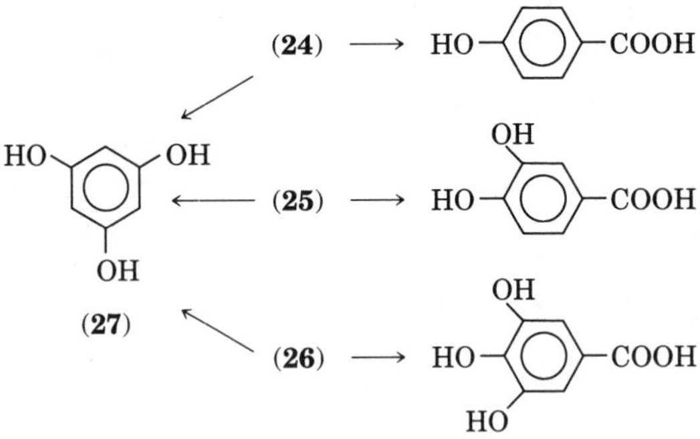

Anthocyanidins were first synthesized by reaction of an aryl Grignard reagent with a coumarin (93).

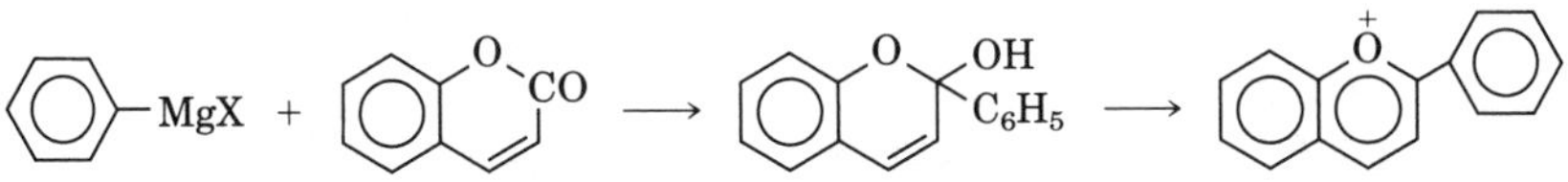

A more convenient method for synthesizing anthocyanidins involves the condensation of an *o*-hydroxybenzaldehyde with an acetophenone (94).

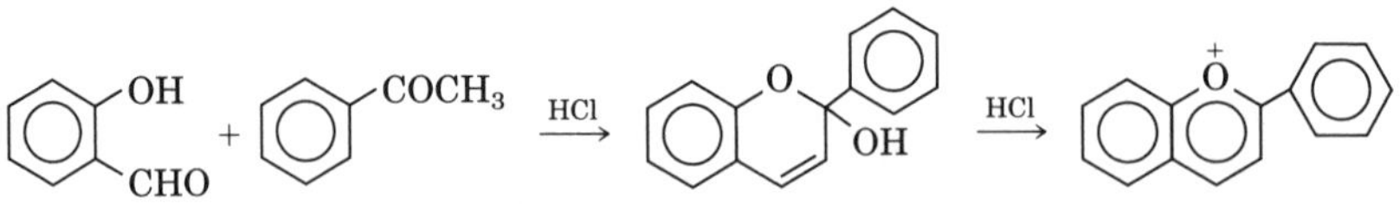

Because of their pH sensitivity, anthocyanins have found little use as industrial dyes. However, a few having the quinoidal form of anthocyanidin (28) were formerly used as dyes. Two of these were carajurin [*491-93-0*] (29) and dracorhodin [*643-56-1*] (30):

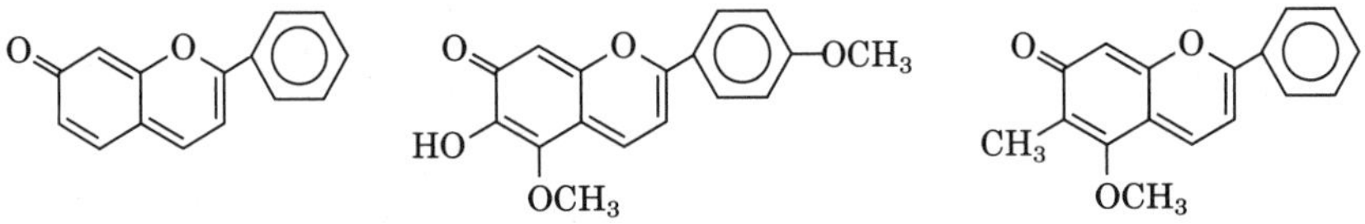

Indigoid Dyes

Tyrian Purple.   The ancient kingdom of Tyre owed its fame and fortune to the purple dye produced from the lowly mollusks found on its shores. From early Egyptian and pre-Roman times, about 1600 BC, it was known that these shellfish produced a secretion which, on exposure to light and air, produced a beautiful and fast purple dye. The dye became known as Tyrian Purple [*19201-58-7*] (CI 75800) reflecting its place of origin. These mollusks belong to the Muricidae family and the genera *Murex* and *Purpura* which include *M. brandaris* and *M. trunculus*, the principal sources of the dye.

There are a number of myths relating to the discovery of Tyrian Purple. One such goes as follows: a Tyrian god was walking along the shore accompanied by his dog and a nymph. Suddenly the dog bit into a shellfish, whereupon his mouth became stained a beautiful purple. Seeing this, the nymph begged the god to have a dress made for her dyed with this new dye. He granted her wish and won her everlasting favor.

Tyrian Purple was the most expensive and rare dye of the ancient world principally because only a small amount of dye could be obtained from each mollusk, roughly 0.12 mg (95). It was always considered a color of distinction and restricted to regal and ecclesiastical uses; in the Eastern Roman Empire, the heir to the throne at Byzantium bore the proud name *Porphyro-Genitur*, born to the purple. The Hebrews used purple in many decorations of the Tabernacle (23).

Pliny described the manufacture of Tyrian Purple, which was dependent on the species of mollusk used. In the case of the large mollusk, *M. brandaris*, a single drop of glanular secretion was extracted from a gland adjacent to the respiratory cavity. In the case of the smaller mollusk, *M. trunculus*, the entire mollusk was crushed and used as such. Irrespective of the method used, the next step consisted in salting the material for about three days, followed by boiling the entire mass for 10 days (96). Sometimes other dyes, such as kermisic acid, were added in order to alter the shade of the dye. Around 1453, the Turks overran the Eastern Roman Empire and destroyed the Tyrian dye plants. From then on, extracting purple dye from shellfish practically ceased.

Toward the latter part of the seventeenth century it was observed that the natives of Ireland were dyeing linen and silk with a secretion from shellfish found along the Somerset and Welsh coast. Cloth dyed with this secretion, when exposed to sunlight, underwent a series of color changes ultimately resulting in a bright crimson color on the fiber. These observations awakened general interest in this dye extracted from mollusks (97). In the period that followed, a number of investigators made valuable contributions about the dye (98). For example, from about 12,000 mollusks ~1.4 g of dye were isolated, and it was determined that the dye was 6,6'-bromoindigotin [*19201-53-7*] (**31**) (99). This structure was confirmed by synthesis (100).

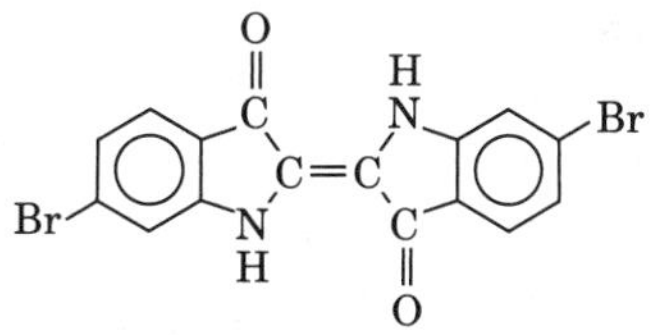

(**31**)

The next important investigation of Tyrian Purple was the determination of the precursor of the dye. Several investigators critically examined the composition of the hypobranchial gland and established that it varied from one species of mollusk to another. In general, the glands were extracted with either alcohol or ether or a combination of both, the extracts were purified by gel and thin-layer chromatography, and the structure of the precursors determined by elemental analysis and spectroscopic methods. Some species of mollusk contained more than one precursor, for example, *Dicathais orbita* contains one (101), whereas *M. trunculus* has four (102). The principal precursor isolated from the glands was sodium-6-bromo-2-methylthioindoxyl sulfate (**32**) (tyrindoxyl sulfate). Also isolated was a quinhydrone mixture consisting of 6-bromo-2-methylthioindoxyl (**33**) and 6-bromo-2-methylthioindoleninone (**34**) (tyriverdin). Both tyrindoxyl sulfate [*74626-31-6*] and tyriverdin produce purple colors. A summary of this work has been reported (103).

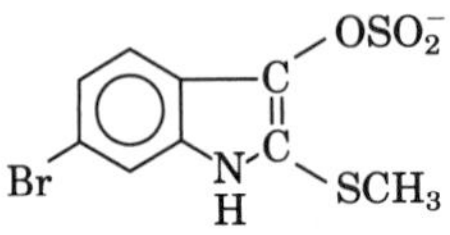

(32) 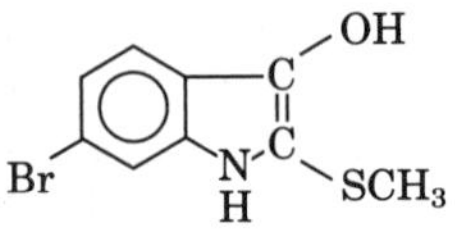   (33) 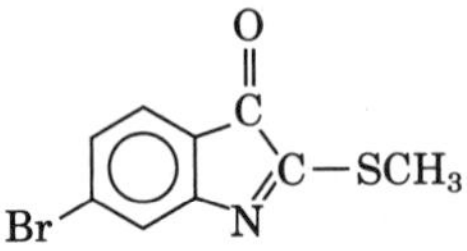   (34)

Indigotin.   The blue dye of the ancient world was derived from indigo and woad. Which plant is the oldest is a matter of conjecture. That indigo was known at least four thousand years ago is evident from ancient Sanskrit writings. Cloth dyed with indigotin [*482-89-3*] (CI Natural Blue) (CI 75780) has been found in Egyptian tombs and in the graves of the Incas in South America. The history of indigo is better documented than that of woad because indigo was indigenous to such wide areas as Asia, Java, Japan, and Central America (104). Woad, on the other hand, was found mainly in Europe. During their conquests in the New World, the Spaniards found Indians who painted their bodies with indigotin, dyed fabrics with it, and used it to paint ceramic vessels.

Indigo belongs to the legume family. Over three hundred species belong to this family, many yielding indigotin in varying quantities. The two most important species are *Indigo tinctoria* and *I. suffruticosa*, found in India and the Americas, respectively. Unaware of the true nature of indigotin, the Romans called it *Indicum* meaning a product from India. Ancient dyers who used indigotin called it *nil*, which ultimately led to the Arabic word *al-nil* meaning blue, and to our word aniline.

Because of the long overland route used to bring indigotin from India to Europe, and because of the small amount of indigotin that was present in the leaves, about 2–4%, indigotin ranked among the most expensive of the ancient dyes (105).

Indigotin was a very profitable item because of cheap labor, and many nations set up indigo plantations in various parts of the world: Spain in Guatemala, France in its Caribbean colonies, and England in the West Indies and India (106). In India, England, represented by its East India Company, had problems arising mainly from its poor economical treatment of the ryots who toiled in the indigo fields. This led to the so-called Blue Mutiny of 1859–1862 (107). As early as 1649, half-hearted attempts were made to grow indigo in America. It was not until 1739 that an indigo plantation was established in South Carolina as a result of the untiring efforts of a Miss Eliza Lucas who later became Mrs. E. Pinckney (108). So successful were her efforts that by 1773 over 500 t of indigo had been exported to England. Ultimately, because of the American Revolution, indigo lost its importance.

The leaves of the indigo plant do not contain the dye as such, but in the form of its precursor, a glycoside known as indican (109). Indican [487-60-5] is the dextrose derivative (**35**) of indoxyl [480-93-3] (110). Indoxyl occurs also in the urine of humans as the potassium salt of indoxyl sulfonic acid (111).

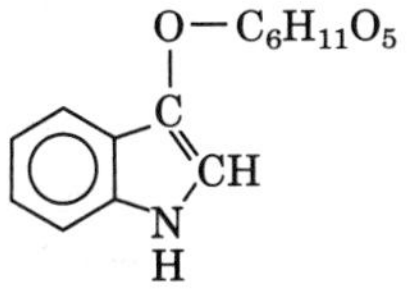

$$O—C_6H_{11}O_5$$

(**35**)

The process used to manufacture indigotin from the plant remained unchanged throughout hundreds of years. An old seventeenth century print shows the equipment used to prepare indigotin (112). A series of tanks are arranged step-wise one above the other. Into the uppermost one, the fermentation tank, water and the freshly cut plants are placed. Here indican is hydrolyzed into indoxyl and glucose by an enzyme known as indimulsin (113). During the fermentation, a vigorous evolution of carbon dioxide occurs along with a gradual color change in the broth. After nine to fourteen hours, the yellowish liquor is drained from the top tank into the next lower one via a spigot. In this tank the indoxyl is air-oxidized to indigotin by agitating the liquor with paddles. As it gradually forms, indigotin settles to the bottom of the tank. The supernatant liquor is siphoned off and the indigotin removed to another tank where it is heated to prevent further fermentation. Then it is filtered, placed into trays and air-dried to a thick paste. The paste is cut into bricks and sold as such. Indigotin was the first of the so-called vat dyes because the entire operation was carried out in tanks or vats.

Indigotin did not appear in Europe until the twelfth century. Venice, because of its strategic land and sea position, was the first European city to receive and use indigotin. However, the further spread of foreign indigotin throughout Europe was strongly opposed by the Europeans because it would compete with indigotin from woad, an ancient and widely cultivated plant. When Caesar invaded England in 55 BC, he encountered a race of people who stained their bodies blue with indigotin from woad (114).

Cultivation and use of woad was a primary industry throughout Europe. Indigotin from abroad would have had disasterous consequences on the economy and lives of many people. Farmers, merchants, and dyers banded together in order to convince the authorities that foreign indigotin was the devil's dye and corrosive to cloth dyed with it. To strengthen their argument, deception was employed whereby corrosive salts were added to the dye bath. Ultimately laws were passed that blocked the use of foreign indigotin in Europe. For the next three or four hundred years, only indigotin from woad was used in Europe.

Woad, *Isatis tinctoris*, belongs to a genus that comprises some thirty species. The plant is widely distributed in many parts of Europe, especially France, England, Germany, and Holland. Woad was probably the first blue dye plant cultivated in America. Undoubtedly it was brought over from England by the early colonists (115). The name woad is derived from the Anglo-Saxon *wad* or *waad*. Because of its weedy nature, its original German name was *weedt* or *weeda*, which may be the origin of our word weed (116).

The method for preparing indigotin from woad differed radically from that used for making indigotin from indigo, possibly because of differences in the dye precursor (117). The fresh woad leaves were ground into a pulp, which was then made into piles for draining. The piles were formed into balls and these dried on racks for one to four weeks. The dried balls were ground to a powder and the powder then spread into layers about one-half to one meter deep. The layers were wet with water, constantly stirred, and left to ferment for about nine weeks. After this time, the dark, clay-like mass was shaped into balls and sold in this form (117). Woad indigotin was about 10% weaker in dye strength than imported indigotin. Ultimately, dyers preferred indigotin from indigo, so that by the seventeenth century woad indigotin had practically disappeared from the marketplace.

Many investigators attempted to unravel the structure of indigotin, including von Baeyer who developed a number of syntheses for the dye. Some of these he believed had potential for the manufacture of the dye. One of these syntheses is shown (118):

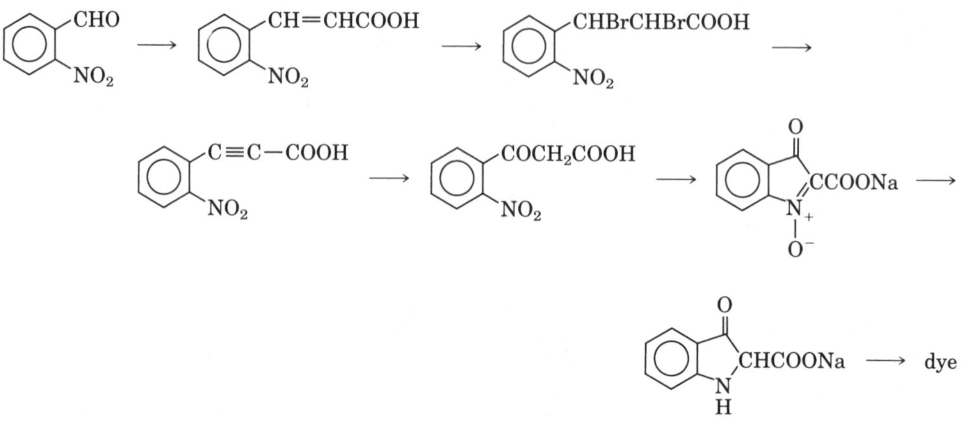

This process was sold to BASF in 1897 for $100,000, but never achieved commercial success.

Baeyer assigned a cis-form to indigotin, but x-ray crystallographic studies indicated that the dye molecule has a center of symmetry that is only possible if the molecule has a trans-configuration (119). Many derivatives of indigotin have been prepared that would not have been possible if indigotin had a cis-structure, eg, (**36**) (120).

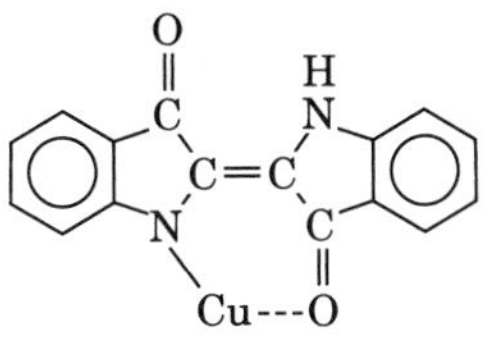

(**36**)

Although there is complete agreement that indigotin has a trans-configuration, many questions remain regarding some of its physical properties (121).

Baeyer's investigation of indigotin spanned a period of almost 20 years. In 1905, he received the Nobel Prize in recognition of his accomplishments.

In 1890, it was observed that treatment of ω-bromoacetanilide with alkali produced oxindole [59-48-3] (122) (**37**):

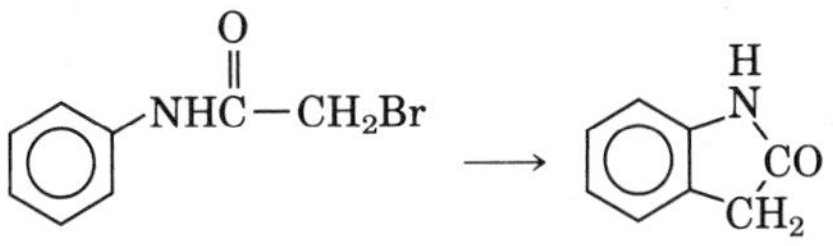

(**37**)

Based on this observation, K. Heumann treated *N*-phenylglycine [103-01-5] with alkali and obtained indoxyl (**38**) (keto form), which on aerial oxidation converted to indigotin:

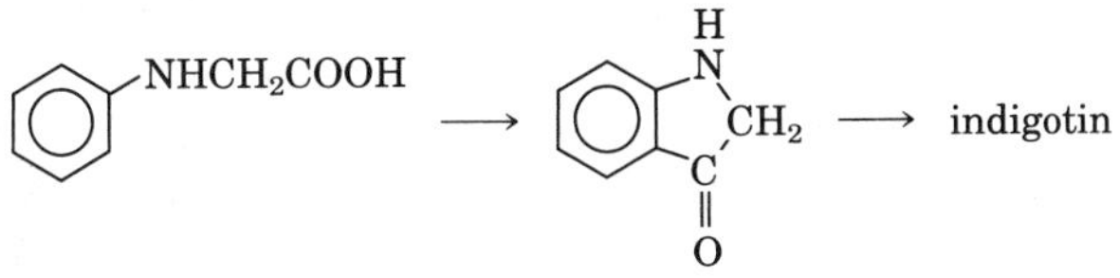

(**38**)

This was the first practical approach to the manufacture of the dye (123). The patent to this process was shared jointly by BASF and Hoechst (124). The yields by this process were so bad that Heumann developed another process involving the use of anthranilic acid [118-92-3] (**39**) (made from naphthalene).

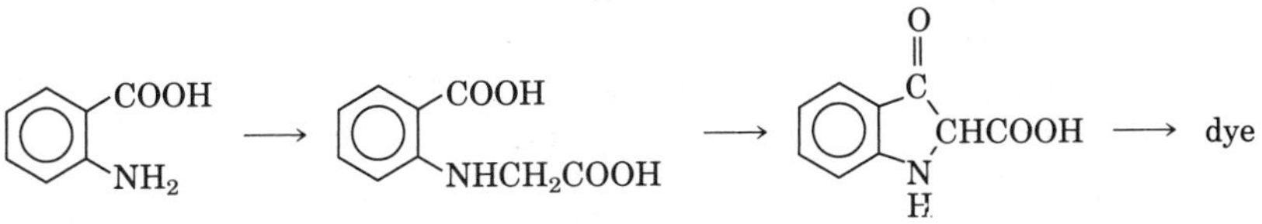

(**39**)

This was the process used by BASF and Hoechst for about 30 years. Later, a variation of the original Heumann process was made: aniline, formaldehyde, and hydrogen cyanide react to form phenylglycinonitrile (**40**) which is hydrolyzed to phenylglycine. This is the most widely used process for manufacturing indigotin.

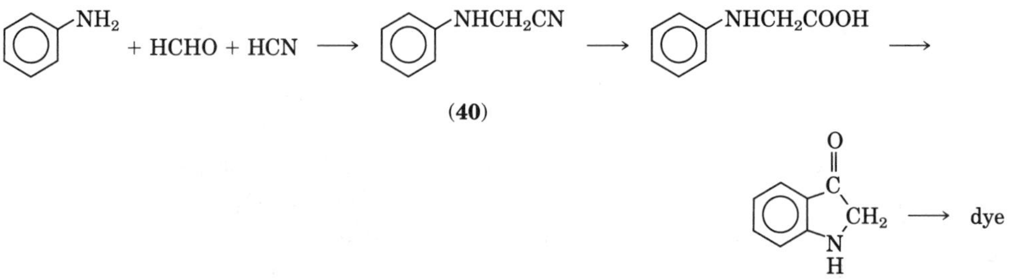

(**40**)

The greatest improvement in the manufacture of indigotin came when sodamide was used with alkali in the conversion of phenylglycine to indoxyl (125). Not only was the fusion temperature lowered from about 300°C to 200°C, but also the reaction was made practically anhydrous by the sodamide reacting with any water present. The result was an almost quantitative yield of dye.

After the second World War, German firms manufacturing indigotin faced serious competition from English and American dyestuff companies. To counteract this, the Germans developed continuous operations for manufacturing the dye. However, because of the complexity of the equipment and the operations (126), the batch process is still the preferred manufacturing method.

Although there is still demand for indigotin for dyeing blue jeans, it has lost a good part of the market to other blue dyes with better dyeing properties. At present, practically all the indigotin consumed in the United States comes from abroad.

Indigotin is available as a 100% pure powder and as a 20% solution. As of February, 1992, the powder form sold for about $20/kg in 80-kg lots. Most of indigotin is sold in solution form.

Natural Food Colors

The use of natural dyes as food colorants has a long and not always an admirable history. Pliny, the Roman scholar, records the use of various vegetable extracts to give young red wines the appearance of mature claret (127). As late as the latter part of the nineteenth century, the juices of the red beet and pokeberry were added to red wine. This stemmed from the fact that the price paid for red wine depended on the richness of its color. In 1892, the use of pokeberry extract was prohibited since it contains an emetic and a purgatory substance (128). Other foods besides wine were artificially colored. In 1935, the government of Paris forbade the coloring of butter since this was often done to conceal an inferior product. For the most part, colorants used for foods were fairly innocuous, but in the early part of the nineteenth century sweets were often colored with metallic salts such as lead chromate or copper arsenite. Obviously some form of legislation was called for to protect people from the improper and harmful use of food additives. Largely through the efforts of Dr. Harvey W. Wiley, who was chief chemist of the agriculture's bureau of chemistry, the Foods and Drug Act of 1906 was initiated (see COLORANTS FOR FOODS, DRUGS, COSMETICS, AND MEDICAL DEVICES). In the 1970s, decertification of the important food colors FD&C Reds 2 and 4 caused much concern among manufacturers of food dyes. With the possibility that other synthetic dyes would be banned, attention was turned to the use of natural dyes as food colorants. Many such dyes had been in use for hundreds of years until they were replaced by synthetic dyes.

The yellow dye curcumin, [458-37-7] (CI Natural Yellow 3; CI 75300) (41), also known as tumeric, occurs in the roots of the plant *Curcuma tinctoria* found growing wild in Asia. The dye was well known to the ancient Romans and Greeks who used it to dye wool, cotton, and silk. The dye is an oil-soluble bright yellow material, and is the only natural yellow dye that requires no mordant. It finds use as a colorant for baked goods such as cakes.

(41)

Carmine [1390-65-4] is the trade name for the aluminum lake of the red anthraquinone dye carminic acid obtained from the cochineal bug. The dye is obtained from the powdery form of cochineal by extraction with hot water, the extracts treated with aluminum salts, and the dye precipitated from the solution by the addition of ethanol. This water-soluble bright red dye is used for coloring shrimp, pork sausages, pharmaceuticals, and cosmetics. It is the only animal-derived dye approved as a colorant for foods and other products.

Carotenoids. The carotenoids are a group of widely distributed, highly colored, fat-insoluble, naturally occurring organic compounds. They owe their color to the four repeating isopyrene units found in the molecule and may, therefore, be classified as tetraterpenoids. The carotenoids may be divided into two principal groups, the carotenes, which are strictly hydrocarbons, and the xanthophylls, which contain oxygen. Carotenoids are found in almost all fruits and vegetables, egg yolk, dairy products, and sea foods. The chemistry of the carotenoids has been described in a number of reviews (129).

Although the carotenoids can be obtained from natural sources, it is far more economical to manufacture them for commercial use (130). Three have been manufactured for many years: β-carotene [7235-40-7] (42), canthaxanthin [514-78-3] (43), and β-apo-8'-carotenal [1107-26-2] (44) (131). Their structures are shown in Figure 1.

In general, the low solubility of the carotenoids creates a problem when they are applied as food dyes. β-Carotene, for example, has poor solubility in fats. To overcome this, a microcrystalline powder form has been prepared. This is then mixed into an edible fat. In this form it finds use for coloring margarine, butter, cake mixtures, and other fat-containing foods. β-Carotene is available also as an emulsion, in a water-dispersed form, and as a liquid suspension. Canthaxanthin is commercially available as a 10% water-dispersable beadlet or spray-dried pow-

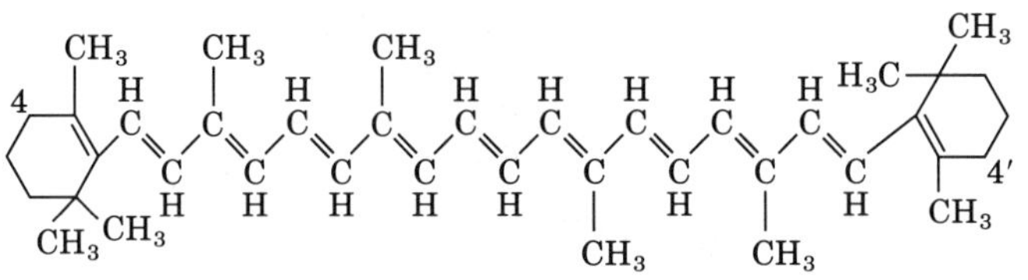

(42)

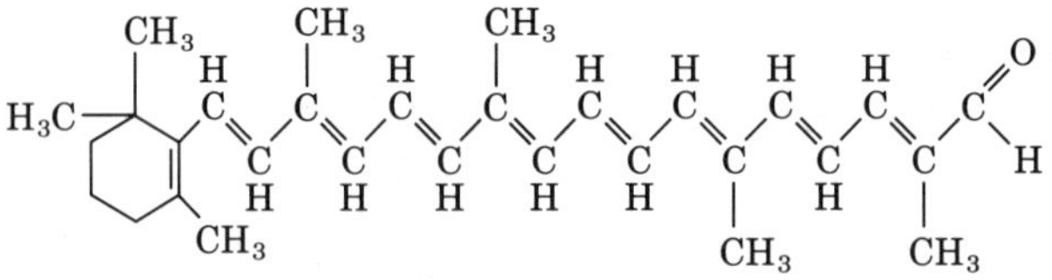

(44)

Fig. 1. Carotenoid pigments: β-carotene (**42**), β-apo-8'-carotenal (**44**), and canthaxanthin (**43**) = structure (**42**) with ketone groups at the 4 and 4' positions.

der. Because of its exceptionally good tomato color-enhancing properties, it finds use in tomato-based products such as pizza and spaghetti sauce. It is useful in water-based foods such as peach ice cream and pink grapefruit beverages. β-Apo-8'-carotenal has high tinctorial strength; because of this, it is marketed in several different strengths, usually as a dispersion in vegetable oils. Its main use is for coloring process cheese and French dressing.

Carotenoids have two general characteristics of importance to the food industry: they are not pH sensitive in the normal 2–7 range found in foods, and they are not affected by vitamin C, making them especially important for beverages. They are more expensive than synthetic food dyes and have a limited color range. In their natural environment they are quite stable, but they become more labile when heated or when they are in solution. Under those conditions, there is a tendency for the trans-double bonds to isomerize to the cis-structure with a subsequent loss of color intensity. The results of controlled tolerance and toxicity tests, using pure carotenoids, indicate that they are perfectly safe as food colors (132).

Bixin [6983-79-5] (CI Natural Orange 4; CI75120) (**45**) is found in the seeds of the plant *Bixa orellana* native to India.

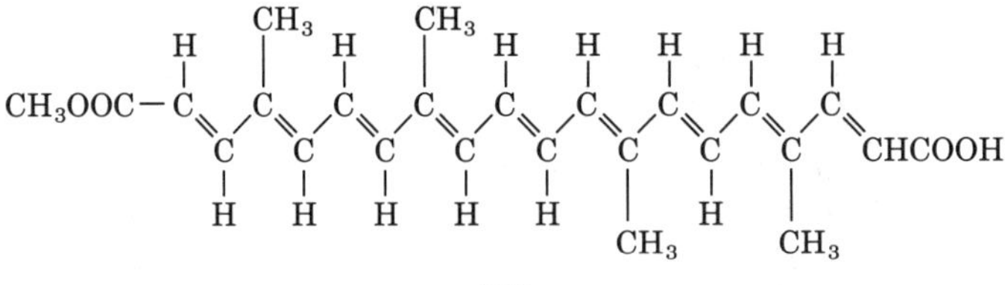

(45)

Later it was found growing in South America where the Indians used the red dye from the seeds as a body paint. An extract of the seeds appears on the market as annatto. This extract is used in coloring butter, margarine, and cheese such as Leicester cheese. In Mexican and South American cuisine, it finds special use as a flavor and coloring matter. The seeds are sold under the name achiote in many Latin grocery stores and markets. Annato is available as an aqueous solution, as an oleaginous dispersion, and a spray-dried powder.

Crocetin [27876-94-4] (CI Natural Yellow 6; CI 75100) occurs in saffron as crocin [42553-65-1] (**46**), the digentiobiose ester of crocetin (**47**).

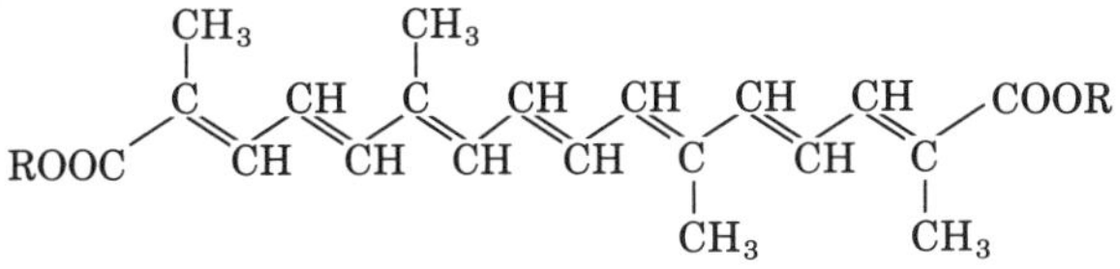

(**46**), (**47**)

R = H for crocetin (**47**) and R = gentiobiose for crocin (**46**)

Saffron is found in the pistils of the plant *Crocus sativus*. Saffron is often confused with safflower, sometimes known as bastard saffron. The name of the plant, *Crocus sativus*, comes from the Arabic word *za faran*, meaning yellow. The Romans and the Greeks used saffron not only as a dye but also as a spice. In the early days of Greece, yellow was the official color, and Grecian women were especially fond of clothes dyed with saffron. Because of its scarcity, saffron ranked among the most expensive dyes of the ancient world.

Betalaines. In 1968, the term betalaines was used to describe collectively two groups of plant pigments: the red betacyanins and the yellow betaxanthins. The red and yellow dyes found in beets, *Beta vulgaris*, fall into this category. An interesting history has been written about these dyes (133).

In 1918, a crude sample of the dye was prepared, as the glycoside betanin, and named betacyane (134). Alkaline hydrolysis of betanin yielded glucose, but the sensitive aglucone could not be isolated. Betanin contains nitrogen and, in some respects, is like an anthocyanidin. This led some investigators to believe that betanin was a nitrogenous anthocyanidin (135). In the meantime, a pure crystalline sample of betanin had been prepared electrophoretically (136). The pure betanin was cleaved enzymatically and yielded the aglucone betanidin (137). Alkaline degradation of betanidin produced 5,6-dihydroxyindol-2-carboxyl acid [4790-08-3] (**48**) and 4-methylpyridine-2,6-dicarboxylic acid [75475-96-6] (**49**) and ammonia (138):

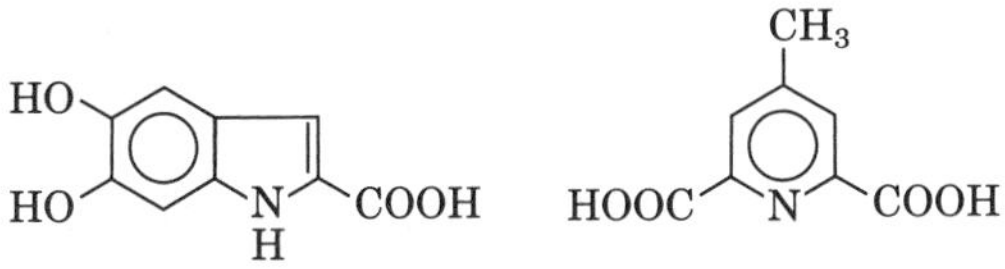

(**48**)   (**49**)

The study of these fragments, coupled with the results of additional research (139), established the structure of betanin [7659-95-2] (**50**). Betanidin [37279-84-8] was synthesized in 1975 (140). A small amount of two yellow pigments is present also in beets (**51**). These have no value as food dyes.

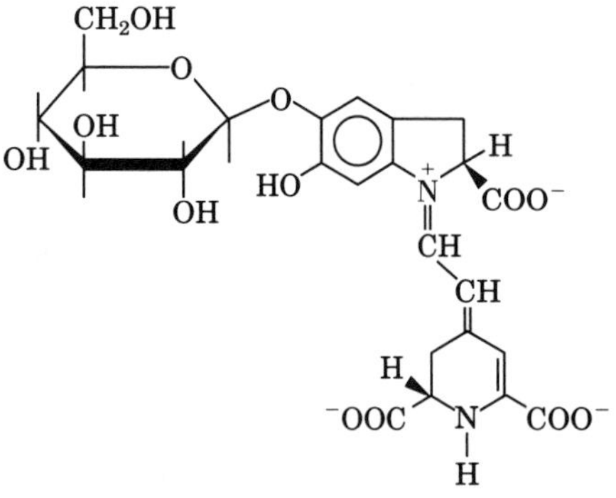

CH2OH
O
O
OH
OH
OH
HO
H
N
COO⁻
CH
CH
H
⁻OOC
N
COO⁻
H

(**50**)

R
O=C
CH2
CH2
CHCOOH
+NH
HOOC
N
COOH
H

where R = NH_2 or OH

(**51**)

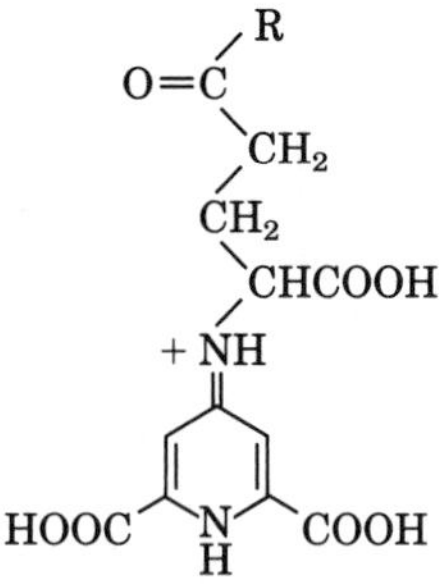

The color of betalaines is barely affected by the pH range normally found in foods. However, the dyes are heat sensitive, which places some limitations on their use as food dyes.

Beet juice contains about 80% of fermentable carbohydrates and nitrogenous compounds. To remove these compounds, a yeast fermentation utilizing *Candida utillis* has been suggested (141). By so doing, a more concentrated form of the dye becomes available. The red dye from beets is sold as beet juice concentrate, as dehydrated beet root, and as a dried powder.

Chlorophyll. The determination of the structure of chlorophyll involved the efforts of many famous chemists, notably R. Willstätter (Nobel Prize 1915) who, during the period 1906–1914, laid the foundation for the future investigation of chlorophyll (142). He not only prepared relatively pure chlorophyll for the first time, but also isolated two different modifications of the molecule. These he designated as a- and b-chlorophyll. H. Fischer (Nobel Prize 1930) suggested the structural formula for chlorophyll based on his work with hemin. In 1960, R. B. Woodward (Nobel Prize 1965) and his co-workers synthesized chlorophyll and hemin (143). Chemically pure chlorophyll is difficult to prepare, since it occurs mixed with other colored substances such as carotenoids. Commercially it is solvent extracted from the dried leaves of various plants such as broccoli or spinach (144). Chlorophyll is water-insoluble. It has none of the characteristics of a dye in that it has no affinity for the usual fibers such as cotton or wool. Chlorophyll is properly classified as a pigment [*8049-84-1*] (CI Natural Green 3; CI 75810). As such, it finds use for coloring soaps, waxes, inks, fats, or oils. Chlorophyll is an ester composed of an acidic part, chlorophyllin, esterified by an aliphatic alcohol known as phytol (**52**):

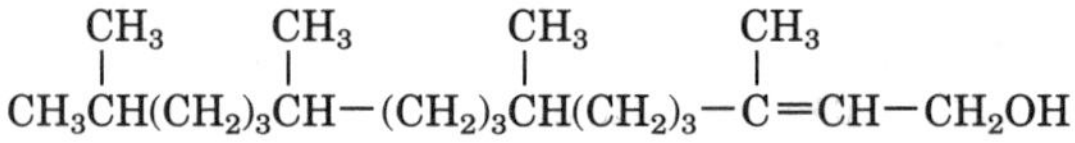

CH3
CH3
CH3
CH3
CH3CH(CH2)3CH−(CH2)3CH(CH2)3−C=CH−CH2OH

(**52**)

Hydrolysis of chlorophyll using sodium hydroxide produces the moderately water-soluble sodium salts of chlorophyllin, phytol and methanol (145). The magnesium in chlorophyllin may be replaced by copper. The sodium copper chlorophyllin salt is heat stable, and is ideal for coloring foods where heat is involved, such as in canning (146).

Health, Safety, and Environmental Factors of Natural Dyes

Natural dyes comprise those colors derived from plant or animal matter without chemical processing. They have been known and used for thousands of years without any reports showing that they are harmful. Modern tests have verified the safety of natural dyes as food colorants; many of these dyes are on the FDA's list of approved food dyes. The FDA no longer looks upon long usage of natural dyes as a criterion of safety. Any new natural dye must conform to the rules and tests established by the FDA. There are no new reports indicating that natural dyes cause health problems, although some fruit juices, colored with natural dyes, have caused diarrhea among children who drank excessive amounts of such juices.

Natural dyes processed for the market do not undergo any chemical operations. Those operations involved are purely physical, such as grinding, spray or vacuum drying, and water or solvent extractions. None of these operations create any great environmental problems.

The use of natural dyes as food colorants evolved over a period spanning thousands of years. During that period, by trial and error, some dyes were found to be safe while others were not. By comparison, the development of synthetic dyes as food colorants has taken place over a comparatively short time. During that period, some synthetic dyes considered safe by existing health standards were used as food colors. Later, with increased knowledge, these were found to create health problems and were removed from the marketplace. The manufacture of synthetic dyes for use on foods creates more of a health and environmental problem than natural dyes, but offers greater variety and stability of color (see DYES, ENVIRONMENTAL CHEMISTRY).

BIBLIOGRAPHY

"Chlorophyll" in *ECT* 1st ed., Vol. 3, pp. 871–882, by P. Rothemund, Charles F. Kettering Foundation; "Natural Dyes" under "Dyes" in *ECT* 1st ed., Vol. 5, pp. 345–354, by G. E. Goheen and J. Werner, General Aniline & Film Corp.; "Plant Derivatives" under "Tints, Hair Dyes and Bleaches" in *ECT* 1st ed., Vol. 14, pp. 169–177, by F. E. Wall, Consulting Chemist; "Chlorophyll" in *ECT* 2nd ed., Vol. 5, pp. 339–356, by P. Rothemund, Consulting Chemical Engineer; "Dyes, Natural" in *ECT* 2nd ed., Vol. 7, pp. 614–629, by A. J. Cofrancesco, General Aniline & Film Corp.; in *ECT* 3rd ed., Vol. 8, pp. 351–373, by R. E. Farris, Sandoz Colors & Chemicals.

1. R. H. Thomson, *The Naturally Occurring Quinones*, 1st ed., Academic Press, Inc., New York, 1957, p. 162.
2. R. Wijnsma and R. Verpoorte, *Progress in the Chemistry of Organic Natural Products*, Springer-Verlag, Wien, 1986, p. 83.
3. H. R. Schweizer, *Künstliche Organische Farbstoffe und ihre Zwischenprodukte*, Springer-Verlag, Berlin, 1964, p. 303.
4. K. Meyers, *Dyes From Nature*, Brooklyn Botanic Garden, New York, 1990, p. 11.
5. S. Robinson, *A History of Dyed Textiles*, MIT Press, Cambridge, Mass., 1969, p. 16.

6. B. Brown, *Text. Color.* **65,** 143 (1943).

7. C. H. Thomson, *Naturally Occurring Quinones,* 3rd ed., Chapman and Hall, New York, 1987, p. 358; H. M. Fox and G. Vevers, *The Nature of Animal Colours,* Sidgwick and Jackson Ltd., London, 1960, p. 140.

8. A. G. Perkin and A. E. Everest, *The Natural Organic Colouring Matters,* Longmans, Green and Co., New York, 1981, p. 32.

9. J. Storey, *Dyes and Fabrics,* Thames and Hudson, Ltd., London, 1978, p. 73.

10. R. J. Adrosko, *Natural Dyes and Home Dyeing,* Dover Publications, Inc., New York, 1971, p. 21.

11. A. G. Perkin and G. F. Attree, *J. Chem. Soc.,* 146 (1931).

12. B. P. Geyer and G. M. Smith, *J. Am. Chem. Soc.* **64,** 1649 (1942).

13. H. E. Fierz-David and M. Rutishauser, *Helv. Chim. Acta* **23,** 1298 (1940).

14. H. L. Haller, *Helv. Chim. Acta* **23,** 466 (1940); K. Venkataraman, *The Chemistry of Synthetic Dyes,* Vol. 1, Academic Press, Inc., New York, 1952, p. 279, and Ref. 9, p. 72.

15. C. Schunck, *Ann. Chem.* **66,** 174 (1848).

16. F. Rochleder, *Ann. Chem.* **80** (1851).

17. D. Richter, *J. Chem. Soc.,* 1701, (1936); G. Zemplen and R. Bognar, *Ber. Deut. Chem. Ges.* **72,** 913 (1939).

18. L. F. Fieser, *J. Chem. Edu.* **7,** 2609 (1930).

19. *FIAT Report 1313* **2,** 58 (1948).

20. U.S. Pat. 1,446,163 (Feb. 20, 1923), A. H. Davies (to Scottish Dyes); U.S. Pat. 1,744,815 (Jan. 28, 1930), J. Thomas and H. W. Hereward (to Scottish Dyes).

21. F. Mayer, *Chemie der Organischen Farbstoffe,* Julius Springer, Berlin, 1934, p. 103.

22. J. Wood in Ref. 4, p. 26.

23. W. F. Leggett, *Ancient and Medieval Dyes,* Chemical Publishing Co., New York, 1964, p. 69.

24. R. Heise, *Arbeit. Kaiserl. Gesund.* **11,** 513 (1895).

25. D. W. Cameron and co-workers, *Aust. J. Chem.* **34,** 2401 (1981).

26. G. Roberge and P. Brassard, *J. Chem. Soc., Perkin Trans.* 1041 (1978).

27. F. L. C. Baranyovits, *Endeavour* **2,** 85 (1978); Ref. 23, p. 82.

28. P. J. Pelletier and J. B. Caventou, *Ann. Chim. Phys.* **8,** 250 (1818).

29. P. Schutzenberger, *Ann. Chim. Phys.* **54,** 52 (1858).

30. R. Furth, *Ber. Deut. Chem. Ges.* **16,** 2169 (1883).

31. C. Liebermann and H. Voswinkel, *Ber. Deut. Chem. Ges.* **30,** 688 and 1731 (1897); O. Dimroth, *Ber. Deut. Chem. Ges.* **43,** 1387 (1910).

32. A. Oppenheim and S. Pfaff, *Ber. Deut. Chem. Ges.* **7,** 929 (1874).

33. O. Dimroth and H. Kammerer, *Ber. Deut. Chem. Ges.* **53,** 471 (1920).

34. M. A. Ali and L. J. Haynes, *J. Chem. Soc.,* 1033 (1959).

35. J. C. Overeem and G. J. M. van der Kerk, *Rec. Trav. Chim.* **83,** 1023 (1964).

36. H. Mühlemann, *Pharm. Acta Helv.* **26,** 204 (1951).

37. S. B. Bratia and K. Venkataramann, *Indian J. Chem.* **3,** 92 (1964).

38. R. H. Thomson, *Naturally Occurring Quinones,* 3rd ed., Academic Press, Inc., New York, 1987, p. 469.

39. R. E. Schmidt, *Ber. Deut. Chem. Ges.* **20,** 1285 (1887).

40. O. Dimroth and S. Goldschmidt, *Ann. Chem.* **399,** 62 (1913).

41. S. Coffey, *Chemistry of Carbon Compounds,* Vol. 3b, Elsevier, Amsterdam, 1956, p. 1421.

42. E. D. Pandhare and co-workers, *Tetrahedron, Suppl.* **8,** 229 (1966).

43. R. Burwood and co-workers, *J. Chem. Soc.,* 6067 (1965).

44. G. Tommasi, *Gazz. Chim. Ital.* **59,** 263 (1920).

45. J. B. La and S. Dutt, *J. Indian Chem. Soc.* **10** (1933).

46. L. F. Fieser, *J. Am. Chem. Soc.* **70,** 3165 (1948); *Reagents for Organic Synthesis*, John Wiley & Sons, Inc., New York, 1967, p. 71.

47. W. Stein, *J. Prakt. Chem.* **99,** 1 (1866).

48. E. Paternó, *Gass. Chim. Ital.* **4,** 505 (1879); **12,** 337 and 622 (1882).

49. S. C. Hooker, *J. Chem. Soc.* **69,** 1355 (1896).

50. A. Vogel and C. Reischauer, *Buchner Neues Rep. fur Pharm* **5,** 106 (1856).

51. R. Buchanan, *A Weaver's Garden*, Interweave Press, Loveland, Colo., 1987, p. 97; R. J. Adrosko, *Natural Dyes in the United States*, Smithsonian Institution Press, Washington, D.C., 1968, p. 39.

52. A. Bernthsen and A. Semper, *Ber. Deut. Chem. Ges.* **18,** 203 (1885).

53. *Ibid.* **20,** 934 (1887).

54. F. Mylius, *Ber. Deut. Chem. Ges.* **17,** 2411 (1884); **18,** 2567 (1885).

55. N. F. Hayes and R. H. Thomson, *J. Chem. Soc.*, 904 (1955).

56. Ref. 53, p. 930.

57. H. J. Teuber and N. Gotz, *Ber. Deut. Chem. Ges.* **87,** 1236 (1954).

58. D. J. Crouse, M. M. Wheeler, and M. Goemann, *J. Organ. Chem.* **46,** 1814 (1981).

59. R. Buchanan in Ref. 51, p. 87.

60. H. Gunther, *The Greek Herbal of Dioscorides*, Hafner, New York, 1959, p. 42.

61. V. P. Papageorgiou, *Experientia* **34,** 1499 (1978).

62. A. Lohmann, *Ber. Deut. Chem. Ges.* **68,** 1487 (1935).

63. Ref. 1, p. 111.

64. H. Brockmann, *Ann. Chem.* **521,** 1 (1936); H. Brockmann and H. Roth, *Naturwissenschaften* **23,** 246 (1935).

65. K. W. Bentley, *The Natural Pigments*, Interscience Publishers, New York, 1960, p. 204.

66. H. Brockman, *Ann. Chem.* **521,** 1 (1936); H. Brockmann and K. Muller, *Ann. Chem.* **540,** 51 (1939).

67. H. Fukui, M. Tabata, and N. Yoshikawa, *Phytochemistry* **22,** 451 (1983).

68. A. Terada and co-workers, *J. Chem. Soc. Comm.*, 987 (1983).

69. S. von Kostanecki and J. Jambor, *Ber. Deut. Chem. Ges.* **28,** 2302 (1895).

70. G. F. Attree and A. G. Perkin, *J. Chem. Soc.*, 234 (1927).

71. C. Kuroda, *J. Chem. Soc.*, 752 (1930).

72. J. Piccard, *Ber. Deut. Chem. Ges.* **6,** 884 (1873); **7,** 888 (1874); **10,** 176 (1877).

73. S. von Kostanecki, *Ber. Deut. Chem. Ges.* **26,** 2901 (1893).

74. J. Czajkowski, S. von Kostanecki, and J. Tambor, *Ber. Deut. Chem. Ges.* **33,** 1988 (1900).

75. T. Emilewica, S. von Kostanecki, and J. Tambor, *Ber. Deut. Chem. Ges.* **32,** 2448 (1899).

76. J. Allan and R. Robinson, *J. Chem. Soc.* **125,** 2334 (1926).

77. J. Tambor, *Ber. Deut. Chem. Ges.* **45,** 1701 (1913).

78. H. M. Wickens in Ref. 4, p. 68.

79. R. Buchanan in Ref. 51, p. 99.

80. *Colour Index*, Vol. 3, 3rd ed., Society of Dyers and Colourists, Bradford, UK, 1971, pp. 3225–3255.

81. M. E. Chevreul, *Ann. Chim. Phys.* **81,** 128 (1812); **82,** 53 and 126 (1812).

82. A. Werner and P. Pfeiffer, *Chemiker-Ztg.* **3,** 388 and 420 (1920).

83. W. H. Perkin, Jr. and R. Robinson, *J. Chem. Soc.* **93,** 489 (1920); F. Mayer, *The Chemistry of Natural Colouring Matters, ACS Monograph No. 89*, trans. and rev. by A. H. Cook, Reinhold Publishing Corp., New York, 1943, p. 241.

84. C. D. Mell, *Text. Color.* **51,** 820 (1929).

85. L. C. Marquart, *Eine Chemisch-Physiol.*, Abhandlung, Bonn, 1835.

86. R. Willstätter and A. E. Everest, *Ann. Chem.* **401,** 189 (1913).

87. D. D. Pratt and R. Robinson, *J. Chem. Soc.* **121,** 1577 (1922).

88. R. Willstätter and co-workers, *Ann. Chem.* **408** 1, 15, 42, 61, 83, 110, and 147 (1915); **412,** 113 (1916); *Ber. Deut. Chem. Ges.* **47,** 2865 (1914).

89. P. Karrer in Klein, ed., *Handbuch der Pflanzenanalyse*, Vol. 3, Springer-Verlag, Vienna, 1933, pp. 941–984; P. Karrer and co-workers, *Helv. Chim. Acta* **10,** 67 and 729 (1927); **12,** 292 (1929); **15,** 507 (1932).

90. R. Robinson, *J. Soc. Chem. Ind.* **52,** 737 (1933); *Nature* **132,** 625 (1933); **135,** 732 (1935); **137,** 94 (1936); *Ber. Deut. Chem. Ges.* **67A,** 85 (1934).

91. T. A. Geissman, ed., *The Chemistry of Flavonoid Compounds*, Macmillan Co., New York, 1962, p. 274.

92. J. C. Bell, R. Robinson, and A. R. Todd, *J. Chem. Soc.*, 806 (1934); R. Robinson and A. R. Todd, *J. Chem. Soc.*, 2299 (1932).

93. R. Willstätter and L. Zechmeister, *Sitzber. Preuss. Akad., Chem. Zeit.* **11,** 1359 (1914).

94. D. D. Pratt and R. Robinson, *J. Chem. Soc.* **121,** 1577 (1922).

95. P. Friedlander, *Ann. Chem.* **351,** 390 (1906); *Ber. Deut. Chem. Ges.* **41,** 765 (1909); *Angew. Chem.* **22,** 990 and 2494 (1909).

96. E. Ploss, *BASF Dig.* **1,** 16 (1963).

97. W. Cole, *Phil. Trans.* **15,** 1278 (1685); H. Munro Fox and G. Vevers, *The Nature of Animal Colours*, Sidgwick and Jackson Ltd., London, 1960, p. 61.

98. G. Bizio, *Ber. Deut. Chem. Ges.* **6,** 142 (1873); A. and G. de Negri, *Ber. Deut. Chem. Ges.* **9,** 84 (1876); E. Schunck, *Ber. Deut. Chem. Ges.* **12,** 1358 (1879); **13,** 2087 (1880); R. Dubois, *Arch. Zool. Exp. Gen.* **2**(5), 471 (1909).

99. P. Friedlander, *Monatsh.* **28,** 991 (1907).

100. *Ibid.*, **30,** 247 (1909).

101. J. T. Baker and M. D. Sutherland, *Tetrahedron Lett.* **1** (1968).

102. H. Fouquet and H. J. Bielig, *Ange. Chem.* (international ed.) **10,** 816 (1971); J. T. Baker and C. C. Duke, *Aust. J. Chem.* **26,** 2153 (1973).

103. J. T. Baker, *Endeavour* **33,** 11 (1974).

104. J. Beckmann, *Text. Color.* **43,** 35 (1921).

105. W. C. Sumpter and F. M. Miller, *Heterocyclic Compounds with Indol and Carbazole Systems*, Interscience Publishers, New York, 1954, p. 173.

106. R. Buchanan in Ref. 51, p. 106.

107. B. B. Kling, *The Blue Mutiny*, University of Pennsylvania Press, Philadelphia, 1966.

108. R. J. Adrosko, *Natural Dyes in the United States*, Smithsonian Institution Press, Washington, D.C., 1968, p. 15; M. Bonta in Ref. 4, p. 33.

109. L. Marchiewski and L. G. Radcliffe, *J. Soc. Chem. Ind.* **17,** 430 (1898).

110. A. G. Perkin and F. Thomas, *J. Chem. Soc.* **95,** 795 (1900); A. Robertson, *J. Chem. Soc.*, 1937 (1927).

111. K. Venkataraman, *The Chemistry of Synthetic Dyes*, Vol. 2, Academic Press, Inc., New York, 1952, p. 1007.

112. Ref. 4, p. 16.

113. N. V. Sidgwick, *The Organic Chemistry of Nitrogen*, Clarendon Press, Oxford, 1937, p. 507.

114. Ref. 6, p. 336.

115. Ref. 6, p. 18.

116. Ref. 51, p. 111.

117. *Dye Plants and Dyeing*, Brooklyn Botanic Garden, New York, p. 34.

118. Ger. Pat. 11,857 (Mar. 1886), A. Baeyer (to BASF); A. von Baeyer, *Ber. Deut. Chem. Ges.* **13,** 3254 (1880); *Frdl.* **1,** 127 (1888).

119. A. Reis and W. Schneider, *Z. Krist.* **68,** 543 (1928); Chem. Abst. **23,** 2083 (1929).

120. K. Kunz, *Ber. Deut. Chem. Ges.* **55,** 3688 (1922).

121. Ref. 111, p. 1010.

122. W. Flimm, *Ber. Deut. Chem. Ges.* **23,** 57 (1890).
123. K. Heumann, *Ber. Deut. Chem. Ges.* **23,** 3048 and 3431 (1890).
124. Ger. Pat. 54,626 (May 1890), (to BASF and Hoechst).
125. Ger. Pat. 137,955 (Feb. 2, 1901), J. Pfleger (to Gold-and Silber-scheider-Anst).
126. J. G. Kern and M. Stenerson, *Chem. Eng. News* **24,** 3164 (1964).
127. R. M. Schaffner in T. E. Furia, ed., *Current Aspects of Food Colorants*, CRC Press, Inc., Cleveland, Ohio, 1977, p. 85.
128. A. S. Dreidine in W. D. Ollis, ed., *The Chemistry of Natural Phenolic Compounds*, Pergamon Press, Inc., Elmsford, N.Y., 1961, p. 194.
129. A. Winterstein, *Angew. Chem.* **72,** 902 (1960); L. Zechmeister, *Fortschritte der Chemie der Organischer Naturstoffe*, Vol. 18, Verlag von Julius Springer, Wien, 1960, p. 223.
130. O. Isler and co-workers, *Helv. Chim. Acta.* **39,** 249 (1956); A. Businger and co-workers, *J. Sci. Ind. Res. (India)* **17A,** 502 (1958).
131. O. Isler, R. Ruegg, and P. Schudel, *Chima* **15,** 208 (1961).
132. R. E. Bagdon, G. Zhinden, and A. Studer, *Toxicol. Appl. Pharmacol.* **2,** 225 (1960).
133. a. S. Dreidine in Ref. 128.
134. H. Wylerand and A. S. Dreiding, *Experentia* **17,** 23 (1961); G. Schudel, *Diss. Zurich-ETH,* (1918).
135. G. M. Robinson and R. Robinson, *J. Chem. Soc.,* 1439 (1932).
136. H. Wyler and A. S. Dreiding, *Helv. Chim. Acta* **40,** 191 (1957).
137. M. Piatteli, L. Minale, and G. Prota, *Annali Chim.* **54,** 955 (1964).
138. H. Wyler and A. S. Dreiding, *Helv. Chim. Acta* **42,** 1699 (1959).
139. *Ibid.,* **45,** 640 (1962); H. Wyler, T. J. Mabry, and A. S. Dreiding, *Helv. Chim. Acta* **46,** 1745 (1963).
140. K. Hermann and A. S. Dreiding, *Helv. Chim. Acta* **58,** 1805 (1975).
141. Ref. 127, p. 37.
142. R. Willstätter and A. Stoll, *Investigation of Chlorophyll*, trans. F. M. Scherta and A. R. Merz, Science Press, Lancaster, Pa., 1928.
143. R. B. Woodward and co-workers, *J. Am. Chem. Soc.* **82,** 3800 (1960).
144. F. Ullmann, *Enzykopaedie der Techischen Chemie*, 3rd ed., Auflage, Urban und Schwrzenberg, Berlin, 1956, p. 138.
145. H. G. Petering, P. W. Morcal, and E. J. Mueller, *Ind. Eng. Chem.* **33,** 1428 (1941).
146. W. H. Shearon, Jr. and O. F. Gee, *Ind. Eng. Chem.* **41,** 218 (1949).

A. J. Cofrancesco
Consultant

DYES, REACTIVE

Reactive dyes are those dyes containing electrophilic functional groups capable of reacting with a nucleophile to form a covalent bond either through addition or displacement. Nucleophiles within fibers that typically react with dyes are hydroxyl groups in cellulose, amino, hydroxyl, and thiol groups in wool, and amino groups in polyamide. The outstanding characteristic feature of reactive dyes is their high wetfastness properties attributed to covalent bonding, an advantage over those dyes fixed through adsorption or mechanical entrapment. Unlike large bulky direct dyes, since reactive dyes are chemically bonded to the substrate, smaller molecules giving brighter colors are possible. The principal use of reactive dyes is far and away greatest for cellulose (cotton and rayon), followed by wool, with polyamide (nylon) a very distant third. Applications to silk and leather represent only very minor uses (see DYES, APPLICATION AND EVALUATION).

History

The basic concept of chemical combination between a dye and cellulose has been credited to Cross and Bevan, who in 1895 esterified cellulose with benzoyl chloride and used the benzene ring for nitration, reduction, diazotization, and coupling to give dyes (1). The first wool reactive dye was introduced in 1930. It was an acid dye (1) with a chloroacetyl reactive group, an azo dye made by coupling aniline to chloroacetyl J-acid (2). It was not realized at that time that the observed high fastness to wash was due to chemical bonding.

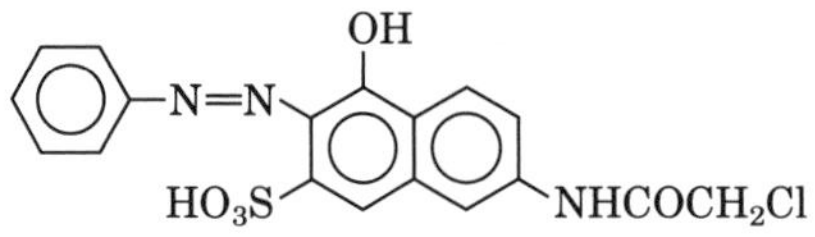

(1)

In 1956, ICI commercialized the first dyes promoted as reactive dyes for cotton. These dyes were marketed under the trade name Procion and contained the dichlorotriazine reactive group bridged to dye through an amino group.

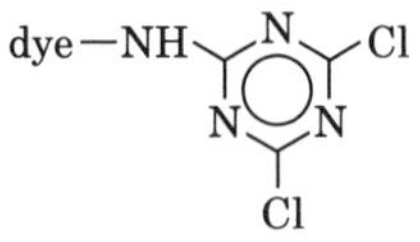

Then in 1957 ICI and CIBA introduced monochlorotriazine.

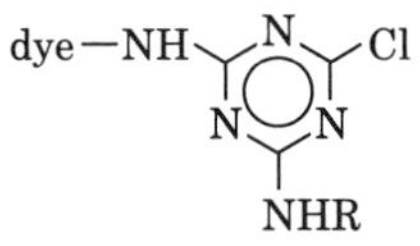

In 1958, Hoechst introduced the vinyl sulfone reactive dyes, offered as the sulfato-ester from which the reactive vinyl sulfone group was generated in the alkaline dye bath.

$$\text{dye} - SO_2CH_2CH_2OSO_3H \longrightarrow \text{dye} - SO_2CH = CH_2$$

It appears that several dye manufacturers then began looking for reactive groups outside those patented. Geigy and Sandoz introduced 2,4,5-trichloro-pyrimidine in 1959.

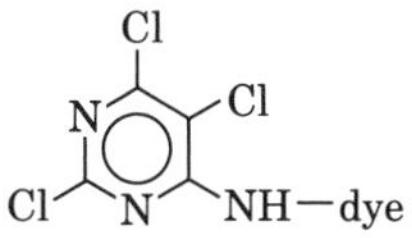

Bayer introduced 2-sulfatoesters of *N*-ethyl sulfonamide in 1958, 2,3-dichloro-quinoxalines in 1961, and 2-methylsulfonyl-4-methyl-6-chloropyridine in 1966.

dye—SO$_2$NHCH$_2$CH$_2$—OSO$_3$H

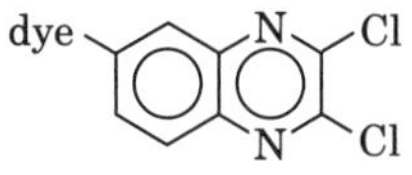

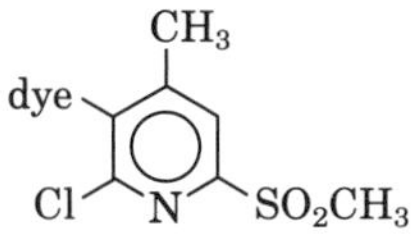

BASF made its entry in 1964 with dichloropyridazone.

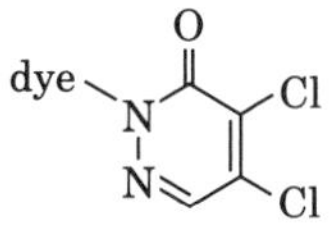

More than 40 reactive groups have been reported since 1956 (3). Interestingly neither Du Pont, GAF, Allied, American Cyanamide, nor American Aniline successfully introduced a line of fiber-reactive dyes, and, except for GAF, left relatively little evidence of activity in the area. ICI and CIBA were the first to introduce reactive dyes designed specifically for wool and polyamide. ICI introduced metal complex dyes containing an acrylamide reactive group, dye—NHCOCH=CH$_2$, in 1964, and CIBA introduced acid dyes containing the α-bromoacrylamide group dye—NHCOCBr=CH$_2$, in 1966. Both Bayer and Sandoz introduced reactive dyes for wool in 1970 containing 2,4-difluoro-5-chloropyrimidine groups. Other reactive groups have been reviewed (4).

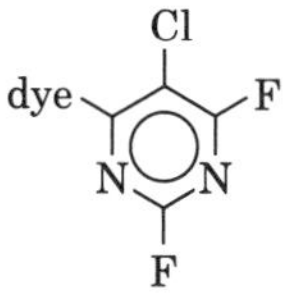

Dyes with phosphonic acid reactive groups were commercialized by ICI under the trade name Procion T (5) and those with monofluorotriazine by CIBA-GEIGY as Cibacron F.

dye—PO(OH)$_2$

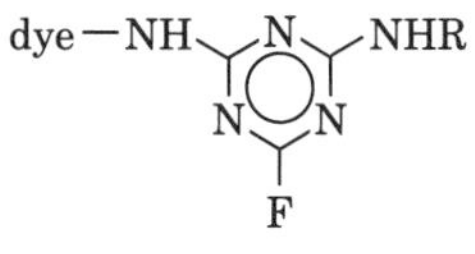

The phosphonic acid reactive dyes were applied to cellulose under slightly acid pH rather than alkaline pH required for other cellulosic reactive dyes. This feature made them especially attractive for one bath application with disperse dyes to cotton/polyester blends. A review of these dyes appears in Reference 5.

Table 1. Yellow and Orange Reactive Dyes

Name	Structure number	CAS Registry Number	CI number	Structure
Reactive Yellow 5	(2)	[56275-25-3]	11859	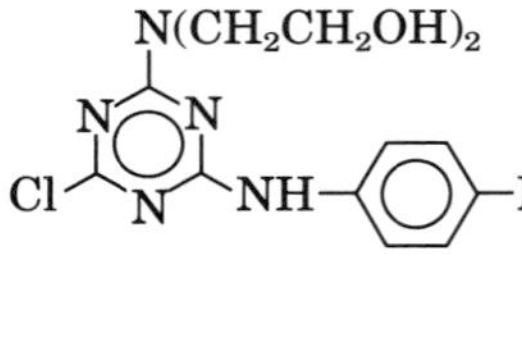
Reactive Yellow 13	(3)	[12769-09-4]	18990	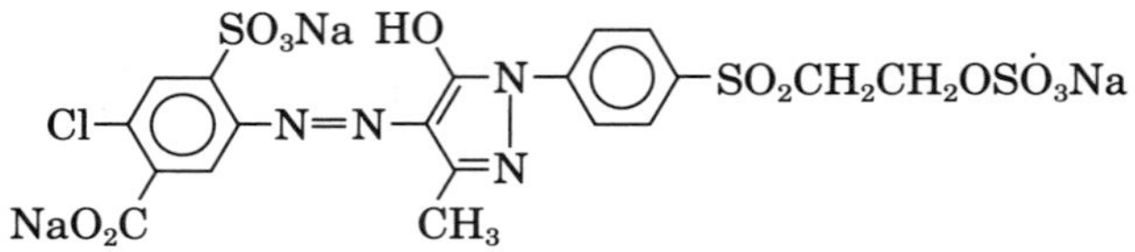
Reactive Yellow 14	(4)	[18976-74-4]	19036	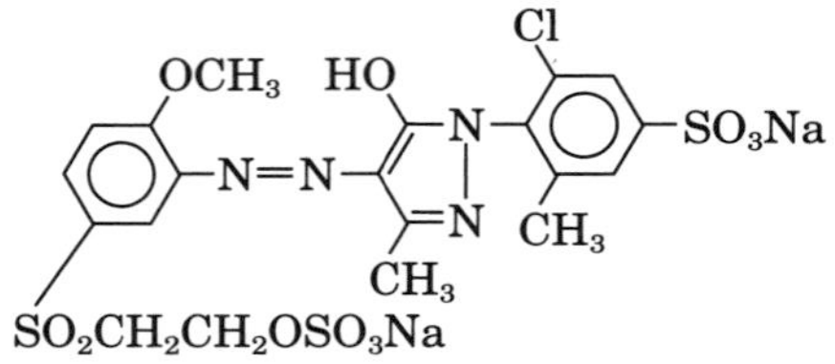
Reactive Yellow 95	(5)	[89923-43-3]		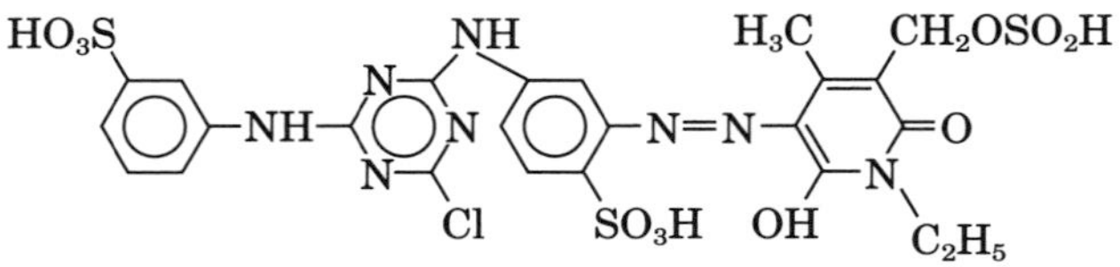
Reactive Orange 14	(6)	[12225-86-4]	19138	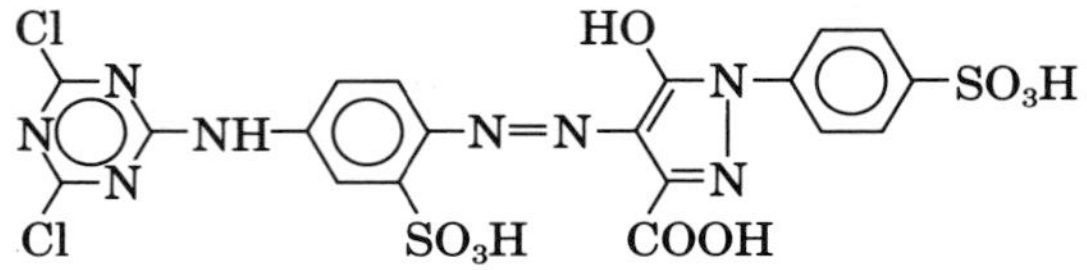

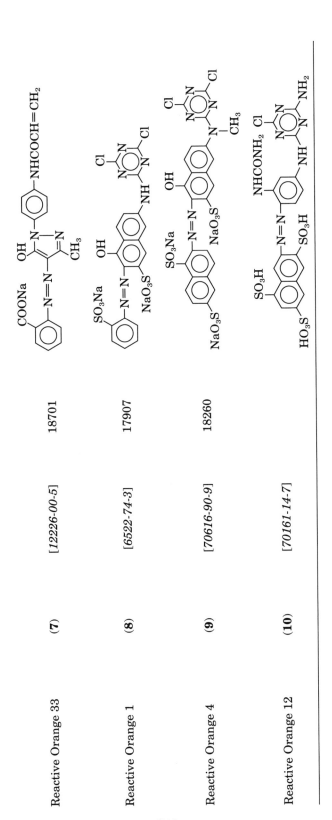

Reactive Orange 33	(7)	[12226-00-5]	18701
Reactive Orange 1	(8)	[6522-74-3]	17907
Reactive Orange 4	(9)	[70616-90-9]	18260
Reactive Orange 12	(10)	[70161-14-7]	

Table 2. Red and Violet Reactive Dyes

Name	Structure number	CAS Registry Number	CI number	Structure
Reactive Red 2	(12)	[17804-49-8]	18200	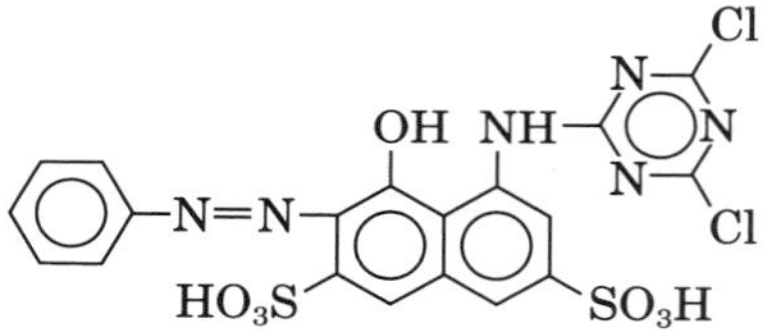
Reactive Red 24	(13)	[72829-25-5]		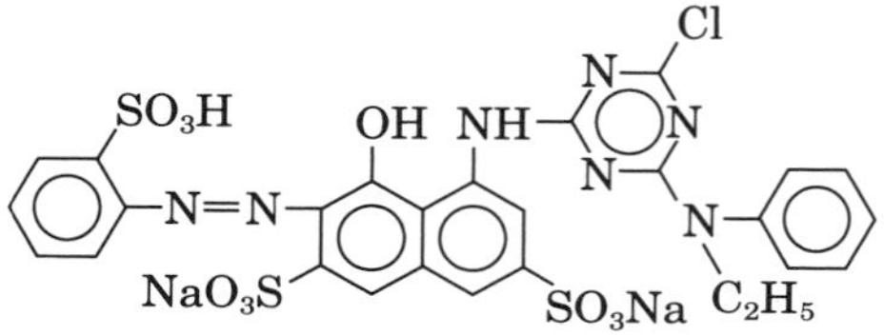
Reactive Red 33	(14)	[12237-01-3]	18280	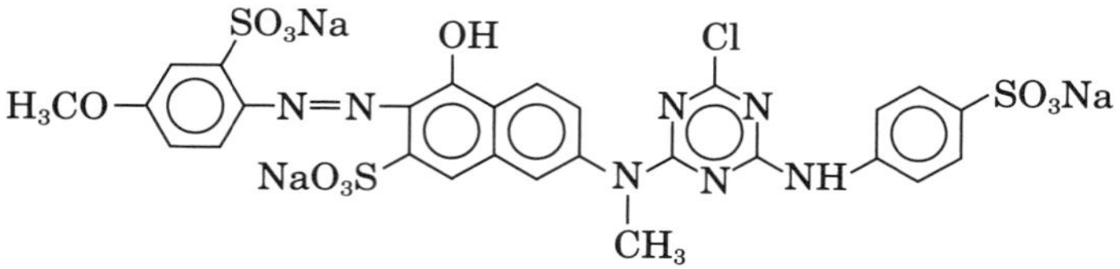
Reactive Red 218	(15)	[84045-65-8]		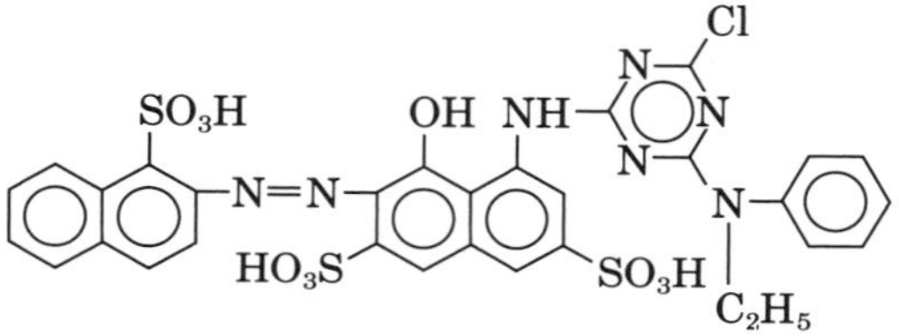

814

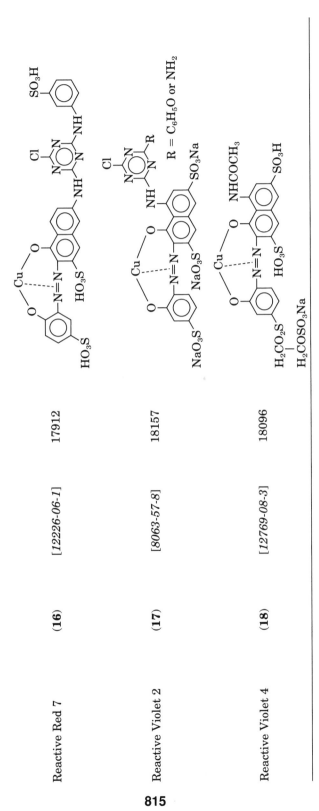

Reactive Red 7	(16)	17912	[12226-06-1]
Reactive Violet 2	(17)	18157	[8063-57-8]
Reactive Violet 4	(18)	18096	[12769-08-3]

Reactive Dye Structure

Reactive dyes consist basically of three components: a dye, a bridging group (B), and the reactive group (R), dye—B—R. The reactive group may be considered in two parts as a carrying group and the reactive component.

Theoretically, the dye or chromogen can be any colored species. Of course, requirements for fastness, solubility, tinctorial value, ecology, and economy must be met. Most commonly used chromophores parallel those of other dye classes. Azo dyes (qv) represent the largest number with anthraquinone and phthalocyanine making up most of the difference. Metallized azo and formazan dyes are important and have gained in importance as a chromophore for blue dyes during recent years (6) (see DYES AND DYE INTERMEDIATES).

Yellow dyes are generally monoazo, and most are pyrazolone or pyridone couplings (7). Orange dyes are generally monoazo derived from couplings to pyrazolones or of slightly substituted phenyl and naphthyl groups (7) (Table 1).

Many red dyes are based on H-acid [90-20-0] (11), eg, Reactive Reds 2, 24, and 218. Others are substituted phenyl and naphthyl or metallized systems (7) (Table 2). Violet dyes are also metallized monoazo dyes (7).

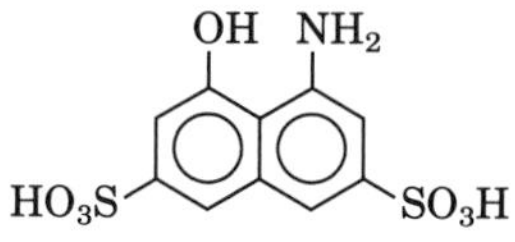

(11)

Blue dyes are derived from anthraquinone, phthalocyanine, or metallized formazan (7) (see DYES, ANTHRAQUINONE) (Figs. 1 and 2). There are also oxazine and thiazine dyes reported (13) (see AZINE DYES) (Fig. 2).

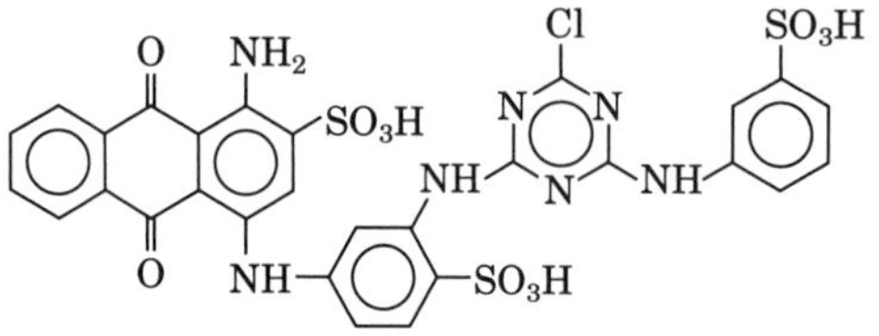

(19)

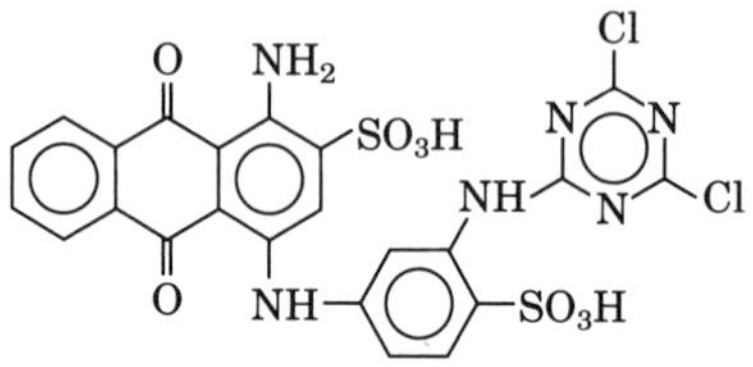

(20)

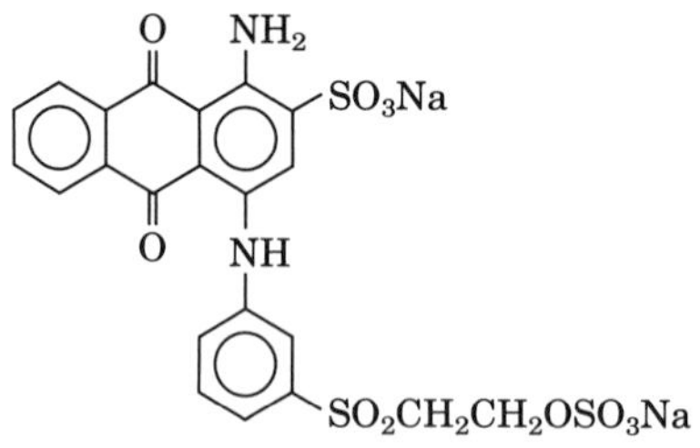

(21)

Fig. 1. Blue reactive dyes from anthraquinone. Also see Table 6. Reactive Blue 5 [16823-51-1] (CI 61210) (19), Reactive Blue 4 [13324-20-4] (CI 61205) (20), eg, Procion Blue MX-R, Reactive Blue 19 [2580-78-1] (CI 61200) (21), eg, Remazol Brilliant Blue R.

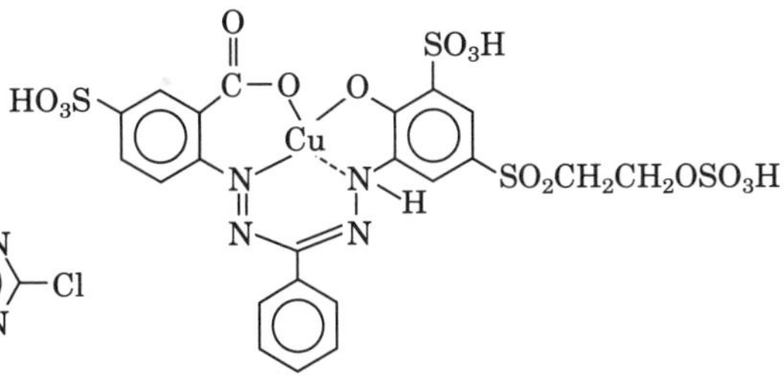

Fig. 2. Phthalocyanine (**22**), and metallized formazan (**23, 24**) and azine (**25**) blue reactive dyes. Reactive Blue 15 [*12225-39-7*] (CI 74459) (**22**); Reactive Blues (**23**) [*78709-74-7*] (8) and (**24**) (9); Reactive Blue 204 [*109125-56-6*] (**25**). Other blue oxazine dyes are other fluorotriazines [*97140-65-3*] (10,11), and a dichlorotriazine [*58104-86-2*] (12).

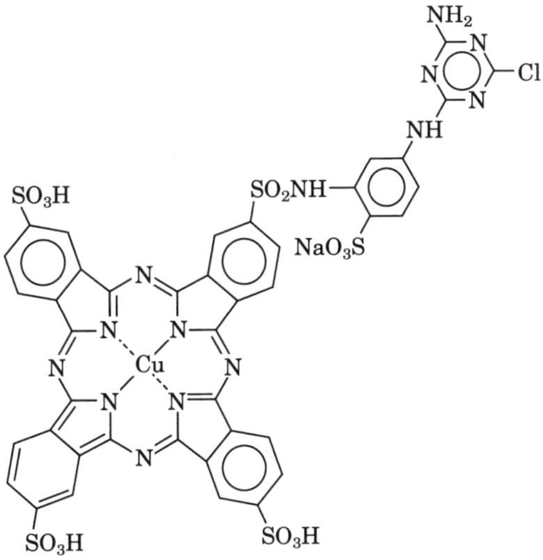

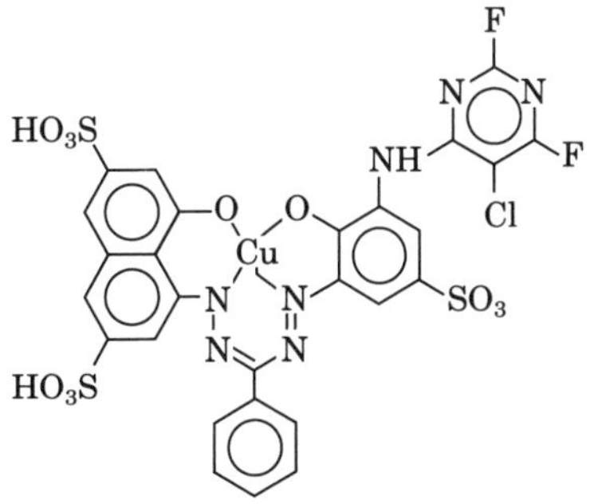

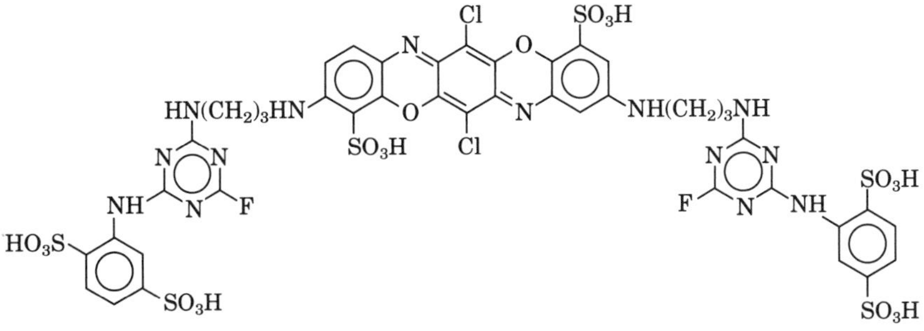

Brown and black dyes are generally disazo with exceptions for metallized or polycyclic structures (7). Two disazo dyes are Reactive Brown 11 [*70161-17-0*] (**26**) and Reactive Black 5 [*17095-24-8*] (CI 20505) (**27**).

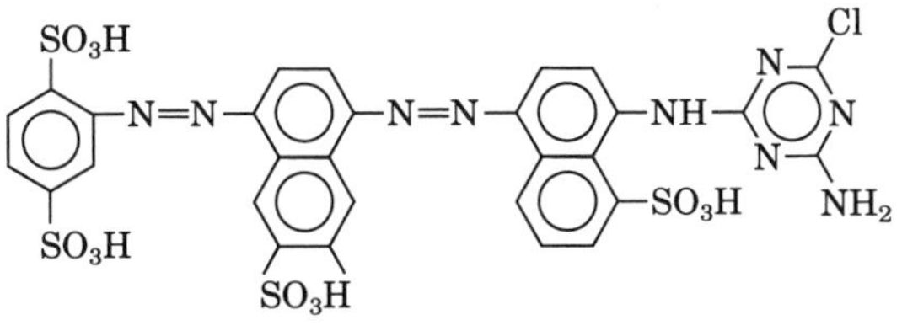

50:50   (**26**)

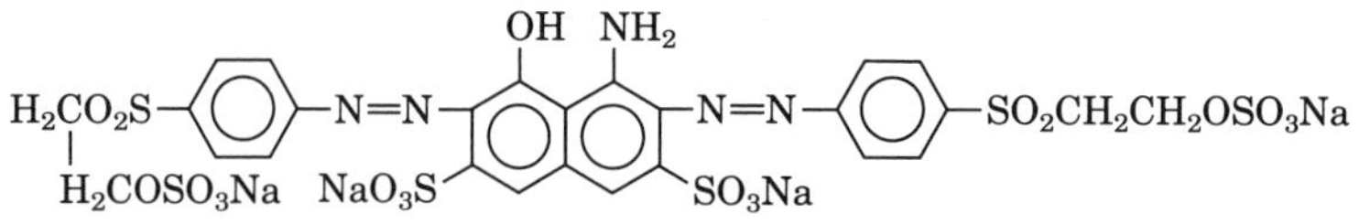

(27)

Green dyes are obtained by bridging an anthraquinone blue chromogen with a yellow chromogen, as in the following reactive green (**28**) [72090-52-9] (14) or from phthalocyanine (7).

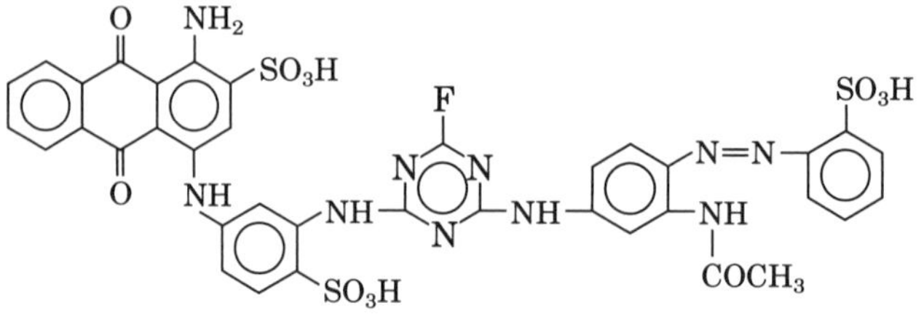

(28)

Properties

Reactive Groups. Although fastness of reactive dyes to washing is good, once fixed on the substrate, hydrolysis during application leads to low dye fixation rate, ie, dye that hydrolyzes in the application process is not fixed on the fiber. Reactive dyes are highly sulfonated, and as such are very water-soluble. Fairly large amounts of salt are required to force the dye into the substrate during application. The hydrolyzed dye is even more soluble than the dye, and presents concerns for the dyer because of effluent color and low biodegradability as well as economic losses. A very large part of the development effort spent on reactive dyes during recent years has been toward improving fixation on the substrate. Approaches taken have been to find different reactive groups as well as to include more than one reactive group in the dye structure and select reactive groups less sensitive to hydrolysis or more reactive to the substrate under application conditions. Sumitomo was the first to commercialize dyes (as Sumifix Supra dyes) with two different reactive groups, a medium and a high energy group (5). However, both ICI and Hoechst had patented bifunctional systems nearly 20 years earlier (5).

During the 1980s, there was a revived interest in bireactive dyes. Every principal dye manufacturer has introduced dyes with more than one reactive group. The ones offering highest fixation have two or more reactive groups with different rates of reaction. Different rates of reaction may be due to selection of different groups, eg, dichlorotriazine and trichloropyrimidine, or by varying substituents on the triazine ring; electron-donating groups decrease reactivity, and electron-withdrawing groups increase reactivity.

Triazines substituted with amines are generally less reactive as the amine becomes more basic, and more reactive as the amine becomes less basic. The

Table 3. Reactive Groups in Fiber-Reactive Dyes

Reactive group[a]	Released	Commercial name	Reactivity[b]	Applications[c] (preferred)
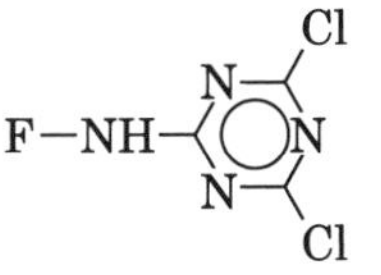 (structure) dichlorotriazine (DCT) F—NH—triazine with Cl, Cl	1956	Procion MX[d] Basilen M[e]	5	exhaust (40°C) pad-batch
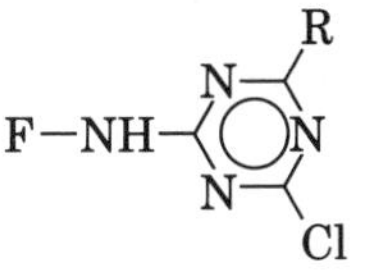 (structure) monochlorotriazine (MCT) F—NH—triazine with R, Cl	1957	Cibacron[f] Cibacron E[f] Procion H, HE, SP[d] Basilen E, P[e] Drimaren P[g]	2	exhaust (80°C) pad-dry–pad-steam pad-thermofix printing (1 phase)
$F—SO_2CH_2CH_2OSO_2H$[h] sulfatoethyl-sulfone (VS)	1957 1958	Remazol[i] Sumifix[j]	3	pad-batch pad-dry–pad-steam exhaust (60°C) printing (2 phase)
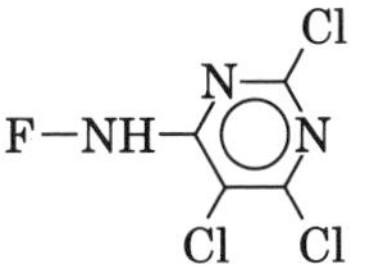 (structure) trichloropyrimidine (TCP) F—NH—pyrimidine with Cl, Cl, Cl	1959 1960	Drimaren Z[g] Cibacron T-E[f]	1	exhaust (80°C)
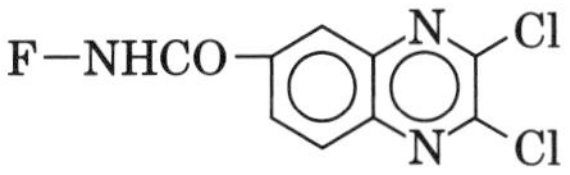 (structure) dichloroquinoxaline (DCQ) F—NHCO—quinoxaline with Cl, Cl	1961	Levafix E[k]	4	exhaust (40°C) pad-steam pad-batch
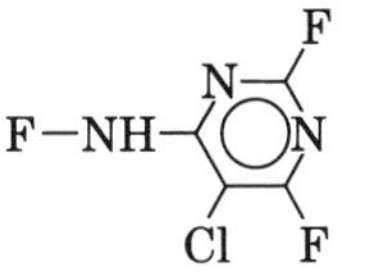 (structure) difluorochloropyrimidine (DFCP) F—NH—pyrimidine with F, Cl, F	1970 1971	Levafix E-A[k] Drimaren K[k]	4	exhaust (40°C) pad-batch
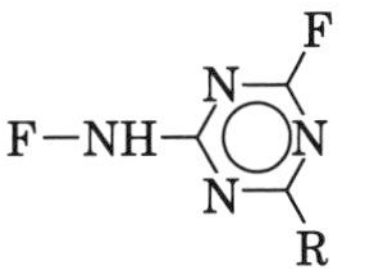 (structure) monofluorotriazine (FT) F—NH—triazine with F, R	1978	Cibacron F[f] Levafix E-N[k]	4	exhaust (40°C) pad-batch pad-steam
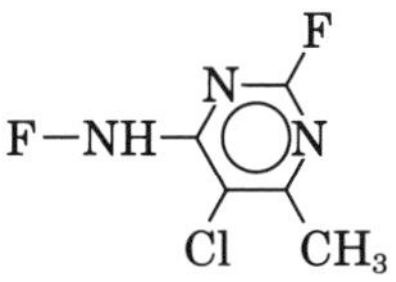 (structure) fluorochloromethyl pyrimidine F—NH—pyrimidine with F, Cl, CH$_3$	1981	Levafix PN[k]	2	printing (1 phase) pad-thermofix

[a]Fixation is by an $S_{N}2$ mechanism unless otherwise noted.   [b]Scale is 1 low to 5 high.
[c]See DYES, APPLICATION AND EVALUATION for a discussion of exhaust dyeing, textile printing, and the various pad processes.   [d]ICI.   [e]BASF.   [f]CIBA-GEIGY.   [g]Sandoz.
[h]Fixation by an addition reaction.   [i]Hoechst.   [j]Sumitomo.   [k]Bayer.

chromophore may be sufficiently electron withdrawing or donating to affect reactivity of the dye.

Reactive groups have minimal auxochrome effect on color intensity, and color yield per molecular weight decreases with increasing numbers of reactive groups. Increased dye fixation and reduced environmental impact of hydrolyzed dye more than compensate for color reduction of additional reactive groups.

The fluorotriazine reactive group was reported in the mid-1970s, and has been the subject of many literature references since that time (15). Monofluorotriazine dyes are especially attractive because of their high color yield at low temperatures (below 40°C) (3).

The more important reactive groups are shown in Table 3 by commercial name with some application information.

Substrates. The principal use for reactive dyes is dyeing textiles; cellulosics constitute the largest use. Cotton is very nearly pure cellulose, and as such has three free hydroxyl groups per glucoside unit that act as dye sites for reactive dyes.

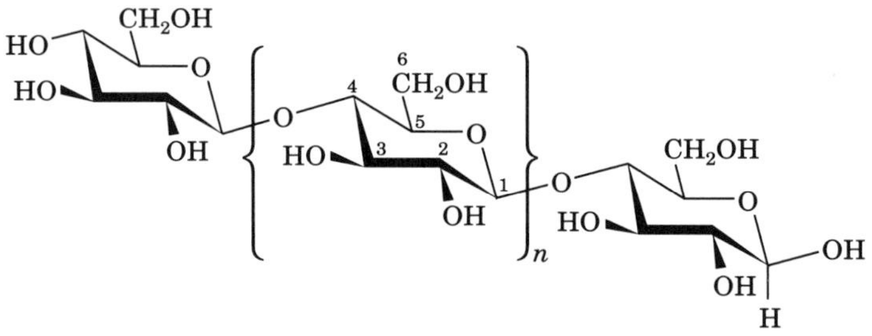

Wool, a natural polyamide, consists of several amino acids including lysine, arginine, and histidine having reactive amine groups; serine, threonine, and tyrosine having reactive hydroxyl groups; and cystine and cysteine capable of forming or having reactive thiol groups. Synthetic polyamides have free amino reactive groups.

Reactivity in the Dyebath

The most important discovery in dyeing cellulose with reactive dyes was the application of Schotten-Baumaun principles. Reaction of alcohols proceeds more readily and completely in the presence of dilute alkali, and the cellulose anion (cell–O$^-$) is considerably more nucleophilic than is the hydroxide ion. Thus the fixation reaction (eq. 1) competes favorably with hydrolysis of the dye (eq. 2).

$$\text{dye—B—R} + \text{cell–O}^- \longrightarrow \text{dye—B—O—cell} \qquad (1)$$

$$\text{dye—B—R} + \text{OH}^- \longrightarrow \text{dye—B—OH} \qquad (2)$$

Ratios of cellulose anion versus hydroxide ion are shown in Table 4.

The most important reactive groups today are monochlorotriazine (high energy), vinyl sulfone (medium energy), and monofluorotriazine (low to medium energy), although dichlorotriazine, 2,3-dichloroquinoxaline and 2,4-difluoro-

Table 4. Ionization of Cellulose

pH	$[OH^-]$ in the dyebath	$[cell\text{--}O^-]$ in the fiber	$\dfrac{[cell\text{--}O^-]}{[OH^-]}$ Ratio
7	10^{-7}	3×10^{-6}	30
8	10^{-6}	3×10^{-5}	30
9	10^{-5}	3×10^{-4}	30
10	10^{-4}	3×10^{-3}	30
11	10^{-3}	2.8×10^{-2}	28
12	10^{-2}	2.2×10^{-1}	22
13	10^{-1}	1.1	11

5-chloropyrimidine have a significant presence. α-Bromoacrylamide and vinyl sulfone from *N*-methyltaurine are the most important reactive dyes for wool. Table 5 gives some examples of dye fixation pathways.

Table 5. Dye–Substrate Reactions

Dye	Substrate	Dyed substrate
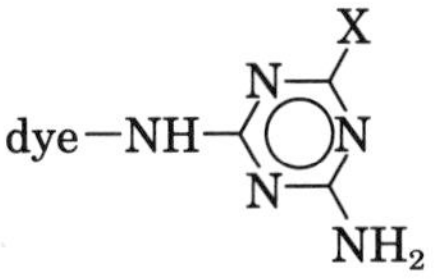 dye—NH—(triazine ring: X top, NH$_2$ bottom) X = Cl or F	cell–OH	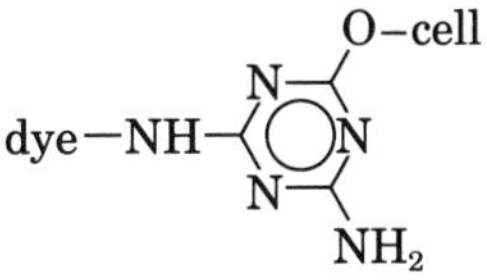 dye—NH—(triazine ring: O–cell top, NH$_2$ bottom)
dye—SO$_2$CH$_2$CH$_2$OSO$_3$H or dye—SO$_2$CH=CH$_2$	cell–OH	dye—SO$_2$—CH$_2$—CH$_2$—O–cell
dye—PO(OH)$_2$	cell–OH	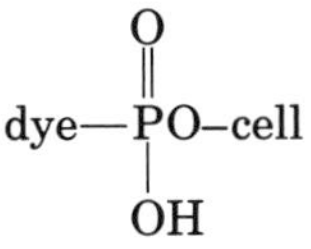 dye—PO-cell (with =O and OH)
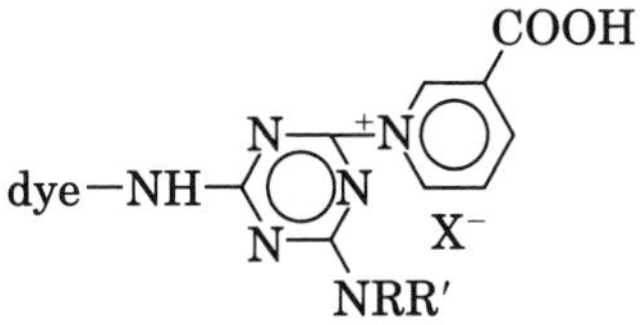 dye—NH—(triazine ring: COOH-phenyl at $^+$N, X$^-$, NRR′)	cell–OH	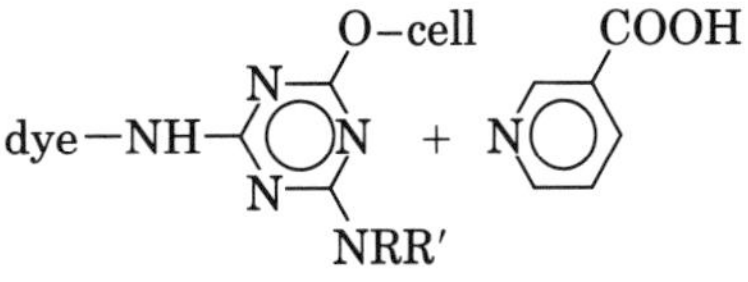 dye—NH—(triazine ring: O–cell) + N-phenyl(COOH) NRR′
dye—NHCOCH=CH$_2$	cell–OH or wool–SH (NH)	dye—NHCOCH$_2$CH$_2$O–cell dye—NHCOCH$_2$CH$_2$S–wool
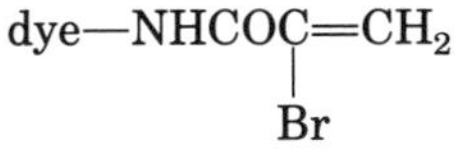 dye—NHCOC=CH$_2$ │ Br	wool–SH	 dye—NHCOC=CH$_2$ │ S–wool or dye—NHCOCHCH$_2$S–wool │ OH 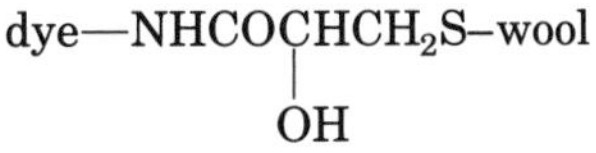

^{a}There is a large variety of chloro or fluoro substituted heterocyclic rings which undergo X$^-$ displacement by cell–O$^-$.

Vinyl sulfones are usually generated under alkaline conditions from β-sulfatoethylsulfones,

$$\text{dye}-SO_2CH_2CH_2OSO_3H + 2NaOH \longrightarrow \text{dye}-SO_2CH=CH_2 + Na_2SO_4 + 2H_2O$$

or in some cases from β-chloroethylsulfones,

$$\text{dye}-SO_2CH_2CH_2Cl + NaOH \longrightarrow \text{dye}-SO_2CH=CH_2 + NaCl + H_2O$$

and in some cases the dye is supplied by the manufacturer as the vinyl sulfone.

An interesting concept is that of generating vinyl sulfones from β-(β-sulfatoethyl)sulfonylpropionamides (4):

$$\text{dye}-\overset{\overset{\displaystyle O}{\|}}{N}HCCH_2CH_2SO_2CH_2CH_2OSO_3H + 2NaOH \longrightarrow$$

$$\text{dye}-\overset{\overset{\displaystyle O}{\|}}{N}HCCH_2CH_2SO_2CH=CH_2 + Na_2SO_4 + 2H_2O$$

This can react with cellulose or hydrolyze.

$$\text{dye}-\overset{\overset{\displaystyle O}{\|}}{N}HCCH_2CH_2SO_2CH_2CH_2O\text{–cell} \quad \text{or} \quad \text{dye}-\overset{\overset{\displaystyle O}{\|}}{N}HCCH_2CH_2SO_2CH_2CH_2OH$$

Either of these structures can react further with caustic to give the acrylamide reactive group, dye—NHCOCH=CH$_2$, which bonds with cellulose to give the more stable β-propionamide derivative.

A reactive dye–cellulose bond is subject to some slight hydrolysis during washing under alkaline conditions.

$$\text{dye}-B-O-\text{cell} + NaOH \longrightarrow \text{dye}-B-OH + HO-\text{cell}$$

Triazinyl reactive dyes show less tendency toward hydrolysis during washing than do the vinyl sulfone type.

To overcome hydrolysis of vinyl sulfone dyes during application under neutral dyeing of wool, Hoechst introduced dyes with the N-methyltaurine group.

$$\text{dye}-SO_2CH_2CH_2\overset{\overset{\displaystyle CH_3}{|}}{N}CH_2CH_2SO_3Na$$

The reactive vinyl sulfone is generated slowly under neutral or slightly acidic conditions, and gives level exhaustion dyeings on wool. Reactive groups based on acrylamide react more slowly and are less subject to hydrolysis than vinyl sulfone; they are also more suitable for wool dyeing. Relative reactivity of the more important groups is shown in Figure 3.

Bifunctional Dyes. There are many examples of dyes with two or more reactive groups, including many mixed reactive systems. Dye fixation is increased

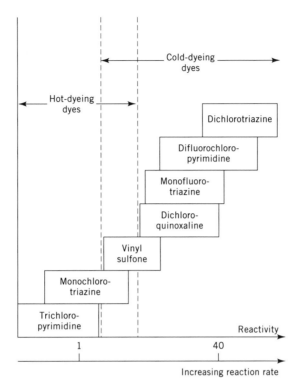

Fig. 3. Reactivity of various reactive dye groups. Hot dyeing is done at 80°C, cold dyeing at 40°C.

significantly with increasing number of reactive groups as shown by the hypothetical situation in Figure 4. Theoretically, fixation in this example could increase from mono- to di- to trireactive from 60 to 84 to 93.6%. Some multiple reactive dyes are claimed to have as high as 95% fixation.

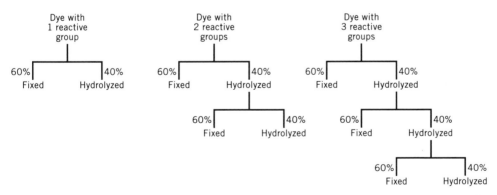

Fig. 4. Dye fixation is increased for dyes with >1 reactive group. One group may be hydrolyzed, but that still leaves the other group(s) available for fixation.

Table 6. Reactive Anthraquinone Dyes[a]

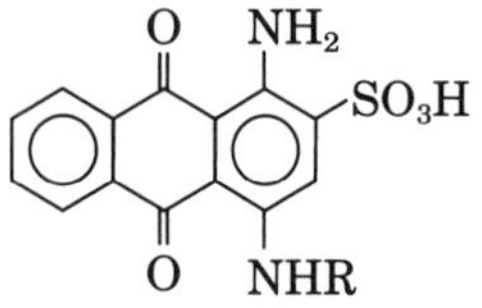

O   NH$_2$
       SO$_3$H

O   NHR

Structure number	CAS Registry Number	R Group	Reference
(29)	[62155-76-4]	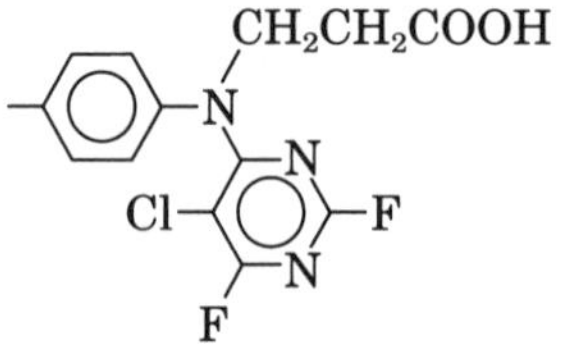 CH$_2$CH$_2$COOH; N; Cl, N, F, N, F	16
(30)	[62155-77-5]	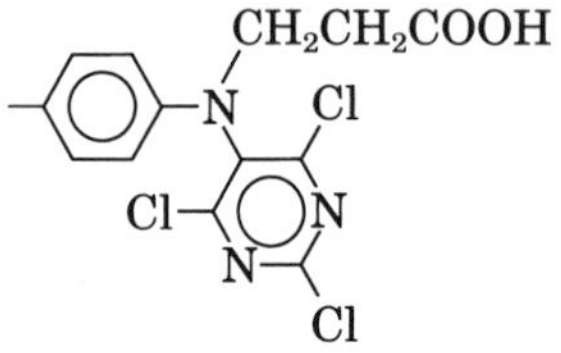 CH$_2$CH$_2$COOH; N; Cl, Cl, N, N, Cl	16
(31)	[64135-02-0]	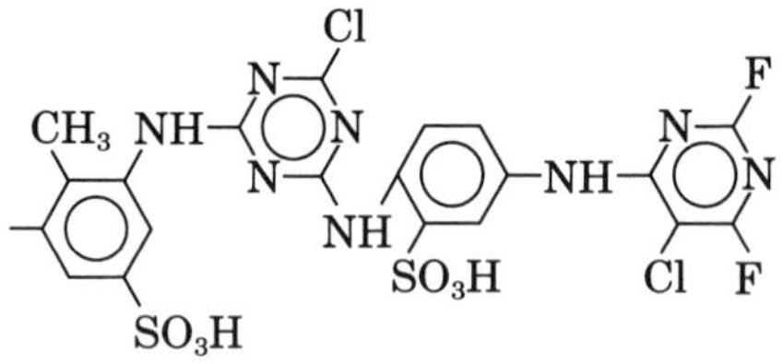 Cl; CH$_3$ NH—N,N; N; NH; SO$_3$H; SO$_3$H; NH; F, N, N, Cl, F	17
(32)	[69658-32-8]	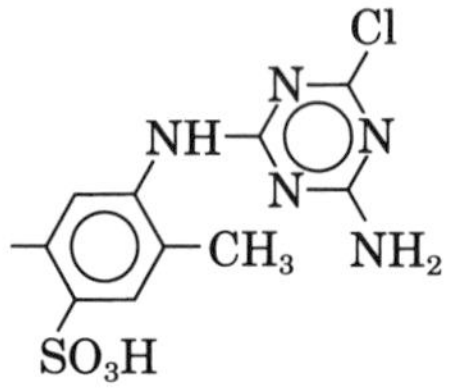 Cl; NH—N,N; N; CH$_3$ NH$_2$; SO$_3$H	18
(33)	[72631-11-9]	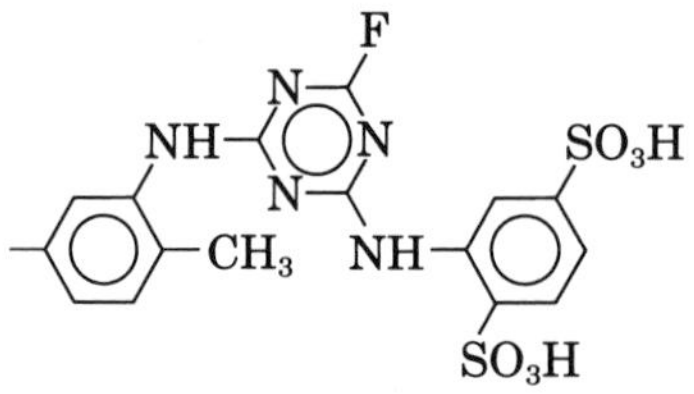 F; NH—N,N; N; CH$_3$ NH; SO$_3$H; SO$_3$H	19
(34)	[72645-84-2]	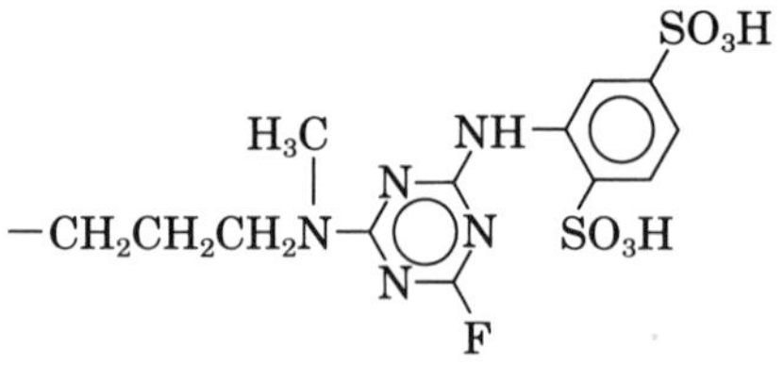 SO$_3$H; H$_3$C NH; —CH$_2$CH$_2$CH$_2$N—N,N; N; SO$_3$H; F	20

824

Table 6. (*Continued*)

Structure number	CAS Registry Number	R Group	Reference
(**35**)	[*75127-47-8*]	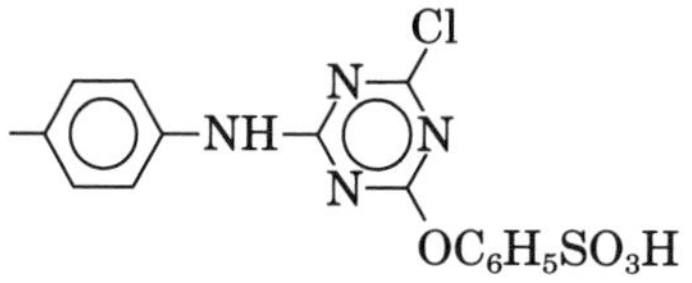	21
(**36**)	[*76619-28-8*]	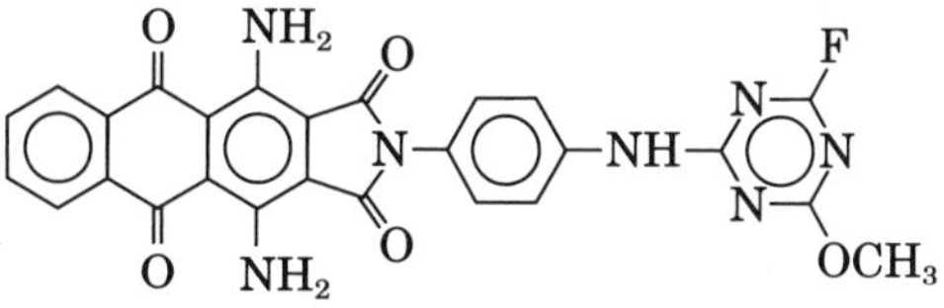	22
(**37**)[b]	[*76655-70-4*]	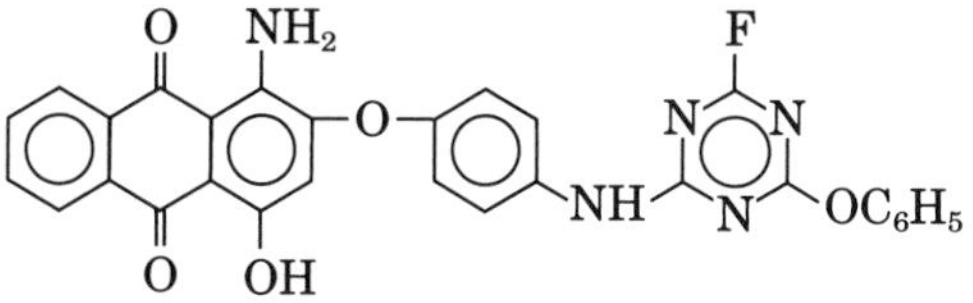	23
(**38**)	[*74878-78-7*]	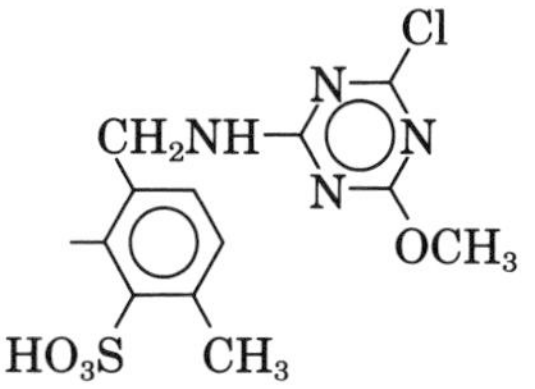	24

[a]These dyes are blue unless otherwise noted.
[b]Pink.

Methods of Synthesis

Reactive dyes are synthesized by (*1*) condensation of an amine function in the chromogen molecule with a reactive group, eg,

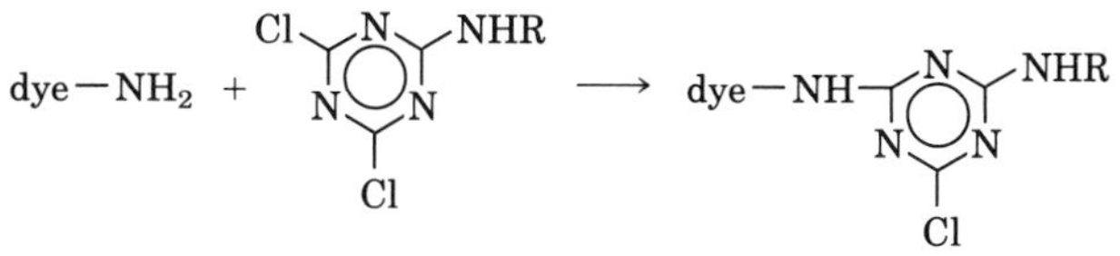

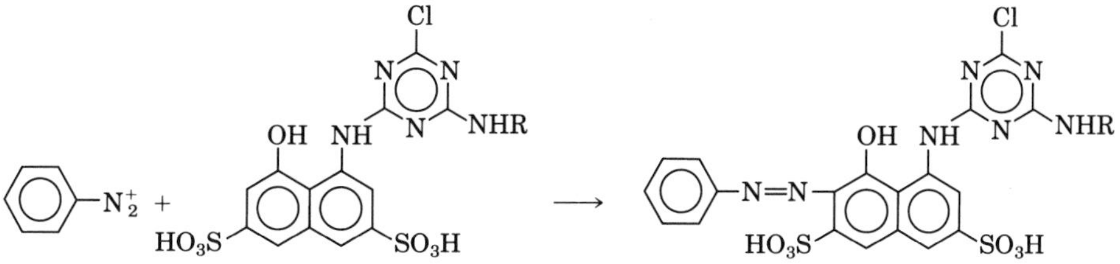

(*2*) by coupling a diazonium salt with a coupling component that has a reactive group

or by coupling a diazonium salt containing a reactive group with a coupler

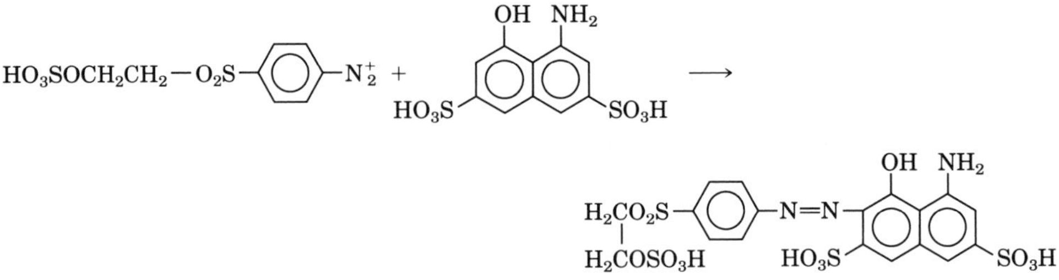

Table 7. Formagan Reactive Dyes

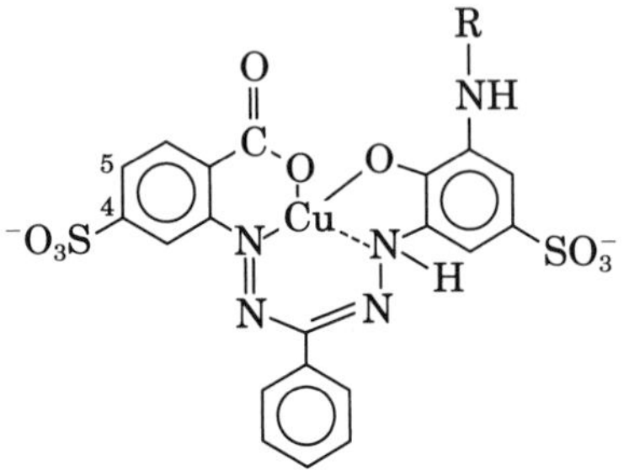

Structure number	CAS Registry Number	R Group	Reference
(39)	[68912-12-9]	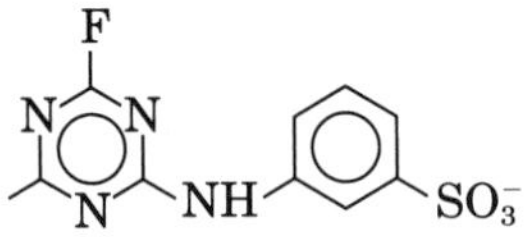	
(40)	[77743-15-8]	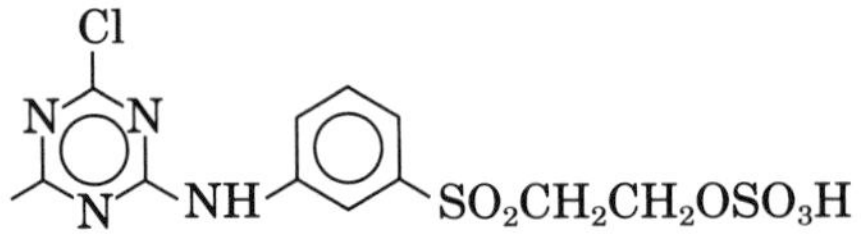	25
(41)[a]	[60265-86-3]	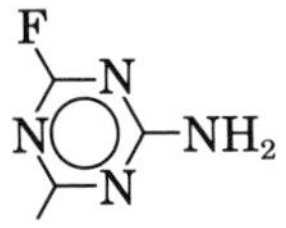	26
(42)		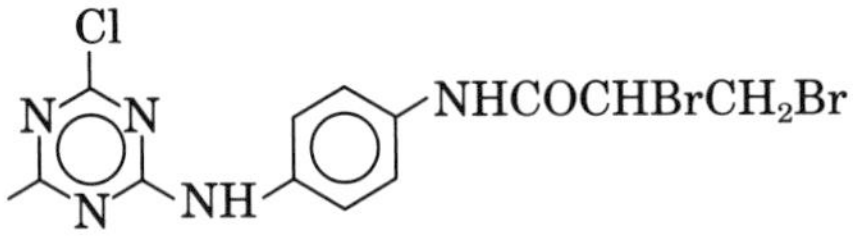	27
(43)		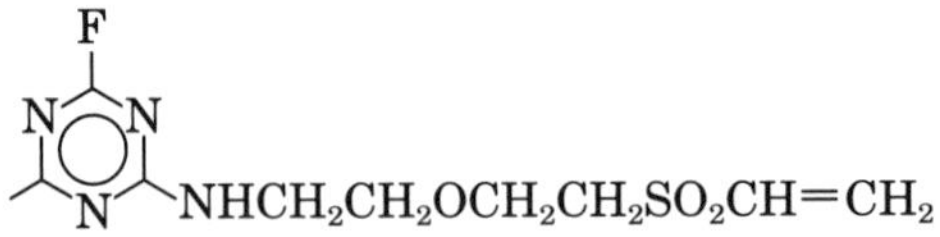	28

[a] In this dye, there is a sulfonic acid group at position 5 rather than position 4.

or (*3*) in the case of copper phthalocyanine (CPC), condensing CPC sulfonyl chloride with an amino-containing bridging group attached to a reactive group

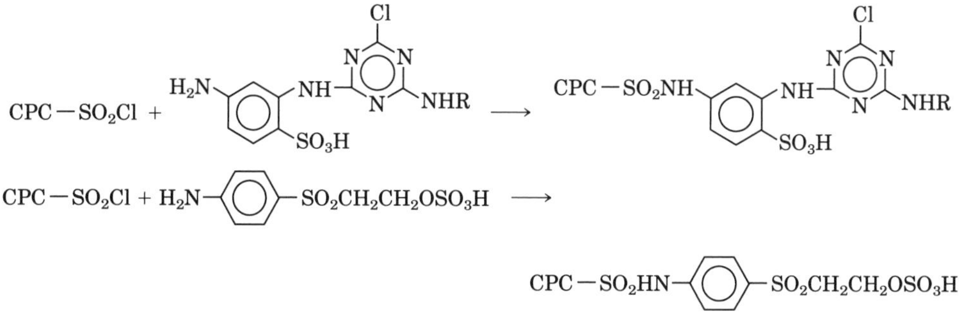

Table 8. Reactive Dyes Based on 3,6-Disubstituted H-Acid (11)[a]

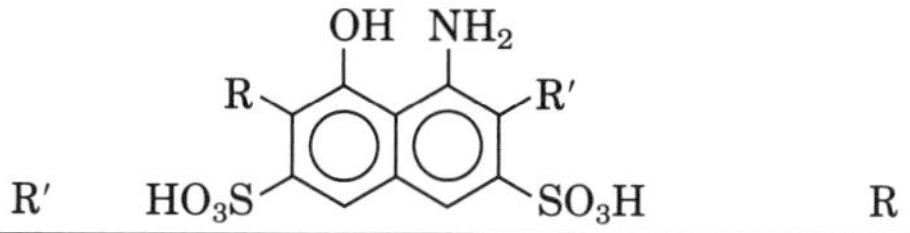

CAS Registry Number	R′	R
[73203-80-2][b]	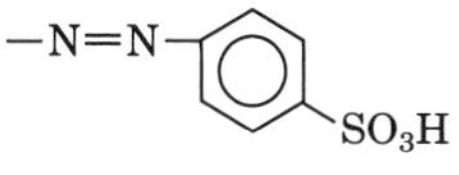 —N=N—⟨⟩—SO₃H	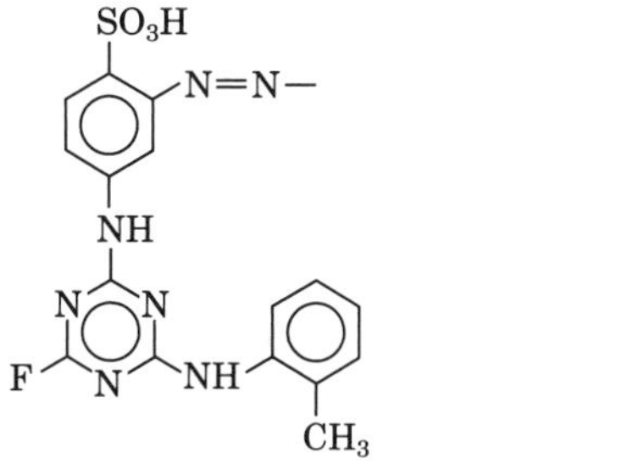 (see structure)
c	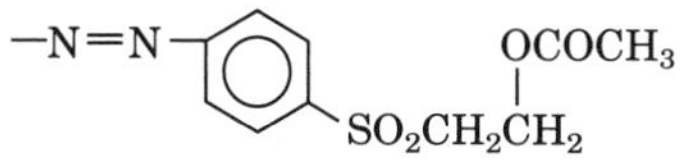 —N=N—⟨⟩ OCOCH₃ / SO₂CH₂CH₂	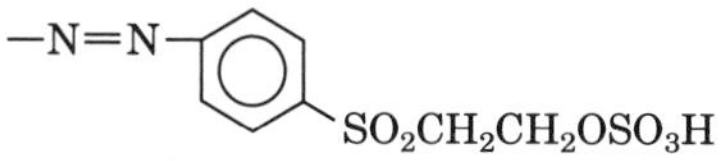 —N=N—⟨⟩—SO₂CH₂CH₂OSO₃H
[73903-43-2][d]	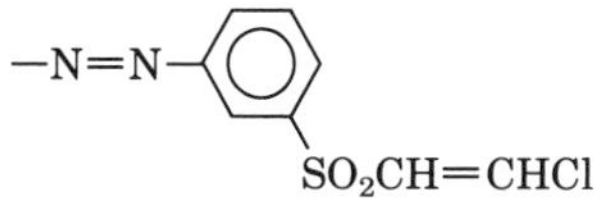 —N=N—⟨⟩—SO₂CH=CHCl	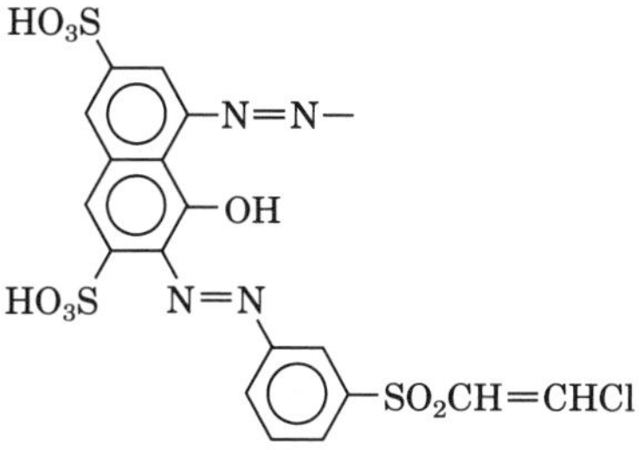 (see structure)
[77365-88-9][e]	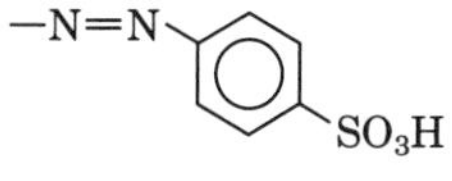 —N=N—⟨⟩—SO₃H	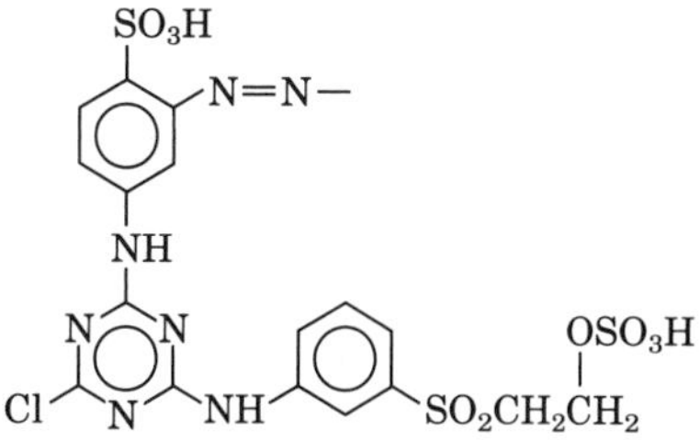 (see structure)

[a]Other blue and navy dyes are described in Refs. 42–47. [b]Navy (48). [c]Black (49). [d]Brown (50). [e]Blue (51).

Table 9. Reactive Dyes Based on 4,6-Disubstituted H-Acid[a] (11)

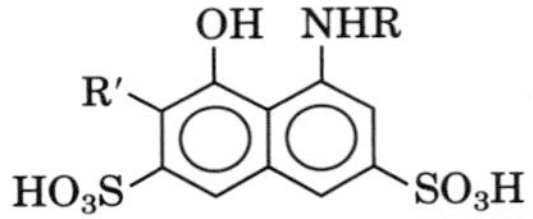

R'	R	CAS Registry Number	Reference
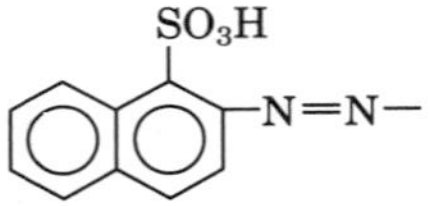	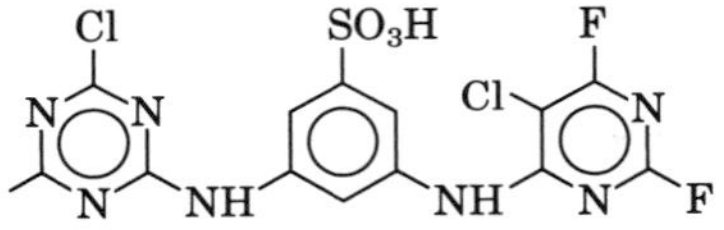	[64329-52-8]	29
	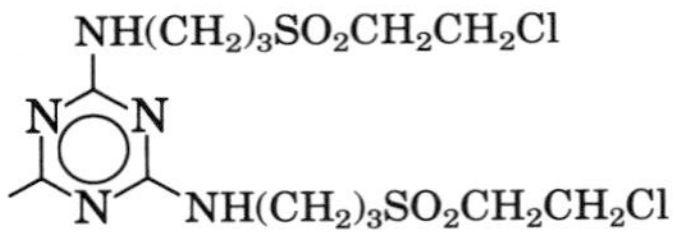 NH(CH₂)₃SO₂CH₂CH₂Cl		30
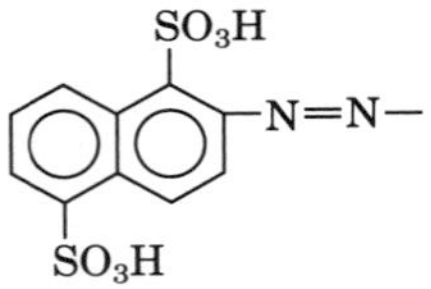	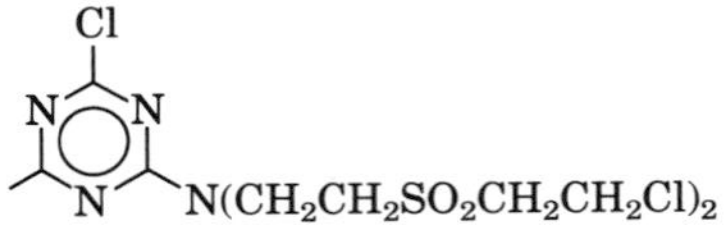 N(CH₂CH₂SO₂CH₂CH₂Cl)₂	[65180-62-3]	31
	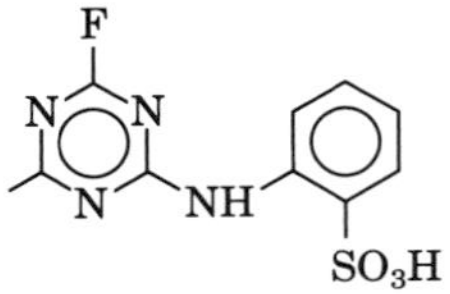	[70817-83-3][b]	32
	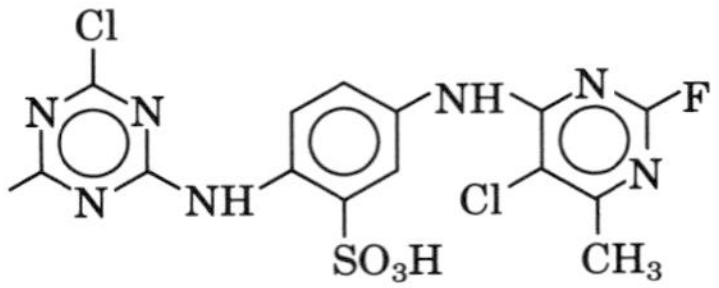	[74526-40-2][c]	33
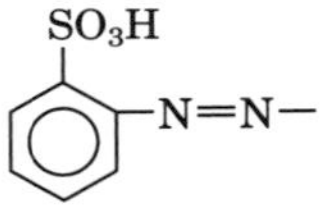	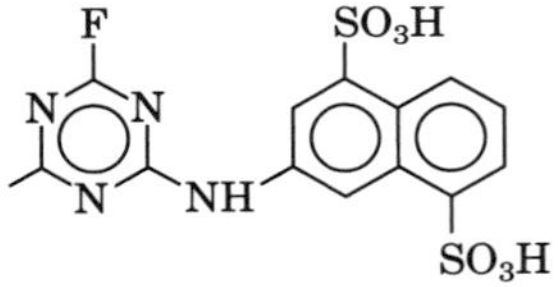	[69774-79-4][b]	34

828

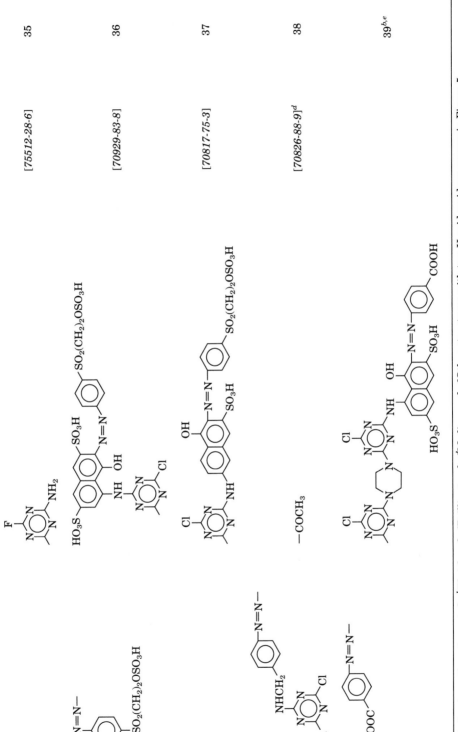

35   [75512-28-6]

36   [70929-83-8]

37   [70817-75-3]

38   [70826-88-9]^d

39^{b,e}

[a]Red unless otherwise noted. [b]Bluish red. [c]Brilliant red. [d]Medium red. [e]Other structures with two H-acid residues appear in Figure 5.

829

Economic Aspects

Development in reactive dyes over the past 15 years has clearly been the most active of any class of dyes. More than 1300 references to reactive dyes were made in *Chemical Abstracts* during the period 1976 to 1991. Representative structures ummarizing this period are listed in Tables 6–13 and Figure 5.

Table 10. Metallized Reactive Dyes Derived from H-Acid

Color	CAS Registry Number	Dye	Reference
red	[69853-33-4]	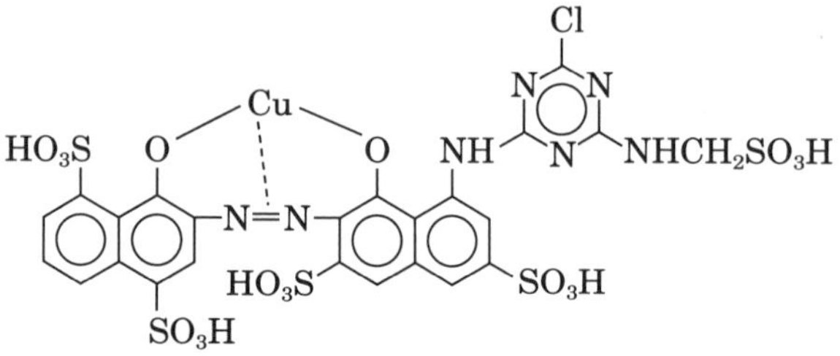	52
violet	[71396-24-2]	R = —COCH₃	53
red	[77980-56-4]	R = —H	54
black	[70867-00-4]	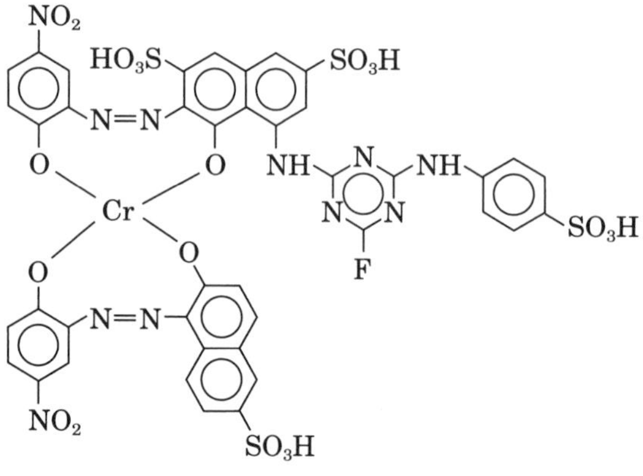	55
violet	[71498-30-1]	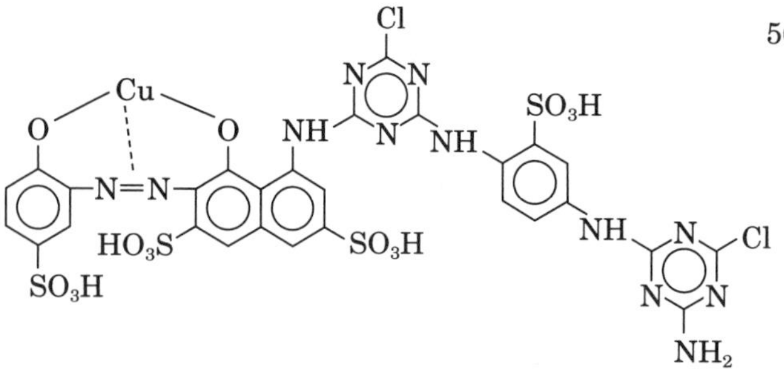	56

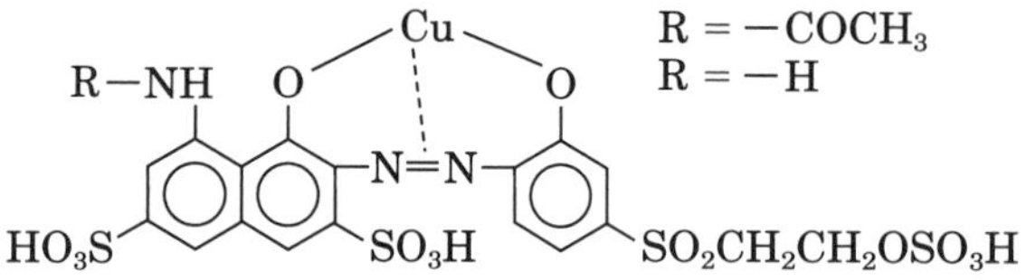

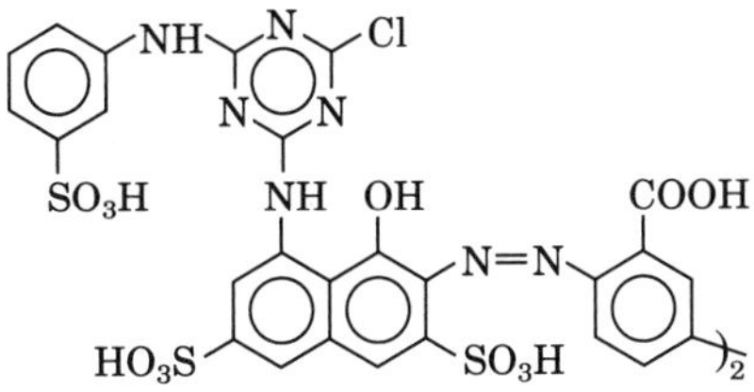

Fig. 5. "Dimeric" azo dyes derived from H-acid. (**44**) is brown (**40**), and (**45**) [70817-79-7] is red (**41**).

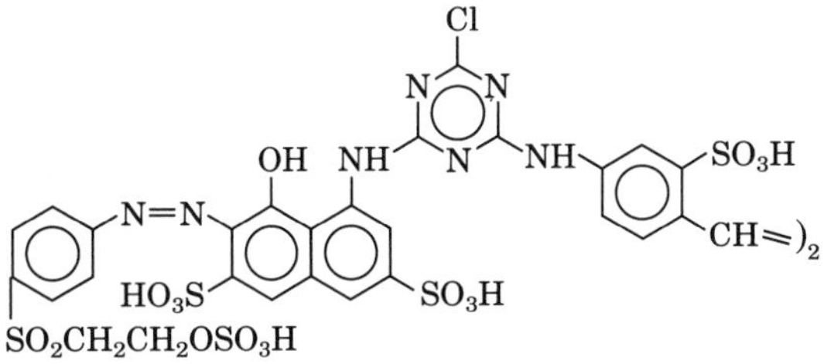

Table 11. Phenyl Pyrazolone Reactive Dyes[a]

Dye	CAS Registry Number	Reference
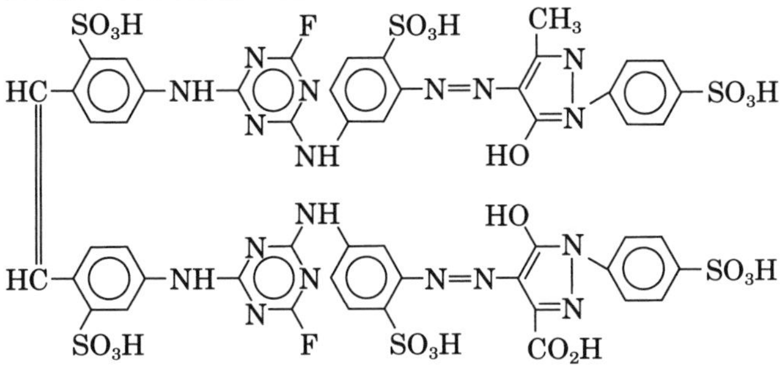	[63400-65-7]	59
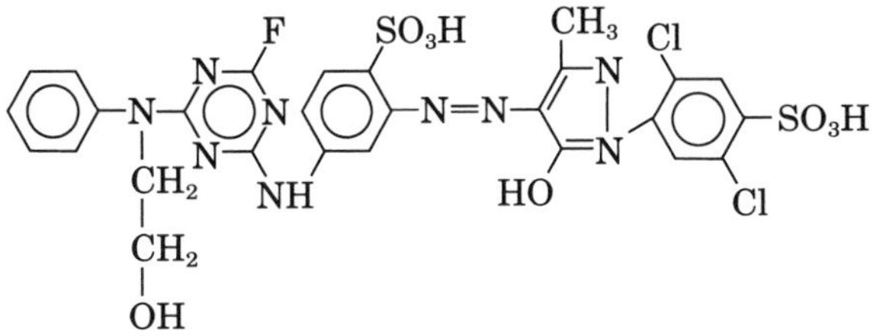	[75487-71-7]	60
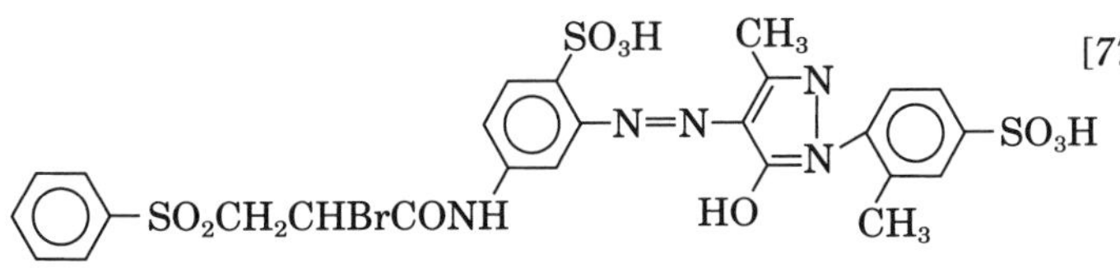	[77850-37-4]	61
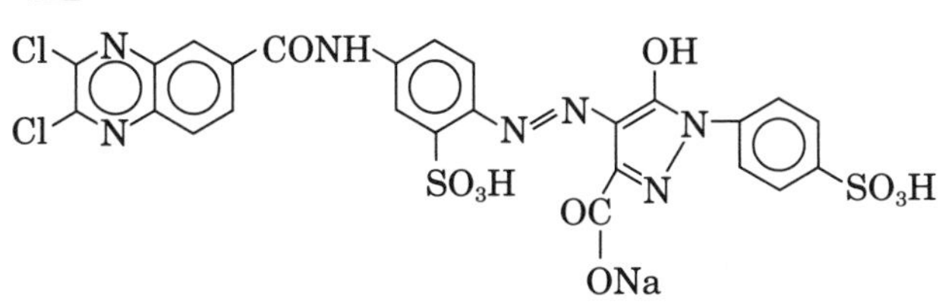	[75199-00-7]	62

[a]These dyes are all yellow. Other yellow dyes of this type are described in Refs. 57 and 58.

Table 12. Dyes Containing Pyridone[a]

Dye	CAS Registry Number	Reference
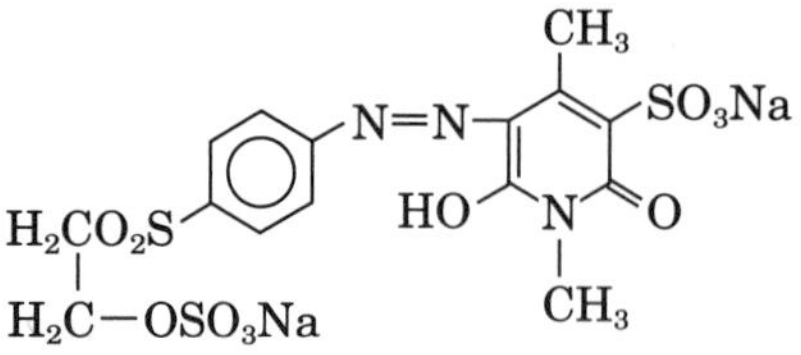	[70911-38-5]	63
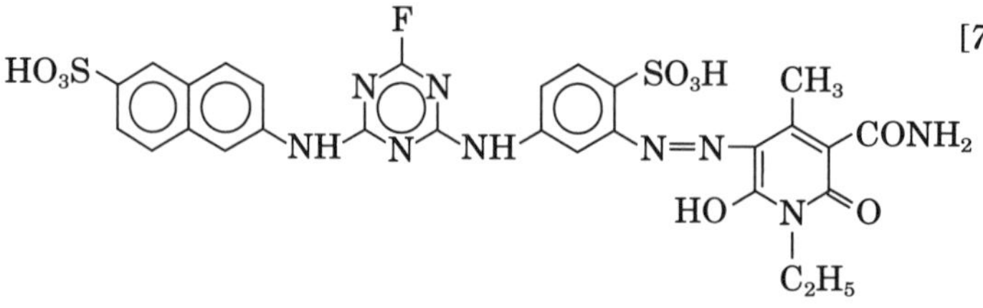	[70868-06-3]	64
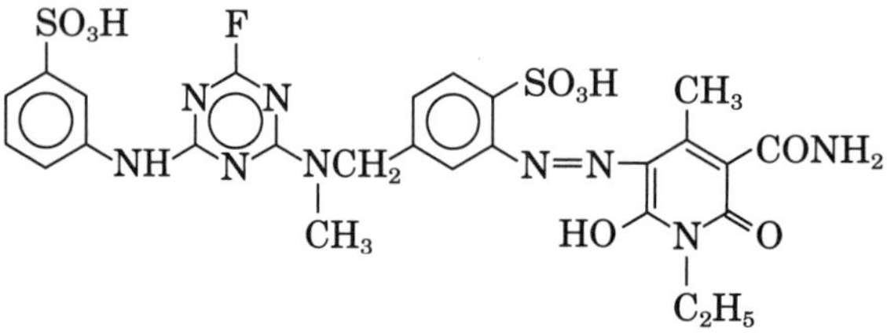	[71363-80-9]	65
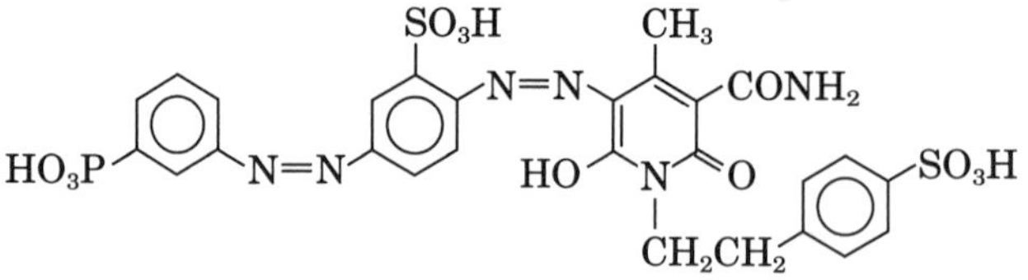	[70911-43-2]	66
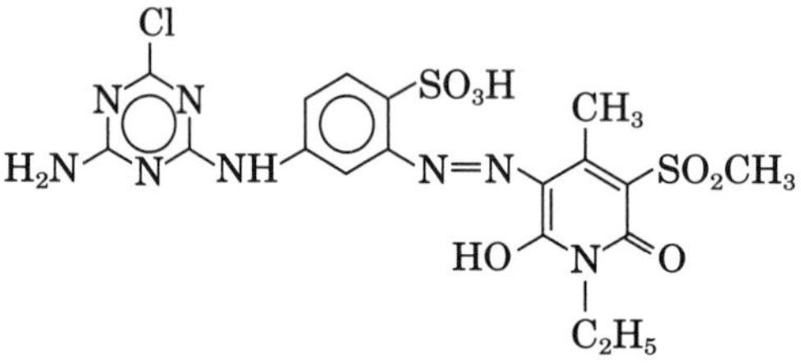	[71552-80-2]	67
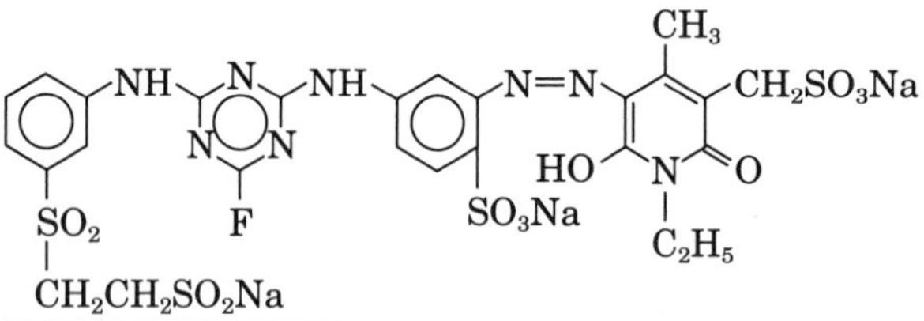	[73570-54-4][b]	69

[a]These dyes are all yellow.
[b]Also see Ref. 68.

832

Table 13. Miscellaneous Reactive Dyes

Color	CAS Registry Number	Dye	Reference
fluorescent red	[62479-97-4]	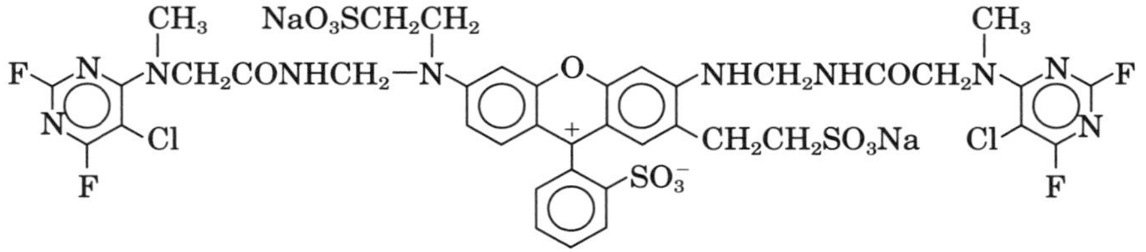	70
yellow	[35168-45-7]	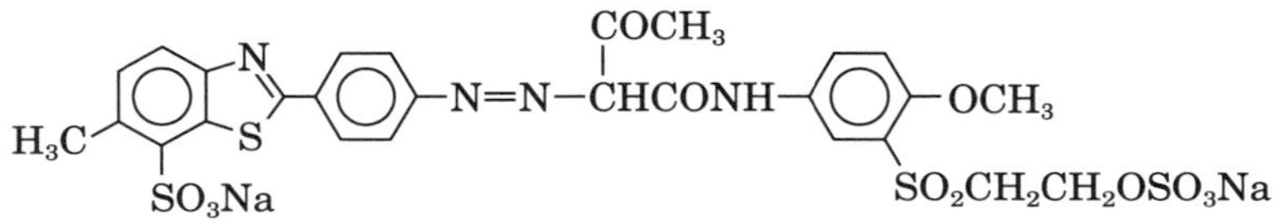	71
red brown[a]	[74920-88-0]	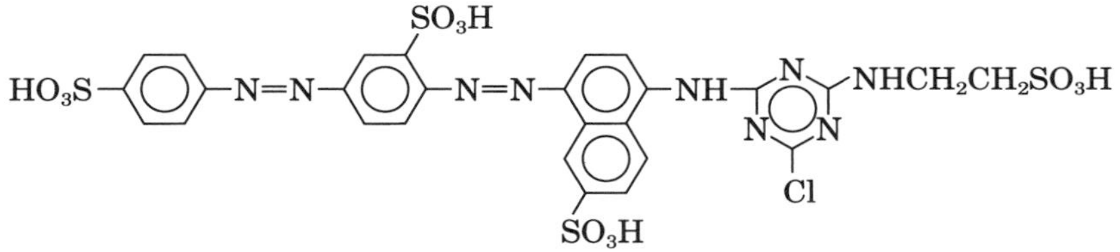	73
orange		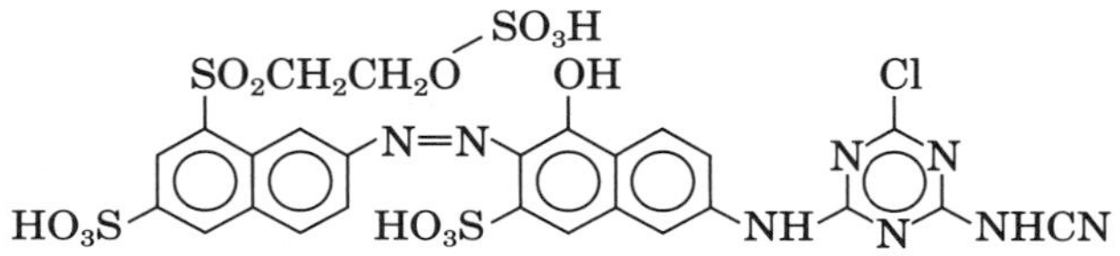	74

Table 13. (*Continued*)

Color	CAS Registry Number	Dye	Reference
navy	[73785-57-6]	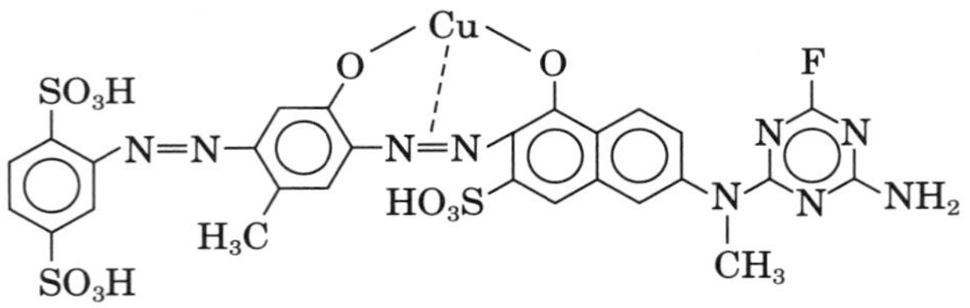	75
yellow-red	[74441-16-0]	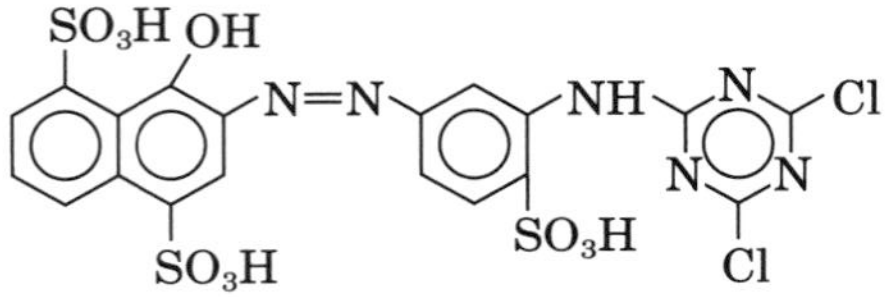	76
navy	[77244-51-0]	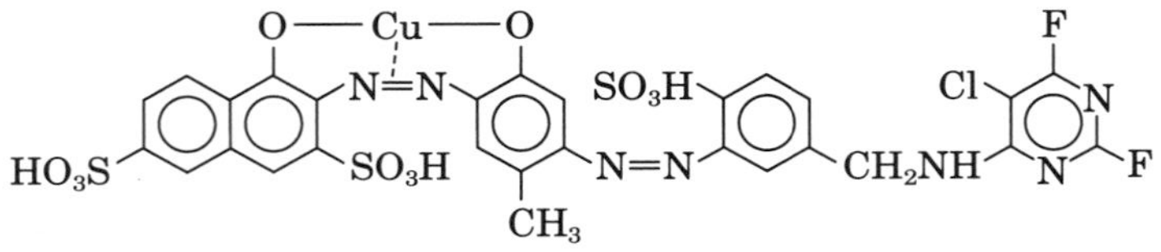	77
red	*b*	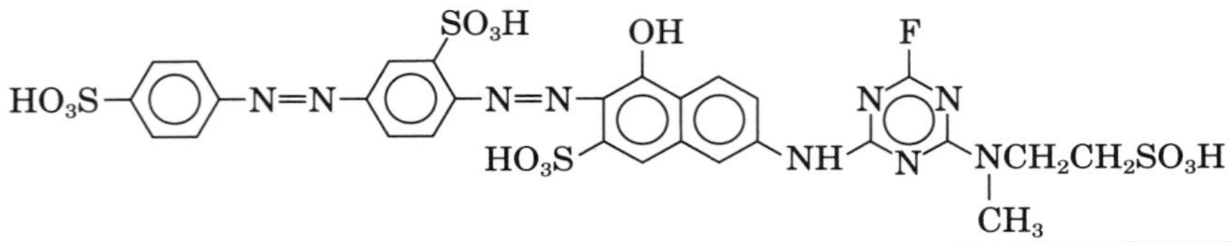	82

[a]The monochlorotriazine reactive group is also found in a metallized brown dye [77934-00-0] (72).

[b]Many other azo dyes with fluorotriazine reactive groups have been synthesized, eg, scarlet [70255-11-7] (78), orange [70239-79-1] (79), yellow [71942-72-8] (80), and orange [73900-71-7] (81).

834

Every significant dye manufacturer is now offering reactive dyes. More than 200 offerings are listed in the 1991 AATCC Buyers Guide. Fifty of those dyes were not listed in 1987. Reactive dyes are offered commercially both as dry powders and buffered liquid forms (83).

BIBLIOGRAPHY

"Dyes, Reactive" in *ECT* 2nd ed., Vol. 7, pp. 630–641, by D. W. Bannister, J. Elliott, and A. D. Olin, Toms River Chemical Corporation; in *ECT* 3rd ed., Vol. 8, pp. 374–391, by John Elliott and Patrick P. Yeung, Toms River Chemical Corporation.

1. C. F. Cross and E. J. Bevan, *Res. Cellulose* **1**, 34 (1895).
2. E. Siegel in K. Venkataraman, *The Chemistry of Synthetic Dyes and Pigments*, Vol. 6, Academic Press, Inc., New York, 1972, p. 4.
3. H. Zollinger, *Text. Chem. Color.* **23**, 12 (1991).
4. Ref. 2, p. 12.
5. A. H. M. Renfrew and J. A. Taylor, *Rev. Prog. Color. Relat. Top.* **20** (1990).
6. Eur. Pat. Appl. 315,045 (May 10, 1989), K. Pandl and M. Patsch (to BASF AG); Ger. Pat. 3,737,537 (Nov. 5, 1987), K. Pandl and M. Patsch (to BASF AG); Ger. Pat. 3,737,536 (Nov. 5, 1987), K. Pandl and M. Patsch (to BASF AG); Ger. Pat. 3,718,397 (Dec. 22, 1988), H. D. Riefenrath and J. Gruetze (to Bayer AG); Ger. Pat. 3,326,638 (Feb. 7, 1985), G. Schwaiger (to Hoechst AG); Ger. Pat. 3,434,818 (Apr. 4, 1985), S. Yamamura, E. Ogawa, and T. Shirasaki (to Nippon Kayaku Co.); Ger. Pat. 2,557,141 (July 1, 1976), G. Hegar and H. Sieler (to CIBA-GEIGY Corp.); Eur. Pat. Appl. 21,351 (Jan. 7, 1981), T. Omura, Y. Tezuka, and M. Sunomi (to Sumitomo Chemical Co.); Ger. Pat. 2,945,537 (May 21, 1981), G. Schwaiger and E. Hoyer (to Hoechst AG); Eur. Pat. Appl. 338,310 (Oct. 25, 1989), H. Henk, K. Herd, and F. Stohr (to Bayer AG); Eur. Pat. Appl. 352,222 (Jan. 24, 1990), U. Lehmann (to CIBA-GEIGY Corp.); Eur. Pat. Appl. 402,318 (Dec. 12, 1990), C. Brinkman (to CIBA-GEIGY Corp.).
7. *The Colour Index*, Vol. 3, 3rd ed., Society of Dyers and Colorists, Bradford, Yorkshire, UK, 1971, pp. 3395–3558.
8. Ger. Pat. 2,945,537 (May 21, 1981), G. Schwaiger and E. Hoyer (to Hoechst AG).
9. Eur. Pat. Appl. 338,310 (Oct. 25, 1989), H. Henk, K. Herd, and F. Stohr (to Bayer AG).
10. Ger. Pat. 2,600,490 (July 15, 1976), J. L. Leng and D. W. Shaw (to ICI).
11. Ger. Pat. 3,330,547 (Oct. 21, 1985), W. Harms, G. Franke, and K. Wunderlich (to Bayer AG).
12. Ger. Pat. 2,503,611 (Aug. 14, 1975), J. L. Leng, B. Parton, and D. Ridyard (to ICI).
13. Brit. Pat. Appl. 2,059,985 (Apr. 29, 1981), R. J. Marklow (to ICI); Ger. Pat. 3,045,471 (Dec. 2, 1980), W. Harms and K. Wunderlich (to Bayer AG); Eur. Pat. Appl. 84,718 (Aug. 3, 1983); Brit. Pat. 8,325,409 (Sept. 22, 1983), R. D. McClelland and A. H. M. Renfrew (to ICI); Ger. Pat. 3,344,253 (Apr. 18, 1985), H. Jaeger (to Bayer AG); Ger. Pat. 3,410,236 (Oct. 3, 1985), H. Jaeger (to Bayer AG); Ger. Pat. 3,409,439 (Sept. 26, 1985), W. Harms, K. Wunderlich, and H. Jaeger (to Bayer AG); Ger. Pat. 3,412,333 (Oct. 10, 1985), W. Harms and K. Wunderlich (to Bayer AG); Ger. Pat. 3,439,755 (Oct. 17, 1985), H. Jaeger, K. Langheinrich, and K. J. Herd (to Bayer AG); Ger. Pat. 3,510,612 (Sept. 25, 1986), K. Wunderlich and co-workers (to Bayer AG); Ger. Pat. 3,537,629 (Apr. 30, 1986), R. Pedrozzi (to Sandoz); Ger. Pat. 3,520,391 (Dec. 11, 1986), W. Harms and K. Wunderlich (to Bayer AG); Ger. Pat. 3,521,358 (Dec. 18, 1986), W. Harms and K. Wunderlich (to Bayer AG); Ger. Pat. 3,426,727 (Jan. 23, 1986), H. Fuchs, H. Springer, and G. Schwaiger (to Hoechst AG); Ger. Pat. 3,530,830 (Mar. 5, 1987), H. Springer, G. Schwaiger, and W. Helmling (to Hoechst AG); Ger. Pat. 3,625,347 (Mar.

26, 1987), G. Schwaiger, H. Springer, and W. Helmling (to Hoechst AG); Ger. Pat. 3,544,982 (June 25, 1987), H. Springer, G. Schwaiger, and W. Helmling (to Hoechst AG); Ger. Pat. 3,625,346 (Jan. 28, 1988), G. Schwaiger, H. Springer, and W. Helmling (to Hoechst AG); Ger. Pat. 3,628,084 (Mar. 3, 1988), H. Springer, W. Helmling, and G. Schwaiger (to Hoechst AG); Eur. Pat. Appl. 260,227 (Mar. 16, 1988), K. Seitz (to CIBA-GEIGY Corp.); Ger. Pat. 3,627,458 (Feb. 25, 1988), H. Springer, W. Helmling, and G. Schwaiger, (to Hoechst AG); Ger. Pat. 3,635,312 (Apr. 21, 1988), W. Harms (to Bayer AG); Eur. Pat. Appl. 281,799 (Sept. 14, 1988), H. Sawamoto, N. Haroda, and T. Omura (to Sumitomo Chemical Co.); Brazilian Pat. 8,800,126 (Aug. 23, 1988), A. H. M. Renfrew, R. A. Denis, and B. Lamble (to ICI); Jpn. Pat. 87 163,166 (June 30, 1987), T. Fugita and co-workers (to Mitsubishi); Ger. Pat. 3,723,459 (July 16, 1987), W. Harmes and K. J. Herd (to Bayer AG); Ger. Pat. 3,740,978 (Dec. 3, 1987), M. Ruske and M. Patsch (to BASF AG); Ger. Pat. 3,828,824 (Aug. 25, 1988), H. Buech and H. Springer (to Hoechst AG); Eur. Pat. Appl. 356,014 (Feb. 28, 1990), D. R. Ridyard and P. Smith (to ICI); Ger. Pat. 3,827,530 (Aug. 13, 1988), W. Harms (to Bayer AG); Ger. Pat. 3,822,850 (Jan. 11, 1990), H. Jaeger and J. Wolff (to Bayer AG); Eur. Pat. Appl. 365,478 (Apr. 25, 1990), A. Tzikas and P. Aeschlimann (to CIBA-GEIGY Corp.); Jpn. Pat. 02 209,969 (Aug. 21, 1990), T. Miyamoto and co-workers (to Sumitomo Chemical Co.); Jpn. Pat. 02 209,980 (Aug. 21, 1990), T. Miyamoto and co-workers (to Sumitomo Chemical Co.).

14. Ger. Pat. 2,812,634 (Sept. 29, 1979), W. Harms, K. von Oertzen, and K. Wunderlich (to Bayer AG).

15. Eur. Pat. Appl. 85,654 (Aug. 10, 1983), J. Markert and H. Seiler (to CIBA-GEIGY Corp.); Eur. Pat. Appl. 302,006 (Feb. 1, 1989), A. Tzikas (to CIBA-GEIGY Corp.); Ger. Pat. 3,825,658 (Feb. 1, 1990), H. Springer and co-workers (to Hoechst AG); Pol. Pat. 101,487 (Mar. 31, 1979), J. Gmaj, C. Sosnowski, and M. Zator.

16. Belg. Pat. 83,696 (Jan. 3, 1977), R. Mislin, W. Schoenover, and K. U. Steiner (to Sandoz).

17. Ger. Pat. 2,603,670 (Aug. 4, 1977), H. S. Bien, D. Hildebrand, and W. Harms (to Bayer AG).

18. Ger. Pat. 2,729,497 (Jan. 4, 1979), H. Jaeger and W. Harms (to Bayer AG).

19. Ger. Pat. 2,817,733 (Oct. 31, 1979), W. Harms, K. Wunderlich, and K. von Oertzen (to Bayer AG).

20. Ger. Pat. 2,817,781 (Oct. 31, 1979), W. Harms, K. Wunderlich, and K. von Oertzen (to Bayer AG).

21. Ger. Pat. 2,852,672 (June 19, 1980), K. von Oertzen, W. Harms, and K. Wunderlich (to Bayer AG).

22. Ger. Pat. 2,854,481 (June 26, 1980), W. Harms, K. Wunderlich, and K. von Oertzen (to Bayer AG).

23. Jpn. Pat. 80 133,456 (Oct. 17, 1980), (to Mitsubishi).

24. Ger. Pat. 2,918,881 (Nov. 20, 1980), W. Harms and co-workers (to Bayer AG).

25. Eur. Pat. Appl. 21,351 (Jan. 7, 1981), T. Omura, Y. Tezuka, and M. Sunomi (to Sumitomo Chemical Co.).

26. Ger. Pat. 2,557,141 (July 1, 1976), G. Hegar and H. Sieler (to CIBA-GEIGY Corp.).

27. Eur. Pat. Appl. 352,222 (Jan. 24, 1990), U. Lehmann (to CIBA-GEIGY Corp.).

28. Eur. Pat. Appl. 402,318 (Dec. 12, 1990), C. Brinkman (to CIBA-GEIGY Corp.).

29. Ger. Pat. 2,607,028 (Aug. 25, 1977), H. S. Bien and D. Hilderbrand (to Bayer AG).

30. Ger. Pat. 3,942,039 (July 5, 1990), H. M. Buech, W. H. Russ, and H. Tappe (to Hoechst AG).

31. Ger. Pat. 2,614,550 (Oct. 27, 1977), R. Mueller and F. Alderbert (to Cassella).

32. Ger. Pat. 2,847,173 (May 31, 1979), K. Seitz (to CIBA-GEIGY Corp.).

33. Ger. Pat. 2,847,938 (May 4, 1980), K. Schundehutte and co-workers (to Bayer AG).

34. Ger. Pat. 2,731,617 (Feb. 1, 1979), K. Wunderlich and W. Harms (to Bayer AG).
35. Ger. Pat. 2,903,594 (Aug. 14, 1980), K. Wunderlich and W. Harms (to Bayer AG).
36. Ger. Pat. 2,748,965 (May 3, 1979), E. Hoyer, F. Meininger, and R. Fass (to Hoechst AG).
37. Ger. Pat. 2,748,966 (May 3, 1979), E. Hoyer and co-workers (to Hoechst AG).
38. Ger. Pat. 2,845,846 (May 10, 1979), B. Anderson and E. Young (to ICI).
39. Brit. Pat. 2,221,471 (Feb. 7, 1990), M. Gisler (to Sandoz).
40. Jpn. Pat. 7,778,926 (July 2, 1977), S. Takahashi and Y. Aizawa (to Nippon Kayaku Co.).
41. Ger. Pat. 2,748,975 (May 3, 1979), E. Hoyer and F. Meininger (Hoechst AG).
42. Ger. Pat. 3,825,658 (Feb, 1, 1990), H. Springer and co-workers (to Hoechst AG).
43. Ger. Pat. 3,843,014 (June 28, 1990), H. Tappe and co-workers (to Hoechst AG).
44. Ger. Pat. 3,843,605 (June 28, 1990), H. Springer and co-workers (to Hoechst AG).
45. Ger. Pat. 3,906,778 (Sept. 6, 1990), R. Haehnle (to Hoechst AG).
46. Eur. Pat. Appl. 400,647 (Dec. 5, 1990), T. Miyamoto and co-workers (to Sumitomo Chemical Co.); Eur. Pat. Appl. 400,648 (Dec. 5, 1990), T. Miyamoto and co-workers (to Sumitomo Chemical Co.).
47. U. S. Pat. 4,968,783 (Nov. 6, 1990), K. Seitz and G. Hegar (to CIBA-GEIGY Corp.).
48. Ger. Pat. 2,924,228 (Dec. 20, 1979), K. Seitz and G. Hegar (to CIBA-GEIGY Corp.).
49. Ger. Pat. 2,929,107 (Jan. 31, 1980), N. Nishimura and co-workers (to Sumitomo Chemical Co.).
50. Pol. Pat. 105,615 (Sept. 1976), S. Stefaniak, E. Stefaniak, and E. Lauer.
51. Eur. Pat. Appl. 22,265 (Nov. 14, 1981), T. Omura, Y. Tezuka, and M. Sunami (to Sumitomo Chemical Co.).
52. Pol. Pat. 94,219 (Dec. 31, 1977), J. Gmaj and C. Sosnowski.
53. Czech. Pat. 177,680 (Mar. 15, 1979), J. Marek, M. Beranek, and V. Hanousek.
54. Czech. Pat. 184,683 (Aug. 15, 1980), J. Marek.
55. Ger. Pat. 2,749,647 (May 10, 1979), W. Scholl (to Bayer AG).
56. Ger. Pat. 2,900,267 (July 19, 1979), K. Seitz (to CIBA-GEIGY Corp.).
57. Ger. Pat. 2,442,553 (Jan. 22, 1976), E. Ungermann (to Hoechst AG).
58. Ger. Pat. 2,850,131 (June 4, 1980), H. Baumann and H. Kaack (to BASF AG).
59. Brit. Pat. 1,461,125 (Jan. 13, 1977), I. K. Barben and C. V. Stead (to ICI).
60. Ger. Pat. 2,901,546 (July 24, 1980), H. Henk, K. Wunderlich, and P. Wild (to Bayer AG).
61. Ger. Pat. 2,935,681 (Mar. 12, 1981), O. Schallner and K. H. Schuendehutte (to Bayer AG).
62. Ger. Pat. 2,948,292 (June 4, 1981), J. Koll and co-workers (to Bayer AG).
63. Belg. Pat. 871,135 (Apr. 10, 1979), (to Hoechst AG).
64. Ger. Pat. 2,847,658 (May 10, 1979), R. Begrich (to CIBA-GEIGY Corp.).
65. Ger. Pat. 2,751,785 (May 23, 1979), H. Jaeger (to CIBA-GEIGY Corp.).
66. Jpn. Pat. 79 71,117 (June 7, 1979), (to ICI).
67. Ger. Pat. 2,902,486 (July 26, 1979), K. Seitz and D. Maeusezahl (to CIBA-GEIGY Corp.).
68. Ger. Pat. 3,900,535 (July 12, 1990), K. H. Schuendehuette (to Bayer AG).
69. Ger. Pat. 2,927,102 (Jan. 17, 1980), H. Seiler (to CIBA-GEIGY Corp.).
70. Ger. Pat. 2,535,077 (Feb. 17, 1977), A. Friedrich, H. Harnisch, and R. Raue (to Bayer AG).
71. Jpn. Pat. 77 0833 (Mar. 8, 1977), T. Ikeda and co-workers (to Sumitomo Chemical Co.).
72. Swiss Pat. 621,358 (Jan. 30, 1981), H. Riat (to CIBA-GEIGY Corp.).
73. Jpn. Pat. 80 62,968 (May 12, 1980), (to Sumitomo Chemical Co.).
74. Ger. Pat. 3,915,305 (Nov. 15, 1990), H. Springer and H. M. Beuch (to Hoechst AG).
75. Brit. Pat. 2,024,236 (Jan. 9, 1990), G. Hegar (to CIBA-GEIGY Corp.).

76. Ger. Pat. 2,846,201 (May 8, 1980), H. Baumann and H. Kaack (to BASF AG).
77. Ger. Pat. 2,925,942 (June 27, 1979), H. Jaeger (to Bayer AG).
78. Ger. Pat. 2,838,540 (Mar. 8, 1979), H. Seiler (to CIBA-GEIGY Corp.).
79. Ger. Pat. 2,839,209 (Mar. 15, 1979), H. Seiler and G. Heger (to CIBA-GEIGY Corp.).
80. Ger. Pat. 2,809,200 (Sept. 6, 1979), E. Kraemer (to Bayer AG).
81. Ger. Pat. 2,838,608 (Mar. 20, 1980), H. Jaeger and K. Wunderlich (to Bayer AG).
82. Ger. Pat. 3,826,060 (Feb. 1, 1990), H. Jaeger (to Hoechst AG).
83. U.S. Pat. 4,448,583 (May 15, 1984), A. J. Corso (to American Hoechst Corp.); Eur. Pat. Appl. 114,031 (July 25, 1984), R. Lacroix (to CIBA-GEIGY Corp.); Eur. Pat. Appl. 144,093 (June 12, 1985), P. Scheibili, A. Kaenzig, and A. Shaub (CIBA-GEIGY Corp.); Ger. Pat. 3,400,412 (July 18, 1985), J. Wolff, K. Wolf, and P. Wegner (to Bayer AG); Ger. Pat. 3,403,662 (Aug. 8, 1985), J. Wolff and K. Wolf (to Bayer AG); Ger. Pat. 3,424,145 (Jan. 9, 1986), J. Wolff, K. Wolf, and G. Seipt (to Bayer AG); Jpn. Pat. 6,155,157 (Mar. 19, 1986), K. Imada and T. Tokieda (to Sumitomo Chemical Co.); Ger. Pat. 3,424,506 (Jan. 9, 1986), F. Meininger, K. Opitz, and J. Semel (to Hoechst AG); Ger. Pat. 3,515,407 (Oct. 30, 1986), M. Haenke and P. Canora (to Hoechst AG); Eur. Pat. Appl. 208,968 (Jan. 21, 1987), N. Yamauchi and K. Imada (to Sumitomo Chemical Co.); Jpn. Pat. 61 204,276 (Sept. 10, 1986), N. Yamonaka and T. Sunoga (to Nippon Kayaku Co.); Ger. Pat. 3,534,729 (Apr. 9, 1987), J. Wolff and K. Wolf (to Bayer AG); Jpn. Pat. 6,243,466 (Feb. 25, 1987), T. Morimitsu, T. Omura, and A. Takeshita (to Sumitomo Chemical Co.); Ger. Pat. 3,702,942 (Aug. 13, 1987), W. Groehke, H. Ochsner, and H. Gilgen (to Sandoz); Eur. Pat. Appl. 234,573 (Sept. 2, 1987), N. Yamauchi, K. Imada, and S. Ikeou (to Sumitomo Chemical Co.); Jpn. Pat. 62 256,869 (Nov. 9, 1987), N. Yamauchi, K. Imada, and Y. Kashiwane (to Sumitomo Chemical Co.); Ger. Pat. 3,703,738 (Aug. 18, 1988), J. Koll, J. Wolff, and D. Szeymies (to Bayer AG); Ger. Pat. 3,703,732 (Aug. 18, 1988), W. Steinbeck and W. Harms (to Bayer AG); Jpn. Pat. 63 213,574 (Sept. 6, 1988), Y. Yamamoto and co-workers (to Sumitomo Chemical Co.); Jpn. Pat. 5,925,838 (Feb. 9, 1984), (to Hoechst AG).

Roy E. Smith
CIBA-GEIGY Corporation

DYES, SENSITIZING

Spectral sensitizing dyes extend the wavelengths of light to which inorganic semi-conductors, organic semiconductors, and chemical (biological) reactions can be photosensitized (see Photochemical technology). Spectral sensitizers are needed for the blue, green, and red portions of the visible spectrum (Fig. 1) (see Color photography). For infrared photographic, electrophotographic, and biological applications, sensitizing dyes are also needed to match output wavelengths of solid-state lasers (optical data storage, laser printing, range finding, and data transmission), to provide color-selective infrared photography (an effective environmental survey method), or to match transmission wavelengths of body tissue.

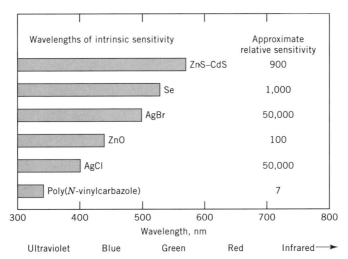

Fig. 1. Spectral sensitivity ranges for undyed semiconductors. Lack of sensitivity in the visible and infrared regions necessitates the use of spectral sensitizing dyes.

Spectral sensitizing dyes are considered "functional" dyes (1) to distinguish them from conventional colorants. The absorption of radiation by a functional dye causes some additional function(s) to occur, and in many cases, the sensitizing dyes can exhibit more than one type of sensitization reaction. For example, spectral sensitization of silver halides in photography (qv) is most common by conduction-band electron generation (n-type photoconduction), and trapped positive holes contribute less to the photoconductivity. However, changes in sensitizer structure can provide dyes that will sensitize with the photohole as the primary carrier (p-type photoconduction), and the photoelectrons are effectively trapped by the dyes (see ELECTROPHOTOGRAPHY; LIGHT GENERATION; OPTICAL DISPLAYS). Many texts and reviews provide extensive details about the diversity of spectrally sensitized processes (2–19).

The detection of spectral sensitizing action often depends on amplification methods such as photographic or electrophotographic development or, alternatively, on chemical or biochemical detection of reaction products. Separation of the photosensitization reaction from the detection step or the chemical reaction allows selection of the most effective spectral sensitizers. Prime considerations for spectral sensitizing dyes include the range of wavelengths needed for sensitization and the absolute efficiency of the spectrally sensitized process. Because both sensitization wavelength and efficiency are important, optimum sensitizers vary considerably in their structures and properties.

Sensitization Wavelength and Efficiency

The wavelengths for which useful spectral sensitization can be achieved are historically best illustrated by the spectral sensitivity of commercial photographic plates and films (20). As shown in Figure 2, spectral sensitivity can be extended

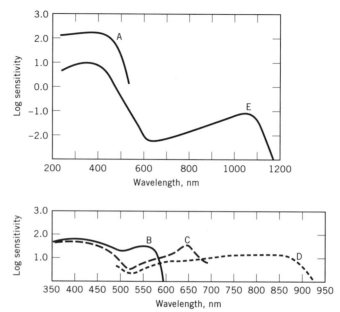

Fig. 2. Spectral sensitivity curves for dyed silver halide plates and films. A, uv-blue; B, green; C, red; D and E, infrared.

beyond the natural ultraviolet and blue-light photosensitivity of silver bromide crystals (Fig. 2, curve A) to include green light (B), red light (C), and infrared wavelengths (D and E). Some commercial products, like the Kodak Spectroscopic Plate, Type 103a–F, are sensitized to have a maximum sensitivity near specific wavelengths, such as the 645-nm line emitted by the sun's hydrogen atoms. Color films sensitive in the visible or infrared regions depend on three spectrally sensitized layers such as B, C, and D in the same film, where the separate amplification reaction produces three differently colored images (one per layer), which are the usual cyan, magenta, and yellow dyes of color photographs.

Commercial products like those in Figure 2 employ the best possible spectral sensitizers for each wavelength region. For example, the sensitivity curve B (Fig. 2) shows that spectral sensitization near 550 can be essentially as efficient as the intrinsic sensitivity of the silver halide itself (350–400 nm). However, in the infrared region, even the chemical properties of well-designed dyes (stability, ease of oxidation or reduction, ease of protonation) cause practical limitations to obtaining highly efficient sensitizers. Both curves D and E (Fig. 2) indicate less than optimum sensitivity in the spectrally sensitized regions, and the chemical properties of these sensitizers diminish the intrinsic silver halide sensitivity as well.

Fortunately, the demand for spectral sensitizers, especially those in the infrared, has extended to systems having somewhat fewer and quite different limitations than for photography. For infrared dyes as examples (1), their photoelectrical effects useful in photography are also important to electrophotography, photopolymerization, and voltage-sensitive biomedical probes. Photochemical reactivity provides especially useful opportunities in photodynamic therapy, at the infrared wavelengths where body tissue has high transparency. Light-induced polarization can be designed to yield infrared nonlinear optical materials. Light absorption and emission, combined with suitable stability and solubility properties, gives dyes used in lasers, medical diagnostics, optical data recording (photothermal), laser welding surgery, and Q-switches. Photosensitizers absorbing in the visible region are used in many applications, including the industrial scale photooxidation of (−)-citronellol to (−)-rose oxide (8), photoisomerization of *cis*-to *trans*-vitamin A acetate (8), and dye-sensitized neurobiology (9).

Structural Classes of Spectral Sensitizers

A useful classification of sensitizing dyes is the one adopted to describe patents in image technology. In Table 1, the Image Technology Patent Information System (ITPAIS), dye classes and representative patent citations from the ITPAIS file are listed as a function of significant dye class. From these citations it is clear that preferred sensitizers for silver halides are polymethine dyes (cyanine, mero-

Table 1. Dyes and Semiconductors, Patent Citations[a]

ITPAIS classification	Silver halide	ZnO	TiO$_2$	ZnS	Se	CdS, CdSe	Misc. organic semiconductor
dyes, cyanine[b]	420	33	27	10	45	16[c]	11
dyes, merocyanine[b]	118	2	1	0	27	2	0
all other polymethine dyes[b]	277	5	4	1	34	3	4
dyes, acridine	0	2	0	0	1	0	0
dyes, azine[d]	4	4	3	0	7	5	0
dyes, azo	107	18	28	3	28	10	4
dyes, arylmethane	3	3	1	2	7	2	0
dyes, quinone type[e]	48	10	16	1	21	2	1
dyes, porphine[f]	12	8	17	1	24	2	1
dyes, xanthene[g]	6	14	11	3	17	3[h]	3
dyes, pyrylium[i]	1	0	1	0	42	3[h]	4

[a]ITPAIS, the Image Technology Patent Information System was developed between 1975–1985 by Eastman Kodak Co., Agfa-Gevaert (Antwerp/Leverkusen), and Fuji Photo Film Co., Ltd., and encompasses selected patents and literature references related principally to the chemical aspects of image technology. Search terms used for this table were the same as in the previous edition, and the Derwent patent database was used for the search data presented here.
[b]Dyes, polymethine: used for dyes having at least one electron donor and one electron acceptor group linked by methine groups or aza analogues; allopolar cyanine, dye bases, complex cyanine, hemicyanine, merocyanine, oxonol, streptocyanine, and styryl. Supersensitization has been reported for these types—18 cites for cyanines, 3 for merocyanine, and 6 for all other polymethine types.
[c]Also 3 citations for misc. inorganic semiconductors.
[d]Dyes, azine: used for azines, thiazines, and oxazines (see AZINE DYES).
[e]Dyes, quinone type: used for anthraquinones, indamines, indoanilines, indophenols, and miscellaneous quinones (see DYES, ANTHRAQUINONE).
[f]Dyes, porphine derivative: used for chlorophyll, phthalocyanines, and hemin.
[g]Dyes, xanthene: used for eosin, fluorescein-type phthaleins, rhodamines, and rose bengal.
[h]Also 1 cite for misc. inorganic semiconductors.
[i]Dyes pyrylium: also used for thiapyrylium and benzothiapyrylium.

cyanine, etc), whereas other semiconductors have more evenly distributed citations. Zinc oxide, for example, is frequently sensitized by xanthene dyes (qv) or triarylmethane dyes (see TRIPHENYLMETHANE AND RELATED DYES) as well as cyanines and merocyanines (see CYANINE DYES).

Spectral Sensitization of Silver Halides

The large number of patents for spectral sensitizers of silver halides indicates their extensive use in photographic plates, films, and papers. Color films that give good color reproduction under a variety of illuminants (daylight, electronic flash, tungsten) and exposure times (short for electronic flash, long for available light) have required more extensive tailoring of spectral sensitizers than imaging systems based on other semiconductors. Many reviews and several books cover aspects of spectral sensitizing dyes specifically for silver halides. These include synthetic methods (2,16,17), general sensitization mechanisms (21), electrochemical potentials and dye efficiency (21,22), dye absorption and aggregation (23), and supersensitization of dyes (24).

Photographic products are often subjected to wide variations in temperature, humidity, and time before processing. Consequently, high priorities are also given to spectral sensitizers that do not degrade either the film or image quality, as well as provide efficient spectral sensitization at the desired wavelengths. Although there are many available dyes and pigments, commercial silver halide films, papers, and plates are efficiently sensitized by just a few types of chromophores in the cyanine and merocyanine dye classes. The spectral sensitizers designated BN, GN, and RN (Fig. 3) spectrally sensitize the photoelectrons as primary carriers in silver halides. Efficient infrared sensitizers are derived from symmetrical dyes: IRN ($n = 1-3$). The merocyanine sensitizer MN and its derivatives ($n = 0$ and 2) sensitize by photoelectron production in silver halides.

The suitability of these chromophores rests in large measure in being able to simultaneously optimize three properties (electrochemical potentials, J-aggregation, and solubility) by choosing various substituent and heteroatom combinations. For example, most commercial silver halides are sensitized by photoelectron carriers. The efficiency for producing dye-sensitized photoelectrons is empirically related to dye reduction potentials. The relationships are diagrammed in Figure 4. Dyes with reduction potentials more negative than -1.0 are difficult to reduce and provide efficient spectral sensitizers where latent image is formed by photoelectron carriers. In Figure 4 the symbols for the typical sensitizing chromophores (BN, GN, etc) are included on the abscissa to indicate the usual electrochemical potential ranges exhibited by these various classes of dyes. Significant changes in these potentials and the efficiency of spectral sensitization result from structural modifications of the heterocyclic groups in the dyes.

Aggregation (self-association) of spectral sensitizers into ordered arrays is of prime importance in silver halide color films and papers (21). Narrow absorption bands, required for accurate color reproduction, do occur for appropriately substituted chromophores like those in Figure 3. The tendency of dyes to aggregate in ordered arrays is not only a function of the type of substituent but also the length of the chromophore and the surface of the solid substrate. J-aggregation

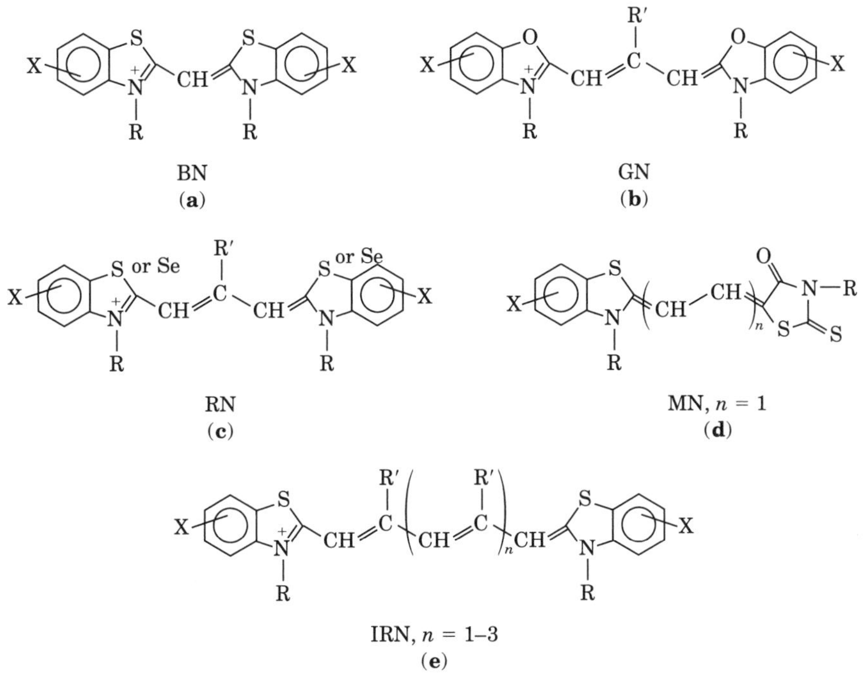

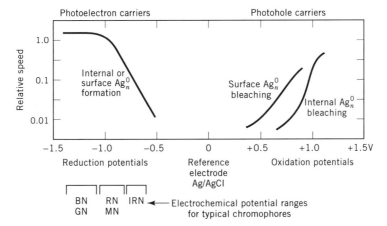

Fig. 3. Spectral sensitizing dyes for silver halides. (**a**) Blue sensitizers (400–500 nm) are designated BN; (**b**) green sensitizers (500–600 nm) are designated GN (the ring oxygen may be replaced by N(R)); (**c**) red sensitizers (600–700 nm) are designated RN; (**d**) MN designates a merocyanine dye; and (**e**), infrared sensitizers (>700 nm) are designated IRN.

Fig. 4. Spectrally sensitized speed for dyed silver halides as influenced by dye reduction or oxidation potential.

843

leads to absorption bands at longer wavelength and narrower than the monomer (Fig. 5) and highly wavelength-selective spectral sensitization. The J-aggregates were first noted by Jelly, and the geometric arrangements of molecules needed for J-aggregate absorption are known for cyanine dyes (qv). Since some J-aggregates are inefficient spectral sensitizers with low relative quantum yields for spectral sensitization, supersensitization of their response by small amounts of added compounds has been extensively studied (13). To improve the relative quantum yields, supersensitizers with suitable oxidation potentials (Fig. 6) are added at about one-tenth of the concentration of J-aggregating dye. These supersensitizers separate the exciton in the aggregate by trapping the positive hole in the supersensitizer and freeing the photoelectron for sensitization. Effects of aggregate size on excited state properties show less efficient sensitization if the aggregate size is large (25). Infrared dyes like IRN (Fig. 3) adsorb to silver halides in many forms and cause spectral sensitization over a broad range of wavelengths. The adsorbed forms of infrared dyes include trans-monomeric dye, isomeric (cis) dyes, and H-aggregates at much shorter wavelengths than the trans-monomeric dye.

Further tailoring of the dye structures (to provide additional improvements in efficiency, spectral response at specific wavelengths, stronger adsorption, and improved solubility) is primarily accomplished by substituent modification. Substituents like methoxy, chloro, and phenyl on the aromatic rings in BN, GN, and RN shift both the solution absorption and spectral sensitization to longer wave-

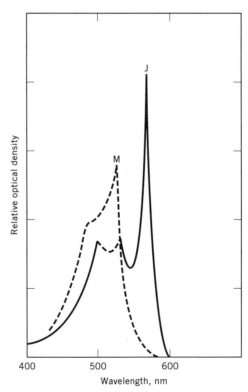

Fig. 5. Effect of aggregation on absorption. M, monomer; J, J-aggregate.

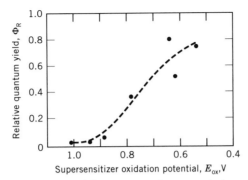

Fig. 6. Effect of added supersensitizers on a J-aggregated spectral sensitizing dye, 1,1′-diethyl-2,2′-quinocyanine chloride [2402-42-8] (**1**), for which $E_R = -1.03$ V and $E_{ox} = 0.99$ V.

(**1**)

lengths and reinforce the dyes' tendencies to form ordered, sharp-absorbing J-aggregates. In the carbocyanines from benzoxazole (GN) and benzothiazole–benzoselenazole (RN), simple out-of-plane substituents like ethyl or phenyl at position R′ strongly increase J-aggregation.

Modification of dye solubility can minimize migration of spectral sensitizers between layers in a multilayer film and improve dye removal by photographic processing solutions. Control of these properties is accomplished primarily by the substituents attached to heterocyclic nitrogen atoms. The structural types of these groups have been summarized (26). Typical groups have included ethyl and other alkyls; sulfoalkyl, $CH_2CH_2CH_2SO_3^-$ unsaturated groups, ie, phenyl and allyl; and carboxyalkyl, phosphonoalkyl, and sulfatoalkyl, $CH_2CH_2CH_2OSO_3^-$. Dyes of the cyanine type have positively charged chromophores and with suitable anions are reasonably soluble in alcohols and other polar solvents. Placing two sulfopropyl groups in the dye structure gives the dye a net anionic charge and improved solubility. Only one sulfopropyl substituent results in a zwitterionic (neutrally charged) dye and considerably lower solubility in polar solvents. Merocyanines (MN in Fig. 3) with alkyl substituents are insoluble in hydroxylic solvents but soluble in pyridine.

Spectral Sensitization of Inorganic and Organic Solids

Relatively few spectral sensitizing dyes continue to serve the large volume color (imaging and printing) photographic markets where silver halide is the dominant semiconductor. Retail photofinishing in the United States, for example, is estimated to grow substantially in the 1990s from 15 billion exposures in 1989 to 23 billion exposures in 1999. Technology to serve both color printing and office copy markets will employ not only silver halides but other inorganic and organic semiconductors as well. Inorganic semiconductors include selenium, germanium, CdS, HgO, HgI_2, ZnO, PbO, Cu_2O, thallium halides, and TiO_2. As noted by patent citations and the published literature, both zinc oxide and titanium dioxide have been extensively investigated (10). Spectrally sensitized photoconduction has also been observed for organic semiconductors, ie, anthracene, poly(N-vinylcarbazole), polyacetylene, copper phenylacetylide, phthalocyanines, and other solid dyes (10) (Fig. 7).

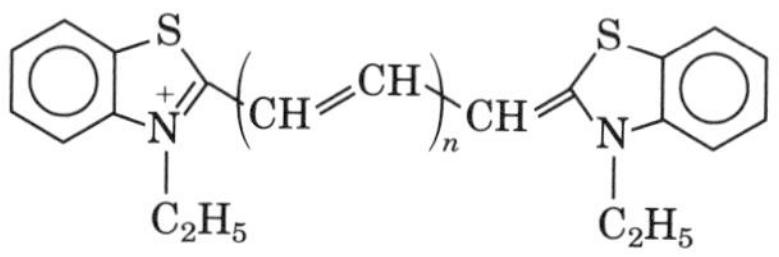

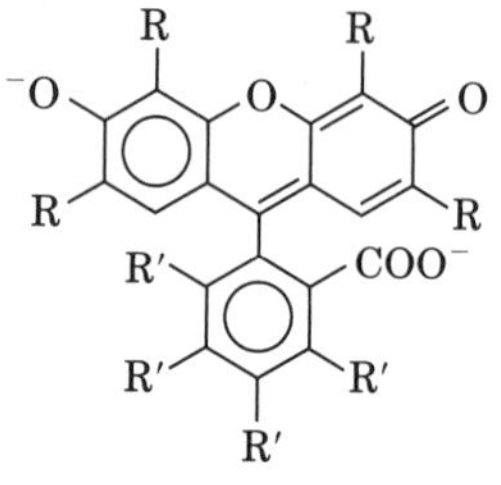

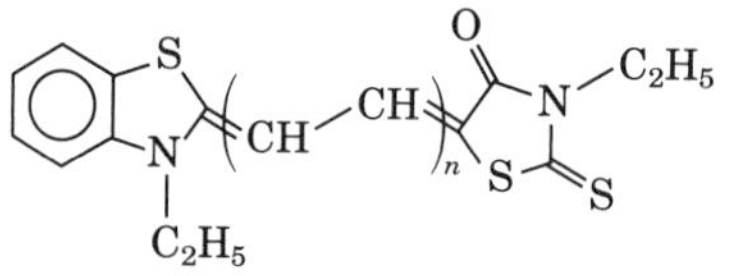

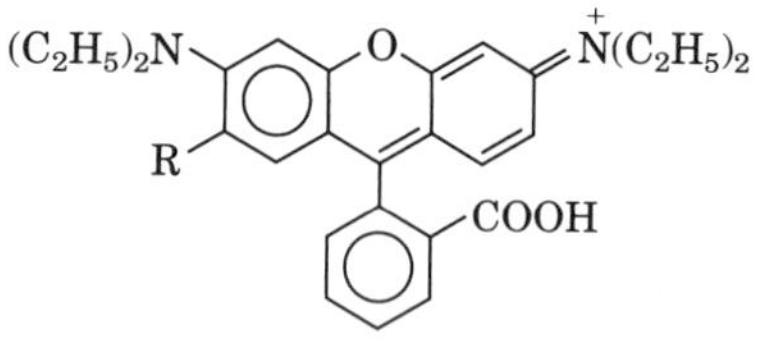

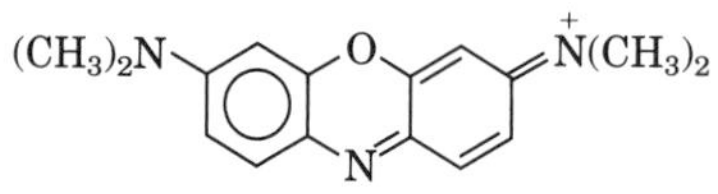

Fig. 7. Structural formulas for spectral sensitizers. (**2**), Vinylogous thiacyanines (thiacarbocyanine, $n = 1$). In (**3**), R = R′ = H, fluorescein; R = Br, R′ =H, eosin; R = I, R′ = H, erythrosin; R = I, and R′ = Cl, rose bengal. (**4**) represents benzothiazolylrhodanines (merocarbocyanine, $n = 1$ [3568-36-3]. (**5**) is rhodamine B and (**6**) is methylene blue. (**7**) ($n = 1$) is benzothiazolyl styryl [3028-97-5] and for ($n = 2$), benzothiazolyl butadienyl [17818-85-8]. (**8**) is the PVK:TNF complex poly(N-vinylcarbazole): 2,4,7-trinitrofluorenone [39613-12-2].

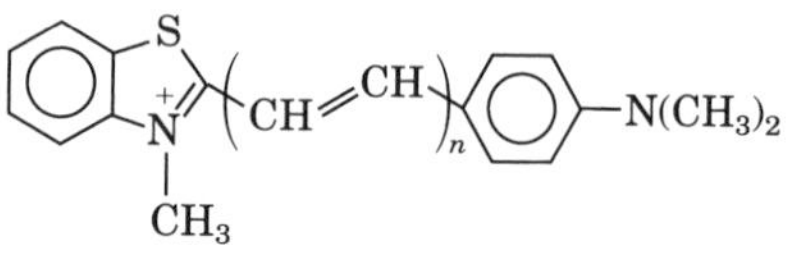

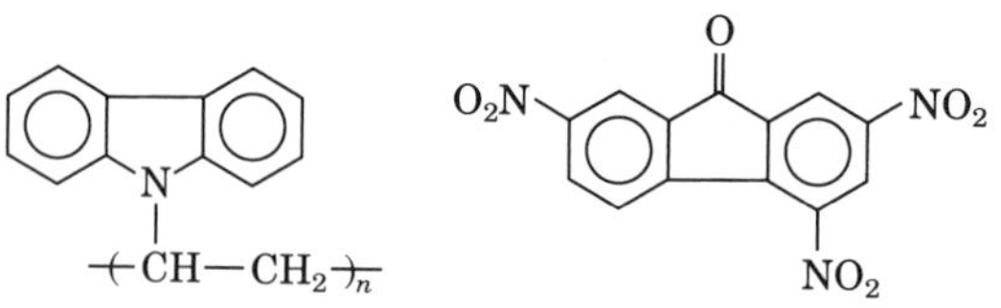

Continued commercial development of nonsilver office copying into color printing and optical data storage to higher densities provides many opportunities for spectral sensitizer improvement among competing technologies. Spectral sensitizers for these other materials are already quite varied (Table 2). Recent reviews focus on both the dye chemistry (4,6) and the imaging systems that utilize dyes (5).

Zinc oxides can be spectrally sensitized by a variety of sensitizing dyes. The Dember effect of pressed ZnO powder was demonstrated early for cyanine dyes, eosin, erythrosin, and rhodamine B (27). ZnO electrophotographic papers are reported to be sensitized with bromophenol blue, auramin, sodium fluorescein, rose bengal, and eosin (see ELECTROPHOTOGRAPHY). In addition, sensitization was observed for cyanine dyes, erythrosin, dihydroxy azo dyes, rhodamine B, and methylene blue. The spectral sensitization is favored by strong dye adsorption (28). Adsorption of sensitizers like rhodamine B and dihydroxyazo dyes results from zinc salt formation with either chromophoric groups or nonchromophoric, charged substituents. Azo dyes (qv) without strong salt-forming ability, eg 4,4'-azodiphenol, are not effective sensitizers. Halide ions significantly increase the spectral sensitization by rhodamine B. Electronegative organic molecules, eg, qui-

Table 2. Spectral Sensitizers for Inorganic and Organic Semiconductors

Semiconductor	Sensitizer	CAS Registry Number[a]
CdS (Cu doped)	thiacarbocyanine	[905-97-5]
	benzothiazolylrhodanines	[3568-36-3]
	rhodamine B	[81-88-9]
ZnO	thiacarbocyanine	[905-97-5]
	erythrosin	[16423-68-0]
	eosin	[17372-87-1]
	phthalocyanine	[574-93-6]
	rhodamine B	[81-88-9]
	rose bengal	[11121-48-5]
	methylene blue	[61-73-4]
	fluorescein	[2321-075]
TiO$_2$	benzothiazolylstyryl dye	[3028-97-5]
	thiacarbocyanine	[905-97-5]
poly(N-vinylcarbazole)	rose bengal	[11121-48-5]
	2,4,7-trinitrofluorenone	[129-79-3]
	thiapyrylium dyes	[25966-12-5][b]
solid dye particles	phthalocyanines	[147-14-8]
photoelectrophoresis	thioindigo	[522-75-8]
	flavanthrone	[475-71-8]
	quinacridone	[1047-16-1]
dye electrodes	benzothiazolylrhodanines	[3568-36-3]
	phthalocyanines	[147-14-8]

[a]Chemical Abstract Service Registry Number for a typical example of the sensitizer listed. See Figure 7 for some structural formulas.
[b]Structure (**10**) where the anion is BF_4^-.

none, also function as cosensitizers for dyed ZnO (28) in analogy to the supersensitization of dyed silver halides.

Spectral sensitization of titanium dioxide is best accomplished by the cationic cyanine dyes. Dyed TiO_2 electrode systems have potential solar cell use and were successfully adapted for the photochemical decomposition of water to hydrogen and oxygen (see PHOTOVOLTAIC CELLS). Typical sensitizers include the vinylogous thiacyanines, styryl and butadienyl dyes, and complex cyanines (29) (see CYANINE DYES). Nonionized merocyanine dyes and many zinc oxide sensitizers (rhodamine B, eosin, erythrosin) are not good sensitizers for TiO_2 emulsions. However, substitution of the cyanine, styryl, merocyanine, and other chromophores with polyhydroxy-containing substituent groups improves the binding of the dyes to TiO_2 and their spectral sensitization. Coverage of the oxide surface with metal ions also improves the effectiveness of certain anionic dyes (30). Thin-film TiO_2 electrodes were sensitized by spectral sensitizers plus supersensitizers (30). For example, detectable photocurrents for oxacarbocyanine were improved by hydroquinone, and rhodamine B was made to sensitize using added 4,5-dihydroxy-*m*-benzenedisulfonic acid. Spectrally sensitized hydrogen generation was achieved in the region 500–630 nm using the TiO_2 films plus supersensitizers and rhodamine B.

Colloidal TiO_2 dyed with Coumarin 343 [*55804-65-4*] (**9**) is an extensively investigated system.

(**9**)

Excitation with visible light causes rapid electron transfer from the dye to TiO_2 and forms nearly 100% yield of the radical cation of the dye. The back electron transfer is several thousand times slower, enabling the light-induced electron transfer state to exist for several microseconds (31). A transparent solar panel to make windows into inexpensive active solar energy collectors is envisioned dye-colloidal-TiO_2 technology (31), where the light-to-electric energy conversion (7–12%) is about as good as in common amorphous solar cells found in calculators. Dye-sensitized electron-transfer processes at semiconductor electrodes apply not only to solar cells but to other technology areas such as waste treatment, materials synthesis and processing, and sensors (11).

Cadmium sulfide can be spectrally sensitized by cyanine, styryl, merocyanine, and xanthene dyes (32,33). In contrast to zinc oxide, which is sensitized by many dyes that desensitize silver halides, cadmium sulfide energy levels require dyes that are also good sensitizers for silver halide. In a comparison of dye energy levels for various photoconductors, vinylogous thiacyanines were noted to sensitize ZnO, AgBr, and CdS, but the silver halide desensitizer methylene blue sensitized only zinc oxide (33). Tin sulfide, dye-sensitized with oxazine 1 [*24796-94-9*], is an excellent model semiconductor surface for kinetic studies of spectrally sensitized electron transfer. From fluorescence quenching studies, the rate of electron transfer from oxazine 1 into the SnS_2 conduction band is 3×10^{13} s^{-1}, corresponding to an electron-transfer time of 40 fs (34).

Hard Copy Systems. Electrophotographic copy processes using organic photoconductors may be spectrally sensitized by many dye classes (35), in sharp contrast to the limited choices for silver halides. The first commercial organic photoreceptor (36) was the charge-transfer complex (**8**) between poly(N-vinylcarbazole) (PVK) and 2,4,7-trinitro-9-fluorenone (TNF) for the IBM Copier I in 1970. Most current photoreceptors are organic dyes or metal-complex dyes, and the useful physical states are dye-aggregates, pigment dispersions, or vacuum-deposited pigments. Aggregates and pigments respond to solvent treatments, the presence of surfactants, and thermal treatments. Considerable detail about the processing steps for electrophotographic organic photoreceptors has been summarized recently (35).

Arylalkane polycarbonates mixed with organic photoconductors can be spectrally sensitized by thiapyrylium and other dyes (Fig. 8), particularly when dye/polycarbonate cocrystalline complexes can be formed (35,37). Relative speeds are improved by vapor-treating the electrophotographic films before exposure to cause either H-aggregation or J-aggregation of the spectral sensitizers. For 4-(dimethylaminophenyl) diphenylthiapyrylium perchlorate, the nonvapor-treated film shows a relative speed of 40 at 580 nm, whereas the vapor-treated coating gives a speed of 1800 at 685 nm (cocrystalline, aggregated dye).

The PVK:TNF and dye-aggregate technologies were useful photoreceptors for high volume, high quality xerographic applications. During the 1980s, manufacture of organic pigments at low cost and large volume was developed for drum-based xerographic applications. Typical aggregate and pigment materials include the dyes shown in Figure 8. All of these respond to similar amounts of incident exposure, $2-5 \times 10^{-7}$J/cm^2. The metal-complex phthalocyanines and related compounds continue to be of interest as charge-generating photoreceptors (38).

Thermal Sensitization. Photothermal actions of dyes are important in sensitizing optical disk data storage layers (12,13,39). For data writing, electronically encoded information controls the output power of a laser, and at high power the sensitizer absorbs enough energy to cause physical deformation (pit formation or layer deformation). Dyes are typically utilized as dye–polymer layers or as thin layers of pure dye. Considerable synthetic effort (6,12,13,39,40) has been expended to match dye absorption with infrared wavelengths from semiconductor lasers while controlling other properties such as crystallization propensity and photodegradation. Dye classes examined for optical disks include pentamethine cyanines (**14**), metal phthalocyanines (or naphthalocyanines), anthraquinone and naphthoquinone dyes, squarilum dyes, and dithiene metal complexes (**15**) (13,39,40) (Fig. 9). All of these dye classes can be designed to provide near-infrared absorbing materials with absorption beyond 1000 nm in some cases (see CYANINE DYES; PHTHALOCYANINES; POLYMETHINE DYES). Additional designing of these long wavelength dyes is pointed toward combining newer heterocycles with groups like squaric acid (1) and substituent effects on metal complex dyes like naphthalocyanines (13) and dithiolene dyes (40).

Biological stains (41) and laser dyes (42) allow wavelength-selective photothermal procedures like laser welding surgery and destruction of dye-stained

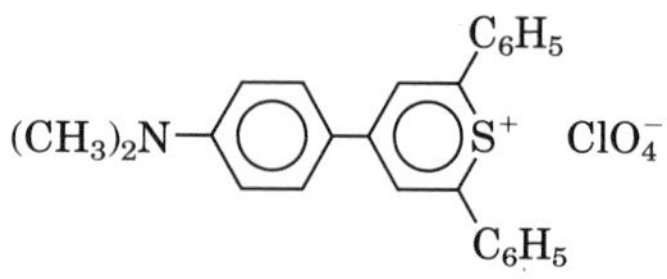

(**10**)

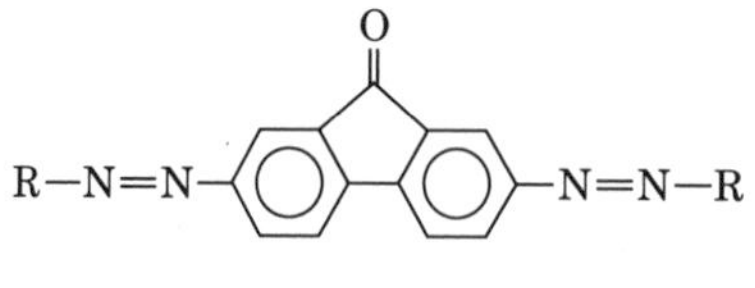

(**11**)

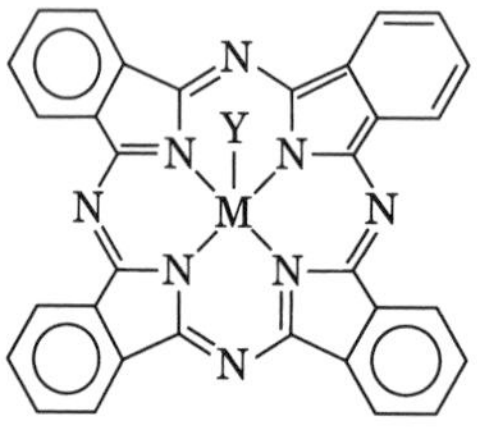

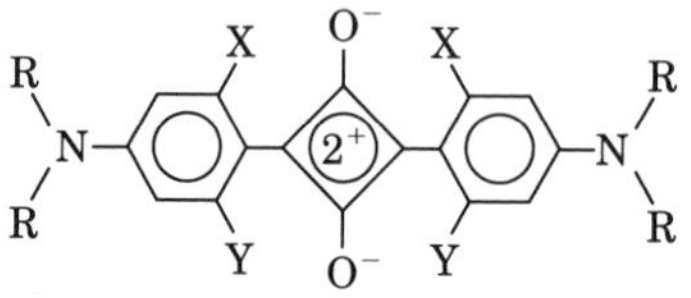

where R =

(**12**)

(**13**)

Fig. 8. Organic photoreceptors for electrophotographic use. (**10**), 4-[4-(Dimethyl-amino)phenyl]-2,6-diphenylthiapyrylium perchlorate [14039-00-0], $E_o = 5 \times 10^{-7}$ J/cm^2; (**11**) is a polyazo dye with $E_o = 2 \times 10^{-7}$ J/cm^2. (**12**) represents phthalocyanines for which $E_o = 2.6 \times 10^{-7}$ J/cm^2 for MY = Cu [147-14-8]. (**13**) is a squaraine dye, $E_o = 3.5 \times 10^{-7}$ J/cm^2. Other sensitizers include naphthalocyanines where M = Si, Ge, Al, Ga, In, or a transition metal and Y = OR, OSiR$_3$, or a polymer.

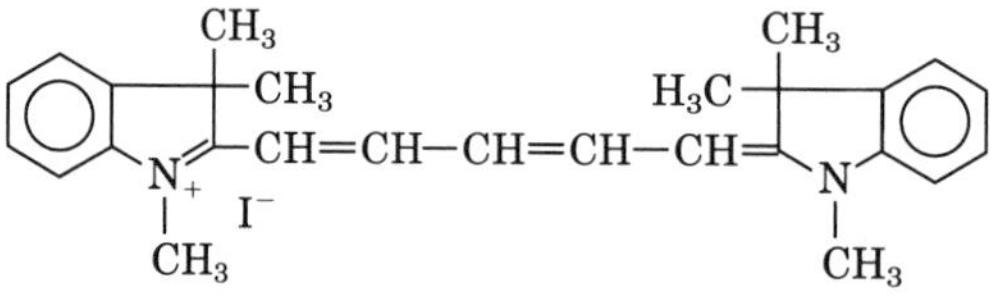

(**14**)

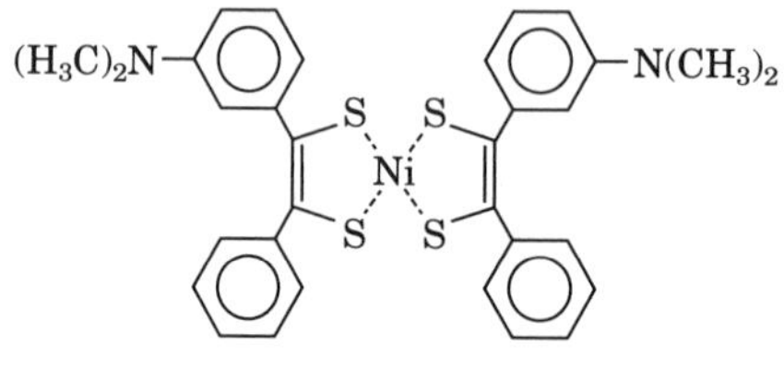

(**15**)

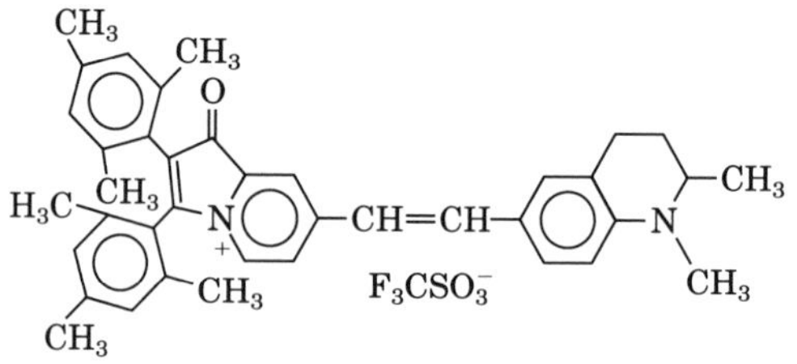

(**16**)

Fig. 9. Examples of dyes used in optical disks. (**14**), pentamethine cyanine [36536-22-8], (**15**), nickel dithiene complex [38465-55-3], and (**16**) IR-810.

tumors. Optimum dyes for photothermal effects on biological cells would, of course, need very short excited state lifetimes and few chemical side reactions, so that dye-absorbed energy would be transferred quickly to nearby cells. A quantitative study with malachite green showed that 10 ns laser pulses at 620 nm caused a 130°C temperature rise within a 10-nm sphere of water around the dye; similar exposures of a dye–enzyme complex inactivated the enzymes (43). Photothermal sensitization by porphyrins and other pigments has been used to make oxide superconductors spectrally sensitized (44). Dye-coated small superconducting junctions showed detectable changes in conductivity when exposed to light absorbed by the dyes.

Spectral Sensitization in Photochemical Technology

Functional dyes (1) of many types are important photochemical sensitizers for oxidation, polymerization, (polymer) degradation, isomerization, and photodynamic therapy. Often, dye structures from several classes of materials can fulfill a similar technological need, and reviewing several dye structures for each technology is a lengthy task. Rather, one or two examples for each photochemical area are cited below.

Photooxidation Reactions in Solution. Excited-state spectral sensitizers may photooxidize nearby substrate or solvent molecules to form radicals. These can undergo further chemistry, including the reaction with ground-state oxygen (type I photooxidation). Direct involvement of oxygen as the primary oxidant (type II photooxidation) occurs through energy transfer from the triplet excited state of the spectral sensitizer giving primarily the $^1\Delta_g$ singlet oxygen state (15,45–47).

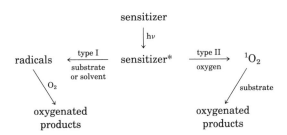

Spectral sensitizers that efficiently sensitize singlet oxygen production must fulfill several requirements: (1) high singlet-to-triplet intersystem crossing yield; (2) sufficient triplet state lifetime to allow sensitizer–oxygen interaction; (3) triplet state energy higher than that of $^1\Delta_g$ state of singlet oxygen (94.2 kJ/mol); (4) spectral absorption in visible, different from substrate to be oxidized; and (5) stability to oxidation by singlet oxygen. Because the $^1\Delta_g$ oxygen energy is relatively low energy, many dyes meet the energetic requirements for singlet oxygen sensitizers. Several examples are shown in Table 3. The dyes include eosin and rose bengal (derivatives of fluorescein); hematoporphyrin [14459-29-1] (17); acridine orange [65-61-2], methylene blue; and psoralen derivatives, eg (18).

Table 3. Singlet Oxygen Yields[a] for Singlet Oxygen Sensitizers

Sensitizer	Structure number	Singlet oxygen yield, ϕ_Δ	Solvent
all-*trans*-retinal		0.66	cyclohexane
		0.20	methanol
rose bengal[b]	(**3**)	0.75	water
erythrosin[b]	(**3**)	0.63	water
eosin Y[b]	(**3**)	0.57	water
fluorescein[b]	(**3**)	0.03	water
hematoporphyrin	(**17**)	0.22	water, pH 7
acridine		1.0	benzene
8-methoxypsoralen		0.0044	benzene
3-carbethoxypsoralen	(**18**)	0.30	benzene

[a] Singlet oxygen yields in the solvents noted and using methods reviewed in Reference 48.
[b] Structures in Figure 7.

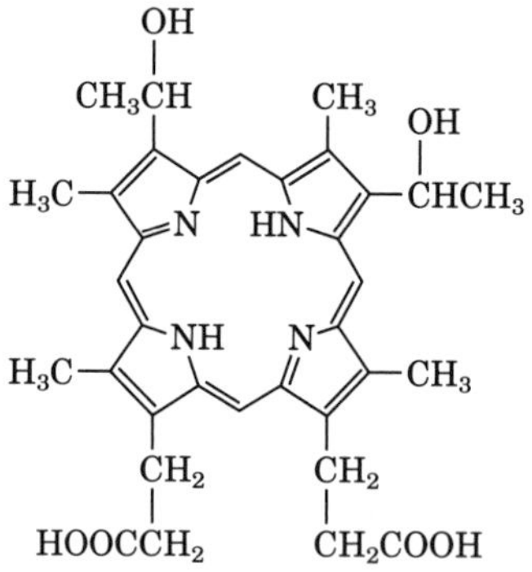

(17) 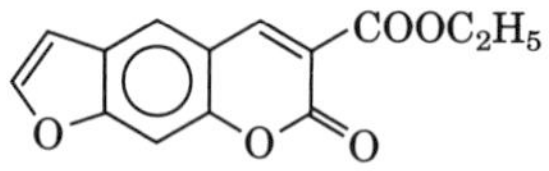   (18)

Characteristic reactions of singlet oxygen lead to 1,2-dioxetane (addition to olefins), hydroperoxides (reaction with allylic hydrogen atom), and endoperoxides (Diels-Alder "4 + 2" cycloaddition). Many specific examples of these spectrally sensitized reactions are found in reviews (45–48), earlier texts (15), and elsewhere in the *Encyclopedia*.

Advances in the technology of photooxidations have also occurred. Product separations, for example, are more convenient when one of several insoluble sensitizers can be used: (*1*) ionic sensitizers like rose bengal on ion-exchange resins; (*2*) sensitizers adsorbed to silica or alumina; or (*3*) polymer-bound sensitizers like rose bengal linked through an ester function to a styrene:divinylbenzene copolymer (49). Efficiency for the insoluble sensitizers can be approximately 50% of the homogeneous reaction (50). Effective methods exist for incorporating sensitizers into latex nanospheres (100–300 nm diameter) (51) or preparing protein conjugates with dyes (52).

Photooxidations are also industrially significant. A widely used treatment for removal of thiols from petroleum distillates is air in the presence of sulfonated phthalocyanines (cobalt or vanadium complexes). Studies of this photoreaction

(53) with the analogous zinc phthalocyanine show a facile photooxidation of thiols, and the rate is enhanced further by cationic surfactants. For the perfume industry, rose oxide is produced in low tonnage quantities by singlet oxygen oxidation of citronellol (54). Rose bengal is the photosensitizer.

Photosensitized Reactions for Polymers. The economic and technical features for photocross-linking, photosolubilization, and photopolymerization reactions have been reviewed (55). The widely used poly(vinyl cinnamates) (PVCN) photocross-link by a photodimerization reaction, forming a cyclobutane from two double bonds in the cinnamoyl groups of the polymer (56). They function as negative working liquid photoresists for integrated circuit manufacture. Triplet sensitizers include Michler's ketone [90-94-8] (**19**), the cyclopentanone analogue [38394-53-5] (**20**) (57), the blue sensitizer (**21**) (56), and coumarin derivatives. Photosolubilization by decomposition of quinone diazides (58) or tetraarylborates (59) provides positive working photoresists. Triplet-state sensitizers do not sensitize the quinone diazides. However, a wide variety of cationic dyes photosolubilize the tetraarylborate anion.

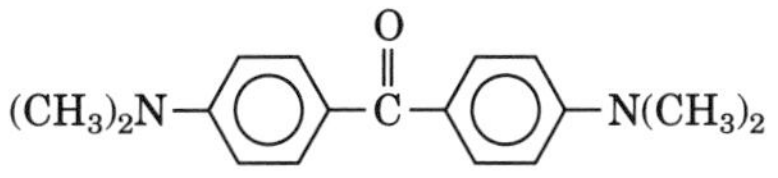

$$(CH_3)_2N-\!\!\left\langle \bigcirc \right\rangle\!\!-\overset{\overset{\displaystyle O}{\|}}{C}-\!\!\left\langle \bigcirc \right\rangle\!\!-N(CH_3)_2$$

(**19**)

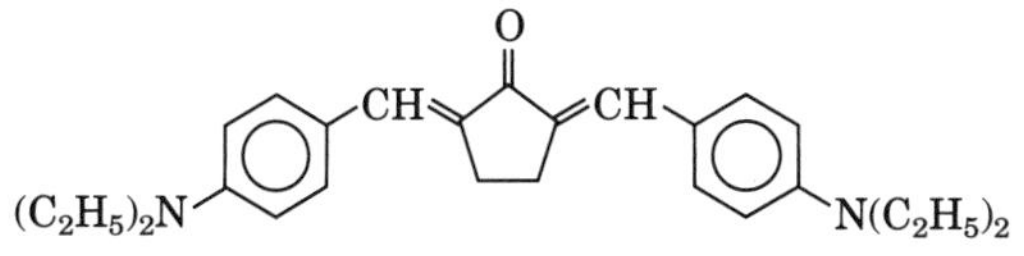

(**20**) 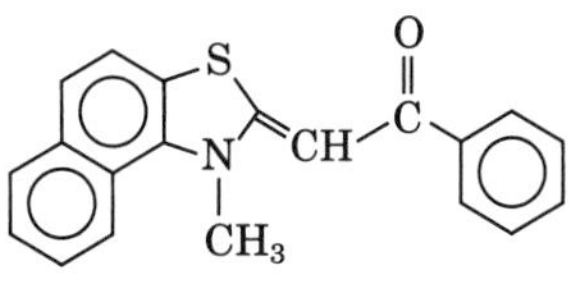   (**21**)

Photopolymerization reactions are widely used for printing and photoresist applications (55). Spectral sensitization of cationic polymerization has utilized electron transfer from heteroaromatics, ketones, or dyes to initiators like iodonium or sulfonium salts (60). However, sensitized free-radical polymerization has been the main technology of choice (55). Spectral sensitizers over the wavelength region 300–700 nm are effective. Acrylic monomer polymerization, for example, is sensitized by xanthene, thiazine, acridine, cyanine, and merocyanine dyes. The required free-radical formation via these dyes may be achieved by hydrogen atom-transfer, electron-transfer, or exciplex formation with other initiator components of the photopolymer system.

Formulation of an aqueous-developable negative working resist used Michler's ketone as the sensitizer, mercaptobenzothiazole as the hydrogen atom donor, and chlorosubstituted hexaarylbiimidazole as a radical generator (61). For a longer wavelength analogue of Michler's ketone, sensitization is reported to occur by electron transfer from the excited singlet state of the ketone to the chloro-hexaarylbiimidazole radical generator (62).

Additional applications of spectrally sensitized charge-transfer processes employ combinations of photosensitizer and relay species in a manner similar to photopolymer formulations. Guidelines for these species include high turnover number and long excited state lifetime for the photosensitizer and high turnover number for the relay species (63). Transition-metal complexes, particularly those with metal-to-ligand charge-transfer excited states, provide quite suitable photosensitizers. Many polypyridine complexes of Ru(II) and other metal ions have been synthesized. The best known and most widely used is the ruthenium bipyridyl complex [*15158-62-0*] (**22**), Ru(bpy)$_3^{2+}$ (λ_{max} = 450 nm). In a system to produce hydrogen, the turnover number for this complex is estimated to range from several hundred to several thousand (64), whereas the turnover number for chlorophyll (65) in natural photosynthetic processes is 10^5. A chemical actinometer based on Ru(bpy)$_3^{2+}$ is also reported (66).

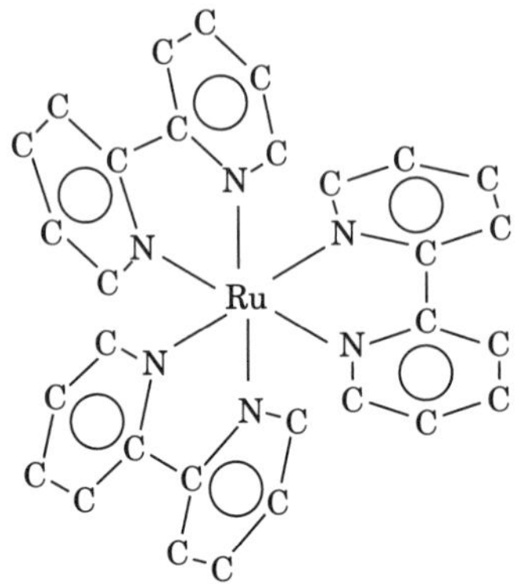

(**22**)

Dye-Sensitized Polymers: Degradation and Imaging.   Colored polymeric materials and natural fibers range from pigmented fabrics to homogeneously dyed thin films. The tendering of cotton (cellulosics) by yellow anthraquinone vat dyes occurs by two pathways: (*1*) hydrogen (or electron) abstraction from the cellulose by the excited dye molecule, and (*2*) dye-sensitized formation of singlet oxygen (67). The anthraquinone dyes also have a phototendering effect on polyamides and silk. Yellow dyes often decrease the stability of other dyes in dye mixtures, presumably by providing an additional sensitizer for singlet oxygen (67,68). Dye effects in polypropylenes and nylon-6,6 films are complex (68,69). In polypropylene fibers phthalocyanine and several other pigments increase stability whereas some azo-condensation pigments decrease it. In nylon-6,6 materials, luminescent carbonyl impurities have been implicated since both polymer and dye stability show parallel trends. Acid azo dyes without a central metal atom cause degradation, whereas those with a metal atom enhance both dye and polymer stability and quench the phosphorescent impurity species in the polymer.

Other dye–polymer systems show utility in digital (70,71) and holographic (71,72) data storage (see HOLOGRAPHY). Write-once optical disks use dyes for infrared (thermal) degradation of polymers. Sensitizers for holographic materials are quite varied (72) and include methylene blue for gelatin and azo dyes for PVA. Spectral hole burning for data storage has utilized the various frequencies (colors) available within the spectral absorption band of dyes, whereas newer applications may utilize time-dependent properties as well (70) (see INFORMATION STORAGE MATERIALS).

Dye-Sensitized Photoisomerization. One technological application of photoisomerization is in the synthesis of vitamin A. In a mixture of vitamin A acetate (all-trans structure) and the 11-cis isomer (**23**), sensitized photoisomerization of the 11-cis to the all-trans molecule occurs using zinc tetraphenylporphyrin, chlorophyll, hematoporphyrin, rose bengal, or erythrosin as sensitizers (73). Another photoisomerization is reported to be responsible for dye laser mode-locking (74). In this example, one metastable isomer of an oxadicarbocyanine dye was formed during flashlamp excitation, and it was the isomer that exhibited mode-locking characteristics.

$$\begin{array}{c}
CH_3 \quad CH{=}CH \\
\diagdown \quad \diagup \quad \diagdown \\
C{=}CH \quad \quad C{=}CH \\
\diagup \quad \quad \diagup \quad \diagdown \\
CH_3 \quad CH{=}CH \quad \quad CH_3 \quad CHO
\end{array}$$

(23)

Natural Sensitizing Dyes and Photodynamic Therapy. The chlorophylls are, of course, among the natural sensitizers for photosynthesis. Considerable interest exists in chlorophyll and related pigments as photosensitizers in biology and medicine (75), isomeric retinal chromophores as visual pigments (76,77), and the use of synthetic photosensitizers in neurobiology (9), hematology (78), and photodynamic therapy (79).

In photosynthesis, the excited singlet state of chlorophyll transfers energy rapidly to nearby chlorophylls until it is trapped by the reaction center of the photosynthetic unit, causing photosynthetic electron flow to occur (75). Relatively little triplet state appears to be produced under ordinary conditions, and protective species like carotenoids are present in high concentrations to quench both triplet species and subsequent reactive oxygen species. In less optimum environments, chlorophyll and its precursors can sensitize the formations of singlet oxygen, superoxide, hydrogen peroxide, and free radicals.

Utilizing this property, various herbicides interfere with the normal photosynthetic chain of reactions, allowing the plant chlorophyll to sensitize via its triplet state (75). Photodynamic tumor therapy with chlorophyll or its porphyrin derivatives (75,79) provides sensitizers with selective adsorption to tumors, effective photosensitization, and rapid clearing from body tissues to avoid long-term photosensitization. Polymer-bound porphyrins, phthalocyanines, and naphthalocyanines, synthetically linked to an uncharged water-soluble polymer, are designed to achieve both longer wavelength absorption bands and more highly selective binding to tumors (79).

Other sensitizing dyes, light, and oxygen can damage and inactivate virtually all classes of organisms through the photooxidation of proteins, polypeptides, individual amino acids, lipids with allylic hydrogens, tocopherols (see VITAMINS), sugars, and cellulose materials. Certain porphyrias in humans are characterized by a sensitivity to light through photosensitizing porphyrins deposited in the skin. Sensitization reactions also have medical value in the treatment of psoriasis and neonatal jaundice. The dependence of photodynamic action on dye structure is generally more marked than for simple singlet oxygen reactions. Thiazine, porphyrin, and xanthene photosensitizers (methylene blue, rose bengal, phenosafranine, eosin, erythrosin, and hematoporphyrin) may operate primarily through singlet oxygen (type II reactions) in the photooxidation of methionine, whereas flavin and anthraquinone sensitizers tend to exhibit mainly type I photooxidation (80). Experimental model studies (81) show that both primary photooxidative mechanisms are operative (Fig. 10). In the oxidation of trypsin by a series of sensitizing dyes with varying triplet yields, the oxidation rate followed the triplet yield and the photooxidation product ratio was constant, indicating a type II reaction involving singlet oxygen. For papain oxidation in a solid gelatin matrix (limited oxygen diffusion), the rate did not follow triplet yields, implicating a type I photooxidation.

The spectral sensitizing pigments for vertebrate vision rods and cones employ 11-*cis*-retinal (76). A diversity in absorption spectra exists among the various pigments, caused primarily by the nature of the protein binding site. For example, 11-*cis*-retinal in rhodopsin peaks near 500 nm, whereas in iodopsin (chicken retina) its absorption is at 562 nm. The longer wavelength absorption is caused by a protein structure similar to rhodopsin plus the presence of chloride ion at the binding site. It has been shown that all four chicken visual pigments have 11-*cis*-retinal as their chromophore, and the broad (415–570 nm) range of absorption maxima among these pigments is attributed to differences in interaction with protein moieties. Bacteriorhodopsin is a photosynthetic pigment in the ancient

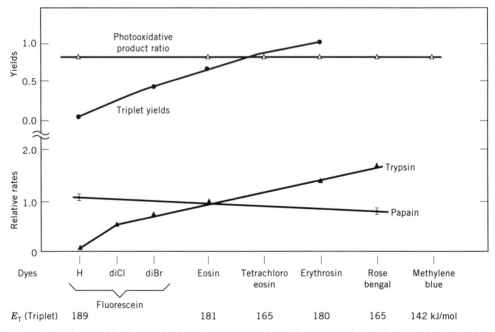

Fig. 10. Photooxidation and photodynamic action. To convert kJ to kcal/mol, divide by 4.184.

Halobacterium halobium, where the all-*trans*-retinal is the spectral sensitizing chromophore (77).

The concepts employed in understanding environmental influences on sensitizer absorption spectra are also incorporated into the design of dye systems. Solvatochromic dye structures are well known and they are used to estimate the polarity environments of such diverse systems as supercritical fluids, alumina chromatographic surfaces, and nonaqueous electrolyte solutions (82). Combining these chromophoric groups with ion-binding macrocyclic ethers (83) has led to ion-sensitive dye chromophores that contain both a solvent sensitive spectral sensitizer and an ion-binding group. For the ion-sensitive solvatochromic dyes (**24,25**), absorption maxima vary from approximately 590–700 nm in the presence of various alkali and alkaline-earth cations.

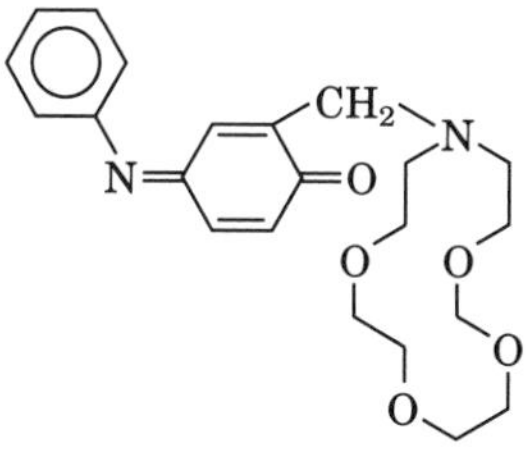

(**24**) 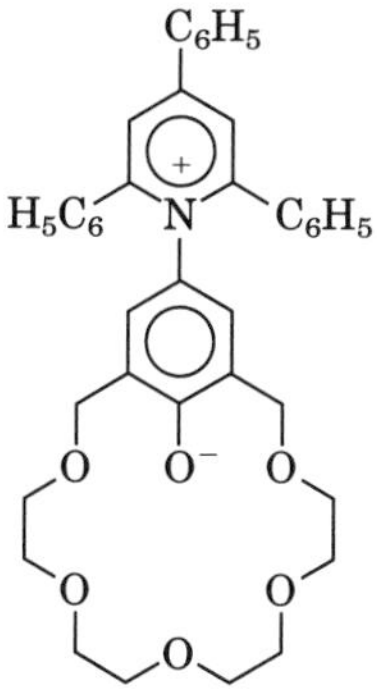　　　　　　　　　　(**25**)

Uses and Suppliers

Sensitizing dyes are used primarily for specialty purposes: photographic sensitizers, electrophotographic sensitizers, laser dyes, infrared (optical disk, etc) imaging, and certain medicinal applications. Because of this, their manufacture is limited to significantly smaller quantities than for fabric dyes or other widely used coloring agents. However, the photographic, laser, and medicinal uses place high demands on the degree of purity required, and the reproducibility of synthetic methods and purification steps is very important. Suppliers of cyanine dyes include manufacturers of other specialty organic and photographic chemicals: Aldrich Chemical Co. (Milwaukee, Wisconsin), Eastman Fine Chemicals (Rochester, New York), Japanese Institute for Photosensitizing Dyes (Okayama, Japan), Molecular Probes (Eugene, Oregon), NK Dyes (Japan), Pfaltz and Bauer (Stamford, Connecticut), Riedel deHaen (Karlsruhe, Germany), and H. W. Sands. More importantly, these firms provide sources of generally useful reagents that, in two or three synthetic steps, lead to many of the commonly used sensitizers.

Sensitizing Dye Toxicology

Nearly every significant class of dyes and pigments has some members that function as sensitizers. Toxicological data are often included in surveys of dyes (84), reviews of toxic substance identification programs (85), and in material safety data sheets provided by manufacturers of dyes. More specific data about toxicological properties of sensitizing dyes are contained in the *Encyclopedia* under the specific dye classes (see CYANINE DYES; POLYMETHINE DYES; XANTHENE DYES).

BIBLIOGRAPHY

"Dyes, Sensitizing" in *ECT* 3rd ed., Vol. 8, pp. 393–408, by D. Sturmer, Eastman Kodak Co.

1. J. Griffiths, *Chimia* **45,** 304 (1991).
2. L. G. S. Brooker, in C. E. K. Mees and T. H. James, eds., *The Theory of the Photographic Process*, 3rd ed., MacMillan Publishing Co, New York, 1966.
3. J. Sturge, V. Walworth, and J. Shepp, eds., *Imaging Materials and Processes* (*Neblette's Eighth Edition*), Van Nostrand-Reinhold, New York, 1987.
4. H. Zollinger, *Color Chemistry*, 2nd rev. ed., VCH Publishers, New York, 1991.
5. A. S. Diamond, ed., *Handbook of Imaging Materials*, Marcel Dekker, Inc., New York, 1991.
6. M. Matsuoka, ed., *Infrared Absorbing Dyes*, Plenum Press, New York, 1990.
7. R. A. Hann and D. Bloor, eds., *Organic Materials for Non-linear Optics II*, Royal Society of Chemistry, Cambridge, UK, 1991.
8. A. M. Braun, M-T Maurette, and E. Oliveros, trans. by D. F. Ollis and N. Serpone, *Photochemical Technology*, John Wiley & Sons, Inc., New York, 1991.
9. J. D. Spikes, *Photochem. Photobiol.* **54,** 1079 (1991).
10. H. Meier, *Spectral Sensitization*, Focal Press, New York, 1968.
11. C. A. Koval and J. N. Howard, *Chem. Rev.* **92,** 411 (1992).
12. K. L. Mittal, *Polymers in Information Storage*, Plenum Press, New York, 1989.
13. P. F. Gordon, in D. R. Waring and G. Hallas, eds., *The Chemistry and Application of Dyes*, Plenum Press, New York, 1990, p. 381.
14. N. Serpone and E. Pelizzetti, eds., *Photocatalysis: Fundamentals and Applications*, John Wiley & Sons, Inc., New York, 1989.
15. A. A. Frimer, *Singlet Oxygen*, CRC Press, Boca Raton, Fla., 1985.
16. F. M. Hamer, in A. Weissberger, ed., *The Cyanine Dyes and Related Compounds, The Chemistry of Heterocyclic Compounds*, Vol. 18, Wiley-Interscience, New York, 1964.
17. D. M. Sturmer, in E. C. Taylor and A. Weissberger, eds., *Special Topics in Heterocyclic Chemistry, The Chemistry of Heterocyclic Compounds*, Vol. 30, Wiley-Interscience, New York, 1977, p. 441.
18. M. Okawara, T. Kitao, T. Hirashima, and M. Matsuoka, *Organic Colorants, Handbook of Selected Dyes for Electro-optical Applications*, Elsevier, Amsterdam, The Netherlands, 1988.
19. P. Gregory, *High-Tech Applications of Organic Colorants*, Plenum Press, New York, 1992; P. Gregory, in A. T. Peters and H. S. Freeman, eds., *Colour Chemistry* (*The Design and Synthesis of Organic Dyes and Pigments*), Elsevier Applied Science, New York, 1991, p. 193.
20. *Scientific Imaging with Kodak Films and Plates*, publication number P-315, Eastman Kodak Co., Rochester, New York, 1987.
21. W. West and P. B. Gilman, in T. H. James, ed., *The Theory of the Photographic Process*, 4th ed., Macmillan Publishing Co., New York, 1977, p. 251.
22. J. R. Lenhard, *J. Imag. Sci.* **30,** 27 (1986); J. R. Lenhard and D. R. Parton, *J. Org. Chem.* **55,** 49(1990); J. Lenhard and A. Cameron, *J. Phys. Chem.* **96** (1993).
23. A. H. Herz, *Adv. Colloid Interface Sci.* **8,** 237 (1977); J. E. Maskasky, *Langmuir* **7,** 407 (1991); T. Matsubara and T. Tanaka, *J. Imag. Sci.* **35,** 274 (1991).

24. J. E. Jones and P. B. Gilman, *Photogr. Sci. Eng.* **17**, 367 (1973).
25. T. Tani, T. Suzumoto, K. Kemmitz, and K. Yoshihara, *J. Phys. Chem.* **96**, 2778 (1992); A. A. Muenter and co-workers, *J. Phys. Chem.* **96**, 2783 (1992).
26. E. J. Poppe, *Z. Wiss. Photogr. Photophys. Photochem.* **63**, 149 (1969).
27. E. K. Putzieko and A. Terenin, *Zh. Fiz. Khim. SSSR* **23**, 676 (1949).
28. K. H. Hauffe, *Photogr. Sci. Eng.* **20**, 124 (1976).
29. R. H. Sprague and J. H. Keller, *Photogr. Sci. Eng.* **14**, 401 (1970).
30. P. D. Fleischhauer and J. K. Allen, *J. Phys. Chem.* **82**, 432 (1978).
31. O. Enea, J. Moser, and M. Graetzel, *J. Electroanal. Chem. Interfacial Electrochem.* **259**, 59 (1989); W. O. Pat. 9116719, (Oct. 31, 1991), M. Nazeeruddi and B. Oregan (to M. Graetzel).
32. P. Yianoulis and R. C. Nelson, *Photogr. Sci. Eng.* **22**, 268 (1978).
33. T. Tani, *Photogr. Sci. Eng.* **17**, 11 (1973).
34. J. M. Lanzafame, R. J. D. Miller, A. A. Muenter, and B. A. Parkinson, *J. Phys. Chem.* **96**, 2820 (1992).
35. P. M. Borsenberger and D. S. Weiss, in Ref. 5, pp. 379–445.
36. R. M. Schaffert, *IBM J. Res. Dev.* **15**, 75 (1971).
37. U.S. Pat. 3,615,414 (Oct. 26, 1971), W. A. Light (to Eastman Kodak Co.).
38. R. O. Loutfy, A. M. Hor, and A. Rucklidge, *J. Imag. Sci.* **31**, 31 (1987); D. E. Nickles, J. E. Kuder, and J. A. Jusuta, *J. Imag. Sci.* **36**, 131 (1992); T. Enokida and R. Hirohashi, *J. Imag. Sci.* **36**, 135 (1992).
39. H. Zollinger, in Ref. 4, pp. 379–383.
40. U. T. Meuller-Westerhoff, B. Vance, and D. I. Yoon, *Tetrahedron* **47**, 909 (1991).
41. F. J. Green, *The Sigma-Aldrich Handbook of Stains, Dyes, and Indicators*, Aldrich Chemical Co., Milwaukee, Wis., 1990.
42. F. J. Duarte, ed., *High-Power Dye Lasers*, Springer-Verlag, New York, 1991.
43. D. G. Jay, *Proc. Natl. Acad. Sci. U.S.A.* **85**, 5454 (1988).
44. J. Zhao, D. Jurbergs, B. Yamazi, and J. T. McDevitt, *J. Am. Chem. Soc.* **114**, 2737 (1992).
45. J. R. Heitz and K. R. Downum, eds., *Light-Activated Pesticides* (*ACS Symposium Series No. 339*), American Chemical Society, Washington, D.C., 1987.
46. J. D. Spikes and J. C. Bommer, in H. Scheer, ed., *Chlorophylls*, CRC Press, Boca Raton, Fla., 1991, pp. 1182–1204; see also Ref. 9.
47. M. Orfanopoulos, in G. Charalambous, ed., "Flavors and Off-Flavors," *Proceedings of the International Flavor Conference*, July, 1989, Elsevier Science Publishers B.V., Amsterdam, The Netherlands, 1989, pp. 865–872.
48. M. A. J. Rodgers, in Ref. 45, p. 76.
49. *Sensitox* Hydron Laboratories, New Brunswick, N.J., 1992; E. Oliveros and co-workers, *Dyes Pigm.* **5**, 457 (1984).
50. E. Gassman and A. M. Braun, *Vortragstag. Fachgruppe Photchem.*, 105 (1983).
51. R. Madison, J. D. Macklis, and C. Thies, *Brain Res.* **522**, 90 (1990).
52. M. Brinkley, *Bioconjugate Chem.* **3**, 2 (1992).
53. D. Woehrle, T. Buck, G. Schneider, G. Schultz-Ekloff, and H. Fischer, *J. Inorg. Organomet. Polym.* **1**, 115 (1991).
54. G. Ohloff, E. Klein, and G. O. Schenck, *Angew. Chem.* **73**, 578 (1961).
55. A. B. Cohen and P. Walker, in Ref. 3, pp. 226–262; J. P. Fouassier, *Photopolym. Dev. Phys. Chem. Apl.* (*SPIE*) **1559**, 76 (1991).
56. J. L. R. Williams, D. P. Specht, and S. Farid, *Polym. Eng. Sci.* **23**, 1022 (1983); A. Reiser and P. L. Egerton, *Photogr. Sci. Eng.* **23**, 144 (1979).
57. M. B. Barnabas, A. Liu, A. D. Trifunac, V. V. Krongauz, and C. T. Chang, *J. Phys. Chem.* **96**, 212 (1992).
58. T. A. Shankoff and A. M. Trozzolo, *Photogr. Sci. Eng.* **19**, 173 (1975).

59. X. Yang, A. Zaitsev, B. Sauerwein, S. Murphy, and G. B. Schuster, *J. Am. Chem. Soc.* **114**, 793 (1992); D. G. Borden, *Photogr. Sci. Eng.* **16**, 300 (1972).
60. H-J. Timpe, *Pure Appl. Chem.* **60**, 1033 (1988).
61. W. J. Chambers and D. F. Eaton, *J. Imag. Sci.* **30**, 230 (1986).
62. A-D. Liu, A. D. Trifunac, and V. V. Krongauz, *J. Phys. Chem.* **96**, 207 (1992).
63. V. Balzani, A. Juris, and F. Scandola, in E. Pelizetti and N. Serpone, eds., *Homogeneous and Heterogeneous Photocatalysis*, D. Reidel Publishing, Boston, Mass., 1985, pp. 1–28.
64. P. A. Lay and co-workers, *Inorg. Chem.* **22**, 2347 (1983).
65. J. R. Bolton, *Science* **202**, 705 (1978).
66. A. Bromberg, K. H. Schmidt, and D. J. Meisel, *J. Am. Chem. Soc.* **107**, 83 (1985).
67. H. Zollinger, in Ref. 4, pp. 321–324; A. A. Krasnovskii and co-workers, *Dokl. Akad. Nauk SSSR* **285**, 654 (1985).
68. N. S. Allen, M. Ledward, and G. W. Follows, *Eur. Polym. J.* **28**, 23 (1992).
69. P. P. Klemchuck, *Polym. Photochem.* **3**, 1 (1983).
70. W. E. Moerner, ed., *Persistent Spectral Hole Burning: Science and Applications*, Springer-Verlag, Heidelberg, Germany, 1988.
71. U. P. Wild, A. Rebane, and A. Renn, *Adv. Mater.* **3**, 453 (1991).
72. S. A. Benton, ed., *Practical Holography V (SPIE)* **1461**, 58, 73 (1991); N. Capolla, C. Carre, J. Lougnot, and R. A. Lessard, *Appl. Opt.* **28**, 4050 (1989).
73. U.S. Pat. 3,838,029 (Sept. 24, 1974), M. Fischer, W. Wiersdorff, A. Nuerenbach, D. Horn, and F. Feichtmayr (to BASF, Germany).
74. D. N. Dempster, T. Morrow, R. Rankin, and G. F. Thompson, *J. Chem. Soc., Faraday Trans. 2* **68**, 1479 (1972).
75. J. D. Spikes and J. C. Bommer, in Ref. 46, pp. 1192–1194.
76. T. Yoshizawa and O. Kuwata, *Photochem. Photobiol.* **54**, 1061 (1991).
77. K. Nakanishi, *Am. Zool.* **31**, 479 (1991).
78. F. Sieber, J. M. O'Brien, and D. K. Gaffney, *Semin. Hematol.* **29**, 79 (1992).
79. D. Woehrle, *Chimia* **45**, 307 (1991).
80. J. D. Spikes and M. L. MacKnight, *Ann. N.Y. Acad. Sci.* **171**, 149 (1970).
81. J. Bourdon and M. Durante', *Ann. N.Y. Acad. Sci.* **171**, 163 (1970).
82. C. Reichardt, *Chimia* **45**, 322 (1991); *Chem. Soc. Rev.* **21**, 147 (1992).
83. F. Voegtle, M. Bauer, C. Thilgen, and P. Knops, *Chimia* **45**, 319 (1991).
84. R. E. Lenga, ed., *The Sigma-Aldrich Library of Chemical Safety Data*, 2nd ed., Sigma-Aldrich Corp., Milwaukee, Wis., 1987.
85. J. Houk, M. J. Dva, M. Dezube, and J. M. Rovinski, in A. T. Peters and H. S. Freeman, eds., *Color Chemistry (The Design and Synthesis of Organic Dyes and Pigments)*, Elsevier Applied Science, New York, 1991, p. 135.

DAVID M. STURMER
Eastman Kodak Company

DYSPROSIUM.   See LANTHANIDES.

ECONOMIC EVALUATION

Economic evaluation is an assessment of the probable benefit or reward of a proposed course of action, relative to other choices. Although the benefit usually takes the form of a financial return, in environmental management, transportation (qv), health care, and other social areas, the benefit may be a social gain instead. Some method is then developed to translate the social gain into a monetary equivalent. The discussion herein is limited to the financial return expected from some type of production or service activity.

This type of evaluation is a primary planning activity from the pilot-plant to the corporate level (see PILOT PLANTS AND MICRO PLANTS), involving research, engineering, manufacturing, finance, sales, marketing, and other business groups (see also MARKET AND MARKETING RESEARCH; RESEARCH AND DEVELOPMENT). Such evaluation is an evolutionary process that is repeated endlessly from the research stage through the manufacturing cycle. Economic studies are employed to assess the expected profitability as the result of various choices such as plant size, location, pricing, product grades, and engineering features, or to make a comparison with other ventures, including the competition. Statistical methodologies can be employed to make use of probabilistic data. The four essential parts of any economic evaluation are problem definition, cost estimation, revenue estimation, and profitability analysis.

Problem Definition

The problem is defined during process development as information becomes available and decisions are made. Initially, the definition is limited, vague, and brief and economic analysis involves a high level of uncertainty. As the project evolves, the definition becomes more complete, more highly specific, and lengthier. At the same time, the economic assessment tends to exhibit less uncertainty.

Whenever an economic evaluation is undertaken, a corresponding problem definition should be provided as the basis on which the evaluation is made. This definition, sometimes called an economic scope, should clearly differentiate between specifications that have actually been selected and features that have been assumed for the evaluation. In a comparison of alternatives, all of the assumptions, data, and conditions must be consistent, realistic, and devoid of bias.

Situations where economic evaluation is involved include equipment selection, process retrofits, commercialization of research, acquisitions, assessment of competition, market strategy development, corporate planning, labor-management negotiations, financial planning, public safety, environmental policy, etc. As a result, there is no general guideline for problem definition. For complete process development cases, good estimates, and as much firm data as possible, about the following items are desirable over the life of the proposed problem scenario: capital investment (amount, schedule); construction time; project lifetime; manufacturing costs; production capacity and availability; selling price and market projection; financial and tax data; regulatory constraints (environmental, safety, fiscal); societal constraints (perceived risks, probabilities, benefits); economic constraints (material costs and availability, utilities); technical constraints (process constraints); legal constraints (laws, regulations, patents); and competitive constraints (processes, products, future).

Cost Estimation

The three general types of cost estimates needed are equipment cost, capital investment cost, and product cost. Equipment cost estimates are needed as part of the capital investment estimate, which indicates the amount of money that is needed to start the venture. Both of these estimates are reflected in the product cost estimate, which is important to both management and marketing groups.

Equipment Costs.   Equipment costs include the purchased cost of process and materials handling equipment, storage facilities, waste treatment equipment, structures, and site service facilities. Installation costs such as insulation, piping, painting and finishing, foundations, process structures, instrumentation, and electrical service connections are estimated or factored separately. Actual quoted prices from suppliers are the best data, but these are not usually available when estimates are made. The quick, inexpensive cost estimates are based largely on personal cost files, internal company cost data, or published cost correlations.

Published Cost Correlations.   Purchased cost of an equipment item, ie, fob at seller's site or other base point, is correlated as a function of one or more equipment–size parameters. A size parameter is some elementary measure of the size or capacity, such as the heat-transfer area for a heat exchanger (see HEAT-EXCHANGE TECHNOLOGY). Historically the cost–size correlations were graphical log–log plots, but the use of arbitrary equation forms for correlation has become quite common. If cost–size equations are used in computer databases, some limit logic must be included so that the equation is not used outside of the applicable size range.

One simple equation form is the exponential equation, which gives a straight line on a log–log plot:

$$C = (C_R)(S/S_R)^n \tag{1}$$

where C is cost, S is size, n is slope on the log–log plot, and the subscript R is a reference cost–size point. The value of n is frequently around 0.6 or 0.7 for many types of equipment. The applicable size range should be provided with any cost correlation. Descriptive information accompanying cost correlations is often quite meager. For example, it is not always clear if driver costs are included in pump costs. When information is incomplete, different correlations should be compared so that the estimate is meaningful. Cost correlations are available for most types of process equipment (1–5).

Time Translation of Cost Data.   Cost data for any particular point in time can be corrected to the present or any other time by means of cost indexes in the relation:

$$\text{COST AT TIME } P = (\text{COST AT TIME } N) \left[\frac{\text{INDEX FOR TIME } P}{\text{INDEX FOR TIME } N} \right] \tag{2}$$

The accuracy of such cost extrapolation tends to decrease with the length of time involved. Various important U.S. cost indexes are the M&S Index, CE Plant Cost Index, Nelson Construction Cost Index, and the ENR Construction Index. Similar indexes are available for other countries.

The Marshall and Swift (M&S) Equipment Cost Index (6), formerly Marshall and Stevens, for installed equipment costs is published monthly in the *Chemical Engineering* journal. The indexes reported are the all-industries, process industries, and several specific industry indexes. The yearly all-industries index, given in Table 1, is based on 47 industrial categories. This is commonly used for the translation of purchased process equipment costs, even though it was developed for installed equipment.

The CE Plant Cost Index (7) is also published monthly in the *Chemical Engineering* journal. Index values are given for various categories of equipment, installation, labor, building, and supervision, as well as a composite plant cost index. The composite index for complete plant costs, tabulated in Table 1, is frequently used for the translation of purchased equipment costs, even though the equipment component of the index would be better.

The Nelson Refinery Construction Index, which appears monthly in the *Oil and Gas Journal*, is a weighted construction materials and labor index. The ENR Construction Cost Index, reported in the *Engineering News Record*, also weights construction materials and labor.

Other Equipment Cost Modifiers.   Temperature, pressure, or corrosive conditions can act as modifiers of the base cost by requiring thicker vessel walls, more expensive alloys, special seals, more expensive fabrication, and special testing procedures. Separate materials and process severity factors for temperature, f_t; pressure, f_p; and material, f_m, multiply the base (mild steel) cost as:

$$\text{COST} = (\text{BASE COST}) f_t f_p f_m \tag{3}$$

Table 1. Equipment Cost Indexes[a]

Year	M&S[b] Equipment Cost Index	CE[c] Plant Cost Index
1970	303	126
1975	444	182
1980	660	261
1985	790	325
1986	798	318
1987	814	324
1988	852	343
1989	895	355
1990	915	358
1991	931	361

[a]Complete yearly listings available (1).
[b]Marshall & Swift, published monthly in *Chemical Engineering*.
[c]*Chemical Engineering*, published monthly.

Typical f factors (8) are given in Table 2.

Table 2. Materials and Process Severity Factors for Equipment Cost

Factor value	Specification
Materials factors, f_m	
1.0	mild carbon steel
1.075	aluminum
1.28–1.5	stainless steel
1.65	monel
Temperature[a] factors, f_t	
$1 + 0.002(\log T) + 0.015(\log T)^2$	10–1000
1.0	0–10
$1 - (1.2 \times 10^{-3})T$	−100–0
Pressure[b] factors, f_p	
$0.8 + 0.0162(\log P) + 0.0191(\log P)^2$	1,400–34,000
1.0	28–1400
$2.00 - 0.692(\log P)$	0.7–28

[a]Temperature in °C.
[b]Pressure, P, in kPa. To convert kPa to psi, multiply by 0.145.

Capital Investment Cost. The capital investment involved in a proposed project is important because it represents the money that must be raised to get the project started, is used in profitability forecasts, and is reflected in the estimated manufacturing cost of a product. The capital investment is classified herein as fixed capital, working capital, and land cost. Sample capital investment estimate forms provide for separate materials (M) and labor (L) categories, or just combined M&L figures.

Fixed Capital. Fixed capital can be classified as direct plant costs, indirect plant costs, and nonplant costs. The direct plant costs include the process equipment, as well as the material and labor costs associated with installation, instrumentation, piping, electrical, buildings, structures, services, and site improvement. Indirect plant costs include engineering, construction site expense, and any other plant items that cannot be charged directly to equipment, materials, or labor accounts. Nonplant costs, which are sometimes classified as indirect costs, include contractor fees, contingency allowances, and occasionally both construction interest and some start-up expenses. The actual cost categories used generally follow the accounting procedures that have been adopted.

Land cost, a part of the direct plant cost, is placed in a separate capital category because it is not depreciable. Land cost is site-specific and highly variable.

The fixed capital estimate depends on the definition of the plant. A grass-roots plant is a complete facility at a new location, including all utilities, services, storage facilities, land, and improvements. If a process plant is located at an existing processing complex, it can usually share some of these auxiliary facilities. A battery-limits plant is defined as the process facility itself, so that the auxiliaries, off-site, and land-related items are excluded from the fixed capital estimation. However, a battery-limits plant may be assigned allocated capital charges for the share of common utility and service facilities used by the plant.

The accuracy of a fixed capital estimate tends to be a function of the design effort involved. As the project definition is refined, the estimates evolve from the various preliminary phases, ie, order of magnitude, predesign, factor estimates, etc, into the more detailed estimates used for budget authorization, project control, and contracts. At the same time, the uncertainty in the estimate decreases from $\pm 50\%$ to as little as $\pm 5\%$.

Order-of-Magnitude Estimates. Unit capital cost data (dollars per annual ton of product) are occasionally reported for chemical plants. These data can be multiplied by a selected plant capacity to estimate a capital cost. This is, however, feasible only if the reference process, conditions, and capacity are similar. The unit cost approach is widely used for quick estimates of the capital cost of utilities, waste treatment facilities, and buildings, where data in $/kW, $/t of waste, $/m^2, etc, are available.

At processing complexes, central utilities and other facilities are shared by several battery-limits process plants. The capital cost of a central utility is sometimes charged to the capital cost of each battery-limits plant as an allocated capital cost based on the unit capital cost of the utility facility and the units of capacity of the utility required by the plant. In this case, the use charge per unit consumed only covers operating expenses. The alternative is to recover utility capital costs, as well as operating expense, in the unit usage charge.

A second order-of-magnitude approach is the ratio method, based on the assumption that capital investment can be correlated with plant capacity in a manner similar to that used for equipment. This gives

$$\left[\frac{\text{CAPITAL FOR CAPACITY } A}{\text{CAPITAL FOR CAPACITY } B}\right] = \left[\frac{\text{CAPACITY } A}{\text{CAPACITY } B}\right]^n \qquad (4)$$

where the plant capacity exponent n averages about 0.7 for many plants. These plant costs can then be transferred to any year using the CE Plant Cost Index. Data are available in the literature for many chemical processes (9). More recent data are given in the "Petrochemical" or "Gas Processing Handbook" issues of the *Hydrocarbon Processing* journal. However, capital cost data found in the literature should be examined to determine what the figure covers. Items such as site development, storage, off-site costs, working capital, capitalized construction interest, and start-up costs might be included. Large size extrapolations should be avoided because the need for multiple units, or changes in equipment type because of size, can affect the apparent capacity exponent.

The ratio method is also particularly useful for quick estimates over a range of capacities after a calculated estimate for one size has been made (10). The calculated estimate can be separated into groups $C_1, C_2, \ldots$, according to individual equipment exponents $n_1, n_2, \ldots$. Then an overall plant capacity exponent n can be calculated from

$$\left(\frac{S_b}{S_a}\right)^n = \left(\frac{C_1}{C}\right)\left(\frac{S_b}{S_a}\right)^{n1} + \left(\frac{C_2}{C}\right)\left(\frac{S_b}{S_a}\right)^{n2} + \cdots \qquad (5)$$

where S_a is the capacity of the calculated plant and S_b is another capacity.

Predesign Estimates. Methods are available to make a capital cost estimate based on a preliminary flow sheet or block diagram, but before material balances, energy balances, or equipment sizes have been calculated (8). Such predesign methods generally attempt to identify the number of units in a tentative flow sheet and then use an average capital cost per unit. This average cost is correlated in some way to process complexity, process conditions, throughput, and materials of construction.

These predesign estimates are empirical methods where accuracy depends strongly on the skill and experience of the user. The value of such approaches has been questioned, but the methods provide cheap estimates without significant design effort. The principal disadvantage is that there is no way to assess the accuracy of the results.

The Viola method is a typical predesign approach based on the number N of principal processing blocks, the production rate P in tons per year, and a complexity factor K. The 1981 battery-limits cost C in dollars can be correlated to K and P:

$$\log\frac{C}{Q} = \left(\frac{\log K}{1.182}\right)^{2.1} \qquad (6)$$

where for $900 < P < 22,500$

$$Q = 6.6 \times 10^3 \, P^{0.6} \qquad (7)$$

The complexity factor K is given by

$$K = NS\phi(1-0.6\,f_s) \tag{8}$$

where f_s is the fraction of the N steps that handle solid–fluid mixtures and should be set less than 0.45; ϕ is $1.0167[(I/O)(1/N)]^{0.1067}$; I/O is the input/output, ie, raw material/product, weight ratio; and S is the special materials and pressure factor. A typical S correlation for stainless steel can be written as

$$S = m[M_f + 0.8] + 0.6 \tag{9}$$

where M_f is the fraction of stainless steel; m is $0.8 + 6.5 \times 10^{-5}p_i + (5.3 \times 10^{-8})$ p_i^2; p_i is the average process pressure.

As an example, the battery-limits capital cost can be estimated for the production of 10,000 t/yr of ethylene (qv) from ethanol (11). Seven processing blocks, ie, vaporizer, reactor, water quench, compressor, dryer, distillation, and energy recovery, can be identified. The highest temperature is 350°C (reactor), and the highest pressure is about 1.7 MPa (17 atm) (compressor, two towers). If a materials-pressure factor, S, of 1.03 is assumed, then for N = 7; $\phi = 0.87$; I/O = 1:64; and $f_s = 0$; $K = 6.3$. This gives the 1981 cost as 4.4×10^6. The 1991 battery-limits investment can be obtained, by updating with the CE Plant Cost Index, as 5.3×10^6.

Overall Factor Estimates. The next level of fixed capital estimate is based on a preliminary design that includes a flow sheet, material balances, energy balances, and enough equipment design to size all of the principal process equipment, including pumps and tanks.

An overall Lange factor, F_L, can be used to relate the battery-limits fixed capital investment I_B to the delivered equipment cost E_D so that $I_B = F_L E_D$. The Lange factor was originally given as 3.10 for solids processing, 3.63 for solids–fluid processing, and 4.74 for fluids processing. More recent work (12) suggests an average value of 3.20 for all types of plants.

A popular overall factor refinement, known as the Hand factor approach, uses a different factor to estimate overall costs for each class of equipment to cover all labor; field materials, eg, piping, insulation, electrical, foundations, structures, and finishes; and indirect costs, but not contingencies. Hand factors range from 4 for fractionating towers down to 2.5 for miscellaneous equipment.

Category Factor Estimates. Various capital categories can be related to total equipment costs by factors, reported as percentages of equipment cost. Both purchased equipment costs, including pumps, tanks (qv), and instruments; and delivered equipment costs, excluding instruments, but including some off-sites, have been used in this approach (1,2).

Module Factor Estimates. All equipment of a given type can be lumped together into a module, such as a heat-exchanger module. Factors are given (9) to relate the various capital cost categories for each module type to the total purchased equipment cost of the module. The capital cost categories are then summed over all module types. This approach offers the advantage that the cost of commodity materials and labor are usually available in module categories to maintain accurate up-to-date module factors. Although the method requires a larger database, it appears to offer greater potential accuracy than overall or category factor methods.

Mid-1984 materials and labor factors have been reported for over 30 equipment types (13). The commodity materials include factors for concrete foundations, piping, steel supports, instrumentation, insulation, electrical, and painting. Installation labor factors are given for each of the commodity material categories, as well as for equipment setting. These data can be used with purchased equipment costs to obtain battery-limits installed equipment costs. Additional factors are given for the balance of direct costs, indirect costs, and other items in the fixed capital investment.

Unproven Technology.   When a project involves new or unproven technology, the capital estimates tend to understate the development, construction, and start-up costs. A quantitative approach to account for the capital cost and performance shortfalls associated with unproven technology has been reported (14), but the data are meager.

Working Capital.   Working capital is the money required for the day-to-day operation of the venture over and above the fixed investment. The amount varies daily, may be cyclical, and can be a significant part of the investment in some cases. In the accounting sense, working capital is the difference between current assets and current liabilities.

The working capital includes the cost of inventories, such as raw materials, materials-in-process, products, etc; as well as supplies, accounts receivable less accounts payable, prepaid expenses, other cash needs such as payroll, and some start-up expenses, eg, materials and wages. Typical inventories can be taken as one month's supply of raw materials, products, and materials-in-process. The materials-in-process can be valued at one month's sales. Other operating cash can be estimated as the actual cash need for one month.

Methods are available for making detailed estimates of working capital (15). Shortcut ratios for estimating working capital are 15–20% of the fixed capital, 15% of the total capital, or 10–30% of annual sales.

Product Cost.   An estimate of total product cost is an important part of economic evaluation and management planning. Total product cost can be viewed as the sum of the manufacturing cost and the general expense.

The manufacturing cost consists of direct, indirect, distribution, and fixed costs. Direct costs are raw materials, operating labor, production supervision, utilities, supplies, repair, and maintenance. Typical indirect costs include payroll overhead, quality control, storge, royalties, and plant overhead, eg, safety, protection, personnel, services, yard, waste, environmental control, and other plant categories. However, environmental control costs are frequently set up as a separate account and calculated directly. The principal distribution costs are packaging and shipping. Fixed costs, which are insensitive to production level, include depreciation, property taxes, rents, insurance, and, in some cases, interest expense.

General expense consists of corporate services such as administration, sales and shipping departments, marketing, financial, technical service, research and development, engineering, legal, accounting, purchasing, public relations, human resources, and communications. Accounting groups are responsible for the consistent allocations of overhead and general expense items. Most companies have product cost estimate forms that give the classifications used.

The product cost can be computed on an annual or daily basis, but is frequently reported on a product unit basis. An estimate of the cost for various levels of production is often needed. This can be done by separating the manufacturing cost into fixed and variable components, with the variable components being those that vary with production level. On occasion, manufacturing costs are computed on an incremental or marginal basis instead of the usual allocated basis. For example, raw material cost discounts might be available at higher production levels. In this case, the cost of raw materials needed for incremental production might be charged at a lower rate.

Joint product costing, when more than one product is manufactured by a single process, is a common situation (16). This can be treated on a weight or a volume basis. If one product is largely a by-product, it might not be assigned any manufacturing cost and is accounted for in terms of a by-product credit based on its sales volume. The suitable allocation of costs should reflect such factors as production level, changing markets, and raw materials.

A related problem is the allocation of costs when a raw material for one process operation is produced internally by another process operation of the same organization (17). The transfer or captive price assigned to the raw material can range from the production cost to a market price that reflects a total profit margin for the material producer, depending on the accounting procedures adopted.

Unit Cost Method. Typical operating cost data in dollars per weight or volume basis of product are frequently available. These data, which usually do not include capital costs, can be scaled directly over a moderate range of annual capacities to give rough estimates of annual operating costs as a function of annual capacity. Capital costs can be annualized using a capital charge factor, which converts the capital cost to equivalent annual costs.

Unit cost data should be carefully assessed to ensure that process type, size, and raw materials are similar to the proposed venture. Operating cost data sometimes are reported for separate categories such as operating labor, maintenance labor, supervision, and utilities (9).

Factor Methods. A more detailed product cost estimation method is to relate manufacturing cost items to a few calculated items, such as raw materials, labor, and utilities by means of simple factors (1,2). Internal accounting groups often develop factors to use with this method. This factor method is very popular.

Raw material costs should be estimated by direct computation from flow rates and material prices. The flow rates are determined from flow sheet material balances. The unit prices are obtained from vendors, company purchasing departments, or the *Chemical Marketing Reporter*. For captive raw materials produced internally, a suitable transfer price must be established. Initial catalyst charges can be treated as a start-up expense, working capital component, or depreciable capital, depending on the expected catalyst life and cost. Makeup catalyst is frequently treated as a raw material.

Utility needs should be calculated directly from the process material and energy balances. Unit costs for the various utilities can be obtained from suppliers or purchasing agents. Although regional variations can be quite large, typical U.S. utility costs in 1992 are tabulated in Table 3.

Table 3. 1992 U.S. Utility Costs

Utility	Cost per unit, $
electricity, kWh	0.09
fuel oil, L[a]	0.225
natural gas, GJ[b]	2.00
refrigeration, GJ[b]	
$-34°C$	25
$-5°C$	10
steam, kg $\times$ 10³	
8720 kPa[c]	3.75
1135 kPa[c]	2.80
450 kPa[c]	2.35
water	
cooling, m³	0.025
demineralized, kg $\times$ 10³	1.70
boiler feedwater, kg $\times$ 10³	2.50
compressed air, standard, m³	0.0053

[a]To convert liter to gallon, divide by 3.77.
[b]To convert J to cal, divide by 4.184.
[c]To convert kPa to psi, multiply by 0.145.

Direct labor costs can be estimated using the flow sheet, typical labor needs (persons/shift) for each piece of process equipment, and the local labor rate. Company files are the best source for labor needs and rates, although some literature data are available (1,2). The hourly cost of labor in the United States can be estimated from the *Monthly Labor Review* of the Bureau of Labor Statistics. Production supervision costs can usually be taken as a factor, such as 15% of the direct labor cost.

Annual plant maintenance and repair costs average about 6% of the fixed capital investment, but should be calculated directly from person–hour per shift data or estimates. The annual cost for supplies can be taken as 15% of the total maintenance and repair cost.

Annual indirect costs are estimated as percentages of the direct labor and fixed capital costs. Typical direct labor percentage ranges are 25–30% for payroll overhead, 15–20% for stores and supplies, 10–20% for control laboratory, 10–20% for security, 10% for yard, and 10–15% for process improvements. That is, total indirect costs are usually 80–115% of the direct labor cost (1).

The most common approach to fixed cost estimation involves the use of a capital recovery factor to give the annual depreciation and return on capital. This factor typically is between 15 and 20% of the total capital investment. Property taxes are taken as 1–5% of the fixed capital and insurance is assumed to be 1–2% of the fixed capital. If annual depreciation is estimated separately, it is assumed to be about 10% of the fixed capital investment. The annual interest expense is sometimes neglected as an expense in preliminary studies. Some economists even believe that interest should be treated as a return on capital and not as part of the manufacturing expense.

General expense can be approximated as 15–20% of the total product cost. Typical category factors, as percentage of total product cost, are 3% for administration, 10% for sales, and 4% for research and development (R&D) (1).

Revenue Estimation

Revenues are money inflows from venture activities. In addition to product sales, revenues might also include earnings of unconsolidated subsidiary ventures, license and royalty fees, gains on foreign currency exchange, investment income, and earnings on various other transactions. Product sales, based on the unit product price ($/unit) and the yearly sales volume (units/yr), are the only type of revenue considered herein. The price and sales volume are related to each other and typically contribute much of the uncertainty in economic forecasts of chemical process systems. Whereas the estimation of price and sales volume is a marketing or sales function, technical factors are frequently part of the estimation process.

Product Price. If a need is not met by any other available product, then price can be set as high as the need can support. However, a very high price tends to limit market growth and encourage the introduction of competitive or substitute products. The preferred strategy is to establish a moderately high initial price, but to plan on future price reductions to help expand the market and meet any competitive pressures. Variations of this strategy are seen in the pricing of pharmaceutical products that meet a unique need or have good patent protection.

If a need is met by a variety of products, then the price should reflect the price of these competitive products and any unique or advantageous features of the product being priced. For example, some plastic parts for automotive applications can be priced higher than corresponding metal parts because of lighter weight, corrosion resistance, or other features.

If the product is essentially identical to that produced by other manufacturers, then the price is determined principally by the commodity market price. However, contract features such as guaranteed delivery schedules can influence price. Examples of commodity pricing are petrochemicals, petroleum feedstocks, petroleum products, and primary metals.

If the product is an engineering service, then the price is determined by the availability, reputation, and price of competitive providers, as well as the uniqueness and scope of the service. Examples include engineering design, technical laboratory services, project management, and construction.

From an internal viewpoint, it is desirable that the price be high enough to generate an adequate rate of return. This is reflected in various cost or margin-based pricing policies, such as constant markup, for new industrial product pricing. The percentage markup is defined as [100(selling price−cost)/selling price]. Pricing policy is discussed in the literature (18,19). Current chemical and commodity prices in the United States are reported in the *Chemical Marketing Reporter*.

Transfer pricing of intermediates produced and used internally in the same organization comes under product costs. Even the choice of name, ie, transfer cost or transfer price, reflects the differing viewpoints of the internal consumer and producer.

Sales Volume.   The quantity of annual sales is often called the sales volume, although the units may be mass quantities instead of volume units. The estimation of the annual sales volume over the expected life of a production facility is extremely difficult. It requires an estimate of the total market, production capabilities and costs of the competition, market share expected, selling strategy to be employed, and future economic conditions.

The expected annual sales volume is important not only for estimating sales revenue, but also for the selection of plant capacity (20) or process type. An economy of scale is typical of many process operations because both investment and some operating costs tend to vary with capacity to a fractional power less than unity.

Profitability Analysis

Profitability analysis involves the generation of criteria that relate to the financial return to be expected from a proposed investment choice, as well as a comparative assessment with other choices. It provides quantitative measures that aid in the subjective decision-making process; this is an art that depends on both experience and luck. The scope and direction of the analysis effort changes at different phases of process evolution. For example, the most elementary concept of profitability at the time of project conception is that sales revenues must be greater than raw materials cost. During manufacturing, the emphasis is on manufacturing costs and profit margins. Between these two, a variety of methods for assessing profitability have been developed.

Essence of Profitability.   If an investor purchases a computer in the morning and sells it in the afternoon for a larger sum, then a return on the investment is realized. This is profit; a reward for the effort (investment) made. A somewhat more difficult situation is the case of three choices having different purchase costs and expected selling prices:

Option	Investment	Sale	Return	Rate of return
A	$2000	$2400	$400	20%
B	$2400	$2950	$550	23%
C	$2800	$3400	$600	21%

Option A has the lowest investment at risk, B has the highest rate of return on investment, and C has the highest return. In general, the objective would be to maximize either the rate of return or the return, within the limits of available investment funds.

Before a decision is made, all three items, ie, investment, return, and rate of return, would be examined, as would the current cash position, perceived risk, other venture opportunities, and a variety of subjective criteria. For this elementary situation, economists would also employ an incremental approach analogous to the above, based on the tenet that each increment of investment should itself make an adequate return. Rarely is there a unique correct decision. Only future events determine the wisdom of the selection; even then, the results that another decision would have produced are rarely known. This is the essence of profitability analysis.

In multiyear process ventures, the money flows are more complicated and must be discounted to a common point in time before they can be combined. However, the three basic parameters, investment, return, and rate of return, should be retained in some logical and consistent manner.

Multiyear Process Ventures. The economics of a proposed process venture, from initial development through final shutdown, should be examined. A venture timetable and schedule of money flows is part of the problem definition. All money flows are typically tabulated as end-of-year flows for simplification. The money flows for any year can be represented by the diagram shown in Figure 1. Investments and other outflows are negative; sales revenues and other inflows are positive. Not all of these occur each year. For example, investment flows are typically preoperational costs; sales revenues do not start until the project is operational.

Money Flows. Estimation of sales revenues and manufacturing expenses has been discussed. Depreciation, part of the manufacturing expense, is treated separately in the money flow, as is interest.

Capital Investment. From the viewpoint of a project, all of the capital that must be raised is external capital. Equity capital is the ownership capital, eg, common and preferred stocks or retained cash, whereas debt capital consists of bonds, mortgages, debentures, and loans. Nearly all investment involves a mixture of both types so as to maximize the return on investment (21). The debt ratio (debt/total capital) for the chemical industry is typically over 30%. Because financial details are not well known during the preliminary phases of project analysis, the investment is viewed simply as the total capital that must be expended to design and build the project.

The investment consists of the fixed capital, eg, equipment, buildings, and facilities; land cost; and working capital. Interest charges during construction are frequently considered part of the fixed capital. This is called capitalization of the construction interest expense. Part of the start-up costs are occasionally treated in the same manner.

The nondepreciable investments, ie, land and working capital, are often assumed to be constant preoperational costs that are fully recoverable at cost when the project terminates. Equipment salvage is another end-of-life item that can represent a significant fraction of the original fixed capital investment. However, salvage occurs at the end of life, can be difficult to forecast, and is partially offset by dismantling costs. For these reasons, a zero salvage assumption is a reasonable approximation in preliminary analysis.

Interest. The interest block on Figure 1 represents the interest expense associated with the debt capital. It is an allowed corporate pretax expense for federal income tax purposes and is actually paid out to the debt holders as their earning.

Tax-Basis Depreciation. Depreciation is a loss in value resulting from use or obsolescence. A depreciation allowance on equipment, buildings, and other facilities is a permitted pretax expense for federal income tax purposes. The tax-basis depreciation allowance reduces the taxable income, but is not an actual out-of-pocket expense. Consequently, this allowance is limited by tax regulations, which define depreciation methods and allowable tax-basis lifetimes for equip-

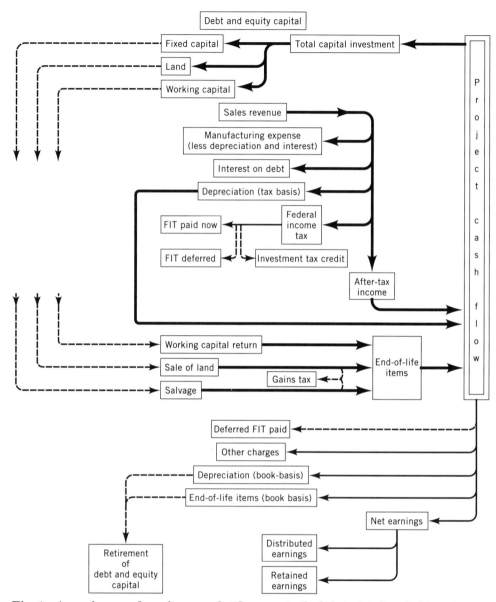

Fig. 1. Annual money flows diagram. Outflows are to the left, ie, total capital investment, and are negative. Inflows are to the right, ie, sales revenues, and are positive. (━), Economic cash flow; (‐ ‐ ‐), money flows; (—), accounting cash flows.

ment. Tax regulations also permit a depletion allowance on exhaustable resources such as oil reservoirs, mineral deposits, and woodlands.

Federal Income Tax. The federal income tax (FIT) is based on the net pretax income. The FIT rate has been below 50% for many years. As of 1992 it was 34%. Although part of the tax is actually deferred and paid in a future year (Fig. 1), it is assumed in profitability studies that any tax due is paid in the current year.

An income tax rate higher than the FIT is sometimes assumed in profitability studies to account for state and other income taxes.

In the past, income tax could be reduced by an investment tax credit. This item, designed to stimulate investment, was a tax credit amounting to some percentage of the new capital investment in certain eligible types of production equipment. It was credited when the investment was made and could be used to offset the tax due, until exhausted, for a prescribed period of years. This credit was eliminated in the United States for most equipment in 1986, but is frequently advocated for investment stimulation.

End-of-Life Items.   The end-of-life items are working capital return, sale of land, and salvage. If there is a capital gain on land sale or salvage, above the remaining tax-basis asset value, then this gain is treated as taxable ordinary income in the United States; historically, capital gains were taxed separately at a lower rate than ordinary income. Although equipment replacement or changing working capital requirements can occur anytime during the project lifetime, both salvage and working capital return are frequently assumed to be end-of-life items in profitability studies.

Book-Basis Depreciation.   The book-basis depreciation is arbitrarily determined by management on a year-to-year basis, subject to acceptable accounting practice. This is not an out-of-pocket expense. It is simply a charge for the recovery of capital in earnings calculations and is available as capital for reinvestment or distribution. Some consistent treatment for recovery of capital must be assumed in profitability analysis.

Distributed Earnings.   The dividends distributed to stockholders provide the earning on equity capital in the same way that interest is the earning on the debt capital. However, the dividends are an after-tax expense and represent an arbitrary management decision.

Retained Earnings.   After the equity earnings are subtracted from the net after-tax earnings, the balance is called the retained earnings and represents an increase in equity. The retained earnings can be visualized theoretically as the new cash generated beyond that needed to provide for a return to investors and an orderly retirement of the investment.

Annual Cash Flow.   The net annual cash flow is the actual cash generated by the project in a given year. This can be defined for any project year as

$$
\begin{matrix} \text{net} \\ \text{annual} \\ \text{cash} \\ \text{flow} \end{matrix} = \begin{matrix} \text{after-tax} \\ \text{income} \end{matrix} + \begin{matrix} \text{depreciation} \\ \text{tax-basis} \end{matrix} + \begin{matrix} \text{end-of-life} \\ \text{items} \end{matrix} - \begin{matrix} \text{yearly} \\ \text{investment} \end{matrix} \qquad (10)
$$

The annual cash flow is an important management consideration and is the starting point for profitability analysis.

Recovery of Capital.   In Figure 1, the annual book depreciation is used to retire the fixed capital investment. Whereas this accounting model does not correspond to the typical money flow, it is one possible model for recovery of capital. This model assumes that the investment is reduced each year by the amount of the annual depreciation. Another model (22) assumes that a uniform yearly book depreciation payment is made to an interest-bearing sinking fund that accumulates to the depreciable fixed capital amount at the end of the venture. Using this second model, the investment is outstanding throughout the lifetime of the project. This also does not correspond to the actual money flow in most cases. Profitability analysis utilizes a third model based on discounted cash flows.

Discounted Cash Flows. Because the flows below the cash flow box in Figure 1 tend to be arbitrary management decisions that are generally difficult to predict, the prediction of profitability is based on the expected cash flows instead of earnings. As a result, some logical assumptions to account for the cost of capital and the recovery of the investment must be made.

If money is borrowed, interest must be paid over the time period; if money is loaned out, interest income is expected to accumulate. In other words, there is a time value associated with the money. Before money flows from different years can be combined, a compound interest factor must be employed to translate all of the flows to a common present time. The present is arbitrarily assumed; often it is either the beginining of the venture or start of production. If future flows are translated backward toward the present, the discount factor is of the form $(1 + i)^{-n}$, where i is the annual discount rate in decimal form ($10\% = 0.10$) and n is the number of years involved in the translation. If past flows are translated in a forward direction, a factor of the same form is used, except that the exponent is positive. Discounting of the cash flows gives equivalent flows at a common time point and provides for the cost of capital.

A logical choice for the discount rate is the average capital cost rate, where capital includes both the equity and debt capital. The estimation of a suitable value for the discount rate is not straightforward (23), but financial specialists always seem ready to provide a number.

Profitability Criteria. *Net Present Value.* Each of the net annual cash flows can be discounted to the present time using a discount factor for the number of years involved. The discounted flows are then all at the same time point and can be combined. The sum of these discounted net flows is called the net present value (NPV), a popular profit criterion. Because the discounted positive flows first offset the negative investment flows in the NPV summation, the investment capital is recovered if the NPV is greater than zero. This early recovery of the investment does not correspond to typical capital recovery patterns, but gives a conservative and systematic assumption for investment recovery.

The NPV represents the present-value net return, because provision has been made for capital recovery and the cost of capital. In other words, the NPV is a discounted net return or profit, analogous to the net return of the example introduced earlier.

Discounted Total Capital. Because the investment can occur over a period of years, the investment flows should be discounted to the same present time and combined to give the discounted total capital (DTC). This present-worth investment parameter is the second criterion; it corresponds to the investment of the earlier example.

Net Return Rate. The NPV can be divided by the DTC to give a measure analogous to the net return on investment over the life of the venture. If this is divided by the venture life, the result is the annual net return on investment, called the net return rate (NRR) defined as (24):

$$\text{NRR}(\%/\text{yr}) = \frac{\text{NPV}}{(\text{DTC})(\text{VENTURE LIFE})} \tag{11}$$

More exactly, this third profitability parameter is an average discounted annual net return rate on the total investment and corresponds to the net return rate of the example. It is a discounted annual rate of profit criterion that relates to the NPV as a discounted profit criterion.

Any positive value of the NRR represents a profitable situation. NRR values for different venture choices can be compared directly. As an alternative, a NRR cutoff level could be selected as the minimum level for acceptability of any venture.

Risk and uncertainty associated with each venture should translate, in theory, into a minimum acceptable net return rate for that venture. Whereas this translation is often accomplished implicitly by an experienced manager, any formal procedure suffers from the lack of an equation relating the NRR to risk, as well as the lack of suitable risk data. A weaker alternative is the selection of a minimum acceptable net return rate averaged for a class of proposed ventures. The needed database, from a collection of previous process ventures, consists of NPV, investment, venture life, inflation, process novelty, decision (acceptance or rejection), and result data.

This gives two choices in interpreting calculated NRR values, ie, a direct comparison of NRR values for different options or a comparison of the NRR value of each option with a previously defined NRR cutoff level for acceptability. The NPV, DTC, and NRR can be interpreted as discounted measures of the return, investment, and return rate, analogous to the parameters of the earlier example. These three parameters characterize a venture over its entire life. Additional parameters can be developed to characterize the cash flow pattern during the early venture years. For example, the net payout time (NPT) is the number of operating years for the cumulative discounted cash flow to sum to zero. This characterizes the early cash flow pattern; it can be viewed as a discounted measure of the expected operating time that the investment is at risk.

Another possibility is the net payout fraction (NPF), defined as the ratio of the NPT to the operating life of the venture. This is the fraction of the expected operating lifetime needed to recover the discounted investment.

Internal Return Rate. Another rate criterion, the internal return rate (IRR) or discounted cash flow rate of return (DCFRR), is a popular ranking criterion for profitability. The IRR is the annual discounting rate that makes the algebraic sum of the discounted annual cash flows equal to zero or, more simply, it is the total return rate at the point of vanishing profitability. This is determined iteratively.

The total annual return rate on investment is the sum of both the capital cost rate, ie, discount rate, and the net return rate (NRR). Any given numerical value can represent a low capital cost rate and a high net return rate, or a high capital cost rate and a low net return rate. The IRR, as the discounting rate that gives a vanishing net return, cannot be related to the total return rate at appropriate discount rates because of the nonlinear nature of the discounting step.

Simplified Profitability Criteria.   Approximate profitability criteria that do not require a detailed year-by-year financial analysis are sometimes employed as simple figure-of-merit measures.

Return on investment (ROI) criteria can be defined by the general form:

$$\text{ROI}(\%/\text{yr}) = \frac{\text{ANNUAL RETURN}}{\text{INVESTMENT}}(100) \tag{12}$$

Possible numerators include the gross income; net pretax income; net after-tax income; gross profit, ie, gross income minus book depreciation; cash flow; or net income. An average return value is selected by defining a typical or mature proof year as the basis of calculation. The denominator can be the original total investment, depreciated book-value investment, lifetime averaged investment, or fixed capital investment.

Payout time (PT), or payback period, is a measure of the time, usually in years, required to recover the investment in a scenario in which the time value of money is neglected. This can be represented by the general form

$$\text{PT} = \frac{\text{INVESTMENT}}{\text{ANNUAL RETURN}} \tag{13}$$

which corresponds to a reciprocal of the ROI. As before, the numerator and denominator can represent a variety of quantities. The customary choice for the annual return term in the ROI is the after-tax profit; for the PT, it is the cash flow. For this reason, these two criteria are not commonly viewed as reciprocals.

A simplified annual cost view is sometimes employed when comparing alternatives. The approach is to convert the capital cost to an equivalent annual expense using a fixed charge factor or an amortization factor. Then the total annual costs or profits of the choices can be compared. This type of analysis is commonly employed for components where one choice might have a higher capital cost, but a lower operating cost. The fixed charge factor is a rate that represents the annual cost of capital and an annual component of capital recovery; typical values range from 12 to 20%.

In energy production ventures, the unit energy costs (cents per kilowatt-hour) are estimated as a measure of venture feasibility. An averaging approach widely used for such projects includes converting the cost items occurring in the various years to an equivalent present worth using a discount factor, and then converting the present worth to an equivalent uniform annual amount using the capital recovery factor

$$\left[\frac{i(1 + i)^n}{(1 + i)^n - 1}\right] \tag{14}$$

where i is the annual discount rate in decimal form and N is the number of years involved in the translation, for the project lifetime. The result is a levelized annual amount.

Incremental Criteria. When mutually exclusive ventures having different levels of investment are compared, an attractive concept is that each increment of investment must itself yield a satisfactory return. This concept has led to a variety of incremental approaches for profitability analysis. Because the risk can vary with investment level and cloud the meaning of satisfactory, any incremental approach to multiyear investment analysis should be viewed with caution.

Cash Flow Examples. Several hypothetical ventures are presented to illustrate cash flow analysis. Venture A exhibits a cash flow pattern typical of process ventures. Other ventures are introduced for comparison and to provide additional insight into cash flow analysis.

Venture A. Input data for Venture A consist of the parameters given at the top of Figure 2 and the first six line-items above the dashed line in the figure. These data estimates must be obtained from the various groups involved in the analysis of the venture, eg, R&D, engineering, purchasing, sales, finance, etc. In line-items 2 and 3, land cost and working capital are assumed to be fully recovered when the venture ends in the year 2004. Cash flow calculations are tabulated in line-items 7–18. When details about the debt structure, ie, debt ratio, interest rates, repayment schedules, are unknown, as in preliminary project analysis, interest can be neglected as a line-item expense (line-item 10) and accounted for in the discount rate. This leads to a small error in the line-item tax and cash flow, but is a popular and conservative approach. If line-item interest is included in line 10, then the interest should be added to the cash flow in line-item 17 before discounting in order to avoid an apparent double return on the debt from interest and discounting. This addition is not part of the accounting cash flow where interest is always treated as a line-item expense and there is no discounting.

For this example, the tax-basis depreciation method in line-item 11 is a straight-line calculation based on the capitalized fixed capital, ie, fixed capital plus interest to the start of operation; any salvage should be subtracted from the capitalized fixed capital and the result divided by the number of expected operating years to obtain the annual tax-basis depreciation.

Values of the DTC, NPV, NRR, NPT, and NPF, ie, profitability results, for Venture A are given at the bottom of Figure 2. These serve as the basis of a profitability analysis.

Profitability Diagrams. Profitability diagrams of the type shown in Figure 3a for Venture A provide insight into venture profitability. Total return rate is defined as the sum of the discount rate and the net return rate (NRR). The discount rate, net return rate, and total return rate are all shown on the diagram as functions of the discount rate. Because the NPV is a nonlinear function of the discount rate, the NRR and total return rate are also nonlinear. The NRR, as a measure of the profitability, correctly decreases as the discount rate increases.

The internal return rate (IRR), a fixed point on the diagram, cannot be viewed as a measure of profitability, which should vary with the cost of capital (discount rate). Because the curvature of the total return curve cannot be predicted from the single IRR point, there is no way that the IRR can be correlated

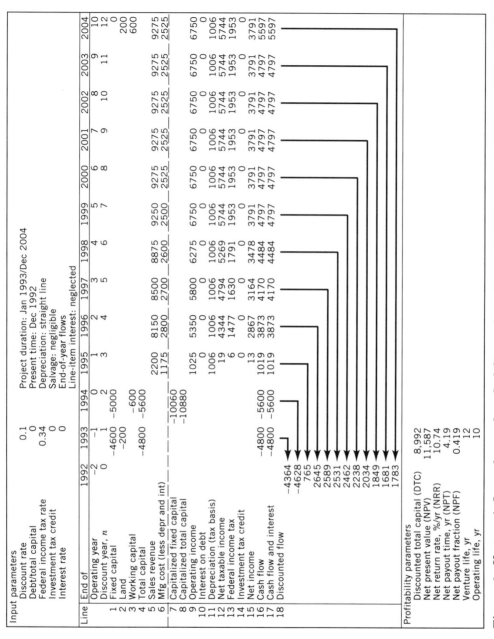

Figure 2. Venture A: Input data and profitability analysis. For profitability analysis, DTC equals line 8 multiplied by the discount factor $[1 + i]^{-n}$ where $i = 0.1$ (given discount rate) and $n = 2$; NPV is equal to the summation of the discounted flow for each year (found under column 1992), derived by multiplying cash flow by the discount factor above, where n = discount year for which the adjustment is made, ie, for 1996 $n = 4$; NRR is defined according to equation 11. NPT and NPF are defined in text.

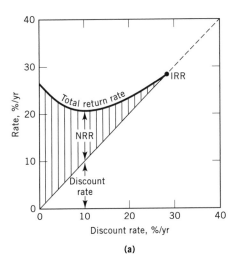

(a)

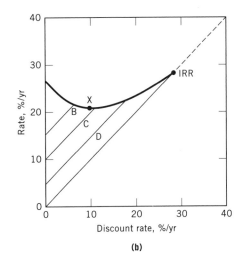

(b)

Fig. 3. Profitability diagram for Venture A. (**a**) Simple diagram. NRR is net return rate; IRR, the internal rate of return, is a given fixed point. (**b**) Three NRR cutoff lines for Venture A where B, C, and D represent NRR values of 15, 10, and 5%/yr, respectively. For example, at a discount rate of 10% per year, the NRR cutoff for Venture A could be as high as 10.74% per year for marginal acceptance (point X). Acceptable levels are to the left of NRR cutoff lines.

with profitability at meaningful discount rates. Even both end points, ie, the IRR and the total return at zero discount rate, are not enough to predict the curvature of the total return curve.

An NRR cutoff level can be selected to reflect the minimum acceptable profitability level. Lines of constant cutoff rate can be drawn on a profitability diagram, as shown in Figure 3**b** for NRR cutoff rates of 5, 10, and 15% per year. The acceptable region changes with discount rate and relative venture risk, ie, cutoff rate.

Different cash flow patterns are used in Figures 4–6 to illustrate advantages of the NRR as a profitability parameter. These ventures have input data identical

Table 4. Input Data for Ventures A–I

Venture	Fixed capital[a]		Sales revenue[b]									
	1993	1994	1995	1996	1997	1998	1999	2000	2001	2002	2003	2004
A	−4600	−5000	2,200	8,150	8,500	8,875	9,250	9,275	9,275	9,275	9,275	9,275
B	same as A		10,675	10,500	10,000	9,000	8,500	6,000	4,500	4,500	4,500	4,500
C	same as A		2,250	3,900	4,139	7,000	8,000	12,000	13,000	13,000	13,500	14,000
D	same as A		7,418	8,000	8,000	7,500	6,500	6,000	6,000	6,000	6,000	5,000
E	same as A		2,270	6,515	8,500	9,000	9,000	9,000	11,000	11,500	11,500	11,500
F	same as A		2,745	4,370	4,270	10,005	14,000	15,720	21,917			
G	−5600	−7500	10,645	9,475	7,975	7,489	6,290	6,300	8,240	8,475	9,000	12,951
H	−5260	−6100	same as A									
I	−4140	−4500	same as A									

[a]This is the same as line-item 1 on Figure 5.
[b]This is the same as line-item 5 on Figure 5.

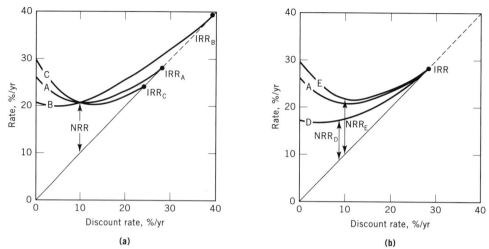

(a)   (b)

Fig. 4. Effect of NPV on profitability where investment and lifetime are the same for all Ventures (see Table 4). (**a**) Sale revenues for Ventures B and C have been selected so that at a discount rate of 10% per year Ventures A, B, and C each have the same NPV and NRR. IRR values are as given and do not relate to NPV, NRR, or total return rate (TRR). The diagram indicates that at discount rates less than 10%, Venture C has the largest NRR, but the IRR indicates Venture B is the choice for all discount rates. (**b**) Sale revenues for Ventures D and E have been selected so that Ventures A, D, and E each have a different NPV at a discount rate of 10% per year, but all three have the same IRR. The diagram indicates that at the selected discount rate of 10%, the NRR is different for each venture.

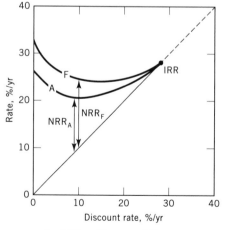

Fig. 5. Effect of lifetime on profitability. Venture F has a shorter operating lifetime than Venture A, but the same investment and IRR (see Table 4); the NPV is the same at the 10% discount rate. The diagram indicates that the profitability of Venture F is higher than that of Venture A at all discount rates; the shorter lifetime leads to a higher annual net return rate (NRR). The IRR rate does not indicate this difference in profitability.

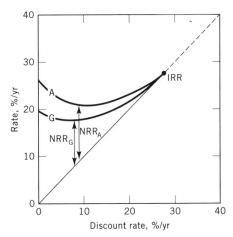

Fig. 6.   Effect of investment on profitability. Venture G has a larger investment and sales revenue than Venture A (see Table 4). NPV at 10%, operating lifetime, and IRR are the same for both ventures. The diagram indicates that the profitability of Venture G, as indicated by NRR, is smaller than that for Venture A. The IRR rate does not indicate this difference in profitability.

to Venture A except for the values given in Table 4; the resulting annual cash flow patterns are quite different.

Sensitivity Analysis.   Sensitivity Diagram.   The changes in NRR, or NPV, as a result of changes in a parameter, can be represented on a sensitivity diagram where the change in the criterion is plotted versus the change in the parameter. This is illustrated in Figure 7**a**, where the changes in NRR resulting from changes in discount rate for Venture A are plotted.

Profitability Diagram.   The sensitivity of profitability criteria to parameter changes or other effects can also be represented on a profitability diagram. For example, plus 10% change in fixed capital investment for Venture A gives the results shown in Figure 7**b**. If several other investment levels were plotted, then interpolation would be a simple task and the sensitivity to investment level could be visualized readily.

Break-Even Charts.   A break-even chart is a visual tool for analyzing operating profitability at various levels of production. In this type of diagram, annual expenses are separated into fixed, variable, and semivariable categories. Fixed expenses do not vary with production level, variable expenses vary linearly with production, and semivariable expenses vary in some nonlinear way with production level. Annual dollars of sales and expenses are plotted as a function of production level, ie, % of design capacity, as shown in Figure 8. Profitable operation is represented by the vertically lined region. This diagram also shows break-even levels of production (Points B and C) that are marginally profitable. Points A and D represent shutdown points where the total loss equals the fixed cost. At these shutdown points it is more economical to cease production, paying only fixed costs, than to operate the facility.

A typical break-even chart is used with production models to predict optimum production levels, break-even points, and shutdown conditions under var-

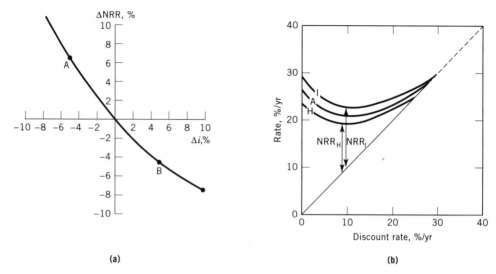

(a)

(b)

Fig. 7. Sensitivity analysis. (**a**) Changes in discount rate as indicated by a sensitivity diagram. At center axis the discount rate $i = 10\%$ and NRR = 10.74%; point A indicates $\Delta i = $ ca -5% with a corresponding ΔNRR of $+6.5\%$; point B indicates $\Delta i = $ ca $+5\%$ with a corresponding ΔNRR of -4.75%. (**b**) Changes in investment as indicated by a profitability diagram, where Venture H has a 10% higher fixed capital investment and Venture I has a 10% lower fixed capital investment than does Venture A (see Table 4). Diagram indicates the NRR for Ventures H and I at a discount rate of 10%.

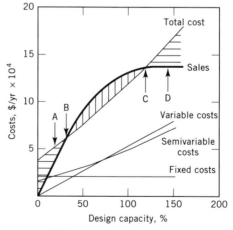

Fig. 8. Break-even chart where ▯ represents profit and ▱ is loss. A and D indicate shut-down points, ie, total loss = fixed cost; B and C are marginally profitable break-even points, ie, zero profit.

884

ious scenarios. These models tend to involve a reasonable amount of approximation. For example, sales revenue as a function of production level involves numerous variables and relationships that are not always well known. Such charts, however, provide useful guides for production operations.

Inflationary Effects. Inflation can have a significant effect on the profitability of a venture. However, the U.S. federal tax laws do not allow for indexing the inflationary effects on depreciation schedules, salvage values, replacement costs, or taxable income. Inflation rates can vary unpredictably with time and can differ for certain revenues or expenditures.

Prevailing interest rates probably tend to reflect an estimate of future inflation and contain a component that can be attributed loosely to inflationary expectations. However, the classical treatment is to assume that an inflation-free interest rate, r_e, and average inflation rate, r_i, over the project lifetime can be identified. A discount factor $(1 + r)^{-n}$ can be modified (25) so that

$$(1 + r)^{-n} = [(1 + r_e)^{-n}(1 + r_i)^{-m}] \tag{15}$$

where m is the number of years to the reference or constant dollar year. It is often assumed that $m = n$. Various treatments of inflationary effects have been reported (26,27).

Statistical Criteria. Sensitivity analysis does not consider the probability of various levels of uncertainty or the risk involved (28). In order to treat probability, statistical measures are employed to characterize the probability distributions. Because most distributions in profitability analysis are not accurately known, the common assumption is that normal distributions are adequate. The distribution of a quantity then can be characterized by two parameters, the expected value and the variance. These usually have to be estimated from meager data.

The two principal approaches of interest for situations involving risk and uncertainty are decision trees and Monte Carlo simulation (26,29,30). These find wide application in general business analysis but are less widely used in engineering economic analysis.

BIBLIOGRAPHY

"Economic Evaluation" in *ECT* 2nd ed., Vol. 7, pp. 642–660, by O. Axtell, Celanese Chemical Co.; in *ECT* 3rd ed., pp. 409–429, by O. Axtell and J. M. Robertson, Celanese Chemical Co.

1. M. S. Peters and K. D. Timmerhaus, *Plant Design and Economics for Chemical Engineers*, 4th ed., McGraw-Hill Book Co., Inc., New York, 1991.
2. K. K. Humphreys, *Jelen's Cost & Optimization Engineering*, 3rd ed., McGraw-Hill Book Co., Inc., New York, 1990.
3. S. M. Walas, *Chemical Process Equipment: Selection and Design*, Butterworths, Boston, 1988.
4. R. S. Hall, W. M. Vatavuk, and J. Matley, *Chem. Eng.* **95**(17), 66 (1988).
5. W. M. Vatavuk, *Chem. Eng.* **97**(5), 126 (1990).
6. J. Matley, *Chem. Eng.* **89**(8), 153 (1982).
7. J. Matley, *Chem. Eng.* **93**(9), 75 (1985).
8. T. J. Ward, in J. J. McKetta, ed., *Encyclopedia of Chemical Processing and Design*, Vol. 38, Marcel Dekker, New York, 1991, p. 172.

9. K. M. Guthrie, *Process Plant Estimating, Evaluation, and Control*, Craftsman, Solana Beach, Calif., 1974.

10. J. D. Chase, *Chem. Eng.* **77**(7) 113 (1970).

11. N. K. Kochar and R. L. Marcell, *Chem. Eng.* **87**(2) 80 (1980).

12. J. Cran, *Chem. Eng.* **88**(7) 65 (1981).

13. I. V. Klumpar and S. T. Slavsky, *Chem. Eng.* **92**(15) 73 (1985); **92**(17) 76 (1985); **92**(19) 85 (1985).

14. R. F. Horvath, *Pioneer Plants Study User's Manual*, Rept. R-2569/1-DOE, Rand Corp., Santa Monica, Calif., June 1983.

15. T. B. Lyda, *Chem. Eng.* **79**(21), 182 (1972).

16. K. E. Lunde, *Chem. Eng.* **91**(25), 89 (1984); **92**(1), 95 (1985).

17. K. E. Lunde, *Chem. Eng.* **92**(3), 85 (1985).

18. T. M. DeVinney, ed., *Issues in Pricing: Theory and Practice*, Lexington Books, Lexington, Mass., 1988.

19. T. Nagle, *The Strategy and Tactics of Pricing*, Prentice-Hall, Englewood Cliffs, N.J., 1987.

20. M. A. Malina, in Ref. 8, p. 166.

21. A. Bohl and F. H. Murphy, *Eng. Econ.* **37**(1), 61 (1991).

22. J. Happel and D. G. Jordan, *Chemical Process Economics*, 2nd ed., Marcel Dekker, New York, 1975.

23. T. E. Powell, *Chem. Eng.* **92**(23), 187 (1985).

24. T. J. Ward, *Chem. Eng.* **96**(3), 151 (1989).

25. F. A. Holland, F. A. Watson, and J. K. Wilkinson, *Introduction to Process Economics*, 2nd ed., John Wiley & Sons, Inc., New York, 1985.

26. D. Young, *Modern Engineering Economy*, John Wiley & Sons, Inc., New York, 1992.

27. R. V. Oakford and A. Salazar, *Eng. Econ.* **27**(2), 127 (1982).

28. J. C. Agarwal and I. V. Klumpar, *Chem. Eng.* **82**(20), 66 (1975).

29. L. E. Bussey and T. G. Eschenbach, *Economic Analysis of Industrial Processes*, 2nd ed., Prentice-Hall, Englewood Cliffs, N.J., 1992.

30. C. S. Park and G. P. Sharp-Bette, *Advanced Engineering Economics*, John Wiley & Sons, Inc., New York, 1990.

General References

Engineering Economics

T. Au and T. P. Au, *Engineering Economics for Capital Investment Analysis*, 2nd ed., Prentice-Hall, Englewood Cliffs, N.J., 1992.

J. Couper and W. Rader, *Applied Finance and Economic Analysis for Scientists and Engineers*, Van Nostrand Reinhold, New York, 1986.

T. G. Eschenbach, *Cases in Engineering Economy*, John Wiley & Sons, Inc., New York, 1989.

W. J. Fabrycky and B. S. Blanchard, *Life Cycle Cost and Economic Analysis*, Prentice-Hall, Englewood Cliffs, N.J., 1991.

L. A. Gordon and G. E. Pinches, *Improving Capital Budgeting: A Decision Support System Approach*, Addison-Wesley, Reading, Mass., 1984.

H. Levy and M. Sarnot, *Capital Investment and Financial Decisions*, Prentice-Hall, Englewood Cliffs, N.J., 1978.

D. G. Newman, *Engineering Economic Analysis*, 2nd ed., Engineering Press, San Jose, Calif., 1983.

J. A. White, M. H. Agee, and K. Case, *Principles of Engineering Economic Analysis*, 3rd ed., John Wiley & Sons, Inc., New York, 1989.

Cost Estimation

F.D. Clark and A. B. Lorenzoni, *Applied Cost Engineering*, 2nd ed., Marcel Dekker, New York, 1978.

K. K. Humphreys and S. Katell, *Basic Cost Engineering*, Marcel Dekker, New York, 1981.

E. M. Malstrom, *What Every Engineer Should Know About Manufacturing Cost Estimating*, Marcel Dekker, New York, 1981.

L. M. Matthews, *Estimating Manufacturing Costs—A Practical Guide For Managers and Estimators*, McGraw-Hill Book Co., Inc., New York, 1983.

C. A. Miller, *Modern Cost Engineering: Methods and Data*, McGraw-Hill Book Co., Inc., New York, 1979.

P. F. Ostwald, *Engineering Cost Estimating*, 3rd ed., Prentice-Hall, Englewood Cliffs, N.J., 1992.

W. M. Vatavuk, *Estimating Costs of Air Pollution Control*, Lewis Publishers/CRC Press, Boca Raton, Fla., 1990.

Chemical Process Economics

O. Axtell and J. Robertson, *Economic Evaluation in the Chemical Process Industries*, John Wiley & Sons, Inc., New York, 1986.

A. Chauvel and co-workers, *Manual of Economic Analysis of Chemical Processes: Feasibility Studies in Refining and Petrochemical Processes*, McGraw-Hill Book Co., Inc., New York, 1980.

F. A. Holland, F. A. Watson, and J. K. Wilkinson, *Introduction to Process Economics*, 2nd ed., John Wiley & Sons, Inc., New York, 1985.

J. F. Valle-Riestra, *Project Evaluation in the Chemical Process Industries*, McGraw-Hill Book Co., Inc., New York, 1983.

J. B. Weaver and H. C. Thorne, eds., *Investment Appraisal for Chemical Engineers*, AIChE Symposium Series No. 285, American Institute of Chemical Engineers, New York, 1991.

J. Wei, T. W. F. Russell, and M. W. Swartzlander, *The Structure of the Chemical Processing Industries*, McGraw-Hill Book Co., Inc., New York, 1979.

D. R. Wood, *Financial Decision Making in the Process Industries*, Prentice-Hall, Englewood Cliffs, N.J., 1975.

THOMAS J. WARD
Clarkson University

EGGS

Eggs are defined herein as eggs from chickens, and refer to both shell eggs and/or egg products. Egg products in liquid, frozen, or dried form contain egg as the principal ingredient. An egg product can be anything from a frozen or dried product made from 100% egg white to a scrambled egg mix that has 51% whole egg. Eggs are primarily used as food. Shell eggs used in the home, restaurants, and institutions are fried, hard-cooked, poached, etc, or are used as ingredients in other foods. Egg products generally are utilized in the food industry.

Eggs contribute important proteins (qv), fats (see FATS AND FATTY OILS), vitamins (qv), and minerals (see MINERAL NUTRIENTS) to the diet. They have many functional properties, eg, binding, whipping, and emulsifying, which make eggs useful in different foods. Comprehensive reviews of the chemistry and biology (1) and marketing (2) of eggs are available.

Properties

Physical Properties. The egg is composed of three basic parts: shell, whites (albumen), and yolk. Each of these components has its own membranes to keep the component intact and separate from the other components. The vitelline membrane surrounds the yolk, which in turn is surrounded by the chalaziferous layer of albumen, keeping the yolk in place. Egg white (albumen) consists of an outer thin layer next to the shell, an outer thick layer near the shell, an inner thin layer, and finally, an inner thick layer next to the yolk. Thick layers of albumen have a higher level of ovomucin in addition to natural proportions of all the other egg white proteins. This ovomucin breaks into shorter fibers when the egg white is blended on a high speed mixer (3), or when the egg white ages. Viscosity is greatly reduced when the egg white is blended in this way.

Table 1 shows the various physical properties for components of eggs (4). Specific gravity of whites, yolks, and whole egg is the same, ie, density is 1035 kg/m^3 (64.6 lb/ft^3 = 8.63 lb/gal) for all three types of egg products shown. The viscosity of blended liquid egg components varies over a wide range of temperatures; at temperatures higher than those indicated in Table 1, the protein starts to denature and coagulate, increasing viscosity.

Chemical Properties. Egg white contains mostly proteins having the physical and chemical characteristics given in Table 2. Some proteins in egg white have biological activities that protect from microbiological growth, eg, lysozyme lyses certain bacteria, conalbumin ties up iron, and avidin binds biotin. Most of these activities are destroyed when the egg white is cooked. pH of liquid egg white

Table 1. Physical Properties of Liquid Egg Products[a]

Property	Whites	Yolks	Whole
solids, %	12.1	44.0	24.5
specific gravity	1.035	1.035	1.035
specific heat	0.940	0.780	0.880
freezing point, °C	−0.4	−0.4	−0.4
specific heat below freezing	0.500	0.500	0.500
latent heat of freezing, kJ/kg[b]	531.4	338.9	451.9
viscosity, mPa($=$cP)			
5°C	12	260	20
50°C	5		
60°C		45	7

[a]Ref. 4.
[b]To convert kJ to kcal, divide by 4.184.

Table 2. Proteins in Egg White[a]

Protein	Amount of albumen, %	pH[b]	Mol wt	Denaturation, °C	Characteristics
ovalbumin	54	4.5	45,000	84.0	phosphoglyco-protein
ovotransferrin[c]	12	6.1	76,000	61.0	binds metallic ions
ovomucoid	11	4.1	28,000	70.0	inhibits trypsin
ovomucin	3.5	4.5–5.0	5.5–8.3×10^6		sialoprotein, viscous
lysozyme[d]	3.4	10.7	14,300	75.0	lyses some bacteria
G_2 globulin	4.0	5.5	3.0–4.5×10^4	92.5	
G_3 globulin	4.0	4.8			
ovoinhibitor	1.5	5.1	49,000		inhibits serine proteases
ficin inhibitor	0.05	5.1	12,700		inhibits thioproteases
ovoglycoprotein	1.0	3.9	24,400		sialoprotein
ovoflavoprotein	0.8	4.0	32,000		binds riboflavin
ovomacroglobulin	0.5	4.5	7.6–9.0×10^5		strongly antigenic
avidin	0.05	10.0	68,300		binds biotin

[a]Refs. 1, 5, and 6.
[b]Isoelectric point.
[c]Also known as conalbumin.
[d]CAS Registry Number is [901-63-2].

is normally about 9.0. However, egg white from freshly laid eggs has a pH of about 7.6. pH increases quite rapidly as carbon dioxide escapes during storage. The high 9.0 pH of natural egg white retards the growth of many bacteria.

The yolk is separated from the white by the vitelline membrane, and is made up of layers that can be seen upon careful examination. Egg yolk is a complex mixture of water, lipids, and proteins. Lipid components include glycerides, 66.2%; phospholipids, 29.6%; and cholesterol [57-88-5], 4.2%. The phospholipids consist of 73% lecithin [8002-43-5], 15% cephalin [5681-36-7], and 12% other phospholipids. Of the fatty acids, 33% are saturated and 67% unsaturated, including 42% oleic acid [112-80-1] and 7% linoleic acid [60-33-3]. Fatty acids can be changed by modifying fatty acids in the laying feed (see CARBOXYLIC ACIDS).

Yolk can be separated into two fractions, granules and plasma (5), by high speed centrifugation (see SEPARATION, CENTRIFUGAL). Granules contain a high percentage of high density lipoproteins (HDL) and lesser amounts of low density lipoproteins (LDL) and water-soluble proteins (phosvitins). The plasma contains water-soluble proteins (livetins) and finely dispersed LDL; most of the glycerides reside in this LDL fraction. The glycerides apparently form the inner core of the LDL, which is surrounded by a phospholipid shell, with protein wrapped around the shell. Half of the water in egg yolk is bound to the proteins and lipoproteins; half is free. The pH of yolk is normally about 6.6, and in freshly laid eggs it is 6.0.

Table 3 indicates the nutritional composition of the three types of egg products, plus the shell egg itself. Eggs, considered to be one of the most nutritious

Table 3. Nutritional Composition of Eggs[a]

Nutrient	Shell,[b] whole	Liquid/frozen[c] Whole	Liquid/frozen[c] White	Liquid/frozen[c] Yolk
		Essential constituents		
solids, g	13.47	24.5	12.1	44.0
calories	84	152	50	313
protein[d]	6.60	12.0	10.2	114.9
total lipids, g	6.00	10.9		27.5
ash, g	0.55	1.00	0.68	1.49
		Lipids		
fatty acids, g				
saturated, total	*2.01*	*3.67*		*9.16*
8:0	0.027	0.05		0.13
10:0	0.082	0.15		0.38
12:0	0.027	0.05		0.12
14:0	0.022	0.04		0.09
16:0	0.137	2.5		6.2
18:0	0.462	0.84		2.14
20:0	0.022	0.04		0.10
monounsaturated, total	*2.53*	*4.60*		*11.80*
14:1	0.005	0.01		0.03
16:1	0.214	0.39		0.97
18:1	2.31	4.2		10.8
polyunsaturated, total	*0.73*	*1.32*		*3.37*
18:2	0.660	1.20		3.07
18:3	0.011	0.02		0.06
20:4	0.055	0.10		0.24
cholesterol, g[e]	0.205	0.36		0.96
lecithin, g	1.27	2.32		5.81
cephalin, g	0.253	0.46		1.15
		Vitamins		
A, IU	264	480		1240
D, IU	27	50		129
E, mg	0.88	1.6		4.1
B_{12}, µg	0.48	0.88		2.27
biotin, µg	11.0	20.0	6.8	40.8
choline, mg	237	430	1.2	1130
folic acid, mg	0.023	0.060	0.016	0.128
inositol, mg	5.94	10.8	4.0	21.4
niacin, mg	0.045	0.082	0.092	0.067
pantothenic acid, mg	0.83	1.52	0.24	3.5
pyridoxine, mg	0.065	0.119	0.021	0.273
riboflavin, mg	0.18	0.33	0.28	0.41
thiamine, mg	0.05	0.09	0.011	0.22

Table 3. (*Continued*)

Nutrient	Shell,[b] whole	Liquid/frozen[c] Whole	Liquid/frozen[c] White	Liquid/frozen[c] Yolk
		Minerals		
calcium, mg	29.2	53	10	121
chlorine, mg	96.0	175	174	176
copper, mg	0.033	0.061	0.023	0.121
iodine, mg	0.026	0.047	0.003	0.114
iron, mg	1.08	1.97	0.14	4.83
magnesium, mg	6.33	11.5	10.8	12.5
manganese, mg	0.021	0.038	0.007	0.09
phosphorus, mg	111	202	22	485
potassium, mg	74	135	150	110
sodium, mg	71	129	165	74
sulfur, mg	90	164	163	165
zinc, mg	0.72	1.30	0.12	3.15

[a] Refs. 1, 7–10.

[b] Per egg; based on 60.9 g shell egg weight with 55.1 g total liquid whole egg, ie, 38.4 g white and 16.7 g yolk.

[c] Per 100 g; based on 24.5% and 12.1% solids, respectively, for whole and white liquid. Yolk contains 44% egg solids, diluted with egg white only.

[d] Protein based on total nitrogen multiplied by 6.25.

[e] Reflects USDA figures for cholesterol in egg and egg products, 22% less than earlier figures.

foods, have the highest quality protein of any food, and are important as a source of minerals and certain vitamins. Lipids in eggs are easily digested, and the amount of unsaturated fatty acids is greater than in most animal products.

Cholesterol has received the most attention of the components in eggs. Considerable controversy surrounds the role of dietary cholesterol in eggs and the part it plays in the development of arteriosclerosis (see CARDIOVASCULAR AGENTS; FAT SUBSTITUTES). The concentration of blood cholesterol is not affected strongly by dietary cholesterol, but rather is dependent on the degree of saturation of dietary triglycerides (10). The USDA has found that the average large egg contains 22% less cholesterol than previously believed (11). Cholesterol is an important part of the animal tissue and cells, eg, it is necessary in the production of Vitamin D, certain hormones (qv), and bile salts. It is carried to the tissues by blood. The body maintains a certain cholesterol level, and synthesizes any additional amount that is not supplied by the diet. There is evidence that high levels of egg in the diet do not present a greater risk of heart disease in a normal individual (9). However, as a precautionary measure, diets low in cholesterol may be advised for persons having higher than normal blood cholesterol levels, and for persons who may be prone to heart disease.

The egg shell is 94% calcium carbonate [471-34-1], $CaCO_3$, 1% calcium phosphate [7758-23-8], and a small amount of magnesium carbonate [546-93-0]. A water-insoluble keratin-type protein is found within the shell and in the outer cuticle coating. The pores of the shell allow carbon dioxide and water to escape during storage. The shell is separated from the egg contents by two protein membranes. The air cell formed by separation of these membranes increases in size because of water loss. The air cell originally forms because of the contraction of the liquid within the egg shell when the temperature changes from the body temperature of the hen at 41.6°C to a storage temperature of the egg at 7.2°C.

Functional Properties. Eggs function in different ways to give food products certain desirable characteristics.

Coagulating and Thickening. Egg protein denatures when heated over a wide range of temperatures (from 55 to 90°C). This denaturation is an important property and is the reason eggs bind or thicken foods such as cakes, custards, omelets, and puddings. The heat coagulating characteristics of egg white protein has been demonstrated in the baking of an angel food cake; no other protein material has been found to substitute for egg whites in an angel food cake.

Whipping or Beating. Eggs incorporate air when beaten with a mechanical device, such as a wire whisk, resulting in the formation of foam. Egg white can be easily whipped into stable foams by itself or when mixed with other ingredients, such as sugar. Proteins of egg white unfold at the surface, ie, surface denature, to give the foam a strong supporting structure.

Whole egg and yolk, which contain a large amount of lipids in a highly emulsified state, also foam but at a lower rate than egg white. If the emulsion is broken, whole egg or yolk loses the ability to foam. The foaming properties of eggs are quite sensitive, and can be adversely affected by certain processing procedures, eg, heating liquid to high temperatures, or drying the products.

Emulsifying. Emulsifying ability in eggs has always been attributed to phospholipids, but other components such as protein also contribute. For example, liquid egg white does not have as good emulsifying properties as liquid egg yolk (12) when substituted on a weight-for-weight liquid basis, for which the egg white would have one-fourth the solids of egg yolk. However, if dried egg white is used on a basis of just one-half the solids of egg yolk, the emulsion is extremely stable and is more like the emulsion formed with egg yolk. It has been indicated that when lipids, including phospholipids, are extracted from whole egg, the remaining portion has good emulsifying properties (13). An excellent example of the emulsifying properties of egg is in the making of mayonnaise, where egg is the only emulsifier. Eggs must support a stable emulsion containing a minimum of 65% vegetable oil (see VEGETABLE OILS). Emulsifying properties are also important in many baked items where fats and oils are present (see also EMULSIONS).

Miscellaneous Functions. Eggs retard crystallization of sugar and contribute to smoothness, moistness, and certain desirable textural characteristics of baked goods and candies. By binding ingredients together, eg, in cakes, eggs offer a barrier against water evaporation. Thus eggs also help retard moisture loss during baking and storage.

Eggs have a distinct flavor that makes them desirable for eating by themselves. They also contribute to the flavor and mouthfeel of baked goods and other food products in which they are used. The natural color in egg yolk comes from xanthophylls and other fat-soluble pigments (qv). Eggs contribute color to the products in which they are used, although this is seldom the primary function. Although egg color can be substituted by other coloring materials, there are some products in which U.S. Federal Standards permit only eggs as the coloring material, eg, in egg noodles.

Shell Eggs

Production. Figure 1 is a schematic diagram of a plant operation handling shell eggs, as well as dried egg products. Most production is carried out on farms having 30,000 hens or more per flock or house. Almost everything in the house is automated, eg, feeding, watering, ventilation, and gathering and sorting of eggs. Eggs are put on filler-flats and placed on racks to be transferred to the processing area. The racks are brought into a refrigerator at about 10°C until they are picked up for transfer to the shell egg or egg products plant. The eggs are usually tempered at 10°C before going to the egg products plant where they are broken and separated into whites, yolks, and mix, ie, standardized whole egg solids.

Grading. Eggs are graded and sorted according to size and to quality factors, which include both shell and interior quality. Historically, all eggs were candled by hand. The egg was placed before a candling light and given a quick twist. Appearance and motion of the yolk and size of the air cell gave an indication of the interior quality. Candling is also used for detecting and subsequently removing eggs with blood spots, and those with checked or cracked shells and other obvious defects. In modern processing, eggs are flash-candled on a continuous conveyor within a short time after being laid (see CONVEYING). Because of their freshness, most eggs have uniformly high interior quality where the proportion of thick to thin egg white is relatively high. The USDA has an egg grading program which is run on a voluntary basis in cooperation with each state. Most states also have their own egg grading laws, usually patterned after USDA regulations. The United States size or weight classifications for eggs (14) are listed below.

Size or weight class	Minimum net weight per dozen, kg (oz)
extra large	0.77 (27)
large	0.68 (24)
medium	0.60 (21)
small	0.51 (18)

Quality Specifications. Eggs are downgraded according to specific conditions of the shell.

Dirty. The shell is unbroken and has adhering dirt or foreign materials, prominent stains, or moderate stains covering more than one-fourth of the shell surface.

Checks. The individual egg has a broken shell or a crack in the shell, but the shell membrane is intact and its contents do not leak.

Leakers. The shell and membrane are broken so that the contents are leaking. USDA regulations prohibit use of this type of egg for human consumption.

Grade AA and A Quality. These eggs are described in Table 4. They are generally recommended for most household uses. However, lower grades can usually be used for many cooking and baking purposes. Higher grade eggs have somewhat better functional properties, such as foam-forming power.

One of the quality changes in eggs during storage is the thinning of the thick whites, which results in a lower grade of eggs. This can be noted through a can-

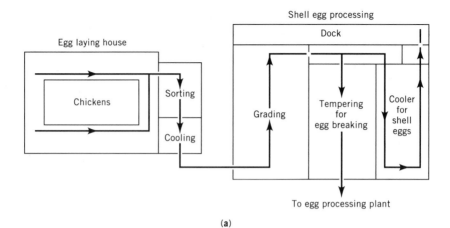

(a)

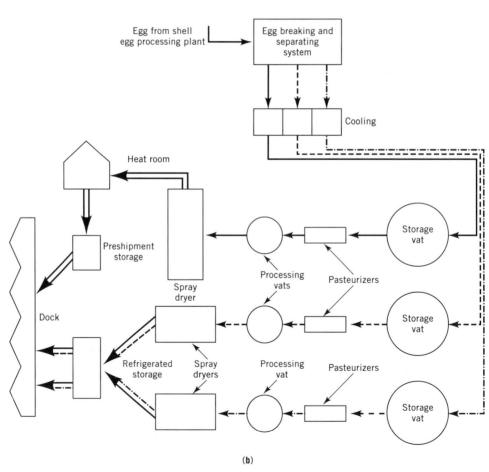

(b)

Fig. 1. (**a**) Shell egg treatment, (**b**) egg processing plant: (——), egg white liquids; (– – –), whole egg liquid; (–·–·–), yolk liquid; (═══), spray-dried egg white; (═ ═ ═), spray-dried whole egg; (═·═·═), spray-dried yolk.

894

Table 4. United States Standards for Quality of Individual Shell Eggs[a]

Factor	AA	A	B
shell	clean, unbroken, practically normal	clean, unbroken, practically normal	clean to slightly stained,[b] unbroken, abnormal
air cell	3.2 mm or less in depth, unlimited movement, and free or bubbly	4.8 mm or less in depth, unlimited movement, and free or bubbly	over 4.8 mm in depth, unlimited movement, and free or bubbly
white	clear, firm	clear, reasonably firm	weak and watery, small blood and meat spots present[c]
yolk	outline slightly defined, practically free from defects	outline fairly well defined, practically free from defects	outline plainly visible, enlarged and flattened, clearly visible germ development but no blood, other serious defects

[a]Ref. 8.
[b]Moderately stained areas permitted, ie, 1/32 of surface if localized, or 1/16 if scattered.
[c]If they are small, ie, aggregating not more than 3.2 mm in diameter.

dling light by a more distinct appearance and greater movement of the yolk. This can also be observed when the egg is broken out on a flat surface. Large percentages of the thick whites of the fresh egg cause the egg to stand high. Eggs of poorer quality spread out and the yolk flattens. Height of the egg when broken out is one way of expressing the quality factor determined by a formula which relates to the egg white height and the weight of the egg.

Eggs also lose quality by bacterial contamination, usually because of improper shell egg cleaning or washing procedures. When eggs are washed at a temperature less than the temperature of the egg, wash solution can be pulled in through the shell because of the contraction of the egg contents. Such eggs are found to be more susceptible to spoilage, particularly if the wash water contains iron. Iron overcomes the ability of conalbumen to inhibit bacterial growth, and wash water contaminated with iron may allow certain bacteria to grow, resulting in spoilage of the egg (15).

Processing. Methods for handling shell eggs are highly automated. This includes collecting, sorting, washing, sanitizing, drying, candling, and packing.

Almost all eggs are washed before they are packed except in countries where washing of eggs is prohibited by law, ie, The Netherlands. They then are dried to prevent bacterial contamination. Washing is usually done in a continuous system where eggs are conveyed through a washing chamber. Temperature of the wash solution should be 32°C or higher, or 11°C warmer than the eggs being washed. This assures that the wash solution is not drawn in through the shell. After washing, the eggs are rinsed and sanitized with either a chlorine or iodine solution.

Eggs are generally packed with the large end up, and are sprayed and coated with mineral oil at this end to retard escape of carbon dioxide and water, thus retaining quality for longer periods. Eggs are packed for either retail or wholesale (commercial) use. For retail use, the usual pack is one dozen eggs per carton. For commercial use, there are 30 dozen eggs per case.

Eggs that do not qualify as AA or A quality, but otherwise have good interior quality, are usually moved to an egg breaking operation for use in egg products. Sometimes eggs are transported directly to the breaking operation from the laying houses, and all eggs are used for the production of egg products.

Egg Products

Manufacturing.   The first step in making egg products is breaking and separating the whites from the yolks or breaking out the eggs as whole egg. Equipment for handling, washing, breaking, and separating the eggs is shown in Figure 2. An automatic loading device picks up eggs from filler flats and deposits them on a conveyor of an egg washer. The eggs that pass through the washer are simultaneously scrubbed by brushes and flushed by wash solution containing a cleaning compound; they are then rinsed and sanitized using a solution of either chlorine or iodine. Next, the eggs pass through a candling inspection area before being fed to the egg breaking/separating machine. This machine cracks and opens the shell, and deposits the contents into a device that separates the whites from the yolks. Single-tier, single-row machines are operated at speeds up to 32,000 eggs per hour, ie, 90 cases per hour or 9 eggs per second. The double-tier, single-row machine (Fig. 2) is rated to operate at twice this capacity. These speeds are made possible by using a yolk detection device to divert any egg white that has yolk in it onto the mix tray. The systems require three persons as operators. A loader presents the shell eggs to the automatic loader and at the same time inspects and removes any eggs that should not be broken on the machine; the inspector of the washed shell eggs removes any eggs that are not properly cleaned or have other obvious defects; and an egg breaking and separating machine operator inspects eggs that have been broken and separated, and controls the operation speed of the entire system.

Fig. 2.   Double-tier egg breaking and separating system. Egg loader, left; washer and inspection, center; and two-tier egg breaking and separating, right. Courtesy of Sanovo Engineering.

As indicated in Figure 1**b**, three components, ie, whites, yolks, and mix, flow away from the breaking machines to small inspection vats. After inspection, the liquids are pumped through filters or centrifuges, and then through cooling plates to a storage vat where they are held for further processing. The solids of yolk and whole egg (or mix) are usually standardized at this point by the addition of whites or yolks.

The egg products are finally processed and spray-dried. Sometimes liquid egg whites are concentrated before spray-drying by ultrafiltration (qv) or reverse osmosis procedures. Table 5 presents the effect of egg quality on the different egg product manufacturing processes.

Pasteurization. All egg products must be pasteurized to render them *Salmonella*-negative. Conventional plate-type pasteurizers having the usual attachments, including holding tubes, flow diversion valve, regeneration cycle, etc, are used (see STERILIZATION). Minimum pasteurization requirements are based on the bacterial kill obtained when heating whole egg to 60°C for a holding time of 3.5 min. Because of viscosity, flow of egg liquid is laminar, as opposed to turbulent, through the holding tubes. Because flow is laminar, the holding time of the fastest particle is only one-half that of the average holding time; the average holding time is 3.5 min, but holding time is actually 1.75 min for the fastest particles. The ultrapasteurization flow through the holding tubes is turbulent flow because of high velocities used.

Table 6 shows the minimum pasteurization conditions required by the USDA. It is necessary to pasteurize all egg products under these conditions, except for egg white which is to be dried.

There are three additional acceptable methods for pasteurizing egg white. In one method, a stabilized liquid egg white is pasteurized at 60°C for a minimum holding time of 3.5 min (17). The white is stabilized by adjusting pH to 7.0 using lactic acid, and adding aluminum sulfate which complexes with conalbumin to give greater heat stability. The adjusted pH stabilizes the other proteins of the egg white, eg, lysozyme, ovomucoid, ovomucin, and ovalbumin.

In another method, hydrogen peroxide can be added to the liquid egg white after it has been heated at 52°C for a holding time of 1 minute (18) to inactivate the natural catalase and to allow the hydrogen peroxide to react against bacteria. Holding time after addition of the hydrogen peroxide is 2.5 min at 50°C. The catalase is then introduced into the egg white to inactivate the hydrogen peroxide and disperse it. A final method involves heat treating dry egg white at greater than 54°C for a minimum holding time of 7 d. Combination liquid heat treatment with the heat treatment of dried egg product kills bacteria and gives greater assurance against the presence of *Salmonella* (19).

Pasteurizing of egg white could cause adverse effects on whipping properties. Whipping aids are presented as an optional ingredient in liquid and frozen egg white. For example, triethylcitrate [77-93-0], $C_{12}H_{20}O_7$, is sometimes added as a whipping aid, as well as a gum to increase viscosity and to improve stability of the egg white foam.

Liquid Egg Products. Liquid egg products include egg white; egg yolk; whole egg; extended shelf life refrigerated liquid egg products, ie, whites, yolks, whole; and concentrated sugared whole egg. These products, generally consumed by large users such as large bakeries who have the necessary handling equipment,

Table 5. Effect of Egg Quality on Products[a]

Property	Shell egg quality		
	Excellent	Average	Poor
Incoming			
shell eggs, kg/case	21.75	21.79	21.68
quality of shell			
checks, %	2.0	5.0	6.5
leakers, %	0.0	1.0	4.0
stucks, %	0.0	1.5	5.2
dirt, %	0.0	3.6	4.2
Outgoing			
edible liquid, kg/case			
egg whites	10.92	10.35	9.45
yolk std. to 43.5%[b]	7.27	6.86	5.93
mix std. to 24.5%[c]	0.0	0.79	1.72
total edible	*18.19*	*18.00*	*17.10*
inedible yield, kg/case			
shells	2.62	2.53	2.47
liquid and shell wringings	.91	1.06	1.74
measured losses, kg/case			
transfer room floor	0.0045	0.0181	0.0363
breaking room floor	0.0045	0.0045	0.0091
egg wash water	0.0091	0.1406	0.2177
breaking machine wash water	0.0091	0.0227	0.0544
Total	*21.75*	*21.79*	*21.64*
Unaccounted for losses			
machine set speed, case/h	90	80	74
effective machine speed, case/h	90	75.5	58.0
egg white solids, %	11.1	11.2	11.72
yolk contents of whites, %	0.008	0.026	0.046
mix solids,[d] %	24.5	24.5	24.5
yolk solids,[d] %	43.5	43.5	43.5

[a]From data supplied by Henningsen Foods, Inc.
[b]Yolk comes from breaking machines at 45% or greater, and this is standardized to 43.5% solids by adding mix and whites to the liquid. USDA standards do not allow going below 43.0% solids for yolk. Solids to which product is standardized depends on customer specifications.
[c]Mix comes off the machines at less than 24% solids and is standardized to 24.5% solids by adding yolk.
[d]After standardizing.

are usually transported by refrigerated tank truck holding approximately 20 t (see TRANSPORTATION). Liquid whole egg and yolk must be held below 5°C; egg white must be held below 7°C. Portable refrigerated vats that hold about 500 kg of product are also used. Bakeries and users must have adequate refrigeration facilities for holding liquid egg products in smaller containers for 30 days maximum.

Table 6. Minimum Pasteurization Requirements for Liquid Egg Products[a]

Product type	Temperature, °C	Holding time, min
albumen[b]	56.7	3.5
	55.5	6.2
whole egg	60	3.5
whole egg blends[c]	61.1	3.5
	60	6.2
fortified whole egg and blends[d]	62.2	3.5
	61.1	6.2
salt whole egg[e]	63.3	3.5
	62.2	6.2
sugar whole egg[f]	61.1	3.5
	60	6.2
plain yolk	61.1	3.5
	60	6.2
sugar yolk[g]	63.3	3.5
	62.2	6.2
salt yolk[h]	63.3	3.5
	62.2	6.2

[a]Ref. 16.
[b]No additives.
[c]Less than 2% added nonegg ingredients.
[d]24–38% egg solids, 2–12% added nonegg ingredients.
[e]2% or more salt added.
[f]2–12% sugar added.
[g]2% or more sugar added.
[h]2–12% salt added.

Liquid egg products must be of excellent microbiological quality with very low total bacteria counts. Pasteurization conditions are more severe than conventional methods for pasteurizing egg products, and aseptic packaging is usually necessary for the success of these products.

Newer liquid egg products are refrigerator shelf-stable for at least 30 days (19) and are aseptically packed in containers holding 13.62 kg of liquid egg product.

A room temperature shelf-stable sugared whole egg with a custard-like consistency has been developed. The concentration of whole egg is great enough, and the sugar level high enough, to give low water activity and good stability at room temperature.

Frozen Egg Products. Frozen egg products include egg white, plain whole egg, whole egg with yolk added (ie, fortified), plain egg yolk, fortified whole egg with corn syrup, sugared egg yolk, salted egg yolk, salted whole egg, and scrambled eggs and omelets. Egg products are frozen in a blast freezer at −40°C for up to 72 h, and then held for storage at −24°C (see REFRIGERATION AND REFRIGERANTS). They are used by large and small bakeries and for other uses.

Whole egg changes consistency during freezing. When thawed, it has a watery, separated appearance. After passing through a strainer or mixing in a vat or container, it appears to be uniform and smooth.

Gelation of egg yolk occurs below $-6°C$. When frozen egg yolk is thawed, it has a gel-like consistency and is difficult to handle, requiring special equipment; water is sometimes added in order to thin the thawed frozen yolk. Frozen yolk products have ingredients such as sugar or salt added to reduce gelation and improve ease of handling.

Dried Egg Products.   Dried egg products are listed as follows.

Egg white	Whole egg	Egg yolk
spray-dried egg white solids (whipping and nonwhipping)	standard whole egg solids	standard egg yolk solids
flake albumen (pan-dried)	stabilized (glucose-free) whole egg solids	stabilized (glucose-free) egg yolk solids
instant egg white with sugar	blends of whole egg with sugar	free-flowing egg yolk solids
	blends of whole egg with corn syrup	blends of egg yolk with sugar
	free-flowing whole egg solids	blends of egg yolk with corn syrup

Most dried egg products are made by spray-drying (qv), which produces a powder form. For almost all egg whites, and some whole egg and yolk products, the natural glucose is removed before spray-drying. This gives dried egg white products excellent stability under almost any storage conditions (20), and dried whole egg and yolk products good stability under room temperature as well as refrigerated storage conditions (21). Glucose causes the browning reaction to occur in dried egg products. The reducing group of the glucose reacts with the amino groups of the protein leading to browning, poor solubility, off-flavor, and off-odor developments.

Dried Egg White.   Glucose may be removed from egg white before drying by bacterial fermentation using a controlled bacterial culture, fermentation using baker's yeast, or oxidation of glucose to gluconic acid using a glucose oxidase/catalase enzyme system. For oxidation, the oxygen is supplied by addition of hydrogen peroxide; reaction rate is controlled by the amount of enzyme, temperature, and rate of hydrogen peroxide addition.

Whipping aids help to preserve the whipping properties of dried egg white. Sodium lauryl sulfate [151-21-3], is preferred, but other approved whipping aids include triethyl citrate, triacetin [102-76-1], and sodium desoxycholate [302-95-4]. These additives are effective at levels of less than 0.02% on a solids basis.

The yolk content of good quality egg white products is very low, less than 0.03% on a liquid basis (22). Fat content on a dry basis is then 0.06%.

Dispersibility of egg white powders into water is relatively poor. The powders tend to clump and form balls that are difficult to disperse. Instant dispersing product is made by mixing the whites with sugar and then agglomerating the particles (23).

Dried Whole Egg and Yolk.   Dried plain whole egg and yolk products are either dried as is, or have the glucose removed to improve stability and shelf life of the product. Glucose is removed before drying by use of glucose oxidase or by yeast fermentation (see YEASTS). Bacterial fermentation is not used because of off-flavor and off-odor development.

Dried whole egg and yolk products with glucose are less stable, and are usually held under refrigeration until used. The whole egg product has more glucose than the egg yolk and therefore is less stable.

Dried whole egg and yolk products should be spray-dried so that a minimum amount of heat is imparted to the product during drying. Sodium silicoaluminate [1344-00-9] or silicon dioxide [7631-86-9] is added at a level of less than 2.0% to give free-flowing and noncaking characteristics.

Dried blends of whole egg and yolk with carbohydrates have sucrose or corn syrup added to the liquids before spray-drying. Such carbohydrates (qv) preserve the whipping properties of whole egg and yolk by keeping the fat in an emulsified state. Corn syrup also gives anticaking characteristics, better flowability, and improved dispersibility in water. Dried blends of egg and carbohydrate function well in emulsified, as well as unemulsified, sponge cakes.

Specialty Dried Egg Products. A dried scrambled egg mix purchased for the U.S. military by USDA is a product having 51% whole egg, 30% skim milk, 15% vegetable oil, 2.5% salt, and 2.5% moisture.

Imitation whole egg having a low cholesterol content contains egg white as a base; nonfat milk and vegetable oil, substituting for egg yolk, are added to give a composition similar to whole egg. These are in frozen, liquid, or dried forms.

Low cholesterol egg products are formed by extraction of cholesterol from the egg. Attempts have been made to extract cholesterol by using hexane or by supercritical CO_2 extraction methods (24,25). A whole egg product in which 80% of the cholesterol is removed by a process using beta-cyclodextrin, a starch derivative, added to egg yolks has been introduced. The cyclodextrin binds up to 80% of the cholesterol, the mixture is centrifuged, and the liquid separated. The cholesterol-reduced yolk is then blended with egg white, pasteurized, and packed in asceptic containers to give a liquid whole egg product having a shelf life of 60 days under refrigeration (see FOOD PACKAGING).

Hard-cooked eggs are usually packed in acid solution, such as vinegar, which contains spice.

Specifications. Typical specifications are indicated in Table 7 for liquid, frozen, and dried egg products. Every lot of egg product is tested for moisture, pH, total bacteria count, coliform, yeast and mold, and *Salmonella*.

Special tests are run for certain customers. Such tests reflect how the product performs in a particular application.

Health and Safety Factors

The interior of shell eggs is mostly sterile at the time of lay. A few eggs may be contaminated inside the shell because of infection of the birds at the time the egg is being formed. However, contamination of the outside of the shell occurs after lay from fecal matter, nesting material, floor litter, dust, etc. Although shell eggs have several physical and chemical barriers that protect the contents from bac-

Table 7. Specifications for Egg Products[a]

| | Liquid or frozen | | | Solids | | | |
| | | | | | | Yolk | |
Specification	White	Yolk[b]	Whole	Whites, spray-dried	Whole, plain	Plain	Free[c]-flowing
moisture, %				<8.0	<5.0	<5.0	<3.0
total solids, %	>11.0	>43.0	>24.2[c]	>80.0	>45.0	>30.0	>30.0
crude protein, %	>10.0	>14.0	>12.0				
total lipids, %	<0.01%	>28.0	>10.5	<0.06	>40.0	>56.0	>56.0
pH	8.9 ± 0.3	6.2 ± 0.1	7.3 ± 0.3	7.0 ± 0.5	8.3 ± 0.3	6.4 ± 0.3	6.4 ± 0.3
carbohydrates[d], %				glu. free	SOP	SOP	SOP
total microbial count, gm[e]	<5,000	<5,000	<5,000	<10,000	<10,000	<10,000	<10,000
granulation[f]				USBS-60	USBS-16	USBS-16	USBS-16

[a]Ref. 8.
[b]Egg yolk contains approximately 17% egg white; natural egg yolk contains about 52% solids.
[c]Free-flowing products contain less than 2% sodium silicoaluminate.
[d]Most egg white solids are desugared. Whole egg and yolk products are desugared if specified on purchase (SOP).
[e]Includes 10 gm maximum each of yeast, mold, and coliform; all products must be *Salmonellae* negative.
[f]USBS-60 corresponds to 60-mesh (~0.25 mm) screen size; USBS-16 corresponds to 16-mesh (~1.19 mm) screen size.

terial contamination, eg, shell membrane and antibacterial factors in egg whites, eggs have been implicated in food poisoning outbreaks.

Salmonella enteriditis is the microorganism that has caused most of the problems in shell eggs. Evidence from the Southeastern Poultry and Egg Association (Decatur, Georgia) suggests that this bacteria is contained within the contents of the shell, a result of the bacteria colonizing within the bird. Only a very few birds are contaminated in this way. In 1990, the incidence of *S. enteriditis* was relatively low. One egg in 250,000 was found to be contaminated. Additionally only 7% of the producers had contaminated eggs. In the case of these producers, one egg in 15,000 was found to be positive. However, because one contaminated egg can contaminate the entire batch of eggs mixed with it, it is recommended that all eggs be cooked thoroughly before serving.

Egg products are relatively free of *Salmonella* because pasteurization and testing for *Salmonella* is required. Since July 1, 1971, the USDA has been responsible for mandatory inspection of all egg products in the United States. These regulations specify the minimum standards for sanitary conditions of plant facilities and equipment, pasteurizing conditions, etc. All egg products must be *Salmonella*-negative using a specified sampling and testing program.

Certain individuals are allergic to eggs and egg products. Most allergies occur as a result of the egg processing plants' transfer rooms where shell eggs are presented to the egg washing machine before the breaking machine. Dust from cases and incidental contact with egg material may cause problems, which are mostly respiratory. Respiratory problems can also occur as a result of the spray-drying of egg white; the finished product is fairly dusty, and can cause problems for those individuals who are allergic to egg white powder. When using egg white powder in large quantities, a dumping station with an exhaust system having a bag collector to facilitate the removal of dust from the air is recommended. Whole egg and yolk products are nondusty and usually do not cause a problem.

Economic Aspects

In the United States in 1990, a total of 230.7 eggs were consumed per person, compared to 272.5 in 1980; 44.0 (19.1%) of those 230.7 eggs were consumed as egg products in 1990 as compared to 35.1 (12.9%) eggs in 1980.

Following are figures, in metric tons, for the amount of liquid, frozen, and dried egg products produced in 1990 and 1991. Dried egg products are given as the liquid equivalent.

Product, t	1990	1991
liquid eggs	213,000	239,000
frozen eggs	184,000	182,000
dried eggs	201,000	227,000

An important aspect of economic consideration is the prevention of egg and egg product loss to the drain or the atmosphere, eg, a checked or cracked egg may be broken in the washer, and the contents go down the drain with the wash water.

Other measurable losses during egg product production are listed in Table 5.

Spray dryers may lose dried egg products out of the stack. Cyclone-type collectors, usually used as the secondary collector, are not properly designed to collect products efficiently. A well-designed cyclone collector recovers only 85–90% of spray-dried egg white, with 10–15% lost out of the stack, and 94–95% of egg products co-dried with carbohydrates. Bag-type collectors are needed for 100% recovery of both these products. Plain whole egg and plain egg yolk products are capable of being collected almost 100% with a cyclone collector, but a bag collector is advised (see AIR POLLUTION CONTROL METHODS).

BIBLIOGRAPHY

"Eggs" in *ECT* 1st ed., Vol. 5, pp. 465–477, by A. L. Romanoff, Cornell University; in *ECT* 2nd ed., Vol. 7, pp. 661–676, by W. W. Marion, Iowa State University; in *ECT* 3rd ed., Vol. 8, pp. 429–445, by D. H. Bergquist, Henningsen Foods, Inc.

1. R. W. Burley and D. V. Vadehra, *The Avian Egg Chemistry and Biology,* John Wiley & Sons, Inc., New York, 1989.
2. W. J. Stadelman and O. J. Cotterill, eds., *Egg Science and Technology,* 3rd ed., Haworth Press, Binghamton, N.Y., 1986.
3. R. H. Forsythe and D. H. Bergquist, *Poultry Sci.* **30,** 302 (1951).
4. USDA, *Egg Pasteurization Manual,* Western Utilization Research and Development Division, Agricultural Research Service, Albany, Calif., 1969.
5. R. E. Feeney in H. W. Schultz and A. F. Anglemeier, eds., *Symposium on Foods: Proteins and Their Reactions,* Haworth Press, Binghamton, N.Y., 1964.
6. W. D. Powrie in Ref. 2, pp. 97–139.
7. L. P. Posati and M. L. Orr, *Composition of Foods, Dairy and Egg Products, Raw-Processed-Dried, Agriculture Handbook No. 8-1,* USDA Research Service, Washington, D.C., Nov., 1976.
8. O. J. Cotterill and W. I. Stadelman, *A Scientist Speaks About Egg Products,* 2nd ed., American Egg Board, Park Ridge, Ill., 1990.
9. F. Cook and G. M. Briggs in Ref. 2, pp. 141–163.
10. *Food, Fat, and Health, Task Force Report No. 118,* Council for Agricultural Science and Technology, Ames, Iowa, 1991.
11. *Eggcyclopedia,* 2nd ed., American Egg Board, Park Ridge, Ill., 1989.
12. B. Lowe, *Experimental Cookery,* 4th ed., John Wiley & Sons, Inc., New York, 1955.
13. R. B. Chapin, *Some Factors Affecting the Emulsifying Properties of Hen's Egg,* Ph.D. dissertation, Iowa State College, Ames, Iowa, 1951.
14. *Regulations Governing the Grading of Shell Egg and United States Standards, Grades, and Weight Classes for Shell Eggs, USDA 7 CFR,* Part 56, U.S. Government Printing Office, Washington, D.C., May 1, 1991.
15. J. A. Garibaldi and H. G. Baynes, *Poultry Sci.* **39,** 1517 (1960).
16. *Regulations Governing the Inspection of Eggs and Egg Products, USDA 7 CFR,* Part 59, U.S. Government Printing Office, Washington, D.C., May 1, 1991.
17. F. E. Cunningham and H. Lineweaver, *Food Technol.* **19,** 1442 (1965).
18. U.S. Pat. 2,776,214 (Jan. 1, 1957), W. E. Lloyd and L. A. Harriman (to Armour and Co.).
19. U.S. Pat. 4,994,291 (Feb. 19, 1991), K. R. Swartzel and co-workers (to North Carolina State University).
20. R. W. Kline and G. F. Stewart, *Ind. Eng. Chem.* **40,** 919 (1948).

21. L. Kline and T. T. Sonoda, *Food Technol.* **5**, 90 (1951).
22. D. H. Bergquist and F. Wells, *Food Technol.* **10**, 48 (1956).
23. U.S. Pat. 4,115,592 (Sept. 19, 1978), D. H. Bergquist, F. E. Cunningham, and R. M. Eggleston (to Henningsen Foods, Inc.).
24. G. W. Froning and co-workers, *J. Food Sci.* **55**, 95–98 (1990).
25. M. W. Warren and co-workers, *Poultry Sci.* **70**, 1991–1997 (1991).

DWIGHT H. BERGQUIST
Henningsen Foods, Inc.

EINSTEINIUM. See ACTINIDES AND TRANSACTINIDES.

ELASTOMERS, SYNTHETIC

SURVEY

The purpose of this article is to provide a brief overview of the materials designated synthetic elastomers and the elastomeric or rubbery state. Subsequent entries describe the individual classes of elastomers in detail. Table 1 provides a fundamental description of the principal classes of synthetic elastomers. Table 2 gives the widely accepted ASTM abbreviations for synthetic rubbers.

Definition of Elastomers

The term elastomer is the modern word to describe a material that exhibits rubbery properties, ie, that can recover most of its original dimensions after extension

Table 1. Elastomers[a] and Their Characteristics

Name	CAS Registry Number	Chemical name	Repeat unit structure	Vulcanizing agent	Stretching crystallization	Gum tensile strength
natural rubber	[9006-04-6]	cis-1,4-polyisoprene (>99%)	b	sulfur	good	good
styrene–butadiene rubber	[9003-55-8]	poly(butadiene-co-styrene)	$-(CH_2-CH=CH-CH_2-)_m-(CH_2-CH-)_n-$ with C_6H_5	sulfur	poor	poor
butadiene rubber	[9003-17-0]	polybutadiene (>97% cis,-1,4)	$-(CH_2-CH=CH-CH_2-)-$	sulfur	poor to fair	poor to fair
isoprene rubber	[9003-31-0]	cis-1,4-polyisoprene (>97%)	$-(CH_2-C(CH_3)=CH-CH_2-)-$	sulfur	good	good
EP(D)M	c,d	poly(ethylene-co-propylene-co-diene)[c]	$-(CH_2-CH_2-)_m-(CH-CH_2-)_n-(\quad)_o$ with CH_3 and $CH=CH_2$	sulfur	poor	poor
butyl rubber	[9010-85-9]	poly(isobutyene-co-isoprene)	$-(CH_2-C(CH_3)_2-)_{50}-(CH_2-C(CH_3)=CH-CH_2-)-$	sulfur	good	good
nitrile rubber	[9003-18-3]	poly(butadiene-co-acrylonitrile)	$-(CH_2-CH=CH-CH_2-)_m-(CH_2-CH-)_n-$ with CN	sulfur	poor	poor

chloroprene rubber	[*9010-98-4*]	polychloroprene (mainly trans)	$-(CH_2-C(Cl)=CH-CH_2)-$	MgO or ZnO	good	good
silicones		polydialkylsiloxane (mainly polydimethyl-siloxane)	$-(Si(R)(R)-O)-$	peroxides	poor	poor
fluorocarbon elastomers		poly(vinylidene fluoride-*co*-hexafluoropropene)	$-(CH_2-CF_2)_x-(CF_2-CF(CF_3))_y-$	diamines	poor	poor
polysulfide rubber		poly(alkylene sulfide)	$-(CH_2-CH_2-S_{2-4})-$	metal oxides	fair	poor
polyurethanes		polyurethanes	$HO-(R-OCONHR'NHCOO)_x-R-OH$	diisocyanate	fair	good

[a]Not inclusive; see also ACRYLIC ELASTOMERS, PHOSPHAZENES, CHLOROSULFONATED POLYETHYLENE, ETHYLENE–ACRYLIC ELASTOMERS, POLYETHERS under the title ELASTOMERS, SYNTHETIC.
[b]See ISOPRENE RUBBER.
[c]*o* = zero for ethylene–propylene rubber; poly(ethylene-*co*-propylene) [*9010-79-1*].
[d][*25038-36-2*] when the diene is norbornene.

Table 2. ASTM Elastomer (Rubber) Designations

Abbreviation	Elastomer
ABR	acrylate–butadiene
BR	butadiene
CR	chloroprene
EPM	ethylene–propylene
EPDM	ethylene–propylene–diene[a]
IR	isoprene
IM	isobutylene
IIR	isobutylene–isoprene or butyl rubber
SBR	styrene–butadiene
SIR	styrene–isoprene
NBR	acrylonitrile–butadiene
NCR	acrylonitrile–chloroprene
PBR	vinylpyridine–butadiene
PSBR	vinylpyridine–styrene–butadiene
XNBR	carboxylic-acrylonitrile–butadiene
XSBR	carboxylic-styrene–butadiene
CIIR	chloroisobutylene–isoprene or chlorobutyl rubber
BIIR	bromoisobutylene–isoprene or bromobutyl rubber
MQ	dialkysiloxane
GPO	poly(propylene oxide)
AU	polyesterurethanes
EU	polyetherurethanes

[a]Nonconjugated diene; residual unsaturation from the diene is in the pendent group.

or compression. Ever since the pioneering work of Staudinger in the early 1900s (1), it has been accepted that such rubbery behavior results from the fact that the material is composed of a tangled mass of long-chain, flexible polymer molecules. When such a material is extended or stretched, the individual long-chain molecules are partially uncoiled, but will retract or coil up again when the force is removed because of the kinetic energy of the segments of the polymer chain. The flexibility of such polymer-chain molecules is actually the result of the ability of the atoms comprising the chain to rotate around the single bonds between them. Theories of rubberlike elasticity are well-developed (2).

The properties of elastomeric materials are also greatly influenced by the presence of strong interchain, ie, intermolecular, forces which can result in the formation of crystalline domains. Thus the elastomeric properties are those of an amorphous material having weak interchain interactions and hence no crystallization. At the other extreme of polymer properties are fiber-forming polymers, such as nylon, which when properly oriented lead to the formation of permanent, crystalline fibers. In between these two extremes is a whole range of polymers, from purely amorphous elastomers to partially crystalline plastics, such as polyethylene, polypropylene, polycarbonates, etc.

A most interesting class of materials is comprised of those amorphous elastomers that show the ability to undergo a temporary crystallization when stretched to a high extension, thus virtually becoming fibers, but that retract to their original dimension when the force is removed. Such crystallizing rubbers can thus demonstrate unusually high tensile strength in the stretched condition, but revert to the amorphous state when the force is relaxed because of relatively weak interchain forces.

Effect of Temperature on Polymer Properties

There are two principal forces that govern the ability of a polymer to crystallize: the interchain attractive forces, which are a function of the chain structure, and the countervailing kinetic energy of the chain segments, which is a function of the temperature. The fact that polymers consist of long-chain molecules also introduces a third parameter, ie, the imposition of a mechanical force, eg, stretching, which can also enhance interchain orientation and favor crystallization.

In addition to the phenomena of crystallization and melting, which both represent a change of state in the material, there is a third transition which plays a strong role in the behavior of polymers, although it is by no means absent in the behavior of simple liquids. This is the glass-transition temperature, T_g. The role of these transitions in the behavior of elastomers is shown in Figure 1, which depicts the volume–temperature relations during the cooling or heating of a sample of natural rubber. The volume is expressed in arbitrary units. This type of measurement is usually performed by noting the shrinkage or expansion of a fluid in a pycnometer containing a sample of rubber in a neutral liquid, which can be heated or cooled.

If a sample of natural rubber is first heated above room temperature, eg, to 30–40°C, and then slowly cooled, the volume shows a linear decrease until the temperature approaches about 20°C, at which point there is a sharp drop in volume. This represents a true change of state, ie, the onset of crystallization or the freezing/melting point of the rubber. On further cooling, the volume continues to

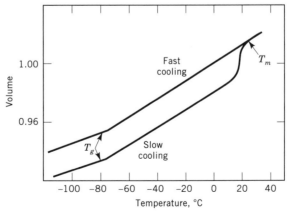

Fig. 1. Phase transition in Hevea rubber. T_g is glass-transition temperature; T_m is melting point.

show a linear decrease, but on a different line than that of the original. This continues until a temperature of $-72°C$ is reached, at which point there is no discontinuity, as in the case of the freezing/melting point, but a change in the slope of the volume–temperature line. This point does not represent a true change of state but a second-order transition known as the glass-transition temperature or "glass point" (3). Thus the freezing/melting point, T_m, of the rubber is the temperature at which crystallization or melting occurs, a true first-order transition, whereas the T_g of the rubber is the point at which the liquid state becomes a glassy state, ie, the molecular chain segments lose their ability to move around.

A further examination of Figure 1 shows that it is possible to avoid the freezing/melting point entirely by quenching the rubber, ie, cooling too fast for crystallization to take place, and shrinkage continues along the original volume–temperature line.

Figure 1 thus helps to define the general features of the physical state of polymers. Polymers fall into two classes, those that are capable of crystallization and those that are not. A noncrystalline (amorphous) polymer is considered a liquid (although a highly viscous one), which becomes a glass at reduced temperatures. Thus, atactic polystyrene is always amorphous, but it is in a "glassy" state at room temperature, since its T_g is about 105°C. Above its T_g it becomes rubbery although its chemical stability in air at that temperature is so poor as to render it useless. On the other hand, polyethylene and isotactic polypropylene are crystalline, to a greater or lesser extent, at room temperature, and hence do not exhibit rubbery behavior.

The class of rubbers that show the ability to crystallize when stretched represent a special class of rubbers. In the listing of the various principal elastomers in Table 1, this unique property is shown to increase the strength of rupture of these elastomers.

Compounding and Vulcanization of Elastomers

In order to "cure" or "vulcanize" an elastomer, ie, cross-link the macromolecular chains (Fig. 2), certain chemical ingredients are mixed or compounded with the rubber, depending on its nature (4,5). The mixing process depends on the type of elastomer: a high viscosity type, eg, natural rubber, requires powerful mixers (such as the Banbury type or rubber mills), while the more liquid polymers can be handled by ordinary rotary mixers, etc (see RUBBER COMPOUNDING).

The principal rubbers, eg, natural, SBR, or polybutadiene, being unsaturated hydrocarbons, are subjected to sulfur vulcanization, and this process requires certain ingredients in the rubber compound, besides the sulfur, eg, accelerator, zinc oxide, and stearic acid. Accelerators are catalysts that accelerate the cross-linking reaction so that reaction time drops from many hours to perhaps 20–30 min at about 130°C. There are a large number of such accelerators, mainly organic compounds, but the most popular are of the thiol or disulfide type. Zinc oxide is required to activate the accelerator by forming zinc salts. Stearic acid, or another fatty acid, helps to solubilize the zinc compounds.

In addition to the ingredients that play a role in the actual vulcanization process, there are other components that make up a typical rubber compound (see

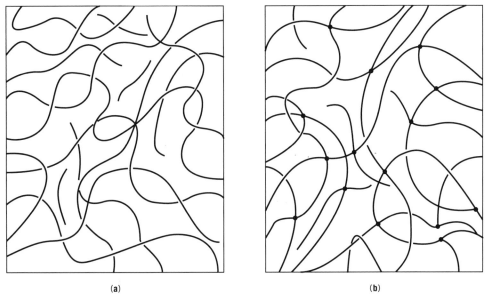

(a)   (b)

Fig. 2. Vulcanization of rubber macromolecules: (**a**), before cross-linking; (**b**), after cross-linking.

RUBBER CHEMICALS). Softeners and extenders, generally inexpensive petroleum oils, help in the mastication and mixing of the compound. Antioxidants (qv) are necessary because the unsaturated rubbers can degrade rapidly unless protected from atmospheric oxygen. They are generally organic compounds of the amine or phenol type (see also ANTIOZONANTS). Reinforcing fillers, eg, carbon black or silica, can help enormously in strengthening the rubber against rupture or abrasion. Nonreinforcing fillers, eg, clay or chalk, are used only as extenders and stiffeners to reduce cost.

General-Purpose Elastomers

The main thrust of the early attempts to create synthetic rubber was aimed at a general-purpose elastomer to displace natural rubber. Such attempts only became possible after Greville Williams, in 1860, showed that natural rubber could be pyrolyzed to yield isoprene (and its homologues) (6). Hence, there were many attempts during the last part of the nineteenth century to "convert" isoprene into a rubber. However, because this involved a polymerization reaction, and because the concept of polymers had not yet been developed, it is not surprising that such efforts were all unsuccessful. Among the more noteworthy work was that of Bouchardat who was able, in 1879, to produce a rubbery substance by heating isoprene, obtained from natural rubber, with hydrochloric acid (7). Tilden, in 1884, showed that the same results could be achieved with isoprene obtained from other sources than rubber, so that there was nothing unusual about the isoprene used by Bouchardat (8).

In the early part of the twentieth century, success began to crown some efforts at the synthesis of rubber. Thus Matthews and Strange, in 1910, showed that metallic sodium can react with isoprene to convert it to a rubbery substance, and Harries, in 1913, showed the same effect with butadiene and sodium (9). It was the latter work that ultimately resulted in the first commercial production of a general-purpose synthetic rubber, sodium polybutadiene. There had been a limited commercial production in Germany during World War I of "methyl rubber" prepared by the thermal polymerization of 2,3-dimethylbutadiene, but this rubber was of such poor quality and required such a difficult and unreliable process that it was hastily dropped at the conclusion of the war. Hence it was the polymerization of butadiene by means of sodium metal that became the first synthetic rubber process, giving rise to the Buna rubbers in Germany, and the SKF rubbers in the USSR during the 1920s.

At the same time, however, considerable research was being done, especially in Germany, on a novel process called emulsion polymerization, in which the monomer was polymerized as an emulsion in the presence of water and soap. This seemed advantageous since the product appeared as a latex, just like natural rubber, leading to low viscosity even at high solids content, while the presence of the water assured better temperature control. The final result, based mainly on work at the I.G. Farbenindustrie (IGF) (10), was the development of a butadiene–styrene copolymer prepared by emulsion polymerization, the forerunner of the present-day leading synthetic rubber, SBR.

The ESSO company had an agreement with IGF during the years prior to World War II for an interchange of information on petroleum products, including synthetic rubber, so that information on the production of SBR was available to the United States during the wartime emergency, which started with the seizure of Far East rubber plantations by Japan in 1941. Thus Germany continued to produce SBR emulsion and the USSR continued with sodium polybutadiene throughout the war, while the United States developed an "instant" synthetic rubber industry in 1942–1945 for the production of emulsion SBR.

Styrene–Butadiene Rubber (SBR).   This is the most important synthetic rubber and represents more than half of all synthetic rubber production (Table 3) (see STYRENE–BUTADIENE RUBBER). It is a copolymer of 1,3-butadiene, $CH_2=CH-CH=CH_2$, and styrene, $C_6H_5CH=CH_2$, and is a descendant of the original Buna S first produced in Germany during the 1930s. The polymerization is carried out in an emulsion system where a mixture of the two monomers is mixed with a soap solution containing the necessary catalysts (initiators). The final product is an emulsion of the copolymer, ie, a fluid latex (see LATEX TECHNOLOGY).

The original recipe adopted by the U.S. Government Synthetic Rubber Program was known as the "Mutual Recipe" and is shown in Table 4. As can be seen, the reaction temperature was set at 50°C, which resulted in 75% conversion to polymer in about 12 h. The reaction was then stopped by addition of a "shortstop," such as 0.1 parts hydroquinone, which destroyed any residual catalyst (persulfate), and generated quinone, which helped inhibit any further polymerization.

An antioxidant such as phenyl-β-naphthylamine, was then added to the latex (1.25 parts) prior to coagulation by salt–acid mixtures and drying of the rubber crumb, which was compressed into 34-kg bales. The SBR produced by this recipe

Table 3. World Synthetic Rubber Consumption[a,b], 10^3 t

Rubber type	1987	1988	1989[c]	1993[c]
SBR solid	2324	2382	2405	2616
SBR latex	241	245	250	265
carboxylated latex	984	1049	1088	1189
polybutadiene	1044	1098	1117	1215
ethylene–propylene	498	540	557	642
polychloroprene	249	251	251	258
nitrile, solid and latex	221	237	240	262
other synthetics[d]	958	1022	1040	1163
Total	*6519*	*6824*	*6948*	*7610*

[a]Ref. 11. Courtesy of the International Institute of Synthetic Rubber Producers.
[b]Excludes Communist countries.
[c]Estimated.
[d]Includes butyl and polyisoprene.

Table 4. The "Mutual Recipe" for Polymerization of SBR

Component	Parts by weight
water	180
butadiene	75
styrene	25
soap	5
potassium persulfate	0.3
n-dodecyl mercaptan	0.5
polymerization temperature	50°C
polymerization time	12 h for 70% conversion

had a Mooney viscosity of about 50 and a number-average molecular weight M_n of 100,000–200,000.

The thiol (n-dodecyl mercaptan) used in this recipe played a prominent role in the quality control of the product. Such thiols are known as chain-transfer agents and help control the molecular weight of the SBR by means of the following reaction where M = monomer, eg, butadiene or styrene; $R(M)_n$ = growing free-radical chain; k_p = propagation-rate constant; k_{tr} = transfer-rate constant; and k'_p = initiation-rate constant.

$$R(M)_{n+1} + C_{12}H_{25}SH \xrightarrow{k_{tr}} R(M)_{n+1}H + R'S$$

$$R'S + M \xrightarrow{k'_p} \xrightarrow{k_p} \xrightarrow{k_p} R'S(M)_n$$

Thus the thiol $C_{12}H_{25}SH$ is capable of terminating a growing chain and also initiating a new chain. If the initiation-rate constant, k'_p is not much slower than the propagation-rate constant, k_p, the net result is the growth of a new chain without any effect on the overall polymerization rate (retardation). That represents a true chain transfer, ie, no effect on the rate but a substantial decrease in molecular weight (12).

The use of these transfer agents, known as modifiers, in SBR polymerization has a profound effect in producing an acceptable synthetic rubber because without them the SBR emulsion polymerization system produces a very high molecular-weight polymer, which can easily become sufficiently cross-linked to result in gel, ie, a rubber insoluble in any solvent. A high gel rubber is also difficult to process with the normal mixing equipment, ie, rubber mills or Banbury mixers. Hence the use of thiol modifiers made it possible to produce processible elastomers. Thus, in the Mutual Recipe of Table 4, the reaction was stopped at about 70% conversion in 12 h to obtain a gel-free SBR, having a Mooney viscosity of about 50. Increasing the conversion above this point would run the risk of gel formation because of depletion of the thiol modifier.

The most significant post-war development in SBR production was the introduction of low temperature polymerization recipes, capable of being used at a temperature near the freezing point of water, ie, 5°C. This "cold rubber" could be produced gel-free even at a Mooney viscosity of 100. Hence the average molecular weight was substantially higher than that of the 50°C SBR, with a smaller fraction of low molecular-weight "tails" that might introduce network defects in the final vulcanizate. The fact that this 100 Mooney cold rubber was more difficult to process during compounding could be easily circumvented by adding up to 25% of an emulsion of petroleum oil to the latex before coagulation. This "oil-extended cold rubber" became the standard SBR for use in tire treads, etc, because of its superior abrasion resistance, resilience, and strength.

Processing and Properties.   Since SBR is an amorphous polymer (irregular chains), it does not exhibit crystallization either on stretching or cooling, and hence exhibits negligible strength unless it is reinforced with a fine-particle carbon black. It is compounded much like natural rubber, although it is not quite as easy to process and is also vulcanized with sulfur. It is not quite as tacky as natural rubber in the tire-building process and also shows a lower "green" strength of the unvulcanized compound.

However, the final properties of the tire show some points of superiority over natural rubber, ie, in higher abrasion resistance of cold SBR. Since the polymer has a T_g of -45°C compared with the -72°C of natural rubber, it shows poorer low temperature properties. Also, since the resilience of SBR is only about 50%, compared to at least 70% for the natural rubber, there is more heat build-up with SBR. In fact, although it is entirely possible to produce an all-synthetic automobile tire, this is not the case for truck tires because their greater mass leads to an unacceptable degree of heat build-up (13).

Polybutadiene.   The homopolymer polybutadiene (PB) is next in importance to SBR, as shown by its world production (Table 3). It is prepared by polymerization of butadiene in solution, using organometallic initiators, either of the Ziegler-Natta type or lithium compounds. It is, therefore, a result of the discovery of stereospecific polymerization during the 1950s and 1960s (see ELASTOMERS, SYNTHETIC–POLYBUTADIENE).

Molecular Structure.   The butadiene units in the chain are of three types: *cis*-and *trans*-1,4 ~CH_2—CH=CH—CH_2~ or 1,2 ~CH_2—CH~   1,2 units are atactic.

$$CH=CH_2.$$

Ziegler-Natta type catalysts can generate a very high cis-1,4 structure (>90%), which is the choice polymer for tires. It is made to specifications similar to SBR, ie, molecular weight average of 100,000–200,000, Mooney viscosity 50, and oil-extended. Lithium catalysts on the other hand yield variable chain structures, depending on the solvent used, ie, mixed structures of cis-1,4 and trans-1,4 and 1,2. These polymers are generally in the lower molecular-weight range and are not used for tire rubber. They are also important in the production of block copolymers for thermoplastic elastomers.

Processing and Properties.   Polybutadiene is compounded similarly to SBR and vulcanized with sulfur. The high cis-1,4 type crystallizes poorly on stretching so it is not suitable as a "gum" stock but requires carbon black reinforcement. It is generally used for automotive tires in mixtures with SBR and natural rubber. Its low T_g ($-95°C$) makes it an excellent choice for low temperature tire traction, and also leads to a high resilience (better than natural rubber) which in turn results in a lower heat build-up. Furthermore, the high *cis*-polybutadiene also has a high abrasion resistance, a plus for better tire tread wear.

Polyisoprene (Synthetic).   Polyisoprene has four possible chain unit geometric isomers: *cis*- and *trans*-1,4-polyisoprene, 1,2-vinyl, and 3,4-vinyl.

~CH_2—C=CH—CH_2~ CH_3 ~CH—CH_2~
          | ~CH_2—C~ |
        CH_3 | CH_3—C=CH_2
                              CH=CH_2

Natural rubber (Hevea) is 100% *cis*-1,4-polyisoprene, whereas another natural product, gutta-percha, a plastic, consists of the trans-1,4 isomer. Up until the mid-1900s, all attempts to polymerize isoprene led to polymers of mixed-chain structure.

The revolutionary development of stereospecific polymerization by the Ziegler-Natta catalysts also resulted in the accomplishment in the 1950s of a 100-year-old goal, the synthesis of *cis*-1,4-polyisoprene (natural rubber). This actually led to the immediate termination of the U.S. Government Synthetic Rubber Program in 1956 because the technical problem of duplicating the molecular structure of natural rubber was thereby solved, and also because the rubber plantations of the Far East were again available.

Polymerization and Molecular Structure.   As in the case of polybutadiene, the Ziegler-Natta type of initiators are used to produce a polymer of high cis-1,4 structure (~98–99%) (natural rubber is 100% cis-1,4). Lithium organometallics can also be used in hydrocarbon solvents to produce relatively high cis-1,4 chain structures but not quite as high as either natural rubber or the Ziegler-Natta polymer. In practice it is the latter polymer that is produced commercially, generally to a high molecular weight (1–2 million), with some proportion of actual gel polymer (see ELASTOMERS, SYNTHETIC–POLYISOPRENE).

Processing and Properties.   This polymer is compounded like natural rubber and has similar ease of processing. It is also vulcanized with sulfur. Like natural rubber, it can crystallize on stretching, leading to high gum strength. However, the slightly lower cis-1,4 content of the high cis polymer compared to natural rubber leads to lower strength. This slight deficiency also leads to lower "green strength" in tire building as well as lower "building tack."

Actually, production of synthetic polyisoprene is relatively small because of the sufficient and increasing supply of natural rubber. It is important, however, in ensuring that *cis*-1,4-polyisoprene (natural rubber), as a strategic material, is less subject to political uncertainties.

Age-Resistant Elastomers

Ethylene–Propylene (Diene) Rubber.   The age-resistant elastomers are based on polymer chains having a very low unsaturation, sufficient for sulfur vulcanization but low enough to reduce oxidative degradation. EPDM can be depicted by the following chain structure:

$$-(CH_2-CH_2)_m-(CH-CH_2)_n-(\underset{CH=CH_2}{\diagdown})_o$$
$$|$$
$$CH_3$$

where $m = {\sim}1500$ (${\sim}60$ mol %), $n = {\sim}975$ (${\sim}39$ mol %), and $o = 25$. The diene comonomer provides one double bond for the copolymerization and a residual double bond for sulfur vulcanization. Ethylene–propylene rubbers (EPR) are also commercial products (see ELASTOMERS, SYNTHETIC—ETHYLENE–PROPYLENE–DIENE RUBBER).

Processing and Properties.   Because of its irregular chain structure, EPDM is amorphous at <60 wt % ethylene and shows no crystallization on stretching (or cooling). Hence it exhibits poor strength and requires the assistance of carbon black reinforcement. At higher ethylene contents, the materials are semicrystalline and exhibit controlled green strength and crystallization on stretching. The presence of unsaturation in the side chain makes it possible to use sulfur vulcanization, but a higher proportion of accelerators must be used because of the low unsaturation. This also makes it difficult to use blends of EPDM with the highly unsaturated elastomers (NR, SBR, and PB), since the unsaturated rubber uses up the sulfur too rapidly. It exhibits poor building tack in tire manufacturing. However, it has excellent aging properties and shows no ozone cracking because of no in-chain low saturation. It also has good low temperature behavior ($T_g{\sim}$ $-60°C$) and resilience. Its excellent aging and low temperature properties make it ideal for use as a sheet rubber for roofing applications, where it has shown a rapid growth.

Butyl Rubber.   Butyl rubber was the first low unsaturation elastomer, and was developed in the United States before World War II by the Standard Oil Co. (now Exxon Chemical). It is a copolymer of isobutylene and isoprene, with just enough of the latter to provide cross-linking sites for sulfur vulcanization. Its molecular structure is depicted in Table 1.

The polymerization system is of the cationic type, using coinitiators such as $AlCl_3$ and water at very low temperatures ($-100°C$) and leading to an almost instantaneous polymerization (see ELASTOMERS, SYNTHETIC–BUTYL RUBBER).

Molecular Structure. The chain structure is as shown in Table 1 and molecular weights of 300,000–500,000 are achieved. The Mooney viscosities are in the range of 40–70 leading to a soft elastomer, which requires carbon black reinforcement for higher modulus.

Processing and Properties. Sulfur vulcanization is generally used with higher accelerator levels, as in the case of EPDM. Unlike EPDM, this elastomer has sufficient chain regularity to permit crystallization on stretching so it can exhibit high gum strength. However, its low modulus requires carbon black for stiffening, not for strength. Butyl rubber has a T_g of $-72°C$, but has a broad loss peak; thus it shows low resilience at room temperature, with a high hysteresis loss. It is therefore a useful damping rubber. Butyl rubber is remarkable for its excellent impermeability to gases, which led to its widespread use for inner tubes for tires, and more recently for the barrier in tubeless tires. It cannot be blended with high unsaturation rubbers in sulfur vulcanization. However, its halogenated derivatives, chlorobutyl and bromobutyl, can be covulcanized with the unsaturated elastomers, eg, SBR in tubeless tires, because of the interaction with the zinc oxide in the compound.

Solvent-Resistant Elastomers

Nitrile Rubber (NBR). This is the most solvent-resistant of the synthetic elastomers, except for Thiokol, which, however, has rather severe limitations. NBR was developed both in Germany and the United States by private industry prior to World War II. It is a copolymer of butadiene, $CH_2=CH—CH=CH_2$, and acrylonitrile, $CH_2=CHCN$, corresponding to the molecular structure shown in Table 1.

Molecular weights are in the same range as for SBR ($\sim$100,000), Mooney 20–90. Vulcanization is carried out with sulfur, as for SBR (see ELASTOMERS, SYNTHETIC–NITRILE RUBBER). It is the nitrile group, $—C\equiv N$, that confers oil resistance to this polymer, and the nitrile content can vary from 10 to 40%, leading to increasing solvent resistance.

Polymerization System. This elastomer is prepared by emulsion polymerization, similar to that used for SBR, but generally carried out to virtually 100% conversion. As for SBR, the chain irregularity leads to a noncrystallizing rubber, so that this polymer requires carbon black reinforcement for strength.

Processing and Properties. Nitrile rubber requires carbon black for strength. It exhibits excellent resistance to hydrocarbon solvents, but not as much to polar solvents. Because of its high T_g (strong interaction between nitrile groups) of -20 to $-40°C$, it has poor low temperature properties. Hence, for example, the desirability of a high nitrile content for solvent resistance, eg, gasoline hose, has to be balanced against low temperature stiffening.

Hydrogenated nitrile is a more recent development, in which the unsaturation of the polymer is markedly reduced leading to better aging.

Acrylic Elastomers. These materials are based principally on an acrylate chain structure, as follows:

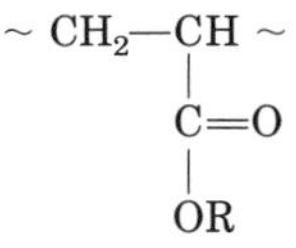

$$\sim CH_2-CH \sim$$
$$|$$
$$C=O$$
$$|$$
$$OR$$

where R is generally an alkyl group. Because of the absence of unsaturation, vulcanization is carried out by amine compounds instead of sulfur, but this absence of unsaturation also confers good aging properties. This rubber also shows good resistance to hydrocarbon solvents because of the polar acrylic ester groups present (see ELASTOMERS, SYNTHETIC—ACRYLIC ELASTOMERS; ETHYLENE–ACRYLIC ELASTOMERS).

Chloroprene Rubber.   Polychloroprene can be represented by the formula:

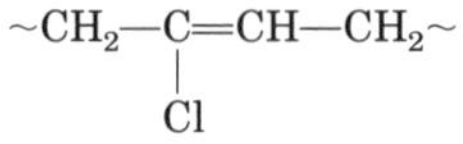

$$\sim CH_2-C=CH-CH_2\sim$$
$$|$$
$$Cl$$

The trans-1,4 configuration predominates but there are also small amounts of the three other isomeric structures, ie, cis-1,4, -1,2, and -3,4. It was one of the first synthetic rubbers produced in the United States as early as 1931, by the Du Pont Co., because of its good solvent resistance. Its Du Pont trade name is neoprene. Even though it is not as solvent resistant as nitrile rubber, it has many other advantageous properties (see ELASTOMERS, SYNTHETIC—POLYCHLOROPRENE).

Polymerization.   The polymer is prepared by emulsion polymerization of chloroprene. Since *trans*-1,4-polychloroprene is capable of crystallizing on stretching (or cooling) it can exhibit a high gum tensile strength without carbon black just like natural rubber. In fact, it is the only elastomer, other than natural rubber, to be used for latex dipped goods. Hence, just like natural rubber latex, neoprene latex is used in many applications, eg, dipping (rubber gloves), coating, and impregnating. It is generally produced as a homopolymer, since copolymerization is difficult to achieve because of the high reactivity of chloroprene. Some grades of neoprene contain a few percent of a comonomer in order to reduce the tendency to crystallize (harden) on storage. This tendency results from the fact that the melting point (T_m) of neoprene crystals ($\sim50°C$) is substantially higher than that of natural rubber ($\sim20°C$).

Vulcanization.   Some of the chlorine atoms along the chain (1,2 units) are very labile and reactive, and provide excellent sites for cross-linking. Hence neoprene is not vulcanized by sulfur but by metal oxides, eg, magnesium and zinc oxides, although sulfur is generally included in the compound to control the rate of vulcanization.

Processing and Properties.   Neoprene has a variety of uses, both in latex and dry rubber form. The uses of the latex for dipping and coating have already been indicated. The dry rubber can be handled in the usual equipment, ie, rubber mills and Banbury mixers, to prepare various compounds. In addition to its excellent solvent resistance, polychloroprene is also much more resistant to oxidation or ozone attack than natural rubber. It is also more resistant to chemicals and has the additional property of flame resistance from the chlorine atoms. It exhibits good resilience at room temperature, but has poor low temperature properties (crystallization). An interesting feature is its high density (1.23) resulting from the presence of chlorine in the chain; this increases the price on a volume basis.

Temperature-Resistant Elastomers

Silicone Rubber. These polymers are based on chains of silicon rather than carbon atoms, and owe their temperature properties to their unique structure. The most common types of silicone rubbers are specifically and almost exclusively the polysiloxanes. The Si–O–Si bonds can rotate much more freely than the C–C bond, or even the C–O bond, so the silicone chain is much more flexible and less affected by temperature (see SILICON COMPOUNDS, SILICONES).

Polymerization. The polymer chain is formed by a ring-opening reaction caused by the action of alkalies on the monomer, a cyclic siloxane:

$$\left(\!\!\begin{array}{c} CH_3 \\ | \\ Si-O \\ | \\ CH_3 \end{array}\!\!\right)_{\!\!4}$$

Molecular weights are about 500,000 but the polymer is still a very viscous liquid and requires a silica filler to become an elastomer.

Vulcanization. Generally this is carried out by the action of peroxides, which can cross-link the chains by abstracting hydrogen atoms from the methyl groups and allowing the resulting free radicals to couple into a cross-link. Some varieties of polysiloxanes contain some vinylmethylsiloxane units, which permit sulfur vulcanization at the double bonds. Some liquid (short-chain) silicones can form networks at room temperature by interaction between their active end groups.

Processing and Properties. Silicones are soft, weak rubbers, even with silica fillers (carbon black does not work well), leading to tensile strength of about 5–8 MPa (700–1200 psi). However, they offer excellent resistance to stiffening at very low temperatures (dry ice), as well as softening at elevated temperatures, thus retaining their properties because of the great flexibility of the polymer chains. However, they cannot be considered high temperature stable elastomers because they can degrade (3 months at 250°C).

Fluorocarbon Elastomers. These elastomers were developed by both the Du Pont and 3M companies during the 1950s. They are the most resistant elastomers to heat, chemicals, and solvents known, but they are also the most expensive, ie, between \$22 and \$35 per kg. The most common types are copolymers of vinylidene fluoride and hexafluoropropene, thus:

$$\left(\!CH_2-CF_2\!\right)_{\!x}\!\left(\!\begin{array}{c} CF_2-CF \\ | \\ CF_3 \end{array}\!\right)_{\!y} \sim$$

The fluorine atom confers chemical inertness, but some hydrogen atoms must be in the chain to maintain rubbery properties. Some fluorinated silicones are also available where superior low temperature properties are required (see ELASTO-MERS, SYNTHETIC—FLUOROCARBON ELASTOMERS).

Polymerization.   Emulsion polymerization is used, but the latex is too un-stable for use and all the latex is coagulated to dry rubber. The molecular weight range is 100,000–200,000 with a Mooney viscosity of 50–70.

Vulcanization.   Diamines are used to form cross-links by reacting with the fluorine atoms. Peroxides are also added to assist in the cross-linking and carbon black is used for reinforcement.

Processing and Properties.   These elastomers are generally designed for high temperature use, with mechanical properties as a secondary consideration. Thus tensile strengths of only 12–15 MPa (1700–2200 psi) are common at ambient temperatures, but they change very little after exposure to high temperatures. Thus these elastomers have an indefinite life at 200°C and can be heated to 315°C up to 48 h. However, they have poor low temperature properties, reaching a brittle point at -30°C, compared to nitrile rubber at -40°C.

Liquid Rubber Technology

An entirely new concept was introduced into rubber technology with the idea of "castable" elastomers, ie, the use of liquid, low molecular-weight polymers that could be linked together (chain-extended) and cross-linked into rubbery networks. This was an appealing idea because it avoided the use of heavy machinery to masticate and mix a high viscosity rubber prior to molding and vulcanization. In this development three types of polymers have played a dominant role, ie, poly-urethanes, polysulfides, and thermoplastic elastomers.

Polyurethanes.   The polyurethanes were developed in Germany and Great Britain just prior to World War II, largely because of the work of Otto Bayer in Germany (14). This was based on the reaction of liquid, low molecular-weight polyesters or polyethers, having hydroxyl end groups, with organic diisocyanates to yield solid elastomers, as follows, where HO–R–OH is a low molecular-weight polyester or polyether and OCN–R–NCO is an organic diisocyanate.

$$\text{HO–R–OH + OCN–R'–NCO} \longrightarrow \text{HO} \underset{x}{(\text{R–O}\overset{\overset{\text{O}}{\|}}{\text{C}}\text{NHR'NH}\overset{\overset{\text{O}}{\|}}{\text{C}}\text{O})} \text{R–OH}$$

This process is based on the very high reactivity of the isocyanate group toward hydrogen present in hydroxyl groups, amines, water, etc, so that the chain exten-sion reaction can proceed to 90% yield or better. Thus when a linear polymer is formed by chain extension of a polyester or polyether of molecular weight 1000–3000, the final polyurethane may have a molecular weight of 100,000 or higher (see URETHANE POLYMERS).

In addition to linear chain extension, excess diisocyanate leads to cross-link-ing into a network because the diisocyanate groups can also react with the hy-drogen atoms of the –NH– groups in the chains. Furthermore, the well-known polyurethane foam rubber can be made by adding water to the mixture because the isocyanate groups react vigorously with water to liberate carbon dioxide gas as follows:

$$-NCO + H_2O \longrightarrow -NH_2 + CO_2$$

The amine groups thus formed can also react vigorously with the isocyanate groups to continue the chain extension and cross-linking reactions. Hence, in the systems there are simultaneous foaming, polymerization, and cross-linking reactions, which produce foam elastomers (or plastics).

Properties.   Polyurethane elastomers generally exhibit good resilience and low temperature properties, excellent abrasion resistance, moderate solvent resistance, and poor hydrolytic stability and poor high temperature resistance. As castable rubber, polyurethanes enjoy a variety of uses, eg, footwear, toys, solid tires, and foam rubber.

Polysulfides.   The polysulfide elastomer, best known under the trade name Thiokol, represents the earliest commercially developed synthetic rubber, developed in 1930 by J. C. Patrick as a highly solvent and age-resistant elastomer (15). It is still considered the most solvent-resistant rubber, but its poor mechanical properties provide a serious disadvantage (see POLYMERS CONTAINING SULFUR).

The polymerization involves the reaction of sodium polysulfide with ethylene dichloride in aqueous media at 70°C for 2–6 h, yielding an aqueous dispersion (latex) of the polysulfide rubber.

$$NaS_{(2-4)}Na + ClCH_2CH_2Cl \longrightarrow +CH_2-CH_2-S_{(2-4)}\!\!\rightarrow_n$$

It is now generally used in the form of a liquid, low molecular-weight polymer, having reactive end groups obtained by degradation of the base polymer, as follows:

$$+CH_2-CH_2-S_{(2-4)}\!\!\rightarrow_n \xrightarrow{\text{NaSH or NaHSO}_3} H+S-CH_2-CH_2-S\!\!\rightarrow_n H$$

$$\text{liquid low mol wt polymer}$$

The liquid polymer is then compounded with metal oxides or peroxides, as well as fillers (carbon black) and can undergo cold vulcanization, ie, chain extension and cross-linking into a solid matrix. It is largely used as a sealant and gasket material for windows, automobile windshields, etc.

Thermoplastic Elastomers.   These represent a whole class of synthetic elastomers, developed since the 1960s, that are permanently and reversibly thermoplastic, but behave as cross-linked networks at ambient temperature. One of the first was the triblock copolymer of the polystyrene–polybutadiene–polystyrene type (Shell's Kraton) prepared by anionic polymerization with organolithium initiator. The structure and morphology is shown schematically in Figure 3. The incompatibility of the polystyrene and polybutadiene blocks leads to a dispersion of the spherical polystyrene domains (ca 20–30 nm) in the rubbery matrix of polybutadiene. Since each polybutadiene chain is anchored at both ends to a polystyrene domain, a network results. However, at elevated temperatures where the polystyrene softens, the elastomer can be molded like any thermoplastic, yet behaves much like a vulcanized rubber on cooling (see ELASTOMERS, SYNTHETIC–THERMOPLASTIC ELASTOMERS).

The particular type of thermoplastic elastomer (TPE) shown in Figure 3 exhibits excellent tensile strength of 20 MPa (2900 psi) and elongation at break

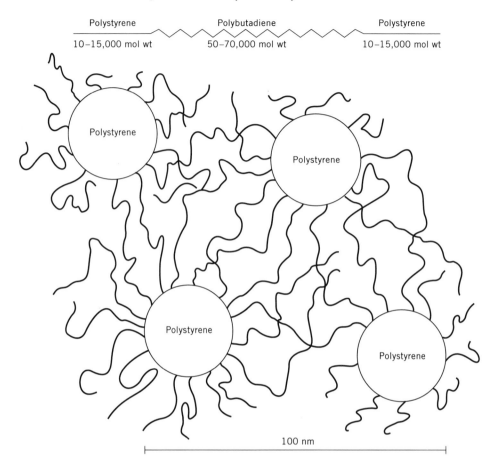

Fig. 3.　Morphology of thermoplastic elastomers.

of 800–900%, but high compression set because of distortion of the polystyrene domains under stress. These TPEs are generally transparent because of the small size of the polystyrene domains, but can be colored or pigmented with various fillers. As expected, this type of thermoplastic elastomer is not suitable for use at elevated temperatures (>60°C) or in a solvent environment. Since the advent of these styrenic thermoplastic elastomers, there has been a rapid development of TPEs based on other molecular structures, with a view to extending their use to more severe temperature and solvent environments.

The styrenic triblock copolymers shown in Figure 3 are prepared by anionic "living" polymerization with organolithium initiator, where excellent control can be exercised over the molecular weight of each block or segment in the polymer chain. These methods can only be used with the styrene and diene-type monomers. In order to prepare thermoplastic elastomers based on other molecular structures, suitable chemical reactions must be used for each particular case to obtain alternating blocks of hard and soft segments which can generate the type of morphology shown in Figure 3.

BIBLIOGRAPHY

"Rubber, Synthetic" in *ECT* 1st ed., Vol. 11, pp. 827–852, by J. D. D'Ianni, the Goodyear Tire & Rubber Co.; "Elastomers, Synthetic" in *ECT* 2nd ed., Vol. 7, pp. 676–705, by W. M. Saltman, The Goodyear Tire & Rubber Co.; "Elastomers, Synthetic (Survey)" in *ECT* 3rd ed., Vol. 8, pp. 446–459, by J. E. McGrath, Virginia Polytechnic Institute and State University.

1. H. Staudinger, *Ber.* **53,** 1073 (1920).
2. J. P. Queslel and J. E. Mark, in J. I. Kroschwitz, ed., *Encyclopedia of Polymer Science and Engineering*, 2nd ed., Vol. 5, John Wiley & Sons, Inc., New York, 1986, pp. 365–408.
3. R-J. Roe, in Ref. 2, Vol. 7, pp. 531–544.
4. J. K. Gillham, in Ref. 2, Vol. 4, pp. 519–524.
5. A. Y. Coran, in Ref. 2, Vol. 17, pp. 666–698.
6. G. Williams, *Proc. Roy. Soc.* **10,** 516 (1860).
7. G. Bouchardat, *Compt. Rend.* **89,** 1117 (1879).
8. W. A. Tilden, *J. Chem. Soc.* **45,** 411 (1884).
9. Brit. Pat. 24, 790 (Oct. 25, 1910), F. E. Matthews and E. H. Strange; C. Harries, *Ann.* **395,** 211 (1913).
10. Ger. Pat. 511,145 (1927), W. Bock and E. Tschunkur (to I. G. Farbenindustrie A.G.); Ger. Pat. 570,980 (Feb. 27, 1933), W. Bock and E. Tschunkur (to I. G. Farbenindustrie A.G.).
11. *Chem. Eng.*, (Apr. 17, 1989).
12. In Ref. 2, Vol. 3, pp. 288–290.
13. R. S. Bhakuni and co-workers, in Ref. 2, Vol. 16, pp. 834–861.
14. O. Bayer, E. Mueller, S. Peterson, H. F. Piepenbrink, and E. Windemuth, *Angew. Chem.* **62,** 57 (1960).
15. U.S. Pat. 1,890,191 (1932), C. J. Patrick.

MAURICE MORTON
The University of Akron

ACRYLIC ELASTOMERS

Acrylic elastomers have the ASTM designation ACM (1) for polymers of ethyl acrylate and other acrylates. Conventionally, the M indicates a polymer having a saturated chain of the polymethylene type. The repeat structure of this polymer family shows both the presence of a saturated backbone, which is responsible for the high heat and oxidation resistance, and the ester side groups, which contribute to the marked polarity of this elastomer chain. These two main properties are not present in general-purpose rubbers. As a result, ACMs are designated as specialty elastomers, and are extensively used in highly demanding applications such as automotive underhood components.

$$-(CH_2-CH)_n$$
$$|$$
$$C=O$$
$$|$$
$$OR$$

The earliest study describing vulcanized polymers of esters of acrylic acid was carried out in Germany by Rohm (2) before World War I. The first commercial acrylic elastomers were produced in the United States in the 1940s (3–5). They were homopolymers and copolymers of ethyl acrylate and other alkyl acrylates, with a preference for poly(ethyl acrylate) [9003-32-1], due to its superior balance of properties. The main drawback of these products was the vulcanization. The fully saturated chemical structure of the polymeric backbone in fact is inactive toward the classical accelerators and curing systems. As a consequence they required the use of aggressive and not versatile compounds such as strong bases, eg, sodium metasilicate pentahydrate. To overcome this limitation, monomers containing a reactive moiety were incorporated in the polymer backbone by copolymerization with the usual alkyl acrylates.

Manufacture

Acrylate esters can be polymerized in a variety of ways. Among these is ionic polymerization, which although possible (6–9), has not found industrial application, and practically all commercial acrylic elastomers are produced by free-radical polymerization. Of the four methods available, ie, bulk, solution, suspension, and emulsion polymerization, only aqueous suspension and emulsion polymerization are used to produce the ACMs present in the market. Bulk polymerization of acrylate monomers is hazardous because it does not allow efficient heat exchange, required by the extremely exothermic reaction.

Solution Polymerization.   This method is not commercially important, although it is convenient and practical, because it provides viscous cements that are difficult to handle. Also, the choice of the solvent is a key parameter due to the high solvent chain-transfer constants for acrylates.

Suspension Polymerization.   This method (10) might be considered as a number of bulk polymerizations carried out simultaneously in the monomer droplets with water acting as a heat-transfer medium. A monomer-soluble initiator, eg, a peroxide or azo compound, and a protective colloid like poly(vinyl alcohol) or bentonite, are required. After completion of the polymerization, the excess of monomer(s) is steam stripped, and the beads of polymer are collected and washed on a centrifuge or filter and dried on a vibrating screen or by means of an expeller–extruder.

Emulsion Polymerization.   In this method, polymerization is initiated by a water-soluble catalyst, eg, a persulfate or a redox system, within the micelles formed by an emulsifying agent (11). The choice of the emulsifier is important because acrylates are readily hydrolyzed under basic conditions (11). As a consequence, the commonly used salts of fatty acids (soaps) are preferably substituted by salts of long-chain sulfonic acids, since they operate well under neutral and acid conditions (12). After polymerization is complete the excess monomer is steam-stripped, and the polymer is coagulated with a salt solution; the crumbs are washed, dried, and finally baled.

Residual monomers exhibit a characteristic sharp odor even in subtoxic concentration, due to the very low olfactory threshold. Modern requirements in terms of environmental safeguard have led to significant improvements in the control of polymerization effluents, driving off gases, and residual monomer in the raw polymer. Consequently, the acrylic elastomers of the 1990s are practically odor-free, and represent a significant improvement over the products of the past.

Monomers

Two kinds of monomers are present in acrylic elastomers: backbone monomers and cure-site monomers. Backbone monomers are acrylic esters that constitute the majority of the polymer chain (up to 99%), and determine the physical and chemical properties of the polymer and the performance of the vulcanizates. Cure-site monomers simultaneously present a double bond available for polymerization with acrylates and a moiety reactive with specific compounds in order to facilitate the vulcanization process.

Backbone Monomers. The most important monomer both from a historical and commercial point of view is ethyl acrylate [140-88-5]. Substantially, homo-polymers of ethyl acrylate with little amounts of proper cure-site monomer(s) con-stitute the base grade of the ACMs present on the market (13). These grades offer excellent oil resistance and a very satisfactory thermal resistance, but limited low temperature behavior, because the glass-transition temperature (T_g) is about $-15°C$ (14). Alkyl acrylates with longer side chains exhibit lower T_g (15). On the other hand, the longer alkyl group reduces oil resistance (16) so that a proper balance has to be found. Experimental and theoretical considerations based on the Group Contribution Theory (GCT) (17) are useful to predict both the low tem-perature flexibility, expressed as T_g, and the most important oil resistance related parameter δ as a function of the side-group chain length of poly(n-alkyl acrylates) (18). Figure 1, which correlates the predicted glass-transition temperatures (T_g) and the solubility parameters (δ) indicates that apart from poly(ethyl acrylate) and poly(propyl acrylate) [24979-82-6] no other poly(alkyl acrylate) reaches a sat-isfactory balance of property requirements. Thus butyl acrylate [141-32-2] is only used as a comonomer, usually with ethyl acrylate.

To improve the performance of acrylic elastomers, side chains are required where the δ value is higher than with n-alkyl groups. Thus the use of polar groups, for instance heteroatoms, is suggested. The general formula for these acrylate monomers may be portrayed as follows:

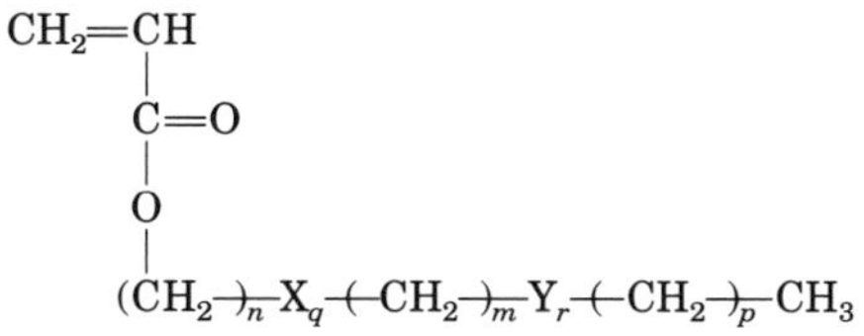

$$CH_2{=}CH$$
$$|$$
$$C{=}O$$
$$|$$
$$O$$
$$|$$
$$(CH_2)_n X_q (CH_2)_m Y_r (CH_2)_p CH_3$$

If $q = 1$, r and $p = 0$ and X = oxygen, the family of alkoxyalkyl acrylates is obtained. The improvement of the solubility parameter for this family compared to poly(n-alkyl acrylates) is shown in Figure 2.

The most important member of this class is 2-methoxyethyl acrylate (MEA) [3121-61-7] (16), which along with ethyl acrylate and butyl acrylate constitutes the building blocks of current acrylic elastomers.

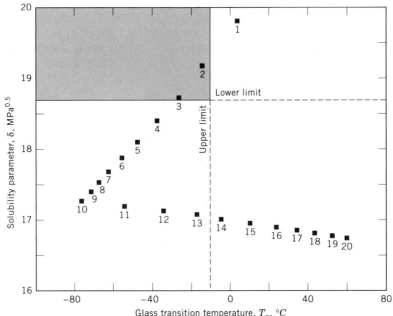

Fig. 1. T_g vs δ for poly(alkyl acrylates). ■ = number of C atoms in alkyl side chain. To convert $MPa^{0.5}$ to $(cal/cm^3)^{0.5}$, divide by 2.05.

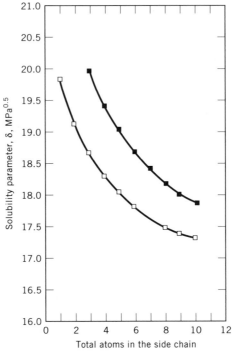

Fig. 2. Improvement in δ achieved by alkoxy substituents. □ = only carbon atoms; ■ = with one oxygen atom. To convert $MPa^{0.5}$ to $(cal/cm^3)^{0.5}$, divide by 2.05.

Cure-Site Monomers. A large variety of cure-site monomers has been proposed, but only a few have achieved commercial significance. Two of the most important classes are labile chlorine containing monomers and epoxy/carboxyl containing monomers.

Labile Chlorine Containing Monomers. Chlorine is introduced in the acrylic elastomer chain by analogy to polychloroprene (19). The monomers are characterized by the simultaneous presence of a double bond available for polymerization with acrylates and a chlorine atom ready to react easily during the vulcanization step. The general formula is as follows where R is a group that might enhance the reactivity of the double bond and/or of the vicinal chlorine atom.

$$CH_2{=}CH{-}(R){-}\overset{|}{\underset{|}{C}}{-}Cl$$

The historical evolution of R producing ever more reactive chlorine and/or double bonds is summarized in the following by way of examples of representative monomers.

Year	R	Monomer	CAS Registry Number	Reference
1950	—O—CH$_2$—	2-chloroethyl vinyl ether	[*110-75-8*]	19
1960	—O—C— (‖ O)	vinyl chloroacetate	[*2549-51-1*]	20
1970	 —⟨O⟩—	*p*-vinylbenzyl chloride	[*30030-25-2*]	21

In order to enhance the reactivity of the chlorine atom, a second reactive monomer can be adopted giving dual cure sites. According to the literature, the second monomer can contain carboxyl (22–24), cyanoalkyl (25), hydroxypropyl (26), or epoxy groups (27,28).

Epoxy/Carboxy Cure Sites. Epoxy/carboxy cure sites probably represent the most important alternative to labile chlorine containing monomers. There has been increasing interest in them due to the discovery of the highly efficient quaternary ammonium salt-based accelerators (29–34). The reaction between the epoxy ring and carboxylic acid can happen in the following three ways:

Reactive sites in the polymer	Curing system
epoxy containing reactive monomer	poly(carboxylic acid) + catalyst
carboxy containing reactive monomer	polyepoxide compound + catalyst
carboxy containing reactive monomer + epoxy containing reactive monomer	catalyst

Double-Bond Cure Sites.   The effectiveness of this kind of reactive site is obvious. It allows vulcanization with conventional organic accelerators and sulfur-based curing systems, besides vulcanization by peroxides. Fast and controllable vulcanizations are expected so double-bond cure sites represent a chance to avoid post-curing. Furthermore, blending with other diene elastomers, such as nitrile rubber [9003-18-3], is greatly facilitated.

The idea of using polyunsaturated monomers is rooted in the early history of acrylic elastomers. The first monomers used were butadiene [106-99-0] (35), isoprene [78-79-5] (36), and allyl maleate [999-21-3] (37), but they did not find commercial success because during polymerization large portions of polymer were cross-linked. Other monomers have been proposed more recently: tetrahydrobenzyl acrylate (38), dicyclopentenyl acrylate [2542-30-2] (39), and 5-ethylidene-2-norbornene [16219-75-3] (40). The market potential, at least for the more recent ones, is still to be determined.

Structure–Property Relationships

The modern approach to the development of new elastomers is to satisfy specific application requirements. Acrylic elastomers are very powerful in this respect, because they can be tailor-made to meet certain performance requirements. Even though the structure–property studies are proprietary knowledge of each acrylic elastomer manufacturer, some significant information can be found in the literature (18,41). Figure 3a shows the predicted T_g, according to GCT, and the volume swell in reference fluid, ASTM No. 3 oil (42), related to each monomer composition. Figure 3b shows thermal aging resistance of acrylic elastomers as a function of backbone monomer composition.

The cure sites present in the polymer also significantly influence the expected properties. Labile chlorine cure sites generally give good elongation at break retention after heat aging, and show low sensitivity to acidic compounds in the mix recipe. Epoxy/carboxyl cure sites give good compression set and good hydrolysis resistance. Dual cure site-based polymers vulcanized by quaternary ammonium salts may behave in different ways (43,44) compared to conventional cure sites. These considerations and Figure 3a and 3b taken together offer an example of a molecular engineering approach to the design of acrylic elastomers tailor-made with specific requirements in mind.

Applications

Specialty rubbers, to which ACM belongs, are selected on the basis of their performance. As a consequence, many elastomeric families may compete for the same application. A first selection can be carried out on the basis of oil and temperature resistance, according to ASTM D2000 (Fig. 4).

Selection of the right elastomer has to take into account not only the low temperature resistance, the elastic properties, and mechanical properties, but also price, which can vary widely in specialty elastomers.

On these bases ACMs have found commercial applications mainly in automotive nontire applications. Some significant examples are as seals: lip, pinion,

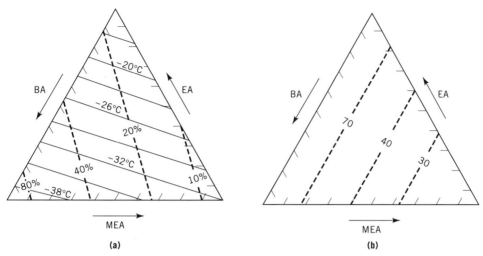

Fig. 3. Elastomer properties as a function of monomer composition, butyl acrylate (BA), ethyl acrylate (EA), and methoxyethyl acrylate (MEA). (**a**), (—) glass-transition temperature; (– – –) swelling in ASTM No. 3 oil; (**b**), (– – –) residual elongation at break, %, after heat aging.

shaft, transmission, clutch, O-rings, bearing dust seals; as gaskets: oil pan, emission control, cover valve, cork-bound, sponge gaskets; in electrical insulation (ignition cable, spark plug boots); valve stem oil deflectors; hose and hose linings; miscellaneous mechanical goods; wire and cable jacketing; noise control and vibration damping; and plastic modification.

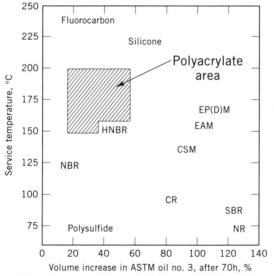

Fig. 4. Oil and temperature resistance of various elastomers.

Compounding

A rubber compound typically consists of many additives and ingredients, although in the case of acrylic rubber compound, the recipes are usually simpler. General recipes include the following:

Ingredient	phr
polymer	100
stearic acid	1–3
processing aid	1–3
antioxidant	1–2
filler (black and/or white)	20–100
plasticizer	0–5
cure system	0.5–8

The compound recipe is a function both of cure site(s) present in the polymer and of application requirements.

As a general rule zinc oxide is not used because it is not involved in the vulcanization mechanism of acrylic rubber. Zinc salts, eg, zinc stearate, have to be avoided because they may adversely affect the performance of the compound and/or the vulcanizate.

Plasticizers (qv) are usually present at lower concentrations compared to general-purpose rubber-based compounds, because of their volatility at typical ACM service temperatures and/or their partial extractability by the aggressive fluids where acrylic elastomers are employed. Other additives are therefore required to improve processibility. These processing aids act as lubricating agents and enhance the release characteristics of the acrylic compound and/or reduce compound viscosity.

Antioxidants (qv) usually provide only a marginal improvement in heat aging resistance because of the inherent oxidation resistance of the saturated backbone of ACM. The antioxidants most frequently used are nonvolatile amine compounds.

Carbon blacks are the preferred reinforcing fillers because they provide a better overall balance of vulcanizate properties than the mineral white fillers. As with other elastomers, smaller particle size blacks provide the highest reinforcement, whereas higher structure types contribute to processing characteristics. As a general rule, relatively low amounts of the ASTM N300 series of blacks are suggested for injection molding applications, and higher amounts of medium large particle size blacks (N500 and N700) are preferred in extrusion (see CARBON, CARBON BLACK).

Where the application requires effective electrical insulation or color coding, white filler reinforcing agents have to be used. High reinforcing silica tends to give compounds that harden during storage. Aluminum silicate and silicoaluminate are usually preferred. To improve filler–polymer interaction and vulcanization behavior silane-modified fillers are often employed.

Graphites (natural or synthetic) are black semireinforcing fillers used almost exclusively in rotary shaft seal applications where surface lubricity and abrasion resistance must be enhanced (see CARBON, NATURAL GRAPHITE).

Vulcanization Systems

Because of the different vulcanization chemistry involved in each commercial ACM, a vulcanization system specific to the cure site present has to be adopted. Many cure systems for labile chlorine containing ACM have been proposed (45). Among these the alkali metal carboxylate–sulfur cure system, or soap–sulfur as it is called in the United States, became the mainstay of acrylic elastomer technology in the early 1960s (46), and continues to be widely used.

New efficient vulcanization systems have been introduced in the market based on quaternary ammonium salts initially developed in Italy (29–33) and later adopted in Japan (34) to vulcanize epoxy/carboxyl cure sites. They have been found effective in chlorine containing ACM dual cure site with carboxyl monomer (43). This accelerator system together with a retarder (or scorch inhibitor) based on stearic acid (43) and/or guanidine (29–33) can eliminate post-curing. More recently (47,48), in the United States a proprietary vulcanization package based on zinc diethyldithiocarbamate [*14324-55-1*] (49) has been reported that offers a superior balance between cure rate at molding temperatures and safety at process temperatures. It is proposed as a curative system that does not require post-vulcanization.

More detailed compounding suggestions and additives information can be found in the technical bulletins published by ACM and additives suppliers (50–54).

Economic Aspects

Acrylic rubbers, as is the case for most specialty elastomers, are characterized by higher price and smaller consumption compared to general-purpose rubbers. The total rubber consumption in 1991 was forecast (55) at 15.7 million t worldwide with a 66% share for synthetic elastomers (10.4×10^6 t). Acrylic elastomers consumption, as a minor amount of the total synthetic rubbers consumption, can hardly be estimated. As a first approximation, the ACM consumption is estimated to be 7000 t distributed among the United States, Western Europe, and Japan/Far East, where automotive production is significantly present.

The price of ACM is approximately four to five times greater than that of general-purpose rubbers like SBR or NR, and about three times the price of standard engineering rubbers such as NBR and EPDM. However, they cost about four to six times less than other specialty elastomers like fluorocarbon elastomers FKM [*25790-89-0*] or HNBR.

Some significant information about trade names, producers, plant location, and capacity are reported in Table 1.

Table 1. Production Capacities of Acrylic Elastomers

Company	Trade name	Location	Capacity, t/yr
EniChem Elastomeri	Europrene AR Cyanacryl[a]	Ravenna (Italy)	2000
Nippon Zeon	HyTemp[b]	Louisville, Ky.	4000
Nippon Zeon	Nipol	Kawasaki (Japan)	500[c]
Toa Paint	Acron	Osaka (Japan)	1300
NOK	Noxtite	na (Japan)	na
Toa Resin	Acron	Taipei (Taiwan)	600

[a]Formerly produced by American Cyanamid in Bound Brook, N.J. (U.S.).
[b]Formerly trademark Hycar, produced by B.F. Goodrich.
[c]Estimated capacity.

Health and Safety

Acrylic elastomers are normally stable and not reactive with water. The material must be preheated before ignition can occur, and fire conditions offer no hazard beyond that of ordinary combustible material (56). Above 300°C these elastomers may pyrolize to release ethyl acrylate and other alkyl acrylates. Otherwise, thermal decomposition or combustion may produce carbon monoxide, carbon dioxide, and hydrogen chloride, and/or other chlorinated compounds if chlorine containing monomers are present in the polymer.

Overexposure to acrylic rubbers is not likely to cause significant acute toxic effects. ACM however may contain residual monomers, mainly acrylate monomers, vapors of which are known to cause eye and/or skin irritation.

According to ACGIH (57) the Threshold Limit Values (TLV) for a normal 8-h workday and 40-h workweek exposure (TWA) are 5 ppm (20 mg/m^3) for ethyl acrylate, and 10 ppm (55 mg/m^3) for butyl acrylate. Furthermore these monomers are codified A2 (Suspect Human Carcinogen). Therefore, according to good practice of industrial hygiene, these types of rubber should be used in a well-ventilated area. The use of eye protection and gloves is recommended when handling acrylic elastomers.

BIBLIOGRAPHY

"Elastomers, Synthetic (Acrylic)" in *ECT* 3rd ed., Vol. 8, pp. 459–469, by T. M. Vial, American Cyanamid Co.

1. *ASTM D1418-85*, American Society for Testing and Materials, Philadelphia, Pa., 1989.
2. Ger. Pat. 262,707 (Jan. 31, 1912), O. Rohm.
3. C. H. Fisher, G. S. Whitby, and E. M. Beavers in G. S. Whitby, ed., *Synthetic Rubber*, John Wiley & Sons, Inc., New York, 1954.
4. W. C. Mast and co-workers, *Ind. Eng. Chem.* **36**, 1022 (1944).

5. U.S. Pat. 2,509,513 (May 30, 1950), W. C. Mast, C. H. Fisher (to U.S. Department of Agriculture).
6. J. Furukawa, *Polymer* **3,** 487 (1963).
7. U.S. Pat. 3,476,722 (Nov. 4, 1969), (to B. F. Goodrich).
8. J. Furukawa and co-workers, *Makromol. Chem.* **42,** 165 (1960).
9. E. A. H. Hopkins and M. L. Miller, *Polymer* **4,** 75 (1963).
10. E. H. Riddle, *Monomeric Acrylic Esters*, Reinhold Publishing Corp., New York, 1954.
11. W. C. Mast and C. H. Fisher, *Ind. Eng. Chem.* **41,** 790 (1949).
12. H. A. Tucker and A. H. Jorgensen in J. P. Kennedy and E. G. Tornquist, eds., *Polymer Chemistry of Synthetic Elastomers*, Part 1, Wiley-Interscience, New York, 1968, p. 250.
13. *The Synthetic Rubber Manual*, 11th ed., IISRP, Inc., Houston, Tex., 1989.
14. P. H. Starmer, *Prog. Rubber Plastics Technol.* **3**(1) (1987).
15. C. E. Rehberg and C. H. Fisher, *J. Am. Chem. Soc.* **66,** 1203 (1944).
16. R. D. De Marco, *Rubber Chem. Technol.* **52,** 173 (1979).
17. D. W. van Krevelen, *Properties of Polymers*, 2nd ed., Elsevier Science Publishing Co., New York, 1976.
18. A. L. Spelta, G. Cantalupo, and L. Gargani, paper presented at *The German Rubber Conference*, Nuremberg, July 4–7, 1988.
19. W. C. Mast and C. H. Fisher, *Ind. Eng. Chem.* **40,** 107 (1949).
20. U.S. Pat. 3,201,373 (Aug. 15, 1965), (to American Cyanamid).
21. U.S. Pat. 3,763,119 (Oct. 20, 1973), (to B.F. Goodrich).
22. U.S. Pat. 3,875,092 (Apr. 1, 1975), (to B.F. Goodrich).
23. U.S. Pat. 3,912,672 (Oct. 14, 1975), (to B.F. Goodrich).
24. U.S. Pat. 3,919,143 (Nov. 11, 1975), (to B.F. Goodrich).
25. U.S. Pat. 3,925,281 (Dec. 9, 1975), (to B.F. Goodrich).
26. Jpn. Pat. 80 112,212 (Aug. 29, 1980) (to Nippon Oil Seal Industry).
27. U.S. Pat. 3,510,442 (May 5, 1970), (to Polysar).
28. U.S. Pat. 4,237,258 (Dec. 2, 1980), (to Montedison).
29. E. Crespi and L. Fiore, *Rubber World* **10** (1980).
30. E. Giannetti and co-workers, *Rubber Chem. Technol.* **56** (1983).
31. E. Lauretti and co-workers, paper presented at *PRI Rubber Conference*, Birmingham, U.K., Mar. 21, 1984.
32. L. Gargani and co-workers, paper presented at *SCR Scandinavian Rubber Conference*, Copenhagen, June, 1985.
33. Bozzetto EniChem Synthesys, *Arax B18 MB50 Technical Bulletin*, Jan. 1986.
34. T. Nakagawa and co-workers, in J. Lal and J. E. Mark, eds., *Advantages in Elastomers and Rubber Elasticity*, Plenum Publishing Corp., New York, 1986.
35. C. H. Fisher and co-workers, *Ind. Eng. Chem.* **36,** 1032 (1944).
36. W. C. Mast, L. T. Smith, and C. H. Fisher, *Ind. Eng. Chem.* **36,** 1027 (1944).
37. U.S. Pat. 3,402,158 (Sept. 17, 1968), (to Thiokol Chemical Co.).
38. Fr. Pat. 1,511,011 (Jan. 26, 1968), (to Ugine Kuhlman).
39. D. D. Berry and co-workers, *Rubber World* **170,** 42 (1974).
40. N. Nakajima and R. A. Miller, "Processing Ease and Rubber Carbon Black Interaction," paper presented at *ACS meeting*, Montreal, 1987.
41. A. L. Spelta, G. Cantalupo, and L. Gargani, paper presented at *The Arctic Conference*, Tampere, Finland, Jan. 30–Feb. 2, 1989.
42. *ASTM D471-79*, American Society for Testing and Materials, Philadelphia, Pa., 1989.
43. R. M. Montague, *Rubber World* **199**(3), 20 (1988).
44. B. F. Goodrich, *Hycar 4050 ACM Non-Post-Cure Technical Data Bulletin PA-87-1*, 1987.
45. P. H. Starmer and F. R. Wolf in J. I. Kroschwitz, ed., *Encyclopedia of Polymer Science and Engineering*, Vol. 1, John Wiley & Sons, Inc., New York, 1985, pp. 306–325.

46. H. W. Holly, F. F. Mihal, and I. Starer, *Rubber Age* **96**(4), 565 (1965).
47. E. Chang and E. Mazzone, "A New Non-Postcure Curative Package for Polyacrylate Elastomers", paper presented at *ACS Rubber Division*, Detroit, Mich., Oct. 17–20, 1989.
48. E. Chang and G. E. Dunn, "Injection Molding of Polyacrylic Gasketing Requiring No Postcure", *SAE Technical Paper Series 900202,* Detroit, Mich., Feb. 26–Mar. 2, 1990.
49. American Cyanamid, *Material Safety Data, Ezcure 2X (MSDS 5742-01) and Ezcure 2Y (MSDS 5737-01),* Dec. 18, 1989.
50. American Cyanamid, *Cyanacryl Acrylic Elastomers, Technical Bulletin EPT-037C,* 1986.
51. B. F. Goodrich, *Hycar Polyacrylic Rubbers, bulletin HPA-1,* 1986.
52. Zeon Chemicals, *HyTemp Acrylic Elastomers, bulletin HPA-1A,* Oct., 1989.
53. Nippon Zeon Co., *Nipol AR 53,* 1987.
54. EniChem Elastomeri, *Europrene AR Acrylic Elastomers, technical bulletin,* Milano, Italy, 1988.
55. *Worldwide Rubber Statistic 1991,* IISRP, Inc., Houston, Tex., 1991.
56. EniChem Elastomeri, *Material Safety Data, Cyanacryl R (MSDS 1030),* Milano, Italy, Mar. 1992.
57. *1991–1992 Threshold Limit Values for Chemical Substances and Physical Agents and Biological Exposure Indices,* ACGIH, Cincinnati, Ohio, 1991.

A. L. Spelta
EniChem Elastomeri

BUTYL RUBBER

Butyl rubber and other isobutylene polymers of technological importance include various homopolymers and isobutylene copolymers containing unsaturation achieved by copolymerization with isoprene. Bromination or chlorination of the unsaturated site is practiced commercially, and other modifications are being investigated.

Isobutylene was first polymerized in 1873. High molecular weight polymer was later synthesized at I. G. Farben by decreasing the polymerization temperature to − 75°C, but the saturated, unreactive polymer could not be cross-linked into a useful synthetic elastomer. It was not until 1937 that poly(isobutylene-*co*-isoprene) [*9010-85-9*] or butyl rubber was invented at the Standard Oil Development Co. (now Exxon Chemical Co.) laboratories (1).

The first sulfur-curable copolymer was prepared in ethyl chloride using AlCl₃ coinitiator and 1,3-butadiene as comonomer; however, it was soon found that isoprene was a better diene comonomer and methyl chloride was a better polymerization diluent. With the advent of World War II, there was a critical need to produce synthetic elastomers in North America because the supply of natural rubber was drastically curtailed. This resulted in an enormous scientific and engineering effort that resulted in commercial production of butyl rubber in 1943.

Prior to butyl rubber, the known natural and synthetic elastomers had reactive sites at every monomer unit. Unlike natural rubber, polychloroprene, and polybutadiene, butyl rubber had widely spaced olefin sites with allylic hydrogens. This led to the principle of limited functionality synthetic elastomers that was later applied to other synthetic elastomers, eg, chlorosulfonated polyethylene, silicone rubber, and ethylene–propylene terpolymers.

The first published information on the halogenation of butyl rubber was provided by B. F. Goodrich Co. (2). Brominating agents such as *N*-bromosuccinimide were used; the bromination occurred in a bulk reaction. This technology was commercialized in 1954, but withdrawn in 1969 (3). Exxon Chemical researchers pursued the chlorination of butyl rubber in hexane solution using elemental chlorine, and a continuous process was commercialized in 1961 (4). Currently, both chlorination and bromination are carried out in continuous-solution processes.

The first use for butyl rubber was in inner tubes, the air-retention characteristics of which contributed significantly to the safety and convenience of tires. Good weathering, ozone resistance, and oxidative stability have led to applications in mechanical goods and elastomeric sheeting. Automobile tires were manufactured for a brief period from butyl rubber, but poor abrasion resistance restricted this development at the time.

Halogenated butyl rubber greatly extended the usefulness of butyl rubber by providing much higher vulcanization rates and improving the compatibility with highly unsaturated elastomers. Moreover, the halogenated elastomers can undergo vulcanization by different mechanisms than those for SBR or NR. These properties permitted the production of tubeless tires with chlorinated or brominated butyl inner liners. Tire durability was extended by the retention of air pressure (5) and low intercarcass pressure (6).

Polyisobutylene is produced in a range of mol wts, and has found a host of uses. The low mol wt liquid polybutenes have applications as adhesives, sealants, coatings, lubricants, and plasticizers, and for the impregnation of electrical cables (7). Moderate mol wt polyisobutylene was one of the first viscosity-index modifiers for lubricants (8). High mol wt polyisobutylene is used to make uncured rubbery compounds, and as an impact additive for thermoplastics.

Synthesis

Monomers for manufacture of butyl rubber are 2-methylpropene [*115-11-7*] (isobutylene) and 2-methyl-1,3-butadiene [*78-79-5*] (isoprene) (see OLEFINS). Polybutenes are copolymers of isobutylene and *n*-butenes from mixed-C_4 olefin-containing streams. For the production of high mol wt butyl rubber, isobutylene must be of >99.5 wt % purity, and isoprene of >98 wt % purity is used. Water and oxygenated organic compounds interfere with the cationic polymerization mechanism, and are minimized by feed purification systems.

ISOBUTYLENE POLYMERIZATION MECHANISM

The mechanism of cationic polymerization of isobutylene and copolymerization of isobutylene with isoprene with Lewis acids is highly complex (8,9). Friedel-Crafts type Lewis acids at low temperature give an extremely high polymerization rate in hydrocarbon or halogenated hydrocarbon solvents. In the first step, the initiation reaction, a carbenium ion is formed between a Lewis acid coinitiator, an initiator, and the isobutylene. Typical Lewis acid coinitiators include $AlCl_3$, (alkyl)$AlCl_2$, BF_3, $SnCl_4$, $TiCl_4$ etc, whereas initiators are Brønsted acids such as HCl, RCOOH, H_2O, or alkyl halides, eg, $(CH_3)_3CCl$, $C_6H_5(CH_3)_2CCl$. Initiation is followed by propagation where monomer units add to the carbenium ion forming the chain. The isoprene monomer mainly adds to the growing polymerization chain by *trans* 1,4-addition (>90%), and to a lesser extent 1,2-addition (<10%). These reactions are fast and highly exothermic. The chemistry of the propagation is affected by the temperature, polarity of the solvent, and counterions. The propagation proceeds until chain transfer or termination occurs.

In the chain-transfer reaction, the carbenium ion chain end reacts with isobutylene, isoprene, or a species with an unshared electron pair, ie, RX, solvents, and olefins. Since the activation energy of the chain-transfer reaction is much larger than that of the propagation reaction, the molecular weight of the copolymer strongly depends on the polymerization temperature and isoprene comonomer content (10,11). Termination reactions are the irreversible destruction of the propagating carbenium ion, and discontinuance of the kinetic chain. The termination reactions include: the collapse of the propagating carbenium ion–counterion pair, hydride extraction from isoprene, and formation of dormant or stable allylic carbenium ion and destruction of the carbenium ion by nucleophiles, eg, alcohols, amines, or alkoxides. The reactivity ratios are strongly affected by polymerization conditions (Table 1). A laboratory procedure for the preparation of butyl rubber is described in Reference 21.

A living cationic polymerization of isobutylene and copolymerization of isobutylene and isoprene has been demonstrated (22,23). Living copolymerizations, which proceed in the absence of chain transfer and termination reactions, yield the random copolymer with narrow mol wt distribution and well-defined structure, and possibly at a higher polymerization temperature than the current commercial process. The isobutylene–isoprene copolymers are prepared by using cumyl acetate: BCl_3 complex in CH_3Cl or CH_2Cl_2 at $-30°C$. The copolymer contains $1 \sim 8$ mol % *trans* 1,4-isoprene units, and has $M_n = 2,000 \sim 12,000$ with $M_w/M_n \approx 1.8$.

MODIFICATION OF BUTYL RUBBERS

Halobutyls. Chloro- and bromobutyls are commercially the most important butyl rubber derivatives. The halogenation reaction is carried out in hydrocarbon solution using elemental chlorine or bromine (equimolar ratio with enchained isoprene). The halogenation is fast, and proceeds mainly by an ionic mechanism. The structures that may form include the following:

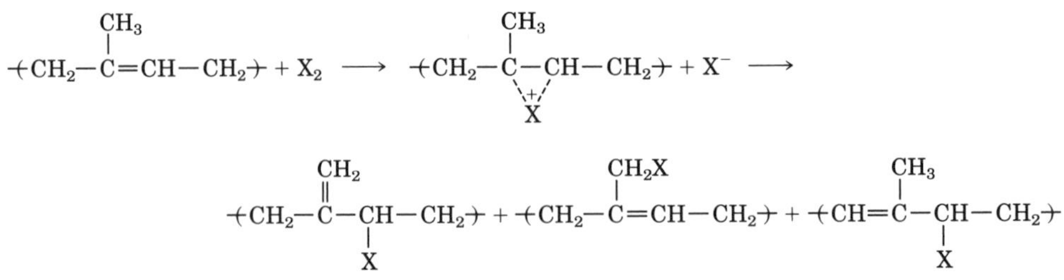

(1)   (2)   (3)

Normally, the structure (**1**) is the dominant structure (>80%) (24,25). More than one halogen atom per isoprene unit can also be introduced. However, the reaction rates for excess halogens are lower, and the reaction is complicated by chain fragmentation (26).

Conjugated-Diene Butyl. CDB can be obtained by the controlled dehydrohalogenation of halogenated butyl rubber (27). This product concept remains in the development stage.

$$-CH_2-\underset{\underset{CH_3}{|}}{\overset{\overset{CH_3}{|}}{C}}-CH_2-\underset{\underset{X}{|}}{\overset{\overset{CH_2}{\|}}{C}}-CH-CH_2- \quad \xrightarrow{-HX} \quad -CH_2-\underset{\underset{CH_3}{|}}{\overset{\overset{CH_3}{|}}{C}}-CH_2-\overset{\overset{CH_2}{\|}}{C}-CH=CH-$$

The conjugated diene butyl chain can be cross-linked with peroxide or radiation exposure. Free radicals also are used to graft cure with vinyl monomers, eg, methacrylic acid or styrene, which lead to transparent rubber exhibiting a T_g of about $-59°C$.

Other Derivatives. Various derivatives have appeared on the market or reached the market development stage. A carboxy-terminated polyisobutylene (28) was prepared by ozonization of high mol wt butyl rubber or poly(isobutylene-co-piperylene) [26335-67-1] in the presence of pyridine. The resulting polymers were viscous liquids with viscosity average mol wt about 2,000 to 4,000, and an average functionality of ~ 1.8 COOH groups per chain. This predominantly diacid could be converted to networks by various reactions, eg, with epoxides or aziridine. High mol wt isobutylene–cyclopentadiene rubbers containing up to 40% cyclopentadiene were produced (29) and developed by Exxon. In addition to these derivatives, butyl rubber is available in latex form. The latexes are manufactured by emulsifying solutions of standard butyl grades with about 60% solids (30).

ISOBUTYLENE–ISOPRENE–DIVINYLBENZENE TERPOLYMER

A partially cross-linked isobutylene–isoprene–divinylbenzene terpolymer containing some unreacted substituted vinylbenzene appendages is commercially available from Polysar Division, Bayer AG. Because of the residual reactive functionality, it can be cross-linked by peroxides that degrade conventional butyl rubbers. It is employed primarily in the manufacture of sealant tapes and caulking compounds (31).

LIQUID BUTYL RUBBER

This material is commercially produced by degradation of a conventional high mol wt butyl rubber, most likely via extrusion at high shear rate and temperature. Liquid butyl rubber is a viscous liquid of about 20,000 to 30,000 viscosity average mol wt (M_v). The principal areas of use are in sealant, caulks, potting compounds, and coatings, where advantage can be taken of its relatively low viscosity in formulating high solids compounds that can be poured, sprayed, troweled, or spread.

NEW MATERIALS

Grades of polyisobutylene, butyl rubber, halogenated butyl rubber, and partially cross-linked isobutylene–isoprene–divinylbenzene terpolymer have been developed to meet specific processing and property needs. Recently, two new poly-isobutylene-based elastomers have been developed. One is now available commercially as Exxon SB Butyl Polymers (32) and the other is under market development as Exxon bromo XP-50.

Star-Branched (SB) Butyl. Butyl rubbers have unique processing characteristics due to their viscoelastic properties and lack of crystallization of compounds on extension. They exhibit both low green strength and low creep resistance as a consequence of high mol wt between entanglements. To enhance green strength, traditional butyl rubber requires a relatively high mol wt. Increasing mol wt also causes an increase in relaxation time along with high viscosity. In such situations it is usually helpful to broaden the mol wt distribution, but this is difficult to accomplish in conventional butyl rubber polymerization. Physical blending of low and high mol wt polymer can also provide broader mol wt distributions, but results in other processing problems such as high extrudate swelling in flow-through shaping dies.

Star-branched butyl has a bimodal mol wt distribution with a high mol wt branched mode and a low mol wt linear component. The polymer is prepared by a conventional cationic copolymerization of isobutylene and isoprene in methyl chloride below $-80°C$ with a Friedel-Crafts coinitiator, eg, $AlCl_3$, in the presence of a polymeric branching agent. The high mol wt branched molecules are formed during the polymerization via a graft-from or a graft-onto mechanism. A graft-from reaction takes place if a macroinitiator/macrotransfer reagent, such as hydrochlorinated poly(styrene-co-isoprene) or chlorinated polystyrene is used. A graft-onto reaction takes place if a multifunctional terminating agent, eg, poly(styrene-co-butadiene) is employed as the branching agent. In general, the star-branched butyl has 10–20% high mol wt branched molecules which have a random comb-like structure with 20–40 butyl branches. Although this is not a true star topology, it approaches a star structure since the branching agent is relatively short and the branching density is relatively high, ie, the mol wt between branching points is low compared to the segment length of the butyl branches. Star-branched butyl rubbers offer a unique balance of viscoelastic properties resulting in significant processability improvements. Dispersion in mixing and mixing rates are improved. Compound extrusion rates are higher, die swell is lower, shrinkage is reduced, and surface quality is improved. The balance between green strength and stress relaxation at ambient temperature is improved, making shaping operations such as tire building easier. Several grades of Exxon SB Butyl Polymers including copolymer, chlorinated, and brominated copolymers are now commercially available.

Brominated Poly(Isobutylene-*co*-*p*-Methylstyrene). *para*-Methylstyrene [*622-97-9*] (PMS) can be readily copolymerized with isobutylene via classical cationic copolymerization using strong Lewis acid, eg, $AlCl_3$ or alkyl aluminum in methyl chloride at low temperature. The copolymer composition is very similar to the feed monomer ratio because of the similar copolymerization reactivity ratios, ie, $r_1 = 1$ and $r_2 = 1.4$. These new high mol wt copolymers encompass an enormous range of properties from polyisobutylene-like elastomers to poly(*p*-methylstyrene)-like tough hard plastic materials with T_gs above 100°C, depending on monomer ratio. A highly reactive and versatile benzyl bromide functionality, $C_6H_5CH_2Br$, can be introduced by the selective free-radical bromination of the benzyl group in the copolymer. The brominated copolymer can be cross-linked with a variety of cross-linking systems. The benzyl bromide in the brominated copolymer can also be easily converted by nucleophilic substitution reactions to a variety of other functional groups and graft copolymers as desired for specific properties and applications. This new functionalized copolymer preserves polyisobutylene properties, low permeability, and unique dynamic response while adding the behavior of inertness to ozone, like ethylene–propylene rubbers. Copolymers with PMS below 10 mol % are most useful for elastomeric applications because the T_gs are near $-60°C$. The brominated copolymer is under development as bromo XP-50 for various tire and nontire applications (33).

Manufacturing

The bulk of the world production of butyl rubber is made by a precipitation (slurry) polymerization process in which isobutylene and a minor amount of isoprene are copolymerized using aluminum chloride in methyl chloride at -100 to $-90°C$. General descriptions of the process and patent information are found in the literature (34–56). An alternative solution process, developed in Russia, uses a C_5–C_7 hydrocarbon as solvent and an aluminum alkyl halide. The polymerization is conducted in scraped surface reactors at -90 to $-50°C$. The solution process avoids the use of methyl chloride, and is an advantage when butyl rubber is to be halogenated. However, the energy costs are higher than that for the slurry process, and because of continuing technical advances, the well-established slurry process is unlikely to be displaced. Halogenated butyl rubbers are produced commercially by dissolving butyl rubber in hydrocarbon solvent and contacting the solution with elemental halogens.

Monomer Purification. A number of commercial processes are available for production of the required high purity isobutylene. An extraction process based on sulfuric acid has been developed by Exxon Chemical Co. (57), Compagnie Francaise de Raffinage (58), and Petro-Tex Chemical Co. Significant quantities of isobutylene are also produced by dehydration of *tert*-butyl alcohol developed by BASF (59). The newest isobutylene recovery route is based on the selective reaction of dilute isobutylene in a C_4 stream with methanol over an acid ion-exchange resin, eg, Amberlyst 15, to form methyl *t*-butyl ether [1634-04-4] (MTBE). This ether is produced mainly as a high octane blending component for low lead gasoline (see ETHERS). Catalytic decomposition at 170–200°C and 600 kPa (5.9 atm) over a fixed-bed acid catalyst, eg, $SiO_2Al_2O_3$ or Amberlyst 15, produces high purity isobutylene (60–62). Typical isobutylene, as supplied to the butyl rubber process, has a purity in the range of 95–99%, and includes varying amounts of propene, 1-butene, 2-butene, isobutylene dimer, and *tert*-butyl alcohol, and trace quantities of a variety of oxygen-containing compounds, depending on the process employed.

Polymerization Process. A simplified flow diagram of the polymerization section of a slurry process is shown in Figure 1. Isobutylene is dried and fractionated to 2-butenes and high boiling components. The feed blend contains 25–40 wt % of isobutylene and 0.4–1.4 wt % of isoprene, depending on the grade of butyl rubber to be produced; the remainder is recycled methyl chloride. Coinitiator solution is produced by passing pure methyl chloride through packed beds of gran-

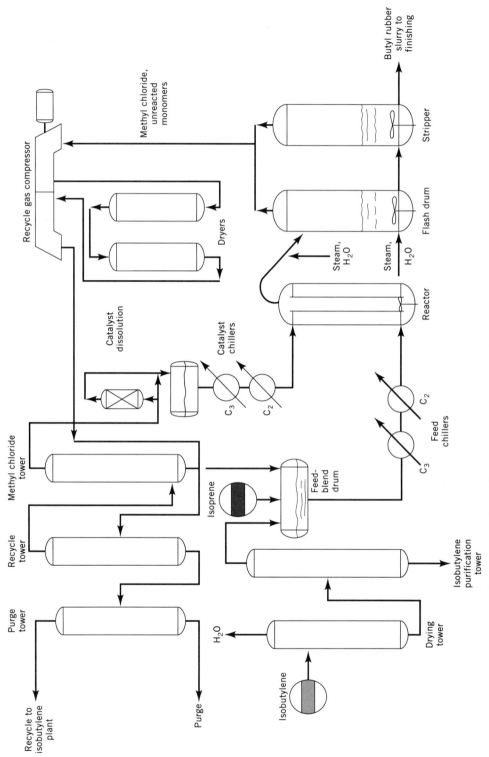

Fig. 1. Slurry butyl rubber polymerization; C_2, C_3, and C_4 are unsaturated hydrocarbon streams.

ular aluminum chloride at 30–45°C. The concentrated solution formed is diluted with additional methyl chloride, to which the initiator(s) is added and the solution later stored. The feed blend and initiator/coinitiator solutions are chilled to −100 to −90°C before entry to the reactor; the latter comprises a central vertical draft tube surrounded by concentric rows of cooling tubes. The reactor is mixed by an axial flow pump located at the bottom of the draft tube that circulates slurry through the cooling tubes. The copolymerization reaction is exothermic, releasing approximately 0.82 MJ/kg of polymer (350 Btu/lb). This heat is removed by exchange to boiling ethylene supplied as liquid to jackets that enclose the tube section of the reactor. The reactor is constructed with 3.5 or 9 wt % nickel steel, or alloys that have adequate impact strength at the low temperature of the polymerization reaction.

The production rate is 2–4 t/h, depending on the feed rate, monomer concentration in the feed, and conversion. The conversion of isobutylene and isoprene typically ranges from 75–95% and 45–85%, respectively, depending on the grade of butyl rubber being produced. The composition and mol wt of the polymer formed depend on the concentration of the monomers in the reactor liquid phase and the amount of chain transfer and terminating species present. The liquid-phase composition is a function of the feed composition and the extent of monomer conversion. In practice, the principal operating variable is the flow rate of the initiator/coinitiator solution to the reactor; residence time is normally 30–60 minutes.

Significant fouling occurs during polymerization. Rubber film deposits on heat-transfer surfaces and agglomerates block cooling tubes and impair slurry circulation. A cyclical operation is required in which some reactors operate while others are washed with hot solvent to remove deposited rubber. Typical runs last from 18 to 60 h, depending on production rate, feed purity, and slurry concentration. Patents that describe the use of block copolymer as stabilizing agents to prevent rubber slurry agglomeration are available (63).

Butyl slurry at 25–35 wt % rubber continuously overflows from the reactor through a transfer line to an agitated flash drum operating at 140–160 kPa (1.4–1.6 atm) and 55–70°C. Steam and hot water are mixed with the slurry in a nozzle as it enters the drum to vaporize methyl chloride and unreacted monomers that pass overhead to a recovery system. The vapor stream is compressed, dried over alumina, and fractionated to yield a recycle stream of methyl chloride and isobutylene. Pure methyl chloride is recovered for the coinitiator ($AlCl_3$) preparation. In the flash drum, the polymer agglomerates as a coarse crumb in water. Metal stearate, eg, aluminum, calcium, or zinc stearate, is added to control the crumb size. Other additives, such as antioxidants, can also be introduced at this point. The polymer crumb at 8–12 wt % in water flows from the flash drum to a stripping vessel operated under high vacuum to remove remaining methyl chloride and monomers. The stripped rubber crumb is pumped to intermediate storage tanks and successively dewatered in a series of drying extruders to a final moisture content of < 0.5 wt %. The dried crumb product is compressed into bales and packaged in containers for shipping. High mol wt polyisobutylene is manufactured similarly.

Halogenated Butyl Rubber. The halogenation is carried out in hydrocarbon solution using elemental chlorine or bromine in a 1:1 molar ratio with enchained isoprene. The reactions are fast; chlorination is faster. Both chlorinated and brominated butyl rubbers can be produced in the same plant in blocked operation. However, there are some differences in equipment and reaction conditions. A longer reaction time is required for bromination. Separate facilities are needed to store and meter individual halogens to the reactor. Additional facilities are required because of the complexity of stabilizing brominated butyl rubber.

The halogenation process begins with the preparation of a hydrocarbon (eg, hexane) solution of butyl rubber with desired mol wt and enchained isoprene. A slurry from a butyl polymerization reactor or solid butyl rubber is dissolved (64–66). In the former case, the cold butyl slurry in methyl chloride flows from the reactor into a drum containing hot liquid hexane that rapidly dissolves the fine slurry particles. Hexane vapor then is added to flash methyl chloride and unreacted monomers overhead for recovery and recycle. The hexane solution passes to a stripping column where the remaining methyl chloride and monomers are removed. The hot solution is then concentrated to 20–25 wt % for halogenation in an adiabatic flash step. Alternatively, bales of butyl rubber are chopped or ground to small pieces and conveyed to a series of agitated dissolving vessels or to a large vessel divided into multiple stages. Solutions of 15–20 wt % can be prepared in 1–4 h depending on temperature, agitation, and size of rubber particles. This method has the advantage of being independent from the butyl polymerization process. However, two finishing operations are required: one to produce dry butyl rubber feedstock, the second to finish the halogenated butyl products. Investment and energy costs for the two dissolving processes are similar.

In the halogenation process shown in Figure 2, the butyl rubber solution reacts with chlorine or bromine in one or more highly agitated reaction vessels at 40–60°C. For safety reasons, chlorine is introduced as a vapor or in dilute solution because liquid chlorine reacts violently with butyl rubber solution. However, bromine may be used in liquid or vapor form because of its lower reaction rate. The halogenation by-product, HCl or HBr, is neutralized with dilute aqueous caustic solution in high intensity mixers. Antioxidants and stabilizers such as calcium stearate and epoxized soybean oil are added. The solution is sent to a multivessel solvent-removal system where steam and water vaporize the solvent and produce crumb-like rubber particles in water. The final solvent content and the steam usage for solvent removal depends on the conditions in each vessel. Typically, the lead flash drum is operated at 105–120°C and 200–300 kPa (2–3 atm). Conditions in the final stripping stage are 101°C and 105 kPa (1.04 atm). The hexane can be reduced to 0.5–1.0 wt % with steam usage of 2.0–2.5 kg/kg rubber. The stripped rubber in water is sent to a finishing unit similar to the butyl rubber process. Extrusion temperatures must be kept low to prevent dehydrohalogenation of halogenated butyl rubbers.

Several patents describe solvent-free bulk-phase halogenation (67–69). Dry solid butyl rubber is fed into a specially designed extruder reactor and contacted with chlorine or bromine vapor. The by-product HCl or HBr are vented directly without a separate neutralization step. Halogenated butyl rubbers produced are essentially comparable in composition and properties to commercial products made by the solution process.

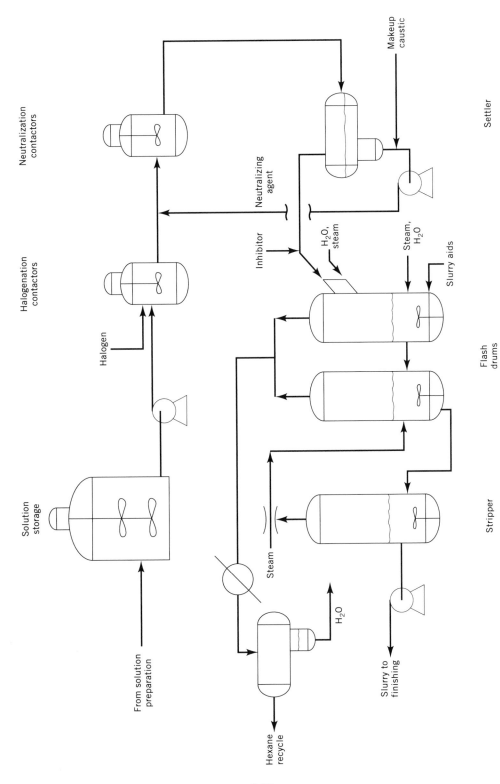

Fig. 2. Halogenated butyl rubber process.

943

Polymer Properties

MOLECULAR STRUCTURE

Polyisobutylene. Isobutylene polymerizes in a regular head-to-tail sequence to produce a polymer [*9003-27-4*] having no asymmetric carbon atoms. The glass-transition temperature is about $-70°C$ (70). In the unstrained state the polymer is amorphous. The high mol wt chains crystallize on extension at room temperature, and the helical chain conformation in the crystalline regions repeats every eight monomer units (71). Because of severe steric overlap of two methyl groups on alternate carbon atoms, a planar zigzag model is ruled out. Steric restrictions cause the bond angle at the methylene carbon to be distorted to ca 123° compared to 109.5° for a tetrahedral carbon (72).

One chain-end is typically unsaturated due to chain transfer and termination mechanisms. Mol wts can range from several hundred to several million. There is no long-chain branching unless special synthesis methods are employed. The mol wt distribution is commonly the most probable, $M_w/M_n = 2$.

Polybutenes. Copolymerization of mixed isobutylene and 1-butene containing streams with a Lewis acid catalyst system yields low mol wt (several hundred to a few thousand) copolymers that are clear, colorless, viscous liquids. The chain-ends are unsaturated, and they are often chemically modified through this functionality (7,73).

Butyl Rubber. In butyl rubber, isoprene is enchained by 1,4-addition in the trans configuration (74).

$$x\ CH_3-\underset{\underset{CH_3}{|}}{C}=CH_2 + y\ CH_2=\underset{\underset{}{|}}{\overset{CH_3}{C}}-CH=CH_2 \longrightarrow \left(CH_2-\underset{\underset{CH_3}{|}}{\overset{CH_3}{C}}\right)_{\!x}\!\!\left(CH_2-\overset{CH_3}{C}=CH-CH_2\right)_{\!y}$$

Depending on the grade, the unsaturation is between 0.5 and 2.5 mol per 100 mol of monomer. The low content of isoprene and a reactivity ratio product near unity (see r_1 and r_2 in Table 1) indicate a random distribution of unsaturation along the chain. The mol wt distribution of butyl rubber depends on the grade, but many products have a M_w/M_n of 3–5.

Halogenated Butyl Rubber. Halogenation at the isoprene site in butyl rubber proceeds by a halonium ion mechanism leading to a double-bond shift and formation of an exomethylene alkyl halide. Both chlorinated and brominated rubber show the predominate structure (**1**) (>80%), by nmr, as described earlier (33,34). Halogenation of the unsaturation has no apparent effect on the isobutylene backbone chains. Cross-linked samples do not crystallize on extension due to the chain irregularities introduced by the halogenated isoprene units.

CHEMICAL PROPERTIES

Polyisobutylene has the chemical properties of a saturated hydrocarbon. The unsaturated end groups undergo reactions typical of a hindered olefin and are used,

Table 1. Reactivity Ratios of Isobutylene Copolymerization[a]

Monomer 2	Initiating system	Solvent	Temperature, °C	r_1	r_2	References
1,3-butadiene	$AlC_2H_5Cl_2$	CH_3Cl	−100	43	0	12
	$AlCl_3$	CH_3Cl	−103	115 ± 15	0.01 ± 0.01	13
cyclopentadiene	$BF_3O(C_2H_5)_2$	toluene	−78	0.60 ± 0.15	4.5 ± 0.5	14
	$SnCl_4TCA^b$	toluene	−78	0.21 ± 0.02	6.3 ± 0.6	14
cyclohexadiene	$SnCl_4TCA$.	toluene	−78	0.77 ± 0.20	1.05 ± 0.25	14
	$SnCl_4TCA$	CH_2Cl_2	−78	0.98 ± 0.20	1.07 ± 0.45	14
2,3-dimethylbutadiene	$AlC_2H_5Cl_2$	pentane: CH_3Cl = 50:50	−70	3.20 ± 1.10	0.98 ± 0.64	15
isoprene	$AlCl_3$	CH_3Cl	−103	2.5 ± 0.5	0.4 ± 0.1	13
	$AlC_2H_5Cl_2$	hexane: CH_3Cl = 100:0	−80	0.80	1.28	16
	$AlC_2H_5Cl_2$	hexane: CH_3Cl = 88:12	−80	1.17	1.08	16
	$AlC_2H_5Cl_2$	hexane: CH_3Cl = 50:50	−80	1.90	1.05	16
	$AlC_2H_5Cl_2$	hexane: CH_3Cl = 12:88	−80	2.15	1.03	16
styrene	$AlC_2H_5Cl_2$ + Cl_2	CH_3Cl	−35	2.5 ± 0.5		17
	$AlC_2H_5Cl_2$	pentane: CH_2Cl_2 = 50:50	−70	1.56 ± 0.19	0.95 ± 0.17	15
	$AlCl_3$	CH_3Cl	−30	2.36 ± 0.06	0.76 ± 0.13	18
	$AlCl_3$	CH_3Cl	−90	1.66 ± 0.02	0.42 ± 0.02	18
	$AlCl_3$	CH_3Cl	−103	3 ± 1	0.6 ± 0.3	19
p-methylstyrene	$AlCl_3$	CH_3Cl	−90	1	1.4	20

[a]Ref. 8. Isobutylene is monomer 1 and has reactivity r_1.
[b]TCA = trichloroacetic acid.

particularly in the case of low mol wt materials, as a route to modification; eg, the introduction of amine groups to produce dispersants for lubricating oils. The in-chain unsaturation in butyl rubber is attacked by atmospheric ozone, and unless protected can lead to cracking of strained vulcanizates. Oxidative degradation, which leads to chain cleavage, is slow, and the polymers are protected by antioxidants (75).

Cross-linking reactions for the polyisobutylene-type polymers depend on adding a reactive site, usually an allylic hydrogen or halogen. These reactive sites allow vulcanization with sulfur and accelerators or metal oxides (76,77).

SOLUTION PROPERTIES

Polyisobutylene is readily soluble in nonpolar liquids. The polymer–solvent interaction parameter χ is a good indication of solubility. Values of 0.5 or less for a polymer–solvent system indicate good solubility; values above 0.5 indicate poor solubility. Values of χ for several solvents are shown in Table 2 (78). The solution properties of polyisobutylene, butyl rubber, and halogenated butyl rubber are very similar. Cyclohexane is an excellent solvent, benzene a moderate solvent, and dioxane a nonsolvent for polyisobutylene polymers.

Table 2. Polymer–Solvent Interaction Parameters for Polyisobutylene and Butyl Rubber

Solvent	Polymer[a]	Method[b]	χ
cyclohexane	PIB	VP	0.42
cyclohexane	B	SE	0.44
n-heptane	PIB	glc	0.57
n-heptane	B	SE	0.48
benzene	PIB	Op	0.66
n-pentane	PIB	VP	0.57
methylene chloride	B	SE	0.58
pyridine	B	SE	1.42

[a]PIB = polyisobutylene; B = butyl rubber.
[b]VP = vapor pressure of solvent; glc = gas–liquid chromatography; SE = swelling equilibria; Op = osmotic pressure.

The relationship between intrinsic viscosity $[\eta]$ and mol wt for isobutylene polymers in diisobutylene is (79):

$$[\eta]_{20°C} = 3.6 \times 10^{-4} M^{0.64} \quad (10^4 \leq M)$$

$$[\eta]_{20°C} = 1.78 \times 10^{-2} M^{0.465} \quad (M \leq 10^4)$$

Data on other solvents have been tabulated (80).

The bulk viscosity at zero shear rate η_0 also depends on mol wt (81):

$$\log \nu_0 = 2.4 \log M + 5.6 \times 10^5/T^2 - 15.85 \quad (\text{for } M > 17,000)$$

and

$$\log \eta_0 = 1.75 \log M + 5.6 \times 10^5 \, (1/T^2 - 1/490^2) \, \exp^{-163/M} \quad (\text{for } M < 17{,}000)$$

PHYSICAL PROPERTIES

The most important physical properties of the elastomeric polyisobutylenes are those exhibited by vulcanized compounds, especially high loss modulus and low permeability air.

A discussion of and values for the storage and loss moduli and the elongational relaxation modulus for polyisobutylene can be found in References 82 and 83. The rotational restriction of the polyisobutylene backbone results in a high monomeric friction coefficient (84) and unique William-Landel-Ferry constants (85) compared to hydrocarbon rubbers of similar glass-transition temperature, eg, natural rubber. Polyisobutylene-based elastomers are highly damping at 25°C, even though the T_g is ca $-70°C$. At elevated temperatures ($> 100°C$) damping properties are low. The chains also display a low plateau modulus, and thus in the uncross-linked state exhibit a high degree of tack or self-adhesion.

Polyisobutylene and similar copolymers appear to "pack" well (density of 0.917 g/cm^3) (86) and have fractional free volumes of 0.026 (vs 0.071 for polydimethylsiloxane). The efficient packing in PIB is attributed to the unoccupied volume in the system being largely at the intermolecular interfaces, and thus a polymer chain surface phenomenon. The thicker cross section of PIB chains results in less surface area per carbon atom.

Sluggish chain mobility and low free volume result in low diffusion constants, and when combined with low solubility of gases lead to very low permeability. The diffusivity of several gases in butyl rubber and natural rubber are shown in Table 3 (82) (see BARRIER POLYMERS).

Table 3. Diffusivity of Gases in Butyl Rubber and Natural Rubber at 25°C, (cm^2/s) $\times$ 10^6

Gas	Butyl rubber	Natural rubber
He	5.93	21.6
H_2	1.52	10.2
O_2	0.081	1.58
N_2	0.045	1.10
CO_2	0.058	1.10

ELASTOMERIC VULCANIZATES

As with almost all rubbers, the final properties are determined by compounding and subsequent vulcanization or cross-linking. Various fillers, processing aids, plasticizers, tackifiers, cure systems, and antidegradants are used.

Fillers.   Of the compounding ingredients, fillers (qv) most significantly influence stress–strain and dynamic properties (88). Polymer chains interact with the surface of reinforcing fillers altering chain dynamics. Carbon black (qv) because of its high surface area (small particle size) enhances tensile properties and abrasion resistance. Butyl rubber has a low affinity for carbon black when compared to highly unsaturated rubbers such as SBR (89,90). This can be an important consideration in elastomeric blends since the carbon black can accumulate in one phase. The effect of carbon black on butyl and halobutyl rubbers is similar to that on other elastomers. Particle size and structure determine the reinforcing characteristics of the carbon black and, hence, the processing and cured properties of the polymer. Increasing the reinforcing characteristics, ie, higher surface area and structure, enhances viscosity, hardness, modulus, and damping (91). Tensile strength increases to a maximum with increasing carbon black content and decreases at higher loading. For butyl rubber, the maximum tensile strength is obtained at ca 50–60 phr of carbon black.

Mineral fillers are used for light-colored compounds. Talc has a small particle size and is a semireinforcing filler. It reduces air permeability and has little effect on cure systems. Calcined clay is used for halobutyl stoppers in pharmaceutical applications. Nonreinforcing fillers, such as calcium carbonate and titanium dioxide, have large particle sizes and are added to reduce cost and viscosity. Hydrated silicas give dry, stiff compounds, and their acidity reduces cure rate; hence, their content should be minimized.

Plasticizers and Processing Aids.   Petroleum-based oils are commonly used as plasticizers. Compound viscosity is reduced, and mixing, processing, and low temperature properties are improved. Air permeability is increased by adding extender oils. Plasticizers are selected for their compatibility and low temperature properties. Butyl rubber has a solubility parameter of ca 15.3 $(J/cm^3)^{1/2}$ [7.5 $(cal/cm^3)^{1/2}$], similar to paraffinic and naphthenic oils. Polybutenes, paraffin waxes, and low mol wt polyethylene can also be used as plasticizers (qv). Alkyl adipates and sebacates reduce the glass-transition temperature and improve low temperature properties. Process aids, eg, mineral rubber and Struktol 40 MS, improve filler dispersion and cured adhesion to high unsaturated rubber substrates.

Other Ingredients.   Other compounding ingredients include tackifiers, flame retarders, odorants, and lubricants. Various resins with a T_g higher than the elastomer act as tackifiers by altering the dynamic mechanical properties of the unvulcanized elastomer (92). Hydrocarbon resins (qv) improve tack without side reactions. On the other hand, phenol–formaldehyde resins contain reactive methylol groups that can react with halobutyl to give premature cure. Hindered phenolic antioxidants (qv) added during manufacture prevent autoxidation during finishing steps, storage, and compounding. Antiozonants (qv) improve the resistance of butyl and halobutyl rubber vulcanizates to ozone cracking. The low segmental mobility in butyl rubber apparently limits the crack growth rate; however, antiozonants, such as N,N'-dioctyl-p-phenylenediamine, are effective at high concentrations of plasticizer (93). Halogenated butyl rubber must be stabilized against dehydrohalogenation (94,12) at elevated temperature. To achieve this, calcium stearate is used for chlorinated butyl; brominated butyl requires a mixture of calcium stearate and an epoxy compound, eg, epoxidized soybean oil.

Vulcanization

Vulcanization or curing is accomplished via chemical cross-linking reactions involving allylic hydrogen or halogen sites along the polymer backbone to produce a polymer network. Some typical vulcanization systems for butyl and halobutyls are shown in Table 4. Most technical vulcanizates have about 1×10^{-4} mols of cross-links/cm^3, or about 250 backbone carbon atoms between chemical cross-links.

Table 4. Some Typical Vulcanization Systems[a]

Property	Butyl rubber		Halobutyl rubber[b]				
	Sulfur/ accelerator	Resin	Sulfur/ accelerator	Resin	Amine	Zinc oxide	Peroxide
Ingredients							
ZnO	5	5	5	3		5	
PbO$_2$							
stearic acid	2	1			1	1	1
sulfur	2		0.5				
MBTS[c]	0.5		1.5				
TMTD[d]	1.0		0.25				
Maglite D			0.5		3		
hexamethylene diamine					1		
SP-1045				5			
SP-1055 resin		12					
Dicup 40C							2
HVA-2[e]							1
Conditions							
T, °C	155	180	160	160	160	160	155
t, min	20	80	20	15	15	15	20

[a]Concentrations are in parts per 100 parts of rubber.
[b]Can also be cured at room temperature (25°C) with 5 phr ZnO, 2 phr SnCl$_2$, 2 phr ZnCl$_2$.
[c]Bisbenzothiazolyl sulfide.
[d]Tetramethylthiuram disulfide.
[e]m-Phenylene bismaleimide.

Sulfur Cures. Accelerated sulfur cures are widely used for butyl rubber and ultra-accelerators are used to obtain acceptable rates at the low allylic hydrogen level. The allylic halogens in the modified butyls give faster rates and the ability to co-cure with high unsaturation elastomers. Sulfur cross-links rearrange at elevated temperature because of their low bond energy, and this results in creep and permanent set in strained elastomers if held for long times at high temperatures. Physical properties obtained with sulfur cures are excellent.

Resin Cure. Resin cure systems yield carbon–carbon cross-links and, consequently, thermally stable materials. Butyl rubber vulcanized with resins are used as tire-curing bladders, and have a life of 300–700 curing cycles at steam temperature of 175°C at about 20 m/cycle. Alkylphenol–formaldehyde resins, eg,

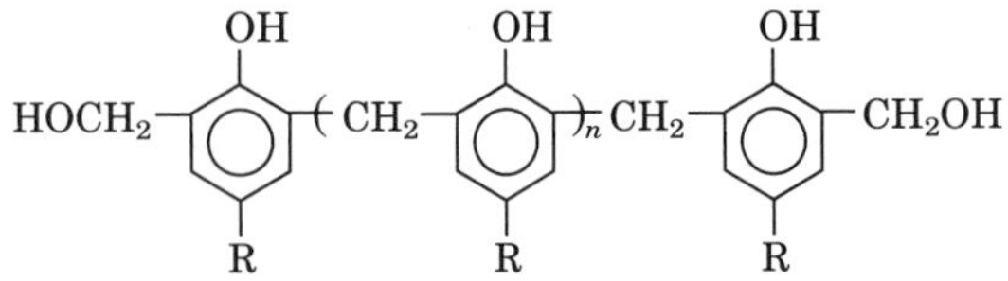

$$HOCH_2 - \underset{R}{\underset{|}{\overset{OH}{\overset{|}{\bigcirc}}}} - (CH_2 - \underset{R}{\underset{|}{\overset{OH}{\overset{|}{\bigcirc}}}})_n CH_2 - \underset{R}{\underset{|}{\overset{OH}{\overset{|}{\bigcirc}}}} - CH_2OH$$

produce the cross-links through the two phenol methylol groups, usually with a Lewis acid catalyst (see PHENOLIC RESINS).

Halobutyl Cures. Halogenated butyls cure faster in sulfur-accelerator systems than butyl; bromobutyl is generally faster than chlorobutyl. Zinc oxide-based cure systems result in C–C bonds formed by alkylation through dehydrohalogenation of the halobutyl to form a zinc chloride catalyst (94,95). Cure rate is increased by stearic acid, but there is a competitive reaction of substitution at the halogen site. Because of this, stearic acid can reduce the overall state of cure (number of cross-links). Water is a strong retarder because it forms complexes with the reactive intermediates. Amine cure may be represented as follows:

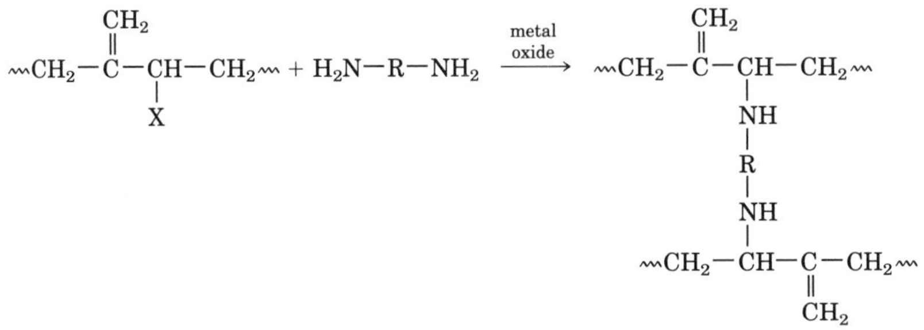

$$\mathsf{mCH_2 - \overset{\overset{\displaystyle CH_2}{\|}}{C} - \underset{\underset{\displaystyle X}{|}}{CH} - CH_2 m + H_2N - R - NH_2 \xrightarrow{\overset{\text{metal}}{\text{oxide}}} mCH_2 - \overset{\overset{\displaystyle CH_2}{\|}}{C} - \underset{\underset{\underset{\underset{\underset{\displaystyle CH_2}{\|}}{C - CH_2m}}{NH}}{R}}{\underset{\displaystyle NH}{|}} CH}}$$

Whereas polyisobutylene and butyl rubber exhibit chain cleavage on free-radical attack, halobutyls, particularly bromobutyl and CDB, are capable of being cross-linked with organic peroxides. The best cure rate and optimal properties are achieved using a suitable co-agent, such as *m*-phenylene bismaleimide. This cure is used where high temperature and steam resistance is required.

Uses

The polyisobutylene portion of the butyl chain imparts chemical and physical characteristics that make it highly useful in a wide variety of applications. The low degree of permeability to gases accounts for the largest uses of butyl and halobutyl rubbers, namely inner tubes and tire inner liners. These same properties are also of importance in air cushions, pneumatic springs, air bellows, accumulator bags, and pharmaceutical closures (96). The thermal stability of butyl rubber makes it ideal for rubber tire-curing bladders, high temperature service hoses, and conveyor belting for hot-materials handling.

Butyl-type polymers exhibit high damping, and the viscous part of the dynamic modulus is uniquely broad as a function of frequency or temperature. Molded rubber parts for damping and shock absorption find wide application in automotive suspension bumpers, auto exhaust hangers, and body mounts.

Blends of halogenated butyl rubber are used in tire sidewalls and tread compounds (97). In sidewalls, ozone resistance, crack cut growth, and appearance are critical to performance. Properly formulated blends with natural rubber (and triblends with ethylene propylene terpolymers) yield excellent sidewalls. The property balance for tire tread compounds can be enhanced by the incorporation of a more damping halobutyl rubber phase. Improvements in wet skid resistance and dry traction (with some compromise in abrasion resistance and rolling resistance) for high performance tires is accomplished by using up to 30 phr of bromo- or chlorobutyl in the compounds.

Blends of isobutylene polymers with thermoplastic resins are used for toughening these compounds. High density polyethylene and isotactic polypropylene are often modified with 5 to 30 wt % polyisobutylene. At higher elastomer concentration the blends of butyl-type polymers with polyolefins become more rubbery in nature, and these compositions are used as thermoplastic elastomers (98). In some cases, a halobutyl phase is cross-linked as it is dispersed in the polyolefin to produce a highly elastic compound that is processible in thermoplastic molding equipment (99) (see ELASTOMERS, SYNTHETIC–THERMOPLASTIC).

Polybutenes enjoy extensive use as adhesives, caulks, sealants, and glazing compounds. They are used as plasticizers in rubber formulations with butyl rubber, SBR, and natural rubber. In linear low density polyethylene (LLDPE) blends they induce cling to stretch-wrap films. Polybutenes when modified at their unsaturated end groups with polar functionality are widely employed in lubricants as dispersants. Blends of polybutene with polyolefins produce semisolid gels that can be used as potting and electrical cable filling materials.

Proper compounding and formulation are critical to the successful uses of most elastomeric materials. The suppliers should be contacted for information.

Health and Safety Factors

Polyisobutylene and isobutylene–isoprene copolymers are considered to have no chronic hazard associated with exposure under normal industrial use. Some grades can be used in chewing-gum base, and are regulated by the FDA in 21 CFR 172.615. Vulcanized products prepared from butyl rubber or halogenated butyl rubber contain small amounts of toxic materials as a result of the particular vulcanization chemistry. Although many vulcanizates are inert, eg, zinc oxide cured chlorobutyl is used extensively in pharmaceutical stoppers, specific recommendations should be sought from suppliers.

Economic Aspects

Butyl and halogenated butyl rubbers are manufactured by Exxon Chemical Co., Bayer AG (formed Polysar Ltd.), and in the former USSR. Table 5 provides a list of plant locations and capacity estimates. The 1993 U.S. listed sales prices for isobutylene polymers are given in Table 6. Polybutenes are manufactured by 10 companies throughout the world: Amoco Chemical Co., (Texas City, Texas; Whiting, Indiana; Antwerp, Belgium); Chevron Chemical Co., (Richmond, California); Cosden Petroleum Co., (Big Spring, Texas); Exxon Chemical Co., (Baytown,

Table 5. Butyl and Halobutyl Producers

Producer	Products	Est. capacity, 10^3t
Exxon Chemical Co.		
Baton Rouge, La.	halobutyl rubbers	109
Baytown, Tex.	butyl rubber	121
Notre Dame de Gravenchon, France	butyl rubber	53
Fawley, UK	butyl, halobutyl rubber	62
Japan Butyl Co., Ltd.[a]		
Kawasaki, Japan	butyl rubber	75
Kashima, Japan	halobutyl rubber	25
Bayer, AG		
Sarnia, Canada	butyl, halobutyl rubber	120
Zwijndrecht, Belgium	butyl, halobutyl rubber	95
former USSR		
Togliatti	butyl rubber	35
Nizhnekamsk	butyl rubber	50

[a]Jointly owned with Japan Synthetic Rubber Co., Ltd.

Table 6. 1993 U.S. Sales Prices for Isobutylene Polymers

Polymer	$/kg
butyl rubber	2.20–2.50
chlorobutyl rubber	2.50
bromobutyl rubber	2.50
polyisobutylene	
high mol wt	2.80
low mol wt	3.00
polybutene	1.00

Texas; Bayway, New Jersey; Köln, Germany); Lubrizol, (Deer Park, Texas; Rouen, France); Petrofina, (Montreal, Canada); Polybutenos Argentinos, (Ensenada, Argentina); BASF, (Ludwigshafen, Germany); British Petroleum Co., (Grangemouth, UK; Naphtachemie, France), and Nipon Sekiyu Kagaku, Ltd. (Kawasaki, Japan).

BIBLIOGRAPHY

"Rubber, Halogenated Butyl" in *ECT* 1st ed., 2nd Suppl. Vol., pp. 716–734 by F. P. Baldwin and I. Kuntz, Esso Research and Engineering Co.; "Elastomers, Synthetic" in *ECT* 2nd ed., Vol. 7, pp. 676–705 by W. M. Saltman, The Goodyear Tire & Rubber Co.; "Elastomers, Synthetic (Butyl Rubber)" in *ECT* 3rd ed., Vol. 8, pp. 470–484 by F. P. Baldwin and R. H. Schatz, Exxon Chemical Co.

  1. U.S. Pat. 2,356,127 (Dec. 29, 1937), R. M. Thomas and W. J. Sparks (to Standard Oil Development Co.).

2. U.S. Pats. 2,681,899 (June, 1959); 2,689,041 (Dec., 1954); 2,720,479 (Oct., 1955), R. A. Crawford and R. T. Morrissey (to B.F. Goodrich Co.).

3. R. T. Morrissey, *Rubber World* **138,** 725 (1955).

4. F. P. Baldwin, *Rubber Chem. Technol.* **52,** 677 (1979); J. V. Fusco and R. H. Dudley, *Rubber World* **144,** 67 (1961).

5. D. Coddington, *Rubber Chem. Technol.* **59,** 905 (1979).

6. S. A. Banks, F. Brzenk, and C. S. Hua, *Rubber Chem. Technol.* **38,** 153 (1965).

7. *Bulletin 12-M,* Amoco Chemicals Corp., 1990.

8. J. P. Kennedy and M. Marechal, *Carbocationic Polymerization,* John Wiley & Sons, Inc., New York, 1982.

9. P. H. Plesch and A. Gandini, *The Chemistry of Polymerization Processes, Monograph 20,* Society of Chemical Industry, London, 1966.

10. J. P. Kennedy and R. M. Thomas, *Polymerization and Polycondensation Processes, Advances in Chemistry Series No. 34,* American Chemical Society, Washington, D.C., 1962, p. 111.

11. *Ibid.,* p. 326.

12. J. P. Kennedy and N. H. J. Canter, *J. Polym. Sci., Part A1* **5,** 2455 (1967).

13. U.S. Pat. 2,356,128 (Aug. 22, 1944), R. M. Thomas and W. J. Sparks (to HASCO).

14. Y. Imanishi and co-workers, *Chem. High Pol. Jpn.* **23,** 152 (1966).

15. C. Corno and co-workers, *Macromolecules* **17,** 37 (1984).

16. W. A. Thaler and J. D. Buckley, Jr., *Rubber Chem. Technol.* **49,** 960 (1976).

17. S. Cesca and co-workers, *Makromol. Chem.* **176,** 2339 (1975).

18. J. Rehner, R. L. Zapp, and W. J. Sparks, *J. Polym. Sci.* **11,** 21 (1953).

19. U.S. Pat. 2,743,993 (June 30, 1953), B. R. Tegge (to Esso Research & Engineering Co.).

20. H-C. Wang and K. W. Powers, *Elastomerics,* Jan. 1992.

21. E. N. Kresge, R. H. Schatz, and H-C Wang, in J. I. Kroschwitz, ed., *Encyclopedia of Polymer Science and Engineering,* Vol. 8, 2nd ed., John Wiley & Sons, Inc., New York, 1987, pp. 423–448.

22. B. Ivan and J. P. Kennedy, *Designed Polymers by Carbocationic Macromolecular Engineering: Theory and Practice,* Hanser Publishers, Munich, 1991.

23. R. Faust, A. Fehervari, and J. P. Kennedy, *Br. Polym. J.* **19,** 379 (1987).

24. R. Vukov, *Rubber Chem. Technol.* **57,** 275 (1984).

25. A. Van Tongerloo and R. Vukov, *Proceedings of the International Rubber Conference,* Venice, 1979, p. 70.

26. F. P. Baldwin and co-workers, *Rubber and Plastic Age* **42,** 500 (1960).

27. F. P. Baldwin and I. J. Gardner, *Chemistry and Properties of Crosslinked Polymers,* Academic Press, Inc., New York, 1977, p. 273.

28. F. P. Baldwin and co-workers, *Adv. Chem. Ser.* **91,** 448 (1969).

29. W. A. Thaler and D. J. Buckley, Sr., *Rubber Chem. Tech.* **49,** 960 (1976).

30. *Enjay Butyl Latex 80-21, Bulletin 012,* Enjay Chemical Co., 1968.

31. J. Walker, G. J. Wilson, and K. J. Kumbhani, *J. Inst. Rubber Ind.* **8,** 64 (1974).

32. U.S. Pat. 5,071,913 (Dec. 10, 1991), K. W. Powers, H-C. Wang, D. C. Handy, and J. V. Fusco (to Exxon Chemical Co.); H-C. Wang, K. W. Powers, and J. V. Fusco, paper presented at *The ACS Rubber Division Meeting,* May 9–12, Mexico City, 1989; I. Dudevani, L. Gursky, and I. J. Gardner, paper presented at *The ACS Rubber Division Meeting,* May 9–12, Mexico City, 1989; L. Gursky and co-workers, *Rubber World* **202,** 41 (1990).

33. U.S. Pat. 5,162,445 (Nov. 10, 1992), K. W. Powers and H-C. Wang (to Exxon Chemical Co.); H-C. Wang and K. W. Powers, *Elastomerics,* Jan. and Feb. 1992.

34. R. M. Thomas and W. J. Sparks in G. S. Whitby, ed., *Synthetic Rubber,* John Wiley & Sons, Inc., New York, 1954, Chapt. 24.

35. A. M. Chatterjee in J. I. Kroschwitz, ed., *Encyclopedia of Polymer Science and Engineering,* Vol. 2, 2nd ed., John Wiley & Sons, Inc., New York, 1985, p. 590.

36. C. E. Schildknecht, *Vinyl and Related Polymers*, John Wiley & Sons, Inc., New York, 1952, p. 571.
37. R. J. Adams and E. J. Buckler, *Trans. Inst. Rubber Ind.* **29,** 17 (1953).
38. R. M. Thomas, *India Rubber World* **130,** 203 (1954).
39. C. E. Schildknecht in *Polymer Processes*, Vol. 10, Interscience Publishers, New York, 1956, p. 208.
40. J. Walker, *Rubber J.* **131,** 39 (1956).
41. R. A. Labine, *Chem. Eng.* **66**(24), 60 (1959).
42. H. S. Pylant, *Oil Gas J.* **61,** (Dec. 2, 1963).
43. R. Dolez, *Genie Chim.* **93**(2), 41 (1965).
44. U.S. Pat. 2,243,658 (May 27, 1941), R. M. Thomas and O. C. Slotterbeck (to Standard Oil Development Co.).
45. U.S. Pat. 2,356,129 (Aug. 22, 1944) W. J. Sparks and R. M. Thomas (to JASCO).
46. U.S. Pat. 2,399,672 (May 7, 1946), A. D. Green, E. T. Marshall, and S. Lane (to Standard Oil Development Co.).
47. U.S. Pat. 2,401,754 (June 11, 1946), A. D. Green (to Standard Oil Development Co.).
48. U.S. Pat. 2,462,124 (Feb. 22, 1949), J. F. Nelson (to Standard Oil Development Co.).
49. U.S. Pat. 2,463,866 (Mar. 8, 1949), A. D. Green (to Standard Oil Development Co.).
50. U.S. Pat. 2,474,592 (June 28, 1949), F. A. Palmer (to Standard Oil Development Co.).
51. U.S. Pat. 2,523,289 (Sept. 26, 1950), P. K. Frolich (to JASCO).
52. U.S. Pat. 2,529,318 (Nov. 7, 1950), B. R. Tegge (to Standard Oil Development Co.).
53. U.S. Pat. 2,530,129 (Nov. 14, 1950), J. H. McAteer and co-workers (to Standard Oil Development Co.).
54. U.S. Pat. 2,581,147 (Jan. 1, 1952), H. G. Schultze (to Standard Oil Development Co.).
55. U.S. Pat. 2,999,084 (Sept. 5, 1961), H. K. Arnold and E. R. Gurtler (to Esso Research and Engineering Co.).
56. U.S. Pat. 3,005,808 (Oct. 24, 1961), R. T. Kelley, J. E. Walker, and B. R. Tegge (to Esso Research and Engineering Co.).
57. G. T. Baumann and M. R. Smith, *Hydrocarbon Process. Pet. Refiner.* **33**(5), 156 (1954).
58. A. M. Valet and co-workers, *Hydrocarbon Process. Pet. Refiner.* **41**(5), 119 (1962).
59. H. Kroper, K. Schlomer, and H. M. Weitz, *Hydrocarbon Process. Pet. Refiner.* **48**(9), 195 (1969).
60. U.S. Pat. 3,665,048 (Jan. 14, 1970), H. R. Grane and I. E. Katz (to Atlantic Richfield Co.).
61. V. Fattore and co-workers, *Hydrocarbon Process.* **60**(8), 101 (1981).
62. Belg. Pat. 882,387 (July 16, 1980), L. A. Smith, Jr. (to Chemicals Research and Licensing Co.); Ger. Pat. 2,534,544 (Feb. 12, 1976), R. Tesei, V. Fattore, and F. Buonomo (to SNAM Piogetti S.p.A.).
63. U.S. Pat. 4,252,710 (Feb. 24, 1981), K. W. Powers and R. H. Schatz (to Exxon Research & Engineering Co.); U.S. Pat. 4,358,560 (Nov. 9, 1982), K. W. Powers and R. H. Schatz (to Exxon Research & Engineering Co.); U.S. Pat. 4,474,924 (Oct. 2, 1984), K. W. Powers and H-C. Wang (to Exxon Research & Engineering Co.).
64. U.S. Pat. 3,023,191 (Feb. 27, 1962), B. R. Tegge, F. P. Baldwin, and G. E. Serniuk (to Esso Research & Engineering Co.).
65. U.S. Pat. 2,940,960 (June 14, 1960), B. R. Tegge and co-workers (to Esso Research & Engineering Co.).
66. U.S. Pat. 3,257,349 (June 21, 1966), J. A. Johnson, Jr. and E. D. Luallin (to Esso Research & Engineering Co.).
67. U.S. Pat. 4,384,072 (May 17, 1983), N. F. Newman and R. C. Kowalski (to Exxon Research & Engineering Co.).
68. U.S. Pat. 4,573,116 (Apr. 23, 1985), R. C. Kowalski, W. M. Davis, and L. Erwin (to Exxon Research & Engineering Co.).

69. U.S. Pat. 4,548,995 (Oct. 22, 1985), R. C. Kowalski and co-workers (to Exxon Research & Engineering Co.).
70. L. A. Wood, *Rubber Chem. Technol.* **49,** 189 (1976).
71. C. S. Fuller, C. J. Frosch, and H. R. Pape, *J. Am. Chem. Soc.* **62,** 1950 (1940).
72. R. H. Boyd and S. M. Breitling, *Macromolecules* **5,** 1 (1972); U. W. Suter, E. Saiz, and P. J. Flory, *Macromolecules* **16,** 1317 (1983).
73. I. Puskas, E. M. Banas, and A. G. Nerhein, *J. Polym. Sci., Plym. Sympos.* **56,** 191 (1976).
74. H. Y. Che and J. E. Field, *J. Polym. Sci., Part B* **5,** 501 (1957).
75. J. F. S. Yu, J. L. Zakin, and G. K. Patterson, *J. Appl. Polym. Sci.* **23,** 2493 (1979).
76. J. V. Fusco and P. Hous in R. F. Ohm, ed., *Vanderbilt Rubber Handbook*, R. T. Vanderbilt Co., Norwalk, Conn., 1990.
77. D. C. Edwards, W. Hopkins, and K. J. Kumbhani, *Poly. News* **14,** 136 (1989).
78. R. A. Orwoll, *Rubber Chem. Technol.* **50,** 451 (1977).
79. T. Fox and P. J. Flory, *J. Phys. Colloid Chem.* **53,** 197 (1949); T. Matsumoto, N. Niskoka, and H. Fujita, *J. Polym. Sci. Part A2* **10,** 23 (1972).
80. J. Brandrup and H. Immergut, eds., *Polymer Handbook*, Vol. 9, 2nd ed., John Wiley & Sons, Inc., New York, 1975.
81. T. Fox and P. J. Flory, *J. Phys. Colloid Chem.* **55,** 221 (1951).
82. D. Ferry, *Viscoelastic Properties of Polymers*, 3rd ed., John Wiley & Sons, Inc., New York, 1980, p. 606.
83. *Ibid.*, p. 608.
84. *Ibid.*, p. 341.
85. *Ibid.*, p. 277.
86. R. H. Boyd and P. V. Krishna Pant, *Macromolecules* **24,** 6325 (1991).
87. V. Stannett in J. Crank and G. S. Park, eds., *Diffusion in Polymers*, Academic Press, Inc., Orlando, Fla., 1968, Chapt. 2.
88. G. Kraus in F. R. Eirich, ed., *Science and Technology of Rubber*, Academic Press, Inc., Orlando, Fla., 1978, p. 297.
89. E. M. Dannenberg, *Rubber Chem. Technol.* **48,** 410 (1975).
90. A. M. Gessler, *Rubber Chem. Technol.* **41,** 1494 (1969).
91. A. R. Payne, *J. Appl. Polym. Sci.* **7,** 873 (1963).
92. G. R. Hamed, *Rubber Chem. Technol.* **54,** 576 (1981).
93. J. R. Dunn in G. Scott, ed., *Developments in Polymer Stabilization 4*, Applied Science Publications, Ltd., London, 1981.
94. F. P. Baldwin and co-workers, *Rubber Plast. Age* **42,** 500 (1961).
95. I. Kuntz, R. L. Zapp, and P. J. Pancirov, *Rubber Chem. Technol.* **57,** 813 (1984).
96. S. Newman, "Use of Butyl Elastomers in the Medical Industry," presented at *ACS Rubber Division Meeting*, Las Vegas, Nev., May 29, 1990.
97. E. T. McDonel, K. C. Baranwal, and J. C. Andries in D. R. Paul and S. Newman, eds., *Polymer Blends*, Vol. 2, Academic Press, Inc., New York, 1978, Chapt. 19.
98. E. N. Kresge, *Rubber Chem. Technol.* **64,** 469 (1991).
99. R. C. Puydak and D. R. Hazelton, *Plast. Eng.* **44,** 37 (1988).

EDWARD KRESGE
H-C. WANG
Exxon Chemical Company

CHLOROSULFONATED POLYETHYLENE

Chlorosulfonated polyethylene, CSM, [68037-39-8], as described according to ASTM D1418, represents a family of cureable polymers, ranging from soft and elastomeric to hard and plastic, containing pendent chlorine and sulfonyl chloride groups. Chlorosulfonated base resins, other than polyethylene, are closely related and are therefore considered a part of this family. Addition of chlorine and sulfonyl chloride groups onto these base resins enhances solubility in common solvents and, when properly compounded and cured, gives resistance to light discoloration, oil, flame, and oxidizing chemicals. Resistance to thermal degradation and ozone attack result from the absence of unsaturation in the polymer backbone. The additional functionality of comonomers and grafts contribute to adhesion and polymer mechanical reinforcement. This combination of value-added properties promotes special end use applications in coatings, adhesives, roofing membranes, electrical wiring insulation, automotive and industrial hose, tubing and belts, and in molded goods.

In the 1940s, Du Pont chemists began a study of chlorinated low density (branched) polyethylene, in response to wartime needs for new synthetic rubbers. The initial result was a rubbery product at about 30% chlorine, called S-1, with vulcanizates having good oil, chemical, and heat resistance properties (1–3). Vulcanization could only be accomplished with peroxides (4), however, which greatly limited its service. Subsequently, simultaneous chlorination and chlorosulfonation of these resins produced polymers with a sulfonyl chloride cure site which allowed cross-linking with safe curatives known in the industry (5). This product, with a molecular weight of about 20,000, was initially called S-2, but was later commercialized in 1952 as Hypalon 20. It was followed in 1957 by commercialization of a lower molecular-weight version called Hypalon 30, targeted for solution applications.

Linear polyethylene (high density) was introduced in the late 1950s, with the development of coordination catalysts. Chlorosulfonation of these base resins gave products that were superior to the earlier, low density types in both chemical resistance and mechanical properties and with distinct advantages in rubber processibility (6,7).

Introduction of linear low density polyethylene in the 1970s and 1980s offered yet another design parameter, giving chlorosulfonated products with the advantages of linear types but with improved low temperature performance (8).

Extension of the chlorosulfonation technology to base resins other than polyethylene, where value can be added, seems a logical next step. Polypropylene and ethylene copolymers containing additional functionality, ie, maleic anhydride graft, vinyl acetate, acrylic acid, etc, have been chlorinated and chlorosulfonated to broaden the application base, particularly in coatings and adhesives (9,10).

The combined worldwide market for this entire family of elastomers had grown to about 48,000 metric tons per year in 1991.

Physical Properties

The rubbery character of the chlorosulfonated polyethylene is derived from the flexibility of the polyethylene chain in the absence of crystallinity. Introduction of chlorine onto the polymer chain provides sufficient molecular irregularity to prevent crystallinity in the relaxed state, but glass-transition temperature, T_g, increases with chlorine content. The degree of rubberiness is, therefore, a function of the balance of residual crystallinity and T_g. The optimum chlorine content for most general rubber applications is the minimum amount required to completely destroy the crystallinity. This chlorine amount varies with the amount of crystallinity in the starting resin. As the chlorine content increases beyond this optimum level, the glass-transition temperature becomes more dominating and the polymers become more plastic and eventually hard and brittle (Table 1).

These values assume chlorination in carbon tetrachloride solution under homogeneous conditions favoring random distribution of chlorine atoms along the chain. Viscous reaction conditions, faster chlorine addition rates, lower temperature conditions, etc, can lead to higher ΔH at equivalent chlorine levels because of more blocky chlorine distribution on the polymer chain.

The sulfonyl chloride groups provide cross-linking sites for nonperoxide curing procedures. Although concentrations as high as 4–5% for specific applications may be achieved, the optimum level for most general rubber applications is normally about 1–1.5 wt %, measured as sulfur, or one sulfonyl chloride group for every 85–110 carbon atoms on the chain. The sulfonyl chloride group, because of its bulk, is more responsible for interference in crystallinity than chlorine atoms on an equimolar basis, but is less important because of its low level.

CSM products may be divided into three groups depending on the type of precursor resin: low density (LDPE), high density (HDPE), and linear low density (LLDPE). LDPE is made by a high pressure free-radical process, while HDPE and LLDPE are made via low pressure, metal coordination catalyst processes (12) (see OLEFIN POLYMERS).

The uncured physical properties of polymers within each group depend on the molecular weight, molecular-weight distribution, and the extent and distribution of chlorination and chlorosulfonation. The molecular weight, molecular-weight distribution (MWD), and chain branching are generally set by the choice of parent polyethylene resin, ie, neither chain scission nor cross-linking take place during chlorosulfonation. This is illustrated by a comparison of gel-permeation

Table 1. Effect of Chlorine Content of CSM on Crystallinity and T_g[a]

Chlorine content, %	Linear ΔH, J/g[b]	T_g, °C	Branched ΔH, J/g[b]	T_g, °C
20	43	−43	9.2	−46
25	24	−26	2.5	−31
30	9	−13	0.2	−19
35	0.1	+4	0.0	+10
40	0.0	+9	0.0	+13
45	0.0	+19	0.0	+27
52			0.0	+65
64	0.0	+95		

[a]Ref. 11.
[b]Heat of fusion of the polymer. To convert J to cal, divide by 4.184.

chromatography (gpc) data of unchlorinated and chlorinated high density polyethylene, which show nearly identical molecular-weight distribution shape functions differing only in the chlorine content (Table 2) (13). The agreement of measured inherent viscosity values and those calculated from gpc indicate no significant long-chain branching in either the unchlorinated or chlorinated polymer.

Table 2. Comparison[a] of Mol Wt Properties of HDPE and Chlorinated HDPE Chlorosulfonated Product

Properties	Polyethylene	Chlorinated[b] product
melt index[c], g/10 min	5.6	
M_n	23,000	36,800
M_w	78,000	117,293
$\eta_{inh}{}^d$, dL/g (calc)	1.18	1.48
$\eta_{inh}{}^d$, dL/g (meas)	1.16	1.50

[a]Gpc data.

[b]35% Cl.

[c]Amount of polymer flowing through a std capillary viscosimeter 2.095-mm dia and 8-mm long at 190°C and a load of 2.16 kg for 10 min.

[d]Inherent viscosity $= \dfrac{\ln \eta_r}{c}$.

These conclusions are further supported by expected physical properties of dried film of chlorosulfonated polyethylene from the different types of polyethylene (Table 3).

These values are given for polymers of narrow molecular-weight distribution, with number-average molecular weights (M_n) of about 20,000 prior to chlorination. Chlorination reactions are carried out under homogeneous conditions in CCl_4 solutions at temperatures between 90 and 110°C with viscosities at about 5 Pa (50 P).

The uncured property most often used for CSM in dry applications is Mooney viscosity, a low shear bulk viscosity (ca 1.6 s^{-1}) determined at 100°C. Mooney viscosity is a rubber industry standard used to predict raw rubber and compound processibility, ie, mixing, extrusion, molding, etc.

Table 3. Properties of Uncured CSM[a]

Property	CSM HDPE		CSM LLDPE		CSM LDPE	
Cl, wt %	21.5	27.7	22	30	20.8	29
sulfur, wt %	1.7	0.51	1.0	0.98	1.3	1.5
tensile strength, MPa[b]	13.4	10.9	9.3	12.9	5.4	0.48
elongation at break, %	420	880	1259	1935	2100	25
ΔH^c, J/g[d]	38	14	12	1.5	11	2

[a]Ref. 14.

[b]To convert MPa to psi, multiply by 145.

[c]Heat of fusion.

[d]To convert J to cal, divide by 4.184.

The bulk viscosity control parameter for CSM, as with other elastomers, is molecular weight (M_w) and molecular-weight distribution (MWD). Mooney viscosity for CSM is determined by selection of the polyethylene precursor.

The melt index (MI) of polyethylene is also a low shear measurement used in the plastics industry to predict processibility and properties. Both MI and Mooney viscosity are related to zero shear viscosity and, therefore, the weight average molecular weight (M_w). Mooney viscosity is directly proportional to M_w, and MI is inversely proportional to M_w. For this reason, polyethylene MI is usually specified for prediction of the CSM Mooney viscosity. Melt index, however, is not always a good measure of M_w, because in most cases it is measured in the non-Newtonian region of the flow curve. The shape and slope in this region are strongly affected by the MWD of the resin. To correct for this, a second MI measurement of polyethylene at a higher shear rate is often measured. The slope of the curve between these points is a function of MWD and may be used, in conjunction with MI, to more accurately predict the CSM viscosity. Increasing the chlorine content of the polymer backbone also increases Mooney viscosity because of increased chain entanglement and intermolecular hydrogen bonding. Generally the Mooney viscosity of the polymer doubles for each 10% chlorine increase at equivalent chain length.

Bulk viscosity for all elastomeric polymers decreases with increasing temperature. CSM viscosity decreases more rapidly with temperature than for most other elastomers because hydrogen bonding and increasing chain entanglement result in high apparent bulk viscosity at low temperatures. At higher temperatures, the hydrogen bonding disappears lowering the bulk viscosity. Figure 1 compares the temperature viscosity relationship of Hypalon 40 and SBR 1500, as well as the interrelationship of all CSM types (15). Among the various CSM types, the more rubbery grades are less temperature-sensitive than those whose chlorine levels place them outside this range because of the absence of crystallinity or glassy regions. The viscosity/shear rate relationship is as expected, ie, polymers with narrow MWD are less shear-sensitive than those with broad MWD (16).

Chemical Properties

The known chemistry of the functional groups in CSM, ie, chlorine and sulfonyl chloride groups, make reactions predictable from their functions in low molecular-weight substances. Acid, ester, and amide derivatives of the sulfonyl chloride group have been prepared and their infrared spectra have been studied (17). The chlorine content, at equivalent cross-link density and base resin type, is the principal factor in determining the chemical properties of the vulcanized CSM polymers. Figure 2 illustrates diagramatically the general trends of several product properties with chlorine content. Actual values for these properties are not given because they depend heavily on compounding, the curing system, and cure states. At low chlorine levels, the polymers retain some of the characteristics of the polyethylene. They are stiffer and harder because of residual polyethylene crystallinity. They also show good electrical properties, heat resistance, and low temperature flexibility compared to other CSM types. With increasing chlorine content, the polymers and vulcanizates become increasingly rubbery. The softest and most

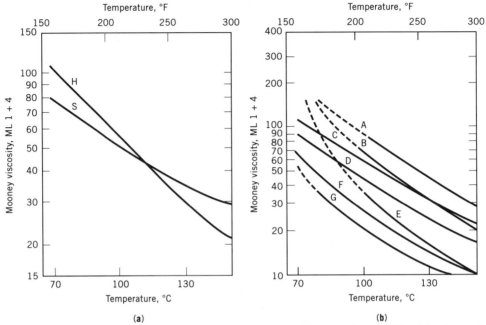

Fig. 1. Viscosity–temperature relationship of (**a**) Hypalon 40 (H) and SBR 1500 (S) and (**b**) various Hypalon polymers: A, Hypalon 4085; B, Hypalon 48; C, Hypalon 40; D, Hypalon 40S; E, Hypalon 45; F, LD999; and G, Hypalon 623.

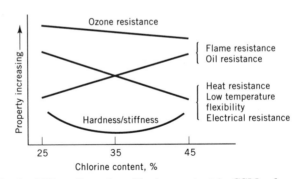

Fig. 2. Effect of varying chlorine content in CSM polymers.

rubbery region is at the minimum chlorine level where the crystallinity disappears, which depends on the crystallinity of the original polyethylene. For branched polymers, ie, LDPE or LLDPE, at a density of 0.92, this chlorine level is about 30% or one chlorine atom for every seven carbon atoms. For HDPE, at 0.96 density, the optimum is about 35% or about one chlorine atom for every five carbon atoms. As the chlorine level is increased beyond this point, the polymers become stiffer again as the glass-transition temperature approaches ambient. These high chlorine types are characterized by excellent oil resistance, flame retardance, and solvent solubility, but have poor resilience and low temperature flexibility.

The distribution of chlorine atoms along the polymer chain has been studied in great detail. The distribution in various functional types is shown in Table 4 (18). High density polyethylene chlorosulfonated to 35% Cl and 1% S has been found to contain only 1.7% highly active chlorines, ie, reactive to weak bases. All of these are attributed to the chlorine in the sulfonyl chloride group and those in beta position to SO_2Cl. No vicinal chlorides groups were found (19).

Model studies with 1-chlorobutane and 1-chlorohexane show that further substitution is directed away from the chlorine substituent in decreasing intensity to four carbon atoms (20). However, the reaction conditions, solvent type, and chlorosulfonating agent can significantly affect the chlorine distribution. Thus CSM polymers prepared in carbon tetrachloride solution using chlorine and sulfur dioxide gas were found to match predictions for random substitution at 75°C (21), whereas a selective distribution was found when the reaction was carried out in chlorobenzene at its boiling point using sulfuryl chloride as the chlorosulfonating agent. Use of gaseous chlorine creates high local concentrations at the bubble interface in the reaction mass. A possible explanation is that the chlorination rate is faster than the diffusion rate, into the reaction mass, causing irregular distribution, resulting in blocky chlorine distribution and higher crystallinity. The melting point and heat of fusion of the polymer depend on the number and length of such blocks (22).

The sulfonyl chloride group is the cure site for CSM and determines the rate and state of cure along with the compound recipe. It is less stable than the Cl groups and therefore often determines the ceiling temperature for processing. The optimum level of sulfonyl chloride to provide a balance of cured properties and processibility is about 2 mol % or 1–1.5 wt % sulfur at 35% Cl. It also undergoes normal acid chloride reactions with amines, alcohols, etc, to make useful derivatives (17).

Table 4. Chlorine Distribution in Chlorosulfonated Polyethylene

	% of Total chlorine	
Structure type	Branched PE	Linear PE
RCH_2—Cl	2.7	0.3
RCHCl—$(CH_2)_n$—CHCl—R	71	83
($n>2$)		
	18	11
($n<2$)		
RR'R''C—Cl	2–3.5	0.0
RCH_2SO_2Cl	0.08	0.0
R—CHCl—CH(SO_2Cl)—R	0.3	0.7
R—CH_2—CH(SO_2Cl)—CH_2—R	4.2	3.8

Vulcanization

Most end use applications of CSM, as with other thermoset polymers, involve mixing with various fillers, plasticizers, processing aids, curatives, etc, then shaping and cross-linking in its final form. Acid acceptors are required in CSM compounds because acidic by-products of curing reactions interfere with the cure and cause equipment corrosion. Acceptable acid acceptors include magnesia, litharge, organically bound lead oxide, calcium hydroxide, synthetic hydrotalcite, and epoxy resins. Zinc oxide or other zinc-containing process aids, ie, zinc stearate used for other elastomers, should be avoided because the zinc chloride formed during curing causes severe polymer degradation.

Compound formulations are chosen to accentuate desired end product properties, ie, heat resistance, water resistance, low temperature flex, or combinations of properties for specific applications, while maintaining cost goals (21). For some applications, ie, nuclear power cable, high filler loadings are undesirable because of their contribution to radiation leakage. Unlike most other elastomers, large quantities of reinforcing fillers are not required to achieve good mechanical properties for high density CSM vulcanizates (23). Carbon black fillers give the best reinforcement of physical properties and the best resistance to chemical degradation. Small amounts of carbon black can also give significant improvement in weatherability. Mineral fillers are used to take advantage of CSM's nondiscoloring characteristics. Clays, silicas, and calcium carbonate augment flame and heat resistance. White compounds include titanium dioxide for light stability. Plasticizers are added to reduce viscosity during processing and increase low temperature flexibility of the final product. Petroleum oils are widely used because of low cost. Ester plasticizers provide the best combination of low temperature flex, heat resistance, and mechanical property retention.

Early recommendations for cross-linking CSM involved the use of divalent metal oxides to form metal sulfonate cross-links (24). The mechanism involves the hydrolysis of the sulfonyl chloride group with a carboxylic acid, ie, stearic acid, which produces water at curing temperatures.

$$RSO_2Cl + H_2O \longrightarrow RSO_3H + HCl$$

$$2\ RSO_3H + MO \longrightarrow RSO_3^-M^{2+-}SO_3R + H_2O$$

$$MO + 2\ HCl \longrightarrow MCl_2 + H_2O$$

These cures, characterized by their ability to proceed at low temperatures, are accelerated by moisture and develop high modulus.

Three different covalent cure systems are commonly used: sulfur-based or sulfur donor, peroxide, and maleimide. These systems rely on a cross-linking agent and one or more accelerators to develop high cross-link density.

Sulfur-based cures give the widest flexibility in choice of compounding ingredients and are the most widely used. The cure involves the decomposition of sulfonyl chloride groups to form polymer radicals that react with active forms of activator decomposition products. The most common activators are dipentamethylenethiuram hexasulfide [971-15-3] (Tetrone A), tetramethylthiuram disulfide [137-26-8] (TMTD), and bisbenzothiazolyl disulfide [120-78-5] (MBTS). Litharge or a substitute is usually necessary to give high states of cure. The litharge probably reacts at cure temperature with the sulfur donor, eg, Tetrone A, to form dithiocarbamate and some active form of sulfur. The carbamate causes decomposition of the sulfonyl chloride group which then reacts with the activated sulfur to form a cross-link.

The best heat resistance is obtained when nickel dibutyldithiocarbamate [*13927-77-0*] (NBC) is incorporated into the compound. NBC contributes to the heat resistance by causing the elimination of unused sulfonyl chloride groups which are then unavailable for additional cross-linking during heat aging. The presence of large amounts of litharge probably also result in some ionic cross-link formation.

In the maleimide cure, the cross-linking agent is N,N'-m-phenylenedimaleimide [*3006-93-7*], HVA-2. This system has two significant advantages: litharge is not required for high cross-link density and low compression set may be obtained. The accelerators are weak bases, ie, N,N'diphenylethylenediamine. The cure mechanism probably involves an amine-catalyzed decomposition of the sulfonyl chloride group or a path of radical anions. The cross-link probably involves the HVA-2. Calcium hydroxide or other SO_2 absorbers must be included for development of good mechanical properties.

Peroxide curing systems are generally the same for CSM as for other elastomers but large amounts of acid acceptor must be present to complete the cure. A small amount of a polyfunctional alcohol, ie, pentaerythritol (PER) in the compound significantly reduces the amount of base required by acting as a solubilizer. Triallyl cyanurate [*101-37-1*] is an additional cure promoter and leads to higher cross-link density.

A comparison of compound recipes and physical properties for the various cure systems is given in Table 5.

Uses

Commercial CSM polymers are currently made by E. I. du Pont de Nemours & Co., Inc. in Northern Ireland and the United States and by Toyo Soda Mfg. Ltd. in Japan. Commercial grades of CSM made by Du Pont under trade names Hypalon and Acsium are shown in Table 6. Similar grades are made by Toyo Soda under CSM TS and CSM CP trade names.

Solution-Grade Polymers   Hypalon 20 is made from a highly branched low density polyethylene (LDPE). It was the original Hypalon elastomer and was formerly used in the manufacture of molded and extruded goods. It was also used in premium-grade white sidewall tires for many years to resist ozone cracking. The base polyethylene contains about four lower alkyl branches and one long-chain alkyl branch for every 100 carbon atoms. This structure gives very rough extrudates unless heavily plasticized. At a chlorine content of 29%, it has slightly poorer oil resistance than CR. The branched-polymer structure, however, makes it readily soluble in common solvents and gives relatively low solution viscosity for its molecular weight. Thus it is useful in roof coatings and the manufacture of tarpaulins and colored awnings.

Hypalon 30 is made from a much lower molecular-weight polyethylene than Hypalon 20, and is also highly branched. The higher chlorine content (43%) gives a glass-transition temperature above ambient and produces hard glossy films with good oil and flame resistance. Its higher chlorine content also makes it more soluble in common solvents. It has the lowest solution viscosity of any commercial CSM and can therefore be used at high solids in solvent paints.

Table 5. Comparison of CSM Cure Systems

Compound, phr	Sulfur cure	Maleimide	Optimum heat resistance	Peroxide
Hypalon 40	100	100	100	100
SRF Black	40	40	40	40
Litharge (PbO)	22		22	
Magnesia (MgO)	10	4	10	5
NBC	3		3	
Tetrone A			0.75	
benzthiazyl disulfide	0.5		0.5	
sulfur	0.56			
PER-200				3
calcium hydroxide		4		
N,N′-m-phenylethylenediamine		2	1	
HVA-2		3		
Varox powder				6
triallyl cyanurate				0.4

Curing conditions

	Sulfur cure	Maleimide	Optimum heat resistance	Peroxide
temperature, °C	153	153	153	160
time, min	30	30	30	20

Vulcanizate properties

	Sulfur cure	Maleimide	Optimum heat resistance	Peroxide
original modulus, 100%, MPa[a]	8.1	6.8	8.5	9.5
tensile strength, MPa[a]	27.5	23.1	31.4	27.6
elongation, %	245	300	285	320
compression set, %, 22 h at 70°C	30	19	23	13

Heat-aged[b] properties

	Sulfur cure	Maleimide	Optimum heat resistance	Peroxide
tensile strength, MPa[a]	26.2	21.3	29.2	26.7
elongation, %	65	60	92	83
vol increase after 72 h at 100°C in ASTM #1 oil	12	14	5	7

[a]To convert MPa to psi, multiply by 145.
[b]After 7 days and 150°C.

General-Purpose Grades. The five grades of Hypalon products based on HDPE at around 35% chlorine may be considered general-purpose elastomers. They are used in hose, tubing, electrical wiring, industrial rolls and belts, molded and sheet goods, as well as many other extruded applications. The linear structure provides good mechanical properties and processibility. Heat and oil resistance is better than the branched types because of the absence of tertiary chlorine atoms and higher chlorine content in the amorphous region. They have good ozone, flame, abrasion resistance, and insulating properties, but their principal uses result from their excellent combination of heat and oil resistance with a balance of good mechanical and processing properties. Figure 3 shows the general position of these types in relation to other general-purpose elastomers (25). Hypalon 40 is

Table 6. Commercial CSM Grades

CSM grade	Chlorine content,%	Sulfur content,%	Mooney viscosity	Principal use
From LDPE				
Hypalons				
20	29	1.4	28	solutions
30	43	1.3	30	solutions
From HDPE				
LD999	34	1.0	30	general-purpose
40S	35	1.0	45	general-purpose
40	35	1.0	56	general-purpose
4085	36	1.0	97	general-purpose
610	35	1.0	126	general-purpose
48	43	1.0	62	oil-resistant hose and freon hose
45	24	1.0	37	membranes
623	24	1.0	21	membranes
From LLDPE[a]				
Acsiums				
6367	26	1.0	40	auto belts
6932	30	1.0	60	auto belts
6983	26	1.0	90	auto belts and hose

[a]Density = 0.92.

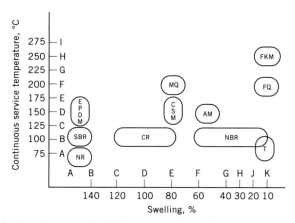

Fig. 3. Heat and oil resistance of CSM compared to other elastomers by ASTM D2000. A–K indicate grades of CSM. The other ASTM designations are as follows: AM, acrylic elastomers; CR, chloroprene rubber; EPDM, ethylene–propylene–diene rubber; FKM, fluorocarbon elastomers; FQ, fluorosilicones; MQ, silicone rubber; NBR, nitrile–butadiene rubber; NR, natural rubber; SBR, styrene–butadiene rubber; and T, thiokol rubber.

a medium viscosity type and is the most versatile of the group. Hypalon 40S and LD999 have lower viscosities than Hypalon 40 and are often used to customize compound viscosity to improve processibility. Hypalon 4085 and Hypalon 610 are higher in molecular weight and are useful where higher uncured integrity, during processing, is needed, ie, pan-cured hose. They are made from very narrow MWD polyethylene which allows high extension with fillers and plasticizers to reduce costs while retaining good mechanical properties in the cured product.

High Chlorine Grades.   Hypalon 48 is a high chlorine grade with a high glass-transition temperature, making it a more plastic material than the general-purpose grades. Oil resistance is equivalent to low grades of nitrile rubber, but heat resistance and low temperature flexibility, although better than nitrile, are poorer than other CSM types. It is made from HDPE, which imparts toughness making it useful in some uncured applications, but its most important uses are in automotive and air conditioning hose, where it has very low permeability to Freon fluorocarbons. Low solution viscosity also makes it useful in some coatings applications where its dry slick finish is advantageous, ie, auto door and window strip coatings.

Low Chlorine Grades.   Hypalon 45 and 623 are used mostly in unsupported roofing membranes and pond liners. The low chlorine level along with special control of reaction conditions provide high levels of polymethylene blocks which produce high crystallinity to give high strength uncured compounds. They can be compounded with white or light-colored pigments and calendered into uncured sheet with sufficient strength to be applied as a sheet roofing or pond liner. Since it is uncured, the overlapped seams can be solvent welded or heat-sealed to give a continuous film. A small amount of divalent metal oxide added to the compound provides a slow ionic cure when the membrane is in place and exposed to moisture, adding to long-term strength. Hypalon 623 is a low viscosity variant designed for easier processing. The CAS Registry Number for Hypalon 623 [*103170-38-3*] indicates a 1-octene polymer with ethene, chlorosulfonated. The octane content of the base resins, however, is very low (0.5–1%) having a density of about 0.95 vs 0.92 for LLDPE. Both Hypalon 45 and 623 contain a chlorine content balanced to give excellent weatherability and low temperature flexibility. When properly compounded, they can be calendered into smooth sheets and rolled for easy application. Hypalon 45 roofs have been found to have good integrity after 20 years weather exposure. Their weather resistance in light pastel colors has made them especially useful in architectural roofing.

Acsium Grades.   Alkyl CSM, ACSM, or Acsium polymers, based on linear low density polyethylene, represent a new class of CSM specifically designed for applications requiring excellent dynamic properties over a broad temperature range, ie, auto timing and accessory drive belts. Hypalon 40 or LD999 have adequate heat and oil resistance for these applications, but have neither low temperature flexibility nor dynamic property requirements. For CSM these properties are dependent almost entirely on chlorine content and state of cure. Introduction of pendent, short alkyl groups on the polymer backbone allows reductions of chlorine content without interference from crystallinity (25,26). By adjusting reaction conditions for selective chlorine distribution and using LLDPE (0.92 density), as a source of pendent alkyl groups, it is possible to lower the chlorine content to about 26% without significant crystallinity. ACSIUM 6367 and 6983 are optimized by low chlorine content for low hysteresis and high resilience. The low T_g also gives good low temperature flexibility. The crystallinity of these polymers is very low ($\Delta H < 4$ J/g). ACSIUM 6932, at 30% Cl, has a sufficiently high chlorine content to give oil resistance equivalent to CR at the expense of slightly higher hysteresis. Acsium polymers can be compounded to give greatly improved low temperature fracture resistance, resilience, damping, and heat resistance over standard CSM grades. Whereas the PE base for Hypalon 623 contains <1% (usually <0.5%) octene comonomer, Ascium grades may contain comonomers other than octene, ie, hexene, butene, or combinations. Actual compositions are considered proprietary by Du Pont.

Hypalon CP Grades. These represent a family of polymers closely related to CSM in that addition of chlorine to precursors other than polyethylene add value as modifiers for adhesives, coatings, and inks because of increased solubility and compatibility.

Hypalon CP-700. In response to the need to reduce the amount of volatile organic solvents used in the application of many coatings and adhesives, Du Pont introduced a line of low molecular-weight chlorosulfonated polyethylene resins. Coded Hypalon CP-700, these resins maintain the same characteristics as the high molecular-weight Hypalon resins they are designed to replace (Hypalon 20 and 30), eg, flexibility, solvent, and chemical resistance and wet adhesion, but can be formulated into considerably higher solids formulations. This significantly reduces the amount of solvent required for their application. Since these are low molecular-weight materials ($M_w \sim 3000$), cure systems needed to be developed to assure that the resins would be able to perform equivalent to their higher molecular-weight predecessors. Several cure chemistries have been developed for Hypalon CP-700 resins including amino-alkoxy silanes, polyamine and polyepoxy, polyol and polyisocyanate, polyamine and polyisocyanate, and redox acrylics. The performance of the cured Hypalon CP-700 resins has been shown to equal or exceed that of similar coatings prepared with the high molecular-weight Hypalon resins alone.

Hypalon CP 826. This is a chlorinated, maleic anhydride modified polypropylene having a chlorine content of about 25% and maleic anhydride content of about 0.8%, developed to promote adhesion of inks and coatings to polypropylene or blends containing polypropylene. It has a solution viscosity of 125 mPa·s (= cP) at 20% solids in xylene and can be used in dilute solutions as a wash primer or a tie layer between materials that are difficult to adhere. CP 827 is a higher molecular-weight analogue with a solution viscosity of 280 mPa·s (= cP) at 20% solids in 80/20 xylene/methyl isobutyl ketone.

Manufacture

Unlike most elastomeric polymers, which are made by direct polymerization of monomers or comonomers, chlorosulfonated polyethylene, as the name implies, is made by chemical modification of a preformed thermoplastic polymer. The chlorination and chlorosulfonation reactions are usually carried out simultaneously but may be carried out in stages.

A simplified form of the reaction may be represented by

$$-(CH_2)_x + (y + z)Cl_2 + (z)SO_2 \xrightarrow{\text{initiator}} -(CH_2)_{x-y-z}-(CH)_y-(CH)_z + (y + z)\,HCl$$
$$\qquad\qquad\qquad\qquad\qquad\qquad\qquad\qquad\qquad\quad |\qquad\quad |$$
$$\qquad\qquad\qquad\qquad\qquad\qquad\qquad\qquad\qquad\quad Cl\qquad SO_2Cl$$

or

$$-(CH_2)_x + (y + z)SO_2Cl_2 \xrightarrow[\text{catalyst}]{\text{initiator +}} -(CH_2)_x-(CH)_y-(CH)_z + (y + z)\,HCl + y\,SO_2$$
$$\qquad\qquad\qquad\qquad\qquad\qquad\qquad\qquad\qquad\quad |\qquad\quad |$$
$$\qquad\qquad\qquad\qquad\qquad\qquad\qquad\qquad\qquad\quad Cl\qquad SO_2Cl$$

The initiator is usually an azo compound and the sulfonation catalyst is usually a tertiary amine, ie, pyridine. Sulfuryl chloride and chlorine may be used without the sulfonation catalyst to produce the same product. Values for x, y, and z are variables and help determine the various product grades.

A number of preparation schemes for CSM have been demonstrated on a laboratory or pilot-plant scale, including fluidized-bed (27), stirred dry powder chlorosulfonation (28), dispersions of fine powders in a nonaqueous medium (28), melt/extrusion (29), falling films or powders (30), and various solution processes both batch and continuous reaction (30–32). The only process of current commercial significance is a solution batch process operating at pressures of 1–350 kPa (0–50 psig). For this process, the solvent must dissolve both the polyethylene and the final product, be essentially inert toward chlorinating agents, have a boiling point such that very high pressures are not required to maintain single-phase reaction, and such that it can be removed below the decomposition temperature of the CSM. Many solvents have been used in laboratory studies, ie, chlorobenzene (33,34), 1,1,2,2-tetrachloroethane, trichlorofluoromethane, and chloroform, but only carbon tetrachloride (33–35) or mixtures with other solvents (36) have been of significant commercial importance in batch units.

Carbon tetrachloride is toxic and an experimental carcinogen requiring stringent environmental control measures. It was placed on the Montreal Protocol list of potential ozone depleting agents in 1989, and its manufacture is scheduled to be completely phased out by the year 1996. It is likely to be replaced as a chlorosulfonation medium by chloroform, not currently considered an ozone depleting agent, or other nonozone-depleting agents which may also be nontoxic noncarcinogens.

The chlorosulfonation reaction is carried out with a mixture of sulfur dioxide and chlorine (37), sulfuryl chloride and chlorine (38), or sulfuryl chloride and a catalytic amount of an organic base (35,37).

The commercial manufacture consists of charging polyethylene pellets into a glass-lined stirred reactor kettle fitted with an agitator and at least one condenser. Staged condensers are sometimes used for improved reaction temperature control. The reactor is then closed and pressurized to about 200 kPa (2 atm). Superheated (150°C) dry CCl_4 is added under pressure, with agitation, such that the polyethylene is dissolved at the end of solvent addition. For high density polyethylene, the dissolving temperature is at least 98°C. The chlorosulfonating agents and free-radical initiator, usually 2,2'-azobisisobutyronitrile, are fed continuously during reaction. The heat of reaction, about 54 J/g (13 cal/g), is removed by condensing solvent vapors and returning the cooled reflux to the reaction vessel. When sulfuryl chloride is the chlorosulfonating agent, the total heat balance is negative resulting in a decrease in temperature during the reaction cycle. The by-product HCl and SO_2 are vented through the condenser and neutralized or optionally mixed with make-up Cl_2, equimolar to the SO_2 content, and passed through an activated carbon reactor to produce sulfuryl chloride for further use as a chlorosulfonating agent. Control of the reaction temperature is important in determining the polymer product chemical composition. Low reaction temperatures aid in increasing polymer S to Cl ratio. When the desired composition is reached as determined by ir analysis, the by-product HCl and SO_2 are removed by decreasing the reaction pressure to atmospheric or a slight vacuum. An epoxy resin stabilizer is then added to scavenge remaining amounts of HCl and to provide storage stability for the final product.

Solvent removal has been accomplished on a commercial scale by steam distillation (39), extraction/extrusion (40), or drum drying (41). The most common isolation procedure, drum drying, involves evaporation on steam heated double-drum surfaces at temperatures of 150–170°C. The counter-rotating drums are set to form a boiling pool of polymer solution above the drums. A thin film of polymer-rich solution is deposited on the drum surface and the remaining solvent is evaporated as the drum rotates through the bite. The dried film, containing <0.2% solvent is removed with a doctor blade to form a sheet that is gathered into a rope, conveyed, and cut into small pieces called chips. The chips are dusted with antimassing agent and packaged in reinforced and polyethylene-lined 25-kg paper bags. The packaged product is shipped by boxcar or trucks.

Process and environmental air is compressed and passed through activated beds to reduce air emission levels to <5 ppm. Process wastewater is air stripped to remove CCl_4. The solvent containing air is also passed through the activated carbon beds. The total air flow through the beds averages about 3965 m^3/min (140,000 SCFM).

Economic Aspects

Production of chlorosulfonated polyethylene products on a worldwide basis is estimated to be approximately 50,000 t/yr. The Du Pont Co. is the primary manufacturer with one plant in the United States having a capacity of about 33,000 t and one plant in Northern Ireland with about 13,000 t capacity. The remaining world capacity is provided by Toyo Soda Manufacturing Ltd. in Japan. The Du Pont Co. manufactures all CSM types under the trade names of Hypalon and Acsium Synthetic Rubber at each of its plant sites. Toyo Soda makes closely related products under the trade name CSM-CP, or Ts. Since the precursor material is primarily ethylene, materials costs are related to petroleum prices. Costs of environmental control procedures surrounding the use of carbon tetrachloride solvent have escalated extensively because it was placed on the Montreal Protocol of potential ozone depleting agents in 1989. Because carbon tetrachloride is to be completely phased out by the year 1996, significant investment in new equipment designed to handle replacement solvents is anticipated. General-purpose grades of CSM, ie, Hypalon 40, have a selling price of $2.80–3.00/kg whereas special grades, ie, Acsium, sell for about $7.70/kg. Average selling prices of all CSM grades is about $3.50/kg.

Health and Safety Factors

Hypalon may contain small amounts of carbon tetrachloride residue and a much lesser amount of chloroform. These chemicals are toxic and carcinogenic with TWA exposure limits of 5 ppm. Both are regulated as air contaminants in the United States under the Occupational Safety and Health Act (OSHA) (42). When large quantities of raw polymer are stored or processed, it is advisable for protection of personnel to provide adequate ventilation to keep employee exposure below regulated levels. Significant amounts of sulfur dioxide and hydrogen chloride may also be evolved during mixing or processing.

Hypalon raw polymer compounds or cured product may be disposed of in an approved landfill. Incineration is not recommended because of the evolution of toxic gases. Additional information is available from Du Pont concerning these and other potential health hazards when handling Hypalon compounds, finished products, thermal decomposition products, or waste disposal (43).

BIBLIOGRAPHY

"Elastomers, Synthetic (Chlorosulfonated Polyethylene)" in *ECT* 3rd ed., Vol. 8, pp. 484–491, by P. R. Johnson, E. I. du Pont de Nemours & Co., Inc.

1. U.S. Pat. 2,503,253 (Apr. 11, 1950), M. L. Ernsberger and P. S. Pinkney (to E. I. du Pont de Nemours & Co., Inc.).
2. U.S. Pat. 2,212,786 (Aug. 27, 1940), D. M. McQueen (to E. I. du Pont de Nemours & Co., Inc.).
3. U.S. Pat. 2,416,016 (Feb. 18, 1947), A. M. McAlevy, D. E. Strain, and F. S. Chance (to E. I. du Pont de Nemours & Co., Inc.).
4. R. E. Brooks, D. E. Strain, and A. M. McAlevy, *Rubber World* **127**, 791 (1953).
5. A. M. Neal, *Report 55-3*, Rubber Chemicals Division, E. I. du Pont de Nemours & Co., Inc., Wilmington, Del., 1955.
6. U.S. Pat. 2,982,759 (May 2, 1961), R. O. Heuse (to E. I. du Pont de Nemours & Co., Inc.).
7. U.S. Pat. 2,972,604 (Feb. 21, 1961), W. B. Reynolds and P. J. Canterino (to Phillips Petroleum Co.).
8. R. L. Dawson, *Hypalon Bulletin HP-2573*, E. I. du Pont de Nemours & Co., Inc., Wilmington, Del., 1991.
9. R. J. Pomije and G. R. McClure, *Am. Paint Coat. J.* **61** (Apr. 30, 1990).
10. E. G. Brugel, *Hypalon preliminary data sheet*, E. I. du Pont de Nemours & Co., Inc., Wilmington, Del., 1991, CP 826 and CP 827.
11. *Report No. 91D-3*, Elastomer Chemicals Division, E. I. du Pont de Nemours & Co., Inc., Wilmington, Del., 1991.
12. J. A. Neumann and F. J. Bockhoff, *Mod. Plast.* **32**(12) (1955).
13. R. L. Dawson, *Hypalon Bulletin HP 276.12*, E. I. du Pont de Nemours & Co., Inc., Wilmington, Del., 1984.
14. P. J. Canterino, in N. M. Bikales, ed., *Encyclopedia of Polymer Science and Technology*, Vol. 6, John Wiley & Sons, Inc., New York, 1967, p. 442.

15. I. C. DuPuis, *Hypalon Bulletin HP201.1 (R1)*, E. I. du Pont de Nemours & Co., Inc., Wilmington, Del., 1962.
16. N. G. Kumar, *J. Polym Sci. Macromol. Rev.* **15,** 255 (1980).
17. M. A. Smook, E. T. Pieski, and C. F. Hammer, *Ind. Eng. Chem.* **45,** 2731 (1953).
18. F. Devlin and T. F. Folk, *Rubber Tech.*, 1098 (Nov./Dec. 1984).
19. E. G. Brehme, *Hypalon Technical Bulletin HP 301 F*, E. I. du Pont de Nemours & Co., Inc., Wilmington, Del., 1987.
20. H. K. Frensdorff and O. Eikener, *J. Polym. Sci. Part A-2* **5,** 1157 (1967).
21. E. G. Brame, *J. Polym. Sci. A-1* **9,** 2051 (1971).
22. C. Zhikuan, S. Liange, and R. N. Sheppard, *Polymers* **25**(3), 369 (1984).
23. *Vanderbilt Rubber Handbook*, R. T. Vanderbilt Co., Norwalk, Conn., 1987, pp. 185–189.
24. S. K. Ostrowski, *Hypalon Bulletin HP-210.1 (R2)*, E. I. du Pont de Nemours & Co., Inc., Wilmington, Del.
25. J. E. A. Williams, paper presented at *The International Rubber Conference*, Paris, June 2–4, 1982.
26. R. E. Ennis and J. G. Pillow, *Materiaux Tecniques* **78**(11–12), 17 (1990).
27. U.S. Pat. 4,560,731 (Dec. 24, 1985), M. R. Rifi (to Union Carbide Co.).
28. Ger. Pat. 1,068,012 (Oct. 29, 1959) (to Farbewerke Hoechst A. G.).
29. U.S. Pat. 3,347,835 (Oct. 17, 1967), J. C. Lorenz (to E. I. du Pont de Nemours & Co., Inc.).
30. U.S. Pat. 3,542,747 (Nov. 20, 1970), R. E. Ennis and J. W. Scott (to E. I. du Pont de Nemours & Co., Inc.).
31. U.S. Pat. 3,296,222 (Jan. 3, 1967), S. Dixon and R. E. Ennis (to E. I. du Pont de Nemours & Co., Inc.).
32. U.S. Pat. 4,871,815 (Oct. 3, 1989), T. Nakagawa and co-workers (to Toyo Soda Mfg. Ltd., Japan).
33. U.S. Pat. 3,314,925 (Apr. 15, 1967), K. F. King (to E. I. du Pont de Nemours & Co., Inc.).
34. G. M. Rankin, *Int. Polym. Sci. Technol.* **8**(2), T/33 (1981).
35. Jpn. Pat. 3-210,305 (Sept. 13, 1991), Y. Fujii and co-workers (to Toyo Soda Mfg. Ltd., Japan).
36. U.S. Pat. 4,544,709 (Oct. 1, 1985), Narui and co-workers (to Toyo Sota Mfg. Ltd., Japan).
37. M. S. Kharasch and A. T. Read, *J. Am. Chem. Soc.* **61,** 3089 (1939).
38. U.S. Pat. 3,299,014 (Jan. 17, 1967), J. Kalil (to E. I. du Pont de Nemours & Co., Inc.).
39. U.S. Pat. 2,592,814 (Apr. 15, 1952), J. L. Ludlow (to E. I. du Pont de Nemours & Co., Inc.).
40. U.S. Pat. 3,387,292 (Mar. 22, 1968), S. G. Smith (to E. I. du Pont de Nemours & Co., Inc.).
41. U.S. Pat. 2,923,979 (Feb. 9, 1960), J. Kalil (to E. I. du Pont de Nemours & Co., Inc.).
42. *Code of Federal Regulations,* Title 29, Government Printing Office, Washington, D.C., Pt 1910-1000.
43. I. C. DuPuis, *Hypalon Bulletin HP-110.1*, E. I. du Pont de Nemours & Co., Inc., Wilmington, Del., 1984.

ROYCE ENNIS
Dupont-Beaumont Works

ETHYLENE–ACRYLIC ELASTOMERS

Ethylene–acrylic is a term used to describe a family of acrylic elastomers which was first introduced commercially in 1975 under the trademark VAMAC by Du Pont Co. The original elastomer was the result of an intensive research effort to develop an oil-resistant polymer with greater heat resistance than nitrile or polychloroprene rubbers, at a moderate cost well below that of silicones and fluoroelastomers. Ethylene–acrylic elastomers are best known for their excellent heat and oil resistance, but they also possess a good balance of compression set resistance, flex resistance, physical strength, low temperature flexibility, and weathering resistance. Special compounded attributes include uniquely temperature-stable vibrational damping properties and the ability to produce flame-resistant compounds with combustion products having an exceptionally low order of smoke density, toxicity, and corrosiveness. Because of this balance of properties, ethylene–acrylic elastomers have found ready acceptance in many high performance applications, especially in the automotive market (1–3).

Polymer Properties

Polymer Composition.   Ethylene–acrylic elastomer terpolymers are manufactured by the addition copolymerization of ethylene [74-85-1] and methyl acrylate [96-33-3], in the presence of a small amount of an alkenoic acid to provide sites for cross-linking with diamines (4).

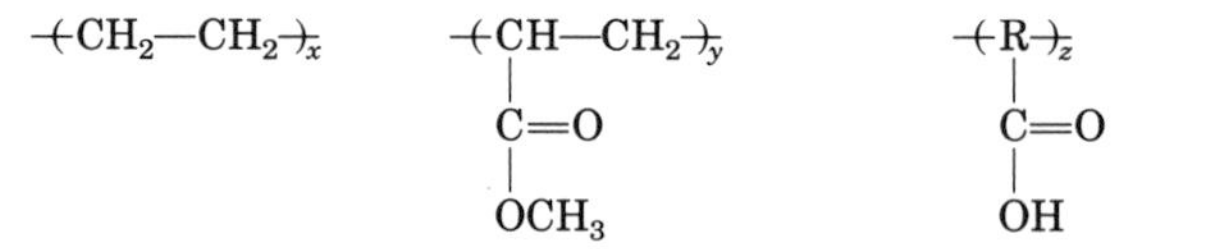

$$\begin{array}{ccc}
-\!\!\!\!+\!CH_2\!-\!CH_2\!\!+\!\!\!-_{\overline{x}} & -\!\!\!\!+\!CH\!-\!CH_2\!\!+\!\!\!-_{\overline{y}} & -\!\!\!\!+\!R\!\!+\!\!\!-_{\overline{z}} \\
 & | & | \\
 & C\!=\!O & C\!=\!O \\
 & | & | \\
 & OCH_3 & OH \\
\text{ethylene} & \text{methyl acrylate} & \text{cure-site monomer}
\end{array}$$

More recently, Du Pont Co. has commercialized a new family of copolymers of just ethylene and methyl acrylate, where the cure-site monomer has been removed from the polymer backbone.

The process yields a random, completely soluble polymer that shows no evidence of crystallinity of the polyethylene type down to $-60°C$. The polymer backbone is fully saturated, making it highly resistant to ozone attack even in the absence of antiozonant additives. The fluid resistance and low temperature properties of ethylene–acrylic elastomers are largely a function of the methyl acrylate to ethylene ratio. At higher methyl acrylate levels, the increased polarity augments resistance to hydrocarbon oils. However, the decreased chain mobility associated with this change results in less flexibility at low temperatures.

Commercial Forms.   Four different base polymers of VAMAC ethylene–acrylic elastomer are commercially available (Table 1). Until 1990, existing grades of ethylene–acrylic elastomers were based on a single-gum polymer, VAMAC G, defined as a terpolymer of 55% methyl acrylate, ethylene, and a cure-site monomer (5). In 1991, a higher methyl acrylate terpolymer, VAMAC LS, was intro-

Table 1. VAMAC Ethylene–Acrylic Elastomer Polymers

Commercial designation	Monomers[a]	Methyl acrylate level	Type of cure system
VAMAC G	E/MA/CS	average	amine
VAMAC LS	E/MA/CS	high	amine
VAMAC D	E/MA	average	peroxide
VAMAC DLS	E/MA	high	peroxide

[a]E is ethylene; MA, methyl acrylate; and CS, proprietary cure-site monomer.

duced. The composition of this polymer was specifically chosen because it significantly increases the oil resistance of the polymer while minimizing losses in low temperature flexibility (6).

A new family of peroxide-cured dipolymers was introduced in 1991. The peroxide cure provides copolymers that cure faster and exhibit good compression set properties without a postcure. The removal of the cure-site has also made the polymer less susceptible to attack from amine-based additives. By varying the methyl acrylate level in the dipolymer, two offerings in this family have been synthesized, VAMAC D and its more oil-resistant counterpart, VAMAC DLS (6).

Curing. Carboxyl cure sites are incorporated in the ethylene–acrylic terpolymer to permit cross-linking with primary diamines (1,7). Guanidines are added to accelerate the cure. Peroxides may also be used as curing agents in the terpolymer, but generally give inferior properties to vulcanizates based on diamine systems (8). Dipolymers are cured only with peroxides.

Aging Properties. The main features of ethylene–acrylic elastomers are heat (177°C) and oil resistance. At elevated temperatures, ethylene–acrylic elastomers age by an oxidative cross-linking mechanism, resulting in eventual embrittlement, rather than reversion. A general heat resistance profile is as follows:

Temperature of continuous exposure, °C	Approximate useful life >50% abs elongation
121	24 months
150	6 months
170	6 weeks
177	4 weeks
191	10 days
200	7 days

As shown, ethylene–acrylic elastomers will function for greater than 24 months at 121°C, or 6 weeks at 170°C continuous service. Exposures up to 190–200°C can be tolerated, although service life at these temperatures are measured in days rather than weeks.

Ethylene–acrylic elastomers are highly resistant to the damaging aspects of weather, ie, sun, water, oxygen, and ozone. Vulcanizates have shown little change in tensile properties and no visible signs of surface deterioration after exposure to the elements in Florida for 10 years. Samples under 20% tensile strain (static) displayed no cracks after one week's exposure to 100 ppm ozone in air, a concentration 100 times greater than is usually specified in qualifying tests.

Fluid Resistance. Ethylene–acrylic elastomers are well suited for applications requiring continuous exposure to hot aliphatic hydrocarbons, a class that includes most lubricants derived from petroleum (9). Volume swell data in Table 2 illustrate the good resistance of ethylene–acrylic elastomers to most common automotive lubricants and hydraulic fluids and to ASTM oils at elevated temperatures. The higher methyl acrylate polymer, VAMAC LS, exhibits lower volume swell in these fluids than does VAMAC G. Low swell in motor oils and transmission fluids indicates usefulness for service as various seals, gaskets, and cooler hoses for transmission and engine oil and seals for wheel and crankshaft bearings. Although resistance to water and glycol is excellent, some antifreeze-additive packages can cause excessive stiffening of the terpolymer vulcanizates. Ethylene–acrylic elastomers should not be selected for service in continuous contact with gasoline, brake fluid, highly aromatic fluids, or polar solvents such as esters and ketones.

Table 2. Chemical Resistance

Medium	Time, h/temp., °C	Volume swell, %	
		VAMAC G	VAMAC LS
ASTM #1 oil	70/150	+5	+2
ASTM #2 oil	70/150	+24	+10
ASTM #3 oil	70/150	+50	+25
motor oil-5W/30	168/150	+19	+9
motor oil-5W/30	1008/150	+20	+10
ATF fluid	168/150	+25	+14
ATF fluid	3000/150	+32	+20
power steering fluid	168/150	+20	+11
fuel B	168/23	+73	+57
kerosene	168/23	+31	
unleaded gasoline	168/23	+68	
water	504/100	+6	
ethylene glycol	504/100	+4	

Mechanical Properties. Typical properties of ethylene–acrylic elastomers, like those of other compounded rubbers, vary widely with formulation and also polymer grade. Among compounding ingredients, reinforcing fillers and plasticizers as well as type and amount of curing agents exert the greatest influence. A typical compound based on VAMAC G and VAMAC LS with pertinent vulcanizate properties is shown in Table 3. Note that whereas VAMAC LS has improved oil resistance versus VAMAC G, about 2–5°C in low temperature flexibility is lost.

Low Temperature Properties. Medium hardness compounds of average methyl acrylate, ie, VAMAC G, without a plasticizer typically survive 180° flex tests at −40°C. Such performance is good for a heat-resistant polymer. Low temperature properties can be greatly enhanced by the use of ester plasticizers (10). Careful selection of the plasticizer is necessary to preserve the heat resistance

Table 3. Vulcanizate Properties of VAMACs G and LS in Black Loaded Compound[a,b]

Property	VAMAC G	VAMAC LS
Physical properties at RT		
100% modulus, MPa[c]	6.1	7.6
tensile strength, MPa[c]	11.9	12.7
elongation at break, %	235	225
hardness, Durometer A	76	80
tear die B, kN/m[d]	59	55
tear die C, kN/m[d]	32	32
Low temperature properties		
glass-transition by dsc, °C	−35	−33
Clashberg torsional stiffness,		
T-69 MPa, °C	−32	−27
TR-10, °C	−27	−22
Fluid resistance, 70 h at 150°C		
volume change, %		
in ASTM #1 oil	2	−3
in ASTM #2 oil	24	10
in ASTM #3 oil	49	25
Compression set, Method B, plied		
70 h at 150°C	16	18
336 h at 150°C	31	33

[a]Compound parts: polymer, 100; Naugard 445 (substituted diphenylamine), 2; Armeen 18D (octadecyl amine), 0.5; stearic acid, 2; Vanfre VAM (complex organic alkyl acid), 0.5; SRF Carbon Black (N774), 100; DIAK #1 (hexamethylenediamine), 1.25; and di-o-tolylguanidine, 4.
[b]Press cure is 10 min at 177°C. Postcure is 4 h at 177°C.
[c]To convert MPa to psi, multiply by 145.
[d]To convert kN/m to ppi, divide by 0.175.

performance of the polymer. Plasticized high methyl acrylate grades lose only a few °C in flexibility, compared to grades with average methyl acrylate levels.

Flame Resistance and Smoke Suppression. Ethylene–acrylic elastomers are not inherently resistant to burning. Through compounding the rate of burning can be retarded and the amount of smoke generated can be suppressed. An important feature of ethylene–acrylic elastomers is their ability to respond to the addition of hydrated alumina (11). This polymer/filler combination provides vulcanizates with good flame resistance, freedom from corrosive gases, and most importantly in many judgments, an unusually low smoke density.

Dynamic Mechanical Properties. Ethylene–acrylic elastomers have a high capacity for damping that is uniquely insensitive to temperature changes between −30 and 160°C. Damping characteristics at room temperature, as indicated by loss tangent (tan ∂), are similar to those of butyl rubber, which is noted for its damping properties. Ethylene–acrylic elastomers differ from butyl and other elastomers, however, by their ability to maintain a high loss tangent as temperature is raised to 160°C. This loss tangent remains virtually unchanged after six months aging in air at 150°C. Damping properties of ethylene–acrylic elastomers are also relatively insensitive to compound variations (12,13).

Processing

Mixing. Ethylene–acrylic elastomers are processed in the same manner as other elastomers. An internal mixer is used for large-scale production and a rubber mill for smaller scales. In either case, it is important to keep the compound as cool as possible and to avoid overmixing. Ethylene–acrylic elastomers require no breakdown period prior to addition of ingredients. Mixing cycles for a one-pass mix are short, typically 2.5–3.5 min. When compounds are mixed on a rubber mill, care should be taken to add the processing aids as soon as possible, after the polymer has been banded on the mill. Normal mill mixing procedures are followed otherwise.

Extrusion and Calendering. Most compounds of ethylene–acrylic elastomers have low nerve and yield smooth extrusions or calendered sheets. To improve collapse resistance, compounding techniques should be used to maximize compound viscosity, eg, VAMAC HG, a higher viscosity version of VAMAC G, and a higher structure carbon black or fumed silica. The extruder temperatures should be kept quite low. A suggested starting temperature gradient would go from 30 to 65°C, with 75°C at the die. Extruded hose is usually vulcanized by exposure to high pressure steam in an autoclave. Other applications, such as wire insulation and jacketing, are subjected to fast, continuous, high pressure steam vulcanization. Vulcanization at atmospheric pressure produces highly porous vulcanizates and is not recommended.

Molding. Parts can be produced from ethylene–acrylic elastomers using compression, transfer, and injection molding techniques. Because the viscosity of these polymers is usually lower than the typical rubber used in the industry, compounds of ethylene–acrylic elastomers have a tendency to trap air during molding, especially in a compression mold. This situation can be avoided with adequate venting of the mold, the use of an effective mold lubricant, the use of compounding techniques to maximize compound viscosity, good preforming techniques, and proper mold temperatures. The low viscosity of ethylene–acrylic elastomers makes them especially good for injection molding.

Mold temperatures vary between 150–200°C, depending on the molding methods and part size. Parts can be molded in 1.5–10 min depending on the configuration and thickness of the part, the mold temperature, and the desired state of cure at demolding. Since most ethylene–acrylic parts are postcured, it is sometimes possible to demold partly cured articles and complete vulcanization in the postcuring oven.

Post-Curing. Whenever production techniques or economics permit, it is recommended that compounds based on terpolymer grades be post-cured. Relatively short press cures can be continued with an oven cure in order to develop full physical properties and maximum resistance to compression set. Various combinations of time and temperature may be used, but a cycle of 4 h at 175°C is the most common. The post-cure increases modulus, greatly improves compresson set performance, and stabilizes the initial stress/strain properties, as chemically the polymer goes from an amide formation to a more stable imide formation. Peroxide-cured dipolymer compounds need not be post-cured.

Adhesion. Commercially available 1- or 2-coat adhesive systems produce rubber failure in bonds between ethylene–acrylic elastomer and metal (14). Adhesion to nylon, polyester, or aramid fiber cord or fabric is greatest when the cord or fabric have been treated with carboxylated nitrile rubber latex.

Additional information on processing compounds of ethylene–acrylic elastomers can be found in References 15–17.

Economic Aspects

The market for ethylene–acrylic elastomers was greater than 2300 t/yr in 1992. The growth rate for ethylene–acrylic elastomers has been greater than 10% since the late 1980s. Over 50% of ethylene–acrylic elastomers are sold in Europe. The price for ethylene–acrylic elastomers in 1992 varied from \$5.59/kg for VAMAC G to \$6.60/kg for VAMAC LS.

Uses

The favorable balance of properties of ethylene–acrylic elastomers has gained commercial acceptance for these elastomers in a number of demanding applications, especially in the automotive industry and in wire and cable jacketing.

Approximately 60% of ethylene–acrylic elastomers is used in automotive applications with hose and tubing as the single largest end use. The recent steady increases in automobile operating temperatures make ethylene–acrylic a prime candidate for under-the-hood applications. Applications include oil and transmission seals, O-rings and gaskets, high velocity CVJ boots, spark plug boots, dampers, and extruded sponges (18,19). Hose applications include transmission and engine oil cooler hose, radiator hose, and turbo charger hose.

Industrial applications include pipe seals, hydraulic system seals, dampers for machinery and high speed printers, and motor lead wire insulation. The fact that the polymer contains no halogens along with certain unique compounding techniques for flame resistance prompts the selection of ethylene–acrylic as jacketing material on certain transportation/military electrical cables and in floor tiles.

BIBLIOGRAPHY

"Acrylic Elastomers" under "Elastomers, Synthetic" in *ECT* 3rd ed., Vol. 8, pp. 459–469, by T. M. Vial, American Cyanamid Co.

1. J. F. Hagman, R. E. Fuller, W. K. Witsiepe, and R. N. Greene, *Rubber Age* **108**(5), 29 (1976).

2. R. G. Peck, *Ethylene/Acrylic Elastomer—Meeting The Challenges of a Demanding Market*, Bulletin EA-020.0185, Du Pont Polymers, Stow, Ohio, Jan. 24, 1985.
3. T. M. Dobel, *Auto. Polym. Des.* **9**(6), 26 (1990).
4. U.S. Pat. 3,883,472 (May 13, 1975), R. N. Greene and K. J. Lewis (to Du Pont Co.).
5. J. F. Hagman, *VAMAC G Gum Ethylene/Acrylic Elastomer*, Bulletin E-52096, Du Pont Polymers, Stow, Ohio, Sept. 1978.
6. T. M. Dobel, *New Development in Ethylene/Acrylic Elastomers*, Paper No. 28, American Chemical Society Rubber Division, Detroit, Mich., Oct. 1991.
7. U.S. Pat. 3,904,588 (Sept. 4, 1975), R. N. Greene (to Du Pont Co.).
8. J. F. Hagman, *Curing Mechanisms of VAMAC*, Bulletin EA-030.0684, Du Pont Polymers, Stow, Ohio, 1980.
9. W. M. Stahl, *Fluid Resistance of VAMAC*, Bulletin H-02366, Du Pont Polymers, Stow, Ohio, Aug. 1988.
10. J. F. Hagman, *Compounding VAMAC For Low-Temperature Performance*, Bulletin E-10770, Du Pont Polymers, Stow, Ohio, Sept. 1978.
11. R. J. Boyce, *Flame Retardance in Mineral-Filled Compounds of VAMAC*, Bulletin E-17762, Du Pont Polymers, Stow, Ohio, Jan. 1978.
12. A. E. Hirsch and R. J. Boyce, *Dynamic Properties of Ethylene/Acrylic Elastomers: A New Heat Resistant Rubber*, Bulletin EA-530.604, Du Pont Polymers, Stow, Ohio, May 1977.
13. A. E. Hirsch, *Dynamic Characteristics of Fluoroelastomers and Ethylene/Acrylic Copolymers*, Bulletin EA-530.602, Du Pont Polymers, Stow, Ohio, 1980.
14. J. F. Hagman, *Bonding Systems for VAMAC*, Bulletin E-36062, Du Pont Polymers, Stow, Ohio, July 1980.
15. J. F. Hagman, *Processing VAMAC*, Bulletin E-10755, Du Pont Polymers, Stow, Ohio, August 1976.
16. C. Williams, *VAMAC Ethylene/Acrylic Elastomer, A Survey of Properties, Compounding and Processing*, Bulletin H-34753, Du Pont Polymers, Stow, Ohio, Jan. 1992.
17. J. W. Crary, *Ethylene/Acrylic Elastomer—Basic Principles of Compounding and Processing*, Bulletin EA-030.0482, Du Pont Polymers, Stow, Ohio, Apr. 1982.
18. R. G. Peck, *Auto. Eng.* **95**(7), 37 (1987).
19. R. E. Vaiden, *Elastomeric Materials for Engine and Transmission Gaskets*, Paper No. 920132, Society of Automotive Engineers, Detroit, Mich., Feb. 1992.

THERESA M. DOBEL
Du Pont Chemical Company

ETHYLENE–PROPYLENE–DIENE RUBBER

Copolymers of ethylene and propylene (EPM) and terpolymers of ethylene, propylene, and a diene (EPDM) as manufactured today are rubbers based on the early work of G. Natta and co-workers (1). A generic formula for EPM and EPDM may be given as follows, where $m = \sim 1500$ (~ 60 mol %), $n = \sim 975$ (~ 39 mol %), $o = \sim 25$ for EPDM (~ 1 mol %), and zero for EPM in an average amorphous molecule, and the comonomers are statistically distributed along the molecular chain.

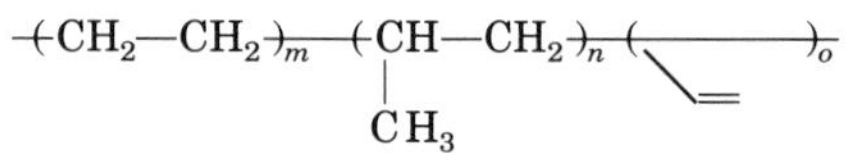

$$-(CH_2-CH_2)_m-(CH-CH_2)_n-(\underline{})_o$$
$$|$$
$$CH_3$$

EPM can be vulcanized radically by means of peroxides. A small amount of built-in third diene monomer in EPDM permits conventional vulcanization with sulfur at the pendent sites of unsaturation.

Among the variety of synthetic rubbers, EPM and EPDM are the fastest growing elastomers, particularly by virtue of their excellent ozone resistance in comparison with natural rubber (cis-1,4-polyisoprene) and its synthetic counterparts IR (isoprene rubber), SBR (styrene–butadiene rubber), and BR (butadiene rubber). Secondly, EPDM rubber can be extended with fillers and plasticizers to an extremely high level in comparison with the other elastomers mentioned, and still give good processibility and properties in end articles. This gives it a price advantage.

Even though EPM and EPDM rubbers have been commercially available for more than 30 years, the technology concerning these products, both their production and their applications, is still very much under development.

Polymer Properties

The properties of EPM copolymers are dependent on a number of structural parameters of the copolymer chains: the relative content of comonomer units in the copolymer chain, the way the comonomers are distributed in the chain, the variation in the comonomer composition of different chains, average mol wt, and mol wt distribution. In the case of EPDM terpolymers there are additional structural features to be considered: amount and type of unsaturation introduced by the third monomer, the way the third monomer is distributed (more or less randomly) along the chain, and long-chain branching. These structural parameters can be regulated via the operating conditions during polymerization and the chemical composition of the catalyst.

Although the rubbery properties of ethylene–propylene copolymers are exhibited over a broad range of compositions, weight percentages of commercial products generally range from 50:50 to 75:25 ethylene:propylene.

In addition to the ethylene:propylene ratio, the average mol wt of the rubber is controlled by polymerization variables. Whereas the polymer chemist generally measures the average mol wt by gel permeation chromatography or intrinsic viscosity, the rubber compounder uses Mooney viscosity for practical purposes. The ethylene–propylene rubbers are controlled within the range of raw polymer Mooney viscosities that has been found to fit the various processing and applications requirements of the rubber industry and includes most other commercial synthetic rubbers. Mooney viscosity of EPM and EPDM is preferably measured four minutes after a one-minute warm-up at 125°C (2). The measurement is expressed as ML (1 + 4) at 125°C and ranges between ca 10 and 90. Grades with higher mol wt are also produced, but are generally extended with either paraffinic or naphthenic oil to reduce the Mooney viscosity for processing purposes.

The structure of EPM shows it to be a saturated synthetic rubber. There are no double bonds in the polymer chain as there are in the case of natural rubber and most of the common commercial synthetic rubbers. The main-chain unsaturation in these latter materials introduces points of weakness. When exposed to the degrading influences of light, heat, oxygen, and ozone, the unsaturated rubbers tend to degrade through mechanisms of chain scission and cross-linking at the points of carbon–carbon unsaturation. Since EPM does not contain any carbon–carbon unsaturation, it is inherently resistant to degradation by heat, light, oxygen, and, in particular, ozone.

The double bonds in natural rubber and the common polydiene synthetic rubbers are essential to their curing into useful rubber products using conventional chemical accelerators and sulfur. As a saturated elastomer, EPM cannot be cured or cross-linked using the long-established manufacturing practices and chemicals pertinent to the unsaturated rubbers. EPDM is a more commercially attractive product that retains the outstanding performance features, ie, heat, oxygen, and ozone resistance, and includes some carbon–carbon unsaturation from a small amount of an appropriate diene monomer to accommodate it to conventional sulfur vulcanization chemistry. A great variety of dienes have been investigated as third monomers (3), of which only three are used commercially at present. A characteristic of the structure of commercially used third monomers is that the two double bonds are nonconjugated. They are either straight-chain diolefins or cyclic and bicyclic dienes with a bridged-ring system.

The most commonly used third monomer is 5-ethylidene-2-norbornene [*16219-75-3*], or ENB:

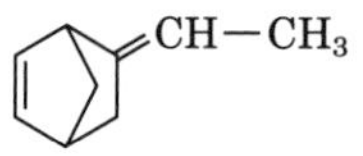

which is polymerized into the ethylene–propylene chain to give poly(ethylene-*co*-propylene-*co*-ENB) [*25038-36-2*] (**1**). The double bond in the bridged, or strained, ring is the more active with respect to polymerization, and the five-membered ring with its double bond is left as a pendent substituent to the main polymer chain.

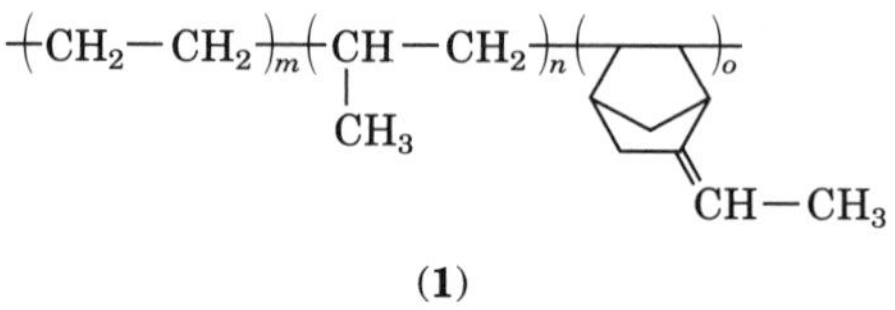

(**1**)

Less commonly used as third monomer is dicyclopentadiene [*77-73-6*], or DCPD, for which, due to its symmetrical shape, the tendency of the second double bond to take part in the polymerization process is more pronounced than for ENB. This is one of the reasons for the formation of long-chain branches. The resulting product is poly(ethylene-*co*-propylene-*co*-DCPD) [*25034-71-3*].

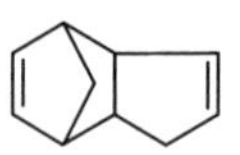

One manufacturer of EPDM uses 1,4-hexadiene [*592-45-0*], or HD:

$$CH_3CH{=}CHCH_2CH{=}CH_2$$

The terminal double bond is active with respect to polymerization, whereas the internal unsaturation remains in the resulting terpolymer as a pendent location for sulfur vulcanization. The polymer is poly(ethylene-*co*-propylene-*co*-1,4-hexadiene) [*25038-37-3*].

Combinations of more than one third monomer are also occasionally applied. The amount of third monomer in general-purpose grades is about 1 mol % or ca 4 wt %. For faster curing grades this amount may be as high as 2 mol % or ca 8 wt %, and there is a tendency to go to even higher amounts. At equal amounts of a third monomer (in mol %) the vulcanization speed of ENB and HD is about equal. DCPD as a third monomer leads to polymers that require about twice as long a curing time for sulfur vulcanization at equal mol % (4). Both EPDM and EPM show outstanding resistance to heat, light, oxygen, and ozone because one double bond is lost when the diene enters the polymer, and the remaining double bond is not in the polymer backbone but external to it. The properties of typical EPDM rubbers are shown in Table 1.

Table 1. Properties of Raw Ethylene–Propylene–Diene Co- and Terpolymers

Property	Value
specific gravity	0.86–0.87
appearance	glassy white
ethylene/propylene ratio by wt	
amorphous types	50/50
crystalline or sequential types	75/25
onset of crystallinity, °C	
amorphous types	below −50
crystalline types	below ca 30
glass-transition temperature,[a] °C	−45 to −60
heat capacity, kJ/(kg·K)[b]	2.18
thermal conductivity, W/(m·K)	0.335
thermal diffusivity, m/s	1.9×10^{-5}
thermal coefficient of linear expansion per °C	1.8×10^{-4}
Mooney viscosity, ML (1 + 4) 125°C[c]	10–90

[a]All types dependent on third monomer content.
[b]To convert kJ to kcal, divide by 4.184.
[c]Oil extended grades, when viscosity >100% of the raw polymer.

Manufacture

The two principal raw materials for EPM and EPDM, ethylene [*74-85-1*] and propylene [*115-07-1*], both gases, are available in abundance at high purity. Propylene is commonly stored and transported as a liquid under pressure. Although ethylene can also be handled as a liquid, usually at cryogenic temperatures, it is generally transported in pipelines as a gas. Of the third monomers, DCPD is also available in abundance (see CYCLOPENTADIENE AND DICYCLOPENTADIENE). ENB is produced by Diels-Alder reaction of cyclopentadiene (in equilibrium with

DCPD) and butadiene. The resulting product vinylnorbornene is rearranged to ethylidene norbornene via proprietary processes. HD is captively produced by the single company using it for EPDM.

EPM and EPDM rubbers are produced in continuous processes. Most widely used are solution processes, in which the polymer produced is in the dissolved state in a hydrocarbon solvent (eg, hexane). These processes can be grouped into those in which the reactor is completely filled with the liquid phase, and those in which the reactor contents consist partly of gas and partly of a liquid phase. In the first case the heat of reaction, ca 2500 kJ (598 kcal)/kg EPDM, is removed by means of cooling systems, either external cooling of the reactor wall or deep-cooling of the reactor feed. In the second case the evaporation heat from unreacted monomers also removes most of the heat of reaction. In other processes using liquid propylene as a dispersing agent, the polymer is present in the reactor as a suspension. In this case the heat of polymerization is removed mainly by monomer evaporation.

All EPDM manufacturing processes are highly proprietary and differ greatly between various suppliers. A great number of patents cover the many details of various processes.

Using a solution process, the choice of catalyst system is determined, among other things, by the nature of the third monomer and factors such as the width of the mol wt distribution to be realized in the product. A number of articles review the influence of catalyst systems on the structural features of the products obtained (3,5–7). The catalyst comprises two main components: first, a transition-metal halide, such as $TiCl_4$, VCl_4, $VOCl_3$, etc, of which $VOCl_3$ is the most widely used; second, a metal alkyl component such as $(C_2H_5)_2AlCl$ diethylaluminum chloride, or monoethylaluminum dichloride, $(C_2H_5)AlCl_2$, or most commonly a mixture of the two, ie, ethylaluminum sesquichloride, $[(C_2H_5)_3Al_2Cl_3]$.

Under polymerization conditions, the active center of the transition-metal halide is reduced to a lower valence state, ultimately to V^{2+}, which is unable to polymerize monomers other than ethylene. The ratio V^{3+}/V^{2+}, in particular, under reactor conditions is the determining factor for catalyst activity to produce EPM and EPDM species. This ratio V^{3+}/V^{2+} can be upgraded by adding to the reaction mixture a promoter, which causes oxidation of V^{2+} to V^{3+}. Examples of promoters in the earlier literature were carbon tetrachloride, hexachlorocyclopentadiene, trichloroacetic ester, and benzotrichloride (8). Later, butyl perchlorocrotonate and other proprietary compounds were introduced (9,10).

For EPDM, the alkylaluminum halide and the V^{3+}/V^{2+} ratio, and more specifically the Lewis acidity (11), influence the degree of long-chain branching by the third monomer. Particularly, ENB tends to cationically couple with another ENB molecule built into another polymer chain due to Lewis acidity, thereby creating another form of long-chain branches, contrary to the first form with DCPD, due to the tendency of its second double bond to take part in the polymerization. In the presence of Lewis bases, the amount of branching is markedly reduced. An increase in reaction temperature enhances the decay of V^{3+}/V^{2+} and the cationic side reactions, thereby increasing the tendency toward long-chain branching. Extended forms of long-chain branching finally lead to gel formation, which damage the processibility of the products in the final application. The coordination Ziegler-Natta catalysts are extremely sensitive to water and other polar materials, as they decompose the catalyst to Lewis acids. Only a few parts per million of water are allowed in any of the feed streams.

In the most widely used solution process, dry solvent, ethylene, propylene, diene, and catalyst and cocatalyst solutions are continuously and proportionately fed to one or a series of polymerization vessels. Polymerization of individual molecules, or chains, is extremely fast, and a few seconds at most is the average life of a single growing polymer molecule from initiation to termination. The polymerization is highly exothermic. This heat must be removed, since the polymerization temperature (ca 35°C) has to be kept within narrow limits to ensure a product with the desired mol wt and mol wt distribution. Most commonly, hydrogen is used as a chain-transfer agent to regulate the average molecular weight.

As the polymer molecules form and dissociate from the catalyst, they remain in solution. The viscosity of the solution increases with increasing polymer concentration. The practical upper limit of solution viscosity is dictated by considerations of heat transfer, mass transfer, and fluid flow. At a rubber solids concentration of 8–10%, a further increase in the solution viscosity becomes impractical, and the polymerization is stopped by killing the catalyst. This is usually done by vigorously stirring the solution with water. If this is not done quickly, the unkilled catalyst continues to react, leading to uncontrolled side reactions, resulting in an increase in Mooney viscosity called Mooney Jumping.

The reactivity of ethylene is high, whereas that of propylene is low and the various dienes have different polymerization reactivities. The viscous rubber solution contains some unpolymerized ethylene, propylene, unpolymerized diene, and about 10% EPDM, all in homogeneous solution. This solution is passed continuously into a flash tank, where reduced pressure causes most of the unpolymerized monomers to escape as gases, which are collected and recycled.

Catalyst residues, particularly vanadium and aluminum, have to be removed as soluble salts in a water-washing and decanting operation. Vanadium residues in the finished product are kept to a few ppm. If oil-extended EPDM is the product, a metered flow of oil is added at this point. In addition, antioxidant, typically of the hindered phenol type, is added at this point.

The rubber is then separated from its solvent by steam stripping. The viscous cement is pumped into a violently agitated vessel partly full of boiling water. The hexane flashes off and, together with water vapor, passes overhead to a condenser and to a decanter for recovery and reuse after drying. Residual unpolymerized ethylene and propylene appear at the hexane condenser as noncondensibles, and are recovered for reuse after drying. The polymer, freed from its carrier solvent, falls into the water in the form of crumb.

The rubber crumb, now a slurry in hot water, is pumped over a shaker screen to remove excess water. The dewatered crumb is fed to the first stage of a mechanical-screw dewatering + drying press. Here, in an action similar to a rubber extruder, all but 3–6% of the water is expressed as the rubber is pushed through a perforated plate by the action of the screw. The cohesive, essentially dry rubber then passes into the second-stage press. This is similar to the first-stage dewatering machine, except that the mechanical action of the screw causes the rubber in the barrel to heat up to temperatures as high as 150°C. This rubber is extruded through a perforated die plate at the end of the machine, the small amount of remaining water is flashed off as a vapor, and the nearly dry rubber crumb is finally subjected to air-drying in a fluid-bed or tunnel drier at temperatures of ca 110°C to reduce the level of remaining volatile matter to <0.3 wt %. This EPDM crumb is then continuously weighed, pressed into bales, and packaged for storage and shipment. Highly crystalline or sequential types are sufficiently form-stable to be produced in the form of pellets by a direct extruder operation, if necessary with application of a small amount of a separating agent.

Noteworthy developments in the field of EPDM-manufacture include the development of highly active catalyst species, achieved by supporting titanium halide on a magnesium chloride carrier. The amount of catalyst consumed per kg of EPDM is below a few ppm. In this case the catalyst-removal operation can be omitted (12). Another development is replacement of the steam-stripping operation by a progressive series of degassing operations, vessels, and extruders, for cost reasons and to do away with water in the whole operation (13). Special reactor designs with multiple feeding locations to achieve special molecular structures for specific purposes have been developed (14). Gas-phase polymerization of EPDM is possible as an extension of the well-known gas-phase processes for polyethylene and polypropylene (15).

Production capacities of EPM/EPDM of the largest manufacturers are listed in Table 2.

Table 2. EPM/EPDM Rubber Production Capacities[a]

Manufacturer	Country	Capacity, t/yr
Bayer/Polysar	U.S.	35,000
Bunawerke Hüls GmbH	Germany	41,000
DSM Elastomers	Netherlands, U.S., Japan	161,000
E.I. du Pont de Nemours & Co., Inc.	U.S.	77,000
EniChem Elastomeri	Italy	85,000
Exxon Chemical	U.S., France	165,000
Japan Synthetic Rubber	Japan	40,000
Mitsui Petrochemical	Japan	60,000
Sumitomo Chemical	Japan	35,000
Uniroyal Chemical Co.	U.S.	90,000
others		30,000

[a]Ref. 16.

Compounding

EPM/EPDM grades have to be compounded with reinforcing fillers if high levels of mechanical properties are required. EPM/EPDM grades with a high Mooney viscosity and a high ethylene content are particularly capable of accepting high loadings of filler, eg, 200 to 400 parts per hundred parts rubber (phr), and plasticizer, eg, 100 to 200 phr, and still giving useful vulcanizates.

Carbon blacks are usually used as fillers. The semi-reinforcing types, such as FEF (Fast Extrusion Furnace) and SRF (Semi-Reinforcing Furnace) give the best performance (see CARBON, CARBON BLACK). To lower the cost and improve the processibility of light compounds, or to lower the cost of black compounds, calcined clay or fine-particle calcium carbonate are used.

The most widely used plasticizers are paraffinic oils. For applications that specify high use temperatures, or for peroxide cures, paraffinic oils of low volatility are definitely recommended. However, since paraffinic oils exude at low temperatures from EPDM vulcanizates, or from high ethylene EPDMs, they are often blended with naphthenic oils. On the other hand, naphthenic oils interfere with peroxide cures. Aromatic oils reduce the mechanical properties of vulcanizates, and they also interfere with peroxide cures. Therefore, they are not recommended for EPM/EPDM.

Although EPM can only be cross-linked with peroxides, peroxide or sulfur plus accelerators or even other vulcanization systems like resins can be used for EPDM. The choice of chemicals used in an EPDM vulcanizate depends on many factors, such as mixing equipment, mechanical properties, cost, safety, and compatibility. In sulfur vulcanization, ENB-containing EPDM is about twice as fast as DCPD-containing EPDM. If peroxide cures are required for better heat stability, DCPD-containing EPDM gives higher cure states than EPM. The reactivity of ENB–EPDM is lower in peroxide cures. For peroxide cures of EPM and to a lesser degree of DCPD–EPDM, activators such as sulfur, acrylates, or maleimides are also needed.

Modern legislation puts much emphasis on the prevention of the formation of carcinogenic secondary nitrosamines as by-products of sulfur vulcanization. This requires the choice of specific accelerators.

In Table 3 a few examples are given of typical EPDM recipes with pertinent cured properties.

Processing

Only compounds of low Mooney EPM or EPDM grades can be mixed on open mills. EPM/EPDM compounds are therefore almost exclusively mixed in internal mixers. In the latter case, the cycles and dump temperatures are about the same as would be used for SBR. The mechanisms, involved in the mixing of EP(D)M, as a function of the structural parameters, in particular mol wt, mol wt distribution, and branching have been studied (17). It turns out that the speed of carbon black dispersion is greatly dependent on these parameters within practical mixing cycles. The mol wt distribution of the EP(D)M rubber should be suitably chosen, depending on the desired degree of carbon black dispersion in the application concerned: a narrow distribution for average carbon black dispersions, and a broad distribution for applications requiring excellent carbon black dispersions.

For EPM/EPDM grades with high ethylene contents, the disintegration of the bales is a dominating factor. If the polymer is in the form of a compact bale, it is difficult to disintegrate the rubber sufficiently to form a fine dispersion with the other compounding ingredients, so that the shearing action of the mixer can disperse the ingredients evenly. In such cases particulate forms of the rubber such as crumbs, pellets, and friable bales shorten the mixing cycle considerably.

Table 3. EPM/EPDM Compounding Recipes for a 60° Shore A Solid Application

Components	Sulfur cure[a]	Low nitrosamine sulfur cure[a]	Peroxide cure[b]
EPDM, 4 wt % ENB, Keltan[c] 4802	100	100	
EPDM, 4 wt % DCPD, Keltan[c] 720			100
zinc oxide	5	5	
stearic acid	1	2	0.5
carbon black, N 683	105		
carbon black, N 765		95	
carbon black, N 550			110
CaCO$_3$ whiting	50		80
oil, Sunpar[d] 2280	70	50	
oil, Sunpar[d] 150			75
CaO, 75% paste	8	6	6
poly(ethylene glycol) 4000			2
N-cyclohexylbenzothiazole-2-sulfenamide, 70%	2.1		
zinc dibutyldithiocarbamate, 80%	2.5		
ethylthiourea	1.0		
tellurium diethyl dithiocarbamate, 75%	0.63		
tetramethylthiuram disulfide, 50%	1.0		
zinc dialkyl dithiophosphate, 67%		2.5	
dithiodicaprolactam, 80%		1.0	
zinc dibenzyldithiocarbamate, 70%		1.0	
2-mercaptobenzothiazole, 80%		1.9	
sulfur	1.5	1.5	
1,3-bis(t-butylperoxy-isopropyl)benzene, 40%			6
1,1-di-t-butylperoxy-3,3,5-trimethylcyclohexane, 40%			3
trimethylolpropane trimethacrylate			1

Properties

compound Mooney[e] ML (1 + 4) 100°C	83	92	44
profile vulcanization	UHF/hot air	UHF/hot air	LCM
temperature, °C/time, s	250/160	250/215	240/60
hardness[f], IRHD	64	60	61
tensile strength, N/mm^2[g,h]	10.0	13.7	7.7
elongation at break[h], %	400	460	330
compression set[i], 22 h/70°C, %	13	19	
compression set[i], 22 h/100°C, %			16

[a]Automotive profile, phr.
[b]Building profile, phr.
[c]Keltan is a registered trademark of DSM.
[d]Sunpar is a registered trademark of Sun-Oil Co.
[e]ISO 289.
[f]ISO 48.
[g]To convert N/mm^2 to psi, multiply by 145.
[h]ISO 37.
[i]ISO 815.

In general, EPM/EPDM compounds can be extruded easily on all commercial rubber extruders. Furthermore, EPDM compounds can be calendered both as unsupported sheeting and onto a cloth substrate.

EPM/EPDM compounds are cured on all of the common rubber-factory equipment: press cure, transfer molding, steam cure, hot-air cure, and injection molding are all practical. Where profile extrusion is the most important shaping technique, molten-salt (LCM) cure and ultra high frequency (UHF) electromagnetic heating followed by hot-air are the most common vulcanization techniques. UHF-heating, however, is only applicable for black-filled compounds. Further, in the case of peroxide curing, open-air vulcanization in the presence of free oxygen in UHF and hot air is problematic, because of surface degradation due to peroxide radicals in combination with the oxygen. This, however, is an intrinsic problem of these curing techniques, occurring with all other rubbers as well.

Properties of EPM and EPDM Vulcanizates

Mechanical properties depend considerably on the structural characteristics of the EPM/EPDM and the type and amount of fillers in the compound. A wide range of hardnesses can be obtained with EPM/EPDM vulcanizates. The elastic properties are by far superior to those of many other synthetic rubber vulcanizates, particularly of butyl rubber, but they do not reach the level obtained with NR or SBR vulcanizates. The resistance to compression set is surprisingly good, in particular for EPDM with a high ENB content.

The resistance to heat and aging of optimized EPM/EPDM vulcanizates is better than that of SBR and NR. Peroxide-cured EPM can, for instance, be exposed for 1000 h at 150°C without significant hardening. Particularly noteworthy is the ozone resistance of EPM/EPDM vulcanizates. Even after exposure for many months to ozone-rich air of 100 pphm, the vulcanizates will not be seriously harmed. EPM/EPDM vulcanizates have an excellent resistance to chemicals, such as dilute acids, alkalies, alcohol, etc. This is in contrast to the resistance to aliphatic, aromatic, or chlorinated hydrocarbons. EPM/EPDM vulcanizates swell considerably in these nonpolar media.

The electrical-insulating and dielectric properties of the pure EPM/EPDM are excellent, but in compounds they are also strongly dependent on the proper choice of fillers. The electrical properties of vulcanizates are also good at high temperatures and after heat-aging. Because EPM/EPDM vulcanizates absorb little moisture, their good electrical properties suffer minimally when they are submerged in water.

Health and Safety Factors

EP(D)M is not classified as hazardous. It is not considered carcinogenic according to OSHA Hazard Communications Standard and IARC Monographs. Commonly used paraffinic extender oils contain less than 0.1 wt % polynuclear aromatics PNAs.

In handling EP(D)M, normal industrial hygienic procedures should be followed. It is advisable to minimize skin contact. The use of EP(D)M is permitted for food contact under the conditions given in the respective FDA-paragraphs: §177.1520 for Olefin polymers, and §177.2600 for rubber articles intended for repeated use.

Uses

Expressed as percentages of total annual synthetic rubber consumption world-wide, EPM and EPDM have increased from 0% in 1964 to 10.6% in 1990, as shown in Table 4. Contrary to the general-purpose elastomers such as NR, SBR, and BR, EPM and EPDM still show a steady growth over the years. Part of this growth still comes from replacement of these commodity rubbers by virtue of their better ozone and thermal resistance.

The main uses of EPM of EPDM are in automotive applications as profiles, (radiator) hoses, and seals; in building and construction as profiles, roofing foil, and seals; in cable and wire as cable insulation and jacketing; and in appliances as a wide variety of mostly molded articles.

Another important application for EPDM is in blends with general-purpose rubbers. Ozone resistance is thus provided, with the host rubber comprising the principal portion of the blend. This technique has been applied in enhancing the ozone and weathering resistance of tire sidewalls and cover strips. This use accounts for essentially all EPDM consumption in tires, as the dynamic and wear properties of EPDM do not favor its use for the carcass and tread parts of the tire. EPDM compounds are, moreover, nontacky, whereas tackiness is a prerequisite for building tires. Although this can be solved to some extent, there are no all-EPDM tires currently being produced. Economic factors favor the use of natural and general-purpose synthetic rubbers in tires.

Considerable amounts of EPM and EPDM are also used in blends with thermoplastics, eg, as impact modifier in quantities up to ca 25% wt/wt for polyamides, polystyrenes, and particularly polypropylene. The latter products are used in many exterior automotive applications such as bumpers and body panels. In blends with polypropylene, wherein the EPDM component may be increased to become the larger portion, a thermoplastic elastomer is obtained, provided the EPDM phase is vulcanized during the mixing with polypropylene (dynamic vul-

Table 4. Worldwide EPM and EPDM Consumption as Percentage of Total Rubber Consumption[a]

Year	EPM/EPDM, t/yr	EPM/EPDM consumption, %[b]
1964		0.0
1970	89,000	
1980	328,000	
1990	609,000	10.6

[a] Ref. 16.
[b] Of total synthetic rubber consumption.

canization) to suppress the flow of the EPDM phase and give the end product sufficient set.

Substantial amounts of EPM are also used as additives to lubrication oils because of its excellent heat and shear stability under the operating conditions of automobile engines.

BIBLIOGRAPHY

"Elastomers, Synthetic (Ethylene–Propylene)" in *ECT* 3rd ed., pp. 492–500, by E. L. Borg, Uniroyal Chemical Co.

1. G. Natta, *Chim. Ind.* **39**, 733–743 (1957).
2. Evaluation procedures, *ISO 4097 Rubber, ethylene–propylene–diene (EPDM)*, Non-oil extended raw general-purpose types, 1992.
3. S. Cesca, *Macromol. Rev.* **10**, 1–230 (1975).
4. C. A. van Gunst, J. W. F. van 't Wout, and H. J. G. Paulen, *Gummi Asbest Kunststoffe* **29**, 184–189 (1976).
5. J. Boor, *Ziegler-Natta Catalysts and Polymerizations*, Academic Press, Inc., New York, 1979, p. 563.
6. C. Cozewith and G. VerStrate, *Macromol.* **4**, 482 (1971).
7. G. Natta, A. Valvasori, and G. Satori in J. P. Kennedy and E. G. M. Törnquist, eds., *Polymer Chemistry of Synthetic Elastomers*, John Wiley & Sons, Inc., New York, 1969, p. 687.
8. D. L. Christman, *J. Polym. Sci. Part A1* **10**, 471 (1972).
9. H. Emde, *Angewandte Makromol. Chem.* **60/61**, 1–20 (1977).
10. U.S. Pats. 4,420,595 (Dec. 13, 1983) and 4,435,552 (Mar. 6, 1984), G. Evens and co-workers (to Stamicarbon).
11. C. Cozewith, W. W. Graessley, and G. Verstrate, *Chem. Eng. Sci.* **34**, 245–248 (1979); *J. Appl. Polym. Sci.* **25**, 59–62 (1980).
12. U.S. Pat. 3,789,036 (Jan. 29, 1974), P. Longi and co-workers (to Montecatini Edison S.p.A).
13. U.S. Pat. 3,726,843 (Apr. 10, 1973), C. Anolick and co-workers (E.I. du Pont de Nemours & Co., Inc.).
14. U.S. Pats. 4,540,753 (Sept. 10, 1985) and 4,716,207 (Dec. 29, 1987), C. Cozewith, S. Ju, and G. W. Verstrate (to Exxon Research and Engineering Co.).
15. U.S. Pat. 4,710,538 (Dec. 1, 1987), R. J. Jorgensen (to Union Carbide Corp.).
16. *Worldwide Rubber Statistics 1991*, International Institute of Synthetic Rubber Producers Inc., Houston, Tex.
17. J. W. M. Noordermeer and M. Wilms, *Kautschuk + Gummi. Kunststoffe*, **41**, 558–563 (1988); **44**, 679–683 (1991).

General References

F. P. Baldwin and G. VerStrate, *Rubber Chem. Technol.* **45**, 709–881 (1972).

JACOBUS W. M. NOORDERMEER
DSM Elastomers Europe

FLUOROCARBON ELASTOMERS

Fluorocarbon elastomers are synthetic, noncrystalline polymers that exhibit elastomeric properties when cross-linked. They are designed for demanding service applications in hostile environments characterized by broad temperature ranges and/or contact with chemicals, oils, or fuels.

Military interest in the development of fuel and thermal resistant elastomers for low temperature service created a need for fluorinated elastomers. In the early 1950s, the M. W. Kellogg Co. in a joint project with the U.S. Army Quartermaster Corps, and 3M in a joint project with the U.S. Air Force, developed two commercial fluorocarbon elastomers. The copolymers of vinylidene fluoride, $CF_2{=}CH_2$, and chlorotrifluoroethylene, $CF_2{=}CFCl$, became available from Kellogg in 1955 under the trademark of Kel-F (1–3) (see FLUORINE COMPOUNDS, ORGANIC—POLYCHLOROTRIFLUOROETHYLENE; POLY(VINYLIDENE) FLUORIDE). In 1956, 3M introduced a polymer based on poly(1,1-dihydroperfluorobutyl acrylate) trademarked 3M Brand Fluororubber 1F4 (4). The poor balance of acid, steam, and heat resistance of the latter elastomer limited its commercial use.

In the late 1950s, the copolymers of vinylidene fluoride and hexafluoropropylene, $CF_2{=}CFCF_3$, were developed on a commercial scale by 3M (Fluorel) and by Du Pont (Viton) (5–8). In the 1960s, terpolymers of vinylidene fluoride, hexafluoropropylene, and tetrafluoroethylene, $CF_2{=}CF_2$, were developed (9) and were commercialized by Du Pont as Viton B. At about the same time, Montedison developed copolymers of vinylidene fluoride and 1-hydropentafluoropropylene as well as terpolymers of these monomers with tetrafluoroethylene, marketed as Tecnoflon polymers (10,11).

In the 1960s and 1970s, additional elastomers were developed by Du Pont under the Viton and Kalrez trademarks for improved low temperature and chemical resistance properties using perfluoro(methyl vinyl ether), $CF_2{=}CFOCF_3$, as a comonomer with vinylidene fluoride and/or tetrafluoroethylene (12,13) (see FLUORINE COMPOUNDS, ORGANIC—TETRAFLUOROETHYLENE POLYMERS AND COPOLYMERS).

Bromine- and iodine-containing fluoroolefins have been copolymerized with the above monomers in order to allow peroxide cure (14–21). The peroxide cure system does not require dehydrofluorination of the polymer backbone, resulting in an elastomer that shows improved properties after heat and fluid aging.

Copolymers of propylene and tetrafluoroethylene, which are sold under the Aflas trademark by 3M, have been added to the fluorocarbon elastomer family (21–26). Also 3M has introduced an incorporated cure copolymer of vinylidene fluoride, tetrafluoroethylene and propylene under the trademark Fluorel II (27). These two polymers (Aflas and Fluorel II) do not contain hexafluoropropylene. The substitution of hexafluoropropylene with propylene is the main reason why these polymers show excellent resistance toward high pH environments (28). Table 1 lists the principal commercial fluorocarbon elastomers in 1993.

Properties

Table 2 summarizes general characteristics of vulcanizates prepared from commercially available fluorocarbon elastomer gumstocks.

Table 1. Commercial Fluorocarbon Elastomers

Copolymer	CAS Registry Number	Trademark	Supplier
poly(vinylidene fluoride-*co*-hexafluoropropylene)	[*9011-17-0*]	Dai-el	Daikin
		Fluorel	3M
		Tecnoflon	Ausimont
		Viton	Du Pont
plus cure-site monomer[a]		Fluorel	3M
poly(vinylidene fluoride-*co*-hexafluoropropylene-*co*-tetrafluoroethylene) with and without cure-site monomer[a]	[*25190-89-1*]	Dai-el	Daikin
		Fluorel	3M
		Tecnoflon	Ausimont
		Viton	Du Pont
poly(vinylidene fluoride-*co*-tetrafluoroethylene-*co*-perfluorovinyl ether) plus cure-site monomer[a]		Viton	Du Pont
		Tecnoflon	Ausimont
poly(tetrafluoroethylene-*co*-perfluoro(methyl vinyl ether) plus cure-site monomer[b]		Kalrez	Du Pont
poly(tetrafluoroethylene-*co*-propylene)[a]	[*27029-05-6*]	Aflas	Asahi Glass
		Aflas	3M
poly(vinylidene fluoride-*co*-chlorotrifluoroethylene)[a]	[*9010-75-7*]	Kel-F	3M
poly(vinylidene fluoride-*co*-tetrafluoroethylene-*co*-propylene)[b]	[*54675-89-7*]	Fluorel II	3M
		Aflas	Asahi Glass
plus cure site		Aflas	3M

[a] Peroxide curable.
[b] Proprietary cure system.

Thermal Stability. The retention of elongation after thermal aging of fluorocarbon elastomers is an indication of their thermal stability. Figure 1 is a plot of percent retention of initial elongation vs days exposure to dry heat (150°C) for a number of oil-resistant elastomers (29), and shows that fluorocarbon elastomers are far superior to hydrocarbon elastomers. A more severe test at 205°C shows that a typical fluorocarbon molded goods compound retains 95% of initial elongation after one year. Retention of tensile strength is another important characteristic of fluorocarbon elastomers. Figure 2 shows the results of long-term heat aging on a typical O-ring compound made from vinylidene fluoride/hexafluoropropylene copolymer. Fifty percent of the initial tensile strength is retained after a period of one year at 205°C or after more than two months at 260°C.

Chemical Resistance. Fluorocarbon elastomer compounds show excellent resistance to automotive fuels and oils, hydrocarbon solvents, aircraft fuels and oils, hydraulic fluids, and certain chlorinated solvents, and may be used without reservation.

They show good to excellent resistance to highly aromatic solvents, polar solvents, water and salt solutions, aqueous acids, dilute alkaline solutions, oxidative environments, amines, and methyl alcohol. Care must be taken in choice of proper gum and compound. Hexafluoropropylene-containing polymers are not recommended for use in contact with: ammonia, strong caustic (50% sodium hydroxide above 70°C), and certain polar solvents such as methyl ethyl ketone and

Table 2. Fluorocarbon Elastomers Physical Property Ranges

Property	Value
Physical properties	
tensile strength, MPa[a]	7.00–20.00
100% modulus, MPa	2.00–16.00
elongation at break, %	100–500
hardness range, Shore A	50–95
compression-set[b]	
70 h at 25°C	9–16
70 h at 200°C	10–30
1000 h at 200°C	50–70
specific gravity (gumstock)	1.54–1.88
low temperature flexibility, °C[c]	0 to −30
brittle point (ASTM D746), °C	0 to −50
thermal degradation temperature, °C	400 to 550
General characteristics	
gas permeability	very low
flammability	self-extinguishing or nonburning (when properly formulated)
radiation resistance	good to fair
abrasion resistance	good and satisfactory for most uses
weatherability and ozone resistance	outstanding (unaffected after 200 h exposure to 150 ppm ozone)

[a]To convert MPa to psi, multiply by 145.
[b]ASTM method B, 3.5 mm O-ring.
[c]Highly dependent on grade of material used.

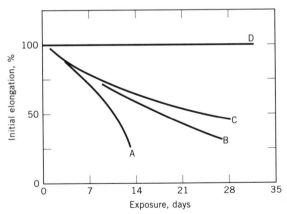

Fig. 1. Retention of elongation of vulcanized elastomers at 150°C. A, nitrile rubber, NBR; B, ethylene–propylene–diene rubber, EPDM; C, acrylic elastomer, AM; D, fluorocarbon elastomer.

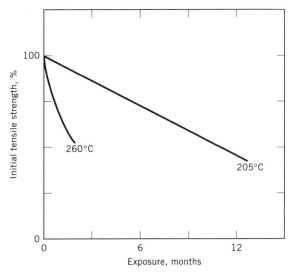

Fig. 2. Tensile strength retention, continuous service, for fluorocarbon elastomers, Compound I (see Table 4).

low molecular weight esters. However, perfluoroelastomers can withstand these fluids. Propylene-containing fluorocarbon polymers can tolerate strong caustic.

Recent innovations in fluorocarbon elastomer development have led to more highly fluorinated materials that possess greater solvent resistance. Included in this class of highly fluorinated materials are Fluorel FLS 2530 and 2650, Viton GF and GFLT, Dai-el G-912, and Kalrez. Figure 3 demonstrates the effect of high fluorine incorporation on volume swell resistance. These highly solvent resistant materials are expected to find wide applicability in the automotive, pollution control, and petrochemical markets.

Compression Set Resistant.   One property of fluorocarbon elastomers that makes them uniquely valuable to the sealing industry is their extreme resistance to compression set. Figure 4 plots compression set vs time for compounds prepared especially for compression set resistance (O-ring grades).

Manufacture and Processing

Manufacture of Fluorocarbon Elastomers.   Elastomers listed in Table 1 are typically prepared by high pressure, free-radical, aqueous emulsion polymerization techniques (30–33). The initiators (qv) can be organic or inorganic peroxy compounds such as ammonium persulfate [7727-54-0]. The emulsifying agent is usually a fluorinated acid soap, and the temperature and pressure of polymerization ranges from 30 to 125°C and 0.35 to 10.4 MPa (50–1500 psi). The molecular weight of the resultant polymers is controlled by the ratios of initiator to monomer, by the choice of chain-transfer reagents, or both. Typical chain-transfer

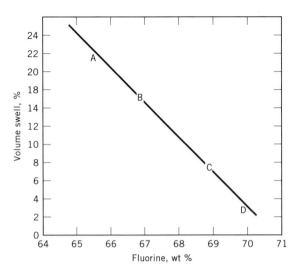

Fig. 3. The percent volume swell in benzene after seven days at 21°C compared with the wt % of fluorine on standard recommended compounds. A, copolymers of vinylidene fluoride–hexafluoropropylene; B, terpolymers of vinylidene fluoride–hexafluoropropylene–tetrafluoroethylene; C, terpolymers of vinylidene fluoride–hexafluoropropylene–tetrafluoroethylene-cure site monomer; D, copolymer of tetrafluoroethylene–perfluoro(methyl vinyl ether)-cure site monomer.

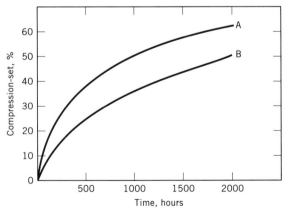

Fig. 4. Compression-sets of fluorocarbon elastomers at 200°C, 3.5 mm O-rings; A, Compound I (see Table 4); B, Compound II (see Table 4).

agents are 2-propanol, methanol, acetone, diethyl malonate [*105-53-3*], and do-decylmercaptan (34–36). A typical polymerization recipe is as follows.

Component	Amount, g
vinylidene fluoride	61
hexafluoropropylene	39
diethyl malonate	0.1
ammonium persulfate	0.5
ammonium perfluorooctanate	0.12
potassium phosphate, dibasic	1.0
water	304

The aqueous emulsion polymerization can be conducted by a batch, semibatch, or continuous process (Fig. 5). In a simple batch process, all the ingredients are charged to the reactor, the temperature is raised, and the polymerization is run to completion. In a semibatch process, all ingredients are charged except the monomers. The monomers are then added continuously to maintain a constant pressure. Once the desired solids level of the latex is reached (typically 20–40% solids) the monomer stream is halted, excess monomer is recovered and the latex is isolated. In a continuous process (37), feeding of the ingredients and removal of the polymer latex is continuous through a pressure control or relief valve.

The polymer latex is then coagulated by addition of salt or acid, a combination of both, or by a freeze–thaw process. The crumb is washed, dewatered, and dried. Since most fluorocarbon elastomer gums are sold with incorporated cure systems, the final step in the process involves incorporation of the curatives. This can be done on a two-roll mill, in an internal mixer, or in a mixing extruder.

Cross-Linking Chemistry. Like other thermosetting elastomers, fluorocarbon elastomers must be cured in order to get useful properties. Three distinct cross-linking systems have been developed to achieve this goal: diamine, bisphenol–onium, and peroxide curing agents (Table 3). Over the years, the bisphenol–onium cure system, which is the most practical in terms of processing latitude and cured properties, has become the most widely used.

These three cure systems have in common the need for a two-step cure cycle to generate the best cured properties. The first step is the application of heat and pressure in a mold to shape the article (press cure). The second step is a high temperature oven cycle at atmospheric pressure to obtain the final cured properties.

The manufacture of the majority of fluorocarbon elastomer gums includes the addition of an incorporated cure system comprising an organic onium cure accelerator, such as triphenylbenzylphosphonium chloride [*1100-88-5*], and a bisphenol cross-linking agent, such as hexafluoroisopropylidenediphenol [*1478-61-1*]. These incorporated cure systems offer improved compression set performance, processing safety, and fast cure cycles to fabricators, who need add only metal oxides as acid acceptors and reinforcing fillers for a complete formulation (38–43).

The chemistry of this cure system has been the subject of several studies (44–47). It is now generally accepted that the cure mechanism involves dehydro-

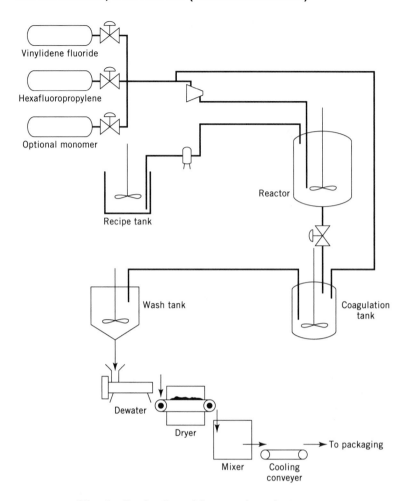

Fig. 5. Production of fluorocarbon elastomers.

fluorination adjacent to hexafluoropropylene monomer units. The subsequent fluoroolefin is highly reactive toward nucleophilic attack by a variety of curatives (eg, diamines, diphenols).

In addition to the incorporated cure gums, there are also raw gums that contain no curative, to which the fabricator adds cure ingredients, such as diamines, bisphenols, or peroxides (48), in addition to formulation (compound) ingredients. Although peroxide cure systems historically have suffered a poor reputation with respect to processibility (mold sticking and mold fouling), recent advancements in these areas have yielded greatly improved products. The use of iodine–bromine containing fluorinated chain-transfer agents has resulted in high fluorine grade polymers with iodine–bromine end groups. The rheology of these materials is characterized by very fast cure rates. In addition, these materials exhibit much better mold release properties than earlier grades (49–51). The advantages and disadvantages of the cure systems are given in Table 3.

Table 3. Cure Systems for Fluorocarbon Elastomers

Property	Diamine	Dihydroxy cross-linker/ accelerator	Peroxide/coagent
rheology			
cure	slow	excellent	good
scorch	poor	excellent	good
compression set	poor	excellent	intermediate
chemical resistance	typical	typical	typical[a]
thermal resistance	excellent	excellent	good
processing	poor	excellent	poor to good[b]
other	FDA approved good bondability	easy to modify	HAV[c] curable

[a]Improved steam and base resistance.
[b]Special grades required.
[c]HAV = hot air vulcanization.

Compounding. Owing to the number of ingredients required in a conventional rubber recipe, fluorocarbon elastomer compounding seems simple compared to typical hydrocarbon elastomer recipes. However, the apparent simplicity of such formulations makes a selection of appropriate ingredients especially important in order to obtain the excellent properties inherent in available gumstocks. A typical recipe in parts per hundred of rubber (phr) by weight is as follows:

Component	Amount, phr
rubber	100
inorganic base: magnesium oxide, calcium hydroxide	6–20
filler (reinforcing or nonreinforcing)	0–60
accelerators or curatives (if not included in base rubber)	0–6
process aids	0–2

With a clear idea of use requirements and rubber response to specific additives, a formulation may be selected. Uses generally fall into one of three classes: O-rings, molded goods, and extruded forms.

O-Rings. In O-ring applications, the primary consideration is resistance to compression set. A fluorocarbon elastomer gum is chosen for O-ring applications based on its gum viscosity, cross-link density, cure system, and chemical resistance so that the best combination of processibility and use performance is obtained. Sample formulations for such uses are given in Table 4.

Long-term compression set resistance is described in Figure 4. Lower set values are achievable by use of higher viscosity gumstock at comparable cross-link densities. Compression set resistance is also very dependent on the cure system chosen. The bisphenol cure system offers the best compression set resistance available today, as shown in Table 5.

Molded Goods. In molded goods compounding, the most important physical property in the final vulcanizate is usually elongation to break, with compression

Table 4. Fluorocarbon Elastomer O-Ring Compounds

Typical formulation	I[a]	II[b]	III[c]
Compound ingredients, phr			
MT black (N-990)	30	30	25
calcium hydroxide	6	6	
magnesium oxide	3	3	
hexafluoroisopropylidenediphenol	2.1	2.1	
triphenylbenzylphosphonium chloride	0.45	0.45	
triallyl isocyanurate			5
α,α-bis(*t*-butylperoxy)diisopropylbenzene			1
sodium stearate			1
Physical properties[d]			
tensile strength, MPa[e]	15.0	15.0	16.0
elongation at break, %	200	200	300
hardness, Shore A	75	75	71
compression-set (3.5 mm O-rings), %			
for 70 h at 200°C, (ASTM D395)	15	10	35
specific gravity	1.8	1.8	1.6

[a]100 phr FKM2230 (ML 1 + 10 at 121°C = 40) where FKM2230 is poly(vinylidene fluoride-*co*-hexafluoropropylene); available from 3M.
[b]100 phr FKM2178 (ML 1 + 10 at 121°C = 100) where FKM2178 is poly(vinylidene fluoride-*co*-hexafluoropropylene); available from 3M.
[c]100 phr FKM100S (ML 1 + 10 at 121°C = 90) where FKM100S is poly(tetrafluoroethylene-*co*-propylene); available from 3M.
[d]Press cure: 5 min at 177°C; post-cure: 24 h at 230°C.
[e]To convert MPa to psi, multiply by 145.

set being a secondary consideration. Since complex shapes are often required, compound flow is also an important parameter. These objectives generally are best met by beginning with gum of the lowest initial viscosity that is consistent with good physical properties. This gum is then cured to lower cross-link density, compared to O-ring formulations, to permit high elongation.

Comparison of starting viscosities and properties of a molded goods compound (Table 6) with an O-ring compound (see Table 4) shows differences in elongation as a result of lower cross-link density. Even higher elongations are achievable with special formulations, or products designed for exceptionally high tear strengths in the press cured state.

The effects of specific fillers in molded goods applications are known (52), and are of special importance for water- and acid-resistant compounds. It is good policy to contact the suppliers for specific recommendations to meet the balance of properties required.

Extruded Articles.   In extruded article compounding, the most important parameters are scorch safety and flow characteristics (53). The bisphenol cure system again offers the best scorch resistance of the available fluorocarbon elastomer cure systems. Good flow characteristics can be achieved through proper selection of gum viscosities. Also, the addition of process aids to the formulation can en-

Table 5. Effect of Cure System on Processing Safety and Compression Set Resistance for FKM2260[a]

Formulation	Amine	Bisphenol	Peroxide
Compound ingredients, phr			
MT black (N-990)	30	30	35
magnesium oxide	10	3	
calcium hydroxide		6	3
N,N'-dicinnamylidene-1,6-hexanediamine	2.5		
hexafluoroisopropylidenediphenol		2.1	
triphenylbenzylphosphonium chloride		0.45	
triallyl isocyanurate			2.5
2,5-dimethyl-2,5-di-t-butylperoxyhexane			2.5
Properties			
Mooney scorch[b] at 121°C			
minimum	68	66	48
point rise (25 min)	76	2	6
compression-set, %[c]	48	13	25

[a] 100 phr of this material, which has ML 1 + 10 at 121°C = 60 and is poly(vinylidene fluoride-*co*-hexafluoropropylene) plus cure site monomer; available from 3M.
[b] ASTM D1646.
[c] 3.5 mm O-rings for 70 h at 200°C (ASTM D395).

hance the flow characteristics. Typical formulations for extrusion grade fluorocarbon elastomers are given in Table 7.

Formulation Parameters. Gum viscosity is of primary importance to the determination of processibility, as this factor affects vulcanizate properties, especially compression set. Gums are available with Mooney Viscosity (ML1 + 10 at 121°C) values of 5–160; a range of 20–60 is preferred for the optimum combination of flow and physical properties. Higher viscosities can cause excessive heat buildup during mixing without a compensatory gain in physical strength. Compound viscosity depends on gum viscosity and on filler selection (type and loading). A preferred range, as measured by Mooney scorch (MS at 121°C), is 25–60.

Compound stability and safety must also be considered when determining processibility, as they are strongly affected by compounding ingredients and cure systems. The data in Table 5 clearly show the effect that choice of cure system can have on scorch stability and processing minimums as determined by standard testing techniques (MS at 121°C). The most workable formulations are compounded with raw gums containing the bisphenol or incorporated cure systems. These elastomers offer the processor the best starting point for maximum processibility.

Mixing. Fluorocarbon elastomer formulations can be compounded by any standard rubber mixing technique. Open mill mixing can be used since most commercial gums mix well. Exceptions to this are very low viscosity gums that have a tendency to stick to the rolls, and very high viscosity gums that are excessively tough.

Table 6. Fluorocarbon Elastomer Molded Goods Compound

Typical formulation	I[a]	II[b]
Compound ingredients, phr		
MT black (N-990)	30	25
calcium hydroxide	6	
magnesium oxide	3	
triphenylbenzylphosphonium chloride	0.4	
hexafluoroisopropylidenediphenol	1.7	
triallyl isocyanurate		5
α,α-bis(t-butylperoxy)diisopropylbenzene		1
sodium stearate		1
Post-cured physical properties[c]		
tensile strength, MPa[d]	14.4	13.5
elongation at break, %	265	315
hardness, Shore A	74	71
compression set, (1.27 cm disk), %		
for 70h at 200°C (ASTM D395)	20	40
specific gravity	1.8	1.6

[a]100 phr FKM2230 (ML 1 + 10 at 121°C = 40) where FKM2230 is poly(vinylidene fluoride-*co*-hexafluoropropylene); available from 3M.
[b]100 phr FKM 150 P (ML 1 + 10 at 121°C = 75) where FKM 150 P is poly(tetrafluoroethylene-*co*-propylene); available from 3M.
[c]Press cure: 5 min at 177°C; post-cure: 24 h at 230°C.
[d]To convert MPa to psi, multiply by 145.

Internal mixing is widely used with fluorocarbon elastomers. Gumstocks and compounds that are particularly successful fall in the viscosity ranges discussed earlier, and use both incorporated bisphenol-type and peroxide cure systems. A typical internal mix cycle runs 6–8 min with a drop temperature of 90–120°C. The typical formulations in Tables 4 and 7 are readily mixed in an internal mixer.

Preforming. Extrusion preforming is easily accomplished if relatively cool barrel temperatures are used with either a screw or piston type extruder (Barwell). It is important that the gums be used in the appropriate viscosity ranges, and that scorching be avoided.

Calendering operations are done routinely, and warm rolls (40–90°C) are recommended for optimum sheet smoothness. A process aid, such as low molecular weight polyethylene wax, is often used. Sheet thicknesses of 0.5–1.3 mm (20–50 mils) can normally be produced.

Molding. Compression molding is generally used when it is desirable to conserve material, and when a molding operation is set up to allow preparation of large numbers of preforms with minimum labor costs. Flow requirements are minimal and high viscosity gums may be used.

Transfer molding minimizes preforming, and is usually used for the production of very small parts; however, this technique may generate excessive amounts

Table 7. Fluorocarbon Elastomer Extrusion Grade Compound

Formulation	I[a]	II[a]
Compound ingredients, phr		
hexafluoroisopropylidenediphenol	1.9	1.9
triphenylbenzylphosphonium chloride	0.45	0.45
MT black (N-990)	35	15
HAF black (N-326)		5
SRF black (N-762)		7
magnesium oxide	3	9
calcium hydroxide	6	
carnauba wax	1	1
Physical properties[b]		
tensile strength, MPa[c]	7.6	12.4
elongation at break, %	280	330
hardness, Shore A	75	75

[a]100 phr FKM2145 (ML 1 + 10 at 121°C = 30) where FKM2145 is poly(vinylidene fluoride-*co*-hexafluoropropylene); available from 3M.
[b]Press cured 45 min at 160°C.
[c]To convert MPa to psi, multiply by 145.

of scrap material. Flow requirements can be quite high, but fluorocarbon elastomers are available that are effective in this application.

Injection molding is finding expanded usage in the rubber industry. Fluorocarbon elastomers can be successfully molded via this technique. Selection of the proper viscosity and cure rheology are very important due to the occurrence of high shear and fast cures.

All types of molding may be carried out at 150 to 200°C. This allows molding times of five minutes or less for most fluorocarbon elastomer parts, but this time is dependent on part size.

Extrusion. Extrusion techniques are used in the preparation of tubing, hose, O-ring cord, preforms and shaped gaskets. Typical extrusion conditions are 70 to 85°C for the barrel temperature and 95 to 110°C for the head temperature. The extruded forms are normally cured in a steam autoclave at 150 to 165°C. Some special grades of peroxide curable fluorocarbon elastomers can be hot air vulcanized.

Post-Curing. Post-curing at elevated temperatures develops maximum physical properties (tensile strength and compression-set resistance) in fluorocarbon elastomers. General post-cure conditions are 16 to 24 h at 200 to 260°C.

Economic Factors

Annual worldwide fluorocarbon elastomer usage totals about 7300 metric tons. Approximately 40% of this usage is in the United States, 30% in Europe, and 20% in Japan. Prices in 1991 were $30–110/kg.

Specifications

Commercially available fluorocarbon elastomers meet automotive specifications in the HK section of ASTM D2000 and SAE J-200. ASTM D1418 specifies designations of composition, eg, fluorocarbon elastomers are designated CFM, FKM, or FFKM. Commercially available fluorocarbon elastomers offer a balance of those properties needed to meet the major O-ring specifications, such as AMS 7276, AMS 7280A, AMS 7259, MIL 83248 Amendment 1 Type II, Class I and II.

Certain grades and formulations of the fluorocarbon elastomers are qualified under the code of Federal Regulations, 21, Food and Drugs, Part 177.2600 for use as rubber articles whose intended applications require repeated or continuous contact with food. Elastomer suppliers will provide assistance in formulating for specified uses.

Test Methods

The fluorocarbon elastomer raw gums provided for rubber molding are tested for Mooney viscosity (ASTM D1646) and for specific gravity (ASTM D297). When compounded as described above, the stocks are tested for Mooney cure (ASTM D1646), Mooney scorch (ASTM D1646), and oscillating-disk rheometer cure rate (ASTM D2084). The vulcanizates are evaluated regarding original physical properties (ASTM D412, D2240, and D1414), aged physical properties (ASTM D573), compression set (ASTM D395), and fluid aging (ASTM D471). Low temperature properties are measured by low temperature retraction TR10 (ASTM D1329) and brittle point (ASTM D2137) tests.

Health and Safety Factors

In general, under normal handling conditions, the fluorocarbon elastomers have been found to be low in toxicity and irritation potential. Specific toxicological, health, and safe handling procedures are provided by the manufacturer of each fluorocarbon elastomer product upon request.

Uses

About 60% of the United States usage is in ground transportation. Typical components include engine oil seals, fuel system components such as hoses and O-rings, and a variety of drive train seals. Growth in this area is expected to continue with the general strength of the U.S. automotive industry coupled with increased demands from higher underhood temperatures, alcohol containing fuels, and more aggressive lubricants. Other major U.S. segments include petroleum/ petrochemical, industrial pollution control, and industrial hydraulic and pneumatic applications. These areas will be more dependent upon general industrial production and overall energy demands, and will show slower growth than the automotive segment.

The usage pattern in Europe and Japan is more dependent upon the automotive industry. However, with the recent concern about acid rain, the European and U.S. markets should show increased interest in fluorocarbon elastomers for pollution control applications. On the other hand, the Japanese market has a sizable outlet in electrical and general machinery manufacturing (eg, copiers). Petroleum applications are of little interest outside the United States.

The principal original use of fluorocarbon elastomers in the aircraft industry now accounts for less than 10% of the total fluorocarbon elastomer consumption.

BIBLIOGRAPHY

"Elastomers, Synthetic (Fluorinated)" in *ECT* 3rd ed., Vol. 8, pp. 500–515, by Arthur C. West and Allan G. Holcomb, 3M Co.

1. M. E. Conroy and co-workers, *Rubber Age* **76,** 543 (1955).
2. C. B. Griffis and J. C. Montermoso, *Rubber Age* **77,** 559 (1955).
3. W. W. Jackson and D. Hale, *Rubber Age* **77,** 865 (1955).
4. F. A. Bovey and co-workers, *J. Polym. Sci.* **15,** 520 (1955).
5. U.S. Pat. 3,051,677 (Aug. 28, 1962), D. R. Rexford (to E. I. du Pont de Nemours & Co., Inc.).
6. S. Dixon, D. R. Rexford, and J. S. Rugg, *Ind. Eng. Chem.* **49,** 1687 (1957).
7. U.S. Pat. 3,318,854 (May 9, 1967) F. J. Honn and W. M. Sims (to 3M Co.).
8. J. S. Rugg and A. C. Stevenson, *Rubber Age* **82,** 102 (1957).
9. U.S. Pat. 2,968,649 (Jan. 17, 1961), J. P. Pailthrop and H. E. Schroeder (to E. I. du Pont de Nemours & Co., Inc.).
10. U.S. Pat. 3,331,823 (July 18, 1967), D. Sianesi, G. Bernardi, and A. Regio (to Montedison).
11. U.S. Pat. 3,335,106 (Aug. 8, 1967), D. Sianesi, G. C. Bernardi, and G. Diotalleri (to Montedison).
12. U.S. Pat. 3,235,537 (Feb. 15, 1966), J. R. Albin and G. A. Gallagher (to E. I. du Pont de Nemours & Co., Inc.).
13. Ger. Offen. 2,457,102 (Aug. 7, 1975), R. Baird and J. D. MacLachlan (to E. I. du Pont de Nemours & Co., Inc.).
14. U.S. Pat. 4,035,565 (July 12, 1977), D. Apotheker and P. J. Krusic (to E. I. du Pont de Nemours & Co., Inc.).
15. U.S. Pat. 4,418,186 (Nov. 29, 1983), M. Yamabe and co-workers, (to Asahi Glass Co., Ltd.).
16. A. L. Barney, G. H. Kalb, and A. A. Kahn, *Rubber Chem. Technol.* **44,** 660 (1971).
17. A. L. Barney, W. J. Keller, and N. M. Van Gulick, *J. Polym. Sci. A-1* **8,** 1091 (1970).
18. G. H. Kalb, A. L. Barney, and A. A. Kahn, *Am. Chem. Soc. Dev. Polym. Chem.* **13,** 490 (1972).
19. S. M. Ogintz, *Lubric. Eng.* **34,** 327 (1978).
20. U.S. Pat. 4,251,399 (Feb. 17, 1981), M. Tomoda and Y. Ueta (to Daikin Kogyo Co., Ltd.).
21. U.S. Pat. 4,263,414 (Apr. 21, 1981), A. C. West (to 3M Co.).
22. Y. Tabata, K. Ishigure, and H. Sobue, *J. Polym. Sci.* **A2,** 2235 (1964).
23. G. Kojima and Y. Tabata, *J. Macromol. Sci. Chem.* **A5**(6), 1087 (1971).
24. K. Ishigure, Y. Tabata, and K. Oshima, *Macromolecules* **6,** 584 (1973).

25. G. Kojima and Y. Tabata, *J. Macromol. Sci. Chem.* **A6**(3), 417 (1972).
26. G. Kojima, H. Kojima, and Y. Tabata, *Rubber Chem. Technol.* **50,** 403 (1977).
27. U.S. Pat. 4,882,390 (Nov. 21, 1989), W. Grootaert and R. E. Kolb (to 3M Co.).
28. W. M. Grootaert, R. E. Kolb, and A. T. Worm, *Rubber Chem. and Technol.* **63**(4), 516–522 (1990).
29. J. R. Dunn and H. A. Pfisterer, *Rubber Chem. Technol.* **48,** 356 (1976).
30. U.S. Pat. 3,051,677 (Aug. 28, 1962), D. R. Rexford (to E. I. du Pont de Nemours & Co., Inc.).
31. U.S. Pat. 3,053,818 (Sept. 11, 1962), F. J. Honn and S. M. Hoyt (to 3M Co.).
32. U.S. Pat. 2,968,649 (Jan. 17, 1961), J. P. Pallthorp and H. E. Schroeder (to E. I. du Pont de Nemours & Co., Inc.).
33. A. L. Logothetis, *Prog. Polym. Sci.* **14,** 251–296 (1989).
34. U.S. Pat. 3,069,401 (Dec. 18, 1962), G. A. Gallagher (to E. I. du Pont de Nemours & Co., Inc.).
35. U.S. Pat. 3,080,347 (Mar. 5, 1963), C. L. Sandberg (to 3M Co.).
36. U.S. Pat. 3,707,529 (Dec. 26, 1972), E. K. Gladding and J. C. Wyce (to E. I. du Pont de Nemours & Co., Inc.).
37. U.S. Pat. 3,845,024 (Oct. 29, 1974), S. D. Weaver (to E. I. du Pont de Nemours & Co., Inc.).
38. U.S. Pat. 3,655,727 (Apr. 11, 1972), K. U. Patel and J. E. Maier (to 3M Co.).
39. U.S. Pat. 3,712,877 (Jan. 23, 1973), K. U. Patel and J. E. Maier (to 3M Co.).
40. U.S. Pat. 3,752,787 (Aug. 14, 1973), M. R. deBrunner (to E. I. du Pont de Nemours & Co., Inc.).
41. U.S. Pat. 3,857,807 (Dec. 31, 1974), Y. Kometani and co-workers (to Daikin Kogyo Co.).
42. U.S. Pat. 3,864,298 (Feb. 4, 1975), Y. Kometani and co-workers (to Daikin Kogyo Co.).
43. U.S. Pat. 3,920,620 (Nov. 18, 1975), C. Ceccato, S. Geri, and L. Calombo (to Montedison).
44. W. W. Schmiegel, *Kautsch. Gummikunst.* **31,** 137 (1971).
45. W. W. Schmiegel, *Angew. Makromolek. Chem.* **76/77,** 39 (1979).
46. P. Venkateswarlv and co-workers, *Paper 123 presented at the 136th ACS Rubber Division Meeting*, Detroit, Mich., Oct. 17–20, 1989.
47. V. Ardella and co-workers, *Paper 57 presented at the 140th ACS Rubber Division Meeting*, Detroit, Mich., Oct. 8–11, 1991.
48. J. E. Alexander and H. Omura, *Elastomerics* **2,** 19 (1978).
49. U.S. Pat. 4,243,770 (Jan. 6, 1981), M. Tatemoto and co-workers (to Daikin Kogyo Co.).
50. U.S. Pat. 4,501,869 (Feb. 26, 1985), M. Tatemoto and co-workers (to Daikin Kogyo Co.).
51. U.S. Pat. 4,745,165 (May 17, 1988), V. Arcella and co-workers (to Ausimont S.p.A.).
52. 3M Company product brochure, *FLUOREL 2170-Compounding with Various Fillers*, available from Commercial Chemicals Division, St. Paul, Minn., June 1984.
53. R. Christy, *Rubber World* **184**(6), 38 (1981).

General References

K. J. L. Paciorek and L. A. Wall, eds., *High Polymers*, Vol. 25, Wiley-Interscience, New York, 1972, pp. 291–313.

L. E. Creneshaw and D. L. Tabb, *The Vanderbilt Rubber Handbook*, 13th ed., R. T. Vanderbilt Comp., Inc., Norwalk, Conn., 1990, pp. 211–222.

M. Morton, ed., *Rubber Technology*, 3rd ed., Van Nostrand Reinhold Co., New York, 1987, pp. 410–437.

R. G. Arnold, A. L. Barney, and D. C. Thompson, *Rubber Chem. Technol.* **46,** 619 (1973).

J. C. Montermoso, *Rubber Chem. Technol.* **37,** 1521 (1961).

W. W. Schmiegel, *Makromekulare Chemie* **76/77,** 39 (1979).

D. Apotheker and co-workers, *Rubber Chemistry and Technology*, 1004 (1982).

J. C. Arthur, Jr., ed., *ACS Symposium Series 260, Polymers for Fibers and Elastomers*, American Chemical Society, Washington, D.C., 1984.

A. L. Logothetis, *Prog. Polym. Sci.* **14**, 251–296 (1989).

WERNER M. GROOTAERT
GEORGE H. MILLET
ALLAN T. WORM
3M Company

NITRILE RUBBER

Nitrile rubber is a synthetic polymer made from 1,3-butadiene and acrylonitrile using emulsion polymerization techniques. The ASTM designation (D1418) for nitrile rubber is NBR and the Chemical Abstracts Service (CAS) registry number for poly(acrylonitrile-*co*-1,3-butadiene) is [9003-18-3]. Nitrile rubber was first developed in Germany during the early 1930s (1). Domestic production of NBR started during the 1940s. Nitrile rubber is classified as a specialty rubber and is well known for its resistance to various oils, fuels, and chemicals. After mixing with other ingredients (fillers, plasticizers, antidegradants, curatives) and curing, NBR compounds commonly see use in various seals, gaskets, hose, and roll applications. Nitrile rubber has been chemically modified by a solution hydrogenation process in order to extend its high temperature performance. Hydrogenated nitrile rubber (HNBR) [88254-10-8] was developed and commercialized during the early 1980s. The saturated backbone of the HNBR polymer leads to improved heat resistance of the rubber, while retaining excellent oil/chemical resistance.

Chemical Properties

Nitrile rubbers are high molecular-weight amorphous copolymers of 1,3-butadiene [106-99-0], $CH_2=CH-CH=CH_2$, and acrylonitrile [107-13-1], $CH_2=CH-CN$. Average molecular weights of commercial products have been reported to be between 250,000 and 600,000, with a wide distribution around the average in any single product (2). Structure of the polymer can vary widely from largely linear to highly branched to cross-linked, depending on the conditions of polymerization (3).

Distribution of the monomer units in the polymer is dictated by the reactivity ratios of the two monomers. In emulsion polymerization, which is the only commercially significant process, reactivity ratios have been reported (4). If M_1 = butadiene and M_2 = acrylonitrile, then $r_1 = 0.28$, and $r_2 = 0.02$ at 5°C. At 50°C, $r_1 = 0.42$ and $r_2 = 0.04$. As would be expected for a combination where $r_1 r_2 =$ near zero, this monomer pair has a strong tendency toward alternation. The degree of alternation of the two monomers increases as the composition of the polymer approaches the 50/50 molar ratio that alternation dictates (5,6). Another complicating factor in defining chemical structure is the fact that butadiene can enter the polymer chains in the cis (**1**), trans (**2**), or vinyl(1,2) (**3**) configuration:

$$\underset{(1)}{\overset{\displaystyle \underset{\text{www}CH_2}{\nearrow}\overset{\displaystyle CH{=}CH}{}\underset{CH_2\text{www}}{\searrow}}{}}\qquad\underset{(2)}{\overset{\displaystyle \text{www}CH_2{\searrow}\ \ CH{=}CH{\searrow}\ CH_2\text{www}}{}}\qquad\underset{(3)}{\overset{\displaystyle \text{www}CH_2{-}CH\text{www}}{}}$$

In a copolymer of 33% acrylonitrile, the most common composition for commercial products, the butadiene occurs in the approximate ratio of 90% trans, 8% vinyl, and 2% cis. At higher acrylonitrile content the cis configuration disappears, and at lower levels it increases to about 5%; the vinyl configuration remains approximately constant (6,7). Since actual compositions of commercial nitrile rubbers are between 15 and 50% acrylonitrile, they also vary somewhat in sequence distribution and in the content of the three isomeric butadiene configurations.

This combination of monomers is unique in that the two are very different chemically, and in their character in a polymer. Polybutadiene homopolymer has a low glass-transition temperature, remaining rubbery as low as $-85°C$, and is a very nonpolar substance with little resistance to hydrocarbon fluids such as oil or gasoline. Polyacrylonitrile, on the other hand, has a glass temperature of about 110°C, and is very polar and resistant to hydrocarbon fluids (see ACRYLONITRILE POLYMERS). As a result, copolymerization of the two monomers at different ratios provides a wide choice of combinations of properties. In addition to providing the rubbery nature to the copolymer, butadiene also provides residual unsaturation, both in the main chain in the case of 1,4, or in a side chain in the case of 1,2 polymerization. This residual unsaturation is useful as a cure site for vulcanization by sulfur or by peroxides, but is also a weak point for chemical attack, such as oxidation, especially at elevated temperatures. As a result, all commercial NBR products contain small amounts ($\sim$0.5–2.5%) of antioxidant to protect the polymer during its manufacture, storage, and use.

Chemically modified nitrile rubbers are also produced commercially with the objective of changing their chemistry in such a way that specific properties are enhanced (8). The oldest of the chemically modified NBRs are the carboxylated varieties, made by copolymerizing methacrylic or acrylic acid with the butadiene and acrylonitrile. The resultant products, which typically contain 2–6% by weight of the acid monomer, can be vulcanized with polyvalent metals in addition to the normal sulfur or peroxide cure systems (9). The most outstanding result of this modification is a large improvement in abrasion resistance. Another chemical modification currently in use is the copolymerization of a monomer which causes an antioxidant structure to be attached to the polymer (10). The result is improved resistance to oxidation, particularly after the rubber has been exposed to a hydrocarbon fluid that would normally extract a conventional antioxidant from the polymer, leaving it unprotected. The most recent and potentially the most important chemical modification is hydrogenation of the polymer so that little of the unsaturation remains (11–15). This results in a product with much improved resistance to oxidation and weathering, but with little or no sacrifice in other useful properties. The usual procedure is to hydrogenate in a solution process until just enough unsaturation is left to provide sites for curing, using catalysts based on ruthenium, rhodium, or palladium. A latex reduction process has been reported using hydrazine and an oxidizer such as hydrogen peroxide (16,17).

Other chemical modifications such as attachment of isocyanate or hydroxyl functionality have been reported (18), but have not become commercially significant.

Physical Properties

Nitrile rubbers are produced over a wide range of monomer ratios and molecular weights, so their physical constants and basic polymer properties also cover a range of values. Some of the more widely used properties are listed in Table 1.

Manufacturing

Virtually all nitrile elastomers are manufactured by emulsion polymerization, according to technology that evolved from efforts in Germany in the 1930s and in the United States in the 1940s to find a substitute for natural rubber. The primary steps in the process include polymerization into a latex form, coagulation of the latex into a wet crumb, and then drying into a final product. In practice, latex is usually collected in large blend tanks where it is sampled and tested before "finishing" to a dry product. As a result, a description of the manufacturing of nitrile rubber can be conveniently divided into two areas: polymerization and finishing.

   Polymerization.   The two different methods of carrying out emulsion polymerization on a commercial scale are shown in Figures 1 and 2. In the batch

Table 1. Physical Properties of Nitrile Rubber

Property	Acrylonitrile content, %	Value
specific gravity	15	0.94
	20	0.95
	35	0.99
	45	1.02
	50	1.03
T_g, °C	15	−49
	22	−40
	30	−30
	40	−19
	50	−9
thermal conductivity[a],kJ/(m·h·°C)[b]	28	0.90
	33	0.90
	38	0.92
thermal expansion coefficient[a] $\times 10^6$/°C	28	175
	33	170
	38	150
specific heat[a], J/(g·°C)[b]	40	$0.00283\,T + 1.126$

[a]Ref. 4.
[b]To convert J to cal, divide by 4.184.

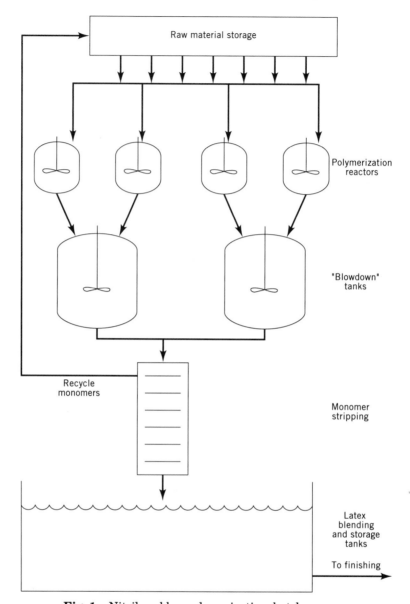

Fig. 1. Nitrile rubber polymerization, batch process.

process, each reactor operates independently to carry out a complete polymerization reaction. Raw materials are charged to each reactor and the polymerization proceeds to the desired end point in each reactor, where it is either stopped or dropped to a "blowdown" tank and stopped. When large quantities of a single product are being made, many individual polymerizations are carried out, and the latices from them are blended together. In the continuous process, a number of reactors are connected in series, usually three or more. Raw materials are charged to the first reactor where polymerization is started. The reacting mass

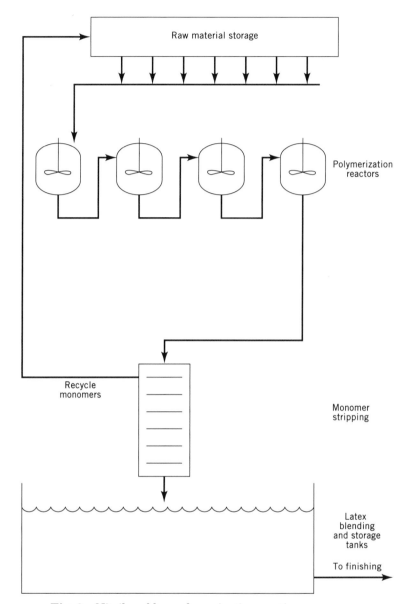

Fig. 2.   Nitrile rubber polymerization, continuous process.

then flows from one reactor to the next continuously, until it reaches the desired end point where polymerization is stopped.

The batch process is most useful when relatively short runs of a large number of products are planned. A number of products can be made simultaneously and collected in separate storage tanks for testing and later finishing. The continuous process is more efficient when large quantities of a few products are desired. Reactors remain in operation a larger fraction of the time because no time is lost for emptying and recharging such as in the batch mode. In both cases, the progress of polymerization is closely monitored to follow the conversion of monomer to polymer. The conversion is important for two reasons: some recipes call for additions of ingredients at specific points during polymerization, and all recipes have a specific final conversion at which they must be stopped so that desired product properties are achieved.

Simplified nitrile rubber polymerization recipes are shown in Table 2 for "cold" and "hot" polymerization. Typically, cold polymerization is carried out at 5°C and hot at 30°C. The original technology for emulsion polymerization was similar to the 30°C recipe, and the redox initiator system that allowed polymerization at lower temperature was developed shortly after World War II. The latter uses a reducing agent to activate the hydroperoxide initiator and soluble iron to reactivate the system by a reduction–oxidation mechanism as the iron cycles between its ferrous and ferric states.

In the recipes shown in Table 2, the amount of water can vary widely, depending on the available heat-transfer capacity of the reactor and the rate of polymerization. Each of the monomers has a heat of polymerization of about 75 kJ/mol (18 kcal/mol), so removing the heat of polymerization to control temperature is often the limiting factor on rate of polymerization.

The ratio of monomers can be varied from zero acrylonitrile up to approximately 60% to produce copolymers of zero to about 50% acrylonitrile. This is, of course, a crucial factor in determining the properties of the final material. Rubbers with low acrylonitrile content have extremely low glass temperatures but little resistance to solvents and oils. The opposite is true at high acrylonitrile levels.

The choice of emulsifiers is dependent on cost to some degree, but is also chosen on the basis of ease of coagulation in the finishing process and on the influence of residual emulsifier in the product on properties such as water sensitivity and cure rate. Ideally, an emulsifier is chosen which is inexpensive, results in latex that is very stable to shear during stripping and pumping, easy to coag-

Table 2. Typical Nitrile Rubber Polymerization Recipes

Materials	Recipe, parts by weight, active	
	5°C Polymerization	30°C Polymerization
soft or demineralized water	200	200
butadiene	67	67
acrylonitrile	33	33
primary emulsifier	2.5	2.5
secondary emulsifier	0.5	0.5
electrolyte	0.3	0.3
modifier (mol wt regulator)	0.4	0.4
iron chelate	0.005	
reducing agent	0.03	
hydroperoxide	0.04	
potassium persulfate		0.25
oxygen scavenger (if needed)	0.01	0.01
shortstop (at desired final conversion)	0.2	0.2

ulate, and inert in the final product. All the practical polymerization recipes use an anionic primary emulsifier such as fatty soaps or synthetic detergents like sodium alkyl benzenesulfonate. The secondary emulsifier, when used, can be a nonionic or a wetting agent like sodium naphthalenesulfonate.

Electrolyte is added to the recipe to help control particle size of the latex produced. Materials like sodium carbonate or trisodium phosphate are commonly used.

One of the most important ingredients is the modifier added to control molecular weight. These operate by stopping the polymerization of a growing polymer chain and transferring the free radical to start a new copolymer molecule. For that reason, they are also called chain-transfer agents. Without this control, emulsion polymerized butadiene copolymers would be extremely high in molecular weight and would be highly branched. They would be almost impossible to process in subsequent mixing, extrusion, and molding operations. The most commonly used modifiers are alkyl mercaptans, especially *tert*-dodecyl mercaptan.

The initiator ingredients are designed to generate free radicals at a moderate rate at the temperature of the polymerization. Sodium, potassium, or ammonium persulfate are commonly used for hot polymers. The initiator ingredients in a cold recipe are more complex and a wide variety of materials can be used. The iron is usually added in a chelated form so it is soluble, but not as reactive as free ions in solution. A typical reducing agent is sodium formaldehyde sulfoxylate and *p*-menthane hydroperoxide is often used to provide the necessary free-radical source. Usually, a high acrylonitrile recipe requires less initiator than a low acrylonitrile recipe. All of these systems are sensitive to oxygen, so in many cases a small amount of a material like sodium hydrosulfite is used in case purging of air before polymerization was incomplete.

The choice of initiator system depends on the polymerization temperature, which is an important factor in determining final product properties. Cold polymers are generally easier to process than hot polymers and in conventional cured rubber parts have superior properties. The hot polymers are more highly branched and have some advantages in solution applications such as adhesives, where the branching results in lower solution viscosity and better cohesion in the final adhesive bond.

All ingredients in the polymerization recipe are not always added at the beginning of the process. For example, better latex stability can sometimes be achieved by starting with only part of the emulsifier, saving the rest for later addition. Sometimes a portion of the modifier is held out for late addition to allow higher final conversion without premature consumption of all of it. Occasionally, if a low acrylonitrile product is the objective, part of the acrylonitrile monomer will be saved for late addition so that a chemically more uniform copolymer is produced, which can sometimes enhance properties in critical applications.

A shortstop is necessary to control the final conversion of monomer to polymer. If allowed to run to very high conversion (>95%), the resultant product is difficult to process, usually because the modifier has been consumed, and because free radicals begin to react with polymer instead of additional monomer, which at high conversion is in short supply. Uncontrolled molecular weight and branching begin to occur. In practice, a high conversion nitrile rubber polymerization is allowed to run to about 90% conversion and results in a product with excellent physical properties, but is difficult for the user to mix and mold. A low conversion polymer is shortstopped at 65–80% conversion and is much easier to process in later operations. Effective shortstops are destroyers of free radicals and include materials that liberate free hydroxylamine, hydroquinone and its derivatives, and carbamates such as sodium diethyldithiocarbamate [148-18-5].

After polymerization is complete, especially in the case of low conversion products, residual monomer must be removed and recovered for recycling. Butadiene can be removed by warming the latex under vacuum, since it boils at a very low temperature. The acrylonitrile and remaining small amount of butadiene monomer are usually removed by steam stripping in a column. This process requires that the latex be mechanically stable to avoid loss of product and consequent fouling of the stripping equipment. For reasons of cost, environmental concern, and product safety, the stripping process must be designed to remove residual monomer down to low ppm levels.

Polymerizer design variables such as length/diameter ratio of the polymerizer, type and speed of agitation, and materials of construction, have important effects on operation of the process but little effect on the product itself. Polymerization reactors are usually jacketed or contain cooling coils (or both) for temperature control. Stainless steel is common in modern reactors and agitation is usually designed for low shear and high pumping rates, the objective being to allow good heat transfer while avoiding the production of small particle size latex.

Finishing.   After manufacturing a polymer in latex form by either batch or continuous polymerization, it is collected in large tanks for further processing to a dry rubber form. The most common finishing process involves addition of stabilizer to the latex, coagulation of the latex to form a slurry of rubber crumb in water, washing of the wet material, and dewatering and drying to a finished product. A typical finishing process is described in Figure 3. In modern facilities this is always a continuous process, although it could be carried out in a batch mode.

Because nitrile rubber is an unsaturated copolymer it is sensitive to oxidative attack and addition of an antioxidant is necessary. The most common practice is to add an emulsion or dispersion of antioxidant or stabilizer to the latex before coagulation. This is sometimes done batchwise to the latex in the blend tank, and sometimes is added continuously to the latex as it is pumped toward further processing. Phenolic, amine, and organic phosphite materials are used. Examples are di-*tert*-butylcatechol, octylated diphenylamine, and tris(nonylphenyl) phosphite [26523-78-4]. All are meant to protect the product from oxidation during drying at elevated temperature and during storage until final use. Most rubber processors add additional antioxidant to their compounds when the NBR is mixed with fillers and curatives in order to extend the life of the final rubber part.

The choice of coagulant for breaking of the emulsion at the start of the finishing process is dependent on many factors. Salts such as calcium chloride, aluminum sulfate, and sodium chloride are often used. Frequently, pH and temperature must be controlled to ensure efficient coagulation. The objectives are to leave no uncoagulated latex, to produce a crumb that can easily be dewatered, to avoid fines that could be lost, and to control the residual materials left in the product so that damage to properties is kept at a minimum. For example, if a significant amount of a hydrophilic emulsifier residue is left in the polymer, water resistance

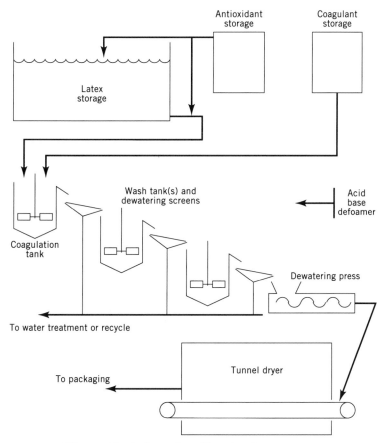

Fig. 3. Typical nitrile rubber finishing process.

of final product suffers, and if the residue left is acidic in nature, it usually contributes to slow cure rate.

Once the latex is coagulated, the resultant slurry is filtered through a screen and sent to a washing tank or tanks. The purpose is to remove water-soluble solids, sometimes at controlled pH, by addition of fresh water. Whether this is carried out in one tank or in two steps in series, water is usually recycled, and screens are used for separating the wet rubber crumb from the free water. The final operation in the washing process in modern plants is to dewater the crumb in a press. These are commonly extruder presses in which the screw forces the wet material through a restriction, and which have a slotted barrel to allow free water to drain away. Coagulated crumb of NBR is typically about 60% by weight water when fed to an extruder press, and is dewatered to about 10% moisture by this process. The benefits are that a minimum of water has to be removed by evaporation in the dryer (an energy savings), and that the water squeezed out carries much of the remaining soluble impurities with it.

After dewatering the crumb it is fed to the drying process which is usually carried out in a continuous tunnel dryer. The crumb is spread on a perforated stainless steel bed through which hot air is passed to evaporate the remaining water. Typically, in the first portion of the dryer, air at 110–140°C is used, with lower temperatures being used as the product approaches dryness. A typical target for final moisture content is 0.5% or less. At the exit of the dryer the product is cooled and conveyed to a baler which shapes it into bales for packaging and shipment.

For special products or product forms, NBR is sometimes finished by other methods. A powdered form can be produced by feeding latex directly to a spray dryer, where it is atomized and dried in a hot air stream. An anticaking material such as talc must be added to preserve the powder form during storage and shipment. Some products are ground to particulates of various sizes; they are carried out of the tunnel dryer to a grinding process where size is reduced and anticake is added. Finally, low molecular-weight NBR is a molasses-like liquid and is produced largely for use as a plasticizer for other polymers. It must be dried in an evaporator under vacuum and packaged in drums.

Economic Aspects

Late 1992 pricing for NBR was approximately $2.2/kg, with specialty grades slightly higher. Nitrile rubber is generally considered a mature product and growth of production and sales have been relatively low as reflected in the production history and long-term consumption forecast figures provided by the International Institute of Synthetic Rubber Producers (IISRP) (Tables 3 and 4) (19).

There are five nitrile rubber producers in North America. Details of production facilities for nitrile rubber as provided by the IISRP are shown in Table 5.

Grades of Nitrile Rubber

There are many grades of nitrile rubber available on the market today; for example, Zeon Chemicals offers approximately 75 different grades of NBR. The principal variables that are commonly changed include the following:

Table 3. Worldwide Nitrile Rubber Production History[a]

Year	Production, 10^3 t
1982	175
1983	189
1984	238
1985	213
1986	214
1987	231
1988	262
1989	243
1990	246
1991	244

[a]Ref. 19; includes latex and dry rubber.

Table 4. Long-Term Rubber Consumption Forecast by Area[a], 10^3 t

Geographic area	1992	1996
North America	108	111
Western Europe	87	92
Asia	75	87
Latin America	12	15
Africa/Middle East	4	4
Total	286	309[b]

[a] Ref. 19; includes latex and dry rubber.
[b] +8.0%

Acrylonitrile Content. Standard grades available in the market contain between 15 to 50% acrylonitrile. The acrylonitrile content of nitrile rubber has a significant effect on two properties: chemical resistance and low temperature performance. As the acrylonitrile content of the polymer is increased, the chemical resistance is improved whereas the low temperature properties are diminished.

Mooney Viscosity. This is a measurement of the viscosity of the polymer that is commonly used in the rubber industry. Mooney viscosity values typically range from 25 to 100. Mooney viscosity generally relates to polymer molecular weight, with the lower Mooney viscosity polymers providing improved flow and processing characteristics and the higher Mooney NBRs providing improved physical properties.

Emulsifier Type. The manufacturers of NBR use a variety of emulsifiers (most commonly anionic) for the emulsion polymerization of nitrile rubber. When the latex is coagulated and dried, some of the emulsifier and coagulant remains with the rubber and affects the properties attained with the rubber compound. Water resistance is one property in particular that is dependent on the type and amount of residual emulsifier. Residual emulsifer also affects the cure properties and mold fouling characteristics of the rubber.

Antioxidant Type. Producers of nitrile rubber add stabilizers to the nitrile rubber latex prior to coagulation and drying of the NBR. These antioxidants are required because the polymer contains residual double bonds that can be oxidized during the elevated temperature drying process. The mechanisms of oxidation of diene elastomers (including nitrile rubber) have been covered in a recent review (20). Many different types of antioxidants are used, including amine and phenolic derivatives. Of course these antioxidants affect the properties of the nitrile rubber in the final rubber compound. Choice of antioxidant is especially important when food contact is planned for the rubber. Grades of nitrile rubber are available to meet various regulations and requirements from the Food and Drug Administration (FDA) and the National Sanitation Foundation.

Third Monomers. In order to achieve certain property improvements, nitrile rubber producers add a third monomer to the emulsion polymerization process. When methacrylic acid is added to the polymer structure, a carboxylated nitrile rubber with greatly enhanced abrasion properties is achieved (9). Carboxylated nitrile rubber carries the ASTM designation of XNBR. Cross-linking monomers,

Table 5. World Production Facilities for Nitrile Rubber

Producer	Plant location	Trade name	Capacity, 10^3 t
	North America		
Zeon Chemicals	Louisville, Ky.	Nipol	35
Goodyear Tire & Rubber	Houston, Tex.	Chemigum	28
Uniroyal Chemical Co.	Painesville, Ohio	Paracril	20
Polysar/Bayer	Sarnia, Ontario	Krynac	20
Copolymer Rubber/DSM	Baton Rouge, La.	Nysyn	10
Total North America			*113*
	Western Europe		
Bayer	Germany	Perbunan	35
Polysar/Bayer	France	Krynac	33
Enichem Elastomeri	Italy	Europrene N	30
Goodyear Chemicals Europe	France	Chemigum	11
Zeon Chemicals Europe	South Wales	Breon	10
Buna A.G.	Germany		5
Total Western Europe			*124*
	Asia		
Japan Synthetic Rubber	Japan	JSR N	40
Nippon Zeon	Japan	Nipol	20
Nantex	Taiwan	Nancar	12
Korea Kumho	Korea		10
Synthetics and Chemicals	India		2
Takeda Chemical	Japan		1
Total Asia			*85*
	Latin America		
Nitriflex	Brazil	Nitriflex	11
Hules Mexicanos	Mexico	Humex	6
Pasa	Argentina	Arnipol	2
Total Latin America			*19*
Total global capacity			*341*

eg, divinylbenzene or ethylene glycol dimethacrylate, produce precross-linked rubbers with low nerve and die swell. To avoid extraction losses of antioxidant as a result of contact with fluids during service, grades of NBR are available that have utilized a special third monomer that contains an antioxidant moiety (10). Finally, terpolymers prepared from 1,3-butadiene, acrylonitrile, and isoprene are also commercially available.

Preplasticized. A plasticizer can be incorporated into the rubber during the manufacturing process. By adding a phthalate plasticizer to the rubber latex prior to coagulation and drying, the NBR producer can prepare a rubber that is particularly well suited to low hardness applications such as rolls (21).

Blends with PVC. Nitrile rubber may be blended with poly(vinyl chloride) (PVC) by the polymer producer by two different techniques: (1) blending of NBR latex with PVC latex followed by co-coagulation and drying, or (2) physically mixing the solid NBR and PVC powder in mixing equipment such as an internal mixer. NBR–PVC polymer blends are well known for the good ozone resistance that is imparted by the PVC.

Masterbatches with Carbon Black. Two producers (Zeon Chemicals and Co-polymer) offer nitrile rubber–carbon black masterbatches. These grades are prepared by mixing the carbon black with the rubber latex prior to the coagulation and drying process.

Physical Form. Nitrile rubber is available in several different forms, ie, solid bale, particulate/powder, and high viscosity liquid. The most common form is a 50-lb (22.7-kg) bale. Particulate and powder grades of nitrile rubber are produced by either grinding of the solid rubber or spray drying of the nitrile rubber latex. Liquid grades are produced by using high levels of mercaptan molecular-weight modifiers during the polymerization process to produce a relatively low molecular-weight/high viscosity liquid, eg, Brookfield viscosity of Nipol 1312LV is 13,000 mPa·s(=cP) at 50°C.

Storage of nitrile rubber is recommended in a cool, dry place away from direct light. Producers of nitrile rubber add an antioxidant to the rubber for stability during the rubber drying operation and subsequent warehouse storage. As such, assurances of one-year warehouse storage stability are typically provided. Special problems exist with the storage of powdered NBRs because they tend to agglomerate when stored for long periods of time at elevated temperatures.

Rubber Analysis and Quality Control Testing

The Rubber Manufacturer's Association (RMA) is a cooperative manufacturing trade association. Recently, the RMA has issued a technical bulletin to standardize the reporting of key analytical/quality control data on nitrile rubber (22). The various tests commonly run on nitrile rubber include raw polymer Mooney viscosity, ML1' + 4 at 100°C, ASTM D1646; cure profile using ASTM D3187 compound, ASTM D2084; total volatile matter, total ash, and total extractables, ASTM D1416; and bound ACN.

Health and Safety Factors

Nitrile rubber presents no unusual hazards when processed and handled in areas with good housekeeping and ventilation. Like other elastomers, nitrile rubber emits fumes and vapors when heated to high temperatures during processing and curing. Acrylonitrile and 1,3-butadiene monomers are known or suspected carcinogens, thus producers of NBR must follow strict Occupational Safety and Health Administration (OSHA) standards that regulate permissible exposure of personnel to these materials during the nitrile rubber manufacturing process. In order to minimize the residual acrylonitrile and 1,3-butadiene in the final product, nitrile rubber producers physically strip or remove the unreacted monomers from the in-process latex prior to the coagulation and drying of the final product. As a result, typical acrylonitrile and 1,3-butadiene residues in the final NBR are on the order of a few parts per million of each of the two monomers for most NBRs. This is important to minimize exposure of the consumer of the NBR to acrylonitrile and 1,3-butadiene during the mixing, processing, and curing of any nitrile rubber article.

Acrylonitrile (qv), listed as a carcinogen by several government agencies, is covered by a specific OSHA standard (29CFR1910.1045) that regulates exposure to two parts acrylonitrile per million parts of air (2 ppm) as an 8-h time-weighted average. The regulation also states that no employee shall be exposed to an airborne concentration of acrylonitrile in excess of 10 ppm over any 15-min period (ceiling limit). The current permissible exposure limit for airborne 1,3-butadiene is 1000 ppm as an 8-h time-weighted average. However, in Aug. 1990 OSHA drafted a proposed rule that would lower the permissible exposure limit to 2 ppm as an 8-h time-weighted average. In Sept. 1991 a new risk assessment for 1,3-butadiene was submitted to OSHA by the National Institute for Occupational Safety and Health (NIOSH). This new risk assessment states that the cancer risk for workers exposed to 1,3-butadiene is greater than believed when OSHA drafted its proposed rule. Thus NIOSH recommended that worker exposure to 1,3-butadiene be kept to the lowest feasible level.

Uses

Most nitrile rubber is consumed in applications utilizing vulcanized rubber compounds. Rubber compounds are mixed with a wide variety of ingredients, including various fillers, plasticizers, processing aids, stabilizers, and curatives (23). After mixing the rubber compound, various vulcanization techniques, such as compression molding, transfer molding, and injection molding, can be used to prepare the final cured rubber article. Common applications for nitrile rubber take advantage of the chemical resistance of the rubber and are as follows:

seals	belts
O-rings	wire and cable insulation
gaskets	hose tubes/covers
oil field parts	rolls
diaphragms	weather stripping
printing supplies	footwear/shoe products
gloves	milking inflations
pump stators	miscellaneous molded rubber goods

Nitrile rubber is also used in many applications that do not involve the mixing of a traditional rubber compound as described above. Some of these areas include the following:

Modification of Plastics. Many plastics, such as PVC, ABS, polypropylene, and nylon, are blended with nitrile rubber to improve flexibility, toughness, or appearance. An oil-resistant thermoplastic elastomer has been prepared by blending nitrile rubber and polypropylene (24).

Adhesives/Cements/Sealants/Coatings. Excellent adhesives of high strength and high oil resistance can be prepared using nitrile rubber (25). Many references have discussed the use of nitrile rubber–phenolic and nitrile rubber–epoxy adhesives for printed circuit boards.

Friction Materials. Nitrile rubber has been used as a modifier in various friction materials such as brake lining and clutch pads.

Hydrogenated Nitrile Rubber

Hydrogenated nitrile rubber is a high performance polymer with significantly better high temperature properties compared to nitrile rubber because of the largely saturated backbone of the rubber (11–15). Hydrogenated nitrile rubber was first developed during the late 1970s and commercialized during the early 1980s. It is produced by a solution process that leads to a selective hydrogenation of the butadiene unsaturation in the polymer. Late 1992 prices varied from approximately \$22 to \$31/kg. Estimates of production capacities for the two commercial producers of hydrogenated nitrile rubber in the world today are shown in Table 6.

Table 6. World Production Facilities for Hydrogenated Nitrile Rubber[a]

Producer	Plant location	Trade name	Capacity, 10^3 t
Zeon Chemicals	Bayport, Tex.	Zetpol	1500
Nippon Zeon	Japan	Zetpol	1800
Polysar/Bayer	Orange, Tex.	Tornac/Therban	1600

[a]Ref. 19.

As in nitrile rubber there are different grades of hydrogenated nitrile rubber that vary in acrylonitrile content and Mooney viscosity. In addition, various grades with different extents of hydrogenation are available. Finally, grades of hydrogenated nitrile rubber are available containing zinc oxide and methacrylic acid. Extremely high tensile strength, up to 58 MPa (8400 psi), and excellent abrasion resistance have been reported for hydrogenated nitrile rubber compounds containing zinc oxide and methacrylic acid (26).

Applications for hydrogenated nitrile rubber are similar to those of nitrile rubber in that they take advantage of the excellent chemical and oil resistance of the polymer. However, the increased high temperature performance of the hydrogenated nitrile rubber makes it far superior to standard grades of nitrile rubber. Examples of applications for hydrogenated nitrile rubber are shown in Table 7 (27,28).

Table 7. Applications for Hydrogenated Nitrile Rubber[a]

Oil-field applications	Automotive applications	Industrial applications
blowout preventers	rotating shaft seals	hydraulic hose
downhole packers	lip seals	hydraulic packings
drill pipe protectors	bearing/wheel seals	cable jacket for power station
mud pump pistons	O-rings	laminating rolls
well head seals	valve cover gaskets	printing rolls
O-rings	fuel pump diaphragm	textile rolls
valve seals	fuel pump isolators	paper mill rolls
cable jackets	fuel hose tubes	military components ·
drilling hoses	timing belts	chemical plant diaphragms
pressure accumulators	valve stem seals	
wipers	water pump seals	
swab cups		
pump stators		

[a] Refs. 27 and 28.

BIBLIOGRAPHY

"Rubber, Synthetic" in *ECT* 1st ed., Vol. 11, pp. 827–852, by J. D. D'Ianni, The Goodyear Tire & Rubber Co.; "Elastomers, Synthetic" in *ECT* 2nd ed., Vol. 7, pp. 676–705, by W. M. Saltman, The Goodyear Tire & Rubber Co.; "Elastomers, Synthetic (Nitrile Rubber)" in *ECT* 3rd ed., Vol. 8, pp,. 534–546, by H. W. Robinson, Uniroyal, Inc.

1. Ger. Pat. 658,172 (Apr. 26, 1930); Brit. Pat. 360,821 (May 30, 1930); Fr. Pat. 710,901 (Feb. 4, 1931); U.S. Pat. 1,973,000 (Sept. 11, 1934), E. Konrad and E. Tschunkur (to I. G. Farbenindustrie A.G.).
2. D. A. Seil, "How New Nitrile Rubbers Improve Product Quality," *Rubber Plas. News* (Sept. 23, 1985).
3. N. Nakajima, E. R. Harrel, P. R. Kumler, D. A. Seil, and A. H. Jorgensen, *Adv. Polym. Technol.* **4**(3/4), 267 (1983).
4. W. Hoffmann, *Rubber Chem. Technol.* **37**(2), part 2 (April–June 1964).
5. D. F. Kates and H. B. Evans, in Clara D. Craver, ed., *Polymer Characterization: Interdisciplinary Approaches*, Plenum Press, New York, 1971, p. 213.
6. B. G. Willoughby, *Polym. Test.* **8,** 45 (1989).
7. R. Schmolke and W. Kimmer, *Plaste Kautschuk* **21**(9), 651 (1974).
8. P. W. Milner, *Dev. Rubber Technol.* **4,** 57 (1987).
9. P. H. Starmer, *Plast. Rubber Proc. Appl.* **9**(4), 209 (1988).
10. J. W. Horvath, "Bound Antioxidant Stabilized NBR in Automotive Applications", *Annual Meeting Proceedings of the IISRP*, Paper No. 11, 1979, p. 20.
11. S. Hayashi, M. Oyama, K. Hashimoto, and T. Nakagawa, "New Improved Low Temperature Hydrogenated Nitrile Rubber (HNBR)," paper presented at the *A.C.S. Rubber Division Meeting*, Detroit, Mich., Oct. 8–11, 1991.
12. K. Hashimoto and Y. Todani, in A. Bhownick and H. Stephens, eds., *Handbook of Elastomers*, Marcel Dekker, Inc., New York, 1988, Chapt. 24, p. 741.
13. R. C. Klingender, in M. Morton, ed., *Rubber Technology*, 3rd ed., Van Nostrand Reinhold, New York, 1987, p. 488.
14. N. A. Mohammadi and G. L. Rempel, *Macromolecules* **20**, 2362 (1987).
15. Y. Kubo, K. Hashimoto, and N. Watanabe, "Structure and Properties of Highly Saturated Nitrile Elastomers," paper presented at the *A.C.S. Rubber Division Meeting*, New York, Apr. 7–11, 1986.

16. Int. Pat. Appl. (PCT) 91/06579 (May 16, 1991), H. W. Schiessl and F. W. Migliaro (to Olin Corp.).
17. U.S. Pat. 4,452,950 (June 5, 1984), L. G. Wideman (to The Goodyear Tire & Rubber Co.).
18. D. K. Parker, H. A. Colvin, A. H. Weinstein, and S. L. Chen, *Rubber Chem. Technol.* **63**(4), 582 (1990).
19. *Worldwide Rubber Statistics* 1992, International Institute of Synthetic Rubber Producers, Inc., Houston, Tex., 1992, pp. 10, 13, 68–70.
20. C. Adam, J. Lacoste, and J. Lemaire, *Actual. Chim.* (2), 85 (1991).
21. T. L. Jablonowski, *Rubber World* **200**(4), 25 (1989).
22. Rubber Manufacturers' Association, *RMA Form SP-241*, Washington, D.C., 1991.
23. H. L. Stephens, in Ref. 13, p. 20.
24. U.S. Pat. 4,409,365 (Oct. 11, 1983), A. Y. Coran and R. Patel (to Monsanto Co.).
25. D. E. Mackey and C. E. Weil, in Irving Skeist, ed., *The Handbook of Adhesives*, 3rd ed., Van Nostrand Reinhold, New York, 1990, p. 206.
26. Y. Saito, A. Fujino, and A. Ikeda, *High Strength Elastomers of Hydrogenated NBR Containing Zinc Oxide and Methacrylic Acid*, Society of Automotive Engineers (SAE) Technical Paper Series, No. 890359, Detroit, Mich., Feb. 27–Mar. 3, 1989.
27. K. Hashimoto, Y. Otawa and Y. Todani, "Applications for HSN Elastomers," *Autom. Polym. Des.* **8**(3) (Feb. 1989).
28. K. Hashimoto and co-workers, "Highly Saturated Nitrile Elastomer, A Review," paper presented at the *A.C.S. Energy Rubber Group*, Dallas, Tex., Jan. 19, 1989.

General References

J. R. Purdon, in *The Vanderbilt Rubber Handook*, 13th ed., R. T. Vanderbilt Co., Inc., New York, 1990, p. 166.
D. A. Seil and F. R. Wolf, in Ref. 13, Chapt. 11, p. 332.
P. W. Milner, *Dev. Rubber Technol.* **4,** 57 (1987).
J. R. Dunn, *Kautsch. Gummi Kunstst.* **36**(11), 966 (1983).
H. H. Bertram, *Dev. Rubber Technol.* **2,** 51 (1981).
J. R. Dunn, *Rubber Chem. Technol.* **51**(3), 389 (1978).
A. H. Jorgensen, in J. McKetta and W. Cunningham, eds., *The Encyclopedia Chemical Processes and Design*, Vol. 1, Marcel Dekker, New York, 1976, p. 439.
C. H. Lufter, "Vulcanization of Nitrile Rubber," in *Vulcanization of Elastomers*, Reinhold Publishing Corp., New York, 1964.

DONALD MACKEY
AUGUST H. JORGENSEN
Zeon Chemicals, USA

PHOSPHAZENES

Polyphosphazenes have a backbone of alternating nitrogen and phosphorus atoms with two substituents on each phosphorus atom. The backbone is isoelectronic with that of silicones; these polymer backbones share the characteristics of thermal stability and high flexibility. The normal synthesis route provides a large range of possible derivatives. Coatings, fluids, elastomers, and thermoplastics can be produced by varying the polymer molecular weight and the substituents on the phosphorus. These materials have been suggested for use in a wide variety of areas including solid electrolytes for advanced batteries, and biomedical devices, including implants and drug carriers. Other reviews are available covering various aspects of polyphosphazene chemistry and applications (1,2) (see INORGANIC POLYMERS).

Two elastomers have been commercialized with unique property profiles. One has fluoroalkoxy substituents that provide resistance to many fluids, especially to hydrocarbons. This material also has a broad use temperature range and useful dynamic properties. Aryloxy substituents provide flame retardant materials without halogens.

Synthesis

Phosphazene polymers are normally made in a two-step process. First, hexachlorocyclotriphosphazene [940-71-6], trimer (**1**), is polymerized in bulk to poly(dichlorophosphazene) [26085-02-9], chloropolymer (**2**). The chloropolymer is then dissolved and reprecipitated to remove unreacted trimer. After redissolving, nucleophilic substitution on (**2**) with alkyl or aryloxides provides the desired product (3).

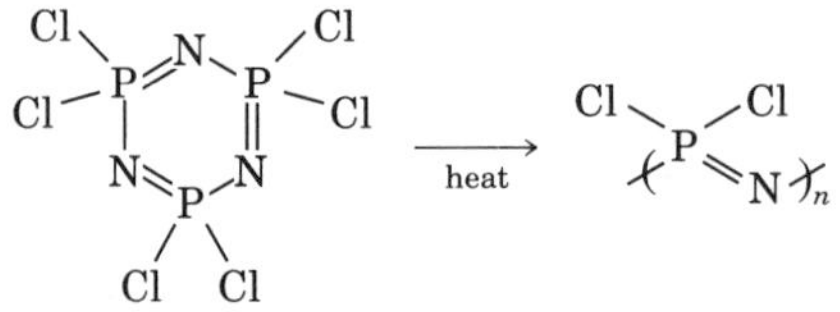

$$(1) \qquad\qquad (2)$$

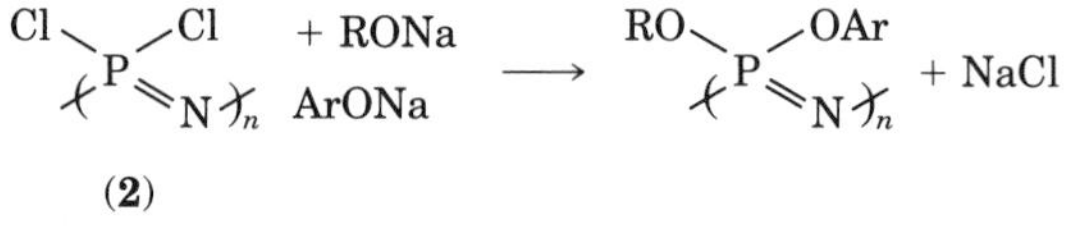

$$(2)$$

Production of a high purity linear chloropolymer of appropriate molecular weight is a demanding task. The original approach, a ring opening polymerization of cyclic trimer, can be done without deliberate addition of catalyst, although it is now accepted that ultrapure trimer will not polymerize without some catalyst. This thermal polymerization leads to high molecular weight (typically over one million) and broad molecular weight distributions. Temperatures of 250°C for 24 to 48 hours are frequently employed to obtain conversions of 30 to 50%. Conversions are limited to avoid gelation. Use of a catalyst provides control of molecular weight, and gives significantly higher conversions with soluble (substantially linear) product. With catalysts of appropriate activity, lower polymerization temperatures can be utilized. A large number of catalysts have been studied and catalysts are used commercially in preparing chloropolymer. A patent was issued claiming aprotic Lewis acids as catalysts (4).

The standard route to chloropolymer requires high purity cyclic trimer, which is expensive. Other potentially less expensive routes have been investigated which may be considered polycondensation reactions. One approach is a multistep procedure involving phosphorus pentachloride and ammonium chloride (5). Another route involves polycondensation of P-trichloro-N-dichlorophosphoryl monophosphazene (3) or the thio analogue, $Cl_3P{=}N{-}P(S)Cl_2$ (6). This approach is being pursued by Atochem. All these syntheses require the processing of extremely corrosive materials at temperatures of 200–300°C.

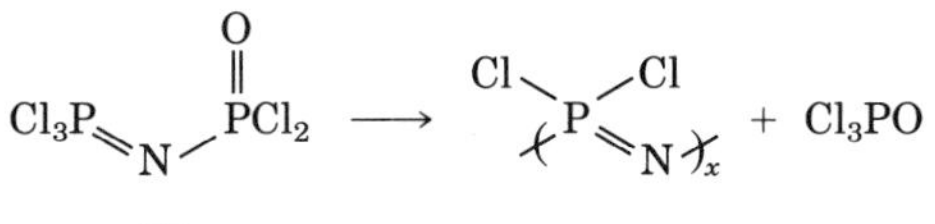

$$Cl_3P{\diagdown}_N{\diagup}\overset{\overset{\displaystyle O}{\|}}{PCl_2} \longrightarrow \underset{x}{\left[\overset{Cl}{\diagdown}\underset{\diagdown N\diagup}{P}\overset{Cl}{\diagup}\right]} + Cl_3PO$$

(3)

Substitution of chloropolymer is possible using a variety of nucleophiles. The most common are sodium salts of alcohols and phenols. Thermoplastics are obtained using a single substituent, whereas multiple substituents of sufficiently different size lead to elastomers (2). Liquid crystal behavior similar to polysiloxanes has been noted in most homopolymers. The homopolymer formed using trifluoroethanol as a substituent has received a fair amount of academic scrutiny (7).

Elastomers formed using alkoxide salts of trifluoroethanol and higher fluoroalcohols have been found to have an interesting range of thermal and fluid resistance. These materials were developed under U.S. Army sponsorship, and initially commercialized by Firestone using the name PNF, later sold by Ethyl Corp. as EYPEL-F elastomer. ASTM has reserved the designation FZ for elastomers of this class which have the nominal structure given as (4).

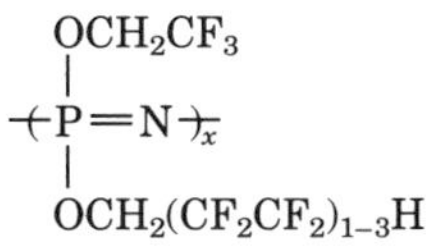

$$\underset{(4)}{\overset{OCH_2CF_3}{\underset{OCH_2(CF_2CF_2)_{1-3}H}{\left[P{=}N\right]_x}}}$$

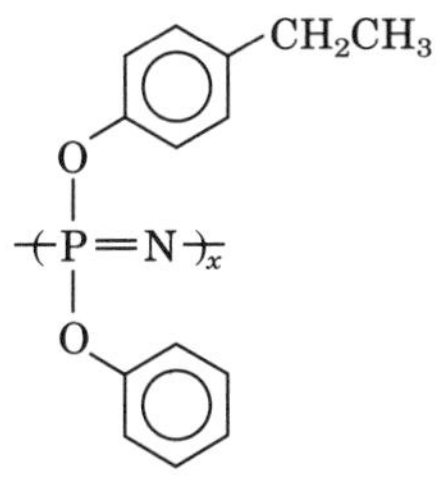

(5)

Roughly 65% of the substituents are trifluoroethoxy, and 35% are telomer alcohols prepared from tetrafluoroethylene and methanol. About 0.5 mol % of an allylic substituent is used as a cross-link site. The substituent pattern is believed to be strictly statistical.

Another elastomer to find use is the substitution product of phenol and *p*-ethylphenol along with an allylic monomer to provide cross-link sites (**5**). This is trademarked EYPEL-A elastomer [*66805-77-4*] by Ethyl Corp., and designated PZ by ASTM. The substitution ratio is roughly 52 mol % phenol, 43% *p*-ethylphenol, and 5% allylic substituent.

Routes to prepare substituted polymer directly were pioneered with the polymerization of *N*-trimethylsilylphosphoranamines to form low to moderate molecular weight polyphosphazenes (**6**) where R is alkyl or aryl (8).

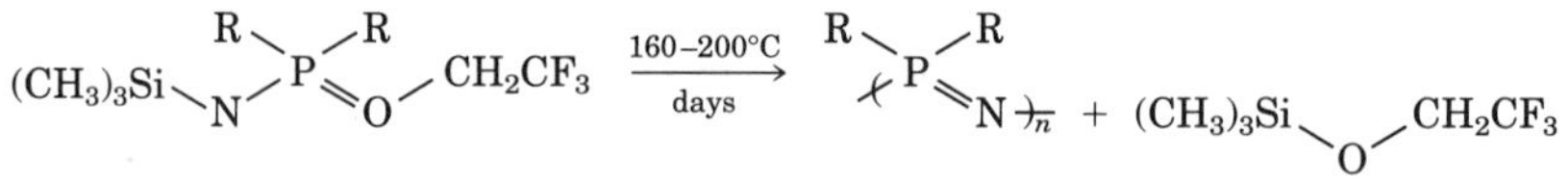

$$(CH_3)_3Si\diagdown_{N} \diagup^{R} \diagdown_{P} \diagup^{R} \diagdown_{O} \diagup CH_2CF_3 \xrightarrow[\text{days}]{160-200°C} \diagup^{R} \diagdown_{P} \diagup^{R} \diagdown_{N}\!\!\big)_{\!\!\overline{n}} + (CH_3)_3Si\diagdown_{O} \diagup CH_2CF_3$$

(**6**)

These polymers differ from those prepared in the traditional two-step routes in that there is a direct carbon–phosphorus bond instead of a potentially weaker O, N, or S linkage to phosphorus. A related approach has been discovered using a catalyzed polymerization of tris(trifluoroethoxy)-*N*-(trimethylsilyl)phosphoranimine (9). This allows greater synthetic variation while still giving the control of a one-step polymer synthesis. Neither of these approaches are practiced commercially. The monomers required are costly, and frequently the polymer molecular weights are low.

Toxicology

Hexachlorocyclotriphosphazene (cyclic trimer) is a respiratory irritant. Nausea has also been noted on exposure (10). Intravenous and intraperitoneal toxicity measurements were made on mice. The highest nonlethal dose (LD_0) was measured as 20 mg/kg (11). Linear chloropolymer is also believed to be toxic (10). Upon organic substitution, the high molecular weight linear polymers have been shown to be inert. Rat implants of eight different polyphosphazene homopolymers indicated low levels of tissue toxicity (12). FZ has been found to be reasonably compatible with blood (13), and has lower lipid absorption than fluorosilicone.

Fluoroalkoxyphosphazene Elastomers

FZ Characterization. FZ elastomer is a translucent pale brown gum with a glass-transition temperature, T_g, of -68 to $-72°C$. The gum can be cross-linked using peroxides such as dicumyl peroxide and α,α'-bis(*t*-butylperoxy)diisopropylbenzene. It is provided in compounded form for specific applications, or in masterbatch form to allow custom compounding. Surface treated silica and semireinforcing carbon black are preferred fillers (qv). Acidic fillers such as precipitated or fumed silicas and small particle carbon blacks inhibit the peroxide cure and lead to poorer thermal stability and compression set. Clays, silicates, and nonreinforcing blacks can be used as extending fillers. Compounding is performed

using an internal mixer, and completed using a roll mill. Gum physical properties
are as follows (14):

T_g, °C	−68
Mooney viscosity, ML 1 + 4, 100°C	20
specific gravity	1.75
refractive index	1.41

The gum is soluble in lower ketones and esters, amide solvents, methanol, and
tetrahydrofuran.

Compounds can be processed using most conventional rubber processing
equipment. Roll milling of low durometer stocks is easiest using a warm roll.
Mooney viscosities of these compounds range from 30 to 140 at 100°C. This typically leads to easy flow in molding, although significant orientation is often noted.
Selected compounds are suitable for calendering and extrusion of profiles. These
contain at least some carbon black as filler.

Most applications have been parts molded using compression molding, although transfer molding has been used and injection molding has been demonstrated. Parts made have varied from O-rings weighing under a gram to complex
seals weighing over 20 kg. Good adhesive bonds to metal can be achieved during
the cure cycle using silane-type primers. Molding cycles of 20–30 min at 160°C
are typical using dicumyl peroxide. Shorter cycles are obtainable at higher temperatures, but curing above 200°C is not recommended. Linear mold shrinkage of
roughly 3% is typical for 70 durometer compounds molded at 170°C. Post-cure in
a 175°C oven for four hours provides optimal modulus and compression set values.
The range of typical compound property values is given in Table 1.

The decreases in tensile strength and elongation values as the temperature
is elevated are modest when compared to specialty elastomers having higher glass

Table 1. Properties of FZ Compounds

Property	Value
density, g/mL	1.75–1.85
durometer hardness, Shore A	35–90
tensile strength, MPa[a]	6.9–13.8
100% modulus, MPa[a]	2.8–14.8
ultimate elongation, %	75–250
compression set, 70 h at 150°C, %	15–55
tear resistance, die B, kN/m[b]	to 26
retraction temperature, TR-10, °C	−56
brittle point, °C	−68
weatherability and ozone resistance	excellent
flame resistance, LOX[c]	39–46

[a]To convert MPa to psi, multiply by 145.
[b]To convert kN/m to ppi, divide by 0.175.
[c]Limiting oxygen index.

transitions (Fig. 1). This aids in actual use as well as in molding and other processing. A continuous use temperature of 175°C is based on retention of 50% of both tensile strength and elongation after a thousand hours of aging at this temperature (Fig. 2).

The excellent low temperature properties of FZ have been indicated in Table 1. Modulus curves were obtained using dynamic mechanical spectroscopy to compare several elastomer types at a constant 75 durometer hardness. These curves indicate the low temperature flexibility of FZ is similar to fluorosilicone and in great contrast to that of a fluorocarbon elastomer (vinylidene fluoride copolymer) (Fig. 3) (15).

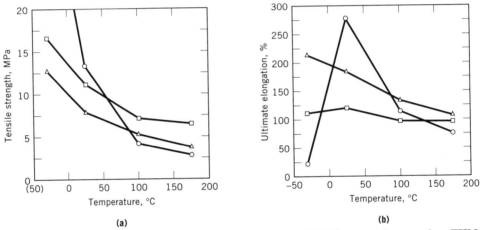

Fig. 1. Comparison of fluoroalkoxy FZ, □; fluorosilicone FVMQ, △; and fluorocarbon FKM, ○, elastomers. (**a**) Tensile strength; (**b**) elongation. To convert MPa to psi, multiply by 145.

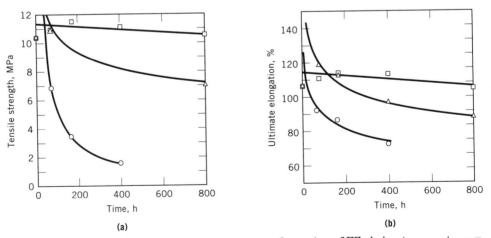

Fig. 2 Tensile strength and elongation retention after aging of FZ. Aging temperature, □, 150°C; △, 175°C; ○, 200°C. To convert MPa to psi, multiply by 145.

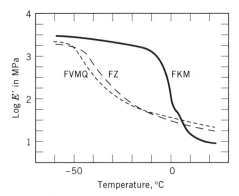

Fig. 3. Modulus comparison at low temperature of FZ, FVMQ, and FKM elastomers.

Dynamic mechanical analysis uses a small sinusoidal mechanical deformation to measure the modulus of the material. Modulus values obtained in the tensile mode are designated with the symbol E, whereas G is used for shear moduli. In viscoelastic materials, the deformation is not in phase with the force applied. The modulus can be resolved into a storage modulus, E' (elastic response) and a loss modulus, E'' (viscous response). The ratio of loss to storage moduli, E''/E', also known as the loss tangent or tan δ, peaks at transitions in the mechanical behavior of the material. These are related to polymer morphology changes, but are also a function of the applied oscillation frequency. Highly resilient materials have low tan δ values, while damping materials have high values in the range of temperature and frequency of interest.

Properly formulated FZ compounds have demonstrated excellent fatigue resistance. One measure of this is the Monsanto fatigue-to-failure test (ASTM D4482) which emphasizes crack propagation. Superior fatigue life was observed at elongation ratios under 2, ie, elongations under 100% (Fig. 4), in a comparison of 50 durometer specialty elastomers. This property is utilized in diaphragms, both in applications involving contact with hydrocarbons and in those without fluid contact.

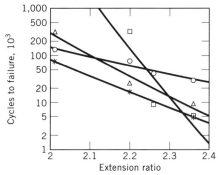

Fig. 4. Tensile fatigue, the Monsanto fatigue-to-failure FTF test, for several elastomers: □, FZ; △, ethylene–acrylic; ○, nitrile (NBR); and ∗, FVMQ.

Shock isolation is also possible using the damping characteristics of FZ elastomer. Dynamic mechanical analysis indicates multiple transitions and a broad damping peak. This damping can be enhanced using formulations containing both silica and carbon black fillers.

FZ elastomers have excellent resistance to hydrocarbons and inorganic acids as expected for a fluorinated elastomer. They are strongly affected by polar solvents, but are more resistant to amines than most other fluorinated elastomers as shown in Table 2.

Economic Factors.   Annual usage is under 10 metric tons, and is largely consumed in the United States. Minor quantities are used in both Europe and Japan. Prices vary by application but exceed $200/kg. Ethyl Corp. has announced its intent to divest or exit this business. At present they are the sole supplier.

Specifications.   FZ compounds meet class EK in the SAE automotive rubber classification J200. Line callouts have been developed for 50, 70, and 80 durometer materials. Specific compounds meet requirements for O-rings for military aerospace applications identified in MIL-P-87175, AMS-7261/1-3, and AMS-7284.

Applications.   Initial applications have been largely in military and aerospace areas. These include hydraulic seals for military aircraft and fuel seals and diaphragms for both military and civilian aircraft. Shock mounts for FZ are used

Table 2. Fluid Resistance of FZ

Fluid	Rating[a]
Aqueous solutions	
inorganic acids	G–E
organic acids	P–F
bases including amines	F–G
ethylene glycol	G
Fuels	
unleaded gasoline	G–E
gasohol	F–G
diesel, jet fuel	E
Hydraulic fluids	
hydrocarbon	E
phosphate ester (Skydrol)	P
ester	G
silicone	E
Organic solvents	
hydrocarbons	E
chlorinated hydrocarbons	G
chlorofluorocarbons	P
carbonyl compounds (acids, ketones, esters), ethers, methanol	P
other alcohols	G

[a]E, Excellent; G, good; F, fair; P, poor.

on aircraft engines. Large fabric-reinforced boot seals are used in the air intake system on the M-1 tank. The material's useful temperature range, fuel and fatigue resistance, and fire resistance were determining factors in this application.

There are also a number of nonaerospace civilian applications for this material. Small diaphragms have found limited use in automotive and related applications. These are used because of the material's excellent fatigue life and hydrocarbon resistance. A compound containing methacrylates functions as a soft liner for dentures (16), providing extended life, shock isolation, and resistance to microbial attack.

Aryloxyphosphazene Elastomers

Properties and Applications. Aryloxyphosphazene elastomers using phenoxy and p-ethylphenoxy substituents have found interest in a number of applications involving fire safety. This elastomer has a limiting oxygen index of 28 and contains essentially no halogens. It may be cured using either peroxide or sulfur. Peroxide cures do not require the allylic cure monomer. Gum physical properties are as follows (17):

T_g, °C	-18
Mooney viscosity, ML 1 + 4, 100°C	40–70
specific gravity	1.25

The gum is soluble in tetrahydrofuran, toluene, and cyclohexane.

Initial uses of aryloxyphosphazenes have focused on areas where fire safety is a paramount concern. Cross-linked aryloxyphosphazene networks form an intumescent char on heating. The gum accepts fairly high filler loadings. Nonreinforcing fillers like alumina trihydrate can be used to make compounds with limiting oxygen indexes above 40. Smoke density of these compounds is also low as measured in the NIST procedure (maximum in flaming mode can be <100). Early research on toxicity of combustion products to small animals rated PZ compounds as roughly 1.5 times as toxic as Douglas fir. Other common synthetics rated from 6 to 25 times as toxic as fir (18).

These advantages in fire safety have been pursued in several product areas. Flexible closed cell foams have been produced with typical densities of 72 kg/m^3 (4.5 lb/ft^3). Thermal conductivity is only 0.037 W/(m·K) and water vapor permeability is low. This product is used in hull and pipe insulation on naval ships and submarines, and has elicited interest for use on aircraft. These foams meet the requirements of specification MIL-I-24703. Dense sheeting made using high gravity fillers is also used in acoustic damping applications by the U.S. Navy. In both these applications, the glass and higher transitions contribute to appreciable damping at ambient temperature and acoustic frequencies.

Wire and cable insulation based on vulcanizates of PZ has also been studied. Again, low fire risk was the target property, and this was achieved. The need to vulcanize the coating, somewhat modest tensile properties, tensile strength of 5.2 to 12.2 MPa (760 to 1770 psi), and high dielectric constant (4–5 at 10,000 Hz) limited interest in this application (19).

Economic Considerations.   Gum production by Ethyl Corp. is under 20 metric tons per year. Products are sold primarily for use in U.S. and N.A.T.O. naval applications. The materials are sold as finished goods, and thus elastomer pricing is not appropriate. Ethyl Corp. has announced that it will either divest or exit this business. Atochem has developed their technology and has constructed a pilot plant to produce aryloxyphosphazenes.

BIBLIOGRAPHY

"Fluoroalkoxyphosphazenes" under "Fluorine Compounds, Organic," in *ECT* 3rd ed., Vol. 10, pp. 936–947, by D. P. Tate and T. A. Antkowiak, Firestone Tire and Rubber Co.

1. H. R. Allcock, *Chem. Eng. News*, 22 (Mar. 18, 1985) (good simple introduction); J. C. van de Grampel and B. De Ruiter, *Organophosphorus Chemistry* **15,** 260 (1985) (emphasizes cyclics, some biomedical applications); D. F. Lohr and H. R. Penton in A. K. Bhowmick and H. L. Stephens, eds., *Handbook of Elastomers*, Marcel Dekker, Inc., New York, 1988, p. 535 (elastomers); S. V. Vinogradova, D. R. Tur, and I. I. Minosyants, *Russian Chemical Reviews* **53,** 49 (1984); R. E. Singler and M. J. Bieberich in R. L. Shubkin, ed., *Synthetic Lubricants and High-Performance Functional Fluids*, Marcel Dekker, Inc., New York, 1992, p. 215 (fluids).
2. R. E. Singler, G. L. Hagnauer, and R. W. Sicka in J. E. Mark and J. Lal, eds., *Elastomers and Rubber Elasticity*, ACS Symposium Series, Vol. 193, Washington, D.C., 1985, p. 229 (good general review).
3. H. R. Allcock, *Phosphorus–Nitrogen Compounds*, Academic Press, Inc., New York, 1972.
4. U.S. Pat. 5,006,324 (Apr. 9, 1991) C. H. Kolich, B. R. Meltsner, and H. R. Braxton (to Ethyl Corp.).
5. U.S. Pat. 4,551,317 (Nov. 5, 1985) H. M. Li (to Ethyl Corp.).
6. R. DeJaeger and co-workers, *Makro. Chemie* **183,** 1137 (1982); R. DeJaeger and co-workers, *Macromolecules* **25,** 1254 (1992). U.S. Pat. 4,554,113 (Nov. 19, 1985) T. A. Chakra and R. DeJaeger (to Societe Nationale Elf Aquitaine).
7. M. Kojima and J. H. Magill, *Polymer* **26,** 1971 (1985).
8. R. H. Nielson and P. Wisian-Neilson, *Chem. Rev.* **88,** 541 (1988).
9. R. A. Montague and K. Matyjaszewski, *J. Am. Chem. Soc.*, **112,** 6721 (1990).
10. L. F. Audrieth, R. Steinman, and A. D. F. Toy, *Chem. Rev.* **32,** 109 (1943).
11. J. F. Labarre and co-workers, *Europ. J. Cancer* **15,** 637 (1979).
12. C. W. R. Wade and co-workers, in C. E. Carraher, Jr., J. E. Sheets, and C. O. Pittman, eds., *Biocompatibility of Eight Poly(organophosphazene) Organometallic Polymers*, Academic Press, Inc., New York, 1978, pp. 289–299.
13. W. M. Reichert, F. E. Filisko, and S. A. Barenberg, *J. Biomed. Mat. Res.* **16,** 301 (1982).
14. Ethyl Corp., *EYPEL-F Performance Elastomers Technical Data*, 1987.
15. Ethyl Corp., unpublished data, 1988.
16. U.S. Pats. 4,543,379 (Sept. 24, 1985) and 4,432,730 (Feb. 21, 1984), L. Gettleman and co-workers (to Gulf South Research Inst.).
17. Ethyl Corp., data sheets, 1989.
18. P. J. Lieu, J. H. Magill, and Y. C. Alarie, *J. Combust. Toxicol.* **7,** 143 (1980).
19. J. T. Books, D. M. Indyke, and W. O. Muenchinger, *Connec. Technol.* **45,** July 1985.

JEFFREY T. BOOKS
Ethyl Corporation

POLYBUTADIENE

1,3-Butadiene [*106-99-0*] can be polymerized to produce various resinous and elastomeric polymers. The basic microstructural units in polybutadiene [*9003-17-2*] include *cis*-1,4, *trans*-1,4, and 1,2 units. A variety of polymers with different properties can be produced by changing the ratio of these units and their tacticity. For example, polybutadienes with high 1,2 contents include syndiotactic polybutadiene with various melting points; isotactic polybutadiene with a melting temperature, T_m, of 170°C; and amorphous atactic polybutadiene ($T_g = +5$°C). High *trans*-1,4-polybutadiene is a resinous crystalline material with three crystal structure modifications having T_ms of 55, 150, and 175°C. High *trans*-1,4-polybutadiene [*40022-02-4*] is partially amorphous when it contains either *cis*-1,4 or 1,2 units which disrupt its crystal structure and reduce its melting point by as much as 20–60°C. It has a glass-transition temperature of −80°C. Other useful elastomers made from 1,3-butadiene include high (99%) *cis*-1,4-polybutadiene [*40022-03-5*] with a T_m of −13°C and a T_g of −100°C, and a mixed microstructure polybutadiene composed of various ratios of *cis*-1,4, *trans*-1,4, and 1,2 units.

The preparation and characterization of 1,3-butadiene monomer is discussed extensively elsewhere (1–4) (see BUTADIENE). Butadiene monomer can be purified by a variety of techniques. The technique used depends on the source of the butadiene and on the polymerization technique to be employed. Emulsion polymerization, which is used to make amorphous *trans*-1,4-polybutadiene (75% *trans*-1,4; 5% *cis*-1,4; 20% 1,2), is unaffected by impurities during polymerization. However, both anionic and Ziegler polymerizations, which are used to prepare *cis*-1,4-polybutadiene, mixed *cis*-1,4 and *trans*-1,4-polybutadiene, and high 1,2-polybutadiene [*26160-98-5*], are greatly affected by impurities. During polymerization, oxygenated compounds such as alcohols, aldehydes or ketones and acidic compounds such as organic acids must be avoided.

Between the 1920s when the initial commercial development of rubbery elastomers based on 1,3-dienes began (5–7), and 1955 when transition metal catalysts were first used to prepare synthetic polyisoprene, researchers in the U.S. and Europe developed emulsion polybutadiene and styrene–butadiene copolymers as substitutes for natural rubber. However, the tire properties of these polymers were inferior to natural rubber compounds. In seeking to improve the synthetic material properties, research was conducted in many laboratories worldwide, especially in the U.S. under the Rubber Reserve Program.

The discovery by Ziegler that ethylene and propylene can be polymerized with transition-metal salts reduced with trialkylaluminum gave impetus to investigations of the polymerization of conjugated dienes (7–9). In 1955, synthetic polyisoprene (90–97% *cis*-1,4) was prepared using two new catalysts. A transition-metal catalyst was developed at B. F. Goodrich (10) and an alkali metal catalyst was developed at the Firestone Tire & Rubber Co. (11). Both catalysts were used to prepare *cis*-1,4-polyisoprene on a commercial scale (9–19).

The Firestone group also polymerized 1,3-butadiene to give an extremely high mol wt polybutadiene of 70% *cis*-1,4 structure. In their research, they purposefully avoided the preparation of vinyl structures in both polyisoprene and polybutadiene since it was believed that vinyl groups adversely affected tire performance. Since natural rubber was 99.9% *cis*-1,4 structure and had superior properties, they believed that a 1,4 structure was necessary for acceptable physical properties. The addition of polar compounds to the lithium-catalyzed polymerization of butadiene changes the microstructure from the 90% *cis*-1,4 structure to a mixed *cis*-1,4 and *trans*-1,4 microstructure.

Microstructures of Polybutadiene

The conjugated structure of 1,3-butadiene gives it the ability to accept nucleophiles at both ends and distribute charge at both carbon 2 and 4. The initial addition of nucleophiles leads to transition states of π-allyl complexes in both anionic and transition-metal polymerizations.

It has been postulated that the syn π-allyl structure yields the *trans*-1,4 polymer, and the anti π-allyl structure yields the *cis*-1,4 polymer. Both the syn and anti π-allyl structures yield 1,2 units. In the formation of 1,2-polybutadiene, it is believed that the syn π-allyl form yields the syndiotactic structure, while the anti π-allyl form yields the isotactic structure. The equilibrium mixture of syn and anti π-allyl structures yields heterotactic polybutadiene. It has been shown (20–26) that the syndiotactic stereoisomers of 1,2-polybutadiene units can be made with transition-metal catalysts, and the pure 99.99% 1,2-polybutadiene (heterotactic polybutadiene) [26160-98-5] can be made by using organolithium compounds modified with bis-piperidinoethane (27). At present, the two stereoisomers of 1,2-polybutadiene that are most used commercially are the syndiotactic and the heterotactic structures.

HIGH VINYL POLYBUTADIENE

These 1,2 addition products are categorized into three main groups: syndiotactic, isotactic, and atactic 1,2-polybutadiene. The 1,2 vinyl products follow a stereospecific addition in which the chiral carbon carrying the pendent vinyl group leads to the formation of ldl, lll, or ddd configurations. The ldl structure is called syndiotactic polybutadiene, and the lll or ddd structures are called isotactic polybutadiene. A mixed structure (dldlldldl) is called atactic polybutadiene. Each of these structures gives polymers with unique physical, mechanical, and rheological properties.

Syndiotactic Polybutadiene. Syndiotactic polybutadiene is a unique material that combines the properties of plastic and rubber. It melts at high (150–220°C) temperatures, depending on the degree of crystallinity in the sample, and it can be molded into thin films that are flexible and have high elongation. The unique feature of this plastic-like material is that it can be blended with natural rubber. *cis*-1,4-Polybutadiene and the resulting blends exhibit a compatible formulation that combines the properties of plastic and rubber.

Syndiotactic polybutadiene was first made by Natta in 1955 (28) with a melting point of 154°C. Syndiotactic polybutadiene [31567-90-5] can be prepared with various melting points depending on its vinyl content and degree of crystallinity. The physical, mechanical, and rheological properties of the polymer are greatly affected by these parameters.

syndiotactic PB

Preparation. There are several methods described in the literature using various cobalt catalysts to prepare syndiotactic polybutadiene (29–41). Many of these methods have been experimentally verified; others, for example, soluble organoaluminum compounds with cobalt compounds, are difficult to reproduce (30). A cobalt compound coupled with triphenylphosphine aluminum alkyls water complex was reported by Japan Synthetic Rubber Co., Ltd. (JSR) to give a low melting point (T_m = 75–90°C), low crystallinity (20–30%) syndiotactic polybutadiene (32). This polymer is commercially available.

The addition of dienophile to cobalt compounds reduced with alkyl aluminum compounds has been reported (31); diethyl fumarate was used as modifier, and a high melting point syndiotactic polybutadiene (T_m = 150°C) was obtained. Preparation of a highly crystalline and high melting point syndiotactic polybutadiene has been reported by many investigators (32–36). The present authors have prepared a syndiotactic polybutadiene with a T_m of 200–210°C by adding carbon disulfide to a cobalt–aluminum alkyl catalyst. Increased yield and a highly crystalline material was developed at Firestone (38) using sulfur, nitrogen ligand, and cobalt compounds. Moreover, the use of carbon oxysulfide was found to give the same type of highly crystalline polymer (38,42). The catalyst, based on cobalt complexed with polar compounds in the presence of reducing agent (trialkylaluminum) and carbon disulfide (CS_2), produced a highly crystalline syndiotactic polybutadiene. This process has been commercialized by Ube Industries Co., Ltd. (29). The Ube group proposed a mechanism for this polymerization which consists of a side-on coordination of the CS_2 to cobalt, anti-π-allyl growing end cisoid bidentate coordination of butadiene. This mechanism was based on an aluminum-free catalyst, $Co(C_4H_6)(C_8H_{13})$–CS_2.

An unusual method for the preparation of syndiotactic polybutadiene was reported by The Goodyear Tire & Rubber Co. (43); a preformed cobalt-type catalyst prepared under anhydrous conditions was found to polymerize 1,3-butadiene in an emulsion-type recipe to give syndiotactic polybutadienes of various melting points (120–190°C). These polymers were characterized by infrared spectroscopy and nuclear magnetic resonance (44–46). Both the Ube Industries catalyst mentioned previously and the Goodyear catalyst were further modified to control the molecular weight and melting point of syndio-polybutadiene by the addition of various modifiers such as alcohols, nitriles, aldehydes, ketones, ethers, and cyano compounds. The use of water as a co-catalyst in Ziegler-type polymerizations was first introduced in 1962 (47). The reaction kinetics and crystallinity of the resulting polymers measured by x-ray scattering has been studied (48–51).

Physical Properties. By using different catalysts and polymerization techniques, syndiotactic polybutadiene can be prepared with various melting points. An extensive review of high melting syndiotactic polybutadiene has been published (51). Two types of syndiotactic polybutadiene are most relevant to industrial applications; a high melting syndio-polybutadiene (T_m = 190–216°C) and a low melting syndio-polybutadiene (T_m = 70–90°C). The low melting type is commercially available from the Japan Synthetic Rubber Co. (JSR).

The physical properties of syndiotactic polybutadiene are controlled by its melting point, degree of crystallinity, and molecular weight. Typical differential scanning calorimetry (dsc) scans for various low melting syndiotactic polybutadienes are shown in Figure 1. Even though the 1,2 content of the polybutadienes differs by only 3%, the T_m varies from 71 to 105°C. The T_gs are all about −10°C. This implies that the melting point of the syndio-polybutadiene is dependent on the amount of amorphous structure embodied in the crystalline matrix rather than the vinyl content. Molecular weight also influences the melting point and glass-transition temperature. The molecular weight of the JSR material varies between 30,000 and 100,000. The higher molecular weight material has the highest melting point.

The physical properties of low melting point (60–105°C) syndiotactic polybutadienes commercially available from JSR are shown in Table 1. The modulus, tensile strength, hardness, and impact strength all increase with melting point. These properties are typical of the polymer made with a cobalt catalyst modified with triphenylphosphine ligand.

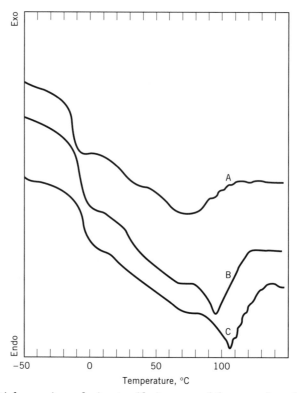

Fig. 1. Differential scanning calorimeter (dsc) curves of three grades of low melting syndiotactic 1,2-polybutadiene. A, 90% 1,2: T_m = 71°C; B, 92% 1,2: T_m = 95°C; C, 93% 1,2: T_m = 105°C. Heating rate: 20°C/min. Courtesy of the Japan Synthetic Rubber Co., RB product brochure.

Table 1. Physical Properties of Low Melting Syndiotactic 1,2-Polybutadiene

Properties	Test method	Measured value JSR[a] RB805	JSR[a] RB810	JSR[a] RB820	JSR[a] RB830
density, g/cm³	ASTM D1505 modify (at 20°C)	0.898	0.901	0.906	0.909
microstructure, % of 1,2 bonds	infrared spectrum	90	90	92	93
refractive index	ASTM D542	1.508	1.510	1.512	1.517
thermal properties, °C					
vicat softening point	ASTM D1525		39	52	68
brittle point	ASTM D746	−42	−40	−37	−35
melting point	ASTM D3418	59	71	95	105
tensile properties,	ASTM D412				
300% modulus, MPa[b]		2.4	3.9	5.9	7.8
strength at break, MPa[b]		4.9	6.4	10.3	12.7
elongation at break, %		800	750	700	670
hardness	ASTM D2240				
Shore D		19	32	40	47
Shore A		70	82	95	99
izod impact strength, notched, (N·cm)/cm[c]	ASTM D256	21.6–22.6	——does not break——		
transmittance of parallel light[d], %	ASTM D1003	93	91	89	82
haze value[d], %	ASTM D1003	2.7	2.6	3.4	8.0
mold shrinkage[e], %	JSR method	2.3–3.3	0.7–0.9	0.3–0.5	0.3–0.6

[a]Japan Synthetic Rubber Co.; noncross-linked; melt flow index, 150°C, 2.160 g is 3 g/10 min for all grades.
[b]To convert MPa to kgf/cm², multiply by 10.2; to psi, multiply by 145.
[c]To convert (N·cm)/cm(J/m) to ftlbf/in., divide by 53.38.
[d]Values measured with 2 mm thickness sheet molded by injection molding machine set at cylinder temperature 150°C, mold temperature 20°C.
[e]130°C medium speed injection molding; mold temperature 20°C; ASTM No. 1 dumb bells.

The stress–strain behavior of several low melting syndio-polybutadienes is similar to that of low density polyethylene. Increasing the melting point and crystallinity of the syndio-polybutadienes increases its modulus and brittle point. The polybutadienes with higher crystallinity and T_ms have higher moduli. The storage modulus of the low melting material is very similar to ethylene–vinyl acetate polymer, but lower than low density polyethylene. At low temperatures, syndio-polybutadiene has a higher storage modulus than either EVA films or LDPE (39).

High melting syndio-polybutadiene (T_m between 190–210°C) made with cobalt catalyst has different physical properties than the low melting material described previously. A typical dsc curve shows a sharp melting peak at 207°C. Annealing the polymer at 205°C for 12 h slightly increased the T_m to 212°C. The annealing did not significantly change the melting temperature, suggesting that the crystallites in the original sample were well ordered.

Typical polymerization conditions used to produce a syndiotactic 1,2-polybutadiene with a melting point of 206°C have been given (52). The dynamic storage modulus E' and loss modulus E'' have been plotted as a function of temperature for this high melting polymer ($T_m = 206°C$) (52). The peak in the E'' curve indicates a glass-transition temperature of $+40°C$. This means that this syndiotactic polybutadiene is highly crystalline and has a high vinyl content (>95%).

Measurements of the mechanical properties of high melting syndiotactic polybutadiene compared to isotactic polypropylene indicate that syndio-polybutadiene has higher tensile strength at the breaking point, lower initial modulus, and lower distortion temperature than polypropylene. In addition, syndiotactic polybutadiene can be more easily cross-linked, functionalized, and cyclized due to its side-chain unsaturation as compared to polypropylene. Moreover, syndio-polybutadiene can be easily loaded with carbon fibers and graphite and at higher loadings.

Amorphous High 1,2-Polybutadiene. The increased emphasis on energy conservation puts pressure on the tire industry to produce a tire with low rolling resistance and better fuel economy. Tire designers, by changing the tire geometry and tread patterns, have produced only limited success. The compounding technologist using existing polymers such as natural rubber, emulsion styrene–butadiene rubber, and cis-1,4-polybutadiene has been unable to simultaneously reduce rolling resistance and increase traction. Typically, a polymer with low rolling resistance also has low traction. Developing tread formulations for optimum traction and high resilience is paramount. However, in doing so many other tire properties are drastically affected. High tread modulus and hysteresis affect tire performance in traction, tread wear, and high resilience. A single-tread compound having single-hardness/hysteresis combinations cannot provide maximum performance in all tire-related properties simultaneously. High resilient (low hysteresis) tread rubber is desirable for reducing the lower rolling resistance tread compounds. The amount of energy which is lost or connected to heat is minimized thereby reducing tire rolling resistance. Whereas highly resilient tread compounds are desirable for reducing rolling resistance, they generally result in lower wet traction and poor tread resistance.

A goal for the synthetic polymer chemist is to molecularly engineer a polymer chain such as high vinyl polybutadiene to give low rolling resistance without compromising on wet grip and abrasion. The introduction of vinyl units into polybutadiene significantly changes its glass-transition temperature and physical properties. In this manner, the synthetic polymer chemist may be able to balance wet grip and fuel economy.

In recent years, high vinyl polybutadiene has become increasingly important to the tire industry. On a mid-size car, 7–10% of the fuel consumption can be attributed to rolling resistance losses from the tire. More than 50% of this tire energy loss comes from the tread. This makes a low hysteretic tread rubber very desirable as long as it maintains other important properties such as wet grip and tread wear. Amorphous, high vinyl polybutadiene is useful in tread formulations developed for energy conservation due to its low hysteresis and good wet grip characteristics. An amorphous, high 1,2-polybutadiene has been commercialized by the Nippon Zeon Co. under the trade names BR1240 and BR1245. These polymers showed better wet grip/rolling resistance behavior of butadiene rubbers having different vinyl contents. Rubber with 72% vinyl displayed a better compromise between rolling resistance and wet grip than customary rubbers due to the shift and rotation of the loss modulus pattern as a function of frequency.

Preparation.   The preparation of amorphous high (99%) 1,2-polybutadiene was first reported in 1981 (27). The use of a heterocyclic chelating diamine such as dipiperidine ethane in the polymerization gave an amorphous elastomeric polymer of 99.9% 1,2 units and a glass-transition temperature of +5°C. In a previous description (53,54) of the use of a chelating diamine such as N,N,N',N'-tetramethylethylene diamine, an 80% 1,2-polybutadiene with a glass-transition temperature of −30°C was produced.

Several reports in the literature describe the preparation and characterization of low, medium, and high vinyl polybutadienes (55–69). Each of these references used polar modifiers including chelating diamines, oxygenated ether compounds, acetals, ketals, and compounds of similar structures (56–64).

The random spatial arrangement of vinyl groups (ie, lddldllldl) in atactic polybutadiene results in an amorphous, rubbery polymer whose glass-transition temperature is a function of vinyl content (Fig. 2). The vinyl content of polybutadiene is controlled by the ratio of the polar modifier to the active lithium catalyst as well as the polymerization temperature (70–72). Raising the modifier/lithium ratio increases the vinyl content and T_g of the polybutadiene (Fig. 3). Above a ratio of about 2:1, additional modifier has no effect on the polymer microstructure. At a 2:1 modifier:lithium ratio, the vinyl content decreases with increasing polymerization temperature up to about 60°C (Fig. 4). Increasing the polymerization temperature above 60°C does not further reduce the vinyl content. By controlling the polymerization temperature and modifier ratio, the synthetic polymer chemist can prepare a variety of polymer microstructures.

Properties of Amorphous High Vinyl Polybutadiene.   The microstructural control described is possible only with living anionic polymerizations of conjugated diene monomers such as 1,3-butadiene, isoprene, and 2,3-dimethylbutadiene, and

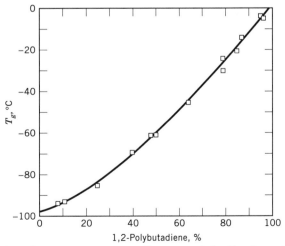

Fig. 2. T_g of polybutadiene vs 1,2 content. Courtesy of the Goodyear Tire & Rubber Co.

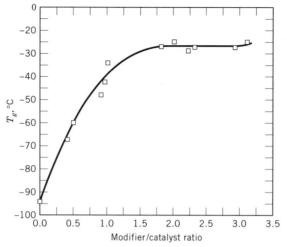

Fig. 3.   Effect of modifier ratio on polymer T_g. Courtesy of the Goodyear Tire & Rubber Co.

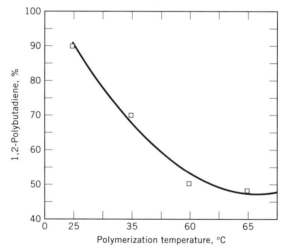

Fig. 4.   Effect of polymerization temperature on vinyl content at a 2:1 modifier:lithium ratio. Courtesy of the Goodyear Tire & Rubber Co.

in copolymerizations of conjugated diene monomers and vinylaromatics to prepare solution styrene–butadiene copolymers. This microstructural variation can give polymers with a wide range of glass-transition temperatures. Using this method, one can develop a single polymer in which the desired combination of physical properties can be obtained. For example, tire properties such as wet grip, hysteresis, and wear can be controlled by the introduction of vinyl groups to polybutadiene. A low vinyl polybutadiene (10–30% 1,2; T_g between -70 and $-85°C$) has good wear and excellent fuel economy but poor traction. The high vinyl polymer

(80–95% 1,2; T_g between -10 and $-30°C$) has excellent traction and fuel economy but poor wear properties. By manipulating the polymer structure, one can prepare a 70% vinyl polybutadiene with a T_g of $-40°C$ that has the optimum balance of traction, wear, and rolling resistance. The effect of polymer microstructure on physical properties is depicted in Figure 5.

Based on this variety of properties, amorphous polybutadiene has found a niche in the rubber industry. Moreover, it appears that the anionically prepared polymer is the only polymer that can be functionalized by polar groups. The functionalization is done by using aromatic substituted aldehydes and ketones or esters. Functionalization has been reported to dramatically improve polymer-filler interaction and reduce tread hysteresis (70–73).

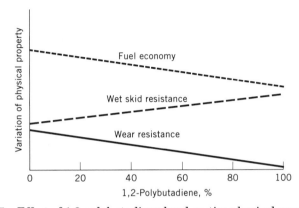

Fig. 5. Effect of 1,2-polybutadiene level on tire physical properties.

cis-1,4-POLYBUTADIENE

There are numerous references in the literature on the preparation of *cis*-1,4-polybutadiene (74–83). These authors have used transition metals in the presence of a reducing agent such as organoaluminum compounds or its chloride or hydride derivatives. In many cases nickel, cobalt, iron, or vanadium have been used in combination with organometallics of Groups I, II, or III. The resulting polybutadienes usually have *cis*-1,4 contents between 90–98%. The molecular weights of these polymers are very high, and usually a transfer agent such as internal or terminal olefins is employed (75–80).

For example, *cis*-1,4-polybutadiene has been made using a titanium halide such as $TiCl_4$, $TiBr_4$, or TiI_4. The most useful combination is $TiCl_4/(C_2H_5)_3Al$. The highest *cis*-1,4 content ($>90\%$) is obtained from TiI_4. The combination of $TiI_4/(C_2H_5)_3Al$ has been successfully scaled up to give a gel-free, $> 90\%$ *cis*-1,4-polybutadiene. This catalyst is heterogeneous, and requires special equipment to make a fine catalyst dispersion. Another system that has also been scaled up commercially is the cobalt system. This heterogeneous catalyst is homogeneously prepared using the cobalt salt of octanoate or napthenates, or complexed with pyridine. The cobalt system employs both aluminum alkyls and aluminum sesquichlorides (80–86). Cobalt systems usually produce polybutadienes with very high molecular weights. A chain-transfer agent such as ethylene, hydrogen, or methyl-allene is usually added to control the molecular weight.

Another transition-metal system that gives greater than 90% *cis*-1,4 content is based on nickel. This nickel system is currently used commercially to produce over 450,000 kg per day of *cis*-1,4-polybutadiene. The catalyst was first developed based on the BF_3 etherate system reduced with trialkylaluminum. The BF_3 etherate system became fully commercial in the late 1970s, and was used until the early 1980s. At that time, the system was completely modified using hydrogen fluoride to give a gel-free, 98% *cis*-1,4-polybutadiene with a controlled molecular weight and broad molecular weight distribution (87–94).

The unique advantage of the nickel system is that it can produce either structures of *cis*-1,4-polybutadiene, *trans*-1,4-polybutadiene, or a mixture of both depending on the reducing agent and the co-catalyst used. For example, chloride catalyst yields *cis*-1,4-polybutadiene, whereas bromide or iodide yields *trans*-1,4-polybutadiene. The counterion also has an effect on the polymer microstructure. A 50/50 *cis*-1,4/*trans*-1,4-polybutadiene has been prepared using a carboxylic counterion (95–105).

trans-1,4-POLYBUTADIENE

trans-1,4-Polybutadiene can be prepared using transition-metal catalysts or a nontransition-metal catalyst system based on Group I and II metals. The transition metals used include titanium, vanadium, chromium, rhodium, iridium, cobalt, and nickel (106–111).

The *trans*-1,4-polybutadiene made by transition-metal catalysis (112,113) is a resin-like material that has two melting temperatures, 50 and 150°C. This solid resinous material has not found much application because it is difficult to stabilize. However, the same type of polymer was prepared using a Group I metal alkoxide in combination with organometallics of the same group. For example, previous work (114,115) showed that reduction of potassium alkoxide with either organolithium or organomagnesium gave a 90–99% *trans*-1,4 structure with melting points of 50 and 175°C. This resinous, crystalline polymer was insoluble in hexane, and was isolated in quantitative yield. It can be isomerized to give a rubbery material.

A *trans*-1,4-polybutadiene that is useful as a tire rubber and can be stabilized and processed using conventional equipment has been made using an alkoxide of group II reduced with organolithium or organomagnesium compounds in the presence of lithium alkoxide salts and aluminum alkyls (116–123). This homogeneous catalyst system gives a rubbery, *trans*-1,4-polybutadiene with controlled molecular weight and a T_m of about $+30$ to $+40$°C. The polymer crystallizes upon stretching, and shows 30% crystallinity by wide angle x-ray scattering analysis. The broad crystalline melt peak observed by dsc suggests that a wide distribution of crystallite sizes exist in a rubbery matrix. The microstructure of this polymer based on ^{13}C-nmr shows 80% *trans*-1,4, 17% *cis*-1,4, and 3% 1,2. This polymer is different than the ultrahigh mol wt polybutadiene made with Alfin catalyst, since it is soluble in hydrocarbon solvents such as hexane (125–131). The polymer made by the well-known Alfin catalyst (132–134) gives an ultrahigh mol wt polybutadiene with a broad T_m at $+40°C$, and a microstructure of 70% *trans*-1,4 and 30% 1,2 units. X-ray diffraction analysis shows about 25–30% crystallinity.

MIXED MICROSTRUCTURE POLYBUTADIENE

An amorphous polybutadiene of 10% vinyl, 35% *cis*-1,4, and 55% *trans*-1,4 structure can be made with a living anionic catalyst (135–139). This polymer is currently sold by Firestone under the name of Diene, and by Asahi under the name of Taktene. It has poor green strength, but shows excellent physical properties when compounded in a tire tread. This tread rubber, made in a living polymerization using a lithium catalyst, shows excellent tread wear as well as low hysteresis. Moreover, this type of polymer can be functionalized since the end of the polymer chain carries an active carbon metal linkage. This carbon-bound metal has been functionalized with ketones, aldehydes, acid chlorides, and metal halides of silicon and tin. Using silicon tetrachloride to couple the polybutadiene gives a polymer with a molecular weight four times that of the original polymer, and improves its hysteretic properties. Moreover, it has been claimed that it also shows better tread wear due to improved carbon black dispersion without a loss in hysteresis. There are claims in the literature that tin-coupled polybutadiene has lower hysteresis due to the formation of carbon–tin bonds that help break up the carbon black agglomerates into smaller aggregates; this results in improved carbon black dispersion (140–146).

BIBLIOGRAPHY

"Elastomers, Synthetic (Polybutadiene)" in *ECT* 3rd. ed, Vol. 8, pp. 546–567, by L. J. Kuzma and W. J. Kelly, The Goodyear Tire & Rubber Co.

1. H. E. Armstrong and A. K. Miller, *J. Chem. Soc.* **49**, 80 (1986).
2. E. G. M. Torqvist in J. P. Kennedy and E. G. M. Torqvist, eds., *High Polymers, A Series of Monographs on the Chemistry, Physics and Technology of High Polymeric Substances*, Wiley-Interscience, New York, 1968, p. 57.
3. C. D. Harries, *Ann.* **383**, 157 (1911); *Ibid.*, 406 (1914); *Ibid.*, **395**, 211 (1913).
4. E. Caventou, *Ann.* **127**, 93 (1963).
5. K. Ziegler and co-workers, *Angew. Chem.* **67**, 541 (1955).
6. U.S. Pat. 1,938,731 (Dec. 12, 1933), W. Brock and E. Tschunker (to I. G. Farbenindustries).
7. U.S. Pat. 1,973,000 (Sept. 11, 1934), E. Korhard and E. Tschunker (to I. G. Farbenindustries).
8. K. Ziegler, *Angew. Chem.* **49**, 499 (1936).
9. K. Zeigler, *Rubber Chem. Technol.* **38**, xxiii (1965).
10. S. E. Horne and co-workers, *Ind. Eng. Chem.* **48**, 784 (1956).
11. F. W. Stavely and co-workers, *Ind. Eng. Chem.* **48**, 778 (1956).
12. K. Ziegler, F. Dersch, and H. Wollthan, *Ann.* **511**, 13 (1934).
13. K. Ziegler and L. Jakob, *Ann.* **511**, 45 (1934).
14. K. Ziegler and co-workers, *Ann.* **511**, 64 (1934).
15. K. Ziegler, *Angew. Chem.* **49**, 499 (1936).
16. F. C. Forster and J. L. Binder, *Advances in Chemistry Series*, No. 19, American Chemical Society, Washington, D.C., 1957, p. 26.
17. R. S. Stearns and L. E. Forman, *J. Poly. Sci.* **41**, 381 (1959).

18. M. Szwarc and J. Smid in G. Porter, ed., *Progress in Reaction Kinetics*, Vol. 2, McMillan, New York, 1964, p. 267.
19. S. G. Horne, Jr., *Rubber Chem. Technol.* **53**(3), 671 (1980).
20. G. Natta and L. Porri in J. P. Kennedy and E. G. M. Torqvist, eds., *Polymer Chemistry of Synthetic Elastomers*, Part 2, Wiley-Interscience, New York, 1969, p. 597.
21. W. Cooper, *Ind. Eng. Chem. Prod. Res. Dev.* **9**, 457 (1972).
22. P. Teyssie and F. Dawans, *Ind. Eng. Chem. Res. Dev.* **10**, 261 (1971).
23. B. A. Dolgoplosk, *Polymer Sci. USSR* **13**, 367 (1971).
24. P. Teyssie and F. Dawans in N. Saltman, ed., *The Stereo Rubbers*, John Wiley & Sons, Inc., New York, 1977, Chapt. 3.
25. J. Boor, Jr., *Zeigler-Natta Catalysts and Polymerizations*, Academic Press, Inc., New York, 1979.
26. J. Furukawa, *Acc. Chem. Res.* **13**, 1 (1980).
27. A. F. Halasa, D. F. Lohr, and J. E. Hall, *J. Poly. Sci., Poly. Chem. Ed.* **19**, 1347 (1981).
28. G. Natta, *Makromol. Chem.* **16**, 213 (1955).
29. G. Natta and P. Corradin, *J. Poly. Sci.* **20**, 251 (1956).
30. E. Susa, *J. Poly. Sci., Part C* **4**, 399 (1964).
31. C. Longiave and R. Castelli, *J. Poly. Sci., Part C* **4**, 387 (1964).
32. S. Otsuka, K. Mori, and M. Kawakanni, *Kogyo Kagaku Zasshi* **67**, 1652 (1964).
33. K. Matsuzaki and T. Yasukawa, *J. Poly. Sci., A1* **5**, 511 (1967).
34. T. Saito and co-workers, *Kogyo Kagaku Zasshi* **66**, 1099 (1963).
35. J. Zymonas, E. R. Santee, Jr., and H. J. Harwood, *Macromolecules* **6**, 129 (1973).
36. Y. Takeuchi, A. Sekimoto, and M. Abe, *Am. Chem. Soc. Symp. Ser. 4*, American Chemical Society, Washington, D.C., 1974, pp. 15–26.
37. Y. Takeuchi, M. Ichikawa, and K. Mori, *Polym. Prepr. Jpn.* **15**, 423 (1966).
38. U.S. Pat. 4,051,308 (1975), A. F. Halasa (to the Firestone Tire & Rubber Co.).
39. N. K. Mori, T. Taketomi, and F. Imaizumi, *Nippon Kagaku Kaishi* **11**, 1982 (1975).
40. S. Sugiura and co-workers, *Jpn. Kokuku* **72-19**, 892 (1969).
41. M. Takayanagi, paper presented to *The 19th Annual Meeting of the International Institute of Synthetic Rubber Producers*, Hong Kong, 1978.
42. U.S. Pat. 4,104,465 (Aug. 1, 1978), A. F. Halasa (to the Firestone Tire and Rubber Co.); U.S. Pat. 3,914,210 (Oct. 21, 1975), A. F. Halasa (to the Firestone Tire and Rubber Co.).
43. U.S. Pat. 4,429,085 (Aug. 12, 1984), 4,506,031 (Nov. 12, 1988), and 5,011,890 (Feb. 22, 1991), A. F. Halasa (to the Firestone Tire and Rubber Co.).
44. V. D. Mochel, *J. Poly. Sci., A1* **10**, 1009 (1972).
45. Y. Araki and co-workers, *Kobunshi Kagaku* **29**, 397 (1972).
46. J. L. Binder, *J. Poly. Sci., A-1* **1**, 47 (1963).
47. M. Gippin and G. Michell, *Ind. Eng. Chem. Prod. Res. Dev.* **1**, 32 (1962); **4**, 160 (1965).
48. K. F. Elgert, G. Quak, and R. Stutzel, *Makromol. Chem* **175**, 1955 (1974); T. Kagiya, M. Izu, and K. Fukui, *Bull. Chem. Soc. Jpn.* **10**, 1045 (1967).
49. Y. Doi and co-workers, *J. Poly. Sci., Poly. Chem. Ed.* **13**, 2491 (1975).
50. H. Okamoto and co-workers, *J. Poly. Sci., Poly. Chem. Ed.* **17**, 1267–1279 (1959).
51. H. Ashitaka, K. Junda, and H. Ueno, *J. Poly. Sci., Poly. Chem. Ed.* **21**, 1989 (1983); **21**, 1951 (1983); **21**, 1973 (1983).
52. *Ibid.*, **21**, 1853 (1983).
53. A. W. Langer, Jr., *Trans. NY Acad. Sci.*, 741 (1965).
54. A. W. Langer, Jr., *Am. Chem. Soc., Div. Chem. Preprint No. 1* **7**, 132 (1966).
55. V. A. Kropachev, B. A. Dolgoplash, and N. I. Mikolaev, *Dokl. Akad. Nauk USSR* **115**, 51 (1957).
56. A. V. Tobolsky, D. I. Kelley, and H. J. Hsieh, *J. Poly. Sci.* **26**, 240 (1957).

57. A. A. Korothkov, paper presented at *IUPAC Meeting*, Prague, 1957; *Angew Chem.* **70,** 85 (1958).

58. A. A. Yakubovich and S. S. Medvedev, *Dokl. Akad. Nauk USSR* **159,** 1305·F(1964).

59. J. L. Binder, *Anal. Chem.* **26,** 1877 (1954).

60. A. A. Korotkov, S. P. Mitsengendler, and K. M. Aleyev, *Poly. Sci., USSR* **2,** 487 (1961).

61. C. Screttas and J. Eastham, *J. Am. Chem. Soc.* **87,** 3276 (1965).

62. I. J. Kuntz, *J. Poly. Sci.* **54,** 569 (1961).

63. S. Bywater, D. Worsfold, and P. Black, *Makromol. Chem. Suppl.* **15,** 31 (1989).

64. V. Frolov and co-workers, *Prom, Ohraztsy Tovaryne Zanki* **10,** 116 (1982).

65. M. Hiraoka, *Kagaku Kyoiku* **24**(5), 383 (1976).

66. U.S. Pat. 3,879,367 (Apr. 1975), A. F. Halasa (to the Firestone Tire and Rubber Co.).

67. E. Duck, *Annual Meeting, Inst. Synth. Rubber Prod.* **5,** 11 (1974).

68. U.S. Pat. 3,663,480 (Apr. 1972), R. Zelinskui and R. Sonnenfeld (to Phillip's Petroleum Co.).

69. U.S. Pat. 4,696,986 (Aug. 1987), A. F. Halasa and R. E. Cunningham (to The Goodyear Tire & Rubber Co.).

70. A. F. Halasa, V. D. Mochel, and G. Frankeal, *Amer. Chem. Soc., Div. Poly. Chem., Preprint* **21,** 1 (1980).

71. T. A. Antkowiak and co-workers, *J. Poly. Sci., A1* **10,** 1319 (1972).

72. F. C. Forster and J. L. Binder, *Adv. in Chem. Series* **19,** 26 F(1957).

73. Y. Yang, *Thany Xueming, Geofengi Tongxum*, (4), 312–316 (1992).

74. G. Natta and co-workers, *Chim. Ind. (Milan)* **41,** 398 (1959).

75. G. Natta, *Chim. Ind. (Milan)* **42,** 1207 (1960).

76. W. Franks, *Kautsch. Gummi Kunstst.* **11,** 254 (1958).

77. I. J. Poddubonyi, V. A. Grechanovsky, and E. G. Ehrenburg, *Makromol. Chem.* **94,** 268 (1966).

78. M. H. Lehr and P. H. Moyer, *J. Poly. Sci.* **A3,** 217 and 753 (1965).

79. W. Marconi and co-workers, *Chim. Ind. (Milan)* **46,** 245 (1967).

80. M. Gippin, *Ind. Eng. Chem. Prod. Res. Dev.* **1,** 32 (1962); **4,** 160 (1965).

81. A. I. Diaconescu and S. S. Medvedev, *J. Poly. Sci.* **A3,** 31 (1965).

82. W. M. Saltman and L. J. Kuzman, *Rubber Chem. Technol.* **46,** 1066 (1973).

83. V. N. Zgonnik and co-workers, *Vysolomol. Soldin* **4,** 1000 (1962).

84. B. A. Dolgoplosk and co-workers, *Dokl. Akad. Nauk USSR* **135,** 847 (1960).

85. F Engel and co-workers, *Rubber Plast. Age* **45,** 1499 (1964).

86. H. Scott and co-workers, *J. Poly. Sci.* **A3,** 3233 (1964).

87. U.S. Pats. 3,170,904; 3,170,905; 3,170,906; and 3,170,907 (Feb., 1965) (to The Bridgestone Tire and Rubber Co.).

88. U.S. Pat. 3,178,403 (Apr. 1965) (to The Bridgestone Tire and Rubber Co.).

89. S. Kitagawa and Z. Harada, *Jpn. Chem. Q* **IV-1,** 41 (1968).

90. M. C. Throckmorton and F. S. Farson, *Rubber Chem. Technol.* **45,** 268 (1972).

91. C. Dixon and co-workers, *Eur. Poly. J.* **5,** 1359 (1970).

92. E. W. Duck and co-workers, *Eur. Poly. J.* **6,** 1359 (1970).

93. G. Wilke and co-workers, *Angew Chem. Int. Ed.* **5,** 151 (1966).

94. G. Wilke, *Angew Chem. Int. Ed.* **2,** 105 (1963).

95. L. Porri, G. Natta, and M. C. Gallazzi, *J. Poly. Sci., C* **16,** 2525 (1971).

96. F. Dawans and P. Teyssie, *Ind. Eng. Chem. Prod. Res. Dev.* **10,** 261 (1971).

97. F. Dawans and P. Tyessie, *C. R. Acad. Sci. Paris* **263C,** 1512 (1966).

98. B. A. Dolgoplosk and co-workers, *Bull. Acad. Sci. USSR, Chem. Ser.* **9,** 429 (1967).

99. F. Dawans and P. Teyssie, *J. Poly. Sci.* **B7,** 111 (1969).

100. J. P. Durand, F. Dawans, and P. Teyssie, *J. Poly. Sci.* **B5,** 785 (1967).

101. E. Tinyakova and co-workers, *J. Poly. Sci.* **C16,** 2625 (1967).

102. B. D. Babitskii and co-workers, *Vysokomol Soedin* **6,** 2202 (1964).

103. J. Furukawa, *Acc. Chem. Res.* **13**(6), 1 (1980).
104. J. Furukawa, E. Kobayashi, and T. Kawagoe, *Poly. J.* **5**(3), 231 (1973).
105. J. Furukawa and co-workers, *Poly. J.* **2**(3), 371 (1971).
106. G. Natta, L. Porri, and A. Carbonaro, *Rend. Accad. Naz. Lincei* **31**(8), 189 (1961).
107. G. Natta and co-workers, *Chim. Ind.* **40**, 362 (1958).
108. G. Natta, L. Porri, and A. Mazzei, *Chim. Ind.* **41**, 116 (1959).
109. G. Natta and co-workers, *Chim. Ind.* **41**, 398 (1959).
110. G. J. Van Amorongen, *Adv. Chem. Ser.* **52**, 36 (1966).
111. G. Natta and co-workers, *Gazz. Chim. Ital.* **89**, 761 (1959).
112. D. K. Jenkins, *Polymer* **26**, 147 (1985).
113. J. Boor, Jr., *Ziegler–Natta Catalysts and Polymerizations*, Academic Press, Inc., New York, 1979, Chapts. 5–6.
114. D. Patterson and A. F. Halasa, *Macromolecules* **24**, 4489 (1991).
115. D. Patterson and A. F. Halasa, *Macromolecules* **24**, 1583 (1991).
116. B. I. Nakhmanovich and A. A. Arest-Yakubovich, *Vysokomol. Soedin., Ser. B* **26**(6), 476 (1984).
117. E. V. Kristal'nyi, A. A. Arest-Yakubovich, and E. F. Koloskova, *Vysokomol. Soedin., Ser. B* **19**(10), 767 (1977).
118. R. V. Basova, G. A. Kabalina, and A. A. Arest-Yakubovich, *Vysokomol. Soedin., Ser. B* **18**(12), 910 (1976).
119. Z. M. Baidakova, B. I. Nakhmanovich, and A. A. Arest-Yakubovich, *Dokl. Acad. Nauk. USSR* **230**(1), 114 (1976).
120. L. N. Moskalenko and co-workers, *Izobert Otkrytia, Prom. Obratztsy, Tovernye Zanki* **52**(46), 173 (1975).
121. Z. M. Baidakova, L. N. Moskalenko, and A. A. Arest-Yakubovich, *Vysokomol. Soedin., Ser. A* **16**(10), 226 (1974).
122. A. A. Arest-Yakubovich and L. N. Moskalenko, *Vysokomol. Soedin., Ser. A* **13**(6), 1242 (1971).
123. P. Maleki and B. Francois, *Makromol. Chem.* **156**, 31 (1972).
124. U.S. Pat. 4,996,273 (Feb. 26, 1991), A. A. Van der Haizen (to Shell Oil Co.).
125. U.S. Pats. 4,355,156 (Oct. 19, 1982); 4,302,568 (Oct. 19, 1982); 4,033,900 (July, 1977); 3,992,561 (July, 1975), R. E. Binghan and co-workers, (to General Tire Corp.).
126. S. L. Aggrawal, I. G. Hargis, and R. A. Livigni, *Paper presented at ACS Rubber Division Meeting*, Houston, Tex., Oct., 1983.
127. S. L. Aggrawal and co-workers, in J. E. Mark, ed., *Advances in Elastomer and Rubber Elasticity*, Plenum Press, New York, 1968, p. 16.
128. U.S. Pt. 4,933,401 (June 12, 1990), I. Hattori and co-workers, (to the Japan Synthetic Rubber Co.).
129. P. G. Wang and A. E. Woodward, *Macromolecules* **20**, 1818, 1823, and 2718 (1987).
130. S. Tseng and A. E. Woodward, *Macromolecules* **15**, 343 (1982).
131. U.S. Pat. 4,225,690 (Sept. 30, 1980), A. F. Halasa and J. E. Hall (to Firestone Tire and Rubber Co.).
132. A. A. Morton, *Ind. Eng. Chem.* **42**, 1488 (1950).
133. A. A. Morton, J. Nelidow, and E. Shoenberg, *Rubber Chem. Technol.* **30**, 326 (1957).
134. A. A. Morton and E. J. Lanpher, *J. Poly. Sci.* **44**, 233 (1960).
135. M. Morton and L. J. Fetters, *Rubber Chem. Technol.* **48**, 359 (1975).
136. A. A. Korotkov and N. N. Chesnokova, *Polymer Sci. USSR* **2**, 284 (1960).
137. F. Tsutsumi, M. Sakakibara, and N. Oshima, *Paper presented at ACS Rubber Division Meeting*, Cincinnati, Ohio, Oct. 18, 1988.
138. C. Uvaneck and J. N. Short, *J. Appl. Polymer Sci.* **14**, 1421 (1970).
139. K. H. Nordsiek and K. M Kiepert, *Proceedings of the I.R.C.*, Venice, Oct. 1979.
140. L. H. Krol, *IISRP 23rd Annual Meeting*, Apr. 1982.

141. J. P. Ferry, *Viscoelastic Properties of Polymers*, 3rd ed., John Wiley & Sons, Inc., New York, 1980, Chapt. 10.

142. A. F. Johnson and D. J. Worsfold, *Macromol. Chem.* **85,** 273 (1965).

143. C. F. Wofford and H. L. Hsieh, *J. Poly. Sci. Part A-1* **7,** 461 (1969).

144. R. Zelmiski and C. W. Childers, *Rubber Chem. and Technol.* **41,** 161 (1968).

145. A. F. Halasa and co-workers, in F. G. A. Stone and R. West, eds., *Advances in Organometallic Chemistry*, Vol. 18, Academic Press, Inc., New York, 1980.

146. A. W. Meyer, R. R. Hampton, and J. A. Davison, *J. Am. Chem. Soc.* **74,** 2294 (1952).

ADEL F. HALASA
J. M. MASSIE
Goodyear Tire & Rubber Company

POLYCHLOROPRENE

Polychloroprene [*9010-98-4*] was the first commercially successful synthetic elastomer. It was discovered in 1930 (1) and introduced in 1932 by Du Pont under the name DuPrene. The name Neoprene was adopted a short time later, and is now in general use, but the commercial material is now correctly designated as CR, or chloroprene rubber. Commercial acceptance of CR was based on the material having an unusually good combination of environmental resistance and toughness, particularly in dynamic applications involving heat buildup and resistance to flex cracking. The unusual chemistry of chloroprene monomer and the versatility of emulsion polymerization have led to a broad spectrum of products, often optimized for particular end use applications. Currently, worldwide annual consumption is about 350,000 metric tons (2) with a value approaching $1.4 billion.

Polymerization

Chloroprene [*126-99-8*] monomer undergoes dimerization and autopolymerization when stored at ordinary temperatures. These reactions are parallel and independent. The dimerization reaction is second order, and is unaffected by free-radical initiators (3–6). Chloroprene monomer is highly reactive in free-radical polymerization because of the influence of the electron-rich chlorine atom, which facilitates free-radical addition to the monomer. Even specially purified monomer undergoes spontaneous polymerization without added catalyst. Over the temperature range of 20–80°C, this thermal reaction is close to first order and has an activation energy of 82 kJ/mol (19.6 kcal/mol) (3). The rate of bulk polymerization is strongly catalyzed by chloroprene peroxides (4,7) formed either by deliberate or adventitious exposure of the monomer to oxygen. Bulk and solution polymerization rate varies with the amount of added free radical (3,8–10), and is slowed by typical radical inhibitors. The heat of polymerization is 68–75 kJ/mol (16–18 kcal/mol) of monomer (11,12).

Commercial chloroprene polymerization is most often carried out in aqueous emulsion using an anionic soap system. This technique provides a relatively concentrated polymerization mass having low viscosity and good transfer of the heat of polymerization. A water-soluble redox catalyst is normally used to provide high reaction rate at relatively low polymerization temperatures.

In practice, monomer is added to an aqueous phase containing a suitable surface active agent and subjected to high shear to reduce monomer droplets to a few micrometers in size. The soap and monomer that are present distribute between the monomer droplets and micelles. The micelles are perhaps 100 times smaller than emulsion droplets, and possess considerably more surface area. When free radicals are generated in the aqueous phase, they initiate polymerization and migrate to the soap micelles where further growth occurs. The mechanism for emulsion polymerization of chloroprene is complex, but seems generally consistent with the Smith-Ewart Model developed based on styrene polymerization (13–15) (see EMULSIONS; LATEX TECHNOLOGY).

Reaction of polymerizing chloroprene and vinyl acetate with ^{14}C labeled model compounds was used to determine the relative kinetics of chain-transfer and double-bond addition reactions with polymer. Chain transfer to allylic hydrogen or chlorine atoms on the polymer chain leads to long-chain branching. Addition of the growing chain across a polymer double bond can lead directly to gel formation. Either reaction affects polymer rheology and processibility. Vinyl acetate radicals, as a model for initiator fragments, were found to be 570 to 2,427 times faster in chain-transfer reactions than chloroprene radicals. In all cases studied, the rate of double-bond addition reactions were significantly faster than transfer reactions (16).

Emulsions stabilized with a nonionic surfactant and catalyzed with a monomer soluble initiator were found to follow kinetics dependent on initiator concentration (17).

There are only a limited number of commercial applications of bulk or solution polymerization of chloroprene. These involve graft polymerization of adhesives and production of liquid polymers.

Polymer Structure and Properties

The emulsion polymerization process enables considerable variation in the properties of polychloroprene, and provides an opportunity to tailor polymers for a wide variety of uses.

Branching and Gel Formation. Long-chain branching and gel formation are caused by chain-transfer and double-bond addition reactions with polymer. The rate of these reactions increases with increasing temperature, lack of a chain-transfer agent, and increasing conversion. In the absence of a chain-transfer agent, up to 90% of the polymer formed at 30% monomer conversion will be gelled (18). A quantitative method for measuring long-chain branching and incipient gel formation involves comparison of the intrinsic viscosity of a sample with its molecular weight by size exclusion chromatography. The method is calibrated with a series of linear polymers of increasing molecular weight. A branching parameter, γ, related to the frequency of long-branches on the polymer chain can then be determined for unknowns. When γ equals 0, no branches are present and γ equals 1 at the point of incipient gel formation. This parameter should be useful for predicting processibility (19,20). Either dodecyl mercaptan or diethylxanthogen disulfate inhibit long-chain branching and prevent gel formation until 60–70% conversion (20) at normal molecular weight. A difunctional monomer, such as ethylene glycol bismethacrylate, can be used to increase gel formation (21).

The extent of the long-chain branching of a series of ethyl xanthogen disulfide modified polymers carried to increasing conversion is shown in Table 1. No branches are found until 56% conversion. The gel point is a little over 82%. Polymer rheology deteriorates between 56 and 82% conversion.

The relative rates of propagation and cross-linking reactions in the absence of a chain-transfer agent have been estimated from swelling ratios. The dependence of volume of the monomer droplets, surface tension, and cross-linking density on conversion correspond to the Harkins and Smith-Ewart Model (15). Qualitative data on long-chain branching can be obtained from size exclusion chromatography using either a light scattering or a viscosity detector (22). An analytical ultracentrifuge may be used, in place of swell ratio, to determine degrees of swelling and cross-linking on very small particles (23).

Molecular Weight Control. Polymer bulk viscosity, eg, Mooney viscosity, is widely used to predict processability. It is usually controlled by addition of a chain-transfer agent to the polymerization system. Normally a mercaptan is used, but other materials such as xanthogen disulfides (24), iodoform (25), and aromatic disulfides (26) have also been used. The chain-transfer agent additionally enables polymerizations to be run to higher conversion without gel polymer formation. The chain-transfer reaction of mercaptans involves hydrogen transfer to the growing polymer chain with the formation of a thiyl radical. The thiyl radical can then initiate a new chain. This series of reactions was proven for a high pH polymerization by use of radiosulfur tagged dodecyl mercaptan (27). Use of a mixture of two chain-transfer agents with dissimilar reactivity can result in more uniform molecular weight control from the beginning to end of the polymerization (28). Mercaptans produce a polymer with unreactive end groups. A disulfide may produce end groups that either participate during cure or are otherwise reactive.

Sulfur Copolymers. Molecular weight may also be controlled by copolymerization of the monomer with sulfur. The copolymer contains polysulfide units along the polymer chain that are subsequently cleaved chemically to control molecular weight. These polysulfide units (29) were found to be either 2 or 8 sulfur atoms long with only a few units containing 3–7 sulfur atoms. Sulfur initially adds to the polymer chain by opening the normal 8-member sulfur ring, forming a chain end containing 8 sulfur atoms. After addition of another chloroprene unit, some of the chains depropagate by a backbiting mechanism to provide a chain

Table 1. Polychloroprene Branching[a]

Parameter	Value					
conversion, %	11.7	33.6	55.9	62.8	71.9	82.3
$M_n \times 10^{-5}$	1.44	1.44	1.19	1.07	1.05	1.26
M_w/M_n	2.3	2.3	3.4	4.2	5.2	4.9
branches/100,000 mol wt			0.15	0.23	0.36	0.52
branching parameter	0	0	0.38	0.50	0.66	0.76

[a]Emulsion polymerization at 40°C using xanthate modification (20).

end having 2 sulfur atoms. Continued polymerization at this chain end results in incorporation of a disulfide linkage. The process also enriches the amount of cis-1,4 monomer units attached to sulfur.

The initial sulfur copolymer that is formed is often high conversion and gelled. Molecular weight is reduced to the required level by cleaving some of the polysulfide linkages, usually with tetraethylthiuram disulfide. An alkali metal or ammonium salt (30) of the dithiocarbamate, an alkali metal salt of mercaptobenzothiazole (31), and a secondary amine (32) have all been used as catalysts. The peptization reaction results in reactive chain ends. Polymer peptized with diphenyl tetrasulfide was reported to have improved viscosity stability (33).

Both a cyclic sulfur compound, 1,2,3,4-tetrasulfocyclohexane (34), and a polymeric disulfide (35), have been used as alternatives to free sulfur. The resulting polymers were reported to have superior heat resistance.

Copolymerization.   Over the years, almost every vinyl and diene monomer has been tested with chloroprene in free-radical polymerization. Reactivity ratios for a few of these are given in Table 2. Few monomers are sufficiently reactive to compete with chloroprene. As a result, copolymers usually contain only a limited amount of random comonomer with much unreacted comonomer. If the polymerization is driven to completion, the second monomer may form a combination of homopolymer and graft polymer. Sometimes, however, a minor amount of comonomer provides beneficial functional groups, eg, carboxylated adhesives (37), where the comonomer promotes adhesion and cohesive strength. An exception to the reactivity problem is 2,3-dichlorobutadiene, which is even more reactive than chloroprene. Its use as an additive for crystallization resistance (38) can be optimized by gradual addition of the second monomer throughout the polymerization cycle (39), rather than all in the initial batch makeup. Dichlorobutadiene is also unusual in that it has a very high homopolymer melting point and, at higher concentration, provides very tough crystalline copolymers with chloroprene (40). A large number of graft polymers of polychloroprene have been described, but the only ones of commercial significance are those made with acrylates and methac-

Table 2. Chloroprene (M1) Reactivity Ratios[a]

Comonomer, M2	$r1$	$r2$
methyl acrylate	10.40	0.06
acrylonitrile	5.38	0.056
butadiene	3.41	0.06
butadiene		
2,3-dichloro-	0.31	1.98
1-(2-hydroxyethylthio)-	1.00	0.20
2-fluoro-	3.70	0.22
diethyl fumarate	6.51	0.02
tetrachlorohexatriene	3.60	0.20
isoprene	2.82	0.06
methyl methacrylate	6.33	0.08
methacrylic acid	2.7	0.15

[a]From more extensive data in Ref. 36.

rylates. These are particularly useful for adhesion to plasticized vinyl (41,42). The graft polymers may be made either in solution or emulsion polymerization (43).

Interpenetrating networks have been made by co-curing polychloroprene with copolymers of 1-chloro-1,3-butadiene [627-22-5]. The 1-chloro-1,3-butadiene serves as a cure site monomer, providing a cure site similar to that already in polychloroprene. The butadiene copolymer with 1-chloro-1,3-butadiene (44) and an octyl acrylate copolymer (45) improved the low temperature brittleness of polychloroprene. The acrylate also improved oil resistance and heat resistance.

Finally, block copolymers have been made in a two-step process. First a mixture of chloroprene and p-xylenebis-N,N'-diethyldithiocarbamate is photopolymerized to form a dithiocarbamate terminated polymer which is then photopolymerized with styrene to give the block copolymer. The block copolymer has the expected morphology, spheres of polystyrene domains in a polychloroprene matrix (46).

Microstructure. Whereas the predominate structure of polychloroprene is the head to tail *trans*-1,4-chloroprene unit (1), other structural units (2,3,4) are also present. The effects of these various structural units on the chemical and physical properties of the polymer have been determined. The high concentration of structure (1) is responsible for crystallization of polychloroprene and for the ability of the material to crystallize under stress. Structure (3) is quite important in providing a cure site for vulcanization, but on the other hand reduces the thermal stability of the polymer. Structures (3),(4), and especially (2) limit crystallization of the polymer.

trans-1,4-unit	*cis*-1,4-unit	1,2-unit	3,4-unit
(1)	(2)	(3)	(4)

A further structural irregularity was found to be inverted alignment of the structural units to give tail to tail and head to head arrangements in addition to the predominant head to tail arrangement (47). There have been numerous analytical studies employing infrared and Raman spectroscopy (48,49), ^{1}H and ^{13}C nmr (50,51), and nqr (52,53). The total amount of structures (2),(3), and (4) varies with polymerization temperature from 5% at $-40°C$ to about 30% at 100°C. The amounts produced of the various structures at -150 to $+90°C$ have been determined by ^{13}C nmr (Table 3) (51). A more recent study using similar techniques found a greater amount of the head to tail *trans*-1,4 structure with reduced amounts of the irregular structures at a given temperature. The thermodynamics of the various modes of unit addition have been discussed (50). Other structural studies have involved 2,3-dichlorobutadiene homopolymer (54), the free-radical random copolymer with methyl methacrylate (55) with chloroprene, and the alternating copolymer of sulfur dioxide with chloroprene (56).

When the polymerization is run at less than about 20°C, the resulting polymer is hard and tough and has valuable properties as an adhesive. Polymer made

Table 3. Microstructure of Polychloroprene by ^{13}C nmr[a]

Polymerization temperature, °C	1,4 Addition			1,2 Addition	Isomerized 1,2	3,4 Addition
	trans	inverted	cis			
+90	85.4	10.3	7.8	2.3	4.1	0.6
+40	90.8	9.2	5.2	1.7	1.4	0.8
+20	92.7	8.0	3.3	1.5	0.9	0.9
0	95.9	5.5	1.8	1.2	0.5	1.0
−20	97.1	4.3	0.8	0.9	0.5	0.6
−40	97.4	4.2	0.7	0.8	0.5	0.6
−150	~100	2.0	<0.2	<0.2	>0.2	<0.2

[a]Ref. 51.

at higher temperature, with more chain irregularities, tends to be much slower crystallizing, and is more suitable for mechanical goods applications.

Stereoregular poly-*trans*-1,4-chloroprene (57) and poly-*cis*-1,4-chloroprene (58) have been synthesized. The cis structure has a higher glass-transition temperature, and despite the large number of inverted units, a relatively high melting point.

Other Types of Polymerization. Nonradical polymerization has not produced commercially useful products, although a large variety of polymerization systems have been tested. The structural factors that activate chloroprene toward radical polymerization often retard polymerization by other mechanisms.

Cationic polymerization with Lewis acids yields resinous homopolymers containing cyclic structures and reduced unsaturation (58–60). Polymerization with triethylaluminum and titanium tetrachloride gave a product thought to have a cyclic ladder structure (61). Anionic polymerization with lithium metal initiators gave a low yield of a rubbery product. The material had good freeze resistance compared with conventional polychloroprene (62).

Alternating copolymers of chloroprene have been prepared from a number of donor acceptor complexes in the presence of metal halides. Frequently this enables preparation of copolymers from monomers having unfavorable reactivity ratios in radical polymerization. Ethylaluminum sesquichloride with a vanadium oxychloride cocatalyst yielded alternating copolymers of chloroprene with acrylonitrile, methyl acrylate, and methyl methacrylate when equimolar amounts of the two monomers were used (63). Polymer composition tends to follow the composition of the monomer mixture (64). The chloroprene units were shown to be in the *trans*-1,4-configuration (64) on the basis of infrared spectra. Variables affecting the acrylonitrile copolymerization were studied in detail (65). The alternating copolymer of acrylonitrile and chloroprene is resinous. A copolymer containing 36 mol % acrylonitrile was a soft, oil resistant elastomer (65). Stability constants have been determined for complexes of acrylic monomers with ethylaluminum sesquichloride, and related to the kinetics of copolymerization with chloroprene (66). Kinetic data have been determined for polymerization in the presence of a manganese cocatalyst (67).

A series of graft polymers on polychloroprene were made with isobutylene, *i*-butyl vinyl ether, and α-methylstyrene by cationic polymerization in solution.

The efficiency of the grafting reaction was improved by use of a proton trap, eg, 2,6-di-*t*-butylpyridine (68).

Popcorn Polymerization ω-Polymerization, frequently referred to as popcorn polymerization because of the appearance of the product, can be a dangerous side reaction if not carefully controlled. The polymerization appears to proceed without external initiation (69–71), and is catalyzed by the tightly gelled polymer seeds that are a product of the polymerization. Once seeds are present and immersed either in the liquid or vapor phase of monomer, their weight increases exponentially with time.

Fresh radicals are formed continuously by mechanical rupture of polymer chains being swollen by imbibed monomer (69,72). Termination of polymer radicals, in turn, is inhibited by the rigidity of the polymer network. The reaction is temperature sensitive, and can be minimized with adequate cooling (69). On the other hand, heat transfer may be impaired as the mass of material grows. Polymerization continues until the available monomer is consumed or gross amounts of inhibitor are added to the system. A number of inhibitors, eg, organic nitrites, nitroso compounds (69), oxides of nitrogen (73), alkali metal mercaptides (74), or nitrogen tetroxide adducts with unsaturates (75,76) have been recommended. The best control, however, is routine inspection and clean out of equipment to eliminate seeds.

Manufacture of Chloroprene Rubber

Chloroprene rubber is usually manufactured by either batch or continuous emulsion polymerization and isolated either by freeze coagulation or drum drying of a polymer film. Figure 1 is a schematic flow sheet of this process.

The process for manufacture of a chloroprene sulfur copolymer, Du Pont type GN, illustrates the principles of the batch process (77,78). In this case, sulfur is used to control polymer molecular weight. The copolymer formed initially is carried to fairly high conversion, gelled, and must be treated with a peptizing agent to provide a final product of the proper viscosity. Key control parameters are the temperature of polymerization, the conversion of monomer and the amount/type of modifier used.

To start the process, appropriate amounts of sulfur and a pine tar resin are dissolved in chloroprene to make a monomer solution. A water solution is also made up containing sodium hydroxide, and the sodium salt of a naphthalenesulfonic acid–formaldehyde condensation product. The two solutions are emulsified together, giving an oil-in-water emulsion. A sodium rosinate emulsifier is formed *in situ* on mixing of the aqueous and organic solutions. The formaldehyde condensation product is present to colloidally stabilize the latex at the low pH during the polymer isolation procedure. Emulsification is achieved by recirculating the two liquid phases through a centrifugal pump to give a particle size of about 3 μm. When emulsification is complete, the mixture is pumped to the polymerizer. This is a jacketed, glass-lined reaction vessel having a glass-coated agitator. An aqueous solution of potassium persulfate is added as required to initiate and maintain the polymerization. The temperature is kept at ~40°C by circulating brine through the kettle jacket and by changing the agitator speed.

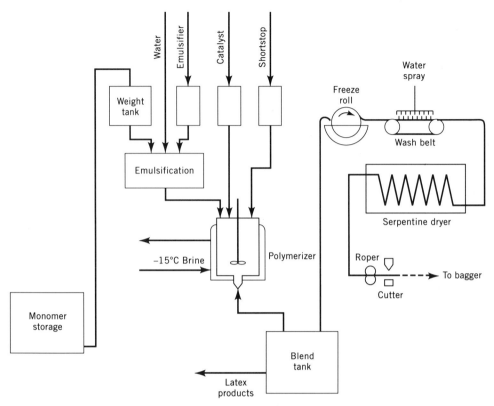

Fig. 1. Schematic flow sheet for the polymerization and isolation of polychloroprene; the polymerization section is fully functional for the manufacture of polychloroprene latex products. Courtesy of *Rubber Chemistry and Technology.*

Monomer conversion (79) is followed by measuring the specific gravity of the emulsion. The polymerization is stopped at 91% conversion (sp gr 1.069) by adding a xylene solution of tetraethylthiuram disulfide. The emulsion is cooled to 20°C and aged at this temperature for about 8 hours to peptize the polymer. During this process, the disulfide reacts with and cleaves polysulfide chain segments. Thiuram disulfide also serves to retard formation of gel polymer in the finished dry product. After aging, the alkaline latex is acidified to pH 5.5–5.8 with 10% acetic acid. This effectively stops the peptization reaction and neutralizes the rosin soap (80).

The acidified latex containing only the acid stable dispersing agent is fed to a pan in which a stainless steel drum rotates partly submersed in latex (81). The drum is chilled to about −15°C by recirculating brine. The polymer in the frozen film is coagulated in less than a full revolution of the drum and skived off the roll by a stationary knife. The film is dropped onto a woven stainless steel belt where it is thawed and washed by sprays of water on the top of the film that are forced through the film by reduced pressure applied underneath. Water and remaining serum are squeezed out of the film by rotating rolls to reduce the water content of the film to about 25%. The film is dried in a current of air at 120°C as it is carried continuously through a multicompartment dryer by an endless conveyer of cloth covered aluminum girts. The conveyer travels slightly slower than the squeeze roll to accommodate shrinkage during drying. The last two dryer compartments operate at 50°C to cool the film. The dried film is cooled further by passing it over a water cooled roll, and is then gathered into a rope, chopped into small pieces, and bagged.

The success of the freeze roll isolation process depends on the latex being stable enough to prevent premature coagulum formation, but so unstable that it will coagulate within a revolution of the freeze roll. The film must be tough enough to remain intact through the washing and drying operations, but thin enough to enable removal of salts during the washing step.

Continuous polymerization in a staged series of reactors is a variation of this process (82). In one example, a mixture of chloroprene, 2,3-dichloro-1,3-butadiene, dodecyl mercaptan, and phenothiazine (15 ppm) is fed to the first of a cascade of 7 reactors together with a water solution containing disproportionated potassium abietate, potassium hydroxide, and formamidine sulfinic acid catalyst. Residence time in each reactor is 25 min at 45°C for a total conversion of 66%. Potassium ion is used in place of sodium to minimize coagulum formation. In other examples, it was judged best to feed catalyst to each reactor in the cascade (83).

Alternative processes for polymer isolation have involved direct drum drying of latex (84), extrusion isolation of coagulated crumb (85), and precipitation/drying or spray-drying of the rubber as a powder (86). The powder can be processed directly in continuous compounding equipment (87). The manufacture and use of powdered CR has been reviewed (88).

These general polymerization processes have continued in use with occasional incremental changes to improve process control and to reduce environmental emissions. Recipe changes have also been made to take advantage of new process information. The original type introduced in 1932 was dropped in 1939 to introduce the G family of sulfur modified polymers (89). The mercaptan modified W family was introduced in 1949 followed by a gel grade, WB, for dry blends to improve the processibility of W family polymers. The T family was introduced in 1970 to provide a further improvement in processibility together with improved physical properties.

Properties of CR

General Vulcanizate Properties. Table 4 summarizes the properties of CR vulcanizates that are important to the designers of rubber parts, and compares these properties to those of two competitive materials. The comparison is general and varies depending on compound design (90). The obvious conclusion from the table is that EPDM has exceptional heat/ozone resistance, and that NBR has excellent oil resistance, but that both are weak in other areas. CR is not exceptional in those categories, but does have a good balance of toughness and environmental resistance. The high strength of the gum (ie, without reinforcing filler) CR vulcanizate, is due to its tendency to crystallize under stress. Few elastomers, other than natural rubber, do this.

Crystallization. The rate of crystallization and the melting point of polychloroprene both increase as the polymerization temperature decreases. This is shown in Figure 2 (91) and is related to the decreasing content of structural ir-

Table 4. General Vulcanizate Properties[a,b]

Polymer type	EPDM	CR	NBR
density	0.86	1.23	1.0
hardness, Shore A	40–95	40–90	45–100
tensile strength, MPa[c]			
gum stock	VL	21	VL
black stock	>21	21	17
service temperature, °C	−50 to +150	−40 to +120	−20 to +120
heat resistance	H	FH	M
cold resistance	M	FH	L–M
T_{10}, °C	−45	−45	−20
tan delta at 20°C		0.09	0.1–0.18
set resistance	VH	FH	M–FH
tear resistance	M	H	M
aging general	VH	H	M
ozone/corona resistance	VH	FH–H	M
aliphatic oils resistance	L	M–H	H
electrical resistance	H	M	M
bonding to substrates	M	VH	VH
useful properties	excellent aging; medium strength	low damping; high weather, heat, flame resistance; moderate oil resistance; good flex and tack	resists oil, chemicals; moderate heat aging
limiting properties	self tack, poor building of composites	moderate tensile strength	poor cold resistance

[a]Ref. 90.
[b]VH = very high; H = high; FH = fairly high; M = moderate; L = low; VL = very low.
[c]To convert MPa to psi, multiply by 145.

regularities as the temperature of polymerization is decreased. The melting point also varies with the temperature of crystallization and the time and temperature of any subsequent annealing (57,92). Copolymerization with 2,3-dichlorobutadiene, butadiene, methyl methacrylate or styrene, can be used to reduce the rate and extent of crystallization. To a limited extent certain structural irregularities such as inverted *trans*-1,4 units and dichlorobutadiene units, can be accommodated in the crystalline lattice. Other units, such as *cis*-1,4-chloroprene, 1,2-chloroprene, styrene, and methyl methacrylate, cannot (92). The effect of comonomers on microstructure and crystallization has been investigated (52).

X-ray diffraction studies have shown that the polychloroprene unit cell is orthorhombic, $a = 0.88$ nm, $b = 1.02$ nm, and $c = 0.48$ nm (93).

Crystallization kinetics have been studied by differential thermal analysis (92,94,95). The heat of fusion of the crystalline phase is approximately 96 kJ/kg (23 kcal/mol), and the activation energy for crystallization is 104 kJ/mol (25 kcal/mol). The extent of crystallinity may be calculated from the density of amor-

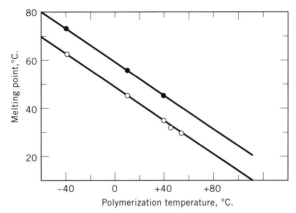

Fig. 2. Effect of polymerization temperature on the crystalline melting point of chloroprene rubbers produced by emulsion polymerization: ●, highest observed value; ○, lowest observed value (91).

phous polymer (d = 1.23), and the crystalline density (d = 1.35). Using this method, polymer prepared at $-40°C$ melts at 73°C and is 38% crystalline. Polymer made at $+40°C$ melts at 45°C and is about 12% crystalline.

The tendency of polychloroprene to crystallize enhances its value as an adhesive (97). The cured or uncured polymer can crystallize on stretching thereby increasing the strength of gum vulcanizates. Elastomers that cannot crystallize have poor gum vulcanizate properties (98).

The crystallization resistance of vulcanizates can be measured by following hardness or compression set at low temperature over a period of time. The stress in a compression set test accelerates crystallization. Often the curve of compression set with time has an S shape, exhibiting a period of nucleation followed by rapid crystallization (Fig. 3). The mercaptan modified homopolymer, Du Pont Type W, is the fastest crystallizing, a sulfur modified homopolymer, GN, somewhat slower, and a sulfur modified low 2,3-dichlorobutadiene copolymer, GRT, and a mercaptan modified high dichlorobutadiene copolymer, WRT, are the slowest. The test is often run near the temperature of maximum crystallization rate of $-12°C$ (99). Crystallization is accelerated by polyester plasticizers and delayed with hydrocarbon oil plasticizers. Blending with hydrocarbon diene rubbers may retard crystallization and improve low temperature brittleness (100).

Studies have considered the effect of crystallinity on the performance of CR adhesives (97), on segmental mobility as determined by nqr studies (101), on strain induced property changes (102), and on relaxation processes (103).

Heat Aging and Degradation. The resistance of polychloroprene to air oxidation, ozone attack, and weathering is enhanced, compared to other diene elastomers, by the presence of an electronegative chlorine atom on the repeat unit double bond. In contrast with natural rubber, which softens on aging, CR gradually increases in modulus and loses elongation until it becomes hard and brittle. Although the aging characteristics of the polymer are basically determined by the *trans*-1,4-chloroprene structure, irregularities that provide reactive tertiary chlorine or hydrogen atoms weaken the polymer's resistance to aging.

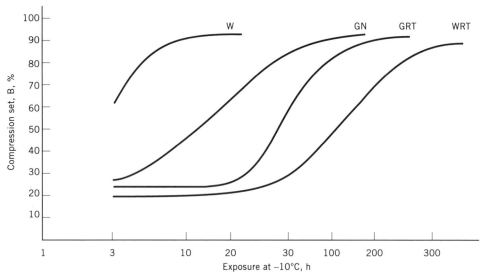

Fig. 3. Crystallization rate of polychloroprene vulcanizates under compressive stress (99).

Raw polychloroprene, made by emulsion polymerization and isolated by co-agulation and freeze drying, does not contain the isomerized 1,2-units found in polymer dried at elevated temperature (Table 3) (51). After one hour at 130°C, the original 1,2-units undergo an allylic rearrangement to form a primary allylic chlorine. Structure (**5**) is a polymer-chain sequence containing an isomerized 1,2-unit and a 1,4-unit. As the polymer is heated to higher temperature, eg, to 150°C for an hour, the isomerized units are consumed with the evolution of hydrogen chloride. On the basis of nmr evidence, it was concluded that the isomerized units undergo a combination of intra- and intermolecular reactions (104). This is not the simple loss of hydrogen chloride as originally postulated (93). Rather, the reaction involves the abstraction of allylic chlorine with the formation of a carbon to carbon bond at a neighboring 1,4-unit. In addition to the intramolecular reaction leading to ring formation, eg, structure (**6**), intermolecular reaction results in the formation of cross-links between polymer molecules.

Polymer units formed by either 3,4- or 1,4-addition do not evolve hydrogen chloride under these conditions in the absence of oxygen. Also, chain scission resulting in the formation of lower molecular weight polymers does not occur (104).

At higher temperatures under nitrogen, the polymer is reduced to coke with the evolution of hydrogen chloride and organic liquids such as chloroprene dimer. At temperatures below 275°C, polymers prepared at low temperature, with less 1,2- and 3,4-addition, are less reactive. Dehydrochlorination under nitrogen is not a radical chain process below about 275°C (105).

The practical service ceiling for polychloroprene is around 120°C under dynamic conditions in the presence of air depending on precise polymer structure. Several studies involving infrared and a variety of other techniques concluded that autoxidation begins at tertiary hydrogen or chlorine formed by either 1,2- or 3,4-addition during polymerization (106–108). It was also found that the induction period preceding rapid autoxidation of polychloroprene increased with decreasing temperature of polymerization (109). The presumption is that initial autoxidation at chain irregularities is the catalyst for reaction of the more inert sequences formed by 1,4-addition. Heat resistant mercaptan and sulfur modified polymers, that make use of low temperature polymerization to reduce the amount of these chain irregularities, have been described, and do in fact have improved hot air aging resistance (110).

If tertiary chlorine atoms are indeed critical to heat resistance, then reactions that consume them should improve polymer stability. This is indeed the case. Post-reaction of polychloroprene with dodecyl mercaptan (111), use of higher levels of ethylene thiourea for curing (112), and inclusion of reactive thiols such as mercaptobenzimidazole in cure systems (113) all improve heat resistance. This latter technique is especially effective in improving the heat resistance of mercaptan modified polychloroprene.

Compounding and Processing Chloroprene Rubber

Compound processibility is a key factor in the optimization of new polychloroprene types. As a result, commercial compounds can be mixed, shaped, and cured by virtually all the methods used in the rubber industry. A typical polychloroprene compound includes a variety of additives designed to improve compound rheology, cure rate, and vulcanizate properties.

CR type	selected according to application
processing aid	mill release, reduce viscosity, improve dispersion
magnesium oxide	acid acceptor, moderate cure
antidegradant	retard autoxidation, ozone, and flex cracking
reinforcing filler	to reinforce, and to reduce cost
plasticizer	reduce viscosity and hardness, improve low temperature properties
accelerator	increase rate and extent of cure, optional for G family
retarder	avoid premature cure
zinc oxide	curative

Curing Systems. Polychloroprene can be cured with many combinations of metallic oxides, organic accelerators, and retarders (114). The G family of polymers, containing residual thiuram disulfide, can be cured with metallic oxides alone, although certain properties, for example compression set, can be enhanced by addition of an organic accelerator. The W, T, and xanthate modified families require addition of an organic accelerator, often in combination with a cure retarder, for practical cures.

The organic accelerator most frequently used is ethylene thiourea. However, concerns about the toxicity of this material have led to its use as a dispersion in an inert elastomer, or to replacement by other materials.

The chemistry of cures with a combination of zinc oxide and magnesium oxide was originally thought to involve formation of ether linkages (115) but more recent ^{1}H-nmr studies have shown that the cure chemistry parallels degradation reactions involving monomer units formed by 1,2-addition. The cure reaction is ionic and catalyzed by zinc chloride that is formed during the process. The function of magnesium oxide is to moderate the reaction by consuming a portion of the zinc chloride being formed and converting it to a less active mixture of zinc oxychloride and magnesium chloride. Magnesium oxide will also function as an acid acceptor to protect the vulcanizate. As with degradation, the first reaction is isomerization of the 1,2-units followed by formation of an allyl carbonium ion. The ion can form ring structures by an intramolecular reaction, or cross-links by an intermolecular reaction (116). Only a single cure site is needed to form a cross-link, but many of the cure sites are consumed by the intramolecular cyclization reaction. As a result, the state of cure is relatively low.

Ethylenethiourea reacts to form monosulfide cross-links (117). A number of alternative curatives have been proposed to avoid use of ethylene thiourea. These include polyhydric phenols (118), hydroxyphenyl and mercapto substituted triazoles (119), thiolactams (120), thiazolidinethiones as Vulkacit CRV (121), and alkanethioamides (122). Among these, Vulkacit CRV is the most widely used. An accelerator is ordinarily used in combination with a retarder to control premature cross-linking. Tetramethylthiuram disulfide [137-26-8] is ordinarily used for this purpose when the accelerator is either ethylenethiourea [96-45-7] or a thiazolidinethione.

Antidegradants. Although CR is more resistant to autoxidation than other diene elastomers, it will autoxidize readily at ordinary temperature in the absence of a small amount of antioxidant (114). Degradation of CR at elevated temperatures is clearly related to air exposure. Degradation is 15 times slower under nitrogen than air (123). The addition of even a small amount of a free-radical inhibitor has a significant effect on the resistance of vulcanizates to oxidative attack. The inhibitor can be an aromatic amine for dark stocks or a hindered phenolic for light colored stocks (see ANTIOXIDANTS). Octylated diphenylamine is often selected as an inhibitor since it is nonstaining, does not affect the cure, and is among the most effective materials (124). Its activity can often be improved further by addition of an organic monosulfide or phosphite (125). A small amount of added polybutadiene or unsaturated oil, eg, rapeseed oil, is also effective in scavenging sulfur and oxygen (126).

The best oxidation inhibitors are not usually the best antiozonants (qv). A disubstituted *para*-phenylenediamine such as *N*-isopropyl-*N*'-phenyl-*p*-phenylenediamine is often selected for that purpose. *p*-Phenylenediamine derivatives interfere with cure chemistry and scorchiness, and can stain objects in contact with the vulcanizate (114). On balance, *N*-(1,3-dimethylbutyl)-*N*'-phenyl-*p*-phenylenediamine and phenyl/tolyl-*p*-phenylenediamines have the best combination of properties. They are less scorchy and provide excellent ozone and heat resistance. Additional protection is gained in blends with a small amount of EPDM rubber (126).

In some applications, eg, CVJ boots, stabilizers are extracted from the part by grease or oil, leading to early failure. *N*-Isopropyl-*N*'-phenyl-*p*-phenylenediamine reacts with CR during cure so that it is no longer extractable. Reaction is between allylic chlorine on the polymer and the isopropyl amino group of the antiozonant. Although bonded, it continues to act as a stabilizer (127). The reaction product of a liquid polychloroprene with *p*-phenylenediamine derivatives has been tested as an antidegradant (128). Additives are also available to protect against photodegradation (129) and radiation (130).

Reinforcing Fillers. Polychloroprene and natural rubber are both self-reinforcing rubbers, and do not require reinforcing fillers to achieve high tensile strength. However, reinforcing fillers are normally used to improve certain properties, such as hardness and modulus, to improve processability and to reduce cost. Carbon black (qv) is the most widely used filler. A considerable range of carbon black types is available varying in particle size and structure. For maximum reinforcement as measured by tensile strength, high abrasion furnace blacks (SAF,N110 and HAF,N330) are commonly used. Where processability is of more concern and maximum strength not required, semireinforcing furnace (SRF,N762), or general-purpose furnace (GPF,N660) blacks are used. Fast extruding furnace (FEF,N550) is used in extrusion applications. SRF and GPF provide an adequate degree of reinforcement together with economical price and, hence, are widely used. The principle nonblack mineral fillers (qv) are hard clay, silica, and calcium silicate. Hard clay provides a good degree of reinforcement and tear strength, and is used to extend more expensive black compounds. Calcined clays are used for best electrical properties. Hydrated alumina is used for improved flame resistance, and platy talcs for good extrusion and electrical properties (131).

Plasticizers. These are used to improve compound processability, modify vulcanizate properties, and reduce cost. For many applications, where cost and processability are the objective, naphthenic and aromatic oils are preferred. They are inexpensive yet effective in improving processability at high filler levels. The compatibility of the naphthenic oils is limited to about 20 parts per hundred rubber. Aromatic oils are more compatible and can be used at higher levels (132).

If flammability is an issue, liquid chloroprene polymers (eg, Du Pont FB or Denki LCR-H-050) can be used. They cocure and, for that reason, are nonvolatile and nonextractable. They are particularly useful in hard compounds where they do not detract from physical properties as much as nonreactive plasticizers (132,133). Methacrylate esters have been used as reactive plasticizers (qv). For example, hexa(oxypropylene)glycol monomethacrylate can be used as a reactive plasticizer to enhance flex life without increasing hardness (134).

Polyester plasticizers are much more effective than hydrocarbon oils in reducing the brittle point and the glass-transition temperature of vulcanizates.

There is a trade off, however. The plasticizers that lower the brittle point the most also promote crystallization (135). The trade off can be adjusted by using blends of dioctyl sebacate with aromatic oil or by going to an unsaturated vegetable oil such as rapeseed oil. Ester plasticizers are best used with more freeze-resistant CR grades such as Du Pont WRT, Baypren 110 VSC (78,135), or preferably Du Pont Showa SND-35.

Mixing. CR may be mixed with the various types of equipment used for general-purpose elastomers. The key factor is to minimize mix time and temperature to avoid scorch. This is often best done in an internal mixer. In a conventional mixing cycle, the addition of polymer, magnesium oxide, and antioxidants is followed by the addition of process aids and peptizing agents. The magnesium oxide helps to retard the cure. Next, fillers and plasticizers are added. To obtain the best dispersion, these last two ingredients should be added separately. Zinc oxide and any accelerators are best added on the sheetoff mill or in a second operation just before the stock is needed (131,136).

Calendering and Extrusion. Friction compounds are used to build up composite structures of fabric and rubber. The surface of the calendered fabric must have good green tack, and for that reason compounds to be calendered are best made from a slow crystallizing peptizable polymer such as Du Pont GRT or Baypren 610. The relative temperatures of the fabric and various rolls must be carefully controlled so that the polymer does not stick to the rolls yet penetrates the fabric (137).

Extrusion compounds use FEF carbon black to obtain good die definition, low nerve, and a good level of physical properties. Gel-Sol blends are commonly used for extrusion because they contribute to the favorable rheology of the compound and also give resistance to collapse to the uncured structure during vulcanization. The screw of the extruder should have a constant diameter root with increasing pitch. Heat history of the compound should be minimized with only a brief warm-up before extrusion. The barrel and screw should be run cool, 50°C, and die hot, 95°C (131,136).

Molding. Molding is used widely for fabricating CR into belting, hose, sponge, and a variety of industrial products. All of the standard molding techniques have been used successfully with commercially available equipment. Molding methods include compression, transfer and injection, blow molding, vacuum molding, and tubing mandrel wrap. An optimum cure cycle, time–temperature relationship, must be selected based on the curing characteristics of the compound and the suitability of existing equipment. Mold design must take into consideration easy, rapid removal of the cured part without damage to it (138).

Commercial Polymer Types

The *Synthetic Rubber Manual* lists 93 chloroprene rubber dry polymer types manufactured by six producers, not including those types produced in China or those for adhesive applications (113). For the most part, the polymers fall into the categories described in Table 5. These categories differ according to the method used for molecular weight control, the means used to control crystallization resistance, and blending with gel polymer. Each polymer category has its own set of advan-

Table 5. Distinguishing Features of CR Grades[a]

Standard types	Advantages	Limitations
mercaptan modified	heat resistance	cure rate
W types	compression set	flex resistance
	shelf stability	
	nonpeptizable	
xanthate modified types	tensile properties	heat resistance
	cure rate	flex resistance
	tan delta	
	compression set	
	rheology	
sulfur copolymers	peptizable	heat resistance
G types	building tack	compression set
	flex resistance	shelf stability
	tan delta	
	cure chemistry	
precross-linked grades	low nerve	tensile properties
T types	fast extrusion	
	low die swell	
	collapse resistance	
freeze-resistant copolymers		
2,3-dichlorobutadiene	slow crystallization	slightly higher T_g
styrene		
polymers made at higher	slow crystallization	heat resistance
temperatures		

[a]Ref. 139.

tages and disadvantages (139). Lists of commercial types from various manufacturers have been compiled (Table 6) (140). Although general characteristics are similar, each manufacturer has optimized properties for particular customers and manufacturing capabilities.

The various methods used for molecular weight control also affect the end group chemistry of the polymer, and in the case of sulfur, change the chemistry of the main chain. Mercaptan modification gives polymers that are stable in storage and resistant to breakdown during processing. Their molecular weight and molecular weight distribution are determined during polymerization. If they have a deficiency, it is that polymer chain ends are not reactive, leading to loss of loose chain ends from the elastic network formed during cure. As a result, vulcanizates from low molecular weight polymers are not quite so tough as vulcanizates from high molecular weight polymers (141).

Xanthogen disulfide modified polymers are a close relative of the mercaptan modified polymers in that polymer molecular weight is determined during polymerization. They differ in that polymer modified with a xanthogen disulfide has cure reactive end groups (142,143). As a result, vulcanizates from xanthate modified polymers have better utilization of chain ends and a generally tighter cure. This results in improved tensile properties and reduced hysteresis loss and creep. The downside is that sulfur residues from this chemistry impair the heat resist-

Table 6. Commercial Polychloroprene Grades[a]

					Company (trade name)			
Grade	Du Pont (neoprene)	Bayer (Baypren)	Miles (Baypren)	Distugil (Butaclor)	Denki Kagaku (chloroprene)	Tosoh (Skyprene)	Du Pont Showa (neoprene)
slow crystallizing	WX	110	M3.25	MC-10	DCR-70	TSR51	WRT
	WRT	110VSC			DCR-35	B-5	SND-35
medium crystallizing	W	210	M1	MC-30	M-40	B-30	
	WHV	230	M2	MH-30	M-120	Y-30	WHV
fast crystallizing	AC	331	AT-M	MA-41K	TA-95	G-41K	AC
	AD	320	A3	MA-40S	A-90	G-40T	AD
precross-linked	TW	215		DE-302	MT-40	E-33	TW
	TRT	115	EM-1	DE-102	ES-40	E-20	TRT
sulfur modified	GW	510	SM	SC-202	DCR-46		GRT
	GRT	611	S3	SC-10	PS-40	R-10	GW

[a] Ref. 140.

ance of vulcanizates. Depending on chain-transfer kinetics for particular mercaptans and xanthogen disulfides, the xanthate modified polymer may have a broader molecular weight distribution, and, for that reason, improved processibility (110,143).

Molecular weight control, by means of sulfur copolymerization, is a two-step process involving polymerization of a high molecular weight, often gelled polymer, which is subsequently peptized to a lower molecular weight sol polymer. The Du Pont G family of polymers is representative of this class. Chain ends are cure reactive (144), and the polymer chains continue to have some polysulfide sequences. Depending on several factors, the polymer may either peptize further or cross-link during storage and processing.

The ability of sulfur modified polymers to break down during processing enables softer stocks at high filler level without the use of plasticizers. In addition, compounds calender smoothly to give stocks that adhere well to each other for manufacture of rubber/fabric structures (building tack). Since chain ends are cure reactive, vulcanizate toughness does not normally vary with molecular weight. Properties vary with the amount of copolymerized sulfur. Polymers described as having low sulfur content have higher viscosity but improved heat resistance and compression set. Properties are intermediate between those of the G and W families of polymers (145–147).

The precross-linked grades of polychloroprene, such as the Du Pont T family, contain a highly cross-linked gel form of polychloroprene that acts as an internal processing aid (148,149). The blend has many of the same characteristics as the sol component, but different compound rheology. This results in smoother, faster extrusion, and calendering with reduced die swell and shrinkage for a given filler level. Extrudates also retain their shape better with good definition until vulcanized. The down side is that vulcanizates may not develop tensile properties equivalent to those of the sol polymer. This has been attributed to the cross-link density of the two phases being dissimilar, thus causing uneven stress under load (150). Various means have been suggested for reducing this problem (151,152). Sol–gel blends of sulfur modified sol and gel polymers have also been described (153).

Freeze-Resistant Polymers. Chloroprene homopolymers made at conventional polymerization temperatures of 40–50°C are not sufficiently freeze resistant for some applications. In particular, automotive parts such as belts, boots, and air springs are used in dynamic applications and need substantial freeze resistance during cold weather. Certain polychloroprene manufacturers make use of 2,3-dichloro-1,3-butadiene as a comonomer to retard crystallization. The Du Pont RT polymers are representative of this type of material. In using the RT polymers, there is a small trade off in that the glass-transition temperature is increased slightly. Styrene is apparently used in some Armenian types, while other manufacturers without dichlorobutadiene capability polymerize at increased temperature.

Applications

Dry Polymer Type. Polychloroprene has relatively high compound cost both because of its high manufacturing cost and relatively high density. It is not used in applications where less expensive materials can perform satisfactorily. To off-set the cost factor, it has a well-balanced combination of properties including processibility, strength, flex and tear resistance, weatherability, flame resistance, and adhesion together with sufficient heat and ozone resistance for most applications. Some of its uses are outlined in Table 7 together with estimates of market volume in Japan (154). About half the volume is in automotive applications. Much of the automotive and general industrial usage is in applications where the polymer is under dynamic load, frequently at elevated temperature. Typical applications involve power transmission and timing belts, automotive boots, airsprings, and truck engine mounts (139). Other applications involve the proven long-term dependability of the material. Examples involve all sorts of things from bridge bearing pads to automotive air bags that must remain tough and flexible years after manufacture.

Table 7. Applications of Polychloroprene in Japan*ᵃ*, t

Application	Year		
	1986	1987	1988
automotive	15,000	19,500	20,300
boots, belts, hoses, etc			
general industrial	16,700	13,800	14,600
belts, hoses, cellular rubber, sheet stock, gaskets, weather seals, etc			
adhesives	5,200	5,200	5,200
wire and cable	1,800	1,700	1,900
latex	700	700	700
paper treatment, foam, dipped goods, coated fabric			
Total	*39,400*	*40,900*	*43,200*

*ᵃ*Ref. 154.

Solvent Adhesive Applications. U.S. consumption of chloroprene rubber for adhesive applications was 13.6×10^6 dry kg in 1990. This included both solvent-based applications (155–157) and latex-based applications (158,159). A large number of CR types are involved, both standard and specially designed, for adhesive uses. The special polymer types involved are designed to have high un-cured cohesive strength either through polymer crystallization or by ionomer formation. When properly compounded and applied to substrates, they provide almost instantaneous bonding as a pair of treated surfaces are brought together. This quick grab characteristic is almost unique among adhesives, and has led to numerous applications.

The type of chloroprene polymers used is perhaps best illustrated by the variety of special products, designed for adhesive applications, that Du Pont has developed. These are described in Table 8. Standard polymer grades are also often used, especially to modify adhesive properties and to reduce cost.

Polymers such as AC and AD are normally compounded according to the generic recipe given in Table 9. The oxides and inhibitor may or may not be mill

Table 8. Neoprene Adhesive Polymer Types[a]

Type	Crystallization rate	Characteristics
AC	very fast	fast bond strength development
AD	very fast	similar to AC, more stable for graft
AD-G	very fast	polymer adhesives
AF	slow	methacrylic acid copolymer, superior hot bond strength, forms ionomer
AG	very slow	dispersable, high gel polymer
WHV-A	moderate	high viscosity W family polymer, modified for adhesive use

[a]Ref. 155.

Table 9. CR Contact Bond Adhesive Compounding[a]

Ingredient	Amount, phr	Function
polymer	100	cohesive strength
magnesium oxide	4–8	acid acceptor, retarder
antioxidant	2	retard degradation
zinc oxide	5	curative, acid acceptor
t-butylphenolic resin	40	promote adhesion
water	1	help MgO–rosin rxn
solvent	as needed	

[a]Ref. 155.

mixed into the polymer prior to the mixture being dissolved in the solvent. The t-butylphenolic resin improves specific adhesion and auto-adhesion, increases tack time, and increases hot cohesive strength. Its quality is critical to the performance of the bond. Although the cohesive strength of the bond primarily depends on the crystallinity of the neoprene, reaction of the phenolic resin with the magnesium oxide and polymer contributes additional strength. The solvents ordinarily used are a blend of aromatic, aliphatic, and oxygenated solvents.

Carboxylated polymers such as AF use similar but not identical compounds. The higher strength, especially hot bond strength, is due to the interaction of the carboxyl groups on the polymer chain with the metal oxides. The crystallization rate of AF is low and does not contribute to bond strength. Manufacture of adhesive compounds from AF is more demanding than manufacture of those from AD.

Gel polymers such as AG are used as part or all of the polychloroprene component in an adhesive. Its principal value is in providing relatively high solids, low viscosity solvent dispersions. This is particularly desirable in sprayable adhesive compounds. WHV-A can be used as an adhesive in its own right, or can be blended with crystalline sol polymers when reduced crystallinity is desired.

Type AD-G is used in an entirely different sort of formulation. The polymer is designed for graft polymerization with methyl methacrylate. Typically, equal amounts of AD-G and methyl methacrylate are dissolved together in toluene, and the reaction driven to completion with a free-radical catalyst, such as benzoyl peroxide. The graft polymer is usually mixed with an isocyanate just prior to use. It is not normally compounded with resin. The resulting adhesive has very good adhesion to plasticized vinyl, EVA sponge, thermoplastic rubber, and other difficult to bond substrates, and is of particular importance to the shoe industry (42,43).

Latex Adhesive Applications.   Polychloroprene latex adhesives have a long history of use in foil laminating adhesives, facing adhesives, and construction mastics. Increasingly stringent restrictions on the emission of photoreactive solvents has heightened interest in latex compounds for broader applications, particularly contact bond adhesives. Table 10 makes a general comparison of solvent and latex contact bond adhesives (158).

On a practical basis, two different types of polymers are used for latex and solvent adhesives. The solvent system uses a tough, crystalline polymer that contains little or no gel. In the usual latex system, the polymer used is soft and slow crystallizing, but is partly gelled. The stronger bonding of the solvent system melts out at about 52°C, and the bond loses most of its strength. For example, a solvent bond from a crystalline polymer develops a film strength of 25 MPa (3600 psi) after 7 days, but most of this strength is lost at 52°C. The strength of an air dried film of a gel polymer ranges from 14–20 MPa (2000–2900 psi) with much of this retained at higher temperature. Both types of polymers continue to form cross-links, and improve with time in the presence of the zinc oxide normally included in adhesive compounds (158).

Compounding is quite different for the two systems. The solvent base system is dependent on magnesium oxide and a *t*-butylphenolic resin in the formulation to provide specific adhesion, tack, and added strength. Neither of these materials have proven useful in latex adhesive formulations due to colloidal incompatibility. In addition, zinc oxide slowly reacts with carboxylated latexes and reduces their tack. Zinc oxide is an acceptable additive to anionic latex, however. Other tackifying resins, such as rosin acids and esters, must be used with anionic latexes to provide sufficient tack and open time.

Table 10. Comparison of Latex and Solvent Contact Bond Adhesives[a]

Adhesive type	Latex	Solvent
solids content	high	low
compound rheology	easy	difficult
sprayability	good	moderate
drying rate	very slow	rapid
rate of bond development	moderate	fast
adhesion to porous surfaces	good	good
wetting of nonporous surfaces	poor	good
tack time	moderate	good
hot bond strength	good	moderate
water resistance	initially poor	good

[a]Ref. 158.

The carboxylated latexes are formulated to use a reduced amount of a less reactive zinc complex. Special resin blends provide an optimum balance of film tack and strength, and are colloidally compatible with the carboxylated latexes (158). Epon resins may also be used as an acid acceptor in place of zinc oxide (160).

The use of polychloroprene in solvent and latex adhesives has been the subject of a recent review (155).

Polychloroprene Latexes

Polychloroprene latexes can often be used where either dry or solvent-based processes would be impractical. The water base systems have high solids, minimal solvent emissions, are unaffected by polymer rheology, and have extremely small particle size. The main problem areas are the slow evaporation rate of water and the need to ensure that drying conditions favor film formation. The factors that control the properties of dry polymers also control the properties of latex polymers. However, long-chain branching and gel formation, which are usually unacceptable with dry polymers, are useful ways to increase the cohesive strength of latex polymers. This is an important difference from dry polymer technology. Polymer films dried from latex can have high cohesive strength without further cure. Gel content is only limited when it prevents film formation or reduces tack. An added factor in the complexity of latex manufacture has to do with routinely reproducing the colloidal properties of the latex and maintaining the highest possible solids level for certain applications. Polychloroprene with a density of 1.23 is at a disadvantage compared with natural rubber (density = 1) in terms of the volume fraction of rubber in the latex. This can be an important factor, for example, in manufacture of foam rubber. On the other hand, the particle size of neoprene latex is about 0.07 to 0.25 μm, thus much smaller and with a narrower distribution than natural rubber. This can be a key factor in saturation of paperboard, for example, where the latex must penetrate the web of the paperboard. Finally, the latex compound must have sufficient mechanical stability to run smoothly without coagulum (161) formation.

Latex Types. Latexes are differentiated both by the nature of the colloidal system and by the type of polymer present. Nearly all of the colloidal systems are similar to those used in the manufacture of dry types. That is, they are anionic and contain either a sodium or potassium salt of a rosin acid or derivative. In addition, they may also contain a strong acid soap to provide additional stability. Those having polymer solids around 60% contain a very finely tuned soap system to avoid excessive emulsion viscosity during polymerization (162–164). Du Pont also offers a carboxylated nonionic latex stabilized with poly(vinyl alcohol). This latex type is especially resistant to flocculation by electrolytes, heat, and mechanical shear, surviving conditions which would easily flocculate ionic latexes. The differences between anionic and nonionic latexes are outlined in Table 11.

A few commercial cationic latexes are offered, eg, by Du Pont Showa and Denki. These are stabilized with quaternary ammonium surfactants.

As with the solid polymers, the useful properties of the elastomers contained in the latex vary with polymerization conditions. Conversion is limited with the dry types to maintain good dry and solution rheology. This is usually not an im-

Table 11. Comparison of Anionic and Nonionic Latexes[a]

Latex type	Anionic	Nonionic–carboxylated
emulsifier	sodium or potassium resinates	poly(vinyl alcohol)
colloidal stabilization	ionic	entropic
stability		
mechanical	good	exceptional
electrolytic	good	excellent
to ionic contaminants	fair	excellent
storage	good	fair[b]
pH	12+	7
adjusted with	caustic	diethanolamine
solids, %	38–60	47
surface tension, mN/m		
(= dyn/cm)	39	58
Brookfield viscosity, mPa·s (= cP)	5–500	500
average particle size, μm	0.1	0.3

[a]Ref. 158.
[b]Excellent colloidal stability but gel formation may interfere with tack.

portant factor with latex types, and some of the latex types are driven to high conversion to improve cohesive strength. Thus the gel content of latex polymers runs from approximately 0 to 90% depending on type compared with only a few percentages for dry types. Polymer properties such as cohesive strength, modulus, and resilience increase with increasing gel content. Elongation, dry tack, and oil swell decrease. Polymer crystallization affects polymer properties in a generally similar way except that the effects of crystallinity disappear at about 52°C. Figure 4 (158) compares the crystallinity and gel contents of a number of Du Pont latex and dry types. Note that the conventional dry types are all in the low gel area, the latex (numbered) types extend to the very high gel region. Two types are of special interest. Latex 115 is a carboxylated polymer in nonionic dispersion. It is not crystalline, but it is strengthened by ionomer formation. Latex 400 is a copolymer with relatively high 2,3-dichlorobutadiene content. The dichlorobutadiene content is high enough for those polymer units to crystallize, with the result that the polymer is quite tough and acts as though it contained high gel (158).

Latex Compounding.   Latex compounding must take into account the stability of the latex both before and after compounding. Where consideration of solids concentration permits, the additives are best predispersed in a compatible aqueous surfactant before addition to the latex. The volume of additives, especially if clay fillers are involved, may easily be enough to starve the system for soaps and flocculate the compound. On the other hand, dry powders or molten resins may often be added directly to the nonionic latex.

All latex compounds should contain at least 5 phr of zinc oxide. This is needed to absorb evolved hydrochloric acid either in the compound or finished part. A larger amount should be considered if the part contains or is in contact with acid-sensitive materials such as cotton cloth. Magnesium oxide may destabilize anionic soap systems, and is avoided for that reason. The compound should

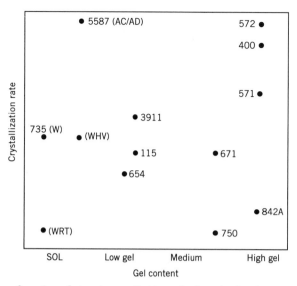

Fig. 4 Properties of various latex types. Letters designate dry types; numbers indicate latex (158).

also contain two parts of an antioxidant, and consideration should be given to the need for a uv screen in light-colored products.

Accelerators may be added to improve the physical properties of the polymer when needed. Where high modulus or low oil swell is required, thiocarbanilide is the preferred accelerator, with a cure time of 60 min at 100°C. Tetraethyl thiuram disulfide and sodium dibutyl dithiocarbamate are preferred for high tensile strength and cured at 121°C.

Since polychloroprene crystallizes under stress, fine particle size carbon black is not ordinarily needed or used to enhance tensile strength. More frequently, mineral fillers, for example clay, can be added to reduce cost. A light process oil, free of polycyclic aromatics, can be used to improve the flexibility or hand of films. On the other hand, an ester plasticizer can be used to improve low temperature properties (161).

Processing and Applications.   Latexes are used in a variety of special ways to provide useful products. The following items illustrate both uses and processes for latexes (165).

Binders. CR latex compounds are used to bond a variety of particulates. The particulates may be cellulosic fibers such as shoe board or ground rubber, used on an athletic track. In the first instance the compound is used to saturate a cellulosic web moving through a paper-making machine at high speed. The web is then dried on-line with drum dryers. Another application involves mixing ground rubber with a specially formulated anionic latex compound and a coagulant mixture of Portland cement in oil. The latex is slowly coagulated as calcium ions from the oil dispersion migrate into the latex. The resulting surface is tough and resilient.

Coatings. CR has been used to coat a variety of substrates, from cloth for rainwear to concrete decks for protection against salt water. A sol-type latex is preferred to ensure good adhesion to concrete decks. A crystalline polymer latex is preferred where added durability is needed. The compound includes a nonionic surfactant to improve its chemical stability. A number of thin coatings are applied to the surface to allow better coverage and facilitate drying. A similar formulation could be used to coat the interior of tanks, but an accelerator is needed to improve toughness.

Dipped Goods. Gloves with and without cloth support are a typical example. Unsupported applications involve specialty medical gloves and weather balloons. For CR to compete with less expensive natural rubber, it must provide added value. Natural rubber medical examination gloves contain proteins that can cause an allergic reaction in both patients and medical personnel. Polychloroprene gloves avoid the problem for those individuals, but are too expensive for general use. Natural rubber weather balloons have less lift than CR balloons and are rapidly attacked by ozone in the upper atmosphere.

Dipped goods are manufactured by dipping a coagulant coated form into a latex compound. The object is then washed, dried, and cured on the form. If a cloth support is used, the colloidal properties of the compound must be such that rubber will adhere to the lining but not penetrate it (166).

Elasticizers. CR latex is mixed with melted asphalt to be used in the construction of roads surfacing material. At a 2% level, the blend has increased ductility and toughness together with a higher softening point and viscosity (167). An optimum latex composition has been developed by means of a statistical design experiment (168). Polychloroprene may also be used to toughen films deposited from aqueous emulsion (166).

Foam. Polychloroprene latex has been used to manufacture foam for carpet underlayment and cushioning. It is manufactured by whipping air into a latex compound to make a froth. The froth is gelled by either heat or chemical coagulation. The chemical gelling agent is sodium silicofluoride. This material hydrolyzes to generate acid at a predictable rate. The acid coagulates the froth before it has a chance to collapse. The gelled foam is then dried and cured. In laboratory testing, properly compounded CR foams have shown reduced flame spread. Combustion of properly compounded polychloroprene needs a strong ignition source and maintenance of significant temperature if flame propagation is to occur. The chemical structure of CR causes it to char rather than melt, and spread, as is the case with other elastomeric foams. These advantages are obtained at the expense of smoke formation and evolution of hydrogen chloride (169).

Economic Aspects

CR has now been used commercially for around 60 years. Over that period of time, it has established a reputation for reliability, and has become the material of choice for innumerable applications. Its annual volume has approached around 350,000–400,000 metric tons. It is currently manufactured in five countries in Western Europe, North America, and Japan with a combined capacity of 385,000 metric tons. Du Pont is the largest supplier with 49% of the total capacity and plants in the United States, UK, and Japan. Other suppliers are Bayer/Miles (22.6%) in Germany and the United States, Denki Kagaku Kogyo (11.7%) in Japan, DISTUGIL (10.4%) in France, and TOSOH (6.2%) in Japan (2).

Polychloroprene consumption worldwide, except for eastern European countries and China, has plateaued at about 250,000 metric tons per year with some continued slow growth expected. Annual production averaged 307,000 metric tons during the 1980s with at least part of the difference being exported to formerly Socialist countries. Production in Armenia has been limited to a fraction of its capacity of 60 metric tons by environmental problems and, in fact, is currently shut down. The People's Republic of China has three plants with a combined capacity of 20 metric tons (2).

Although CR continues to have about the same production volume, its share of total elastomers production has dropped from around 5% in 1975 to 3% in 1989 (2). In part, this is related to the introduction of new materials that may not match CR's balance of properties but work quite well in specific areas and are less expensive. For a typical compound, CR's cost is about twice that of EPDM. Applications that require only weatherability and heat resistance tend to move to the less expensive product. Use of polyetheresters, eg, HYTREL, in CV-J boots replacing CR, is based on a combination of an easier process and improved quality (impact resistance) with a more modern and much more expensive material. Another difficulty facing CR has to do with changes in product requirements. Longer automotive guarantee periods together with higher underhood temperatures will place more stringent requirements on the heat resistance of all underhood parts. This may result in partial replacement of CR in some underhood applications.

On the other hand, CR is a versatile material that is able to take advantage of new opportunities. Right now, these are in dynamic applications such as air bags, air springs, and heat-resistant engine mounts.

Quality Management

Quality control systems look not only at the analytical results, but also are able to track the material back through the process to the conditions of manufacture and the quality of the lots of incoming raw materials. ISO standard 9002 lays out the requirements for such a quality management system. Basically, the participating companies must write down in detail 18 aspects of their quality management system covering such things as control and testing of raw materials, process control procedures, physical testing procedures, instrument calibration and maintenance procedures, warehousing, etc (170). The program also involves third party audits of the systems to determine if participating companies are in fact doing what they say they will do. Where compliance with this standard is voluntary, companies that have a first-rate program have a competitive edge.

In addition to this drive to look beyond manufacturing to specifications, new analytical methods such as molecular weight distribution, Mooney relaxation, and other measures of polymer processability are being explored.

Health, Safety, and Environmental Factors

Fire and uncontrolled polymerization are a concern in the handling of chloroprene monomer. The refined monomer is ordinarily stored refrigerated under nitrogen and inhibited. This is supported by routine monitoring for polymer formation and vessel temperature. Tanks and polymerization vessels are equipped for emergency inhibitor addition. Formalized process hazard studies, which look beyond the plant fence to potential for community involvement, are routine for most chemical processes.

After polymerization, excess monomer is stripped and recycled. The residual monomer content of the stripped emulsion does not represent an acute hazard. Worker exposure to monomer is monitored, and sources of exposure identified and corrected.

Neoprene latexes contain 0.5 to 0.02% residual chloroprene depending on the specific latex type. The amount of free alkali in the water phase of latexes varies from 0.1 to 0.08% depending on type and age of the material. Eye protection and appropriate skin protection have been recommended for use in situations where splashes or spills are possible. Toxicity and safe handling practices have been recommended for Du Pont types (171). Since compositions may vary with other manufacturers, specific information should be obtained for other products.

Neoprene Type TW was shown to have low oral toxicity in rats. The LD_{50} was found to be in excess of 20,000 mg/kg. Human patch tests with Types GN, W, WRT, and WHV showed no skin reactions (169). The FDA status of Du Pont Neoprene polymers is described (172). Although polychloroprene itself has not been shown to have potential health problems, it should be understood that many rubber chemicals that may be used with CR can be dangerous if not handled properly. This is particularly true of ethylenethiourea curatives and, perhaps, secondary amine precursors often contained in sulfur modified polychloroprene types. Material safety data sheets should be consulted for specific information on products to be handled.

The principal atmospheric emissions from a chloroprene polymerization facility result from gaseous venting during monomer transfer, from incomplete condensation of monomer after distillation and stripping operations, and from the polymer dryer vents. Back-up condensers are used to capture monomer venting from monomer-transfer operations. Chilled brine cooling is used to maximize condenser efficiency. Recipes have been modified to minimize dryer emissions. Equipment and emissions control procedures are ordinarily arrived at by agreement with appropriate governmental agencies.

The effluent from the isolation wash belt is the principal wastewater stream from the polymerization process. It contains highly diluted acetic acid and a surfactant that is not biodegradable. The wastewater streams are sent to sewage treatment plants where BOD is reduced to acceptable levels. Alternative biodegradable surfactants have been reported in the literature (173).

Polymer that does not meet specifications is sold at reduced price. What cannot be sold is buried in nonhazardous landfill operations. Bags used for shipping polychloroprene are considered to be hazardous waste in some countries because of the use of talc as a parting agent. Even without that complication, the bags are a waste disposal problem for rubber manufacturers. A bag that can be dispersed in the rubber mix has been described and tested (174). Part of the vulcanizate scrap from rubber parts manufacture is ground and returned to the process. Not all can be handled this way, however, so the rest of it must be landfilled. The problem is especially acute in the grinding operation during manufacture of serpentine belts. Various means have been proposed to work off the material in fresh compound. The amount of scrap that can be recycled is increased if the scrap is treated with liquid rubber during compounding (175).

BIBLIOGRAPHY

"Elastomers, Synthetic–Neoprene" in *ECT* 3rd ed., Vol. 8, pp. 515–534, by P. R. Johnson, E. I. du Pont de Nemours & Co., Inc.

1. A. M. Collins, *Rubber Chem. Technol.* **46,** G48 (1978).
2. *Worldwide Rubber Statistics 1991,* Int. Institute of Synthetic Rubber Producers, Houston, Tex., 1991.
3. L. J. Op, P. J. Hwan, and B. H. On, *Hwahak Kwa Hwahak Kongop* **20,** 119 (1977).
4. W. Kern and co-workers, *Macromol. Chem.* **3,** 223–246 (1950).
5. C. A. Stewart, Jr., *J. Am. Chem. Soc.* **93,** 4815 (1971); **94,** 635 (1972).
6. F. Hrabak and J. Webr, *J. Macromol Chem.* **104,** 275 (1967).
7. K. A. Nersesyan, R. O. Chaltykyan, and N. M. Beileryan, *Arm. Khim. Zh.* **40**(2), 92–95 (1987).
8. F. Hrabak and M. Bezdek, *Collect. Czech. Chem. Commun.* **33,** 278–285 (1968).
9. E. S. Voskanyan, L. E. Gasparyan, and S. M. Gasparyan, *Arm. Khim. Zh.* **36**(7), 467–476 (1983).
10. Ya. Kh. Bieshev and co-workers, *Dokl. Akad. Nauk. USSR* **301**(2), 362 (1988).
11. I. Williams, *Ind. Eng. Chem.* **31,** 1204 (1939).
12. S. Ekegren and co-workers, *Acta Chem. Scand.* **4,** 126 (1950).
13. W. V. Smith, *J. Am. Chem. Soc.* **70,** 3695 (1948); W. V. Smith and R. H. Ewart, *J. Chem. Phys.* **16,** 592 (1968); W. D. Harkins, *J. Polym. Sci.* **5,** 217 (1950).
14. F. Hrabak and Ya. Zachoval, *J. Polym. Sci.* **52,** 134 (1961); F. Hrabak, V. Hynkova, and M. Bezdec, *J. Polym. Sci, A1* **7,** 111 (1969); F. Hrabak and co-workers, *J. Polymer Sci., Part C* **16,** 1345 (1967); F. Hrabak and M. Bezdec, *Collect. Czech. Chem. Commun.* **33,** 278 (1968).
15. K. Itoyama and co-workers, *Polym. J. (Tokyo)* **23,** 859 (1991).
16. F. Hrabak and M. Bezdec, *Makromol. Chem.* **115,** 43–55 (1968); M. Bezdec and co-workers, *Makromol. Chem.* **147,** 1 (1971).
17. O. F. Kodenko and co-workers, *Tr. Mosk. Inst. Tonkoi Khim. Technol.* **5**(2), 99 (1975); **5**(1), 78 (1975).
18. W. E. Mochel, *J. Polym. Sci.* **8,** 583 (1949).
19. E. E. Drott and R. A. Mendelson, *J. Polym. Sci., A-2* **8,** 1361 (1970); G. Kraus and C. J. Stacy, *J. Polym. Sci., Polym. Phys. Ed.* **10,** 657 (1972); *J. Polym. Sci. Polym. Symp.* **43,** 329 (1973).
20. M. M. Coleman and R. E. Fuller, *J. Macromol. Sci., Phys.* **11**(3), 419 (1975).
21. Ger. Offen. DE 3,105,339 (Feb. 13, 1981), U. Eisele and co-workers (to Bayer AG).
22. R. C. Jordan and M. L. McConnell, *ACS Symposium Ser.* **138,** 107 (1980).
23. H. Lange, *Colloid Polym. Sci.* **264**(6), 488–493 (1986).
24. Eur. Pat. Appl. EP 53,319 (June 9, 1982), G. Aren and co-workers (to Bayer AG).
25. U.S. Pat. 2,481,044 (Aug. 15 1950), G. W. Scott (to E. I. du Pont de Nemours & Co., Inc.).
26. Eur. Pat. Appl. EP 336,824 (Mar. 31, 1989), F. Sauterey, P. Branlard, and P. Poullet (to Societe Distugil).
27. W. E. Mochel, *J. Am. Chem. Soc.* **71,** 1426 (1949).
28. Fr. Demande FP 2,556,730 (June 21, 1985), P. Branlard and F. Sauterey (to Distugil SA).
29. Y. Miyata and S. Matsunaga, *Polymer* **28,** 2233 (1987); Y. Miyata and M. Sawada, *Polymer* **29,** 1495 and 1683 (1988).

30. Ger. Offen. 2,755,074 (June 15, 1978), K. L. Miller (to E. I. du Pont de Nemours & Co., Inc.).

31. Ger. Offen. DE 3,344,065 (June 7, 1984), E. M. Banta and K. D. Fitzgerald (to Denka Chemical Corp.).

32. Ger. Offen DE 3,246,748 (June 20, 1984), R. Musch and co-workers (to Bayer AG).

33. Ger. Offen. 2,924,660 (Jan. 15 1981), R. Musch and co-workers (to Bayer AG).

34. Eur. Pat. Appl. EP 146,131 (June 26, 1985), N. Emura, T. Ariyoshi, and T. Kato (to Toyo Soda Mfg. Co., Ltd.).

35. Jpn. Kokai Tokkyo Koho JP 61,238,808 (Oct. 24, 1986), M. Kamezawa, T. Ariyoshi, and Y. Sakanaka (to Toyo Soda Mfg. Co., Ltd.).

36. J. Brandrup and E. H. Immergut, eds., *Polymer Handbook,* 3rd ed., John Wiley & Sons, Inc., New York, 1989.

37. J. F. Smith, *Adhesives Age* **13,** 21 (1970).

38. M. M. Coleman, R. J. Petcavich, and P. C. Painter, *Polymer* **19,** 1243 (1978).

39. Jpn. Pat. Appl. 60 65,011 (Apr. 13, 1985), (to Toyo Soda Mfg. Co., Ltd.); Jpn. Kokai Tokkyo Koho JP 01 185,309 (July 24, 1989), Y. Masuka and M. Akimoto (to Denki Kagaku Kogyo).

40. Jpn. Kokai Tokkyo Koho 81 20,010 (Feb. 25, 1981), (to Denki Kagaku Kogyo).

41. C. R. Cuervo and A. J. Maldonado, *Solution Adhesives Based on Graft Polymers of Neoprene and Methyl Methacrylate,* Du Pont Informal Bulletin, Wilmington, Del., Oct. 1984; K. Itoyama, M. Dohi, and K. Ichikawa, *Nippon Setchaku Kyokaishi* **20,** 268 (1984).

42. K. S. V. Srinivasan, N. Radhakrishnan, and M. K. Pillai, *J. Appl. Polym. Sci.* **37,** 1551 (1989).

43. Jpn. Kokai Tokkyo Koho 01 284,544 (Nov. 15, 1989), S. Takayoshi, Y. Denda, and K. Ichikawa (to Denki Kagaku Kogyo).

44. Jpn. Kokai Tokkyo Koho JP 59 056,440 (Mar. 31, 1984), (to Toyo Soda Mfg. Co., Ltd.).

45. Jpn. Kokai Tokkyo Koho JP 61 118,440 (June 5, 1986), Kato and co-workers, (to Toyo Soda Co., Ltd.).

46. Eur. Pat. Appl. EP 421,149 (Apr. 10, 1991) S. Ozoe and H. Yamakawa (to Tosoh Corp.).

47. R. C. Ferguson, *J. Polym. Sci.* **A2,** 4735 (1964).

48. M. M. Coleman and co-workers, *J. Polym. Sci. Polym. Letters Ed.* **12,** 577 (1974); D. L. Tabb, J. L. Koenig, and M. M. Coleman, *J. Polym. Sci. Phys. Ed* **13,** 1145 (1975).

49. M. M. Coleman, P. C. Painter, and J. L. Koenig, *J. Raman Spect.* **5,** 417 (1976); M. M. Coleman, R. J. Petcavich, and P. C. Painter, *Polymer* **19,** 1243 (1978).

50. R. Petiaud and Q. R. Pham, *J. Polym. Sci. Polym. Ed.* **23** 1333–1334 (1985).

51. M. M. Coleman, D. L. Tabb, and E. G. Brame, *Rubber Chem. Technol.* **51,** 49 (1978); M. M. Coleman and E. G. Brame, *Rubber Chem. Technol.* **51,** 668 (1978).

52. T. A. Babushkina, L. N. Gvozdeva, and G. K. Semin, *J. Mol. Struct.* **73,** 215 (1981).

53. T. A. Babushkina and co-workers, *Vysokomol. Soedin., Ser. A* **23**(8), 1810 (1981).

54. R. J. Petcavich and M. M. Coleman, *J. Macromol. Sci.-Phys.* **B18,** 47–71 (1980).

55. J. R. Ebdon, *Polymer* **15,** 782 (1974); D. J. T. Hill, J. H. O'Donnell, and P. W. O'Sullivan, *Polymer* **25,** 569 (1984); A. A. Kahn and E. G. Brame, Jr., *Rubber Chem. Technol.* **50,** 272 (1977).

56. R. E. Cais and G. J. Stuk, *Macromolecules* **13,** 415 (1980).

57. R. R. Garrett, C. A. Hargeaves II, and D. N. Robinson, *J. Macromol. Sci. Chem.* **A4,** 1679 (1970); T. Okada and T. Ikushige, *J. Polym. Sci., Polym. Chem. Ed.,* 2059 (1976).

58. C. A. Aufdermarsh, *J. Org. Chem.* **29,** 194 (1964); C. A. Aufdermarsh and R. Pariser, *J. Polym. Sci.* **A2,** 4727 (1964); R. C. Ferguson, *J. Polym. Sci.* **A2,** 4735 (1964).

59. N. G. Gaylord and co-workers, *J. Polym. Sci.* **A2,** 3969 (1966).

60. N. G. Gaylord and co-workers, *J. Polym. Sci. A-1* **4,** 2493 (1966).

61. N. G. Gaylord and co-workers, *J. Am. Chem. Soc.* **85**, 641 (1963).
62. U.S. Pat. 3,004,011 (Feb. 5, 1958), H. L. Jackson (to E. I. du Pont de Nemours & Co., Inc.); U.S. Pat. 3,004,012 (Feb. 5, 1958), K. L. Seligman and H. L. Jackson (to E. I. du Pont de Nemours & Co., Inc.).
63. N. G. Gaylord and B. K. Patnik, *J. Polym. Sci.* **13**, 837 (1975); A. Masaki, M. Yasui, and I. Yamashita, *J. Macromol. Sci.-Chem.* **A6**, 1285 (1972).
64. T. Okada and M. Oysuru, *J. Appl. Polym. Sci.* **23**, 2215 (1979).
65. K. Irako and co-workers, *Nippon Kagaku Kaishi*, (4), 670 (1976); K. Irako, H. Koyama, and A. Shisra, *Kogyo Kagaku Zasshi* **74**(10), 2210 (1971).
66. X. Han, F. Guo, and D. Wang, *Yingyong Huaxue* **1**(3), 41 (1984).
67. X. Han and E. Chen, *Gaofenzi Tongxum*, (6), 435 (1985).
68. J. P. Kennedy and S. C. Guhaniyogi, *J. Macromol. Sci., Chem.*, **A18**(1), 103 (1982).
69. L. J. Op, P. J. Hwan, and B. H On, *Hwahak Kwa Hwahak Kongop* **20**(6), 299 (1977).
70. H. K. Banock, R. S. Lehrle, and J. C. Robb, *J. Polym. Sci., Part C* **4**, 1165 (1964).
71. G. H. Miller, G. P. Chock, and E. P. Chock, *J. Polym. Sci., Part A* **3**, 3353 (1965).
72. A. N. Pravednikov and S. S. Medvedev, *Dokl. Akad. Nauk., USSR* **109**, 579 (1956).
73. M. F. Margantova and M. P. Zverev, *Ukr. Khim. Zh.* **23**, 734 (1957).
74. U.S. Pat. 2,942,037 and 2,942,038 (June 21, 1960), P. A. Jenkins (to Distillers Co., Ltd.).
75. U.S. Pat. 2,770,657 (Nov. 13, 1956), J. R. Hively (to E. I. du Pont de Nemours & Co., Inc.).
76. U.S. Pat. 3,175,012 (Mar. 23, 1965), G. P. Colbert (to E. I. du Pont de Nemours & Co., Inc.).
77. U.S. Pat. 2,264,173 (Nov. 25, 1941), A. M. Collins (to E. I. du Pont de Nemours & Co., Inc.).
78. C. A. Hargreaves II in J. P. Kennedy and E. G. Tornqvist, eds., *Polymer Chemistry of Synthetic Elastomers,* Wiley-Interscience, New York, 1986, pp. 227–252.
79. R. S. Barrows and G. W. Scott, *Ind. Eng. Chem.* **40**, 2193 (1948).
80. M. A. Youker, *Chem. Eng. Prog.* **41**, 391 (1947).
81. U.S. Pat. 2,187,146 (Jan. 16, 1940), W. S. Calcott and H. W. Starkweather (to E. I. du Pont de Nemours & Co., Inc.).
82. U.S. Pat. 2,831,842 (Apr. 22, 1958), C. E. Aho (to E. I. DuPont de Nemours & Co., Inc.).
83. Ger. Offen. 3,002,711 (July 30, 1981), R. Musch and co-workers (to Bayer AG); Ger. Offen. DE 3,605,331 (Aug. 20, 1987), R. Musch and co-workers, (to Bayer AG).
84. U.S. Pat. 2,914,497 (Nov. 24, 1959), W. J. Keller (to E. I. du Pont de Nemours & Co., Inc.).
85. Eur. Pat. Appl. EP 426,023 (May 8, 1991) F. Y. Kafka, A. R. Bice, and D. K. Burchett (to E. I. du Pont de Nemours & Co., Inc.).
86. Jpn. Kokai Tokkyo Koho JP 58,180,501 (Oct. 22, 1983), (to Denki Kagaku Kogyo); Jpn. Kokai Tokkyo Koho JP 58,189,202 (Nov. 4, 1983), (to Denki Kagaku Kogyo).
87. *Toyo Soda Kenkya Hokoku* **26**(2), 93 (1982).
88. *Toyo Soda Kenkyu Hokoku* **24**, 83 (1979).
89. L. M. White and co-workers, *Ind. Eng. Chem.* **37**, 770 (1943).
90. P. K. Freakley and A. R. Payne, *Theory and Practice of Engineering with Rubber,* Applied Science Publishers, Ltd., London, 1978.
91. J. T. Maynard and W. E. Mochel, *J. Polym. Sci* **13**, 242 (1954).
92. G. I. Tsereteli, I. V. Sochava, and A. Buka, *Vesta Leningr. Univ., Fiz. Khim.* **1975**, 67 (1975).
93. C. W. Bunn, *Proc. R. Soc. London, Ser. A.* **180**, 40 (1942).
94. M. M. Coleman, R. J. Petcavich, and P. C. Painter, *Polymer* **19**, 1253 (1978).

95. B. Y. Teitelbaum and N. P. Anoshina, *Vysokol. Soedin.* **7**, 978 (1965); B. Y. Teitelbaum and co-workers, *Dokl. USSR Phys. Chem. Sec., (English)* **150**, 463 (1963).
96. M. Hanok and I. N. Cooperman, *Proc. of the Int. Rubber Conference, Preprints Paper,* Washington, D.C., 1952, p. 582.
97. F. Riva, A. Forte, and C. D. Monica, *Colloid Polym Sci.* **259**, 606 (1981); C. Della Monica and co-workers, *Conv. Ital. Sci. Macromol.,* [*Atti*] **2**, (6) 179 (1983); A. Forte and co-workers, *Conv. Ital. Sci. Macromol.,* [*Atti*], (5) 381.
98. W. R. Krigbaum and R. J. Roe, *J. Polym. Sci., Part A,* **2**, 4391 (1964); W. R. Krigbaum, Y. I. Bata, and G. H. Via, *Polymer* **7**, 61 (1966); A. K. Bhowmick and A. K. Gent, *Rubber Chem. Technol* **56**, 845 (1983).
99. R. M. Murray and J. D. Detenber, *Rubber Chem. Technol* **34**, 668 (1961).
100. E. Rhode, H. Bechon, and M. Mezger, *Scandinavian Rubber Conference, SCR/89,* Tampere, Finland, 1989.
101. O. F. Belyaev, V. Z. Aloev, and Yu. V. Zelenev, *Vysokomol. Soedin. Ser. A* **30**, 2382 (1988); O. F. Belyaev, V. Z. Aloev, and Yu. V. Zelenev, *Acta Polym.* **39**, 590 (1988).
102. H. W. Siesler, *Makromol. Chem., Rapid Commun.* **6**(10), 699 (1985).
103. A. M. Mashuryan, K. A. Gasparyan, and G. T. Ovanesov, *Vysokomol. Soedin., Ser. A,* **27**, 1660 (1985); G. T. Ovanesov and co-workers, *Vysokomol. Soedin., Ser. A* **28**, 1052 (1986).
104. Y. Miyata and M. Atsumi, *J. Polym. Sci., Part A* **26**, 2561 (1988).
105. D. L. Gardner and I. C. McNeill, *Eur. Polym. J.* **7**, 569, 593, and 603 (1971).
106. R. J. Petcavich, P. C. Painter, and M. M. Coleman, *Polymer* **19**, 1249 (1978).
107. V. S. Davtyan and co-workers *Arm. Khin. Zh.* **42**, 7 (1989).
108. K. Itoyama and S. Mastunaga, *Preprints of the ACS 122nd Rubber Division Meeting,* Chicago, Oct. 4, 1982, American Chemical Society, Washington, D.C.
109. H. C. Bailey, *Rev. Gen. Caoutch. Plast.* **44**, 1495 (1971).
110. U.S. Pat. 3,988,506 (Oct. 26, 1976), M. Dohi, T. Sumida, and K. Yokobori (to Denki Kagaku Kabushiki); Eur. Pat. Appl. EP 223,149 (May 27, 1987), T. Takeshita (to E. I. du Pont de Nemours & Co., Inc.); Eur. Pat. Appl. EP 175,245 (Mar. 26, 1986), T. Takeshita (to E. I. du Pont de Nemours & Co., Inc.).
111. K. L. Seligman and P. A. Roussel, *Rubber Chem. and Technol.* **34**, 869 (1961).
112. T. E. Schroer, *Approaches to Improved Heat Resistance for Dynamic Applications,* Du Pont Elastomers Bulletin C-NP-510.0584, E. I. du Pont de Nemours & Co., Inc., Wilmington, Del., 1984.
113. *The Synthetic Rubber Manual,* 11th ed., Int. Institute of Synthetic Rubber Producers, Inc. Houston, Tex., 1989.
114. W. Schmidt, *Curing Systems for Neoprene,* Du Pont Bulletin NP-330.1, 1982; J. C. Bament, *Neoprene Compounding and Processing Guide Plus Formulary,* E. I. du Pont de Nemours & Co., Inc., 1987; R. M. Murray and D. C. Thompson, *The Neoprenes,* E. I. du Pont de Nemours & Co., Inc., Wilmington, Del., 1963.
115. N. Kawasaki and T. Hashimoto, *J. Polym. Sci., Polym. Chem. Ed.* **11**, 671 (1973).
116. Y. Miyata and M. Atsumi, *Rubber Chem. Technol.* **62**, 1 (1989).
117. R. Pariser, *Kunstst.* **50**, 623 (1960).
118. H. Kato and H. Fujita, *Rubber Chem. Technol* **55**, 949 (1981); Jpn. Kokai Tokkyo Koho JP 02,28,532 (Dec. 10, 90) N. Yamamoto and co-workers (to Ouchi Shinko Chem. Ind. Co., Ltd.); N. Yamamoto and co-workers, *Nippon Gomu Kyokaishi* **64**, 48 (1991).
119. W. Schunk and co-workers, *Gummi, Fasern, Kunstst.* **43**, 617 (1990); F. M. Helay and co-workers, *Elastomerics,* **121**(9), 26 (1989).
120. Y. Sakuramoto and co-workers, *Nippon Gomu Kyokaishi* **56**,(4), 218 (1983).
121. U. Eholzer and T. Kempermann, *Rubber Plast News,* 49 (May 25, 1981); W. Warrach and Don Tsou, *Elastomerics,* **116**, 26 (1984).

122. Jpn. Kokai Tokkyo Koho JP 58,091,737 (May 31, 1983), (Ouchi Shinko Chemical Industrial Co., Ltd.).
123. R. M. Murray and D. C. Thompson, *The Neoprenes,* E. I. du Pont de Nemours & Co., Inc., Wilmington, Del., 1963.
124. D. C. H. Brown and J. Thompson, *Rubber World* **185**(2), 32–35 (1981).
125. U.S. Pat. 4,829,115 (May 9, 1989) K. S. Cottman (to Goodyear Tire and Rubber Co.); Jpn. Kokai Tokkyo Koho JP 62,205,142 (Sept 9, 1987), (to NOK Corp.).
126. Fr. Demande FR 2,459,266 (Jan 9, 1981) J. E. Vostovich (to General Electric Co.).
127. M. S. Al-Mehdawe and J. E. Stucky, *Rubber Chem. Technol.* **62**, 13 (1989); M. S. Al-Mehdawe and co-workers, *J. Pet. Chem. Res.* **7**(2), 99 (1988).
128. Jpn. Kokai Tokkyo Koho, JP 56,125,440 (Mar. 10, 1980), (to Toyoda Gosei Co., Ltd.).
129. E. A. Abdel-Razik, *J. Polym. Sci., Part A: Polym. Chem.* **26**, 2359 (1988).
130. R. L. Clough and K. T. Gillen, *Polym. Degrad. Stab.* **30**, 309 (1990); M. Ito, S. Ohada, and I. Kuriyama, *J. Mater. Sci.* **16**, 10 (1981).
131. J. C. Bament, *Neoprene Compounding and Processing Guide Plus Formulary,* E. I. du Pont de Nemours & Co., Inc., Wilmington, Del., 1978.
132. J. W. Graham, *Selecting a Plasticizer,* Du Pont Bulletin NP-320.1, E. I. du Pont de Nemours & Co., Inc., Wilmington, Del., 1981.
133. Jpn. Kokai Tokkyo Koho, JP 56,106,934 (Jan. 25, 1981), (to Dainichi Nippon Cables, Ltd.); Jpn. Kokai Tokkyo Koho, JP 62,267,394) (Nov. 20, 1987), (to Furukawa Electric Co., Ltd.).
134. *UK Res. Discl.* **211**, 403 (Du Pont (UK) Ltd.).
135. E. Ronde, H. Bechen, and M. Metzger, *Kautsch. Gummi, Kunstst.* **42**, 1121 (1989); E. Ronde. H. Bechen, and M. Mezger, *Polychloroprene Grades and Compounding for Long Term Flexibility at Low Environmental Temperatures,* Scandinavian Rubber Conference SCR/1989, Tampere, Finland, Jan. 30, 1989.
136. S. W. Schmidt, *Extrusion of Neoprene,* Du Pont Elastomers Bulletin NP-430.1, E. I. du Pont de Nemours & Co., Inc., Wilmington, Del., 1987.
137. S. W. Schmidt, *Calendering Compounds of Neoprene,* Du Pont Elastomers Bulletin NP-440.1, E. I. du Pont de Nemours & Co., Inc., Wilmington, Del., 1987.
138. S. W. Schmidt, *Molding Neoprene,* Du Pont Elastomers Bulletin NP-450.1, E. I. du Pont de Nemours & Co., Inc., Wilmington, Del., 1987.
139. J. W. Graham, *Neoprene: An Overview of Properties, Mixing and Compounding,* Du Pont Elastomers Informal Bulletin, C-NP-050.044, E. I. du Pont de Nemours & Co., Inc., Wilmington, Del., 1982.
140. *Correlator for Baypren Polychloroprene,* Miles Inc. Technical Literature, 1992.
141. S. W. Schmidt, *The W Family,* Du Pont Technical Bulletin NP-230.1, 1977.
142. Ger. Pat. 1,186,215 (Jan. 28, 1965), K. L. Miller (to E. I. du Pont de Nemours & Co., Inc.); (U.S. Pat. 3,655,827 (Apr. 11, 1972), J. B. Finlay (to E. I. du Pont de Nemours & Co., Inc.).
143. R. Musch and U. Eisle, *Preprints of the ACS Rubber Division Meeting,* Detroit, 1989, The American Chemical Society, Washington, D.C., 1989.
144. Y. Miyasita, S. Matsunaga, and M. Mitani, *Int. Rubber Conference Kyoto,* 16A-12.217, 1985.
145. Ger. Offen. DE 2,755,047 (June 15, 1978), K. L. Miller (to E. I. du Pont de Nemours & Co., Inc.)
146. G. Herzog and E. Rohde, *Advantages of New Baypren Grades,* Kaukuk Deregi, 1st Rubber Fair, Istanbul, June 7–11, 1989.
147. S. W. Schmidt and J. P. Dowd II, *The Neoprene G Family,* Du Pont technical bulletin NP-220, E. I. du Pont de Nemours & Co., Inc., Wilmington, Del., 1983.
148. U.S. Pat. 3,042,652 (July 3, 1962) R. Pariser and R. D. Soufie (to E. I. du Pont de Nemours & Co., Inc.).

149. S. W. Schmidt, *The Neoprene T Family,* Du Pont technical bulletin NP-240.1, E. I. du Pont de Nemours & Co., Inc., Wilmington, Del., 1977.

150. P. Mueller, U. Eisele, and G. Pampus, *Angew. Makromol. Chem.* **98**, 97 (1981).

151. U.S. Pat. 3,655,827 (Feb. 28, 1969), J. B. Finlay and J. F. Hagman (to E. I. du Pont de Nemours & Co., Inc.).

152. Eur. Pat. Appl. EP 65,718 (Dec 1, 1981), R. Musch and co-workers, (to Bayer AG); Ger. Offen. DE 3,105,339 (Sept. 2, 1982) U. Eisele and co-workers (Bayer AG).

153. Jpn. Kokai 73 102,149 (Dec. 22, 1973), T. Kadowaki, K. Takahashi, and M. Dohi (to Denki Kagaku Kogyo).

154. K. Sandow and H. Murato, *Nippon Gomu Kyokaishi* **63**, 331–340 (1990).

155. S. K. Guggenberger in I. Skiest ed., *Handbook of Adhesives,* 3rd ed. Van Nostrand Rheinhold Co., Inc., New York, 1990.

156. A. J. Maldonado, *Adhesives Based on Neoprene,* Du Pont elastomers informal bulletin C-ADH-101.0584, 1984.

157. P. Branlard, F. Sauterey, and P. Poullet, *New Chloroprene Rubber Types for Solvent-Based Adhesives,* Butaclor Symposium, Societe Distugil, Paris, 1988.

158. D. G. Coe, *Neoprene Latex Based Adhesives,* du Pont de Nemours Int. SA Geneva, 1991.

159. C. M. Matulewicz and A. M. Snow, Jr., *Adhes. Age* **24**(3), 40 (1980); *Neoprene Latex Pressure Sensitive Adhesives,* Tape Council June 16, 1980, E. I. du Pont de Nemours & Co., Inc. 1980.

160. U.S. Pat. 4,342,843 (Aug. 3, 1982), W. Perlinski, I. J. Davis, and J. F. Romanick (to National Starch and Chemical Corp.).

161. C. H. Gelbert, *Selection Guide for Neoprene Latexes,* Du Pont elastomers bulletin NL-020.1, 1985; J. C. Carl, *Neoprene Latex,* E. I. du Pont de Nemours & Co., Inc., Wilmington, Del, 1962.

162. Ger. Offen. DE 2,944,152 (May 14, 1981), W. Nolte, W. Keller, and H. Esser (to Bayer AG).

163. Ger. Offen. DE 3,111,138 (Sept. 30, 1982), W. Nolte and H. Esser (to Bayer AG).

164. Ger. Offen. 2,047,449 (Apr. 8, 1971), A. M. Snow, Jr. (to E. I. du Pont de Nemours & Co., Inc.); Ger. Offen. 2,047,450 (Apr. 1, 1971), A. M. Snow, Jr (to E. I. du Pont de Nemours & Co., Inc.).

165. F. L. McMillian, *Neoprene Latexes and Their Applications,* Du Pont elastomers informal bulletin, E. I. du Pont de Nemours & Co., Inc., Wilmington, Del, 1991.

166. C. H. Gelbert and H. E. Berkheimer, *Neoprene Latex and Its Applications with Emphasis on Manufacture of Dipped Goods,* Du Pont elastomers informal literature, E. I. du Pont de Nemours & Co., Inc., Wilmington, Del, 1987.

167. S. S. Tremelin, *Neoprene Modified Asphalt,* Du Pont elastomers bulletin C-NL-541.055, E. I. du Pont de Nemours & Co., Inc., Wilmington, Del, 1982.

168. Can. Pat. Appl. CA 2,020,179 (Dec. 31, 1990), L. A. Christel (to E. I. du Pont de Nemours & Co., Inc.).

169. C. W. Stewart, Sr., R. L. Dawson, and P. R. Johnson, *Effect of Compounding Variables on the Rate of Heat and Smoke Release from Polychloroprene Foam,* Du Pont elastomer bulletin C-NL-550.871, 1974.

170. J. H. Startton, *Quarterly Progress,* 67 (Jan. 1992).

171. *Toxicity and Safe Handling of Neoprene Latexes,* E. I. du Pont de Nemours & Co., Inc., *Bulletin NL-110.1,* 1984.

172. FDA Status of Du Pont Neoprene Solid Polymers and Latexes, bulletin NP-120.1, E. I. du Pont de Nemours & Co., Inc., Wilmington, Del., 1984.

173. U.S. Pat. 4,283,510 (Aug. 11, 1981) N. L. Turner (to Denka Chemical Corp.).

174. Fr. Demande FR 2,437,348 (Apr. 25, 1980), A. C. Adam (E. I. du Pont de Nemours & Co., Inc.).

175. F. J. Stark, Jr., *Polychloroprene as an Extender for Recycled Rubber, Preprints of the ACS Rubber Division Meeting,* 1983, The American Chemical Society, Washington, D.C., 1989.

General References

C. A. Stewart, Jr., T. Takeshita, and M. L. Coleman in J. I. Kroschwitz, ed., *Encyclopedia of Polymer Science and Engineering,* Vol. 3, 3rd ed., John Wiley & Sons, Inc., New York, 1985, pp. 441–462.
P. R. Johnson, *Rubber Chem Technol.* **49**, 650–702 (1976).

W. K. WITSIEPE
E. I. du Pont de Nemours & Co., Inc.

POLYETHERS

Polyether elastomers commercially available today are the result of key developments in the 1950s. It had been predicted that high mol wt polyethers would exhibit interesting physical properties, due to the presence of the oxygen atoms in the polymer backbone, which would impart more flexibility to the polymer chain (1). However, conventional catalysts had been effective only in synthesis of low mol wt polymers of epoxides. When these low mol wt polyethers are synthesized with hydroxyl end groups and the resulting polymer allowed to react with a diisocyanate, a polyether urethane product results (2). Since that time, numerous block copolymers incorporating polyethers with functionalized end groups have been synthesized. These compounds are beyond the scope of this article; herein only high mol wt, commercially important, polyether rubber compounds are discussed.

In 1957, it was discovered that organometallic catalysts gave high mol wt polymers from epoxides (3). The commercially important, largely amorphous polyether elastomers developed as a result of this early work are polyepichlorohydrin (ECH) (4,5), ECH–ethylene oxide (EO) copolymer (6), ECH–allyl glycidyl ether (AGE) copolymer (7,8), ECH–EO–AGE terpolymer (8), ECH–propylene oxide (PO)–AGE terpolymer (8,9), and PO–AGE copolymer (10,11). The American Society for Testing and Materials (ASTM) has designated these polymers as follows:

Polymer	ASTM designation
ECH	CO
ECH–EO	ECO
ECH–AGE	GCO
ECH–EO–AGE	GECO
ECH–PO–AGE	GPCO
PO–AGE	GPO

The catalysts for these compounds are based on aluminum (Nippon Zeon) or tin (Daiso) (12,13). The preferred catalysts are trialkylaluminum–water combinations used with or without a chelating agent such as acetylacetone. Except for minor variations, few changes in catalyst composition have been made since it was first formulated.

The early technology for the high mol wt polymers was developed by Hercules, Inc., and licensed to the B. F. Goodrich Co. which began marketing ECH homopolymer, ECH–EO copolymer, and ECH–EO–AGE terpolymer under the trade name Hydrin in 1965. Hercules began marketing similar elastomers in 1966 under the trade name Herclor. Hercules also produced PO–AGE copolymer, Parel 58. Two Japanese companies have been in the polyether elastomer business for many years; Daiso Co., Ltd., (formerly Osaka Soda Co., Ltd.) manufactures three chlorinated, amorphous polyethers (ECH, ECH–EO, and ECH–EO–AGE) under the trade name Epichlomer, and Nippon Zeon Co., Ltd., markets ECH–EO–AGE and ECH–AGE under the trade name Gechron. Nippon Zeon and B. F. Goodrich had been associated in an information exchange program since the early 1950s. Nippon Zeon has also begun producing amorphous ECH homopolymer and ECH–EO copolymer, selling these chlorinated elastomers under the Gechron trade name. The company also markets ECH–PO–AGE terpolymers under the Zeospan trademark.

Hercules and B. F. Goodrich are no longer in the polyether manufacturing business. In 1986, Hercules sold its polyether elastomer operation to B. F. Goodrich, which, after operating it for several years, in turn sold it and their whole specialty elastomer division to Zeon Chemicals USA, Inc., in 1989 (14). Zeon Chemicals USA, Inc., is a subsidiary of the Nippon Zeon Co., Ltd. At the present time, manufacture of polyethers is done by Zeon Chemicals in Hattiesburg, Mississippi, Nippon Zeon in Tokuyama, Japan, and Daiso in Mizushima, Japan. Total production is estimated to be 13,000 to 15,000 tons per year.

Structure

Epichlorohydrin Elastomers without AGE. ECH homopolymer, polyepichlorohydrin [24969-06-0] (**1**), and ECH–EO copolymer, poly(epichlorohydrin-*co*-ethylene oxide) [24969-10-6] (**2**), are linear and amorphous. Because it is unsymmetrical, ECH monomer can polymerize in the head-to-head, tail-to-tail, or head-to-tail fashion. The commercial polymer is 97–99% head-to-tail, and has been shown to be stereorandom and atactic (15–17). Only low degrees of crystallinity are present in commercial ECH homopolymers; the amorphous product is preferred.

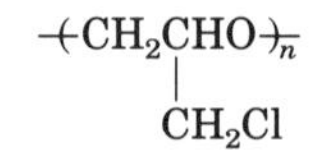

$$+CH_2CHO \xrightarrow{}_{n} \qquad\qquad +CH_2CHO \xrightarrow{}_{m} +CH_2CH_2O \xrightarrow{}_{n}$$
$$\underset{CH_2Cl}{|} \qquad\qquad\qquad \underset{CH_2Cl}{|}$$

$$(\mathbf{1}) \qquad\qquad\qquad\qquad (\mathbf{2})$$

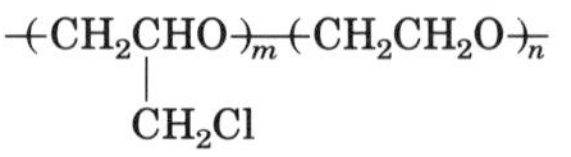

Crystallinity in ECH and ECH–EO finished products increases over time, and may be detected by x-ray analysis or differential scanning calorimetry. In synthesizing ECH–EO, the process is designed to maximize random monomer sequence and minimize crystallinity. The ECH–EO molecular ratio in these products ranges from approximately 3:1 to 1:1.

AGE-Containing Elastomers. ECH–AGE, poly(epichlorohydrin-*co*-allyl glycidyl ether) [*24969-09-3*] (**3**), ECH–EO–AGE, poly(epichlorohydrin-*co*-ethylene oxide-*co*-allyl glycidyl ether) [*26587-37-1*] (**4**), ECH–PO–AGE, and PO–AGE are also amorphous polymers.

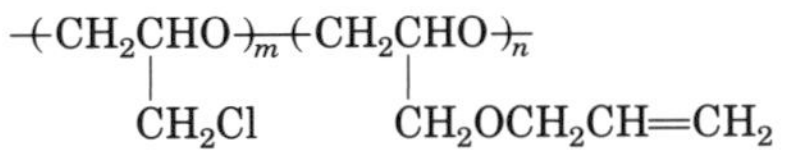

$$\text{+CH}_2\text{CHO+}_m\text{+CH}_2\text{CHO+}_n \qquad\qquad \text{+CH}_2\text{CHO+}_m\text{+CH}_2\text{CH}_2\text{O+}_n\text{+CH}_2\text{CHO+}_o$$
$$\text{CH}_2\text{Cl} \quad \text{CH}_2\text{OCH}_2\text{CH=CH}_2 \qquad \text{CH}_2\text{Cl} \qquad\qquad \text{CH}_2\text{OCH}_2\text{CH=CH}_2$$

(3) 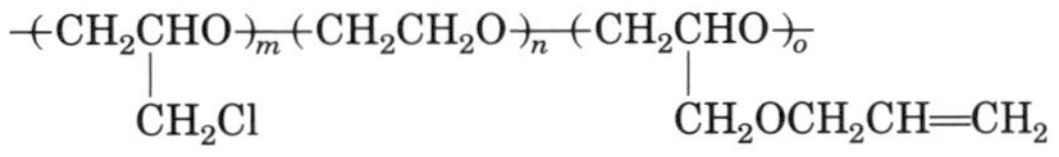   (4)

Crystallinity is low; the pendent allyl group contributes to the amorphous state of these polymers. Propylene oxide homopolymer itself has not been developed commercially because it cannot be cross-linked by current methods (18). The copolymerization of PO with unsaturated epoxide monomers gives vulcanizable products (19,20). In ECH–PO–AGE, poly(propylene oxide-*co*-epichlorohydrin-*co*-allyl glycidyl ether) [*25213-15-4*] (**5**), and PO–AGE, poly(propylene oxide-*co*-allyl glycidyl ether) [*25104-27-2*] (**6**), the molar composition of PO ranges from approximately 65 to 90%.

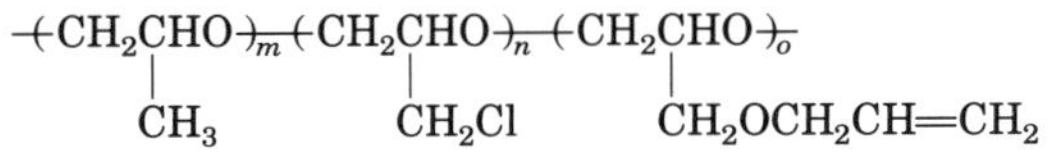

$$\text{+CH}_2\text{CHO+}_m\text{+CH}_2\text{CHO+}_n\text{+CH}_2\text{CHO+}_o \qquad\qquad \text{+CH}_2\text{CHO+}_m\text{+CH}_2\text{CHO+}_n$$
$$\text{CH}_3 \quad\ \text{CH}_2\text{Cl} \quad\ \text{CH}_2\text{OCH}_2\text{CH=CH}_2 \qquad\qquad \text{CH}_3 \quad\ \text{CH}_2\text{OCH}_2\text{CH=CH}_2$$

(5) 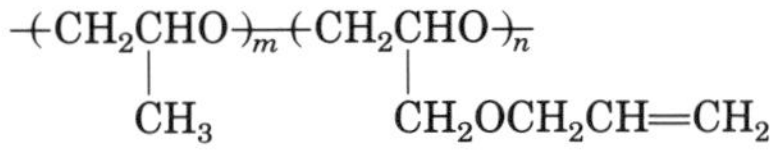   (6)

Properties

Properties of the uncompounded elastomers are listed in Table 1 and properties of the compounded polymers in Table 2.

Epichlorohydrin Elastomers without AGE. The ECH homopolymer has very low gas permeability (two to three times greater than that of butyl rubber),

Table 1. Commercial Polyether Elastomers

Elastomer	ECH, %	Chlorine, %	Ethylene oxide, %	CAS Registry Number	Specific gravity	ML[a]	T_g, °C
ECH	100	38	0	[*24969-06-0*]	1.36	40–80	−22
ECH–EO	68	26	32	[*24969-10-6*]	1.27	40–130	−40
ECH–AGE	92	35	0	[*24969-09-3*]	1.24	60	−25
ECH–EO–AGE	48–70	24–29	20–50	[*26587-37-1*]	1.27	50–100	−38
ECH–PO–AGE	40	15	0	[*25213-15-4*]	1.12	60–80	−48
PO–AGE	0	0	0	[*25104-27-2*]	1.01	[b]	−62

[a]Mooney viscosity at 100°C, ASTM 1646.
[b]Oscillating disk rheometer (ODR) viscosity is 21–26.

Table 2. Vulcanizate Properties of Polyether Elastomers

Properties	CO	ECO	GCO	GECO	GPCO	GPO
Mooney viscosity at 100°C						
polymer	78	80	62	80	78	72
compound	73	93	60	86	75	83
originals cured at 170°C						
100% modulus, MPa[a]	7.6	5.4	7.2	5.5	5.6	3.4
tensile strength, MPa[a]	16.5	15.0	15.5	15.1	11.4	10.2
elongation, %	225	275	205	280	210	365
hardness, Shore A	74	74	74	73	77	68
air oven aged 70 h at 125°C						
100% modulus, MPa[a]	7.6	5.9	7.5	5.7	0	6.2
tensile strength, MPa[a]	15.3	14.4	13.6	14.7	11.8	10.6
elongation, %	195	240	165	250	80	180
hardness, Shore A	75	72	74	73	90	76
compression set						
70 h at 100°C	13	19	18	19	39	39
70 h at 150°C	58	65	62	66	100	66
brittleness, temp[b] °C	−20	−40	−24	−40	−46	−61
tear strength, kN/m[c]	32	33	26	35	46	51
ozone resistance, 100 ppm at 49°C, h	>168	>168	>168	>168	>168	>168
volume change 70 h, %						
ASTM fuel A at std[d] T	0	0	0	0	12	48
ASTM fuel B at std T	20	17	22	18	65	134
ASTM fuel C at std T	34	32	38	34	106	175
ASTM oil #1 at 125°C	−5	−4	−5	−5	0	11
ASTM oil #1 at 150°C	−4	−5	−4	−5	2	16
ASTM oil #3 at 125°C	5	5	6	5	40	92
ASTM oil #3 at 150°C	6	6	7	6	56	102
water at 23°C	1	6	1	6	3	4
water at 100°C	8	2	10	14	14	10

[a]To convert MPa to psi, multiply by 145.
[b]ASTM 2137.
[c]To convert kN/m to lb/in., multiply by 5.71.
[d]Std T is 23°C.

outstanding ozone resistance, good building tack, and low heat buildup. The polymer is flame retardant as a result of its high chlorine content. It has poor resilience at room temperature, but this improves upon heating. The ECH–EO copolymer is less flame retardant due to its lower chlorine content. It has some impermeability to gases similar to that of the medium high acrylonitrile rubbers. It has low temperature flexibility to −40°C and exhibits good heat resistance. In contrast with ECH homopolymer, ECH–EO has poor tack.

Vulcanizates of ECH homopolymer and ECH–EO copolymer are resistant to ASTM oils, aliphatic solvents, and aromatic-containing fuels, showing low swell after exposure. The polymers do not harden after exposure to these fluids, although plasticizer may be extracted. Overall, these polymers offer a good balance of heat, ozone, and fuel resistance over a broad temperature range.

AGE-Containing Elastomers. ECH–AGE copolymer shows excellent ozone resistance and good resistance to softening upon heat aging. The ECH–EO–AGE terpolymers display good ozone resistance with ozone resistance increasing with increasing AGE content. Resistance to softening upon heat aging also increases with increasing AGE content. Like ECH and ECH–EO, vulcanizates of ECH–AGE and ECH–EO–AGE are resistant to ASTM oils, aliphatic solvents, and aromatic-containing fuels. As AGE content of the ECH–EO–AGE terpolymer increases, tensile strength and moduli decrease. Compounding with Group IIA oxides or hydroxides gives the vulcanizates of this terpolymer nitrogen oxide resistance (21). ECH, ECH–EO, ECH–AGE, and ECH–EO–AGE elastomers may be blended; the properties of the blended product are proportional to the amount of each polymer.

Two propylene oxide elastomers have been commercialized, PO–AGE and ECH–PO–AGE. These polymers show excellent low temperature flexibility and low gas permeability. After compounding, PO–AGE copolymer is highly resilient, and shows excellent flex life and flexibility at extremely low temperatures (ca −65°C). It is slightly better than natural rubber in these characteristics. Resistance to oil, fuels, and solvents is moderate to poor. Wear resistance is also poor. Unlike natural rubber, PO–AGE is ozone resistant and resistant to aging at high temperatures. The properties of compounded ECH–PO–AGE lie somewhere between those of ECH–EO copolymer and PO–AGE copolymer (22). As the ECH content of the terpolymer increases, fuel resistance increases while low temperature flexibility decreases. Heat resistance is similar to ECH–EO; fuel resistance is similar to polychloroprene. The uncured rubber is soluble in aromatic solvents and ketones.

Molecular Weight Determination and Solution Behavior. Molecular weight determinations using dilute solution viscosity measurements have been reported for ECH (23,24). Intrinsic viscosity is related to molecular weight by the Mark-Houwink equation: $[\eta] = KM^a$, where K and a are measured experimentally. The molecular weight is a viscosity average molecular weight, but with the use of the required correction factors, the equation may be used to obtain the number average molecular weight, $\overline{M}_n$, and the weight average molecular weight, $\overline{M}_w$, (25). Constants have been determined for amorphous ECH at 100°C in α-chloronaphthalene, and $\overline{M}_w$ may be determined from this relationship: $[\eta] = 8.93 \times 10^{-5}$ $\overline{M}_w^{0.73}$ (23). A later study of this polymer in α-chloronaphthalene at 25°C gave values of $a = 0.71$ and $K = 8.23 \times 10^{-5}$ dL/g; in tetrahydrofuran at 25°C, $a = 0.80$ and $K = 3.35 \times 10^{-5}$ dL/g (26). When low angle laser light scattering photometry (lalls) is used in conjunction with gel permeation chromatograph (gpc) techniques, M_i values in the eluate may be determined continuously (27,28). The values obtained using this technique agree with those obtained by dilute solution viscometry (24). Equations relating the sedimentation constant, S, and intrinsic viscosity, $[\eta]$, to molecular weight have been determined for dilute solutions of polyepichlorohydrin in benzene (29):

$$S = 0.145 \, M^{0.33} \text{ and } [\eta] = 0.155 \times 10^{-4} \, M^{0.89}$$

When viscometric measurements of ECH homopolymer fractions were obtained in benzene, the nonperturbed dimensions and the steric hindrance parameter were calculated (24). From experimental data collected on polymer solubility in 39 solvents and intrinsic viscosity measurements in 19 solvents, Hansen (30) model parameters, δ_d and δ_a, could be determined (24). The notation δ_d symbolizes the dispersion forces or nonpolar interactions; δ_a, a representation of the sum of δ_p (polar interactions); and δ_h (hydrogen bonding interactions). The homopolymer is soluble in solvents that have solubility parameters $\delta_d \geq 7.9$, $\delta_p \geq 5.5$, and $0.2 \leq \delta_h \leq 5.0$ (31). Solubility was also determined using a method (32) in which δ represents the solubility parameter and γ the hydrogen bonding parameter. When δ is between 8.6 and 9.8 and the squared hydrogen bonding parameter, γ^2, is between 0.05 and 0.90, the polymer is soluble (31).

Molecular weight determinations of ECH–EO, ECH–AGE, ECH–EO–AGE, ECH–PO–AGE, and PO–AGE have not been reported. Some solution studies have been done on poly(propylene oxide), and these may approximate solution behavior of the PO–AGE copolymer (33,34).

Polymer Preparation

Epichlorohydrin Elastomers without AGE. Polymerization on a commercial scale is done as either a solution or slurry process at 40–130°C in an aromatic, aliphatic, or ether solvent. Typical solvents are toluene, benzene, heptane, and diethyl ether. Trialkylaluminum–water and trialkylaluminum–water–acetylacetone catalysts are employed. A cationic, coordination mechanism is proposed for chain propagation. The product is isolated by steam coagulation. Polymerization is done as a continuous process in which the solvent, catalyst, and monomer are fed to a back-mixed reactor. Final product composition of ECH–EO is determined by careful control of the unreacted, or background, monomer in the reactor. In the manufacture of copolymers, the relative reactivity ratios must be considered. The reactivity ratio of EO to ECH has been estimated to be approximately 7 (35–37).

The molecular weight of the polymers is controlled by temperature (for the homopolymer), or by the addition of organic acid anhydrides and acid halides (37). Although most of the product is made in the first reactor, the background monomer continues to react in a second reactor which is placed in series with the first. When the reaction is complete, a hindered phenolic or metal antioxidant is added to improve shelf life and processibility. The catalyst is deactivated during steam coagulation, which also removes solvent and unreacted monomer. The crumbs of water-swollen product are dried and pressed into bale form. This is the only form in which the rubber is commercially available. The rubber may be converted into a latex form, but this has not found commercial application (38).

AGE-Containing Elastomers. The manufacturing process for ECH–AGE, ECH–EO–AGE, ECH–PO–AGE, and PO–AGE is similar to that described for the ECH and ECH–EO elastomers. Solution polymerization is carried out in aromatic solvents. Slurry systems have been reported for PO–AGE (39,40). When monomer reactivity ratios are compared, AGE (and PO) are approximately 1.5 times more reactive than ECH. Since ECH is slightly less reactive than PO and AGE and considerably less reactive than EO, background monomer concentration must be controlled in ECH–AGE, ECH–EO–AGE, and ECH–PO–AGE synthesis in order to obtain a uniform product of the desired monomer composition. This is not necessary for the PO–AGE elastomer, as a copolymer of the same composition as the monomer charge is produced. AGE content of all these polymers is fairly low, less than 10%. Methods of molecular weight control, antioxidant addition, and product work-up are similar to those used for the ECH polymers described.

Processing and Fabrication

All of the polyether elastomers, like other vulcanizable elastomers, can be compounded with processing aids, fillers, plasticizers, stabilizers, and vulcanizing agents to make useful rubber products. A typical compounding recipe for epichlorohydrin elastomer is as follows:

Ingredients	phr
polymer	100
carbon black	70
plasticizer	5
antioxidant	1
processing aid	1
acid acceptor	5
vulcanizing agent	1.5

More specific recipes appear in Table 3. The ingredients are added to the elastomers on standard two-roll mills or in internal mixers. Finished compounds are readily extruded, calendered, or molded in standard equipment. Vulcanization of extrudates is accomplished in live steam autoclaves, liquid salt baths, fluidized beds, and microwave equipment.

Table 3. Recipes for Table 2 Vulcanizates, phr[a]

Ingredient	CO	ECO	GCO	GECO	GPCO	GPO
N-762[b]	70	70	70	70	80	80
DOP[c]	5	5	5	5	5	5
stearic acid	1	1	1	1	1	1
NBC[d], 70%	1.4	1.4	1.4	1.4	1.4	1.4
Pb_3O_4, 90%	5.5	5.5	5.5	5.5		
zinc oxide					5	5
ethylenethiourea, 75%	1.5	1.5	1.5	1.5		
MBT[e]					1.5	1.5
TMTM[f]					1.5	1.5
sulfur					1.3	1.3
Total[a]	*184.4*	*184.4*	*184.4*	*184.4*	*196.7*	*196.7*

[a]Column headings indicate the elastomer at 100 phr.
[b]A carbon black designation (see CARBON, CARBON BLACK).
[c]Dioctyl phthalate.
[d]Nickel dibutyl dithiocarbamate.
[e]2-Mercapto benzothiazole.
[f]Tetramethylthiuram monosulfide.

For two-roll mixing, a roll temperature between 60–80°C is recommended. Lower temperatures increase the probability of roll sticking, whereas higher temperatures induce scorch and undesirable precure. The elastomer is banded on the mill with the addition of a processing aid. The fillers are added in increments followed by the remaining solid ingredients except the vulcanizing agent. Plasticizers are added after the solids, and the vulcanizing agent is added last with care being taken to ensure that the compound temperature does not rise high enough to cause scorch.

With internal mixers, the elastomer, processing aid, and one-third to one-half of the fillers are mixed, followed by the remainder of the fillers and the other solid ingredients except the vulcanizing agent. Plasticizers should then be added. This first pass mix may be mixed and dropped as high as 175°C to remove moisture. The vulcanizing agent should be added during a second pass with the temperature not exceeding 100°C.

The polyethers are shear sensitive and undergo molecular weight reduction during both mill mixing and internal mixing. This breakdown can be accelerated in the PO–AGE, ECH, and ECH–EO polymers by the addition of an organic peroxide, such as benzoyl peroxide. Since the other polyethers are cross-linkable with peroxides, this technique will not work. The thermoplasticity of the polyethers increases as the mixing temperature increases. This reduces shear and reduces the polymer breakdown. It is extremely important to control the amount of polymer breakdown from batch to batch to control the compound Mooney viscosity and eliminate flow-related problems during subsequent processing.

Compounding Ingredients.   Some common rubber additives can cause problems with the ECH and ECH–EO elastomers. Peroxides cause backbone scission. Acidic materials interfere with most cure systems and can increase polymer degradation during aging. Epoxidized materials, for example, plasticizers, interfere with lead–ethylene thiourea cure systems. Zinc-containing materials may form the Lewis acid zinc chloride, which causes backbone scission during aging of these elastomers as well as the other polyethers containing ECH.

Processing Aids.   Stearic acid [57-11-4] or other fatty acids and/or metal soaps of fatty acids are added to reduce shear degradation and mill sticking during mixing. Sorbitan monostearate (ICI's Span 60) is one of the best processing aids to reduce mill sticking.

During some molding and extrusion operations, knit line failures, incomplete mold fill, die drag, and excessive heat buildup, ie, scorch, are problems. Many of these problems are reduced or eliminated by the addition of internal lubricants such as low mol wt polyethylene or Vanfre AP-2 Special, a product of R. T. Vanderbilt.

Fillers.   The physical properties of unfilled polyether elastomer vulcanizates are typical of amorphous synthetic elastomers. Tensile strength and hardness tend to be low. Properties are improved by reinforcing fillers (qv) such as carbon blacks, aluminas, and silicas. Reinforcement also is typical. For example, high structure carbon blacks at low loading and silicas yield high tensile strength. Nonreinforcing fillers, calcium carbonate, talc, ground coal, and nonacidic clay are used.

Plasticizers.   Addition of plasticizers (qv) to polyether elastomers alters physical properties, improves processing, and can improve low temperature flexibility. Plasticizers also reduce vulcanizate costs by allowing the use of higher levels of less expensive fillers.

The polarity of the polyethers makes them incompatible with hydrocarbon-type plasticizers, which tend to bleed. Effective plasticizers are ethers such as di(butoxyethoxyethyl)formal [143-29-3] (Thiokol's TP-90B), esters such as di(2-ethylhexyl) phthalate [117-81-7] dioctyl phthalate (DOP), polyesters such as Paraplex G50 (Rohm and Haas), and ether–esters such as di(butoxyethoxyethyl) adipate [114-17-3] (Thiokol's TP-95). The lower mol wt plasticizers, DOP, TP-90B, and TP-95 improve vulcanizate low temperature performance. The polymeric plasticizers maintain higher temperature and long-term aging properties. Epoxidized plasticizers should be avoided because they interfere with vulcanization.

Stabilizers and Antioxidants. During the manufacture of the polymers, a hindered phenolic antioxidant is added for protection during storage. Additional antioxidants (qv), such as nickel dimethyldithiocarbamate [15521-65-0], nickel dibutyldithiocarbamate [13927-77-0] (NBC), or 2-mercaptobenzimidazole [149-30-4] are added during compounding to improve air aging and ozone resistance. The hindered phenolics, bis(3,5-di-*tert*-butyl-4-hydroxyhydrocinnamoyl)hydrazine, and calcium bis-*O*-ethyl(3,5-di-*tert*-butyl-4-hydroxybenzyl) phosphonate provide protection against sour (peroxidized) gasoline. These hindered phenolic materials are ineffective in combination with the nickel dithiocarbamates. Overall aging characteristics of compounds containing the hindered phenolic sour gasoline stabilizers are improved with the addition of AgeRite White.

The principal mechanism of polymer degradation during aging is the acid-catalyzed cleavage of the ether linkage in the backbone. The acid acceptor, a cure activator added as part of the cure system, protects against this mechanism of degradation (41). Some of the more common heat stabilizing acid acceptors are red and white lead oxides, calcium and magnesium oxides, and calcium and barium carbonates (see HEAT STABILIZERS).

Curing Systems. The most commonly used vulcanizing agent for the polyethers not containing AGE, that is, ECH and ECH–EO, is 2-mercaptoimidazoline, also called ethylenethiourea [96-45-7]. Other commercially applied curing agents include derivatives of 2,5-dimercapto-1,3,4-thiadiazole, trithiocyanuric acid and derivatives, bisphenols, diamines, and other substituted thioureas.

In addition to the vulcanizing agent, an acid scavenger that acts as an activator is required. For most of the curatives, a lead-containing compound, such as red lead oxide, litharge, lead phthalate, and lead phosphite, is used most commonly. However, lead-containing materials cause mold fouling during molding operations. This has led to the use of other activators, such as calcium oxide or hydroxide and magnesium oxide for the thiourea curatives. These, and weaker bases such as calcium carbonate and barium carbonate, are used with the dinucleophilic curatives, the dimercaptothiadiazoles and derivatives and trithiocyanuric acid and derivatives. The function of these activators during curing is to scavenge HCl to drive the cross-linking reaction and prevent the HCl from causing backbone cleavage of the polymer.

Although these curative systems may also be used with the polyepichlorohydrin elastomers containing AGE, the polymers were developed to be cured with conventional rubber curatives, sulfur, and peroxides. These polymers containing the pendent allyl group are readily cured with a typical sulfur cure system such as zinc oxide, and sulfur along with the activators, tetramethylthiuram monosulfide [97-74-5] (TMTM) and bis(2,2'-benzothiazolyl)disulfide [120-78-5] (MBTS). A typical peroxide cure system for these elastomers contains sodium or potassium stearate, a peroxide such as dicumyl peroxide, and an unsaturated coagent such as trimethylolpropane trimethacrylate.

Some of the terpolymers containing high levels of AGE give superior sour gasoline and ozone resistance, particularly dynamic ozone resistance. Since the unsaturation is not in the polymer backbone, it can be, and apparently is, sacrificed under sour gasoline or ozone aging. This protection scheme is limited with the peroxide and sulfur cure systems as they involve the allyl functionality of the polymer. The protection is maximized when a dinucleophilic curative, such as trithiocyanurate, is used.

There are no known practical peroxide cure systems for the PO–AGE polymers. Apparently the peroxide attacks the polymer backbone at a rate that is unfavorably competitive with the cross-linking rate. A typical sulfur cure system consists of zinc oxide [1314-13-2], tetramethylthiuram monosulfide (TMTM), 2-mercaptobenzothiazole [149-30-4] (MBT), and sulfur. A sulfur donor cure system is zinc oxide, di-o-tolylguanidine [97-39-2] (DOTG) and tetramethylthiuram hexasulfide.

Economic Aspects

Polyether elastomers are moderately costly when compared with other synthetic rubbers because of their specialty, small-volume nature (Table 4).

Table 4. 1992 Prices for Polyether Elastomers

Elastomer	Cost per kg, $
CO	5.50
ECO	5.40
GCO	5.50
GECO	5.80
GPCO	5.80
GPO	5.10

Health and Safety Factors; Toxicity

Monomers.   The monomers used for commercial production of these elastomers are suspected carcinogens. The International Agency for Research on Cancer (IARC) has rated ECH, EO, and PO as probable human carcinogens. Evidence for human carcinogenicity is inadequate for ECH and PO and limited for EO. However, evidence from animal studies is considered sufficient to identify these monomers as potential human carcinogens. The National Toxicology Program (NTP) has found some evidence for carcinogenesis in animal studies involving the inhalation of AGE (42). Studies in human subjects are inconclusive because workers exposed to only one chemical, such as ECH, are not available. Frequently, subjects have been exposed to many chemicals in the course of their employment, and physiological aberrations cannot be attributed to only one factor (43). These monomers, in addition to being potential human carcinogens, also produce acute toxic effects upon exposure. They are harmful when ingested or inhaled, or when skin is contacted. Vapors are extremely irritating to the respiratory tract. Burns result from skin exposure, and EO, ECH, and AGE may cause skin sensitization. AGE, ECH, and PO are absorbed through the skin and may cause systemic toxic effects.

The American Conference of Governmental Industrial Hygenists (ACGIH) has set threshold limit values (TLVs) for airborne concentrations in the workplace (ppm, time-weighted average, 8-h day) as follows: AGE, 5; ECH, 2; EO, 1; and PO, 20 (44). However, the ACGIH has reexamined data on ECH, and has proposed lowering the TLV to 0.1 ppm; it has categorized this monomer as Group A2, a suspected human carcinogen (45).

Polymers. Studies to determine possible exposure of workers to residual epichlorohydrin and ethylene oxide monomers in the polymers have been done. Tests of warehouse air where Hydrin H and Hydrin C are stored showed epichlorohydrin levels below 0.5 ppm. Air samples taken above laboratory mixing equipment (Banbury mixer and 6″ × 12″ mill) when compounds of Hydrin H or C were mixed gave epichlorohydrin levels below detectable limits, and ethylene oxide levels less than 0.2 ppm, well below permissible exposure limits (46). A subacute vapor inhalation toxicity study in which animals were exposed to emission products from compounded Parel 58 suggests that no significant health effects would be expected in workers periodically exposed to these vapors (47).

Compounding Ingredients. Ethylene thiourea (ETU), the most commonly used curing agent for epichlorohydrin elastomers, has been determined to be carcinogenic, teratogenic, and goitrogenic in animal studies. One study showed no evidence of cancer or birth defects in humans (48). A study on workers who mixed ETU into masterbatch rubber showed these subjects to have significantly lower levels of thyroxine than process workers or matched controls (49). To minimize respiratory and skin exposure, ETU is, and should be, used as a dispersion of the powdered ETU in a polymeric binder. Suppliers of dispersions of ETU include Polymerics, Inc., Cuyahoga Falls, Ohio; Rhein Chemie Corp., Trenton, New Jersey; and Synthetic Products Co., Cleveland, Ohio.

The common acid acceptors, red lead oxide and barium carbonate, are both toxic when inhaled or ingested. They are, and should be, used in industry as dispersions in EPDM and ECO. Suppliers of red lead oxide include Polymerics, Inc., Rhein Chemie Corp., and Akrochem Co., Akron, Ohio. Barium carbonate in an ECO binder is available from Rhein Chemie Corp. and Synthetic Products Co.

Uses

Epichlorohydrin Elastomers without AGE. Vulcanizates of ECH homopolymer and ECH–EO copolymer have outstanding ozone and gas permeability resistance. They also retain their flexibility at low temperatures and are fuel resistant. This combination of properties makes them important in automotive applications such as fuel, air, and vacuum hoses, vibration mounts, and adhesives. Homopolymer is used in adhesives because of its natural tack. Used as an additive, ECH–EO copolymer can also impart antistatic properties to plastics. Other industrial applications of these polymers include drive and conveyor belts; hoses, tubing, and diaphragms; pump parts including inner coatings, seals, and gaskets; printing rolls and blankets; fabric coatings for protective clothing; pond liners, and membranes in roofing material. In oil well drilling equipment, uses include drill-pipe protectors, packers, pipe scrubbers, and submersible power-cable jacketing.

AGE-Containing Elastomers. Modification by addition of a small amount of AGE to the two polymers mentioned above allows for sulfur curing while still maintaining the desirable properties of the original polymers. As ECH levels decrease, fuel resistance also decreases, but this does not present a problem as only small (<10 wt %) levels of AGE are used. ECH–EO–AGE vulcanizates offer excellent heat and ozone resistance, and also excellent compresion set (50). This rubber finds applications in constant velocity boots, mounting isolators, and hose and wire covers. The presence of propylene oxide imparts even greater low temperature flexibility and gas permeation resistance. Thus, ECH–PO–AGE terpolymer and PO–AGE copolymer have outstanding high and low temperature resistance and are used in high flex applications. Both elastomers have good ozone resistance. Although polychloroprene rubbers have good fuel resistance, they do not offer the low temperature flexibility, high temperature resistance, and ozone resistance of ECH–PO–AGE vulcanizates (51). Because of its ECH content, ECH–PO–AGE terpolymer still retains some fuel resistance, whereas the performance of PO–AGE in this area is only fair. Automotive applications of ECH–PO–AGE include dust and fuel hose covers and rubber boots for suspension and transmission systems. This polymer is also used in some automotive anti-vibration applications as well as covering for cable. PO–AGE has vibration damping properties similar to those of natural rubber, but superior heat resistance. It is used in automotive under-the-hood applications such as motor mounts and suspension bushings, where high temperatures preclude the use of natural rubber. Addition of PO–AGE to polycarbonate gives a product that is both impact and chemical resistant (52). When styrene is polymerized with PO–AGE, a product is obtained with better weather resistance than high impact polystyrene (53).

Epichlorohydrin Elastomer Derivatives

The principal route of chemical modification of epichlorohydrin elastomers is nucleophilic substitution on the pendent chloromethyl group. Reported nucleophilic substitution products include acetate, glycolate, hydrazine, α-pyrrolidonate (36), azide (54), phosphinyl (55), thiosulfate (56), α-mercaptoacetic acid, thiosalicylic acid (57), thioethers (58), dithiocarbamate (59), isothiouronium (60,61), imides (62), α,β-unsaturated carboxylic acids (63), cinnamate (64,65), carbazole (66), and methoxide (67). Preparations of quaternary ammonium salts have also been reported (36,62,68–77). Substitution reactions of the pendent chloromethyl group effected by the use of a phase-transfer catalyst have been reported (78). Applications of these modified polyethers include water thickeners (68,70,72,74), breaking emulsions (69), flocculating agents (68), drainage aids in paper manufacture (71), selectively permeable membranes (75,79), photosensitive material (64,65,80), flame retardants (36), and shrinkproofing wool (60).

Modifications of epichlorohydrin elastomers by radical-induced graft polymerization have been reported. Incorporated monomers include styrene and acrylonitrile, styrene, maleic anhydride, vinyl acetate, methyl methacrylate, and vinylidene chloride (81), acrylic acid (82), and vinyl chloride (81,83,84). When the vinyl chloride-modified epichlorohydrin polymers were used as additives to PVC, impact strength was improved (83,84).

BIBLIOGRAPHY

"Elastomers, Synthetic (Polyethers)" in *ECT* 3rd ed., Vol. 8, pp. 568–582, by E. J. Vandenberg, Hercules Inc.

1. C. C. Price, *Chemist* **38**, 131 (1961).
2. U.S. Pat. 2,866,774 (Dec. 30, 1958), C. C. Price (to University of Notre Dame).
3. E. J. Vandenberg, *J. Polym. Sci.* **47**, 486 (1960).
4. U.S. Pat. 3,158,580 (Nov. 24, 1964), E. J. Vandenberg (to Hercules, Inc.).
5. E. J. Vandenberg, *Rubber Plast. Age* **46**, 1139 (1965).
6. U.S. Pat. 3,158,581 (Nov. 24, 1964), E. J. Vandenberg (to Hercules, Inc.).
7. A. Maeda, *Nippon Gomu Kyokaishi* **53**, 341 (1980).
8. U.S. Pat. 3,158,591 (Nov. 24, 1964), E. J. Vandenberg (to Hercules, Inc.).
9. A. Maeda, *Nippon Gomu Kyokaishi* **61**, 169 (1988).
10. U.S. Pat. 3,728,320 (Apr. 17, 1973), E. J. Vandenberg (to Hercules, Inc.).
11. U.S. Pat. 3,728,321 (Apr. 17, 1973), E. J. Vandenberg (to Hercules, Inc.).
12. T. Nakata, *Nikkakyo Geppo* **41**, 13 (1988).
13. U.S. Pat. 3,766,101 (Oct. 16, 1973), H. Komai and co-workers (to Japanese Geon Co., Ltd. and Osaka Soda Co., Ltd.).
14. E. J. Vandenberg, *Rubber Chem. Technol.* **64**, G56 (1991).
15. K. E. Steller in E. J. Vandenberg, ed., *Polyethers*, American Chemical Society, Washington, D.C., 1975, p. 136.
16. K. E. Steller, *ACS Symp. Ser. 6*, American Chemical Society, Washington, D.C., 1975.
17. H. N. Cheng and D. A. Smith, *J. Appl. Polym. Sci.* **34**, 909 (1987).
18. C. C. Price and M. Osgan, *J. Am. Chem. Soc.* **78**, 4787 (1956).
19. E. E. Gruber, and co-workers *Ind. Eng. Chem. Prod. Res. Dev.* **2**, 199 (1963).
20. E. J. Vandenberg and A. E. Robinson in E. J. Vandenberg, ed., *Polyethers*, American Chemical Society, Washington, D.C., 1975, p. 101.
21. Jpn. Kokai Tokkyo Koho JP 58 002,344 (Jan. 7, 1983), (to Osaka Soda Co., Ltd.).
22. A. Maeda and M. Inagami, *Rubber & Plastics News*, 42 (1987).
23. E. J. Vandenberg in W. J. Bailey, ed., *Macromolecular Synthesis*, Vol. 4, John Wiley & Sons, Inc., New York, 1972, p. 49.
24. H. Balcar and V. Bohackova, *Collect. Czech. Chem. Commun.* **42**, 2145 (1977).
25. E. Schröder, G. Müller, and K.-F. Arndt, *Polymer Characterization*, Hanser Publisher, New York, 1989, p. 94.
26. H. Balcar and J. Polacek, *Collect. Czech. Chem. Commun.* **41**, 2519 (1976).
27. H. Li and co-workers, *Eur. Polym. J.* **25**, 1065 (1989).
28. D. Chen and co-workers, *Certif. Ref. Mater., Proc. ISCRM 89*, Int. Acad. Publ., Beijing, People's Republic of China, 1989, p. 183.
29. L. S. Yasenkova and co-workers, *Vysokomol. Soedin. Ser. B* **13**, 366 (1971).

30. C. M. Hansen, *J. Paint Technol.* **39**, 104 (1967).
31. J. Kozakewicz and P. Penczek, *Angew. Makromol. Chem.* **50**, 67 (1976).
32. E. P. Lieberman, *Off. Dig. Fed. Paint Var. Prod. Clubs* **34**, 32 (1976).
33. G. Allen, C. Booth, and M. N. Jones, *Polymer* **5**, 195 (1964).
34. G. Allen, C. Booth, and C. Price, *Polymer* **8**, 397 (1967).
35. E. J. Vandenberg, *Pure Appl. Chem.* **48**, 295 (1976).
36. T. N. Kuren'gina, L. V. Alferova, and V. A. Kropachev, *Vyskomol. Soedin. Ser. A* **11**, 1985 (1969).
37. U.S. Pat. 3,313,743 (Apr. 11, 1967), L. J. Filar and E. J. Vandenberg (to Hercules, Inc.).
38. U.S. Pat. 3,634,303 (Jan. 11, 1972), E. J. Vandenberg (to Hercules, Inc.).
39. U.S. Pat. 3,776,863 (Dec. 4, 1973), K. Shibatami and S. Nagata (to Kuraray Co. Ltd.).
40. U.S. Pat. 3,957,697 (May 18, 1976), R. K. Schlatzer (to B. F. Goodrich Co.).
41. J. Day and W. W. Wright, *Br. Polym. J.* **9**, 66 (1977).
42. K. B. Chansky, ed., *Suspect Chemicals Sourcebook: A Guide to Industrial Chemicals Covered Under Major Federal Regulatory and Advisory Programs*, Roytech Publications, Inc., Burlingame, Calif., 1991.
43. L. Hagmar and co-workers, *Int. Arch. Occup. Environ. Health* **60**, 437 (1988).
44. *1991–1992 Threshold Limit Values for Chemical Substances and Physical Agents and Biological Exposure Indices*, American Conference of Governmental Industrial Hygenists (ACGIH), Cincinnati, Ohio, 1991.
45. *Appl. Occup. Environ. Hyg.* **5**, 629 (1990).
46. *Form Number 4, Material Safety Data Sheet, Hydrin Elastomers, MSDS Number Z0812.89*, Zeon Chemicals USA, Inc., Louisville, Ky., 1989.
47. *Form Number 8, Material Safety Data Sheet, Parel Elastomers, MSDS Number Z1112.89*, Zeon Chemicals USA, Inc., Louisville, Ky., 1989.
48. D. M. Smith, *J. Soc. Occupa. Med.* **26**, 92 (1976).
49. D. M. Smith, *Br. J. Ind. Med.* **41**, 362 (1984).
50. K. Hashimoto and co-workers, *Elastomerics* **119**, 12 (1987).
51. A. Maeda, K. Hashimoto, and M. Inagami, *SAE Technical Paper Series*, 870194, Conf. Code 09796, SAE, Warrendale, Pa., 1987.
52. U.S. Pat. 4,929,676 (May 29, 1990), M. K. Laughner (to The Dow Chemical Company).
53. J. T. Oetzel, *Rubber World* **172**, 55 (1975).
54. U.S. Pat. 3,645,917 (Feb. 29, 1972), E. J. Vandenberg (to Hercules, Inc.).
55. U.S. Pat. 3,660,314 (May 2, 1972), E. J. Vandenberg (to Hercules, Inc.).
56. U.S. Pat. 3,706,706 (Dec. 19, 1972), E. J. Vandenberg (to Hercules, Inc.).
57. U.S. Pat. 3,417,036 (Dec. 17, 1968), E. J. Vandenberg (to Hercules, Inc.).
58. U.S. Pat. 3,417,060 (Dec. 17, 1968), D. S. Breslow (to Hercules, Inc.).
59. T. Nakai and M. Okawara, *Bull. Chem. Soc. Jpn.* **41**, 707 (1968).
60. U.S. Pat. 3,694,258 (Sept. 26, 1972), E. J. Vandenberg and W. D. Willis (to Hercules, Inc.).
61. U.S. Pat. 3,594,355 (July 20, 1971), E. J. Vandenberg and W. D. Willis (to Hercules, Inc.).
62. T. V. Markman, N. A. Mukhitdinov, and M. A. Askarov, *Deposited Doc.* 1974, VINITI 451-74; *Chem. Abstr.* **86**, 121980d (1977).
63. U.S. Pat. 3,415,902 (Dec. 20,1967), R. A. Hickner and H. A. Farber (to The Dow Chemical Company).
64. Brit. Pat. 1,404,927 (Sept. 3, 1975), (to Mitsubishi Chemical Industries Co., Ltd.).
65. Ger. Offen. 2,124,686 (Dec. 2, 1971), H. Fukutomi and H. Ohotani (to Dainippon Ink and Chemicals, Inc.).
66. T. D. N'Guyen, A. Deffieux, and S. Boileau, *Polymer* **19**, 423 (1978).
67. J. Kelly, W. M. MacKenzie, and D. C. Sherrington, *Polymer* **20**, 1048 (1979).
68. U.S. Pat. 3,403,114 (Sept. 24, 1968), E. J. Vandenberg (to Hercules, Inc.).

69. U.S. Pat. 3,591,520 (July 6, 1971), M. T. McDonald (to Nalco Chemical Co.).
70. C. K. Riew, *Polym. Prepr. Am. Chem. Soc. Div. Polym. Chem.* **14**, 940 (1973).
71. U.S. Pat. 3,746,678 (July 17, 1973), C. R. Dick and E. L. Ward (to The Dow Chemical Company).
72. U.S. Pat. 3,864,288 (Feb. 4, 1975), C. K. Riew and R. K. Schlatzer (to B. F. Goodrich Co.).
73. E. Bortel and R. Lamot, *Rocz. Chem.* **50**, 1765 (1976).
74. Ger. Offen. 2,540,310 (Apr. 8, 1976), R. K. Schlatzer, Jr., and H. A. Tucker (to B. F. Goodrich Co.).
75. U.S. Pat. 4,005,012 (Jan. 25, 1977), W. Wrasidlo (to the U.S. Dept. of the Interior).
76. E. Pulkkinen and T. Peteja, *Finn. Chem. Lett.* **1**, 22 (1977).
77. J. Stamberg, *Coll. Czech.Commun.* **29**, 478 (1964).
78. T. Nishikubo and co-workers, *J. Polym. Sci., Part A: Polymer Chemistry,* **26**, 2881 (1988) and references contained therein.
79. U.S. Pat. 3,567,631 (Mar. 2, 1971), C. A. Lukach, E. J. Vandenberg, and W. L. Young III (to Hercules, Inc.).
80. U.S. Pat. 3,923,703 (Dec. 2, 1975), H. Fukutana and co-workers (to Mitsubishi Chemical Industries, Ltd.).
81. U.S. Pat. 3,632,840 (Jan. 4, 1972), E. J. Vandenberg (to Hercules, Inc.).
82. U.S. Pat. 3,546,321 (Dec. 8, 1970), H. Jabloner and E. J. Vandenberg (to Hercules, Inc.).
83. Ger. Offen. 2,147,290 (Apr. 13, 1972), F. Wollrab, F. Declerk, and P. Georlette (to Solvay et Cie).
84. F. Wollrab and co-workers, *Polym. Prepr.* **13**, 499 (1972).

General References

D. A. Berta and E. J. Vandenberg in A. K. Bhowmick and H. L. Stephens, eds., *Handbook of Elastomers: New Developments and Technology*, Marcel Dekker, Inc., New York, 1988, p. 643.
R. W. Body and V. L. Kyllingstad in J. I. Kroschwitz, ed., *Encyclopedia of Polymer Science and Engineering*, Vol. 6, John Wiley & Sons, Inc., New York, 1986, p. 307.

KATHRYN OWENS
VERNON L. KYLLINGSTAD
Zeon Chemicals USA, Incorporated